Biomaterials and Bioengineering Handbook

Biomaterials and Bioengineering Handbook

edited by
Donald L. Wise
Cambridge Scientific, Inc.
Belmont, Massachusetts

MARCEL DEKKER
MARCEL DEKKER, INC.
NEW YORK • BASEL

ISBN: 0-8247-0318-9

This book is printed on acid-free paper.

Headquarters
Marcel Dekker, Inc.
270 Madison Avenue, New York, NY 10016
tel: 212-696-9000; fax: 212-685-4540

Eastern Hemisphere Distribution
Marcel Dekker AG
Hutgasse 4, Postfach 812, CH-4001 Basel, Switzerland
tel: 41-61-261-8482; fax: 41-61-261-8896

World Wide Web
http://www.dekker.com

The publisher offers discounts on this book when ordered in bulk quantities. For more information, write to Special Sales/Professional Marketing at the headquarters address above.

Current printing (last digit):
10 9 8 7 6 5 4 3 2 1

PRINTED IN THE UNITED STATES OF AMERICA

Preface

The reference book *Biomaterials and Bioengineering Handbook* contains cutting-edge chapters by leading practitioners who deal with critical issues concerning biocompatible materials—polymers, metals, and other substances—used in or on the human body. The topics range from biopolymers used in controlled-release drug delivery systems and synthetic burn-wound dressings to biopolymeric plates used in bone repair and specific orthopedic devices. Each application-oriented chapter integrates basic science, engineering, and medical experience. Most chapters offer discussions of quality control and regulatory approval.

This book offers a wealth of valuable data and experience that will be of use to all bioengineers, materials scientists, and practicing physicians concerned with the properties, performance, and use of materials—from research engineers faced with selecting materials for given tasks to physicians and surgeons interested in materials' biocompatibility, behavior, and toxicology. The many chapters, some of which include case studies, provide rich insights into our experiences today with a broad spectrum of modern biomaterials applications. The book features discussions of the following:

- Basic science, engineering, and medical experience with many biomaterials
- The properties, performance, and use of biomaterials
- Issues of biomaterials behavior, toxicology, and biocompatibility
- Maintaining quality control using a variety of instruments

The medical device and drug industry is consistently among the strongest technological performers, and materials are a key ingredient in this dynamic growth. Development of these materials is in a constant state of activity, as old materials fail to withstand the tests of time and modern applications demand new materials. This book focuses on materials used in or on the human body—materials that define the world of "biomaterials." The book covers the range of biomaterials from polymers to metals to ceramics. The diversity of the field necessitated careful integration of basic science, engineering, and practical medical experience in a variety of applied disciplines. As a result, scientists, engineers, and physicians are among the authors. They provide a full and detailed accounting of the state of the art in this rapidly growing area, reflecting the diversity of the field.

Clearly, the *Biomaterials and Bioengineering Handbook* focuses on materials development and characterization. Chapters deal with issues ranging from biocompatibility and biostability to structure–function relationships. Chapters also focus on the use of specific biomaterials based on their physicochemical and mechanical characterization. Integral to these chapters are discussions of standards in analytical methodology and quality control.

The users of this book will represent a broad base of backgrounds ranging from the basic sciences (e.g., polymer chemistry and biochemistry) to the more applied disciplines (e.g., mechanical and chemical engineering, orthopedics, and pharmaceutics). To meet varied needs, each chapter provides clear and fully detailed discussions. This in-depth but practical coverage should also assist recent inductees to the biomaterials circle. This volume conveys the intensity of this fast-moving field in an enthusiastic presentation.

"Biomaterials" range from biopolymers used in controlled-release drug delivery systems to biopolymeric plates used in bone repair to specific orthopedic devices described in case studies. Basic science, engineering, and medical experience are each necessarily integrated into the contents.

The book is organized by the type of biomaterial application, including: selected biomaterials and biocompatibility studies; case studies of orthopedic biomaterials; specific biomedical applications of biomaterials; and functional biomaterials. This organized format allows the reader to focus on a specific application of interest.

As noted above, this book addresses polymers as well as metals and other materials that are used in or on the human body. For example, chapters on biopolymers include controlled-release drug delivery systems, bone repair cements and resorbable fixtures, synthetic burn-wound coverings/dressings, and so on. Integral to chapters on these applications of biomaterials are discussions on quality control using a variety of instrumentation, such as gel permeation chromatography, thermal gravimetric analysis and differential scanning calorimetry, and high-performance liquid chromatography.

The readers of *Biomaterials and Bioengineering Handbook* are expected to have broad and varied specialty backgrounds, but to be focused in their own applied work. Materials scientists and materials engineers will be interested in applications concerned with properties, performance, and use. Design criteria are emphasized because the research engineer will be concerned with the selection process for given tasks. The academic physician and practicing surgeon will be concerned with materials behavior, toxicology, and biocompatibility and will want to see these aspects presented in a readily accessible form. As a result, each chapter provides details satisfying to professional colleagues interested in biomaterials.

Donald L. Wise

Contents

Biomaterials and Bioengineering Handbook

1

Biocompatibility: Review of the Concept and Its Relevance to Clinical Practice

J. H. Boss
Bnai Zion-Medical Center, Haifa, Israel

I. INTRODUCTION

Based on current ideas and past experience, as reflected in a towering literature, clinicians entrust certain biomaterials with the virtue of *biocompatibility* and condemn others with the infamy of *bioincompatibility*. Not having the edge of personal bedside skills, materials scientists are likely to miss what has been learned in the course of decades of clinical practice of implantation. They choose to classify biomaterials by their performance in the highly artificial niches of test tubes or, at best, in the experimental animal. Such ratings coincide with explicitly preconceived assumptions. Not surprisingly, the same biomaterial may be censured in the tranquility of the laboratory and prized in the bustle of clinics, and vice versa. Bio(in)compatibility is an innate property of a material. Property is an essential quality or capability of a material or a quality common to a whole class of materials [1,2]. The quintessence of a material cannot be formulated by a pair of opposites. The paradoxically conflicting viewpoints of researchers are dissipated by the concept of the *relativity of biocompatibility* [3]. Having established that the biological behavior cannot be discoursed out of a working context, deviating conduct traits of a biomaterial under disparate circumstances are, logically, contingent on particular environmental situations.

II. DEFINITIONS AND CONCEPTS

Biomaterials are substances, other than food or drugs, introduced in therapeutic or diagnostic systems which are in contact with the tissues or biological fluids [4]. They constitute a heterogeneous group of living or inanimate, natural or synthetic, inorganic or organic compounds. They are inserted into or placed onto the body with the aim of improving the function of or replacing a diseased, damaged or lost tissue or a whole organ. Biocompatibility describes the interactions between the living system and the material introduced into this system. A material is *bioinert* when the material-tissue interface is stable, i.e., constituents of the tissue and material neither react chemically with each other nor dissolve into each other. When the interface is not in equilibrium, that is, when the interface is unstable, host-material relationships characterized by irritation, inflammation, damage, immunogenicity, pyrogenicity, toxicity, mutagenic-ity or carcinogenicity evince a state of bioincompatibility while a state of biocompatibility exists in their absence. Biomaterials which cannot be eliminated by the body are *biotolerant* when they are fibrotically encapsulated and *bioactive* when they evoke a critical host reaction. Bioactive hydroxyapatite is a case in point in as much as it stimulates proliferation and differentiation of fibroblasts and osteoblasts and induces collagen synthesis by these cells. Types III and V collagen are synthesized in the immediate postimplantation period. Their substitution by type I collagen is a distinctive attribute of the response to a biocompatible material. Constituting operative interruption of the tissue integrity, any implantation inevitably elicits an immediate inflammatory reaction. An implant evokes a specific response if the ensuing inflammatory, granulomatous and fibrosing processes are over and above those expected to occur in the course of simple wound healing. The patterns of wound healing and the response to nonirritating materials, including the fleeting infiltration of neutrophils and the action of tissue debris-resorbing enzymes, parallel one another during the earliest phases, but deviate from each other at the later stages. Implants are initially enclosed by leukocytes, the density of which lessens with time and, compared with other white blood cells, macrophages preferentially adhere to the surface of the foreign bodies. Some recruited monokaryonic macrophages fuse to form polykaryonic macrophages—the so-called foreign body giant cells. The mono- and polykaryonic macrophages abut on the implant's surface and phagocytose small breakdown particles, if such are scattered in periprosthetic tissues. From the early to the late stages, the evolution of the periprosthetic processes depend, among others, on the implant-tissue interaction-driven activation of the leukocytes. This in turn determines production of integrins, adhesion molecules and intermediary substances of inflammation. It is self-evident that sufficient and undisturbed incorporation into the living environs and rapid functional takeover are imperative for an effectual reconstructive surgery. Locally toxic chemicals provoke unrelentingly ongoing inflammatory and necrotizing reactions. Investigators are aware that the results of local toxicity testing are incidental on the animal species evaluated. To set an example, neutrophils encircle polyurethane particles in the mouse's subcutis for many months, but in rabbits they are absent during the late phases around the particles in the subcutis. In addition, to get a faithful view of in vivo biocompatibility, qualitative parameters of local reactions ought to be amended by quantitative scores, e.g., the intensity and extent of tissue damage, thickness of an encapsulating fibrous membrane or density and differential count of an infiltrating cell population. As opposed to the local effect of a toxic substance, systemic toxicity, due to either leached components or poisonous degradation products, is hazardous to the recipient's health and may even be fatal [5–22].

Remnants of fabrication-associated additives or solvents as well as impurities are of concern in view of their possible noxious effects on the tissues. For instance, unless removed prior to use, the dross (fragmented fibrils) included in commercially available carbon filaments turns into an implantation debris. Likewise, release of Al_2O_3, Ta_3O_5, ZrO_2 and phosphates from ceramics inhibits mineralization of newly formed bone subjacent to the implant. Also worrisome is the prospect of formation of toxic compounds by sterilization of biomaterials, e.g., with ethylene oxide or gamma irradiation. Breakdown particles of otherwise biocompatible materials incite disastrous reactions when deposited at host-implant interfaces. They may initiate annoying, at times serious side effects when migrating to distant sites. It is important to keep in mind that fabrication-related shortcomings and an implant's poorly chosen configuration tend to increase generation of breakdown particles. It is an attractive proposition to convert inactive into active biomaterials by incorporating pharmaceutical agents (e.g., drugs and growth factors) into the bulk or surface of the implants [23–29].

Patterns of the tissular responses depend on the chemical as well as the physical properties of the implant and on the host's reaction potentials. The inventory of critical factors encompass toxicity, chemical composition, hydrophobicity/hydrophilicity, size, shape, weight per area, surface texture, surface free energy, surface charge, surface-adsorbed proteins, wettability, solubility, porosity, pore size, rate of degradation, degradation products, particulate burden, implantation site and species-specific reactivity. Implants induce foreign body reactions. With rare exceptions, they are sequestered from the tissue by a two-tiered interfacial membrane. The inner coat consists of mono- and polykaryonic macrophages admixed with lymphocytes and, as the case may be, neutrophils, eosinophils, plasma cells and mast cells. The outer coat, merging with the surrounding tissues, is composed of concentrically oriented fibrocytes and collagen as well as reticulin fibers. Macrophages abut upon the implant's surface as early as the third postoperative day. Foreign-body-activated macrophages synthesize monokines and growth factors, which in turn stimulate proliferation and differentiation of mesenchymal cells, multiplication of fibroblasts (or other cells, as the case may be) and synthesis of collagen fibers (or other matrical elements, as the case may be). Fibroblast-stimulating activities are amplified in vitro by preadsorption of a material with proteins, among which albumin, gammaglobulin, fibrinogen, and fibronectin are of a particular interest in the context of periprosthetic matrical molecules. Incidental to the implant's nature, the cellular composition, content of the extracellular matrix and thickness of the interfacial membrane change postoperatively. Fibrous encapsulation is typical of the reaction to biocompatible, nondegradable biomaterials. Equal things being equal, the capsular thickness parallels the amplitude of relative motion at the tissue-implant interface. Thick fibrous interfacial membranes may impair an implant's functional capacity. Bioincompatible substances generally elicit intense inflammatory and necrotizing responses. Deductively, the histologic characteristics* of the interfacial membrane are the measure of the local biocompatibility of an implanted material [26–38].

Safety and efficacy are the most important considerations in implantology. By and large, up-to-date biomaterials are neither poisonous nor tumorigenic in man. Yet, the prospect that their breakdown products may be detrimental to the patient's well-being is of concern. In view of literature reports, foreign body carcinogenesis (tumorigenesis contingent on physical—as opposed to chemical—properties of a foreign body) is a possibility to be reckoned with. Although not cancer initiators, implants may act the role of promoters. Indeed, sarcomas arise sporadically near patients' implants many years postoperatively. They also develop now and then in rats with a subcutaneous or intramuscular, metallic or polymeric (nonresorbable or resorbable) implant. In rats, various biomaterials are tumorigenic. The rate at which neoplasms develop is mainly material dependent. At least some additives used in the manufacture of biomaterials are endowed with tumor-promoting activity. The tumor-promoting activity is influenced by surface characteristics of as well as by an antioxidant added to a polyethylene film. Extrapolations from the experimental to the clinical settings are fraught with uncertainties, yet, the clinical case reports and experimental data is worrisome. That the rate of implant-induced tumors is incidental to the species as well as the strain of the animals studied is what makes investigative endeavors so difficult [39–43].

Prior to addressing the subject matter of this chapter, it is proper to make some prefacing comments for the benefit of the noninitiated reader. Biocompatibility relates to the exploitation by a material of the proteins and cells of the body to meet specific performance goals. In addition to biological aspects, biocompatibility further pertains to the harmonious operation of the implant-tissue complex. Unless planned for short-term service, the time dimension plays a

---

*Some researchers have communicated their quantitative and statistical methods for assessment of the interfacial membranes. No consensus having been reached, and in this author's judgment one technique is as gratifying as any other, the subject is not further dealt with herein.

crucial role since biocompatibility, biostability, biodurability and biofunctionality must last for the duration of the material's programmed function in the body. Biostability is of concern since leaching and degradation of constituents of a material may alter the implicit biocompatibility [44–46].

All implants degrade with time but at different rates. The degradation is governed by diverse mechanical stresses. Hence, biostability is a relative term too. Acceptedly, materials degrading tardily, e.g., polyethylene, are labeled as biostable. Employment of a device of poor biodurability cannot be justified if its degradation, with or without mediation of enzyme-catalyzed processes, results in release of toxic products. Some of the well-established effects of cellular activities on polymeric surfaces are fissuring, cracking and rupture. Duration of an implant's service is a function of its biodurability. For instance, the duration of service is reduced when an implant calcifies such that it loses its flexibility critical for the performance of a polymeric cardiac valve prosthesis. The rate and extent of calcification of natural (e.g., collagenous) and synthetic (e.g., polyurethane) devices depend on adherence of cellular debris, local stress concentrations, absorption of calcium-binding serum proteins, surface defects and inherent affinity sites for calcium ions. Finally, biofunctionality pertains to the mechanical and physical properties enabling an implant to perform its planned functions. Because the forces and stresses acting on the bone regulate its growth and remodeling, differences in mechanical properties can hinder or uphold the success of an implant irrespective of the materials used in its fabrication [29,45–53].

Biocompatibility is modified by a material's surface texture. A microporous surface optimizes attachment of fibroblasts coupled with enhanced synthesis of extracellular matrix and is linked to a reduction of the inflammatory reaction. An implant of polyvinyl chloride-polyacrylonitride copolymer with 1- to 2-μm-sized pores encourages the close adherence of fibrous tissue to the surface, whereas an implant of the same material but with smaller or larger pores incites a granulomatous interfacial response [54,55].

Biomaterials ought to be evaluated in vivo under nonloaded as well as loaded conditions. Assessment of the response to a nonloaded implant appraises the tissular compatibility but only analysis of a loaded implant yields the functional compatibility as well. For example, friction at tissue-host interfaces perpetuates inflammatory reactions. In comparison with restrained implants, compliance of actively loaded devices upsurges the formation of both fibrous and granulation tissue [56–58].

A highly sensitive technique of biocompatibility testing within the bony environs quantitatively grades implant-mediated disturbances of primary matrical mineralization via the matrix vesicles and calcifying globules [59].

Hypersensitivity to one or more constituents reverses the rewarding payoff of an otherwise well-tolerated biomaterial in a susceptible individual. Similarly, an implant's ingredient capable of directly activating the complement cascade may boost the tissular response to an in vitro innocuous biomaterial [60–62].

The exemplified eventualities are mentioned in an attempt to persuade readers that neither in vitro nor simplistically planned in vivo experiments suffice to unmask the gamut of untoward effects an implant may generate in patients. As with cytotoxicity and carcinogenicity tests in general, it is not passable to merely explore one compound or another under artificial laboratory settings. Comprehensive analyses must be adjusted to the proposed clinical application of the device and must comprise any and all of the biomaterials under consideration. Indeed, rats react differently to such a seemingly trivial difference as whether expanded tetrafluoroethylene discs are implanted into the animals' subcutaneous or epididymal fat tissue [63,64].

Two congenial indices of the host response to a biomaterial are the thickness of the membrane insulating it from its environment and the intensity of the inflammatory infiltrate within the membrane. It should be understood, however, that features of the interfacial membrane, as observed under the microscope at any given time, depict but a stage in a drawn-out process. Formation of the interfacial membrane begins with the acute postoperative inflammatory infiltration. It continues with the reaction to chemical substance(s) of the implant's surface and to those leached from it as well as to proteins of the biological fluids adhering thereto. It winds up with the response to the physique of the surface and to interfacial movements. Reflecting the conventional foreign body reaction, encapsulation by fibrous tissue—more often than not invaded by at least some mononuclear and giant cells—does not address the issue of biocompatibility. It is rather the reaction severity which is the measure of bio(in)compatibility [65–68].

The well-balanced interfacial situation occasionally deteriorates, upon which cell-derived chemoattractants recruit ever more leukocytes, and growth factors boost fibroblastic proliferation and production of collagen and enzymes and debase the implant's surface. Leukocytes, including the mono- and polykaryonic macrophages, are short-lived cells. Hence, the perpetual presence of leukocytes at the interface signals continuing control of the local scene by the implant's chemical and physical qualities. Inoffensive foreign body responses being consistent with well-functioning implants, a thin interfacial membrane is a compelling criterion of biocompatibility. As a rule, the thickness of inflamed interfacial membranes increases with the in vitro gauged cytotoxicity of a biomaterial. Deductively, the absence of an interfacial membrane implies an outstanding biocompatibility. Osseointegration specifies lack of a soft tissular interfacial coating within the bony environs. It is defined as the direct contact, at the light microscopical level, of viable osseous tissue with the surface of the implant [27,68–72].

Definitions of biocompatibility are many, their premises depending to a large measure on the authors' professional training, background and interest. The objective of this survey is to highlight the perspectives which are deterministic in the success or failure of the implants, particularly as witnessed in everyday practice of reconstructive surgery. Admittedly, the viewpoints brought up are colored by the morphologist's retrospective aptitude of exploration. The purpose of biocompatibility research is to aid the physician in her or his care of the sick. Short of biofunctionality tests, microscopic study of the implant-tissue interface and its surroundings is an informative practice to assess an implant's in vivo behavior, notably when such an evaluation is carried out on samples retrieved after lengthy and effective service periods. Repercussions on the functional and morphological aspects of distant organs are not to be ignored, but the crucial parameters of biocompatibility are the qualitative and quantitative morphological aspects of the host-implant interface and its contiguous reaction zone [73–76].

III. NONBIODEGRADABLE POLYMERS

In vitro techniques have been worked out to gauge the biocompatibility of nonbiodegradable polymers based on tests ranking cell adhesion, cell migration, and cell growth on the surfaces. It is critical to concurrently determine cell viability in the culture media to which extractables leached from the pertinent biomaterial have been added. This is of special import in view of potential cytotoxicity of catalysts used in manufacturing processes. Future research is anticipated to clarify whether the results of current tests do indeed match the clinical observations in patients with implants as well as the data obtained on histological analysis of retrieved implants [77–80].

Macrophages are stimulated by interacting in vitro with polymers. The intensity of their activation varies from one material to another. Their adhesion to the surface, via receptor molecules, is modulated by adsorbed matrical and humoral constituents. The activation and incitement of the macrophages to synthesize and secrete cytokines and growth factors are qualitatively and quantitatively regulated by the classes of in vivo available proteins. Spreading and differentiation of surface-attached fibroblasts and osteoblasts are modified not only by the compositional and physical properties of the polymeric exterior but also by surface-adsorbed matrical molecules such as immunoglobulins, fibrinogen, fibronectin, vitronectin and laminin. An amorphous extracellular material as well as cellular debris also participate in making up the surface coat. Being controlled by macrophage-derived growth factors, proliferation of fibroblasts and synthesis of collagen are enhanced concurrently. The surface is generally enveloped by mono- and polykaryonic macrophages. Giant cells abound, especially when the macrophages face rough-surfaced polymers. Fibrous tissue usually abuts smooth-surfaced polymers. The density of adherent monokaryonic macrophages and the rate of their fusion into giant cells depend on the composition, hardness and hydrophobicity of the surface. Around filamentous polymers, macrophages spread with finger-like cytoplasmic projections. Within the interstices of polymeric devices, the formation of connective tissue deviates from what happens during conventional wound healing: In addition to its complement of mono- and polykaryonic macrophages, the newly formed fibrous tissue contains a surplus of acid mucopolysaccharides but relatively few and improperly arrayed collagen and reticulin fibers. This newly formed fibrous tissue does not progress with time to resemble its well-ordered, oriented and bundled, native counterpart. Notwithstanding the chemical differences between the various polymers (e.g., polyethylene, polysulfone, polyethylene terephthalate, polytetrafluoroethylene, polystyrene and polypropylene), the kindred behavior of macrophages in contact with these disparate moieties favor the notion that intrinsically similar laws govern the cell-polymer interactions. Correspondingly, alike organizational patterns occur, for instance, within and around prosthetic ligaments fabricated from diverse polymers [10,81–90].

To the degree that most nondegradable polymers are being separated from their bony bed by an interfacial membrane, they poorly qualify for a majority of intraosseous applications. Exceptions to this rule are exemplified by 5714-F1 polyether urethane, polytetramethylene terephthalate segmented polyether polyester copolymer and polyethylene oxide/polybutylene terephthalate copolymer. These bioactive polymers not only become osseointegrated but they regularly turn bone-bonded. The bone-bonding potential of bioactive polymers appears to be contingent on their undergoing calcification (facilitating direct bone formation), cement line-type proteins (osteopontin, osteocalcin and glycosaminoglycans) coprecipitating with carbonate-apatite at the implant-bone interface [91–93].*

The intricacies of cellular and enzymatic interactions at the host-implant interface should be taken into account when discoursing results of in vitro biocompatibility tests. The fluidity of the inflammatory, granulamatous and fibrosing reactions is governed not only by chemical and physical characteristics of the implants but also by the variable amounts and diverse classes of regulatory substances which are produced upon confrontation of macrophages and other leukocytes with the implants and especially with their breakdown particles. Notwithstanding the biocompatibility of modern nondegradable polymers in the context of currently employed usages, their particulates do excite a macrophages-, lymphocytes- and plasma cells-rich, progressively expanding inflammatory reaction [31,94–95].

A. Polyethylene and Polyacetal

Does death of cells growing in the vicinity of a biomaterial or thereon, either in vitro or in vivo, of necessity imply toxicity? May it be that parameters besides cell death or damage

*The morphological aspects of osseointegration and bone-bonding are described in some detail in the chapters dealing with titanium and hydroxyapatite.

should be considered when evaluating biocompatibility? The widely used polyethylene is a case in point. Myriad joint replacements being carried out annually, orthopedic surgeons are evidently convinced of the biocompatibility of the ultrahigh molecular weight polyethylene, making up an articulating surface of the artificial joints. Indeed, smooth surfaced polyethylene blocks are well accepted by the tissues as expressed by the formation of a thin, bland, fibrous capsule, an apt marker of optimum biocompatibility. Devices such as the Rashel polyethylene net (employed in the treatment of musculofascial defects) and the high-tensile polyethylene filamentous prosthesis (used experimentally in the replacement of the anterior cruciate ligament) incite a mild foreign body granulomatous response and are adequately organized by tissular ingrowth and ongrowth (Figs. 1, 2). Phenotype preservation of highly differentiated cell types in culture is precarious. That the chondrocytes growing within the pores of a polyethylene substrate maintain collagen type II synthesis for several weeks, validates this polymer's in vitro biocompatibility [3,97–100].

Alas, there is another side to the coin. Cells die when cultured on polyethylene in a saline medium (and, by the way, on some other polymers as well). Hence, the polyethylene is, at least conditionally, toxic to the cells; ergo, it is bioincompatible. Conceded, physiological saline does not simulate the milieu interne. This toxic effect is averted, however, when the cells are seeded on the polyethylene in serum-containing medium or if the polymer is preadsorbed in a proteinaceous solution (albumin, lgG or fibronectin). Saline incubated with the polyethylene is innocuous to cells. Hence, the saline is devoid of detectable (leached) toxic traces. Therefore,

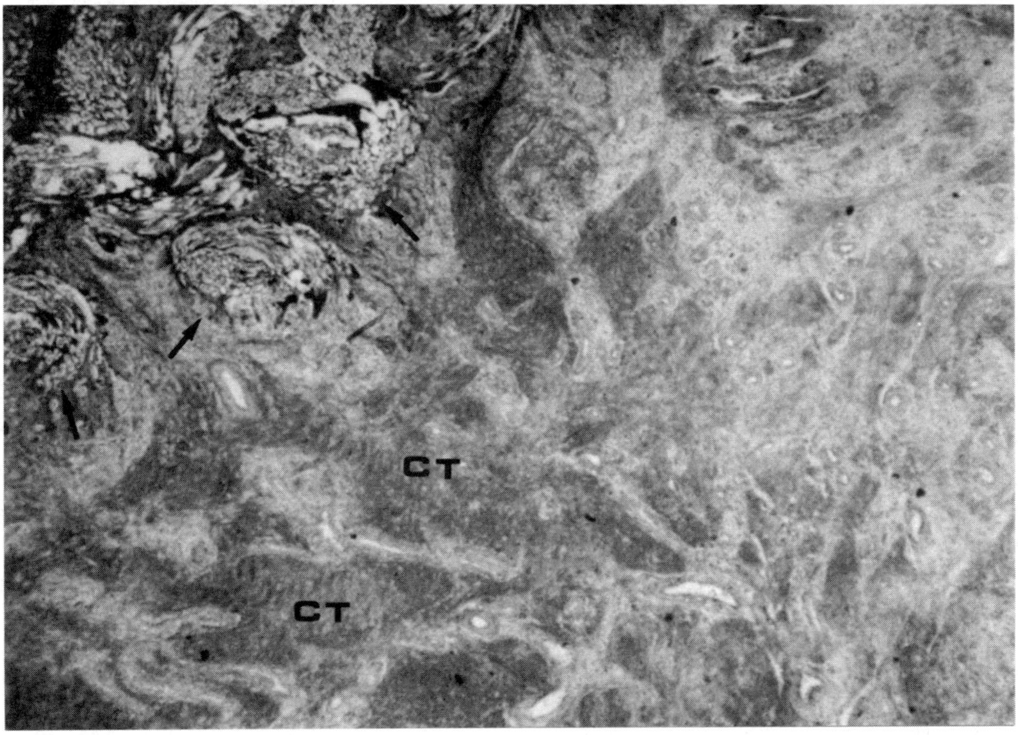

Figure 1 Abundantly formed, well-aligned, bundled collagenous tissue (CT) close to bundled polyethylene filaments (arrows) of a Rashel net which was retrieved a decade after its implantation for augmentation of a large postoperative musculofacial defect. Giemsa stain. ×30.

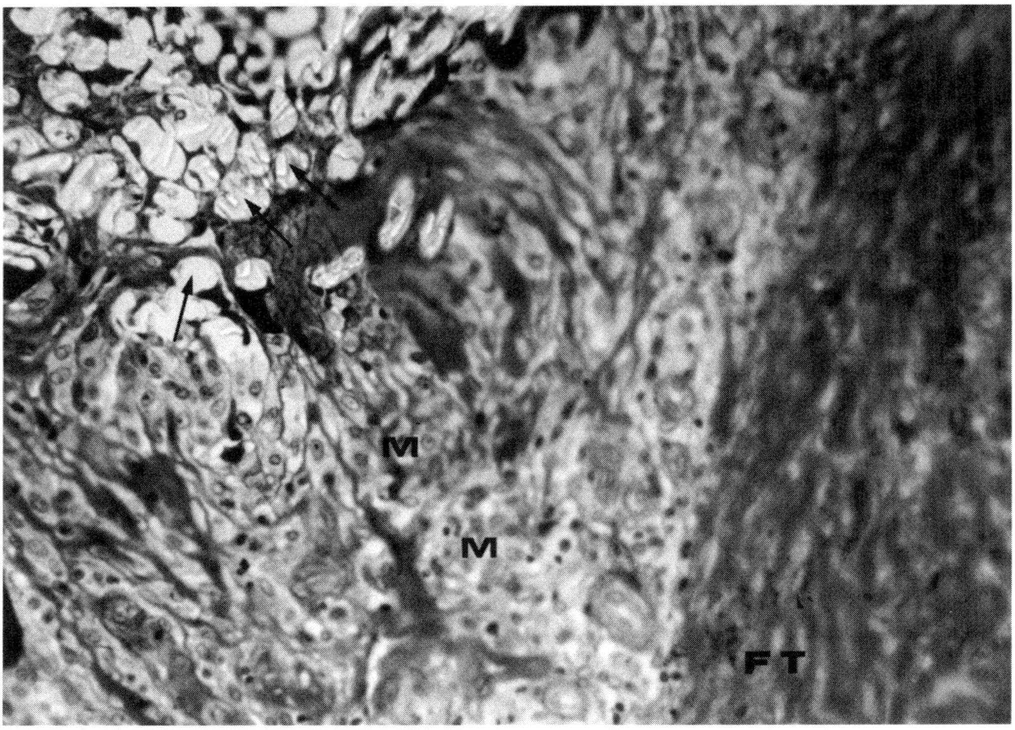

Figure 2 The bundle of polyethylene filaments of the Rashel net is poorly organized, scarce cells having penetrated in between the individual polymeric fibrils (arrows). The bundle is surrounded by mono- and polykaryonic macrophages (M) intermixed with a few lymphocytes. Newly formed fibrous tissue (FT) abounds around the bundle. Giemsa stain. ×200.

the noxious effect of the polyethylene in a saline medium is ascribed to the substrate's physical properties. Cell attachment to the polymer is prerequisite for this intrinsic surface toxicity to occur. Critique is justly directed at assays in which a material is tested in a culture medium without direct contact of the material with the cells. Without close cell-material contact, test results show whether or not the polymer leaches toxic compounds. Such tests gauge the material's (non)toxicity but not its bio(in)compatibility. Likewise, the results of in vitro cytotoxicity tests depend to a large measure on the procedures used by the investigators in the preparation of their materials. In conclusion, the outcome of in vitro tests is ever so often an unreliable indicator of the in vivo conduct of biomaterials. As persuasively gleaned from the analyses of both patients' and animals' well-functioning retrieved implants, the purportedly "toxic" polyethylene is innocuous. On the other hand, the R-4 latex, which in vitro proves devoid of any cytotoxic effects, provokes an intense tissular reaction when inserted into a rat's muscle [101–104].

The morphological assessment of specimens retrieved postmortem from patients with well-functioning total joint replacements endorses the biocompatibility of polyethylene. Yet, not so long ago, aseptic loosening of artificial joints was equated with "polyethylene disease." It was early implied and later corroborated that an erstwhile harmonious bone-implant coexistence deteriorates in the wake of an expansive polyethylene-induced granulomatous response to wear particles shed from the articulating surface and accumulated at the interface. In a rabbit tibial model, polyethylene in bulk form is separated from its osseous bed by a thin, dense, fibrous interfacial membrane with infrequent mono- and polykaryonic macrophages at surface irregularities. On the other hand, polyethylene particles are being separated from their bony bed by a thick layer of florid granulomatous and fibrous tissues. The clinical and experimental data leaves little doubt but that in osseous environs, as at other locales, polyethylene as such is perfectly biocompatible but its breakdown products elicit tissular reactions which are irreconcilable with long-term successful function of implants [105–107].

The rate of generation of polyethylene wear debris is remarkable: Forty million to 40 billion particles are yearly shed into the joint cavity of patients with a Charnley total hip prosthesis. Disintegration proceeds even more briskly when a polyethylene component is implanted without cement. Bone does not grow up to the polyethylene surface. For the first three postoperative years, uncemented polyethylene tibial inserts become separated from the bone by a synovial-like membrane in the presence of but a scant macrophagic response and in the absence of significant osteolysis. Interfacial micromotion-related generation of particles from uncemented polyethylene acetabular cups induces a granulomatous reaction which is associated with alarmingly high rates of aseptically loosened sockets. At the 10-year follow-up, nearly half of all patients with uncemented threaded polyethylene acetabular sockets require revision operations. This unacceptably high failure rate has alerted surgeons to shun a direct polyethylene-bone contact. It turns out that the failure is related to a faulty technique and in no way reflects on the polyethylene's biocompatibility [108–110].

Periprosthetic tissues of aseptically loosened artificial joints display inflammatory, granulomatous and fibrosing changes which are mainly induced by the submicron- to micron-sized polyethylene wear products emanating from the articulating surfaces of the artificial joints. The tiny polyethylene particles are phagocytosed by large, polyhedral, deeply eosinophilic or foamy macrophages (Fig. 3), as evidenced by the very fine birefringence and oil red O-affinity of their cytoplasm. Intra-articularly impacted acrylic fragments may abrade larger polyethylene particles. The forbidding role of biomechanically poor designs in leading to catastrophes is exemplified by precipitous generation of a great deal of up to 130-μm-long polyethylene shards when a patient's knee joint is replaced with a cementless prosthesis manufactured of porous-coated components having too small contact areas between the articulating surfaces [111–113].

The chain of events finalizing failure of artificial joints is initiated by the dispersal of prosthetic debris in the so-called effective joint space. In addition to mono- and polykaryonic macrophages, lymphocytes also participate in the infiltration of the periprosthetic tissues. Depending on the composition of the alloplastic components, a T-cell- or B-cell-predominant or a mixed T- and B-cell reaction is aroused. The periprosthetic tissues close to osteolytic cavities in patients with a loose cemented artificial joint hold many more T-lymphocytes than the periprosthetic tissues of patients with either a well-fixated or a loose cemented arthroplasty without localized osteolysis. The biocompatibility of any one of the ingredients of joint replacements is suspect if this finding is in fact indicative of an immune process. Activated lymphocytes synthesize lymphokines. The detritus-laden macrophages produce monokines. The osteolysis-enhancing factors are of especial interest in as much as they recruit and stimulate the osteoclasts. The stimulated macrophages may, however, also directly resorb the bone, a process referred to as surface resorption, distinguishing it from lacunar resorption performed by the osteoclasts. The degree of ensuing diffuse and localized osteolysis correlates, as a rule, with the number of particles-laden macrophages. Localized osteolysis (occurring even aside stable components) may be related to cement mantle defects providing a route through which debris from the joint cavity reaches the endosteal bony aspect via a cement-metal gap. The high rate of 40% localized acetabular osteolysis around well-fixated sockets of cementless hip arthroplasties is worrisome. Once fixation of a component has deteriorated in the wake of osteolysis, interfacial motion intensifies not only generation of wear particles but promotes osteo-

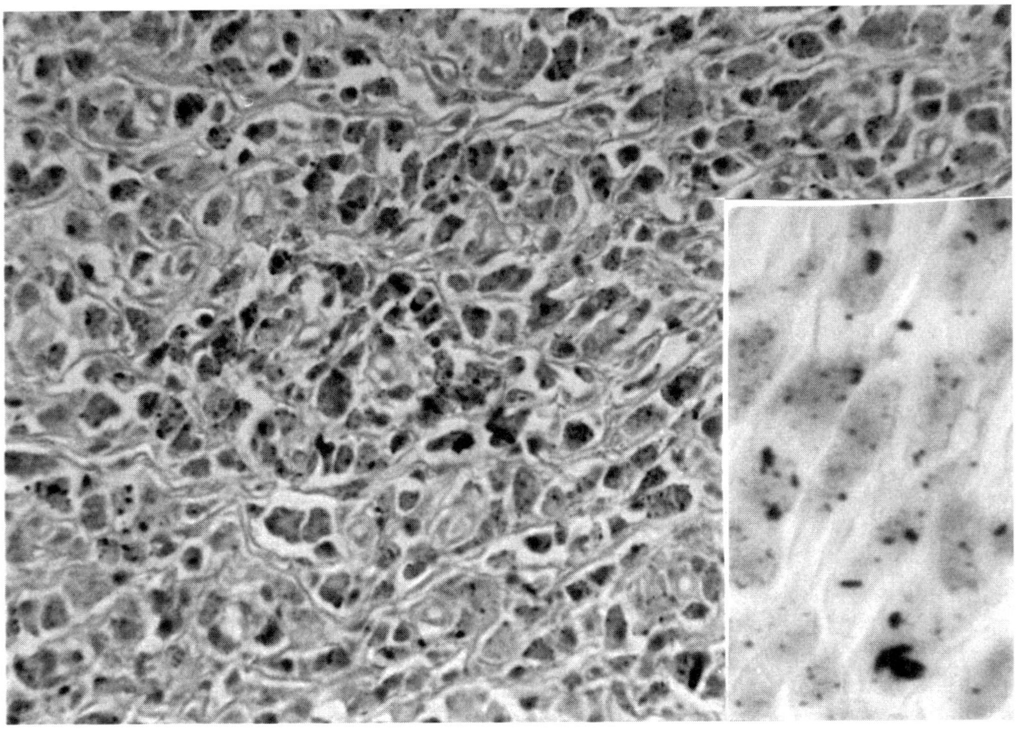

Figure 3 Sheet of large, polyhedral, foamy, submicron- to micron-sized polyethylene-particles-laden macrophages in the interfacial membrane of an aseptically loosened, cemented total hip joint replacement. Several macrophages have also phagocytosed metallic particles. (Inset) The large, foamy, polyethylene and metallic-particles-laden cells are CD68-immunoreactive, evincing their macrophagic nature. Hematoxylin and eosin. ×80. Avidin-biotin immunostain for monocyte/macrophage-associated CD68 antigen. ×320.

clastic activity as well. Many years postoperatively, the dominance of activated macrophages over activated fibroblasts at the interface of loosened arthroplasties is a sign of insufficient clearance mechanisms, implying that the phagocytes are overwhelmed in their capacity to dispose of the large amounts of generated prosthetic and necrotic debris. Activation of the macrophages by the implant may be a blessing in disguise. The activated macrophage-derived transforming growth factor-β along with the basic fibroblast growth factor encourage ingrowth of tissue into porosities but they also further the progressive enlargement of the interfacial membranes. Also, other macrophage-derived factors cause tissue damage and bone resorption. Small to sizable foci of coagulation necrosis accompany extensive polyethylene deposits. This vicious circle results in intractable loosening and failure of artificial joints. It is independent of the reaction to acrylic cement as such. Usually, the reaction to the polyethylene debris is more prominent than the reaction to metallic particles. It stands to reason that the adverse response to particulate polyethylene is the most forceful factor in aseptic loosening of artificial joints, equally condemning cementless and cemented arthroplasties to failure [23,114–128].

A discussion of the tribological aspects of particle generation from a polyethylene surface sliding over a hard counterface is beyond the scope of this chapter. Suffice to say that the amount of the breakdown particles depends on several, only partially controllable factors, prime among which are the contact stress and the geometry as well as yield strength of the component. Polyethylenes of an inferior quality, manufacturing processes adversely affecting wear resistance, lack of congruency, undue properties of the counterface, improper modularity, incorrect thickness of a polyethylene insert, malalignment of the components, hapless skeletal geometry, third-body wear, shift or tilt of a component, and the patients' weight, height and level of activity have all been implicated in excessive debris generation. Having analyzed their cases of tibial osteolysis in patients with uncemented total knee arthroplasties, Peters et al. view particles-associated cavitation as being related to mechanical factors, primarily to destruction of a thin tibial polyethylene insert, excessive abrasion of a prominent polyethylene tibial eminence and failure of the metal-backed patellar component, and secondarily to corrosion between the titanium screws and the cobalt-chromium baseplate. Length of service likewise taking its toll, it is not surprising that polyethylene particles incrementally accumulate in the joint cavity, articular capsule and implant-bone interface. The reality of a profuse deposition of particles, rather than their innate properties—toxicity versus inertness—fore and foremost accounts for exorbitant tissular reactions. Hence, what happens at the implant-bone interface of failed arthroplasties does not reflect on the biocompatibility of the polyethylene, or any other biomaterial for that matter, but rather on the strenuous mechanical conditions of the artificial joint. Component modifications promising improvements while still on the drawing board, such as a carbon filaments-reinforced polyethylene, have not justified themselves, the clinical results with the "bettered" material being worse than the "primeval" one [129–138].

Polyethylene is not unique in the untoward effects of its breakdown products on the integrity of the bone-implant boundary. The microscopical features of the response to polyacetal and polyethylene particles are indistinguishable from each other. However, polyacetal particles are released in extra large amounts from the femoral stem of the so-called isoelastic cementless and the acetabular socket of the Christiansen hip prostheses. Unsurprisingly, the massive inflammatory, granulomatous, fibrosing, and necrotizing reactions in the periprosthetic tissues of artificial hip joints with polyacetal-based components are associated with high failure rates so much so that orthopedic surgeons mostly avoid employing this polymer [139–142].

Principles similar to those governing incorporation of carbon filamentous implants (vide infra) account for deficient bony fixation of the high tensile polyethylene filamentous braid used for replacement of the anterior cruciate ligament. Deposition of wear products is limited. A thin granulomatous layer at the perimeter of the braid and a thick fibrous mantle around it evolve, but there is only scant ingrowth of tissue into its core. A disastrous expansion of the femoral and tibial tunnels ensues within three months of implantation into the goat's stifle. This hardly approximates the outcome expected of a biocompatible biomaterial. Yet when implanted into the subcutis, the high tensile polyethylene filamentous braid is amply ingrown by fibrous tissue and enveloped by a thin fibrous film. It manifests superb biocompatibility in the subcutis. Behavioral disparities of one and the same device—the high tensile polyethylene filamentous braid in the cited example—in different sites of the body and under unequal mechanical settings emphasize once more the importance of reviewing the "unique circumstances" in biocompatibility assays. The chemical and physical properties of the polyethylene braids are obviously identical within the osseous tunnels and in the fat tissue. Locomotion-related motion of the prosthesis within the bony tunnels prevent not only an adequate tissular ingrowth and organization but a secure bony anchorage as well. The adverse effects of the biomechanical factors offset the salutary acceptance of the implant by the tissues, spuriously making the high tensile polyethylene filaments appear bioincompatible when in the stifle [98].

B. Polymethylmethacrylate

Cemented artificial joints attain good long-term clinical results in many patients. Cemented joint replacement has enjoyed wide popularity until orthopaedic surgeons and researchers alike have made the polymethylmethacrylate the scapegoat for many prosthetic failures.

The acrylic cement is cytotoxic in vitro. Its nanoparticles provoke degranulation of neutrophils. Surface and lacunar osteolysis is triggered in vitro by macrophages that have been isolated from subcutaneous polymethylmethacrylate-induced granulomas. The scene in the interfacial membranes of aseptically loosened cemented arthroplasties has been interpreted—wrongly as we are aware today–as being dominated by polymethylmethacrylate "pearls" and the macrophagic response thereto, sometimes to the exclusion of polyethylene debris. The polymethylmethacrylate has been, simplistically, blamed for prosthetic failure and aseptic loosening of cemented artificial joints has been labeled "cement disease." Surgeons' frustration with high rates of prosthetic failure has rationalized the quest for cementless fixation techniques. Basic scientists' encouraging findings and clinicians' early enthusiastic accounts notwithstanding, the analyses of retrieved uncemented artificial joints usually do not evince the anticipated widely spread fixation by bony ongrowth and ingrowth into the porous-coated surfaces but rather an expansive formation of an interfacial membrane. With introduction of the bone-bonding ceramics, orthopedic surgeons have anticipated earlier and stronger fixation of hydroxyapatite-coated surfaces by more extensive osseous ongrowth and ingrowth. The concept of cementless implant fixation was taken one step further. Alas, press-fit interference, porous coating and hydroxyapatite coating of the alloplastic components have not alleviated the loosening predicament. There is more to prosthetic failure than cursorily meets the eye: Indeed, in some surgeons' experience, the incidence of aseptic loosening of noncemented artificial joints is higher than that of conventional cemented arthroplasties. Mechanical overload rather than the discoursed "implant" incompatibilities may be the culprit. This has to be kept in mind when discussing the biocompatibility of the acrylic cement. Polymethylmethacrylate is not an adhesive; i.e., it does not glue opposing sidings to each other. Alloplastic components loosen not only, as ordinarily conceived, at the bone-cement interface, but also at the metal-cement interface. The limited strength of the drawn-out cement projections interdigitating with the rugged bony surface colludes with the limited strength of the bony trabeculae buttressing the implant. The fatigue failure of both the cement and osseous projections under cyclic loading leads to instability and interfacial motion of the cemented (and uncemented) arthroplasties alike [143–156].

The distinct material properties of the polymethylmethacrylate, that is, high fatigue rate, fracture propensity and physical as well as chemical deterioration account for the loss of mechanical strength with time. Its breakdown particles accumulate at the bone-implant interface. The resulting inflammatory-granulomatous reaction, like that due to any other particulates, is associated with production of mediatory substances, importantly with the macrophage-derived prostaglandin E_2, which in turn activate the osteoclasts to enhanced osteolysis. The notion that the formation of interfacial membranes and osteolytic cavities is brought about by a reaction to polymethylmethacrylate particles ("pearls") is supported, or so it seems, by anecdotal instances in which a response to acrylic products occurs in the absence of all other prosthetic detritus. If this were the case, it would be a lone occurrence. In certain published microphotographs, but few cement granulomas appear in a spacious sea of macrophages. These large and finely granular macrophages are laden with submicron- to micron-sized polyethylene debris, which fact is likely to be missed by microscopists unless they exploit oil red O staining and polarization microscopical techniques. There is no difference between types and amounts of the mediatory substances, i.e., collagenase, prostaglandin E_2, interleukin-1 and others, produced by inflammatory cells in the interfacial membranes of loosened cemented and uncemented arthroplasties. Being deposited interfacially, acrylic debris contributes to but, by itself, it hardly accounts for aseptic loosening. On the other hand, fragmentation of the cement mantle leads to prosthetic failure. The methylmethacrylate and *N*,*N*-dimethylel-*p*-toluidine (a catalyst added to the methylmethacrylate monomer) have experimentally been proven to be toxic to tissues.

In clinical practice, osseous, fat and hematopoietic tissues aside recently cemented components looks, histologically, healthy so that participation of the methylmethacrylate and *N,N*-dimethylel-*p*-toluidine in the loosening process of patients' prostheses is largely contentious. Heat produced during methylmethacrylate polymerization has been accused of destroying a 0.5-mm-deep zone of the bone. Coagulative bone necrosis takes place at 72°C, a temperature above that recorded at the surface of curing methylmethacrylate. Even within the small tubular bones, the maximum rise of 11.7°C during the polymerization process does not suffice to cause adverse thermal effects. In view of the cement's bad reputation, it has been recommended to exclude polymethylmethacrylate from the junction of host bone and allograft bone. When this point is explored experimentally by evaluating the effect of intramedullary cement on the incorporation of intercalary bone allografts, it turns out that the polymethylmethacrylate does not affect the healing processes. Cementation has evidently been falsely condemned for intraoperative bone death, prosthetic failure and manifold other mishaps [116,149,157–169].

Deductions apropos an intractable in vivo effect of polymethylmethacrylate on the attainment of a balanced host-implant relationship rest, at least partly, on the study of the periprosthetic tissues removed at revision operations for failed arthroplasties. Conjectures based on such studies fail to discriminate causes from effects. It is simplistic to argue that polymethylmethacrylate is bioincompatible because macrophagic reactions to cement and its fragments occur in the interfacial membrane of failed arthroplasties. In fact, polymethylmethacrylate in bulk form is without any cytotoxic potential in in vitro tests. Osteoblasts adhere to and proliferate on its surface, synthesizing plentiful collagen. On the other hand, macrophages exposed in vitro to pulverized acrylic cement release intermediates (including the bone-resorbing factors) and, along with other cell types, are lethally damaged. A bolus of polymethylmethacrylate is enclosed by a thin fibrous layer within the bone of animals, but particulate cement provokes a giant celled granulomatous response. On the practical side, the reaction to the cement particles is enhanced by the admixed barium sulfate (Fig. 4). When growth hormone is incorporated, the cement bulk is covered by osteoid rather than being encircled by fibrous tissue. Osteoid seams commonly abut on the cement mantle of osseointegrated components of patients' artificial joints (Fig. 5). This author has time and again observed roundish polymethylmethacrylate slabs incorporated in the periprosthetic fibrous tissues without an accompanying inflammatory-granulomatous response (Fig. 6). Yet, at other times, the acrylic chunks are surrounded by a two- to three-cells-thick, uninterrupted film of mono- and polykaryonic macrophages.

When the opportunity lends itself to assess samples of clinically sufficient artificial joints retrieved postmortem, the acrylic mantle is found to be separated from the bone by a thin, bland fibrous membrane or osseointegrated (Figs. 7, 8). Cement mantles of well-fixated tibial inserts often rest on fibrocartilage, a tissue which hardly withstands the rigor of a toxic neighbor. Observers may ponder whatever happened to the purportedly extensive necrosis of the bone and hematopoietic tissue contended to be omnipresent in samples of the cemented implants collected in the immediate postoperative period. All in all, the degree of tissular injury may be sparse. Cement in bulk form does not cause necrosis in the canine bone, it becomes incorporated and displays nothing more than the formation of a thin film of bland fibrous tissue. Whatever in vitro tests may imply, the flourishing osseous and hematopoietic tissues near the cemented components refute earlier claimed inferences about the cytotoxicity of polymethylmethacrylate. In appraising the body's faculty to "disallow" or "countenance" its presence, the cement's physique rather than its chemistry determines the reaction pattern. While methylmethacrylate hypersensitivity certainly inconveniences the occasional surgeon, scientifically well-founded basis for its role in the loosening event is scarce. Dissenting researchers point out that cellular responses to cement particles are more expansive and in vitro cytokine production of the lymphocytes in the presence of polymethylmethacrylate is higher in patients

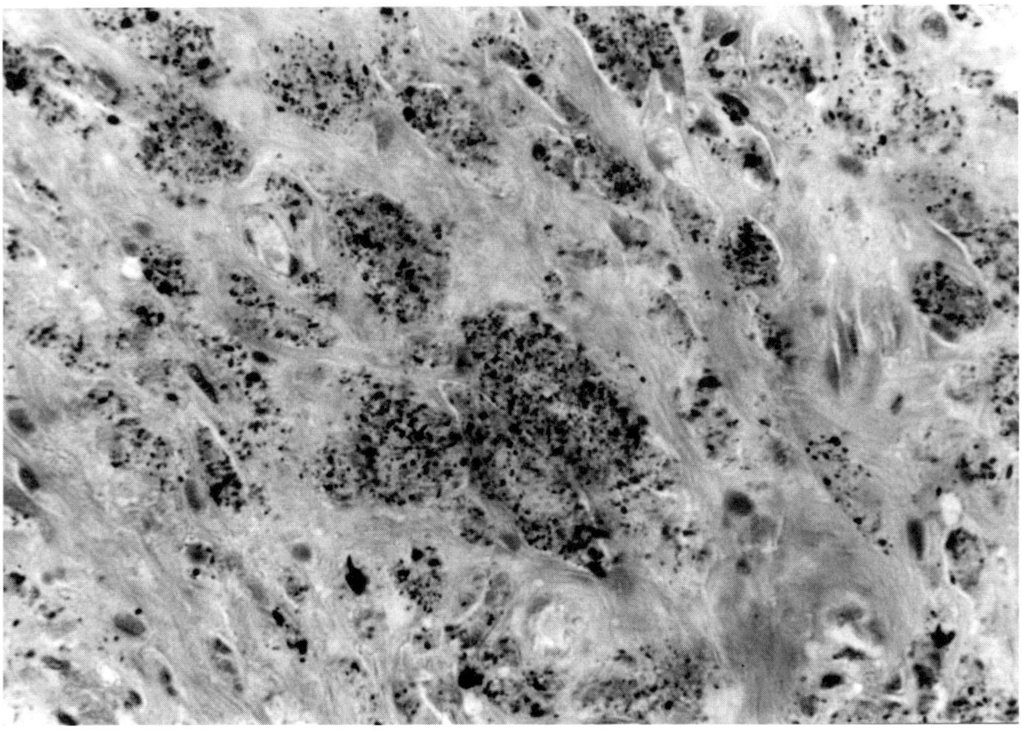

Figure 4 Barium-sulfate-particles-packed mono- and polykaryonic macrophages are strewn in the vicinity of acrylic chunks in the periprosthetic tissues of an aseptically loosened, cemented artificial joint. Hematoxylin and eosin. ×400.

whose artificial joints have been revised for loosening than for mechanical failure or sepsis. The quasi contradiction between the luxurious interfacial membrane near a loosened component, on the one hand, and osseointegrated cement or a thin fibrous interfacial membrane near a stable component, on the other hand, does require reappraisal of the controversial issue of the bioincompatibility of the acrylic cement [121,159,160,165–187].

Ling has made a forceful point in stating that what happens at the bone-implant interface is "substantially dependent on mechanical factors, given that the implant itself is basically nonreactive." When employed for purposes other than fixation of artificial joints, e.g., as a means of securing osteochondral fragments of the canine stifle, polymethylmethacrylate neither interrupts bone repair not does it induce an inflammatory response. Investigators reiterate the role of the "aggressive," synovial-like interfacial membrane, the hallmark of aseptically loosened artificial joints, in periprosthetic osteolysis. The membrane exhibits a roughly three-layered construct: Abutting on the cement, there are plump, palisading macrophages; the midzone is composed of prosthetic debris-induced giant cell granulomas and sheets of finely granulated macrophages; the usually mildly inflamed, predominately fibrous outer layer merges with the underlying remodeled bone. Shoji et al. appropriately maintain that given the granulomatous features of the membrane, there is nothing unique in its morphological nature. Such constructs have long been accepted to constitute the expression of the body's defense mechanisms against living and inanimate "invaders." That loosened cementless arthroplasties are likewise endowed with similar membranes denies the cement any esoterically exclusive role in the formation of the synovial-like interfacial membrane. In a comely experimental model of aseptic loosening,

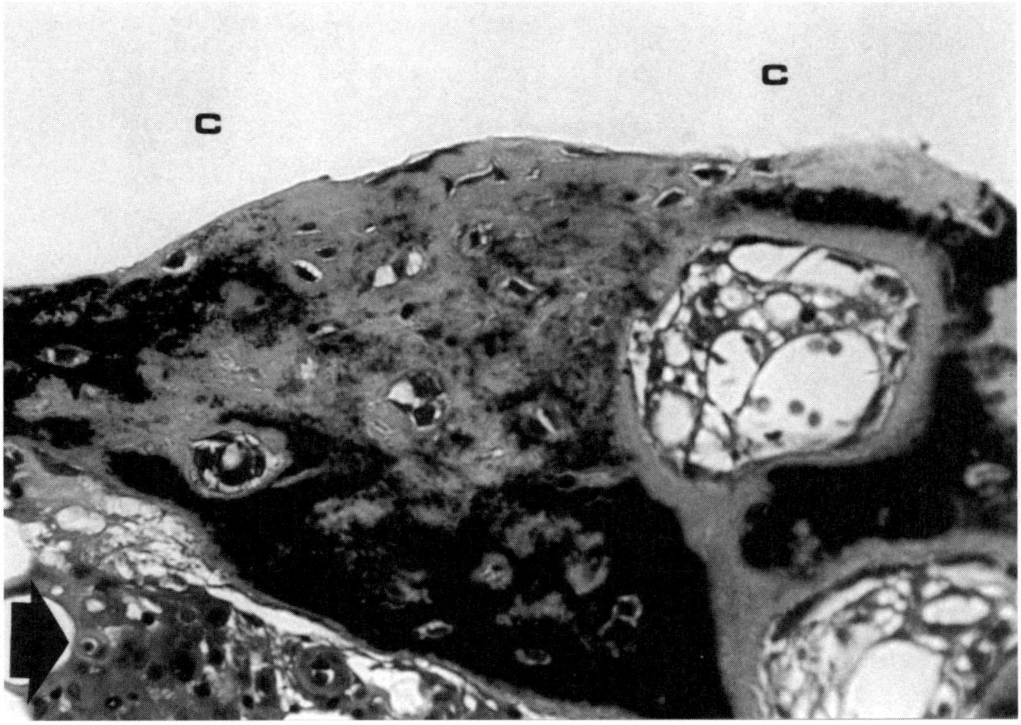

Figure 5 Sizable osteoid seams of bone abutting on an osseointegrated cement mantle of the femoral stem of a Charnley total hip arthroplasty. Notwithstanding the proximity of cement (C) on one side of the osseous trabeculae and polyethylene-laden macrophages (arrow) on their other side, there is ongoing active mineralization of the osteoid. Von Kossa stain with McNeal counterstain. ×250.

it has been evinced that implant instability rather than its mode of fixation, cemented versus uncemented, correlates with the intensity of the inflammatory response and the amounts of released mediators of inflammation [139,188–197].

The early postoperative interlock of the cement with the bony trabeculae persists around successful artificial joints but is lost in the wake of poor mechanical (rather than biological) circumstances. Stable cemented components are well accepted by their osseous environs and are, at least, segmentally osseointegrated, whereas loose ones are sequestered from their bony bed by the synovial-like interfacial membrane. Given that formation of the latter is independent of the cementation per se but dependent on interfacial accumulation of wear particles, prosthetic failure may, if anything, be viewed as a "small particles disease." Its typical expression is the late failure of the acetabular socket of hip arthroplasties, where loosening begins circumferentially at intra-articular margins, the polyethylene particles-driven granulomatous membrane interposing itself in between the bone and the cement and progressing with time toward the dome of the implant [193,198,199].

To conclude, polymethylmethacrylate as such is not pathogenic. In the context of aseptic loosening of total arthroplasties, the epithet "cement disease" is a misnomer. The literature's testimony apropos the bioincompatibility of acrylic cement is pathetic. Alike other species of natural and man-made foreign bodies, acrylic particles provoke a granulomatous response. They partake in the induction of the "aggressive" interfacial membrane. Yet, cement granulomas in periprosthetic tissues of failed artificial joints are quantitatively less conspicuous than

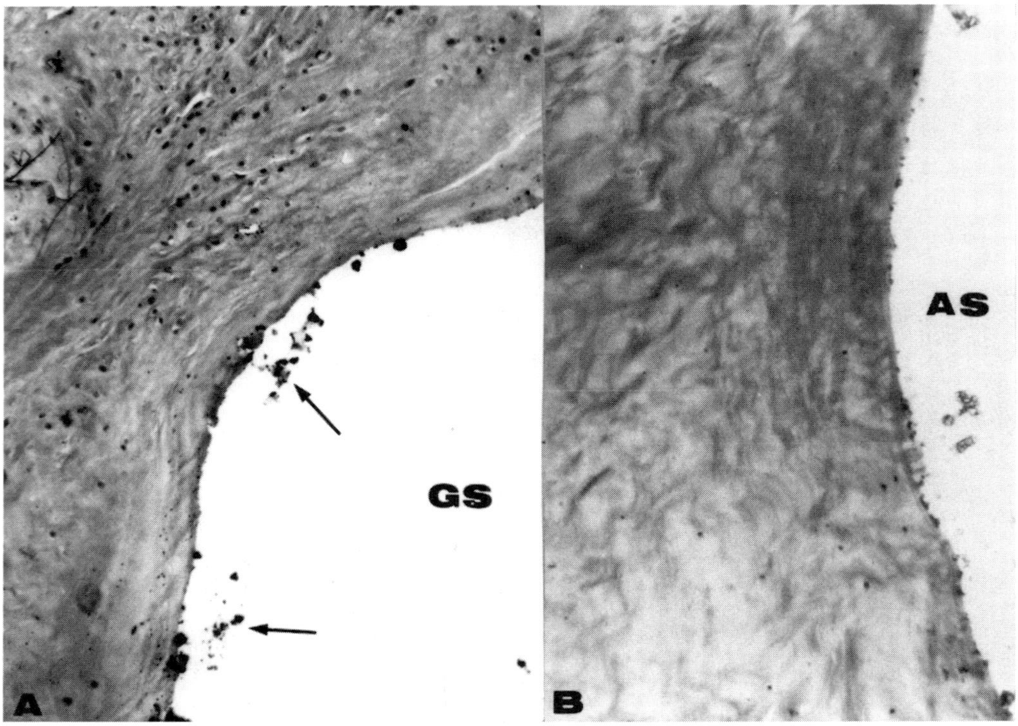

Figure 6 (A) Globular cement slab (GS) is incorporated in a hypocellular and densely textured fibrous tissue without an interposed layer of cellular response. This and many other similar slabs were present in the pseudocapsule of an artificial hip joint which loosened almost 20 years postoperatively. Some $BaSO_4$ granules (arrows) persist in the "cavity" appearing after dissolution of the cement during preparation of the section. Goldner stain. ×80. (B) The bonding of the acrylic cement slab (AS) with the fibrous tissue of the pseudocapsule without mediation of a reactive macrophagic film bears witness to the biocompatibility of the polymethylmethacrylate. Hematoxylin and eosin. ×320.

are the polyethylene or metallic debris-induced granulomas (Figs. 3, 9). The currently in vogue paradigm attributes prosthetic failure to a multifactorial mechanism [200,201]. The obstacles to successful prosthetic joint replacements are primarily mechanical factors. The processes usually responsible for the loosening process are initiated and aggravated by friction at the joint articulation and micromotion at the bone-implant interfaces. As the case may be, the breakdown products of any other alloplastic component, such as polyacetal or hydroxyapatite, may supplant or may be added to the biomaterials specified in the scheme summarizing the loosening paradigm (Fig. 10), which specifically pertains to the commonly implanted Charnley prosthesis.

C. Other Nonresorbable Polymers

Reconstruction of the anterior cruciate ligament has been attempted with a variety of polymeric devices such as polyethylene, polytetrafluoroethylene, polypropylene and terylene. Whether used as a substitute, a tissue ingrowth scaffold or an augmentation device for allogeneic or autogeneic grafts, the outcome in the long run is disappointing and the peril of debilitating detritic synovitis is ever present [202–207]. An anatomically proper and functionally effective

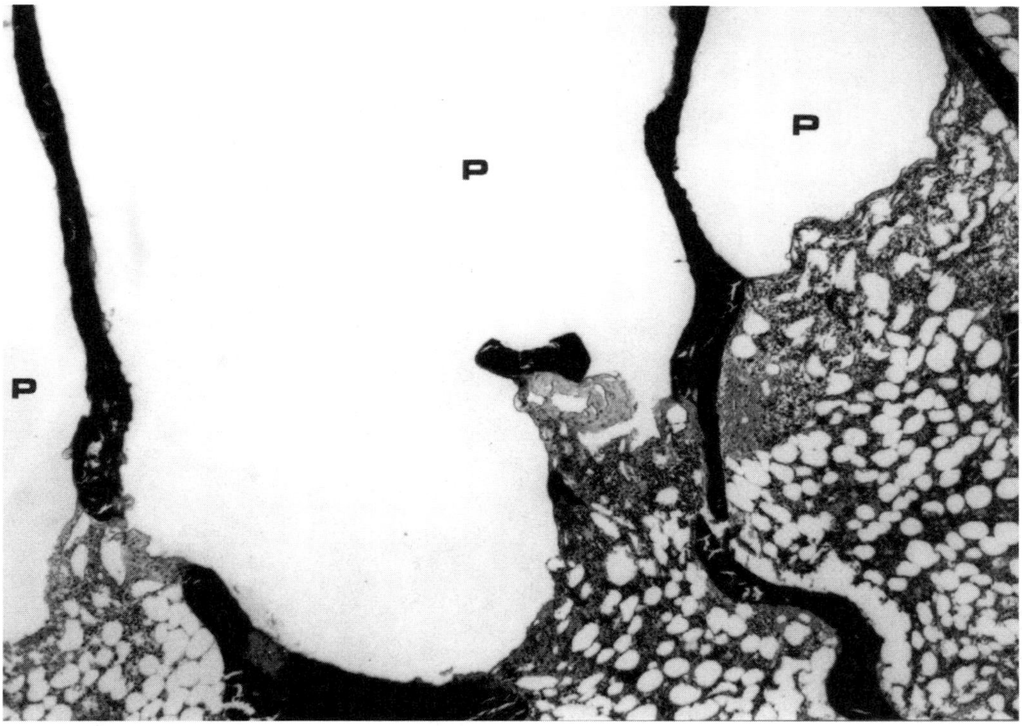

Figure 7 Excellent bone-cement interdigitation at the acetabular socket of a patient who died 16 years postoperatively with a functionally efficacious, cemented, total hip arthroplasty. The projections of the acrylic cement mantle (P) are osseointegrated; that is, they adhere to the bone without mediation of a soft tissular layer. Goldner stain. ×30.

load-transferring implant-bone joining does not ensue postoperatively.* A fibrous interfacial membrane separates the bone from the intraosseous segments of the devices. It is not surprising that a majority of replacements fail at the tibial or femoral insertion site; only a minority display midsubstance rupture. The polymeric implants are conducive to ingrowth and ongrowth of tissue, although it is erroneous to speak of the development of a "neoligament" in the absence of axial orientation of the fibroblasts and collagen fibers within the devices' interstices. Furthermore, the implants are subject to attrition by abrasion and rupture. Particles-induced granulomas are omnipresent and, being responsible for a painful detritic synovitis, may seal the fate of the reconstruction. Replacement of the canine anterior cruciate ligament by tendon autografts is promptly ingrown by well-vascularized tissue. On the other hand, tissular regeneration is defective and revascularization is delayed within tendon autograft replacements augmented with polymeric filaments. After the third postoperative month, implant-augmented grafts have no mechanical advantage over conventional autografts [208–213].

Load-resistance-conferring interdigitation of the polymeric filaments and collagen fibers alternates with granulation tissue in organized polytetrafluoroethylene implants. Bulk polytetra-

*The functional significance of the enthesis in the context of physiological load transfer from a ligament or tendon to the bone is discussed at some length in the section on collagen-based implants.

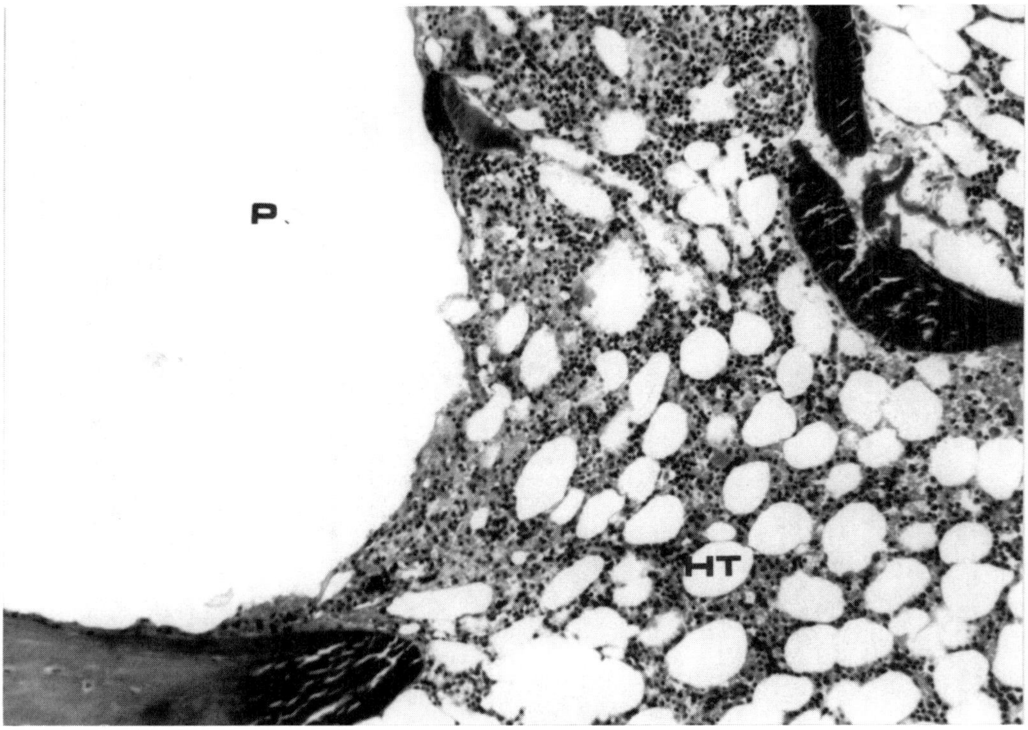

Figure 8 The adipocytes and the normally active hematopoietic tissue (HT) present alongside, and also abutting on, a projection of the acrylic mantle (P) of an acetabular socket of a cemented total hip joint replacement attest to the biocompatibility of the polymethylmethacrylate. Goldner stain. ×80.

fluoroethylene being relatively inert, at least in comparison to most other polymers, the inflammatory response evidently stems from an unfavorable mechanical situation. The microstructure plays a pivotal role in the cell-biomaterial relationships. Thus, fibroblasts adhere to polymeric filaments with striated surfaces. Rough surfaces, on the other hand, provoke a macrophagic response [214–216].

Polymeric filamentous prostheses are prone to mechanically induced adversities in animals' stifles. Half of all terylene implants replacing the anterior cruciate ligament of sheep rupture within a year of operation, whereas none of those substituting medial collateral ligaments do. Reportedly, well aligned, mainly type I collagen fibers admixed with elastic fibers are produced by the fibroblasts growing into the terylene prostheses; these findings are exceptional and should be confirmed. More often than not, normally structured elastic fibers are absent, or at best scarce, in reconstructed soft tissues, the nature of the fibrous tissue shaping in the interstices of polymeric filamentous devices is more akin to a scar tissue than to the native connective tissue [217,218].

Surgeons have prudently abandoned these reconstructive modalities in favor of tendon grafting, not because of biocompatibility predicaments but because of technical hurdles which are presently beyond an expedient solution. The polymeric filamentous implants are subjected to tension and bending, abrasion at bony surfaces, as well as interfilamentous friction. The prospect of preventing generation of wear debris and the ensuing detritic synovitis is an unrealistic goal. Particles, matching in size and shape wear products retrieved from patients' tissues, were produced by grinding terylene, polypropylene, polytetrafluoroethylene, carbon and alloge-

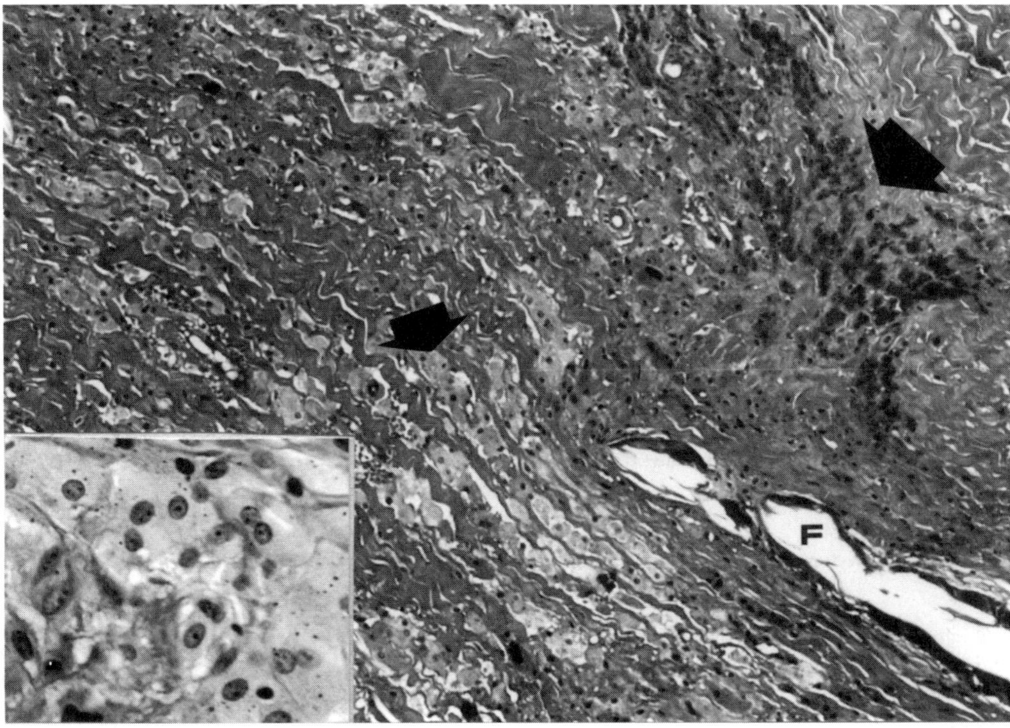

Figure 9 The distinctive perspective of the interfacial membranes of the aseptically loosened arthroplasties, be they cemented or cementless, is the macrophagic reaction to submicron- to micron-sized prosthetic debris. This field is representative of the membranes of loosened, cemented tibial components. In this field, there is no evidence of a tissular response to acrylic detritus, which is generally sparsely present. Polyethylene-particles-laden foamy macrophage (small arrow) and metallic-particles-packed macrophages (large arrow) are scattered within a background of a densely textured fibrous tissue. A few, large polyethylene fibers (F), which are hardly discernible unless viewed under polarized light or in oil red O-stained sections, are surrounded by mono- and polykaryonic macrophages. (Inset) At higher magnification, the foamy macrophages are seen to contain some metallic particles as well. Goldner stain. ×80 and (inset) ×320.

neic or xenogeneic tendon. These particles activated synoviocytes in vitro and their injection incited a "small particles disease" in vivo, the severity of which was dose dependent. Patients and their physicians alike perceive operative interventions as unwarranted in face of the prospect that osteoarthritis or a particles-migration-related lymphadenitis may evolve in the wake of replacement of an anterior cruciate ligament by one polymeric prosthesis or another. The history of the quest for ideal intra-articular ligament substitutes once more attests to the predominance of wear debris over the biomaterials' chemical symbols in settling the fortunes of reconstructive surgery [219,220].

The regenerative potential of hyaline cartilage is poor. As a rule, diseased or lost parts of the cartilage are not replenished. The literature is replete with descriptions on endeavors to promote repair of cartilage defects. Placement of polytetrafluoroethylene felts into osteochondral defects of a rabbit's stifle causes bone to amply grow from the base into the implants' porosities, but a subchondral bone plate does not reform; the alloplastic matrix is not incorporated laterally; ingrowing tissue rarely extends centrally; chondroid differentiation is limited

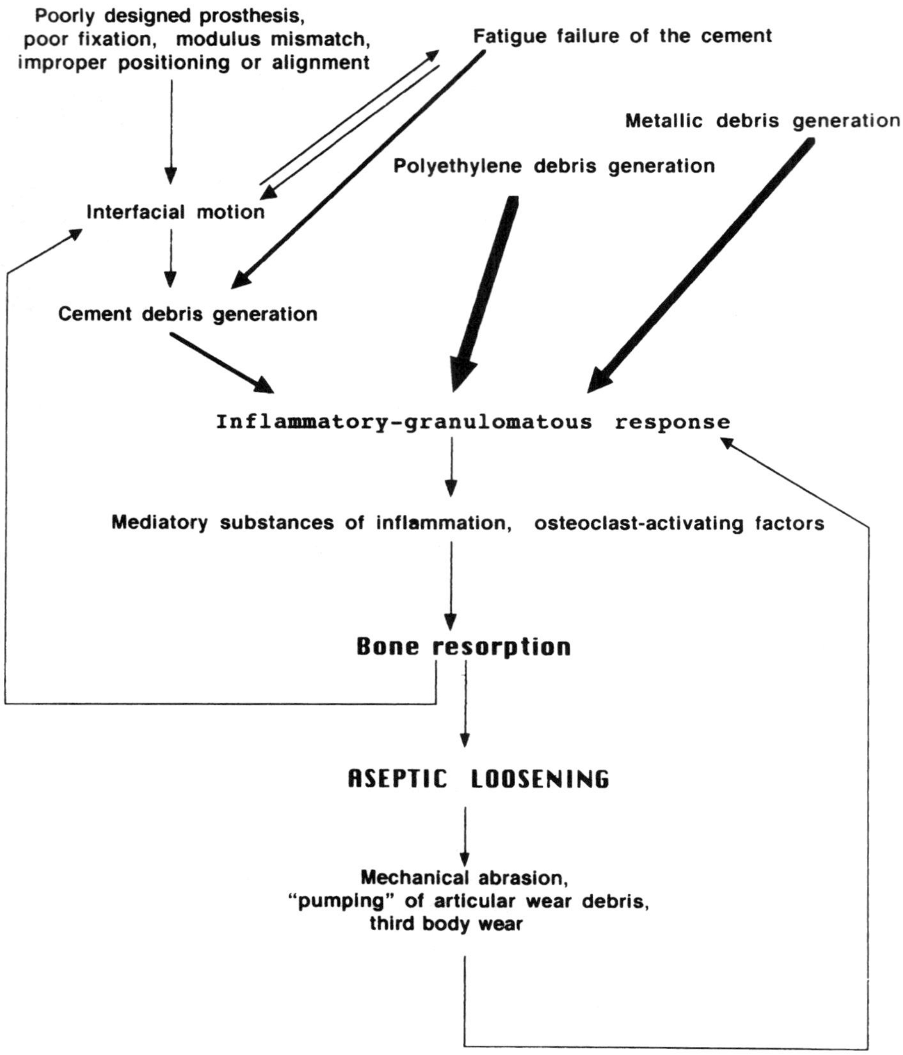

Figure 10 The today on vogue paradigm which rationalizes the cardinal role of the mechanical adversities in aseptic loosening of total arthroplasties. There is no place in this scheme for bioincompatibility of the alloplastic components of the artificial joints. Generation of submicron- to micron-sized debris of the components is the outcome of the harsh mechanical and physical circumstances at the host-implant interfaces and the articulating surfaces. Prosthetic debris induces inflammatory and granulomatous responses. The thereby stimulated macrophages and other leukocytes, as the case may be, release mediatory substances, prime among which are osteoclast-activating factors. Thereupon, bone resorption leads to loss of anchorage of the components in their osseous bed and, thus, to prosthetic failure.

and imperfect; the felts are regularly exposed to the joint cavity; and detritic synovitis is common. The researchers' claim that such synthetic felts "can be used to fill full-thickness cartilage defects" is remarkable [221], skeptics arguing that such reconstructive surgery is a recipe for development of osteoarthritis. This is an illustrative example of how "pernicious" effects of a biomaterial may be the outcome of an unsuitable application rather than biocompatibility issues.

A variety of synthetic, biostable polymeric devices serves the augmentation of soft tissues. Bundled filamentous implants are well accepted by both subcutaneous and muscular tissues (Figs. 14 and 15). Mono- and polykaryonic macrophages enclose on the bundles and filaments (Fig. 16). There is no critical disintegration of these implants such that, despite some interfacial movements, injurious granulomatous reactions do not ensue. With time, the bundles become encircled and separated from one another by ample fibrous tissue, which is initially hypercellular and loosely textured and later matures into a densely packed, scar-like collagenous tissue (Fig. 17). Single filaments and fragments thereof detach from the bundles and are present amidst the newly formed collagen fibers without an accompanying crucial inflammatory response. The mild foreign body reaction and brisk fibrous tissue formation conform with the conjecture that in vitro cytocompatible alloplastic materials, such as polytetrafluoroethylene and polyethylene terephthalate, ensure a functional success of surgical interventions where a "small particles disease" is evaded because of existent biomechanical circumstances [222,223].

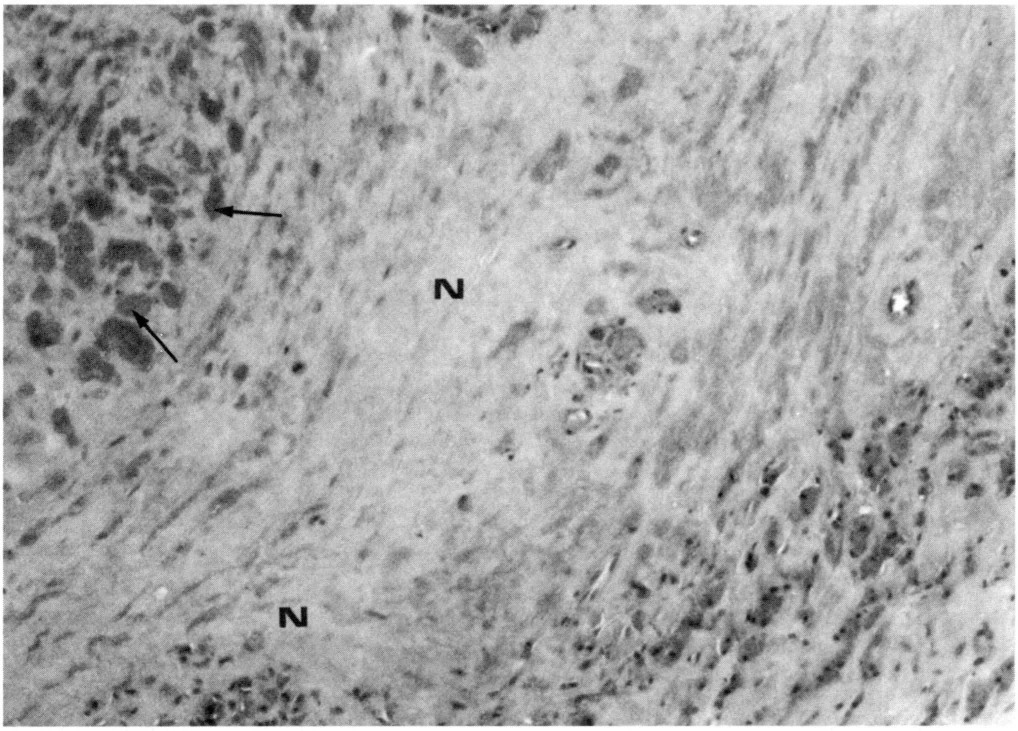

Figure 11 A large focus of coagulation necrosis (N) in the midst of a sheet of metallic and polymeric detritus-laden macrophages. The necrotizing process involves matrical fibrous tissue and macrophages, which appear as "ghost-like" cells (arrows, the viable macrophages possess well-stained nuclei). This focus, the age of which is indeterminate, discloses no sign of undergoing organization by ingrowing granulation tissue. Von Kossa stain with McNeal counterstain. ×80.

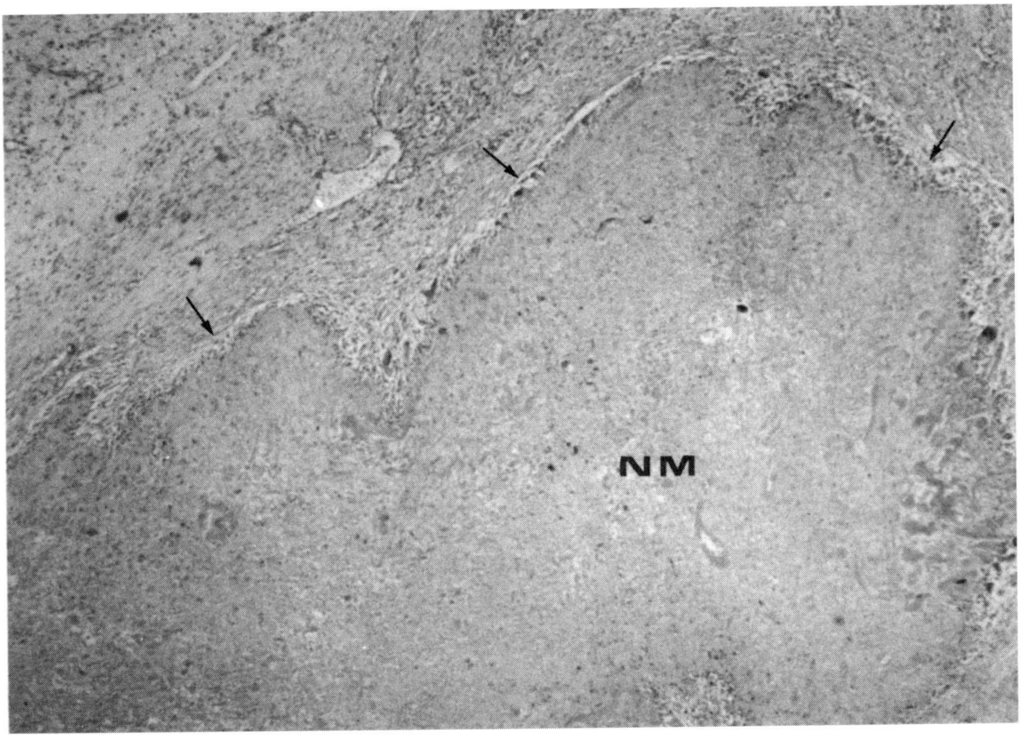

Figure 12 Large giant-celled necrotizing granuloma in the interfacial membrane of a cemented artificial joint. The central necrotic mass (NM) of the granuloma consists of an eosinophilic, amorphous and granular material and is surrounded by a giant-celled granulomatous fence (arrows). Hematoxylin and eosin. ×30.

Because of its good tissue compatibility, hemocompatibility, pliability and material characteristics, polytetrafluoroethylene has explicit advantages over other synthetic as well as natural materials in cardiovascular surgery. The histological patterns of incorporation of polytetrafluoroethylene implants in rats' adipose tissue are illuminating in that they sketch the dependence of tissular reactivities on parameters distinct from the chemical composition of the device. Polytetrafluoroethylene patches with an internodal distance of 100 μm provoke a stronger inflammatory response and are less conducive to vasoneogenesis than polytetrafluoroethylene patches with an internodal distance 30 μm. In addition, both kinds of implants cause a much greater fibroblastic response and stimulate less growth of blood vessels in the subcutaneous than epididymal fat tissue. Polytetrafluoroethylene patches with an internodal distance of 60 μm stimulate the least fibrous tissue and the most rapid formation of blood vessels at both locales. These observations are instructive when one comes to generalize and discover the ruling principles in implantology. They "indicate that both the polymer structure and the site of polymer implantation have dramatic effects on the relative nonbiocompatibility." The authors of this commendable research were, however, not evaluating the biocompatibility of a material used in vascular reconstruction but were scoring the degrees to which physical, as opposed to chemical, characteristics determine the host response to a polymeric device at different body sites [224].

Polyurethane and polypropylene are also widely used in vascular surgery. In a canine model, myofibroblasts and macrophages colonize the inner siding of the prostheses within

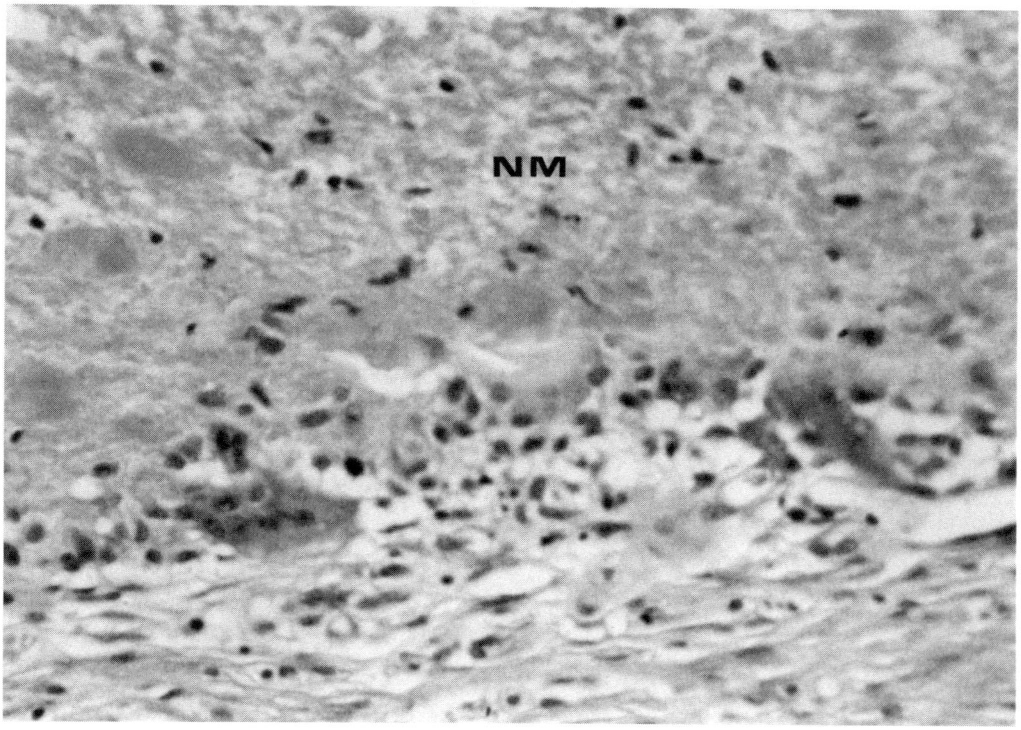

Figure 13 Higher magnification of Fig. 12 showing the mono- and polykaryonic macrophages, intermixed with lymphocytes, at the boundary of the necrotic mass (NM) of the granuloma with the surrounding fibrous tissue. Hematoxylin and eosin. ×250.

two weeks, confluent endothelialization is achieved within one month and a well-collagenized pseudointima is complete within four months. Fibrous tissue grows into the porosities up to a degree. Anchorage of such vascular grafts to adjoining mural structures is segmentally optimal, but other interfacial segments may disclose an inflammatory infiltration or a giant-celled granulomatous response. Some prostheses undergo focal mineralization. In addition to thrombotic or tissular occlusion of the graft's lumen, long-term failure is often related to aneurysmal dilatation due to an inflammatory-granulomatous reaction-induced weakening of the wall of the reconstructed blood vessel. In patients with limb ischemia, placement of a vein autograft provides a two-year salvage rate of almost 90% compared with a less than 60% salvage rate in patients with a polytetrafluoroethylene prosthesis. The use of polymeric grafts in vascular surgery is yet problematic. The gamut of applications of nonresorbable polymers in reconstructive surgery has not been exhaustively covered herein. To name but two examples, polytetrafluoroethylene and polypropylene devices have been employed with success in the construction of a bilioenteric bypass and in the repair of hernias. Patients undergoing repair of abdominal incisional hernias with a mono- or double-filamented polypropylene mesh or expanded polytetrafluoroethylene patch are usually cured, while fistula formation, infections or a recurrent hernia complicates the postoperative course of many patients whose hernia has been fixed with a multifilamented polyester mesh. Techniques of placement having no influence on the outcome, polyester cannot be recommended for the repair of incisional hernias. Polyester manifests no cytotoxicity and only weak cell activation. It well supports the growth of the sensitive endothe-

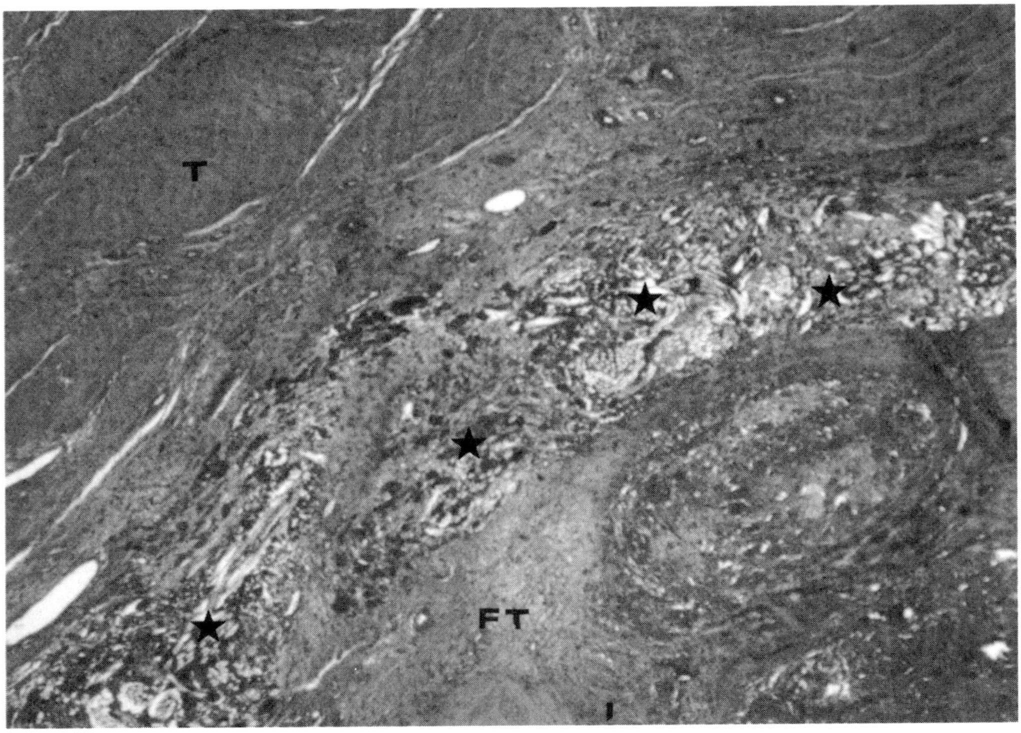

Figure 14 Bundled filamentous tetrafluoroethylene implant (asterisks) inserted aside a tendon (T) has induced a limited granulomatous response and stimulated formation of a glut fibrous tissue (FT), as expected at the time of the reconstructive surgery. Von Kossa stain with McNeal counterstain. ×30.

lial cells. Despite its otherwise laudable biocompatibility, polyester is evidently unsuitable for one surgical intent or another [225–234].

Nonresorbable polymers also support the surgical treatment of eye disorders, for instance, operations for pathological myopia. There is no intention to delve herein into this discipline of the ophthalmologists save to say that several, coated and uncoated, porous and solid materials have been evaluated. Taking the fibrous capsule thickness (the smaller the better), the amount of deposited collagen (the more the better) and the number of macrophages (the lower the better) surrounding the implants as a measure of biocompatibility, polytetrafluoroethylene and polyurethane have been testified to be the most suitable compounds for operations on the sclera [235].

IV. METALS

The biocompatibility of the metals is controlled by chemical, actually, by electrochemical interactions which result in release of metallic ions. The biocompatibility is a function of the toxic effects of those ions gaining access to the tissues. The interaction is adverse when corrosion products injure the immediate environment, distant organs or both. Hence, the biocompatibility of metals correlates with their corrosion resistance. Because of the spontaneous formation of an inert oxide layer on their surface, titanium and titanium alloys are currently the most corrosion-resistant metals in use in medical practice. Having analyzed a large body of toxicological data, Wapner has concluded that there is no unequivocal clinical evidence to ascribe patients' systemic ailments to the effects of metal ions leaked from the implants serving their useful purpose for many years [236–240].

Heavy metals, of which cobalt, chromium and titanium are of particular concern in the context of artificial joints, have been found to exert toxic effects in tissue cultures. Also, they are carcinogenic in experimental animals. The outcome of in vitro cytocompatibility assays relates to the type of the metal tested but also to its concentration and the state of its surface. Cellular functions, for instance, are less impaired by a smooth-surfaced than by a rough-surfaced cobalt-chromium compound. As few as 8×10^{-6} v/v of cobalt particles are toxic for cultured bovine chondrocytes. On the other hand, as many as 8×10^{-1} v/v of chromium or titanium particles are required to affect the cells' metabolic activity. Signs of cytotoxicity are apparent in cultured bone marrow stromal cells exposed to metal ions at concentrations of 50^{-5} to 50^{-8}. Investigators disagree with each other when it comes to grade metals' cytotoxicity. Some argue that chromium is highly toxic, cobalt, molybdenum, iron and nickel are moderately toxic and, finally, titanium, aluminum, vanadium and magnesium are minimally toxic. Others assert that cobalt is most poorly tolerated: At concentrations of 0.01–0.1 mg/ml, the degree to which particulate cobalt damages osteoblast-like cells in vitro and inhibits synthesis of collagen type I, osteocalcin and alkaline phosphatase is not matched by chromium and cobalt-chromium alloy; at a concentration of 1 mg/ml, chromium and cobalt-chromium alloy exert an inhibitory effect as well. Osteogenic cells are irreversibly injured after exposure times as short as 3 to 6 h. One claim to clinical relevance of these findings is that the concentrations of the in vitro tested metal ions and particles equal those measured in patients' periprosthetic tissues. It awaits to be established whether metallic particles and corrosion products modulate osteoblastic activity not only in the artificial culture medium but also at the implant-bone interface. In addition, the role of a cellular immunity in patients with metallic implants, in general, and with an artificial joint, in particular, deserves prospective studies. Allegedly, the cellular response to metallic particles is more expansive and in vitro cytokine production of the lymphocytes in the presence of the alloy is higher in patients whose artificial joints have been revised for loosening than for mechanical failure or sepsis. Except for aluminum, which causes erosions of osseous tissues, other metals are well tolerated in bulk form in the bony environment of man and experimental animals. Bone easily grows into the porosities of a titanium implant but not into those of cobalt-chromium implants. It may be that the osteoprogenitor cells discern in vivo material differences which have as yet eluded the researchers but which parallel the in vitro observations. Long-term in vitro exposure of osteogenic cells are desirable to pursue the argument that ions released from metallic surfaces play a role in osteolysis. After a four-week exposure period to ions released from titanium alloys (but not to ions freed from cobalt alloys), the cultures of osteogenic cells show suppression of osteocalcin secretion and matrix mineralization. The validity of the in vitro assays may go unchallenged but not the notion that this data indicates that periprosthetic osteolysis is related to interference with osteoblastic differentiation by ions released from titanium-based components of an artificial joint. The uptake of ^{99m}Tc and ^{32}P by the newly formed bone early after ablation of the medullary cavity of rats is decreased nearby titanium or stainless steel implants. Whatever drawbacks titanium and its alloys may have in clinical practice, impaired osteogenesis does not seem to be one of them. The inconsistencies between the results of the experimental studies and the excellent clinical experience with metal-based implants remain to be explained [186,238–250].

Results obtained during in vitro testing of neodymium-iron-boron magnets serve to alert materials scientists to the vagaries of their calling. Parylene-coated demagnetized magnets are cytotoxic for human but not murine fibroblasts. Uncoated magnetized, uncoated demagnetized and parylene-coated magnetized magnets are cytotoxic for both human and murine fibroblasts.

Lastly, the corrosion products, and not the bulk material, prove to be cytotoxic. Are the neodymium-iron-boron magnets biocompatible or not? Is the parylene coating cytotoxic? What about the effects of magnetism? There are references to the untoward effects of the magnetizing force on cells [251].

Many studies have established the locally benign traits of stainless steel, cobalt, titanium and their alloys. Under perfect conditions, metallic screws within the osseous environment are enclosed by an hypocellular, densely textured fibrous tissue without an accompanying inflammatory reaction. An acellular, thin, proteinaceous film is often seen to abut on the metallic surface without the mediation of macrophages (Fig. 18). To cite another example, stainless steel compression plates, used for fixation of bone fractures, are generally isolated from the bone by bland interfacial membranes composed of well-aligned collagenous tissue (Fig. 19). In other patients, however, necrotic material, nearby the plate, overlays a nonspecific granulation tissue (Fig. 20). In such interfacial membranes, variously sized, partly large, metallic particles are often seen within and without mono- and polykaryonic macrophages (Fig. 21). Variations on this theme of tissular damage associated with the generation of breakdown products of an implant are repeatedly referred to in this chapter.

The clinical consequences of raised metal levels in the serum of patients with a loosened artificial joint cannot be discounted off hand. Rats with permanent intramuscular implants of chromium-cobalt-molybdenum fail to gain weight; when compared to a control group, more animals with an implant die or develop interstitial pneumonitis. In other authors' experience,

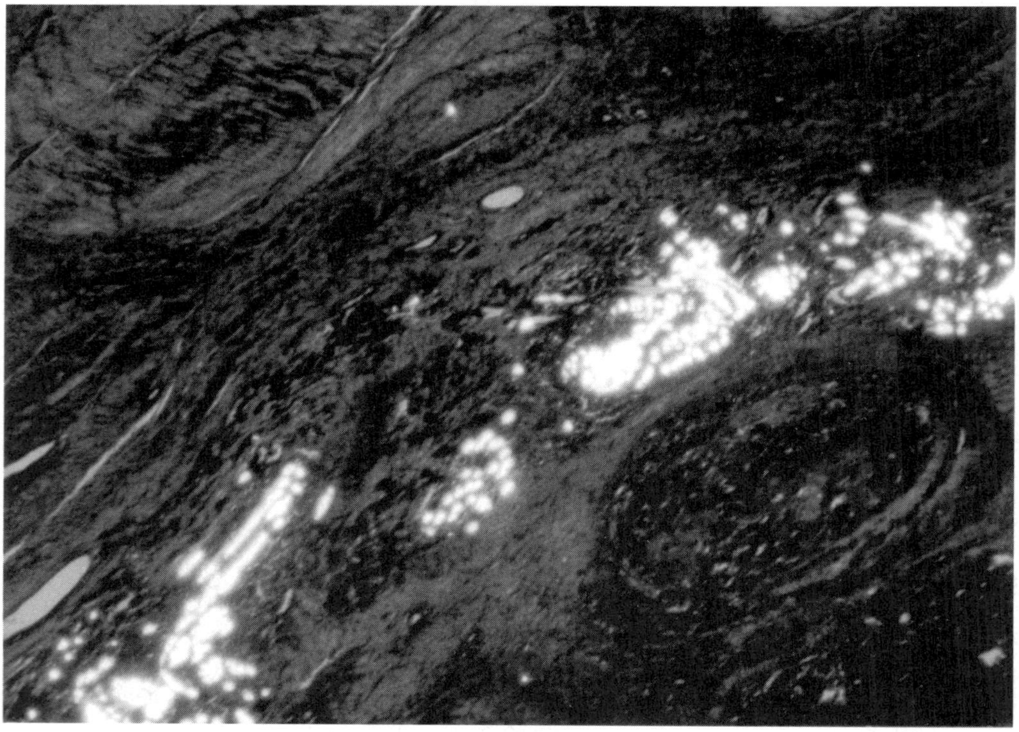

Figure 15 Polarization optical details of the field illustrated in Fig. 14. An injurious granulomatous reaction has not ensued since tetrafluoroethylene filaments barely disintegrate when implanted into the soft tissues. Collagen fibers have been produced by the stimulated fibroblasts in abundance. Von Kossa stain with McNeal counterstain, photographed under polarized light. ×30.

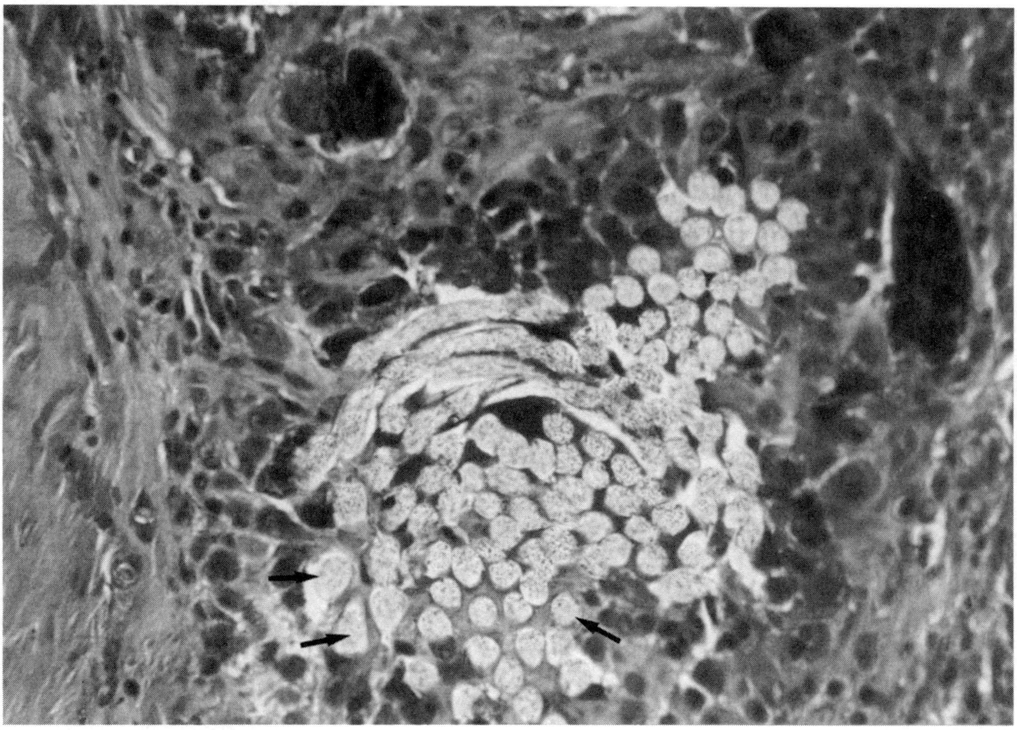

Figure 16 Mono- and polykaryonic macrophages have encircled the bundled tetrafluoroethylene filaments (arrows) but have not penetrated in between individual fibrils although this augmentation device was retrieved many years postoperatively. Von Kossa stain with McNeal counterstain. ×320.

the health of the animals does not fail after the injection of diverse metals in particulate form into different body sites, though an inflammatory-granulomatous reaction ensues locally. These reactions are brisk for iron, medium for cobalt and modest for titanium. It is intriguing and yet unexplained why disposal of the metallic granules from their place of deposition decreases in the same order. With the exception of storage of the cleared metallic particles in the lymph nodes, other organs are unaffected. The same is premised to hold true for the metallic particles deposited in the periprosthetic tissues of patients' arthroplasties [252–256].

The concentration of titanium in the serum and urine of rabbits with intraosseous titanium fiber felts is not raised in comparison with the level in control animals. In the absence of wear of the implant, the metal is evidently released in small amounts and is preferentially retained in the local tissues. It is assumed that raised titanium, and other metal, concentrations in patients' biological fluids result from mechanically induced or assisted release phenomena [257].

Titanium has for some time been the orthopedic surgeons' favorite selection in view of its superior corrosion resistance and its intimate contact with bone. Under optimal experimental conditions, chromium-alloy components are segmentally osseointegrated as well. In the bones of miniature pigs, rough-surfaced titanium cylinders are 60% to 70% osseointegrated, whereas smooth-surfaced cylinders are 75% to 80% apposed by fibrous tissue. Osseointegration is contingent on lack of interfacial motion. Experimentally, just 20 diurnal cycles of interfacial movements (of 0.5-mm amplitude) encourage fibrous tissue rather than bony ongrowth onto titanium implants. Osseointegration is the ultimate affirmation of the biocompatibility of a material. It is a light microscopical image. Light microscopically osseointegrated titanium screws do not,

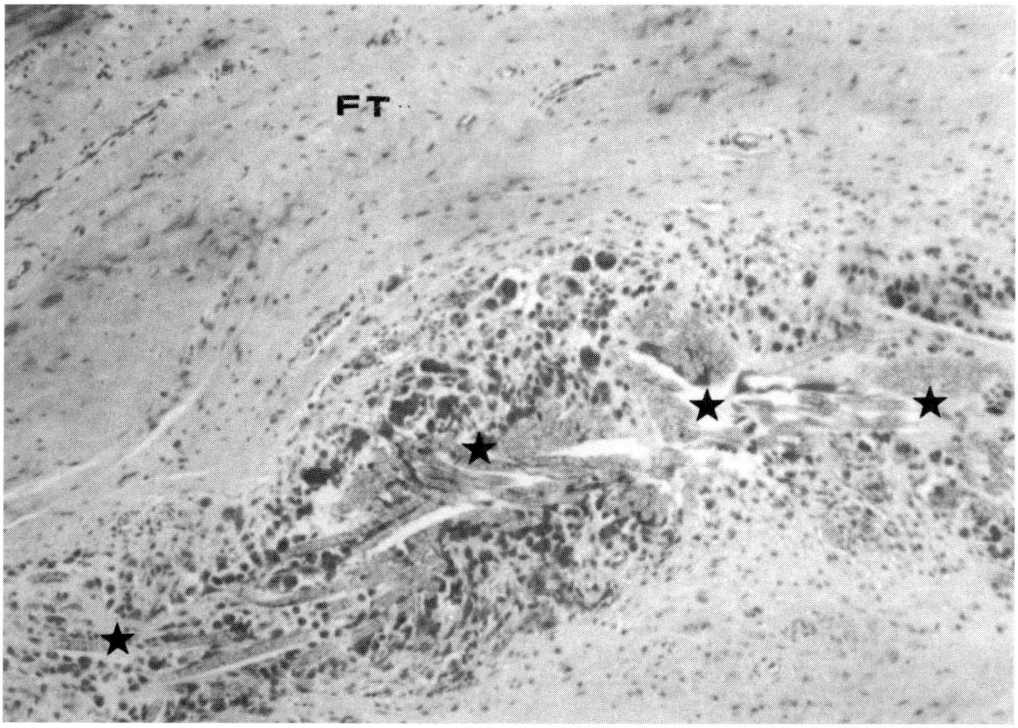

Figure 17 The granulomatous reaction is limited to the vicinity of the bundles of the tetrafluoroethylene filaments (asterisks). Well-aligned collagen fibers of the newly formed fibrous tissue (FT) abound around the augmentation device. Goldner stain. ×80.

ultrastructurally, abut directly on the bone but are rather separated from it by a 100- to 200-nm-wide amorphous film and a 100-nm-wide lamina limitans-like layer [258–265].

The behavior of undifferentiated mesenchymal cells encountering metallic surfaces is affected by physicochemical interactions of the outermost metal oxide layer with the nearby matrical extracellular milieu. In this environs, the proteoglycans and collagen fibers play a major role. Surface oxidation of metals within the tissues is an ongoing process. Calcium and phosphate precipitate within the oxide layer, thus controlling the differentiation of the osteoprogenitor cells, which in turn mature into osteoblasts and lay down a bony matrix onto the oxidized surfaces. Titanium and other metals delay maturation of the extracellular matrix vesicles (the submicroscopic organelles involved in the calcification of osseous matrices). The oxides, leached from the 5- to 100-Å-thick oxide film coating the metallic implants, apparently exert untoward effects on the cells engaged in the synthesis of the matrix vesicles. Yet, a concurrent increase in the production of the vesicles enhances the primary mineralization of the newly formed osteoid. Compared with pure titanium, the inferior biocompatibility of its alloys is possibly related to the release of aluminum ions, which inhibit mineralization. It is not surprising that intense osteolysis in the face of scant osteogenesis occurs around an intraosseous lesion filled up with titanium alloy debris-laden macrophages. Materials scientists have of late explored options of preimplantation modification of the surface properties, e.g., by heating or sol-gel coating, in order to promote not only bone-metal apposition but also bone-bonding. Within the soft tissues, titanium and titanium alloy implants are enclosed by a thin and ever-dwindling fluid space, which contains a few inflammatory cells, and an ever-regress-

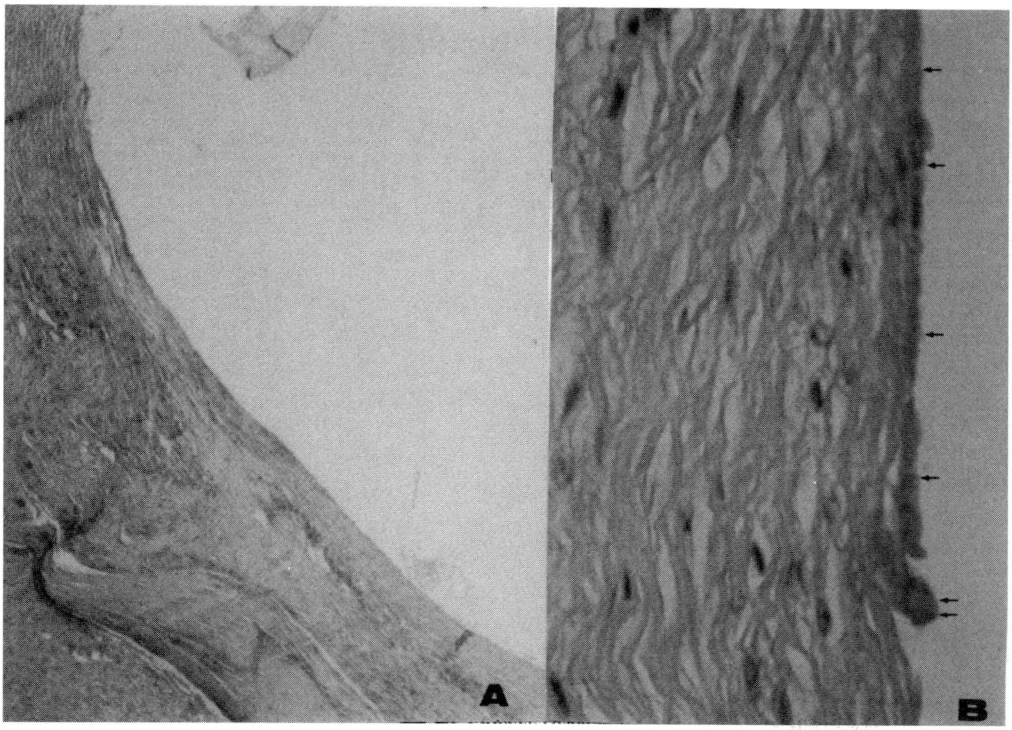

Figure 18 (A) Encapsulation of a metallic screw by densely textured and hypocellular fibrous tissue without an accompanying inflammatory response. Hematoxylin and eosin. ×40. (B) At higher magnification, a fine film of an amorphous material (arrows) is seen to abut on the surface of the screw. There are no CD68-immunoreactive macrophages at the metal-tissue junction, hence, this screw is solely enclosed by fibrocytes and collagen fibers. Hematoxylin and eosin. ×320.

ing fibrous capsule. Mainly scattered close to the surfaces, some mono- and polykaryonic macrophages, lymphocytes, a few mast cells and neutrophils infiltrate the fibrous capsule [258,266–272].

Autoptically retrieved, clinically effective, uncemented total arthroplasties exhibit ingrowth—to varying degrees—of bland fibro-osseous tissue into the porosities of the metallic parts. Where bony ingrowth falls short, "ligamentoid" fibrous tissues affixes the metallic surface to the bone, one more compelling argument endorsing the biocompatibility of the commonly used metals. Anchorage to the bone by Sharpey-like fibers forms at the surface of diverse nonosseointegrated biocompatible implants. Grafting of autogeneic bone chips overcomes the scarce bony ingrowth into a porous-coated component. The deficient bony ingrowth, therefore, pertains to technical difficulties of achieving a close bone-implant fit and does not reflect on properties of the materials. The recently popularized modular artificial hip joints are greatly vulnerable to corrosion as well as to generation of extra large amounts of both metallic and polyethylene wear particles. The inhibition of bony ingrowth into porous-coated components has to do with engineering rather than biological issues of modularity [97,273–276].

Metallic implants are subject to service-related deterioration. Their wear products evoke an injurious macrophagic response. Metallic particles are sometimes deposited in the periprosthetic tissues in such massive amounts as to bring about a blackish discoloration of the tissues, so-called metallosis. Histologically, broad sheets of crowded, large, round to polyhedral, granular to foamy, mono- and polykaryonic macrophages are strewn in the chronically inflamed interfacial membrane and pseudocapsule. The macrophages are packed with black, barely visible to 1- to 3-μm-sized, roundish to rod-shaped or jagged metallic particles. Some larger metallic fragments, up to 100 μm in length, may be present intra- and extracellularly (Figs. 3, 9). If coagulation necrosis ensues, as it often does, the sheets of macrophages are interrupted by variously sized foci of an amorphous, metal-dust-dotted matter, which inexplicitly discloses no or just scant signs of undergoing organization by invading granulation tissue (Fig. 19). The metallic debris may abound within the fibrinous exudate on the synovial surface of the pseudocapsule. Metal-laden macrophages usually pervade the fibrotic bone marrow of the subjacent bone. Periprosthetic tissues of long-term surviving (over 15 years) artificial hip joints with an alumina ceramic femoral head may be burdened by extensive aggregates of macrophages filled to capacity with metallic dust but nearly devoid of either polyethylene or cement. Metallosis, evolving a few months postoperatively, is ascribed to interfacial motion-related fretting-corrosion of an unstable, porous-coated, uncemented femoral stem. Florid macrophagic responses to metallic debris originating from stainless steel or cobalt-chrome cables, used for fixation of trochanteric osteotomies, may bring about acetabular socket loosening. The sometimes aired statements that "microscopic metal particles (are) seen only in tissues from infected hips" or that "metal particles (are) less important in the inflammatory reactions and loosening event than cement and polyethylene particles" are unsubstantiated. True, bacterial metabolic activities may be responsible for a notably high content of metal in the tissues around infected components. But metallic debris-laden macrophages abound as well in the periprosthetic tissues of

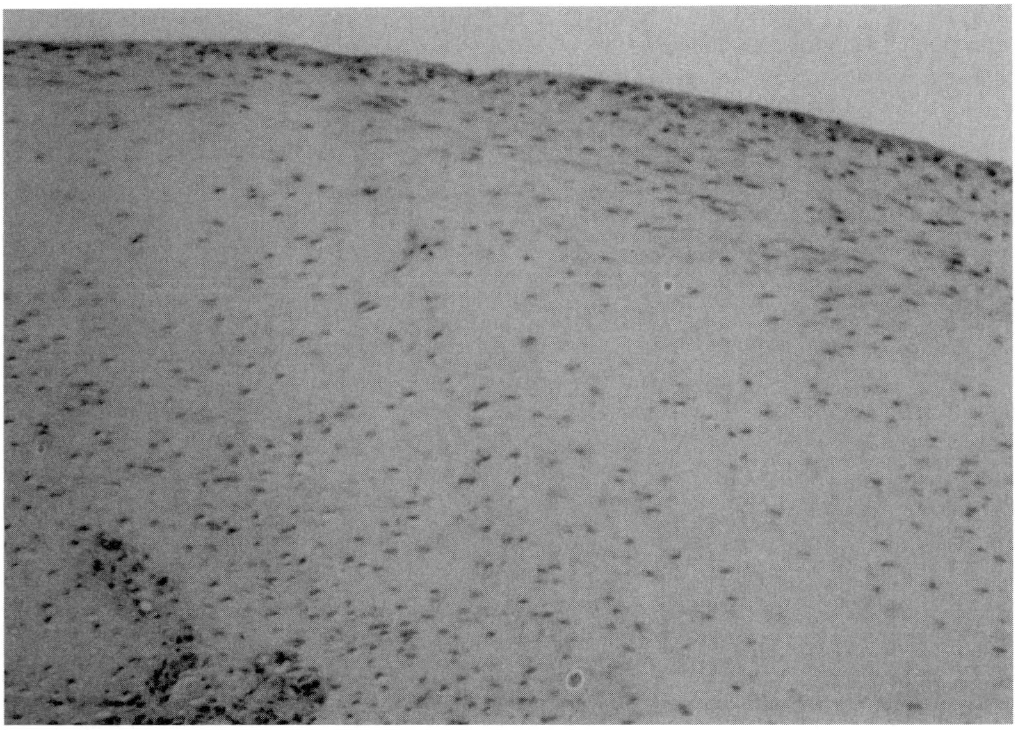

Figure 19 An interfacial membrane of a stainless steel compression plate composed of well-aligned, loosely textured fibrous tissue without an associated inflammatory or granulomatous response. Hematoxylin and eosin. ×100.

aseptically loosened joint replacements and may be, albeit infrequently, the most weighty morphological finding [277–285].

Metallosis may be a dominant or contributing cause of painful detritic synovitis as well as of localized osteolysis and aseptic loosening of an artificial joint with either a cobalt- or a titanium-based component. The cardinal role of the reaction to the metallic debris in bone resorption was already known in the mid-1980s when Amstutz et al. reported on titanium-particles-filled macrophages in bony cysts of dogs with uncemented porous-coated resurfacing hip prostheses. Metallic particles activate macrophages and fibroblasts in vitro. Deterioration of implant fixation and synovitis progress when a metallic surface crumbles. The increased vulnerability to wear and third-body-induced abrasion with consequent shedding of extra large amounts of particles explains the prevalence of early metallosis in patients with titanium- or titanium-alloy-based prostheses. Industrial techniques utilized to produce a matte or a satin finish, intended to promote cemented or cementless fixation, leave processing materials embedded within the surface, apparently resulting in enhanced corrosion and abrasion. Metallosis-related pain necessitating a revision operation of a well-fixated artificial joint is a truly frustrating ordeal. Titanium alloys have been counted on to be "both safe and efficacious in clinical use." However, an almost 20% incidence of aseptic loosening within just 26 months of operation hardly reflects an encouraging clinical outcome. Compared to cobalt-chromium-alloy-based components, the titanium and its alloys have not only a 10 times higher wear rate but they cause more polyethylene wear as well. Titanium-based prostheses can hardly be labeled "inoffensive implants" when a patient with an arthroplasty presents with inguinal lymphadenopathy due to migrating particles [277,280–292].

The cell population near titanium fracture fixation plates or near metallic particles in the periprosthetic tissues of artificial joints consists of helper/inducer T-lymphocytes, cytotoxic/suppressor T-lymphocytes, activated macrophages, a few B-lymphocytes (if any), rare plasma cells and mast cells. The nature of this infiltrate indicts cell-mediated hypersensitivity as being the major pathogenic factor in eliciting the reaction. The response is more intense around stainless steel screws inserted into the bones of nickel-sensitive rabbits than into the bone of normal rabbits, incriminating cellular immunity in the induction of the inflammatory reaction close to the metallic implants or breakdown products thereof. Besides, epithelioid and giant-celled necrotizing granulomas arise, albeit rarely, in the periprosthetic tissues of patients with a cemented or an uncemented arthroplasty (Figs. 20, 21). Polymeric and metallic debris is widely dispersed in the necrotic core, stroma and macrophages of the granulomas. The necrotizing granulomas may be accompanied by necrotizing arteritis and tissue eosinophilia. It is well known that tissue eosinophilia, necrotizing arteritis and granulomas characterize patterns of damage mediated by cellular immune responses to insoluble antigens or haptens. Polyethylene being nonantigenic, allergic-pathergic reactions to metal released from the implants seems to be pathogenically involved in the evolution of these and similar manifestations. Understanding the body's reactivities to the metals may explain some unusual complications in metal-hypersensitive patients with failed uncemented cobalt-chromium joint prostheses. To exemplify, a metal-hypersensitive patient with a loosened uncemented cobalt-chromium femoral stem had a para-articular pseudotumor featuring necrotizing arteritis and epithelioid cell granulomas on the background of an eosinophils-rich inflammatory infiltrate. Tissue eosinophilia and/or necrotizing arteritis have repeatedly been observed around corroded cobalt-chromium implants. Expectedly, not all researchers subscribe to this perspective. To name just one contradictory argument, screws inserted into tibia of nickel-, chromium- or cobalt-sensitized guinea pigs stay well fixated for several months, being separated from the bone by a thin fibrous interfacial membrane [125,273,293–306].

The often portrayed osseointegration of titanium and its alloys is conditional on the

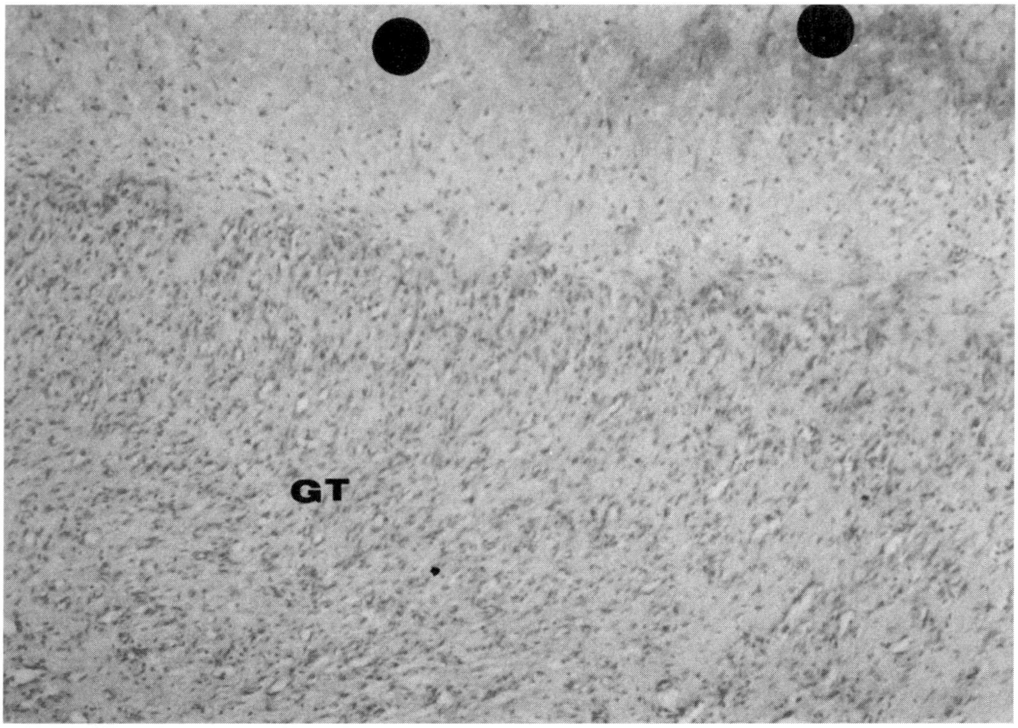

Figure 20 An interfacial membrane of a stainless steel compression plate composed of a florid, non-specific granulation tissue (GT) which is covered by necrotic material (circles) abutting on the metallic surface. Hematoxylin and eosin. ×80.

physique of the implant's surface. Rough-surfaced implants undergo rapid osseointegration, while similarly constructed but smooth-surfaced implants become mostly separated from their bony bed by an interfacial membrane. The arrangement of fibroblasts alongside the implants is mediated via an amorphous electron-dense zone. The orientation of the fibroblasts is a function of the surface topography as well. Irrespective of textural features, wherever a close bone-metal apposition has not materialized, mono- and polykaryonic macrophages abut upon the surface of the implant while osteoclasts resorb the adjacent bone. Dependent on different modes of stress transfer, unlike designs of titanium-based devices (e.g., threaded versus fully or partially porous coated designs) lead to either bone buildup or bone loss. When compared to the segmental and limited osseointegration of artificial joints, it is noteworthy and impressive that titanium-based oral implants are over 80% osseointegrated. This difference in the degrees of osseointegration instructively depicts the crucial technique- and biomechanics-dependent behavioral traits, as opposed to the chemical constitution of the implants. Design, surface topography and load pattern decisively alter genres of cellular activity such that new bone formation prevails under favorable settings, whereas inflammatory infiltration, fibroblastic proliferation and bone resorption monopolize the scene under adverse conditions. Unfortunately, the advantages which titanium and its alloys have over other metals, in that they prompt surface precipitation of apatite, is offset by leaching of injurious ions and voluminous particles generation [259,303–308].

Fractures of the long tubular bones are commonly fixated with variously designed nails fabricated of the biocompatible 316L stainless steel. Pitting, crevice and other types of corro-

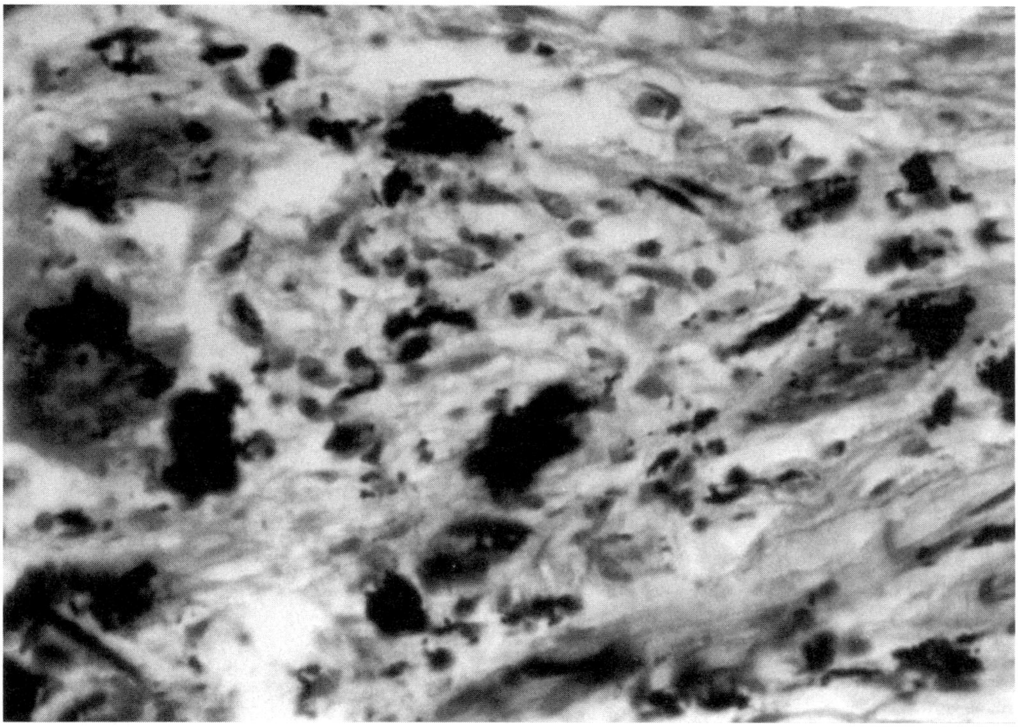

Figure 21 Variously sized metallic particles are scattered extracellularly while others have been phagocytosed by the mono- and polykaryonic macrophages infiltrating the interfacial membrane of a stainless steel compression plate. The intractable effects of metallic breakdown products explain the "conflicting" perspectives of a bland (Fig. 19) versus a highly inflamed (Fig. 20) interfacial membrane, both of which face chemically and texturally identical surfaces of the stainless steel compression plate. Hematoxylin and eosin. ×320.

sive reactions occur as early as six weeks after implantation of stainless steel plugs into the hard or soft tissues of rats, where they progressively increase with time. Is stainless steel perhaps not as innocuous as is generally assumed? The interfacial membranes surrounding stainless steel nails consist of bland fibrous tissue in but a minority of patients. More often, the membranes comprise a chronically inflamed fibrous tissue strewn with clustered, metallic particles-laden mono- and polykaryonic macrophages. The metallic particles are generated not only by corrosion but also by motion-related surface abrasion. The extensive and intensive response to the stainless steel particles suffices to lead to bone resorption. By the way, tracts of pins, inserted for the external fixation of fractures, show similar appearances. Overt metallosis is infrequently found in patients with stainless steel nails. The iron concentrations are high in the metallosis-discolored periprosthetic tissues of implants incorporating stainless steel components. In the experimental animal, both bone resorption and formation are trivial around tight pins, but loose pins are accompanied by a granulomatous interfacial membrane, profuse osteolysis, abundant osteogenesis and, rarely, bone necrosis. Metal ions in excessive amounts are toxic. The focal bone necrosis is ascribed to a direct noxious effect, an immune-mediated reaction or to both. Metal ions being potent haptens, T-lymphocytes are anticipated to be and are, indeed, dominant within the inflammatory infiltrates associated with the ferric deposits* in the membranes of nails. However, necrosis also happens where extra large amounts of the nonantigenic and nontoxic polyethylene detritus collects. Thus, it cannot be excluded that cell death in the vicinity of a stainless steel device is solely due to the overburdening of the tissues by massive amounts of the metallic particles [294,307–316].

What happens at the bone-metal boundary is influenced, inter alia, by release of ions into the tissue, the "corrosion resistant" implants not being exempt. The leached ions appear to play a major role in the formation of the interfacial membrane. Osseointegration expresses the healing potential of the bone. Osseointegration of nonloaded stainless steel pins is a well known occurrence but only on condition that a stable interfacial setting is attained. In contrast to the scene at the nail-bone interface, which fits in with a variation on the theme of the "small particle disease," stainless steel meshes in soft tissues induce a giant-celled granulomatous response in the absence of breakdown particles [234,317].

V. CERAMICS

In vitro biocompatibility tests specify cytotoxic effects of the ceramics. The scope of the cytotoxic effects depends on the material as well as on its surface texture. The manufacturing techniques may as well influence a ceramics' cytotoxicity. Such effects are, fortunately, minimized in the in vivo setting. True, intraperitoneally injected high-dosed ceramic powder is lethal to mice and rats. Since death of the animals is caused by renal failure due to silicon release from the ceramics, this observation of a highly laboratory artifact is peripheral to the issue of biocompatibility in the context of reconstructive surgery. Most calcium phosphate ceramics, in particular the widely employed hydroxyapatite, are highly compatible with the soft and hard tissues, causing no undue inflammatory reactions. Acknowledging that the thinner the interfacial membrane the more congenial an implant's constituents, hydroxyapatite-coated titanium-based composites constitute the most biocompatible moiety of the ceramic alternatives within the bone. Yet other ceramics, e.g., magnesium-whitlockite, do elicit adverse reactions as they induce macrophagic responses and formation of poorly organized fibrous tissue. Whether and to what extent the diverse ceramics are removed by cellular activity is a moot question. Recent studies suggest that, depending on composition and porosity of the implant, the ceramics are resorbed, albeit to unequal degrees, by both osteoclast-type and foreign-body-type giant cells and are then replaced by bone [318–326].

Interest in the ceramics derives from their bioactivity. Hydroxyapatite ceramics are widely used in orthopedic and dental surgery, by reason of their osteoconductive potential, i.e., they direct bone formation in osseous environs. Hydroxyapatite surfaces bond to bone via a mixed mineral phase in which bone-restricted proteins coprecipitate with carbon-apatite. Surface-bound fibronectin benefits anchorage and attachment of osteogenic progenitor cells to the surface of the ceramics. It also supports the proliferation and spreading of the preosteoblasts onto the ceramics. The gradual modification of the hydroxyapatite exterior during ion exchange with the surrounding fluid, dissolution into and precipitation from the fluid allow for the incorporation of the ceramics into the maturing osseous tissue. Significantly, bone-bonding occurs even in the face of cyclic loading. Osseous ingrowth into the porosities of the hydroxyapatite-

*It is acknowledged that reference to iron deposits is an oversimplification, since the 316L stainless steel comprises Fe, Mo, Mn, Cr and Ni and inclusions of their oxides are detected by chemical analysis of the retrieved implants [318].

coated metallic implants takes place earlier, deeper and is more extensive than the ingrowth into the porosities of noncoated implants. Granular hydroxyapatite may also be incorporated into the bone. The emerging ceramo-osseous regeneration complex exhibits an adequate orientation of its newly synthesized collagen fibrils, in alignment with the preexisting collagen fibers. Finally, even under locomotion, the initially laid down woven-fibered bone remodels into lamellar-fibered bone. Because of insufficient strength and toughness of the ceramics and their susceptibility to fatigue failure, hydroxyapatite-coated metallic components are employed in the manufacture of artificial joints. In a canine model, uncoated stems of hip prostheses are loosely attached to the bone by fibrous tissue, while hydroxyapatite-coated stems of the same design become rigidly fixated by bone-bonding within three weeks. Also, osseous tissue fills defects of up to 2 mm within six weeks. Uncoated polymeric implants, which within the bone are invariably enclosed in an interfacial membrane, become osseointegrated when coated with hydroxyapatite. Ducheyne and Cuckler warn that an hydroxyapatite coating may adversely affect the bone-implant coupling if relative motion between the coating and the substrate generates particles and electrochemically increases the wear processes. Because of the ceramics' poor fracture toughness, a hydroxyapatite coating may shear off after extended loading periods [10,321–343].

The mesenchymal cells in the interfacial fibrous tissue may undergo osteoblastic differentiation after the postoperative inflammatory phase subsides. The osteoblasts rest either directly on the hydroxyapatite or they line the newly formed bone apposing the hydroxyapatite. Membranous ossification takes place within the interfacial fibrous tissue and gives rise to osseous bridges, joining the endosteal and hydroxyapatite-apposing bone with each other. Further attesting to the biocompatibility of the hydroxyapatite are the absence of inflammatory infiltrates and preservation of adjacent normal hematopoietic marrow. In the experimental animal, newly produced bone forms along ~10% of hydroxyapatite-coated femoral stems three weeks postoperatively, and along ~50% after 12 weeks, with clearcut evidence of bidirectional bone remodeling [148,334,341,344–346].

Notwithstanding these envisioned advantages of the ceramics, failure of hydroxyapatite-coated artificial joints is yet commonplace in clinical practice. To materialize, osseointegration requires a close initial fit. Osseointegration is forestalled by a gap of just 350 μm in the rabbit model. An implant-bone gap of less than 50 μm is essential to assure success in man. When interfacial motion ensues, calcium phosphate implants become enclosed by fibrous tissue and osteoclast-type giant cells and macrophages assemble at the interface and phagocytose the hydroxyapatite. The metal-ceramic interface being a weak link, hydroxyapatite particles may debond and accumulate interfacially. Once mechanical attachment of the coat to the bone has been achieved, the hydroxyapatite-coating is liable to readily detach from its substrate. The loosened hydroxyapatite-coated press-fit prostheses display a stereotypical synovial-like interfacial membrane replete with ceramic-filled macrophages (Fig. 22). Alike the circumstances leading to aseptic loosening of cemented arthroplasties, a progressive granulomatous reaction to the particulate hydroxyapatite debris also coincides with periprosthetic bone loss [54,259,347–351].

The erstwhile held belief in the inoffensive bearing of particulate hydroxyapatite, purportedly sequestrated intracellularly, turns out to be erroneous. Alike phagocytosis in general, ingestion of hydroxyapatite particles results in macrophagic activation and synthesis of cytokines. Rather than exhibiting the anticipated direct bone-implant apposition, hydroxyapatite-coated titanium-based components are often associated with overt bone loss and are surrounded by an interfacial membrane. Periprosthetic tissues retrieved at revision operations are found to contain macrophages laden with myriad 1- to 15-μm-sized and some larger hydroxyapatite particles. Besides, 0.1- to 75-μm-sized particles may migrate to the joint cavity and become

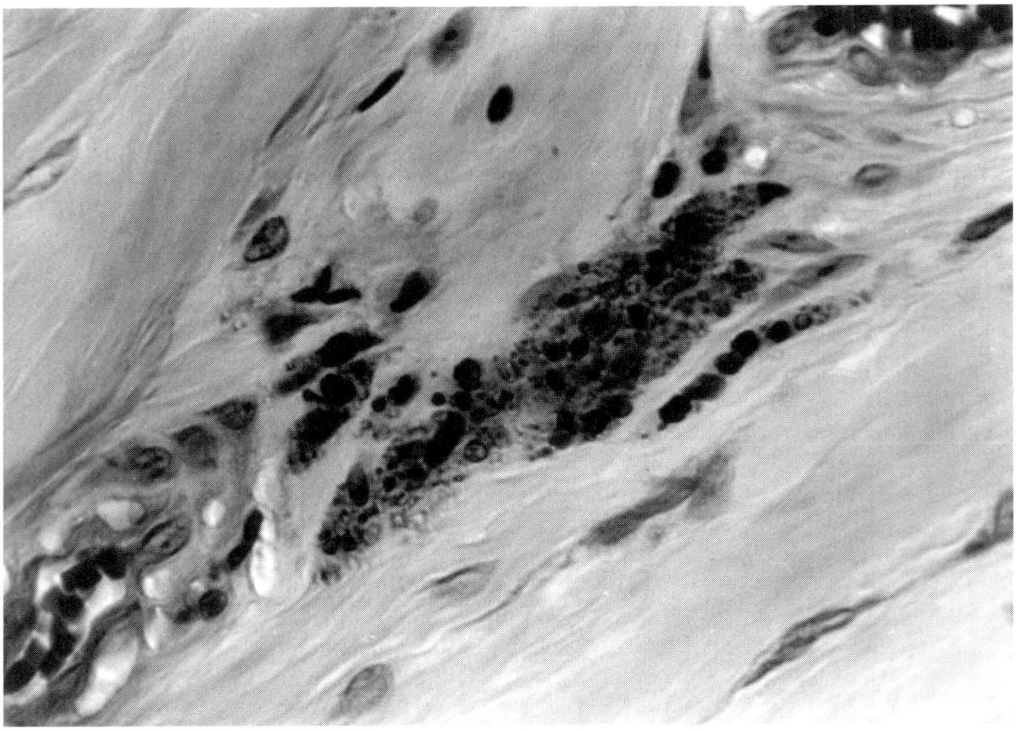

Figure 22 Hydroxyapatite particles-laden macrophages in the interfacial membrane of a loosened, ceramic-coated, press-fit total hip arthroplasty. (Other prosthetic debris was present elsewhere in this membrane.) Hematoxylin and eosin. ×400.

embedded within the polyethylene surface. In the wake of the thereby enhanced third-body wear, the alloplastic femoral head becomes burnished and scratched and the polyethylene covered by cracks, scratches and grooves. Roughening of the articulating surfaces of an artificial joint further speeds generation of particulates [259,352–359].

The alumina ceramic femoral head, introduced in 1970 for total hip replacements, has the inferred advantages of good biocompatibility, fitting hydrophilicity, low friction coefficient, high wear resistance and high mechanical strength. On the other hand, the ceramics are brittle, have poor resistance to tensile loading and are notch sensitive. Alumina in bulk form induces a low-grade reaction within the soft tissues so much so that it was initially considered to be bioinert but it was later shown to actually be slightly bioactive. Particulate alumina provokes a granulomatous response, the intensity of which is less than that to particulate polyethylene and metals. Alumina particles smaller than 5 μm are ingested by macrophages, those larger than 10 μm remain in the extracellular compartment. Alumina becomes osseointegrated but does not bond chemically to the bone. Ingrowth of bone into the porous-coated surfaces is, therefore, required to achieve a strong mechanical union. Inducing formation of fibrous membranes, interfacial micromotions cause separation of the alumina from its osseous bed. Indeed, direct contact of alumina ceramic with the bone is a rare occurrence. In most authors' experience, osseointegration fails to develop. Irrespective of the absence or presence of some wear-particles-laden macrophages and despite component stability, an interfacial membrane is consistently present around alumina-based implants. The interfacial membrane is thin as long as artificial joints remain well fixated. But following loosening, alumina debris-laden mono- and polykaryonic macrophages aggregate throughout a thick synovial-like interfacial membrane and pseudocapsule, lymphocytes and plasma cells abound and the attracted osteoclasts resorb the adjacent bone. Much extracellular ceramic debris may amass. The macrophagic clusters sometimes undergo necrosis. As a rule, the tissular reaction to alumina particles is less extensive and less intense than that to particulate polyethylene, metal and cement. All in all, the performance of total hip arthroplasties with a ceramic femoral ball, whether combined with a polyethylene or a ceramic acetabular insert, share the ordeals of conventional total hip replacements. At the seven-year follow-up evaluation, the revision rate is as high as 30%. Retrieved components exhibit signs of significant wear and dislodged loose grains cause third-body wear [57,108,360–370].

VI. SILICONE

Polydimethylsiloxane, known by its generic name of "silicone," is used in the manufacture of oils, gels and rubbers, often in combination with a variety of additives. It is present in consumer goods (cosmetics and toiletries), processed foods, household products and is widely used in medical practice (as a lubricant in tubing and syringes and in some implantable devices). It was formerly judged to be nontoxic and bioinert. Polyurethane, used in the manufacture of silicone breast prostheses, is characterized by its superb biocompatibility, assertedly superior to that of many other commercially available polymers. Overt signs of toxicity have not been observed in mice followed for about half a year after subcutaneous injections of silicone fluid or gel or subcutaneous implantation of silicone elastomer or polyurethane. Immunoglobulin-producing cells in the spleen may be slightly reduced but the antibody response to sheep erythrocytes is not impaired. The T-lymphocytes are unaffected but natural killer cell activity is depressed in the silicone-treated mice as well as in animals with a polyurethane implant. The animals are not overly susceptible to infections and they normally reject transplanted tumors. As with the other biomaterials, fragments of an implant induce locally a tissular reaction and, having been ingested by macrophages, the silicone particles may be carried to distant sites, e.g., from the peritoneal cavity to the spleen [371–376].

The clinical results of replacement of phalangeal and carpal joints with a silicone rubber prosthesis are satisfactory. Except for painful detritic synovitis, the findings at the follow-up examinations are unremarkable in most patients. However, focal, even widespread, bone destruction occasionally ensues in the wake of a foreign body reaction to particulate silastic and the migrating debris may cause a granulomatous lymphadenitis. Similarly, osteolysis may occur after replacement of the disk of the temporomandibular joint by Dacron-reinforced silicone implants. As with all other intraosseous devices, breakdown products, in this case the particulates generated on abrasion of the silastic prostheses, are responsible for the bone-resorbing potential of the debris-induced giant celled granulation tissue. In the absence of critical breakdown, silicone-based devices serve as commendable templates for tissue regeneration, provoking but a minimal foreign body response [35,377–383].

Silicone rubber stimulates platelets aggregation. Sophisticated experimentation has brought out surface-entrapped air microbubbles, rather than intrinsic material properties, to be responsible, at least in vitro, for this untoward effect of the silicone [384].

The lay and medical media have focused the public's attention on the ostensible health hazards of silicone breast prostheses, in use to cosmetically restore a woman's postmastectomy look, improve the appearance of a disfigured breast or augment an otherwise normal breast. The high complication rate of the procedure is disquieting. Reoperation is frequently required. Implant shift, formation of a fibrous capsule, often followed by mineralization of the capsule, "silicone-bleeding"-associated reactions and immune-mediated diseases, in crescendo order of seriousness, are hazardous to the patients' well-being. The incidence of the complications increases with service time of the implants. Calcium phosphate is the main constituent of the intracapsular mineral deposits. It is accompanied by a small amount of zinc phosphate, the zinc being an ingredient of the glue used in sealing the prostheses [385,386].

The silicone rubber envelope of the breast prostheses bounds a saline or silicone gel core. The envelope is coated with polyurethane foam in some implant types. The commercially available models differ from each other by minor structural and chemical features. Axillary lymphadenopathy becomes clinically overt, at variable postoperative intervals, in a minority of women with an implant. The enlarged lymph nodes disclose either massive sinus histiocytosis or epithelioid celled granulomas. Both these macrophagic reactions are due to lymphatic migration of an implant's leaked constituents to the draining lymph nodes. The prostheses themselves are confined within a host capsule. The latter may consist of a densely textured, bland, hypocellular and hyalinized fibrous tissue. The silica added during the manufacturing process exerts a fibrogenic effect. In other patients, the capsules comprise an inflamed and, at least, focally loosely textured fibrous tissue containing lymphocytes admixed with plasma cells, foamy macrophages and, often, mono- and polykaryonic macrophages. The pericapsular fat tissue is sometimes necrotic. As with other alloplastic materials, the junction with a silicone-gel-filled, saline-filled and polyurethane-coated implants often displays a synovial-like aspect, palisading macrophages abutting the prosthesis. The tissular reaction patterns appear to correlate with the implant type used. The host capsules adjacent to smooth-surfaced implants are of regular structure and uniform thickness. Those subjacent to textured implants vary in thickness, are disrupted along their length and are festooned with 0.25- to 0.5-mm-sized knob-like projections. These features are likened to the detritic synovitis caused by breakdown products of artificial joints. Silicone particles are continually shed at the implant-tissue boundary, a scene equaling that ensuing in the wake of liquid silicone injections. The leading causes of silicone gel escape are seepage via breaches in of the silicone rubber envelope, whereupon fragments scatter throughout the surroundings. Droplets of liquid silicone, rectilinear silicone rubber fragments, and in patients with a coated model also geometric crystalline polyurethane fragments, are embedded in the fibrous tissue or engulfed by mono- and polykaryonic macrophages, which ingest the particulate foreign bodies. Examination of capsules retrieved after variable time intervals indicates that separation of the polyurethane coating begins a few months postoperatively. Fragmenting, slowly disintegrating, iron-encrustated polyurethane particles, partly associated with silicone flakes, seem to be overly active in inducing the granulomatous response (Figs. 23–25). Gross envelope rupture may trigger a follicular mastitis. The adult respiratory distress syndrome may develop when silicone gains access to the vasculature and disseminates hematogenously [387–391].

Little is known about long-range effects silicone breast implants, attention having mostly been focused on the connective tissue diseases. Epidemiologic studies have suggested that the risk of breast carcinoma might actually be reduced among women with a prosthesis but clinical data hint at a potential risk of sarcomas and hematologic malignancies. If this turns out to be true, then the bioincompatibility of silicone breast implants is legible. The type of implants may be pivotal since those with polyurethane foam covers can leak toluene diamine, a demonstrated carcinogen in animals [392].

Some authors view the periprosthetic inflammatory process to solely pertain to a foreign body reaction. Other investigators postulate the involvement of a cell-mediated immune mechanism in the host response. The patients infrequently present with a systemic immune disorder or an adjuvant disease-like state. Silicone-protein complexes are immunogenic in experimental animals, exciting humoral and cellular immunity. It is noteworthy in this context that the

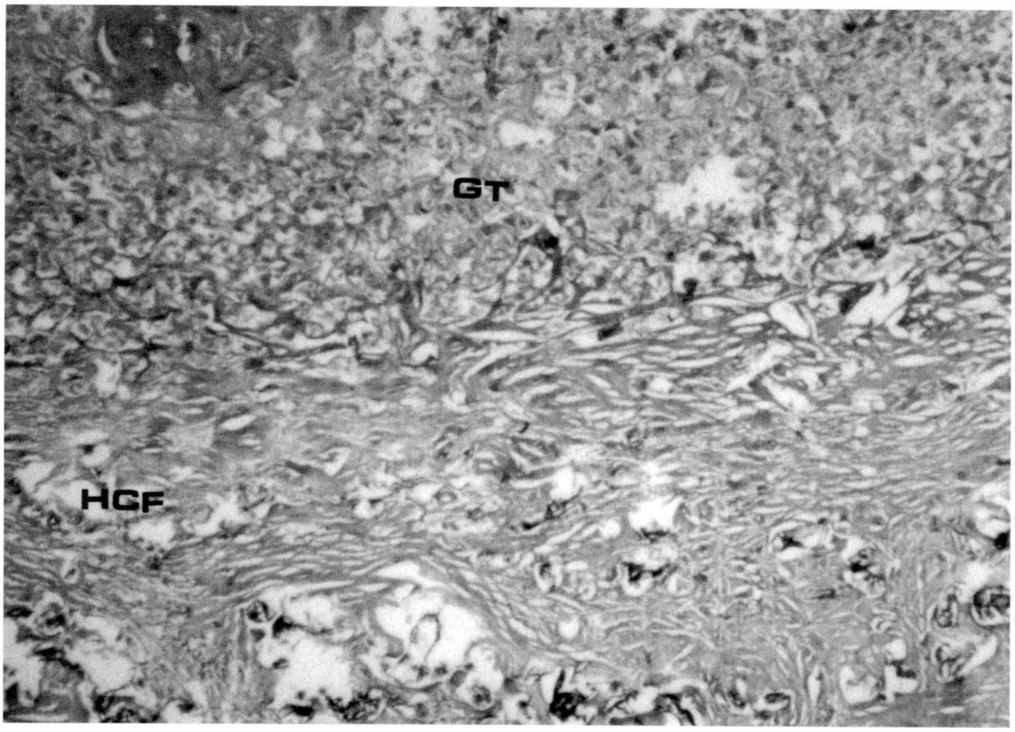

Figure 23 The capsule of a long-term surviving breast prosthesis of a patient with a debilitating neurological syndrome. The capsule comprises hyalinized collagen fibers (HCF), oriented parallel to the envelope of the prosthesis, and a foreign-body-induced granulation tissue (GT) with silicone and polyurethane particles. Masson stain. ×30.

silicone and the silica (vide supra) manifest adjuvant potentials. A variety of immune aberrations are identified after exposure to silicone oil, gel or rubber. It is theorized that the silicone's adjuvanticity is linked to adsorbed native molecules which, having undergone conformational changes, stimulate the formation of antibodies to interstitial and cellular antigens, both kinds of antibodies having been detected in the serum of the patients. Antisilicone antibodies are also common, their titres being highest in women with either ruptured or leaking implants. Kossovsky and Freiman propose a model of silicone immunogenicity that ascribes, in analogy with the traditional vaccinations, adjuvanticity to freed silicone and immunogenicity to surface-adsorbed, denatured native macromolecules. This prototypical pathway is acceptedly operative in experimental autoimmune diseases. Those who, based on epidemiological data, doubt the reality of silicone's capacity to cause systemic disorders, suggest that the clinically observed collagen-vascular diseases happen in women who are predisposed to autoimmunity. Physicians caring for women with augmented breasts should be cognizant of the incidence of implant-associated connective-tissue diseases which is 1 per 3801 implant-years. In a 14-year-ongoing follow-up study of a cohort of 87,501 women, Sanchez-Guerrero et al. have established that the relative risk of connective tissue disorders is 0.6 among women with any type of implant but just 0.3 in women with a silicone-gel-filled prosthesis. The authors, therefore, negate an association between silicone breast implants and connective-tissue diseases. In Bar-Meir and his co-workers' series, autoantibodies to 15 (out of 20 tested) autoantigens were significantly more prevalent in sera of women with breast implants than in a control population; the patients

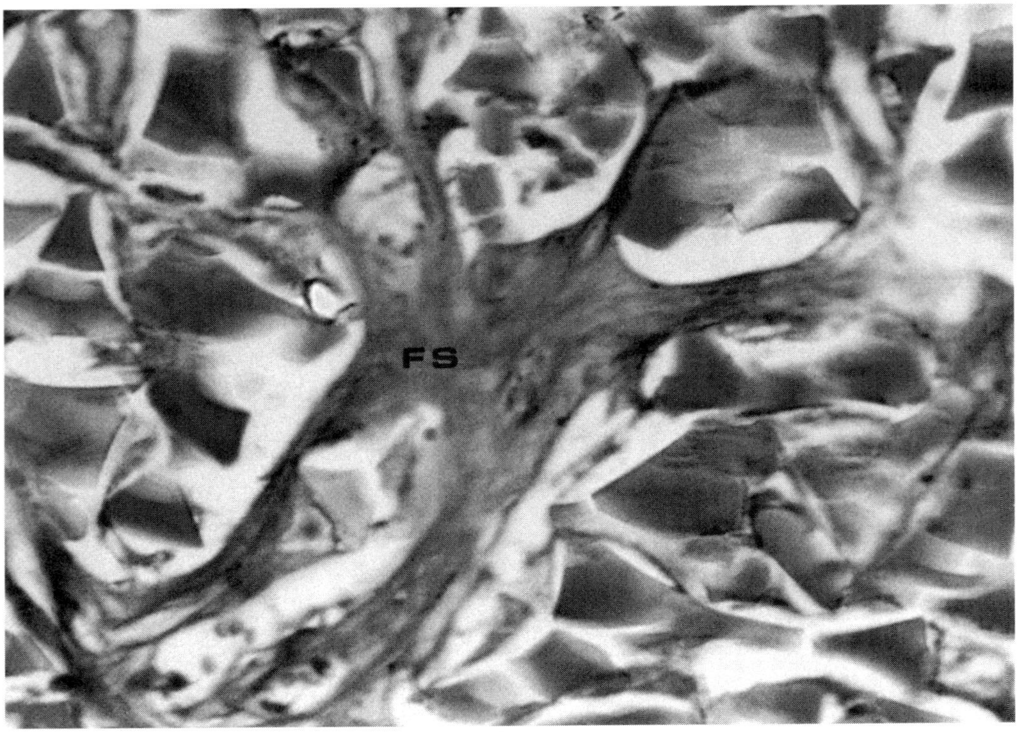

Figure 24 Acellular, hyalinized, fibrous septa (FS) in between the rectangular and triangular polyurethane particles of the ancient capsule of the breast prosthesis illustrated in Fig. 23. Masson stain. ×320.

presented with such nonspecific complaints as polyarthralgias, fatigue, myalgias, morning stiffness and decreased memory. The authors suggest that silicone or its by-products have an adjuvant action on the immune system [393–404]. Conceding that silicone and polyurethane are biocompatible in bulk form, surgeons ought nonetheless to profess that the safety of silicone-based breast augmentation devices is still contentious in view of the grievous locoregional and, albeit infrequent, distant effects.

Silicone tubular devices are employed with success in the reconstruction of blood vessels and the construction of arteriovenous shunts. Modifications of current devices are being explored, e.g., silicone-polymer or silicone-collagen composites. Endothelial or muscular cells seeded in vitro onto silicone rubber tubes survive and stay attached to the luminal surface even after having been subjected to pulsatile and cyclic strain. Endothelial cells proliferate in vivo on the luminal aspect of the silicone tubes. Circumferentially encircling fibrous tissue flourishes and grows inwardly into the interstices of the silastic network. It is accompanied by some mono- and polykaryonic macrophages. This granulomatous reaction is more lively at the abluminal side of the reconstructed blood vessel. Myocytic or myofibroblastic differentiation of the cells infiltrating the interstices of the prosthesis does not come about. Thrombotic occlusion of the conduit and aneurysmal dilatation of the conjoining vascular segment are two forbidding complications. The intra- and extramural granulomatous reactions accompany silicone spallation, the absence of myocytes and elastic fibers in the newly formed tissues and the densely textured, partially myxoid and hyalinized, collagenous scar-like nature of the tissue occupying the interstices of the implant readily account for the weakness of the shaping wall

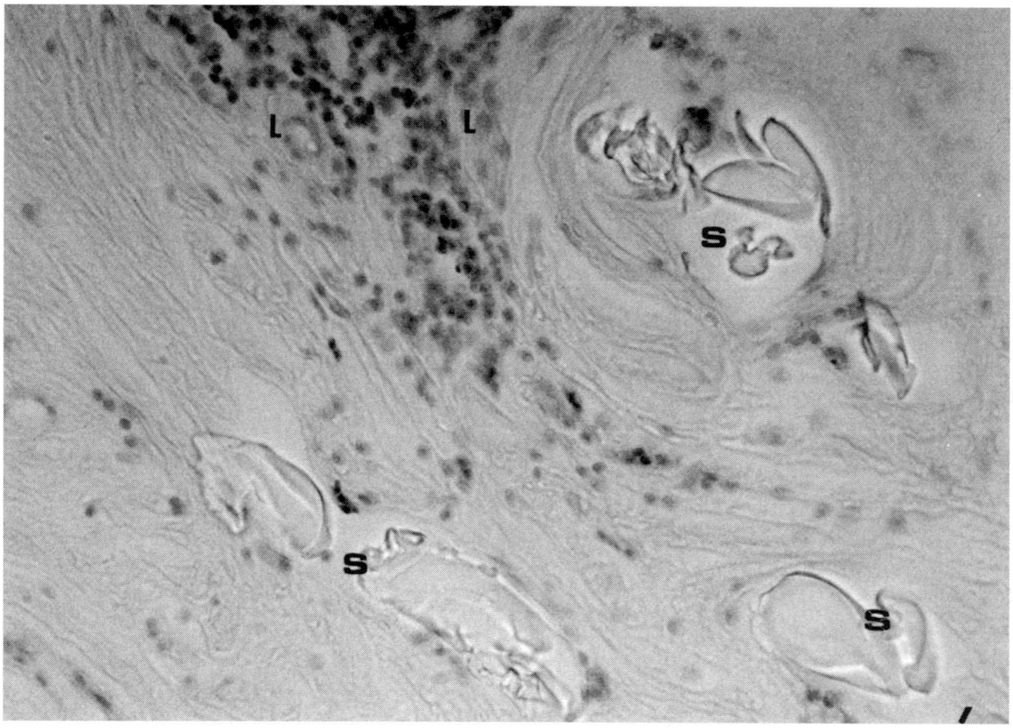

Figure 25 Unstained silicone flakes (S), incorporated in hypocellular fibrous tissue, are associated with focal lymphocytic infiltrates (L). Hematoxylin and eosin. ×250.

and, thus, the resulting aneurysmal dilatation. The functionally deficient scar tissue within and around the implant evolves in the wake of rampant inflammatory and granulomatous changes which are induced by the widespread dispersal of tiny to sizable silicone breakdown particles. The silicone breaks down into such small specks that they are ingested in vast numbers by a single polykaryonic macrophage [225,405,406].

Scientists who have studied the in vitro and animal models applaud the biocompatibility of silicone. They stress its nearly bioinert and nontoxic properties. The mild foreign body reaction provoked by silicone in bulk form does not impede the function of silicone-based implants, which perform, at least initially, as has been intended at the time of their conception. Practitioners who perceive the many and manifold, now and then life-threatening complications arising in patients with silicone-based implants are likely to distrust the judgment of the laboratory workers. A close look portrays a recurrent chain of events characterized by an initial physical damage of the device, a host response to released constituents and lastly the expression of a disease. A clinically consequential exception concerns the hemocompatibility of silicone, thrombi forming on the surface of tubular devices irrespective of what happens at the tissue-implant level. Otherwise, silicone's unsavory behavior is to be understood in the framework of variations on the theme of the "small particles disease." Troubles do not become overt prior to the point in time at which the implant comes apart and polydimethylsiloxane disperses in the tissues. The reactions to breakdown products, albeit those emanating from a particular chemical compound, dictate the steps which directly or indirectly, i.e., via immune mechanisms, exert adverse effects on the individual's health. Be this as it may, silicone is hardly the biocompatible material as formerly assumed [405,406].

VII. CARBON

Carbon in bulk form is thoroughly biocompatible in vitro and in vivo. Experimental assessment has assured that carbon filaments are strong and inert. Still, the literature is replete with reports on the futile attempts to replace patients' or animals' cruciate knee ligaments with carbon filamentous braids or tows. Allegedly, fibrous tissue penetrates rapidly throughout the interstices of carbon filamentous implants. The fibroblasts supposedly produce collagen fibers which intimately cohere to and are oriented along the carbon filaments. In the perfect surroundings, and especially in extra-articular sites, an implant may indeed stimulate the formation of a "neotendon" or a "neoligament," in which fibrocytes and well-aligned collagen fibers are oriented in parallel to the carbon filaments (Fig. 26). Such an organizational pattern is illusory in most cases in as much as the proliferating fibroblasts usually fall short of deeply penetrating in between the filaments such that the implants remain mostly unorganized within a sleeve of fibrous tissue [407–411].

Early enthusiastic documentation notwithstanding, reconstruction of the anterior cruciate ligament of the knee with a carbon filamentous prosthesis is confounded by a high rate of mishaps. Coating of the carbon filaments with absorbable polymers does not alleviate the poor clinical outcome. The carbon filaments have high tensile strength but they are very brittle. Shear and bending forces fracture them into myriad particles. Within a few years to a decade or so of the operation, the patients usually present with a ruptured implant, immensely painful synovitis or both. Whether or not the implant had been intraoperatively ensheathed in a synovial sleeve, and regardless of the degree of tissular growth on and into the prosthesis, the filaments break and their fragments disperse within the joint cavity, become incorporated in the synovial membrane and cause a calamitous detritic synovitis. When exposed to carbon particles small enough to be internalized, cultured synoviocytes richly produce cell-activating

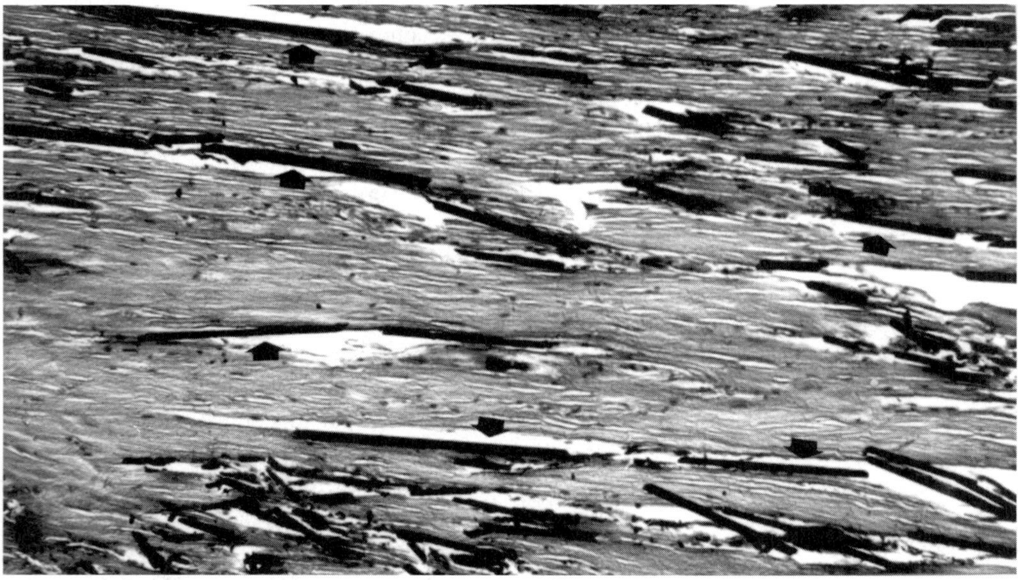

Figure 26 The mature, successfully organized carbon filamentous implant discloses fibrocytes and well-aligned collagen fibers oriented in parallel to the alloplastic fibrils (arrows). Neither inflammatory nor granulomatous changes are evident. Hematoxylin and eosin. ×80.

factors and collagenase. The cells are apparently activated when their membranes make multiple independent contacts with the asperities of the material. In fact, the potency of carbon exceeds that of polymers in this respect [219,408–414].

The centrally located bundles of carbon filaments of an anterior cruciate ligament prosthesis are often just partially organized by fibrous tissue years postoperatively. At the same time, the peripherally located parts of the implant are amply organized by an ongrowing and ingrowing fibrous tissue. The implant is also encased in a sleeve of fibrous tissue. Typically, the individual carbon filaments are enclosed by concentrically layered fibroblasts and macrophages as well as a tube of longitudinally oriented collagen and reticulin fibers (Figs. 27, 28). However, this image too is deceptive since collagen synthesis within the implants is deficient, and the tensile strength of the newly formed fibrous tissue does not near that of the native anterior cruciate ligament. For as long as a prosthesis is in service, mono- and polykaryonic macrophages abut on the filaments and ingest tiny carbon particles. The filaments disintegrate to some extent and there is no effective mechanism of disposal of the debris. Both the filamentous and particulate carbon continually activate an inflammatory-granulomatous reaction, thereby modifying the metabolic activities of the fibroblasts which synthesize more type III and V collagen than type I collagen all of the time. Hence, even a decade or more postoperatively, the nature of the interfilamentous fibrous tissue is that of functionally inferior young scar tissue. In contrast, fibrous tissue in and around biocompatible biomaterials is in a steady state after type I collagen has supplanted types III and V collagen with the lapse of time. Additionally, the carbon filamentous implants are, as a rule, separated from their bed within the intraosseous tunnels by thick interfacial membranes, the formation of which is likely to

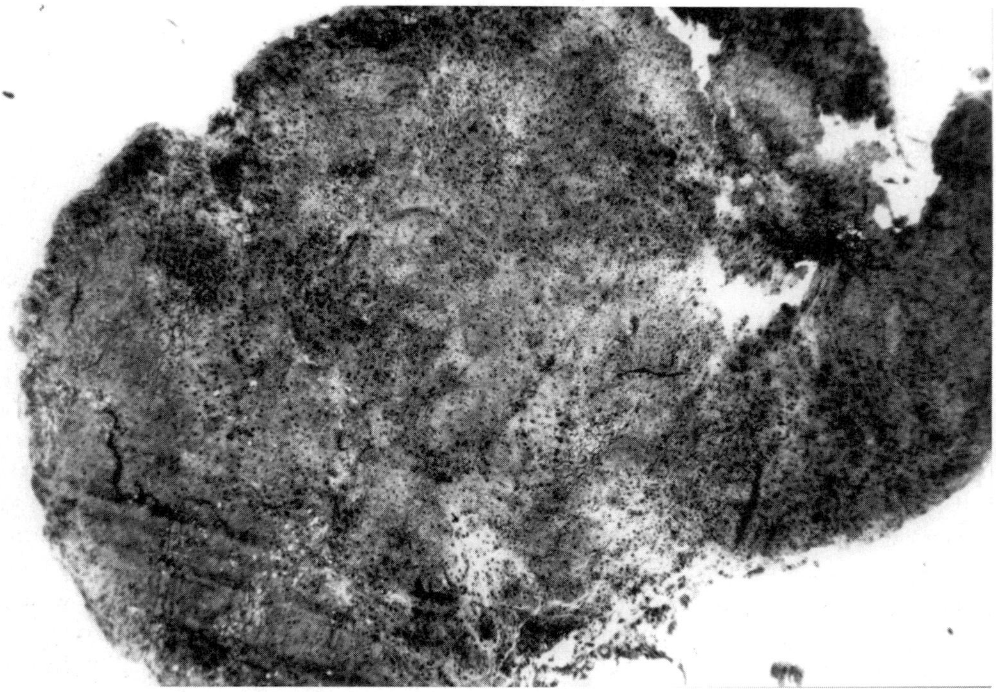

Figure 27 Well-organized carbon filamentous prosthesis used for substitution of the anterior cruciate ligament. Both the peripheral and the central portions of this implant contain much more tissue than carbon filaments (black dots in this cross section). Hematoxylin and eosin. ×10.

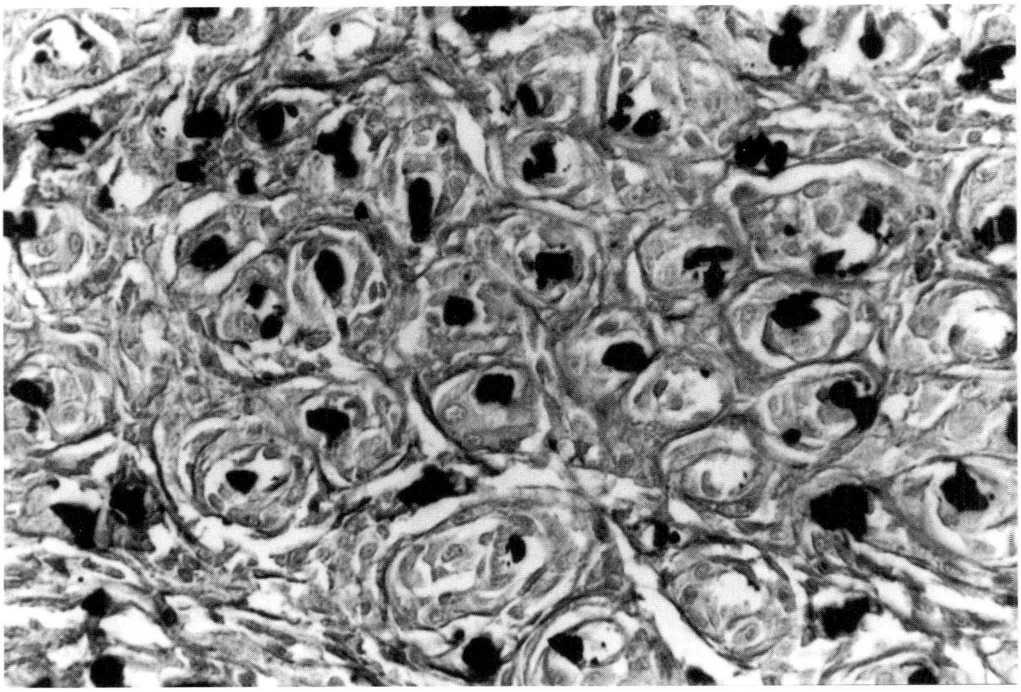

Figure 28 The characteristic peculiarity of the organized carbon filamentous implant is the so-called composite unit. Each unit is composed of the centrally located carbon filament enclosed by concentrically arranged fibroblasts and macrophages and a tube of longitudinally oriented collagen and reticulin fibers. Hematoxylin and eosin. ×400.

have been initiated and then maintained by interfacial motion. In the unusual event of an implant's postoperative stability, a carbon filamentous prosthesis may undergo osseointegration (Fig. 29, 30): The implant becomes anchored to its bony bed by the osseous trabeculae which grow in between single and groups of carbon filaments [35,208,415–419].

In contrast to the intra-articular locale, carbon filamentous devices do not induce excessive tissular reactions in other sites. The carbon-filaments-reinforced plastic plate has been applied successfully in the fixation of fractures and treatment of nonunion of fractures. These very strong but flexible and light plates accelerate the formation of an abundant external callus. Experimentally and clinically, the thin fibrous tissue layer at the plate-bone interface displays just mild inflammatory and granulomatous reactions, notably where isolated carbon filaments or particles collect (the debris mainly concentrates around screw holes); the healing processes under these circumstances are not interfered with. Carbon filamentous constructs designed for anchorage of polymeric implants and enhanced healing of patellar bone defects elicit a mild tissular response, are well accepted in the bone and perform satisfactorily. Fibrous and fibrocartilaginous tissues develop in experimentally produced osteochondral defects packed with carbon filamentous pads. Cartilage does not arise in and around the carbon filamentous pads inserted into surgically created articular defects in patients' patella. The ambition of the resurfacing operations being repopulation of the defects by chondrocytes producing cartilaginous matrix, the woven carbon filamentous pads do not fulfill this expectation. In the nonconducive environment of an ongoing granulomatous reaction, provoked by the carbon filaments (Fig. 31), stem cells infiltrating the pads and their surroundings appear to be incapable of maturing

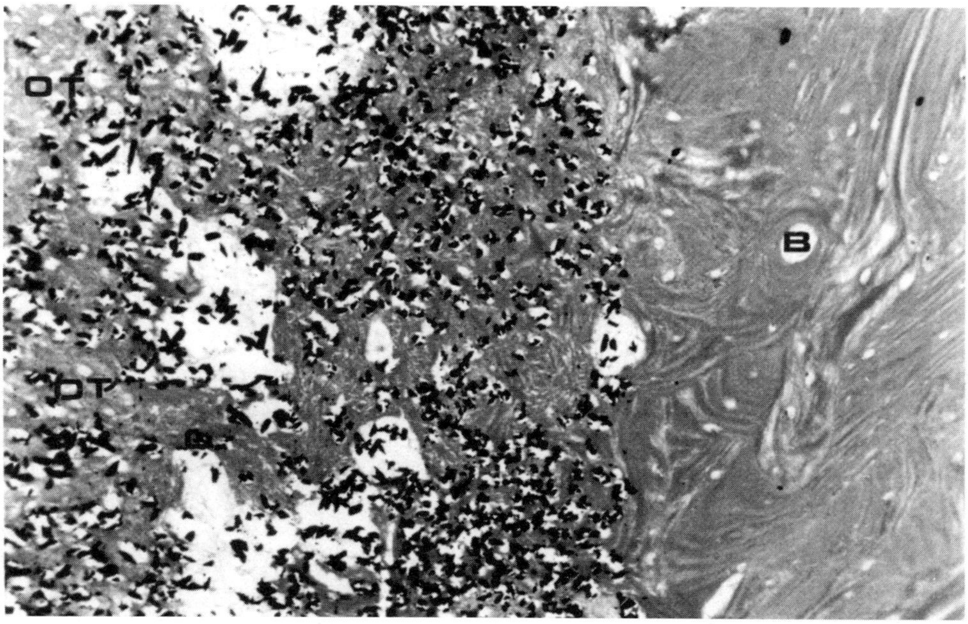

Figure 29 The inner aspect of the bony bed (B) of the intraosseous tunnel and outer aspect of a carbon filamentous braid implanted to replace a patient's ruptured anterior cruciate ligament. Single and groups of carbon filaments are encompassed by bony trabeculae (OT). Neither fibrous nor granulation tissue insulates the bony trabeculae from the carbon filaments. Hematoxylin and eosin. ×200.

into highly differentiated cells (chondrocytes) synthesizing a highly complex (cartilaginous) matrix, instead forming a poorly organized scar-like fibrous tissue (Fig. 32). Animals' meniscal lesions repaired with either bundled carbon filamentous implants or carbon filaments–organic polymer composites heal by ingrowth of fibrous tissue and formation of fibrocartilage without a granulomatous response. A mild and harmless inflammatory response accompanies orderly fibrous tissue ingrowth into the woven carbon filamentous mesh used for the repair of defects in the rabbits' lumbar fascia. It is informative to highlight the differential organizational patterns of different constituents of a device inserted at a certain body site: Replacing a rabbit's meniscus, the concentrically stacked hoops of carbon filaments ensheathed by terylene filaments reveal abundant fibrous tissue in between the polymeric filaments while the carbon portion of the implant remains virtually unorganized [420–426].

The purported inconsistencies in tissular reactions to carbon filamentous devices cannot be ascribed to investigators' bias nor to conflicting interpretations. Ostensibly opposing patterns of host behavior are explained away when one takes into account the diverse biomechanical settings operative at different sites for different applications. In addition to species-specific handling of foreign bodies, the anatomic site, implant stability and interfacial motion determine the tissular response. Extra-articular reconstructive surgery is often successful. Intra-articular operations are usually ineffectual. These facts validate the pivotal roles played by biomechanical, anatomophysiological and other factors in establishing the morphological and functional outcome of reconstructive surgery employing alloplastic materials. The organizational pattern by growth of tissue into a carbon filamentous device replacing an Achilles tendon or a collateral ligament is fundamentally identical with that of its intra-articular counterpart. Yet, fibroblastic proliferation and collagenization within and without the implant progress more favor-

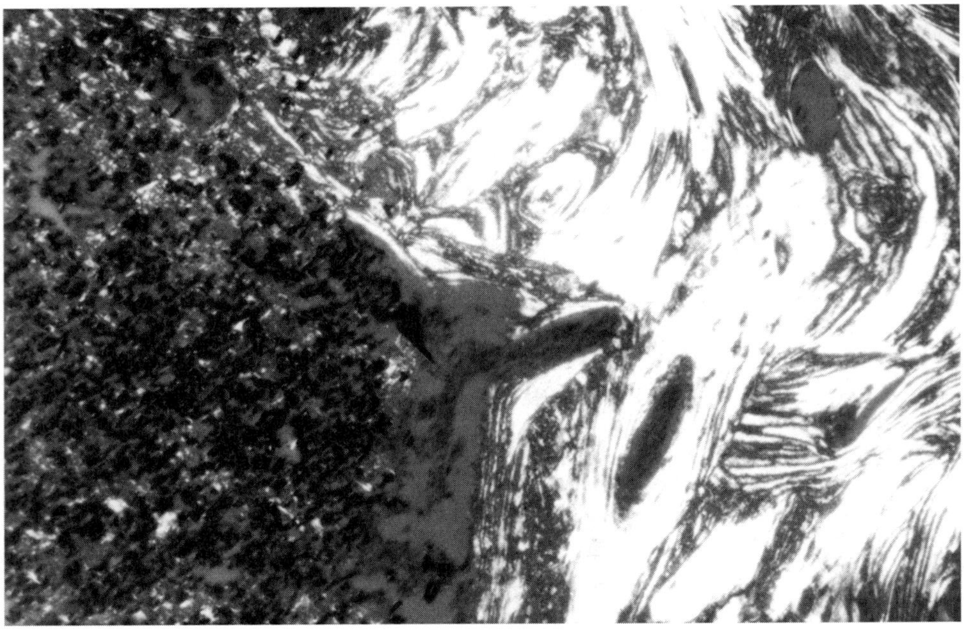

Figure 30 The same field as in Fig. 29 photographed under polarized light. The innermost layer of the implant's osseous bed and the ingrowing osseous trabeculae are composed of woven-fibered bone. The bulk of the bony bed consists of lamellar-fibered bone. Hematoxylin and eosin. ×200.

ably in extra-articular sites such that the implant's diameter increases several fold within a few months. Ingrowth of fibrous tissue into an anterior cruciate ligament prosthesis is slower than into a collateral ligament implant, as dramatically illustrated by a comparison of patients' biopsies simultaneously obtained from both sites. Since wear debris is not generated in significant amounts within collateral ligament and Achilles tendon substitutes, there is no room for the infamous "small particles disease." Also and in sharp contrast to the intra-articular site, undifferentiated mesenchymal cells are available in abundance and revascularization is expansive in extra-articular locales. No wonder that near physiologic tensile strength is rapidly regained postoperatively and, consequently, the replacements successfully substitute for extra-articular tendons and ligaments. Still, when tested in vitro, the constructs fail at the bone-implant junction, because a functionally adequate osseous anchorage mechanism fails to materialize [208,426–429].*

VIII. BIODEGRADABLE SYNTHETIC POLYMERS

Biodegradable polymers have obvious advantages when function is required for a limited time span only, e.g., for bone fracture fixation, bone defect filling, bridging of severed tendons or blood vessels and drug delivery. In addition to possessing proper mechanical strength, these

*The functional significance of the enthesis in the context of physiological load transfer from a ligament or tendon to the bone is discussed at some length in the chapter on collagen-based implants.

Figure 31 Giant-celled granulomatous reaction aside several bundles of a poorly organized carbon filamentous pad inserted in the articular aspect of a patient's patella seven years prior to retrieval. There is hardly any penetration of cells in between individual carbon filaments. Goldner stain. ×200.

polymers must be nontoxic, nonantigenic and cause least morbidity. The aliphatic polyesters are the most desirable biomaterials in this context. Polylactic acid, polyglycolic acid, poly-ε-caprolactone and polydioxanone as well as their copolymers have been extensively studied. These and similar biomaterials have been collectively referred to as biodegradable polymers. In the strict sense, however, they pertain to a heterogeneous group of materials: The *bioabsorbable* polymers dissolve in the body fluids without either a chain cleavage or a molecular mass decrease. The *biodegradable* polymers break down into macromolecules which, though not eliminated from the body, disperse locally or remotely. The *bioresorbable* polymers degrade into products which are eliminated by filtration, either directly or after having been metabolized via the natural pathways. Finally, the *bioerodible* polymers are subject to surface degradation. These subtle distinctions between elimination modes are often disregarded. The prefix "bio" reflects phenomena contingent on a material's contact with tissues, cells or enzymes of the body fluids. The terms (bio)degradability and (bio)resorbability are interchangeably spoken of in the literature. The ideal "biodegradable" polymer provides the least inflammation and its gradual disappearance is synchronized with the remodeling of the ingrowing tissue in the desired direction without leaving a trace of the preceding intervention. In vitro, the surface of a biocompatible synthetic polymer ought to be conducive to cellular growth, attachment, spreading and migration. In vivo, it ought to display an adequate conductive potential for both soft and hard tissues. In practice, the rate of biodegradation is all important because an implant

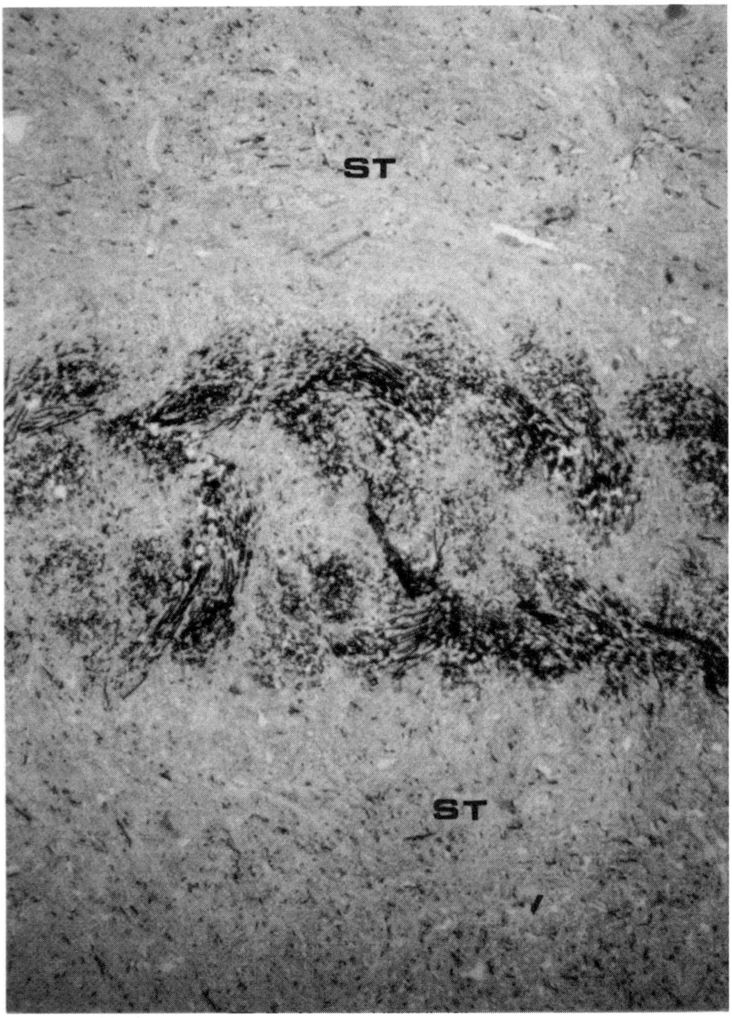

Figure 32 General view of a woven carbon filamentous pad inserted in the articular aspect of a patient's patella. The interstices of the bundled carbon filaments have scarcely been permeated by cells after seven year on situ. Where there are no bundled carbon filaments, a poorly aligned, scar-like fibrous tissue (ST) incorporates solitary carbon filaments. The hoped for growth of cartilaginous tissue into the articular defect has not materialized. Goldner stain. ×80.

should retain its mechanical properties for a specified time period. Apropos the biomechanics of biodegradable devices, the self-reinforcing technologies have been introduced so as to provide the surgeons with implements of adequate properties and strength retention. Polymers' innate or manufacturing-related capacity to activate the complement cascade may elicit unforeseen severe inflammatory reactions in spite of their affirmed in vitro biocompatibility—the results of routine in vitro tests being ill-adapted to predict this and similar unexpected events unless they are specifically searched for [430–434].

The biodegradable synthetic polymers in bulk form, especially the poly-α-hydroxy esters, are well tolerated. In vitro, they favorably sustain the growth of rat osteoblasts, the cells attaching to the surface, early exceeding confluency numbers and adequately producing alkaline phosphatase. However, their degradation products are chemically irritative in vivo. It is not surprising, therefore, that devices manufactured from biodegradable synthetic polymers induce inflammatory, granulomatous and fibrosing reactions. For as long as the implants persist in the host, their external and internal surfaces are insulated from the tissues by mono- and increasingly also polykaryonic macrophages. Following their implantation into soft tissues, the majority of the commonly used biodegradable polymers do not discernibly damage adjacent or remote structures. A mononuclear-celled inflammatory infiltration, with recruitment of macrophages, vasoneogenesis and collagen synthesis by proliferating fibroblasts make up the prototypical response. Degradation of the implants is accompanied by enhanced deposition of collagen fibers in widening spaces arising in the wake of the removal of polymeric fragments. Different biodegradable synthetic polymers differ from one another in the severity of the tissular responses they occasion. This holds true for polymers of the same family but unequal molecular weight and also for those with or without leachable impurities or soluble low molecular weight parts. The severity of the reaction of some polymers is site specific. For example, while the phosphate-fiber-reinforced polycaprolactone/lactide composition is nontoxic in vitro and well accepted in osseous environs, it causes necrosis when inserted into the muscle of rabbits [435–437].

The degradation kinetics of chemically dissimilar and physically differently designed polymers diverge from each other. The degradation kinetics are as well altered by additives. Enzymatic or nonenzymatic hydrolysis and bulk or surface degradation participate in the bioresorption of the biodegradable polymers. The degradation may take place with or without cooperation of the macrophages. It may or may not be associated with oxidation and mechanical perturbation of the polymer. These disparate pathways operative in bioresorption depend on the particular biomaterials studied. The widely used α-polyesters—polylactide and polyglycolide—are degraded principally by hydrolysis. Following deesterefication, the fragments are incorporated into metabolic pathways and the end products are excreted in the urine and feces or exhaled by the lung as carbon dioxide. The biodegradable polymers are not known to exert adverse systemic side effects [7,70,438–449].

Degradation products of a biocompatible polymer may be toxic, especially if they accumulate and are concentrated at the site of their formation. As a rule, monomers inhibit cell proliferation to a greater degree than their polymers. The rate of bioresorption is decisive in determining the extent and intensity of the inflammatory response to a degrading polymer. The rate is not only a function of the material itself but also of the specifics of the implanted device, such as its size, shape, density, surface area and porosity. In vitro tests portray a scale of toxicity of the monomers, in which the ε-caproic and lactic acids have the dubious honor of heading the list (maybe unexpectedly in view of the prestige enjoyed by their respective polymers), followed by glycolic acid, cyclohexane dimethanol and propionic acid and winding up with some of the less well known compounds. The results of Suganuma and Alexander's exquisite experiments catalog the reaction of bone to diverse coupons inserted into the intramedullary cavity of canine femora. More bone grows in between polymeric coupons than in between stainless steel coupons for the first six weeks. Once degradation of the polymer (polylactide) commences, granulomatous reactions to the small breakdown particles usher in extensive bone resorption. Suganuma and Alexander relate the biocompatibility of polylactide to the postoperative time interval. The biocompatibility of polymeric devices appears to decrease dramatically with time of residence in the body while the biocompatibility of metallic implants stays constant [448–452].

Incidental to the surface textural features, large interfaces and plentiful release of particulates, the tissular response is stronger to a porous and degradable implant than to a compact and nondegradable one. Hence, porous and degradable devices seem to be less biocompatible than compact and nondegradable implants. The molecular weight of the polymer used for the manufacture of the implant plays a role as well. The rapid liquefaction of the low molecular weight polymers, e.g., polylactide, causes an exuberant macrophagic response. The excessive phagocytosis of polymeric particles is associated, by itself and of itself, with cell damage and cell death. Theoretically, only the slowly degrading and high molecular weight moieties of the biodegradable polymers behave as genuinely biocompatible materials within the osseous environment as they provoke but mild reactions and the abundantly formed bone directly apposes their surfaces [70,453–456].

The instantaneous influx of neutrophils and eosinophils after implantation of synthetic biodegradable polymers is followed by a moderate macrophagic response, an incremental fibroblastic proliferation and, eventually, a reduction in the cell population accompanied by an enhanced collagen synthesis. Where degradation has sufficiently advanced, macrophages remove polymeric microparticles not only at the perimeter of the implant but also from its centermost region. Compared with the scenery enfolding after insertion of stable materials, the great upsurge of the cellular catalyzing enzymes suggests active attack on the biodegradable material. Attendant on the high rate of disintegration of the polymer, the tissular response and giant-celled reaction are intense. Rapidly degrading implants, e.g., glycine and lactic acid copolymer, elicit more extensive responses than slowly resorbing ones, e.g., polylactide. Finally, the larger the biodegradable implant the more expansive is the tissular reaction [441, 457–460].

Composites tailored to specific requirements combine several biomaterials. They are likely to open new venues in implantation surgery. To cite an example, the filamentous poly(2-hydroxyethylmethacrylate)/poly(caprolactone) blend hydrogel matrix reinforced with polylactic acid mimics structurally and mechanically native tendons. In view of its proficient bearing as a load transmitter, it is intended to temporarily bridge tendon gaps. Unfortunately, ingrowth of tissue in between the individual filaments of the composite does not ensue within two months of the operation. Also, the filament bundles are surrounded by crude scar tissue rather than the expected, axially aligned collagen fibers. This result is diametric to the theorized betterment. Any implant is unsuitable to serve in tendon repair if it results in the formation of intratendinous scars after its resorption, no matter that it is composed of otherwise praiseworthy biomaterials on account of their biodegradable and biocompatible properties. This implant is incongruous, by implication, with the objectives of reconstructive surgery. Research on artificial devices has to target attainment of refashioned tendons and ligaments which are anatomically and functionally adequate following resorption of the polymers [461,462].

Biodegradable rods, pins, screws and plates have been used with success for the internal fixation of fractured bones. They ought to retain their initial biocompatibility, strength, stiffness and ductility—without causing side effects—for several weeks to months prior to undergoing complete bioresorption. In addition, the strength, stiffness and ductility of the biodegradable devices should nearly equal those of stainless steel. Although polylactide and other polyester-based implants promote intramedullary bone formation, are osteoconductive and become osseointegrated, modern biodegradable polymers do not measure up to the requirements. With strict indications for their application, effective healing of bone fractures is attained, as reported by European investigators. Biomechanically, the outcome of fixation of rabbits' osteotomized femora with self-reinforced polylactide or polyglycolide rods, on the one hand, and metallic wires, on the other hand, is equivalent. Initially, osteogenesis proceeds undisturbed. This is the anticipated eventuality, since the inflammatory-granulomatous response does not begin until degradation and liquefaction of the polyesters reach the final phase, which trails bone healing by a long time. The implants are resorbed and replaced by fibrous, osseous and hematopoietic tissues. The traits of composites are altered by the ratio of their components. Thus, by raising calcification of the copolymer, increasing the ratio of poly(ethylene oxide) to poly(butylene terephthalate) enhances the implant's bone-bonding potential. On the debit side, it should not go unnoticed that polylactide-based osteosynthesis devices are said to be cancerogenic in rats [442,463–474].

In clinical practice, the devices fabricated from polylactide, polyglycolide and their copolymers, especially the self-reinforced constructs, measure up to the requirements demanded for fixation of patients' bone fractures. The implants induce a moderate and transitory inflammatory reaction. So far so good. Only that internal fracture fixation with biodegradable rods (notably those made of polyglycolide or lactide-glycolide copolymer) is beset with an unpleasant complication. Swelling appears during the second to fourth postoperative month at the site of the fixated bone in close to 1 of every 10 patients. Spontaneous drainage, leading to formation of a discharging sinus, and surgical incision produce copious amounts of a sterile, pus-like secretion, comprising the residues of the degrading polymer. Biocompatibility of any material depending, inter alia, on its specific application, it is instructive to note that such secretions occur almost five times oftener after fixation of a fractured distal radius than of a fractured ankle. Analysis of the patients' biopsies and animals' retrieved specimens unfolds a chain of events spanning the gamut of plausible courses, with adequate union of the fractured bone fragments at one end of the spectrum and osteolysis and tissue necrosis at the other end. Bone healing does take its ordinary path after fraction fixation with these stiff polymeric devices. The implants are initially well accepted, though they are fenced off from the bone by a synovial-like, giant-celled granulomatous interfacial membrane. Every so often, breakdown products, toxic compounds, pH change, and all collectively damage the tissues, cause necrosis and provoke a brisk neutrophilic infiltration and an expansive inflammatory-granulomatous response. Macrophages ingest the polymeric detritus. Contingent on their activation, the macrophages release osteoclast-activating factors. The recruited and activated osteoclasts resorb the surrounding bone. In some patients and animals, the osteolytic cavity expands considerably, but the process later resolves, probably coincident with complete resorption of the residues. Because fewer than 10% of the patients treated for a fractured bone with self-reinforced polyglycolide rods develop accumulation of sterile fluid requiring drainage, the fixation within cancellous bone is good and no hardware has ever to be removed, biodegradable rods are, according to some authors, best suited for the fixation for certain fractures [64,65,440,475].

Distinct biodegradable synthetic polymers must be individually tested for each proposed application. Take the interference screws. Because of a short degradation time, high concentrations of by-products build up and cause joint effusion and cavitary osteolytic lesions in some patients whose anterior cruciate ligament autograft is fixated to the bone with poly-DL-lactide-glycolide copolymer screws. These difficulties may be avoided if poly-DL-lactide screws are used. While they do not degrade within the first six postoperative weeks, they have cleared completely after 10 months such that there is no foreign body granulomatous response to remnants of the screws [476].

The extent and intensity of the adverse reactions to intraosseous biodegradable polymeric implants differ not only between animals of different species but also between animals of the same species. Even so, when a material undergoes a metamorphosis from the beneficial to the harmful state, Williams' biocompatibility dictum is not satisfied: The compositional changes of the polymer, occasioned by activities of the milieu interne, are accompanied by an excessive and unruly, neither qualitatively nor quantitatively appropriate tissular response. A patient's and physician's apprehension may also be raised by a progressively expanding lymphadenopathy due to a macrophagic response to migrating crystalline remnants of, say, polylactide plugs. This setting, too, does not fit in with the profile of a biocompatible biomaterial. Why do not other polylactide-based implants disclose such adverse behavior at other locales? Thus, the bilayered regeneration template with a porous polylactide undersurface readily incorporates into the subcutis and muscle, the inflammatory response abates with time and fibrous tissue amply grows into the porosities of the device [70,471,477].

Synthetic biodegradable polypeptides constitute a novel class of biomaterials. They deserve only brief mention, because potential clinical applications have not been popularized. The in vivo reactions to the synthetic polypeptides, e.g., polyglutamic acid ester derivatives, are kindred to those described above in the context of the synthetic biodegradable polymers. Following an acute inflammatory infiltration and the fragmentation of the inserted polypeptide, the interstices are invaded by granulation tissue and the implant is fibrotically encapsulated. That neighboring parenchymal tissues are not overtly damaged evidences the polypeptide's biocompatibility [478–481].

IX. COLLAGEN-BASED IMPLANTS

Organ and tissue transplants constitute a form of biomaterial in the widest sense. They are not dealt with herein. On the other hand, collagen-based implants are briefly reviewed. For practical reasons, surgeons avoid the utilization of fresh, allogeneic or xenogeneic collagen-based preparations. In addition to the chance of transmission of infectious diseases and the obstacles posed by the histocompatibility-antigens barrier other hurdles must be overcome prior to the widespread application of the chemically modified collagen-based implants. Their adept processing is decisive in order to attain adequate tissular incorporation. Glutaraldehyde-treated collagen is not permissive of fibroblastic repopulation and induces an ongoing inflammatory reaction. The principal cause of clinical failure of stented glutaraldehyde-treated porcine aortic bioprosthetic valves is calcification. The biocompatibility of glutaraldehyde-treated collagen as such is understandably distrusted [482,483].

The resorbable collagen-based implants may be used as temporary scaffolds for tissue regeneration in the treatment of soft tissue defects, bone fractures and diseased or lost tendons, ligaments and blood vessels. The allogeneic or xenogeneic, frozen or freeze-dried, natural or modified collagen-based devices are rather well tolerated in the cutaneous and subcutaneous sites, the host's reaction being mild. Nonetheless, the macrophagic response to a collagen-based implant, such as fibrin-collagen paste, and the succeeding fibroblastic proliferation may be so profuse as to suppress the desired osteoneogenesis [484].

Cross-linked purified collagen is preferred over other collagen moieties because of its congenial physicochemical properties and minimal immunogenicity. Cross-linked collagen fibers support cell adhesion, locomotion and function. They optimally suit to serve as temporary porous scaffolds for cellular ingrowth and allow for an effective tensioning by the ingrowing myofibroblasts. Their degradation products continually recruit more fibroblasts, an advantage in surgical repair of fibrous structures. A crucial limitation of cross-linked collagen-based devices is a rapid loss of mechanical strength. The degradation rate of reconstituted collagen fibers is critical. Prolonged degradation may inhibit remodeling of the implant by ingrowth of and foster its encapsulation by fibrous tissue. In this context, the stability of the cross-links is of concern. In spite of extensive and lengthy rinsing of the preparations, the glutaraldehyde-cross-linked collagen still releases enough toxic chemicals to kill the fibroblasts in vitro. Failure is a distressing ordeal when the anchoring tissue of a life-sustaining collagen-based prosthesis, e.g., a cardiac valvular substitute, suddenly withers away after a protracted interim of effective clinical performance [485–492].

Following implantation of collagen-based devices, the early neutrophilic infiltrate is promptly replaced by mono- and polykaryonic macrophages and ingrowing fibrous tissue. The foreign collagen and calcific material deposited therein are resorbed within a short time. Concurrently, the foreign collagen is substituted by the host's own fibrous tissue. Yet, the biocompatibility of collagen-based implants is contentious. Knowledge concerning the issue of immune reactions is fragmentary. Anticollagen antibodies circulate in the serum of animals and patients with a with collagen-based implant. Their presence is not associated with any detrimental clinical sequelae. It may solely reflect an epiphenomenon, perhaps announcing elevation of the titers of natural antibodies, which are not harmful to the organism [493–497].

The diverse cross-linked dermal collagen preparations differ from one another as well as from native dermal collagen, of sheep provenance for instance, in their biological and, specifically, in their cytotoxic effects. The mode and the extent of cross-linking have a profound influence on the rate of biodegradation and tissular response. In comparison with the cross-linked collagen-sepiolite complex, the native collagen-sepiolite complex degrades within a short while, provoking a copious granulomatous response. In contrast to glutaraldehyde-treated collagen fibers, collagen fibers cross-linked with cyanamide better support fibroblastic growth in vitro. This and more. Glutaraldehyde-treated fibers are not conducive to invasion of fibroblasts into and collagen synthesis within the implants. But they do elicit a vigorous acute inflammatory reaction, stimulate fibrous encapsulation, cause morphologic anomalies of the neighboring tissues and initiate matrical calcifications. Acyl azide-cross-linked dermal sheep collagen-based implants provoke an extensive macrophagic response. Hexamethylenediisocyanate-cross-linked dermal sheep collagen induces only a mild inflammatory infiltration and stimulates an orderly ingrowth of fibroblasts and ample collagen fiber formation. Better results have been achieved with the reconstituted carbodiimide-cross-linked collagen-based devices. Their structural and mechanical properties approximate those of native collagenous structures. They degrade in vivo at a similar rate to that of an implanted autogeneic tendon graft. Severed rabbits' tendons repaired with a carbodiimide-cross-linked collagenous graft have recovered 10 weeks postoperatively, the implant having been resorbed and replaced by fibrous tissue akin to the collagen fibers native to the site. The reconstructed structures have recovered the normal strength and modulus of the replaced tendons 20 weeks postoperatively. Tendon ruptures heal spontaneously by the formation of a poorly organized scar tissue. This contrasts with the refashioned, normally textured and crimp patterned neotendon observed one year after a large gap in a tendon has been repaired with a carbodiimide-cross-linked collagen-based implant. This accomplishment in the treatment of a notoriously problematic setting in orthopedic surgery speaks of the biocompatibility of carbodiimide-cross-linked collagen fibers and, contrariwise, highlights the bioincompatibility of glutaraldehyde-cross-linked collagen fibers [458,495–498].

The enthesis—the aligned collagen fibers-fibrocartilage-calcified fibrocartilage-bone complex—constitutes the physiological coupling of the ligaments and tendons with the bone. The enthesis is nature's solution to the demanding target of force transfer without undue concentration of stress and shear at the interface of two very different tissues. A leading cause of failure of replacements is the body's inability to refashion an enthesis at the junction a carbon or polymeric filamentous prosthesis with the bone. With rare exceptions, synthetic filaments attach to the bone via a granulomatous or fibrous interfacial membrane, a biomechanically inadequate linking mechanism. It is not surprising, therefore, that failure on loading a ligament or a tendon replaced with synthetic filaments generally occurs at this unphysiological anchorage site. In contrast, an enthesis-like junction is remodeled after substitution of tendons or ligaments, including the anterior cruciate ligament, with tendon autografts. The suitability of collagen-based devices to functionally reduplicate these autografts is plainly elucidated by the fact that enthesis-like junctions are refashioned after intraosseous insertion of carbodiimide- or cyanamide-cross-linked collagen fibers embedded in a collagen matrix [499–501].

Implants of carbodiimide- and cyanamide-cross-linked collagen fibers are initially infiltrated by inflammatory cells and, then, they are rapidly degraded and replaced by the host's well-aligned fibrous tissue. Their replacement by autochtonous connective tissue coincides with formation of refashioned enthesis-shaped structures. On pullout testing a long-term carbodiimide- or cyanamide-cross-linked collagen-fibers-based implant, failure usually occurs within the bulk of the construct rather than at an attachment site. Similarly crafted but glutaraldehyde-treated devices are ill-suited as tendon or ligament replacements, since parts of their original collagen fibers are still intact at the 20th postoperative week. Solvent preservation and preoperative irradiation of collagen fibers-based devices may bring about changes in their chemical and morphological properties. Because of meager and deficient tissular ingrowth, their expediency to perform as effective substitutes is impaired. A hydroxyapatite-tricalcium phosphate-collagen composite has been evaluated for the treatment of bone loss accompanying fractures of the canine femur. One year postoperatively, bone occupies 40–70% of the former defect. The fibrillar collagenous component of the implant has been completely resorbed and its place taken by a newly formed, mostly lamellar-fibered bone, directly apposing the ceramic particles. The composite is blandly incorporated within the bone. Except polarization optically, the newly formed and native bone can hardly be distinguished from each other. Thriving hematopoietic bone marrow within the remodeled implant evinces the biocompatibility of the composite. However, wherever they come into contact with the adjacent soft tissues, scattered ceramic particles elicit an encapsulating giant-celled granulomatous reaction [202,218,502–505].

X. BIOCOMPATIBILITY OF DIALYSIS MEMBRANES AND FLUIDS

Concern has been voiced with respect to the quality of membranes utilized in hemodialysis treatment. Exposure of blood to poorly biocompatible membranes has a detrimental effect on the patients' clinical course even though the renal disease itself is unaffected. Clinical practice suggests that bioincompatible membranes exert systemic toxic effects. Experts in this field have schooling in the principles of hemocompatibility, a subspeciality of biocompatibility. Issues which are mostly peripheral in reconstructive surgery are central in hemodialysis, such as activation of the complement cascade by dialysis membranes. Researchers' goal is to develop systems which minimize changes of the cells and proteins of the blood. Patients dialyzed with the low-complement-activating membranes have a survival advantage over those treated with membranes having a high complement activation potential. The compatibility profile of cellulose-based membranes appears to be inferior to the profile of membranes manufactured from polysulfone, acrylonitrile copolymer or cuprophane; lately, polyacrylonitrile has been evaluated. There are no significant differences in the hemocompatibility of these compounds. On the other hand, the postdialysis neutropenia, thrombocytopenia and hypocomplemetemia are limited after dialysis with a polymeric membrane when compared with the deviations from norm after a cellulose-based membrane is used. At any rate, dialysis with polysulfone membranes results in leukopenia and an increase in the levels of neutrophil elastase and C_{3a} and C_{5a} in the serum. The β-2-microglobulin serum level increases in dialyzed patients. Being larger with bioincompatible membranes, the increments are lowest in patients dialyzed with a polysulfone-based membrane. This deserve special awareness since an elevated serum β-2-microglobulin is associated with amyloidosis. Not just the membrane makeup but also the sterilization methods have to be considered. All alterations are less substantial after employment of steam-sterilized membranes rather than ethylene-oxide-sterilized ones. Allegedly, steam sterilization "improves" the biocompatibility of the membranes. Early hypersensitivity symptoms, intradialysis eosinophilia and high-ethylene-oxide-specific IgE serum levels tend to diminish when steam-sterilized membranes are employed. Alike the biomaterials used for reconstruction of blood vessels, dialysis membranes should be nonthrombogenic. It is self-evident that man-made polymers do not equal the endothelial cells in preventing surface-induced thrombosis and maintaining hemostasis. Current research focuses, among others, on modifying the polymeric surfaces by immobilizing heparin onto them or coating them with a polymer-heparin composite. The nephrologist's dilemma is that proinflammatory pathways are universally activated during all dialysis modalities. Purportedly, the side effects are especially intense in patients dialyzed with "bioincompatible" membranes. Clinically relevant but of uncertain nature is the role played by the interaction between biocompatibility and membrane flux [506–514].

Hemodialysis patients have increased erythrocyte osmotic fragility compared to controls. The structural integrity of the erythrocytes benefit from dialysis with the polyacrylonitrile- rather than the cuprophane-based membranes. Also, platelet-neutrophil coaggregates and neutrophil hydrogen peroxide products increase after dialysis with cuprophane- but not with polyacrylonitrile-based membranes. There is little doubt that, at least as far as hemocompatibility is concerned, the material properties of polyacrylonitrile surpass those of cuprophane [515,516].

The biocompatibility of the conduit as well as of the fluid is of paramount import in the treatment of patients with end-stage kidney disease by peritoneal dialysis. The investigators seek, for instance, to find out whether an amino-acid- or a glucose-based peritoneal dialysis fluid is preferable. To exemplify, the lipopolysaccharide-stimulated macrophages isolated from the latter produce more interleukin-1β and interleukin-8 than cells present in the former. The phagocytosis capacity, as a measure for macrophagic function, is better preserved after exposure to the amino-acid-containing fluid while the higher release of the interleukins suggests a brisk intra-abdominal activation of the macrophages by the glucose-based fluid, which acts as a chemical inflammatory agent. Which of these and other parameters is most relevant for the biocompatibility of the peritoneal dialysis fluid [517]?

In patients undergoing peritoneal dialysis, local effects of the dialysis solution may be a compelling factor in biocompatibility considerations. Reflecting tissular injury, the morphological and functional aftermath of the dialyzing fluid and catheter is complex and involves permeability abnormalities of the peritoneal membrane as well as changes of the cellular and noncellular contents. Sclerosing peritonitis, a life-threatening disorder, develops in about 0.5% of patients undergoing continuous ambulatory peritoneal dialysis for four or more years. The etiopathogenesis is debatable; if bacterial infections are responsible for sclerosing peritonitis, as has been suggested by some investigators, then the disease has nothing to do with biocompatibility issues. Otherwise, dialyzing solutions and catheters should be painstakingly probed for their compatibility with the tissues of the abdominal cavity [518–520].

XI. BONE ALLOGRAFTS AND BONE-DERIVED FACTORS

Implantation of bone allografts, allogeneic or xenogeneic devitalized, mineralized or demineralized bone and bone morphogenetic protein may provoke an inflammatory and granulomatous reaction in the soft and hard tissues. This reaction inhibits induction of bone within the osseous environs. Autogeneic bone grafting is the only means to consistently assure flawless healing of bony defects by osseous tissue. Predictably, other materials would be recognized as nonself by the body. This biologically rational sounding perception is disputed by authors whose esoteric xenogeneic bone grafts do not induce inflammatory responses, and are, therefore, allegedly biocompatible. Minor and major histocompatibility antigen differences between a donor and host as well as different preparation modalities of the grafts verily affect the outcome of bone grafting. Osseoplasty with bone grafts dates back to the Stone Age. The history of implantation of bone grafts and bone-graft substitutes and present-day mastery of the methodologies make interesting reading. Being a detour in this path of this chapter, attention is called to the surveys of Damien, Enneking, Stevenson and others on this subject. Suffice it to say in the context of biocompatibility that a loss of template function is witnessed with each additional, preoperative processing step of the bone grafts, e.g., from freezing to freeze-drying to chemical modification [321,486,521–530].

Implantation of demineralized bone matrix is sometimes succeeded by excessive osteolysis rather than the anticipated expansive osteogenesis. Implantation of a composite preparation incorporating the bone-inducing extract within a carrier may upgrade the low osteoinductive potential of the xenogeneic demineralized bone matrix, most likely because the inflammatory reaction is thereby abrogated. At other times, however, composite implants fabricated from demineralized bone matrix and a biodegradable or a biostable polymer incite inflammatory responses which suffice to repress the activity of the osteoinductive agent [23,531–537].

The pernicious consequences of the inflammatory and granulomatous responses on osteoneogenesis is pictured by "an animal's own bone becoming bioincompatible," a preposterous, if not farcical, sounding proposition. Still, the granulomatous reaction provoked by powdery, irregularly shaped, edgy, micron-sized autogeneic bony particles (Fig. 33) is severe enough to inhibit new bone formation in a gap-healing model in the rabbit. The reaction being outrightly inappropriate in the setting of the bone-healing process, the autogeneic bony debris behaves toward its environment as if it were a bioincompatible matter. The deterministic drive deciding reactivities of the body is not solely the basic chemical nature of the material (provided of course that it is not toxic) but also its innate physical traits such as size, shape and surface topography [538].

XII. THE LESSONS LEARNED

Clinical failure of up-to-date prostheses utilized in reconstructive surgery ought to be divorced from the issue of biocompatibility of the materials employed in fabrication of the implants. With rare exceptions, prosthetic failure relates to untoward mechanical settings. Prosthetic failure cannot be construed to reflect on the biocompatibility topics straightaway. In the context of total joint replacements, a plethora of articles records the mishaps to which the cement is vulnerable, on the one hand, and the excellence of titanium and its alloys, on the other hand. Yet, less than 40% of cemented acetabular cups of total hip arthroplasties have failed after 11 years of service, but close to 40% of uncemented threaded titanium-based acetabular cups are already loose after 6 years. Logically, the stigma of aseptic loosening relates more often than not to factors other than biocompatibility of the alloplastic components, say, flawed design-related biomechanics. Catastrophic events, such as a fracture or rupture of an implant and excessive generation of breakdown products, do not fall within the domain of biological reactivities. Years of studious observation have resaid time and again that *the cells tend to be indiscriminate in their response to small particles*. When it comes to these infamous small particles, the body's defenses do not distinguish between rubble of their own provenance and litter of outside origin (Figs. 34, 35). This fact, among others, accounts for the monotony of the morphological and biochemical scene near diverse failed devices, as is forthwith evident

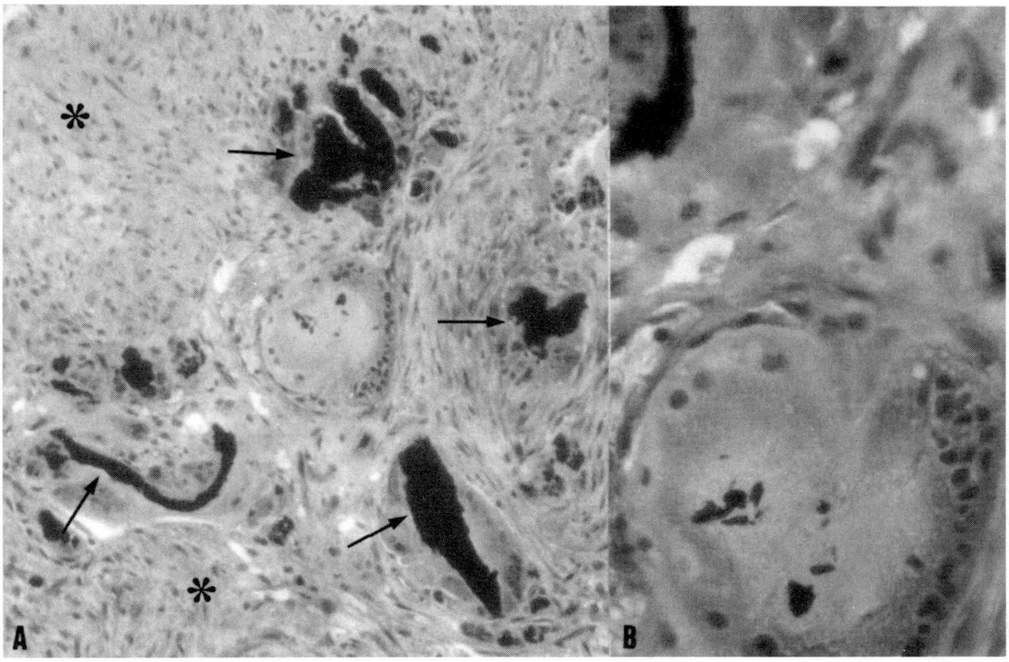

Figure 33 Giant-celled granulomatous reaction to autogeneic bony debris in the tibia of a rabbit. The drilled cortical defect was polluted with the animal's own micron-sized bony particles. Healing of the defect was delayed by the ensuing foreign-body-induced granulation tissue. (A) Mono- and polykaryonic macrophages are phagocytosing and engulfing variously size, bizarrely shaped, calcific, bony particles (arrows). The macrophages are scattered within a densely textured fibrous tissue (asterisk). (B) Small bony particles within the cytoplasm of a polykaryonic macrophage. Von Kossa stain with McNeal counterstain. (A) ×280 and (B) ×400.

when recalling the retrieval studies and experimental models described above. It is timely to put away the impulse to accuse the polymethylmethacrylate, as if it were playing an unrivaled role in periprosthetic bone resorption. Otherwise authors will go on declaring that localized osteolysis aside a stable, cementless femoral stem is "surprising since it occurred in the absence of bone cement," when in fact the intraosseous cavity was filled with polyethylene and metallic debris. It is recalled that localized, and particularly the balloon-like, pelvic osteolysis is more often encountered in patients with an uncemented acetabular cup than in those with a cemented one. It is well established that design features and reconstruction specifics* are principally responsible for initiation of the circumferential invasion of the polyethylene debris-rich granulation tissue into the bone-implant interface of acetabular cups. Having reevaluated the scientific basis of fixation of alloplastic components, Freeman, Hozack and co-workers have aptly concluded that, equivalence of the clinical results being overt, there is neither an overwhelming advantage nor a disadvantage for the cemented versus the cementless techniques. The primacy of biomechanics, rather than of the biocompatibility of the replacing biomaterials, should be

*Type of femoral component, design, position and technique of insertion of the acetabular component, presence of fixation screws or pegs, initial degree of porous surface-bone contact, volumetric wear rate of the polyethylene and stress shielding.

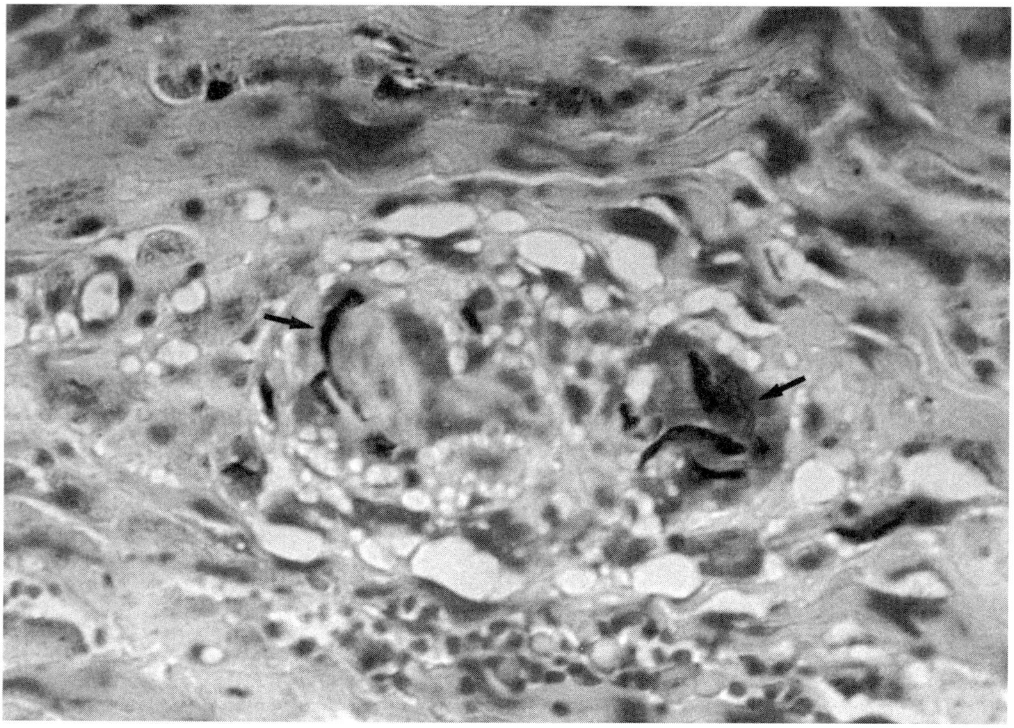

Figure 34 Metallic-particles-induced giant-celled granuloma in the pseudocapsule of an aseptically loosened, uncemented total hip arthroplasty. Some large titanium fragments (arrows) have been ingested by the polykaryonic macrophages. Movat stain. ×320.

acknowledged also in the failure of most intra-articular nonbiodegradable, synthetic, filamentous implants. The latter are incapable of successfully substituting for anterior cruciate or other intra-articular ligaments because they do not emulate the physiologically demanding and complex biomechanical functions of the native structures [408,539–542].

In the context of load transfer between an implant and the bone, osseointegration constitutes functionally the optimal host-implant relationships. Clinically, osseointegration is defined as the process whereby an asymptomatic rigid fixation of alloplastic materials is achieved and maintained in bone during functional loading. Morphologically, it is defined as the direct contact of the bone and implant at the light microscopic level. Ultrastructurally, osseointegrated surfaces of the implants are separated from the bone by a few hundred nanometers thick proteoglycan layer; e.g., this layer measures 300–1000 nm at the surfaces of osseointegrated cemented alloplastic components. Bone-bonding, which is defined as the establishment by physicochemical processes of continuity between an implant's surface and the bone matrix, is the favored state of osseointegrated parts, be they metals, polymers or ceramics. Naturally, just thoroughly biocompatible materials evince bone-bonding. Comparative analysis has clarified that bone-bonding ceramics do not adversely alter mineralization of the bone matrix while the non-bone-bonding ceramics exert harmful effects on the synthesis and maturation of the matrix vesicles. The dominating role of the surface texture is unmistakable. Contingent on whether the undifferentiated mesenchymal cells face a rough- or a smooth-surfaced implant, an osseoin-

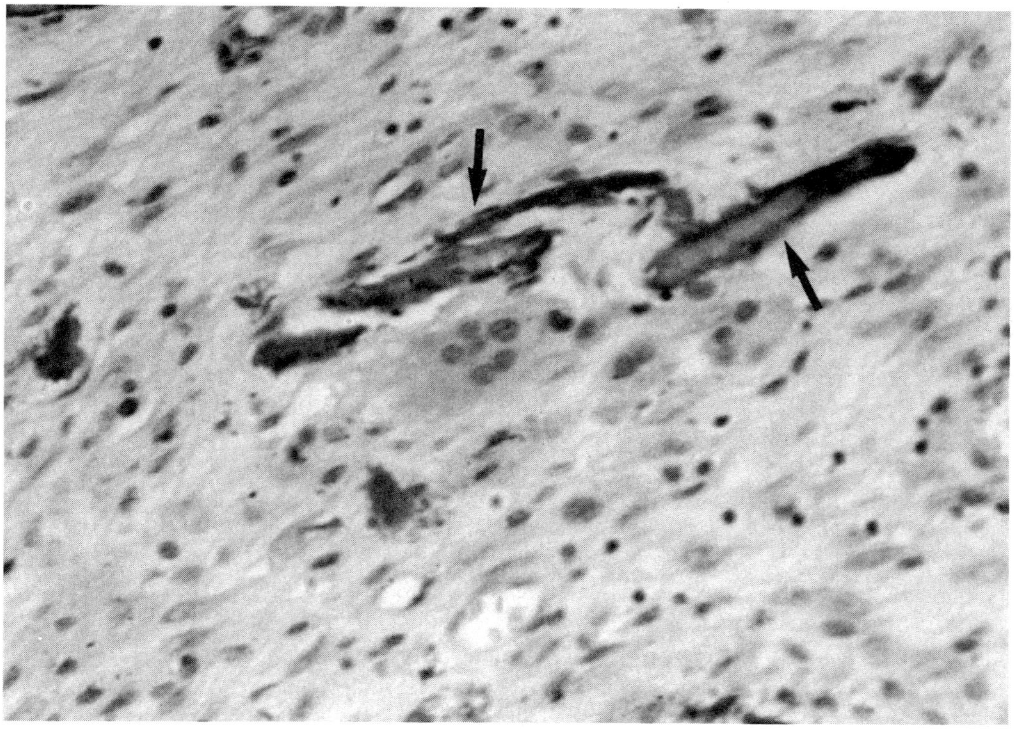

Figure 35 Bony-debris-induced giant-celled granulomatous reaction in the interfacial membrane of an aseptically loosened, uncemented total hip arthroplasty. A few large calcific fragments (arrows) are scattered in a giant-celled granulation tissue in which other than bone-derived foreign bodies are not detectable. Hematoxylin and eosin. ×320.

tegrating or granulomatous setting prevails. The common property of all bone-bonding biomaterials (titanium and its alloys, bioactive ceramics) is the formation of a calcium phosphate-rich, apatite-like surface layer. The reaction of cortical bone to titanium and hydroxyapatite-coated titanium is similar but in addition to promoting more bone formation, hydroxyapatite-coated surfaces also trigger bidirectional bone growth and replacement of a motion-induced fibrous interfacial membrane by bone [266,270,331,334,441–551].

Yet, both titanium and hydroxyapatite particles, generated when a flawless bone-implant apposition does not ensue, provoke a granulomatous reaction. The surmised gap-healing potential of hydroxyapatite does not alleviate the obstacles to the faultless bone-implant apposition. On the contrary, bone contact of hydroxyapatite-coated surfaces decreases, whereas bone contact of the surfaces of uncoated titanium increases with time. The granulomatous reaction to particles formed on delamination of the hydroxyapatite coating and the osteolysis resulting therefrom, particularly in a loaded setting, are responsible for the untoward effect of the coat fragmentation in patients with orthopaedic or dental implants. The course of hydroxyapatite-coated total hip replacements goes from bad to worse if third-body wear, caused by particulate hydroxyapatite entrapped within the joint space, increases breakdown of the polyethylene acetabular insert [259,356,552–557].

The practitioner's predicaments cannot be resolved by the basic scientists. In the reconstructive operations, the barrier to success is not that suitable and biocompatible materials are not available. The hurdles that the surgeon has to overcome in order to attain long-lasting stable fixation of artificial joints are harsh mechanical and physical circumstances at the interfaces and articulating surfaces. Effects of the physiologically operative forces are amplified by anatomical constraints, design faults, malalignment, components placed "onto" rather than "into" the bone, teeter-totter phenomena with tilting caused by asymmetrical loading, abnormal strain patterns, stress shielding, microfracture of bony trabeculae supporting the constructs and gross polyethylene failure. In aseptic loosening of cemented arthroplasties, additional factors playing a role are a poor cementing technique, debonding at the cement-metal interface, fracture of the cement mantle, upsets at the bone-cement unit and fatigue cracks in the cement. Bone remodeling stemming from an altered stress-strain field combines with these events to occasion generation of excessively large amounts of prosthetic detritus. The ensuing inflammatory-granulomatous response causes that extensive osteolysis which leads to prosthetic failure. To sum, mechanical and biological adversities conflict with effective fixation of implants in the long term. In the final analysis, the incidents at and about a prosthesis raise the rate of osteolysis and lower the rate of osteogenesis. Hence, all joint replacement systems, whether cemented or uncemented, are destined to loosen some time. Within the presently popular paradigm, mechanical adversities are solely condemned for the loosening of artificial joints (Fig. 10). In this scheme there is no place for concern about bioincompatibility of the employed ingredients, not one of which is known to provoke osteolysis when facing bone in its unadulterated bulk form. That the so-called interfacial membrane adversely affects the functional integrity of orthopaedic implants has been amply dealt with in the literature and repeatedly pointed out herein. In fact, there is nothing unique to the interfacial membranes within the framework of the body's defense options. They express the formation of protective buffer zones fencing off inanimate and animate invaders. For instance, they bound a polytetrafluoroethylene-based nephrostomy tube (Figs. 36, 37), but they as well surround an echinoccocus cyst. They perform laudable functions under most pathophysiological circumstances. Their unlucky repercussions on the well-being of some patients with implants, notably intraosseous ones, are a payoff modern medicine offers for the outstanding achievements in other patients [116,558–561].

The biological behavior of biomaterials, which under disparate situations evince opposing effects, have been elucidated at some length in an attempt to impress on the reader, be she or he convinced, lately converted or skeptical, that current denotations of biocompatibility are unsatisfactory. Materials deemed biocompatible by basic scientists are made the perpetrators of prosthetic failure by practitioners. Irrespective of the outcome of test tube and animal experiments, definitions ought to be uncommitted and separated in everyday practice from questionably relevant in vitro and unfittingly made in vivo experimental situations. The list of observations stressing this point is long.

1. Implants are bathed in a replenishable, proteinaceous, enzyme-laden and cell-rich body fluid. They are surrounded by a reacting and remodeling tissue. Transient pathophysiological processes regulate and continuously adjust the interfacial environment. Injurious agents, unless accumulating in extra large quantities, are likely to be totally or partially removed by the flowing fluids and surveiling cells. These events cannot be satisfactorily mimicked in vitro experiments.
2. Secondary to unlike handling of biomaterials, in vitro tests often fail to recognize subtle differences in the surface chemistry of materials.
3. Ordinarily, the in vitro tests do not take in account the isolation procedure-effected changes in cell behavior. It is well known that isolated cells respond more readily than explanted cells to changes in the surface chemistry of a substratum they colonize.

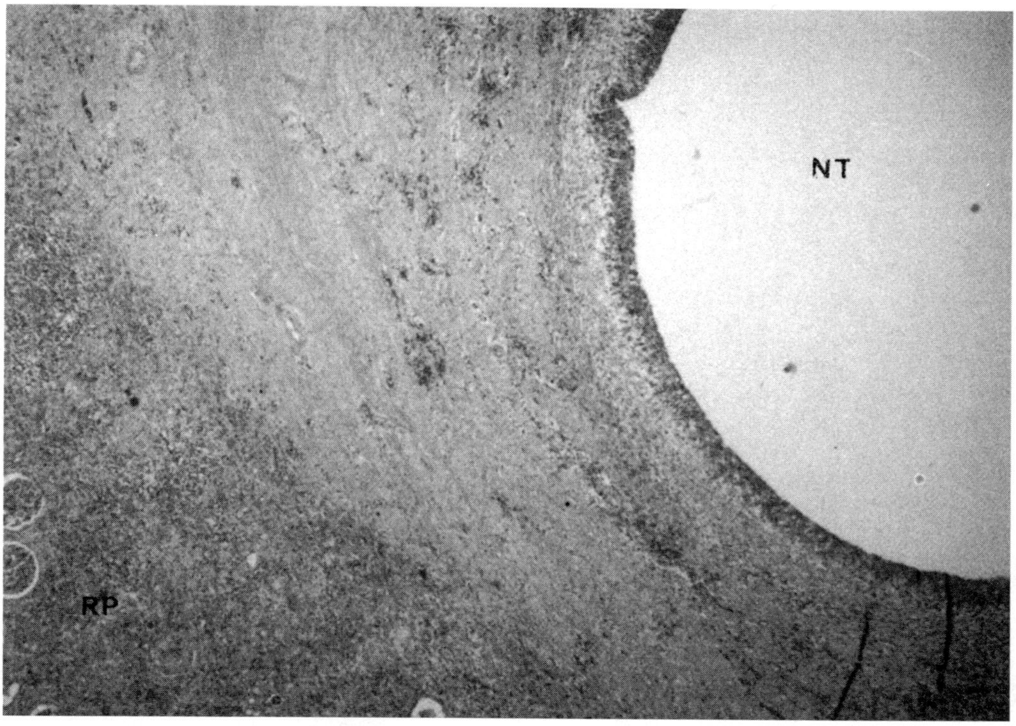

Figure 36 Synovial-like aspect of an interfacial membrane surrounding a polymeric nephrostomy tube (NT). In the absence of interfacial motion, generation of alloplastic debris does not take place and, consequently, the membrane solely comprises fibrous tissue, without those foreign body granulomas which are a characteristic landmark of a failed orthopedic implant. (RP = renal parenchyma). Masson stain. ×30.

4. The in vitro tests miss the in vivo formed, conditioning, surface-coating film, which regulates attachment of cells to implants. Indeed, the thickness of this film is one measure of a material's biocompatibility.
5. The in vitro tests ignore the part played by surface-adsorbed elements of the biological fluids in modulating enzymatic and cellular activities. This explains why the cellular reactions to fresh and retrieved particles of acrylic cement differ from one another.
6. In vitro biocompatibility tests disregard the effects of degradation products arising in vivo after contact of a material with interstitial, and humoral or cellular enzymes.
7. The nonloaded implants in animals do not reduplicate patients' loaded prostheses.
8. The different bone anatomy, dissimilar kinetics of locomotion and human-specific age-related bone changes rationalize the unmatched biomechanical conditions in biped man and quadriped animals. Interspecies extrapolations are to be viewed with skepticism. Many an esoteric alloplastic implant, devised to experimentally explore tissular reaction patterns, are functionally successful in the animal model.
9. Given that the periprosthetic tissues of a patient's failed arthroplasty retain many billions of 200- to 500-nm-sized wear particles per gram, Kossovsky et al. rightfully rationalize that, irrespective of other considerations, "particulates smaller than 500 nm may have an effect on material biocompatibility."

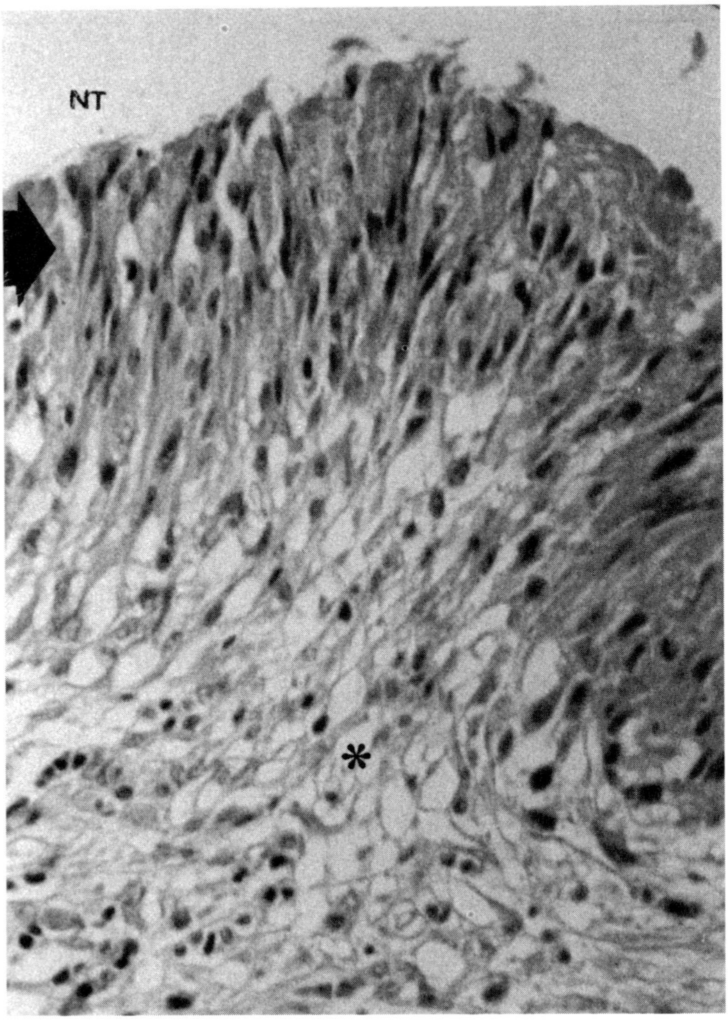

Figure 37 The inner aspect of the interfacial membrane facing the nephrostomy tube (NT) consists of stratified, palisading macrophages (arrow) overlying a loosely texture fibrous tissue (asterisk), thereby imparting a synovial-like appearance which is akin to that commonly seen at the implant-tissue junction of orthopedic devices. Masson stain. ×320.

10. Time, extent and intensity scales of healing processes in animals are so unlike those in man that extrapolations based on data obtained at analyses of the implant-tissue interface in, say, the rabbit or dog are at best suspect and as a rule invalid for the clinical setting.
11. Reactivities to implants are site specific. Tissular behavior to a certain foreign body may deviate from the pattern a researcher expected in view of her or his experience with that material implanted at a different body site.
12. Reactivities to foreign bodies differ not only between species and strains of the animals used in the experiments but even between individuals of the same species [99,478,562–564].

XIII. DEFINITION OF BIOCOMPATIBILITY

Biomaterials are presently in widespread use in all specialties of medicine so that any one observer has personal knowledge and experience with no more than a small fraction of the entire scope. A recent MEDLINE-based literature perusal has disclosed about 3000 papers on biocompatibility, but there are many more articles in which the body's reactions to biomaterials are as well dealt with. The motifs brought up in this chapter have been chosen to suit this author's expertise. The biocompatibility of the materials preferentially applied in operations on the central and peripheral nervous systems or the eye, hemocompatibility and biocompatibility of dialysis membranes and drug delivery systems, to mention just a few of the gamut of themes, have been completely or somewhat neglected. It is premised that the principles deduced after morphologically studying periprosthetic tissues retrieved from patients and animals with certain types of prostheses and after reviewing other authors' masterful reports qualify to outline the crucial and general laws which hold true for the subject matter of biocompatibility in its entirety.

1. Biocompatibility implies a lack of untoward effects of a biomaterial on the adjacent tissues and distant tissues or organs as well as a lack of compositional changes of the material due to the chemical and physical activities of its environs which may impede the functional competence of the implant comprising that material.

Implant-related osteolysis consequent on the production of mediatory substances by a particles-induced granulation tissue is the prototype of deleterious consequences of biomaterials, albeit in a special physical state, on adjoining tissues. Lymphadenopathy due to florid sinus histiocytosis, with cobalt or titanium alloy- and polyethylene particles-laden macrophages distending the nodal sinuses, typifies a harmful effect on a distant tissue by far migrating prosthetic debris. Likewise, debris wandering afar is blamed for pathological fractures through a polyethylene or metallic-particles-induced osteolytic cavity in a femoral shaft. Vice versa, cracks turning up under foreign body giant cells adherent to polyetherurethane surfaces exemplify the damaging effects of cellular activities on a biomaterial [97,561–567].

2. Biocompatibility expresses the ability of a biomaterial composing an implant to perform with an appropriate host response in the specific application originally planned for that particular implant. Foreign matter, whether deliberately introduced or accidentally inserted into or pathologically formed within an organism, incites a response that is intended to attack, destroy and remove it. Williams has rightfully stated his reservations as to simplistic definitions of biocompatibility. Not the response by itself but rather its intensity and extent set limit to the biofunctionality of the implant. Ergo, any definition should be amended by quantitative parameters. Biofunctionality specifies the mechanical and physical properties which permit an implanted device to perform its expected functions. More precisely, *the (in)appropriateness of the tissular response of the host determines the bio(in)compatibility of a material.* It is too broad a generalization to define biocompatibility solely as the state of "passive" mutual coexistence of a material and its environs, neither actuating an undesirable effect on the other. *It is imperative to evaluate bio(in)compatibility in the context of the precise application of the biomaterial in question.* Illustratively, the fact that the physeal plates of growing rabbits' bones are irreversibly damaged by penetrating polyglactin (copolymer of lactide and glycolide) or polydioxanone rods is immaterial to the issue of biocompatibility. Whatever the rods are made of, such treatments are doomed to fail if the diameter of the device is large enough to cause a permanent injury to the physeal plate. Indeed, the growth disturbance of the bone is but brief if the diameter of the rods is kept small [54,409,568–572].

Definitions of biocompatibility do injustice to factual in vivo situations by ascribing behavioral traits to materials while ignoring their physical conformation. Biocompatibility depends, albeit among other items, on the surface texture of the biomaterial and the type of tissue harboring the implant containing that material. Heedful reasoning mandates to make size, shape and surface topography of a material, in addition to its chemical composition and breakdown products, responsible for the extent and intensity of the inflammatory-granulomatous response induced by devices containing the compound. The pivotal commands of texture on the behavior of cells at the tissue-implant interface has been known for quite some time. As substantiated by morphological studies of patients' and experimental animals' retrieved orthopedic implants, prosthesis failure is more often than not brought about by exuberant inflammatory-granulomatous reactions. Undoubtedly, other quantitatively measurable parameters have as well to be considered when formulating the concept of biocompatibility. Just to mention the as yet inexplicable observation of variance of the tissular reaction at different regions of the interface of one and the same implant [89,260,262,573].

Capitalizing on the discussed propositions, one and the same biomaterial will be declared biocompatible under one set of circumstances but professed bioincompatible under other sets of circumstances. In the context of arthroplasties, the granulomatous response to prosthetic debris deposited at component-bone interfaces is inappropriate because it is responsible for aseptic loosening. Deductively, small polyethylene particles are bioincompatible. The polyethylene Rashel net is surrounded and penetrated by fibrous tissue, the fibroblasts of which abundantly produce crimp-fashioned collagen fibers. Macrophages, accompanied by some leukocytes, invade the net. In the wake of organizational activity, the polyethylene filaments are internalized in composite units, consisting of the individual filaments encircled by concentrically aligned macrophages, fibroblasts, collagen fibers and reticulin fibers. In spite of the macrophages' frustrated phagocytosis, attended by exocytosis of cytoplasmic enzymes, injurious effects on the environs are morphologically conspicuous by their absence. The Rashel net endures some interfacial motions during daily activities, but there is minimal generation of wear products such that discrete foreign body granulomas are rarely evident in the adjacent connective and muscular tissues. Since movability and compliance of implants cause enhanced tissular reactions, factors other than just the chemical and textural attributes control the intensity of the response. Not conflicting with the device's planned function, the host response to the Rashel net is appropriate to the circumstances, and it follows that the filamentous polyethylene of the net is biocompatible [3,97,455,574–576].

In addition to the factors brought up above, chemical modifications of a device as well control the extent of the host response and, consequently, the (in)appropriateness of the reaction and, therefore, also the bio(in)compatibility. The fact that just slightly modifying—not refashioning—a device may significantly shift the scene at its surroundings underscores Williams' behest to evaluate implants in the context of their precise applications [570,571].

XIV. CONCLUSIONS

A property designates an attribute. It defines a characteristic which is unrelated to other things, that is, its reality is not conditional on the circumstances. With respect to our perceptions of the concept of biocompatibility, a paradigm shift may be desirable at the present time. Biocompatibility is neither an absolute quantity nor a fixed attribute, that is, an all-or-none asset. *Bio(in)compatibility is to be perceived relative to the events existent at the time of our inquiry.*

To the degree that this concept is sound, polyethylene making up intact smooth surfaced motifs, filamentous nets and tibial or acetabular components of an artificial joint is biocompatible. In contrast, the polymer appears to be bioincompatible when death is the fate of cells grown on it in a saline solution and when its wear particles induce the formation of a sizable interfacial membrane. Following long-term and useful service, ongoing and intensifying inflammatory-granulomatous reactions associated with otherwise intrinsically biocompatible materials, like carbon, polyethylene, polytetrafluoroethylene and titanium, leads to prosthetic failure. To reconcile the seemingly contradictory extremes of the behavioral traits of one and the same material under different circumstances, the concept of the "relativity of biocompatibility" has been introduced. The concept implies that this attribute of biomaterials is subject to the conditions under which biocompatibility is measured and observed. Patients and their physicians may rightfully take it for granted that implanted devices are "safe." Materials scientists and engineers assume that safety of implants depends, at least in part, on the biocompatibility of the materials from which they are fabricated. Black emphatically points out that a biomaterial, as a component of an implant claimed to be safe, does not become, by that association, categorically safe: "In the same way that *safe is not an absolute attribute of a device, it cannot become an absolute attribute of any biomaterial that the device may contain*." In addition to displaying their safety in experimental animals and humans, biomaterials should successfully pass the test of performance standards in specific applications. Black adds that after examining the clinical experience, all that can be said is that a certain biomaterial, in a certain setting, is either biocompatible or not biocompatible. Judging a biomaterial is a subjective affair. It requires the evaluator to weigh the evidenced local, systemic and remote host responses against the implied therapeutic achievements of the device. Host reactions do not have to be entirely benign nor do the materials have to be inert. Evaluators will render a favorable ruling also when the host response is appropriate to the specific application [3,569,570,577].

What happens at tissue-biomaterial interfaces depends not only on the chemical composition of the implants but also on their textural features as well as on the stress and micromotion to which they are subjected. There is more to the explanation of the properties setting the biocompatibility of an implant than just cataloging the chemical essentials of the materials contained in that implant. The study of the implant-activated modifications of the cellular behavior with permutations of the adhesion molecules and the microfilamentous cytoskeleton when confronting a biomaterial is apt to open new vistas in biocompatibility research [3,578–580].

Wear-debris-induced osteolysis around stable hip arthroplasties has been more common in recent years than it was a decade or so ago. Surely, there was no setback in the manufacture of raw materials. The rising prevalence of osteolytic lesions is likely due to design alterations, some contemporary femoral bearings embodying significant mismatches in the surface finish and sphericity of the mating polyethylene and metallic components such that wear particles are shed in extra large amounts. This and further examples affirm the dictum that catastrophic events are more often occasioned by an improper mechanical situation than by bioincompatibility of the ingredients used in the manufacture of a prosthesis. The stature of biocompatibility should be separated from the image of prosthetic failure. With the rare exception, prosthetic failure reflects wear debris- or mechanics-related disasters. Hence, one should question the legitimacy of a phrase such as the "biocompatibility of THR prostheses." Not the implant as such but load-related physical predicaments of the construct and its breakdown products stood between the satisfactory outcome the authors hoped for and the fiasco they witnessed. Mallory did not speak in vain when he forewarned that all prostheses were apt to fail sometime, the interfacial events kicking off "a race between the life of the patient and the life of the prosthesis." One should keep in mind that, except for the biodegradable synthetic and collagen-based compounds, there are no known enzymatic or metabolic pathways whereby intracellularly ingested man-made materials used in joint and other tissue replacements are effectively digested [108,581,582].

Biocompatibility is not subject to an all-or-none law. It is not a property subsidiary to particular chemical formulations. Biocompatibility is first and foremost dependent on circumstances under which the material in question meets the tissue at the interface. In this context, it is deserving to reiterate the forceful role played by biomechanical factors in guiding tissular responses to a biomaterial in a salutary or adverse direction. The patterns of tissular reactions are site specific. This reactivity distinctness is not just the result of distinct functional demands at different body locales. Anatomical specifics are decisive in setting the degree of mismatch of the properties of the implant and tissue, e.g., a mismatch in the elasticity module. Physiological specifics determine metabolic climates particular to every body site. It is recalled that, with the exception of diamonds, all implanted materials cause at least some degree of tissue injury. For the time being, concoction of biomaterials personifying absolute biostability is a goal beyond reach. The researchers' challenge is to discover how and to what degree biocompatibility is swayed by the degradation of the implant, the rate of the degradation and the nature of the degradation products (i.e., those derived from the bulk of the implant, impurities or additives). Qualitatively ascertainable or quantitatively measurable injury of adjacent tissues or distant organs by the degradation products makes use of the biomaterial a safety issue. The interface is the place where interactions occur between two systems. The perspectives of biocompatibility are projected relative to other facets of the events occurring in the immediate domain of the implant or remote from the implantation site, given the spectrum of feasible, local and systemic interfacial reactivities to one and the same biomaterial. Cognition of this principle is crucial to materials scientists, chemists, engineers and medical practitioners who explore the options of restoring a diseased, damaged or lost tissue or organ by placing a viable or nonviable, natural or synthetic, organic or inorganic implant in its stead [46,70,356,442,583,584].

Biomaterials have a bright future. At the outset, biomaterials were not designed for specific applications. They were off-the-shelf supplies found useful in solving an urgent problem. Based on familiarity with the success and the failure of contemporary implants, physicians practicing in the 21st century will predictably have at their disposal bioengineered materials specifically devised to satisfy mechanical properties and surface characteristics required to synchronize the capabilities of the implant with the potentials of the tissues. Surface modifications of and incorporation of "bioactive principles" into a biomaterial, e.g., cell adhesion sequences, and the novel family of "bioartificial polymeric materials," combining synthetic polymers and natural molecules with one another, may enable implants and cells to actively cooperate with each other so as to establish and conserve appropriate interfacial conditions [4,585–587].

Williams' interpretation of biocompatibility is inept in Ratner's view because "the ability of a material to perform with an appropriate host response in a specific application provides no clue as to what biocompatibility really is and offers no useful insight into what materials are biocompatible, how they work or how to design improved materials." Ratner understands biocompatibility to mean "the exploitation by materials of the proteins and cells of the body to meet a specific performance goal." This definition accommodates the many "faces" of biocompatibility by stressing the central role of interfacial proteins, cellular recognition processes and the part played by the biomaterial itself. Scientists are biased by the customs of their disciplines and the accidental paths of their education, wherefore the Ratner-Williams controversy is likely to persevere as it mirrors opposite philosophies. A novel, uncommitted and neutral epithet, say *biosuitability*, may negotiate these disagreements. Biocompatibility specifies fundamental effects of the surface of biomaterials and leached cytoreactive substances on cells and tissues. It also describes tissue-material interactions. Biosuitability defines the performance of biomaterials in the context of clinically relevant settings. It addresses the practical side of implantology. Investigators as well as practitioners embrace the axiom that prostheses ought to perform in the mode programmed at the time of their design on the drawing board. To exemplify the concept of biosuitability, take carbon and polyethylene. Both carbon and polyethylene are called biocompatible by experimenters. Surgeons, however, consider them biounsuitable in the filamentous form for the replacement of intra-articular ligaments. No matter how biocompatible their constituents may be, lack of scaffolding efficacy of the implants and their premature degradation frustrate attempts at reconstructive surgery. The intricately adapted anatomy of the organs to the precise functions, e.g., the three-bundled human anterior cruciate ligament, cannot be duplicated by industrial appliances. Would it not be naive to expect biomechanical sufficiency from unsophisticated devices? Adverse reactions to motion at the interface of optimally conceptualized but unstable constructs or to prosthetic debris should be acknowledged for what they are. Prosthetic failure should not be explained away by indiscriminately ascribing the property of bioincompatibility to one type of an implant or another. A material is biocompatible or bioincompatible. An implant is biosuitable or bioinsuitable. Finally, one should not overlook that rare case in which an individual's unpredictable idiosyncratic reaction may make a biocompatible material look injurious at the bedside [7,9,14,36, 45,588,589].

It is taken for granted that biocompatible materials do not damage distant organs. Yet, there has not been a major investigative endeavor to find out the impacts, if any, of implants on the general health of the patients, though it is well known that particulate debris disseminates systemically in the body. If the higher than expected incidence of lymphomas and leukemias in patients with a total hip arthroplasty is causally related to the oncogenic property of chromium or cobalt, be it an initiating or promoting potential, the unconditional biocompatibility of at least some alloplastic prosthetic components is suspect if not worse. It behooves to climax this overview of the subject with Williams' dictum that *no material is universally biocompatible* [236,359,590,591].

REFERENCES

1. Webster's New World Dictionary of the American Language, DB Guralnik, Ed, Simon and Schuster, New York, 1979.
2. The Concise Oxford Dictionary of Current English, HW Fowler and FG Fowler, Eds, Clarendon Press, Oxford, 1960.
3. Boss, JH, Shajrawi, Aunullah, Mendes DG. The relativity of biocompatibility. A critique of the concept of biocompatibility. Isr J Med Sci 1995;31:203–209.
4. Peppas NA, Langer R. New challenges in biomaterials. Science 1994;263:1715–1720.
5. Krawczynski J, Ondracek G. Biomaterials—a research concept. In: Progress Report on Biomaterials, J Krawczynski and G Ondracek, Eds, WEKA-Druck GmbH, Linnich, 1993;5–17.
6. Geesink RG. Hydroxyapatite-coated total hip prostheses. Two-year clinical and roentgenographic results of 100 cases. Clin Orthop 1990;261:39–58.
7. Jacob-LaBarre JT, Assouline M, Byrd T, McDonald M. Synthetic scleral reinforcement materials: I. Development and in vivo tissue biocompatibility response. J Biomed Mater Res 1994;28:699–712.
8. Kotani S, Fujita Y, Kitsugi T, Nakamura T, Yamamuro T, Ohtsuki C, Kokubo T. Bone bonding mechanism of β-tricalcium phosphate. J Biomed Mater Res 1991;25:1303–1315.
9. Søballe K, Hansen ES, Rasmussen HB, Jorgensen RH, Buenger C. Tissue ingrowth into titanium and hydroxyapatite-coated implants during stable and unstable mechanical conditions. J Orthop Res 1992;10:285–299.
10. Ducheyne P, Cuckler JM. Bioactive ceramic prosthetic coatings. Clin Orthop 1992;276:102–114.
11. Rae T. The macrophage response to implant materials, with special reference to those used in orthopedics. CRC Crit Rev Biocomp 1986;2:97–126.

12. Ross R. Wound healing: Recent progress—future directions. J Dent Res 1971;50:312–317.
13. Coleman DL, King RN, Andrade JD. The foreign body reaction: A chronic inflammatory response. J Biomed Mater Res 1974;8:198–211.
14. Bakker D, van Blitterswijk CA, Hesseling SC, Koerten HK, Kuijpers W, Grote JJ. Biocompatibility of a polyether urethane, polypropylene oxide, and a polyether polyester copolymer. A qualitative and quantitative study of three alloplastic tympanic membrane materials in the rat middle ear. J Biomed Mater Res 1990;24:489–515.
15. Krause TJ, Robertson FM, Greco RS. Measurement of intracellular hydrogen peroxide induced by biomaterials implanted in a rodent air pouch. J Biomed Mater Res 1993;27:65–69.
16. Marchant R, Miller KM, Anderson J. In vivo biocompatibility studies. V: In vivo leukocyte interactions with Biomer. J Biomater Mater Res 1984;18:1169–1188.
17. Wolter D, Burri C, Kinzl L, Müller A. Die Veraenderungen der physikalischen Eigenschaften von Polyacetalharz (Hostaform C), Polyester (Hostadur), Polyäthylen (Hostalen) und Teflon (Hostaflon) nach tierischer Implantation und mehrfachen Autoklavieren. Arch orthop Unfall-Chir 1976; 86:291–302.
18. Ridgon RH. Local tissue reaction to polyurethane—a comparative study in the mouse, rat and rabbit. J Biomed Mater Res 1973;7:79–93.
19. Christel P, Meunier A. A histomorphometric comparison of the muscular tissue reactions to high-density polyethylene in rats and rabbits. J Biomed Mater Res 1989;23:1169–1182.
20. Gombotz WR, Pankey SC, Bouchard LS, Phan DH, Puolakkainen. Stimulation of bone healing by transforming growth factor $beta_1$ released from polymeric or ceramic implants. J Appl Biomater 1994;5:141–150.
21. Demmer P, Fowler M, Marino AA. Use of carbon fibers in the reconstruction of knee ligaments. Clin Orthop 1991;271:225–232.
22. Gross U, Strunz V. The interface of various glasses and glass ceramics with a bony implantation bed. J Biomed Mater Res 1985;19:251–271.
23. Zenni GC, Ellinger J, Lam TM, Greisler HP. Biomaterial-induced macrophage activation and monokine release. J Invest Surg 1994;7:135–141.
24. Bloebaum RD, Nelson K, Dorr LD, Hofmann AA, Lyman DJ. Investigation of early surface delamination observed in retrieved heat-pressed tibial inserts. Clin Orthop 1991;269:120–127.
25. Kilgus DJ, Funahashi TT, Campbell PA. Massive femoral osteolysis and early disintegration of a polyethylene-bearing surface of a total knee replacement. J Bone Joint Surg 1992;74A:770–774.
26. Damien CJ, Parsons, JR, Benedict JJ, Weisman DS. Investigation of a hydroxy-apatite and calcium sulfate composite supplemented with an osteoinductive factor. J Biomed Mater Res 1990;24:639–654.
27. Christel P, Meunier A, Therin M. A method of analysis of cellular distribution in the membrane encapsulating surgically implanted biomaterials. J Appl Biomater 1990;1:205–214.
28. Kao WJ, Zhao GH, Hiltner A, Anderson JM. Theoretical analysis of in vivo macrophage adhesion and foreign body giant cell formation on polydimethylsiloxane, low density polyethylene, and polyetherurethanes. J Biomed Mater Res 1994;28:73–79.
29. Lam KH, Nieuwenhuis P, Molenhaar I, Esselbrugge H, Feijen J, Dijkstra PJ, Schakenraad JM. Biodegradation of porous versus non-porous poly(L-lactic acid) films. J Mater Sci: Mater Med 1994;5:181–189.
30. Rogers SD, Pearcy MJ, Hay SJ, Haynes DR, Bramley A, Howie DW. A method for production and characterization of metal prosthesis wear particles. J Orthop Res 1993;11:856–864.
31. Kossovsky N, Millett D, Juma S, Little N, Briggs PC, Ruz S, Berg E. In vivo characterization of the inflammatory properties of poly(tetrafluoroethylene) particulates. J Biomed Mater Res 1991; 25:1287–1301.
32. Murphy E. History and philosophy of attachment of prostheses to the musculoskeletal system and of passage through the skin with inert materials. J Biomed Mater Res 1973;4:275–295.
33. Bonfield TL, Colton E, Anderson JM. Fibroblast stimulation by monocytes cultured on protein absorbed biomedical polymers. I: Biomer and polydimethylsiloxane. J Biomed Mater Res 1991; 25:165–175.

34. Bonfield TL, Colton E, Anderson JM. Protein adsorption of biomedical polymers influences activated monocytes to produce fibroblast stimulating factors. J Biomed Mater Res 1992;26:457–465.
35. Henderson JD Jr, Mullarky RH, Ryan DE. Tissue biocompatibility of kevlar aramid fibers and polymethylmethacrylate, composites in rabbits. J Biomed Mater Res 1987;21:59–64.
36. Moore GE. Foreign body carcinogenesis. Cancer 199;67:2731–2732.
37. Maekawa A, Ogiu T, Onodera H, Furuta K, Matsuoka C, Ohno Y, Tanigawa H, Salmo GS, Matsuyama M, Hayashi Y. Foreign-body tumorigenesis in rats by various kinds of plastics induction of malignant fibrous histiocytomas. J Toxicol Sci 1984;9:263–272.
38. Troop JK, Mallory TH, Fisher DA, Vaughn BK. Malignant fibrous histiocytoma after total hip arthroplasty. A case report. Clin Orthop 1990;253:297–300.
39. Brien WW, Salvati EA, Healey JH, Bansal M, Ghelman B, Betts F. Osteogenic sarcoma arising in the area of a total hip replacement. A case report. J Bone Joint Surg 1990;72:1097–1099.
40. Black J, Oppenheimer P, Morris DM, Peduto Am, Clark CCJ. Release of corrosion products by F-75 cobalt base alloy in the rat. III: Effects of a carbon surface coating. J Biomed Mater Res 1987;21:1213–1230.
41. Nakamura T, Shimizu Y, Okumura N, Matsui T, Hyon SH, Shimamoto T. Tumorigenicity of poly-L-lactide (PLLA) plates compared with medical-grade polyethylene. J Biomed Mater Res 1994; 28:17–25.
42. Tsuchiya T, Ikarashi Y, Nakamura A. Studies on the tumor-promoting activities of additives in biomaterials: Inhibition of metabolic cooperation by additives such as pigments and phenolic antioxidants. J Long-Term Eff Med Impl 1995;5:243–252.
43. Nakaoka R, Tsuchiya T, Nakamura A. Studies on the tumor-promoting activity of polyethylene: Inhibitory activity of metabolic cooperation of polyethylene films containing an antioxidant. J Long-Term Eff Med Impl 1995;5:253–262.
44. Ratner BD. New ideas in biomaterials science—a path to engineered biomaterials. J Biomed Mater Res 1993;27:837–850.
45. Bruck SD. Identification of biodegradation products of medical implants in adjacent tissues: An overview and recommendations. J Long-Term Eff Med Impl 1991;1:357–364.
46. Renier M, Anderson JM, Hiltner A, Lodoen GA, Payet CR. Infrared spectral analysis of extractables from poly(etherurethane urea) (PEUU) elastomers J Biomater Sci Polym Ed 1993;5:231–244.
47. Remagen W, Morscher E. Histological results with cement-free implanted hip joint sockets of polyethylene. Arch Orthop Trauma Surg 1984;103:145–151.
48. Zhao Q, Agger MP, Fitzpatrick M, Anderson JM, Hiltner A, Stokes K, Urbanski P. Cellular interactions with biomaterials: in vivo cracking of pre-stressed Pellethane 2363-80A. J Biomed Mater Res 1990;24:621–637.
49. Han DK, Park KD, Jeong SY, Kim YH, Kim UY, Min BG. In vivo biostability and calcification-resistance of surface-modified PU-PEO-SO3. J Biomed Mater Res 1993;27:1063–1073.
50. Golomb G. Calcification of polyurethane-based biomaterials implanted subcutaneously in rats: Role of porosity and fluid absorption in the mechanism of mineralization. J Mater Sci, Mater Med 1992;3:272–277.
51. Bean TA, Zhuang WC, Tong PY, Eick JD, Yourtee DM. Effect of esterase on methacrylates and methacrylate polymers in an enzyme simulator for biodurability and biocompatibility testing. J Biomed Mater Res 1994;28:59–63.
52. Lin TW, Corvelli AA, Frondoza CG, Roberts JC, Hungerford DS. Glass peek composite promotes proliferation and osteocalcin production of human osteoblastic cells. J Biomed Mater Res 1997; 36:137–144.
53. Klein CP, Patka P, van der Lubbe HB, Wolke JG, de Groot K. Plasma-sprayed coatings of tetracalciumphosphate, hydroxyl-apatite, and alpha-TCP on titanium alloy: an interface study. J Biomed Mater Res 1991;25:53–65.
54. Campbell CE, von Recum AF. Microtopography and soft tissue response. J Invest Surg 1989;2: 51–74.
55. Von Recum AF. New aspects of biocompatibility: Motion at the interface. In: Clinical Implant

Materials, G Heimke, U Soltesz and AJC Lee, Eds, Elsevier Science Publishers, Amsterdam, 1990, p. 332.
56. Christel PS. Biocompatibility of surgical-grade dense polycrystalline alumina. Clin Orthop 1992; 282:10–18.
57. Müller-Mai C, Schmitz HJ, Strunz V, Fuhrmann G, Fritz T, Gross UM. Tissues at the surface of the new composite material titanium/glass-ceramic for replacement of bone and teeth. J Biomed Mater Res 1989;23:1149–1168.
58. Tramaglini D, Powers DL, Black J. Failure of percutaneous devices due to excessive interfacial stresses at the implant-tissue interface. J Appl Biomater 1993;4:183–191.
59. Sela J, Bab I. The mechanisms of "primary mineralization" in the reaction of bone to injury and administration of implant. J Biomed Mater Res 1985;19:225–231.
60. Rostoker G, Robin J, Binet O, Blamoutier J, Paupe J, Lessana-Leibowitch M, Bedouelle J. Sonneck JM, Garrel JB, Millet P. Dermatitis due to orthopaedic implants. A review of the literature and report of three cases. J Bone Joint Surg 1987;69A:1408–1412.
61. DeLustro F, Dasch J, Keefe J, Ellingsworth L. Immune responses to allogeneic and xenogeneic implants of collagen and collagen derivatives. Clin Orthop 1990;260:263–279.
62. Labarre D, Montdargent B, Carrenno M-P, Maillet F. Strategy for in vitro evaluation of the interactions between biomaterials and complement system. J Appl Biomater 1993;4:231–240.
63. Schmalz G, Hiller K-A, Dörter-Aslan F. New developments in the filter test system for cytotoxicity testing. J Mater Sci, Mater Med 1994;5:43–48.
64. Wan H, Williams R, Doherty P, Williams DF. A study of the reproducibility of the MTT test. J Mater Sci, Mater Med 1994;5:154–160.
65. Keogh JR, Velander FF, Eaton JW. Albumin-binding surfaces for implantable devices. J Biomed Mater Res 1992;26:441–456.
66. Quantitative characterization of cells at the interface of long-term implants of selected polymers. J Biomed Mater Res 1986;20:653–666.
67. Kinzl L, Burri C, Mohr W, Paulini K, Wolter D. Biocompatibility of different plastics (polyoxymethylene-copolymer, polyethylensterephthalate, polyethylene, polytetrafluorethylene). Z Orthop 1976;114:777–784.
68. Ikarashi Y, Toyoda K, Ohsawa N, Uchima T, Tsuchiya T, Kaniwa M, Sato M, Takahashi M, Nakamura A. Comparative studies by cell culture and in vivo implantation test on the toxicity of natural rubber latex materials. J Biomed Mater Res 1992;26:339–356.
69. Beumer GJ, van Blitterswijk CA, Ponec M. Degradative behaviour of polymeric matrices in (sub)-dermal and muscle tissue of the rat: A quantitative study. Biomaterials 1994;15:551–559.
70. Papadimitriou JM. Endocytosis and formation of macrophage polykarya: an ultrastructural study. J Path 1978;126:215–219.
71. Laing PG, Ferguson AB Jr, Hodge ES. Tissue reaction in rabbit muscle exposed to metallic implants. J Biomed Mater Res 1967;1:135–149.
72. Linder L, Carlsson A, Marsal L, Bjursten LM, Brånemark PI. Clinical aspects of osseointegration in joint replacement. A histological study of titanium implants. J Bone Joint Surg 1988;70B: 550–555.
73. Ducheyne P. Bioceramics: Material characteristics versus in vivo behavior. J Biomed Mater Res 1987;21(A2 Suppl):219–236.
74. Crowninshield R. An overview of prosthetic materials for fixation. Clin Orthop 1988;235:166–172.
75. De Lange GL, De Putter C, De Wijs FL. Histological and ultrastructural appearance of the hydroxyapatite-bone interface. J Biomed Mater Res 1990;24:829–845.
76. Jansen JA, Dhert WJ, van der Waerden JP, von Recum AF. Semi-quantitative and qualitative histologic analysis method for the evaluation of implant biocompatibility. J Invest Surg 1994;7: 123–134.
77. Sigot-Luizard MF, Sigot M, Guidoin R, King M, von Maltzahn WW, Kowligi R, Eberhart RC. A novel microporous polyurethane blood conduit: biocompatibility assessment of the UTA arterial prosthesis by an organo-typic culture technique. J Invest Surg 1993;6:251–271.
78. Guidoin R, Sigot M, King M, Sigot-Luizard MF. Biocompatibility of the Vascugraft: evaluation

of a novel polyester urethane vascular substitute by an organotypic culture technique. Biomaterials 1992;13:281–288.
79. Tanzi MC, Verderio P, Lampugnani MG, Resnati M, Dejanan E, Sturani E. Cytotoxicity of some catalysts commonly used in the synthesis of copolymers for biomedical use. J Mater Sci, Mater Med 1994;5:393–398.
80. Bonfield TL, Colton E, Marchant RE, Anderson JM. Cytokine and growth factor production by monocytes/macrophages on protein preadsorbed polymers. J Biomed Mater Res 1992;26:837–850.
81. Miller KM, Rose-Caprara V, Anderson JM. Generation of IL-1-like activity in response to biomedical polymer implants: a comparison of in vitro and in vivo models. J Biomed Mater Res 1989; 23:1007–1026.
82. Chignier E, Guidollet J, Freyria AM, Ardail D, McGregor JL, Louisot P. Dacron vascular biomaterial triggers macrophage ectoenzyme activity without change in cell membrane fluidity. J Biomed Mater Res 1993;27:1087–1094.
83. Schreuders PD, Salthouse TN, von Recum AF. Normal wound healing compared to healing within porous Dacron implants. J Biomed Mater Res 1988;22:121–135.
84. Behling CA, Spector M. Characterization of the in vivo cellular response to selected polymers. In: Biological and Biochemical Performance of Biomaterials, P Christel, A Meunier and AJC Lee, Eds, Elsevier Science Publishers, Amsterdam, 1986;327–330.
85. Steele JG, Johnson G, Underwood PA. Role of serum vitronectin and fibrinogen in adhesion of fibroblasts following seeding onto tissue culture polystyrene. J Biomed Mater Res 1992;26: 861–884.
86. Steele JG, McFarland C, Dalton BA, Johnson G, Evans MDM, Howlett CR, Underwood PA. Attachment of human bone cells to tissue culture polystyrene and to unmodified polystyrene: The effect of surface chemistry upon initial cell attachment. J Biomater Sci Polymer Edn 1993;5: 245–257.
87. Salthouse T. Some aspects of behavior at the implant interface. J Biomed Mater Res 1984;18: 395–401.
88. Rich A, Harris AK. Anomalous preference of cultured macrophages for hydrophobic and roughened substrata. J Cell Sci 1981;50:1–7.
89. Miller KM, Anderson JM. In vitro stimulation of fibroblast activity by factors generated from human monocytes activated by biomedical polymers. J Biomed Mater Res 1989;23:911–930.
90. Bakker D, van Blitterswijk CA, Hesseling SC, Daems WTh, Grote JJ. Tissue/biomaterial interface characteristics of four elastomers. A transmission electron microscopical study. J Biomed Mater Res 1990;24:277–293.
91. Radder AM, Davies JE, Leenders H, van Blitterswijk CA. Interfacial behavior of PEO/PBT copolymers (Polyactive) in a calvarial system: An in vitro study. J Biomed Mater Res 1994;28:269–277.
92. Radder AM, van Blitterswijk CA, Leenders H, Inoue K, Okumura M, Ohgushi H. Gene expression and protein activity in bone-bonding and non-bonding PEO/PBT copolymers. J Mater Sci: Mater Med 1994;5:582–586.
93. Ohlin A, Johnell O, Lerner UH. The pathogenesis of loosening of total hip arthroplasties. The production of factors by periprosthetic tissues that stimulate in vitro bone resorption. Clin Orthop 1990;253:287–296.
94. Charosky CB, Bullough PG, Wilson PD, Jr. Total hip replacement failures. A histological evaluation. J Bone Joint Surg 1973;55A:49–59.
95. Mendes DG, Figarola F, Bullough PG, Loudis P. High density polyethylene prosthetic femoral head replacement in the dog. Clin Orthop 1975;111:274–283.
96. Boss JH, Shajrawi I, Mendes DG. Histological patterns of the tissue reaction to polymers in orthopaedic surgery. Clin Mater 1993;13:11–17.
97. Leininger RI, Bigg DM. Polymers. In: Handbook of Biomaterials Evaluation, AF von Recum, Ed, Macmillan, New York, 1986, p. 34.
98. Boss JH, Shajrawi I, Soudry M, Anullah J, Solomon H, Mendes DG. Studies on a novel anterior cruciate ligament polyethylene fiber prosthesis: The histomorphological pattern of organization

and bony anchorage of a polyethylene fiber prosthesis in the stifle of the goat. Clin Mater 1994; 15:61–67.
99. LiVecchi AB, Tombes RM, LaBerge M. In vitro chondrocyte collagen deposition within porous HDPE: Substrate microstructure and wettability effects. J Biomed Mater Res 1994;28:839–850.
100. Ertel SI, Ratner BD, Kaul A, Schway MB, Horbett TA. In vitro study of the intrinsic toxicity of synthetic surfaces to cells. J Biomed Mater Res 1994;28:667–675.
101. Burridge K, Fath K, Kelly T, Nuckolls G, Turner D. Focal adhesions: Transmembrane junctions between the extracellular matrix and the cytoskeleton. Ann Rev Cell Biol 1988;4:487–525.
102. Wan H, Williams RL, Doherty PJ, Williams DF. The cytotoxicity evaluation of Kevlar and silicon carbide by MTT assay. J Mater Res: Mater Med 1994;5:441–445.
103. Bouet T, Toyoda K, Ikarashi Y, Uchima T, Nakamura A, Tsuchiya T, Takahashi M, Eloy R. Evaluation of biocompatibility, based on quantitative determination of the vascular response induced by material implantation. J Biomed Mater Res 1991;25:1507–1521.
104. Müller ME. Lessons of 30 years of total hip arthroplasty. Clin Orthop 1992;274:12–21.
105. Wroblewski BM. Wear of high-density polyethylene on bone and cartilage. J Bone Joint Surg 1979;61B:498–500.
106. Goodman SB, Fornasier VL, Kei J. The effects of bulk versus particulate ultrahigh-molecular-weight polyethylene on bone. J Arthroplasty 1988;4(Suppl):S41–S46.
107. Clarke IC. Role of ceramic implants. Design and clinical success with total hip prosthetic ceramic-to-ceramic bearings. Clin Orthop 1992;282:19–30.
108. Blaha JD, Insler HP, Freeman MAR, Revell PA, Todd RC. The fixation of a proximal tibial polyethylene prosthesis without cement. J Bone Joint Surg 1982;64B:326–335.
109. Wilson-MacDonald J, Morscher E, Masar Z. Cementless uncoated polyethylene acetabular components in total hip replacement. Review of five- to 10-year results. J Bone Joint Surg 1990;72B: 423–430.
110. Krugluger J, Eyb R. Bone reaction to uncemented threaded polyethylene acetabular components. Intern Orthop 1993;17:259–265.
111. Guttmann D, Schmalzried TP, Jasty M, Harris WH. Light microscopic identification of submicron polyethylene wear debris. J Appl Biomater 1993;4:303–307.
112. Jones SMG, Pinder IM, Moran CG, Malcolm AJ. Polyethylene wear in uncemented knee replacements. J Bone Joint Surg 1992;74B:18–22.
113. Boss JH, Shajrawi I, Soudry M, Mendes DG. Histomorphological reaction patterns of the bone to diverse particulate implant materials in man and experimental animals. In: Particulate Debris from Medical Implants: Mechanisms of Formation and Biological Consequences, ASTM STP 1144, KR St John, Ed, American Society for Testing and Materials, Philadelphia 1992;90–108.
114. Boss JH, Shajrawi I, Mendes DG. The nature of bone-implant interface: The lessons learned from implant retrieval and analysis in man and experimental animal. Med Progr Through Technol 1994; 20:119–142.
115. Goodman SB, Huie P, Song Y, Lee K, Doshi A, Rushdieh B, Woolson S, Maloney W, Schurman D, Sibiey R. Loosening and osteolysis of cemented joint arthroplasties. Clin Orthop 1997;337: 149–163.
116. Schmalzried TP. Jasty M, Harris WH. Periprosthetic bone loss in total hip arthroplasty: The role of polyethylene wear debris and the concept of the effective joint space. J Bone Joint Surg 1992; 74A:849–863.
117. Kaufman RL, Tong I, Beardmore TD. Prosthetic synovitis: Clinical and histologic characteristics. J Rheumatol 1985;12:1066–1074.
118. Vaes G. Cellular biology and biochemical mechanics of bone resorption. A review of recent developments on the formation, activation, and mode of action of osteoclasts. Clin Orthop 1988;231: 239–271.
119. Huffer WE. Morphology and biochemistry of bone remodeling: Possible control by vitamin D, parathyroid hormone, and other substances. Lab Invest 1988;59:418–442.
120. Herman JH, Sowder WB, Anderson D, Appel AM, Hopson CN. Polymethylmethacrylate-induced release of bone-resorbing factors. J Bone Joint Surg 1989;71A:1530–1541.

121. Bataille R, Klein B, Jourdan M, Rossi JF, Durie BGM. Spontaneous secretion of tumor necrosis factor-beta by human melanoma cell lines. Cancer 1989;63:877–879.
122. Morscher EW. Current status of acetabular fixation in primary total hip arthroplasty. Clin Orthop 1992;274:172–193.
123. Hedley AK, Clarke IC, Bloebaum RD, Moreland J, Gruen T, Coster I, Amstutz HC. Viability and cement fixation of the femoral head in canine hip surface replacement. The Hip. Proc 7th Open Scientific Meeting Hip Society, C V Mosby St. Louis, 1979;160–187.
124. Santavirta S, Nordström D, Metsärinne K, Konttinen YT. Biocompatibility of polyethylene and host response to loosening of cementless total hip replacement. Clin Orthop 1993;297:100–110.
125. Anthony PP, Gie GA, Howie CR, Ling RSM. Localised endosteal bone lysis in relation to the femoral components of cemented total hip arthroplasties. J Bone Joint Surg 1990;72B:971–979.
126. Buechel FF, Drucker D, Jasty M, Jiranek W, Harris WH. Osteolysis around uncemented acetabular components of cobalt-chrome surface replacement hip arthroplasty. Clin Orthop 1994;298:202–211.
127. Santavirta S, Konttinen YT, Bergroth V, Eskola A, Tallroth K, Lindholm S. Aggressive granulomatous lesions associated with arthroplasty. Immunopathological studies. J Bone Joint Surg 1990; 72A:252–258.
128. Wroblewski BM, Lynch M, Atkinson JR, Dowson D, Issac GH. External wear of the polyethylene socket in cemented total hip arthroplasty. J Bone Joint Surg 1987;69B:61–63.
129. Cooper JR, Dowson D, Fisher J. Birefringent studies of polyethylene wear specimens and acetabular cups. Wear 1991;151:391–402.
130. Cooper JR, Dowson D, Fisher J. Macroscopic and microscopic wear mechanisms in ultra-high molecular weight polyethylene. Wear 1993;162–164:378–384.
131. Collier JP, McNamara JL, Surprenant VA, Jensen RE. Surprenant HP. All-polyethylene patellar components are not the answer. Clin Orthop 1991;273:198–203.
132. Kilgus DJ, Moreland JR, Finerman GAM, Funahashi TT, Lipton JS. Catastrophic wear of tibial polyethylene inserts. Clin Orthop 1991;273:223–231.
133. Engh GA, Dwyer KA, Hanes CK. Polyethylene wear of metal-backed tibial components in total and unicompartmental knee prostheses. J Bone Joint Surg 1992;74B;9–17.
134. Heck DA, Clingman JK, Kettelkamp DG. Gross polyethylene failure in total knee arthroplasty. Orthopedics 1992;15:23–28.
135. Lee JG, Keating M, Ritter MA, Faris PM. Review of the all-polyethylene tibial component in total knee arthroplasty. A minimum seven-year follow-up period. Clin Orthop 1990;260:87–92.
136. Peters PC, Engh GA, Dwyer KA, Vinh TN. Osteolysis after total knee arthroplasty without cement. J Bone Joint Surg 1992;74A:864–876.
137. Wright TM, Astion DJ, Bansal M, Rimnac CM, Green T, Install JN, Robinson RP. Failure of carbon fiber-reinforced polyethylene total knee replacement components. A report of two cases. J Bone Joint Surg 1988;70A:926–932.
138. Willert HG, Semlitsch M, Buchhorn G, Kriete U. Materialverschleiss und Gewebereaktion bei klinischen Gelenken. Orthopäde 1978;7:62–83.
139. Mathiesen EB, Lindgren JU, Reinholt FP, Sudmann E. Tissue reactions to wear products from polyacetal (Delrin) and UHMW polyethylene in total hip replacement. J Biomed Mater Res 1987; 21:459–466.
140. Boss JH, Shajrawi M, Soudry M, Mendes DG. Histological features of the interface membrane of failed isoelastic cementless prostheses. Intern Orthop 1990;14:399–403.
141. Izquiedro RJ, Northmore-Ball MD. Long-term results of revision hip arthroplasty. Survival analysis with special reference to the femoral component. J Bone Joint Surg 1994;76B:34–39.
142. Mulroy RD, Jr, Harris WH. The effect of improved cementing techniques on component loosening in total joint replacement: An 11-year radiographic study. J Bone Joint Surg 1990;72B:757–760.
143. Quinn J, Joner C, Triffitt JT, Athanasou NA. Polymethylmethacrylate-induced inflammatory macrophages resorb bone. J Bone Joint Surg 1992;74B:652–658.
144. Boyd AD, Goldberg VM, Miller LM, Jr, Malemud CJ. Effects of polymethylmethacrylate on rabbit articular chondrocytes in monolayer culture. Clin Orthop 1984;189:279–293.

145. Papatheofanis FJ, Barmada R. Polymorphonuclear leukocyte degranulation with exposure to polymethylmethacrylate nanoparticles. J Biomed Mater Res 1991;25:761–771.
146. Jones LC, Hungerford DS. Cement disease. Clin Orthop 1987;225:192–206.
147. Cook SD, Thomas KA, Haddad RJ, Jr. Histologic analysis of retrieved human porous-coated total joint components. Clin Orthop 1988;234:225–90–101.
148. Williams RP, McQueen DA. A histopathologic study of late aseptic loosening of cemented total hip prostheses. Clin Orthop 1992;275:174–179.
149. Bauer TW, Geesink RCT, Zimmerman R, McMahon JT. Hydroxyapatite-coated femoral stems: Histological analysis of components retrieved at autopsy. J Bone Joint Surg 1991;73A:1439–1452.
150. Hardy DCR, Frayssinet P, Guilhem A, Lafontaine MA, Delince PE. Bonding of hydroxyapatite-coated femoral prostheses. Histopathology of specimens from four cases. J Bone Joint Surg 1991; 73B:732–740.
151. Geesink RGT, de Groot K, Klein CPAT. Bonding of bone to apatite-coated implants. J Bone Joint Surg 1988;70B:17–22.
152. Søballe K, Hansen ES, Brockstedt-Rasmmussen H, Pedersen CM, Buenger C. Hydroxyapatite coating enhances fixation of porous coated implants: A comparison in dogs between press-fit and noninterference fit. Acta Orthop Scand 1990;61:299–396.
153. Maric Z, Karpman RR. Early failure of noncemented porous coated anatomic total hip arthroplasty. Clin Orthop 1992;278:116–120.
154. Mann KA, Bartel DL, Wright TM, Ingraffea AR. Mechanical characteristics of the stem-cement interface. J Orthop Res 1991;9:798–808.
155. Spector M, Shortkroff S, Hsu H-P, Lane N, Sledge CB, Thronhill TS. Tissue changes around loose prostheses. A canine model to investigate the effects of an anti-inflammatory agent. Clin Orthop 1990;261:140–152.
156. Saha W-J, Pal S. Mechanical properties of bone cement: A review. J Biomed Mater Res 1984;18: 435–462.
157. Seuderi GR, Insall JN, Windsor RE, Moran MC. Survivorship of cemented knee replacement. J Bone Joint Surg 1989;71B:798–803.
158. Johnson JA, Provan JW, Krygier JJ, Chan KH, Miller J. Fatigue of acrylic bone cement. Effect of frequency and environment. J Biomed Mater Res 1989;23:819–831.
159. Goodman SB, Chin RC. Prostaglandin E_2 levels in the membrane surrounding bulk and particulate polymethylmethacrylate in the rabbit tibia. A preliminary study. Clin Orthop 1990;257:305–309.
160. Willert H-G, Bertram H, Buchhorn GH. Osteolysis in alloarthroplasty of the hip. The role of bone cement fragmentation. Clin Orthop 1990;258:108–121.
161. Schmalzried TP, Jasty M, Rosenberg A, Harris WH. Histologic identification of polyethylene debris using oil red O stain. J Appl Biomater 1993;4:119–125.
162. Kim KJ, Rubash HE, Wilson SC, D'Antonio JA, McClain EJ. A histologic and biochemical comparison of the interface tissues in cementless and cemented hip prostheses. Clin Orthop 1993;287: 142–152.
163. Feith R. Side-effects of acrylic cement implanted into bone. Acta Orthop Scand 1975;56 Suppl: 161–201.
164. Linder L. The tissue response to bone cement. In: Biocompatibility of Orthopedic Implants, D F Williams, Ed, CRC Press, Boca Racon, 1982, pp. 1–21.
165. Boesch P, Harms H, Lintner F. Nachweis des Katalysatorbestandteiles Dimethylparatoluidin in Knochenzement nach mehrjaehriger Implantation. Arch Toxicol 1982;51:157–162.
166. Horowitz SM, Gautsch TL, Frondoza CG, Riley L, Jr. Macrophage exposure to polymethyl methacrylate leads to mediator release and injury. J Orthop Res 1991;9:406–413.
167. Schultz RJ, Johnston AD, Krishnamurthy S. Thermal effects of polymerization of methyl-methacrylate on small tubular bone. Int Orthop 1987;11:277–282.
168. Radin EL, Rubin CT, Thrasher EL, Lanyon LE, Crugnola AM, Schiller AS, Paul IL, Rose RM. Changes in the bone-cement interface after total hip replacement. An in vivo animal study. J Bone Joint Surg 1982;64A:1188–1200.
169. Kobayashi S, Takaoka K, Saito N, Hisa K. Factors affecting aseptic failure of fixation after pri-

mary Charnley total hip arthroplasty. Multivariate survival analysis. J Bone Joint Surg Am 1997; 79:1618–1627.

170. Ishiguro N, Kojima T, Ito T, Saga S, Anma H, Kurokouchi K, Iwahori Y, Iwase T, Iwata H. Macrophage activation and migration in interface tissue around loosening total hip arthroplasty components. J Biomed Mater Res 1997;35:399–406.
171. Oakeshott RD, Morgan DAF, Zukor DJ, Rudan JF, Brooks PJ, Gross AE. Revision total hip arthroplasty with osseous allograft reconstruction. Clin Orthop 1987;225:37–61.
172. Straw RC, Powers BE, Withrow SJ, Cooper MF, Turner AS. The effect of intramedullary polymethylmethacrylate on healing of intercalary cortical allografts in a canine model. J Orthop Res 1992;10:434–439.
173. Goodman SB, Fornasier VL. Clinical and experimental studies in the biology of aseptic loosening of joint arthroplasties and the role of polymer particles. In: Particulate Debris from Medical Implants: Mechanisms of Formation and Biological Consequences, ASTM STP 1144, KR St John, Ed, American Society for Testing and Materials, Philadelphia, 1992;27–37.
174. Kobayashi S, Takaoka K, Tsukada A, Ueno M. Polyethylene wear from femoral bipolar neck-cup impingement as a cause of femoral prosthetic loosening. Arch Orthop Trauma Surg 1998;117: 390–391.
175. Puleo DA, Holleran LA, Doremus RH, Bizios R. Osteoblast response to orthopedic implant materials in vitro. J Biomed Mater Res 1991;25:711–723.
176. Downes S, Wood DJ, Malcolm AJ, All SY. Growth hormone in polymethylmethacrylate cement. Clin Orthop 1990;252:294–298.
177. Boss JH, Shajrawi I, Dekel S, Mendes DG. The bone-cement interface: Histological observations on the interface of cemented arthroplasties within the immediate and late phases. J Biomater Sci Polymer Edn 1993;3:221–230.
178. Malefijt JDW. Early features of the bone-implant interface in hip arthroplasty. A comparative study in the proximal femur of the goat after implantation of a cemented versus an uncemented endoprosthesis. Doctoral Dissertation. Drukkerij Leijn te Nijmegen. CIP-Data Koninklijke Biobliotheek, Den Haag 1988.
179. Lazarus MD, Cuckler JM, Mitchell J, Schumacher HR, Baker DG, Ducheyne P, Imonitie V. Biocompatibility of polymethyl methacrylate with and without barium sulfate in the rat subcutaneous air pouch model. In: Particulate Debris from Medical Implants: Mechanisms of Formation and Biological Consequences, ASTM STP 1144, K R St John, Ed., American Society for Testing and Materials, Philadelphia, 1992;127–135.
180. Goodman SB, Fornasier VL, Kei J. The effects of bulk versus particulate polymethylmethacrylate on bone. Clin Orthop 1988;232:255–262.
181. Davis RG, Goodman SB, Smith RL, Lerman JA, Williams RJ, III. The effects of bone cement powder on human adherent monocytes/macrophages in vitro. J Biomed Mater Res 1993;27:1039–1046.
182. Malawer MM, Marks MR, McChensey D, Piasio M, Gunther SF, Schmookler BM. The effect of cryosurgery and polymethylmethacrylate in dogs with experimental bone defects comparable to tumor defects. Clin Orthop 1988;226:299–310.
183. Tibrewal SB, Grant KA, Goodfellow JW. The radiolucent line beneath the tibial components of the Oxford meniscal knee. J Bone Joint Surg 1984;66B:523–528.
184. Schenk RK, Wehrli U. Zur Reaktion des Knochens auf eine zementfreie SL-Femur-Revisionsprotheses. Orthopaede 1989;18:454–462.
185. Fries IB, Fisher AA, Salvati EA. Contact dermatitis in surgeons from methylmethacrylate bone cement. J Bone Joint Surg 1975;57A:547–549.
186. Rutherford CS, Cardea JA, Jesse SD. Polymethylmethacrylate fixation of osteochondral fragments in dog knees. Clin Orthop 1987;223:287–295.
187. Wooley PH, Petersen S, Song Z, Nasser S. Cellular immune responses to orthopaedic implant materials following cemented total joint replacement. J Orthop Res 1997;15:874–880.
188. Ling RSM. Observations on the fixation of implants to the bony skeleton. Clin Orthop 1986;210: 80–96.

189. Mirra JM, Marder RA, Amstutz HC. The pathology of failed total joint arthroplasty. Clin Orthop 1982;170:175–185.
190. Linder L, Lindeberg L, Carlsson A. Aseptic loosening of hip prostheses. A histologic and enzyme histochemical study. Clin Orthop 1983;175:93–104.
191. Goldring SR, Schiller AL, Roelke M, Bourke CM, O'Neill DA, Harris WH. The synovial-like membrane at the bone-cement interface in loose total hip replacements and its proposed role in bone lysis. J Bone Joint Surg 1983;65A:575–584.
192. Goldring SR, Jasty M, Roelke MS, Rourke CM, Bringhurst FR, Harris WH. Formation of a synovial-like membrane at the bone-cement interface. Its role in bone resorption and implant loosening after total hip replacement. Arthr Rheum 1986;29:836–842.
193. Johanson NA, Bullough PG, Wilssson PD, Jr, Salvati EA, Ranawat CS. The microscopic anatomy of the bone-cement interface in failed total hip arthroplasties. Clin Orthop 1987;218:123–135.
194. Goodman SB. Polymer particles potentiate loosening of joint arthroplasties. Orthopaedics Intern Edn 1993;1:316–321.
195. Williams P, MacQueen DA. A histopathologic study of late aseptic loosening of cemented hip prostheses. Clin Orthop 1992;275:174–179.
196. Shoji H, Karube S, D'ambrosia RD, Dabezies EJ, Miller DR. Biochemical features of pseudomembrane at the bone-cement interface of loosened total hip prostheses. J Biomed Mater Res 1983;17: 669–678.
197. Goodman SB, Kang T, Smith RL. Lysosomal enzyme production at the interface surrounding loose and well-fixed cemented tibial hemiarthroplasties in the rabbit knee. J Invest Surg 1993;6: 413–418.
198. Dorr LD, Bloebaum R, Emmanuel J, Meldrum J. Histological, biochemical and ion analysis of tissue and fluids retrieved during total hip arthroplasty. Clin Orthop 1990;261:82–95.
199. Schmalzried TP, Kwong LM, Jasty M, Sedlacek DC, Haire TC, O'Conner DO, Dragdon CR, Kabo M, Malcolm AJ, Harris WH. The mechanisms of loosening of cemented acetabular components in total hip arthroplasty. Analysis of specimens retrieved at autopsy. Clin Orthop 1992;274:60–78.
200. Maguire JK, Jr, Coscia MF, Lynch MH. Foreign body reaction to polymeric debris following total hip arthroplasty. Clin Orthop 1987;216:213–223.
201. Amstutz HC, Campbell P, Kossovsky N, Clarke IC. Mechanism and clinical significance of wear debris-induced osteolysis. Clin Orthop 1992;276:7–18.
202. Boss JH, Shajrawi I, Aunullah J, Solomon H, Mendes DG, Schaefer R, Soltesz U, Ulrich D. Motion-induced inhibition of tissular organization and bony anchorage of a polyethylene anterior cruciate ligament prosthesis in the stifle of the goat. In: Biomedical Engineering Recent Development, J Vossoughi, Ed, University of the District of Columbia, Washington, 1994;1125–1128.
203. Shajrawi I, Aunullah J, Soudry M, Solomon H, Mendes DG, Boss JH. Quantification of the tissue response to a polyethylene prosthesis of the ACL in the goat: A histomorphometric study. Orthopaedics Intern Edn 1993;1:455–460.
204. Bolton CW, Bruchman WC. The GORETEX polytetrafluoroethylene prosthetic ligament. An in vitro and in vivo evaluation. Clin Orthop 1985;196:202.
205. Roth JH, Kennedy JC, Lockstadt H, McCallum CL, Cunning LA. Polypropylene braid augmented and nonaugmented intraarticular anterior cruciate ligament reconstruction. Am J Sports Med 1985; 13:321–336.
206. Engstroem B, Wredmark T, Westblad P. Patellar tendon or Leeds-Keio graft in the surgical treatment of anterior cruciate ligament ruptures. Intermediate results. Clin Orthop 1993;295:190–197.
207. James SL, Woods GW, Homsy CA, Prewitt JM, Slocum DB. Cruciate ligament stents in reconstruction of the unstable knee. A preliminary report. Clin Orthop 1979;143:90–96.
208. Mendes DG, Boss JH. The morphological patterns of organization of synthetic filamentous prostheses in reconstructive orthopaedic surgery. In: Encyclopedic Handbook of Biomaterials and Bioengineering, DL Wise, DJ Trantolo, DE Altobelli, MJ Yaszemski, JD Gresser and ER Schwartz, Eds, Marcel Dekker, New York, 1995;383–414.
209. Klein W, Jensen K-U. Synovitis and artificial ligaments. J Arthroscopic Related Surg 1992;8: 116–124.

210. Panni AS, Denti M, Franzese S, Monteleone M. The bone-ligament junction: A comparison between biological and artificial ACL reconstruction. Knee Surg Sports Traumatol Arthroscopy 1993;1:9–12.
211. Lopez-Vazquez E, Juan JA, Vila E, Debon J. Reconstruction of the anterior cruciate ligament with a Dacron prosthesis. J Bone Joint Surg 1991;73A:1294–1299.
212. Macnicol MF, Penny I, Sheppard L. Early results of the Leed-Keio anterior cruciate ligament replacement. J Bone Joint Surg 1991;73B:377–380.
213. Kurosaka M, Yoshiya S, Andrish JT. Dacron-augmented anterior cruciate ligament reconstruction in dogs. A 6-month follow-up study. Am J Knee Surg 1993;6:61–66.
214. Müller-Mai CM, Gross UM. Histological and ultrastructural observations at the interface of expanded polytetrafluoroethylene anterior cruciate ligament implant. J Appl Biomater 1991;2:29–35.
215. Ricci JL, Gona AG, Alexander H, Parsons JR. Morphological characteristics of tendon cells cultured on synthetic fibers. J Biomed Mater Res 1984;18:1073–1087.
216. Taylor SR, Gibbons DF. Effects of surface texture on the soft tissue response to polymer implants. J Biomed Mater Res 1983;17:205–227.
217. Claes L, Duerselen L, Kiefer H, Mohr W. The combined anterior cruciate and medial collateral ligament replacement by various materials: A comparative animal study. J Biomed Mater Res 1987;21:319–343.
218. Marcacci M, Gubellini P, Buda R, De Pasquale V, Strocchi R. Molgora AP, Zaffagnini S, Guizzardi S, Buggeri A. Histologic and ultrastructural findings of tissue ingrowth. The Leeds-Keio prosthetic anterior cruciate ligament. Clin Orthop 1991;267:115–121.
219. Olson EJ, Kang JD, Fu FH, Georgescu HI, Mason GC, Evans CH. The biochemical and histological effects of artificial ligament wear particles: In vitro and in vivo studies. Am J Sports Med 1988;16:558–570.
220. Macon ND, Lemons JE, Niemann KMW. Polymer particles in vivo: Distribution in the knee, migration to lymph nodes, and associated cellular response following anterior cruciate ligament replacement. In: Particulate Debris from Medical Implants: Mechanisms of Formation and Biological Consequences, ASTM STP 1144, KR St John, Ed, American Society for Testing and Materials, Philadelphia, 1992;189–199.
221. Messmer K, Gillquist J. Synthetic implants for the repair of osteochondral defects of the medial femoral condyle: A biomechanical and histological evaluation in the rabbit knee. Biomaterials 1993;14:513–521.
222. Adams JS. Facial augmentation with solid alloplastic implants: A rational approach to material selection. In: Applications of Biomaterials in Facial Plastic Surgery, AI Glasgold and FH Silver, Eds, CRC Press, Boca Raton, 1991; 297.
223. Bordenave L, Bareille R, Lefebvre F, Bacque C. A comparison between ^{51}Chromium and LDH release to measure cell membrane integrity: Interest for cytocompatibility studies with biomaterials. J Appl Biomater 1993;4:309–315.
224. Salzmann DL, Kleinert LB, Berman SS, Williams SK. The effects of porosity on endothelialization of ePTFE implanted in subcutaneous and adipose tissue. J Biomed Mater Res 1997;34:463–476.
225. Ehler WJ, Cissik JH. Smith VC, Hubbard GB. Evaluation of Gore-Tex graft material in the repair of right ventricular outflow tract defect. J Invest Surg 1990;3:119–127.
226. Firlit CF, Canning JR, Cross RR. Cause of occlusion of arteriovenous cannulas. Surg Gynecol Obstet 1972;135:556–560.
227. Greisler HP, Tattersall CW, Henderson SC, Cabusao EA, Garfield JD, Kim DU. Polypropylene small-diameter vascular grafts. J Biomed Mater Res 1992;26:1383–1394.
228. Griffith CD, Callum KB. Limb salvage in a district general hospital: Factors affecting outcome. Ann R Coll Surg Engl 1988;70:95–98.
229. Greisler HP, Tattersall CW, Henderson SC, Cabusao EA, Garfield JD, Kim DU. Polypropylene small-diameter vascular grafts. J Biomed Mater Res 1992;26:1383–1394.
230. Sanchez-Montes I, Deysine M. Spigelian hernias: a new repair technique using preshaped polypropylene umbrella plugs. Arch Surg 1998;133:670–672.

231. Leber GE, Garb JL, Alexander AI, Reed WP. Long-term complications associated with prosthetic repair of incisional hernias. Arch Surg 1998;133:378–382.
232. Marois Y, Guidoin R, Roy R, Vidovsky T, Jakubiec B, Sigot-Luizard MF, Braybrook J, Mehri Y, Laroche G, King M. Selecting valid in vitro biocompatibility tests that predict the in vivo healing response of synthetic vascular prostheses. Biomaterials 1996;17:1835–1842.
233. Schein M, Assalia A, Hashmoni M. Vascular prosthetic bypass grafting in obstructive jaundice. Experimental and clinical perspectives. Int Surg 1994;79:65–67.
234. Begin GF. Laparoscopic extraperitoneal treatment of inguinal hernias in adults. A series of 200 cases. Endosc Surg Applied Technol 1993;1:204–206.
235. Jacob-LaBarre JT, Assouline M, Byrd T, McDonald M. Synthetic scleral reinforcement materials. I: Development and in vivo tissue biocompatibility response. J Biomed Mater Res 1994;28:699–712.
236. Williams DF. Biomaterials and biocompatibility, an introduction. In: Fundamental Aspects of Biocompatibility, DF Williams, Ed, CRC Press, Boca Raton, 1981; 2–3.
237. Zitter H, Plenk H, Jr. The electrochemical behavior of metallic implant materials as an indicator of their biocompatibility. J Biomed Mater Res 1987;21:881–896.
238. Rae T. The haemolytic action of particulate metal (Cd, Cr, Co, Fe, Mo, Ni, Ta, Ti, Co-Cr alloy). J Path 1978;125:81–89.
239. Heath JC, Freeman MAR, Swanson SAV. Carcinogenic properties of wear particles from prostheses made in cobalt-chromium alloy. Lancet 1971;I:564–566.
240. Williams DF. Titanium—epitome of biocompatibility or cause for concern. J Bone Joint Surg 1994;76B:348–349.
241. Wapner KL. Implications of metallic corrosion in total knee arthroplasty. Clin Orthop 1991;271: 12–20.
242. Maloney WJ, Castro F, Schurman DJ, Smith RL. Effects of metallic debris on adult bovine articular chondrocyte metabolism in vitro. J Appl Biomater 1994;5:109–115.
243. Naji A, Harmand M-F. Study of the effect of the surface state on the cytocompatibility of a Co-Cr alloy using human osteoblasts and fibroblasts. J Biomed Mater Res 1990;24:861–871.
244. Wapner KL, Morris DM, Black J. Release of corrosion products by F-75 cobalt base alloy in the rat. II: Morbidity apparently associated with chromium release in vivo: A 120-day study. J Biomed Mater Res 1986;20:219–233.
245. Buchhorn GH, Willert H-G, Semlitsch M, Schoen, Steinemann S, Schmidt M. Preparation, characterization, and animal testing for biocompatibility of metal particles of iron-, cobalt-, and titanium-based implant alloys. In: Particulate Debris from Medical Implants: Mechanisms of Formation and Biological Consequences, ASTM STP 1144, KR St John, Ed, American Society for Testing and Materials, Philadelphia 1992;177–188.
246. Puleo DA, Huh WW. Acute toxicity of metal ions in cultures of osteogenic cells derived from bone marrow stromal cells. J Appl Biomater 1995;6:109–116.
247. Thompson GJ, Puleo DA. Effects of sublethal metal ion concentrations on osteogenic cells derived from bone marrow stromal cells. J Appl Biomater 1995;6:249–258.
248. Allen MJ, Myer BJ, Millett PJ, Rushton N. The effects of particulate cobalt, chromium and cobalt-chromium alloy on human osteoblast-like cells in vitro. J Bone Joint Surg 1997;79B:475–482.
249. Sela J, Shani J, Kohavi D, Soskolne WA, Katzir I, Boyan B, Schwartz Z. Uptake of ^{99}TcMD^{32}P during endosteal healing around titanium, stainless steel and hydroxyapatite implants in rat tibial bone. Biomaterials 1995;16:1372–1380.
250. Donohue VE, McDonald F, Evans R. In vitro testing of neodymium-iron-boron magnets. J Appl Biomater 1995;6:69–74.
251. Camilleri S, McDonald F. Static magnetic field effects on the sagittal suture in *Rattus norvegicus*. Am J Orthod Dentofac Orthop 1993;103:240–246.
252. Szivek JA, Johnson EM, Magee EP, Emmanual J, Poser R, Koeneman JB. Bone remodeling and in vivo strain analysis of intact and implanted greyhound proximal femora. J Invest Surg 1994;7: 213–233.
253. Bloebaum RD, Rhodes DM, Rubman MH, Hofmann AA. Bilateral tibial components of different

cementless designs and materials. Microradiographic, backscattered imaging, and histologic analysis. Clin Orthop 1991;268:179–187.
254. Willert H-G, Bertram H, Buchhorn G. Osteolysis in alloarthroplasty of the Hip. The role of UHMW polyethylene wear. Clin Orthop 1990;258:95–107.
255. Bothe RT, Beaton LE, Davenport HA. Reaction of bone to multiple metallic implants. Surg Gynecol Obstet 1940;71:598–602.
256. Clarke EGC, Hickman J. An investigation into the correlation between the electrical potential of metals and their behavior in biological fluids. J Bone Joint Surg 1953;35B:467–473.
257. Bianco PD, Ducheyne P, Cuckler JM. Titanium serum and urine levels in rabbits with a titanium implant in the absence of wear. Biomaterials 1996;17:1937–1942.
258. Hayashi K, Matsuguchi K, Uenoyama T, Kanemaru T, Sugioka Y. Evaluation of metal implants coated with several types of ceramics as biomaterials. J Biomed Mater Res 1989;23:1247–1259.
259. Carlsson L, Regner L, Johansson C, Gottlander M, Herberts P. Bone response to hydroxyapatite-coated and commercially pure titanium implants in the human arthritic knee. J Orthop Res 1994; 12:274–285.
260. Buser D, Schenk RK, Steinemann S, Fiorellini JP, Fox CH, Stich H. Influence of surface characteristics on bone integration of titanium implants. A histomorphometric study in miniature pigs. J Biomed Mater Res 1991;25:889–902.
261. Jacobs JJ, Skipor AK, Black J, Urban RM, Galante JO. Release and excretion of metal in patients who have a total hip-replacement component made of titanium-base alloy. J Bone Joint Surg 1991; 73A:14754–1486.
262. Hazan R, Oron U. Bone growth around a metal implant inserted to the medullary canal: A new experimental model in the rat. Clin Mater 1991;7:45–49.
263. Johansson CB, Alberktsson, Ericson LE. A quantitative comparison of the cell response to commercially pure titanium and Ti-6Al-4V implants in the abdominal wall of rats. J Mater Sci, Mater Med 1992;3:126–136.
264. Sennerby L, Thomsen P, Ericson LE. Ultrastructure of the bone-titanium interface in rabbits. J Mater Sci: Mater Med 1992;3:262–271.
265. Goodman S, Aspenberg P. Effect of amplitude of micromotion on bone ingrowth into titanium chambers implanted in the rabbit tibia. Biomaterials 1992;13:944–948.
266. Albrektsson T, Jacobsson M. Bone-metal interface in osseointegration. J Prosthet Dent 1987;57: 597–607.
267. Sundgren JE, Bodo P, Lundstrom I. Auger electron spectroscopic studies of the interface between human tissue and implants of titanium and stainless steel. J Colloid Interface Sci 1986;110:9–20.
268. Kohave D, Schwartz Z, Amir D, Müller-Mai C, Gross U, Sela J. Effects of titanium implants on primary mineralization following 6 and 14 days of rat tibial healing. Biomaterials 1982;13: 255–260.
269. Nasser S, Campbell PA, Kilgus D, Kossovsky N, Amstutz HC. Cementless total joint arthroplasty prostheses with titanium-alloy articular surfaces. A human retrieval analysis. Clin Orthop 1990; 261:171–185.
270. Hazan R, Brener R, Oron U. Bone growth to metal implants is regulated by their surface chemical properties. Biomaterials 1993;14:570–574.
271. Li P, de Groot K, Calcium phosphate formation within sol-gel prepared titania in vitro and in vivo. J Biomed Mater Res 1993;27:1495–1500.
272. Goodman SB, Davidson JA, Fornasier VL, Mishra AK. Histological response to cylinders of a low modulus titanium alloy (Ti-13Nb-13Zr) and a wear resistant zirconium alloy (Zr-2.5Nb) implanted in the rabbit tibia. J Appl Biomater 1993;4:331–339.
273. Vigorita VJ, Minkowitz B, Dichiara JF, Higham PA. A histomorphometric and histologic analysis of the implant interface in five successful, autopsy-retrieved, noncemented porous-coated knee arthroplasties. Clin Orthop 1993;293:211–218.
274. Bloebaum RD, Rubman MH, Hofmann AA. Bone ingrowth into porous-coated tibial components implanted with autograft bone chips. Analysis of ten consecutively retrieved implants. J Arthroplasty 1992;7:483–493.

275. Cook SD, Barrack RL, Baffes GC, Clemow AJT, Serekian P, Dong N, Kester MA. Wear and corrosion of modular interfaces in total hip replacements. Clin Orthop 1994;298:80–88.
276. Bobyn JD, Tanzer M, Ktygier JJ, Dujovne AR, Brooks E. Concerns with modularity in total hip arthroplasty. Clin Orthop 1994;298:27–36.
277. Witt JD, Swann M. Metal wear and tissue response in failed titanium alloy total hip replacements. J Bone Joint Surg 1991;73B:559–563.
278. Black J, Sherk H, Bonini J, Rostoker WR, Schajowicz F, Galante JO. Metallosis associated with a stable titanium-alloy femoral component in total hip replacement. J Bone Joint Surg 1990;72A: 126–130.
279. Bullough PG. Metallosis. J Bone Joint Surg 1994;76B:687–678.
280. Kelley SS, Johnston RC. Debris from cobalt-chrome cable may cause acetabular loosening. Clin Orthop 1992;285:140–146.
281. Buchert PK, Vaughn BK, Mallory TH, Engh CA, Bobyn D. Excessive metal release due to loosening and fretting of sintered particles on porous-coated hip prostheses. J Bone Joint Surg 1986; 68A:606–609.
282. Goodman SB, Fornasier, Lee J, Kei J. The effects of bulk versus particulate titanium and cobalt chrome alloy implanted into the rabbit tibia. J Biomed Mater Res 1990;24:1539–1549.
283. Betts F, Wright T, Salvati EA. Boskey A, Bansal M. Cobalt-alloy metal debris in periarticular tissues from total hip revision arthroplasties. Metal contents and associated histologic features. Clin Orthop 1992;276:75–82.
284. Huo MH, Betts F, Bogumill GP, Kenmore PI, Hayek RJ, Martinelli TJ. Metallic wear debris in acetabular osteolysis in a mechanically stable cementless total hip replacement: Report of a case. 1993;16:1277–1281.
285. Agins HJ, Alcock NW, Bansal M, Salvati EA, Wilson PD, Jr, Pellicci PM, Bullough PG. Metallic wear in failed titanium-alloy total hip replacements. A histological and quantitative analysis. J Bone Joint Surg 1988;70A;347–356.
286. Amstutz HC, Kim WC, O'Carroll PF, Kabo M. Canine porous resurfacing hip arthroplasty. Long-term results. Clin Orthop 1986;207:270–289.
287. Weissman BN, Scott RD, Brick GW, Corson JM. Radiographic detection of metal-induced synovitis as a complication of arthroplasty of the knee. J Bone Joint Surg 1991;73A:1002–1007.
288. Goldring SR, Bennett NE, Jasty MJ, Wang J-T. In vitro activation of monocyte macrophages and fibroblasts by metal particles. In: Particulate Debris from Medical Implants: Mechanisms of Formation and Biological Consequences, ASTM STP 1144, KR St John, Ed, American Society for Testing and Materials, Philadelphia 1992;136–142.
289. Amstutz HC, Kim WC, O'Carroll PF, Kabo M. Canine porous resurfacing hip arthroplasty. Long-term results. Clin Orthop 1986;207:270–289.
290. Scales JT. Black staining around titanium alloy prostheses—an orthopaedic enigma. J Bone Joint Surg 1991;73B:534–536.
291. Davidson JA. Characteristics of metal and ceramic total hip bearing surfaces and their effect on long-term ultra high molecular weight polyethylene wear. Clin Orthop 1993;294:361–378.
292. Shinto Y, Uchida A, Yoshikawa H, Araki N, Kato T, Ono K. Inguinal lymphadenopathy due to metal release from a prosthesis. A case report. J Bone Joint Surg 1993;75B:266–269.
293. Lalor PA, Revell PA, Gray AB, Wright S, Railton GT, Freeman MAR. Sensitivity to titanium: A cause of implant failure? J Bone Joint Surg 1991;73B:25–28.
294. Lalor PA, Revell PA. T-lymphocytes and titanium aluminum vanadium (TiAlV) alloy: Evidence for immunological events associated with debris deposition. Clin Mater 1993;12:57–62.
295. Hunt JA, Williams DF, Ungersboeck A, Perrin S. The effect of titanium debris on soft tissue response. J Mater Sci: Mater Med 1994;5:381–383.
296. Merritt K, Brown SA. Tissue reaction and metal sensitivity. An animal study. Acta Orthop Scand 1980;51:403–411.
297. Cook SD, McCluskey LC, Martin PC, Haddad RJ, Jr. Inflammatory response in retrieved noncemented porous-coated implants. Clin Orthop 1991;264:209–222.
298. Boss JH. Shajrawi I, Luria M, Mendes DG. Necrotizing granulomas within the periprosthetic

tissues of cemented and cementless total arthroplasties. Veter Comp Orthop Traumatol 1995;8: 107–113.
299. Mathiesen EB, Lindgren JU, Blomgren GA, Reinholt FP. Corrosion of modular hip prostheses. J Bone Joint Surg 1991;73B:569–575.
300. Svensson O, Mathiesen EB, Reinholt FP, Blomgren G. Formation of a fulminant soft-tissue pseudotumor after uncemented hip arthroplasty. J Bone Joint Surg 1988;70A:1238–1242.
301. Pizzoferrato A, Savarino L, Stea S, Tarabusi C. Results of histological grading on 100 cases of hip prosthesis failure. Biomaterials 1988;9:314–318.
302. Lewin J, Lindgren JU, Wahlberg JE. Apparent absence of local response to bone screws in guinea pigs with contact allergy. J Orthop Res 1987;5:604–608.
303. Albrektsson T, Brånemark P-I, Hansson H-A, Lindstroem J. Osseointegrated titanium implants: Requirements for ensuring a long-lasting, direct bone-to-implant anchorage in man. Acta Orthop Scand 1981;52:155–170.
304. Chehroudi B, Gould TRL, Brunette DM. A light and electron microscopic study of the effects surface topography on the behavior of cells attached to titanium-coated percutaneous implants. J Biomed Mater Res 1991;25:387–405.
305. Pilliar RM, Deporter DA, Watson PA, Valiquette N. Dental implant design—effect on bone remodeling. J Biomed Mater Res 1991;25:467–483.
306. Albrektsson T, Eriksson AR, Friberg B, Lekholm U, Lindahl L, Nevins M, Oikarinen, Roos J, Sennerby L, Astrand P. Histologic investigations on 33 retrieved Nobel-pharma implants. Clin Mater 1993;12:1–9.
307. Blumenthal NC, Cosma V. Inhibition of apatite formation by titanium and vanadium ions. J Biomed Mater Res 1989;23:13–22.
308. Shirkhanzadeh M. Electrochemical preparation of protective oxide coatings on titanium surgical alloys. J Mater Sci: Mater Med 1992;3:322–325.
309. Disegi JA, Wyss H. Implant materials for fracture fixation: A clinical perspective. Orthopedics 1989;12:75–79.
310. Jansen JA, van der Waerden JPCM, Paquay YCGJ, Histologic evaluation of the soft tissue response to sintered austenitic stainless steel fibre structures. J Mater Sci: Mater Med 1994;5:284–290.
311. Oron U, Alter A. Corrosion in metal implants embedded in various locations of the body in rats. Clin Orthop 1984;185:295–300.
312. Winckler S, Brug E, Baranowski D. Bundle nailing of forearm fractures. Indications and results. Unfallchirurg 1991;94:335–341.
313. Von Lüdinghausen M, Meister P, Probst J. Osteosynthese und Metallose. Med Welt 1970;21: 1913–1916.
314. Boss JH, Behar J, Misselevich I, Mendes DG. The histologic features of the interfacial membrane of intramedullary nails. J Long-Term Eff Med Impl 1995;5:169–183.
315. Matsuda Y, Yamamuro Kotoura Y, Fukumoto A. Severe metallosis observed 17 years after replacement of the knee with a tumor prosthesis: A case report. J Long-Term Eff Med Impl 1992; 1:295–303.
316. Hardy N, Burny F, Deutsch GA. Pin tract histological study. Preliminary clinical investigations. Orthopedics 1984;7:616–618.
317. Pettine KA, Chao EYS, Kelly PJ. Analysis of the external fixator pin-bone interface. Clin Orthop 1993;293:18–27.
318. Sutow EJ, Pollack SR. In Biocompatibility of Clinical Implant Materials, DF Williams, Ed, CRC Press, Boca Raton, 1981, p. 45.
319. Steinemann S. In Evaluation of Biomaterials, GD Winter, JL Leray and K de Groot, Eds, Wiley, New York, 1980, p 1.
320. Hyakuna K, Yamamuro T, Kotoura Y, Kakutani Y, Kitsugi T, Takagi H, Oka M, Kokubo T. The influence of calcium phosphate ceramics and glass-ceramics on cultured cells and their surrounding media. J Biomed Mater Res 1989;23:1049–1066.
321. Kawanabe K, Yamamuro T, Kotani S, Nakamura T. Acute nephrotoxicity as an adverse effect

after intraperitoneal injection of massive amounts of bioactive ceramic powders in mice and rats. J Biomed Mater Res 1992;26:209–219.
322. Ellies LG, Carter JM, Natiellla JR, Featherstone JDB, Nelson DGA. Quantitative analysis of early in vivo tissue response to synthetic apatite implants. J Biomed Mater Res 1988;22:137–148.
323. Hayashi K, Matsuguchi N, Uenoyama K, Kanemaru T, Sugioka Y. Evaluation of metal implants coated with several types of ceramics as biomaterials. J Biomed Mater Res 1989;23:1247–1259.
324. Dhert WJA, Klein CPAT, Jansen JA, van der Velde EA, Vriesde RC, Rozing PM, de Groot K. A histological and histomorphometrical investigation of fluorapatite, magnesium-whitlockite, and hydroxylapatite plasma-sprayed coatings in goats. J Biomed Mater Res 1993;27:127–138.
325. Daculsi G, Passuti N, Martin S, Deudon C, Legeros RZ, Raher S. Macroporous calcium phosphate ceramic for long bone surgery in humans and dogs. Clinical and histological study. J Biomed Mater Res 1990;24:379–396.
326. Basle MF, Rebel A, Grizon F, Daculsi G, Passuti N, Filmon R. Cellular response to calcium phosphate ceramics implanted in rabbit bone. J Mater Sci: Mater Med 1993;4:273–280.
327. Ducheyne P, Kim CS, Pollack SR. The effect of phase differences on the time-dependent variation of the zeta potential of hydroxyapatite. J Biomed Mater Res 1992;26:147–168.
328. Bagambisa FB, Joos U. Preliminary studies on the phenomenological behaviour of osteoblasts cultured on hydroxyapatite ceramics. Biomaterials 1990;11:50–56.
329. Tisdel CL, Goldberg VM, Parr JA, Bensusan JS, Staikoff LS, Stevenson S. The influence of a hydroxyapatite and tricalcium-phosphate coating on bone growth into titanium fiber-metal implants. J Bone Joint Surg 1994;76A:159–171.
330. Driessens FCM, Ramselaar MMA, Schaeken HG, Stols ALH, van Mullem PJ, de Wijn JR. Chemical reactions of calcium phosphate implants after implantation in vivo. J Mater Sci: Mater Med 1992;3:413–417.
331. Moore DC, Chapman MW, Manske D. The evaluation of a biphasic calcium phosphate ceramic for use in grafting long-bone diaphyseal defects. J Orthop Res 1987;5:356–365.
332. Cooke FW. Ceramics in orthopedic surgery. Clin Orthop 1992;276:135–146.
333. Heise U, Osborn JF, Duwe F. Hydroxyapatite ceramic as a bone substitute. Int Orthop 1990;14: 329–338.
334. Søballe K, Gotfredsen K, Brockstedt-Rasmussen H, Nielsen PT, Rechnagel K. Histologic analysis of a retrieved hydroxyapatite-coated femoral prosthesis. Clin Orthop 1991;272:255–258.
335. Kang JD, McKernan DJ, Kruger M, Mutschler T, Thompson WH, Rubash HE. Ingrowth and formation of bone in defects in an uncemented fiber-metal total in hip-replacement model in dogs. J Bone Joint Surg 1991;73A:93–105.
336. Spivak JM, Ricci JL, Blumenthal NC, Alexander H. A new canine model to evaluate the biological response of intramedullary bone to implant materials and surfaces. J Biomed Mater Res 1990;24: 1121–1149.
337. Walters MA, Blumenthal NC, Leung Y, Wang Y, Ricci JL, Spivak JM. Molecular structure at the bone-implant interface: A vibrational spectroscopic study. Calcif Tiss Int 1991;48:368–369.
338. Oonishi H, Noda T, Ito S, Kohda A, Ishimaru H, Yamamoto M, Tsuji E. Effect of hydroxyapatite coating on bone growth into porous titanium alloy implants under loaded conditions. J Appl Biomater 1994;5:23–37.
339. Bloebaum RD, Bachus KN, Rubman MC, Dorr LD. Postmortem comparative analysis of titanium and hydroxyapatite porous-coated femoral implants retrieved from the same patient. A case study. J Arthroplasty 1993;8:203–211.
340. Meenen NM, Osborn JF, Dallek M, Donath K. Hydroxyapatite-ceramic for juxtaarticular implantation. J Mater Sci: Mater Med 1992;3:345–351.
341. Geesink RGT, de Groot K, Klein CPAT. Chemical implant fixation using hydroxyl-apatite coatings. The development of a human total hip prosthesis for chemical fixation to bone using hydroxyl-apatite coatings on titanium substrates. Clin Orthop 1987;225:147–170.
342. Boone PS, Zimmerman MC, Gutteling E, Lee CK, Parsons JR, Langrana N. Bone attachment to hydroxyapatite coated polymers. J Biomed Mater Res 1989;23:183–199.

343. Piotrowski G, Hench LL, Allen WC, Miller GC. Mechanical studies of the bone bioglass interfacial bond. J Biomed Mater Res 1975;6:47–61.
344. Geesink RGT. Hydroxyapatite-coated total hip prostheses. Two-year clinical and roentgenographic results of 100 cases. Clin Orthop 1990;261:39–58.
345. D'Antonio JA, Capello WN, Jaffe WL. Hydroxylapatite-coated hip implants. Multicenter three-year clinical and roentgenographic results. Clin Orthop 1992;285:102–115.
346. Bloebaum RD, Merrell M, Gustke K, Simmons M. Retrieval analysis of a hydroxyapatite-coated hip prosthesis. Clin Orthop 1991;267:97–102.
347. Carlsson L, Roestlund T, Albrektssson B, Albrektssson T. Implant fixation improved by close fit: Cylindrical implant-bone interface studied in rabbits. Acta Orthop Scand 1988;59:272–275.
348. Hofmann AA, Bachus KN, Bloebaum RD. Comparative study of human cancellous bone remodeling to titanium and hydroxyapatite-coated implants. J Arthroplasty 1993;8:157–166.
349. Munting E, Mirtchi AA, Lemaitre J. Bone repair of defects filled with a phosphocalcic hydraulic cement: An in vivo study. J Mater Sci: Mater Med 1993;4:337–344.
350. Filiaggi MJ, Pilliar RM, Coombs NA. Post-plasma-spraying heat treatment of the HA coating/Ti-6Al-4V implant system. J Biomed Mater Res 1993;27:191–198.
351. Lennox DW, Schofield BH, McDonald DF, Riley LH, Jr. A histologic comparison of aseptic loosening of cemented, press-fit, and biologic ingrowth prostheses. Clin Orthop 1987;225:171–191.
352. Furlong RJ, Osborn JF. Fixation of hip prostheses by hydroxyapatite ceramic coatings. J Bone Joint Surg 1991;73B:741–745.
353. Hardy DCR, Frayssinet P, Guilhem A. Bonding of hydroxyapatite-coated femoral prostheses: Histopathology of specimens from four cases. J Bone Joint Surg 1991;73B:732–738.
354. Cheung HS, Halverson PB, McCarty DJ. Phagocytosis of hydroxyapatite or calcium pyrophosphate dihydrate crystals by rabbit articulate chondrocytes stimulates release of collagenase, neutral protease, and prostaglandins E_2 and F_{2a}. Proc Soc Exp Biol Med 1983;173:181–189.
355. Alwan WH, Dieppe PA, Elson CJ, Bradfield JWB. Hydroxyapatite and urate crystal induced cytokine release by macrophages. Ann Rheum Dis 1989;48:476–482.
356. Hong L, Xu H, de Groot K. Tensile strength of the interface between hydroxyapatite and bone. J Biomed Mater Res 1992;26:7–18.
357. Bloebaum RD, Dupont JA. Osteolysis from a press-fit hydroxyapatite-coated implant. J Arthroplasty 1993;8:195–202.
358. Bloebaum RD, Beeks D, Dorr LD, Savory CG, Dupont JA, Hofmann AA. Complications with hydroxyapatite particulate separation in total hip arthroplasty. Clin Orthop 1994;298:19–26.
359. Heimke G, Griss P. Five years experience with ceramic-metal-composite hip endoprostheses. II: Mechanical evaluations and improvements. Arch Orthop Traumatol Surg 1981;98:165–171.
360. Mahoney OM, Dimon JH, III. Unsatisfactory results with a ceramic total hip prosthesis. J Bone Joint Surg 1990;72A:663–671.
361. Saito M, Saito S, Ohzono K, Takaoka K, Ono K. Efficacy of alumina ceramic heads for cemented total hip arthroplasty. Clin Orthop 1992;283:171–177.
362. Takagi H, Yamamuro T, Hyakuna K, Nakamura T, Kotoura Y, Oka M. Bone bonding behavior of bead-coated alumina ceramic under load-bearing conditions. J Biomed Mater Res 1989;23:161–181.
363. Harms J, Mausle E. Tissue reaction to ceramic implant material. J Biomed Mater Res 1979;13:67–87.
364. Williams DF. A model for biocompatibility and its evaluation. J Biomed Eng 1989;11:185–193.
365. Predecki P, Stephan JE, Auslaender BA, Mooney VL, Kirkland K. Kinetics of bone growth into cylindrical channels in aluminum oxide and titanium. J Biomed Mater Res 1972;6:375–400.
366. Griss P, von Andrian-Werburg J, Krempien B, Heimke G. Biological activity and histocompatibility of dense AL203/MgO ceramic implants in rats. J Biomed Mater Res 1973;4:453–462.
367. Griss P, Heimke G. Biocompatibility of high density alumina and its application in orthopaedic surgery. In: Biocompatibility of Clinical Implant Materials, DF Williams, Ed, CRC Press, Boca Raton, 1981; 155–198.

368. Plenk H, Jr. Zur direkten Knochenverankerung von zementfreien Hüften-prothesenschäften. Med Orthop Tech 1991;111:65–70.
369. Willert HG, Semlitsch M. Reactions of the articular capsule to wear products of artificial joint prostheses. J Biomed Mater Res 1977;11:157–164.
370. Dorlot J-M. Long-term effects of alumina components in total hip prostheses. Clin Orthop 1992; 282:47–52.
371. Picha GJ, Goldstein JA. Analysis of the soft tissue response to components used in the manufacture of breast implants: Rat animal model. Plast Reconstr Surg 1991;87:490–500.
372. Habal M. The biological basis for the clinical application of the silicones. Arch Surg 1984;119: 843–848.
373. Bakker D, van Blitterswijk CA, Daems WTh, Grote JJ, Biocompatibility of six elastomers in vitro. J Biomed Mater Res 1988;22:423–439.
374. Bradley SG, White KL Jr, McCay JA, Brown RD, Musgrove DL, Wilson S, Stern M, Luster MI, Munson AE. Immunotoxicity of 180 day exposure to polydimethylsiloxane (silicone) fluid, gel and elastomer and polyurethane disks in female B6C3F1 mice. Drug Chem Toxicol 1994;17: 221–269.
375. Guo W, Willén R, Liu X, Odelius R, Carlén B. Splenic response to silicon drain material following intraperitoneal implantation. J Biomed Mater Res 1994;28:1433–1438.
376. Worsing RA, Engber WD, Lange TA. Reactive synovitis from particulate silastic. J Bone Joint Surg 1982;64A:581–585.
377. Eiken O, Ekerot L, Lindstroem C, Jonsson K. Silicone carpal implants: Risk or benefit? Scand J Plast Reconstr Surg 1985;19:295–304.
378. Verhaar J, Vermeulen A, Bulstra S, Walenkamp G. Bone reaction to silicone metatarsophalangeal joint-1 hemiprosthesis. Clin Orthop 1989;245:228–232.
379. Westesson P-L, Eriksson L, Lindstroem C. Destructive lesions of the mandibular condyle following diskectomy with temporary silicone implant. Oral Surg, Oral med, Oral Path 1987;63:143–150.
380. Gordon M, Bullough PG. Synovial and osseous inflammation in failed silicone-rubber prostheses. J Bone Joint Surg 1982;64A:574–580.
381. Manes HR. Foreign body granuloma of bone secondary to silicone prosthesis. Clin Orthop 1985; 199:239–241.
382. Natsume T, Ike O, Okada T, Takimoto N, Shimizu Y, Ikada Y. Porous collagen sponge for esophageal replacement. J Biomed Mater Res 1993;27:867–875.
383. Kalman PG, Ward CA, McKeowen NB, McCullough D, Romaschin AD. Improved biocompatibility of silicone rubber by removal of surface entrapped air nuclei. J Biomed Mater Res 1991;25: 199–211.
384. Williams JE. Experience with a large series of silastic breast implants. Plast Reconst Surg 1972; 49:483–487.
385. Rolland C, Guidoin R, Marceau D, Ledoux R. Nondestructive investigations on ninety-seven surgically excised mammary J Biomed Mater Res 1989;23:285–298.
386. Domanskis EJ, Owsley JQ. Histological investigation of the etiology of capsule contracture following augmentation mammaplasty. Plast Reconstr Surg 1976;56:689–693.
387. Smahel J. Tissue reactions to breast implants coated with polyurethane. Plast Reconstr Surg 1978; 61:80–85.
388. Luke JL, Kalasinsky VF, Turnicky RP, Centeno JA, Johnson FB, Mullick FG. Pathological and biophysical findings associated with silicone breast implants: a study of capsular tissues from 86 cases. Plast Reconstr Surg 1997;100:1558–1565.
389. Hameed MR, Erlandson R, Rosen PP. Capsular synovial-like hyperplasia around mammary implants similar to detritic synovitis. A morphologic and immunohistochemical study of 15 cases. Am J Surg Pathol 1995;19:433–438.
390. Kasper CS. Histologic features of breast capsules reflect surface configuration and composition of silicone bag implants. Am J Clin Pathol 1994;102:655–659.
391. Brinton LA, Brown SL. Breast implants and cancer. J Natl Cancer Inst 1997;89:1341–1349.
393. Corrin B. Silicone lymphadenopathy. J Clin Path 1982;35:901–902.

394. Heggers JP, Kossovsky N, Parsons RW, Robson MC, Pelley RP, Raine TJ. Biocompatibility of silicone implants. Ann Plast Surg 1983;11:38–45.
395. Sergott TJ, Limoli JP, Baldwin CM, Laub DR. Human adjuvant disease, possible autoimmune disease after silicone implantation: A review of the literature, case studies, and speculation for the future. Plast Reconstr Surg 1986;78:104–114.
396. Kossovsky N, Heggers JP, Robson MC. Experimental demonstration of the immunogenicity of silicone-protein complexes. J Biomed Mater Res 1987;21:1125–1133.
397. Kossovsky N, Zeidler M, Chun G, Papasian N, Nguyen A, Rajguru S, Stassi J, Gelman A, Sponsler E. Surface dependent antigens identified by high binding avidity of serum antibodies in a subpopulation of patients with breast prostheses. J Appl Biomater 1993;4:281–288.
398. Celli B, Textor S, Kovnat DM. Adult respiratory distress syndrome following mammary augmentation. Am J Med Sci 1978;275:81–85.
399. Truong L, Cartwright J, Goodman D, Woznicki D. Silicone lymphadenopathy associated with augmentation mammoplasty. Morphologic features of nine cases. Am J Surg Path 1988;12:484–491.
400. Kossovsky N, Frieman CJ. Silicone breast implant pathology. Clinical data and immunologic considerations. Arch Pathol Lab Med 1994;118:686–693.
401. Silver FH, Olson RM. Silicone implants and autoimmune disease: Fact or fantasy? In: Biomedical Engineering Recent Development, J Vossoughi, Ed, University of the District of Columbia, Washington, 1994; 1113–1116.
402. Handel N, Jensen JA, Black Q, Waisman JR, Silverstein MJ. The fate of breast implants: A critical analysis of complications and outcomes. Plast Reconstr Surg 1995;96:1521–1533.
403. Sanchez-Guerrero J, Colditz GA, Karlson EW, Hunter DJ, Speizer FE, Liang MH. Silicone breast implants and the risk of connective-tissue diseases and symptoms. N Engl J Med 1995;332:1666–1670.
404. Bar-Meir E, Teuber SS, Lin HC, Alosacie I, Goddard G, Terybery J, Barka N, Shen B, Peter JB, Blank M. Multiple autoantibodies in patients with silicone breast implants. J Autoimmun 1995;8: 267–277.
405. Orkin RW, Abbott WM. A compliant tubular device to study the influence of wall strain and fluid shear stress on cells of the vascular wall. J Vasc Surg 1994;20:184–194.
406. Joos KM, Sandra A. Microarterial synthetic graft repair: Interstitial cellular components. Microsurgery 1990;11:268–277.
407. Mäkisalo S, Skutnabb K, Holmström T, Grönblad M, Paasolainen P. Reconstruction of anterior cruciate ligament with carbon fiber. An experimental study on pigs. Am J Sports Med 1988;16: 589–593.
408. Alexander H, Weiss AB, Parsons JR, Ligament repair and reconstruction with an absorbable polymer coated carbon fiber stent. J Orthop Surg Tech 1987;3:1–14.
409. Jenkins DHR. The repair of cruciate ligaments with flexible carbon fibre: A longer term study of the induction of new ligaments and of the fate of the implanted carbon. J Bone Joint Surg 1978; 60B:520–522.
410. Amis AA, Kempson SA, Campbell JR, Miller JH. Anterior cruciate ligament replacement. Biocompatibility and biomechanics of polyester and carbon fibre in rabbits. J Bone Joint Surg 1988; 70B:628–634.
411. Zarnett R, Velazquez R, Salter RB. The effect of continuous passive motion on knee ligament reconstruction with carbon fibre. An experimental investigation. J Bone Joint Surg 1991;73B: 47–52.
412. Forster JW, Ralis ZA, McKibbin B, Jenkins DHR. Biological reaction to carbon fiber implants. The formation and structure of a carbon-induced "neotendon." Clin Orthop 1978;131:299–307.
413. Jenkins DHR, McKibbin B. The role of flexible carbon-fiber implants as tendon and ligament substitutes in clinical practice. A preliminary report. J Bone Joint Surg 1980;62B:497–499.
414. Greis PE, Georgescu HI, Fu FH, Evans CH. Particle-induced synthesis of collagenase by synovial fibroblasts: An immunocytochemical study. J Orthop Res 1994;12:286–293.

415. Mendes DG. An overview on the use of carbon fibres in ligament reconstruction. J Orthop Surg Tech 1987;3:53–65.
416. Maekisalo S, Paavolainen P, Holmstroem T, Skutnabb K. Carbon fiber as a prosthetic anterior cruciate ligament. A biochemical and histological analysis in pigs. Am J Sports Med 1989;17: 459–462.
417. Mendes DG, Soudry M, Levine E, Silbermann M, Shoshan S, Gross U, Shajrawi I, Boss JH. The collagen composition of the carbon fiber composite ligament: An electromicroscopical, biochemical and immunohistological study. J Long-Term Eff Med Impl 1992;1:305–319.
418. Soudry M, Roffman M, Boss JH, Rotem A, Mendes DG. Fixation of prosthetic patella by ingrowth of fibrous tissue. Trans 27th Meet Orthop Res Soc, 1981, p. 323.
419. Boss JH, Shajrawi I, Mendes DG. Intraosseous anchorage of the carbon fiber composite ligament by bony ingrowth. J Appl Biomater 1991;2:241–242.
420. Minns RJ, Muckle DS. Mechanical and histological response of carbon fibre pads implanted in the rabbit patella. Biomaterials 1989;10:273–276.
421. Minns RJ, Muckle DS, Betts JA. Biological resurfacing using carbon fiber. Orthopedics Int Edn 1993;1:414–424.
422. Veth RPH, den Heeten GJ, Jansen HWB, Nielsen HKL. An experimental study of reconstructive procedures in lesions of the meniscus. Use of synovial flaps and carbon fiber implants for artificially made lesions in the meniscus of the rabbit. Clin Orthop 1983;181:250–254.
423. Veth RPH, Jansen HWB, Leenslag JW, Pennings AJ, Hartel RM, Nielsen HKL. Experimental meniscal lesions reconstructed with a carbon fiber-polyurethane poly(L-lactide) graft. Clin Orthop 1986;202:286–293.
424. Wood DJ, Minns RJ, Strove A. Replacement of the rabbit medial meniscus with a polyester-carbon fibre bioprosthesis. Biomaterials 1990;11:13–16.
425. Ward B, Minns RJ. Woven carbon-fibre patch versus Dacron mesh in the repair of experimental defects in the lumbar fascia of rabbits. Biomaterials 1989;10:425–428.
426. Gleason TF, Barmada R, Ghosh L. Can carbon fiber implants substitute for collateral ligament? Clin Orthop 1984;191:274–280.
427. Amis AA, Campbell JR, Miller JH. Strength of carbon and polyester fibre tendon replacements: Variation after operation in rabbits. J Bone Joint Surg 1985;67B:829–832.
428. Forster IW, Shuttleworth A. Tissue reaction to intraarticular carbon fibre implants in the knee. J Bone Joint Surg 1984;66B:282–288.
429. Leyshon RL, Channon GM, Jenkins DHR, Ralis ZA. Flexible carbon fibre in late ligamentous reconstruction for instability of the knee. J Bone Joint Surg 1984:66B:196–200.
430. Vert M, Li SM, Spenlehaur G, Guerin P. Bioresorbability and biocompatibility of aliphatic esters. J Mater Sci: Mater Med 1992;3:432–446.
431. Suuronen S, Pohjonen T, Taurio R, Törmälä P, Wessman L, Roenkkoe K, Vainionpää S. Strength retention of self-reinforced poly-L-lactide screws and plates: An in vivo and in vitro study. J Mater Sci: Mater Med 1992;3:426–431.
432. Hollinger JO. Preliminary report on the osteogenic potential of a biodegradable copolymer of polylactide (PLA) and polyglycolide (PGA). J Biomed Mater Res 1983;17:71–82.
433. Saltzman WM, Parsons-Wingerter P, Leong KW, Lin S. Fibroblast and hepatocyte behavior on synthetic polymer surfaces. J Biomed Mater Res 1991;25:741–759.
434. Kalman PG, McCullough DA, Ward CA. Evacuation of microscopic air bubbles from Dacron reduces complement activation and platelet aggregation. J Vasc Surg 1990;11:591–598.
435. Ishaug SL, Jaszemski MJ, Bizios R, Mikos AG. Osteoblast function of synthetic biodegradable polymers. J Biomed Mater Res 1994;28:1445–1453.
436. Gogolewski S, Jovanovic M, Perren SM, Dillon JG, Highes MK. Tissue response and in vivo degradation of selected polyhydroxyacidsa: Polylactides (PLA), poly(3-hydroxybutyrate) (PHB), and poly(3-hydroxybutyrate-co-3-hydroxyvalerate) (PHB/VA). J Biomed Mater Res 1993;27: 1135–1148.
437. Andriano KP, Daniels AU, Smutz P, Wyatt WB. Preliminary biocompatibility screening of several biodegradable phosphate fiber reinforced polymers. J Biomed Mater Res 1993;4:1–12.

438. Saitoh H, Takata T, Nikai H, Shintani H, Hyon S-H, Ikada Y. Tissue compatibility of polylactic acids in the skeletal site. J Mater Sci: Mater Med 1994;5:194–199.
439. Pitt CG, Gratzlk MM, Kimmel GL, Surles J, Schindler A. The degradation of poly(DL lactide), poly(ε-caprolactone), and their copolymers M-V in vitro. Biomaterials 1981;2:215–220.
440. Majola A, Vainionpaa S, Vihtonen K, Mero M, Vasenius J, Törmälä P, Rokkanen P. Absorption, biocompatibility, and fixation properties of polylactic acid in bone tissue: An experimental study in rats. Clin Orthop 1991;268:260–269.
441. Schliephake H, Klosa D, Rahlff M. Determination of the 3-D morphology of degradable biopolymer implants undergoing in vivo resorption. J Biomed Mater Res 1993;27:991–998.
442. Van Blitterswijk CA, van der Brink J, Leenders H, Bakker D. The effect of PEO ratio on degradation, calcification and bone bonding of PEO/PBT copolymer (polyactive). Cells Mater 1993;3: 23–36.
443. Wu Y, Zhao O, Anderson JM, Hiltner A, Lodoen GA, Payet CR. Effect of some additives on the biostability of a poly(etherurethane) elastomer. J Biomed Mater Res 1991;25:725–739.
444. Daniels AU, Andriano KP, Paul Smutz W, Chang MKO, Heller J. Evaluation of absorbable poly-(ortho esters) for use in surgical implants. J Appl Biomater 1994;5:51–64.
445. Hollinger JO, Battistone GC. Biodegradable bone repair materials. Synthetic polymers and ceramics. Clin Orthop 1986;207:290–305.
446. Williams DF. Enzymatic hydrolysis of polylactic acid. Eng Med 1981;10:5–7.
447. Williams DF. Mechanisms of biodegradation of implantable polymers. Clin Mater 1992;10:9–12.
448. Vasenius J, Vainionpaa S, Vihtonen K, Maekela A, Rokkanen P, Mero M, Törmälä P. Comparison of in vitro hydrolysis, subcutaneous and intramedullary implantation to evaluate the strength retention of absorbable osteosynthesis implants. Biomaterials 1990;11:501–504.
449. Taylor MS, Daniels AU, Andriano KP, Heller J. Six bioabsorbable polymer: In vitro acute toxicity of accumulated degradation products. J Appl Biomater 1994;5:151–157.
450. Van Sliedgregt A, Radder AM, de Groot K, van Blitterswijk CA. In vitro biocompatibility testing of polylactides. Part I. Proliferation of different cell types. J Mater Sci: Mater Med 1992;3:365–370.
451. Suganuma J, Alexander H. Biological response of intramedullary bone to poly-L-lactic acid. J Appl Biomater 1993;4:13–27.
452. Silver FH, Marks M, Li C, Pulapura S, Kohn J. Tissue compatibility of tyrosine-derived polycarbonates and polyiminocarbonates: An initial evaluation. J Long-Term Eff Med Impl 1992;1:329–346.
453. Lam KH, Schakenraad JM, Esselbrugge H, Feijen J, Nieuwenhuis P. The effect of phagocytosis of poly(L-lactic acid) fragments on cellular morphology and viability. J Biomed Mater Res 1993; 27:1569–1577.
454. Merrit K, Shafer JW, Brown SA. Implant site infection rates with porous and dense materials. J Biomed Mater Res 1979;13:101–108.
455. Schakenraad JM, Dijkstra PJ. Biocompatibility of poly(DL-lactic acid/glycine) copolymers. Clin Mater 1991;7:253–269.
456. Saitoh H, Takata T, Nikai H, Shintani H, Hyon S-S, Ikada Y. Tissue compatibility of polylactic acids in skeletal site. J Mater Sci: Mater Med 1994;5:194–199.
457. Schakenraad JM, Oosterbaan JA, Nieuwenhuis P, Molenaar I, Olijslager J, Potman W, Eenink MJD, Feijen J. Biodegradable hollow fibers for the controlled release of drugs. Biomaterials 1988; 9:116–120.
458. Schakenraad JM, Hardonk MJ, Feijen J, Molenaar I, Nieuwenhuis P. Enzymatic activity toward poly(L-lactic acid) implants. J Biomed Mater Res 1990;24:529–545.
459. Schakenraad JM, Nieuwenhuis P, Molenaar I, Helder J, Dijkstra PJ, Feijen J. In vivo and in vitro degradation of glycine/DL-lactic acid copolymers. J Biomed Mater Res 1989;23:1271–1288.
460. Matlaga BF, Salthouse TN. Ultrastructural observations of cells at the interface of a biodegradable polymer: Polyglactin 910. J Biomed Mater Res 1983;17:185–197.
461. Davis PA, Huang SJ, Ambrosio L, Ronca D, Nicolais L. A biodegradable composite artificial tendon. J Mater Sci: Mater Med 1991;3;359–364.
462. Kato YP, Dunn MG, Zawadsky JP, Tria AJ, Silver FH. Regeneration of Achilles tendon with a

collegen tendon prosthesis. Results of a one-year implantation study. J Bone Joint Surg 1991;73A: 561–574.
463. Daniels AU, Chang MKO, Andriano KP, Heller J. Mechanical properties of biodegradable polymers and composites proposed for internal fixation of bone. J Appl Biomater 1990;1:57–78.
464. Böstman OM. Absorbable implants for the fixation of fractures. J Bone Joint Surg 1991;73A: 148–153.
465. Mäkelä EA. Fixation properties and biodegradation of absorbable implants in growing bone. An experimental and clinical study. Academic Dissertation. Helsinki, 1989.
466. Gombotz WR, Pankey SC, Bouchard LS, Ranchalis J, Puolakkainen P. Controlled release of TGF-β_1 from a biodegradable matrix for bone regeneration. J Biomater Sci Polymer Edn 1993;5:49–63.
467. Otto TE, Patka P, Haarman HJTM, Klein CPAT, Vriesde R. Intramedullary bone formation after polylactic acid wire implantation. J Mater Sci: Mater Med 1994;5:407–410.
468. Matsusue Y, Yamamuro T, Yoshil S, Oka M, Ikada Y, Hyon S-H, Shikinami Y. Biodegradable screw fixation of rabbit tibia proximal osteotomies. J Appl Biomater 1991;2:1–12.
469. Böstman OM, Paeivaerinta U, Partio E, Manninen M, Vasenius J, Majola A, Rokkanen P. The tissue-implant interface during degradation of absorbable polyglycolide fracture fixation screws in the rabbit femur. Clin Orthop 1992;285:263–272.
470. Ibnabddjalil M, Loh IH, Hu CC, Blumenthal N, Alexander H, Turner D. Effect of surface plasma treatment on the chemical, physical, morphological, and mechanical properties of totally absorbable bone internal fixation devices. J Biomed Mater Res 1994;28:289–301.
471. Manninen MJ, Päivärinta U, Paetiälä, Rokkanen P, Taurio R, Tamminmäki M, Törmälä. Shear strength of cancellous bone after osteotomy fixed with absorbable self-reinforced polyglycolic acid and poly-L-lactic acid rods. J Mater Sci: Mater Med 1992;3:245–251.
472. Verheyen CC, de Wijn JR, van Blitterswijk CA, Rozing PM, de Groot K. Examination of efferent lymph nodes after 2 years of transcortical implantation of poly(L-lactide) containing plugs: A case report. J Biomed Mater Res 1993;27:1115–1118.
473. Pihlajamäki H, Böstman OM, Manninen M, Päivaerinta U, Törmälä P, Rokkanen P. Absorbable plugs of self-reinforced poly-L-lactic acid in the internal fixation of rabbit distal femoral osteotomies. Clin Orthop 1994;298:277–285.
474. Hoppert T, Pistner H, Stolte M, Muehling J. Sarkomauslösung durch resorbierbares Osteosynthesematerial bei der Ratte. Eine vorläufige Mitteilung. Z Orthop 1992;130:244–253.
475. Böstman O, Hirvensalo E, Vainionpää S, Vihtonen K, Törmälä P, Rokkanen P. Degradable polyglycolide rods for the internal fixation of displaced malleolar fractures. Int Orthop 1990;14;1–8.
476. Stähelin AC, Weiler A, Rüfenacht H, Hoffmann R, Geissmann A, Feinstein R. Clinical degradation and biocompatibility of different bioabsorbable interference screws. A report of six cases. Arthroscopy, J Arthrosc Rel Surg 1997;13:238–244.
477. Böstman O, Hirvensalo E, Vainionpää S, Mäkelä A, Vihtonen K, Törmälä P, Rokkanen P. Ankle fractures treated using biodegradable internal fixation. Clin Orthop 1989;238:195–203.
478. Böstman O, Päivärinta U, Partio E, Vasenius J, Manninen M, Rokkanen P. Degradation and tissue replacement of an absorbable polyglycolic screw in the fixation of rabbit femoral osteotomies. J Bone Joint Surg 1992;74A:1021–1031.
479. Böstman OM. Intense granulomatous inflammatory lesions associated with absorbable internal fixation devices made of polyglycolide in ankle fractures. Clin Orthop 1992;278:193–199.
480. Böstman OM. Osteolytic changes accompanying degradation of absorbable fracture fixation implants. J Bone Joint Surg 1991;73B:679–882.
481. Verheyen CCPM, 329. Lescure F, Gurny R, Doelker E, Pelaprat ML, Bichon D, Anderson JM. Acute histopathological response to a new biodegradable polypeptidic polymer for implantable drug delivery system. J Biomed Mater Res 1989;23:1299–1313.
482. Parsons JR. Resorbable materials and composites. New concepts in orthopedic biomaterials. Orthopedics 1985;8:907–915.
483. Turina J, Hess OM, Turina M, Krayenbuehl HP. Cardiac bioprostheses in the 1990s. Circulation 1993;88:775–781.
484. Solheim E, Pinholt EM, Bang G, Sudmann E. Effect of local hemostatics on bone induction in

rats: A comparative study of bone wax, fibrin-collagen paste, and bioerodible polyorthoester with and without gentamicin. J Biomed Mater Res 1992;26:791–800.
485. Bertone AL, Sullins KE, Stashak TS, Norrdin RW. Effect of wound location and the use of topical collagen gel on exuberant granulation tissue formation and wound healing in the horse and pony. Am J Vet Res 1985;46:1438–1444.
486. Chvapil M, Chvapil TA, Owen JA. Reaction of various skin wounds in the rat to collagen sponge dressings. J Surg Res 1986;41:410–418.
487. Nimni ME, Cheung D, Strates B, Kodama M, Sheikh K. Bioprosthesis derived from cross-linked and chemically modified collagenous tissues. In: Collagen, Vol 3: Biotechnology, ME Nimni, Ed, CRC Press, Boca Raton, 1988; 1–37.
488. Doillon CJ, Whyne CE, Brandwein S, Silver FH. Collagen-based wound dressings. J Biomed Mater Res 1986;20:1219–1228.
489. Chvapil M, Speer DP, Holubec H, Chvapil TA, King DH. Collagen fibers as a temporary scaffold for replacement of ACL in goats. J Biomed Mater Res 1993;27:313–325.
490. Dunn MG, Avasarala PN, Zawadsky JP. Optimization of extruded collagen fibers for ACL reconstruction. J Biomed Mater Res 1993;27:1545–1552.
491. Huang-Lee LLH, Cheung DT, Nimni ME. Biochemical changes and cytotoxicity associated with the degradation of polymeric glutaraldehyde derived crosslinks. J Biomed Mater Res 1990;24: 1185–1201.
492. Ferrans VJ, Spray LT, Billingham ME, Roberts WC. Structural changes in glutaraldehyde-treated porcine heterografts used as substitute cardiac valves. Am J Cardiol 1978;41:1159–1184.
493. Anselme K, Bacques C, Charriere G, Hartmann DJ, Herbage D, Garrone R. Tissue reaction to subcutaneous implantation of a collagen sponge. A histological, ultrastructural, and immunological study. J Biomed Mater Res 1990;24:689–703.
494. McCoy JP, Schade W, Siegle RJ, Vanderveen EE, Zachary CB, Waldinger TP, Swanson NA. Immune response to bovine collagen implants. Significance of pretreatment serology. J Am Acad Dermatol 1987;16:955–960.
495. DeLustro F, Condell RA, Nguyen MA, McPherson JM. A comparative study of the biologic and immunologic response to medical devices derived from dermal collagen. J Biomed Mater Res 1986;20:109–120.
496. DeLustro F, Dasch J, Keefe J, Ellingsworth L. Immune responses to allogeneic and xenogeneic implants of collagen and collagen derivatives. Clin Orthop 1990;260;263–279.
497. Boss JH, Silber E, Nelken D. Nonpathogenicity of species homologous antikidney antibodies in the rat. Br J Exp Path 1968;48:1–5.
498. Olmo N, Turnay J, Gavilanes JG, Lizarbe MA, Herrera JI. Subcutaneous and intramuscular implantation of sepiolite-collagen complexes. J Mater Sci: Mater Med 1992;3:239–244.
499. Law JK, Parsons JR, Silver FH, Weiss AB. An evaluation of purified reconstituted type I collagen fibers. J Biomed Mater Res 1989;23:961–977.
500. Van Wachem PB, van Luyn MJA, Olde Damink LHH, Dijkstra PJ, Feijen J, Nieuwnheuis P. Biocompatibility and tissue regenerating capacity of crosslinked dermal sheep collagen. J Biomed Mater Res 1994;28:353–363.
501. Gilbert DL, Kim SW. Macromolecular release from collagen monolithic devices. J Biomed Mater Res 1990;24:1221–1239.
502. Panni AS, Fabbriciani C, Delcogliano A, Franzese S. Bone-ligament interaction in patellar tendon reconstruction of the ACL. Knee Surg Sports Traumatol Arthroscopy 1993;1:4–8.
503. Dunn MG, Tria AJ, Kato P, Bechler JR, Ochner RS, Zawadsky JP, Silver FH. Anterior cruciate ligament reconstruction using a composite collagenous prosthesis. A biomechanical and histologic study in rabbits. Am J Sports Med 1992;20:507–515.
504. Dunn MG, Maxian SH, Zawadsky JP. Intraosseous incorporation of composite collagen prostheses designed for ligament reconstruction. J Orthop Res 1994;12:128–137.
505. Maeda A, Inoue M, Shino K, Nakamura H, Tanaka M, Seguchi Y, Ono K. Effects of solvent preservation with or without gamma irradiation on the material properties of canine tendon allografts. J Orthop Res 1993;11:181–189.

506. Valeri A, Radhakrishnan J, Ryan R, Powell D. Biocompatible dialysis membranes and acute renal failure: A study in post-operative acute tubular necrosis in cadaveric renal transplant recipients. Clin Nephrol 1996;46:402–409.
507. Himmelfarb J, Tolkoff Rubin N, Chandran P, Parker RA, Wingard RL, Hakim R. A multicenter comparison of dialysis membranes in the treatment of acute renal failure requiring dialysis. J Am Soc Nephrol 1998;9:257–266.
508. Hoenich NA, Woffindin C, Stamp S, Roberts SJ, Turnbull J. Synthetically modified cellulose: An alternative to synthetic membranes for use in haemodialysis? Biomaterials 1997 Oct;18(19): 1299–1303.
509. Muller TF, Seitz M, Eckle I, Lange H, Kolb G. Biocompatibility differences with respect to the dialyzer sterilization method. Nephron 1998;78:139–142.
510. Hakim RM, Wingard RL, Husni L, Parker RA, Parker TF III. The effect of membrane biocompatibility on plasma beta 2-microglobulin levels in chronic hemodialysis patients. J Am Soc Nephrol 1996;7:472–478.
511. Santoro A, Ferrari G, Francioso A, Zucchelli P, Duranti E, Sasdelli M, Rosati A, Salvadori M, Sanna GM, Briganti M, Fusaroli M, Lindner G, Stefani A, Borgatti P, Badiali F, Mignani R, Cagnoli L, Aucella F. Stallone C, Massazza M, Borghi M, Gualandris L, Modoni S, Grandone E, Orlandini G. Ethylene-oxide and steam-sterilised polysulfone membrane in dialysis patients with eosinophilia. Int J Artif Organs 1996;19:329–335.
512. Mulvihill J, Crost T, Renaux JL, Cazenave JP. Evaluation of haemodialysis membrane biocompatibility by parallel assessment in an ex vivo model in healthy volunteers. Nephrol Dial Transplant 1997;12:1968–1973.
513. Kim SW, Jacobs H. Design of nonthrombogenic polymer surfaces for blood contacting medical devices. Blood Purif 1996;14:357–372.
514. Locatelli F, Manzoni C. Biocompatibility in haemodialysis: Fact and fiction. Curr Opin Nephrol Hypertens 1997;6:528–532.
515. Martos MR, Hendry BM, Rodriguez-Puyol M, Dwight J, Diez-Marques ML, Rodriguez-Puyol D. Haemodialyser biocompatibility and erythrocyte structure and function. Clin Chim Acta 1997;265: 235–246.
516. Bonomini M, Stuard S, Carreno MP, Settefrati N, Santarelli P, Haeffner-Cavaillon N, Albertazzi A. Neutrophil reactive oxygen species production during hemodialysis: Role of activated platelet adhesion to neutrophils through P-selectin. Nephron 1997;75:402–411.
517. Brulez HF, Dekker HA, Oe PL, Verbeelen D, ter Wee PM, Verbrugh HA. Biocompatibility of a 1.1% amino acid-containing peritoneal dialysis fluid compared to a 2.27% glucose-based peritoneal dialysis fluid. Nephron 1996;74:26–32.
518. Wieczorowska-Tobis K, Korybalska K, Polubinska A, Radkowski M, Breborowicz A, Oreopoulos DG. In vivo model to study the biocompatibility of peritoneal dialysis solutions. Int J Artif Organs 1997;20:673–677.
519. Afthentopoulos IE, Passadakis P, Oreopoulos DG. Sclerosing peritonitis in continuous ambulatory peritoneal dialysis patients: One center's experience and review of the literature. Adv Ren Replace Ther 1998;5:157–167.
520. Bowers VD, Ackermann JR, Richardson W, Carey LC. Sclerosing peritonitis. Clin Transplant 1994;8:369–372.
521. St John KR, Zardiackas LD, Black RJ, Armstrong R. Response of canine bone to a synthetic bone graft material. Clin Mater 1993;12:49–55.
522. Enneking WF, Mindell ER. Observations on massive retrieved human allografts. J Bone Joint Surg 1991;73A:1123–1142.
523. Miyamoto S, Takaoka K, Okada T, Yoshikawa H, Hashimoto J, Suzuku S, Ono K. Polylactic acid-polyethylene glycol block copolymer. A new biodegradable synthetic carrier for bone morphogenetic protein. Clin Orthop 1993;294:333–343.
524. Schwarz N, Schlag G, Thurnher M, Eschberger J, Dinges HP, Redl H. Fresh autogeneic, frozen allogeneic, and decalcified allogeneic bone grafts in dogs. J Bone Joint Surg 1991;73B:787–790.

525. Heckman JD, Boyan BD, Aufdemorte TB, Abbott JT. The use of bone morphogenetic protein in the treatment of non-union in a canine model. J Bone Joint Surg 1991;73A:750–764.
526. Chappard D, Richard S, Audeval-Gerard C, Noc C, Alexandre C. Biocompatibilité chez l'animal d'une xénogreffe osseuse hauteement purifée: Résultats preliminaries. Innov Tech Biol Med 1991; 12:685–693.
527. Bos GD, Goldberg VM, Zika JM, Heiple KG, Powell AE. Immune response of rats to frozen bone allografts. J Bone Joint Surg 1983;65A:239246.
528. Prolo DJ, Oklund SA. The use of bone and alloplastic materials in cranioplasty. Clin Orthop 1991; 268:270–278.
529. Damien CJ, Parsons JR. Bone graft and bone graft substitutes: A review of current technology and applications. J Appl Biomater 1991;2:187–208.
530. Stevenson S, Li XQ, Martin B. The fate of calcellous and cortical bone after transplantation of fresh and frozen tissue-antigen-matched and mismatched osteochondral allografts in dogs. J Bone Joint Surg 1991;73A:1143–1156.
531. Toriumi DM, Larabee WF, Walike JW, Millay DJ, Eisele DW. Demineralized bone, implant resorption with long-term follow up. Arch Otolaryngol Head Neck Surg 1990;116:676–680.
532. Nathan RM, Bentz H, Armstrong RM, Piez KA, Smestad TL, Ellingsworth LR, McPherson JM, Seyedin SM. Osteogenesis in rats with an inductive bovine composite. J Orthop Res 1988;6: 324–334.
533. Uretzky G, Appelbaum J, Sela J. Inhibition of the inductive activity of demineralized bone matrix by different percutaneous implants. Biomaterials 1988;9:195–197.
534. Solheim E, Pinholt EM, Andersen R, Bang G, Sudmann E. The effect of a composite of polyorthoester and demineralized bone on healing of large segmental defects of the radius in rats. J Bone Joint Surg 1992;74A:1456–1463.
535. Fox GM, McBeath AA, Heiner JP. Hip replacement with a threaded acetabular cup. J Bone Joint Surg 1994;76A:195–201.
536. Maloney WJ, Jasty M, Harris WH, Galante JO, Callaghan JJ. Endosteal erosion in association with stable uncemented femoral components. J Bone Joint Surg 1990;72A:1025–1034.
537. Schmalzried TP, Guttmann D, Grecula M, Amstutz HC. The relationship of the design, position, and articular wear of acetabular components inserted without cement and the development of pelvic osteolysis. J Bone Joint Surg 1994;76A:677–688.
538. Boss JH, Shajrawi I, Allperson M, Mendes DG. Inhibition of osteogenesis by foreign body granulomatous response to bony debris. Histological and histomorphometrical comparison of gap callus in experimentally produced cortical defects with or without bony detritus. Orthopaedics Int Edn 1994;2:447–453.
539. Freeman MAR, Tennant R. The scientific basis of cement versus cementless fixation. Clin Orthop 1992;276:19–25.
540. Hozack WJ, Rothman RH, Booth RE, Jr, Balderston RA. Cemented versus cementless total hip arthroplasty. A comparative study of equivalent patient populations. Clin Orthop 1993;289:161–165.
541. Brånemark PI. Introduction to osseointegration. In: Tissue-Integrated Prostheses: Osseointegration in Clinical Dentistry, GA Zarb and T Albrektsson, Eds, Quintessence Pub, Chicago, 1985; 11–76.
542. Zarb G. Osseointegration: A requiem for the periodontal ligament? Int J Periodont Restr Dent 1991;11:88–91.
543. Linder L. Ultrastructure of the bone-cement and the bone-metal interface. Clin Orthop 1992;276: 147–156.
544. Pas R, Walker P, Spector M, Reilly DT, Robertson DD, Sledge CB. Strategies for improving fixation of femoral components in total hip arthroplasty. Clin Orthop 1988;235:181–194.
545. Takeshita F, Morimoto K, Suetsugu T. Tissue reaction to alumina implants inserted into tibiae of rats. J Biomed Mater Res 1993;27:421–428.
546. Radder AM, Leenders H, van Blitterswijk CA. Interface reactions to PEO/PBT copolymers (polyactive) after implantation in cortical bone. J Biomed Mater Res 1994;28:141–151.

547. Schwartz Z, Amir D, Boyan BD, Cochavy D, Mueller Mai C, Swain LD, Gross U, Sela J. Effect of glass ceramic and titanium implants on primary calcification during rat tibial bone healing. Calcif Tissue Int 1991;49:359–364.
548. Fujita Y, Yamamuro T, Nakamura T, Kotani S, Ohtsuki C, Kokubo T. The bonding behavior of calcite to bone. J Biomed Mater Res 1991;25:991–1003.
549. Ohtsuki C, Kushitani H, Kokubo T, Kotani Yamamuro T. Apatite formation on surface of cervital-type glass ceramic in the body. J Biomed Mater Res 1991;25:1363–1370.
550. Søballe K, Hansen ES, Rasmussen HB, Buenger C. Hydroxyapatite coating converts fibrous anchorage to bony fixation during continuous implant loading. Trans 38th Meet, Orthop Res Soc, Washington, 1992, p. 292.
551. Jansen JA, van de Waerden JPCM, Wolke JGC, de Groot K. Histologic evaluation of the osseous adaptation to titanium and hydroxyapatite-coated titanium implants. J Biomed Mater Res 1991;25: 973–989.
552. Thomas KA, Kay JF, Cook SD, Jarcho M. The effect of surface macrotexture and hydroxyapatite coating on the mechanical strengths and histologic profile of titanium implant materials. J Biomed Mater Res 1987;21:1395–1414.
553. Gottlander M, Albrektssson T. Histomorphometric studies of hydroxyapatite-coated and uncoated CP titanium threaded implants in bone. Int J Oral Maxillofac Implants. 1991;4:399–401.
554. Shen W-J, Chung K-C, Wang G-J, MaLaughlin RE. Mechanical failure of hydroxyapatite- and polysulfone-coated titanium rods in a weight-bearing canine models. J Arthroplasty 1992;7:43–49.
555. Albrektsson T, Astrand P, Becker W, Eriksson AR, Lekholm U, Malmquist J, Sennerby L. Histologic studies of failed dental implants: A retrieval analysis of four different implant designs. Clin Mater 1992;10:225–232.
556. Campbell P, MaKellop H, Park SH, Malcolm A. Evidence of abrasive wear by particles from a hydroxyapatite coated hip prosthesis. Trans Orthop Res Soc 1993;18:224.
557. Bauer TW, Taylor SK, Jiang M, Mendendorp SV. An indirect comparison of third-body wear in retrieved hydroxyapatite-coated, porous, and cemented femoral components. Clin Orthop 1994; 298:11–18.
558. Passick JM, Dorr LD. Primary total knee replacement arthroplasty for the 1990s. Tech Orthop 1990;5:57–66.
559. Harris WH, McGann WA. Loosening of the femoral component after use of the medullary-plug cementing technique: Follow-up note with a minimum five-year follow-up. J Bone Joint Surg 1986;68A:1064–1066.
560. Markel MD, Gottsauner-Wolf F, Bogdanske JJ, Wahner HW, Chao EYS. Dual energy X-ray absorptiometry of implanted femora after cemented and press-fit total hip arthroplasty in a canine model. J Orthop Res 1993;11:452–456.
561. Campbell P, Nasser S, Kossovsky N, Amstutz HC. Histopathological effects of ultrahigh-molecular-weight polyethylene and metal wear debris in porous and cemented surface replacements. In: Particulate Debris from Medical Implants: Mechanisms of Formation and Biological Consequences, ASTM STP 1144, KR St John, Ed, American Society for Testing and Materials, Philadelphia, 1992; 38–50.
562. Callen BW, Sodhi RNS, Shelton RM, Davies JE. Behavior of primary bone cells on characterized polystyrene surfaces. J Biomed Mater Res 1993;27:851–859.
563. Bloebaum RD, Ota DT, Skedros JG, Mantas JP. Comparison of human and canine external femoral morphologies in the context of total hip replacement. J Biomed Mater Res 1993;27:1149–1159.
564. Kossovsky N, Liao K, Gelman A, Campbell P, Amstutz HC, Finerman M, Nasser S, Thomas BJ. Photon correlation spectroscopy analysis of the submicrometre particulate fraction in human synovial tissues recovered at arthroplasty or revision. In: Particulate Debris from Medical Implants: Mechanisms of Formation and Biological Consequences, ASTM STP 1144, KR St John, Ed, American Society for Testing and Materials, Philadelphia, 1992; 68–74.
565. Albores-Saavedra J, Vuitch F, Delgado R, Wiley E, Hagler H. Sinus histiocytosis of pelvic lymph

nodes after hip replacement. A histiocytic proliferation induced by cobalt-chromium and titanium. Am J Surg Path 1994;18:83–90.
566. Pazzaglia U, Byers PD. Fractured femoral shaft through an osteolytic lesions resulting from the reaction to a prosthesis. A case report. J Bone Joint Surg 1984;66B:337–339.
567. Zhao Q, Topham N, Anderson JM, Hiltner A, Lodoen G, Payet CR. Foreign-body giant cells and polyurethane biostability: In vivo correlation of cell adhesion and surface cracking. J Biomed Mater Res 1991;25:177–183.
568. Laitinon D, Alitals I, Toivonen T, Vasenius J, Törmälä P, Vainionpaeae S. Tissue response to a braided poly-L-lactide implant in an experimental reconstruction of anterior cruciate ligament. J Mater Sci: Mater Med 1993;4:547–554.
569. Williams DF. The development and control of the host response to implanted polymers. Scientific Abstracts of the International Conference on Biomedical Polymers: From Molecular Design to Clinical Applications, UNESCO, Jerusalem, 1991, p. 11.
570. Williams DF. Introduction. In: Biocompatibility of Orthopaedic Implants. DF Williams, Ed, CRC Press, Boca Raton, 1985; 2–3.
571. Mäkelä EA, Vainionpää, Vihtonen K, Mero M, Laiho J, Törmälä P, Rokkanen P. The effect of a penetrating biodegradable implant on the epiphyseal plate: An experimental study on growing rabbits with special regard to polyglactin 910. J Pediatr Orthop 1987;7:425–420.
572. Mäkelä EA, Vainionpää, Vihtonen K, Mero M, Helevirta P, Törmälä P, Rokkanen P. The effect of a penetrating biodegradable implant on the growth plate: An experimental study on growing rabbits with special reference to polydioxanone. Clin Orthop 1989;239:56–64.
573. Galante JO, Lemons J, Spector M, Wilson PD, Jr, Wright TM. The biologic effects of implant materials. J Orthop Res 1991;9:760–775.
574. Mendes DG, Soudry M, Roffman M, Boss JH. Maturation of composite ligament. Clin Orthop 1988;234:291–295.
575. Hayashi K, Uenoyama N, Matsuguchi N, Sugioka Y. Quantitative analysis of an vivo tissue responses to titanium-oxide- and hydroxyapatite-coated titanium alloy. J Biomed Mater Res 1991; 25:515–523.
576. Spilizewski KL, Marchant RE, Anderson JM, Hiltner A. In vivo leukocyte interactions with the NHLBI-DTB primary reference materials: Polyethylene and silica-free polydimethylsiloxane. Biomaterials 1987;8:12–17.
577. Black J. "Safe" biomaterials. J Bone Joint Surg 1995;29:791–792.
578. Kuhn TS. The Essential Tension: Selected Studies in Scientific Tradition and Change. University of Chicago Press, Chicago, 1997.
579. Meyle J, Gueltig K, Brich M, Haemmerle H, Nisch W. Contact guidance of fibroblasts on biomaterial surfaces. J Mater Sci: Mater Med 1994;5:463–466.
580. Gwynn IAP. Cell biology at interfaces. J Mater Sci: Mater Med 1994;5:357–360.
581. Manley MT, Serekian P. Wear debris. An environmental issue in total joint replacements. Clin Orthop 1994;298:137–146.
582. Mallory TH. Mallory head hip system. In: Current Concepts in Joint Replacement. S Greenwald, Ed, Mt Sinai Medical Center, Cleveland, 1988; tape 6.
583. Swan A, Dularay B, Dieppe P. A comparison of the effects of urate, hydroxyapatite and diamond crystals on polymorphonuclear cells: Relationship of mediator release to the surface area and adsorptive capacity of different particles. J Rheumatol 1990;17:1346–1352.
584. Gotfried Y, Mendes DG, Boss JH. Guest editorial: Biomaterials. Their past, their present and their future. Orthopaedics Intern Edn, 1993;1:293–294.
585. Johnson SD, Anderson JM, Marchant RE. Biocompatibility studies on plasma polymerized interface materials encompassing both hydrophobic and hydrophilic surfaces. J Biomed Mater Res 1992;26:915–935.
586. Nicol A, Gowda C, Urry DW. Cell adhesion and growth on synthetic elastomeric matrices containing ARG-GLY-ASP-SER. J Biomed Mater Res 1992;26:393–413.

587. Soldani G, Mercogliano R. Bioartificial polymeric materials obtained from blends of synthetic polymers with fibrin and collagen. Int J Artif Organs 1991;14:295–303.
588. Gleick J. Chaos, Making a New Science. Little, Brown, London, 1987.
589. Amis AA, Dawkins GPC. Functional anatomy of the anterior cruciate ligament. Fibre bundle actions related to ligament replacements and injuries. J Bone Joint Surg 1991;73B:260–267.
590. Langkamer VG, Case CP, Heap P, Taylor A, Collins C, Pearse M, Solomon L. Systemic distribution of wear debris after hip replacement. J Bone Joint Surg 1992; 74B:831–837.
591. Visuri T, Koskenvuo M. Cancer risk after Mckee-Farrar total hip replacement. Orthopedics 1991; 14:137–142.

2

Tissue Response to Implants: Molecular Interactions and Histological Correlation

M. Cannas, M. Bosetti, M. Santin, and S. Mazzarelli
University of Eastern Piedmont "A. Avogadro," Novara, Italy

I. INTRODUCTION

Biomaterials—materials used for the elaboration of systems designed for human implantation or organ substitutes—can be synthetic (metals, alloys, ceramics and polymers) or from a natural source [1]. Their uses are largely diversified for soft and hard tissue replacement: in fact, biomaterials can be employed in neural prostheses, cardiovascular devices, blood or bone substitutes, controlled drug delivery matrices, dental materials, artificial organs, dialyzers and so on [2].

II. INTERACTIONS BETWEEN IMPLANT SURFACES AND TISSUES

During the first seconds after implantation, only water, dissolved ions, and free biomolecules (but no regional cells) are in the closest proximity of the surface. The composition of the bioliquid then changes continuously as inflammatory and healing processes proceed and, in turn, causes variations in the composition of the adsorbed layer of biomolecules on the implant surface over the first minutes, hours, days, weeks [3].

Cells and tissues approach the surface and, depending on the nature of the adsorbed layer, respond specifically modifying the adsorbed layer of biomolecules [4]. The types of cells approaching the surface and their activities change with time. The final results may be the formation of a more or less pronounced fibrous capsule with a different degree of tissue integration [5, 6]. The interaction at the surface is therefore a coupled series of events developing in space and time which is in a crucial way influenced by the properties of the original biomaterial surface.

An implant can be considered as a source of irritation or as a stimulus to the tissue in terms of the response provoked by traumas or infections. In fact, most features of the tissue response to an implant bear a close similarity to the classical features of wound repair after trauma or of the cellular and humoral response to invading bacteria.

Acute inflammation is the immediate response to injury. Bleeding is followed by a stagnation of blood flow since it becomes more viscous as water is absorbed by the surrounding tissue due to the abnormal permeability of the capillaries mediated by histamine, serotonin, complement, prostaglandins, leading to stasis and increased pressure. The increased availability of blood at the site of injury is followed by clot formation and by migration of leucocytes and plasma proteins through the capillary wall into the surrounding tissue. Cell migration is directional and is mediated by a chemical process known as chemotaxis (Fig. 1).

One of the most important functions of inflammatory cells is phagocytosis, process initiated by the recognition of the foreign body by the surface of the cell (Fig. 2) through proteins called *opsonin*, such as IgG for which the cell surface has receptors. In cases where the foreign body is not a phagocytable size, phagocytes undergo degranulation with release of enzyme free radicals and mediators into the surrounding tissue, thus mediating and promoting inflammatory responses (Fig. 3).

Chronic inflammation is a proliferative rather than exudative response, and the tissue is characterized by fibroblasts associated with collagen-rich repaired tissue and an accumulation of leucocytes. The cells mainly involved in the chronic response are macrophages, cells able to transform themselves into multinucleated foreign body giant cells, lymphocytes and plasma cells.

III. THE RESPONSE OF HARD TISSUES TO IMPLANTATION

Bone response to implants depends on the type of bone, its previous history, the geometrical and morphological relationships between the bone and implant, the method of attachment of one to the other, the mechanical stress system acting on the implant-bone system and the

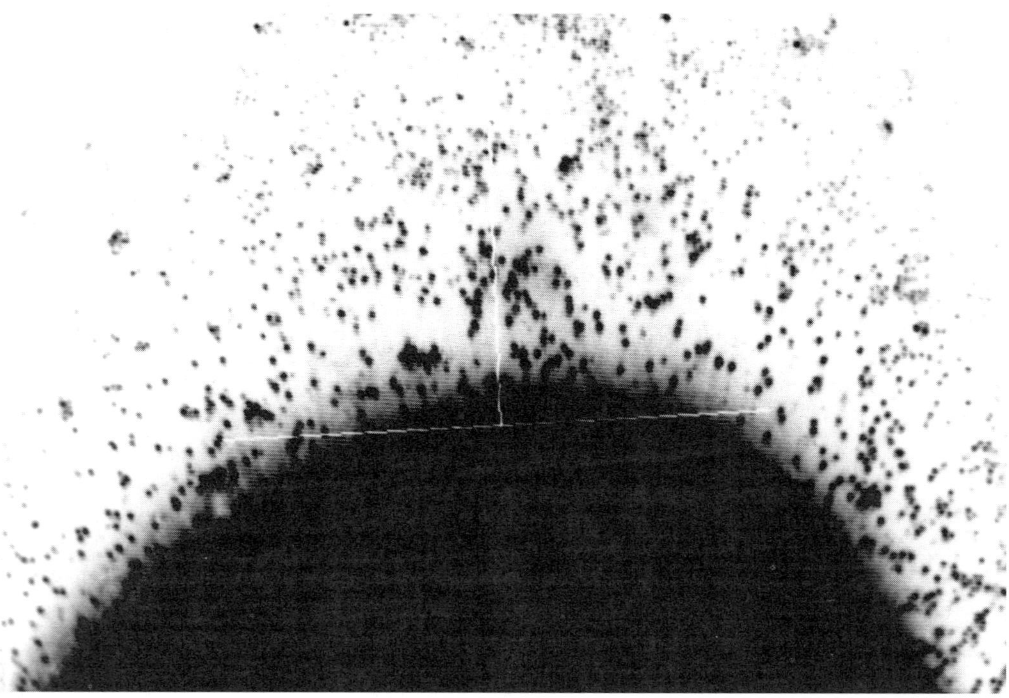

Figure 1 Example of chemotaxis: cells moving toward a biological target.

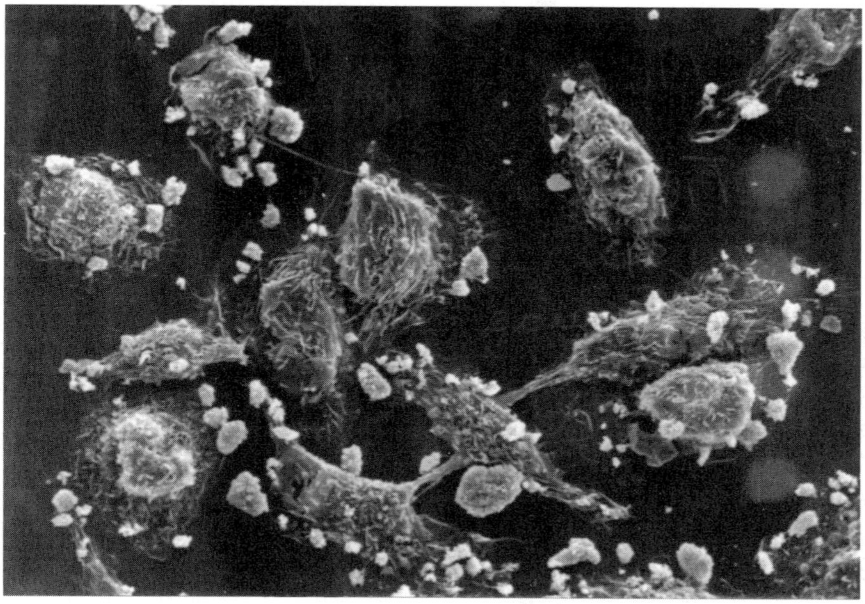

Figure 2 Macrophages exposed to biomaterial (SEM 100x).

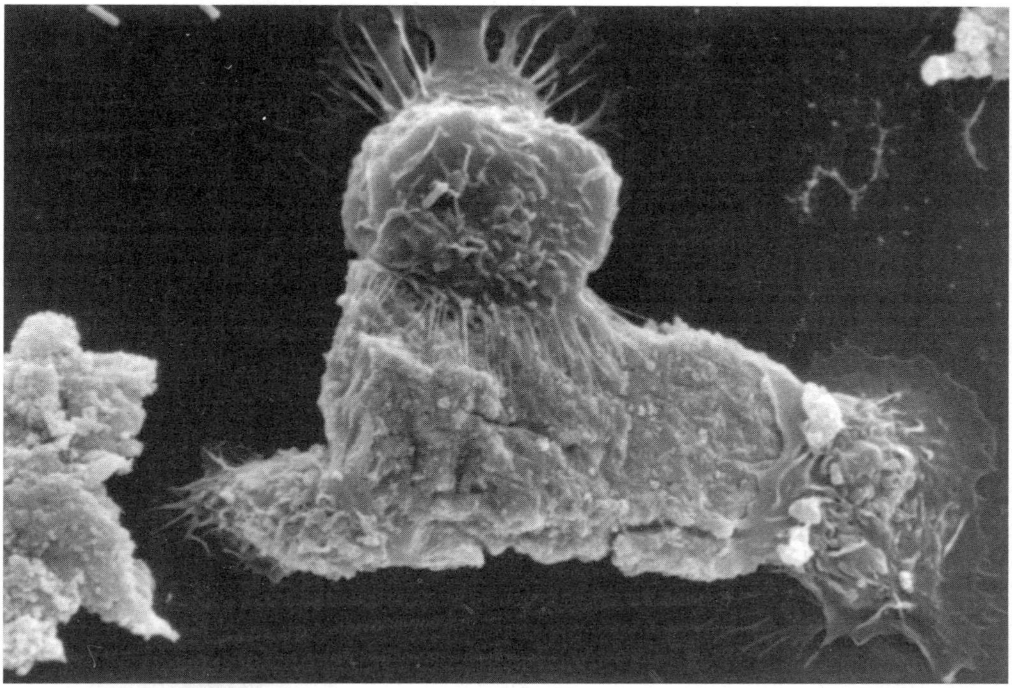

Figure 3 Macrophage activation: the macrophage is unable to phagocytose the foreign body (SEM 2000x).

material chemistry. Biomaterial may contact bone with little direct effect on the bone (plate, nail), since it is not plated within the healing zone, or be vitally important in determining the response of the bone to the implant (joint prostheses, bone screws, dental implants).

In these latter cases the progression of healing in the bony defect is found. Healing can therefore proceed (i) without the aid of any biomaterial (limited to the size of the defect) where defect bridging occurs spontaneously and completely through an initial exudate which slowly reorganizes to form new bone (Fig. 4a); (ii) with a solid object placed in the bony cavity (influenced by mechanical and chemical characteristics of the material) where healing takes place in the space between the implant and the bone. In the latter case, there will be intimate bone-implant contact or a soft fibrous tissue interface between bone and implant with considerable differences in the functional performances of the prosthesis (Fig. 4b).

Orthopedic materials can be classified as (i) biologically inert ("bioinert"), with no positive or negative effects on bone growth; and (ii) "bioactive," which encourages the formation of new bone at the surface. The complex interrelationship between mechanical stresses, surgical technique and characteristics of the implanted material make it difficult to foresee the conditions under which bone-material contact will reach its optimum.

Studies focusing on titanium, PMMA or certain glass-ceramics were found to have thick proteoglycan layers at the interfaces not behaving as natural tissue.

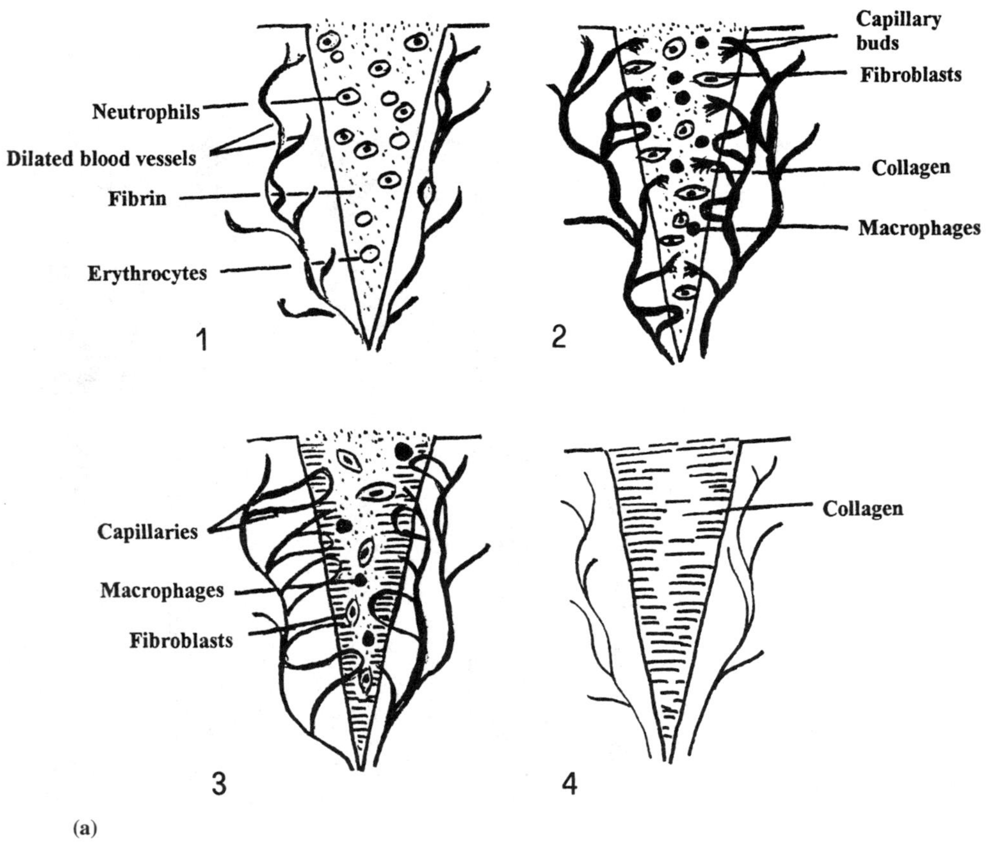

Figure 4a Scheme of process of wound healing in a simple incisional wound.

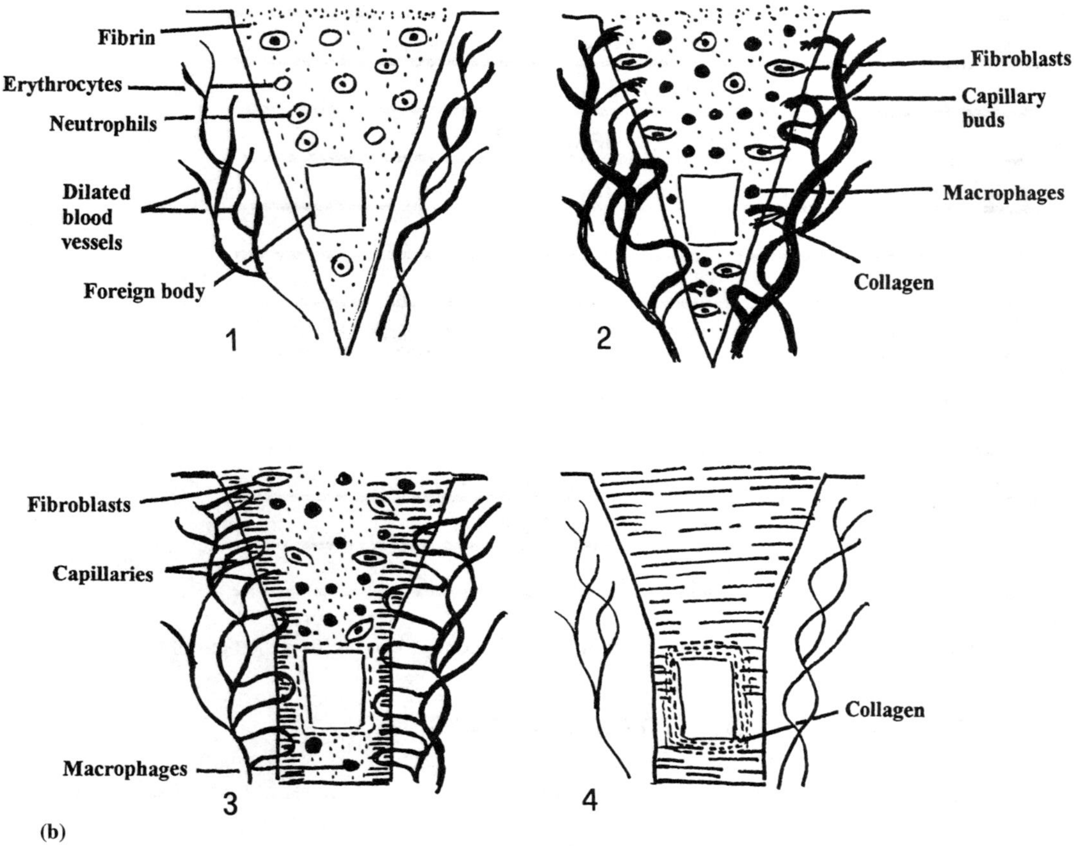

Figure 4b Scheme of process of wound healing in the presence of a foreign body, such as an implant.

About the so-called bioactive materials calcium hydroxylapatite ceramics, certain glasses or glass-ceramics of controlled surface activity and certain porous surfaces are found.

Calcium hydroxylapatite use has been encouraged by its composition which is analogous to the natural mineral phase of the bone. When placed within a bony environment, it becomes well incorporated into the bone, without intervening inflammatory or fibrous tissue elements and with crystals of biological apatites deposited straightforward on its surface.

Glasses or glass-ceramics of controlled surface activity, first developed by Hench [6], based on the SiO_2-CaO-Na_2O phase diagram with a small quantity of P_2O_5, appear to provide a different mechanism (Fig. 5): they have a reactive surface that initially leaches out calcium and phosphate ions, chemically equivalent to the mineral phase of bone. These elements are probably responsible for the formation of new bone at the surface. It has been hypothesized that the surface reaction would be self-limiting because of the protective effect of the silica-rich surface layer that is exposed after preliminary leaching, but the transition zone is of such a character that it is able to incorporate the molecules of the organic matrix of new bone, thus favoring a chemical bond between the glass and the new bone.

Another mechanism, largely independent of chemistry but controlled by the morphology of the material, regulates the bioactivity of porous materials which, placed adjacent to existing bone, may promote the ingrowth of new bone with no osteogenic mechanism.

BONE GROWTH

CRYSTALLIZATION OF MATRIX

GENERATION OF MATRIX

DIFFERENTATION OF STEM CELLS

ATTACHMENT OF STEM CELLS

ACTION OF MACROPHAGES

ADSORPTION OF BIOLOGICAL MOITIES IN HCA LAYER

CRYSTALLIZATION OF HYDROXYL CARBONATE APATITE (HCA)

ADSORPTION OF AMORPHOUS CA+PO_4+CO_3

POLYCONDENSATION OF SIOH+SIOH→ SI-O-SI

FORMATION OF SIOH BONDS

BIOACTIVE GLASS

Figure 5 Sequence of surface reactions in bioactive bonding to bone, from the bottom to the top.

IV. RESPONSE OF BLOOD TO BIOMATERIALS

The principles of the interactions between blood and synthetic materials are described under the headings of the essential characteristics of blood, the mechanism of blood clotting, the effects of materials on clotting proteins and material-platelet interactions.

Blood is a suspension of cells (Table 1) in plasma, an aqueous solution containing a variety of organic and inorganic molecules (Table 2). The plasma proteins include those providing nutrients to the cells (albumin, lipoproteins), those involved in the transport of hormones and chemicals (transferrin, ceruloplasmin, vitamin-binding proteins, steroid-binding proteins) and those involved in defense (immunoglobins, complement, protein of the clotting process).

Among blood cells, platelets are of enormous significance in blood compatibility; under certain conditions, platelets are activated, resulting in significant functional, biochemical and structural alterations to the cell compared to its resting state.

Table 1 Cells of Circulating Blood

Cell	Concentration (N°/mm3)	Normal shape	Volume (%) in blood
Erythrocytes	4 to 6×10^6	Biconcave disk, 8 μm × 1–3 μm	45
Leukocytes			
neutrophils	1.5 to 7.5×10^3	Spherical, 7–22 μm diameter	1
eosinophils	0 to 4×10^2		
basophils	0 to 2×10^2		
lymphocytes	1 to 4.5×10^3		
monocytes	0 to 8×10^2		
Platelets	250 to 500×10^3	Rounded or oval, 2–4 μm	

Table 2 Composition of Human Plasma

Surface	Concentration ($g\ dl^{-1}$)	Molecular weight
Water	90–92	
Proteins		
serum albumin	3.3–4.0	69,000
fibrinogen	0.34–0.43	340,000
α_1-globulins	0.31–0.32	44,000–200,000
α_2-globulins	0.48–0.52	150,000–300,000
β-globulins	0.78–0.81	90,000–1,300,000
γ-globulins	0.66–0.74	160,000–320,000
Cations		
Na^+	0.31–0.34	
K^+	0.016–0.021	
Ca^{2+}	0.009–0.011	
Mg^{2+}	0.002–0.003	
Anions		
Cl^-	0.36–0.39	
HCO_3^-	0.20–0.24	
PO_3^{3-}	0.003–0.004	

The effects of activation are adhesion of the cells to sites of blood vessel wall injury, aggregation and the fusion of granules with the plasma membrane to facilitate the release of granular contents.

The injury to a vessel involves disruption of the endothelium and initiates a sequence of events which allow platelets to adhere to the damaged surface. The first stage is the platelet-collagen interaction, and the stimulated platelet releases a number of substances, principally adenosine diphosphate (ADP), which causes the platelets to aggregate and form a platelet mass.

The second event is the activation of the sequence of events called coagulation cascade in the plasma proteins, which leads to the formation of a thrombus. The coagulation cascade may be initiated by a glycoprotein associated with a phospholipid (extrinsic pathway) or by the exposure of a nonendothelial surface like biomaterials (intrinsic pathway); in both cases there is a localized conversion of inactive molecules to proteolytic enzymes, in a sequenzial pattern, that come together with the conversion of Factor X to Xa and culminates in the conversion of prothrombin to thrombin, which catalyzes the polymerization of fibrinogen to fibrin, as is well known.

The proposed mechanism by which surfaces may influence the factors of the intrinsic pathway is that negatively charged surfaces possess the ability to initiate coagulation: first they introduce a structural change in Factor XII; second the surface promotes an interaction between Factor XII and the inactive molecule prekallikrein; third the surface promotes the activation of Factor XI by surface-bound Factor XIIa.

Much experimental data has also been accumulated concerning the ability of foreign surfaces to adsorb and interact with plasma proteins; it is generally believed that surfaces adsorbing albumin are generally thromboresistant, while those adsorbing fibrinogen, gamma globulin and fibronectin tend to be more thrombogenic. Little is known of the way in which different biomaterials are able to influence these events.

Also the surface characteristics which are responsible for the attraction of platelets have not really been identified. The literature shows that opposing characteristics are equally important (platelet adhesion decreases with decreasing interfacial free energy, and yet, on the other hand, it decreases with increasing interfacial free energy), with the role of adsorbed proteins being very important.

No one theory is able to explain the interaction of platelets with foreign surfaces: it is interesting that there must be substances within the vessel walls that repel the platelets, and one way to prepare nonthrombogenic materials has involved the attachment of such substances (prostacyclin or prostacyclin-like substances) to the surface, and this has indeed shown some success with respect to antiplatelet activity.

It is clear that the future of biomaterials within the cardiovascular system depends on understanding their interactions with blood, expecially those leading to the resistance to clot formation. For many years attempts have been made to define the ideal thromboresistant surface in terms of physicochemical parameters; more recently a more logical biological approach regards antiplatelet substances or coating of polymer surfaces with phospholipids that mimic the cell membrane or with heparin or heparin-like substances that are themselves known to be antithrombogenic.

Other attempts use natural materials, such as glutaraldehyde-treated pericardial tissue derived from cows or pigs, in some heart valves, and finally experiments proposed vessels coated with endothelial cells before implantation.

V. PROTEIN ADSORPTION

The early event that determines the interactions of a surface with the surrounding host tissues is the adsorption of proteins from the body fluid in which they are dissolved: consequently, the biocompatibility of the material is strictly determined by these facts [7]. Since proteins play important roles in the host response processes, the study of this phenomenon is very important for understanding the molecular basis of biocompatibility and for designing novel biomaterials that are able to respond to the functions for which they are needed.

We will give an overview about different aspects of protein adsorption with a major emphasis on (i) the factors governing the process, (ii) the main protein kinds adsorbing onto surfaces, and (iii) their influence on the biological response.

Organic components present in all the main body fluids, such as blood and wound exudate plasmas, urine, cerebrospinal fluid, and peritoneal dialysate, tend to adsorb onto the surfaces of all the most commonly used biomaterials. As a first step, the process of adsorption is mainly conditioned by the rate of diffusion of the molecules from the liquid bulk to the material/fluid interface [7]. It is easily understood that low molecular weight proteins quickly contact the surface due to a rate of diffusion higher than the high molecular weight species. Anyway, the latter are considered favored in establishing interactions with the surface of the material due to their ability to better adapt their conformation to the material surface. Therefore, low molecular weight proteins are usually displaced over time by the high molecular weight ones. The first phase of protein adsorption is, therefore, a race where speed is only one of the prerequisites necessary to win the competition: adaptability to environmental changes appears to be more important in view of durable and biologically significant interactions. That makes the difference between proteins with comparable molecular weights and determines the final protein composition of the conditioning layer deposited on the surface.

At the interface proteins and biomaterial attract and repel each other depending on the affinity of their functional groups [8, 9]. Among all possible chemical interactions (listed in Table 3), van der Waals and charge interactions and hydrogen bonding are considered the most important.

Table 3 Forces Involved in Protein Adsorption onto Surfaces

Kind of interaction
Electrostatic
Charge-dipole
Dipole-dipole
Charge-nonpolar
Dipole-nonpolar
Nonpolar-nonpolar
Hydrogen bond
Covalent bond

It is widely accepted that hydrophobic surfaces adsorb more proteins than hydrophilic ones. The displacement of the water shell surrounding the folded polypeptide chain of the protein by the effect of the material moieties enhances the possibility of stable binding with the surface. Furthermore, van der Waals interactions are very dynamic and allow flexible bounds able to reconstitute upon changes of the microenvironmental conditions.

Due to the importance of this kind of interaction in determining protein adsorption, we have recently set up a novel and simple method to determine the strength of the hydrophobic interactions linking proteins to surfaces [10, 11]. By this protocol biomaterials are treated with body fluids such as plasma or urine and then sequentially washed with distilled water and isopropanol/water solutions at increasing concentration. The supernatants are then concentrated and tested for their protein profiles by electrophoresis. Proteins present in the supernatants of washing solution with no isopropanol or low isopropanol concentration are considered weakly bound to the material surface or belonging to the upper layers of a multilayer (Fig. 6, lanes 1–4). On the other hand, proteins desorbed with high isopropanol concentrations are considered strongly bound to the surface by hydrophobic interactions (Fig. 6, lanes 5–7). This procedure, combined with information about the physicochemical properties of the tested material (i.e., contact angle values), can lead to interesting speculations about protein adsorption, and it represents a tool to foresee the degree of bioinertness (or biointegration) of the material.

Opposite charge interactions are important to the formation of a protein layer on the surface of a biomaterial [7, 8]. This kind of binding takes place at a very fine level due to the possible presence of patches of opposite charge which allow the formation of salt bridges even when the net charge of the molecule is the same of the surface. Hydrogen bonding has also been reported among the interactions occurring at interfaces mainly involving hydroxyl moieties of both the material and protein functional groups [8].

Covalent binding has been found to occur between proteins and surfaces, although rarely. The most remarkable example is represented by the binding and activation of the complement C3 fragment in which a spontaneous reaction produces a thioester bond by the substitution of a nucleophil group of the surface to a disulfide bridge of the protein [12]. Therefore, the material/fluid interface represents a complex microenvironment where the molecular composition dynamically changes, depending on many physicochemical factors.

The pattern of protein deposition can be affected by the coadsorption of different kinds of proteins as well as by the presence of other inorganic and organic components such as salts (i.e., "salting in" and "salting out" species), glycosaminoglycans, and lipids [8]. These molecules can either change the composition of the medium in which proteins are dissolved, favoring their solubilization (or precipitation), or bind the adsorbing proteins, inducing variations in their conformation and therefore influencing their ability to interact with the exposed surface.

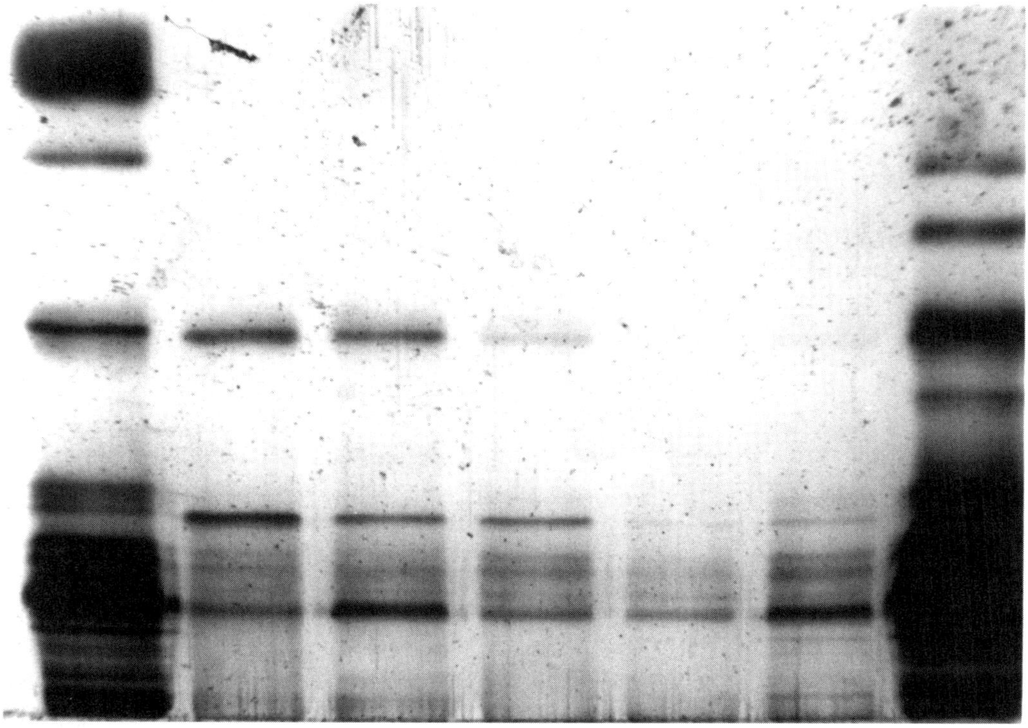

Figure 6 Typical electrophoresis profiles of plasma proteins eluted from the surface of a biomaterial: lanes 1–3—proteins eluted by washing with deionized water; lanes 4–7—proteins eluted by washing with 10%, 30%, 50%, 70% (v/v) isopropanol/water solutions. Protein bands were detected by a silver staining procedure.

Temperature and pH are among the variables able to change the behavior of a given protein at the interface, respectively affecting its thermodynamics and electric charge [8]. In reality, lower pH values, which usually occur at the site of the inflammation, and the pH changes detected in pathological urine can alter the adsorption of the protein, while the temperature range of the body would not change the protein behavior at the interface, since variations of the sorption profiles have been detected at temperatures far from the physiological values [7].

In vitro studies carried out from artificially prepared protein solutions demonstrated the possibility of forming protein multilayers adsorbed onto materials [6, 7]. Although in these models the aggregates were very unstable, in vivo studies demonstrated the developing of thick conditioning layers mainly formed by proteins upon long-term implantation of medical devices [12–16]. This can be explained by hypothesizing the constitution of stable macromolecular complexes by strong intermolecular interactions also involving other organic molecules such as glycosaminoglycans and lipids.

In summary, the deposition of a protein film on the surface of a biomaterial contacting a body fluid has to be considered a highly dynamic process where one species "feels" differently attracted by another depending on the composition of the surrounding environment. For example, a new approaching protein can displace a species previously adsorbed on the material (Vroman effect) by expressing more physicochemical attraction for both the exposed surface and the other components of the deposited film [15].

Proteins can be basically divided into inflammatory and structural proteins, according to their ability to participate in the response toward a foreign body or to their ability to form the framework of the tissues; that is, although many protein kinds adsorb onto the surface of biomaterials, the vast majority of observations from the literature are on the most important proteins concurring to inflammation and tissue regeneration events.

The complement system is undoubtely the most important defense mechanism available to the host organism to resist tissue invasion by foreign bodies [16, 17]. This system accounts for 20 components which are divided into two cascade mechanisms, both of them predisposed to mark the invader as foreign and to contribute to its destruction or isolation. The two cascade mechanisms are called "classical" and "alternative" pathways (Fig. 7, top). In both pathways convertases produce active fragments which have the ability to bind the foreign surfaces and play a receptor role for inflammatory cells. In the classical pathway the binding of a complex constituted by IgG and the C1 fragment of the complement system to the foreign surface is the event triggering the activation of the cascade, which leads to the progressive formation of a protein complex able to destroy the foreign body [18].

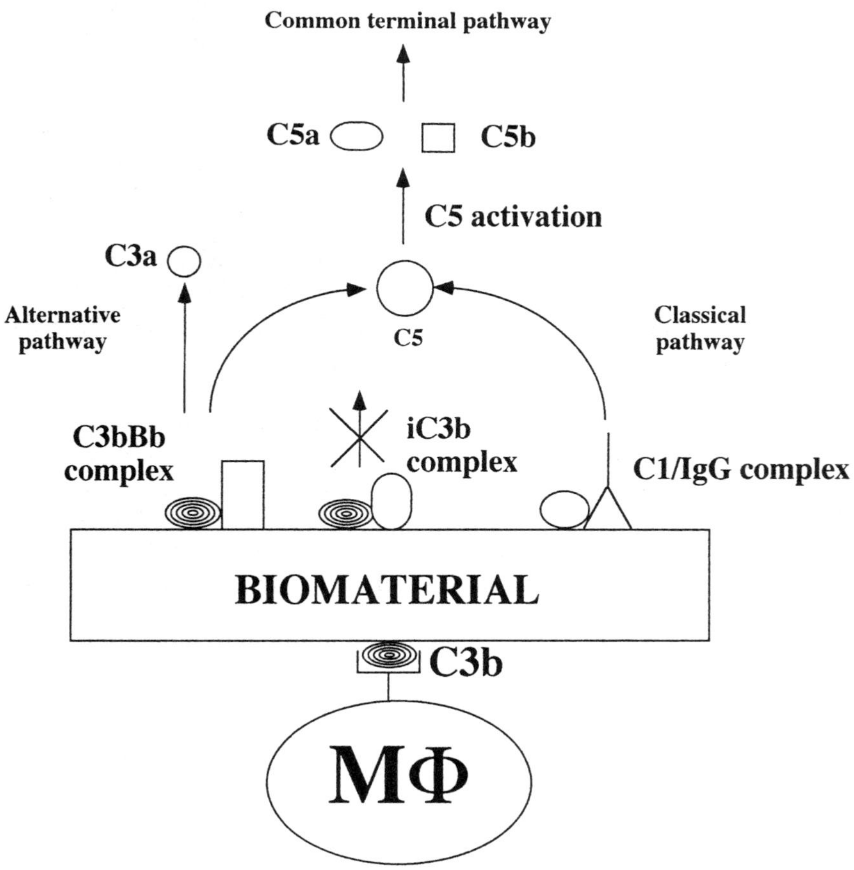

Figure 7 Scheme of the biomaterial-induced complement activation. Top: activation of the classical (right side) and alternative (left side) pathways; Bottom: receptor role of the adsorbed C3b fragment toward macrophages (MΦ) (for details see text).

Mainly devoted to the destruction of undesired cells (i.e., bacterial cells), the classical pathway is not considered important in the evolution of the inflammation upon the implant of a biomaterial [17]. However, the presence of the C1/IgG complex on the surface of several biomaterials means the activation of the "classical" pathway has to be reconsidered in the context of the implant-induced inflammation. This pathway shares with the "alternative" pathway the activation of the C5 fragment of the complement system, which is recognized by the inflammatory cells and is responsible for the amplification of the inflammatory response by agglutinating leukocytes at the inflammed environment [19]. Furthermore, the C1/IgG complex has also been found to produce activated C3a fragments of the alternative pathway [20].

In the field of biomaterials the C3 fragment is the most studied component of the entire complement system [20–24], the reason being its ability to be activated upon adsorption and covalent binding on the surfaces of the biomaterials [21]. In particular, C3 is activated by most of the solid surfaces by spontaneous reduction of a disulfide bond linking its chains. Nucleophil groups present on the surface of biomaterials cleave the disulfide bond, establishing a covalent thioester link with the protein fragment. The formation of separate fragments generates a molecule which in turn can bind Factor B, forming a convertase complex C3Bb able to cleave more C3 into C3a and C3b [22]. C3b binding to the surfaces also favors the adhesion of leukocytes (Fig. 7, bottom), while C3a represents an anaphylatoxin able to recruit cells to the implant/tissue interface [24]. The mechanism of activation of C3 can be either amplified or inhibited depending on the ability of the adsorbed C3b fragment to link, respectively, further Factor B molecules or the inhibitor Factor I [22].

In most cases the correct functionality of a prosthesis or medical device surface depends on its structural interaction with the surrounding tissues. The goal of a good material acceptability is obtained by the production of surfaces with the ability to support cells growing on their surfaces and, therefore, tissue integration.

Although cells have been demonstrated to adhere to materials aspecifically [25], the adsorption on the surface of specific receptors favors the rearrangement of the cytoskeleton toward a pattern similar to that adopted by the cell within the physiological extracellular matrix, thus promoting its appropriate metabolic functions [25–27]. For these reasons, biomaterials able to support the adsorption of extracellular matrix proteins with receptor sites for tissue cells represent materials of choice for those applications in which the integration of the prosthesis within the tissue is required.

Table 4 lists the main structural proteins of the extracellular matrix. These proteins share the presence in their structure of the amino acid sequence Arginine-Glycine-Aspartic acid (RGD), which is responsible for the recognition of most cell membrane integrins [27]. Furthermore, these proteins have relatively high molecular weights and can form macromolecular

Table 4 Main Proteins Promoting Cell Adhesion

Protein kind
Collagen
Fibronectin
Vitronectin
Laminin
Osteopontin
Entactin
Thrombospondin
Fibrinogen

aggregates between them as well as with other components of the connective tissues, thus contributing to the formation of the extracellular matrix network [27].

It must be pointed out that these proteins are also able to support the adhesion of inflammatory cells [28] and bacteria [29], and their binding can occur on the surfaces of prostheses where material inertness is required.

Proteins involved in the clotting system can also offer double-face responses, depending on the site of implantation and on the application of the biomaterial used [30]. The deposition of fibrinogen and its modification to fibrin, for example, can favor the wound healing process consequent to the implant procedure, but can be an undesired substratum for the growing of thrombi onto the lumen surface of vascular grafts.

The study of material biocompatibility also involves the possibility of mineralization processes on the prosthesis surface. Mineralization is a common process in nature, and in the biomaterial field it is approached as both a wanted phenomenon and an undesired one [31–34]. In the case of orthopedic and dental prostheses, surface mineralization is one of the main goals to be pursued to obtain biomaterials able to osteointegrate [31]. On the other hand, mineralization has to be eliminated in applications such as heart valve bioprostheses [33] and urological catheters [34] where the formation of encrustations can lead to the failure of the device.

Mineralization processes in living organisms are always driven by the presence of an organic matrix formed by proteins, lipids, and glycosaminoglycans, which constitute a template where calcium-based crystals can nucleate, grow, and aggregate [35]. This pathway is shared by both physiological (i.e., bone formation) and pathological (i.e., urinary stone formation) events [36, 37].

The proteins involved in these processes are those able to bind calcium and to interact with other proteins and crystals [36, 38]. These features allow them to favor the nucleation of calcium-based crystals and their growing to form macroscopic mineralized structures.

In theory, calcium-binding proteins have been proposed as agents to functionalize the surface of orthopedic biomaterials with the aim of improving their osteointegrative properties, but, to our knowledge, no final product has been synthesized yet on these bases [39].

Few studies have been carried out to asses the levels of adsorption of these kinds of proteins upon biomaterial implantation and their effect especially in pathologic calcifications. For example, the presence of calcium-binding proteins adsorbed on the surface of mineralizing bioprostheses has been demonstrated in rat subcutaneous implants, although no clear relationship has been found with the mineralization process [40].

Recently, we showed changes in the plasma protein adsorption profiles of glutaraldehyde-crosslinked collagen sponges before and after their treatment with calcification inhibitors, suggesting that calcification of the collagen-rich bioprosthesis can be induced by the adsorption of particular protein kinds [41].

VI. HISTOLOGICAL APPROACH TO PERIMPLANTAR TISSUE

Many studies have evidenced that the surgical implantation of biomaterials in bones and soft tissue is followed by the formation of a connective tissue membrane encapsulating the implant which changes with time of implantation and nature of the implanted materials. The fibrous layer is due to a nonspecific response by the tissue to the physical presence of the implant as a solid body [42].

There are many possible causes of tissue changes adjacent to implant not only due to the chemical nature of foreign material that may be approached by histology, as surgical procedures inevitably cause some degree of cell demage that can lead to an end product of fibrous repair such as the regeneration of the fibrous joint capsule following hip arthroplasty. Another important factor, which often occurs after implantation of devices, is the possibility of implant movement that may provide mechanically or physically tissue damage, with histological abnormalities arising from such damage and from the response to it.

The release of particles from the implant may cause a local hypersensitivity reaction that may be evidenced with immunohistochemistry by the presence of particular cells like macrophages migrated near the implant site to phagocyte the foreign particles.

Microscopic examination of biopsy specimens is also a method for assessing the presence of infection; it may show granulocytic infiltration in the subacute form, while the chronic form is characterized by a lymphocyte/plasma cell spread and histiocytic infiltration [43].

Now, the histological approach to peri-implantar tissue may be differentiated into steps, the first being a basic histological staining to morphologically define the tissue, the second and the third are electron microscopy and immunohistochemistry to qualitatively define cell populations, and the last is histomorphometrical analysis to quantitatively define cell distribution.

VII. HISTOLOGICAL REACTION TO MATERIALS IMPLANTED INTO SELECTED TISSUES

Peri-implantar tissue is the body response to implantation of medical devices. As we already said, the site of implantation is an important factor that influences the tissue response to biomaterials. The main differentiation has to be done between soft and hard tissue.

A. Soft Tissue

The tissue response to biomaterials in soft tissue is always characterized by the presence of a fibrous tissue capsule enveloping the implant and by the presence of interfacial cells like macrophages and foreign body giant cells. The thickness of the capsule seems to be related to the biocompatibility of the material and to its chronic inflammatory response. An example of soft tissue response to implant is the one studied in patients with silicone or saline implants as mammary expanders. Histological evaluation of the capsule may be performed after explantation of silicone disks together with surrounding tissue and fixation by 10% buffered formalin solution. Stainings have to be performed using ethylene dichloride instead of xylene to avoid swelling of silicone with the solvent. The magnitude of tissue reactions to implanted silicones may be classified according to the degree of infiltration of inflammatory cells and the appearance of foreign body cells. The thickness of collagenous capsule formed around the implant, measured by optical micrographs of the histological sections after 16 weeks, is around 80–95 μm [44]. The capsule consists of an inner dense membranous zone of thin, collagenous fibers of the same size concentrically arranged around the prosthesis.

Early after implantation granulocytes, lymphocytes and a few fibroblasts were the predominant cell types. Then the population changes progressively to a relatively greater number of histiocytes, a few lymphocytes and many fibroblasts. B and T lymphocytes may be evidenced by immunohistochemical staining [45]. The mature capsule is predominantly acellular and vascularity was well developed.

Hematoxilyn/eosin staining shows the presence of round small cells at the material/capsule interface which decreased toward the physiological tissue border where there is a higher concentration of fibroblasts (Fig. 8).

Along with the collagen fiber acids mucopolysaccharides may be observed, while elastic

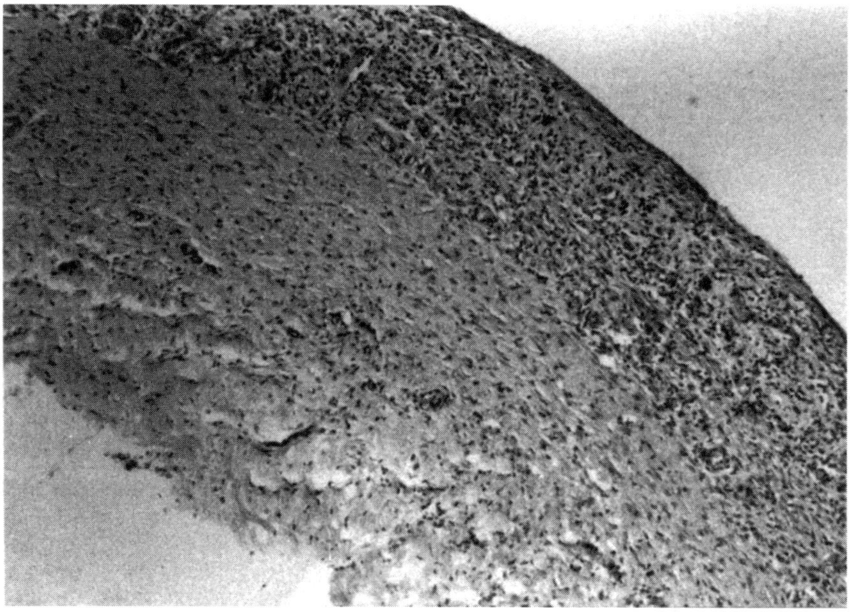

Figure 8 Hematoxylin-eosin staining shows the presence of round small cells decreasing in intensity from the implant interface (right) to the physiological tissue with a higher concentration of fibroblasts (100x).

fibers were not found by Weigert staining with the exception of few cases where they are localized in the dermis (Fig. 9). This capsule represents the final establishment of an equilibrium between the organism and the foreign body implant, and its characteristic depends on the susceptibility of the foreign material to phagocytosis and other cellular mechanisms.

B. Hard Tissue

Histological evaluations of tissue reactions to materials implanted in hard tissues (bone) have identified the direct apposition of new bone to biomaterial. In some cases bone may be found directly apposed to one portion of an implant, while the remaining surface is encapsulated by fibrous tissue similar to that seen in soft tissue. In this case the fibrous tissue capsule takes an additional significance because it reduces the stability of the implant, leading to implant loosening.

Many histological evaluations have been made on this membrane. Since 1975 [43] attention has been focused on two histological parameters: the inflammatory infiltrate and the presence of material debris. The inflammatory reaction is diffused in the peri-implantar tissue: in the acute phase polymorphonuclear cells are predominant, while in the chronic phase lymphocytes and giant cells are more prevalent.

In 1986 Linder [46] described the peri-implantar membrane as composed of three layers: near the prosthesis it has macrophages and giant cells, in the middle layer it is fibrous and poor in cells, and, finally, near the bone it looks like a granulation tissue rich in fibroblasts, macrophages, round cells and blood vessels. Macrophages show in the cytoplasm phagocyted particles of metal and corrosion product.

More recent studies describe the membrane as consisting of an external compact fibrous

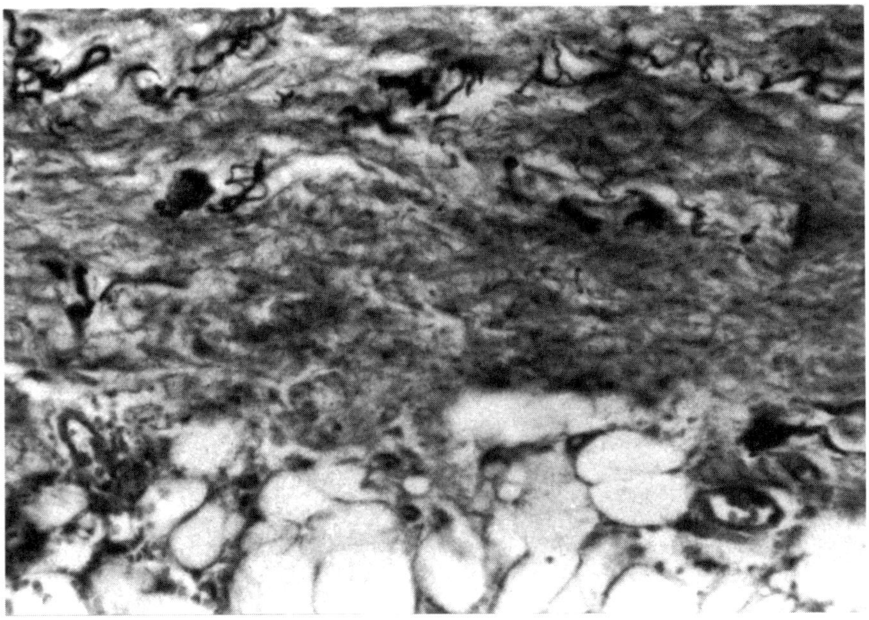

Figure 9 Weigert staining evidences elastic fibers localized on the side of the pseudomembrane (top) through the dermic layer (100x).

layer and an internal layer richer in cells. The external layer has a thickness ranging from 3 to 5 μm formed by orientated fibers. Blood vessels of various size, few fibroblastic cells and small wear debris may be recognized. The internal layer, less than 1 μm thick, is derived from fibroblastic differentiation.

VIII. HISTOLOGICAL TECHNIQUES

The assessment of a tissue reaction toward a biomaterial has traditionally involved the use of staining techniques such as hematoxylin and eosin, Masson staining and others. For histological analysis peri-implantar tissues have to be explanted and fixed in 10% buffered formalin solution. Tissues, dehydrated in increasing ethanol concentrations and clarified in xylol, are embedded in paraffin and cut in semithin (4–6 μm) serial sections using a microtome.

Different stains may be used to qualitatively evaluate the histological characteristics of the tissue around the implant, and often they are used simultaneously to allow the recognization of each constituent of the cell. Hematoxylin/eosin staining is the most diffuse method to evidence nuclei and cell population: nuclear structure is blue while cytoplasm and intercellular substances are reddish. This staining allows observation of numerous vessels in the middle layer.

To evidence the disposition of collagen in the capsule the best technique is Masson staining. Collagen fibers are disposed parallel to the surface of the implant and are more diffused far from the implant (Fig. 10).

Elastic fibers are not frequently observed, but when they are present they are localized near the dermis layer and are easily evidenced by Weigert staining.

Reticular fibers evidenced with silver impregnation staining show them to be concentrated around vessels (Fig. 11) and in the layer near the device.

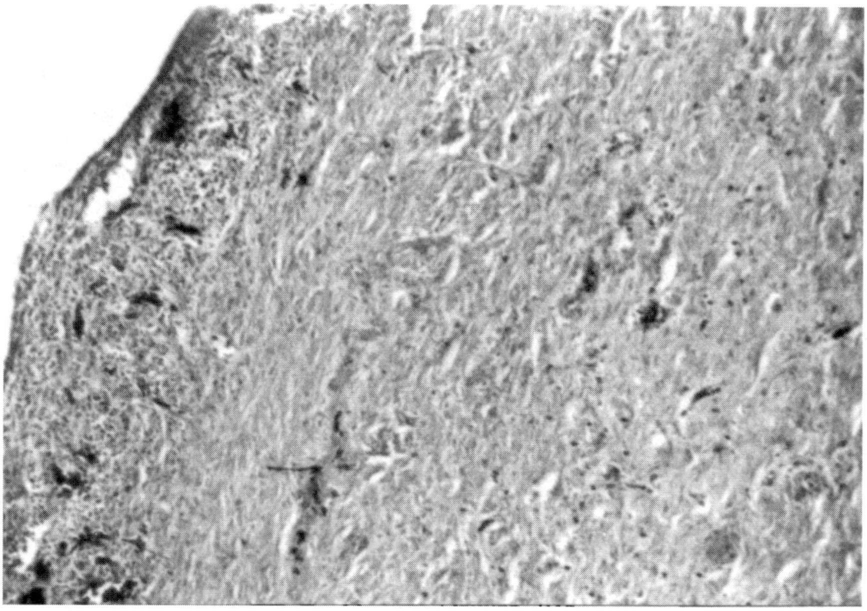

Figure 10 Masson staining shows collagen fibers disposition (100x).

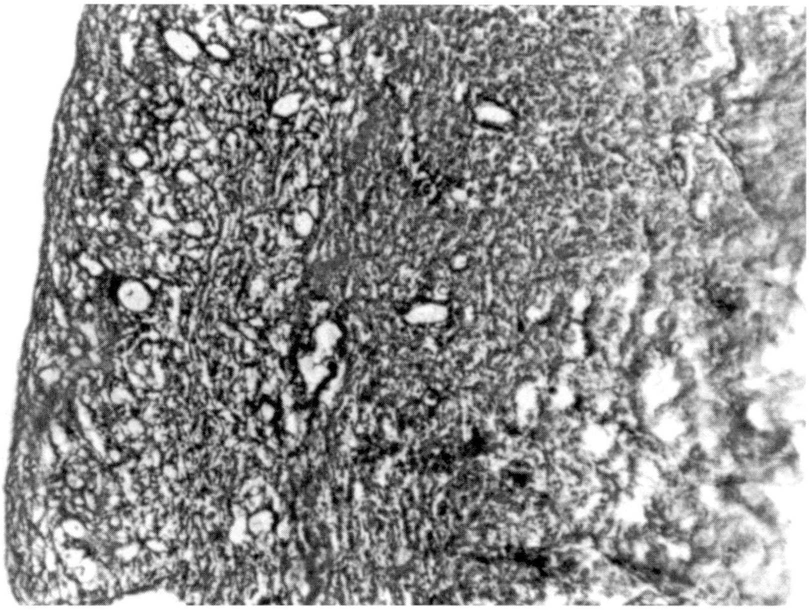

Figure 11 Silver impregnation shows reticulated fibers at the interface-tissue implant (200x).

IX. IMMUNOHISTOCHEMISTRY

Many immunoenzymatic staining methods are used for the localization of tissue antigen and then for the recognization of a cell type. The choice of the appropriate technique depends on the requirements of the operator. The avidin-biotin technique is the most used. It is a three-stage process which utilizes a primary mouse antihuman monoclonal antibody, a biotinylated rabbit antimouse polyclonal secondary antibody and a performed avidin-biotin horseradish peroxidase complex. The process relies on the strong affinity of the avidin-biotin complex binding nonimmunologically to the peroxidase activity, and the position of the monoclonal antibody is then identified using 3,3-diaminobenzidine (DAB).

Studies of peri-implantar tissue may be performed by analysis of the subpopulation of lymphocytes, macrophages, monocytes as a means to characterize immunological events in an inflammatory condition. Numerous investigators have recognized the importance of macrophages in the histological response to implants [47, 48]. While fibroblasts and fibrocollagenous capsules around the implant represent the end stage of repair, the macrophages on the surface are participants in the active reparative stage.

Moreover, macrophages have the ability to mediate many important biological processes. They can secrete a variety of products, including chemotactic agents for other cells, growth factors that stimulate the production of collagen by fibroblasts, and then may fuse to form multinucleated foreign body giant cells (Fig. 12) [42]. The production of chemotactic stimuli for polymorphonuclear neutrophils and monocytes makes them migrate into the wound site where they differentiate into macrophages to attempt their role in phagocytosis of cell, tissue debris and microorganisms.

The detection of reactive macrophages and histiocytic cells may be performed with CD68 monoclonal mouse antihuman macrophage antibodies, clone KP1.

If microorganisms, bacterial fragments and cell debris are completely digested, the reac-

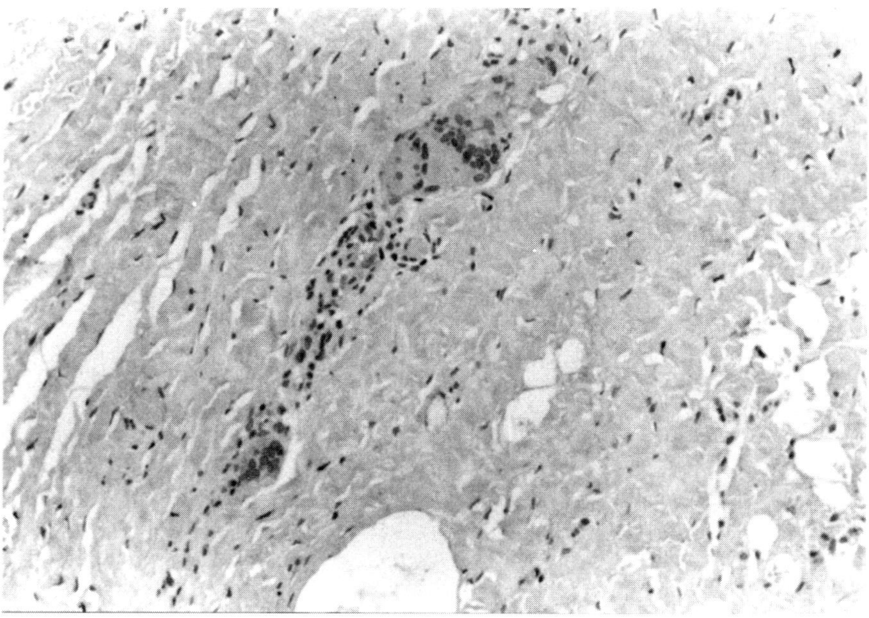

Figure 12 Foreign body giant cells of histiocitic origin (haematoxylin-eosin) (200x).

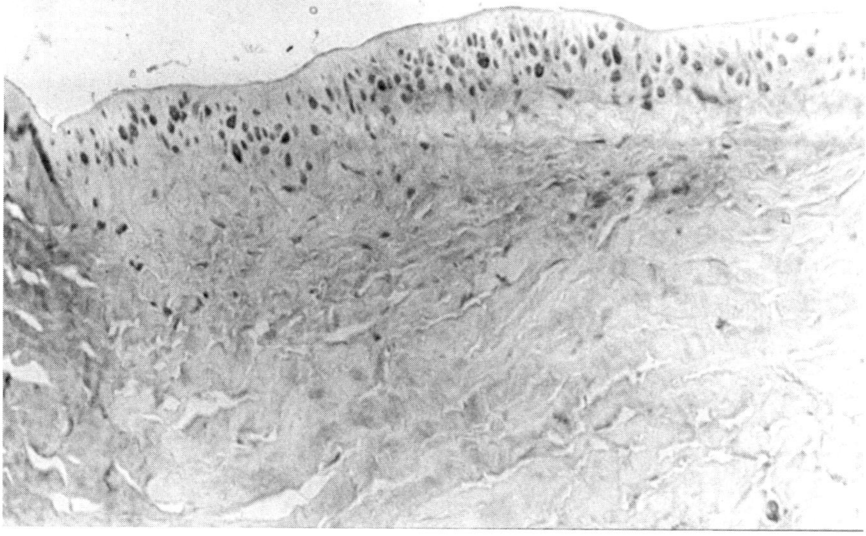

Figure 13 CD3 immunostaining evidences T lymphocyte (200x).

tion is limited to an acute inflammation which tends to exhaustion. But if this is not accomplished, further development of a specific immune reaction starts. The reaction is governed by T cells and B cells, which in cooperation with macrophages and other accessory cells, attack microorganisms using different and sophisticated immune mechanisms [49].

Reactive T cells are detected by CD3 monoclonal mouse antihuman antibodies, clone UCHT1 (Fig. 13), while B cells may be evidenced by CD20 monoclonal mouse antihuman antibodies, clone L26 (Fig. 14).

The detection of the presence of vessels in the analysis of peri-implantar tissue is made by an antibody that evidenced the presence of endothelial cells in the membrane: CD31 monoclonal mouse antihuman endothelial cell antibody (Fig. 15). These antibodies are usually used to define the different responses of tissue to biomaterials.

X. HISTOMORPHOMETRIC TECHNIQUES

The spatial distribution and concentration of distinct cellular elements and the dimension of the reaction zone are of great interest in the evaluation of the biological reaction to implant materials. The histomorphometric evaluation of the soft tissue adjacent to the implant has been carried out for many years by visual control and counting histological structure according to their morphology and their staining behavior [50]. In the last 10 years the use of a computer-aided image analysis system and immunohistochemical techniques have been developed [51]. This image analysis is a versatile technique that has found applications in every wide range of scientific discipline.

A superficial justification for using an image analyzer would be in terms of labor saving because speed of measurement is so great when compared to manual methods. However, the

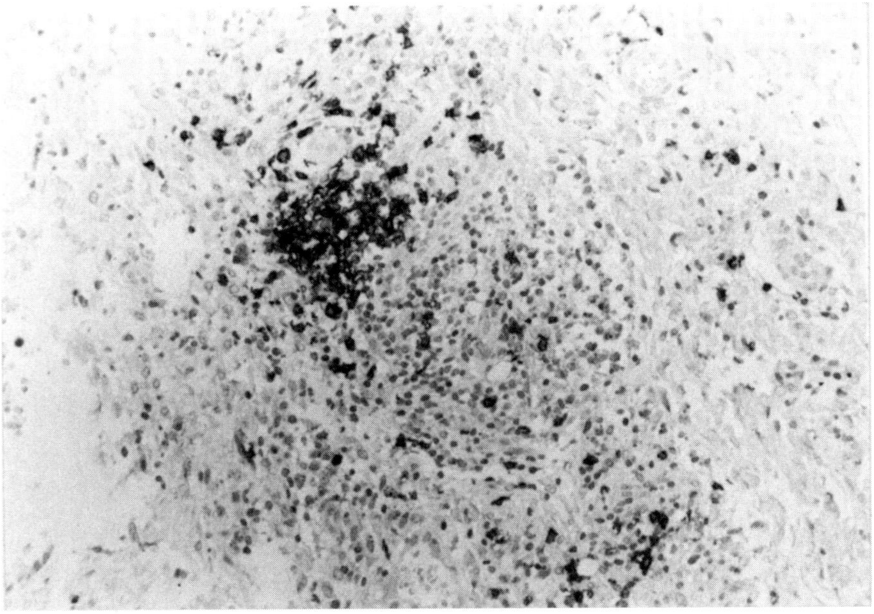

Figure 14 CD20 (L26) immunostaining evidences B lymphocytes (200x): cluster image.

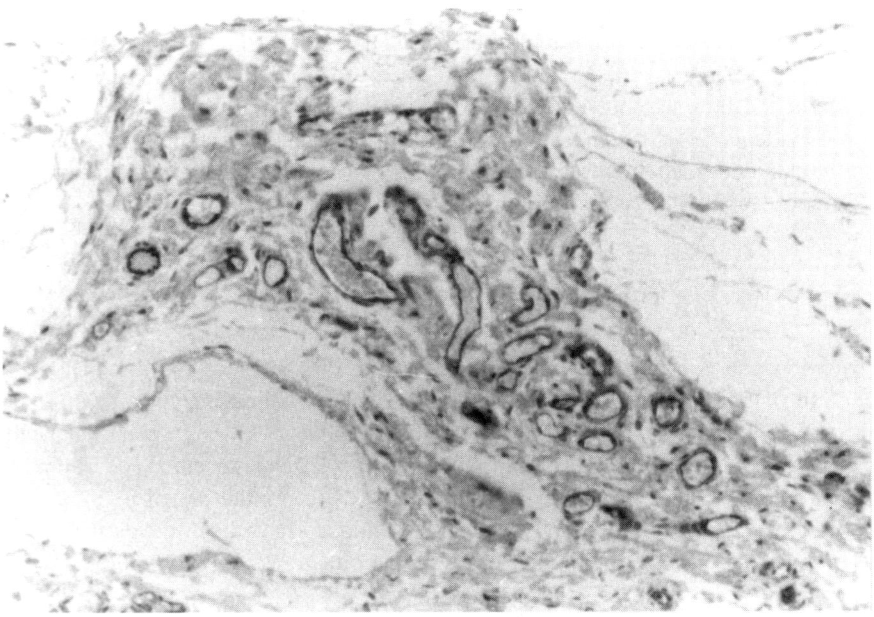

Figure 15 CD31 immunostaining for the localization of vessels.

real justification arises from the great certainty that can be attached to quantitative results obtained by image analysis and to the high reproducibility of the results, even counting different fields on the same slide [51].

The analysis system is composed of a light microscope connected to a computer unit. Microscope objectives are selected so that the image analyzed is magnified 100 or 200 times, and from each section a series of five fields in which the peri-implantar tissue may be divided are captured and stored on the hard drive of the computer.

The quantification is made by automatic selection based on black-and-white or colored image intensity, and a selective deletion provides a means of removing noncellular objects. This is performed on the basis of a set of parameters preventing objects such as fragments of material or cell debris from being included in the measurement stage. The results are expressed in number of cells/fields and may be stored in a data file for analysis.

The quantitative histomorphometric approach to peri-implantar tissue has evidenced that cells are more diffused at the interface, while the connectival component is predominant far from the implant surface [52].

XI. CONCLUSIONS

For a long time the biocompatibility of a medical device surface was intended to be a priori evaluated by in vitro tests. A more recent view of the matter is devoted to the possibility of predicting a possible adverse reaction from the preliminary phase of the production, as in the tissue engineering model; such a possibility requires a complete knowledge of the molecular level of interactions between cells and tissue to the biomaterial. In this sense protein adsorption represents one of the main topics to be addressed to fully understand the variegated aspects of biocompatibility, along with the histological investigations.

ACKNOWLEDGMENTS

The authors are indebted to Chiara Canetta for technical assistance.

REFERENCES

1. Baier, R. E., 1988. Advanced biomaterials development from natural products. J. Biomater. Appl. 4: 615–626.
2. Hanker, J. S. and Giammara, B. L., 1988. Biomaterials and biomedical devices. Science 242, 4880: 885–892.
3. Tang, L. and Eaton, J. W., 1995. Inflammatory response to biomaterials. Am. J. Clin. Pathol. 103, 4: 466–471.
4. Kirkpatrick C. J. et al., 1997. The cell and molecular biological approach to biomaterial research: a perspective. J. Mat. Sci. 8: 131–141.
5. Goldring, S. R. and Schiller, A. R., 1983. The synovial-like membrane at the bone-cement interface in loose total hip replacements and its proposed role in bone lysis. J. Bone Joint Surg. 65A:575–586.
6. Hench, L. L., 1991. Bioactive glasses and composites from gels. J. Am. Ceram. Soc. 74:1487–1510.
7. Wahlgren, M. and Arnebrant, T., 1991. Protein asorption to solid surfaces. TIBTECH 9: 201–208.
8. Norde, W., 1986. Adsorption of proteins from solution at the solid-liquid interface. Adv. Coll. Interface Sci. 25: 267–340.
9. Van Damme, H. S. and Feijen, J., 1991. Protein adsorption at polymer-liquid interfaces using series

of polymers with varying hydrophobicity, charge and chain mobility. In: Modern Aspects of Protein Adsorption on Biomaterials, Y. F. Missirlis and W. Lemm (eds.), Kluwer Academic, Boston; 55–61.

10. Santin, M., Wassall, M. A., Peluso, G. and Denyer, S. P., 1997. Adsorption of α-1-microglobulin from biological fluids onto polymer surfaces. Biomaterials 18: 823–827.
11. Santin, M., Cannas, M., Wassall, M. A. and Denyer, S. P., 1998. Adsorption of serum α-1-microglobulin onto biomaterials. J. Mater. Sci: Mater. Med. 9:135–140.
12. Jayabalan, M., 1993. Biological interactions: causes for risks and failures of biomaterials and devices. J. Biomed. Appl. 8: 64–71.
13. Reid, G., Davidson, R. and Denstedt, J. D., 1994. XPS, SEM and EDX analysis of conditioning film deposition onto ureteral stents. Surf. Interf. Anal. 21: 581–586.
14. Kaene, P. F., Bonner, M. C., Johnston, S. R., Zafar, A. and Gorman, S. P., 1994. Characterization of biofilm and encrustation on ureteric stents in vivo. Br. J. Urol. 73: 687–691.
15. Vroman, L., Adams, A. L., Fischer, G. C. and Munoz, P. C., 1980. Interaction of high molecular weight kininogen, factor XII and fibrinogen in plasma at interfaces. Blood 55: 156–161.
16. Remes, A. and Williams, D. F., 1992. Immune response in biocompatibility. Biomaterials 13: 731–743.
17. Berger, M., Broxup, B. and Sefton, M. V., 1994. Using Elisa to evaluate complement activation by reference biomaterials. J. Mater. Sci.: Mater. Med. 5: 622–627.
18. Porter, R. R. and Reid, K. B. M., 1978. The biochemistry of complement. Nature 275: 699–704.
19. Craddock, P. R., Hammerschmidt, D., White, J. G., Dalmasso A. P. and Jacob, H. S., 1977. Complement (C5a)-induced granulocyte aggregation in vitro. J. Clin. Invest. 60: 260–264.
20. Elwing, H., Nilsson, B., Svensson, K., Askendahl, A., Nilsson, U. R. and Lundstrom, I., 1987. Conformational changes of a model protein (complement factor 3) adsorbed on hydrophilic and hydrophobic solid surfaces. J. Coll. Interf. Sci. 125: 139–145.
21. Payne, M. S. and Horbett, T. A., 1987. Complement activation by hydroxyethylmethacrylate-methylmethacrylate copolymers. J. Biomed. Mater. Res. 21: 843–859.
22. Gemmell, C. H., 1997. A flow cytometric immunoassay to quantify adsorption of complement activation products (iC3b, C3d, SC5b-9) on artificial surfaces. J. Biomed. Mater. Res. 37: 474–480.
23. Allemann, E., Gravel, P., Leroux, J., Balant, L. and Gurny, R., 1997. Kinetics of blood component adsorption on poly(D,L-lactic acid) nanoparticles: Evidence of complement C3 component involvement. J. Biomed. Mater. Res. 37: 229–234.
24. Anderson, J. M., 1988. Inflammatory response to implants. ASAIO 11: 101–107.
25. Healy, K. E., Lom, B. and Hockberger, P. E., 1994. Spatial distribution of mammalian cells dictated by material surface chemistry. Biotech. Bioeng. 43: 792–800
26. Ratner, B. D., 1993. New ideas in biomaterials science—a path to engineered biomaterials. J. Biomed. Mater. Res. 27: 837–850.
27. Massia, S. P. and Hubbell, J. A., 1991. An RGD spacing of 440 nm is sufficient for integrin $\alpha V\beta 3$-mediated fibroblast spreading and 140 nm for focal contact and stress fiber formation. J. Cell Biol. 114: 1089–1100.
28. Bonfield, T. L., Colton, E. and Anderson, J. M., 1992. Protein adsorption of biomedical polymers influneces activated monocytes to produce fibroblast stimulating factors. J. Biomed. Mater. Res. 26: 457–465.
29. Paulsson, M., Kober, M., Frekjlarsson, C., Stollenwerk, M., Wesslen, B. and Ljungh, A., 1993. Adhesion of staphylococci to chemically-modified and native polymers, and the influence of preadsorbed fibronectin, vitronectin and fibrinogen. Biomaterials 14: 845–855.
30. Silver, F. and Doillon, C., 1989. In: Biocompatibility: Interactions of Biological and Implantable Materials, Vol. 1, VHC, 101–106.
31. Reis, R. L., Cunha, A. M., Fernandes, M. H. and Correia, R. N., 1997. Treatments to induce the nucleation and growth of apatite-like layers on polymeric surfaces and foams. J. Mater. Sci: Mater. Med. 8: 897–905.
32. Tanahashi, M. and Matsuda, T., 1997. Surface functional group dependence on apatite formation on self-assembled monolayers in a simulated body fluid. J. Biomed. Mater. Res. 34: 305–315.

33. Nimni, M. E., Myers, D., Ertl, D. and Han, B., 1997. Factors which affect the calcification of tissue-derived bioprostheses. J. Biomed. Mater. Res. 35: 531–537.
34. Tunney, M. M., Keane, P. F. and Gorman, S. P., 1997. Assessment of urinary tract biomaterial encrustation using a modified Robbins device continuous flow model. J. Biomed. Mater. Res. 38: 87–93.
35. Berman, A., Hanson, J., Leiserowitz, L., Koetzle, T. F., Weiner, S. and Addadi, L., 1993. Biological control of crystal texture: a widespread strategy for adapting crystal properties to function. Science 259: 776–779.
36. Buckwalter, J. A., Glimcher, M. J., Cooper, R. R. and Recker, R., 1995. Bone biology part I: Structure, blood supply, cells, matrix, and mineralization. J. Bone Joint Surg. 77A: 1256–1275.
37. McKee, M. D., Nanci, A. and Khan, S. R., 1995. Ultrastructural immunodetection of osteopontin and ostocalcin as major matrix components of renal calculi. J. Bone Miner. Res. 10: 1913–1929.
38. Kelm, R. J., Swords, N. A., Orfeo, T. and Mann, K. G., 1994. Osteonectin in matrix remodeling. J. Biol. Chem. 269: 30147–30153.
39. Bunker, B. C., Rieke, A. A., Tarasevich, P. C., Campbell, B. J., Fryxell, G. E., Graff, G. L., Song, L., Liu, J., Virden, J. W. and McVay, G. L., 1994. Ceramic thin-film formation on functionalized interfaces through biomimetic processing. Science 264: 48–55.
40. Gura, T. A., Wright, K. L., Veis, A. and Webb, C. L., 1997. Identification of specific calcium-binding noncollagenous proteins associated with glutaraldehyde-preserved bovine pericardium in the rat subdermal model. J. Biomed. Mater. Res. 35: 483–495.
41. Santin, M., Motta, A. and Cannas, M., 1998. Changes in serum conditioning profiles on glutaraldehyde-crosslinked collagen sponges after their treatment with calcification inhibitors. J. Biomed. Mater. Res. 40: 434–441
42. Spector, M., Cease, C. and Tong-Li, X., 1989. The local tissue response to biomaterials. Crit. Rev. Biocompat. 5(3): 269–295.
43. Charosky, C. B., Bullog, P. G. and Wilson, P. D., 1978. Total hip replacement failure, a histological evaluation. J. Bone Joint Surg. 55A: 49–56.
44. Okada, T. and Ikada, Y., 1993. Tissue reactions to subcutaneously implanted, surface-modified silicones. J. Biomed. Mat. Res. 27: 1509–1518.
45. Bosetti, M., Navone, R., Rizzo, E. and Cannas, M., 1998. Histochemical and morphometric observations on the new tissue formed around mammary expanders coated with pyrolitic carbon. J. Biomed. Mat. Res. 40: 307–313.
45. Vistnes, L. M., Ksander, G. A. and Kosek, J., 1978. Syudy of encapsulation of silicone rubber implants in animals. A foreign body reaction. Plastic Reconstruct Surg. 26(4): 580–588.
46. Linder, L. and Carlsson, A. S., 1986. The bone-cement interface in hip arthroplasty. A histologic and enzyme study of stable components. Acta Orthop. Scand. 57: 495–500.
47. Anderson, J. M. and Miller, K. M., 1984. Biomaterial biocompatibility and the macrophage. Biomaterials 5: 8.
48. Rae, T., 1986. The macrophage response to implant materials. Crit. Rev. Biocompatibility 2(2): 97–126.
49. Pizzoferrato, A., Ciapetti, G., Stea, S. and Toni, A., 1991. Clin. Mater. 7: 51–81.
50. Ungersbock, A. and Hunt L. A., 1994. Comparison of visually controlled and automatic histomorphometric evaluation of soft tissue. J. Mater. Sci. Mater. Med. 5: 702–704.
51. Hunt, J. H., Vince, D. G. and Williams, D. F., 1993. Image analysis in the evaluation of biomaterials. J. Biomed. Eng. 15: 39–45.
52. Cannas, M., Bosetti, M. and Navone, R., 1995. It. J. Anat. Embryol. 100 (Suppl. 1): 605–612.

3

Biomaterial-Related Infection: Role of Neutrophil Peptides

Sandra S. Kaplan, Robert P. Heine, and Richard L. Simmons
University of Pittsburgh, Pittsburgh, Pennsylvania

I. INTRODUCTION AND BACKGROUND

The use of prosthetic materials (biomaterials) in reconstructive surgery has altered the course of medical practice and has enhanced the lives of many millions of recipients worldwide. Yet despite this phenomenal success and the fact that the incidence of infection actually has been quite low overall (<5% for most kinds of totally implanted devices) (1), infection at the biomaterial surface and surroundings has continued to be a vexing and often intractable problem resulting in a heavy toll in terms of morbidity and expense (1–3). These infections are most often introduced at the time of surgery and, despite the use of prophylactic antibiotics, form the nidus of an infection which may not become manifest for many years (4–8). Some devices, however, especially those which are only partially implanted, have a much higher incidence of infection. Vascular access devices, for example, have an infection rate of up to 37% with a case fatality rate of 20–40% (3). Urinary catheters left in place for longer than a month have an infection rate approaching 100%, but with a lower case fatality rate (1,3,9).

The relationship of foreign bodies to infection has been recognized for at least six centuries. In the 14th century, Guy de Chauliac, a French surgeon, reported that foreign bodies must be removed in order to effect resolution of an infected wound (7), but it was not until the middle of the 20th century that a systematic scientific study of the pathogenesis of this phenomenon began. Elek and Conen are credited with the first scientific study of this issue with their determination that a smaller bacterial inoculum-induced infection in the presence of silk suture material than it would in the absence of the sutures (10). Since that initial study, this observation of enhanced infectivity of foreign materials in general, and prosthetic materials specifically, has been extensively validated as well as generalized to the many different types of materials used for prosthetic devices (11–13). Some of the most important work has investigated the mechanisms of this increased infectivity of biomaterial surfaces. This work may be classified under several different subheadings:

Properties of materials which may enhance infectivity
Properties of bacteria which may facilitate colonization of materials
Role of soluble mediators of inflammation (matrix proteins, cytokines, complement)
Effects of the association of phagocytes with materials on phagocyte-mediated host defense

A. Properties of Materials Which May Enhance Infectivity

It was convincingly demonstrated that bacteria adhered readily to materials and that this action constituted the first step in foreign-body-associated infection (4,14). The properties of materials which may affect this adhesion have therefore been the subject of study. Despite this attention, the structural and chemical properties of the ideal material for implantation have remained elusive and ideas which seemed sound intuitively, and even were supported in some early studies, have not proved to be relevant upon further examination. Whether or not the surface roughness of a material contributed to infectivity has been somewhat controversial. Early studies claimed a direct relationship (15,16), but other workers did not find roughness to be an important factor (17). Under conditions of increased turbulence, however, such as might occur at a rough surface under conditions of flow, increased bacterial adherence was shown to occur (18). In contrast, polymer chemistry in terms of biodegradability did not appear to affect bacterial adherence (15), but surface chemistry as it affects the net surface charge also was found to influence the ability of prosthetic devices to induce cell activation, including complement activation, phagocyte adherence, and cytokine production, although results have been somewhat contradictory (19–21). While it had been assumed that the best possible materials would be those which were most chemically inert, the salutary effects of tissue integration had not been considered. When this aspect was investigated, it was shown that tissue integration, defined as the ability to promote tissue ingrowth, improved biocompatibility (19,22,23).

B. Properties of Bacteria Which May Facilitate Colonization of Materials

Staphylococcus epidermidis colonization accounts for approximately 50% of infections on orthopedic implants and also is an important element in infections of vascular and urinary tract access devices (2,6). *Escherichia coli*, *Peptococcus*, *Pseudomonas aeruginosa*, *Proteus mirabilis*, and beta hemolytic streptococci also are important pathogens associated with prosthetic devices. These organisms have in common their ability to produce a glycocalyx substance known as extracellular slime substance. This substances is believed to impair phagocytic killing by interfering with phagocytosis per se (14,24) as well as interfering with subsequent intralysosomal bacterial killing (25,26). This slime substance also may protect the microorganisms from antibiotics (4). The role of bacterially derived slime in adhesion of organisms to materials has recently been questioned by Higashi et al. who found that a mutant strain of *S. epidermidis*, deficient in both an adhesin and in slime, adhered effectively to polyethylene regardless of the presence of protein even under a range of physiological shear stress conditions (27,28). Organisms such as *S. aureus* also infect biomaterials, but here the element of tissue devitalization may play a contributing role, and a number of matrix proteins (discussed in the next section) act to enhance adherence of this organism to prosthetic devices by means of surface adhesins (12,29–34).

C. Role of Soluble Mediators of Inflammation (Matrix Proteins, Cytokines, Complement)

After implantation, prosthetic devices rapidly become associated with a variety/succession of plasma and matrix proteins, i.e., the Vroman effect (35), which provide receptor sites for bacterial adherence and facilitate the adherence of *S. aureus* to the biomaterial surface (34,36–39). These proteins include fibrinogen, fibronectin, vitronectin, laminin, collagen, and even thrombospondin, but fibrinogen and fibronectin appeared to be the most potent of these for the adherence of *S. aureus* (34). Other workers focused on the attachment apparatus of the *S.*

aureus. This work characterized the genes for the protein which binds to fibrinogen (32) and further determined that mutant staphylococci which were deficient in binding protein failed to bind to materials (40,41).

Fibrinogen associated with prosthetic devices plays another important role having to do with the attachment of inflammatory cells, such as neutrophils and monocytes, to the biomaterial surfaces. This topic has been extensively studied by Tang et al. (42–44) who demonstrated that the spontaneous adsorption of fibrinogen, but not albumin, IgG, or even activated complement components, was responsible for the attachment of phagocytes to the material surface. Other workers, however, not only have shown that fibronectin also plays a role in the adhesion of phagocytes to materials (45), but that other proteins, including the Hageman factor, IgG, and albumin, play a role in the response of adherent phagocytes to produce inflammatory mediators (46). Several groups, however, demonstrated that a number of cytokines (IL-1, IL-6, and TNFα) were produced by macrophages in contact with biomaterials (20,47–53), and this was variably reduced when cells were attached to protein-adsorbed materials. This work suggested that the passivating effect of proteins could be related to whether or not the surface chemistry of the material resulted in enhanced adherence and cytokine production in the first place (20). They further described highly cationic surfaces that were activating and neutral surfaces that were not. Also of interest, these studies demonstrated that an unidentified factor which stimulated thymocyte proliferation was enhanced even as IL-1 production was decreased, suggesting that the role of protein adsorption to materials may be much more complex than anticipated.

That complement is activated at biomaterial surfaces has been appreciated for many years. Complement generally has been believed to be activated by the alternative pathway (50,51). Marosok et al. demonstrated that silicone catheters were considerably more effective at activating complement in this manner than other catheter materials, and that silicone appeared to be more infection-prone than other materials (52). This also resulted in local complement depletion secondary to the augmented activation. Tang et al. also demonstrated that materials varied in their capacity to activate complement, and attributed this variability to the surface chemistry of the material, finding that materials with uncharged, hydroxyl-bearing surfaces activated complement but surfaces derived with amine and carboxyl groups did not (21). These observations would seem to contradict the relationship of surface charge to activation, published by Yun et al. (20), but Yun's studies focused on cell activation while Tang et al. focused on complement activation. Tang et al. further stated that under conditions in which complement was not activated (polyethylene, terephthalate, polyether urethane, polyvinyl chloride, and polyethylene), inflammatory cell accumulation mediated by fibrinogen would occur (21). The role of complement for attachment of phagocytes to biomaterials also is controversial. Kao et al., for example, found that complement was necessary to mediate adhesion, but the complement was sufficient only under static conditions (45). Under conditions of shear stress, other factors such as fibronectin were necessary.

D. Effects of the Association of Phagocytes with Materials on Phagocyte-Mediated Host Defense

Polymorphonuclear neutrophilic leukocytes, known as neutrophils or PMN, and monocyte/macrophage cells are considered to be professional phagocytes with the committed function to effect host defense against microbial and fungal invading organisms by the process of phagocytosis followed by the killing of these ingested microorganisms (53). This cytotoxic function requires that concomitant with phagocytosis there is a marked and rapid production of reactive oxygen intermediates (ROI) such as the superoxide anion (O_2^-) and hydrogen peroxide (H_2O_2)

within the phagocytic vacuole together with the release into this phagocytic vacuole of potent antibacterial constituents from nearby granules (such as myeloperoxidase, lysozyme, and cationic peptides known as defensins) (53,54). Given the importance of phagocytic killing for effective host defense along with the observation that neutrophils rapidly become associated with foreign materials (42–46), it is somewhat curious and paradoxical that infections at biomaterial surfaces are persistent and intractable (1,3,5–7). This apparent paradox, however, becomes understandable in the light of the accumulating evidence that neutrophils in contact with foreign materials degranulate, produce ROI, and then become deactivated and dysfunctional (7,11,22,55–60). Studies by Zimmerli et al. using in vivo and ex vivo models of biomaterial-related infection not only demonstrated that material-associated neutrophils were dysfunctional, but also provided evidence that even the addition of fresh neutrophils could not rescue the infectious process except under highly specific conditions (22,57). These studies demonstrated that in the presence of Teflon, phagocytosis and ROI formation by neutrophils became impaired, granule content became reduced, and phagocytic killing, especially of catalase-positive organisms, was impaired. Microorganisms introduced into the area of the materials were not cleared and infection persisted. Addition of fresh neutrophils improved host defense only if the fresh cells were introduced within 3 h of the bacterial inoculum, or at 20 h but only if fresh serum was also added to the area of infection at this time point. It was suggested that the function of the fresh serum was to provide opsonins, which in previous studies had been shown to be poorly available in the vicinity of an implant (22). Marchant et al., using polyurethane implants, also reported that material-associated neutrophils exhibited decreased phagocytic capacity (58). Later work in our laboratory, using an in vitro model of the biomaterial-neutrophil interaction and a variety of materials used in clinical practice, demonstrated that most materials induced a rapid but transient stimulation of ROI production/release by cells in contact with the material, even in the absence of an additional stimulus. We also observed that this process was passivated by the presence of a serum coat on the material, except in the case of woven Dacron. Here, stimulation was maximum in the presence of serum (59,61). Further investigation to evaluate the mechanisms of this material-associated stimulation demonstrated that the stimulation did not occur via a pertussis toxin-inhibitable G protein, but that protein kinase C was involved since inhibition of protein kinase C by Staurosporine did abrogate the response of neutrophils to polystyrene and polyurethane (but not to woven Dacron). Thus the signaling mechanisms of material-induced ROI production appear to vary among the materials and appear not to be the same (in terms of G protein requirements) as stimulation via a receptor-dependent process (61). Continuing with this work, we demonstrated that material-associated neutrophils (especially cells on woven Dacron) produced a potent chemoattractant (62). Since this would serve to bring a fresh wave of phagocytes to a biomaterial surface, we evaluated the effect of adding fresh neutrophils to cells already material-associated. In support of Zimmerli's findings (57), our work demonstrated that neither ROI production nor phagocytic killing occurred when the fresh cells were added (62). Liu et al. studied the effect of the specific protein adherent to the material on the production of ROI by adherent neutrophils (63). These studies evaluated the effects of albumin, IgG, and fibrinogen since these are the predominant proteins adsorbed to materials, and resulted in the observation that the effects differed according to whether the material was hydrophobic or hydrophilic. IgG coating, for example, was activating to neutrophils, especially when on a hydrophobic material, but fibrinogen and albumin were not. This finding was somewhat unexpected since fibrinogen-coated materials are attractant to neutrophils. It was suggested that protein coating may induce an augmented response to other agents of inflammation such as TNFα, and this indeed was demonstrated by Nathan who observed that many coating materials acted to enhance responses to agents that would have failed to activate cells in the absence of their attachment to a coated surface (64).

Karlsson et al. also found that hydrophobic surfaces were more activating than hydrophilic surfaces, and observed that this was due to complement activation by the former (65). These studies evaluated the effect of the plasma supernatant of cells which had been in static contact with materials, and found that soluble factors had been released which induced metabolic activation in target neutrophils.

II. BIOMATERIAL-ASSOCIATED DYSREGULATION OF FUNCTION

The idea that neutrophils exposed to materials might produce soluble factors, and that these factors might augment the activity of bystander cells or that one or more of these factors might act to attract fresh neutrophils to a material surface, suggested that the function of incoming, second-wave neutrophils deserved further investigation. In contrast to Karlsson et al. (65), however, studies in our laboratory did not find that supernatants of cells exposed to polystyrene, expanded polytetrafluorethylene (ePTFE), or woven Dacron (WD) contained factors that activated fresh neutrophils to produce O_2^-. We did find, however, that substances chemoattractant for neutrophils were released. We therefore determined to characterize the relative amounts of chemoattractant and, more importantly, to evaluate the function of fresh neutrophils added to an inoculum of neutrophils which had been in contact with the material for 60 min. These studies will show that neutrophils on some, but not all, materials produce a chemoattractant for subsequent neutrophil recruitment. We further demonstrated that fresh neutrophils placed together with neutrophils already present on the material failed to become activated to produce ROI and exhibited significantly impaired microbial killing.

A. Methods

1. Cells and Reagents

Neutrophils were isolated from heparinized whole blood using standard density gradient centrifugation techniques originally described by Boyum (66) and modified for these studies as previously described (59,60,62). Neutrophils were suspended in modified Kreb's ringer phosphate buffer, pH 7.4 [16 mM Na_2HP0_4, 123 mM NaCl, 0.5 mM $CaCl_2$, 10.5 mM $MgSO_4$, 5 mM KCl, and 4.4 mM glucose (KRPG) (61)], and in some studies the KRPG included 5% heat-inactivated calf serum or 10% autologous human plasma. Each neutrophil donor provided cells for a single day's experiment, but when fresh neutrophils were added to 24-h-old neutrophils the same donor was used for two consecutive days. Neutrophils were quantitated using Coulter electronic instruments (Hialeah FL) and the purity of the preparation was determined from DiffQuick-stained cytospin smears. Cells were suspended at a concentration of 2×10^6/mL and consisted uniformly of >90% neutrophils. Trypan blue dye exclusion was used to assess viability, which was uniformly >95%. Stock solutions of fMLP were prepared as previously described (59). All other materials were reagent grade and obtained from Sigma Chemical Co. (St. Louis, MO) unless otherwise specified.

2. Biomaterials

The biomaterials were generous gifts and included expanded polytetrafluorethylene (ePTFE) (Gore and Associates Inc., Flagstaff, AZ) and woven Dacron (WD) (Meadox Medicals Inc., Oakland, NJ). Polystyrene (PST) tissue culture dishes (24 well) (Corning CoStar, Corning, NY) were used as a prototype material in some studies. The biomaterials were cut into 13×13 mm squares, sterilized with ethylene oxide as previously described (59), and fit into the wells of 24-well plates. The materials were tested for lipopolysaccharide (LPS) contamination as previously described, and were LPS free (59).

3. Determination of Mediator Release from the Biomaterial-Neutrophil Interactions

Neutrophils, at a concentration of 2×10^5 in KRPG or KRPG with 5% calf serum, were placed on the biomaterials WD, ePTFE, or PST for 30 min or 2 h. The plates were centrifuged and the supernatants were stored at −70°C until tested for their activity as chemoattractants. The supernatants of biomaterials incubated with KRPG in the absence of cells also were tested to determine if materials alone released a chemoattractant.

4. Superoxide Release

Superoxide (O_2^-) was measured as superoxide dismutase-inhibitable reduction of cytochrome c (67). Neutrophils were placed in flat-bottomed 24-well or 96-well plates for varying periods of time. Cytochrome c was added with and without the stimulator formyl-methionyl-leucyl-phenylalanine (fMLP) and the plates were incubated at 37°C for 30 min. Data are expressed as nanomoles $O_2^-/10^6$ cells.

5. Chemotaxis

Chemotaxis was evaluated using 48-well micro-Boyden chambers and 5 μ pore-size polycarbonate filters, as previously described (68). The biomaterial supernatants or fresh KRPG were placed into the lower wells. fMLP was placed into some of the lower wells and served as a positive control. The chambers were assembled and allowed to warm in a 37°C humidified incubator for 10 min prior to placing fresh neutrophils into the upper wells. Incubation was continued for 15 min. The filters were removed, fixed, stained, and mounted on glass slides to determine by microscopy the number of cells that had migrated to the lower surface of the filter.

6. Effect of Adding a Second Inoculum of Fresh Neutrophils to the First Inoculum of Cells

a. Metabolic Studies on Biomaterial Surfaces. Neutrophils were placed on the non-serum-coated materials for 10 or 120 min. Cytochrome c then was added with or without fMLP (10^{-7} M final concentration) and the incubation continued for 30 min. The fresh inoculum of neutrophils was obtained from cells held at room temperature for 120 min which then were added to materials already containing the original inoculum of cells. To control for the effect of aging, these room-temperature-held neutrophils also were added to materials which previously had not been exposed to cells. In both instances, these fresh cells were allowed to settle for 10 min, then cytochrome c was added and incubation continued for 30 min. The samples were centrifuged, and reduced cytochrome c determined in cell-free supernatants. In some studies, the cells that had been overlaid onto the first inoculum were removed to a clean plate and their ability to release O_2^- determined in the fresh plate.

b. Microbial Killing on Biomaterial Surfaces. *S. aureus*, strain D2C, was used as the target organism to determine the ability of the cells to kill bacteria, as previously described (69). In these studies, 5×10^5 neutrophils in 500 μL KRPG were added to biomaterials and after 10 min an inoculum of *S. aureus* in KRPG containing 20% autologous plasma was added (Staph:neutrophil = 5:1). The samples were swirled to mix, and a 10 μL sample was removed immediately after the addition of bacteria, the T-0 sample. The plates were incubated at 37°C on a Kaola-Ty slow variable rotating shaker (Accurate Chemical and Scientific Corp., Westbury, NY) at 90 oscillations per minute. Samples of 10 μL were removed at 5 h and 24 h, and numbers of viable staphylococci determined. At the 24-h time point, a fresh inoculum of neutrophils and staphylococci was added and samples were taken again at the new time 0 and at 2 h, 5 h, and 24 h after the second inoculation. The numbers of viable staphylococci were determined as described above. Neutrophil viability also was determined at times 0, 24 h, and 48 h.

7. Statistical Evaluation

Statistical evaluations were performed where relevant using a StatWorks program and Student's *t*-test or the Wilcoxon signed rank test. Significance is defined as $p \leq 0.05$.

B. Results and Discussion

Figure 1A shows that WD-associated cells produced a potent chemoattractant for fresh neutrophils with a potency nearly as great as 10^{-7} M fMLP, especially when the cells were incubated in the absence of serum, as shown in Fig. 1B. Figure 1A also shows the results with cells associated with PST and ePTFE. With these materials, the presence or absence of serum had little effect. Furthermore, the extent of material-induced O_2^- formation had little effect since PST is highly activating and ePTFE is poorly activating. Figure IB illustrates chemotactic activity using the negative (KRPG) and positive (fMLP) chemoattractant controls. This would be a proinflammatory action with far-reaching implications since it would augment the number of inflammatory cells at the material, thereby potentially enhancing microbial killing at the site of an implant.

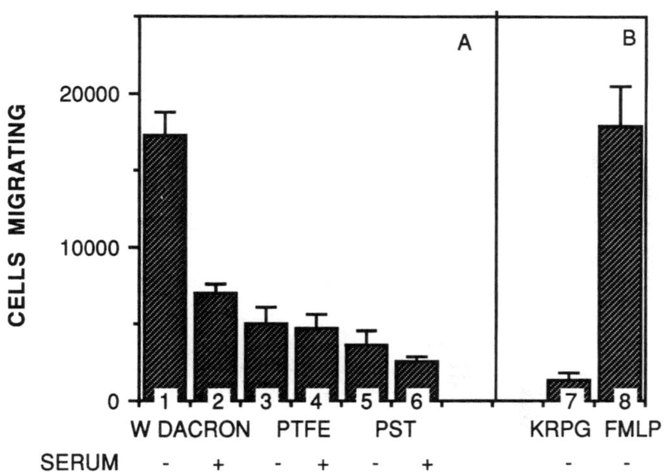

Figure 1 (A) The chemoattractant potency of biomaterial-neutrophil supernatants was determined by placing these preparations in the lower wells of micro-Boyden chambers and fresh neutrophils in the upper wells. The potency is indicated by the number of cells that had migrated through the filter. Woven Dacron-neutrophil supernatants (columns 1 and 2) exhibited potent chemoattraction, especially in the absence of serum (column 1). PTFE-associated neutrophils (columns 3 and 4) released much less chemoattractant under both serum absent (column 3) and serum present (column 4) conditions. PST-associated cells (columns 5 and 6) released slightly less chemoattractant when serum was present (column 6). (B) The chemoattractant potency on the negative control, KRPG (column 7), and the positive control, fMLP (column 8), are shown. (From Ref. 62.)

Because chemoattraction would bring in fresh PMN, we evaluated the ability of a second inoculation of neutrophils overlaid upon previously inoculated neutrophils to produce O_2^- or kill bacteria. Figure 2A shows that after 10 min on PST, the first inoculum of neutrophils produced 38.7 ± 7.2 nmol of $O_2^-/10^6$ neutrophils/30 min, and the addition of fMLP caused the release of an additional 9.0 ± 1.7 nmol. After 2 h, however, very little O_2^- was produced (1.9 ± 0.5) and the response to subsequent exposure to fMLP was somewhat diminished (6.4 ± 1.1), supporting earlier work (59). The addition of a second inoculum of neutrophils, held at room temperature for 2 h, to the neutrophils already associated with the PST resulted in release of only 2.3 ± 0.5 nmol O_2^-/30 min and the response to fMLP was 7.0 ± 0.8 (Fig. 2B, col. A). This was not due to the age of the neutrophil suspension since cells held for 2 h at room temperature and then added to a PST well resulted in 28 ± 6.2 nmol, and responded normally to stimulation with fMLP (Fig. 2B, col. C). We also removed the freshly added cells and evaluated the activity of these cells in a fresh, never inoculated PST well, and these neutrophils produced only 7.7 ± 1.5 nmol of O_2^-. and their response to fMLP was very poor 1.3 ± 0.5 (Fig. 2B, col. B). We also found that the response of these neutrophils to PMA was reduced (17.6 ± 2.7 versus 50–60 nmol/30 min under other conditions). Thus, biomaterial association appears to induce a potent downregulation of ROI production, an action potentially deleterious to host defense, and these results cast doubt, then, as to whether newly recruited neutrophils would function effectively as mediators of antimicrobial activity. We also showed that neutrophils removed to a fresh, noncoated well also exhibited diminished production of ROI. The mechanism of this passivation, however, is unclear.

The various biomaterials affected neutrophils differently. Figures 3A and 3B show that ePTFE, a nonactivating material, had less of an effect on the response of added neutrophils (Fig. 3A) and the response of the second "wave" of neutrophils to fMLP was not impaired (Fig. 3B). WD, in contrast, induced a marked attenuation of the response over 120 min (Fig.

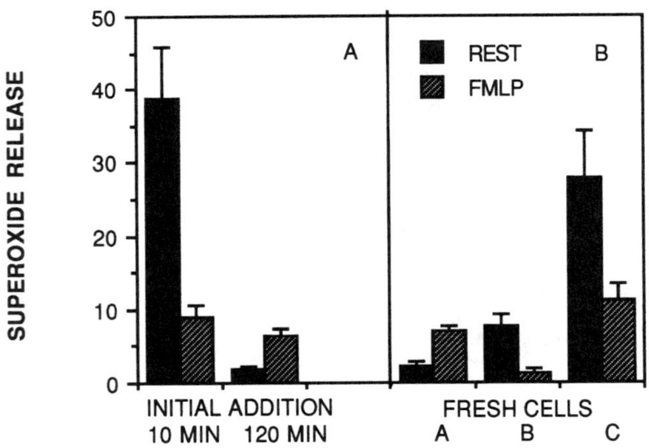

Figure 2 (A) The effect of O_2^- release of incubating neutrophils on PST for 10 min and 120 min in the absence of serum is shown for unstimulated cells (■) and for cells additionally stimulated with fMLP (▨). A marked time related diminution of response is seen. (B) The effect on O_2^- release of adding fresh neutrophils held at room temperature for 2 h to cells already biomaterial associated is shown in the A columns without additional stimulation (■) and after fMLP stimulation (▨). B columns show the effect of removing these cells to a clean plate. C columns show the effect of adding neutrophils held at room temperature for 2 h to a clean plate without previously placing them on plate-associated neutrophils. (From Ref. 62.)

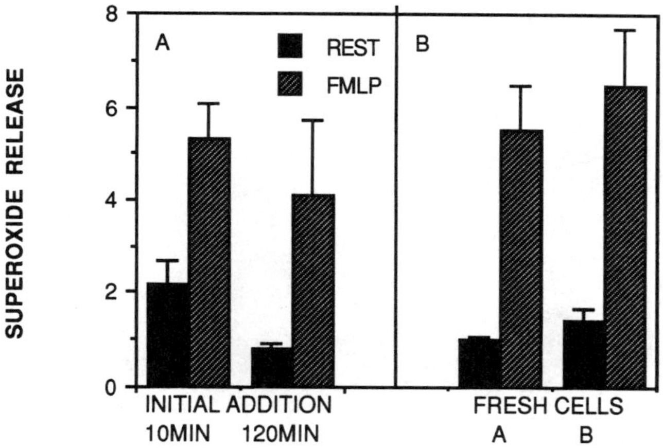

Figure 3 (A) Conditions as for Fig. 2A, when cells are on PTFE. (B) A columns show effect of adding cells held for 120 min at room temperature to materials already containing neutrophils. B columns show effect of adding these cells to a fresh sample of PTFE. (From Ref. 62.)

4A), and the response of fresh neutrophils added to cells already associated with the material was decreased. The response to subsequent fMLP challenge also was decreased (Fig. 4B, col. A). These effects were not due to the age of the neutrophils. This is shown in Fig. 4B, col. B, which shows that neutrophils held for 2 h were capable of normal responsiveness if added to wells that had never contained neutrophils. The effects on WD, however, are less profound than the effects on PST.

The implications of this passivation of ROI release on antimicrobial activity also were of interest. It might be expected that cells which poorly release O_2^- also would exhibit impaired microbial killing, and this, in fact, occurred. A second inoculum of staphylococci was killed very poorly, even though it was added together with fresh neutrophils. Figure 5 shows that over the first 24 h period staphylococci were killed slowly by neutrophils on PST. This second

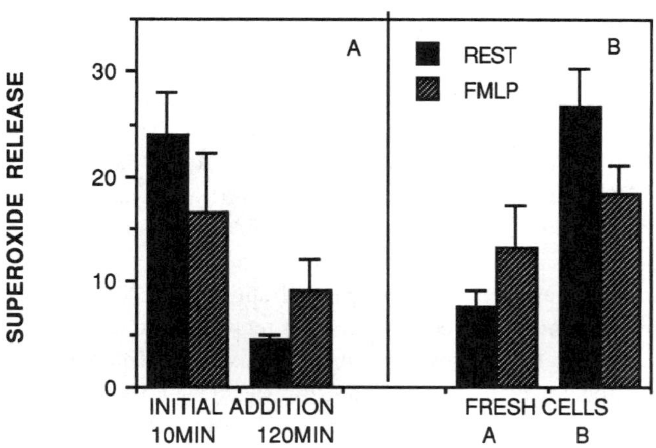

Figure 4 (A) and (B) Conditions as for Figs. 3A and 3B, when the cells are on woven Dacron. (From Ref. 62.)

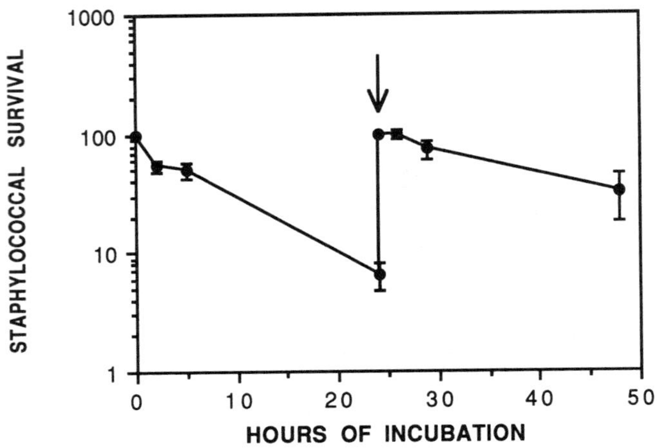

Figure 5 The ability of neutrophils to kill staphylococci during a 48 h incubation on PST is shown. Fresh neutrophils and additional staphylococci are added as indicated by the arrow. (From Ref. 62.)

inoculum was less effectively killed, with (32.5 ± 14.5)% of the inoculum surviving during the second killing period in contrast to (6.5 ± 1.7)% surviving after the first 24 h period ($p \leq 0.05$). The importance of this observation relates to phenomena such as delayed infection at biomaterial sites and the occurrence of infection at the site of sutures remaining from a previous surgical procedure.

The above studies described how neutrophils become activated to produce O_2^- when in contact with biomaterials and that this activation is short-lived. The studies have further shown that the consequences of this biomaterial-induced activation are a dysregulation of oxidative responsiveness and microbial killing by fresh neutrophils. Under normal circumstances, neutrophils process incoming signals from their environment, signals such as formylated peptides elaborated by bacterial growth or the active complement fragment C5a and thereby become activated for host defense. This exquisitely regulated process involves G-protein-mediated signal transduction which initiates a cascade of responses in proportion to the degree of receptor occupancy (70,71). These responses are regulated by the intrinsic capacity of the G protein to become deactivated and then reactivated as part of the same process in which an initial activation occurs (72), as well as by physiologic feedback inhibition by second messengers formed in the process of activation (73). Changes in cell shape occur with low-level stimulation and this facilitates cell movement in the direction of increasing concentrations of chemoattractant. Higher concentrations of ligand, however, inhibit neutrophil movement and instead begin to elicit the metabolic burst leading to ROI formation. Neutrophils, however, are endstage cells. There is little potential for generation of new reactants, there is no DNA produced, and there is only limited protein synthesis. These cells are destined to die as a consequence of their host-defense function. It is not unreasonable, therefore, that unregulated and in fact dysregulated biomaterial-induced activation of a neutrophil might compromise its real function as a phagocytic killer. The hypothesis of dysregulation is supported by our previous work which demonstrated that biomaterial-induced signaling for activation was not transduced via the G-protein mechanism used by physiological ligands and therefore might not be subject to the same control processes.

Despite being endstage cells, neutrophils are known to produce and release substances that modify the inflammatory response, including cytokines such as IL-6, IL-1, TNFα, and IL-

8 (74–78). The regulation of this cytokine expression is largely unknown, but recent studies have shown that IL-10 down-regulates neutrophil expression and release of IL-8 (9,79,80) while ROI, not necessarily derived from NADPH oxidase, appear to cause increased production of IL-8 via increased gene expression (81). Cytokines also may exhibit autocoid properties in which the IL-8 produced by neutrophils which have been stimulated by TNFα and LPS (82), for example, may act to further stimulate these neutrophils (50). Other work, however, has demonstrated that IL-8 also decreases the adherence of neutrophils for endothelial cells and reduces recruitment to sites of inflammation (83–85). Activated neutrophils also produce metabolites of arachidonic acid, including prostaglandins and leukotrienes. This metabolic activity occurs in concert with ROI formation and leads to further stimulation of ROI, including activation of phospholipase D (46). One such metabolite, LTB4, is a potent chemoattractant.

While the studies described above have helped to establish that neutrophil contact with biomaterials acts to downregulate the functional activity of neutrophils that might be attracted to a material surface, the mechanism of this downregulation remains undefined. We and others have shown that material contact induces cytokine release by cells but these cytokines (IL-1 and TNFα) are unlikely to have a downregulating effect on fresh cells. The persistent nature of material-associated infection and the resistance to antibiotic management, however, and the need to remove the material to resolve the infection (7,86,87), suggest that these infections might behave somewhat like an abscess where surgical drainage is needed to resolve the infection.

III. NEUTROPHIL-INDUCED AUTOCOID AND PARACOID TOXICITY

It is well appreciated that neutrophils contain a large number of cytotoxic substances, many of which participate in oxidative and nonoxidative microbial killing, and that these substances also may have a cytotoxic effect on fresh or bystander cells. Among these substances are small and highly cationic peptides (human neutrophil peptides, or HNP) which play a role in the non-ROI-mediated host defense against microbial invaders. These peptides, also known as defensins, are present in the primary granules of neutrophils and comprise approximately 5% of the total cellular protein (88–92). Three peptides, HNP1, HNP2, and HNP3, have been extensively studied and have been shown to consist of 29–35 amino acid residues. Because of their highly cationic charge, they readily form multimeric subunits which insinuate themselves into the cell walls of microorganisms and other cells (91–95) and produce voltage-sensitive channels accompanied by membrane permeabilization (96). These HNP are known to damage eukaryotic cells, as well as the prokaryotic microorganisms, including tumor cells (97–99) and human neutrophils (100), and the NADPH oxidase enzyme system in neutrophils (101). To some extent, the amount of damage depends on the lipid composition of the target cell membrane (102) and the metabolic activity of the target cell (99). Other studies have shown that H_2O_2 is able to act synergistically with HNP to enhance the cytotoxic effect (103). Another cytotoxic effect of HNP, DNA strand breaks (99,104), appears to be due to a direct interaction between the HNP and the DNA.

Because of these cytotoxic effects, especially those to PMN (89,101), and because neutrophils are known to release granule constituents when they are activated by materials (56,105–108), we considered that HNP could be a candidate substance causally related to the biomaterial-associated down-regulation of fresh neutrophils (62). To explore this hypothesis, the effect of crude and purified HNP on the phagocytic killing of neutrophils was evaluated. These studies will demonstrate that there is a dose-related cytotoxicity involved in the process, and this cytotoxicity may be inhibited by specific antibody to HNP.

A. Methods

1. *Cells and Reagents*

Blood was obtained and neutrophils were isolated as described in the previous section. Stock solutions of the substances which activate neutrophils (fMLP), phorbol myristate acetate (PMA), and opsonized zymosan (OZ) were prepared as previously described (59). Semipurified defensins were prepared from granule extracts which had been separated into low molecular weight (<10,000 kD) and larger molecular weight (>10,000 kD) using a filter. Purified human neutrophil defensins (HNP1 and HNP2) were obtained from Sigma Chemical Co., St. Louis, MO, and stored at −20°C until needed. Then the HNP were thawed and suspended in PBS such that the PMN could be incubated with 10–35 μg/mL HNP. Monoclonal antibody to HNP (D1-1 IgG_1, a neutralizing antibody) was a generous gift of T. Ganz, UCLA School of Medicine, Los Angeles, CA, and was prepared as a stock solution of 10,000 μg/mL (10 mg/mL). This was diluted in PBS to the desired concentration just before use. Antibody to VCAM-1, an irrelevant IgG, was a generous gift from T. Carlos, University of Pittsburgh, and was used as a control antibody. It was prepared as a stock solution of 3600 μg/mL (3.6 mg/mL) and was diluted in PBS just before use. All other materials were reagent grade and obtained from Sigma Chemical Co. unless otherwise specified.

2. *Superoxide Release*

These studies were performed essentially as described above, but after preincubation of the neutrophils with semipurified or purified human neutrophil defensins. The cells were then placed into triplicate wells of 24-well plates in KRPG and cytochrome c with and without fMLP, PMA, OZ, or *S. aureus*. These culture dishes were incubated at 37°C for 30 min, following which the reaction was terminated by placing the culture dishes on melting ice and transferring the supernatants to 96-well plates for reading and quantitating of the reduced cytochrome c. As before, the data are expressed as nanomoles $O_2^-/10^6$ cells.

3. *Microbial Killing*

The effects of material association and of monoclonal antibody to HNP and of HNP1 and HNP2 on microbial killing were determined by clonal culture techniques as described in the previous section. The effect of incubating suspended neutrophils incubated with HNP1 and HNP2 on the ability of the PMN to kill staphylococci also was determined. For this study, PMN in KRPG (1×10^6/mL) in 12×75 mm test tubes were incubated with concentrations of HNP1 and HNP2 ranging from 12.5 to 35 μg/mL at 37°C for 1 h, the time demonstrated to be needed for HNP to exert irreversible cytotoxicity for mammalian eukaryotic cells (89,99). These PMN were then washed and transferred to 25-mL Erlenmeyer flasks suspended in KRPG plus 10% plasma, and *S. aureus* was added for a ratio of 10 staph:1 neutrophil. The flasks were incubated in a rotary shaking water bath. A 10-μL sample was removed immediately after the addition of bacteria, the T-0 sample. Samples of 10 μL were removed at 2 and 24 h, and viability of the staphylococci was determined.

The effect on the killing of *S. aureus* by fresh neutrophils added to material-associated neutrophils was also determined (condition 1). The initial incubation was for 1 h at 37°C. The second inoculum was added and the incubation was continued for another hour. To remove these PMN to a fresh well, KRPG containing EDTA was added and the cells were transferred to a tube and washed with KRPG before placing them in a clean culture well. PMN which were stored in a test tube for 1 h at 37°C were also evaluated for the aging phenomenon and then added in complete buffer containing 10% plasma. Viability of the *S. aureus* was determined immediately after their addition as well as after 2 and 24 h (condition 2).

The effect of antibody to HNP (D1-1) on HNP was determined using concentrations from 2 to 200 μg/mL. Antibody to VCAM was at 200 μg/mL. These antibodies were added to the first inoculum of PMN just prior to adding the second inoculum, and the incubation was continued for 1 h more at 37°C. These PMN were washed and transferred to a fresh well, as described above.

4. *Defensin Assay*

Quantitation of PMN defensin content in cells which had been exposed to materials for various time periods was determined using Panyutich's "sandwich"-type enzyme immunoassay method which detects HNP with a range of 0.05 to 10.0 ng/mL (109). In this assay, a monoclonal antibody is used for capture and a biotinylated monoclonal antibody for detection, and cetyltrimethyl ammonium bromide was used to counteract nonspecific binding of HNP to surfaces.

5. *Statistical Evaluation*

Determinations of statistical significance were calculated using Student's *t*-test or the paired *t*-test. Significance was defined as $p \leq 0.05$.

B. Results and Discussion

It is clear that implanted materials can induce an inflammatory reaction and become associated with phagocytic cells which become activated to produce ROI and to release chemoattractant substances (63,66). Nevertheless, and despite this influx of neutrophils, infection at biomaterial surfaces remains problematic, and others have shown that foreign bodies with the greatest capacity to induce an inflammatory response are more easily infected than are materials which induce less of an inflammatory response (8,19). In this manner, the effect of an implanted material appears to be paradoxical since these phagocytic cells appear to be disabled as well as barely capable of resisting and clearing infection.

The effect on bacterial viability of adding fresh cells to material-associated cells (condition 1) is shown in Table 1, compared to staphylococcal viability when staph are added to the PMN without a prior period of PMN/material association. When bacteria are added to a second inoculum of PMN (cells which could have been damaged by releasants from the PMN initially added to the material [condition 1]), fewer of the bacteria are killed, resulting in a larger number of viable organisms at each of the time points when compared to the numbers of viable organisms in condition 2. In condition 1, (38.4 ± 2.8)% of the staphylococci were killed after 2 h and (44.8 ± 4.9)% were killed after 24 h, compared to (63.4 ± 2.3)% and (85.4 ± 2.8)% in

Table 1 Effect of Prior Biomaterial Exposure on Staphylococcal Killing[a] by Neutrophils

	Condition 1[b]	Condition 2[c]
2 h	38.4 ± 2.8	63.4 ± 2.3
24 h	44.8 ± 4.9	85.4 ± 2.8

[a]Data are percent of inoculum killed.
[b]With prior exposure.
[c]Without prior exposure.

Table 2 Effect of Contact with Materials on HNP Content of Exposed Neutrophils

Material exposure	Residual HNP/10^6 PMN
None	3,616 ± 1,098
PST 30 min	*1,313 ± 183
PST 120 min	*886 ± 227
WD 120 min	*453 ± 230
Silastic 120 min	*1,275 ± 215

*$p \le 0.01$ compared to absence of exposure.

condition 2 when PMN and staph were added together ($p \le 0.001$). These results strongly support preliminary data which suggested that PST association for 1 h downregulated microbial killing by the secondary inoculum, suggesting that an environment hostile to antibacterial host defense is created in the vicinity of an implanted material. These data further suggest that the resulting infectivity at biomaterial surfaces may be due, at least in part, to releasant-mediated autocoid and paracoid damage of the attached and bystander PMN themselves, and further suggest that these cytotoxic substances are released when PMN encounter a foreign material.

It has been demonstrated that material-associated PMN degranulate (55,56), and the cytotoxic potential for proteolytic granule constituents such as elastase are well known. Other factors related to cytotoxicity could include metabolic products derived from the production of ROI. In addition, however, PMN are rich in small cationic polypeptides (HNP) and we believe that HNP are an important factor for this downregulation since it has been abundantly demonstrated that HNP are cytotoxic toward both eukaryotic and prokaryotic cells (90,93–95,98,99,104,110). However, when we evaluated the supernatants of PST-associated PMN for the presence of HNP, we found <1000 ng. Based on published data, this concentration would be too low to be cytotoxic. This may be explainable by the strongly cationic nature of HNP resulting in their ready adherence to negatively charged surfaces such as plastic (109) and to cells (111). Another approach would be to determine the residual amount of HNP in control cells and from PST-associated PMN. These data are shown in Table 2. We found a marked reduction in the detectable HNP content of biomaterial-associated cells compared to control cells. A total of 3616 ± 1098 ng/mL/10^6 PMN was extractable from control cells compared to

Table 3 Effect of HNP1 on Staphylocidal Activity of Neutrophils

Concentration of HNP_1 (μg/mL)	Staphylococci killed (%)	
	2 h	24 h
35	31.5 ± 13.4[a,b]	40.2 ± 10.8[a,b]
25	65.3 ± 14.0	73.0 ± 8.0
10	77.7 ± 5.9	88.6 ± 2.2
0	91.2 ± 1.6	85.2 ± 2.5

[a]$p \le 0.01$ compared to 0 or 10 μg/mL.
[b]$p \le 0.03$ compared to 25 μg/mL.

Table 4 Effect of HNP2 on Staphylocidal Activity of Neutrophils

Concentration of HNP_2 (μg/mL)	Staphylococci killed (%)	
	2 h	24 h
25	$6.2 \pm 20.0^{a,b}$	$10.6 \pm 18.0^{a,b}$
15	$43.0 \pm 18.0^{b,c}$	$39.0 \pm 12.0^{b,c}$
10	81.0 ± 6.0	71.0 ± 7.0
0	89.0 ± 1.5	78.0 ± 3.4

[a] $p \leq 0.01$ compared to 10 μg/mL.
[b] $p \leq 0.01$ compared to 0 μg/mL.
[c] $p \leq 0.05$ compared to HNP2 at 10 μg/mL.

1313 ± 183 ng/mL extractable 30 min after PST exposure and 886 ± 277 after 2 h of exposure. The defensin content of woven Dacron and silastic-associated cells was 453 ± 230 and 1275 ± 215, respectively. These results support the notion that PMN will release their granular HNP when these cells are associated with materials such as PST, and further support the notion that the released HNP becomes associated with the PST. The addition of purified HNP1 and HNP2 to PST plates in the absence of any cells and analysis for residual HNP demonstrated that detectable HNP disappeared within 15 min with only 2% remaining in the supernatant, consistent with other studies (102) showing rapid association of HNP with plastic.

We then evaluated the effect of purified HNP on the ability of PMN to kill bacteria. For this study, the PMN were incubated with the HNP for 1 h at 37°C, washed, and resuspended together with *S. aureus*. These data showing the ability of these PMN to kill the *S. aureus* are summarized in Table 3 and demonstrate that 35 μg/mL of HNP1 impaired staphylococcal killing; (31 ± 13)% and (40.2 ± 10.8)% of the inoculum were killed after 2 h and 24 h of incubation, respectively ($p \leq 0.01$ compared with control or with HNP1 at 10 μg/mL; $p \leq 0.03$ compared with HNP1 at 25 μg/mL). HNP1 at 25 μg/mL also impaired staphylocidal activity but inhibition was more variable; HNP2 is a more potent inhibitor of staphylococcal killing by PMN, and its effect is summarized in Table 4. 25 μg/mL was almost completely inhibitory ($p \leq 0.01$ compared to control and to 10 μg/mL), and as little as 15 μg/mL also was inhibitory ($p \leq 0.01$ compared to control; $p \leq 0.05$ compared to HNP2 at 10 μg/mL) when killing was evaluated after 2 and 24 h of incubation. HNP2, therefore, was far more toxic to PMN than HNP1. Table 5 shows that HNP1 at 35 μg/mL also significantly impaired the ability of neutro-

Table 5 Effect of HNP1 on O_2^- Release by Neutrophils

HNP concentration (μg/mL)	O_2^- Release mM/10^6 PMN/30 min			
	Stimulant			
	PMA	fMLP	OZ	Unstimulated
Control/none	80.0 ± 5	12.0 ± 2.7	19.5 ± 3.1	7.5 ± 2.5
17.5	82	17	21.5	8.0
25.0	70	12	16.8	5.3
35.0	14.7 ± 10	1.1 ± 0.1	3.3 ± 1.9	1.2 ± 0.4

Table 6 Effect of Antibody to HNP on Material-Induced Inhibition of Staphylococcal Killing

	Staphylococci killed (%)	
	2 h	24 h
Control	53.4 ± 3.3	81.7 ± 3.3
Condition 1	38.0 ± 3.1	50.5 ± 7.9
Condition 1 + antibody	55.7 ± 6.3*	81.0 ± 8.9*
Condition 1 + irrelevant antibody	44.5 ± 3.5	57.0 ± 13.3

*$p = 0.00008$ compared to Condition 1.

phils to produce O_2^- following stimulation with any of the inducing agents. When similar studies were performed using HNP2, we demonstrated complete inhibition of O_2^- release using 20 μg/mL (data not shown). Thus, we also demonstrated that these HNP impair the ability of PMN to produce ROI. These data suggest a mechanism for the loss of ROI production by material-associated PMN after the period of initial stimulation. This cytotoxicity is consistent with the known effects of HNP on eukaryotic cells (92,96,98,100,104). Despite this cytotoxicity, PMN viability did not appear to be compromised after 5 h of incubation with HNP, as evaluated by trypan blue dye exclusion. After 24 h of association, viability assessed in this was 73%.

The validity of the hypothesis that HNP damage the function of PMN was tested by using a monoclonal antibody to HNP. This was added to PMN on polystyrene prior to the addition of a second inoculum of PMN. This study demonstrated abrogation of the inhibition of function inflicted by material-associated PMN on fresh PMN. Table 6 shows that monoclonal antibody to HNP at a concentration of 200 μg/mL reversed the cytotoxic effect at both the 2 h and 24 h time points ($p \leq 0.05$ at 2 h; $p = 0.0001$ at 24 h) when comparing staphylococcal viability in the absence of antibody. When less antibody was used (2 to 50 μg/mL), there was no effect (data not shown). Similarly, the use of 200 μg/mL of an irrelevant antibody (antibody to VCAM) at 2 and 24 h had no effect.

Defensin-mediated damage should not be construed as the only mechanism of material-associated impairment of phagocytic function. As stated above, neutrophils also release proinflammatory mediators such as oxidative free radicals, IL-8, and metabolites of arachidonic acid, including leukotrienes and prostaglandins, all of which could serve transiently to augment and potentially harm cell function. The finding that antibody to HNP but not irrelevant antibody reverses the inhibition, however, is strong evidence that HNP are a key component of what we observed. Thus, the neutrophils attracted to a material surface appear to become the very instruments of increased infectivity and at least under some conditions help to create this hostile and infection-prone environment.

ACKNOWLEDGMENTS

This work was supported by NIH grant R01-GM-41734. We gratefully acknowledge the assistance of Rebecca E. Pfeifer in the preparation of this chapter.

REFERENCES

1. Sugarman, B., and Young, E. J., 1989. Infections associated with prosthetic devices: magnitude of the problem. *Infect. Dis. Clin. N. Am.* 3:187–198.

2. Gristina, A. G., 1994. Implant failure and the immuno-incompetent fibro-inflammatory zone. *Clin. Ortho. Rel. Res.* 298:106–118.
3. Stickler, D. J., and McLean, R. J. C., 1995. Biomaterial associated infections: the scale of the problem. *Cells Mater.* 5:167–182.
4. Jansen, B., and Peters, G., 1993. Foreign body associated infection. *J. Antimicrob. Chemother.* 32: 69–75.
5. Dougherty, S. H., and Simmons, R. L., 1982. Infections in the bionic man: the pathobiology of infections in prosthetic devices. Part 1. *Curr. Prob. Surg.* 19:221–264.
6. Dougherty, S. H., and Simmons, R. L., 1982. Infections in the bionic man: the pathobiology of infections in prosthetic devices. Part 2. *Curr. Prob. Surg.* 19:265–319.
7. Dougherty, S. H., and Simmons, R. L., 1989. Endogenous factors contributing to prosthetic device infections. *Infect. Dis. Clin. N. Am.* 3:199–209.
8. Dougherty, S. H., 1988. Pathobiology of infection in prosthetic devices. *Rev. Infect. Dis.* 10:1102–1117.
9. Kunin, C. M., 1987. *Detection, Prevention and Management of Urinary Tract Infections*, 4th ed. Lea & Febiger, Philadelphia, PA.
10. Elek, S. D., and Conen, P. E., 1957. The virulence of *Staphylococcus pyogenes* for man: a study of the problems of wound infections. *Br. J. Exp. Pathol.* 38:573–586.
11. Zimmerli, W., Lew, P. D., and Waldvogel, F. A., 1984. Pathogenesis of foreign body infection. Evidence for a local granulocyte defect. *J. Clin. Invest.* 73:1191–1200.
12. Christensen, G. D., Baddour, L. M., Hasty, D. L., Lowrance, J. E. I., and Simpson, W. A., 1989. Microbial and foreign body factors in the pathogenesis of medical device infections. In: Bisno, A. L., and Waldvogel, F. A. (eds). *Infections Associated with Indwelling Medical Devices*, Am. Soc. Microbiol., Washington, DC.
13. Lew, D. P., 1989. Physiopathology of foreign body infections. *Eur. J. Cancer Clin. Oncol.* 25: 1379–1382.
14. Jansen, B., Peters, G., and Pulverer, G., 1988. Mechanisms and clinical relevance of bacterial adhesion to polymers. *J. Biomat. Appl.* 2:520–543.
15. Reid, G., Hawthorn, L. A., Eisen, A., and Beg, H. S., 1989. Adhesion of *Lactobacillus acidophilus*, *Escherichia coli*, and *Staphylococcus epidermidis* to polymer and urinary catheter surfaces. *Colloids Surfaces* 42:299–311.
16. Cheesborough, J. S., Elliott, T. S. J., and Finch, R. C., 1989. A morphological study of bacterial colonization of intravenous cannulae. *J. Med. Micro.* 39:149–157.
17. Barton, A. J., Sagers, R. D., and Pitt, W. G., 1996. Bacterial adhesion to orthopedic implant polymers. *J. Biomed. Mater. Res.* 30:403–410.
18. Wang, I. W., Anderson, J. M., Jacobs, M. R., and Marchant, R. E., 1995. Adhesion of *Staphylococcus epidermidis* to biomedical polymers: contributions of surface thermodynamics and hemodynamic shear conditions. *J. Biomed. Mater. Res.* 29:485–493.
19. Hunt, J. A., Flanagan, B. F., McLaughlin, P. J., Strickland, I., and Williams, D. F., 1996. Effect of biomaterial surface charge on the inflammatory response: evaluation of cellular infiltration and TNF alpha production. *J. Biomed. Mater. Res.* 31:139–144.
20. Yun, J. K., DeFife, K., Colton, E., Stack, S., Azeez, A., Cahalan, L., Verhoeven, M., Cahalan, P., and Anderson, J. M., 1995. Human monocyte/macrophage adhesion and cytokine production on surface-modified poly(tetrafluorethylene/hexafluoropropylene) polymers with and without protein preadsorption. *J. Biomed. Mater. Res.* 29:257–268.
21. Tang, L., Liu, L., and Elwing, H. B., 1998. Complement activation and inflammation triggered by model biomaterial surfaces. *J. Biomed. Mater. Res.* 41:333–340.
22. Gristina, A. G., Giridhar, G., Gabriel, B. L., Naylor, P. T., and Myrvik, Q. N., 1993. Cell biology and molecular mechanisms in artificial device infections. *Int. J. Artif. Organs* 16:755–764.
23. Chinn, J. A., Sauter, J. A., Phillips, R. E., Kao, W. J., and Anderson, J. A., 1998. Blood and tissue compatibility of modified polyester: thrombosis, inflammation, and healing. *J. Biomed. Mater. Res.* 39:130–140.
24. Schumacher-Perdreau, F., Peters, G., Digon-Gusman, F., Jansen, B., and Pulverer, G., 1988. Inci-

dence of coagulase-negative staphylococci in fibrous capsular contracture after augmentation mammoplasty with silicon. In: *Program and Abstracts of the 28th Interscience Conference on Antimicrobial Agents and Chemotherapy*; Los Angeles, Abstract #996, Am. Soc. Microbiol., Washington, DC.

25. Johnson, G. M., Lee, D. A., Regelmann, W. E., Gray, E. D., Peters, G., and Quie, P.G., 1986. Interference with granulocyte function by *Staphylococcus epidermidis* slime. *Infect. Immun.* 54: 13–20.
26. Johnson, G. M., Regelmann, W. W., Gray, E. D., Peters, G., and Quie, P. G., 1987. Staphylococcal slime and host defenses: effects on polymorphonuclear granulocytes. In: Pulverer, G., Quie, P. G., and Peters, G. (eds). *Clinical Significance and Pathogenicity of Coagulase-Negative Staphylococci*, Gustav Fischer Verlag, Stuttgart.
27. Higashi, J. M., Wang, I. W., Shales, D. M., Anderson, J. M., and Marchant, R. E., 1998. Adhesion of *Staphylococcus epidermidis* and transposon mutant strains to hydrophobic polyethylene. *J. Biomed. Mater. Res.* 39:341–350.
28. Wang, I. W., Anderson, J. M., Jacobs, M. R., and Marchant, R. E., 1995. Adhesion of *Staphylococcus epidermidis* to biomedical polymers: contributions of surface thermodynamics and hemodynamic shear conditions. *J. Biomed. Mater. Res.* 29:485–493.
29. Vaudaux, P., Waldvogel, F. A., Morganthaler, J. J., and Nydegger, U. E., 1984. Foreign body infection: role of fibronectin as a ligand for the adherence of *Staphylococcus aureus*. *J. Infect. Dis.* 150:546–553.
30. Falceri, E., Vaudaux, P., Hugggler, E., Lew, P. D., and Waldvogel, F. A., 1987. Role of bacterial exopolymers and host factors on adherence and phagocytosis of *Staphylococcus aureus* in foreign body infection. *J. Infect. Dis.* 155:524–531.
31. Hermann, M., Vaudaux, P. E., Pittet, D., Auckenthaler, R., Lew, P. D., Schumacher-Perdreau, F., Peters, G., and Waldvogel, F. A., 1988. Fibronectin, fibrinogen and laminin act as mediators of adherence of clinical staphylococcal isolates to foreign material. *J. Infect. Dis.* 158:693–701.
32. Espersen, F., Wilkinson, B. J., Gahrn-Hansen, B., Rosdahl, V. T., and Clemmensen, I., 1990. Attachment of staphylococci to silicone catheters in vitro. *Acta Pathol. Microbiol. Immunol. Scand.* 98:471–478.
33. Foster, T. J., and McDevitt, D., 1994. Molecular basis of adherence of staphylococci to biomaterials. In: Bisno, A. L., and Waldvogel, F. A. (eds). *Infections Associated with Indwelling Medical Devices*, 2nd ed., Am. Soc. Microbiol., Washington, DC.
34. Vaudaux, P., Suzuki, R., Waldvogel, F. A., Morgenthaler, J. J., and Nydegger, U. E., 1984. Foreign body infection: role of fibronectin as a ligand for the adherence of *Staphylococcus aureus*. *J. Infect. Dis.* 150:546–553.
35. Vroman, L., Adams, A. L., Fischer, G. C., and Munoz, P. C., 1980. Interaction of high molecular weight kininogen, factor XII, and fibrinogen in plasma at interfaces. *Blood* 55:156–159.
36. Vaudaux, P., Pittet, D., Haeberli, A., Huggler, E., Nydegger, U. E., Lew, D. P., and Waldvogel, F. A., 1989. Host factors selectively increase staphylococcal adherence on inserted catheters: a role for fibronectin and fibrinogen or fibrin. *J. Infect. Dis.* 160:865–875.
37. Vaudaux, P., Pittet, D., Haeberli, A., Lerch, P. G., Morganthaler, J. J., Proctor, R. A., Waldvogel, F. A., and Lew, D. P., 1993. Fibronectin is more active than fibrin or fibrinogen in promoting *Staphylococcus aureus* adherence to inserted intravascular catheters. *J. Infect. Dis.* 167:633–641.
38. Herrmann, M., Suchard, S. J., Boxer, L. A., Waldvogel, F. A., and Lew, D. P., 1991. Thrombospondin binds to *Staphylococcus aureus* and promotes staphylococcal adherence to surfaces. *Infect. Immun.* 59:278–288.
39. Francois, P., Vaudaux, P., and Lew, P. D., 1998. Role of plasma and extracellular matrix proteins in the physiopathology of foreign body infections. *Ann. Vasc. Surg.* 12:34–40.
40. McDevitt, D., Francois, P., Vaudaux, P., and Foster, T. J., 1994. Molecular characterization of the clumping factor (fibrinogen receptor) of *Staphylococcus aureus*. *Mol. Microbiol.* 11:237–248.

41. Vaudaux, P. E., Francois, P., Proctor, R. A., McDevitt, D., Foster, T. J., Albrecht, R. M., Lew, D. P., Wabers, H., and Cooper, S. L., 1995. Use of adhesion-defective mutants of *Staphylococcus aureus* to define the role of specific plasma proteins in promoting bacterial adhesion to canine arteriovenous shunts. *Infect. Immun.* 63:585–590.
42. Tang, L., and Eaton, J. W., 1993. Fibrin(ogen) mediates acute inflammatory responses to biomaterials. *J. Exp. Med.* 178:2147–2156.
43. Tang, L., and Eaton, J. W., 1995. Inflammatory responses to biomaterials. *Am. J. Clin. Pathol.* 103:466–471.
44. Tang, L., Ugarova, T. P., Plow, E. F., and Eaton, J. W., 1996. Molecular determinants of acute inflammatory responses to biomaterials. *J. Clin. Invest.* 97:1329–1344.
45. Kao, W. J., Sapatnekar, S., Hiltner, A., and Anderson, J. M., 1996. Complement-mediated leukocyte adhesion of poly(etherurethane ureas) under shear stress in vitro. *J. Biomed. Mater. Res.* 32: 99–109.
46. Anderson, J. M., Ziats, N. P., Azeez, A., Brunstedt, M. R., Stack, S., and Bonfield, T. L., 1995. Protein adsorption and macrophage activation of polydimethylsiloxane and silicone rubber. *J. Biomat. Sci. Polymer Ed.* 7:159–169.
47. Cardona, M. A., Simmons, R. L., and Kaplan, S. S., 1992. TNF and IL-1 generation by human monocytes in response to biomaterials. *J. Biomed. Mater. Res.* 26:851–859.
48. Bonfield, T. L., Colton, E., Marchant, R. E., and Anderson, J. M., 1992. Cytokine and growth factor production by monocytes/macrophages on protein preadsorbed polymers. *J. Biomed. Mater. Res.* 26:837–850.
49. Bonfield, T. L., and Anderson, J. M., 1993. Functional versus quantitative comparison of IL-1β from monocytes/macrophages on biomedical polymers. *J. Biomed. Mater. Res.* 27:1195–1199.
50. Jahns, G., Haeffner-Cavaillon, N., Nydegger, U. E., Kazatchkine, M. D., 1993. Complement activation and cytokine production as consequences of immunological biocompatibility of extracorporeal circuits. *Clin. Mater.* 14:303–336.
51. Hakim, R. M., 1993. Complement activation by biomaterials. *Cardiovasc. Pathol.* 2:187S–197S.
52. Marosok, R., Washburn, R., Indorf, A., Soloman, D., and Sheretz, R., 1996. Contribution of vascular catheter material to the pathogenesis of infection: depletion of complement by silicone elastomer in vitro. *J. Biomed. Mater. Res.* 30:245–250.
53. Wade, B. H., and Mandell, G. L., 1983. Polymorphonuclear leukocytes: dedicated professional phagocytes. *Am. J. Med.* 74:686–693.
54. Lehrer, R. L., Lichtenstein, A. K., and Ganz, T., 1993. Defensins—antimicrobial and cytotoxic peptides of mammalian cells. *Annu. Rev. Immunol.* 11:105–128.
55. Klock, J. C., and Bainton, D. F., 1976. Degranulation and abnormal bactericidal function of granulocytes procured by reversible adhesion to nylon wool. *Blood* 48:149–161.
56. Wright, D. G., and Gallin, J. I., 1979. Secretory responses of human neutrophils: exocytosis of specific (secondary) granules by human neutrophils during adherence in vitro and during exudation in vivo. *J. Immunol.* 123:285–294.
57. Zimmerli, W., Waldvogel, F. A., Vandaux, P., and Nydegger, V. E., 1982. Pathogenesis of foreign body infection: description and characteristics of an animal model. *J. Infect. Dis.* 146:487–497.
58. Marchant, R. E., Miller, K. M., and Anderson, J. M., 1984. In vivo biocompatibility studies. V. In vitro leukocyte interactions with biomer. *J. Biomed. Mater. Res.* 16:1169–1190.
59. Kaplan, S. S., Basford, R. E., Mora, E., Jeong, M. H., and Simmons, R. L., 1992. Biomaterial-induced alterations of neutrophil superoxide production. *J. Biomed. Mater. Res.* 26:1039–1051.
60. Parsson, H., Nassberger, L., Thorne, J., and Norgren, L., 1995. Metabolic response of granulocytes and platelets to synthetic vascular grafts: preliminary results with an in vitro technique. *J. Biomed. Mater. Res.* 29:519–525.
61. Kaplan, S. S., Basford, R. E., Jeong, M. H., and Simmons, R. L., 1994. Mechanisms of biomaterial-induced superoxide release by neutrophils. *J. Biomed. Mater. Res.* 28:377–386.
62. Kaplan, S. S., Basford, R. E., Jeong, M. H., and Simmons, R. L., 1996. Biomaterial-neutrophil

interactions: dysregulation of oxidative functions of fresh neutrophils induced by prior neutrophil-biomaterial interaction. *J. Biomed. Mater. Res.* 30:67–75.
63. Liu, L., Elwing, H., Karlsson, A., Nimeri, G., and Dahlgren, C., 1997. Surface-related triggering of the neutrophil respiratory burst. Characterization of the response induced by IgG adsorbed to hydrophilic and hydrophobic glass surfaces. *Clin. Exp. Immunol.* 109:204–210.
64. Nathan, C., 1987. Neutrophil activation on biological surfaces. Massive secretion of hydrogen peroxide in response to products of macrophages and lymphocytes. *J. Clin. Invest.* 80:1550–1560.
65. Karlsson, C., Nygren, H., and Braide, M., 1996. Exposure of blood to biomaterial surfaces liberates substances that activate polymorphonuclear granulocytes. *J. Lab. Clin. Med.* 128:496–505.
66. Boyum, A., 1968. Isolation of mononuclear cells and granulocytes from human blood. *Scand. J. Clin. Lab. Invest.* 21:77–89.
67. Kuhns, D. B., Kaplan, S. S., and Basford, R. E., 1986. Hexachlorocyclohexanes, potent stimuli of O_2^- production and calcium release in human polymorphonuclear leukocytes. *Blood* 68:535–540.
68. Kaplan, S. S., Caliguiri, L. A., Basford, R. E., and Zdziarski, U. E., 1987. Transient chemotactic defect in a child with elevated IgE. *Ann. Allergy* 59:213–217.
69. Kaplan, S. S., and Nardi, M., 1977. Impairment of leukocyte function during sickle cell crisis. *J. Reticuloendothel. Soc.* 22:499–506.
70. Sklar, L. A., Hyslop, P. A., Oades, Z. G., Omann, G. M., Jesaites, A. J., Painter, R. G., and Cochrane, C. G., 1985. Signal transduction and ligand-receptor dynamics in the human neutrophil. *J. Biol. Chem.* 260:11461–11467.
71. Snyderman, R., and Uhing, R. J., 1988. Phagocytic cells: stimulus-response coupling mechanisms. In: Gallin, J. I., Goldstein, I. M., and Snyderman, R. (eds). *Inflammation: Basic Principles and Clinical Correlates*, Raven Press, New York.
72. Graziano, M. P., and Gilman, A. G., 1987. Guanine nucleotide-binding proteins: mediators of transmembrane signaling. *Trends Pharmacol. Sci.* 8:478–481.
73. Snyderman, R., Smith, C. D., and Verghese, M. W., 1986. Model for leukocyte regulation by chemoattractant receptors: roles of a guanine nucleotide regulatory protein and polyphosphoinositide metabolism. *J. Leukoc. Biol.* 40:785–800.
74. Lloyd, A. R., and Oppenheim, J. J., 1992. Poly's lament: the neglected role of the polymorphonuclear neutrophil in the afferent limb of the immune response. *Immunol. Today* 13:169–172.
75. Deju, J. Y., Serbousek, D., and Blanchard, D. K., 1990. Release of tumor necrosis factor by human polymorphonuclear leukocytes. *Blood* 76:1405–1409.
76. Cicco, N. A., Lindermann, A., Content, J., Vandenbusske, P., Lubbert, M., Gauss, J., Mertelsmann, R., and Herrmann, F., 1990. Inducible production of interleukin 6 by human polymorphonuclear neutrophils. Role of granulocyte-macrophage colony-stimulating factor and tumor necrosis factor-alpha. *Blood* 75:2049–2052.
77. Baggiolini, M., Walz, A., and Kunkel, S. L., 1989. Neutrophil activating peptide interleukin-8, a novel cytokine that activates neutrophils. *J. Clin. Invest.* 84:1045–1049.
78. Oppenheim, J. J., Zachariae, C. O. C., Mukanda, N., and Matsushina, K., 1991. Properties of the novel proinflammatory supergene "intercrine" cytokine family. *Annu. Rev. Immunol.* 9:617–648.
79. Kasama, T., Strieter, R. M., Lukacs, N. W., Burdick, M. D., and Kunkel, S. L., 1994. Regulation of neutrophil-derived chemokine expression by IL-10. *J. Immunol.* 152:3559–3569.
80. Wang, P., Wu, P., Anthes, J. C., Siegel, M. L., Egan, R. W., and Billah, M. M., 1994. Interleukin-10 inhibits interleukin-8 production in human neutrophils. *Blood* 83:2678–2683.
81. DeForge, L. E., Preston, A. M., Takeuchi, E., Kenney, J., Boxer, L. A., and Remick, D. G., 1993. Regulation of interleukin-8 gene expression by oxidant stress. *J. Biol. Chem.* 268:25568–25576.
82. Strieter, R. M., Kasahara, K., Allen, R. M., Standiford, T. J., Rolfe, M. W., Becker, F. S., Chensue, S. W., and Kunkel, S. L., 1992. Cytokine-induced neutrophil-derived interleukin-8. *Am. J. Pathol.* 141:397–407.
83. Baggiolini, M., and Clark-Lewis, I., 1992. Interleukin-8, a chemotactic and inflammatory cytokine. *FEBS Lett.* 307:97–101.
84. Gimbrone, M. A. Jr., Obin, M. S., Brock, A. F., Luis, E. A., Hass, P. E., Hebert, C. A., Yip, Y. K., Leung, D. W., Lowe, D. G., Kohr, W. J., Darbone, W. C., Bechtol, K. B., and Baker, J. B.,

1989. Endothelial interleukin-8: a novel inhibitor of leukocyte-endothelial interactions. *Science* 246:1601–1603.

85. Ley, K., Baker, J. B., Cybuisky, M. I., Gimbrone, M. A. Jr., and Luscinskas, F. W., 1993. Interleukin-8 inhibits granulocyte emigration from rabbit mesenteric venules without altering L-selectin expression or leukocyte rolling. *J. Immunol.* 151:6347–6357.
86. Greco, R. S., 1997. Body parts: *In vivo veritas*. *J. Biomed. Mater. Res.* 34:409–410.
87. Katz, D. A., and Greco, R. S., 1994. The pathobiology of infections associated with biomaterials. *Prob. Gen. Surg.* April/June:209–226.
88. Lehrer, R. I., and Ganz, T., 1990. Antimicrobial polypeptides of human neutrophils. *Blood* 76: 2169–2181.
89. Lehrer, R. I., Barton, A., Dahrer, K. A., Harwig, S. S. L., Ganz, T., and Selsted, M. E., 1989. Interaction of human defensins with *Escherichia coli*: mechanism of bactericidal activity. *J. Clin. Invest.* 84:553–561.
90. Ganz, T., 1987. Extracellular release of antimicrobial defensins by human polymorphonuclear leukocytes. *Infect. Immun.* 55:568–571.
91. Ganz, T., Selsted, M. E., Szklarek, D., Harwig, S. S., Daher, K., Bainton, D. F., and Lehrer, R. I., 1985. Defensins: natural peptide antibiotics of human neutrophils. *J. Clin. Invest.* 76:1427–1435.
92. Kagan, B. L., Ganz, T., Lehrer, R. I., 1994. Defensins: a family of antimicrobial and cytotoxic peptides. *Toxicology* 87:131–149.
93. Daher, K. A., Selsted, M. E., and Lehrer, R. I., 1986. Direct inactivation of viruses by human granulocyte defensins. *J. Virol.* 60:1068–1074.
94. Lehrer, R. I., Daher, K., Ganz, T., and Selsted, M. E., 1985. Direct inactivation of viruses by MCP-1 and MCP-2, natural peptide antibiotics from rabbit leukocytes. *J. Virol.* 54:467–472.
95. Selsted, M. E., Szklarek, D., Ganz, T., and Lehrer, R. I., 1985. Activity of rabbit leukocyte peptides against *Candida albicans*. *Infect. Immun.* 49:202–206.
96. Kagan, B. L., Selsted, M. E., Ganz, T., and Lehrer, R. I., 1990. Neutrophil antimicrobial peptides (defensins) form voltage-dependent ionic channels in planar lipid bilayer membranes. *Proc. Natl. Acad. Sci. USA* 87:210–214.
97. Lichtenstein, A., 1991. Mechanism of mammalian cell lysis medicated by peptide defensins. *J. Clin. Invest.* 88:93–100.
98. Lichtenstein, A., Ganz, T., Selsted, M. E., and Lehrer, R. I., 1986. *In vitro* tumor cell cytolysis mediated by peptide defensins of human and rabbit granulocytes. *Blood* 68:1407–1410.
99. Lichtenstein, A. K., Ganz, T., Nguyen, T. M., Selsted, M. E., and Lehrer, R. I., 1988. Mechanism of target cytolysis by peptide defensins. Target cell metabolic activities, possibly involving endocytosis, are crucial for expression of cytotoxicity. *J. Immunol.* 140:2686–2694.
100. Yomogida, S., Nagaoka, I., Saito, K., and Yamashita, T., 1996. Evaluation of the effects of defensins on neutrophil functions. *Inflam. Res.* 45:62–67.
101. Tal, T., and Aviram, I., 1993. Defensin interferes with the activation of neutrophil NADPH oxides in a cell-free system. *Biochem. Biophys. Res. Commun.* 196:636–641.
102. Hristova, K., Selsted, M. E., and White, S. H., 1997. Critical role of lipid composition in membrane permeabilization by rabbit neutrophil defensins. *J. Biol. Chem.* 272:24224–24333.
103. Lichtenstein, A. K., Ganz, T., Selsted, M. E., and Lehrer, R. I., 1988. Synergistic cytolysis mediated by hydrogen peroxide combined with peptide defensins. *Cell. Immunol.* 114:104–116.
104. Gera, J. R., and Lichtenstein, A., 1991. Human neutrophil peptide defensins induce single strand DNA breaks in target cells. *Cell. Immunol.* 138:108–120.
105. Borregaard, N., Lollike, K., Kyeldsen, L., Sengelov, H., Bastholm, L., Nielsen, M. H., and Bainton, D. F., 1993. Human neutrophil granules and secretory vesicles. *Eur. J. Haematol.* 51:187–198.
106. Sengelov, H., Follin, P., Kjeldsen, L., Lollike, K., Dahlgren, C., and Borregaard, N., 1995. Mobilization of granules and secretory vesicles during *in vivo* exudation of human neutrophils. *J. Immunol.* 154:4157–4165.
107. Sengelov, H., Kjeldsen, L., and Borregaard, N., 1993. Control of exocytosis in early neutrophil activation. *J. Immunol.* 150:1535–1543.

108. Wright, D. G., Bralove, D. A., and Gallin, J. I., 1977. The differential mobilization of human neutrophil granules. Effects of phorbol myristate acetate and ionophore A23187. *Am. J. Pathol.* 87:273–284.
109. Panyutich, A. V., Voitenok, N. N., Lehrer, R. I., and Ganz, T., 1991. An enzyme immunoassay for human defensins. *J. Immunol. Meth.* 141:149–155.
110. Sheu, M. J., Baldwin, W. W., Brunson, K. W., 1985. Cytotoxicity of rabbit macrophage peptides MCP-1 and MCP-2 for mouse tumor cells. *Antimicrob. Agents Chemother.* 28:626–629.
111. Ganz, T., and Lehrer, R. I., 1995. Defensins. *Pharmacol. Therapeut.* 66:191–205.

4

Biodegradable Biomedical Polymers

Review of Degradation of and In Vivo Responses to Polylactides and Polyhydroxyalkanoates

V. Hasirci
Middle East Technical University, Ankara, Turkey

I. INTRODUCTION

Polymers, regardless of their chemical makeup, degrade under appropriate conditions through breakdown of covalently linked polymer backbone. This is true for all the polymers, whether produced in the laboratory or derived from natural materials. The title nondegradable is, however, still used to indicate polymers that do not degrade during use or for a very long time after use (Gopferich, 1996).

The human body is a very hostile environment for a foreign material, especially for a polymeric one. For biomedical polymers stability is desirable in most clinical applications. There are, however, instances where degradation might be the preferred property. Under those circumstances the control of the degradability of the biomaterial becomes critical for the completion of the assigned function. For example, with material designed for fracture fixation the ideal rate of resorption should not exceed the rate of bone formation, and the reduction in the strength of the implant should closely match the increase in tissue strength (Fini et al., 1995). Otherwise the stresses could be transferred to the healing bone, which obviously would be detrimental for bone healing.

According to Williams (1997) there are three general areas of application for biodegradable polymers in medical devices:

1. Where an implantable device is required to perform a function that is transient and where elimination from the body is essential after the function is no longer necessary
2. When the device cannot function as implanted but can only function during and as a result of the degradation process
3. As biodegradable packaging materials

When biodegradation rates are compared, what basically differs between different polymers is (1) the energy required to break the bond and (2) the location of the bond. Polymers with strong bonds in the backbone and no easily hydrolyzable groups require long times and/or activators or catalysts of some sort to initiate the degradation process. This initiating factor could be heat, electromagnetic radiation like visible light, UV or gamma, chemicals like water, oxygen, ozone and halogenated compounds or any combination of the above. The molecules with such hydrolyzable groups are degraded much more efficiently and rapidly.

The location of the break is the other difference in various degradation routes. Polymers can degrade through the breakage of the final unit on the chain, (unzipping), or through scission of a bond along the length of the polymer backbone, (random scission). Breakage along the chain is random unless cleavage of bond is decided by the enzyme present (which might prefer certain bonds, like certain DNAses do). When it is a result of enzymatic action, the end unit loss can also take place to some extent by reversal of the polymerization process, like distilling off of methylmethacrylate (MMA) upon heating of polymethylmethacrylate (PMMA). From the thermodynamic point of view the reason can be stated as follows. In the equation $\Delta G = \Delta H - T\Delta S$, where ΔG, ΔH, T and ΔS are Gibbs free energy, enthalpy change of reaction, temperature and entropy change, respectively, ΔS is negative for the polymerization process, and with the negative sign in front it contributes a positive value of free energy, rendering it less spontaneous. Enthalpy change is negative (due to energy release during bond formation) for polymerization. Thus, at low temperatures polymerization is favored, but as the temperature is increased polymerization becomes unfavorable and eventually is reversed. When the electronic environments near the reactive sites are different, chain end scission was also reported for polycondensation polymers like cellulose and polyglycine (Shih, 1995).

Backbone breakage is encouraged as the penetration capability of the solvent into the polymeric form is increased. This can be rephrased as "as the hydrophilicity of the polymer is increased so is its biodegradability," because the solvent in the biological media is basically water with quite a high salt content.

Chain scission might not be without side reactions. According to Gogolewski and Mainil-Varlet (1996), upon thermal treatment, polyhydroxyacids can undergo chain scission at the ester bond followed by new bond formation or transesterification. This would cause the creation of molecules with higher chain lengths than the starting materials.

Biodegradability is an asset when the polymeric product is intended for temporary use and removal involves risk of infection, pain and cost. The mode and the extent of degradation for a polymer under a set of conditions have to be known to determine the suitability of the material for a given application. Some researchers even define the range of suitability as 200,000 daltons, below which mechanical properties of the polymer become unacceptable for most purposes (Yasin et al., 1990).

The main parameters which influence the polymer biodegradability are polymer crystallinity, hydrophilicity, composition and form of the product. These will be briefly discussed before examining polyhydroxybutyric acid and its copolymers and polylactide-*co*-glycolides as the potential and the current polymers, respectively, of use in the biodegradable polymer market.

A. Polymer Crystallinity

Polymer crystallinity is a measure of the alignment of polymeric chains along each other. Thus, the polymer which has the highest amount of aligned chains, and the least void or free volume is the most crystalline because it would be the one in whose structure a unit cell or pattern would be repeated along the three axes. The presence of bulky side groups and or branches and of freely mobile atoms (like oxygen in the backbone bonds) adversely influence the alignment of neighboring chains and thus crystallinity. The greater the free volume the higher is the chance of penetration of solvent molecules between the polymer chains and to initiate hydrolysis of the chains.

B. Hydrophilicity

Hydrophilicity is a very important property especially when the major route of polymer degradation is hydrolysis. For the hydrolytic action of the water molecules and accelerators (or catalysts) like acids to be effective, these molecules need to have access to the targeted bonds. This is facilitated if the polymeric structure absorbs water due to hydrophilicity of the backbone or the pendent groups.

C. Composition

The influence of composition on degradation can be observed at two levels: polymer backbone type (C—C backbone or heteroaromatic group containing backbone) and side group type. A typical addition polymer can be constructed to carry various side groups on the same backbone. Various condensation polymers can be prepared with both different backbones and side groups. Due to variations in charge distribution and thus in degree of interaction, as well as due to crystallinity and hydrophilicity changes resulting from these degradation rates of condensation polymers vary.

D. Product Form

Product form is not normally expected to influence the degradation rate of a polymeric product. Initial expectation as influence of form on degradation is that larger surface-to-volume ratios lead to higher degradation rates of hydrophobic polymers due to higher exposure to the lytic molecules, namely water molecules. Thus, thin sheets, disks, small microspheres, thin cylinders, microcapsules or any macro- or microporous structure can be counted among the forms more susceptible to hydrolysis.

II. DEGRADATION OF POLYLACTIDE-GLYCOLIDES AND POLYHYDROXYALKANOATES

In the biomedical field there are two families of polyesters which with their mechanical strengths, pliability, fiber formation capability and above all by their biodegradability have special importance. The rate and mode of degradation influence their service life, mechanical properties during use, and the response of the biological system toward them. For example, a rapid degradation is bound to produce large amounts of either oligomers or monomers, and this is determined by the mode of degradation. The response of the biological system against these, on the other hand, is not the same. Also the mode of degradation decides whether there is a substantial drop in the mechanical properties, determining the suitability of the polymers. It is, therefore, very important to correctly determine the mode of degradation.

A. Detection, Quantification and Interpretation of Degradation Data

Interpretation of degradation mode and rate is very much dependent on the measured indicators of degradation. Among the most commonly followed parameters are mechanical property changes (i.e., tensile or bending strength loss), weight loss, molecular weight change, surface porosity (as judged by reflectance from the surface), surface hydrophilicity (judged by contact angle, IR, ESCA, etc), monomer analog formation, oligomer formation, changes in the medium pH, etc. The sensitivities of these methods differ widely. Their implications do also differ. Therefore, several of these methods should be used simultaneously to study the mode.

How these methods compare to each other will be given below with examples in which PHBV and PLGA were used. In a study, weight loss was deemed to be the least sensitive parameter measured to yield information about the degradation of PHBV20 (in pH 7.4 buffer at 37°C) (Holland et al., 1990). This property was found to remain unchanged for about 400 days, while gloss factor and the polar component of the surface energy were, on the other hand, the most sensitive and started changing immediately after the initiation of the degradation test. Gloss factor showed a decrease while the polar component of surface energy showed an increase, which were interpreted as increased surface roughness and hydrophilicity, respectively. Thus, even though there were changes on the polymer, they were not significant enough for weight loss measurements, while they were for gloss factor and surface energy. Molecular weight, the property which should yield the most information on the degradation mode, showed perceptible changes all throughout the same study, initially being quite insignificant. The molecular weight change could not be followed on a weekly basis as a consequence of insufficient sensitivity because there are no direct measurement systems sensitive enough to detect monomer or oligomer losses of the remaining chain. Thus, one has to wait (weeks) until a change that can be measured takes place. As mentioned earlier it is possible to piece together a degradation pattern using various techniques simultaneously.

It should also be kept in mind during the interpretations of molecular weight data that some authors assume that the number average molecular weight, M_n, values obtained with GPC are less reliable than weight average molecular weight, M_w, due to the errors in accommodating low molecular weight species in calculations (Yasin et al., 1990).

Another important conclusion is that the comparison of biodegradation data for materials prepared by different research groups is difficult unless detailed characterization of the processed samples is carried out and documented. This is because processing conditions and the initial properties of the materials are very much influential on the degradation pattern of the polymers (Yasin et al., 1990). A typical example is the following. Working with a sample of PHBV20 which had a bimodal molecular weight distribution (with a high molecular weight tail) definitely leads to a degradation product distribution different than the one with a more homogeneous starting material. This obviously would add to the confusion in the field unless known before hand.

One final but very important point that has to be taken into consideration is that during the design of in vitro degradation tests the sterility of the degradation media has to be maintained for the duration of the tests, which could be as long as 600 days to avoid any discrepancy in the results (Pouton and Akhtar, 1996) simply because any growth in the medium would alter the result or the data.

The mode of degradation and the response of the biological system against two very commonly used biodegradable structures, poly(lactide-*co*-glycolide) and poly(hydroxybutyrate-*co*-valerate) will now be presented and compared.

III. POLYLACTIDE AND POLYLACTIDE-*co*-GLYCOLIDES

Lactides and glycolides are polymerized to form a family of polyesters called polylactide (PLA) if synthesized of only lactide and polylactide-*co*-glycolide (PLGA), if it is a copolymer. The homopolymer can be an optically active stereoisomer like poly-L-lactide (PLLA) or a racemic mixture of the D and L forms, represented simply by PLA. Polylactide degradation takes a long time especially when it is an optically active stereoisomer of the polymer. According to Bergsma et al. (1995), as polymerized PLLA takes more than 5.6 years for total resorption in the biological system, and the degradation rate becomes slower as the molecular weight becomes higher (Gogolewski et al., 1993).

As for the mode of degradation, a large number of studies point to a random scission route for PLGA degradation. In support of that, Cha and Pitt (1990) state that in vivo and in vitro degradation of PLLA and PLGA occur at the same rate, suggesting no significant contribution by enzymes. They report that the biodegradation of these polyesters proceeds by random hydrolytic chain scission of ester links until the molecular weight has decreased to the point that fragments are small enough to diffuse from the polymer bulk; weight loss then ensues. The significant M_n decrease in solvent cast, compression molded films observed with GPC upon storage in pH 7.4 phosphate buffer which would not have been observed with unzipping is another finding that supports this view (Pitt et al., 1979).

Gopferich (1996) stated that complete degradation of poly(L-lactic acid) takes longer than the loss of tensile strength. This implies that random scission is the main route of degradation because substantial decreases in the molecular weight upon degradation are not possible by removal of monomers from the chain ends, leaving the polymer quite the same size as before; for the same reason, substantial changes in mechanical properties are observed only upon random scission.

The results of Cha and Pitt (1990) indicate a random scission of PLLA (0.6-mm-thick disk) in pH 7.4 buffer as the drop in molecular weight is gradual, starting from day one. The weight loss is not a sudden occurrence initiated on day 20 but rather a gradual one. This supports various data reported in the literature (Li et al., 1990a, 1990b; Grizzi et al., 1993) except that the onset of degradation is much earlier here, possibly due to initial differences in heterogeneity index (HI = $\overline{M_w}/\overline{M_n}$) and thermal and processing history.

Li et al. (1990a), using the same type of polymer, poly(D,L-lactic acid), with Shih (1995), investigated the degradation mode in isotonic saline and pH 7.4 phosphate buffer. They observed the first traces of lactic acid on the fifth week, which also coincided with pH drop, osmolarity increase and weight loss. Size exclusion chromatography (SEC) showed both a uniform peak shift implying a random scission on the seventh week but also a bimodal SEC appearance of the polymers on the surface, implying the formation of very low molecular weight fragments (oligomer) or monomer formation. The rapid, autocatalyzed degradation on the inside appears as a single, distinctly lower molecular weight peak. The continuous increase in water absorption starting from day 1 implies, however, more chain-end presence in the polymeric specimen. Assuming that lactic acid would leach out of the specimen during unzipping while oligomers of random scission would remain behind, water absorption is an indicator of random scission. It thus appears that with sufficiently thick samples both degradation mechanisms take place simultaneously with random scission being dominant in the earlier stages (and especially at the core of the specimen due to autocatalysis) and unzipping in the later stages.

Gogolewski et al. (1993) confirmed the existence of bimodal degradation with PLA during in vivo implantation but the cause as to whether it is an enzymatic or hydrolytic degradation was not clear to them.

Degradation does not lead only to mechanical property changes. An important observation was that degradation leads to an increase in crystallinity as a result of content loss (Li et al., 1990a). This obviously implies that the amorphous regions of the semicrystalline polymer are subjected to degradation earlier than the crystalline regions, leading to an increase in crystallinity. Fini et al. (1995) also reported an increase in crystallinity (2–12%) and a significant drop in the molecular weight (93%) of PLLA rods implanted in the femur of rabbits in 64 weeks.

Another interesting change in the polymer properties upon degradation is the change in the heterogeneity index. Heterogeneity index is an indicator of the narrowness of molecular weight distribution of a polymeric product. An increase in the heterogeneity index upon degradation indicates broadening of the molecular weight distribution possibly resulting from a faster decrease in M_n in comparison to a decrease in M_w.

Alonso et al. (1993) carried out in vitro release studies with tetanus toxoid-loaded PLGA microsheres, and in 10 days a significant decrease in the M_n (75%) and a substantial increase in HI (from 1.2 to 1.6) were observed.

A similar observation was reported by Bergsma et al. (1995). Degradation of PLLA and PLA in distilled water at 100°C led to a substantial increase in the heterogeneity index (M_w/M_n) in only 30 h. This they interpreted to be a result of a random hydrolytic process. This is contrary to what the other researchers have found and could rather be a result of the high temperatures used in the degradation process leading to a substantial degree of chain-end scission or unzipping, too.

One of the earliest studies on PLA and PLGA degradation was carried out by Gilding and Reed (1979). They showed, in a study where the effect of exposure to gamma rays for sterilization was tested, that M_n falls much more rapidly than M_w. This implies that the degradation follows the "unzipping" route. They have implicated this as the primary reason for initial maintenance of strength by PGA through the help of crystalline regions (before implantation) and decrease of tensile strength to zero in 10 days.

Spenlehauer et al. (1989) showed that exposure to 3–7 Mrad caused a molecular weight decrease to 60 days in pH 7.4 phosphate buffer. The mechanism was not given.

Degradation could be induced by other means, leading to other routes of degradation. A study showed that when poly(D,L-lactide) was incubated at 60°C in 0.118 M deuterochloric acid (DCL), substantial amounts of lactic acid was produced, starting from the early stages of storage. A progressive increase in rate was also detected (Shih, 1995). A significant drop in rate without altering the mode of degradation was observed upon replacement of DCL with lactic acid.

In all these studies, unusual conditions like application of gamma irradiation, or incubation at high temperatures, or in acids like deuterochloric acid have led to unzipping, while incubation in buffers of medium pH or water at ambient temperatures or 37°C led to random scission.

With the use of copolymers the changes observed were not in the mode but in the rate of degradation. Poly(lactide-*co*-glycolide) 75 : 25 in distilled water showed racemic lactide weight loss, lactic acid production appeared on the 20th day (Li et al., 1990b). This was faster and earlier (ca. 5–9 days) in pH 7.4 phosphate buffer, but the drop in storage modulus and rate of water absorption (both of which started on the first day) appeared more rapidly upon incubation in distilled water. Molecular weight shift was unimodal on the 17th day both in buffer and in distilled water; however, the shift was to much lower molecular weights with water. Since no lactic acid production and weight loss was observed on the 17th day, these were interpreted as signs of a random-scission degradation which probably starts on day 1. This was supported by storage modulus and water absorption tests. On day 35 a difference between the interior and the exterior appeared for aging in water; the SEC of the polymer on the surface became bimodal, while the bulk yielded a single peak which had a much lower molecular weight, implying the presence of the two modes of degradation on the surface. In buffer, on the other hand, no inner bulk was left while the remainder shell degraded to a much lesser to a much lesser degree than in water.

Spenlehauer et al. (1989) confirmed that as glycolide content of the copolymer increased the degradation rate is also increased. The observation of molecular weight decrease was made much earlier (ca. two weeks) than with copolymers with less glycolide content.

Grizzi et al. (1993) state that form also has an effect on the mode of degradation. For example, out of plates (2 mm thick), films (0.3 mm thick), microspheres (0.125–0.250 mm diam.) and beads (0.5–1.0 mm diam.), only the plates showed a bimodal degradation. The plates are the only ones with appreciable interior where rapid autocatalytic degradation leading to lactic acid formation takes place. Weight loss, water absorption, lactic acid formation, pH change and molecular weight decrease all take place more and earlier with the plates than with the other forms.

In short, random scission was the normal mode of degradation for PLA and PLGA. Thicker samples produced bimodal degradation on the surface and homogeneous degradation in the inside, implying both random scission and unzipping on surface and only random scission in the interior. Also, rather harsh conditions (high acidity, high temperature, or high energy radiation) caused a change in the mode of degradation, mainly from random scission to unzipping.

IV. POLYHYDROXYBUTYRATE AND POLYHYDROXYBUTYRATE-*CO*-VALERATES

Polyhydroxyalkanoates (PHA) are polyesters which are found in the nature as products of a large variety of microorganisms. The most abundant PHA is poly(3-hydroxybutyrate) (P3HB). P3HB is generally found as a copolymer containing varying degrees of 3-hydroxyvalerate and is shown as PHBV7, where the number indicates the molar percentage of the comonomer in the chain. Polyhydroxyalkanoates are known to be degraded in vitro and in vivo as well as by microorganisms. Recently these rates of degradation were compared (Williams, 1997), where PHB was found to degrade in soil, but it was not observed in the more hostile environment of the human body, and the author concluded that PHB has not really found a place as an implantable degradable material.

It is, however, not so clear as to how this degradation process takes place. According to Timmins and Lenz (1994) poly(a-hydroxy acids) are readily hydrolyzed in the body but by a predominantly nonenzymatic route. This is contradicted by Doyle et al. (1991) who report that polyhydroxyalkanoates are not degraded in vivo at all.

However, when patches of PHB were used to close experimentally induced atrial septal defects in calves, the patch was degraded by polynucleated macrophages, not only by simply (Malm et al., 1992). Twelve months postoperatively no polymer material was identifiable at ordinary light microscopy, but polarized light microscopy revealed small particles of polymer. Microscopic observations revealed fragmentation through more rapid hydrolysis of amorphous regions, leaving behind polymer with higher crystallinity.

These contrasting reports do actually result from the slowness of degradation. The rates could be speeded up by external influences. For example, exposure to 2.5 Mrad gamma radiation was reported to cause a substantial deterioration in the properties of PHB when implanted for seven days, while no significant change was observed in the in vitro testing (Miller and Williams, 1987). This is an indication of the involvement of enzymes in the degradation process. These results obviously show the behavior of the polymers in this test and cannot be generalized because PHB is radiation resistant. In a later study, PHBV were shown to lose weight and shape upon exposure to gamma radiation with doses of up to 10 Mrad and then stored in physiological saline (Gürsel and Hasırcı, 1995).

As the following reports show, polyhydroxyalkanoates are degraded in vitro and in vivo, depending on the conditions, very slowly in some cases.

In a study carried out at pH 7.4, in 600 days only a 5% loss took place with PHBV12 and 8% loss took place with PHBV20, showing that at around neutral pH the degradation is very slow (Holland et al., 1990).

In a degradation study by Holland et al. (1990) carried out on apatite nucleated PHBV20 in pH 7.4 buffer at 37°C, the sample weight remained almost unchanged for about 400 days. Gloss factor and the polar component of the surface energy, however, started changing immediately. Gloss factor showed a decrease and the polar component of surface energy showed an increase and these were interpreted as increased surface roughness and hydrophilicity, respectively. Molecular weight showed barely perceptible changes initially but became the most abruptly and distinctly changing parameter at around 250 days, revealing a substantial decrease. Crystallinity, on the other hand, showed an increase starting in the first days.

According to Timmins and Lenz (1994) the crystalline regions are 20 times less susceptible to hydrolytic attack than amorphous regions. This obviously leads to a product with higher crystallinity because the amorphous regions are degraded and removed. Their statement of the insusceptibility of the crystalline regions to degradation and the observation of crystallinity increase by Holland et al. (1990) are definitely supportive of each other. The rapid hydrolysis observation of these researchers is in contrast with what Doyle et al. (1991) reported. In that study, they showed in vitro that sheets of PHB (crystallinity 75%, 2.5 mm thick) lost up to 50% of their modulus within four months, the effect being higher at higher temperatures. A similar trend was seen in bend strength. Since the trends followed a linear path with respect to time, one concludes that the degradation mode is random scission. In vivo implantation of the same material into rabbits, however, did not reveal any signs of extensive structural breakdown, physical degradation or resorption for six months. Since these are gross observations it is possible that degradation could occur without erosion of the material and is probably the case.

Miller and Williams (1987) observed that although in vivo exposure for 182 days did not change the appearance of PHB fibers (under SEM), load at break decreased while strain at break increased. They could not record any weight changes in vitro and in vivo during this time. In vitro tests in PBS showed signs of degradation which were especially higher at elevated temperatures (60°C and 70°C). Within 55 days in PBS, load at break, strain at break, and tensile strength decreased substantially and Young's modulus increased.

In a study by Holland et al. (1990) where the work of Miller and Williams (1987) is quoted, it is stated that as the HV content increases the hydrolytic degradation rate of PHBV copolymers decreases. Their own data involving PHBV copolymer, filled with ≥1% hydroxyapatite, on change of crystallinity with time revealed a gradual increase up to ca. 300 days, which then started decreasing in some samples while the rest maintained the increase. The increases could be due to the removal of amorphous regions which initially let the water in, increasing the crystallinity, which is then followed by degradation in the crystalline regions, causing a decrease in crystallinity.

Introduction of HV into polyester chains led to an increase in the rate of mass loss with respect to PHB, but no direct proportionality between HV content and half-life could be observed (Pouton and Akhtar, 1996). This finding is contrary to Holland et al. (1990) and other researchers where HV content positively influences degradation rate. This controversy can be understood by examining the reasons for crystallinity. HV content alone does not define crystallinity. Sample homogeneity, HI, thermal history, impurities and chain lengths all contribute to crystallinity; therefore materials obtained from other sources with different processing histories or batch numbers would probably not have the same level of crystallinity.

Incorporation of hydrophilic materials into the bulk of PHBV also increased the rate of degradation. Yasin et al. (1989) have shown that blending with materials like sodium alginate, dextrin, amylose and talc substantially increased the rate of weight loss from injection molded plaques in comparison to pure PHBV. For example, 30% sodium alginate loaded PHBV12 lost 10% of its weight in pH 10.6 buffer at 37°C in 2 days and 50% of it in 60 days. The loss was less (25% in 125 days) if the extent of loading with alginate was decreased from 30% to 10%. The weight loss obviously does not directly correlate to a degradative loss. Especially when hydrophilic fillers are present they would initially imbibe water, swell, leach out and lead to erosion, and thus they may enhance hydrolysis due to the presence of more water than there would normally have been. So, it is similar to the effect of low crystallinity induced by high HV content.

So far no mention of the mode of degradation has been made. Data obtained under various conditions carry clues about the mode. In one study, significant amounts of PHB oligomers and small amounts of isocrotonic acid (a PHB monomer analog) were produced upon heating the polymer at around 300°C (Pouton and Akhtar, 1996). This showed that harsh conditions could change the more common degradation route (random scission) and that there is a real possibility of unzipping.

Another study revealed that processing via injection molding (ca. 160°C) decreased the molecular weight (M_w) of the original, unprocessed form from 3.9×10^5 to 1.92×10^5 and from 2.8×10^5 to 1.95×10^5 for PHBV12 and PHBV20, respectively (Yasin et al., 1989). When these injection molded samples were stored in pH 7.4 buffer at 70°C for 120 days, these M_w values were further reduced to 4.27×10^3 and 3.40×10^3, respectively. The M_n values also showed significant decreases. An important observation was that the heterogeneity index decreased from 2.9 to 2.2 to 1.8 and from 4.1 to 2.2 to 1.6 after molding and storage for these two copolymers, respectively. On the other hand, upon storage at a higher pH (pH 10.6) and the same temperature for 46 days the PHBV12 and PHBV20 molecular weights decreased to 1.5×10^4 and 1.22×10^4, respectively. An interesting thing was the very significant increase in the heterogeneity index at this pH; for PHBV12 and PHBV20 samples HI went from 2.2 to 5.5 and from 2.2 to 4.3. Thus, depending on the degradation medium pH, both the rate and the mode of degradation were altered. The increase in heterogeneity index is observed because M_w did not change as much as the M_n, possibly by breaking long chains into still long chains and monomers, causing M_n to decrease faster than M_w. This probably was because chain-end cleavage (unzipping) started becoming more dominant over random scission. Gilding's (1979) observation also supports this explanation. Therefore, the degradation rate was faster in alkaline and the pH increase also led to an increase in HI, contrary to what is observed at pH 7.4.

V. THE COMPARISON OF THE DEGRADATION PATTERNS OF PLGA AND PHBV

As can be deduced from the results presented above, the rates and modes of degradation of these two polyester families appear to differ when studies using only one of them are examined. For example, while PHBV degraded through unzipping under harsh conditions and at high pH values, and with random scission under normal conditions, while PLGA showed unzipping at lower pH values and harsh conditions.

There also appears to be a difference in the rates of degradation of the two copolymer families. Pouton and Akhtar (1996) reported a slower in vitro degradation with PHBV with respect to PLGA. The half-life (based on weight loss) of a thin solvent-cast film of PHB (85 μm thick) in a pH 7.4 buffer at 37°C was estimated as ca. three years. This is a very long time for PLA to degrade to that degree but is relatively rapid for PHB and could be the result of the sample thickness (or thinness) and thus to exposure of more surface to aqueous medium.

The in vitro degradation results were strongly supported by in vivo degradation observations. PHB and PHBV (5–22% HV) were observed to degrade substantially less (0–1.6%) during six months of implantation than PLA (0–50%) (Gogolewski et al., 1993). As expected, PHBV degradation was found to be related to HV content. An important observation was that there was no bimodality with PHBV degradation.

A most interesting report is that under accelerated aging conditions (0.1 N NaOH, 75°C) PHBV showed no molecular weight change upon 87% weight loss (Pouton and Akhtar, 1996). This indicates substantial erosion not involving hydrolysis and is in contrast to PLGA results where mass loss was followed by some molecular weight decrease. It, however, is contrary to the observations where high pH normally leads to unzipping of PHBV.

VI. IN VIVO RESPONSES TO POLYLACTIDE-GLYCOLIDES AND POLYHYDROXYBUTYRATE-*co*-VALERATES

In vivo response of the biological system to a polymer, thus its biocompatibility, is its most important property if its use in the biomedical field is contemplated. PLGA and PHBV are the most important members of the biodegradable biomedical polymers, and thus the response of the biological system is of utmost importance.

Some reports on the biocompatibility of these polymers are, however, not very encouraging. In a study where PLA pins were applied to human subjects, six knees were observed to develop diffuse swelling and a prolonged postoperative course, which could not be attributed to infection (Tegnander et al., 1994). The high frequency of inflammatory signs in the knees treated for OCD and the demonstration of a complement activation potential of PLA pins were thought to warrant further studies on the biocompatibility of the material. The paper was concluded by stating that "until more information is available we do not recommend intraarticular use of these pins." This is, no doubt, a very disturbing note for the users of the PLGA family of polyesters.

In another study porous D,L-PLA disks were used for implantation in rabbit calvariae (with the eventual intention to load with bone morphogenic proteins) and retrieved after one, two, four and six months (Robinson et al., 1995). Disks with the largest pore size (larger than 350 μm) had the greatest bone ingrowth. Multinucleated giant cell response associated with all the implanted disks was, however, found to be disturbing.

When PLLA bone plates were used on four patients with step and straight osteotomy and were followed for 5.5 years, bone healing was uneventful in all the patients, but histological examinations revealed a nonspecific foreign body reaction on highly crystalline PLLA remnants, which were still present (Tams et al., 1996). The researchers concluded that "as polymerized" PLLA was no longer their first choice in maxillofacial surgery. The uncertainty of the degradation duration of the crystalline segments and the possibility of a need for their removal led them to this conclusion.

Spenlahauer et al. (1989) observed at three weeks a subacute inflammatory response in rat livers when they used PLGA. At four weeks the microspheres lost shape, and inflammatory lesions increased with increase in number and size of foreign body Giant cells. At five weeks the microsphere structure was infiltrated by large syncytial Giant cells. At six weeks resorption was complete and intensity of inflammatory response was decreased, too.

Not all studies reported negative findings on biocompatibility. Bergsma et al. (1995) observed no biological response three weeks after the implantation of PLLA samples. Plasma cells, neutrophils or lymphocytes were rarely seen in the explants. They, however, stated that degradation products of PLLA implants which are numerous stable particles of high crystallinity appeared to be related to a subcutaneous swelling in patients three years post-op when used as bone plates, screws, etc. Thus, they suggested that the polymer should be more rapidly degradable and less crystalline to avoid long-lasting presence and these undesirable responses.

Also on a positive line, Fini et al. (1995), in their tests with rabbits, observed noninflammatory cells or fragmentation of bone trabeculae after four weeks of implantation of PLLA rods into the femur. At 64 weeks the bone tissue around the PLLA appeared regularly mineralized with no signs of trabecular resorption, microfractures or altered calcium deposition, as well as good osteointegration without a reactive capsule.

In the sera from the animals which carried PLGA implants loaded with antimycobacterial isoniazid showed no significant differences in renal, hepatic and hematological parameters, and PLGA did not cause local or systemic toxicity in the six week test period (Gangadharam et al., 1991).

When PLGA microparticles were used as a delivery system for measles virus cytotoxic T cell epitope, the microparticles were found to be of excellent biocompatibility except for the pronounced immunostimulatory effect (Partidos et al., 1996).

Ultrahigh strength PLLA pins used in the treatment of osteochondritis and osteochondral fracture of the knee in humans in all cases that were followed for two to seven years and nine months, satisfactory bone union with no detection of inflammatory reaction was reported and the pins were found to be safe and useful (Matsusue et al., 1996).

PLA was also used as a biodegradable barrier in 29 patients with mandibular and maxillary molar defects (Polson et al., 1995). The barriers fragmented and were displaced in three to six weeks, and substantial granulation tissue was sometimes present between the barrier and the root surface. There were, however, clinically and statistically significant improvements in all the other parameters, and the opinion on this barrier material used in Class II furcation defects in humans was very favorable.

With PLLA applied subcutaneously to rats, at three weeks the PLLA particles were found to be surrounded and individually embedded by young fibrous tissue, macrophages and foreign body giant cells (Bergsma et al., 1995). From 16 weeks onward a mature and relatively hypocellular fibrous capsule was obtained. After 70 weeks, fragmentation and formation of smaller particles internalized by phagocytes were observed. With PLA96, however, the situation was slightly different. After three weeks, the particles were embedded in young fibrous tissue which was infiltrated by cells, mainly by macrophages, fibrocytes, foreign body giant cells and some lymphocytes. After 16 weeks the number of giant cells and fibrocytes were diminished, but the number of macrophages that had a foamy appearance and had engulfed polymer particles were increased. At week 80 TEM showed numerous macrophages with large amounts of particles, though no cell damage was observed.

Galgut et al. (1991) found that with PLA membranes implanted to rats using a transcutaneous model on the dorsal surface showed a high incidence of absorption of the polymer and a mainly monocytic infiltrate. On the whole, PLA material was well tolerated.

With PHB and PHBV a more positive picture than that observed with PLA and PLGA emerged.

Saad et al. (1996) tested the effect of a synthetic PHB block copolymer containing short (M_n 2300) crystalline segments on the viability and activation of mouse macrophages (J774), primary rat peritoneal macrophages and mouse fibroblasts (3T3), and their biodegradation or exocytosis in these cells. At concentrations higher than 10 μg/mL, a significant decrease in the number of attached and viable macrophages and a significant increase in tumor necrosis factor-alpha and nitric oxide were observed. At low concentrations, PHB did not induce cytotoxic effects or activate marcophages. Signs of possible biodegradation in macrophages were observed. Fibroblasts showed only limited phagocytosis of PHB and no signs of cellular damage or cell activation (production of collagen types, I and IV and fibronectin).

Rivard et al. (1995) used PHBV9 foams and seeded them with fibroblasts isolated from canine anterior cruciate ligaments. Collagen sponges were used for comparison. After 35 days, PHBV9 sustained a cell proliferation rate similar to that observed in collagen sponges, kept their structural integrity while collagen sponge contracted substantially and total protein production was twice as high as that on collagen, proving porous PHBV to be an adequate substrate for cell culture and tissue engineering.

Polyhydroxybutyrate membranes were used in studies with rats to obtain bone regeneration, and histological analysis demonstrated increased bone fill with the test specimens between days 15–180, whereas only 35–40% of the defect area in the control sites with no PHB membrane was filled with bone after three to six months (Kostopoulos and Karring, 1994a,b).

Another tissue regeneration study carried out with dogs led to a contrasting result. According to Gotfredsen et al. (1994) the PHBV membranes reinforced with polyglactin 910 fibers were frequently surrounded by a fibrous tissue capsule which interfered with the marginal bone healing adjacent to the immediately placed implants. An inflammatory reaction and significantly less marginal bone healing were registered on the membrane side compared with the control side.

Doyle et al. (1991) observed in a study with rabbits that PHB materials produced a consistent favorable bone tissue adaptation response with no indication of an undesirable chronic inflammatory response after implantation up to 12 months. There was no evidence of a Giant cell response within the soft tissue in the earlier stages of implantation. Bone was rapidly formed close to top the material and subsequently became highly organized, with up to 80% of the implant surface lying in direct apposition to new bone. The materials showed no conclusive evidence of extensive structural breakdown in vivo during six months, which could explain the lack of response to the material.

In another type of application Malm et al. (1992) used biodegradable patches of PHB to close experimentally induced atrial septal defects in calves. At implant degradation complete endothelial layers facing the right and left atria were observed, along with a subendothelial layer of collagen and some smooth muscle cells. The patch was degraded by polynucleated macrophages, and 12 months postoperatively no polymer material was identifiable by ordinary light microscopy, but polarized light microscopy revealed small particles of polymer with persistent foreign body reaction. Thus PHB patches implanted in atrial septal defects prompted formation of regenerated tissue that microscopically resembled a native atrial septal wall.

Malm et al. (1994) also used nonwoven patches made of PHB and implanted these transannular patches into the right ventricular outflow tract and pulmonary artery of weanling sheep. The results obtained at 3–24 months revealed no aneurysms. Regeneration of a neointima and a neomedia, comparable to native arterial tissue, was observed in the PHB test samples. In the control where Dacron was used a neointimal layer was present, but no neomedia comparable to native arterial tissue was obtained.

Duvernoy et al. (1995) used biodegradable patches made of polyhydroxybutyrate with success as pericardial substitutes on 59 patients for 24 months.

In order to compare the two families of polyesters better, it might be better to consider the studies in which both polyester types were used. PHBV and PLGA implanted in mice subcutaneously for up to six months were well tolerated (Gogolewski et al., 1993). No signs of acute inflammation, abscess formation or tissue necrosis were observed in the adjacent tissues. Also no tissue reactivity or cellular mobilization was evident at remote sites. Mononuclear macrophages, proliferating fibroblasts and mature vascularized fibrous capsules were the typical tissue response. Polymer degradation was accompanied by collagen deposition.

Among the PLGAs, the D-component led to lesser tissue response. With PHBVs, higher HV led to higher inflammatory cells. Between one and three months PHBV showed more tissue response than PGLA, and this was explained by Gogolewski et al. to be possibly due to leachable impurities and low molecular weight components. Obviously the leachable impurity in PHBV implies improper purification of the microbial products. At six months the extent of tissue response was reported to be the same. During this time 56–99% of the PLGA were reportedly degraded.

The interesting thing here is that while PLGAs degraded almost completely within the experimental period, the PHBVs were reported to be intact. Thus, the biological responses shown by the biological system toward the polyesters should be results of two different response types: in PLGA, toward degradation products, and in PHBV, toward the unchanged foreign mass.

Not all the data involving PHBV are positive. For example, it is reported that when strips of PHBV and PLGA (along with other polymers) deployed on stents were implanted in porcine coronary arteries, after four weeks of implantation PHBV (along with polycaprolactone, polyorthoester, polyurethane and silicone) evoked extensive inflammatory responses and fibrocellular proliferation, while PLGA and polyethyleneoxide/polybutylene terephthalate evoked less but still severe responses (van der Giessen et al., 1996). The authors state that preliminary in vitro tests were not in agreement with these observations and postulated that the responses could be attributable to a combination of factors like polymer biodegradation products and implant geometry. Obviously since PHBV would not degrade within the time frame of the test, the responses were most probably a result of the test conditions (geometry of the stent, site of implantation, skill in implantation, etc.).

VII. CONCLUSION

In short, the difficulty in achieving identical conditions in the preparation and treatment of the test samples makes comparison between various studies hard. It appears, however, that more positive responses occur with PHBVs than with PLAs and PLGAs. The causes for unfavorable responses are

1. Leachable polymerization medium compounds
2. Low molecular weight polymers
3. Particulate degradation products
4. Highly crystalline polymer particles resulting from degradation
5. High concentration of products as a result of rapid degradation

Items 1 and 5 are not observed for PHB and PHBV because they are natural and biotechnologically produced, and their degradation rate is substantially lower than that of PLA and PLGA. One would, therefore, expect polyhydroxyalkanoates to be more biocompatible and find increasing use in the biomedical field.

REFERENCES

Alonso, M. J., S. Cohen, T. G. Park, R. K. Gupta, G. R. Siber, R. Langer, Determinants of release rate of tetanus vaccine from polyester microspheres, *Pharm. Res.* 10(7) 945–953, 1993.

Bergsma, J. E., F. R. Rozema, R. R. M. Bos, G. Boering, W. J. Bruyjn, A. J. Pennings, In vivo degradation and biocompatibility study of in vitro pre-degraded as-polymerized polylactide particles, *Biomaterials* 16, 267–274, 1995.

Cha, Y. and C. G. Pitt, The biodegradability of polyester blends, *Biomaterials* 11, 108–112, 1990.

Doyle, C., E. T. Tanner, W. Bonfield, In vitro and in vivo evaluation of polyhydroxybutyrate and of polyhydroxybutyrate reinforced with hydroxyapatite, *Biomaterials* 12, 841–847, 1991.

Duvernoy, O., T. Malm, J. Ramström, S. Bowald, A biodegradable patch used as a pericardial substitute after cardiac surgery: 6 and 24 month evaluation with CT, *Thorac. Cardiovasc. Surg.* 43(5), 271–274, 1995.

Fini, M., S. Giannini, R. Giordano, G. Giavaresi, M. Grimaldi, N. Nicoli Aldini, L. Orienti, M. Rocca, Resorbable device for fracture fixation: in vivo degradation and mechanical behaviour, *Int. J. Artif. Organs* 18(12), 772–776, 1995.

Galgut, P., R. Pitrola, I. Waite, C. Doyle, R. Smith, Histological evaluation of biodegradable and non-degradable membranes placed transcutaneously in rats, *J. Clin. Periodontal* 18(8), 581–586, 1991.

Gangadharam, P. R., D. R. Ashtekar, D. C. Farhi, D. L. Wise, Sustained release of isoniazid in vivo from a single implant of a biodegradable polymer, *Tubercle* 72(2), 115–122, 1991.

Gilding, D. K., A. M. Reed, Biodegradable polymers for use in surgery—Polyglycolic/poly(lactic acid) homo and copolymers. I, *Polymer* 20, 1459–1464, 1979.

Gogolewski, S., M. Jovanovic, S. M. Perren, J. G. Dillon, M. K. Hughes, Tissue response and in vivo degradation of selected polyhydroxyacids: polylactides (PLA), poly(3-hydroxybutyrate) (PHB), and poly(3-hydroxybutyrate-co-3-hydroxyvalerate) (PHB/VA), *J. Biomed. Mater. Res.* 27(9), 1135–1148, 1993.

Gogolewski, S., P. Mainil-Varlet, The effect of thermal treatment on sterility, molecular and mechanical properties of various polylactides, *Biomaterials* 17, 523–528, 1996.

Gopferich, A., Mechanisms of polymer degradation and erosion, *Biomaterials* 17(2), 103–114, 1996.

Gotfredsen, K., L. Nimb, E. Hjirting-Hansen, Immediate implant placement using a biodegradable barrier, polyhydroxybutyrate-hydroxyvalerate reinforced with polyglactin 910. An experimental study in dogs, *Clin. Oral Implants Res.* 5(2), 83–91, 1994.

Grizzi, I., H. Garreau, S. Li, M. Vert, Hydrolytic degradation of devices based on poly(DL-lactic acid) size-dependence, *Biomaterials* 16, 305–311, 1993.

Gürsel, I., V. Hasırcı, Properties and drug release behaviour of PHB and various P(HB-HV) copolymer microcapsules, *J. Microencapsulation* 12(2) 185–193, 1995.

Holland, S. J., M. Yasin, B. Tighe, Polymers for biodegradable medical devices VII. Hydroxybutyrate-hydroxyvalerate copolymers: degradation of copolymers and their blends with polysaccharides under in vitro physiological conditions, *Biomaterials* 11, 206–215, 1990.

Kostopoulos, L., T. Karring, Guided bone regeneration in mandibular defects in rats using a bioresorbable polymer, *Clin. Oral Implants Res.* 5(2), 66–74, 1994a.

Kostopoulos, L., T. Karring, Augmentation of the rat mandible using guided tissue regeneration, *Clin. Oral Implants Res.* 5(2), 75–82, 1994b.

Li, S. M., H. Garreau, M. Vert, Structure-property relationships in the case of degradation of massive aliphatic poly(α-hydroxy acids) in aqueous media, *J. Mater. Sci.* 1, 123–130, 1990a.

Li, S. M., H. Garreau, M. Vert, Structure-property relationships in the case of degradation of massive poly(α-hydroxyacids) in aqueous media. 2. Degradation of lactide-glycolide copolymers: PLA37.5GA25 and PLA75GA25, *J. Mater. Sci. Mater. Med.* 1, 131–139, 1990b.

Malm, T., S. Bowald, A. Bylock, C. Busch, T. Saldeen, Enlargement of the right ventricular outflow tract and the pulmonary artery with a new biodegradable patch in transannular position, *Eur. Surg. Res.* 26(5), 298–308, 1994.

Malm, T., S. Bowald, S. Karacagil, A. Bylock, C. Busch, A new biodegradable patch for closure of atrial septal defect. An experimental study, *Scand. J. Thorac. Cardiovasc. Surg.* 1, 9–14, 1992.

Matsusue, Y., T. Nakamura, S. Suzuki, R. Iwasaki, Biodegradable pin fixation of osteochondral fragments of the knee, *Clin. Orthop.* 322, 166–173, 1996.

Miller, N. D., D. F. Williams, On the biodegradation of polyhydroxybutyrate (PHB) homopolymer and poly(β-hydroxybutyrate-hydroxyvalerate) copolymers, *Biomaterials* 8, 129–137, 1987.

Partidos, C. D., P. Vohra, C. Aagnostopoulou, D. H. Jones, G. H. Farrar, M. W. Steward, Biodegradable microparticles as a delivery system for measles virus cytotoxic T cell epitopes, *Mol. Immunol.* 33(6), 485–491, 1996.

Pitt, C. G., R. A. Jeffcoat, A. Schindler, R. A. Zweidinger, Sustained drug delivery systems. I. The permeability of poly (ε-caprolactone), poly(DL-lactic acid), and their copolymers, *J. Biomed. Mater. Res.* 13, 497–507, 1979.

Polson, A. M., S. Garrett, N. H. Stoller, G. Greenstein, A. P. Polson, C. Q. Harrold, L. Laster, Guided tissue regeneration in human furcation defects after using a biodegradable barrier: a multi-center feasibility study, *J. Periodontol.* 66(5), 377–385, 1995.

Pouton, C. W., S. Akhtar, Biosynthetic polyhydroxyalkanoates and their potential in drug delivery, *Adv. Drug Delivery Rev.* 18, 133–162, 1996.

Rivard, C. H., C. J. Chaput, E. A. DesRosiers, L. H. Yahia, A. Selmani, Fibroblast seeding and culture in biodegradable porous substrates, *J. Appl. Biomater.* 6(1), 65–68, 1995.

Robinson, B. P., J. O. Holinger, E. H. Szachowicz, J. Brekke, Calvarial bone repair with porous D,L-polylactide, *Otolarynol. Head Neck Surg.* 112(6), 707, 713, 1995.

Saad, B., G. Ciardelli, S. Matter, M. Welti, G. K. Uhlschmid, P. Neuenschwander, U. W. Suter, Characterization of the cell response of cultured macrophages and fibroblasts to particles of short chain poly[(*R*)-3-hydroxybutyric acid], *J. Biomed. Mater. Res.* 30(4), 429–439, 1996.

Shih, C., Chain-end scission in acid catalyzed hydrolysis of poly(D,L-Lactide) in solution, *J. Controlled Release* 34, 9–15, 1995.

Spenlehauer, G., M. Vert, J. P. Benoit, A. Boddaert, In vitro and in vivo degradation of poly(D,L-lactide/glycolide) type microspheres made by solvent evaporation method, *Biomaterials* 10, 557–563, 1989.

Tams, J., F. R. Rozema, R. R. Bos, J. L. Roodenburg, P. G. Nikkels, A. Vermey, Poly(L-lactide) bone plates and screws for internal fixation of mandibular swing-osteotomies, *Int. J. Oral Maxillofac. Surg.* 25(1), 20–24, 1996.

Tegnander, A., L. Engebretsen, K. Bergh, E. Eide, K. J. Holen, O. J. Iversen, Activation of the complement system and adverse effects of biodegradable pins of polylactic acid (Biofix) in osteochondritis dissecans, *Acta Orthop. Scand.* 65(4), 472–475, 1994.

Timmins, M. R., R. W. Lenz, Enzymatic biodegradation of polymers: The polymer chemists' perspective, *Trends Polym. Sci.* 2(1), 15–19, 1994.

van der Giessen, W. J., A. M. Lincoff, R. S. Schwartz, H. M. van Beusekom, P. W. Serruys, D. R. Holmes, S. G. Ellis, E. J. Topol, Marked inflammatory sequalae to implantation of biodegradable and nonbiodegradable polymers in porcine coronary arteries, *Circulation* 94(7), 1690–1697, 1996.

Williams, D. F., Degrading experiences: turning difficulties of polymer degradation to advantage, *Med. Dev. Technol.* April 1997.

Yasin, M., S. J. Holland, A. M. Jolly, B. J. Tighe, Polymers for biodegradable medical devices. VI. Hydroxybutyrate-hydroxyvalerate copolymers: accelerated degradation of blends with polysaccharides, *Biomaterials* 10, 400–413, 1989.

Yasin, M., S. J. Holland, A. M. Jolly, B. J. Tighe, Polymers for biodegradable medical devices. V. Hydroxybutyrate-hydroxyvalerate copolymers: effects of polymer processing on hydrolytic degradation, *Biomaterials* 11, 451–454, 1990.

5

The Role of Free Radicals in Degradation of Biodegradable Biomaterials

C. C. Chu and K. H. Lee
Cornell University, Ithaca, New York

I. INTRODUCTION

Free radical species are essential to many biological systems. These regulatory molecules are the consequences of events in tissue response to invading microorganisms [1]. Usually, these regulatory molecules exist as intermediates or end products from enzyme-catalyzed reactions [2] which involve in the cyclo-oxygenase of the phagocytosis mechanism and lipoxygenase activity of the eicosanoid metabolism [3,4]. During this phagocytosis, the production of superoxide ions, hydrogen peroxide, hydroxyl radical, and singlet oxygen has been well documented [5]. Some investigations have also shown that macrophages that protect us against infection and invading microorganisms are responsible for the cellular degradation of natural polymers like DNA, since these phagocytes undergo both morphological and biochemical transformations that facilitate the degradation of the biopolymers [7]. On the other side, free radicals can become highly destructive to cells and tissues if their production is not tightly controlled. One of the substances involved in the free radical biological processes that are regarded as the cause of several pathological phenomena is superoxide ion, $\cdot O_2^-$. These free radical anion species are involved in many biological processes that can cause serious damage to cells, lipid peroxidation, protein denaturation and DNA destruction [6–8].

Regardless of its toxicity, the chemical properties of the superoxide ion were not widely studied until in early 1970s when four new experimental techniques were used to investigate the chemical behavior of superoxide ion pulse radiolysis [9], flash photolysis in aqueous solution [10], electrochemical reduction of dioxygen [11], and use of KO_2/crown-ether solutions in aprotic media [12].

There are major differences in the reactivity of superoxide in aprotic solvents and in an aqueous solution. In aqueous solution, superoxide ion is easily dismutated to H_2O_2 and O_2, and any reactions initiated by this ion must compete with its rapid disproportionation. At the same time, in aprotic solvents, the superoxide ion acts as an effective nucleophile. A few mechanistic studies on the cleavage of small molecules like alkyl esters of carboxylic acids by $\cdot O_2$ have been reported [13,14]. Acyl halides were also reported to be hydrolyzed by their reaction with superoxide ion to produce carboxylic acid and alcohol [15]. The general mechanism of simple ester hydrolysis by superoxide ion was subsequently proposed [16,17].

Recently, a published review of the biodegradation phenomena of synthetic absorbable sutures suggested that there several extrinsic factors that could affect biodegradation phenomena [18]. However, little information is available about the effect of anionic free radical species on the degradation of synthetic biomedical polymers [19–21]. Williams et al. investigated the degradation mechanisms of biodegradable biomedical polymers by using the free radical species like hydroxyl radicals in an aqueous Fenton solution [22–25]. However, the role of free radicals on the biodegradable glycolic acid/lactic acid copolymer (Vicryl) could not be definitely confirmed, since equal amounts of $\cdot OH$ and OH^- were generated in Fenton reagents; the observed changes in morphological, mechanical and thermal properties of Vicryl could be partially attributed to OH^- ions as well as $\cdot OH$ radicals. In our study of Vicryl by using electron spin resonance (ESR) to determine existence of the α-carbon hydrogen abstraction mechanism of polymer chains in Fenton agent, we concluded, however, that hydroxyl radicals played an important role in the stability of polymers in [26,27].

Due to the extreme reactivity of superoxide ions and hence their stability, it was difficult to examine the effect of superoxide ions on biomaterials in vitro. However, we have used a chemical means to stabilize superoxide ions in degradation media as described in Scheme 1.

In this chapter, we reported on the reactivity of the superoxide ion as an oxygen nucleophile toward a variety of absorbable polymeric biomaterials having an aliphatic polyester structure, such as poly(D,L-lactic acid), PDLLA, poly(L-lactic acid), PLLA, and absorbable suture materials [27a,27b]. In the event of PDLLA and PLLA systems, we used polymer solutions during superoxide ion induced degradation, i.e., homogeneous degradation. In the synthetic absorbable suture materials, however, they remained in a solid state during superoxide ion degradation, i.e., heterogeneous degradation.

PDLLA is an amorphous biodegradable polymer that has been tested as biomaterials for drug delivery systems [28,29] with good compatibility and biodegradability. PLLA, however, is a semicrystalline polymer that has been experimented with for a number of medical devices like vascular grafts [30], surgical sutures [31], drug delivery systems [32], burn wound covering [33], and internal bone fixation for orthopedic applications [34] because of its biodegradability, biocompatibility, processability, good mechanical strength and other useful properties [35]. There has been much research of structure-property relationships for the degradation of PDLLA, PLLA and their copolymers in aqueous media [36–38]. These published studies confirmed that the autocatalytic degradation mechanism was governed and, in the case of semicrystalline polymer, the amorphous domains were more preferentially degraded than crystalline domains.

The absorbable suture materials used were 2/0 size Dexon, coated Vicryl, PDSII, Maxon and Monocryl. Dexon is a braided, undyed multifilament suture that is polymerized from the cyclic dimer of α-hydroxyacetic acid (glycolic acid). Vicryl is a random copolymer of glycolide and L-lactide and has the same braided construction as Dexon, except Vicryl is coated with a 50 : 50 mixture of an amorphous polyglactin 370 (a 65/35 mole ratio of lactide-glycolide

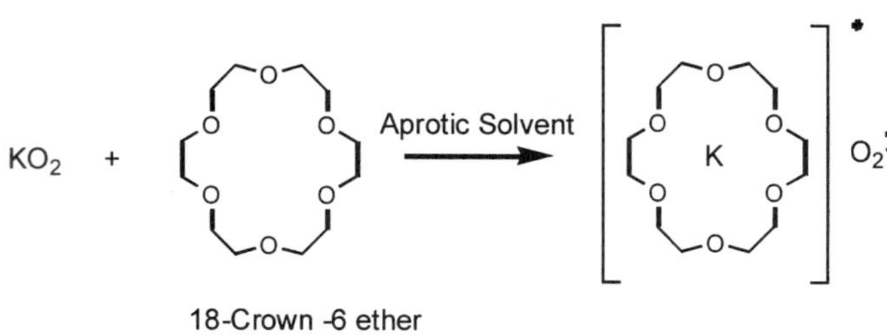

Scheme 1

copolymer) and calcium stearate. PDS is the first commercially available monofilament absorbable suture that derives from *p*-dioxanone monomers. Fiber-grade high molecular weight PDS polymer is made from the ring-opening polymerization of highly purified (>99%) *p*-dioxanone monomers in the presence of organometallic catalysts, such as zirconium acetylacetonate. Recently, an enhanced version of PDS suture (PDS II) has been introduced for improving its flexibility and handling characteristics. PDS II suture differs from PDS in fiber morphology due to different fiber spinning conditions. PDS II suture has a distinctive skin-core morphology and its core has a highly ordered and larger spherulitic crystal structure than its surrounding annular area. The monofilament absorbable suture, Maxon, is made from block copolymer of glycolide and 1,3-dioxan-2-one (trimethylene carbonate). It consists of 32.5% by weight of trimethylene carbonate and 67% by weight of glycolide. The monofilament Monocryl suture is a segmented block copolymer consisting of both soft and hard segments. The soft segment is the copolymer of glycolide and ε-caprolactone, providing good handling property like pliability. This low molecular soft segment prepolymer is further polymerized with glycolides to provide hard segments of polyglycolide for adequate strength. Thus, the total composition of 75% glycolide and 25% ε-caprolactone and high molecular weight were required for sutures with adequate mechanical properties.

II. RESULTS AND DISCUSSION

A. Homogeneous Superoxide-Induced Degradation of PDLLA and PLLA

The common character of the change of the molecular weights of PDLLA and PLLA as a function of hydrolysis time at a constant superoxide ion concentration and reaction temperature is a two-stage reduction in their molecular weight, and a typical example is shown in Fig. 1

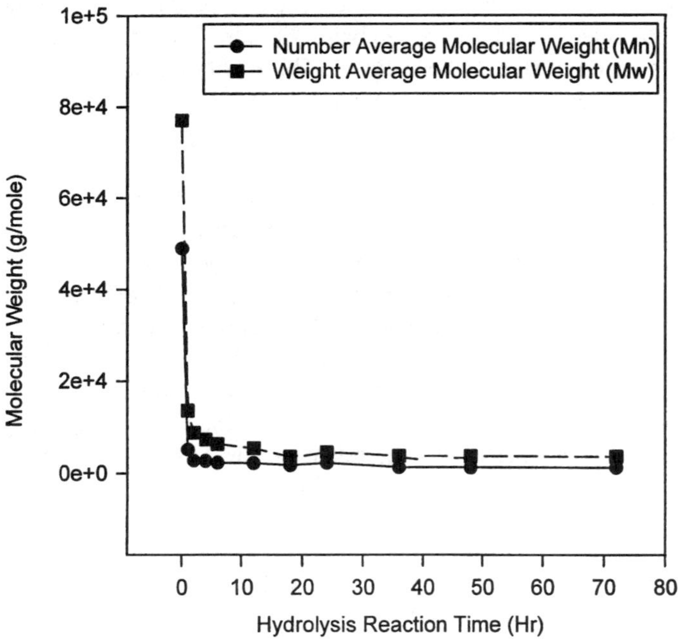

Figure 1 Effect of the hydrolysis reaction time on the molecular weight of PDLLA at 25°C and 0.01 molar superoxide ion concentration.

for PDLLA sample at 0.01 mole superoxide ion concentration and 25°C reaction temperature. This two-stage process of degradation includes an initial dramatic decline in both M_n (the number average molecular weight) and M_n (weight average molecular weight) which occurred within a short period of time followed by a relatively constant M_n and M_w. After 2 h, there was only a 10% and an 18% retention of M_w and M_n of the original PDLLA values, respectively. After this short period, an apparently insignificant decline in molecular weights has been observed up to 72 h. PLLA exhibited the same two-stage superoxide ion-induced degradation but with a smaller effect for the reasons that PLLA is a semicrystalline polymer and methylene chloride was used as the solvent instead of THF as in the case of PDLLA.

This two-stage time-dependent reduction in the molecular weight of biodegradable aliphatic polyesters in a superoxide ion medium can be explained by our modification of the reaction pathway originally proposed by Forrester and Purushotham [16,17]. They suggested that the cleavage of simple organic ester group by a nucleophilic attack of superoxide ion were predominant via S_N2 reaction to yield the corresponding alcohol and carboxylic acid. However, in the case of polyesters, these S_N2 reactions would produce mainly anionic end groups with different chain length, and these active anionic species could subsequently attack the main backbone chains via transesterification with a rapid reduction in molecular weight until thermodynamic equilibrium is reached. The resulting degradation products would eventually lead to a cyclic and/or linear oligomer.

This modified Forrester's mechanism for the hydrolysis of aliphatic polyester by superoxide ion is shown in Scheme 2. Since the nucleophilic attack of the superoxide ion on polymers occurs primarily in steps 1 and 2 which lead to the polyester chain fragmentation and the formation of two different types of species, lower molecular weight macromolecules with peroxy radical chain end group and anionic-catalyzed chain end group, respectively. Unlike the base-catalyzed hydrolysis of simple aliphatic polyesters in an alkaline solution in which the alcoholic anions were stabilized by hydrogen abstraction from the carboxylic end group to produce an alcohol group and a relatively stable carboxylic anion end group, the alcoholate end group from step 2 may subsequently attack main backbone chains via intra- and intermolecular transesterification reactions as illustrated in step 3 of Scheme 1. This additional chain scission in step 3 would not only reduce the molecular weight of biodegradable polymers further but also produce another anionic species that could repeat step 3 again to accelerate the hydrolytic degradation further.

Step 4 is a typical electron transfer reaction from the superoxide ion to the peroxy radical that leads to the formation of the carboxyl anionic intermediates which are expected to be stable. Therefore, their reactivity toward polyester chain as step 3 does might be very low or unlikely. Any excess superoxide ion would react with water shown in step 5 to form a hydroxyl anion and other species. Although it appeared to have apparently competitive reactions between the hydroxyl anion attacking ester linkage of polyester for additional chain scission and the hydroxyl anion reacting with the carboxyl peroxide-anion end group of the polymer chain (step 6) to produce the carboxylic anion without chain scission, it was less feasible for the former (i.e., hydroxyl anion to attack ester linkages) because of the relatively low temperature (25°C). Step 6 was preferable during the final step of termination reaction. Finally, it is possible to exchange potassium counterions with protons via diluted hydrochloric acid solution used for the termination reaction.

Based on the suggested chemical pathway of steps 1 and 2 in Scheme 2, an increase in superoxide ion concentration should also accelerate the hydrolytic degradation of biodegradable polyesters. Figure 2 shows such as effect of concentration of superoxide ion on molecular weight of PDLLA at 25°C and 24 h hydrolysis reaction time. There were two distinctive regions of molecular weight versus concentration effect: a sharp drop of molecular weight at

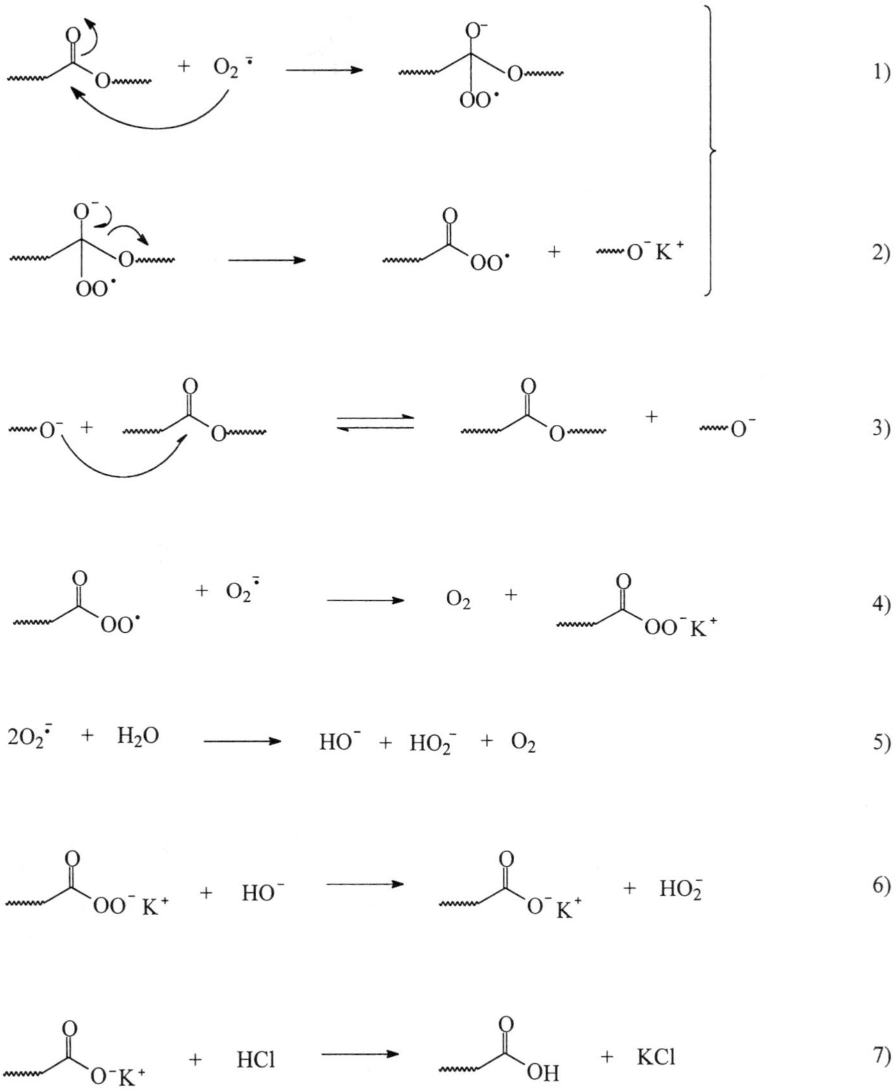

Scheme 2

low concentration of superoxide ion from 0.0005 molar to 0.0025 molar and a plateau region at high concentration of superoxide ion from 0.0025 to 0.01 molar. These two distinctive regions of molecular weight versus concentration effect remained over the entire reaction time (from 1 to 72 h). This common characteristic, however, was the most profound at the longest time of hydrolysis as evident in the difference in slopes of the curves. For example, the 72 h hydrolyzed PDLLA sample showed very little change in M_w as the superoxide ion concentration was doubled from 0.005 to 0.01 molar, while the 1 h hydrolyzed PDLLA sample exhibited a significant reduction in M_w within the same range of superoxide ion concentration. Therefore, the lower superoxide ion concentration samples, particularly at the 0.001 molar, reached the plateau region (i.e., the region with very little change in M_w with hydrolysis time) at a later time than the higher concentration samples.

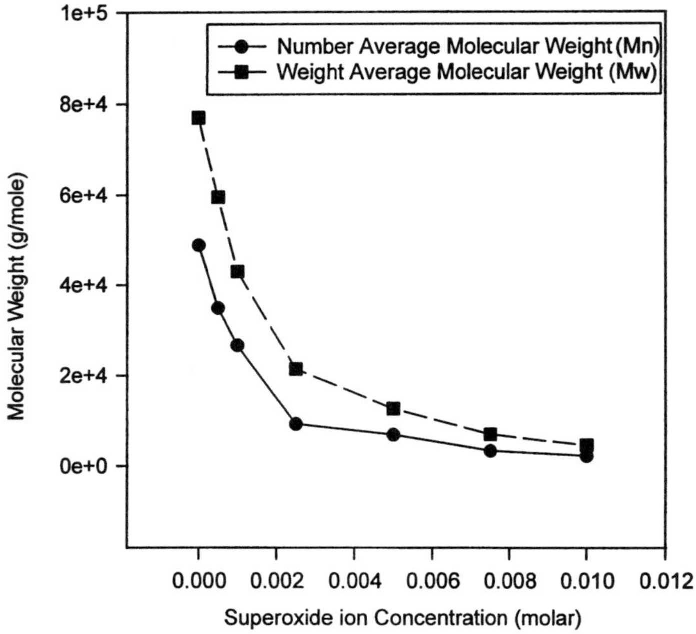

Figure 2 Effect of superoxide ion concentration on the reduction in molecular weight of PDLLA at 25°C and 24 h hydrolysis reaction time.

Because these synthetic biodegradable biomaterials are intended for use in the human body (i.e., at 37°C), the effect of reaction temperature (37°C versus 25°C) on superoxide effect on the hydrolytic degradation of these biodegradable polyesters was also examined at a superoxide concentration (0.01 mole) and reaction time (24 h). The data show that the rate of superoxide-ion-induced hydrolytic degradation of PDLLA was slower at 37°C than at 25°C as shown in Fig. 3. PDLLA at 37°C reaction temperature retained a larger amount of the high molecular weight fractions over the whole range of molecular weight distribution when compared with the same polymer at 25°C. Although both hydrolysis reaction temperatures showed low molecular weight species, the 37°C PDLLA sample had a relatively smaller proportion of the low molecular weight species than the 25°C sample. Both reaction temperatures resulted in low molecular weight oligomers. The less severe chain fragmentation of biodegradable polyesters by superoxide ion at 37°C was later also confirmed by the effect of reaction temperature on T_g of this class of polymers. The deactivation (i.e., stability) of superoxide ions at a higher reaction temperatures might be responsible for this temperature-dependent superoxide effect.

The GPC chromatograms of the superoxide-ion-induced degradation of biodegradable PDLLA and PLLA indicated that there were low molecular weight species due to main chain fragmentation. A combined GPC and chemical tagging method (phenyl isocyanate) of chain end groups of the degradation products, such as hydroxyl group (Scheme 3), were used to characterize the complex polymer-oligomer mixtures, particularly the linear or cyclic nature of the degradation products obtained at 24 h hydrolysis reaction time. Cyclic oligomers could not be tagged by phenyl isocyanate because of the lack of free hydroxyl end groups. This chemical tagging method of using phenyl isocyanate to label the hydroxyl chain end groups of linear oligomers only was used by Manolova et al. for chemical identification [39]. The GPC data

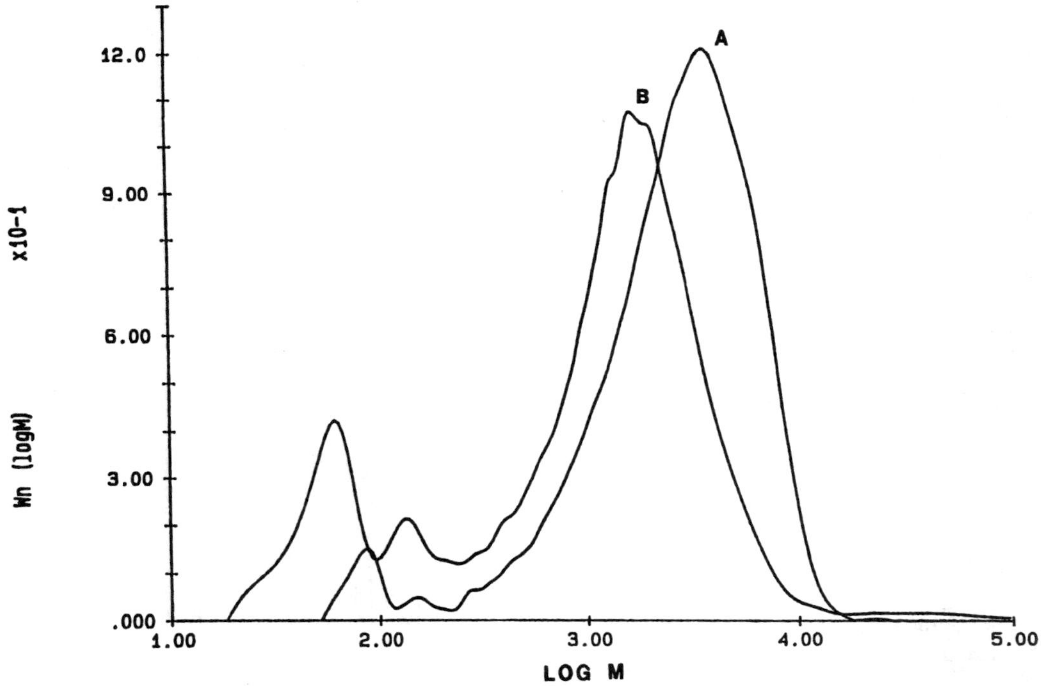

Figure 3 Effect of hydrolysis reaction temperature on the molecular weight distribution of PDLLA at 0.01 molar superoxide ion concentration and 24 h hydrolysis time. (A) 37°C; (B) 25°C.

from the phenyl isocyanate tagged degradation mixtures (Fig. 4B) showed strong UV absorption peaks at the locations corresponding to dimers and trimers (Fig. 4A versus 4B). The strong UV absorption peaks in the GPC chromatogram was due to the presence of unsaturated bonds in the tagged phenyl group. Therefore, by using the combined phenyl isocyanate tagging and GPC data we confirmed that linear oligomer products did form during the superoxide-induced hydrolytic degradation of PDLLA. Consequently, the contribution of step 2 reaction in Scheme 2 to the molecular weight distribution of PDLLA could be regarded as the dominant one.

The observed significant reduction in molecular weight of PDLLA and PLLA due to the superoxide-ion-induced degradation should also be reflected in the change of thermal properties of these polymers, such as the glass transition temperature, T_g. This was because the superoxide-ion-induced fragmentation of PDLLA and PLLA chains would produce more chain ends with shorter chain length. Since the number of polymer chain ends played an important role in determining the free volume of macromolecules, an increase in chain ends by chain fragmentation would lower the glass transition temperature.

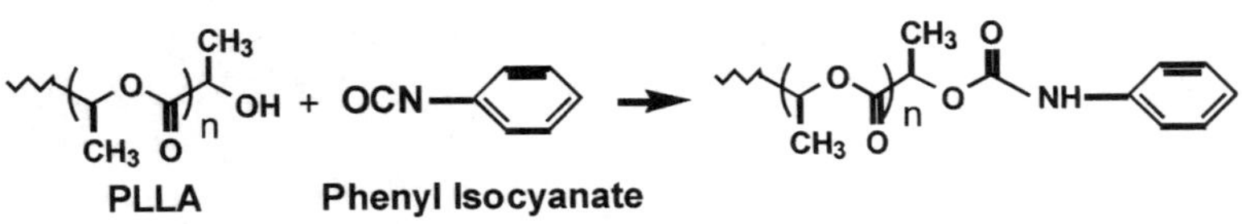

Scheme 3

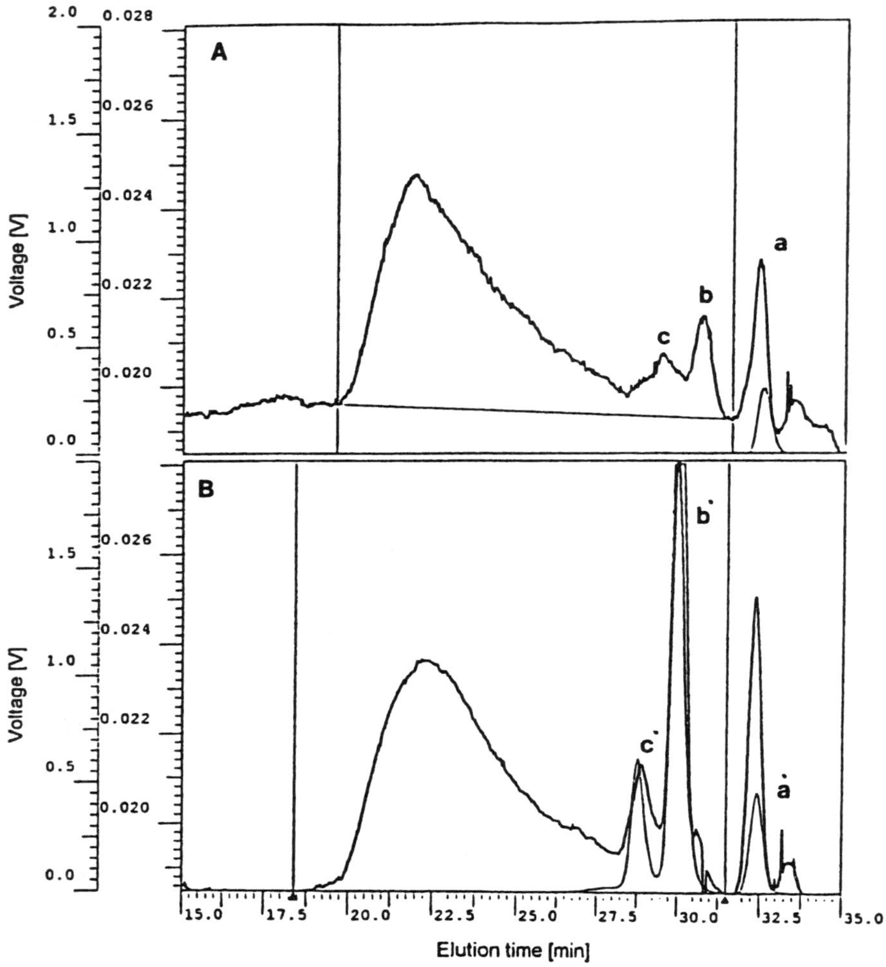

Figure 4 SEC concentration chromatograms of the 24 h hydrolysis reaction mixture from PDLLA before (A) and after (B) reaction with phenyl isocyanate chain end tagging for UV absorption: (a) peak was toluene; (b) peak was dimer and (c) peak was trimer. The appearance of UV peaks at b and c in (B) indicated the presence of linear oligomers of PDLLA having phenyl isocyanate chain end group.

Figure 5 illustrates the effect of superoxide ion reaction time on T_g of PDLLA at 25°C and 0.01 molar concentration of superoxide ion. There were three distinctive regions of change in T_g with reaction time: a significant decrease in T_g within a short period up to 2 h, a gradual but profound reduction in T_g until 24 h, and a region of insignificant change in T_g thereafter. The initial sharp drop of T_g during the first 2 h coincided with the significant reduction in number average molecular weight described previously in Fig. 1. The gradual but profound decrease in T_g from 2 h to 24 h was attributed to both the reduction in molecular weight and the presence of low molecular oligomer species, such as trimers and dimers. These low molecular weight species could act as external plasticizers, an effect which was widely used to improve the flexibility of polymers, and allow polymer chains to have a large amount of extra free volume, i.e., lower T_g.

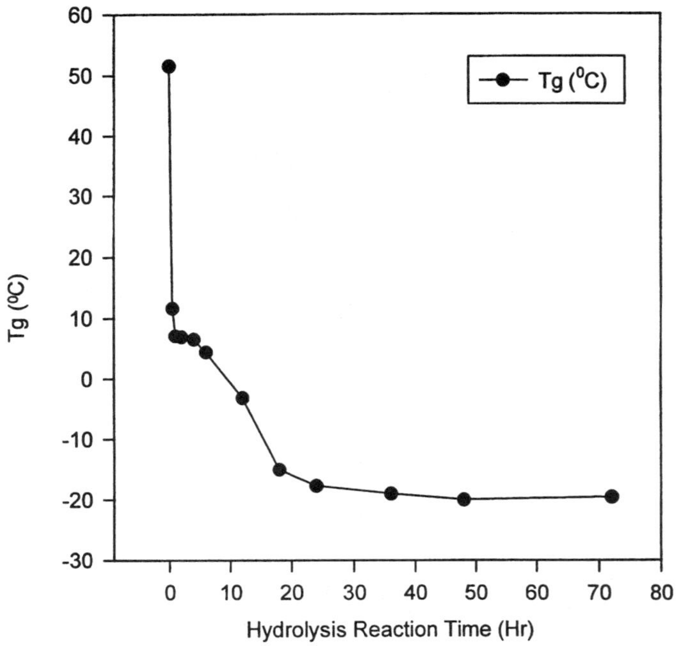

Figure 5 Change of glass transition temperature, T_g, of PDLLA as a function of superoxide ion hydrolysis time at 25°C and 0.01 molar superoxide ion concentration.

Since the superoxide-ion-induced degradation of PDLLA was also temperature dependent, the change of T_g of PDLLA and PLLA due to superoxide ion was also found to be temperature dependent. The magnitude of reduction in T_g of the 37°C PDLLA sample at 0.01 molar superoxide ion concentration over the 24 h hydrolysis time (from 51.58°C to −4.07°C with $\Delta T_g = 55.65°$) was significantly less than the 25°C sample (from 51.58°C to −17.7°C with $\Delta T_g = 69.28°$). These T_g data at different reaction temperatures were consistent with our reaction temperature dependent molecular weight data. Again, the superoxide ion was less stable at 37°C than 25°C, and thus less active superoxide ion was available for degradation at 37°C, i.e., a retention of higher molecular weight and higher T_g.

The effect of superoxide ion concentration on the degradation of PDLLA and PLLA described previously should also be reflected in the thermal properties of PDLLA. As shown in Fig. 6, a close to linear reduction in T_g with superoxide ion concentration was observed and the T_g was reduced from the original 51.58°C at zero superoxide ion concentration to −17.7°C at 0.01 molar of superoxide ion concentration, 24 h hydrolysis time and 25°C. The continuous reduction in T_g beyond 0.0025 molar superoxide ion concentration without significant reduction in M_n (see Fig. 2) could be attributed to the plasticizer effect from the low molecular weight oligomers which would lower T_g of PDLLA further than expected from molecular weight data.

In general, it has been known that T_g was directly related to the number average molecular weight of a polymer as described in the Flory-Fox theory [40,41]:

$$T_g = T_g^{\infty} - \frac{K_g}{M_n} \tag{1}$$

where T_g^{∞} is the limiting T_g at an infinite molecular weight and K_g is a constant. This theory was based on the hypothesis that the local configurational order in a liquid polymer was distrib-

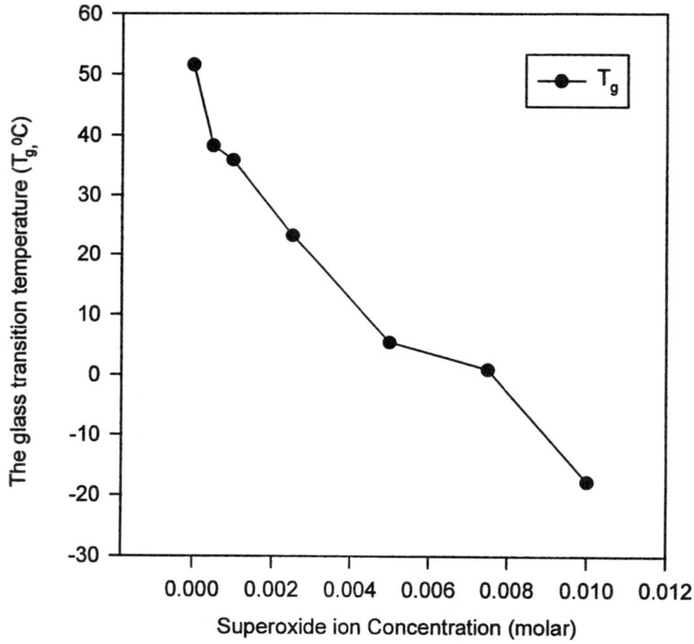

Figure 6 Change of T_g of PDLLA as a function of superoxide ion concentration at 24 h hydrolysis time & 25°C.

uted by the introduction of chain end groups to a degree that was proportional to their number. Boyer [42] reported that the molecular weight dependence of T_g of polystyrene based on Eq. (1) could be represented by three intersecting straight line regions, where each region exhibited a different K_g value. Boyer's K_g was found to decrease with decreasing molecular weight. A number of studies have been reported to detail this relationship [42–46].

It would be interesting to know whether our T_g versus M_n data of PDLLA would fit into the Fox-Flory theory. As shown in Fig. 7, our molecular weight dependence of T_g of PDLLA agreed well with the general characteristic of Fox-Flory relationship of T_g and molecular weight. As the M_n of PDLLA increased, T_g also increased up to an M_n of about 40,000 and the effect of molecular weight on T_g became less pronounced at $M_n > 40{,}000$.

Our experimental K_g values decreased with decreasing molecular weight. This K_g versus molecular weight relationship was suggested by Boyer to be associated, to some degree, with the free volume at chain ends as the reduction in free volume with decreasing molecular weight would lead to a lower K_g value. As shown in Table 1, our experimental K_g values, however, were consistently higher than the Boyer K_g over the molecular weight ranges $M_n > 10^4$ and $10^3 < M_n < 10^4$. This discrepancy in K_g values was attributed to the uniformity or the level of distribution of chain length. The Boyer K_g values were calculated based on a relatively uniform or homogeneous chain length (narrow distribution), while our experimental K_g values were obtained from the mixtures of polymers having broad chain length distribution as shown previously. We suggested previously that the presence of low molecular weight oligomers, trimers and dimers in a polymer-oligomer mixture, could act as external plasticizers to lower T_g more than the Fox-Flory theory predicted. This lower T_g would lead to a higher K_g value, according to Eq. (1), and was indeed found in our experiment.

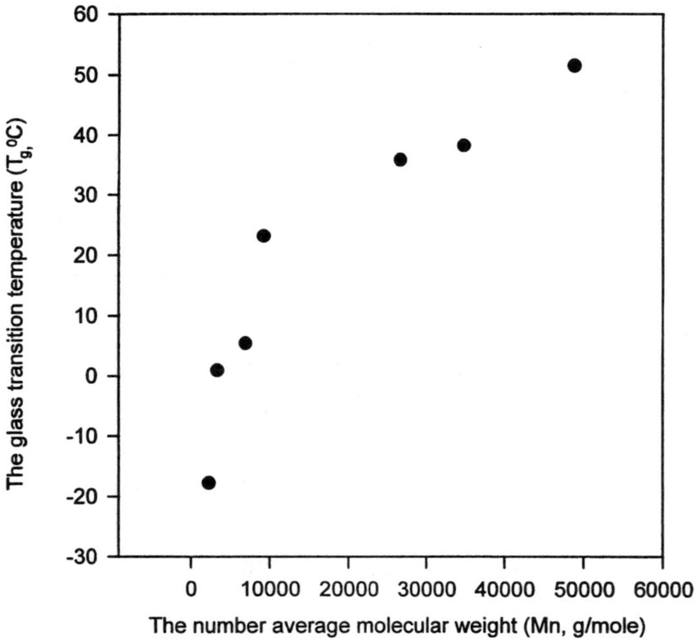

Figure 7 Relationship between number average molecular weight of PDLLA and T_g.

B. Heterogeneous Superoxide-Induced Degradation of Absorbable Suture Materials

The results of the tensile breaking force of the five absorbable sutures as a function of superoxide ion concentration and reaction time are summarized in Table 2 and Fig. 8 for 0.005 molar of superoxide ion concentration. Regardless of reaction time, the measurement of tensile breaking force of these five absorbable sutures at the highest superoxide ion concentration (0.01 molar) was impossible because all absorbable sutures were so degraded that they completely lost their physical integrity as fibers and became almost invisible.

The common characteristic of the reduction in the tensile breaking force of these absorbable sutures with superoxide-ion-induced reaction time at 25°C was an immediate reduction in their tensile breaking forces as early as 2 h reaction time followed by a relatively smaller decrease thereafter. At a superoxide ion concentration of 0.005 molar, its effect on tensile breaking force of the absorbable sutures retained the similar pattern as the lower concentration, but with a significantly larger magnitude. For example, as shown in Fig. 8, a 10-fold increase in superoxide ion concentration from 0.0005 to 0.005 molar resulted in a loss of tensile break-

Table 1 Comparison of the Boyer Constant K_g Values [42] with our Experimental Constant K_g Values

M_n	Boyer constant K_g	Our exp. constant K_g
$M_n > 10^4$	2×10^5	9×10^5
$10^3 < M_n < 10^4$	$\sim 8 \times 10^4$	1×10^5
$M_n < 10^3$	$\sim 2 \times 10^4$	—

Table 2 Tensile Breaking Force (*Kg*) and Its Percentage Retention of Five Absorbable Sutures as a Function of Superoxide Ion Concentration and Hydrolysis Reaction Time at 25°C

	Superoxide ion concentration (molar)						
	0	0.0005		0.005		0.01	
	Hydrolysis time						
	0 h	2 h	24 h	2 h	24 h	2 h	24 h
PDS II	4.75 ± 0.01 (100.0%)[a]	4.59 ± 0.01 (97%)	4.59 ± 0.02 (97%)	4.57 ± 0.01 (96%)	3.09 ± 0.01 (65%)	0	0
Dexon	6.23 ± 0.01 (100.0%)	5.73 ± 0.01 (92%)	5.59 ± 0.01 (90%)	5.32 ± 0.01 (85%)	2.89 ± 0.02 (46%)	0	0
Vicryl	6.26 ± 0.01 (100.0%)	5.22 ± 0.01 (85%)	5.13 ± 0.01 (82%)	4.90 ± 0.01 (78%)	2.39 ± 0.01 (38%)	0	0
Maxon	7.42 ± 0.01 (100.0%)	6.20 ± 0.01 (84%)	5.40 ± 0.02 (73%)	6.16 ± 0.01 (83%)	1.86 ± 0.01 (25%)	0	0
Monocryl	5.66 ± 0.01 (100.0%)	4.40 ± 0.01 (78%)	3.70 ± 0.01 (65%)	3.01 ± 0.01 (53%)	1.14 ± 0.01 (20%)	0	0

[a]Percentage in parentheses indicates retention of original (no superoxide and 0 h) tensile strength.

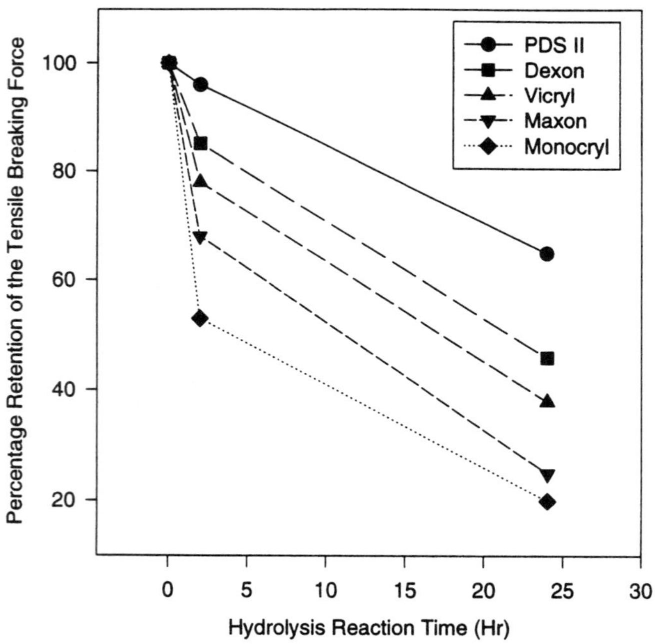

Figure 8 Percentage of retention of tensile breaking force of five synthetic absorbable sutures upon superoxide-ion-induced hydrolytic degradation at 0.005 molar superoxide ion concentration at 25°C.

Table 3 Melting Temperature (T_m, °C) and Heat of Fusion (ΔH_f, cal/g) of Five Absorbable Sutures upon Superoxide-Ion-Induced Hydrolytic Degradation as a Function of Superoxide Ion Concentration at 25°C and 24 h Hydrolysis Reaction Time

	Superoxide ion concentration (molar)							
	0		0.0025			0.005		
	T_m	(ΔH_f)	T_m	(ΔH_f)	ΔT_m	T_m	(ΔH_f)	ΔT_m (°C)
Dexon	217.0 (°C)	(18.3)	209.4 (°C)	(18.9)	7.6	199.3 (°C)	(16.6)	17.7
Vicryl	196.2	(12.4)	—	—		179.8	(2.87)	16.4
Monocryl	162.5	(13.4)	157.6	(13.1)	4.9	149.7	(8.00)	12.8
Maxon	199.7	(9.84)	196.0	(10.9)	3.7	192.0	(4.88)	7.
PDS II	99.3	(21.9)	96.7	(22.6)	2.6	93.0	(18.9)	6.

ing force as large as 47% and 80% for Monocryl suture after 2 and 24 h, respectively. The bulk of the loss of tensile breaking force of Monocryl occurred during the initial 2 h period. Even the most superoxide ion resistant PDS II suture showed an appreciable loss of tensile breaking force at 0.005 molar of superoxide ion concentration. Like PDS II sutures, Dexon, Vicryl and Maxon sutures all showed most of their loss of tensile breaking forces between the 2 and 24 h period. The order of these five absorbable suture materials toward the superoxide ion sensitivity at this relatively higher superoxide ion concentration was the same as the lower superoxide ion concentration case, Monocryl > Maxon > Vicryl > Dexon > PDS II. It is important to know that there would be no change in tensile breaking force of these absorbable sutures in regular saline buffer media at 25°C for many days [18].

Tables 3 and 4 and Figs. 9 and 10 illustrate the changes in T_m and T_g of the five absorbable sutures as a function of superoxide ion concentration at 25°C and 24 h reaction time. Regardless of superoxide ion concentration, T_m of all five absorbable sutures generally decreased with increasing superoxide ion concentration. The order of ΔT_m ($=T_m$ at zero superoxide ion concentration $-T_m$ at a specified superoxide ion concentration) for these five absorbable

Table 4 Glass Transition Temperature (T_g, °C) of Five Absorbable Sutures upon in Vitro Superoxide-Ion-Induced Hydrolytic Degradation at 25°C and 24 h Hydrolysis Reaction Time at Two Different Superoxide ion Concentrations

	Superoxide ion concentration (molar)				
	0	0.0025		0.005	
	T_g	T_g	ΔT_g	T_g	ΔT_g
Dexon	62.7	53.7	9.0	43.2	19.5
Vicryl	65.3	—	—	42.7	12.6
Monocryl	7.8	0.4	7.4	−1.9	9.7
Maxon	26.0	21.9	4.1	17.5	8.5
PDS II	−10.9	−11.7	0.8	−13.1	1.2

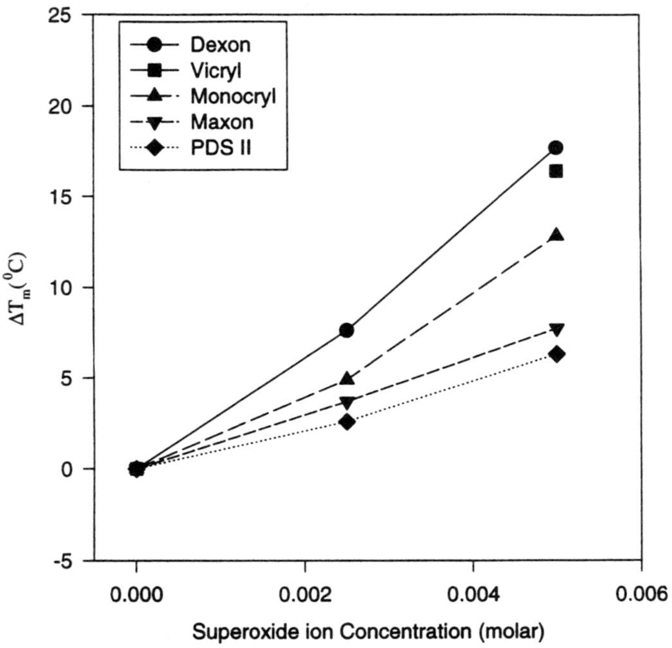

Figure 9 Reduction in melting temperature, T_m, of five synthetic absorbable sutures upon superoxide-ion-induced hydrolytic degradation at 25°C over a superoxide ion concentration ranging from 0.0025 to 0.005 molar. $\Delta T_m = T_m$ (of original at zero superoxide ion concentration) – T_m (of samples at a specified superoxide ion concentration).

sutures was Dexon > Vicryl > Monocryl > Maxon > PDS II as shown in Fig. 9. Dexon suture showed the most dramatic reduction in T_m ($\Delta T_m = 7.6$°C and 17.7°C) at both superoxide ion concentrations, and this suture also has the highest level of crystallinity and T_m among the five absorbable sutures. Among the monofilament absorbable sutures, Monocryl was the most sensitive suture material upon superoxide-ion-induced thermal property changes ($\Delta T_m = 4.9$°C and 12.8°C at 0.0025 molar and 0.005 molar, respectively). This might be due to both solvent affinity and superoxide ion reactivity toward the nature of the chemical structure in Monocryl. The least superoxide-ion-induced change in thermal properties among these five absorbable sutures was PDS II ($\Delta T_m = 2.6$°C, 6.3°C at both superoxide ion concentration) possibly due to its distinctive skin-core morphology in which the core of the PDS II suture has a highly ordered and larger spherulitic crystal structure than the surrounding annular area [47]. Maxon suture showed the next to the most resistant PDS II absorbable suture and its changes in thermal properties due to superoxide ion were relatively small ($\Delta T_m = 3.7$°C and 7.7°C) at both superoxide ion concentrations.

Table 4 summarizes the results of the change in T_g of the five synthetic absorbable sutures as a function of superoxide ion concentration at 25°C and 24 h reaction time. Figure 10 shows the change in T_g and their magnitudes of difference from the controls ($\Delta T_g = T_g$ of zero superoxide concentration $-T_g$ at a specific superoxide concentration) of the five synthetic absorbable sutures. Dexon suture showed the most significant change in T_g at both superoxide ion concentrations ($\Delta T_g = 9.0$°C, 19.5°C) followed by Vicryl ($\Delta T_g = 12.6$°C). Among the monofilament absorbable sutures, Monocryl had the highest T_g reduction ($\Delta T_g = 7.4$°C, 9.7°C), while PDS II was the most resistant to superoxide-ion-induced T_g change with the smallest ΔT_g among all five synthetic absorbable suture materials ($\Delta T_g = 0.8$°C, 1.2°C).

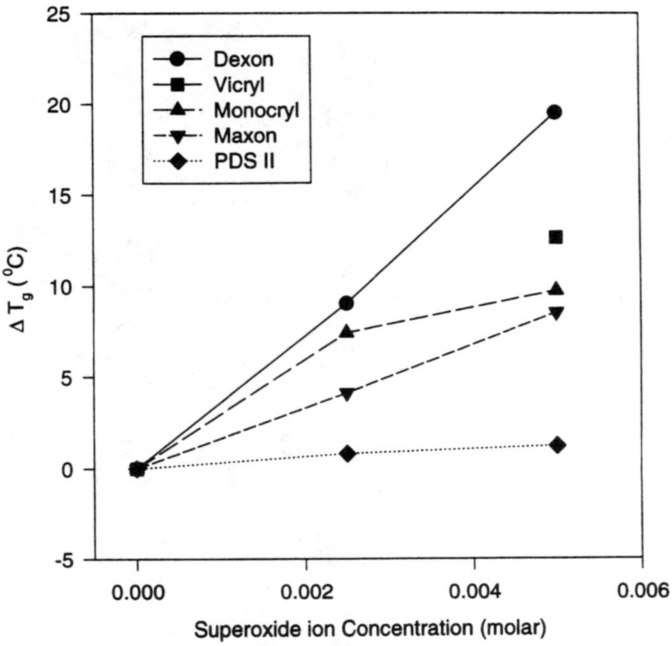

Figure 10 Magnitude of reduction in glass transition temperature, ΔT_g, of five synthetic absorbable sutures upon superoxide-ion-induced hydrolytic degradation at 25°C over a superoxide on concentration ranging from 0.0025 to 0.005 molar. $\Delta T_g = T_g$ (of original at zero superoxide ion concentration) − T_g (of samples at a specified superoxide ion concentration).

The changes in the suture fiber surface morphology upon superoxide-ion-induced degradation were consistent with the observed tensile strength data and were found to depend on the reaction time and superoxide ion concentration. The suture fibers which showed the most unique surface morphological changes are Monocryl and Maxon sutures, particularly at a higher superoxide ion concentration and longer duration of reaction as shown in Figs. 11 (Maxon) and 12 (Monocryl). Both sutures exhibited many moon-crater-shaped impressions of a variety of sizes at a superoxide ion concentration of 0.0025 molar and 24 h reaction.

In the superoxide ion media, some longitudinal and circumferential cracks started to appear on Maxon suture surface at 0.005 molar superoxide concentration. Monocryl sutures had the most unique surface morphology at this highest superoxide ion concentration and 24 h reaction of this study (Fig. 12). Besides the appearance of more moon-crater-shaped impressions, there were many rope-like microfibers of about 10 μm diameter arisen from the underlining Monocryl suture fiber and arranged circumferentially and orderly along the suture surface. This unique surface morphology suggested that there would be a significant loss of mechanical properties of the suture as evident in the observed dramatic reduction in the retention of tensile strength and thermal properties of both Monocryl and Maxon sutures described earlier. Both Maxon and Monocryl sutures lost majority of their size at this relatively high superoxide concentration and at the end of 24 h. At 0.005 molar superoxide ion concentration, the diameter of the Maxon suture decreased to 30% (120 μm) from its original 400 μm diameter, while the Monocryl suture retained only 48% of its original 400 μm diameter (192 μm).

Even the most superoxide-resistant PDS II sutures exhibited few patches of rough spots (PDSII) on the outermost skin of the sutures. The dimensional change of PDS II sutures was

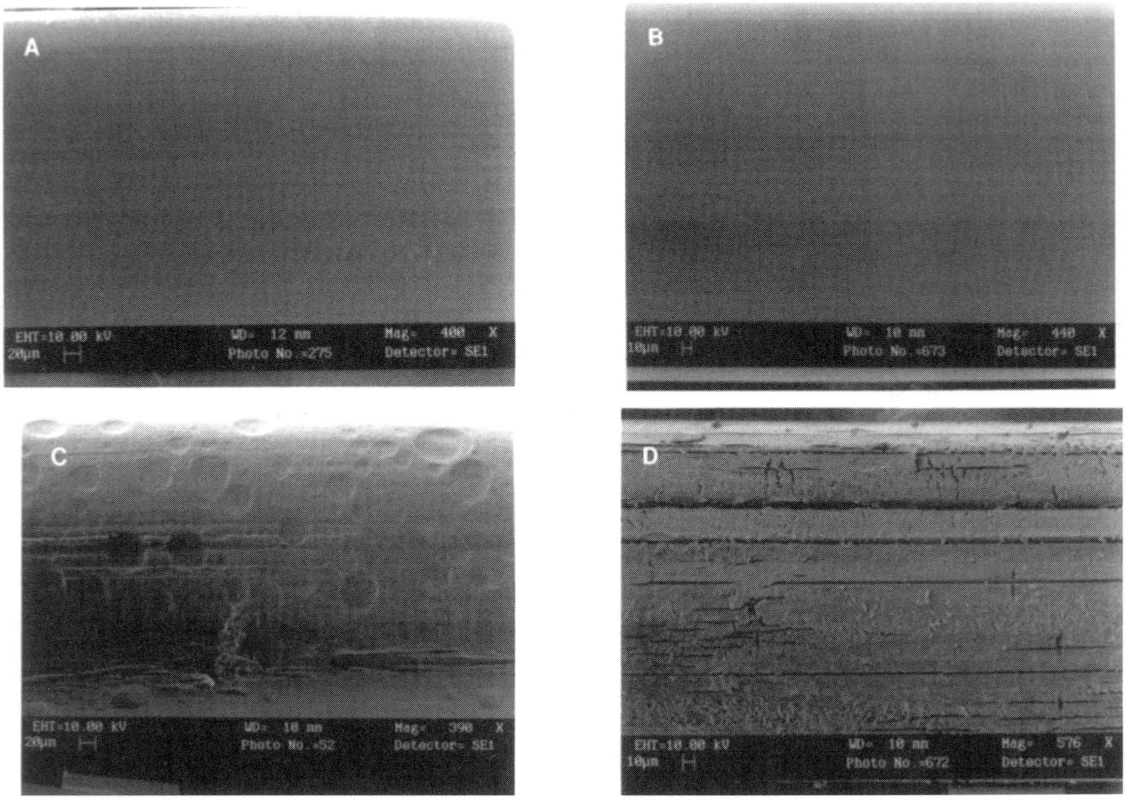

Figure 11 Surface morphology of 2-0 Maxon suture upon superoxide-ion-induced hydrolytic degradation at 25°C for 24 h: (A) original control; (B) 0.0005 molar; (C) 0.0025 molar; (D) 0.005 molar superoxide ion concentration.

significant at the highest superoxide ion concentration (0.005 molar) and the diameter of PDS II sutures decreased to 67% (268 μm) from their original diameter of 400 μm. This diameter reduction of PDSII over a period of 24 h degradation reaction in superoxide ion medium provides clear evidence to support the "peeling onion" mode of superoxide-ion-induced degradation of monofilament sutures, i.e., the exposure of its fresh inner core after the removal of their outermost skin layers during the 24 h superoxide-ion-induced degradation. Since the core of the PDS II suture has a more highly ordered and larger spherulitic crystal structure than the skin of the suture [47], it is expected to be more resistant to degradation than the skin.

The braided multifilament absorbable sutures like Dexon and Vicryl, however, did not show the same dramatic morphological change as the monofilament absorbable sutures did at the higher superoxide ion concentration and longer period of reaction (24 h) at 25°C. In the case of Dexon suture, the disintegration of the braided structure of the suture started to appear at a low superoxide ion concentration (0.0025 molar) and became more severe with the formation of many needle-shaped filament ends as the superoxide ion concentration increased further. The general surface degradation patterns of Vicryl suture were very similar to Dexon suture but were less severe (Fig. 13).

It is important to recognize that none of these and other absorbable sutures show this morphological pattern upon their hydrolytic degradation in normal saline buffer solutions. In

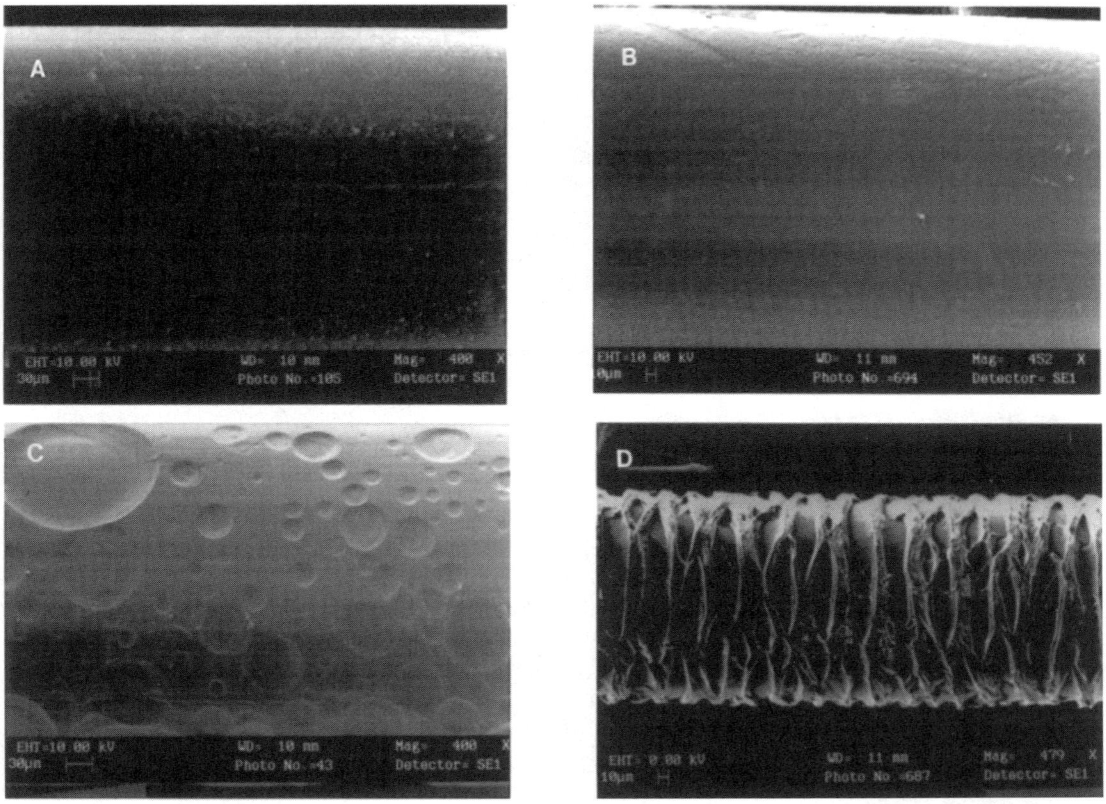

Figure 12 Surface morphology of 2-0 Monocryl suture upon superoxide-ion-induced hyrdrolytic degradation at 25°C for 24 h: (A) original control; (B) 0.0005 molar; (C) 0.0025 molar; (D) 0.005 molar superoxide ion concentration.

the reported morphological studies of all existing absorbable sutures in conventional buffer media [23,24,48–60], the most common surface morphological characteristics upon hydrolytic degradation is the formation of circumferential or/and longitudinal surface cracks that are consistent with the anisotropic characteristic of fibers. It is not fully understood at this stage how superoxide-ion-induced degradation could lead to such unusual surface morphology on Monocryl and Maxon sutures.

III. CONCLUSION

We examined the reactivity of superoxide ion as an oxygen nucleophile agent toward biodegradable polyesters, PDLLA, PLLA and five synthetic absorbable sutures as a function of temperature, time and superoxide ion concentration. The data of molecular weight, mechanical and thermal properties suggested that the hydrolytic cleavage pathway of polymer chain backbone by superoxide ion involved an S_N2 displacement in aprotic solvents. This fragmentation produced various mixture of different chain lengths, such as oligomers, trimers and dimers. A combined GPC and chemical tagging method revealed that the structure of oligomer species in the degradation mixture was of linear characteristic.

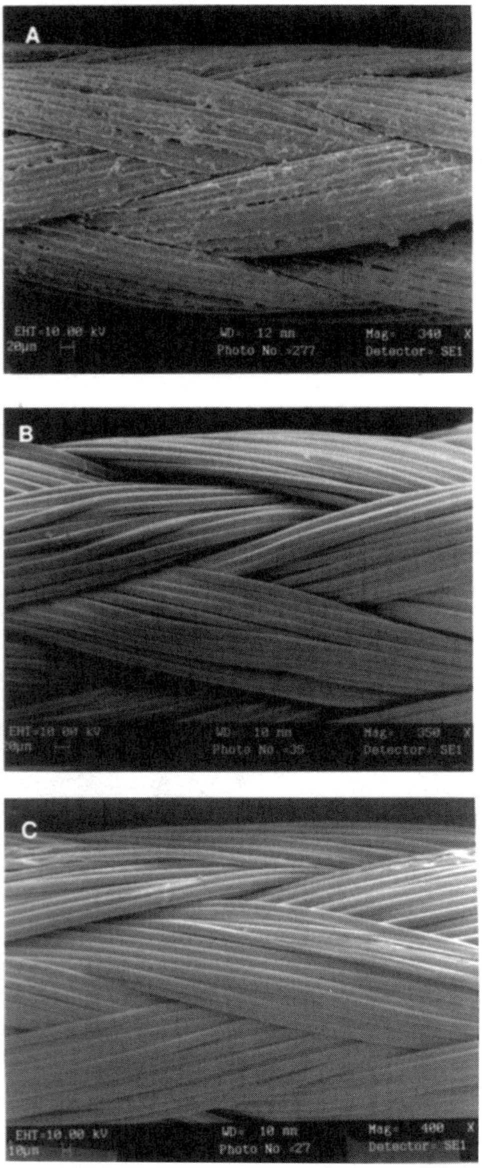

Figure 13 Surface morphology of 2-0 Vicryl suture upon superoxide-ion-induced hydrolytic degradation at 25°C for 24 h: (A) original control; (B) 0.0005 molar; (C) 0.0025 molar superoxide ion concentration.

The reactivity of the superoxide ion toward five commercial 2/0 size synthetic absorbable sutures at 25°C indicated that there was a significant effect of superoxide-ion-induced degradation on the mechanical properties, thermal properties, and surface morphology of these five absorbable sutures. Among the five absorbable sutures and over the concentration range of this study, monofilament Monocryl suture was the most sensitive toward superoxide-ion-induced degradation followed by Maxon, Vicryl, Dexon and PDSII sutures had relatively the least effect of superoxide-ion-induced hydrolytic degradation. The mount of tensile breaking force loss ranged from as low as 3% to as high as 80%, depending on the type of absorbable sutures, reaction time, and superoxide ion concentration. All five absorbable sutures showed significant reductions in both T_m and T_g. Unlike the surface morphological change of the absorbable suture in buffer solutions, the effect of superoxide-ion-induced degradation on the surface morphological change of the five absorbable sutures were unique, particularly the moon-crater-shaped impressions of various sizes and depth found in Monocryl and Maxon sutures that defied the anisotropic characteristic of fibers.

REFERENCES

1. Halliwell, B. and Gutteridge, J. M. C., 1989. Protection against tissue damage in vivo by desferrioxamine: what is its mechanism of action? In: *Free Radicals in Biology and Medicine*, Oxford University Press, New York.
2. Fridovich, I., 1974. Super oxide dismutases. *Advan. Enzymol.*, 41, 35.
3. Babior, B. M., 1978. Oxygen dependent microbial killing by phagocytes. Part 1. *New Engl. J. Med.*, 298, 659–668.
4. Lands, W. E. M., 1979. The biosynthesis and metabolism of prostaglandins. *Annu. Rev. Physiol.*, 41, 633–652.
5. Babior, B. M., Kipnes, R. S. and Curnutte, J. T., 1973. Biological defense mechanisms the production by leukocytes of superoxide a potential bactericidal agent. *J. Clin. Invest.*, 52, 741.
6. Hassan, H. M. and Fridovich, I., 1979. Paraquat and *Escherichia coli* mechanism of production of extracellular superoxide radical. *J. Biol. Chem.*, 254, 10846.
7. Slater, T. F., 1984. Free-radical mechanisms in tissue injury. *Biochem. J.*, 222, 1–15.
8. White, J. R., Vaughn, T. O. and Shiang Yeh, W., 1971. Biochemical pharmacology. II: Oxidation and enzymes. *Fed. Proc., Fed. Amer. Soc. Exp. Bio.*, 30, 1145.
9. Meisel, D. and Czapski, G., 1975. One-electron transfer equilibriums and redox potentials of radicals studied by pulse radiolysis. *J. Phys. Chem.*, 79, 1503.
10. McCord, J. M. and Fridovich, I., 1973. Production of O_2^- in photolyzed water demonstrated through the use of superoxide dismutase. I. *Photochem. Photobiol.*, 117, 115.
11. Sayer, D. T. and Seo, E. T., 1977. One-electron mechanism for the electrochemical reduction of molecular oxygen. *Inorg. Chem.*, 16, 499.
12. Valentine, J. S. and Curtis, A. B., 1975. A convenient preparation of solutions of superoxide anion and the reaction of superoxide anion with a copper II complex. *J. Am. Chem. Soc.*, 97, 224.
13. San Fillipo, Jr., J., Romano, L. J., Chern, C. I. and Valentine, J. S., 1976. Oxidative cleavage of α-keto, α-hydroxy, and α-halo ketones, esters, and carboxylic acids by superoxide. *J. Org. Chem.*, 41, 586.
14. Magno, F. and Bontempelli, G., 1976. On the reaction of kinetics of electrogenerated superoxide ion with aryl benzeates. *J. Electroanal. Chem.*, 68, 337.
15. Johnson, R. A., 1976. Superoxide chemistry. A new, convenient synthesis of diacyl peroxides. *Tetrahedron Lett.*, 331.
16. Forrester, A. R. and Purushotham, V., 1984. Mechanism of hydrolysis of esters by superoxide. *J. Chem. Soc. Chem. Commun.*, 1505.
17. Forrester, A. R. and Purushotham, V., 1987. Reactions of carboxylic acid derivatives with superoxide. *J. Chem. Soc. Perkin Trans.* 1, 945.
18. Chu, C. C., von Fraunhofer, J. A. and Greisler, H. P., 1997. *Wound Closure Biomaterials and Devices*, CRC Press, Boca Raton, FL.
19. Williams, D. F., Smith, R. and Oliver, C., 1985. The degradation of carbon-14 labeled polymers by enzymes. In: *Biological and Biomechanical Performance of Biomaterials*, P. Christel, A. Meunier, and A. J. C. Lee, Eds., Elsevier, Amsterdam, 239.
20. Kossovsky, N., Heggers, J. P. and Robson, M. C., 1987. The bioreactivity of silicone. *CRC Crit. Rev. Biocompat.*, 3, 53.

21. Zhong, S. P. and Williams, D. F., 1991. Are free-radicals involved in biodegradation of implanted polymers. *Adv. Mater.*, 3, 623.
22. Ali, S. A. M., Doherty, P. J. and Williams, D. F., 1993. Mechanisms of polymer degradation in implantable devices. 2. Poly(DL-lactic acid). *J. Biomed. Mater. Res.*, 27, 1409.
23. Ali, S. A. M., Zhong, S. P., Doherty, P. J. and Williams, D. F., 1993. Mechanisms of polymer degradation in implantable devices. 1. Poly(caprolactone). *Biomaterials*, 14, 648.
24. Zhong, S. P., Doherty, P. J. and Williams, D. F., 1994. A preliminary study on the free radical degradation of glycolic acid/lactic acid copolymer. *Plastics, Rubber and Composite Proc. Appl.*, 21, 89.
25. Ali, S. A. M., Doherty, P. J. and Williams, D. F., 1994. The mechanisms of oxidative-degradation of biomedical polymers by free-radicals. *J. Appl. Poly. Sci.*, 51, 1389.
26. Lee, K. H. and Chu, C. C., 1996. Electron spin resonance study of free radical properties of polyglycolic acid upon g-irradiation sterilization, *5th World Biomaterials Congress*, Toronto, May 29–June 2, Vol. 2, 811.
27. Lee, K. H. and Chu, C. C., 1996. The role of free radicals in hydrolytic degradation of absorbable polymeric biomaterials, *5th World Biomaterials Congress*, Toronto, Vol. 2, 91.
27a. Lee, K. H., Won C. Y., Chu, C. C., and Gitsov, I., Hydrolysis of biodegradable polymers by superoxide ions, *J. Polym. Sci. Polym. Chem.* (In Press).
27b. Lee, K. H. and Chu, C. C. The role of superoxide ions in degradation of synthetic absorbable sutures. *J. Biomed. Mater. Res.* (In Press).
28. Van Sliedregt, A., De Groot, K. and Van Blitterswijk, C. A., 1993. *In vitro* biocompatibility testing of polylactides. Part II. Morphologic aspects of different cell types. *J. Mater. Sci. Mater. Med.*, 4, 213.
29. Vert, M., Christel, P., Chabot, F. and Leray, J., 1984. Bioresorbable plastic material for bone surgery. In: G. W. Hastings and P. Ducheyne (Eds.), *Macromolecular Materials*, CRC Press, Boca Raton, FL.
30. Leeslang, J. W., Pennings, A. J., Veth, R. P. H., Nielsen, H. K. L. and Jansen, H. W. B., 1984. A porous composite, based on a biodegradable poly(ι-lactide)-polyurethane matrix and reinforced with carbon fibres, for reconstruction of meniscus lesions. *Makromol. Chem. Rapid Commun.*, 5, 815.
31. Leeslang, J. W., Pennings, A. J., Bos, R. R. M., Rozema, F. R. and Boering, G., 1987. Resorbable materials of poly(L-lactide). VII. In vivo and in vitro degradation. *Biomaterials*, 8, 311.
32. Wakiyama, N., Juni, K. and Nakano, M., 1982. Preparation and evaluation *in vitro* and *in vivo* of polylactic-acid microspheres containing dibucaine. *Chem. Pharm. Null.*, 30, 3719.
33. Shalaby, S. W., 1994. Synthetic absorbable polyesters. In: *Biomedical Polymers: Designed-to-Degrade System*, S. W. Shalaby (Ed.), Hanser, Munich, pp. 1–34.
34. Manninen, M. J., 1993. Self-reinforced poly-L-lactide screws in the fixation of cortical bone osteotomies in rabbits. *J. Mater. Sci. Mater. Med.*, 4, 179.
35. Zhang, X., Wyss, U. P., Pichora, D. and Goosen, M. F. A., 1993. Biodegradable polymers for orthopedic applications-synthesis and processability of poly(L-lactide) and poly(lactide-co-ε-caprolactone). *J. Macromol. Sci. Pure Appl. Chem.*, 30, 933.
36. Li, S. M., Garreau, H. and Vert, M., 1990. Structure property relationships in the case of the degradation of massive aliphatic poly(alpha-hydroxy acids) in aqueous media. 1. Poly(DL-lactic acid). *J. Mater. Sci. Mater. Med.*, 1, 123.
37. Li, S. M., Garreau, H. and Vert, M., 1990. Structure property relationships in the case of the degradation of massive aliphatic poly(alpha-hydroxy acids) in aqueous media. 2. Degradation of lactide-glycolide copolymers PLA37.5GA25 and PLA75GA25. *J. Mater. Sci. Mater. Med.*, 1, 131.
38. Li, S. M., Garreau, H. and Vert, M., 1990. Structure-property relationships in the case of the degradation of massive aliphatic poly-(alpha-hydroxy acids) in aqueous media. 3. Influence of the morphology of poly(L-lactic acid). *J. Mater. Sci. Mater. Med.*, 1, 198.
39. Manolova, N. E., Gitsov, I., Velichkova, R. S. and Rashkov, I. B., 1985. Separation and characterization of ε-caprolactone oligomers by gel permeation chromatography. *Polym. Bull.*, 13, 285.
40. Fox, T. G. and Flory, P. J., 1950. Second-order transition temperatures and related properties of polystyrene. I. Influence of molecular weight. *J. Appl. Phys.*, 21, 581.

41. Fox, T. G. and Flory, P. J., 1954. The glass temperature and related properties of polystyrene. Influence of molecular weight. *J. Polym. Sci.*, 14, 315.
42. Boyer, R. F., 1974. Variation of polymer glass temperatures with molecular weight. *Macromolecules*, 7, 142.
43. Cowie, J. M. G., 1974. Some general features of T_g-M relations for oligomers and amorphous polymers. *Eur. Polym. J.*, 11(4), 297.
44. Kumlmer, P. L., Keinath, S. E. and Boyer, R. F., 1977. ESR studies of polymer transitions. III. Effect of molecular weight and molecular weight distribution of T_g values of polystyrenes as determined by ESR spin-probe studies. *J. Macromol. Sci. Phys.*, B13(4), 631.
45. Turner, D. T., 1978. Glass transition elevation by polymer entanglements. *Polymer*, 19, 789.
46. Pochan, J. M., Beatty, C. L. and Pochan, D. F., 1979. Different approach for the correlation of the T_g of mixed amorphous systems. *Polymer*, 20, 879.
47. Broyer, E., 1994. Thermal treatment of theraplastic filaments for the preparation of surgical sutures. U. S. Patent 5,294,395, Ethicon, Inc.
48. Pratt, L., Chu, C., Auer, J., Chu, A., Kim, J., Zollweg, J. A. and Chu, C. C., 1993. The effect of ionic electrolytes on hydrolytic degradation of biodegradable polymers: Mechanical properties, thermodynamics and molecular modeling. *J. Polym. Sci. Chem. Ed.*, 31, 1759.
49. Zhang, L., Loh, I. H. and Chu, C. C., 1993. A combine γ-irradiation and plasma deposition treatment to achieve the ideal degradation properties of synthetic absorbable polymers. *J. Biomed. Mater. Res.*, 27, 1425.
50. Chu, C. C., Zhang, L. and Coyne, L., 1995. Effect of irradiation temperature on hydrolytic degradation properties of synthetic absorbable sutures and polymers. *J. Appl. Polym. Sci.*, 56, 1275.
51. Chu, C. C. and Williams, D. F., 1983. The effect of γ-irradiation on the enzymatic degradation of polyglycolic acid absorbable sutures. *J. Biomed. Mater. Res.*, 17, 1029.
52. Williams, D. F. and Zhong, S. P., 1991. Are free radicals involved in the biodegradation of implanted polymers. *Adv. Mater.* 3, 623.
53. Loh, I. H., Lin, H. L. and Chu, C. C., 1992. Plasma surface modification of synthetic absorbable sutures. *J. Appl. Biomater.*, 3, 131.
54. Yu, T. J. and Chu, C. C., 1993. Bicomponent vascular grafts consisting of synthetic biodegradable fibers. Part 1. *In vitro* study. *J. Biomed. Mater. Res.*, 27:1329–1339.
55. Lin, H. L., Chu, C. C. and Grubb, D., 1993. Hydrolytic degradation and morphologic study of poly-*p*-dioxanone. *J. Biomed. Mater. Res.*, 27(2):153.
56. Chu, C. C. and Browning, A., 1988. The study of thermal and gross morphologic properties of polyglycolic acid upon annealing and degradation treatments. *J. Biomed. Mater. Res.*, 22(8):699–712.
57. Chu, C. C. and Lecaroz, L. E., 1987. Design and in vitro testing of newly-made bicomponent knitted fabrics for vascular surgery. In: *Advances in Biomedical Polymers*, Charles G. Gebelein (Ed.), *Polymer Science and Technology*, Vol. 35, pp. 185–214, Plenum Press, New York.
58. Chu, C. C. and Campbell, N. D., 1982. Scanning electron microscopic study of the hydrolytic degradation of polyglycolic acid. *J. Biomed. Mater. Res.*, 16(4):417–30.
59. Williams, D. F., Chu, C. C. and Dwyer, J., 1984. The effects of enzymes and gamma irradiation on the tensile strength and morphology of poly(*p*-dioxanone) fibers. *J. Appl. Polym. Sci.*, 29:1865–1877.
60. Chu, C. C., 1991. Recent advancements in suture fibers for wound closure. In: *HiTech Fibrous Materials*, T. R. Vigo and A. F. Turbak (Eds.), American Chemical Society Symposium Ser. #457, American Chemical Book, Washington, DC, Chapter 12, pp. 167–213.

6

In Vitro Testing of Cytotoxicity of Materials

G. Ciapetti, D. Granchi, C. R. Arciola, E. Cenni, L. Savarino, S. Stea, L. Montanaro, and A. Pizzoferrato
Istituti Ortopedici Rizzoli, Bologna, Italy

I. INTRODUCTION

The usefulness of mammalian cell cultures for biocompatibility testing is confirmed by experimental studies which found a good correlation between in vitro and in vivo tests. The rationale is that in vitro models employing human cells to study the interactions between the cell system and the biomaterial/device allow for a reasonable prediction of the performance in vivo of the biomaterial/device.

It has to be acknowledged that isolated cell systems are more sensitive to toxic materials than body tissues. Nevertheless, there is an increasing demand for reliable in vitro methods for two main reasons: (1) cellular mechanisms of toxicity can be described using biochemical assays; (2) valid alternatives to animal models are needed. Using such testing methods the materials can be screened according to their grade of toxicity and discarded, if this is the case, prior to further testing.

A number of characteristics and functions of cells can be verified after challenge with biomaterials: morphology [1], membrane integrity [2], cytoskeleton [3], surface molecules [4], viability, proliferation [5], protein synthesis [6], oxidative response [7], motility [8], secretion [9], response to growth factors and cytokines [10], cell-cell interactions and, recently, gene expression [11,12].

Questions to be answered by such studies may be “Is this material toxic?” or “Which mechanism/function of the cell is hampered/promoted by this material?” Producers of medical devices need an answer to the first one, while biomedical research tries to disclose the second body of knowledge.

II. STANDARDS FOR TESTING CYTOTOXICITY

Preclinical assessment of the toxic potentials of medical devices have to be undertaken, to determine any possible hazard which may be associated with their application. Standards concerning the safety assessment of medical devices are prepared at the International Organization for Standardization level by the ISO Technical Committee 194, and at the European Committee for Standardization level by CEN Technical Committee 206. The European standards EN

30993 are currently being developed on the basis of the corresponding international standards ISO 10993.

III. MATERIALS PREPARATION

The preparation of materials and/or devices for cytotoxicity testing is of utmost importance, as their chemical/physical appearance strongly influences the response of the cells, as well as the whole experimental system. Standards and guidelines for the preparation of materials have been produced by many agencies and associations, including the International Organization for Standardization (ISO), the Ente Nazionale Italiano di Unificazione (UNI), the British Standards Institute (BSI), the Deutsches Institut fur Normung (DIN), the Swiss Association for Standardization (SNN), the Association Francaise de Normalisation (AFNOR), the American Society for Testing and Materials (ASTM), the American Dental Association (ADA). In such documents chemical and physical form of materials, sterilization procedures, shape and size of the samples, and any treatment required to get a sample suitable for cytotoxicity testing are addressed. The adoption of such indications is recommended when interlaboratory comparisons are planned or when a report for a producer has to be written; moreover they guarantee the material is not altered nor masked by the procedure of preparation, as the standards are developed by working groups of experts.

Reference materials should be adopted in all cases: they should serve to check the performance of the method and allow comparison with results from other laboratories. Different types of reference materials to be used as negative and positive controls are suggested in the standard ISO 10993-part 12 [13] and efforts for developing reliable and reproducible materials continue to be made [14].

A. Solid Materials

Solid materials are employed when the surface of the biomaterial/device is known to influence the response of cells definitely. The solid material/cell combination is also the experimental system which more closely reproduces the in vivo situation. Although solid materials are often quite difficult to be inserted in cell culture systems, their interactions with cells can be analyzed following preparation of small-sized samples and their placement into suitable culture dishes [15,16]. Indeed, by varying the ratio of specimen size:media volume, almost any effect from nontoxic to extremely cytotoxic can be obtained.

Standard ISO 10993-part 12 provides indications on (1) volume of cell suspension/size of the sample, (2) volume of cell suspension/container and (3) size of the sample/container.

Adhesion of cells onto biomaterials is considered a positive phenomenon for implant outcome: such process is known to influence cell activities. Cell adhesion onto foreign surfaces is inspected mainly by cell count, while electron microscopic techniques or computer-assisted measurement are used for the assessment of spreading and morphology [17,18]. Tamada [1] showed that not only the rate of adhesion and spreading of cells is conditioned by the presence of a collagen layer onto polymer surface, but the functions of cells are influenced. Ertel et al. [19] were able to demonstrate that the bare surface of some polymers, including PDMS, PE, PMMA and polyurethanes, was able to cause cell death, unless it was covered by a proteinaceous layer.

The ability of surface topography to determine orientation and alignment of cells and to modify bone cell response has been demonstrated in several studies [20,21].

B. Extracts of Materials

When the material cannot be tested as a solid (e.g., when the device is made of a combination of materials to be verified in their sinergy), aqueous extracts are the choice. The specimen to be tested is eluted in an extraction tube containing a precise amount of fluid. The extraction of a material/device is a complex procedure which is influenced by the duration, temperature, material/extracting vehicle ratio, type of extracting vehicle and type of material. The choice of the appropriate duration and temperature for extraction should be made according to the end use of the material/device in vivo [13]. Most labs resort to extracts for cytotoxicity testing, due to their ease of preparation and manipulation [22,23]. However, concern is raised sometimes about the presence of leachables in the extracts and the actual amount of such substances. Bordenave et al. [24] showed that some primary reference materials, including low density polyethylene and cellulose, could become toxic to cells when tested as extracts. Therefore, the concentration of leachables in material/device extracts should be measured wherever possible (1) to correlate cytotoxic responses and (2) to avoid false negative results. When metals are extracted, metal ion concentration in the extracts can be measured by either AAS [25] or other sophisticated techniques, including neutron activation analysis and mass spectrometry. Polymers can be analyzed for their dissolution by high pressure liquid cromatography [26], while ceramics are usually assayed as they are [27].

Standards provide suggestions about (a) medium for extraction, (b) temperature, (c) duration, (d) handling of the extracts. Recommended media include water, saline solution and culture medium with or without serum. In the preparation of the extracts for cytotoxicity testing, it has to be borne in mind that the final destination of the extract is a monolayer of cells. When a large volume of extracting media is required, for example to extract blood bags with a long tubing apparatus, water or saline has to be used for economic reasons. Then the extract can be used to dilute a double concentrated (2×) culture medium and added to the cells: in this case the maximum concentration of extract which can be tested is 50% [28]. If an undiluted extract (100%) has to be tested, the biomaterial/device is placed in the proper culture medium for target cells. We use culture medium without serum for two main reasons: (1) serum substances would be altered by contact with foreign surfaces, so that new serum should be added at the time of testing, with final concentration of serum in the system resulting unknown, and (2) the culture medium is usually a complex mixture of salts and nutrients which can extract efficiently the leachables from the foreign surface. The temperature of extraction ranges from 37°C to 121°C: as a rule the lower the temperature the longer the duration of extraction (from 1 h at 121°C to 120 h at 37°C). The choice is dictated mainly by the sensitivity of the material/device to heat: e.g., low density polymers cannot bear high temperatures without alteration of their properties. Most authors choose 37°C: this is the temperature of the body which the biomaterial/device will really experience after implantation in vivo [24,26].

As far as the handling of extracts is concerned, including centrifugation, filtration and storage, some matters need consideration. The EN-30993-part 12 standard also recommends that each of these steps, if taken, has to be justified. Centrifugation is needed when particulates are present in the extracts and will interfere with subsequent analysis: this is the case of AAS measurement, where false positive or overrange values are obtained if metal particles are contained in the medium assayed. Filtration through .22 μm-pore filter is required to obtain sterile extracts from liquids where nonsterile samples of biomaterial/device were dipped: sterility of solutions is an absolute requirement for cell cultures.

Some preliminary experiments were performed in our lab to look for the effect of centrifugation and filtration procedures on extracts from metal powders. In Table 1 the results from

Table 1 Toxicity of Extracts of Chromium and Cobalt on L929 Cells (mean ± sd of three separate runs)

	Time of exposure (h)		
Extract	24	48	72
Chromium	62 ± 17	91 ± 12	64 ± 59
Cobalt	24 ± 15	2 ± 2	42 ± 36

cytotoxicity testing of "crude" extracts of chromium and cobalt onto L929 cells are presented. The cells were exposed for 24, 48 and 72 h to the undiluted extracts: the experiments were performed separately using the same batch of Cr or Co extract on different days. After the challenge with the extracts the cells were fixed and stained using crystal violet dye, and the absorbance of the samples was compared to that found for the control cells. The results on different days were largely different with both chromium and cobalt at any time of exposure. This was interpreted as an uneven distribution of ions within the extract, so that different aliquots of the same batch contained different amount of Cr or Co. This was confirmed by graphite-furnace atomic absorption spectrometry (GFAAS) analysis, by which the content of several aliquots of the same batch of extract was measured (Table 2). It was concluded that the presence of particulate in the extracts without filtration prior to testing was responsible for the difference of ion content in extract samples from the same batch. Centrifugation and fitration of the extracts before storage provides uniform distribution and low variability of the ion content, as demonstrated by the low standard deviation. Since the cells are very sensitive to minute amounts of foreign substances, we recommend that particles generated by degradation of materials during extraction procedure be removed by centrifugation and filtration.

For testing cytotoxicity the extracts of biomaterials/devices have to be preferred to solids samples for several reasons:

a. Extracts can be added to cells in the amount desired and can be diluted to get a toxicity profile.
b. Many replicates can be tested.
c. Problems related to the size and shape of solids are avoided, as well as any mechanical damage or feeding impairment caused to cells by solids.

IV. METHODS OF EVALUATION

In the introduction of the ISO 10993 standard part 5 document, the crucial parameters for cytotoxicity testing are addressed but not specified, in order to leave the research lab to decide

Table 2 Concentration of Metal Ions in the Extracts Measured by GFAAS

Metal ion	Raw extracts (m ± sd)	Filtered extracts (m ± sd)
Chromium (μg/mL)	22.7 ± 20.8	0.008 ± 0.002
Cobalt (μg/mL)	36.59 ± 25.76	58.20 ± 4.66

steps to be taken with regard to (a) cell type, (b) duration of the exposure, (c) method of evaluation, etc. [29]. An overview of this document has recently been published [30].

Many types of cells have been proposed for use in cytotoxicity testing of materials. The use of cells freshly derived from explants, i.e., primary cells, or established cell lines has been debated since the introduction of cell culture systems in biocompatibility testing and is not yet resolved [31]. If it is assumed that tissue reaction has to be reasonably approximated in cell culture systems, cells derived from explants should be preferred, due to their complex mechanisms of response [32]. But when large numbers of materials have to be assayed for potential toxicity, established cell lines provide many advantages [33]. Suitability of L929 cells or 3T3 fibroblasts for cytotoxicity screening of materials has been experienced over a long time: actually their use is recommended in standard documents. In our opinion the reproducibility of intralaboratory results and the comparison of interlaboratory results are very important: they both are warranted by the use of established cells. The duration of the exposure of cells to materials/devices is usually dictated by the method of assay, but the in vivo persistence of the material/device may also be considered.

The 24-h period is the endpoint commonly used for cytotoxicity testing, but in many instances the exposure may be prolonged or shortened. Tomas et al. tested bone-marrow derived cells for 7, 14, 21 days to look for proliferation and osteogenic differentiation in vitro [34]. On the other hand, the duration of contact with materials may be limited to 1–2 h, as recommended for some types of dental materials which will be applied to oral mucosa for a few minutes [35]. The finding of clear toxicity for a material/device after a 24-h testing in vitro could lead to its discard, due to high sensitivity of cell culture systems, although in vivo it is applied for few minutes. If this material/device has unique properties and optimum performance in vivo with short exposure, the benefit from its use could be higher than the risk for the patient [36].

In the ISO 10993 standard part 5 for cytotoxicity testing it is stated that "It is the intention of this part of ISO 10993 to leave open the choice of type of evaluation." Four categories of evaluation types are listed in the standard, with the first being an assessment of cell morphology and the other measurements of cell parameters. In our opinion the greatest source of variation in cell culture assays is the subjective assessment of cytotoxicity. For this reason we recommend the use of quantitative methods for scoring toxicity. A number of biochemical methods have been described, but also visual methods can be modified to allow for quantitation of results.

For initial screening of materials the methods to be adopted by the lab have to be technically simple, reproducible and broadly applicable and to give predictive values. Cell viability, or conversely cell death, and inhibition of cell growth are common parameters currently used in the initial screening of candidate materials. Using these endpoints the acute toxicity of specimens is measured and materials are ranked relative to their cytotoxic effect. Although cell viability and cell growth are well different characteristics of cells, the distinction becomes less clear when a testing method is used to measure both parameters. This often happens with biochemical assay of enzymatic activities: MTT conversion to formazan (see below) is used as a marker of cell viability/proliferation [34].

For rapid cytotoxicity screening of materials/devices cell cultures in microplates and colorimetric methods to quantify cell parameters in situ are becoming increasingly popular. They offer several advantages including easy manipulation, possibility of large number of replicates and production of quantitative results. Several assay methods providing an index of toxicity have been recently compared for their sensitivity [37].

A. Cell Viability

Historically, the hemolysis test was the very first test for the assessment of the damage induced by biomaterials to red blood cells [38]. The test is simple, but nearly every substance not "physiological" is able to induce massive red blood cell lysis, so positive results are obtained with the most part of materials. ^{51}Chromium release and LDH release have been adopted for the assessment of cell integrity, and they are still in use by some groups, though the use of radioactive tracers is under discussion [2,39].

The uptake of neutral red (NR) and the exclusion of propidium iodide (PI) are used by our lab for routine evaluation of cell viability after challenge with materials or extracts. The uptake of NR, initially developed for toxicological and immunological purposes [40,41], was then modified to meet biocompatibility needs. Although it is suitable for quantitative evaluation through spectrophotometric measurement of the amount of dye inside the cells and in this way is described by an AFNOR standard [42], it is often used as a qualitative method [26].

In our hands NR uptake method is sensitive to any material and its reliability has been assessed in comparison with other methods (Fig. 1). It must be considered that the results may be affected by consistent alteration of medium ph, being NR a sensitive probe for monitoring the intra-cellular pH [43]: actually we found a dramatic enhancement of NR uptake, which was not accompanied by a real increase of the number of cells, using an extract of aluminium. The staining of nucleic acids with PI is commonly used in a number of flow cytometry tests: this dye stains dead cells after crossing the injured membrane [44]. Since these methods rely on different mechanisms and different cell compartments, that is viable and dead cells are stained by NR and PI respectively, the entire cell population is covered and different steps of toxicity mechanism may be detected.

Results from PI and NR assay methods show a good correlation; nevertheless occasional disagreement between the two methods was found with NR detecting toxicity where PI was negative. This can be explained by the fact that a delay of cell growth may decrease NR uptake, but not necessarily result in PI uptake if the cell membrane is intact. Another method is the exclusion of trypan blue, with qualitative or quantitative assessment [19].

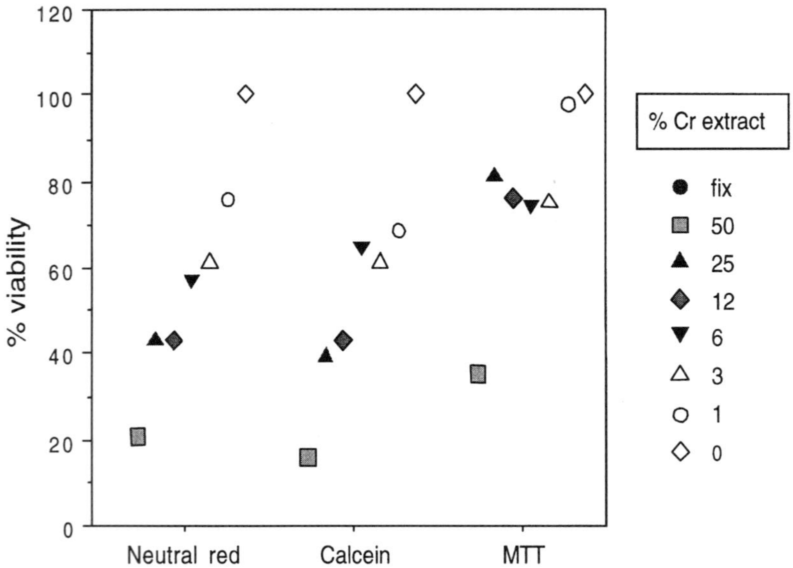

Figure 1 Viability of L929 cells after 24 h challenge with chromium extract at several dilutions measured with the uptake of neutral red (NR), the calcein fluorescence assay (calcein) and the reduction of tetrazolium salts (MTT).

Quite recently the MTT test has become very popular in biocompatibility testing with cell culture systems [26,45]. The conversion of the yellow dye (dimethylthiazolyl)-dyphenyl tetrazolium bromide (MTT) to blue formazan is used as a marker of either cell viability or cell enzymatic activity (mitochondrial) or cell growth. While it is sure that a dead cell is no longer growing, a nondividing cell may well be viable: for this reason we do not consider the results from MTT test a measure of cell growth. Moreover, it has been well documented that the conversion of MTT is driven not only by mitochondrial dehydrogenases, but by other enzymatic and metabolic factors [46], and influenced by glucose consumption [47]. Therefore the MTT assay should be considered a biochemical marker of the metabolic activity of cells.

Such a test is employed by many labs, due to its inherent simplicity and possibility of processing cells in situ: compared to other methods it was found either sensitive [48] or irreproducible [49]. We use it as an additional test for cell viability [50], but it has some drawbacks. Ignatius and Claes [51], using MTT with degradable polymers, found it increased not due to a real increase of cell viability, but probably due to participation of the degradation products to the MTT conversion [51]. In a recent study it was shown that mainly aluminium, but also vanadium, were able to increase MTT reduction by immortalized rat osteoblasts, but a real cell growth was not demonstrated [49]. We also happened to find high levels of MTT conversion, compared to controls, with some extracts of materials (Fig. 2). It is influenced by the pH of the medium, so that a neutral medium has to be provided before MTT conversion takes place; moreover we experienced a poor within-plate precision, as demonstrated by other authors, too [52].

A new microplate-based method has been recently adopted for monitoring cells challenged with materials : the Alamar blue dye changes its color (and its absorbance) with time as oxidoreduction reactions of cells go on [53]. Due to its properties of easy manipulation, nontoxicity to cells and sensitivity, it is now applied for cytotoxicity with a number of cells, including lymphocytes, endothelial cells, neural cells, osteoblasts and fibroblasts. An additional advantage of this dye is that it can be used in conjuction with other methods, so maximizing information from one single assay system.

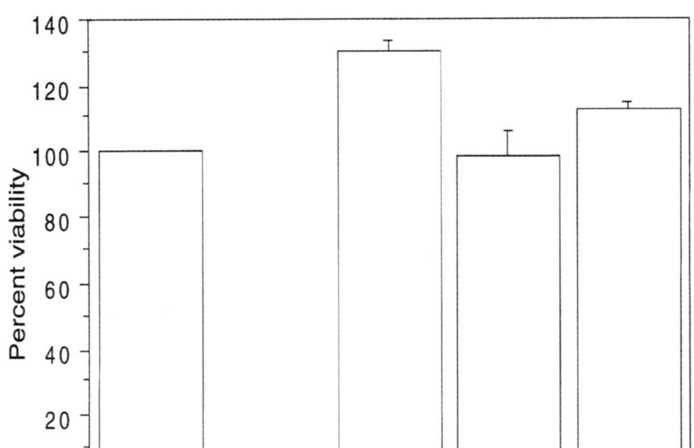

Figure 2 MTT test on lymphocytes challenged for 24 h with extracts of three ceramics (alumina-zirconia and pure alumina): two samples score an index of viability higher than control.

Like the MTT assay, the results from Alamar blue technique are used to describe viability, proliferation, mitochondrial function and metabolic activity of cells [54,55]. A possible limitation of the method is that once color change has occurred due to cell proliferation, the color is stable even if cells, for example, are detached from the plate: thus, the color change does not indicate necessarily the actual status of cells.

In our hands the NR uptake test is the most sensitive in the assessment of cell viability following exposure to biomaterials: Nishi et al. comparing different in vitro methods found a very good sensitivity and a relatively good correlation with the in vivo toxicity estimated by the Draize test [56].

B. Cell Growth

Although time consuming, cell counting with hemocytometer or Coulter counter is still used for cell enumeration [28,57]. The cell counting methods are at risk for inaccuracy mainly because of the steps required for obtaining cells in supension, including repeated washes and cell transfer. Indirect methods to measure the number of cells rely on either labeling of nuclear-DNA or staining of cytoplasmic proteins. The incorporation of radioactive thymidine is a precise and sensitive method for DNA determination [58]: the main drawbacks are the expensive equipment and the problems related to radioactive waste.

The bromodeoxyuridine assay has been adopted for cytotoxicity testing and found reproducible, although not very sensitive [59]; moreover the experimental protocol is quite complicated and flow cytometry equipment is required. In our lab a Hoechst dye is currently used for quantifying DNA of cells cultured in microplates: since there is no need for cell transfer after contact with biomaterials the procedure is really simple [60]. A number of colorimetric methods for cell growth in microplates have been developed. Amido black [61], crystal violet [62] and methylene blue [59] are all dyes which stain cells after fixation: following solubilization of the adsorbed dye from cells the absorbance of the blue color is measured at the proper wavelength on a plate reader. Adsorption of the dye by the material surface may be taken into account and verified when working with cells onto solid samples. The total protein content can be estimated using the Lowry procedure or bicinchoninic acid (BCA): the measurement of total intracellular proteins in cell lysates ensures that any material interference is avoided [37,63]. Moreover, this method could provide evalution of the status of cells nondividing but metabolically active.

The colony forming assay provides the best demonstration of the residual ability of cells to grow in vitro after challenge with biomaterials [64]. The main limitation of this technique is the duration: the results of an experiment may be attained after two to three weeks, since materials which temporarily block cell division may be deemed cytotoxic if cells are processed prematurely. Moreover, some cell lines were not able to grow at the low cell density required for colony formation: the plating efficiency for each cell type should be tested at the very beginning of the experimentation. Despite the number of tests devised for cytotoxicity measurement, the question of which assay provides the best evaluation of biomaterial injury to cells is still unanswered. If the material has an intracellular site of damage the assay methods which measure cell viability by membrane integrity (vital dyes, ^{51}Cr, LDH, etc.) could underestimate the toxicity. A material which requires time for expressing its activity may go unnoticed if the test is short. On the other hand, two to three weeks for getting results could be unacceptable, especially when large batteries of materials have to be assayed. In this case also, cell counting techniques become unsuitable. Indeed, no current test ensures consistency, sensitivity, reproducibility and simplicity all together.

To avoid artefacts or misinterpretation, we advocate the use of a combination of methods with different endpoints to assess cytotoxicity of candidate materials. In our lab NR uptake is coupled with PI uptake for detection of different aspects of cell viability, followed by a method for cell growth, e.g., staining with amido black or crystal violet, total protein content, DNA assay, etc.

The EN30993 part 5 standard inherently suggests the adoption of more than one test, by saying " . . . test results may be invalid if the test specimen releases substances which interfere with the test system or measurement." For example, we found that cells exposed to aluminum ion showed an increased uptake of neutral red not due to an increased viability of cells, but to a specific accumulation of NH_4^+ in the lysosomes: their swelling led to high values of NR uptake.

We are aware that different labs have a tendency to favor a specific assay method rather than others: the advantage is that information on different cell compartments may be collected and an exhaustive picture of cell/material interactions is likely to be obtained. This is the approach of toxicological investigation: only a battery of tests is able to fully mimic the in vivo situation, which is characterized by a number of cell populations with different sensitivities to the agent tested. In conclusion, quantitative methods are imperative for rating cytotoxicity of materials. They are needed in the screening phase of a battery of samples or when a pass/fail answer is required, while morphological or qualitative methods add valuable information but cannot provide decision by themselves: they are the methods of choice for a second-step evaluation or exhaustive characterization of materials.

V. QUANTITATIVE APPROACH FOR THE CYTOTOXICITY EVALUATION

In the study of cytotoxicity of biomaterials the damage should be quantified in order to define a threshold of nontoxicity. The cytotoxic effect is mostly quantified as the proportion of cell survival after contact with biomaterials: using percent values the results from different assay methods can be compared.

No agreement is found on the threshold value for "cytotoxicity." Some authors classify a material as cytotoxic when cell survival is less than 50% or 75%, while others employ statistics to detect significant differences between materials and controls.

Quantification of cytotoxic effects is imperative whenever different materials have to be compared. It may occur that a material is applied clinically in spite of its relative toxicity, due to the lack of valid alternatives. Following graduation of various materials with similar properties in order of toxicity, the less toxic can be selected for clinical application.

Bone cement, widely employed for fixation of prostheses to bone, represented a turning-point in orthopedic prosthetic surgery, in spite of some mechanical and biological drawbacks. The toxicity of bone cement in prosthetic implants has been widely demonstrated. Both methyl-methacrylate monomer (MMA), the main component of the acrylic cement, and its additives can damage the cells irreparably [65,66]. The toxicity of MMA is proportional to the amount of monomer within the composite [67]. After the setting time, usually 15 min, about 3–5% of the monomer is present, to be reduced to 1–2% with time [68]. A part of this residue is eliminated through the bloodstream and breathing; the part which is entrapped in situ exerts a toxic effect. In our lab we analyzed the toxicity in vitro of some bone cements [69]. The cement components were mixed according to the manufacturer's indications and extracted in Iscove's medium at different times after the polymerization. By analyzing the growth inhibition and the cell cycle of osteoblast-like cells, the degree of toxicity was expressed as a percentage of inhibition of cell proliferation (CPI%) according to the formula: CPI% = 1− (cell number of sample/cell number of control) × 100 [70]. Using CPI% we were able to identify four levels of different toxicity, i.e., "high toxicity" (CPI = 50%), "medium toxicity" (CPI < 50% and >25%), "low toxicity" (CPI ≤ 25%) and "no toxicity" (CPI = 0) (Fig. 3).

A quantitative approach in cytotoxicity testing is also needed when a dose/response effect has to be measured: the aqueous extract from a device can show overt toxicity when tested undiluted, while after dilution cells are no more affected. Using the dose/response curve the toxic and nontoxic concentrations of substances released in biological fluids are calculated by mathematical models.

An experimental model for the evaluation of dose/response effects was applied in our lab to measure the cytotoxicity of metal extracts [71]. A lot of biological functions are influenced by metal ions, which are characterized often by a paradoxical behavior in the biological systems: they are essential trace elements but they become toxic when in excess. Often the cutoff between toxic and nontoxic doses is not clear and an essential element can shift to dangerous substance by a little increase in concentration [72,73]. Many in vitro methods to test the toxicity of metals have been applied, but the comparison among studies can be difficult due to experimental differences, including cell lines, target cell number, time of exposure to cations, concentration, chemical state of the metal, and so on. For example, 2.5 μg/ml of Ni^{2+} are enough to inhibit the growth of primary gingival cells, but at least 10 μg/ml are needed to obtain the same effect on 3T3 fibroblast cell line [74].

In our lab the toxicity of metal extracts was investigated in vitro on human lymphocytes: extracts were obtained following the ISO guidelines. Graphite furnace atomic absorption spec-

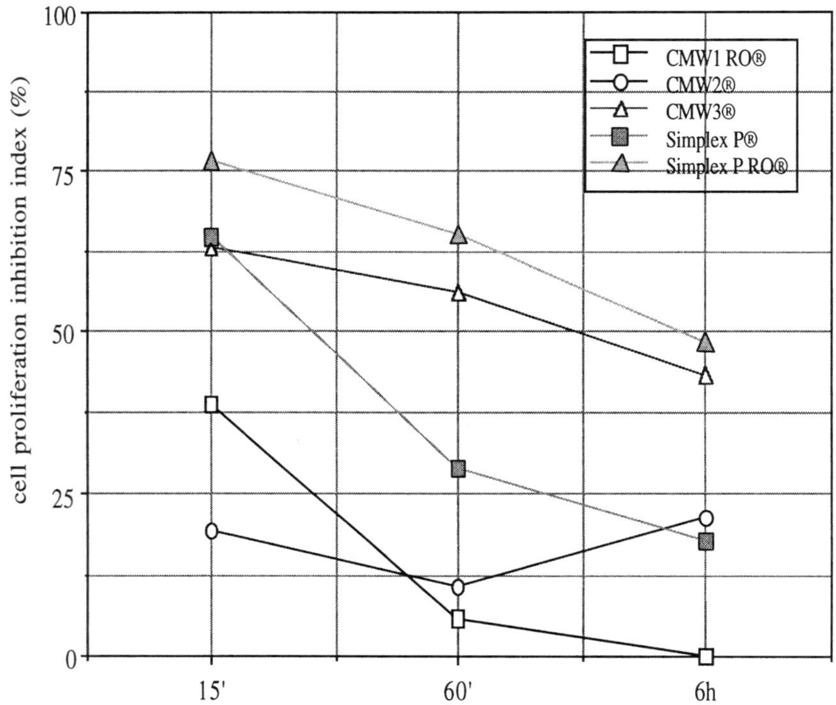

Figure 3 In vitro effects of cements tested onto MG63 cells. The time after polymerization (15 min to 6 h) is plotted versus the index of inhibition of cell proliferation (CPI%), evaluated by cell counting. After 6 h Simplex P RO and CMW3 exerted a "medium toxicity" (CPI: 48.2 ± 11% and 43.3 ± 12%, respectively).

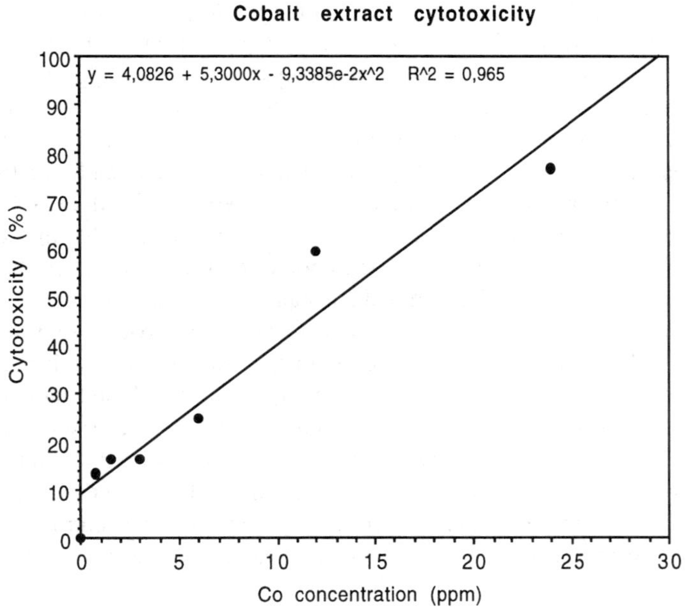

Figure 4 Regression plot of ion concentration and cytotoxic effect of metal extracts.

about 200 base pairs (bp). These DNA fragments can be extracted from apoptotic cells and result in the appearance of "DNA laddering" when the DNA is analyzed by agarose gel electrophoresis; the DNA of nonapoptotic cells which remain largely intact does not display this "laddering," while the denaturated DNA of necrotic cells show a large smear near the deposition area.

3. The coordinated structural changes of the apoptotic process represent the final effect of an intricate network of events. All living cells contain a variety of molecules that following activation participate in these events. The process of programmed cell death is the result of an interaction between initiating stimuli, such as physiological stimuli or injury by toxic substances, and "factors" which determine the susceptibility of the cells to start the cascade of events leading to the structural changes. These "factors" are specific genes coordinating and regulating apoptosis, as well as cell growth and differentiation. For example, bcl-2 is a potent suppressor of apoptosis, while ICE, c-myc and p53 induce apoptosis. Because most substances become cytotoxic depending on the concentration, some authors suggest that the threshold for the onset of cell death can be determined by the expression of genes which promote or suppress apoptosis [80]. The gene expression can be measured by analyzing the PCR products from RNA of cells exposed to toxic substances; moreover, fluorochrome- or enzyme-conjugated monoclonal antibodies are now available: the presence of gene products can be revealed by the flow cytometry and histochemistry.

Flow cytometry can be applied to differentiate the two modes of cell death, and several cytometric methods have been recently described for identifying cells undergoing DNA fragmentation [81]. Using flow-cytometric methods the apoptotic phenomena are detected on a per-cell basis, so that the amount of apoptotic cells is estimated.

The following parameters were found useful to distinguish apoptotic from necrotic cells: (a) Unfixed *dead cells* lose plasma membrane integrity, and the damage is probed by the uptake of propidium iodide that occurs rapidly in necrotic cells, while it is a late event in apoptosis. (b) Ethanol-fixed *apoptotic cells* show a decreased staining with propidium iodide: a hypodiploid peak is found in the histogram, with a DNA index ranging from 0.60 to 0.95 with respect to the diploid peak, as a result of loss of DNA fragments outside the cells.

In vivo apoptotic cells are removed rapidly by phagocytes, while in an in vitro system they remain in situ, lose cell membrane integrity and get the flow-cytometric characteristics of the necrotic cells. In this circumstance it is very difficult to establish whether cell death is due to necrosis or apoptosis and a double analysis on both unfixed and ethanol-fixed cells is needed. In our experience the ratio dead cells/apoptotic cells can be considered a reliable index of the toxic effects of materials [79]. When the ratio is ≤1 the apoptotic events prevail on the dead cells. This means that the toxic effects of the examined substance are linked to apoptotic phenomena. The >1 ratio means that the material under test behaves as a strong toxic substance, causing structural alterations. In Table 4 the results of our experiments on the cytotoxicity of metal extracts, namely chromium, cobalt and nickel, are summarized. Peripheral blood mononuclear cells were exposed to different concentrations of extracts and the characteristics of both apoptosis and necrosis were evaluated by flow-cytometry at different culture endpoints. Our data demonstrated that the toxic effect depends on the type of metal, on the extract concentration and on the time of contact between mononuclear cells and extract; high concentrations of cobalt and nickel seem to determine cell necrosis phenomena, whereas by diminishing the

Table 4 Ratio between Dead Cells and Apoptotic Cells by Flow Cytometry of Mononuclear Cells Cultured with Metal Extracts

	Dead cells/Apoptotic cells		
	24 h	48 h	72 h
Control	0.2 ± 0	0.4 ± 0.1	0.5 ± 0.1
Chromium			
50%	0.3 ± 0.1	0.5 ± 0.1	0.7 ± 0.1
25%	0.5 ± 0.3	0.6 ± 0.1	0.7 ± 0.1
12.5%	0.5 ± 0.2	0.6 ± 0.1	0.7 ± 0.2
6%	1.1 ± 0.5	0.5 ± 0.1	0.7 ± 0.1
3%	1.1 ± 0.6	0.4 ± 0.1	0.7 ± 0.1
1.5%	0.7 ± 0.3	0.6 ± 0.3	0.5 ± 0.1
Cobalt			
50%	0.7 ± 0.4	1.7 ± 0.3	2.9 ± 0.5
25%	0.4 ± 0.2	1.1 ± 0.1	2.3 ± 0.2
12.5%	0.5 ± 0.3	0.4 ± 0.1	0.9 ± 0.1
6%	0.7 ± 0.5	0.4 ± 0.1	0.7 ± 0.1
3%	0.4 ± 0.3	0.4 ± 0.1	0.7 ± 0.2
1.5%	0.7 ± 0.6	0.3 ± 0.1	0.6 ± 0.1
Nickel			
50%	0.5 ± 0.1	0.9 ± 0.1	1.6 ± 0.1
25%	0.3 ± 0.1	0.6 ± 0.1	1 ± 0.1
12.5%	0.3 ± 0.1	0.6 ± 0.1	0.7 ± 0.1
6%	0.2 ± 0.1	0.5 ± 0.1	0.7 ± 0.1
3%	0.2 ± 0	0.4 ± 0.1	0.6 ± 0.1
1.5%	0.2 ± 0	0.2 ± 0.1	0.4 ± 0.1

extract concentration apoptotic phenomena prevail; high chromium concentrations induce cell death by apoptotic phenomena.

From these results some hypotheses on the role of cobalt, chromium and nickel ions released from an implant due to corrosion are put forward. The release of large amounts of nickel and cobalt could favour necrotic phenomena and the inflammatory reaction of the surrounding tissue, as it is observed in the tissue biopsies from loosened prostheses. The release of chromium or of limited amounts of cobalt and nickel could be followed, however, by toxic effects, characterized by apoptotic phenomena: this situation is less severe, because tissue adaptation to the implant is not seriously hampered.

The large number of DNA fragments appearing in apoptotic cells means a multitude of 3′-hydroxyl termini of DNA ends. This property can be used to identify apoptotic cells by labeling the DNA breaks with fluorescent-tagged deoxyuridine triphosphate nucleotides (f-dUTP). The enzyme terminal deoxynucleotidyl transferase (TdT) catalyzes a template-independent addition of F-dUTP to the 3′-hydroxyl DNA ends of double or single stranded DNA: a substantial number of these sites are available in apoptotic cells providing the basis for the flow-cytometric analysis.

Recently a new method for detection of apoptosis by flow cytometry on a per cell basis has been described [82]. In normal blood cells the plasma membrane exhibits significant phospholipid asimmetry, with phosphatidylcholine and sphingomyelin predominantly on the external leaflet, and most of the membrane phosphatidylethanolamine and phosphatidylserine on the inner leaflet. In the early stages of apoptosis the phosphatidylserine translocates from the inner part of the plasma membrane to the outer layer, to be exposed at the external surface of the cell. This event seems to favour recognition and phagocytosis of apoptotic cells by macrophages. It was shown that the anticoagulant annexin V preferentially binds to negatively

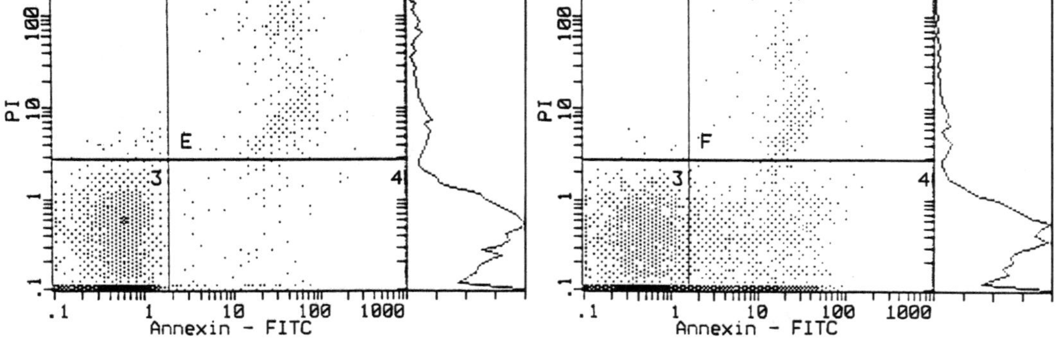

Figure 5 Flow cytometric analysis of apoptotic HL-60 cells after staining with Annexin V-FITC and propidium iodide. Histograms show the effects of 4h cultivation in absence (left panel) or presence (right panel) of 10 μM campthotecin. The areas of apoptotic cells are E4 (2.1%) and F4 (27%), while the necrotic cells are collected in E2 (7.1%) and F2 (6.5%).

charged phospholipids like phosphatidylserine. Thus, fluorescein isothiocyanate (FITC) labelled Annexin V can be used for the quantification of apoptosis. Since necrotic cells also expose phosphatidylserine as they lose membrane integrity, a DNA stain for dye exclusion test, that is propidium iodide, is applied simultaneously. The apoptotic cells become Annexin V-positive after nuclear condensation has started, but before the cell has become permeable to propidium iodide. In Fig. 5 representative histograms of apoptotic HL-60 cells after staining with FITC-Annexin V and propidium iodide are reported.

In conclusion, although theoretically simple, cell culture assays for biocompatibility testing require a number of endpoints to be decided after accurate analysis. As outlined before, main points to be analyzed include material preparation and handling, assay method and quantitative assessment: such items are imperative for materials to be allocated into a class of toxicity. We also suggest that the mechanism of cell killing by materials is ascertained: the result may be explanatory of some tissue reactions observed after in vivo implantation.

New techniques and refinement of the existing methods are the tools for improving the studies in vitro and for connecting the results to the clinical outcome of implants.

REFERENCES

1. Tamada, Y. and Ikada, Y., 1994. Fibroblast growth on polymer surfaces and biosynthesis of collagen. *J. Biomed. Mater. Res.*, 28: 783–789.
2. Bordenave, L., Bareille, R., Lefebvre, F. and Baquey, C., 1993. A comparison between ^{51}chromium release and LDH release to measure cell membrane integrity: interest for cytocompatibility studies with biomaterials. *J. Appl. Biomater.*, 4: 309–315.
3. Naij, A. and Harmand, M. F., 1990. Study on the effect of the surface state on the cytocompatibility of a Co-Cr alloy using human osteoblasts and fibroblasts. *J. Biomed. Mater. Res.*, 24: 861–871.
4. van Kooten, T. G., Klein, C. L., Kohler, H., Kirkpatrick, C. J., Williams, D. F. and Eloy, R., 1997. From cytotoxicity to biocompatibility testing *in vitro*: cell adhesion molecule expression defines a new set of parameters. *J. Mater. Sci.: Mater. Med.*, 8: 835–841.
5. Rudolph, A. S., Stilwell, G., Cliff, R. O., Kahn, B., Spargo, B. J., Rollwagen, F. and Monroy, R. L., 1992. Biocompatibility of lipid microcylinders: effect on cell growth and antigen presentation in culture. *Biomaterials*, 13: 1085–1091.
6. Frondoza, C. G., Tanner, K. T., Jones, L. C. and Hungerford, D. S., 1993. Polymethylmethacrylate particles enhance DNA and protein synthesis of human fibroblasts in vitro. *J. Biomed. Mater. Res.*, 27: 611–617.
7. Shanbhag, A., Yan, J., Lilien, J. and Black, J., 1992. Decreased neutrophil respiratory burst on exposure to cobalt-chrome alloy and polystyrene in vitro. *J. Biomed. Mater. Res.*, 26: 185–195.
8. Remes, A. and Williams, D. F., 1990. Chemotaxis and the inhibition of chemotaxis of human neutrophils in response to metal ions. *J. Mater. Sci.: Mater. Med.*, 1: 26–32.
9. Papatheofanis, F. J. and Barmada, R., 1991. Polymorphonuclear leukocyte degranulation with exposure to polymethylmethacrylate nanoparticles. *J. Biomed. Mater. Res.*, 25: 761–771.
10. Kieswetter, K., Schwartz, Z., Hummert, T. W., Cochran, D. L., Simpson, J., Dean, D. D. and Boyan, B. D., 1996. Surface roughness modulates the local production of growth factors and cytokines by osteoblast-like MG-63 cells. *J. Biomed. Mater. Res.*, 32: 55–63.
11. Sun, Z. L., Wataha, J. C. and Hanks, C. T., 1997. Effects of metal ions on osteoblast-like cell metabolism and differentiation. *J. Biomed. Mater. Res.*, 34: 29–37.
12. Chou, L., Firth, J. D., Uitto, V. and Brunette, D. M., 1998. Effects of titanium substratum and grooved surface topography on metalloproteinase-2 expression in human fibroblasts. *J. Biomed. Mater. Res.*, 39: 437–445.
13. International Organization for Standardization, 1994. Biological evaluation of medical devices. Part 12: Sample preparation and reference materials. ISO/DIS 10993.

14. Tsuchiya, T., 1994. Studies on the standardization of cytotoxicity tests and new standard reference materials useful for evaluating the safety of biomaterials. *J. Biomater. Appl.*, 9: 138–157.
15. Berstein, A., Bernauer, I., Marx, R. and Geurtsen, W., 1992. Human cell culture studies with dental metallic materials. *Biomaterials*, 13: 98–100.
16. Gotoh, Y., Tsukada, M. and Minoura, N., 1998. Effect of the chemical modification of the arginyl residue in Bombyx Mori silk fibroin on the attachment and growth of fibroblast cells. *J. Biomed. Mater. Res.*, 39: 351–357.
17. Jansen, J. A., van der Waerden, J. P. C. M. and de Groot, K., 1991. Fibroblast and epithelial cell interactions with surface-treated implant materials. *Biomaterials*, 12: 25–31.
18. Knabe, C., Grosse-Siestrup, C., Hunder, A. and Ziemann, A., 1997. A computer-assisted in-vitro biomaterials test for percutaneous devices using human keratinocyte cultures. *J. Mater. Sci.: Mater. Med.*, 8: 577–582.
19. Ertel, S. I., Ratner, B. D., Kaul., A., Schway, M. B. and Horbett, T. A., 1994. In vitro study of the intrinsic toxicity of synthetic surfaces to cells. *J. Biomed. Mater. Res.*, 28: 667–675.
20. Den Braber, E. T., de Ruijter, J. E., Smits, H. T. J., Ginsel, L. A., von Recum, A. F. and Jansen, J. A., 1995. Effect of parallel surface microgrooves and surface energy on cell growth. *J. Biomed. Mater. Res.*, 29: 511–518.
21. Boyan, B. D., Batzer, R., Kieswetter, K., Liu, Y., Cochran, D. L., Szmuckler-Moncler, S., Dean, D. D. and Schwartz, Z., 1998. Titanium surface roughness alters responsiveness of MG63 osteoblast-like cells to 1a,25-$(OH)_2D_3$. *J. Biomed. Mater. Res.*, 39: 77–85.
22. Wenz, L. M., Merritt, K., Brown, S. A. and Moet, A., 1990. In vitro biocompatibility of polyetherketone and polysulfone composites. *J. Biomed. Mater. Res.*, 24: 207–215.
23. Morrison, C., Macnair, R., MacDonald, C., Wykman, A., Goldie, I. and Grant, M. H., 1995. In vitro biocompatibility testing of polymers for orthopedic implants using cultured fibroblasts and osteoblasts. *Biomaterials*, 16: 987–992.
24. Bordenave, L., Bareille, R., Lefebvre, F., Caix, J. and Baquey, C., 1992. Cytocompatibility study of NHLBI primary reference materials using human endothelial cells. *J. Biomater. Sci.—Polymer Edn.*, 3: 509–516.
25. Granchi, D., Ciapetti, G., Savarino, L., Cavedagna, D., Donati, M. E. and Pizzoferrato, A., 1996. Assessment of metal extract toxicity on human lymphocytes cultured in vitro. *J. Biomed. Mater. Res.*, 31: 183–191.
26. Dupraz, A. M. P., van der Meer, S. A. T., De Wijn, J. R. and Goedemoed, J. H., 1996. Biocompatibility screening of silane-treated hydroxyapatite powders, for use as fillers in resorbable composites. *J. Mater. Sci.: Mater. Med.*, 7: 731–738.
27. Bruijn, J. D., van der Brink, I., van der Meer, J., Nijweide, P. and Bovell, Y. P., 1995. Bone cell responses to artificially produced surface apatite layers. In: J. Wilson, L. L. Hench and D. Greenspan (Editors), *Bioceramics*, vol. 8, Elsevier Science, Oxford, U.K., pp. 29–34.
28. Ito, A., Okazaki, Y., Tateishi, T. and Ito, Y., 1995. In vitro biocompatibility, mechanical properties, and corrosion resistance of Ti-Zr-Nb-Ta-Pb and Ti-Sn-Nb-Ta-Pd alloys. *J. Biomed. Mater. Res.*, 29: 893–900.
29. European Committee for Standardization, 1994. Biological evaluation of medical devices. Part 5: Tests for cytotoxicity: In vitro methods. EN 30993.
30. Harmand, M. F., 1997. Cytotoxicity. Part I: Toxicological risk evaluation using cell culture. In: J. H. Braybrook (Editor), *Biocompatibility Assessment of Medical Devices and Materials*, Wiley, Chichester, England, pp. 119–124.
31. Lontz, J. F., Nadijcka, M. D. and Holmes, R., 1983. Assessment of biocompatibility of orofacial materials and devices by culturing with human excised donor tissues. In: S. A. Brown (Editor), *Cell-Culture Test Methods*, ASTM Special Technical Publication 810, ASTM 1916 Race Street, Philadelphia, pp. 77–87.
32. Harmand, M. F., Bordenave, L., Bareille, R., Naji, A., Jeandot, R., Rouais, F. and Ducassou, D., 1991. In vitro evaluation of an epoxy resin's cytocompatibility using cell lines and human differentiated cells. *J. Biomat. Sci.—Polymer Edn.*, 2: 67–79.

33. Pizzoferrato, A., Ciapetti, G., Stea, S., Cenni, E., Arciola, C., Granchi, D. and Savarino, L., 1994. Cell culture methods for testing biocompatibility. *Clin. Mater.*, 15: 173–190.
34. Tomas, H., Carvalho, G. S., Fernandes, M. H., Freire, A. P. and Abrantes, L. M., 1997. The use of rat, rabbit or human bone marrow derived cells for cytocompatibility evaluation of metallic elements. *J. Mater. Sci.: Mater. Med.*, 8: 233–238.
35. International Organization for Standardization, 1995. Dentistry—Preclinical evaluation of biocompatibility of medical devices used in dentistry: Test methods. ISO-DIS 7405.
36. Ciapetti, G., Granchi, D., Stea, S., Savarino, L., Verri, E., Gori, A., Savioli, F. and Montanaro, L., 1998. Cytotoxicity testing of materials with limited in vivo exposure is affected by the duration of cell-material contact. *J. Biomed. Mater. Res.*, 42: 485–490.
37. Macnair, R., Rodgers, E. H., Macdonald, C., Wykman, A., Goldie, I. and Grant, M. H., 1997. The response of primary rat and human osteoblast and an immortalized rat osteoblast cell line to orthopaedic materials: comparative sensitivity of several toxicity indices. *J. Mater. Sci.: Mater. Med.*, 8: 105–111.
38. Rae, T., 1978. The hemolytic action of particulate metals (Cd, Cr, Co, Fe, Mo, Ni, Ta, Ti, Zn, Co-Cr alloy). *J. Pathol.*, 125: 81–87.
39. Allen, M., Millet, P., Dawes, E. and Rushton, N., 1994. Lactate dehydrogenase activity as a rapid and sensitive test for the quantification of cell numbers in vitro. *Clin. Mater.*, 16: 189–194.
40. Borenfreund, E. and Puerner, J. A., 1985. Toxicity determined in vitro by morphological alteration and neutral red adsorption. *Toxicol. Lett.*, 24: 119–124.
41. Hansen, M. B., Nielsen, S. E. and Berg, K., 1989. Re-examination and further development of a precise and rapid dye method for measuring cell growth/cell kill. *J. Immunol. Methods*, 119: 203–210.
42. Norme expérimentale par l'Association Française de Normalisation (AFNOR) S 90-702. Matériel médico-chirurgical: Evaluation in vitro de la cytotoxicité des materiaux et dispositifs médicaux, publieé en decembre 1988.
43. Bashford, C. L., 1987. Measurement of the pH of Cellular Compartments. In: C. L. Bashford and D. A. Harris (Editors), *Spectrophotometry and Spectrofluorimetry. A Practical Approach.* IRL Press, Washington, DC, pp. 131–135.
44. Darzynkievicz, Z., 1990. Probing nuclear chromatin by flow cytometry. In: M. R. Melamed, T. Lindmo and M. L. Mendelson (Editors). *Flow Cytometry and Sorting*, Wiley, New York, pp. 291–314.
45. Sgouras, D. and Duncan, R., 1990. Methods for evaluation of biocompatibility of soluble synthetic polymers which have potential for biomedical use: 1. Use of the tetrazolium-based assay (MTT) as a preliminary screen for evaluation of in vitro cytotoxicity. *J. Mater. Sci.*, 1: 61–68.
46. Marshall, N. J., Goodwin, C. J. and Holt, S. J., 1995. A critical assessment of the use of microculture tetrazolium assays to measure cell growth and function. *Growth Regulation*, 5: 69–84.
47. Vistica, D. T., Skehan, P., Scudiero, D., Monks, A., Pittman, A. and Boyd, M. R., 1991. Tetrazolium-based assay for cellular viability: a critical examination of selected parameters affecting formazan production. *Cancer Res.*, 51: 2515–2520.
48. Schweikl, H. and Schmalz, G., 1996. Toxicity parameters for cytotoxicity testing of dental materials in two different mammalian cell lines. *Eur. J. Oral Sci.*, 104: 292–299.
49. McKay, G. C., Macnair, R., MacDonald, C. and Grant, M. H., 1996. Interactions of orthopedic metals with an immortalized rat osteoblast cell line. *Biomaterials*, 17: 1399–1344.
50. Ciapetti, G., Cenni, E., Pratelli, L. and Pizzoferrato, A., 1993. Evaluation of cell/biomaterial interaction by MTT assay. *Biomaterials*, 14: 359–364.
51. Ignatius, A. A. and Claes, L. E., 1996. In vitro biocompatibility of bioresorbable polymers: poly(L,DL-lactide) and poly(L-lactide-co-glycolide). *Biomaterials*, 17: 831–839.
52. Wan, H., Williams, R., Doherty, P. and Williams, D. F., 1994. A study of the reproducibility of the MTT test. *J. Mater. Sci.:Mater. Med.*, 5: 154–159.
53. Benghuzzi, H. A., 1995. The role of various biomedical polymers concentration on the adhesion rate and viability of monocyte and monocyte derived macrophages. *Biomed. Sci. Instrum.*, 31: 121–126.
54. Jonsson, K. B., Frost, A., Larsson, R., Ljunghall, S. and Ljunggren, O., 1997. A new fluorometric

assay for determination of osteoblastic proliferation: effects of glucocorticoids and insulin-like growth factor. I. *Calcif. Tissue Int.*, 60: 30–36.
55. Nociari, M. M., Shale, A., Benias, P. and Russo, C., 1998. A novel one-step, highly sensitive fluorometric assay to evaluate cell-mediated cytotoxicity. *J. Immunol. Methods*, 213: 157–167.
56. Nishi, C., Nakajima, N. and Ikada, Y., 1995. In vitro evaluation of cytotoxicity of diepoxy compounds used for biomaterials modification. *J. Biomed. Mater. Res.*, 29: 829–834.
57. Ryhanen, J., Niemi, E., Serlo, W., Niemela, E., Sandvik, P., Pernu, H. and Salo, T., 1997. Biocompatibility of nickel-titanium shape memory metal and its corrosion behavior in human cell cultures. *J. Biomed. Mater. Res.*, 35: 451–457.
58. Maloney, W. J., Castro, F., Schurman, D. J. and Smith, R. L., 1994. Effects of metallic debris on adult bovine articular chondrocyte metabolism in vitro. *J. Appl. Biomater.*, 5: 109–115.
59. Clifford, C. J. and Downes, S., 1996. A comparative study of the use of colorimetric assays in the assessment of biocompatibility. *J. Mater. Sci.: Mater. Med.*, 7: 637–643.
60. Granchi, D., Verri, E., Ciapetti, G., Savarino, L., Cenni, E., Gori, A. and Pizzoferrato, A., 1998. Effects of chromium extract on cytokine release by mononuclear cells. *Biomaterials*, 19: 283–291.
61. Ciapetti, G., Granchi, D., Verri, E., Savarino, L., Cavedagna, D. and Pizzoferrato, A., 1996. Application of a combination of neutral red and amido black staining for rapid, reliable cytotoxicity testing of biomaterials. *Biomaterials*, 17: 1259–1264.
62. Koyano, T., Minoura, N., Nagura, M. and Kobayashi, K., 1998. Attachment and growth of cultured fibroblast cells on PVA/chitosan-blended hydrogels. *J. Biomed. Mater. Res.*, 39: 486–490.
63. Nichols, K. G. and Puleo, D. A., 1997. Effect of metal ions on the formation and function of osteoclastic cells in vitro. *J. Biomed. Mater. Res.*, 35: 265–271.
64. Tsuchiya, T., Ikarashi, Y., Hata, H., Toyoda, K., Takahashi, M., Uchima, T., Tanaka, N., Sasaki, T. and Nakamura, A., 1993. Comparative studies of the toxicity of standard reference materials in various cytotoxicity tests and in vivo implantation tests. *J. Appl. Biomater.*, 4: 153–156.
65. Ciapetti, G., Stea, S., Granchi, D., Cavedagna, D., Gamberini, S. and Pizzoferrato, A., 1995. The effects of orthopaedic cements on osteoblastic cells cultured in vitro. *Chir. Org. Mov.*, 25: 409–415.
66. Stea, S., Granchi, D., Zolezzi, C., Ciapetti, G., Visentin, M., Cavedagna, D. and Pizzoferrato, A., 1997. High-performance liquid chromatography assay of n,n-dimethyl-p-toluidine released from bone cements: evidence for toxicity. *Biomaterials*, 18: 243–246.
67. Dahl, O. E., Garvick, L. J. and Lyberg, T., 1994. Toxic effects of methylmethacrilate monomer on leukocytes and endothelial cells in vitro. *Acta Orthop. Scand.*, 65: 147–153.
68. Linder, L., 1982. Reaction to bone cement. In: D. F. Williams (Editor), *Biocompatibility of Orthopedic Implants*, Vol. II, CRC Press, Boca Raton, FL, pp. 1–23.
69. Granchi, D., Stea, S., Ciapetti, G., Savarino, L., Cavedagna, D. and Pizzoferrato, A., 1995. Cell cycle of osteoblast-like cells after in vitro exposure to bone cements. *Biomaterials*, 16: 1187–1192.
70. Van Luyn, M. J. A., van Wachem, P. B., Jonkman, M. F. and Nieuwenhuis, P., 1991. Cytotoxicity testing of wound dressings using methylcellulose cell culture. *Biomaterials*, 13: 267–275.
71. Granchi, D., Ciapetti, G., Savarino, L., Cavedagna, D., Donati, M. E. and Pizzoferrato, A., 1996. Assessment of metal extract toxicity on human lymphocytes cultured in vitro. *J. Biomed. Mater. Res.*, 31: 183–191.
72. Rae, T., 1981. Metal enzyme interactions. In: D. F. Williams (Editor), *Systemic Aspects of Biocompatibility*, Vol. I, CRC Press, Boca Raton, FL, pp. 21–38.
73. Takeda, S., Kakiuchi, H., Doi, H. and Nakamura, M., 1989. Cytotoxicity of pure metals. *Shika. Zairyo. Kikai*, 8: 648–652.
74. Wataha, J. C., Hanks, C. T. and Craig, R. G., 1991. The in vitro effects of metal cations on eukaryotic cell metabolism. *J. Biomed. Mater. Res.*, 25: 1133–1149.
75. Tomei, L. D. and Cope, F. O., 1991. Apoptosis: The molecular basis of cell death. *Curr. Commun. Cell Mol. Biol.*, Vol. 3., Cold Spring Arbor, NY.
76. Carracedo, J., Ramirez, R., Martin Malo, A., Rodriguez, M. and Aljama, P., 1998. Nonbiocompatible hemodialysis membranes induce apoptosis in mononuclear cells: the role of G-protein. *J. Am. Soc. Nephrol.*, 9: 46–53.
77. Schedle, A., Samorapoompichit, P., Rausch-Fan, X. H., Franz, A., Füreder, W., Sperr, W. R., Sperr,

W., Ellinger, A., Slavicek, R., Boltz-Nitulescu, G. and Valent, P., 1995. Response of L929 fibroblasts, human gingival fibroblasts and human tissue mast cells to various metal cations. *J. Dent. Res.*, 74: 1513–1515.

78. Schedle, A., Samorapoompichit, P., Füreder, W., Rausch-Fan, X. H., Franz, A., Sperr, W. R., Sperr, W., Slavicek, R., Simak, S., Klepekto, W., Ellinger, A., Ghannadan, M., Baghestanian, M. and Valent, P., 1998. Metal ion-induced toxic histamine release from human basophils and mast cells. *J. Biomed. Mater. Res.*, 39: 560–567.
79. Granchi, D., Cenni, E., Ciapetti, G., Savarino, L., Stea, S., Gamberini, S., Gori, A. and Pizzoferrato, A., 1998. Cell death induced by metal ions: necrosis or apoptosis? *J. Mater. Sci.: Mater. Med.*, 9: 31–37.
80. Hickmann, J. A. and Boyle, C. C., 1997. Apoptosis and cytotoxins. In: A. H. Wyllie (Editor), *British Medical Bullettin*, British Council by the Royal Society of Medicine Press Limited, 53: 632–643.
81. Darzynkievicz, Z., Bruno, S., Del Bino, G., Gorczyca, W., Hotz, M. A., Lassota, P. and Traganos, F., 1992. Features of apoptotic cells measured by flow cytometry. *Cytometry*, 13: 795–808.
82. Vermes, I., Haanen, C., Steffen-Nakken, H. and Reutelingsperger, C., 1995. A novel assay for apoptosis. Flow cytometric detection of phosphatidylserine expression on early apoptotic cells using fluorescein labelled Annexin V. *J. Immunol. Methods*, 184: 39–51.

7

Animal Models for Preclinical Testing of Medical Devices

H. V. Mendenhall
Primedica Corporation, Worcester, Massachusetts

I. INTRODUCTION

The determination of the safety of a material and, subsequently, its efficacy as a device, will, of necessity, require implantation in animals. Such studies, if performed in the correct animal model, have proven to be directly applicable to humans. The phenomena of life-sustaining processes and the adjustments to disease are similar, if not identical, certainly throughout the class *Mammalia*, if not others. Most biological events, wound healing in particular, are also comparable. Only the magnitude and timing of specific features may differ.

The literature is replete with descriptions of animal models that have been used for evaluation of various materials and devices. Unfortunately, it is quite deficient in pathological data as derived from observations gained from material and device implantation in humans. This deficiency is related to the economics of retrieval and histological evaluation of an implanted device when death of the patient was not related to its presence. Equally suspect is the selection of an animal species on the basis of its expense and/or availability, rather than its relevance to the (presumed) human situation.

The decision regarding which test system (animal model) to use for a specific study involving the determination of the safety and/or efficacy of a new device must come from an extensive review of the literature. This review must be related to not only the specifics of the system into which the new device is to be implanted, but also to the physiological and chemical nature of the system to be studied. All this must refer to its final relation and applicability to humans: the target species.

Despite these limitations, a careful review of what literature does exist concerning documented observations following the implantation and subsequent retrieval of devices placed in specific anatomical locations in humans should be related to similar observations of similar implants placed in similar anatomical locations in various animal species. Once this is done, the likelihood of choosing an animal model that will accurately predict the outcome of human implantation of the material and the device is greater. As dictated by the ISO regulations, Phase I testing of a new material is best done in tissue culture. Phase II testing for any application is best done through subcutaneous or intramuscular implantation in rodents, with subsequent histopathological evaluation of the tissue that develops around the test material. If it is intended for orthopedic use, the material should be implanted in the femoral intramedullary cavity of rabbits, with subsequent histopathological examination of the surrounding tissue. The animal model of choice for Phase III testing of the completed device depends upon its eventual final application.

The ideal animal model should have consistently reproducible features that simulate an analogous or homologous condition where the biomaterial would be used in man. The scientific criteria for animal model selection depend upon the intended application of the biomaterial. More specifically, anatomical, biochemical, physiological, pathological, and/or psychological characteristics need to be considered. Over the years it has been found by trial and error—and keen relational observation—that human systems resemble similar systems in various animal species. Delineation of all of the subjective "proof" for these physiological and anatomical similarities and differences, would take up many volumes, and is out of the range of this treatise.

The American College of Laboratory Animal Medicine has sponsored several reference texts on spontaneous animal models of human disease [1–2], and the Armed Forces Institute of Pathology has developed a series of publications that summarize nearly 300 animal models [3]. Several other publications frequently list newly developed animal models, some of which may be applicable to biomaterial technology [4–6].

Beyond the anatomical, pathological, and physiological similarities to man that must be looked for in selection of an animal model, the experimental surgeon may have to limit his or her choice by practical considerations. The size and conformation of the animal may be an important one. Small animals cost less and are easier to handle. Surgical procedures in large animals, however, are more easily performed. Body conformation is especially important where long-term restraint is required. For example, pigs and nonhuman primates both have cardiovascular systems similar to that of man; however, for a study requiring long-term catheterization and continuous monitoring, the primate may in fact be the animal model of choice.

The choice of experimental animal to be used in implant research should be based on considerations other than the risk of infection. There are no reliable data indicating that one species is more susceptible to septic complications than another [7,8].

Realistically, it must be admitted that all the results of an experiment involving the implantation of a material either as the raw material or as the finished device will never exactly duplicate those seen when the material or the device is finally implanted in man. However, it is certainly possible to predict the safety of the material in use, and from there to be able to extrapolate presumed efficacy, if the final use of the implant under study is accurately stated and similar work properly reviewed.

The following is a summary of suggested test systems for study of implants according to the surgical specialty for which the device may be used. This list is by no means complete, and largely represents my own bias, gleaned from over 30 years experience in these areas, and from periodic literature review.

II. ABDOMINAL

The anatomy and physiology of the abdominal contents of the dog are very similar to that of humans, making this species the ideal animal model for any experimental surgery involving the abdomen [9].

Studies involving the development of an artificial pancreas for treatment of diabetes mellitus have primarily been conducted in the dog [3]. Dogs are easily rendered diabetic either by surgical pancreatectomy or by the administration of streptozocin. Implantation of the device into the omentum mimics the venous drainage of the pancreas.

The evaluation of antiadhesion products is usually conducted in rats or rabbits, with definitive studies again being performed in dogs. The animal model of choice is the rat cecal abrasion model or, secondarily, the rabbit uterine horn model. In the former, the cecum and peritoneum are mechanically abraded in a reproducible fashion. The wounded edges are treated with the candidate material and opposed. The degree of adhesions between the two structures is determined at necropsy, usually 14 days after the insult. A similar procedure is performed in the latter model.

III. CARDIOVASCULAR

There are many considerations in cardiovascular research, and nearly all species have been used. The choice of the model depends upon the objective of the experiment. Briefly, arterial healing and vascular grafts are best studied in goats [10,11], thrombogenicity in pigs and nonhuman primates [12], hemodynamics in dogs and secondarily in nonhuman primates [13], heart valves in sheep [3,14,15], and artificial hearts and left ventricular assist devices in calves [16].

A relatively new area of vascular research involving "restenosis," the smooth muscle cell proliferation seen in an artery subsequent to endothelial cell removal and rupture of the internal elastic lamina, as seen following balloon angioplasty, or placement of intraluminal stents, is now being extensively studied. The pig iliofemoral and coronary artery seem to most closely resemble the situation seen in man. Unfortunately, all treatments so far determined to be efficacious in animals have not proven to be so in humans. Other test systems used in this area of research have included the rat internal carotid artery, the rabbit aorta, and the nonhuman primate saphenous artery.

IV. NEUROLOGICAL

The rat sciatic nerve is a good model for preliminary testing of the effects of new treatments designed to improve peripheral nerve regeneration [17]. However, the nonhuman primate should be used for validation of the efficacy of the treatment because of the similarity in complexity of the peripheral nerves in higher animals [18]. The spinal cord and brain of the cat, rat and nonhuman primate are the best-mapped and understood [19]. Hydrocephalus shunts are best tested in dogs, however, because of the relatively large size of the lateral ventricle in this species.

V. OPHTHALMOLOGICAL

The testing of new ophthalmologic devices or pharmacological agents can potentially involve three animal models: rabbits, white domestic geese, and cynomolgus monkeys. The rabbit eye is very sensitive to irritation, and should be the primary source of information regarding the toxicology of a new ophthalmic material [20]. The anterior segment is, however, too large and reactive for the evaluation of intraocular devices. From my own experience, this type of work is best done in the white domestic goose. The lens capsule of the goose is nearly identical in size to that of the human, and the reaction of the eye to intraocular surgery is quite similar, but accelerated. Evaluations of a new intraocular lens haptic design or for a potential treatment and/or prophylaxis of posterior capsular opacity can thus be completed in a relatively short time (12 weeks). For long-term studies, the cynomolgus monkey is the most appropriate, because of its similarity to the human with regard to the size of the anterior segment in general, the reactivity to intraocular surgery, and healing time and processes. Determination of the efficacy of glaucoma shunts and other treatments, medical or surgical, for increased intraocular pressure are best performed in nonhuman primates. The anatomy of the iridocorneal angle in other species, especially rabbits, is not suitable for the creation of closed angle glaucoma by means of laser ablation.

VI. ORTHOPEDICS

The healing of bone and its reaction to biomaterials in the goat has been shown to very closely resemble that seen in the human with regard to cell type, collagen structure, organization of both hard and soft tissue structures, and time. Sheep, on the other hand, tend to calcify soft tissue structures. Beagle dogs tend to be chondrodystrophic [10].

These three animal models have been extensively utilized in the evaluation of artificial ligaments and tendons, bone graft substitutes, aids to spinal fusion, artificial joints, and evaluation of various medical treatments for osteoarthritis. Animal models for osteoarthritis include dogs in which the anterior cruciate ligament has been sectioned, sometimes followed by dorsal root rhizotomy. Surface arthroplasty of the femoral head in dogs, sheep and goats has also been reported to be a useful model for evaluation of drug therapy for osteoarthritis.

Tendon healing is best studied in chickens or turkeys. Apart from management considerations, goats and sheep, as well as rats, tend to develop ectopic ossification in these structures [3], making them poor models for the study of candidate artificial tendons or research into the healing characteristics of these structures.

A test system commonly used in orthopedic research is the rabbit. Despite a large body of literature that indicates the inadequacy of this animal model in predicting the outcome of similar biomaterial implants in humans [3], it is still used extensively. The reasons are probably more historical and economical than scientific.

Dental: Dental implants have been extensively studied in dogs, monkeys, baboons and pigs. Although the specific anatomy is quite different between these species and humans, the tissue response to periodontal disease and gingival recession is quite similar [3]. Beyond that, this area of research resembles that seen in orthopedics.

VII. OTOLOGIC

The microscopic anatomy of the temporal bone and its contents in cats and chinchillas bears a remarkable resemblance to that of the human. This is especially true of the auditory ossicles and the cochlea. These characteristics have made these species the test systems of choice for the evaluation of cochlear implants, artificial stapes, and other otologic prostheses. The development of surgical techniques for the treatment of otosclerosis and otitis media have been developed and refined in these animals, especially the cat [3].

VIII. RESPIRATORY

Again, because of similar anatomical size and healing characteristics, the animal models most frequently used for evaluation of materials to be implanted into the respiratory system include the dog and pig. There is an extensive amount of literature regarding the dog as an animal model for evaluation of candidate artificial trachea [3]. The pig has also been extensively used for testing of tissue adhesives and sealants following lung resection.

IX. UROGENITAL

The evaluation of various urological prosthetic devices has been limited to stent-like devices to help prevent stenosis following ureteral or urethral anastomosis. Primarily because of the size of these structures, the animal models utilized have again included dogs and pigs. The microscopic anatomy and healing characteristics of these structures appears to be quite similar to humans [3]. The development and evaluation of surgical techniques and instrumentation for transurethral prostatectomy were developed in dogs. Intrauterine contraceptive devices have been extensively studied in rabbits. This work revealed similar effectiveness and tissue reaction to that subsequently observed in humans.

X. WOUND HEALING

Wound healing studies designed to study the effectiveness of various treatments on "donor site" wounds are best performed in the young domestic pig. The rate of reepithelialization of partial thickness (<0.5 mm deep) blade (dermatome) wounds or burns is largely dependent upon the concentration per square centimeter of keratinocytes. This concentration in the skin of young pigs is nearly identical to that of humans (11 hair follicles/cm^2) [21]. The use of other test systems for wound healing studies should be limited to studying the effect of materials on the healing of incisional wounds.

REFERENCES

1. Andrews, E. J., Ward, B. C. and Altman, N. H. (eds.). 1979. *Spontaneous Animal Models of Human Diseases*. Vol. I. New York: Academic Press.
2. Andrews, E. J., Ward, B. C., and Altman, N. H. (eds.). 1979. *Spontaneous Animal Models of Human Disease*. Vol. II. New York: Academic Press.
3. *Animal Models of Human Disease*, Vols. 1–26. 1972–1998. The Registry of Comparative Pathology, Armed Forces Institute of Pathology, Washington, DC.
4. *ILAR News*. Institute of Laboratory Animal Resources, National Research Council (published quarterly).
5. *Veterinary Pathology*. American College of Veterinary Pathologists Inc., Washington, DC (published bimonthly).
6. *Comparative Pathology Bulletin*. Registry of Comparative Pathology, Armed Forces Institute of Pathology (published quarterly).
7. Brown, M. J., Pearson, P. T., and Tomson, F. N. 1993. Guidelines for animal surgery in research and teaching. *Am. J. Vet. Res.* 54(9): 1544–1559.
8. Dougherty, S. H. 1986. Implant Infections. In: von Recum, A. F. (ed.), *Handbook of Biomaterials Evaluation*, New York: Macmillan, pp. 276–289.
9. Markowitz, J., Archibald, J. and Downie, H. G. 1964. *Experimental Surgery*, 5th ed. Baltimore: Williams & Wilkins, p. 6.
10. Gibbons, D. F. Personal communications, 1982–1992.
11. Clowes, A. W. 1987. Pathobiology of arterial healing. In: Strandness, D. E., Didisheim, P., Clowes, A. W., and Watson, J. T. (eds.), *Vascular Diseases. Current Research and Clinical Applications*. Orlando: Grune and Stratton, pp. 351–362.

12. Parks, P. J., and Ericson, D. G. 1993. Testing and evaluation of thrombogenicity. *Evaluation and Testing of Cardiovascular Devices. Short Course*. Minneapolis: Society for Biomaterials.
13. Rushmer, R. F. 1961. *Cardiovascular Dynamics*, 2nd ed. Philadelphia: WB Saunders.
14. Levy, R. J., Schoen, F. J., Anderson, H. C., Harasaki, H., Koch, T. H., Brown, W., Lian, J. B., Cumming, R., and Gavin, J. B. 1991. Cardiovascular implant valcification: a survey and update. *Biomaterials* 12:707–714.
15. Levy, R. J. 1993. Calcification models and evaluation of the efficacy of controlled release drug delivery for anticalcification. *Evaluation and Testing of Cardiovascular Devices. Short Course*. Minneapolis: Society for Biomaterials.
16. Schoen, F. J., Anderson, J. M., Didisheim, P., Dobbins, J. J., Gristina, A. G., Harasaki, H., and Simmons, R. L., 1990. Ventricular assist device (VAD) pathology analyses: guidelines for clinical studies. *J. Appl. Biomater.* 1:49–56.
17. Seckel, B. R., Ryan, S. E., Gagne, R. G., Chiu, T. H., and Wadkins, E. 1986. Target specific nerve regeneration through a nerve guide in the rat. *Plast. Reconstr. Surg.* 78:793.
18. Tountas, C. P., Bergman, R. A., Lewis, T. W., Stone, H. E., Pyrek, J. D., and Mendenhall, H. V., 1993. A comparison of peripheral nerve repair using an absorbable tubulization device and conventional suture in primates. *J. Appl. Biomater.* 4:261–268.
19. Nauta, W. J. H. and Karten, H. J. 1970. A general profile of the vertebrate brain, with sidelights on the ancestry of cerebral cortex. In: Schmitt, R. O. (ed.), *The Neurosciences*, New York: Rockefeller University Press, pp. 7–25.
20. Beckley, J. H. 1965. Comparative eye testing: man vs. animal. *Toxicol. Appl. Pharmacol.* 7:93–101.
21. Winter, G. D. 1977. Oxygen and epidermal wound healing. *Adv. Exp. Med. Biol.* 94:673–678.

8

Blood-Biomaterial Interaction

A Review of Some Evaluation Methods

E. Cenni, C. R. Arciola, G. Ciapetti, D. Granchi, L. Savarino, S. Stea, L. Montanaro, and A. Pizzoferrato
Istituti Ortopedici Rizzoli, Bologna, Italy

I. INTRODUCTION

Blood-biomaterial interaction means every interaction between an artificial device and blood, which determines effects on the material on one side and on blood, organs or tissues on the other (1). Such interaction is influenced by the flow conditions, the contact duration, the surface area and the implant site.

The surface of devices hardly has a hemocompatibility degree similar to endothelium, which physiologically releases antiaggregating substances and maintains the balance between the plasma phase of coagulation and the fibrinolytic process.

Even though to a different extent, all the artificial surfaces adsorb plasma proteins; this event is followed by activation of platelets, coagulation factors, complement and leukocytes. The artificial materials may induce hemolysis too.

Platelet aggregation and coagulation factor activation cause thrombus formation. Thrombus determines a reduction of blood flow, ischemia of tissues and embolism.

The risk for thromboembolism caused by blood-contacting implants still represents a serious clinical problem, so that the substitution of small size vessels with artificial prostheses is a serious hazard.

The adhesion of PMNs, mediated by C3a and C5a, favors thrombi formation, too. Besides, the leukocytes secrete enzymes and cytokines which have a proinflammatory effect (2).

Even the immune response can be modified by contact with biomaterials.

Thromboxane A_2, serotonin, fibrinopeptide A, C4a, C3a and C5a determine systemic effects, such as vasoconstriction, fever and bronchial constriction (Fig. 1).

Conversely, the performance of the device may be affected by the adsorption of plasma proteins, lipids and calcium and by the adhesion of platelets, leukocytes and erythrocytes (3).

II. EVENTS FOLLOWING THE BLOOD-MATERIAL CONTACT

The main events following the contact of blood with an artificial surface are first protein adsorption, then changes of blood cells and soluble factors.

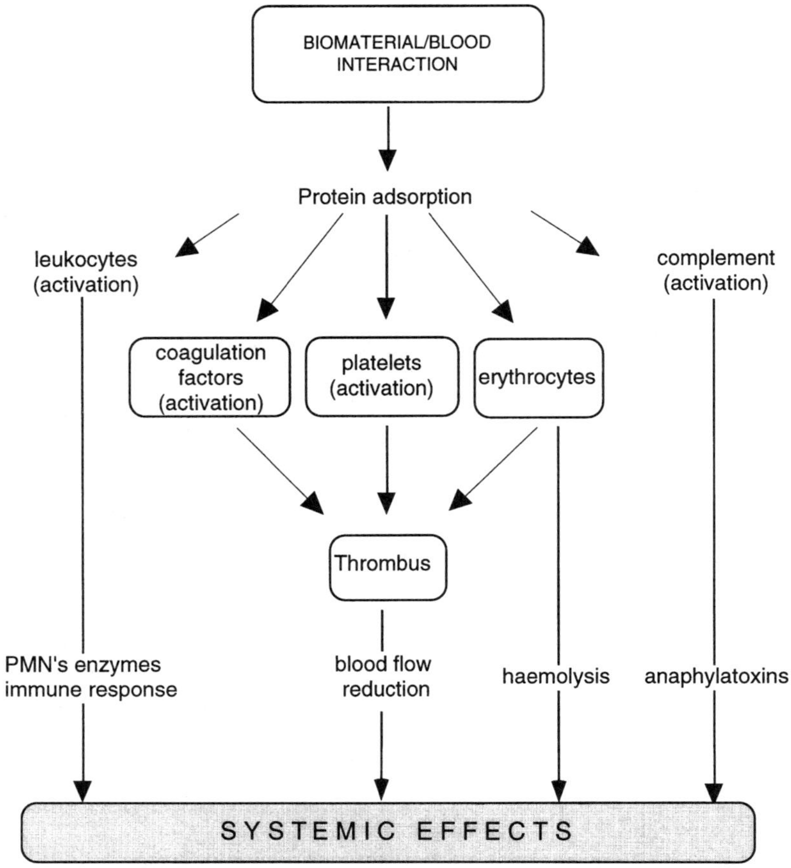

Figure 1 Effects Consequent to the Blood-Biomaterial Interaction in the Living Organism.

A. Protein Adsorption

Proteins are adsorbed when blood gets in contact with an artificial surface, during the short latency time which precedes the adhesion of platelets and blood cells. Inorganic ions and water molecules are adsorbed first, then plasma proteins follow, as to form a layer about 100 nm thick. Adsorption is a dynamic phenomenon, governed by the chemical properties of the material, by the duration of contact with blood and by the surface area exposed. Protein adsorption depends on dynamic conditions too: during nonflow conditions the amount of adsorbed proteins is lower than under flow (4).

Adsorbed albumin inhibits platelet adhesion and activation (5), but on some surfaces, adsorbed albumin is substituted by IgG, which favors coagulation.

Adsorbed fibrinogen activates platelets by interacting with membrane receptors.

On some surfaces, within a few minutes fibrinogen is replaced by high molecular weight kininogen, which favors factor XII activation. Decreasing levels of fibrinogen were found after 15 min on polydimethylsiloxane and suggested that this phenomenon might be due to the displacement of such protein (6).

Also fibronectin, which favors cell adhesion, and von Willebrand factor, which determines platelet adhesion to subendothelial tissues, adhere to foreign surfaces.

B. Platelet Activation

Platelet adhesion is enhanced by prior adsorption of γ-globulin or fibrinogen and is reduced by albumin. Probably, incomplete heterosaccharides of γ-globulin and fibrinogen react with glycosyl transferase groups on the platelet membrane. The inhibitory effect of albumin may be due to the absence of such saccharide chains (7). When platelets contact a fibrinogen-coated surface, their round shape undergoes a transformation to a fully spread form which is flattened against the surface. Platelet spreading is accompanied by a cytoskeletal reorganization. The surface-bound fibrinogen moves toward the center of the spreading platelet on the membrane and is translocated rapidly from the peripheral margin to the cell center and to channels of the open canalicular system. After a 20 min incubation, fibrinogen is localized in vacuole-like and granule-like structures (8).

The cytoskeletal reorganization is accompanied by changes in the levels of biochemical messengers. Calcium ions potentiate the morphological changes.

The final events are platelet release reaction and aggregation. In addition, the surface of an activated platelet is a prerequisite for the formation of clotting enzyme complexes (9).

C. Activation of Coagulation

The contact phase includes enzymatic reactions which take place during the initial phases of blood-material interaction. Factor XII binds to negatively charged surfaces, in particular to carboxylic groups. In fact all the substances activating factor XII, such as glass, collagen, kaolin, celite and ellagic acid, possess carboxylic groups or negative charges; this activity is no longer present upon neutralization of the negative charges. Factor XII may be activated also by contact with aggregated platelets. Following adsorption, factor XII breaks down into the fragments α-FXIIa and β-FXIIa. β-FXIIa converts prekallikrein to kallikrein, which converts factor XIIa to Hageman factor fragment, which in turn induces further activation of prekallikrein to kallikrein.

Factor XIIa activates factor XI, which triggers the intrinsic pathway of blood coagulation.

The extrinsic pathway is triggered by the tissue damage associated with surgical implantation of a device. Also monocytes activated by artificial materials release tissue factor and other substances directly activating factor X and factor II.

A mathematical model of coagulation on artificial surfaces concluded that the contact phase is very fast and depends on both surface reactivity and flow (10). Also on inert surfaces, the presence of trace levels of factor XII leads to activation of coagulation. The common pathway takes place onto platelet membranes and produces levels of thrombin depending on the flow conditions. Even trace levels of factors XIIa and XIa are capable of producing substantial thrombin concentrations, with appropriate flow conditions. On the contrary, if the flow rate is high, also high levels of factor XIa do not necessarily determine high thrombin concentrations.

The inhibitors of coagulation factors, such as antithrombin III, protein C, protein S, heparin cofactor II, the inhibitor of C1-esterase, may be modified by contact with artificial materials.

The blood-biomaterial interaction can influence the fibrinolytic system, both directly and indirectly, though the activation of factor XII and kallikrein, which splits plasminogen into plasmin.

The activation of coagulation and fibrinolysis can lead to thrombus formation on the implant site, but it does not lead to disseminated intravascular coagulation nor to hemorrhagic disorders. In the thrombotic process induced by implantation, phenomena determining local thrombosis often occur. The consumption of coagulation factors is limited and it is balanced by the continuous secretion of new factors. Therefore it can be hypothesized that the blood-material interaction sometimes originates chronic intravascular coagulation.

D. Leukocyte Activation

Understanding the process of leukocyte adhesion to blood vessels and to artificial materials is important in medical devices applications and is an important requisite for successful implants (11). Leukocyte adhesion is induced both by direct contact with the materials, by adsorbed proteins (especially IgG, thrombin and prothrombin) and by C3b and C5a.

Adherence is often followed by activation. Activated granulocytes release superoxide ions and produce prostaglandins, leukotrienes and platelet aggregating factor (PAF). PAF activates platelets, neutrophils and monocytes and constricts smooth muscles. Material-activated monocytes release tissue factor and substances directly activating factor X and factor II. Activated mononuclear cells release also cytokines. In cell cultures put in contact with velour Dacron and woven Dacron, interleukin-1 and tumour necrosis factor α (TNFα) release by human monocytes was demonstrated (12). Also in vivo high TNFα levels were demonstrated during hemodialysis with Cuprophan membranes (13). Cytokine secretion is influenced by the characteristics of the artificial surface and by the proteins adsorbed. Biomaterials may induce the release of IL-1 receptor antagonist, TGF-β and PGE_2. PGE_2 suppresses IL-1 production by monocytes and TGF-β and IL-1 receptor antagonist suppresses the ability of IL-1 to stimulate thymocytes (6).

Some biomaterials may induce the expression of the leukocyte receptors LFA-1, Mac-1/CR3 and LECAM-1. These antigens are expressed by activated leukocytes and promote their adhesion onto endothelium (14). In chronic hemodialysis patients the expression of Mac-1 and LECAM-1 is changed following the migration of leukocytes through materials for dialysis membranes, such as polyacrylonitrile or polysulfone (15). The leukocytes exposed to polyvinyl alcohol hydrogel surface express high levels of CD11b/CD18 antigens and a decrease of LECAM-1 (16). Polyethylene terephthalate induces adhesion of leukocytes onto endothelium (17). Woven Dacron upregulates CD11b and CD11c on both granulocytes and monocytes (18). An alteration in the CD11/CD18 complex was observed on the polymorphonuclear cell surface after incubation with knitted Dacron. In this study it was also demonstrated that CD11/CD18 expression varies among donors for a given vascular prosthesis, and that each donor shows a pattern of responsiveness toward synthetic vascular grafts (19).

E. Complement Activation

Complement activation is one of the causes of failure of blood-contacting devices. It leads to the formation of peptides with biological activity, which activate platelets and induce granulocytes and monocytes to adhere to the endothelium, with subsequent migration into tissues and release of their enzymes. Complement activation therefore contributes to the inflammatory response, the onset of immune reaction and the formation of thrombi.

Usually biomaterials activate the complement through the alternative pathway, even though the involvement of the classic pathway cannot be excluded (20). In fact, classical pathway activation was demonstrated for thiol-modified gold surfaces at a lower serum concentration or after a shorter incubation time, while at a high serum concentration or with longer incubation time the classical pathway was inhibited and the alternative pathway dominated (21,22).

Complement activation induced by materials used in hemodialysis circuits, extracorporeal circulation and apheresis (23) has been particularly investigated, as well as complement activation induced by vascular catheters, as a factor able to affect the onset of infectious complications (24). Also complement activation induced by materials used for vascular prostheses was investigated. It was demonstrated that uncoated PET activates complement (25–27). Such activation, favoring platelet adhesion and intimal hyperplasia, can contribute to vascular implant failure. There is also research on complement activation induced by different coatings of PET. The coating with bovine collagen does not reduce complement activation compared to uncoated PET (28,29).

Fragments with biological activity derive from complement activation. C3a and C5a are chemoattractants for leukocytes and favor their aggregation. Furthermore, C5a induces the granulocytes and monocytes to adhere to the endothelium, to migrate into tissues, to release their enzymes and to produce procoagulant or platelet aggregating substances, like PAF. C3a and the C5b-9 complex can directly activate platelets.

Complement activation on the device surface should not be considered as a strictly local phenomenon, but results also in systemic effects, when activated factors are carried away in the bloodstream.

F. Erythrocyte Damage

Erythrocytes, upon adhesion on the layer of adsorbed proteins, undergo hemolysis and release ADP, so inducing platelet aggregation. Blood red cell adhesion on the artificial surfaces is favored by IgG and, in decreasing order, by albumin, fibronectin, transferrin, and fibrinogen (30). Shear-induced hemolysis is influenced by the shear stress and by the nature of the surface in contact with blood.

III. EXPERIMENTAL MODELS IN THE STUDY OF BLOOD-BIOMATERIAL INTERACTION

A number of experimental models have been developed to study blood-material interactions in vitro, in vivo and ex vivo.

There is no overlapping among different methods. In vitro tests are screening tests and permit the use of human blood; in vivo tests are closer to the clinical situation, but data obtained with animals cannot easily be transferred to humans and ethical problems are involved, too. Ex vivo tests eliminate some problems of in vitro tests (e.g., the use of anticoagulants) and permit many materials to be examined simultaneously in the same animal, but have the same limits reported above for in vivo tests.

Tests should use an appropriate model which simulates the conditions of contact of the device with blood during clinical applications. These include duration of contact, temperature, sterility and flow conditions. An appropriate ratio between the material surface area and the amount of blood should be employed. Positive and negative control materials should be used, including well-characterized reference materials or materials already in clinical use (3).

A. In Vitro Tests

Variables that influence in vitro tests are the blood source, anticoagulant, modality of sample collection, delay between the collection and the beginning of the experiment, the blood component used for the test, the hematocrit, presence of an air-blood interface, material surface to blood volume ratio, temperature, dynamic conditions, and duration of blood-biomaterial contact.

- Human blood should be used wherever possible, because there are differences in the hemostatic system of different animal species.
- The choice of the anticoagulant represents an important variable affecting the results. EDTA reduces platelet activity and heparin hinders the formation of fibrin. Citrate, heparin, ACD or CPD-A are used, but also these anticoagulants may interfere with platelets and coagulation factors.
- Stasis during blood collection should be avoided, because tissue factor can be released from the vascular wall.
- The tests on blood should be performed as soon as possible after collection, because coagulation factors and platelet activity are progressively reduced with time. Three or four hours are the maximum delay allowed if the sample was properly stored.
- Whole blood, platelet-rich plasma (PRP), platelet suspensions and platelet-poor plasma (PPP) may be used in in vitro tests, according to the parameters to be tested. Whole blood or PRP should be used, because platelet factor 3 is important for the activation of plasma factors of coagulation.
- The formation of any air-blood interface should be avoided, as the activation of the intrinsic pathway of coagulation can be favored.
- The experimental results are affected by the material size. An inadequate blood-material ratio can give false negative results. Suggestions for the optimal surface/blood ratio are reported in ISO 10993/12: Biological Evaluation of Medical Devices—Sample Preparation and Reference Materials, 1996 (31).
- Also the geometry of the material and the procedure of sterilization can affect the results. Therefore, the material should be tested in the end-use form.
- The temperature of the tests should be 37°C, because at this temperature the optimal platelet and factor activations take place.
- The hemodynamic parameters influence blood-material interaction: though very difficult, the flow in vitro should reproduce the physiological flow whenever possible.
- The duration of blood-material contact should be accurately calculated, because an inadequate time may affect the reliability of the results.

In vitro tests are simpler, faster and cheaper and more standardizable than in vivo and ex vivo tests; they permit study of the basic mechanism of coagulation activation induced by artificial materials and the results are quantifiable. The shortcoming of in vitro tests is the fact that the experimental conditions widely differ from clinical situation, including the need for anticoagulants. In vitro tests are regarded as useful for the screening of materials. A material which does not yield good results in vitro is likely to perform unsatisfactorily once implanted in vivo. In vitro tests are classified into static and dynamic tests.

1. *Static Tests*

Static tests entail the contact of materials with whole blood, PRP, platelet suspensions or PPP without flow conditions. The advantages of static tests are low cost and, in some tests, the absence of anticoagulant; their shortcoming is the absence of flow.

Static models of in vitro contact between materials and blood may consist of tubes (32), beads (33) or perfusion chambers coated with the test material. The blood-contacting device can be left undisturbed for the experiment duration or it can be mixed at low speed to favor the interaction of blood with the material. After the contact time, blood is collected and some parameters are evaluated.

The perfusion chamber is generally a sandwich assembled with two slides. The chamber is coated with the test material. It is primed with buffer to avoid any blood-air interface, then blood is introduced from one side and allowed to remain in the chamber for a fixed time. In the study of Park et al. (34) the chamber was assembled with a glass slide, a glass coverslip and a silicone rubber gasket as a spacer. The slide and the coverslip were treated with dimethyldichlorosilane and the gasket was treated with silicone grease. Platelets in PBS were introduced in the perfusion chamber and allowed to adhere to the surface. Platelets into the chamber were observed with an inverted light microscope and their spreading was monitored by video-enhanced differential interference contrast microscopy.

In another study flat sheets of the test materials were cut and placed onto polystyrene incubation plates. Blood was placed onto the surface; then lids were placed over the plate to reduce evaporation. The system was incubated statically at 37°C for 30 min (9).

In another study the test materials in form of disks were placed into a 24-well cell culture plate and fixed with a silicone ring. Whole blood or PRP was poured on the disk and allowed to remain for 60 min at 37°C. After removing blood or PRP, the disks were examined by scanning electron microscopy (35).

2. *Dynamic Tests*

Dynamic tests entail the contact of materials with circulating blood, under laminar flow conditions, in order to simulate the clinical situation.

The recirculation system most strictly resembles the clinical situation. In this system, whole blood or PRP or platelet suspensions circulate into a close cycle system, where a pump ensures the laminar flow. The test material is inserted into the system (36,37).

An example of a recirculation system consists of two parallel circuits, one made of the test material, and the control one made of silicone. Heparinized whole human blood was forced through the circuits at 37°C: the speed and pressure were controlled to reproduce those of the femoral artery. During the 1-h recirculation, blood specimens were collected every 15 min for platelet count and aggregation test. At the beginning and at the end of the experiment the blood gases, pH and hemoglobin were evaluated, and the formed blood elements were counted. Moreover, platelet consumption was evaluated by using platelets labeled with ^{111}Indium. At the end of the experiment the materials of the two circuits were fixed and examined by scanning electron microscopy (38).

Another perfusion system consisted of a tubing connected at each end to syringes (9). This arrangement allowed a laminar flow along more than 99% of the length of the tube. The syringe containing blood was attached to a microprocessor-controlled pump and the whole blood was perfused along the tubing at various shear rates at 37°C. The blood was constantly re-perfused back and forth through the tube, allowing a high degree of blood-material contact with a minimal blood-air interface.

Another dynamic test for the assessment of the blood compatibility of flat polymers consisted of a chamber where small amounts of blood contacted a large geometrical test surface. The construction consisted of a spiral-shaped flow channel covered with the test biomaterials from both sides: in the chamber laminar flow conditions were maintained (39).

The blood compatibility of sulfonated polyurethanes was examined through a blood loop analysis (40). Loops made of the test materials were first primed with an electrolyte solution, which was displaced by whole blood from a donor's vein. Heparin was introduced into the loop, then a computer-controlled motor system was used to impart unidirectional pulses precisely defined to the blood-filled loops submerged in a 37°C temperature bath.

Another system consisted of a ring formed by two syringe pumps, fitted with polypropylene syringes, attached to both ends of a tube made with the polymer to be tested. The tube was laid across a flat perspex plate and compressed by a second perspex plate. With this configuration a laminar flow was created in the tube (41).

In the rheological test chamber the test material spins into blood. The material evaluation by this method coincides with the results obtained by the bead column test (42).

For studying the effect of temperature and shear rate on platelet aggregation, a computer-controlled concentric cylinder rotational viscometer is used (43,44): citrated whole blood is subjected to known shear rates at different temperatures.

The cone-and-plate device allows well-defined flow conditions. It consists of a four-well microtiter plate, with each well constituting an independent unit. The plate surface consists of the test material. A platelet–red cell suspension is placed in the wells and the cones are lowered until the cone tip touches the bottom of the well. Then cones are rotated at different speeds. After a fixed time, the rotation is stopped and the material surface is examined (45).

A laminar flow chamber consisted of two conical ends, one for inflow and the other for outflow, which were coupled to a cylinder. In this central part the samples to be tested were placed. The ends of the device were connected to a pump, which assured appropriate flow rate and pressure to the blood (46).

B. In Vivo Tests

In vivo tests do not need anticoagulants and allow the materials to be examined in conditions closely resembling the clinical situation. Blood flow is properly evaluated only in vivo; this evaluation is mandatory as it affects the thrombus formation on artificial surfaces. In vivo tests permit the evaluation of thromboembolization and infarction in organs.

The variables that affect in vivo tests are the animal model, the implant site and duration. The choice of an appropriate animal model may be restricted by size requirements, the availability of certain species and the cost. Similarities and differences in physiology of animals and humans should be considered (3). Animals differ from humans as to (1) hemostatic system, (2) inflammatory and reparative response to implant and (3) anatomical and hemodynamic characteristics of vascular system. Therefore, there is no animal model perfectly suited to the study of blood-contacting devices. The sheep is often used because its hemostatic system, and in particular platelets, is quite similar to human system. Other variables which can interfere with clotting are the age, the animal breed and the anesthetic used.

The implant should be tested in the end-use form. The sterile material should be implanted in the same site where it will be implanted in humans; moreover, flow conditions as close as possible to the final ones should be reproduced.

The arterial implant is generally performed in the carotid artery, femoral artery, or aorta, while the jugular vein, the femoral vein and the vena cava are the elective sites for venous implants. The prosthetic valves can be implanted in the atrioventricular or aortic ring.

Before and during implant, coagulation should be periodically monitored and the implant patency should be evaluated. At the time of explant, thrombi are counted and weighed, and the presence of kidney infarct is ascertained; the implant may also be examined by light or electron microscopy.

1. Main In Vivo Tests

In the past, some in vivo tests have been employed for the screening of cardiovascular materials. Actually, in vivo screening tests should not be used, because many valid in vitro tests are available. According to the most recent opinions, in vivo tests should be employed as second-level tests for the evaluation of materials that have already shown good hemocompatibility in vitro.

A material for the construction of small size arterial prostheses was examined upon implantation as aorta-iliac bilateral bypass in dogs. Each animal received a bypass made of the test material and an expanded polytetrafluoroethylene bypass as control. The implants were left in situ for six months and the patency of the implant periodically was examined by angiography. At the retrieval, the thrombus formation on the bypass surface was quantitatively evaluated by light microscopy connected to an image analysis system (47).

The hemocompatibility of vascular prosthetic materials was also examined by evaluating platelet and fibrinogen survival. Platelet survival is measured with ^{111}In-labeled platelets; fibrinogen survival is determined with ^{131}I-labeled fibrinogen. In calves implanted with total artificial heart and left ventricular assist device platelet and fibrinogen survival were measured and correlated with autopsy findings (48). A shortened platelet survival and thromboembolic complications were found probably due to the continuous platelet activation by the artificial surface.

C. Ex Vivo Tests

In ex vivo tests the blood circulates out of the animal body, through a cannulated vessel, in a shunt or chamber. The variables that affect ex vivo tests are the kind of animal employed, the type of system, the implant duration. The choice of animal presents the same problems as in vivo tests.

The blood circulation system can be either open or closed. In the open system, the blood does not return to the animal, but it is collected in tubes or in a chamber coated with the material to be tested.

In the closed system the blood returns to the animal after a passage through either a perfusion chamber or an arterial-venous or arterio-arterial or veno-venous shunt made of the test material (49–51). In this way several materials are tested simultaneously in the same experimental conditions (52). An example of a closed system is the device constructed by San Roman et al. (46) and described in the section on in vitro tests. The device could be connected also to the femoral artery and the right atrium of a dog for ex vivo studies.

According to the duration, the implant is defined as acute or chronic: in the acute model the implant duration ranges from a few minutes to a few hours, whereas in the chronic model the duration is longer.

Ex vivo tests do not need anticoagulation, blood is assayed under hemodynamic conditions similar to the physiological one and several materials are evaluated in parallel. For these advantages, ex vivo tests are considered as elective tests for the assay of external communicating devices. The most important disadvantage is the difference in the hemostatic systems of animals and humans.

Recently, some ex vivo models using human blood were introduced. These systems permit the evaluation of polymers intended for external communicating devices in contact with circulating blood, such as materials for dialytic membranes (53). Test materials are exposed to flowing human blood under controlled conditions of flow. Venous blood from healthy donors was pumped through minidialyzer modules made with the materials to be tested. This system allowed the use of labeled proteins without contaminating the blood donors. The modules were removed at selected endpoints and the adsorption of proteins was evaluated (54).

IV. BLOOD EVALUATION AFTER CONTACT WITH BIOMATERIALS

The tests described in this section evaluate the changes induced by artificial materials in human blood. The testing methods tend to highlight any change of blood formed elements and proteins. Some procedures have also been conceived for detecting alterations of the material surface. When using animal blood, the normal range for the species employed must be established. In some tests which analyze immunological parameters animal blood cannot be used. Actually, monoclonal antibodies specific for the species more frequently employed for in vivo or ex vivo testing are available.

A. Platelets

Platelet performance is studied using platelet-rich plasma or platelet suspensions. The damage to the membrane, the adhesion onto materials, the morphological changes, the aggregability and the release reaction are focal points to be examined.

- Platelet adhesion is usually assessed by differential counting before and after contact with the test material, using electronic counters or phase contrast microscopy. It is possible also to evaluate platelet adhesion indirectly, measuring LDH activity after platelet lysis (55). The deposition of platelets onto artificial surfaces may be evaluated also by the use of monoclonal antibodies specific for membrane glycoproteins, such as GPIIb/IIIa.
- The changes in the levels of biochemical messengers may be investigated through the determination of intracellular free calcium in platelets adhering to artificial surfaces, using a fluorescent dye (56,57).
- The biomaterial-induced morphological modifications of platelets are assessed by microscopy and image analysis system. The loss of the round shape and development of pseudopodia indicate that activation has been triggered. The spreading platelets may be classified into five morphological forms correspondent to increasing activation: round, dendritic, spread dendritic, spreading and fully spreading (58).
- Modifications of platelet aggregability are evaluated through platelet aggregation test induced by different agents, such as ADP, collagen, thrombin, adrenalin, ristocetin (59).
- The production of arachidonic acid metabolites is evaluated by thromboxane B_2 assay.
- The platelet release reaction induced by artificial materials is evaluated in plasma by measuring the substances released by dense bodies and α-granules, including serotonin, ADP, ATP, β-thromboglobulin and platelet factor 4. As a general rule, the materials which determine a higher platelet adhesion also cause a higher release of β-thromboglobulin and platelet factor 4.
- Activated platelets are detected by flow cytometry or enzyme immunoassay using monoclonal antibodies specific for activation-dependent platelet epitopes (CD62 and PAC-1). CD62 binds to P-selectin (GMP140 or PADGEM), an α-granule glycoprotein that is incorporated into the outer cell membrane during activation. P-selectin is present only on activated platelets and is used as a marker of platelet activation on biomaterials (43,44,60–63). PAC-1 binds to the activated form of GPIIb/IIIa, which, following platelet stimulation by agonists, is converted to an activated state that binds soluble fibrinogen and mediates platelet aggregation (64). Annexin V recognizes platelet activation because it detects procoagulant phospholipid exposure on platelet membrane (65).

Some materials were tested in our laboratory:

- Polyethylene terephthalate (PET) in the form of a woven fabric, supplied by Vascutek Limited (Inchinnan, Renfrewshire, Scotland, UK)
- Polyethylene terephthalate in the form of double velour coated with collagen and pyrolitic carbon (PC), supplied by Sorin (Saluggia, Italy)
- Glow discharge-treated polybutylene terephthalate coated with a hydrophilic polymer made of partially hydrolyzed polyvinylacetate containing polyethylene oxide/polypropylene oxide copolymer block as lateral chains (PVA), supplied by Biofil S.r.l. (Cavezzo, Modena, Italy).

An in vitro method was standardized in our laboratory for testing the materials (66–69). Venous blood from healthy donors was collected in siliconized tubes containing sodium citrate at 3.8%, at a blood/citrate ratio of 9/1. Platelet-rich plasma (PRP) was obtained by centrifugation at 120g for 5 min. PRPs from different donors were combined into a pool. All experiments started within 2 h of blood collection, as the platelet number and activity, as well as the concentration and activity of the coagulation factors, change as time goes on. PRP was added to siliconized tubes containing the test material and PRP not in contact with the materials was used as a negative control. All specimens were rotated at 37°C for 30 min.

This method is very simple, repeatable and versatile. Some materials, such as PET, were tested during years and gave substantially the same results. The method is rapid and quite cheap. It is carried out on human blood. On the other hand, a disadvantage is the need for anticoagulant. We use sodium citrate that influences platelets and factor activity less than EDTA or heparin.

At fixed times, plasma was separated from the test material and analyzed. After 15 min contact, platelet adhesion and CD62 were measured, while after 30 minutes platelet adhesion, CD62, β-thromboglobulin and thromboxane B_2 were assayed.

CD62 was evaluated by flow cytometry using a monoclonal antibody.

For the determination of β-thromboglobulin and thromboxane B_2, a platelet activity inhibiting solution of EDTA and theophylline was added to PRP, immediately after contact with the test materials. The specimens were centrifuged for 30 min at 2000g at 4°C, in order to obtain platelet-poor plasma (PPP): β-thromboglobulin and thromboxane B_2 were determined by radioimmunoassay in PPP. The results are reported in Tables 1–4.

PET induced a highly significant reduction in platelet number, both after 15 and 30 min, but no significant variations in CD62 expression. It also caused a significant increase of β-TG and TXB_2 concentrations after 30 min contact.

PC induced a significant reduction of platelet number only after 30 min. This material

Table 1 Arithmetic Mean, Standard Error and Significance (according to Student's *t* test) of Platelet Counting after Contact with Test Materials

Material	Platelet no. after 15 min ($10^3/\mu l$)	p	Platelet no. after 30 min ($10^3/\mu l$)	p
PRP	410.9 ± 27	—	391.1 ± 23	—
PET	244.1 ± 18	<0.01	164.2 ± 25	<0.01
PC	389.5 ± 28	n.s.	321.8 ± 19	<0.05
PVA	382.5 ± 23	n.s.	353.2 ± 18	n.s.

Table 2 Arithmetic Mean, Standard Error and Significance (according to Student's *t* test) of the Percentage of Platelets Positive for CD62, after Contact with the Test Materials

Material	% positive platelets for CD 62 after 15 min	*p*	% positive platelets for CD62 after 30 min	*p*
PRP	24.2 ± 6	—	17.3 ± 5	—
PET	24.1 ± 6	n.s.	18.6 ± 4	n.s.
PC	42.8 ± 7	<0.05	31.1 ± 6	n.s.
PVA	25.8 ± 7	n.s.	18.3 ± 6	n.s.

determined a remarkable and significant increase in CD62 expression at 15 min and in β-TG and TXB_2 concentrations after 30 min.

PVA induced no significant difference in platelet number and in CD62 expression with respect to the control after 15 or 30 min. This material induced no significant variations of β-TG, but a significant increase of TXB_2.

The discrepancy observed for PET between platelet release reaction and CD62 expression may be due to the fact that platelets which expressed the antigen still adhered to the polymer, as demonstrated by the reduction in platelet number after 15 min.

The increase of CD62 by PC was assumed as a marker of platelet activation at 15 min before adhesion.

PVA did not induce significant variations of platelet number and of CD62 expression. However, a significant release of TXB_2 was detected, confirming the high sensitivity of such a test.

CD62 allows the detection of the activation process, even before platelet adhesion onto the material surface. The sensitivity of this marker is good in the early phase of platelet activation, but scanty when platelets adhere massively to the material, as happens with PET. Probably, this discrepancy is due to the fact that almost all activated platelets adhere to PET and only few platelets in PRP express CD62 antigen.

Among platelet tests we chose the assays specific for platelet activation, because they are more in accordance with clinical phenomena in implanted patients. Therefore we prefer to perform platelet counting and to study the release reaction, while the tests evaluating the damage to membrane or morphological modification seem to be insignificant to activation induced

Table 3 Arithmetic Mean, Standard Error and Significance (according to Student's *t* test) of β-Thromboglobulin (β-TG), after Contact with the Test Materials

Material	β-TG after 30 min (ng/ml)	*p*
PRP	712.7 ± 49	—
PET	5971.6 ± 572	<0.01
PC	2170.6 ± 534	<0.05
PVA	1252.4 ± 388	n.s.

Table 4 Arithmetic Mean, Standard Error and Significance (according to Student's *t* test) of TXB_2, after Contact with the Test Materials

Material	TXB_2 after 30 min (pg/ml)	*p*
PRP	617.3 ± 81	—
PET	9022.9 ± 755	<0.01
PC	4881.2 ± 1193	<0.01
PVA	1842.1 ± 262	<0.01

by materials. Platelet counting alone is not sufficient to assess thoroughly platelet modifications induced by artificial materials: also some products of the release reaction, chosen between those specific for platelets, have to be measured. The determination of membrane antigens specific for platelet activation, such as GMP140, is a promising test to assess platelet changes induced by artificial materials.

B. Plasma Factors of Hemostasis

The activation of plasma factors, induced by artificial materials, can be explored with global methods or with selective tests. The former evaluate the coagulation pathway on the whole and are not able to detect which phase is modified by the biomaterial. By selective testing even the change of a single factor can be detected. As a general rule, the global methods and many selective ones are more sensitive to coagulation deficiencies than to activation. Among selective methods there are also methods which evaluate the concentration of activated factors and fragments which split during activation.

1. *Global Tests*

A global evaluation is obtained by measuring changes of clotting time or thromboelastogram (TEG) induced by materials. As mentioned, these tests do not indicate which phase of hemostasis is involved in the interaction with biomaterials.

Clotting time is obtained by measuring blood clot formation in a tube coated with the test material, with glass or siliconized tubes used as reference values: the longer the clotting time, the less thrombogenic the material. The test is very simple, but insensitive.

A thrombus formation test and a plasma recalcification time were carried out in vitro (70). In the first test, the material was attached to a watch glass. ACD-blood was placed on the sample and incubated at 37°C. The clotting reaction was started by adding 0.1 M $CaCl_2$ solution to the blood: after an incubation time varying from 10 to 40 min, distilled water was added to stop the reaction and the weight of the thrombus formed on the material was determined.

The plasma recalcification time was performed on PPP placed on the sample attached to a watch glass and incubated statically at 37°C; 0.025 M $CaCl_2$ solution was then added and the clotting time was determined.

2. *Selective Tests*

By selective tests the different phases of coagulation are separately evaluated. The activated partial thromboplastin time (APTT) evaluates the intrinsic pathway, while the prothrombin time (PT) evaluates the extrinsic pathway and the thrombin time (TT) measures fibrin formation. These times increase for the deficiency of coagulation factors. These tests are, however, nearly insensitive to activation of coagulation factors which can occur after contact with thrombogenic materials. The nonactivated partial thromboplastin time is recommended, as the activated PTT includes components which mask the acceleration of coagulation by materials (3). Shortening of the PTT following contact with a material under standard conditions indicates activation of the contact phase of coagulation. A prolonged PTT suggests a deficiency in any coagulation factor but VII or XIII.

More promising and meaningful tests for material thrombogenicity include the assays of kallikrein (71–73), activated factors XII, XI, IX, X, thrombin (74–77), fragment 1 + 2, which splits from prothrombin, fibrinopeptide A (FPA), which is split from fibrinogen by thrombin (78), and thrombin-antithrombin III complex (TAT) (79). It is possible also to determine both free plasma factor XIIa and the activity of factor XII attached to biomaterial surface (41,80). These tests, used for clinical diagnosis of thrombosis for some years, could demonstrate factor activation and thrombin formation induced by artificial materials.

In our laboratory different tests are adopted for detecting changes of the plasma phase of coagulation induced by the biomaterials. We examined some materials whose characteristics were described in the previous section.

APTT and the activity of coagulation factors were determined immediately after contact with materials. We demonstrated that PET and PC determined a significant prolongation of APTT. As for PET, such prolongation was associated to a significant reduction of the factors of the intrinsic pathway, with the exception of factor VIII. After plasma challenge with PC a significant decrease in factor XI activity was observed. The reduction of the coagulation factors can be explained by their adsorption onto the biomaterial surface (Table 5).

More recently methods for detection of the activation of coagulation were adopted in our lab. The fragment 1 + 2, marker of thrombin formation, and fibrinopeptide A, marker of the fibrinogen conversion into fibrin, were assayed (81). The following polymers were tested:

- PET + collagen: polyethylene terephthalate in the form of collagen coated knitted Dacron (PET + collagen), supplied by InterVascular Inc. (La Ciotat, France).
- Polybutylene terephthalate 1 (PBT-1): a hydrophobic polymer with linear chain, obtained with "melt blown" technique. It was tested as nonwoven sample, with fibers of 2–3 mm diameter.
- Polybutylene terephthalate 2 (PBT-2): polybutylene terephthalate modified through corona treatment at atmospheric pressure in airflow. This treatment was performed on both surfaces by means of a radiofrequency corona discharge apparatus, and induces the formation of polar groups, with an increase in wettability and in adhesion

Table 5 Arithmetic Mean, Standard Deviation and Significance (according to Student's *t* test) of APTT and Activity of the Intrinsic Pathway Factors after Contact with PET and PC

Materials	APTT (s)	Factor XII (%)	Factor XI (%)	Factor IX (%)	Factor VIII (%)
PRP	28.4 ± 0.9	86.5 ± 10.1	75.4 ± 7.6	98.8 ± 46.0	166.3 ± 61.9
PET	29.5 ± 1.3*	78.6 ± 9.8*	68.2 ± 8.7*	48.0 ± 5.4*	181.6 ± 27.7
PC	30.3 ± 1.1*	85.1 ± 6.7	53.8 ± 16.2*	91.4 ± 43.4	146.9 ± 57.9

*$p < 0.01$.

Table 6 Arithmetic Mean, Standard Deviation and Significance (according to Student's *t* test) of Parameters for the Activation of the Plasma Phase of Coagulation, after Contact with Some Materials

Materials	F1 + 2 (nmol/l)	FPA (ng/ml)
PRP	1.71 ± 0.41	37.30 ± 10.18
PET + collagen	1.75 ± 0.16	49.67 ± 18.01*
PBT-1	1.99 ± 0.24*	45.09 ± 13.77
PBT-2	1.97 ± 0.14*	46.56 ± 15.78

*$p < 0.05$.

forces. The presence of these groups is limited to a very thin surface layer, with a thickness of about 1000 Å.

In order to assay fibrinopeptide A (FPA), after the contact of plasma with biomaterials, a thrombin-inhibiting solution of heparin plus aprotinin was added to the plasma at 9 : 1 plasma:solution ratio. After centrifugation at 2000g for 30 min at 4°C, a bentonite solution was added to the supernatants (1 : 2 plasma:bentonite ratio) in order to adsorb fibrinogen molecules which could interfere with the fibrinopeptide A assay. The bentonite was eliminated by centrifugation at 1000g for 20 min at 4°C. Fibrinopeptide A in the supernatant was then determined by radioimmunoassay.

Prothrombin fragment F 1 + 2 was determined by enzyme immunoassay with rabbit antibodies.

PBT-1 and PBT-2 increased significantly fragment 1 + 2; the other biomaterials determined only insignificant increases. Fibrinopeptide A was significantly increased by collagen-coated PET (Table 6), probably for an effect due either to the collagen itself or to the chemical compound used for the cross-linking procedure.

Thrombin formation rate is very important in the assessment of the activation of the plasma phase of coagulation. The assay of markers of thrombin formation and of activated factors represents a promising way of evaluating material-blood interaction.

C. Leukocyte Adhesion and Activation

Leukocyte adhesion is evaluated by counting the leukocytes before and after contact with the material. Likewise, it can be verified how white cell populations are distributed.

Functional modifications are detected by assaying leukocyte enzymes, such as elastase (82), β-glucuronidase, alkaline and acid phosphatase (83), and cytokine production (84).

Cytokine assay is important because these proteins mediate inflammatory reaction and contribute to the failure of the implant. Cytokines useful for the evaluation of biomaterials are interleukin-1 (IL-1), IL-4, IL-6, tumor necrosis factor-α (TNF-α), γ-interferon (γ-IFN) and transforming growth factor-β (TGF-β) (3): their production is assayed directly by enzyme immunoassay or indirectly demonstrating specific mRNA into the leukocytes. The amount of cytokines can be used as a measure for the evaluation of activation of leucocytes when contacting foreign materials (70). Cytokine secretion may result in systemic effects: they play a major role in regulating the immunological, inflammatory and wound healing responses; they control the growth of fibroblasts, smooth muscle cells and endothelial cells.

The expression of adhesion molecules involved in leukocyte activation could be modified by contact with artificial materials. We studied the expression of LFA-1, Mac-1/CR3 and LECAM-1 after contact between leukocytes and polyethylene terephthalate with and without pyrolytic carbon-coating (respectively PC and PET) (85). LFA-1 (CD11a/CD18) is involved in the adhesion of leukocytes to the endothelium during inflammatory reactions, as well as in T cell responses. Mac-1 (CD11b/CD18) is expressed on neutrophils, monocytes, natural killer cells and macrophages and its main function is to mediate the adherence of both monocytes and neutrophils to vascular endothelium for subsequent extravasation. LECAM-1 (CD62L) is expressed on lymphocytes and monocytes surface. It plays a crucial role in the early phase of the leukocyte extravasation during the inflammatory process.

In order to evaluate the activation of leukocyte, heparinized whole human blood was exposed for 30 min at 37°C to (a) AIM-V (Gibco); (b) PET; (c) PC; (d) a solution of bacterial endotoxin (LPS) (lipopolysaccharide of Escherichia coli 055 : B5 by Sigma) in AIM-V to a final concentration of 10 μg/ml. After incubation the expression of CD11a, CD11b, CD18 and CD62L was determined by flow cytometry. The expression of the adhesion molecules on lymphocytes did not change following exposure to the materials (Table 7). On the contrary, the expression of the monocytic-granulocytic population was strikingly modified (Table 8). Both endotoxin and materials increased significantly the percentage of monocyte-granulocyte positive for the CD11b antigen, decreased the CD62L antigen and did not modify the proportion of the CD11a antigen. The inflammatory stimulus represented by LPS induced a significant increase in the mean channel of fluorescence intensity of CD11a, CD11b, CD18, whereas the exposure to the materials increased significantly only the CD11b intensity.

Our results demonstrated that PET can activate leukocytes by modifying the expression by neutrophils of the molecules involved in the early phase of the inflammatory process, i.e., Mac-1 and LECAM-1. The expression of LFA-1, which is usually involved in the specific immune response, was not altered. The coating of the surface with pyrolytic carbon does not eliminate the property of PET to activate the circulating leukocytes. Similar changes were observed when the cells underwent the in vitro inflammatory stimulus represented by endotoxin. This means that the materials under test are able to induce in leukocytes the modifications which appear in the early phase of an inflammatory process, so as to allow the adhesion of monocytes and granulocytes to the vascular endothelium.

Table 7 Arithmetic Mean, Standard Deviation of Adhesion Molecules on Lymphocytes

	Control	LPS	PET	PC
	Proportion of positive cells			
CD11a	31 ± 8	30.8 ± 12	29.8 ± 11	32.1 ± 9
CD11b	6.3 ± 2	7.3 ± 2	6.8 ± 3	6.1 ± 2
CD18	33.7 ± 10	31.9 ± 8	32.1 ± 8	33.2 ± 10
CD62L	34.4 ± 15	36.3 ± 17	38 ± 14	32 ± 19
	Fluorescence intensity (mean channel)			
CD11a	1.6 ± 0.3	1.7 ± 0.3	1.6 ± 0.3	1.6 ± 0.3
CD11b	1.3 ± 0.2	1.5 ± 0.2	1.3 ± 0.3	1.3 ± 0.3
CD18	1.4 ± 0.3	1.3 ± 0.3	1.3 ± 0.3	1.3 ± 0.4
CD62L	1.3 ± 0.1	1.4 ± 0.3	1.3 ± 0.1	1.3 ± 0.3

Table 8 Arithmetic Mean, Standard Deviation and Significance (according to ANOVA and Bonferroni-Dunn's Multiple Comparison test) of Adhesion Molecules on both Monocytes and Granulocytes

	Control	LPS	PET	PC
	Proportion of positive cells			
CD11a	12.3 ± 6	11.4 ± 5	13.4 ± 5	12.6 ± 4
CD11b	80.2 ± 17	94.9 ± 2*	92.4 ± 4*	94.5 ± 3*
CD18	82 ± 23	95.6 ± 4	91.7 ± 9	88.4 ± 20
CD62L	50.4 ± 9	7.1 ± 3*	27.1 ± 7*	20.9 ± 5*
	Fluorescence intensity (mean channel)			
CD11a	1.7 ± 0.5	2.1 ± 0.3*	1.7 ± 0.4	1.7 ± 0.4
CD11b	5.5 ± 2.9	19.7 ± 2.9*	10.2 ± 3.5*	9 ± 1.5*
CD18	1.5 ± 0.7	3.9 ± 1.3*	1.9 ± 0.5	2.4 ± 1.4
CD62L	1.2 ± 0.2	1.5 ± 0.7	1.1 ± 0.3	1.1 ± 0.3

*$p < 0.01$.

D. Complement Activation

Complement evaluation includes inspection of the classical pathway, the alternative pathway, the common pathway and the membrane attack complex (86,87). The C4a and C4d fragments are markers of the classical pathway. The C4a fragment is one of the complement anaphylatoxins, while the C4d fragment is one of the final degradation products of C4b. The Bb, C3a and iC3b fragments are markers of the alternative pathway. The C5a fragment is a marker of the common complement pathway. This anaphylatoxin derives from the cleavage of C5, by the classical or the alternative pathway.

The membrane attack complex is generated by the assembly of C5 through C9 as a consequence of the complement activation by either the classical or the alternative pathway. Complexes formed in the absence of a target membrane bind to a regulatory serum protein, the S protein. The S protein binds to C5b-9 complexes, forming the soluble, nonlytic SC5b-9 complex.

Evaluation of complement activation induced by medical devices should include both the assessment in serum or plasma and the bound phase, i.e., on the biomaterial surface (7). Radioimmunoassay and enzyme immunoassay have limitations in accurate determination of surface-bound components. In situ ellipsometry-antibody techniques and ellipsometric measurements in air before and after serum and subsequent antibody incubations allow a reliable quantitation of surface-bound complement fractions (22,88).

Some studies investigated complement activation after contact of biomaterials with plasma or serum (89,90), or with whole blood, in order to mimic the complement activation which occurs in vivo.

We compared complement activation, through the classic or alternative pathway, induced by pyrolytic carbon coated PET (PC) and uncoated PET. Blood samples were collected from healthy donors in tubes containing lithium heparinate and stored on ice until tested. The contact between blood and materials was performed within an hour from the collection (91).

Each sample of whole blood was incubated with test materials for 30 min at 37°C. Immediately after incubation, the materials were separated and dysodic EDTA was added to whole blood (final concentration: 5 mM). After centrifugation of blood for 15 min at 2000g at

Table 9 Arithmetic Mean, Standard Error and Significance (according to ANOVA and the Bonferroni-Dunn test) of Complement Fraction C4d after Contact with PET and PC

Material	C4d (μg/ml)	*p*
Control	1.974 ± 0.173	—
PET	2.338 ± 0.215	n.s.
PC	1.764 ± 0.176	n.s.

+4°C, the complement fractions C4d and Bb were determined by enzyme immunoassay (Quidel, San Diego, CA). PET and PC determined unsignificant variations in C4d (Table 9). Contact with PET determined a significant Bb generation compared to the control without materials, while contact with PC did not determine any significant variation. A significant difference between PET and PC with respect to the formation of Bb fragment was observed (Table 10).

Uncoated PET activates the complement through the alternative pathway, but does not affect the classic pathway. Pyrolytic carbon coating significantly reduces the complement activation compared to uncoated PET.

E. Erythrocyte Damage

The assessment of damage to red blood cells is usually evaluated using the hemolysis test. In vitro biomaterial-induced hemolysis is evaluated by spectrophotometrical determination of the hemoglobin released by red blood cells after contact with the material or with its extract. An elevated hemoglobin level may reflect red blood cell membrane fragility in contact with devices (3).

F. Blood Elements Adsorbed on the Artificial Surface

The artificial surface is examined by scanning electron microscopy in order to evaluate the presence of platelets (92), leukocytes and erythrocytes, as well as fibrin strands (93). The number and the spread area of adherent platelets may be measured by inverted light microscopy and image analysis (94). The surface radioactivity after exposure to radioisotope-labeled proteins or platelets can be measured. Also the surface fluorescence after incubation with fluoro-

Table 10 Arithmetic Mean, Standard Error and Significance (according to ANOVA and the Bonferroni-Dunn test) of the Bb Fraction of the Complement after Contact with PET and PET + PC

Material	Bb (μg/ml)	*p*
Control	1.148 ± 0.095	—
PET	1.683 ± 0.069	0.0012 (vs. control) 0.0013 (vs. PET + PC)
PC	1.153 ± 0.132	n.s. (vs. control)

chrome-labeled proteins may be evaluated. Monoclonal antibodies anti-GP IIb/IIIa and anti-GMP 140 were employed to evaluate platelet adhesion and activation on polyurethane elastomers (95). Fluorescent antibodies specific to fibrinogen, complement factor C1q, prothrombin/thrombin and platelet membrane antigens allow to quantitate the surface-adsorbed proteins and cells, by computer-aided image analysis (96). Dynamic adhesion and detachment of platelets from polymeric surfaces in presence of flow may be studied with epifluorescent video microscopy (97).

V. DISCUSSION

In developing medical devices for contact with blood, it is important to understand the mechanism of thrombus formation. The blood compatibility of artificial materials should be evaluated in vitro before implantation: only final verification should be done in vivo.

The International Organization for Standardization (ISO) presented a standard for a selection of tests to evaluate the interaction of medical devices with blood (ISO 10993, Biological Evaluation of Medical Devices—Part 4: Selection of Tests for Interactions with Blood). According to this standard, hemocompatibility should be examined by carrying out different tests on platelets, plasma phase of coagulation, erythrocytes, leukocytes and complement system. In vitro investigations are considered as screening tests for a first evaluation of materials. External communicating devices, such as hemodialysis equipment, should be tested by ex vivo systems. Implant devices, such as artificial heart valves and vascular prostheses, should be evaluated using in vivo systems (98).

The multiparametric approach is mandatory because the materials can differently affect blood constituents, including platelets, coagulation factors, leukocytes, complement, erythrocytes.

Some considerations can be drawn from the outlined review of the models used for the hemocompatibility assessment. The advantages of in vitro experimental models include simplicity, standardization, low costs, usage of human blood and simultaneous evaluation of several materials. However, the experimental conditions differ from the clinical situation and the clearance of activated factors by the reticuloendothelial system is not considered. The in vitro study of blood compatibility requires the consideration of many parameters: static or dynamic contact, flow rate, form of material to be tested, duration of contact, clotting components including platelets, coagulation, complement, leukocytes, erythrocytes.

In vivo experimental models permit evaluating the device performance under conditions very close to clinical implantation; however, they are expensive and ethical problems are raised.

Using ex vivo experimental models the simultaneous evaluation of several materials is allowed; however, they are expensive and the prosthesis long-term performance is not verified.

Unfortunately, the results from in vivo and ex vivo models are not easily transferred to humans because of the difference in the coagulation mechanisms of experimental animals.

Among the different tests, platelet count is widely used, because it is relatively simple and fast to perform. Release reaction can be studied by measuring the substances released by platelets, such as β-thromboglobulin and platelet factor 4 which are correlated to platelet adhesion onto biomaterial surfaces. The flow cytometric demonstration of GMP 140 on platelet external membrane is a promising and fast test for the detection of activated platelets after contact with biomaterials. It was pointed out that the lack of platelet adhesion is not necessarily a criterion for blood compatibility, but there is some agreement that progressive platelet deposition, fibrinogen deposition and complement activation are correlated with thrombogenesis (99).

Platelet may be considered the most significant element affecting the blood response to a material in the short term. For longer duration implanted devices, the effects of other components of blood may also have a significant effect on the biocompatibility (41).

Decrease of the coagulation factors provides an evidence of their adsorption onto materials, but does not demonstrate their activation. In fact, factor consumption should also be considered. Fibrinogen adsorption onto the biomaterial surface limits the amount available for clotting. Also other factors may be consumed, reducing the coagulation rate.

The activation of the phase contact is demonstrated directly by kallikrein or activated factor XII assays. Thrombin formation is demonstrated indirectly by prothrombin fragment 1 + 2 and fibrinopeptide A assays. The quantitative determination of activated factors is important for evaluating the thrombotic potential of biomaterials. The activation of the plasmatic phase is strengthened by platelet adhesion too, because factors adsorb and activate onto platelet surface.

Leukocytes should be studied, as during their activation some substances are produced favoring blood coagulation locally or systemically. During the activation of granulocytes, elastase is also released, which inhibits activated clotting factors. Cytokines are assayed owing to their inflammatory effect.

Complement activation is evaluated by the assays of fragments specific for the classical and the alternative pathway; the terminal complex is evaluated, too.

The damage to erythrocytes should be estimated, because hemolysis is a serious complication of artificial heart valves or extracorporeal circuits.

It should be pointed out that each material or device has specific characteristics which require a well-conceived program for biocompatibility evaluation. This concept is important for new devices designed for deliberate and specific interactions with blood, i.e., heparinized surfaces which should be evaluated by methods examining the role played by surface-bound heparin in modulating the coagulation system; or the binding of adhesion molecules to surfaces for the attachment of cells, which should be evaluated with methods for the phenotypic expression of attached cells. The tests should permit evaluation of the interactions of biomaterials with enzymatic systems (kinin, fibrinolytic, coagulation and complement) and formed elements (platelets, polymorphonuclear granulocytes, eosinophils, lymphocytes and monocytes), which may be activated or inhibited (3).

VI. CONCLUSIONS

Some conclusions are drawn from the review of the methods of testing blood-biomaterial interaction:

1. Blood-biomaterial interaction is a complex event and should be evaluated using a multiparametric approach.
2. Hemocompatibility evaluation is necessary for all devices contacting blood, either temporarily or permanently.
3. In vitro methods are screening tests and should be always carried out for a preliminary evaluation of blood-contacting materials.
4. Ex vivo methods are second level tests for external communicating device evaluation.
5. In vivo methods are second level tests for implant device evaluation.
6. The choice of the parameters to be assayed after blood-material contact should consider the distinct evaluation of the principal blood constituents: platelets, coagulation factors, leukocytes, complement and erythrocytes.

7. For platelets, both adhesion and activation should be tested.
8. For coagulation factors, tests able to demonstrate their activation should be preferred.
9. For complement, the classical and the alternative pathway, and the terminal complex should be evaluated by distinct tests.
10. Comparisons among in vitro, ex vivo and in vivo results are of outmost importance, so as to replace gradually the tests on animals with in vitro tests that show the same ability to predict thrombosis.

ACKNOWLEDGMENT

This work was supported by Istituti Ortopedici Rizzoli, "Ricerca Corrente."

REFERENCES

1. Missirlis, Y. F., 1992. How to deal with the complexity of the blood-polymer interactions. *Clin. Mater.*, 11: 9–12.
2. Remes, A. and Williams, D. F., 1992. Immune response in biocompatibility. *Biomaterials*, 13: 731–743.
3. Anderson, J., 1997. Interaction with blood. In: J. H. Braybrook (Editor), *Biocompatibility Assessment of Medical Devices and Materials*. Wiley, London, pp. 129–151.
4. Ortega-Venuesa, J. L., Tengvall, P., Walivaara, B. and Lundstrom, I., 1998. Stagnant versus dynamic conditions: a comparative adsorption study of blood proteins. *Biomaterials*, 19: 251–262.
5. Amiji, M., Park, H. and Park, K., 1992. Study on the prevention of surface-induced platelet activation by albumin coating. *J. Biomater. Sci. Polym. Ed.*, 3: 375–388.
6. Anderson, J., Ziats, N. P., Azeez, A., Brunstedt, M. R., Stack, S. and Bonfield, T. L., 1995. Protein adsorption and macrophage activation on polydimethylsiloxane and silicone rubber. *J. Biomater. Sci. Polym. Ed.*, 7: 159–169.
7. Banerjee, R., Nageswari, K. and Puniyani R. R., 1997. Hematological aspects of biocompatibility—Review article. *J. Biomater. Appl.*, 12: 57–76.
8. Park, K., Gemeinhart, R. A., and Park, H., 1998. Movement of fibrinogen receptors on the ventral membrane of spreading platelets. *Biomaterials*, 19: 387–395.
9. Rhodes, N. P., Kumary, T. V. and Williams, D. F., 1996. Influence of wall shear rate on parameters of blood compatibility of intravascular catheters. *Biomaterials*, 17: 1995–2002.
10. Basmadjian, D., Sefton, M. V. and Baldwin, S. A., 1997. Coagulation on biomaterials in flowing blood: some theoretical considerations. *Biomaterials*, 18: 1511–1522.
11. Jones, D. A., Smith, C. W. and McIntire, L. V., 1996. Leucocyte adhesion under flow conditions: principles important in tissue engineering. *Biomaterials*, 17: 337–347.
12. Cardona, M. A., Simmons, L. R. and Kaplan, S. S., 1992. TNF and IL-1 generation by human monocytes in response to biomaterials. *J. Biomed. Mater. Res.*, 26: 851–859.
13. Jorres, A., Safak, H., Froese, P., Fischer, C., Muller, C., Gahl, G. M. and Vienken, J., 1992. Systemic levels of Tumor Necrosis Factor Alpha during hemodialysis with cellulosic membranes: no effect of the sterilization procedure. *Artificial Organs*, 16: 559–563.
14. Arnaout, M. A., 1990. Structure and function of the leukocyte adhesion molecules CD11/CD18. *Blood*, 75: 1037–1050.
15. Zachee, P., Daelemans, R., Pollaris, P., Boogaerts, M. A. and Lins, R. L., 1994. Neutrophil adhesion molecules in chronic hemodialysis patients. *Nephron.*, 68: 192–196.
16. Gemmell, C. H., Black, J. P., Yeo, E. L. and Sefton, M. V., 1996. Material-induced up-regulation of leukocyte CD11b during whole blood contact: Material differences and a role for complement. *J. Biomed. Mater. Res.*, 32: 29–35.

17. Parsson, H., Jundzill, W., Jonung, T., Thorne, J. and Norgren, L., 1993. The adhesion of labelled neutrophils on synthetic vascular grafts. An experimental porcine study. *Eur. J. Vasc. Surg.*, 7: 257–262.
18. Swartbol, P., Truedsson, L., Parsson, H. and Norgren, L., 1997. Surface adhesion molecule expression on human blood cells induced by vascular graft materials in vitro. *J. Biomed. Mater. Res.*, 32: 669–676.
19. Jakubiec, B., Roy, R., Isles, M. B., Marois, Y. and Guidoin, R., 1994. Measurement of CD11/CD18 integrin expression on the polymorphonuclear cell surface after incubation with synthetic vascular prostheses. *Am. Soc. Artif. Intern. Ors. J.*, 40: M616–8.
20. Jahns, G., Haeffner-Cavaillon, N., Nyddegger, U. E. and Kazatchkine, M. D., 1993. Complement activation and cytokine production as consequences of immunological bioincompatibility of extracorporeal circuits. *Clin. Mater.*, 14: 303–336.
21. Liu, L. and Elwing, H., 1996. Complement activation on thiol-modified gold surfaces. *J. Biomed. Mater. Res.*, 30: 535–541.
22. Tengvall, P., Askendal, A. and Lundstrom, I., 1996. Complement activation by 3-mercapto-1,2-propanediol immobilized on gold surfaces. *Biomaterials*, 17: 1001–1007.
23. Tridon, A., Palcoux, J. B., Jouanel, P., Bezou, M. J., Coulet, M. and Betail, G., 1992. Complement activation during low-density lipoprotein apheresis. *Artificial Organs*, 16: 577–585.
24. Marosok, R., Washburn, R., Indorf, A., Solomon D. and Sherertz, R., 1996. Contribution of vascular catheter material to the pathogenesis of infection: depletion of complement by silicone elastomer in vitro. *J. Biomed. Mater. Res.*, 30: 245–250.
25. Kalman, P. G., McCullough, D. A. and Ward, C. A., 1990. Evacuation of microscopic air bubbles from Dacron reduces complement activation and platelet aggregation. *J. Vasc. Surg.*, 11: 591–598.
26. Plecha, E. J., Kaufman, B. R., Vincent, C. K., Kwak, Y. S., De Luca, D. J. and Graham, L. M., 1990. Effect of complement and arachidonic acid pathway inhibition on white blood cell count and deposition on vascular grafts. *ASAIO Trans.*, 36: M741–744.
27. Coleman, J. E., McEnroe, C. S., Gelfand, J. A., Connolly, R. J. and Callow, A. D., 1991. Complement activation by vascular sutures both alone and in combination with synthetic vascular prostheses. *Eur. J. Vasc. Surg.*, 5: 287–290.
28. De Mol Van Otterloo, J. C., Van Bockel, J. H., Ponfoort, E. D., Brommer, E. J., Hermans J. and Daha, M. R., 1992. The effects of aortic reconstruction and collagen impregnation of Dacron prostheses on the complement system. *J. Vasc. Surg.*, 16: 774–783.
29. Wang, E. Y., Giclas, P. C., Tu, R. H., Hata C. and Quijano, R. C., 1993. A comparative study of complement activation by Denaflex, Bioflow, and BioPolyMeric vascular grafts. *ASAIO J.*, 39: M691–694.
30. Steinberg, J., Neumann, A. W., Absolom, D. R. and Zingg, W., 1989. Human erythrocyte adhesion and spreading on protein-coated polymer surfaces. *J. Biomed. Mater. Res.*, 23: 591–610.
31. ISO 10993/12: Biological evaluation of medical devices—Sample preparation and reference materials, 1996.
32. Ito, Y., Sisido, M. and Imanishi, Y., 1990. Adsorption of plasma proteins and adhesion of platelets onto novel polyetherurethanes—Relationship between denaturation of adsorbed proteins and platelet adhesion. *J. Biomed. Mater. Res.*, 24: 227–242.
33. Van Delden, C. J., Engbers, G. H. M., Feijen, J., 1996. The effect of protein adsorption on the anticoagulant activity of surface immobilized heparin. *J. Biomater. Sci. Polym. Ed.*, 7: 727–740.
34. Park, K., Mao, F. W. and Park, H., 1990. Morphological characterization of surface-induced platelet activation. *Biomaterials*, 11: 24–31.
35. Ishihara, K., Tanaka, S., Furukawa, N., Kurita, K. and Nakabayashi, N., 1996. Improved blood compatibility of segmented polyurethanes by polymeric additives having phospholipid polar groups. I. Molecular design of polymeric additives and their functions. *J. Biomed. Mater. Res.*, 32: 391–399.
36. Brinkman, E., Poot, A., van der Does, L. and Bantjies, A., 1990. Platelet deposition studies on copolyether urethanes modified with poly(ethylene oxide). *Biomaterials*, 11: 200–205.
37. Feuerstein, I. A. and Ratner, B. D., 1990. Adhesion and aggregation of thrombin prestimulated

human platelets: evaluation of a series of biomaterials characterized by ESCA. *Biomaterials*, 11: 127–132.

38. Sheehan, S. J., Rajah, S. M. and Kester, R. C., 1991. In vitro assessment of prosthetic graft thrombogenic potential. In: C. P. Sharma and M. Szycher (Editors), *Blood Compatible Materials and Devices: Perspectives Towards the 21st Century*, Technomic, Lancaster, PA, pp. 213–220.
39. Groth, T. H., Vassiliedff, C. H. R., Wolf, H., Richter, G. and Foerster, F., 1992. Development of a new dynamic method for quantitative evaluation of in vitro hemocompatibility of biomedical materials. *J. Biomater. Sci. Polym. Ed.*, 3: 285–300.
40. Keogh, J. R., Wolf, M. F., Overend, M. E., Tang, L. and Eaton, J. W., 1996. Biocompatibility of sulphonated polyurethane surfaces. *Biomaterials*, 17: 1987–1994.
41. Bernacca, G. M., Gulbransen, M. J., Wilkinson, R. and Wheatley, D. J., 1998. In vitro blood compatibility of surface-modified polyurethanes. *Biomaterials*, 19: 1151–1165.
42. Lee, J. S., Kaibara, M. and Sasabe, H., 1992. In vitro rheological evaluation of antithrombogenicity or anticoagulability of styrene derivative polymer. *Biomaterials*, 13: 1025–1030.
43. Shortland, A. P., Rhodes, N. P., Rattray, A., Black, R. A. and Williams, D. F., 1997. The effect of temperature and shear rate on platelet aggregation. *MIM*, 8: 887–890.
44. Rhodes, N. P., Shortland, A. P., Rattray, A., Black, R. and Williams, D. F., 1997. Activation status of platelet aggregates and platelet microparticles shed in sheared whole blood. *MIM*, 8: 747–751.
45. Skarja, G. A., Kinlough-Rathbone, R. L., Perry, D. W., Rubens, F. D. and Brash, J. L., 1997. A cone-and-plate device for the investigation of platelet biomaterial interactions. *J. Biomed. Mater. Res.*, 34: 427–438.
46. San Roman, J., Bujan, J., Bellon, J. M., Gallardo, A., Escudero, M. C., Jorge, E., de Haro, J., Alvarez, L. and Castillo-Olivares, J. L., 1996. Experimental study of the antithrombogenic behavior of Dacron vascular grafts coated with hydrophilic acrylic copolymers bearing salicylic acid residues. *J. Biomed. Mater. Res.*, 32, 19–27.
47. Brothers, T. E., Stanley, J. C., Burkel, W. E. and Graham, L. M., 1990. Small-caliber polyurethane and polytetrafluoroethylene grafts: a comparative study in a canine aortoiliac model. *J. Biomed. Mater. Res.*, 24: 761–771.
48. Al-Mondhiry, H., Pae, W. E., Miller, C. A. and Pierce, W. S., 1992. Platelet and fibrinogen survival in calves implanted with artifical heart and ventricular assite device—correlation with autopsy findings. *Thromb. Haemost.*, 67: 413–416.
49. Badimon, L., Badimon, J. J., Turitto, V. T. and Fuster, V., 1990. Platelet interaction to prosthetic materials—Role of Von Willebrand factor in platelet interaction to PTFE. *J. Biomat. Appl.*, 5: 27–48.
50. Wabers, H. D., Hergenrother, R. W., Coury, A. J. and Cooper, S. L., 1992. Thrombus deposition on polyurethanes designed for biomedical applications. *J. Appl. Mat.*, 3: 167–176.
51. McCoy, T. J., Wabers, H. D. and Cooper, S. L., 1990. Series shunt evaluation of polyurethane vascular graft materials in chronically AV-shunted canines. *J. Biomed. Mater. Res.*, 24: 107–129.
52. Lin, H.-B., Hergenrother, R. W., Silver, J. H., Lin, J.-C., Lim, F., Brodhagen, T. W. and Cooper, S. L., 1994. Ex-vivo blood compatibility of silicone-containing biomaterials. *MIM*, 5: 207–213.
53. Pacitti, A., Tetta, C., Mangiarotti, G., Canavese, C. and Segoloni, G. P., 1993. Beta-2-microglobulin serum profiles in different settings of mass transport and fluid pyrogen content. *Kidney Int. Suppl.*, 41: S96–99.
54. Mahiout, A., Matata, B. M., Vienken, J. and Courtney, J. M., 1997. Ex vivo complement protein adsorption on positively and negatively charged cellulose dialyser membranes. *MIM*, 8: 287–296.
55. Kang, I.-K., Kwon O. H., Byun K. H. and Kim, Y. H., 1996. Surface modification of polyetherurethaneureas and their antithrombogenicity. *MIM*, 7: 135–140.
56. Waples, L. M., Olorundare, O. E., Goodman, S. L., Lai, Q. J. and Albrecht, R. M., 1996. Platelet-polymer interactions: morphologic and intracellular free calcium studies of individual human platelets. *J. Biomed. Mater. Res.*, 32: 65–76.
57. Hauch, K. D. and Horbett, T. A., 1996. Platelet activation by biomaterials indicated by oscillations in intracellular free calcium concentration. In: Trans. 5th World Biomaterials Congress, vol. 1, May 29–June 2, 1996, Toronto, p. 334.

58. Goodman, S. L., Tweden, K. S. and Albrecht, R. M., 1996. Platelet interaction with pyrolytic carbon heart-valve leaflets. *J. Biomed. Mater. Res.*, 32: 249–258.
59. Rubens, F. D., Labow, R. S. and Robblee, J. A., 1996. Hematologic effects of the Davol autotransfusion device in re-operative cardiac surgery. In: Trans. 5th World Biomaterials Congress, vol. 2, May 29–June 2, 1996, Toronto, p. 678.
60. Rhodes, N. P., Zuzel, M., Williams, D. F. and Derrick, M. R., 1994. Granule secretion markers on fluid-phase platelets in whole blood perfused through capillary tubing. *J. Biomed. Mater. Res.*, 28: 425–439.
61. Groth, T. H., Campbell, E. J., Herrmann, K. and Seifert, B., 1995. Application of enzyme immunoassays for testing haemocompatibility of biomedical polymers. *Biomaterials*, 16: 1009–1015.
62. Seifert, B., Groth, T. H., Herrmann, K. and Romaniuk, P., 1995. Immobilization of heparin on polylactide for application to degradable biomaterials in contact with blood. *J. Biomater. Sci. Polym. Ed.*, 7: 277–287.
63. Jirouskova, M., Bartunkova, J., Smetana, K. Jr., Lukas, J., Vacik, J. and Dyr, J. E., 1997. Comparative study of human monocyte and platelet adhesion to hydrogels in vitro-effect of polymer structure. *MIM*, 8: 19–23.
64. Ginsberg, M. H., Frelinger, A. L., Lam, S. C., Forsyth, J., McMillan, R., Plow, E. F. and Shattil, S. J., 1990. Analysis of platelet aggregation disorders based on flow cytometric analysis of membrane glycoprotein IIb-IIIa with conformation-specific monoclonal antibodies. *Blood*, 76: 2017–2023.
65. Lindhout, T., Blezer, R., Maassen, C., Heijnen, V. and Reutelingsperger, C. P. M., 1995. Platelet procoagulant surface as an essential parameter for the in vitro evaluation of the blood compatibility of polymers. *MIM*, 6: 367–372.
66. Leake, D. L., Cenni, E., Cavedagna, D., Stea, S., Ciapetti, G. and Pizzoferrato, A., 1989. Comparative study of the thromboresistance of Dacron combined with various polyurethanes. *Biomaterials*, 10: 441–444.
67. Cenni, E., Cavedagna, D., Falsone, G., Mari, G. and Pizzoferrato, A., 1993. Numerical and functional modifications in platelets induced by polyester coated by a hydrophilic polymer. *Biomaterials*, 14: 588–590.
68. Cenni, E., Granchi, D., Verri, E., Cavedagna, D., Gamberini, S., Falsone, G. and Pizzoferrato, A., 1997. CD62, thromboxane B2, and beta-thromboglobulin: A comparison between different markers of platelet activation after contact with biomaterials. *J. Biomed. Mater. Res.*, 36: 289–294.
69. Cenni, E., Arciola, C. R., Ciapetti, G., Granchi, D., Savarino, L., Stea, S., Cavedagna, D., Curti, T., Falsone, G. and Pizzoferrato, A., 1995. Platelet and coagulation factors variations induced in vitro by polyethylene terephthalate (Dacron®) coated with pyrolytic carbon. *Biomaterials*, 16: 973–976.
70. Kang, I. K., Kwon, O. H., Kim, M. K., Lee, Y. M. and Sung, Y. K., 1997. In vitro blood compatibility of functional group-grafted and heparin-immobilized polyurethanes prepared by plasma glow discharge. *Biomaterials*, 18: 1099–1107.
71. van der Kamp, K. W. H. J., Hauch, K. D., Feijen, J. and Horbett, T. A., 1995. Contact activation during incubation of five different polyurethanes or glass in plasma. *J. Biomed. Mater. Res.*, 29: 1303–1306.
72. Guerra, G. D., Barbani, N., Lazzeri, L., Lelli, L., Palla, M. and Rizzo, C., 1993. The activation of human plasma prekallikrein as a hemocompatibility test for biomaterials. II. Contact activation by EVAL and EVAL-SMA copolymers. *J. Biomater. Sci. Polym. Ed.*, 4: 643–652.
73. Yun, Y. H., Turitto, V. T., Daigle, K. P., Kovacs, P., Davidson, J. A. and Slack, S. M., 1996. Initial hemocompatibility studies of titanium and zirconium alloys: prekallikrein activation, fibrinogen adsorption, and their correlation with surface electrochemical properties. *J. Biomed. Mater. Res.*, 32: 77–85.
74. Matata, B. M., Courtney, J. M., Sundaram, S., Wark, S., Bowry, S. K., Vienken, J. and Lowe, G. D. O., 1996. Determination of contact phase activation by the measurement of the activity of supernatant and membrane surface-adsorbed Factor XII (FXII): its relevance as a useful parameter for the in vitro assessment of haemodialysis membranes. *J. Biomed. Mater. Res.*, 31: 63–70.
75. Seifert, B., Groth, T. H., Herrmann, K. and Romaniuk, P., 1995. Immobilization of heparin on

polylactide for application to degradable biomaterials in contact with blood. *J. Biomater. Sci. Polym. Ed.*, 7: 277–288.

76. Aldenhoff, Y. B. J., Blezer, R., Lindhout, T. and Koole, L. H., 1997. Photoimmobilization of dipyridamole (Persantin) at the surface of polyurethane biomaterials: reduction of in vitro thrombogenicity. *Biomaterials*, 18: 167–172.
77. Blezer, R., Fouache, B., Willems, G. M. and Lindhout, T., 1997. Activation of blood coagulation at heparin-coated surfaces. *J. Biomed. Mater. Res.*, 37: 108–113.
78. Skarada, D. J., Erickson, G. R., Warnecke, K. L. and du Laney, T. V., Greenberg, C. S., Ritter, E. F. and Klitzman, B., 1995. Assessments of thrombogenicity by three in vitro techniques. *J. Biomed. Mater. Res.*, 29: 1039–1045.
79. Bienert, H., Thelen, H., Klee, D., Hocker, H., Mittermayer, C. and Jakse, G., 1996. Correlation of hemocompatibility and cytotoxicity parameters with physicochemical surface properties of a plasma etched silicone. In: Trans. 5th World Biomaterials Congress, vol. 2, May 29–June 2, 1996, Toronto, p. 101.
80. Matata, B. M., Courtney, J. M., Sundaram, S., Wark, S., Bowry, S. K., Vienken, J. and Lowe, G. D. O., 1996. Determination of contact phase activation by the measurement of the activity of supernatant and membrane surface-adsorbed factor XII (FXII): Its relevance as a useful parameter for the in vitro assessment of haemodialysis membranes. *J. Biomed. Mater. Res.*, 31: 63–70.
81. Cenni, E., Ciapetti, G., Cervellati, M., Cavedagna, D., Falsone, G., Gamberini, S. and Pizzoferrato, A., 1996. Activation of the plasma coagulation system induced by some biomaterials. *J. Biomed. Mater. Res.*, 31: 145–148.
82. Sundaram, S., Lim, F., Cooper, S. L. and Colman, R. W., 1996. Role of leukocytes in coagulation induced by artificial surfaces: investigation of expression of Mac-1, granulocyte elastase release and leukocyte adhesion on modified polyurethanes. *Biomaterials*, 17: 1041–1047.
83. Brunstedt, M. R., Anderson, J. M., Spilizewski, K. L., Marchant, R. E. and Hiltner, A., 1990. In vivo leukocyte interactions on Pellethane surfaces. *Biomaterials*, 11: 370–378.
84. Plotz, F. B., van Oeveren, W., Hultquist, K. A., Miller, C., Bartlett, R. H. and Wildevuur, C. R. H., 1992. A heparin-coated circuit reduces complement activation and the release of leukocyte inflammatory mediators during extracorporeal circulation in a rabbit. *Artificial Organs*, 16: 366–370.
85. Granchi, D., Cenni, E., Verri, E., Ciapetti, G., Gamberini, S., Gori, A. and Pizzoferrato, A., 1998. Flow-cytometric analysis of leukocyte activation induced by polyethylene-terephthalate with and without pyrolytic carbon coating. *J. Biomed. Mater. Res.*, 39: 549–553.
86. Marosok, R., Washburn, R., Indorf, A., Solomon, D. and Sherertz, R., 1996. Contribution of vascular catheter material to the pathogenesis of infection: depletion of complement by silicone elastomer in vitro. *J. Biomed. Mater. Res.*, 30: 245–250.
87. Gemmell, C. H. and Sefton, M. V., 1996. Quantitative analysis of complement protein adsorption: C3b breakdown products and sC5b-9 on biomaterials. In: Trans. 5th World Biomaterials Congress, vol. 1, May 29–June 2, 1996, Toronto, p. 18.
88. Tengvall, P., Askendal, A. and Lundstrom, I., 1996. Complement activation by IgG immobilized on methylated silicon. *J. Biomed. Mater. Res.*, 31: 305–312.
89. Black, J. P., Yeo, E. L. and Sefton, M. V., 1996. An enzyme-linked immunoflow assay for quantifying complement activation. In: Trans. 5th World Biomaterials Congress, vol. 1, May 29–June 2, 1996, Toronto, p. 332.
90. Kao, W. J., Sapatnekar, S., Hiltner, A. and Anderson, J. M., 1996. Complement-mediated leukocyte adhesion on poly(etherurethane ureas) under shear stress in vitro. *J. Biomed. Mater. Res.*, 32: 99–109.
91. Cenni, E., Granchi, D., Ciapett, G., Stea, S., Verri, E., Gamberini, S., Gori, A., Pizzoferrato, A. and Zucchelli, P., 1997. In vitro complement activation after contact with pyrolytic carbon-coated and uncoated polyethylene terephthalate. *MIM*, 8: 771–774.
92. Kim, S. S., Kim, H. W., Yuk, S. H., Oh, S. Y., Pak, P. K. and Lee, H. B., 1995. Blood and cell compatibility of gelatin-carrageenan mixtures cross-linked by gluteraldehyde. *Biomaterials*, 17: 813–821.
93. Bandekar, J. and Sawyer, A., 1995. FT-IR spectroscopic studies of polyurethanes: IV. Studies of

the effect of the presence of processing aids on the hemocompatibility of polyurethanes. *J. Biomater. Sci. Polym. Ed.*, 7: 485–501.

94. Kohler, A. S., Parks, P. J., Mooradian, D. L., Rao, G. H. R. and Furcht, L. T., 1996. Platelet adhesion to novel phospholipids materials: modified phosphatidylcholine covalently immobilized to silica, polypropylene, and PTFE materials. *J. Biomed. Mater. Res.*, 32: 237–242.
95. Mitzner, E. and Groth, T. H., 1996. Modification of poly(ether urethane)elastomers by incorporation of poly(isobutylene)glycol. Relation between polymer properties and thrombogenicity. *J. Biomater. Sci. Polym. Ed.*, 7: 1105–1118.
96. Nygren, H., Tengvall, P. and Lundstrom, I., 1997. The initial reactions of TiO_2 with blood. *J. Biomed. Mater. Res.*, 34: 487–492.
97. McClung, W. G. and Feuerstein, I. A., 1992. Epifluorescent video microscopy (EVM) for platelet-biomaterial interaction: elimination of photoactivation and dye effects. *Biomaterials*, 13: 871–877.
98. ISO 10993, Biological evaluation of medical devices. Part 4: Selection of tests for interactions with blood, 1992.
99. Ip, W. F. and Sefton, M. V., 1991. Platelet consumption by NHLBI reference materials and silastic. *J. Biomed. Mater. Res.*, 25, 1321–1324.

9

New Optical Characterization Technique for Synthetic Biomaterials

C. C. Chu
Cornell University, Ithaca, New York

I. INTRODUCTION

The interest in biodegradable polymeric biomaterials for biomedical engineering application has increased significantly during the past two decades [1–8]. This is because this class of biomaterials has two major advantages that nonbiodegradable biomaterials do not have. First, biodegradable biomaterials do not elicit permanent chronic foreign-body reaction due to the fact that they would be gradually absorbed and metabolized by the human body and hence do not permanently retain any trace of residual in the implantation sites. Second, some of these biodegradable biomaterials have been found to be able to regenerate tissues in animals, particularly vascular tissues, through the interaction of their biodegradation with immunological cells like macrophages [9–20]. Hence, surgical implants made from biodegradable biomaterials could be used as temporary scaffold for tissue regeneration. This approach toward the reconstruction of injured, diseased or aged tissues is one of the most promising fields in the next century.

The earliest and most commercially significant biodegradable polymeric biomaterials originate from linear aliphatic polyesters like polyglycolide, polylactide and their copolymers. Their most important and successful commercial use in medicine are in wound closure [6]. However, the recent introduction of several new synthetic and natural biodegradable polymeric biomaterials extends the domain beyond this family of linear aliphatic polyesters. These new commercially significant biodegradable polymeric biomaterials include poly(orthoesters), polyanhydrides, polysaccharides, poly(ester-amides), tyrosine-based polyarylates or polyiminocarbonates or polycarbonates, poly(D,L-lactide-urethane), poly[bis(carboxylatophenoxy)phosphazene], poly(ϑ-hydroxybutyrate), poly(M-caprolactone), poly(amino acids), pseudo-poly(amino acids) and copolymers derived from amino acids and non-amino acids.

The next largest biomedical application of biodegradable polymeric biomaterials that are commercially satisfactory is drug control/release devices. Some well-known examples of this application are polyanhydrides, poly(orthoester) and biodegradable hydrogels derived from hydrolyzable linear aliphatic polyesters containing vinyl groups, block copolymers of linear aliphatic polyesters and poly(ethylene glycol), polysaccharide-based with degradable crosslinkers.

Biodegradable polymeric biomaterials, particularly totally resorbable composites, have also been used in the field of orthopedics, mainly as components for internal bone fracture fixation like PDS pins. However, their wide acceptance in other parts of orthopedic implants may be limited due to their inherent mechanical properties and their biodegradation rate.

Besides the commercial uses described above, biodegradable polymeric biomaterials have been experimented with as vascular grafts [9–17], vascular stents, vascular coupler for vessel anastomosis, nerve growth conduits, augmentation of defected bone, ligament/tendon prostheses, intramedullary plug during total hip replacement, anastomosis ring for intestinal surgery, and stents in ureteroureterostomies for accurate suture placement.

All these synthetic biodegradable polymers and fibers have been extensively characterized by conventional standard polymer and fiber techniques like FTIR, NMR, SEM, DSC, GPC, X-ray, optical microscopes, etc. The recent availability of laser confocal microscope (LCM) for visual examination of biological systems, however, would provide a unique visual means to study these polymers and fibers that neither SEM nor optical microscope would be able to achieve. The most important aspect of LCM is its ability to view the interior of a material through optical thin sectioning. This capability would eliminate any artifact from conventional physical section.

In this chapter techniques recently developed in the author's lab, using LCM to characterize synthetic biodegradable biomaterials, are reported. Several examples will be used to illustrate these new LCM applications and their limitations in biomaterials science. A brief background of LCM is also given for those who are not familiar with this instrument.

II. BACKGROUND OF LASER CONFOCAL MICROSCOPY

The major difference between confocal and conventional microscopes is the object field. Instead of viewing the entire object field simultaneously as in the conventional microscope, the laser confocal microscope views only one volume element of an object at a time by a point laser light source for reducing the demand on the viewing device, i.e., no need to collect the signals from other focal planes. The entire object is scanned for each volume element and the point images are then reassembled together by a computer. The optical thin sectioning capability of LCM is achieved due to the image of the focused plane largely free of signal interference from other focal planes. Refer to Wilson's book for a complete description of LCM [21].

The use of LCM is frequently associated with a chemical probe agent that not only can be specifically associated with a targeted object but also can emit radiation upon exposure to laser light source. The most frequently used probe agents are fluorescent dyes like rhodamine B. These fluorescent dyes would either chemically or physically associate with a targeted object. Upon laser irradiation, the dyes would emit light at a wavelength longer than the laser source. This remitted light would then be detected by the LCM equipped with specific filters that only permit the passage of fluorescent signals. The fluorescent signals from the testing object can then be visualized and recorded for analysis. Most of the LCM applications have been in biology [22] because of the availability of a wide range of fluorescent dyes, such as the use of immunofluorescence to study osteoblast growth in biodegradable biomaterial matrix [23] and the assessment of the cell morphology and distribution of keratinocytes within Tissu-Vlies collagen sponges [24].

The use of LCM to probe the morphology and properties of synthetic materials, however, has received very little attention and is still in its infancy. In this chapter this author uses four examples to illustrate how he has used LCM to examine the biodegradation properties and interior pH profiles of synthetic biodegradable biomaterials and the interfacial structure of biodegradable fiber-reinforced composite.

III. EXAMPLE I: USE OF LCM TO PROBE THE DEPTH OF CHEMICAL SURFACE MODIFICATION OF SYNTHETIC BIODEGRADABLE POLYMERS FOR TISSUE REGENERATION [25]

A. Objective and Brief Background

The objective was to visually probe the depth of a unique chemical reaction beneath the surface of synthetic biodegradable biomaterials and the effect of such a chemical reaction on the in vitro hydrolytic degradation phenomena. Chemical surface modification of existing synthetic biodegradable polymers and fibers has always been an attractive and economically sound means to alter their chemical, physical and mechanical and biological properties for specific purposes.

In our search for ideal synthetic biodegradable polymers for tissue regeneration, the author's lab developed a unique chemical means to modify the chemical functional groups located on the surface of synthetic biodegradable polymers for the purpose of generating a new class of synthetic biodegradable polymers that would biodegrade at a faster rate to timely stimulate tissue regeneration without the expenses of a premature loss of their mechanical properties upon this accelerated biodegradation [25]. One of the most important questions that needed to be answered was how deep the new chemical surface modification had proceeded beneath the surface of the synthetic biodegradable biomaterials. This piece of information is very important for us to assess the propensity of the chemically surface modified biomaterials toward stimulated tissue regeneration and any adverse affect of the surface chemical reaction on the mechanical properties of the modified biomaterials. Although both SEM and optical microscope can view the surface of the chemically modified biomaterials, they cannot view the material beneath the surface without physically sectioning the materials. The capability of optical thin sectioning of LCM, however, permits us to view the interior of materials without physically breaking the material apart.

B. Experimental Procedures

In principle, polymeric samples for LCM study were prepared by immersion in a 1.0 mM aqueous solution of rhodamine B for 24 h (a fluorescent dye with its excitation wavelength at 555 nm) followed by blotting the excess dye with filter paper. The purpose of using fluorescent dyes is to assist LCM in locating where the dye was within the biomaterials. Since the chemically modified and unmodified portions of the synthetic biodegradable biomaterials have different density and compactness, the dye would prefer to accumulate in the regions with a lower density and compactness. In addition, synthetic biodegradation biomaterials would become more porous upon hydrolytic degradation due to the dissolution of fragmented chain segments. The microvoid space generated within the biodegradable biomaterials would provide the sites for the dye to accumulate preferentially. Therefore, an optical imaging device like LCM could be used to optically analyze the interior of the biomaterials. By detecting the fluorescence emitted upon exposure to laser light, differences in dye concentration levels can be observed and recorded as a function of depth from the surface of biomaterials. This fluorescence map of a biomaterial could provide a visual physical description of the interior morphology of the material.

After removing the excessive water on the surface of the biomaterial, the sample was then mounted on a glass cover slip in immersion oil and examined by a Biorad MRC-600 LCM equipped with a krypton-argon ion laser that is capable of providing excitation at 488, 568, and 647 nm. The LCM was interfaced with a Zeiss Axiovert 10 microscope (an inverted microscope) and the objective lenses used were oil immersion lenses with magnifications of 40×, 63×, and 100×. The focal plane was adjusted manually in order to determine the region of interest, followed by a scan of this region in which the light intensity was plotted as a function of distance from an arbitrary zero point. During data manipulation, the surface of the sample was taken as the first point where the light intensity was greater than zero. The light intensity of the dye was plotted as a function of depth from the surface of the biomaterial in order to obtain a dye penetration.

C. Results

Figure 1 illustrates the light intensity profiles of the fluorescent dye as a function of depth from the surface of a synthetic biodegradable biomaterial (polyglycolide) that had been surface chemically modified under different reaction time from 4 to 10 h. The dye was seen to penetrate only about 12 mm into the unmodified biomaterial. In contrast, the chemically modified samples permitted the dye to penetrate deeper to 15 to 25 mm, depending on the duration of chemical reaction. The peak intensity increased with the chemical reaction time. The 10 h reaction showed the highest light intensity, but it occurred nearer to the surface, at about 12 mm.

Figure 2 summarizes the maximum dye penetration into polyglycolide as a function of the duration of chemical treatment (hours) and hydrolysis time (days) [25]. It appears that either longer duration of chemical treatment or hydrolysis time, in general, increased the maximum depth of dye penetration (MDDP) in both chemically modified and unmodified polyglycolide with some variations. For example, the MDDP of the unhydrolyzed sample increased from 12 mm in the untreated disk to 25 mm in the 10 h chemically treated sample, a 108% increase in MDDP due to chemical treatment alone. The MDDP of untreated control disks

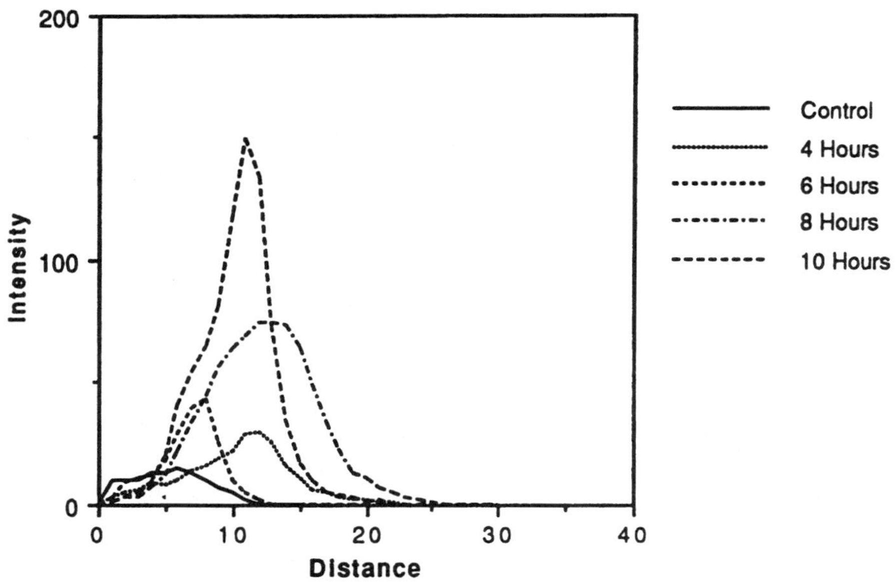

Figure 1 Change in optical density of a polyglycolide disk from a laser confocal microscopic measurement as a function of the depth (mm) of the disk sample that was surface chemically modified by dimethyltitanocene at different treatment time from 0 to 10 h. (From Ref. 25.)

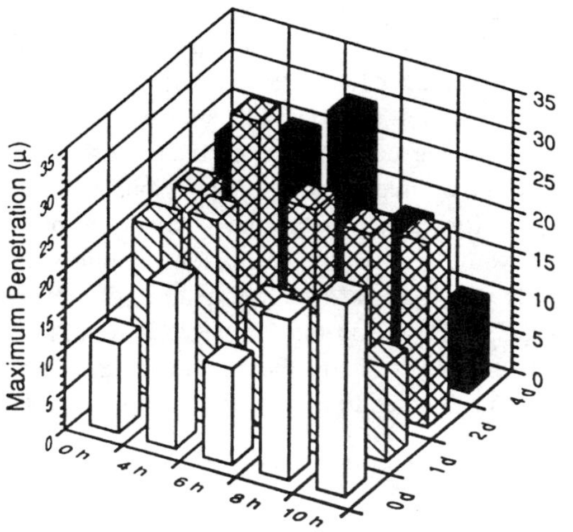

Figure 2 Maximum depth of penetration (μm) of rhodamine B dye into a polyglycolide disk as a function of both the dimethyltitanocene-induced surface chemical reaction time (h) and the hydrolytic degradation time (days). (From Ref. 25.)

continuously increased from about 12 mm at day 0 hydrolysis to about 22 mm at the end of the four day hydrolysis period, an 83% increase in the depth of dye penetration due to hydrolysis alone. It is understandable that hydrolysis of polyglycolide results in more microporous space which allowed dye to penetrate deeper into the interior of the sample.

IV. EXAMPLE II: USE OF LCM TO EVALUATE THE INTERFACIAL ADHESION OF SYNTHETIC BIODEGRADABLE COMPOSITES FOR ORTHOPEDIC APPLICATIONS

A. Objective and Brief Background

The objective was to visually assess the change in interfacial morphology of totally biodegradable fiber-reinforced composites and to correlate the observed change in interfacial morphology to the interfacial adhesion strength obtained by conventional mechanical measurements.

Biodegradable fracture fixation devices made from totally biodegradable composites offer many advantages over conventional metallic implants. For example, there is no need for a second surgery to remove the biodegradable implants after complete healing. Because the strength and modulus of the totally biodegradable fixation devices would decrease with time upon biodegradation, these devices would permit the gradual stress transfer from the device back to the healing bone. Such a gradual transfer would eliminate the stress-protection induced bone atrophy. Thus, the use of high modulus fiber-reinforced biodegradable composite materials is a promising alternative for temporary bone fracture fixation [26–29].

It has been known that the interfacial adhesion strength of a composite dicates the applicability of that composite in a specific application. Frequently, the failure of any fiber-reinforced composite largely initiates at the fiber-matrix interface. A breakdown of the fiber-matrix interface would prevent the load transfer from matrix to fibers, or vice versa. Such a transfer is known to make composites superior to noncomposites. Therefore, a better understanding of the interfacial phenomena is crucial to the design of successful composites.

In the case of totally biodegradable fiber-reinforced composites, their failure mode is far more complicated than nonbiodegradable ones because of the biodegradation nature of both the reinforced fibers and the surrounding polymeric matrix. Since the fiber-matrix interface is considered the weakest point in a composite due to the heterogeneous adhesion between two different phases (fiber and matrix), this interface is also suspected to be the most sensitive to hydrolytic degradation than either the reinforced fiber or matrix component. Consequently, upon hydrolytic degradation, the debonding of the reinforced fiber from the surrounding polymeric matrix in a totally biodegradable composite would lead to the formation of microvoid spaces in the fiber-matrix interface. Such microvoids would allow the accumulation of a probing agent like a fluorescent dye, and detected by LCM. Based on this principle, the author's lab had recently developed a LCM-based visual technique to study the interfacial morphology of totally biodegradable composites that have been subjected to in vitro hydrolytic degradation [30].

B. Experimental Procedures

The continuous single fiber-reinforced composites were prepared according to our published procedures [28,29]. The reinforcing fibers were biodegradable ceramic calcium phosphate (CaP), polyglycolide and chitin and nonbiodegradable AS4 carbon. The only matrix material used was semicrystalline poly(L-lactic acid), PLLA. The probe agent used was rhodamine B, the same fluorescent dye for the chemical surface modification study described in Example I. These biodegradable composites were subject to in vitro hydrolytic degradation in saline buffer solutions at 37°C for predetermined periods. At the end of each hydrolysis period, the sample was removed and dried in a desiccator for at least 24 h. All samples were then immersed in a 5 mM rhodamine B solution at 37°C for additional 24 h before LCM imaging was taken.

These composites were placed on the same BioRad MRC-600 LCM as described in Example I. Figure 3 schematically illustrates the arrangement of such a specimen on the LCM [30]. Two-dimensional images of either *X-Z* or *X-Y* planes were obtained by optical sectioning as shown in Fig. 3. A typical LCM image of a partially hydrolyzed PGA/PLLA composite is shown in Fig. 4 [31]. In a black-and-white LCM image, the lighter region of the LCM image indicates the location of fluorescent dye accumulated and the level of the lightness or gray scale would indicate the relative concentration of the accumulated dye. A whiter region would indicate a higher relative concentration of the dye due to a higher fluorescent signal, while a complete black region indicates no dye accumulation (i.e., no emission detected by the detector). In a color LCM image, the gradient of dye concentration is represented by different colors with white indicating the highest relative dye concentration and dark blue or black indicating no, or insignificant, dye. The data in Fig. 4 indicate clearly that the fluorescent probe agent preferentially locates at the interface of a fiber-reinforced composite. Therefore, Fig. 4 visually demonstrates that the fiber-matrix interface of a composite is its weakest point due to the lack of proper interfacial adhesion.

Quantitative information from the LCM images could also be extracted from measuring the point-to-point distances in the dye accumulation regions (i.e., lighter regions) using the COMOS software designed for the Biorad LCM. The gap between the fiber and the matrix, the so-called interface gap width (IGW) of a composite, was measured at different locations of each of five sampling LCM images for proper statistical average. The IGW values obtained could then be used as one of the new indicators that the author's lab created for assessing the fiber-matrix interfacial adhesion strength of a composite. The changes in IGW values with many different variables like the type of fiber-matrix combination and the hydrolytic degrada-

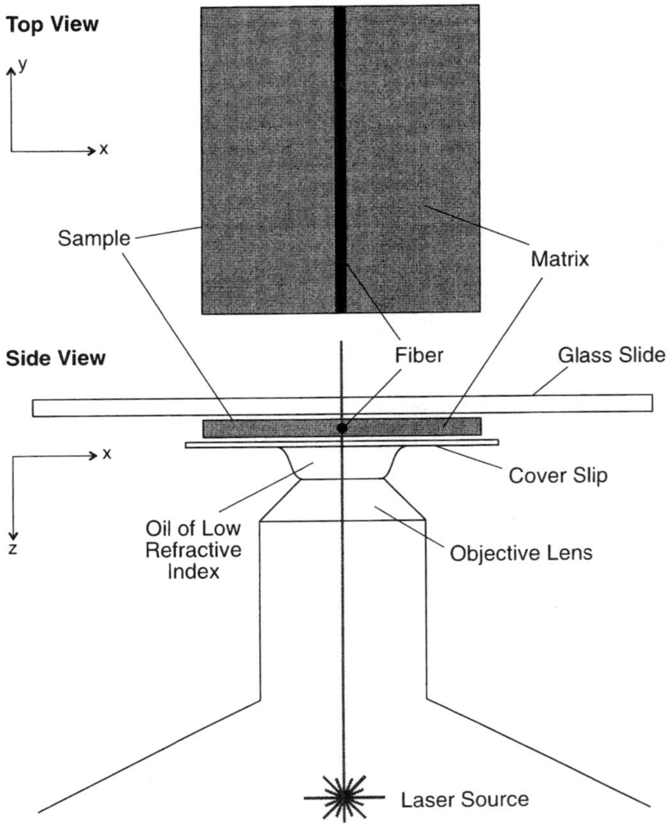

Figure 3 Schematic of the setup of viewing interfacial region of a single fiber composite by laser confocal microscope. The top view (*X-Y* plane) would reveal the longitudinal aspect of the fiber within matrix, while the side view (*X-Z* plane) would reveal the cross-sectional aspect of the fiber within matrix. (From Ref. 30.)

tion time could be used to determine whether a particular combination of fiber-matrix composite at a particular time of hydrolytic degradation would have proper mechanical properties.

C. Results

1. Change of IGW with Time of Hydrolytic Degradation

Because the fiber-matrix interfacial regions are the weakest points in a totally biodegradable composite due to the poor adhesion between the reinforcing fibers and their surrounding matrix, these regions are expected to be degraded preferentially and reflected in an increase in IGW value with time of hydrolysis. Figure 5 shows the LCM images of CaP/PLLA composite at different durations of hydrolysis [31], and its IGW values extracted from the LCM image are given in Fig. 6 [30]. The increase in IGW with time of hydrolysis is both visually evident and quantifiable. The observed continuous increase in IGW of a totally biodegradable composite with hydrolysis time suggests that the interfacial adhesion strength of the composite would decrease with time of hydrolysis. This prediction was verified by our experimental measurement of the interfacial shear strength of a single fiber composite via a microbond test [29].

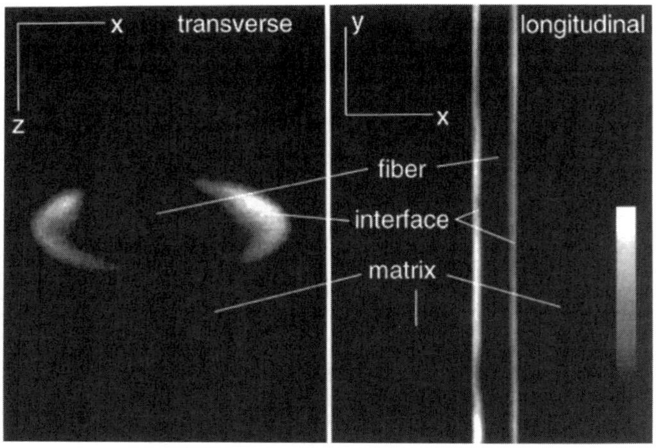

Figure 4 A typical laser confocal microscopic micrograph of a fiber-reinforced composite (polyglycolide fiber/poly-L-lactide matrix) in both *X-Z* view (left) and *X-Y* view (right). The white regions located at the interface between fiber and matrix are where rhodamine dye accumulated. The gray scale at the right lower corner indicates the relative concentration of the dye. The whiter the scale, the higher the concentration of the dye. (From Ref. 31.)

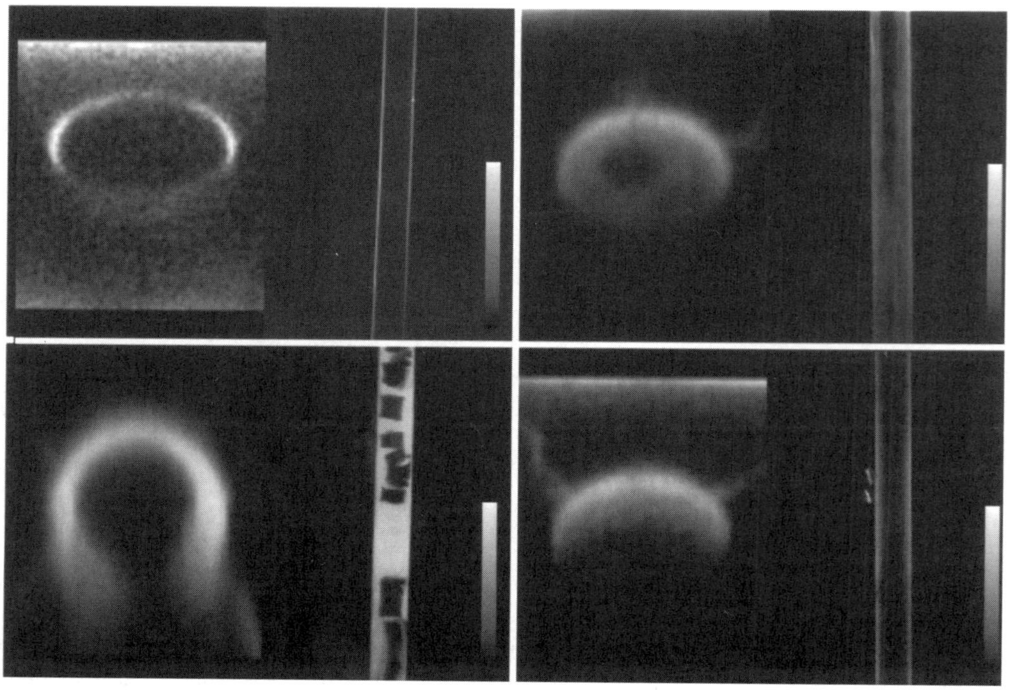

Figure 5 Laser confocal micrographs of CaP/PLLA composite after in vitro hydrolysis. 0 day control (upper left); 5 days (upper right); 15 days (lower left); and 25 days (lower right). The left and right of each micrograph is the *X-Z* cross-sectional and *X-Y* longitudinal views, respectively. (From Refs. 30 and 31.)

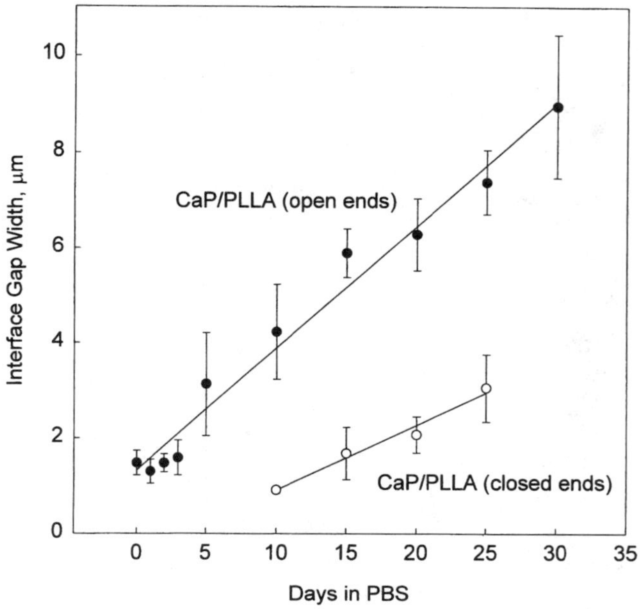

Figure 6 The changes in interfacial gap width (IGW) of a CaP/PLLA composite upon in vitro hydrolysis in phosphate buffer of pH 7.4 at 37°C. The open circle was for open end CaP/PLLA composite; i.e., its CaP fiber ends were exposed directly to the buffer media. The solid circle was for the closed end CaP/PLLA composite whose fiber ends were capped by the PLLA matrix. (From Refs. 30 and 31.)

2. *Failure Modes of a Totally Biodegradable Fiber-Reinforced Composite*

The failure modes of a composite could also be observed by LCM images and included matrix cracking, fiber breakage, and weld line formation as shown in Figs. 7 and 8 [30]. As can be seen from Fig. 7, there is a characteristic pattern of multiple matrix cracking lines which form at a right angle to the perimeter of the fiber. These channels would permit the easy transportation of water outside a composite into its interior for accelerated hydrolytic degradation.

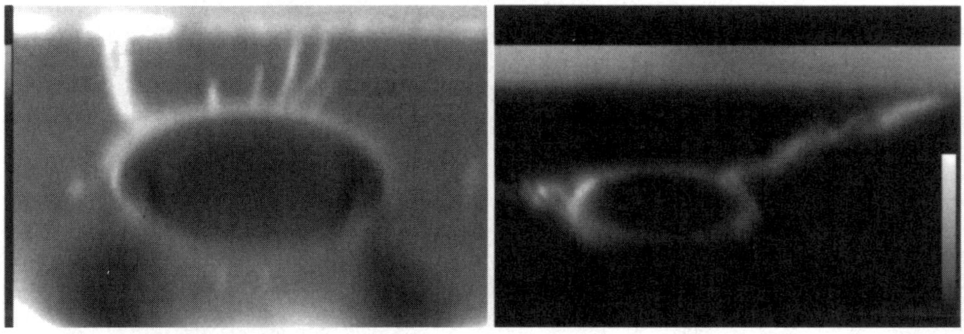

Figure 7 Laser confocal micrographs of the formation of fine channels in composites due to matrix cracking. Those channels were perpendicular to the perimeter of the reinforcing fiber and provided the avenue for degradation media like water to transport into the interior and interface of the composites. CaP/PLLA composite at 25% strained (left); PGA/PLLA composite at 20 days hydrolysis (right). (From Refs. 30 and 31.)

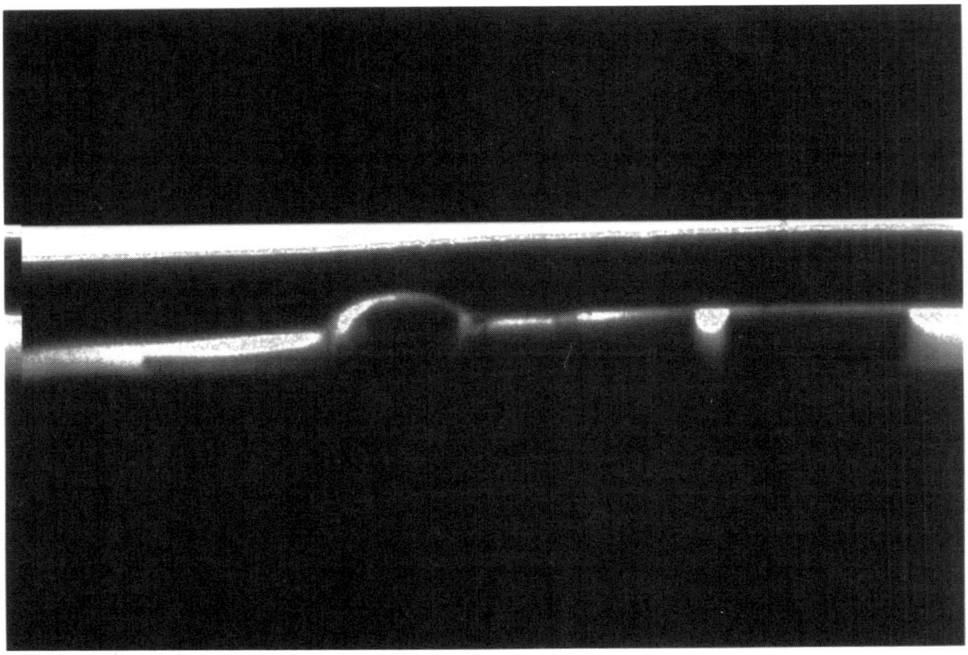

Figure 8 Laser confocal micrographic illustration of the presence of a weld line (white line) in a 4 Mrad irradiated (in air and room temperature) and five days hydrolyzed CaP/PLLA composite. The weld line provides the site for delaminating the composite upon mechanical load. (From Refs. 30 and 31.)

Figure 8 shows the formation of a weld line occurred between layers of the matrix polymer that had been heat-compression-molded together. The presence of the weld line between layers of matrix was due to sample fabrication technique. Obviously, the weld line is a weak location in composites and could lead to delamination mode of failure, a problem which can cause increased flexibility and failure of the composite material. The weld line also provides the preferential site of a biodegradable composite for hydrolytic degradation to initiate.

The fiber breakage within a composite could also be visually observed by LCM as shown in Fig. 9 [30]. The location or space where the fiber broke was filled with fluorescent dyes as

Figure 9 Laser confocal micrographs of the breakage of the reinforcing CaP fiber in CaP/PLLA composites. The broken site between the two broken ends of a fiber was filled with fluorescent dye as evident by the solid white band along the fiber length. (From Refs. 30 and 31.)

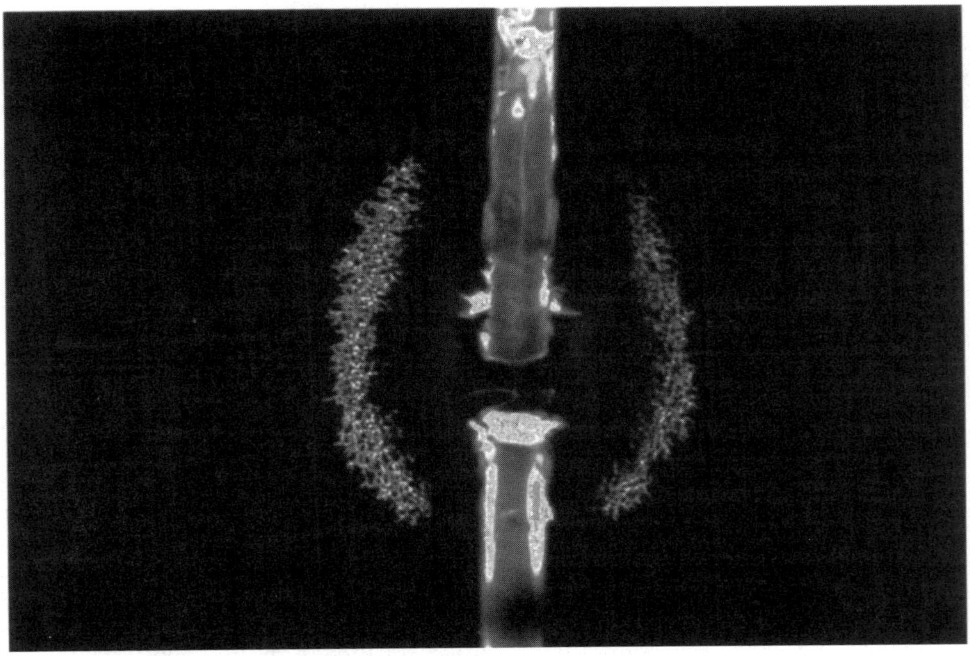

Figure 10 Laser confocal micrographic illustration of the mechanical load transfer between the reinforcing CaP fiber and the PLLA matrix within CaP/PLLA composite. Due to the breakage of the reinforcing fiber, the original load experienced by the fiber was transferred to its surrounding matrix as shown by the two arc white bands surrounding the two broken ends of the fiber. (From Ref. 30.)

indicated by the well-defined solid white bands in the LCM image. Because a fiber-reinforced composite has the capability to transfer load between the reinforcing fibers and their surrounding matrix to improve their mechanical properties that noncomposites could not have, a broken fiber within a composite due to a mechanical load higher than the fiber strength would not cause an immediate failure of the composite because the surrounding matrix could pick up the excessive load. Figure 10 is a LCM visual observation of such a load transfer within a composite and clearly illustrates the mechanical advantage of a composite [30]. The two arc white bands surrounding the two ends of a broken fiber indicate the route where the mechanical load was transferred from the broken fiber to its surrounding matrix. The tight matrix structure located on the route of mechanical load transferring would be disrupted and became loose, i.e., to permit the diffusion and accumulation of the probing fluorescent dye as evident in the arc-shaped white bands.

V. EXAMPLE III: USE OF LCM TO ASSESS THE INTERIOR pH PROFILES OF SYNTHETIC BIODEGRADABLE FIBERS UPON THEIR HYDROLYTIC DEGRADATION [31,43]

A. Objective and Brief Background

The objective of this study is to use LCM technique to measure the interior pH value of synthetic biodegradable fibers upon their hydrolytic degradation without physically sectioning the fibers. Majority of the research activities in biodegradable implants centered on the use of linear aliphatic polyesters like polyglycolide (PGA), poly-L-lactide (PLLA) and their copolymers. One of the most well-known concerns of using this class of biodegradable poly(α-hydroxyl acids) is the acidity of the degradation products released during biodegradation. Several investigators indicated that the glycolic or lactic-acid-rich degradation products have the potential to significantly lower the local pH in a closed region buffered by a small amount of body fluid surrounded by bone [32–37] or in vitro degradation media [38,39]. This is particularly true if the degradation process proceeds with a burst mode (i.e., a sudden and rapid release of degradation products). This acidity tends to cause abnormal bone resorption and/or demineralization. The resulting environment may be cytotoxic [38,40]. Indeed, inflammatory foreign-body reactions with a discharging sinus and osteolytic foci have been encountered in clinical studies [41]. Hollinger and Winet recently confirmed this acidity problem associated with PGA and/or PLLA orthopedic implants [42]. They found that a rapid degradation of a 50 : 50 ratio of glycolide-lactide copolymer in bone chambers of rabbit tibias inhibited bone regeneration. However, they emphasized that the extrapolation of in vitro toxicity to in vivo biocompatibility must consider the microcirculatory capacity surrounding the implant.

An increase in the local acidity due to a faster accumulation of the highly acidic degradation products within biodegradable implants is also known to lead to an accelerated acid-catalyzed hydrolysis in the immediate vicinity of the accumulated acid. This accelerated hydrolysis could lead to a faster premature loss of mechanical property of the implant than expected, particularly in totally biodegradable composites. Although the higher local acidity in the interior of biodegradable implants has been proposed to explain both in vitro and in vivo data, there are no experimental data to directly support this speculation. The lack of proper characterization methods may be the cause for the lack of experimental evidence because there is no known conventional method that could measure the interior pH profile of a solid material without physically opening up the material.

In this example, the author reports a new characterization method which permits scientists to conduct a nondestructive examination of the interior pH profile of biodegradable biomaterials as a function of their hydrolytic degradation [43]. This new characterization method is based upon the capability of "optical" thin-sectioning of a solid material by LCM that was originally developed for examining the interior of biological and electronic circuit board materials.

B. Experimental Procedures

PDSII monofilament biodegradable sutures (size 210 with 260 μm diameter) from Ethicon were used as the model compound for developing this new pH characterization technique. PDSII is made from *p*-dioxanone via ring-opening polymerization [6]. These suture fibers were provided in sterile, sealed packages and were subject to in vitro degradation study in a potassium phosphate buffered solution of 0.1 M, pH 7.44 for predetermined periods (0, 15, 34, 43, 52, and 64 days) at 37°C. The buffer was refreshed once every week.

Two fluorescent dyes were used for the measurement of the interior pH profile of biodegradable fibers based on fluorescence ratio imaging technique. These two fluorescent dyes were Texas Red sulfonyl chloride (TXR) (excitation 568 nm, emission 585 nm) and CL-NERF (excitation 488 nm, emission 525 nm) obtained from Molecular Probes. The TXR dye was the pH insensitive dye, while the CL-NERF was the pH sensitive dye. The purpose of using the pH insensitive dye along with the pH sensitive dye was to factor out the effect of dye concentration on fluorescent intensity. The pK_a for CL-NERF was about 4.5, limiting its range of pH measurement to between about 3.5 and 6. The lower limit of the pH range of CL-NERF appeared adequate for this study because Chu reported that the aqueous degradation media of polyglycolide and poly(glycolide-*co*-lactide) copolymer reached to pH of 3.21–3.48 and 3.12–3.17, respectively, at the end of 70 days in vitro hydrolysis [39].

Before a quantitative analysis of the LCM images of the PDSII fibers could be done, a calibration curve was required to correlate the LCM measured ratio of the two fluorescent dyes to known pH values. This calibration curve of the two fluorescent dyes solutions was constructed in the following manner. First, aqueous solutions with known pH varying from 3 to 6.5 in increments of 0.5 were prepared. Second, an appropriate amounts of the fluorescent dye solution was added to the above-prepared aqueous solution to make the final dye concentration equivalent to the concentration used to soak the PDSII fiber samples. Third, these calibration solutions of varying pHs prepared in the second step were injected into the chambers made from microscope slides and cover slips (Fig. 11) [43] and their LCM images were recorded. The LCM images of both the TXR and CL-NERF emissions were saved as digital graphic files, with each pixel being assigned a value from 0 to 255. This value indicated the intensity of the pixel. The ratio was then taken by dividing the value of the pixel from the CL-NERF image by the same pixel from the TXR image for all pixels in the image. This complex calculation of ratio of signal intensity of CL-NERF to TXR could be easily achieved by using BioRad's COMOS and SOM software. The ratioed images were then analyzed for average intensity using the histogram feature of the COMOS software. These average intensity values were then plotted against their corresponding known pH values and a linear least-squares regression was used to construct a best fit line. From this best linear fit, an equation of the form $y = ax + b$, where x was the pH value and y was the ratio value, was obtained. The values of a and b were calculated from known x (standard pH medium with known pH) and y (ratioed signal intensity), and were found to be 3.30 and −9.51, respectively. Any unknown interior pH value from the testing sample could then be obtained from this best fit linear equation.

After a predetermined period of in vitro hydrolysis in phosphate buffer solutions at 37°C, the PDSII fiber samples were removed, dried, immersed in a fluorescent dye solution consisting of 200 mM equimolar concentration of TXR and CL-NERF fluorescent dyes in deionized water for 24 h and then removed from the dye solution bath for examining the interior pH profiles. The same BioRad MRC 600 LCM interfaced with a Zeiss Axiovert 10 inverted micro-

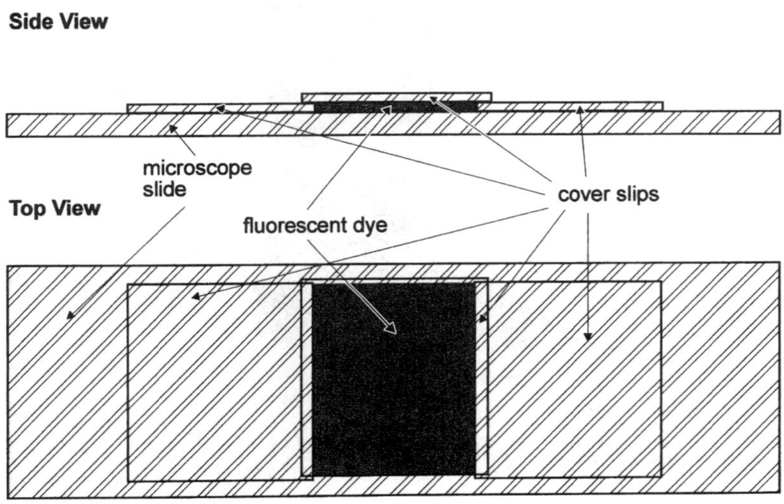

Figure 11 Schematic of the setup for establishing a calibration curve for interior pH measurement. (From Ref. 43.)

scope as the previous examples was used for an interior pH profile examination. The TXR and CL-NERF dyes were excited simultaneously using a krypton-argon ion laser. Two separate filter sets were used to measure the fluorescent intensity from each of the two fluorescent dyes at the same time.

Either the partially hydrolyzed or unhydrolyzed PDSII fiber samples were then mounted between a microscope slide and a cover slip for LCM imaging [43]. LCM images of the PDSII fiber samples were obtained on their surface, and at regular depths toward the interior of the sample and length through optical thin sections. Figure 12 illustrates the optical thin sectioning and the orientation between depth (y) and length (x) of the fiber for LCM imaging. The maximum depth that could be imaged by the present LCM system was about 70 μm beneath the surface of the PDSII fibers which is about 54% of the radius of the fiber. Each image of the PDSII fiber sample was ratioed by using the SOM software as in the construction of the calibration curve described previously.

C. Results

Figure 13 shows a calibration curve for a series of known pH aqueous solutions [43]. The R^2 value of the linear regression plot of the calibration curve was high enough (0.98) for us to conclude that the curve was linear enough for subsequent use.We also found out that heating of the fluorescent dye solutions prepared for the calibration purpose would improve the linearity of the curve due to improved solubility of these two fluorescent dyes in water. For this study, the pH of the dye solution alone was measured to be 5.6. Obviously, it would be beneficial if the lower end of the pH could be extended below pH 3 to broaden the applicability

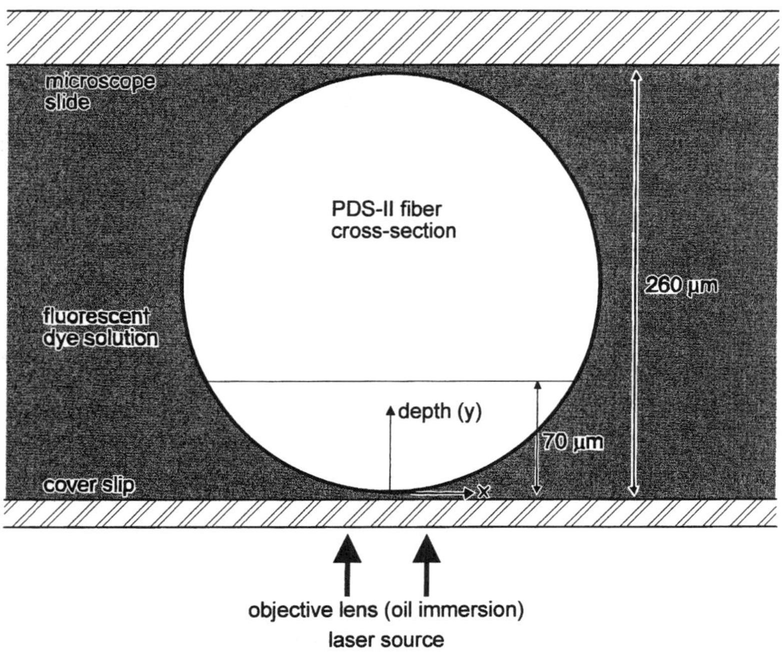

Figure 12 Schematic of the arrangement of depth (y) and distance (x) of LCM image of a PDSII fiber. (From Ref. 43.)

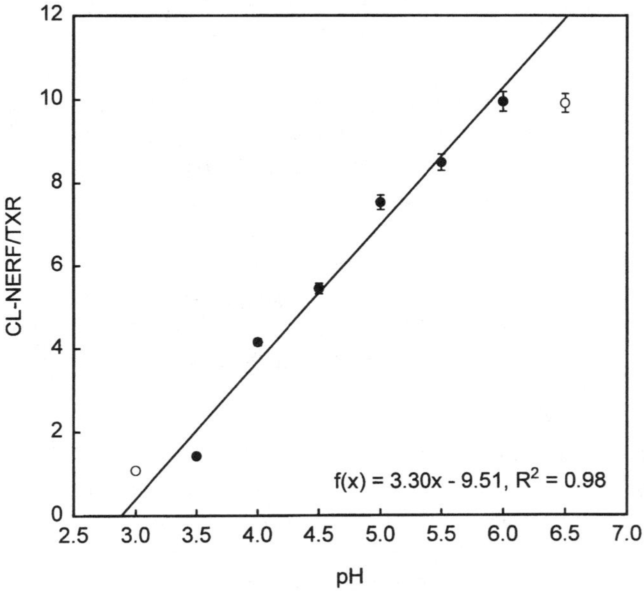

Figure 13 A calibration curve used for determining subsequent interior pH profiles of PDSII fibers. (From Ref. 43.)

of this new characterization technique. Unfortunately, the lack of commercial availability of such fluorescent dyes with lower pH capability limits our study to pH 3.

The results of the interior pH profiles of the 34 days hydrolyzed PDSII fibers as a function of depth (y) and distance (x) of the fibers are shown in Fig. 14 [43]. The data suggest that lower pH values were indeed found at either different depth or length of the fiber. The magnitude of pH reduction, ΔpH, relative to the pH on the PDSII fiber surface reached as high

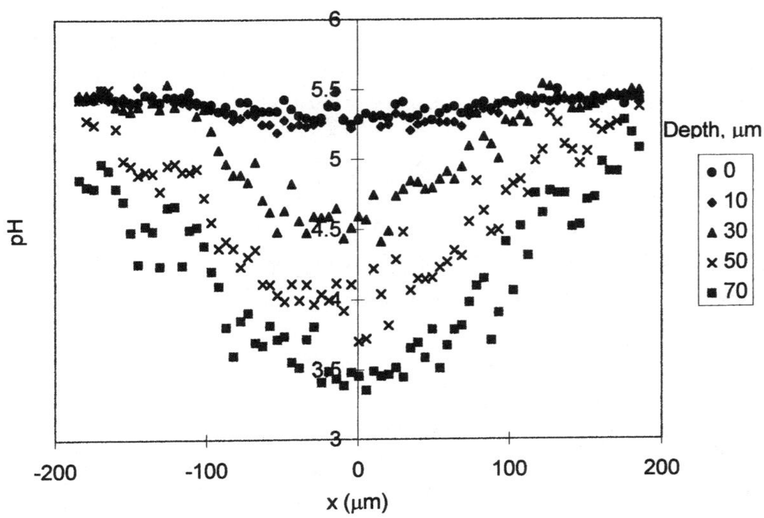

Figure 14 Interior pH profiles of 34 days hydrolyzed 2/0 size PDSII suture fiber control as a function of both the depth (y) and distance (x) of the fiber. Refer to Fig. 12 for x and y. (From Ref. 43.)

as 2 from the surface to 70 μm depth of the fiber. The ΔpH also depended on the duration of hydrolytic degradation. It increased with an increase in the duration of hydrolytic degradation, reached a maximum ΔpH value between 34 and 43 days of hydrolysis and decreased thereafter.

Therefore, the data presented in this example experimentally support the generally speculated theory that solid biodegradable biomaterials could exhibit heterogeneous hydrolytic degradation and the core portion of the biodegradable biomaterial could degrade at a faster rate than the surface layer due to the accumulated acidic degradation products in the core that are difficult to diffuse out of the core fast enough to maintain a constant pH within the core. Such an accumulation of acidic degradation products would lead to acid catalyzed hydrolytic degradation. The level of this acid catalyzed effect, of course, depends on the size or thickness of the solid biodegradable biomaterials, the level of crystallinity and orientation, and their chemical constituents. Vert et al. observed large voids in the interior of linear aliphatic biodegradable polymers upon hydrolytic degradation and suggested that an excessive accumulation of the acidic degradation products within the polymers was the major cause for their findings [44]. Very recently, Gogolewski et al. showed the presence of large void space in the core of poly(D,L-lactide) and a mixture of L- and D,L-polylactide cylindrical rods (138 mm radius which was slightly larger than the radius of PDS-II fiber used in our study) upon in vitro and in vivo aging from one to six months [45]. At the end of six months in vitro, a very large void space about one-half of the total cross-sectional area appeared at the core of the rods. The in vitro aging appeared to lead to a far larger void space than the in vivo mode [45]. Similar findings of the appearance of large void space within the interior of biodegradable biomaterials were also reported by the Loh et al. study of the effect of plasma surface treatment on the degradation of Maxon sutures [46]. They suggested that the hydrophobic plasma surface treatment served as a barrier to reduce the accessibility of acidic degradation products from diffusing out of the core of the suture fiber and hence resulted in a rapid accumulation of acids within the core that led to the observed void formation within the core of the fiber.

As with any new technique, there are some limits of using LCM to examine the interior pH profiles of biodegradable biomaterials. The major limit of this reported new technique will be the depth of the biomaterial that LCM can reach. The depth is limited both by the working distance of the objective lens and the opacity of the sample. The fluorescence dye in the PDSII fibers diffuses, diffracts and refracts the laser light from LCM, which means that the detector sees a lower intensity with increasing depth into the sample. A highly opaque material would not allow enough light intensity emitted from the fluorescent dye within the material back to the detector for proper imaging. Thus, the new technique should work the best with transparent biomaterials.

VI. EXAMPLE IV: USE OF LCM TO ASSESS THE MORPHOLOGY OF SYNTHETIC BIODEGRADABLE SUTURE FIBERS UPON THE EFFECT OF γ-IRRADIATION ON THEIR HYDROLYTIC DEGRADATION

A. Objective and Brief Background

The objective of this study is to explore whether LCM could be used to reveal the effects of γ-irradiation and hydrolytic degradation on the morphology of synthetic biodegradable suture fibers that conventional optical and scanning electron microscopes cannot. The hypothesis to be tested is that the treatments of γ-irradiation and hydrolytic degradation of synthetic biodegradable biomaterials would produce microvoids on and within the biomaterials that are invisible to optical and scanning electron microscopes but could be detected by the preferential accumulation of fluorescent dyes using LCM.

The effect of γ-irradiation on synthetic absorbable biomaterials, particularly wound closure biomaterials, has been extensively studied during the past two decades [46–60]. Due to the predominant chain scission mode across the dosage range studied, the effect of K-irradiation on synthetic absorbable biomaterials is generally undesirable because it leads to an accelerated loss of tensile breaking strength upon biodegradation. Consequently, all existing synthetic absorbable biomaterials cannot be sterilized by a conventional Co^{60} γ-irradiation process. This well-known adverse effect of γ-irradiation on the biodegradation properties of synthetic absorbable sutures, however, has very recently been overcome by Chu and Lee [52]. In that study, Chu and Lee suggested the incorporation of extremely low temperature and vacuum during γ-irradiation would result in biodegradation properties of γ-irradiated synthetic absorbable sutures similar to the corresponding ethylene oxide sterilized absorbable sutures.

Whether the γ-irradiation effect is characterized by predominant chain scission or by crosslinking depends on several factors, including the chemical structure of materials to be irradiated, the amount of dosage, the rate of dosage, the environment of the material during irradiation, and the heat of polymerization [57,58]. The orientation of long chain molecules has also been reported to have some influence on the direction of the overall radiation reaction, and was due to the differences in chain mobility [59]. In the linear saturated polyesters, such as PGA and PGL, D'Alelio et al. [60] reported that the crosslinking reaction would occur if the radicals formed on the carbon atoms of the methylene group in the main chains combine together, whereas the scission process would be expected to occur within the ester moiety.

Although Co^{60} γ-irradiation process deteriorates the mechanical properties of synthetic biodegradable biomaterials and subsequently accelerates their hydrolytic degradation, there is no visible surface morphological change of these biomaterials until they have been subject to an extended period of hydrolytic degradation [47–49,51,55]. In other words, the appearance of the unhydrolyzed unirradiated and irradiated synthetic biodegradable biomaterials are identical under SEM or optical microscopic observation. It would be helpful if there is a simple and nondestructive means that could distinguish them without using mechanical property testing. The LCM method to be described below has been shown to be able to distinguish biomaterials that appear identical by SEM or optical microscope.

B. Experimental Procedures

Basically, similar procedures for Examples I and II were used in this study. The probe agent used was rhodamine B, the same fluorescent dye for the chemical surface modification study described in Examples I and II. Maxon monofilament suture fibers of 2/0 size were used as the model compound for such a study. Maxon is a poly(ester-carbonate) and is made from block copolymer of glycolide and 1,3-dioxan-2-one (trimethylene carbonate or GTMC). It consists of 32.5% by weight (or 36 mole%) of trimethylene carbonate with the repeating unit $[-OCH_2CO-]_{67}[-O(CH_2)_3OCO-]_{33}$ [6]. These biodegradable fibers were subject to Co^{60} γ-irradiation process at two irradiation temperatures (55°C and −78°C) and two radiation dosages (2 and 10 Mrads). Both the irradiated Maxon suture fibers and unirradiated controls were subject to in vitro hydrolytic degradation in saline buffer solutions at 37°C for up to 21 days. At the end of each hydrolysis period, the suture sample was removed and dried in a desiccator for at least 24 h. All samples were then immersed in a 5 mM rhodamine B solution at 37°C for additional 24 h before LCM imaging was performed.

These rhodamine B soaked suture fibers were then placed on the same BioRad MRC-600 LCM as those used in the previous examples. Two-dimensional images of either *X-Z* or *X-Y* planes were obtained by optical sectioning as previously shown in Fig. 3 of Example II.

C. Results

As suggested in the hypothesis, LCM coupled with rhodamine B fluorescent dye indeed could reveal the effects of Co^{60} γ-irradiation and in vitro hydrolytic degradation, particularly in the early stage that conventional SEM or optical microscope cannot as shown in Fig. 15. The data in Fig. 15 clearly demonstrate that 2 Mrad-irradiated/unhydrolyzed Maxon suture exhibited clearly colored (or gray scale in black and white) appearance, while the unirradiated/unhydrolyzed control was invisible under LCM due to its inability to uptake rhodamine B. The intensity of the dye uptake as reflected by changing color (or gray scale in black and white) toward red or white increased with an increase in radiation dosage from 2 to 10 Mrad. The same LCM characteristic was also found in partially hydrolyzed/unirradiated Maxon suture fibers. The color (or gray scale in black and white) appearance of the hydrolyzed Maxon started to show as early as 14 days of hydrolysis and became very intense at 21 days. It is important to recognize that neither SEM nor conventional optical microscope showed any apparent morphological difference between unhydrolyzed/unirradiated and 21 days hydrolyzed/unirradiated Maxon fibers and between unhydrolyzed/unirradiated and unhydrolyzed/2 Mrad- or 10 Mrad-irradiated Maxon suture fibers [6,47,48]. The LCM images of the 10 Mrad-irradiated and partially hydrolyzed Maxon suture fibers exhibited cracks perpendicular to the fiber axis and more cracks were found at a longer period of hydrolysis. The color (gray scale in black and white) intensity of the cracks were more toward red and white than the remaining

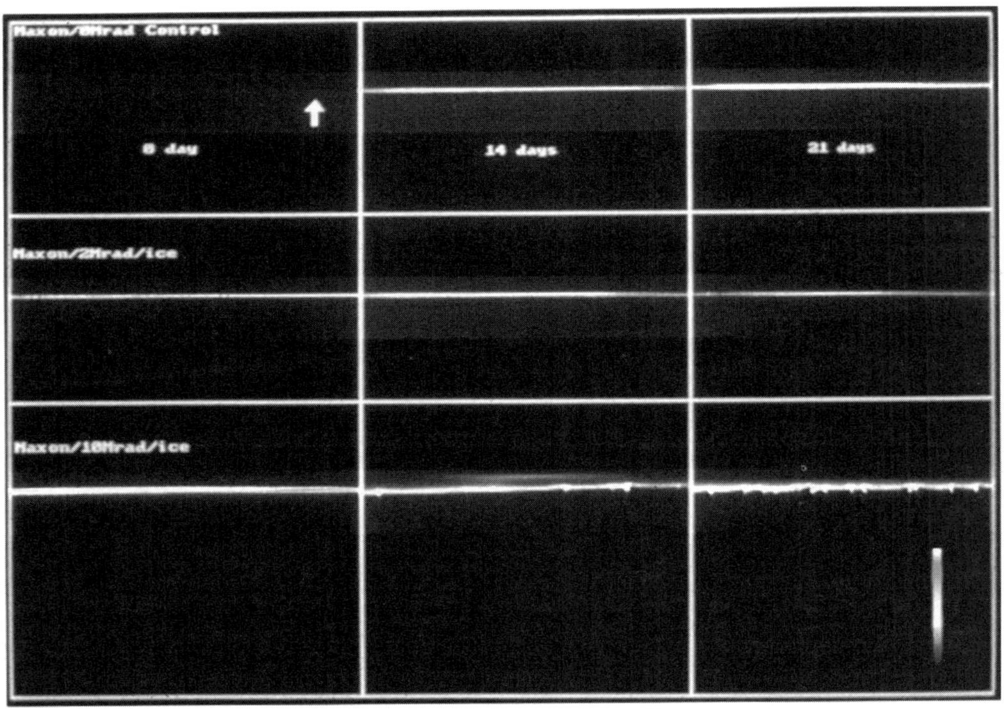

Figure 15 Laser confocal microscopic images of 2/0 size Maxon suture fibers (longitudinal view) that were subject to γ-irradiation at −78°C and in vitro hydrolytic degradation in buffer of pH 7.4 at 37°C. Upper left corner is the unirradiated/unhydrolyzed control Maxon fiber. From left to right indicates hydrolysis time (14 fand 21 days), while from top to bottom indicates increasing irradiation dosage (2 and 10 Mrad).

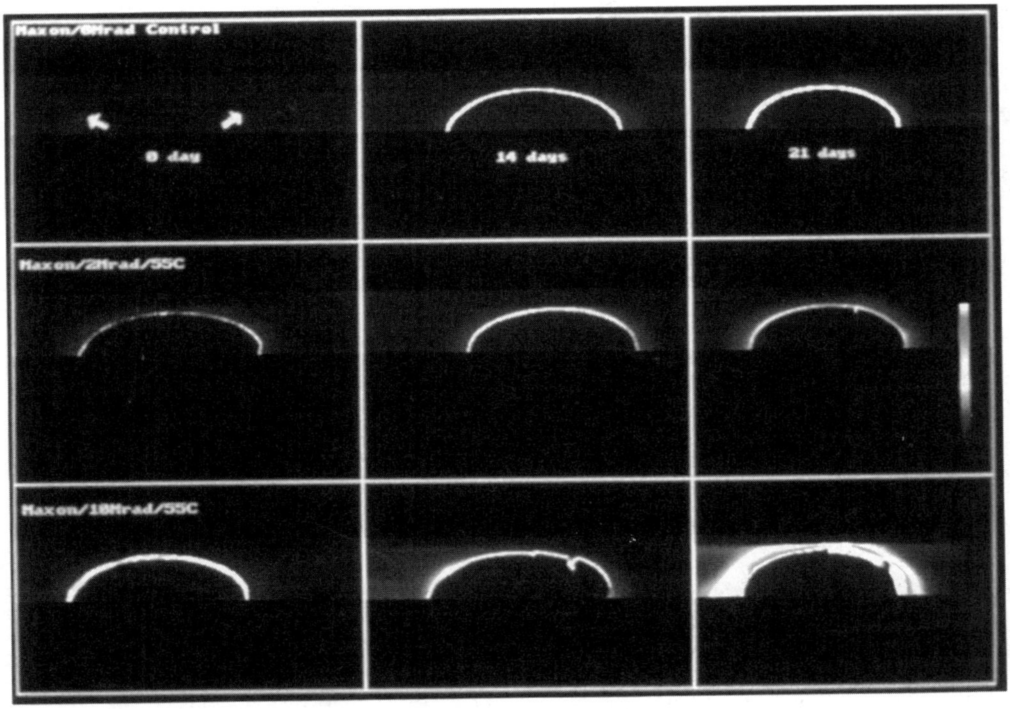

Figure 16 Laser confocal microscopic images of 2/0 size Maxon suture fibers (cross-sectional view) that were subject to γ-irradiation at 55°C and in vitro hydrolytic degradation in buffer of pH 7.4 at 37°C. Upper left corner is the unirradiated/unhydrolyzed control Maxon fiber. From left to right indicates hydrolysis time (14 and 21 days), while from top to bottom indicates increasing irradiation dosage (2 and 10 Mrad).

body of the fibers, an indication that there was far more accumulation of rhodamine B dye within the cracks.

The capability of revealing the effect of irradiation and hydrolysis of synthetic biodegradable biomaterials by LCM image is further demonstrated in Fig. 16, which exhibits the depth of rhodamine B dye penetration into fibers (cross-sectional view) as a function of irradiation dosage and hydrolysis time. The shift of color (or gray scale in black and white) toward red and white with increasing either hydrolysis time or irradiation dosage and the appearance of dents (i.e., cracks) on the fiber cross-sectional contour of the 10 Mrad/14 or 21 days hydrolysis are consistent with the findings described in Fig. 15. The data also showed that the effects of irradiation and early stage of hydrolysis (≤21 days) were generally limited to the outermost surface layer of the Maxon suture fiber or its surface cracks where the dye resided.

ACKNOWLEDGMENT

The author wishes to thank his former graduate students, Dr. Larry Pratt and Mr. Mike Slivka, and former undergraduates, Caryn Chu and John Awad. Their dedication and hard work made this review chapter feasible.

REFERENCES

1. Shalaby, S. W., 1994. *Biomedical Polymers: Designed-to-Degrade Systems*. Hanser, New York.
2. Barrows, T. H., 1986. Degradable implant materials: A review of synthetic biodegradable polymers and their applications. *Clin. Mater.*, 1: 233–257.
3. Kimura, Y., 1993. Biodegradable polymers. In: T. Tsuruta, K. Hayashi, K. Kataoka, K. Ishihara, and Y. Kimura (Editors), *Biomedical Applications of Polymeric Materials*. CRC Press, Boca Raton, FL, pp. 164–190.
4. Chu, C. C., 1996. Biodegradable suture materials: Intrinsic and extrinsic factors affecting biodegradation phenomena. In: D. L. Wise, D. E. Altobelli, E. R. Schwartz, M. Yszemski, J. D. Gresser, and D. J. Trantolo (Editors), *Encyclopedic Handbook of Biomaterials and Applications, Part A. Materials*, 1. Marcel Dekker, New York, pp. 543–688.
5. Chu, C. C., 1995. Biodegradable polymeric biomaterials: An overview. In: J. D. Bronzino (Editor-in-Chief), *Biomedical Engineering Handbook*. CRC Press, Boca Raton, FL. Chapter 44, pp. 611–626.
6. Chu, C. C., von Fraunhofer, J. A., and Greisler, H. P., 1997. *Wound Closure Biomaterials and Devices*. CRC Press, Boca Raton, FL.
7. Hollinger, J. (Editor), 1995. *Biomedical Applications of Synthetic Biodegradable Polymers*. CRC Press, Boca Raton, FL.
8. Vert, M., Feijen, J., Albertsson, A., Scott, G. and Chiellini, E., 1992. *Biodegradable Polymers and Plastics*. Royal Society of Chemistry, Cambridge, England.
9. Yu, T. J., Ho, D. M., and Chu, C. C., 1994. Bicomponent vascular grafts consisting of synthetic biodegradable fibers. Part II. *In vivo* healing response. *J. Investigative Surg.*, 7: 195–211.
10. Yu, T. J., and Chu, C. C., 1993. Bicomponent vascular grafts consisting of synthetic biodegradable fibers. Part I. In vitro study. *J. Biomed. Mater. Res.*, 27:1329–1339.
11. Bowald, S., Busch, C., and Eriksson, I., 1979. Arterial regeneration following polyglactin 910 suture mesh grafting. *Surgery*, 86(5): 722–729.
12. Bowald, S., Busch, C., and Eriksson, I., 1980. Biodegradable material in vascular prosthesis. *Acta. Chir. Scand.*, 146: 391–395.
13. Greisler, H. P, Kim, D. U., Price, J. B., and Voorhees, A. B., 1985. Arterial regenerative activity after prosthetic implantation. *Arch. Surg.*, 120: 315–323.
14. Greisler, H. P., 1982 Arterial regeneration over biodegradable prostheses. *Arch. Surg.*, 117: 1425–1431.
15. Greisler, H. P., Kim, D. U., Dennis, J. W., Klosak, J. J., Widerborg, K. A., Endean, E. D., Raymond, R. M., and Ellinger, J., 1987. Compound polyglactin 910/polypropylene small vessel prostheses. *J. Vasc. Surg.*, 5: 572–583.
16. Greisler, H. P., Ellinger J., Schwarcz T. H., Golan, J., Raymond, R. M., and Kim, D. U., 1987. Arterial regeneration over polydioxanone prostheses in the rabbit. *Arch. Surg.*, 122: 715–721.
17. Greisler, H. P., Endean, E. D., Klosak, J. J., Ellinger, J., Dennis, J. W., Buttle, K., and Kim, D. U., 1988. Polyglactin 910/polydioxanone bicomponent totally resorbable vascular prostheses. *J. Vasc. Surg.*, 7: 697–705.
18. Greisler, H. P., 1988. Macrophage-biomaterial interactions with bioresorbable vascular prostheses. *Tran. ASAIO*, 34(4): 1051–1059.
19. Greisler, H. P., Dennis, J. W., Endean, E. D., and Kim, D. U., 1988. Derivation of neointima of vascular grafts. *Circulation Supp.*, I, 78: I6–I12.
20. Greisler, H. P., Tattersall, C. W., Kloask, J. J., et al., 1991. Partially bioresorbable vascular grafts in dogs. *Surgery*, 110(4): 645–655.
21. Wilson, T., 1990. *Confocal Microscopy*. Academic Press, San Diego, CA.
22. Sarafis, V., 1990. Biological perspectives of confocal microscopy. In: T. Wilson (Editor), *Confocal Microscopy*, Academic Press, San Diego, CA, pp. 325–333.
23. Attawia, M. A., Devin, J. E., and Laurencin, C. T., 1995. Immunofluorescence and confocal laser scanning microscopy studies of osteblast growth and phenotypic expression in three-dimensional degradable synthetic matrices. *J. Biomed. Mater. Res.*, 29: 843–848.

24. Hanthamrongwit, M., Wilkinson, R., Osborne, C., Reid, W. H., and Grant, M. H., 1996. Confocal laser-scanning microscopy for determining the structure of and keratinocyte infiltration through collagen sponges. *J. Biomed. Mater. Res.*, 30: 331–339.
25. Pratt, L., and Chu, C. C., 1994. Dimethyltitanocene induced surface chemical degradation of synthetic biodegradable polyesters. *J. Polym. Sci. Chem. Ed.*, 32: 949–960.
26. Zimmerman, M. C., Parsons, J. R., and Alexander, H., 1987. The design and analysis of a laminated partially degradable composite bone plate for fracture fixation. *J. Biomed. Mater. Res., Appl. Biomater.*, 21: 345–361.
27. Zimmerman, M. C., Alexander, H., Parsons, J. R., and Bajpai, P. K., 1991. The design and analysis of laminated degradable composite bone plates for fracture fixation. In: T. L. Vigo and A. F. Turbak (Editors), *High-Tech Fibrous Materials*, American Chemical Society, Washington, DC, pp. 132–148.
28. Ibnabddjalil, M., Loh, I. H., Chu, C. C., Blumenthol, N., Alexander, H., and Turner, D., 1994. The effect of plasma treatment on the chemical, physical morphological and mechanical properties of biodegradable CaP fibers and polyglycolide acid matrix for use in internal fixation devices. *J. Biomed. Mater. Res.*, 28(3): 289–301.
29. Slivka, M., and Chu, C. C., 1997. Fiber/matrix interface studies on biodegradable composite materials for internal fixation of bone fractures. Part I. Raw material evaluation and measurement of fiber/matrix interfacial adhesion. *J. Biomed. Mater. Res.*, 36(4): 469–477.
30. Slivka, M. and Chu, C. C., 1997. Fiber/matrix interface studies on biobiodegradable composite materials for internal fixation of bone fractures. Part II. A new method using laser scanning confocal microscopy. *J. Biomed. Mater. Res.*, 37(3): 353–362.
31. Slivka, M. A., 1996. Fiber/matrix interface studies on totally and partially biodegradable composite materials for internal fixation of bone fractures: A new visual characterization method using laser confocal microscopy. MS thesis, May 1996, Cornell University, Ithaca, NY.
32. Claes, L. Requirements on resorbable implant materials. 1989. In: G. Heimke, U. Soltesz and A. J. C. Lee (Editors), *Clinical Implant Materials*. Elsevier, Amsterdam, pp. 161–167.
33. Claes, L. Mechanical characterization of biodegradable implants. 1990. In: G. O. Hofmann (Editor), *Biodegradable Implants in Orthopaedic Surgery*, Kommunikation, Berlin, pp. 83–93.
34. Rehm, K. E., Helling, H. J., et al., 1989. Biologisch abbaubare osteosynthesematerialien. In: H. Bunte, T. Junginger and A. Holzgreve (Editor.), *Jahrbuch der Chir*, Biermann, Zulpich, pp. 223–232.
35. Hofmann, G. O., 1993. Biodegradable implants in orthopaedic surgery—A review on the state of the art. *Clin. Mater.*, 10:75.
36. Suganuma, J., and Alexander, H., 1993. Biological response of intramedullary bone to poly-L-lactic acid. *J. Appl. Biomater.*, 4(1): 13.
37. Sugnuma, J., Alexander, H., Traub, J. and Ricci, J. L., 1993. Biological response of intramedullary bone to polylactic acid. In: L. G. Cima and E. S. Ron (Editors), *Tissue-Inducing Biomaterials*. Mater. Res. Soc. Sump. Proc., 252: 339–343.
38. Taylor, M. S., Daniels, A. U., Andriano, K. P. and Heller, J., 1994. Six biobiodegradable polymers: In vitro acute toxicity of accumulated degradation products. *J. Appl. Biomater.*, 5:151–157.
39. Chu, C. C., 1985. Degradation phenomena of two linear aliphatic polyester fibers used in medicine and surgery. *Polymer*, 26: 591–594.
40. Daniels, A. U., Taylor, M. S., Andriano, K. P., and Heller, J., 1992. Toxicity of biodegradable polymers proposed for fracture fixation devices. Trans. 38th Ann. Mtg. Orthop. Res. Soc., 17: 88.
41. Eitenmüller, J., Gerlach, K. L., Schmickal, T., and Muhr, T., 1989. Die Versorgung von Sprung gelenksfrakturen unter Verwendung von Platten und Schrauben aus resorbierbarem. Polymer material. Presented at Jahrestagung der Deutschen Gesellschaft für Unfallheilkunde, November, 1989, Berlin.
42. Winet, H., Hollinger, J. O., 1993. Incorporation of polylactide-polyglycolide in a cortical defect: Neoosteogenesis in a bone chamber. *J. Biomed. Mater. Res.*, 27: 667.
43. Slivka, M. Chu, C. C., and Zhang, Y. L., (Pending). Laser confocal microscopic study of pH profiles of synthetic biodegradable fibers upon in vitro hydrolytic degradation. *J. Mater. Sci. Mat. in Med.*.

44. Vert, M., Li, S. M., Spenlehauer, H., and Guerin, P., 1992. Bioresorbability and biocompatibility of aliphatic polyesters. *J. Mater. Sci., Mater. In Med.*, 3: 423–446.
45. Mainil-Varlet, P., Curtis, R., and Gogolewski, S., 1997. Effect of in vivo and in vitro degradation in molecular and mechanical properties of various low-molecular-weight polylactides. *J. Biomed. Mater. Res.*, 36(3): 360–380.
46. Loh, I. H., Lin, H. L., and Chu, C. C., 1992. Plasma surface modification of synthetic biodegradable sutures. *J. Appl. Biomater.*, 3(2): 131–146.
47. Zhang, L., Loh, I. H., and Chu, C. C., 1993. A combined γ-irradiation and plasma deposition treatment to achieve the ideal degradation properties of synthetic absorbable polymers. *J. Biomed. Mater. Res.*, 27: 1425.
48. Chu, C. C., Zhang, L., and Coyne, L., 1995. Effect of irradiation temperature on hydrolytic degradation properties of synthetic absorbable sutures and polymers. *J. Appl. Polym. Sci.*, 56: 1275.
49. Chu, C. C. and Campbell, N. D., 1982. Scanning electron microscopic study of the hydrolytic degradation of poly(glycolic acid) suture. *J. Biomedical Mater. Res.*, 16: 417.
50. Chu, C. C. and Louie, M., 1985. A chemical means to examine the degradation phenomena of polyglycolic acid fibers. *J. Appl. Polym. Sci.*, 30: 3133.
51. Williams, D. F., Chu, C. C., and Dwyer, J., 1984. The effects of enzymes and gamma irradiation on the tensile strength and morphology of poly(*p*-dioxanone) fibers. *J. Appl. Polym. Sci.*, 29: 1865.
52. Lee, K. H. and Chu, C. C., 1996. Co^{60} γ-irradiation sterilization of biomaterial medical devices or products with improved degradation and mechanical properties. U.S. Patent 5,485,496, January. 16, 1996.
53. Lee, K. H. and Chu, C. C., 1996. Electron spin resonance study of free radical properties of polyglycolic acid upon γ-irradiation sterilization. The 5t World Biomaterials Congress, May 29–June 2, 1996, Toronto.
54. Campbell, N. D. and Chu, C. C., 1981. The effect of γ-irradiation on the biodegradation of polyglycolic acid synthetic sutures—Tensile strength study. 27th Int. Symp. on Macromolecules, II, July 6–9, 1981, Strasbourg, France.
55. Chu, C. C. and Williams, D. F., 1983. The effect of γ-irradiation on the enzymatic degradation of polyglycolic acid absorbable sutures, *J. Biomed. Mater. Res.*, 17: 1029.
56. Reed, A. M. and Gilding, D. K., 1979. Biodegradable polymers for use in surgery—Poly(glycolic)/poly(lactic acid) homo and copolymers: 1. *Polymer*, 20: 1459.
57. Shalaby, S. W., 1993. *Irradiation of Polymeric Materials*, ACS Symposium Series. E. Reichmanis, C. W. Frank, and J. H. O'Donnell (Editors), American Chemical Society, Washington, DC, 527: 315.
58. Charlesby, A., 1960. *Atomic Radiation and Polymers*. Pergamon, Oxford, England.
59. Sehnabel, W., 1978. In: H. H. G. Jellinek (Editor), *Aspects of Degradation and Stabilization of Polymers*. Elsevier, New York, Chap. 4.
60. D'Alelio, G. F., Haberli, R., and Pezdirtz, G. F., 1968. Effect of ionizing radiation on a series of saturated polyester, *J. Macromol. Sci. Chem.*, A2: 501.

10
Polymeric Drug Delivery to the Brain

Zvi H. Israel
Oregon Health Sciences University, Portland, Oregon

Abraham J. Domb
The Hebrew University of Jerusalem, Jerusalem, Israel

I. INTRODUCTION

The blood brain barrier (BBB) has evolved to protect the brain against peripheral neurotransmitters, cytotoxins and microorganisms. However, while providing an excellent homeostatic milieu for the central nervous system (CNS), the BBB also prevents effective therapeutic access to the CNS. Many drugs that have shown promise in the laboratory have not met their clinical potential because of the lack of a delivery method that could effectively bypass the obstacle of the BBB. Furthermore, it has been estimated that the combined effort in academic and industrial laboratories devoted to CNS drug delivery is at least 2 log orders of magnitude less than the comparable effort in CNS drug discovery [1].

The morphological, chemical and metabolic correlates of the BBB have largely been characterized [2–4]. The endothelial cells of brain capillaries are nonfenestrated and the intercellular clefts are closed by circumferential belts of tight junctions. Thus the BBB can be viewed as two separate impermeable cellular membranes, one on the inside of the vessel wall (luminal side) and the other on the outside (abluminal side) separated by 300- to 500-nm-thick cytoplasm [4]. Only small, electrically neutral, lipophilic molecules (molecular weight less than 500 Daltons) can easily traverse this continuous lipid bilayer from the cerebrovascular space to enter the brain interstitial space. Nutrients and peptides enter via carrier-mediated or receptor-mediated transport processes respectively [5].

Pardridge has classified the available strategies for brain drug delivery as being invasive, pharmacological or physiological [1,2,5]. The invasive or neurosurgically based techniques include intraventricular, intrathecal or intraparenchymal drug infusion, polymeric intracerebral slow release implants and osmotic or pharmacological BBB disruption. The pharmacological routes include using lipid carriers or liposomes. The physiological methods look to exploit the normal, endogenous pathways of either carrier mediated transport of nutrients or receptor mediated transport of peptides.

Hydrophilic drugs can be altered to a more lipophilic analog; however, this often changes their therapeutic potential. This approach and other delivery methods have been extensively reviewed elsewhere [1,4,6–16].

The polymeric site-specific pharmacotherapy (PSSP) concept [17,18] aims to selectively deliver a therapeutic agent over a prolonged period to its intended site of action by implanting a preloaded polymer directly to the diseased site. PSSP has been investigated for many organ specific diseases. As polymer science has progressed, our ability to treat or palliate diseases localized in a specific tissue using this approach has considerably increased. This has been especially true for the brain.

Polymeric drug delivery to the brain offers some striking potential advantages. Effective local concentrations of the drug in the brain can be achieved without the sequelae sometimes associated with systemic administration, such as systemic side effects, peripheral drug inactivation, poor drug absorption, serum protein binding, inadequate BBB penetration and poor patient compliance. Drugs or other therapeutic agents with a short half-life can be protected within polymers until their release. Polymeric implants may be placed in the desired brain region preventing exposure of the entire CNS to the active drug. There are disadvantages too. Polymers have to be placed surgically and this means that if discontinuation of therapy is necessary, a further surgical procedure is required. Dose adjustments cannot be made once the polymer is implanted. The only solution for chronic, long-term release is repeated implantations. Furthermore, certain neurological ailments affect the entire CNS, so that local therapy may be inappropriate [18,19].

No single delivery strategy provides a complete solution to the delivery problem; however, some (notably the invasive) neurosurgical techniques have been developed into powerful clinical tools for certain neurological complaints [12,20].

II. POLYMERS FOR DRUG DELIVERY

Polymers to be used for drug delivery must be sufficiently biocompatible and have appropriate physical properties to provide controlled delivery. Specifically, criteria include minimal tissue reaction after implantation, high polymer purity and reproducibility, a reliable drug release profile, and for biodegradable polymers, nontoxic and readily excreted degradation products. For any individual drug to be delivered to a given site an appropriate polymer must be chosen and the composite delivery system designed to provide the desired rate and pattern of drug release. These designs exploit various mechanisms of drug transport through and release from polymer materials [6,17,21].

Biocompatible polymeric systems that have been employed for local controlled drug delivery have been classified into nondegradable polymers, biodegradable polymers, osmotic systems and drug-conjugated polymers [17,19].

Nondegradable polymers were the first polymers to be used as controlled-release devices [22,23]. These polymers are stable in biological systems and drug release is typically diffusion regulated. One system is based on a hollow polymer tube filled with a drug suspension. The drug is released by dissolution into the polymer and then diffused through the polymer wall. Polysilicones and poly(ethylene-vinyl acetate) (EVAc) have been used in this way. Alternatively, a solid matrix may be used which is simpler and potentially safer than a reservoir system [6]. More importantly, at a high enough drug concentrations the polymer matrix can retain drug molecules within a series of interconnecting pores allowing slow delivery of larger and more ionic drug molecules [24]. As drug release is dependent only upon diffusion, drug release kinetics are predictable and reproducible and determined by the size and shape of the polymer and drug particle size and lipophilicity. Nondegradable systems also have drawbacks. Constant rates of release can be difficult to achieve from solid matrix systems [25] and nondegradable polymers must remain as a permanent implant.

Biodegradable polymers dissolve gradually, usually by hydrolytic degradation. Drug incorporated within the polymer matrix remains protected from hydrolysis and enzymatic degradation until released as the polymer disintegrates. Synthetic examples of such polymers include polyesters such as poly(lactide-*co*-glycolide) (PLGA), polyanhydrides such as poly(*p*-carboxyphenoxypropane-*co*-sebacic acid) (PCPP-SA), poly(orthoesters), polyphosphazenes and poly-(phosphate-esters). Naturally existing polymers such as collagen, fibrin and gelatin have also been utilized in this context. The rate of drug release from the polymer is dependent upon the type and rate of polymer degradation. Most degradable polymers display bulk erosion that causes the polymer to dissolve throughout its entire matrix. Constant release rates become difficult to achieve and the patient is exposed to the danger of dose dumping. To prevent the problems associated with bulk erosion and to facilitate continuous drug release it is necessary to utilize a polymer that displays surface erosion whereby the matrix dissolves layer by layer. Geometrically, a long flat slab is one shape that could achieve a constant rate of drug release. Bioerodible polymers can be synthesized to display such surface erosion. The polyanhydrides consist of hydrophobic monomers joined by water-labile bonds; thus, controlled surface erosion occurs while the polymer interior remains anhydrous. The rate of polymer degradation can be modified by changing the proportions of the two monomers [26]. However, increasing the concentration of hydrophilic drugs within biodegradable polymers can result in complex release kinetics as a result in a combination of both polymer degradation and drug diffusion. Furthermore, certain biodegradable polymers can evoke an inflammatory response in the surrounding brain.

While a variety of polymers have been used for brain applications experimentally, the polymer group that has seen the widest clinical application for CNS disorders are the polyanhydrides. The properties of other polymers used in brain and other applications have been extensively reviewed elsewhere.

Several reviews have been published on the chemistry and drug delivery applications of polyanhydrides [27–30]. The majority of the polyanhydrides are prepared by melt polycondensation.

A drug-incorporated matrix can be formulated by either compression or injection molding. The polymer and drug can be ground in a Micro Mill grinder, sieved into a particle size range of 90–120 μm and pressed into circular disks using a Carver press. Alternatively, the drug can be mixed into the molten polymer to form small chips of drug-polymer matrix. These chips are fed into the injection molder to mold the drug/polymer matrix into the desired shape. One must consider the thermal stability of the polymer and potential chemical interaction between drug and polymer at high temperatures of injection molding.

The preferred method of drug delivery, in many instances, is by injection. This requires the development of drug microcapsules or microspheres. Several techniques have been developed for the preparation of microspheres from polyanhydrides and include 'hot-melt' microencapsulation and solvent removal [31].

The copolymer P(CPP:SA) in a 20:80 molar ratio is in use clinically for the delivery of carmustine [1,3-bis(2-chloroethyl)-1-nitrosourea, or BCNU] to the brain parenchyma following glioma resection.

III. DRUG RELEASE FROM POLYMERS

Drug release from a polymer implant can be conceptualized to occur in three stages [19]: (1) drug release from the polymer; (2) drug diffusion into the aqueous environment outside of the polymer matrix; and (3) drug distribution into the surrounding tissue. These stages have been intensively studied for drug delivery to the brain using novel polymers, notably the polyanhydrides developed specifically for use within the CNS. The release rates of drugs from a polymer implant can vary widely and have been predicted with several mathematical models. In general the rate of drug release can be described by three simple types of release kinetics: zero order, first order and square root of time-release kinetics. As the drug reaches the implant surface it must penetrate the unstirred fluid surrounding the implant. Measuring and predicting diffusion through this unstirred layer can be described by Fick's laws. Animal models have been used to predict drug distribution within the brain and have shown that to maximize the extent of drug distribution, the agent should be water-soluble, slowly eliminated and have a high diffusion coefficient in the brain tissue.

IV. EXPERIMENTAL AND CLINICAL APPLICATIONS OF PSSP WITHIN THE BRAIN

To date the only clinical application of polymeric drug delivery to the brain in widespread use is the delivery of the antineoplastic drug carmustine, or BCNU, as part of the multimodal therapy for the malignant brain tumor glioblastoma multiforme (GBM) (Gliadel) [32,33]. We review some of this work to show how the rationale for treating GBM in this way prompted the successful experimental and clinical development of this valuable therapeutic tool. We also briefly summarize some of the other experimental applications for polymeric drug delivery to the brain.

V. POLYMERIC DELIVERY OF BCNU FOR GLIOBLASTOMA

The rationale for using polymeric slow release implants in the treatment of GBM is multifactorial. GBM is a highly malignant primary brain tumor. Approximately 12,000 new cases are diagnosed every year in the United States. Despite great advances in imaging and surgical technique, these tumors remain incurable. The current treatment involves initial surgical resection followed, when possible, by radiation therapy. BCNU exhibits modest but significant clinical effect when added to the treatment protocol [34]. BCNU therapy, while an effective adjunct to surgery and radiation, is limited by systemic toxicity as only a small fraction of an intravenous dose can reach the tumor due to poor passage across the BBB and the blood tumor barrier (BTB). Mean survival after diagnosis remains about nine months and five year survival is less than 5%. GBM rarely metastasizes outside the CNS and macroscopic recurrence is local (within 2 cm of the previous resection) in more than 90% of patients [35]. Although microscopic disease can be detected distant from the tumor even at an early stage, it was hypothesized that aggressive local control could improve the survival of these patients. Henry Brem proposed that introduction of polymeric slow release implants into the surgical resection site to provide high local concentrations of an antineoplastic drug of choice could improve the clinical efficacy of the drug therapy without exposing the patient to the side effects of systemic administration [7,33]. The drug chosen to test this hypothesis was BCNU because of its known activity against GBM.

The polyanhydride polymer P(CPP:SA) was tested for its biocompatibility in rabbit cornea, rat, rabbit and monkey brain. These studies showed that P(CPP:SA) evoked only a mild inflammatory response and was safe and nontoxic and suitable for trials in humans. In vivo release of BCNU from the polymer was analyzed in rat and rabbit brain. These experiments showed that only the implanted hemisphere was exposed to significant levels of the drug over a period of approximately one week with minimal systemic levels of the drug. Efficacy studies were performed in a rat 9L gliosarcoma model. Intracranially polymer delivered BCNU significantly extended rat median survival from the intracranial tumor relative to rats receiving systemic BCNU and even effected some "cures": 42% of the rats in the polymer delivered group survived the experiment until the planned sacrifice date and had no evidence of tumor at autopsy [32].

Based on these experiments, human phase I and phase II trials were undertaken in 21 patients with recurrent GBM using three different concentrations of BCNU within the polymer. Patients were followed by the Karnofsky performance score (KPS, a neurological measure reflecting a patient's ability to function independently), blood analysis, urinalysis, and cranial CT. BCNU-P(CPP:SA) polymers were found to be safe when implanted intracranially in humans.

A phase III efficacy study was performed as a prospective, randomized, double-blind, placebo-controlled trial in 27 centers. A total of 222 patients with recurrent malignant glioma were enrolled. There were no clinically important adverse reactions either in the brain or systemically. Median survival of the treated patients was 31 weeks compared with 23 weeks for the patients who received placebo. This difference was found to be statistically highly significant [20]. On the basis of this work the Food and Drug Administration (FDA) approved the use of Gliadel in the treatment of recurrent GBM. Disappointingly, further follow up of these two groups of patients did not show any difference in long-term survival.

Subsequent to this study it was hypothesized that implanting BCNU polymers at the time of initial surgery could further improve median and possibly long-term survival [36]. This was investigated as a prospective randomized trial in 32 patients [37]. Results were encouraging (including a nonstatistically significant increase in survival at two years); however, because of the relatively small numbers, a larger multinational prospective, randomized, double-blind, placebo-controlled trial aiming to recruit at least 240 patients is currently underway to address this issue. Another question that arose from these studies was whether a higher drug load would have shown greater efficacy. Drug escalation studies are at the Phase II stage of investigation.

Perhaps the greatest criticism of the experimental work and this clinical approach to treating GBM has been the issue of what volume of brain is exposed to the active drug. Critics argue that what may be true for diffusion within the small brain of an experimental animal cannot reflect the case of a 1.5 kg human brain. Brem's group has addressed this issue by investigating the relationship between the rates of drug release from polymer implants and drug concentration within the brain of adult monkeys [38]. High concentrations of drug (BCNU, taxol and cyclophosphamide) were found within the first 3 mm from the polymer implant. Significant concentrations of all three drugs were measured at up to approximately 5 cm from the implant for as long as 30 days after implantation.

Other drugs such as taxol [39], carboplatin [40], methotrexate, cyclophosphamide [41], cisplatin [42], 5-fluorouracil [43] and various drug combinations are being investigated for polymeric delivery to malignant brain tumors. The topoisomerase I inhibitor camptothecin has also been studied. Other experimental therapies include immunotoxins, cytokines, radiosensitizers, steroids and angiogenesis inhibitors. Our own recent focus of interest has been to use polymers to deliver antisense oligonucleotides to the mRNA of various growth factors and intracellular secondary messengers known to be important in experimental gliomas [44].

VI. OTHER CNS APPLICATIONS

One of the most devastating complications in patients with subarachnoid hemorrhage (SAH) is a delayed ischemic neurological deficit (NIND) as a result of cerebral vasospasm, which typically affects 20–30% of patients between six and eight days following the initial hemorrhage. A total of 7% of SAH patients die as a direct result of vasospasm induced ischemic infarction [45]. Pathogenesis is likely multifactorial and the condition is notoriously poorly responsive to therapy. One of the ways that such vasospasm has been treated has been by using the nonspecific vasodilator papaverine either topically at surgery or intra-arterially [46,47]; however, the vasodilator effect seen is typically short lived. Several groups have incorporated papaverine into a slow release polymer [48–50] and have shown that such a preparation is efficacious in an animal model of vasospasm [51]. This has not yet been applied in the clinical setting.

The polymeric delivery of neurotransmitters and neuromodulators has been reviewed by Byrd and Hamilton-Byrd [52]. Potential applications include the neurodegenerative diseases Alzheimer's and Parkinson's, stroke, central and peripheral nervous system trauma and epilepsy. Another exciting potential application is the treatment of neuropathic pain.

With the advent of the applications of molecular biology to modern medicine, including the current era of gene therapy, polymers may offer an attractive mode of delivery for such therapy [44] for a wide gamut of disorders within the CNS and without. Advances in the minimally invasive delivery of polymers to the brain (e.g., by stereotactic injection) will help bring many of today's experimental applications into the clinical arena.

REFERENCES

1. Pardridge WM. Preface: Overview of brain drug delivery. *Adv. Drug Deliv. Rev.* 15:1–3, 1995.
2. Pardridge WM. Transport of small molecules through the blood-brain barrier: biology and methodology. *Adv. Drug Deliv. Rev.* 15: 5–36, 1995.
3. Burke M, Brem H, Langer R. Drug delivery to the central nervous system. *The Encyclopedia of Controlled Drug Delivery*, 1998.
4. Zlokovicz BV, Apuzzo MLJ. Strategies to circumvent vascular barriers of the central nervous system. *Neurosurgery* 43: 877–878, 1998.
5. Pardridge WM. Drug delivery to the brain. *J. Cerebal Blood Flow Metabolism* 17: 713–731, 1997.
6. Fung LK, Saltzman WM. Polymeric implants for cancer chemotherapy. *Adv. Drug Deliv. Rev.* 26: 209–230, 1997.
7. Brem H, Langer R, Polymer-based drug delivery to the brain. *Science & Medicine* 3: 2–11, 1996.
8. Inamura T, Fukui M. Experimental trials of brain tumor therapy through blood-brain barrier/blood tumor barrier. *Crit. Rev. Neurosurg.* 6:351–55, 1996.
9. Lieberman DM, Laske DW, Morrison PF et al. Convection-enhanced distribution of large molecules in gray matter during interstitial drug infusion. *J. Neurosurgery* 82:1021–1029, 1995.
10. Kroll RA, Pagel MA, Muldoon LL et al. Increasing volume of distribution to the brain with interstitial infusion: Dose, rather than convection, might be the most important factor. *Neurosurgery* 38: 746–754, 1996.
11. Kroll RA, Pagel MA, Muldoon LL et al. Improving drug delivery to intracerebral tumor and surrounding brain in a rodent model: A comparison of osmotic versus Bradykinin modification of the blood brain barrier and/or blood-tumor barriers. *Neurosurgery* 43:879–889, 1998.
12. Kroll RA, Neuwelt EA. Outwitting the blood-brain barrier for therapeutic purposes: Osmotic opening and other means. *Neurosurgery* 42:1083–1100, 1998.
13. Ho SY, Barbarese E, D'Arrigo JS et al. Evaluation of lipid-coated microbubbles as a delivery vehicle for taxol in brain tumor therapy. *Neurosurgery* 40:1260–1268, 1997.
14. Siegal T, Horowitz A, Gabizon A. Doxorubicin encapsulated in sterically stabilized liposomes for the treatment of a brain tumor model: biodistribution and therapeutic efficacy. *J. Neurosurgery* 83: 1029–1037, 1995.

15. Sharma US, Sharma A, Chau RI et al. Liposome-mediated therapy of intracranial brain tumors in a rat model. *Pharmaceutical Res.* 14:992–998, 1997.
16. Huwyler J, Wu D, Pardridge WM. Brain drug delivery of small molecules using immunoliposomes. *Proc. Natl. Acad. Sci. USA*, 93:14164–14169, 1996.
17. Langer R. Drug Delivery and targeting. Review. Nature (Supp.), 392:5–10, 1998.
18. Brem H, Walter KA, Tamargo RJ, Olivi A, Langer R. Drug delivery to the brain. In: *Polymeric Site Specific Pharmacotherapy*, Ed. Domb AJ, Wiley, New York, 1994, pp. 117–139.
19. Domb AJ, Ringel I. Polymeric drug carrier systems in the brain. In: *Providing Pharmacological Access to the Brain*, Ed. Flanagan TR et al., *Methods in Neurosciences*, Vol. 21, Academic Press, 1994, pp. 169–183.
20. Brem H, Piantadosi S, Burger PC et al. Placebo-controlled trial of safety and efficacy of intraoperative controlled delivery by biodegradable polymers of chemotherapy for recurrent gliomas. *Lancet*, 345: 1008–1012, 1995.
21. Langer R. New methods of drug delivery. *Science* 249:1527–1533, 1990.
22. Folkman J, Long D. The use of silicone rubber as a carrier for prolonged drug therapy. *J. Surg. Res.* 4:139–42, 1964.
23. Folkman J, Long DM, Rosenbaum R. Silicone rubber: a new diffusion propert useful for general anesthesia. *Science*, 154:18–149, 1966.
24. Langer R, Folkman J. Polymers for the sustained release of proteins and other macromolecules. *Nature* 263:797–800, 1976.
25. Yang MB, Tamargo RJ, Brem H. Controlled delivery of BCNU from ethylene-vinyl acetate copolymer. *Cancer Res.* 49:5103–5107, 1989.
26. Leong KW, Brott BC, Langer R. Bioerodible polyanhydrides as drug carrier matrices: I. Characterization, degradation and release characteristics. *J. Biomed. Mater. Res.* 19:941–955, 1985.
27. Chasin M, Domb A, Ron E et al. In: *Biodegradable Polymers as Drug Delivery Systems*, Eds. M. Chasin and R. Langer, Marcel Dekker, New York, 1990.
28. Domb AJ, Amsalem S, Maniar M. In: *Biopolymers*, Eds. R. Langer and N. Peppas, Springer-Verlag, Heidelberg, 107, 1993, p. 93.
29. Lyman DJ. In: *Polymers in Medicine*, Eds. E. Chiellini, P. Giusti, Plenum Press, New York, 1983.
30. Leong K, Domb AJ, Ron E, Langer R. Polyanhydrides, *Encyclopedia of Biotechnology*, 2nd ed., 1991.
31. Domb AJ, Langer R. Polyanhydrides: I. Preparation of high molecular weight polymers. *J. Polym. Sci.* 25: 3373, 1987.
32. Tamargo RJ, Myseros JS, Epstein JI et al. Interstitial Chemotherapy of the 9L gliosarcoma: Controlled release polymers for drug delivery in the brain. *Cancer Res.* 5:329–333, 1993.
33. Tamargo RJ, Langer R, Brem H. Interstitial drug delivery to the central nervous system using controlled release polymers: Chemotherapy for brain tumors. In: *Providing Pharmacological Access to the Brain*, Ed. Flanagan TR et al., *Methods in Neurosciences*, Academic Press, Vol. 21, 1994, pp. 135–149.
34. Pardridge WM. *Ann Rev. Pharmacol. Toxicol.* 28:25–39, 1988.
35. Hochberg FH, Pruitt A. Assumptions in the radiotherapy of glioblastoma. *Neurology* 30:907–11, 1980.
36. Brem H, Piantadosi S, Burger PC et al. The safety of interstitial chemotherapy with BCNU-loaded polymer followd by radiation therapy in the treatment of newly diagnosed malignant gliomas: Phase I trial. *J. Neurooncol.* 26:111–23, 1995.
37. Valtonen S, Timonen U, Toivanen P et al. Interstitial chemotherapy with carmustine-loaded polymers for high-grade gliomas: A randomized double-blind study. *Neurosurgery* 41:44–49, 1997.
38. Fung LK, Ewend MG, Sills A et al. Pharmacokinetics of interstitial delivery of carmustine, 4-hydroperoxycyclophosphamide and Paclitaxel from a biodegradable polymer implant in the monkey brain. *Cancer Res.* 58:672–684, 1998.
39. Walter KA, Cahan MA, Gur A et al. Interstitial taxol delivered from a biodegradable polymer implant against experimental malignant glioma. *Cancer Res.* 54: 2207–2212, 1994.
40. Olivi A, Ewend MG, Utsuki T et al. Interstitial delivery of carboplatin via biodegradable polymers is effective against experimental glioma in the rat. *Cancer Chemother. Pharmacol.* 39:90–96, 1996.

41. Judy KD, Olivi A, Buahin KG et al. Effectiveness of controlled release of a cyclophosphamide derivative with polymers against rat gliomas. *J. Neurosurg.* 82:481–486, 1995.
42. Lillehei KO, Kong Q, Withrow SJ et al. Efficacy of intralesionally administered cisplatin-impregnated biodegradable polymer for the treatment of 9L gliosarcoma in the rat. *Neurosurgery* 39: 1191–1199, 1996.
43. Menei P, Boisdron-Celle M, Crouè A et al. Effect of stereotactic implantation of biodegradable 5-fluorouracil-loaded microspheres in healthy and C6 glioma-bearing rats. *Neurosurgery* 39:117–124, 1996.
44. Israel Z, Domb AJ. Polymers in gene therapy: Antisense delivery systems. *Polym. Adv. Technol.* 9: 1–7, 1998.
45. Kassell NF, Sasaki T, Colohan ART et al. Cerebral vasospasm following aneurysmal subarachnoid hemorrhage. *Stroke* 16: 562–72, 1985.
46. Kaku Y, Yonekawa Y, Tsukahara T et al. Superselective intra-arterial infusion of papaverine for the treatment of cerebral vasospasm after subarachnoid hemorrhage. *J. Neurosurg.* 77: 842–7, 1992.
47. Kassell NF, Helm G, Simmons N et al. Treatment of cerebral vasospasm with intra-arterial papaverine. *J. Neurosurg.* 77: 848–52, 1992.
48. Oda Y, Konishi T, Okumura T et al. Treatment of vasospasm after ruptured cerebral aneurysms by "pellet-cisternal-drain:PCD" containing sustained-release vasodilators [Japanese]. *No Shinkei geka* 16:965–970, 1988.
49. Oda Y, Okumura T, Morimoto M et al. Topical application of papaverine for cerebral vasospasm using prolonged-release pellet: Basic experiment [Japanese]. *No Shinkei geka* 11: 1027–1034, 1983.
50. Heffez DS, Leong KW. Sustained release of papaverine for the treatment of cerebral vasospasm: In vitro evaluation of release kinetics and biological activity. *J. Neurosurg.* 77: 783–787, 1992.
51. Shiokawa K, Kasuya H, Miyajima M et al. Prophylactic effect of papaverine prolonged-release pellets on cerebral vasospasm in dogs. *Neurosurgery* 42:109–116, 1998.
52. Byrd KE, Hamilton-Byrd EL. Polymeric delivery of neurotransmitters and neuromodulators. In: *Polymeric Site Specific Pharmacotherapy*, Ed. Domb AJ, Wiley, New York, 1994, pp. 141–155.

11

Tissue Adhesives for Growth Factor Drug Delivery

John Bowman
McWhorter School of Pharmacy, Samford University, Birmingham, Alabama

Tom Barker, Barbara Blum, Deepak Kilpadi, and Dale Feldman
University of Alabama at Birmingham, Birmingham, Alabama

I. TISSUE ADHESIVES

The utilization of tissue adhesives in surgery spans more than 50 years. Although fibrin glues or matrices (FM) have been used the most extensively and for the widest variety of applications, the cyanoacrylates (CA), gelatin-resorcinol-formaldehyde (GRF), mussel adhesive proteins (MAP) and others could be used for similar purposes if fully exploited. Tissue adhesives, particularly fibrin, have been used clinically in Europe for several decades for tissue welding, as hemostatic agents to control bleeding, or as fluid and gas barriers for fistulas, vascular grafts, and meningeal and pulmonary punctures. Fibrin was recently approved for clinical use in the United States but limited to surgical hemostasis.

In addition to the traditional uses as adhesives, sealants, and hemostatic agents, researchers have explored the use of tissue adhesives as coatings for implants, as regenerative scaffolds for wound healing, and as drug delivery systems. The various applications of tissue adhesives each have different design requirements for optimal effectiveness. Table 1 shows some of the characteristics that might be optimized differently for different applications. For example, when used for hemostasis and tissue closure of surgical and traumatic injuries, the adhesive should be moderately viscous so that once applied it remains in the field, yet readily spreads and conforms to the wound contours. It should set rapidly, not cause or worsen tissue inflammation, and degrade more quickly than the tissues heal. The adhesive might optimally be degraded and removed by advancing regenerative tissue.

This chapter will focus on the use of tissue adhesives as drug delivery systems and scaffolds for wound healing. For wound healing, the optimal tissue adhesive serves as a scaffold to facilitate the healing response. For example, bone repair might be accomplished using an adhesive that promotes osteointegration with the implant. For skin grafts, the adhesive can stimulate reattachment of the blood supply.

Tissue adhesives have properties that are of particular advantage in wound care. Unlike most wound dressings, they adhere to the wound and conform to its contours. They can adhere two pieces of tissue together with adequate mechanical strength throughout the healing process, but without the uneven stress distribution that results in localized tissue trauma and scarring

Table 1 Desirable Biomaterial Characteristics for Tissue Adhesives

	Biomaterial characteristic						
Application	Viscosity	Setting time	Adhesive strength	Time to degrade	Tissue irritation	Tissue ingrowth	Biofeedback-Controlled degradation
Arterial hemostasis	Moderate	Rapid	High	Rapid	Low	No	Maybe
Venous hemostasis	Moderate	Rapid	Medium	Rapid	Low	No	Maybe
Sealant (air, fluid)	Moderate	Rapid	High	Medium	Low	No	Maybe
Embolization	Low	—	—	Rapid	High	No	Maybe
Fistula occlusion	Low	—	—	Medium	High	No	Maybe
Joining tissues	Moderate	Rapid	Medium	Rapid	Low	Yes	Yes
Coating implants	Low	—	Low	Slow	Low	Maybe	Maybe
Wound healing	Low	—	Medium	Medium	Low	Yes	Yes
Drug delivery	—	—	—	Slow	Low	No	Maybe
Bone repair	High	Slow	High	Slow	Low	Yes	Yes
Skin grafts	Low	Slow	Medium	Rapid	Low	Yes	Yes

as well as the dehiscence and fistulas sometimes seen in sutured wounds. They also can be applied quicker and easier than sutures in most cases and can be used in additional applications such as attaching implants to the host tissue as well as for hemostasis or tissue sealing.

The ability to degrade at an appropriate rate is a critical feature for tissue adhesives. If the adhesive degrades too rapidly, presumably sufficient strength could be lost so that the adhesive fails and the tissues separate prematurely. If it degrades too slowly, it could impede the regeneration process. The ultimate design, therefore, would be a system where the regenerative process controls the degradation, which may be termed biofeedback-controlled degradation. This would allow the system to automatically take into account patient-to-patient variability in healing rate; to adjust the healing rate continually rather than just approximate the overall rate; and to be used in different tissues with different healing rates. Biofeedback-controlled degradation of an implant can occur in a number of ways. Consider the normal physiological process of wound healing: as the cells arrive to repair the tissue they break down the provisional matrix. Some of the natural tissue adhesives, such as fibrin, degrade in this manner as a result of normal enzymatic processes and phagocytosis, while synthetic tissue adhesives typically degrade due to hydrolysis. In systems without biofeedback control the degradation rate has to be controlled to match the healing rate as best as possible by changes in structure or chemistry.

Because a tissue adhesive is often applied to an injured area, it becomes an attractive vehicle for providing healing substances directly to the site of injury. For example, antibiotics can help reduce the risk of infection, and angiogenic agents can speed the wound healing process by stimulating capillary ingrowth. Alternatively, the adhesive may be intended for drug delivery to a localized area, such as providing chemotherapy to a tumor. In this case, the risk of giving a systemic drug is minimized because the medication is applied topically in a smaller dose to a particular area of the body. Furthermore, a tissue adhesive can be designed to release a medication over a period of time to obviate the need for repeated doses. If the release of medication is tied to the degradation rate of the tissue adhesive, it can be said to have biofeedback-controlled release of the therapeutic agent. Identifying appropriate therapeutic agents to incorporate into a biodegradable tissue adhesive depends on the application. For applications in which enhanced wound healing is desired, such as wound closure and skin grafting, growth factors and other biological response modifiers can be used. An angiogenic agent is used for enhancing wound healing, because new capillaries are required to supply nutrients and remove waste products for the cells involved in healing. For cancer treatment, the use of angiogenic inhibitors locally at tumor sites could retard cancerous growth or, if used during cancer surgery, might prevent tumor recurrence [1].

This chapter will discuss the use of tissue adhesives as drug delivery systems for biological response modifiers. Although these systems can and have been used for many applications, the focus will be on the delivery of growth factors, particularly angiogenic agents, for wound healing applications. While there are a number of different tissue adhesives, this chapter will emphasize the natural tissue adhesives that exhibit biofeedback degradation.

A. Fibrin

Fibrin was the first truly successful surgical adhesive in medicine [2]. Its ability to function as an adhesive was noted in 1940 in anastomosing nerves [3], and later in adhering skin grafts [4]. Fibrin adhesive has been used in diverse applications, including vascular surgery, neurosurgery, plastic surgery and skin grafts [5, 6]. Fibrin acts both as a hemostatic barrier, adhering to surrounding tissue and cells, and as a scaffold for migrating fibroblasts.

1. Functions of Fibrin

Fibrinogen, the main structural protein in the blood responsible for forming clots, is converted into fibrin monomer by thrombin [7]. The fibrin monomers assemble into fibrils, eventually forming fibers in a three-dimensional matrix or gel which prevents further loss of blood. Factor XIII is activated by thrombin in the presence of Ca^{2+}. The activated Factor XIII causes covalent bonds to form among the assembled fibrin monomers. This renders the fibrin gel less susceptible to proteolytic digestion by plasmin and increases the overall strength and stiffness of the gel. The fibrin gel adheres to a variety of adherents such as collagen and cell surface receptors, most notably integrins on platelets and other cells. Enzymatic and phagocytic pathways readily resorb fibrin during tissue regeneration.

Inflammatory and repair cells are attracted by fibrin. Fibrinopeptide A, formed during fibrinogen to fibrin conversion, acts as a chemotactic agent for polymorphonuclear leukocytes. Fibronectin is incorporated covalently into the gel structure to act as an attachment site for migrating fibroblasts. Fibrin degradation products, formed by proteolytic digestion of the fibrin network, stimulate the migration of monocytes, which convert to macrophages. These cells in turn phagocytize fibrin. Stimulated fibroblasts migrate through the fibrin gel structure and deposit type III collagen [8]. The infiltrating cells secrete plasminogen activators that further degrade the fibrin network. Plasmin digestion of fibrin is viewed as a continuous process of proteolytic attack on accessible, enzyme-susceptible sites, resulting in soluble degradation products that are removed from the injured site [9]. Neovascularization follows shortly thereafter. This composite of fibrin, macrophages, fibroblasts, collagen and vascular buds comprises the granulation tissue.

Fibroblast accumulation in a cutaneous wound requires phenotypic modulation of fibroblasts [10]. In response to injury, resident fibroblasts in the surrounding tissue proliferate for the first three days and then at day 4 migrate into the wounded site. Once within the wound, they produce type I procollagen as well as other matrix molecules and deposit these extracellular matrix molecules in the local milieu. By day 7, abundant extracellular matrix has accumulated and fibroblasts switch to a myofibroblast phenotype replete with actin bundles along the cytoplasmic face of the plasma membrane. Wound contraction occurs as these myofibroblasts gather in the wound extracellular matrix by extending pseudopodia, attaching to extracellular matrix molecules such as fibronectin and collagen, then retracting the pseudopodia. Once these processes have been accomplished, the fibroblasts appear to undergo apoptosis. Therefore, during cutaneous wound repair, fibroblasts appear to progress through four phenotypes: first proliferating, second migrating, third synthesizing extracellular matrix molecules, and fourth expressing thick actin bundles as myofibroblasts.

Recent evidence suggests that fibrin stimulates angiogenesis. The role of fibrin in the generation of new blood vessels was examined using a wound chamber model [11]. The results support the view that angiogenesis is the product of interactions between endothelial cells and a changing extracellular matrix. Endothelial cells have integrin receptors for fibrin. Furthermore, fibrin appears to be intimately involved in both hemostasis and angiogenesis during the initial phase of wound healing. Thus, fibrin occupies a central position and provides a vital link in the initiation of the cascade of events of wound healing.

2. Mechanical Properties of Fibrin

The coagulation process can be readily engineered into an adhesive system, usually by having the fibrinogen (monomer) and Factor XIII (crosslinker) as one component analogous to the "resin" of a two-part epoxy kit, and thrombin (initiator) in a calcium chloride solution as the second "catalyst" component. The components may be applied sequentially or simultaneously to the repair site by syringe or by spraying. Prior to polymerization, fibrin adhesive acts as a "sticky" liquid and readily adheres to wet surfaces. Once polymerized in situ it becomes a semirigid, hemostatic, fluid-tight adhesive mass, capable of holding tissue or materials in a desired configuration. Sequentially mixed fibrinogen/thrombin material is not evenly polymerized and is inhomogeneous [12]. In order to avoid excess formation of granulation tissue and slow absorption of the adhesive, only thin layers of the two components should be applied [13].

The three-dimensional structure of the fibrin gel may be altered by changing any one of several factors: fibrinogen, thrombin or calcium ion concentrations; ionic strength; pH; and, to a lesser extent, temperature. Varying the ionic strength or pH will produce "fine gels" or "coarse gels." The polymerization rate increases with increasing temperature, reaching a maximum at 37°C after which degradation (denaturation) occurs. The presence of fibronectin does not appear to affect pore size or gelation time, although it does increase fibril diameter [14]. Factor XIII has no effect on gel porosity [15], as it crosslinks adjacent protofibrils, fibrils and fibrillar structures which are already in intimate contact.

Shear loading of fibrin gels has been extensively studied under a wide range of conditions. "Fine" gels, regardless of Factor XIII crosslinking, have a lower shear compliance than "coarse" gels, although they undergo more permanent deformation, i.e., possess more viscous behavior. When compared to gels lacking fibronectin, coarse crosslinked gels demonstrate a large increase in modulus and a decrease in creep compliance, whereas fine gels exhibit a decrease in modulus and a corresponding increase in creep compliance [16]. A decrease in calcium ion concentration below a plateau concentration of 10 mM results in a decrease in shear modulus [17].

Adhesive strength in full thickness incisional wounds in rats is higher for fibrin-sealed wounds than for suture controls until the fourth postoperative day, when both groups are equivalent [18]. However, mechanical strength is delayed during the wound maturation phase. Adhesive strength increases with time from about 50 g/cm^2 at 1 min to 200 g/cm^2 at 120 min [19]. Fibrin adhesive with a fibrinogen concentration of approximately 39 g/L and a thrombin concentration of 200–600 units/mL with no added Factor XIII will result in wounds with significantly increased stress, energy absorption and elasticity values [20]. The breaking strength of 1 cm^2 grafts of human split thickness skin glued with fibrin for 3.5 min is greater than unglued grafts when pulled off with a dynamometer attached with suture [21]. No correlation between fibrinogen concentration and breaking strength was observed.

The tensile properties of fibrin have also been evaluated. Using a commercial pooled source fibrinogen or increasing thrombin or fibrinogen concentrations results in an increase in ultimate tensile strength and Young's modulus [12]. Thus, the volume ratios of fibrinogen and thrombin in solution affect mechanical and mixing properties. The tensile strength is greatest for samples prepared in a 1 : 1 v/v ratio. The mechanical properties increase over time, with no plateau observed up to 100 min after polymerization. At higher fibrinogen concentrations, no further increases in strength are observed [22]. The mechanical properties of fibrin and other tissue adhesives are noted in Table 2.

3. Clinical Uses of Fibrin

Fibrin matrices have found clinical application in cardiovascular surgery, orthopedic surgery, neurosurgery, ophthalmic surgery, general and trauma surgery, plastic surgery, middle ear procedures, and drug delivery. In cardiovascular surgery, they have been used as a hemostatic sealant for vascular graft attachment, cardiovascular patches, and heart valve attachment [23], to preclot porous vascular grafts [24,25], and to seal ventricular septal defects [26]. Woven

Table 2 Mechanical Properties of Tissue Adhesives

Tissue adhesive	Ultimate tensile strength (MPa)	Young's modulus (MPa)	Time to set (min)	Incisional strength (kPa)			Skin graft adhesion (kPa)	
				Post-op	4–7 days	14 days	Tensile	Shear
Methyl-2-CA	28–55	210–340	1	16	9	32		
Propyl-2-CA			1	30	13	30		
Isobutyl-2-CA	65		1					
Heptyl-2-CA			1	32	19	22		
Fibrin	0.1–0.2	0.15	1–120		8	63	20	6–30
Albumin (25)/PEG (15)			2		36	90		37
Albumin (25)/PEG (10)			2		7	28		
Albumin (30)/PEG (10)			2		13	27		
MAP			2–20					
GRF				62				
Sutures	0.7		NA	8	16	53		

Source: From Refs. 7, 264, 265, 269, 395, and 470.

polytetrafluoroethylene (PTFE, Dacron) vascular grafts tend to form a poorly adherent, obstructing layer of pseudointima. Knitted high-porosity grafts allow fibrous and vascular ingrowth and more secure anchoring of the pseudointima, and are also more flexible. However, at the time of insertion, the porosity must be sufficiently low to prevent excessive bleeding in the heparinized patient. Pretreatments with biological sealants are used to temporarily reduce the porosity of knitted Dacron at the time of implantation with subsequent resorption of the sealant. Fibrin seals porous vascular grafts better than whole blood to prevent leakage along the graft. Fibrin also seals Dacron vascular grafts better than regenerated oxidized cellulose [27]. Dacron vascular grafts coated with fibrin and growth factors improve endothelial cell attachment after presurgical seeding compared to untreated grafts [28]. Microvascular anastomoses can be "spot-welded" with fibrin adhesive [29,30]. In addition, fibrin has been used for obliteration of arteriovenous malformations [31,32] and repair of aortic dissections [33].

Tissue adhesives have been used for many applications in neurosurgery. Fibrin has been employed for peripheral nerve reattachment and dural sealing to prevent cerebrospinal fluid leakage with favorable efficacy and clinical utility [34–36]. Fibrin and cyanoacrylate adhesives both sealed experimental dural leaks successfully, but fibrin induced much less tissue inflammation [35]. Although controlled studies are few and the surgical outcomes usually have not been assessed rigorously, fibrin matrix appears to reduce or eliminate the need for sutures in nerve anastomosis, thus reducing the overall procedure time [37–40]. Fibrin matrix forms an adhesive "conduit" around the anastomotic site, preventing ingrowth of epineural tissue, giving mechanical stability, and maintaining apposition of the two ends. A major advantage of fibrin matrix is that once appositioning is achieved, joining with little or no displacement often occurs, a difficult result to achieve with microsuturing. The surgeon must relieve tension across the anastomosis to prevent disruption of the repair.

Fibrin adhesives have been used in ophthalmic procedures, such as retinal reattachment and sealing perforated corneas [41–44]. Fibrin adhesives have been extensively utilized in general surgery and traumatology. Fibrin acts as both a sealant and a hemostatic agent for the repair of traumatic ruptures of the spleen, kidney, and liver [45–50]. Tissue adhesives have been used to successfully repair esophageal perforations [51,52], pulmonary air leaks [53–63] and fistulas in animals and humans [55,63–75]. Packing and hemostasis have been obtained with fibrin matrices in many general surgical applications [6,50,76–91]. Fibrin matrix has also been used to prevent the formation of adhesions [92–103]. The efficacy of fibrin matrix in preventing peritoneal adhesions is somewhat surprising since fibrinogen is implicated in adhesion formation [104].

Fibrin has also been used in many plastic surgery procedures, including blepharoplasty [105–108], face lifts [109,110], and rhinoplasty when mixed with cartilage chips [111]. Fibrin matrices have been employed in a variety of craniofacial reconstruction procedures, including bone restructuring [112,113] and in periodontal tissue grafting [114,115].

Fibrin adhesives have been used extensively in otolaryngology and head and neck surgery [116–146]. As in other microsurgical procedures, the use of tissue adhesives reduces the time to perform the procedure, and improves the ease of securing tissue together. Recently, cartilage has been molded into shapes from chondrocytes in fibrin, offering promise for reconstructive surgery of severely damaged or missing body parts, such as an ear [147–151]. Other applications include obstetrics and gynecological surgery [101,102,152–155].

Skin grafting has been successfully accomplished using fibrin adhesive to adhere the graft, even in areas of complex anatomical contouring such as hands in burn patients [6,21,156–169]. Fibrin forms a stronger attachment than sutures when using autologous fibrin to adhere a skin graft [21]. Contraction of full thickness skin grafts is markedly reduced when a commercial fibrin product (Tissucol) is used instead of staples to attach grafts in a rodent model [159].

This is a result of the increased adherence of the graft to the recipient tissue bed. In general, fibrin matrix provides a close approximation of the graft to the recipient tissue bed, thereby preventing the formation of seromas [170–179], although seroma formation was not prevented in mastectomy cases [180]. Tisseel fibrin adhesive was evaluated in the graft take of cultured human epidermal sheets in an athymic mouse model [181]. On days 4, 10 and 21 postgrafting, human epithelium is present and a basement membrane has developed. Fibrin adhesive enhances the mechanical stability of these fragile grafts, increases the percentage of graft take, and does not harm the epidermal sheets. Keratinocytes cultured on sheets of fibrin matrix (Biocoll) have been used as a burn dressing in humans [182].

A study of diabetic mice noted histologically delayed wound healing with both fibrin adhesive and Opsite semipermeable dressing [183]. However, other investigators have reported positive effects on dermal wound healing. Skin incisional wounds of rats treated with fibrin sealant possess increased maximum tensile strength and relative failure energy after two days of healing [18].

For open porcine ureteric reanastomosis, fibrin glue produces effective anastomoses within less operating time than sutures [184]. Gelatin-resorcinol-formaldehyde (GRF) glue produces adhesion that is not sufficiently flexible to withstand rotation of the anastomosis. Fibrin-glued anastomoses withstand leak pressures equal to those from laser welding with albumin, but are superior in requiring a shorter operating time and in their electron and light microscopic appearances.

Exogenous ^{125}I-labeled fibrinogen is eliminated primarily via urine (>90% by 10 days) in patients undergoing skin grafting [185]. Adverse events with fibrin adhesive are rare, but some patients have experienced anaphylactic reactions and others have developed antibodies to coagulation factors [186–193].

4. *Sources of Fibrin*

Although just commercially introduced into the United States in 1998, fibrin adhesives have been available in Europe for many years. The chief obstacle to U.S. FDA approval has been the potential for viral contamination of pooled source fibrinogen. As a result, procedures for obtaining autologous or single-donor fibrinogen have been developed and utilized for various surgical procedures when its hemostatic and adhesive effects were needed [194]. Cryoprecipitation is the most commonly used method in the United States for autologous fibrinogen procurement today [195]. Citrated plasma is stored for at least 1 h at −20°C or less, and then slowly thawed between 2 and 8°C (usually overnight). The precipitate is isolated by centrifugation and the serum supernatant is decanted.

The clinical use of a commercial pooled-source fibrinogen product (Tisseel, Baxter Corporation) has recently been approved by the FDA, but only for hemostasis applications. In other clinical applications, therefore, fibrinogen must be extracted from the blood of the patient. This requires a lag time between blood extraction and availability of the fibrinogen for surgical use, and requires plasmapheresis of a patient whose health is already compromised. With the marketing of Tisseel, renewed interest in and wider use of fibrin tissue adhesives should occur. In addition, its ready availability will attract more research. A number of reports involving the use of Tisseel have been published since 1983 [41,44,91,119,126,127,146,156,169,181,196–226].

Tisseel is provided as a two-component kit of four separate vials: a sterile, nonpyrogenic, freeze-dried, vapor-heated powder preparation made from pooled human plasma containing fibrinogen and Factor XIII; a bovine source aprotinin solution for reconstituting the fibrinogen powder (aprotinin inhibits proteases such as plasmin that would otherwise degrade the fibrinogen during storage); a sterile, nonpyrogenic, freeze-dried, vapor-heated powder preparation of thrombin made from pooled human plasma; and calcium chloride solution to reconstitute the thrombin [13]. The fibrinogen solution contains 75–115 mg/mL of fibrinogen, 3000 KIU/mL of aprotinin and total protein content of 100–130 mg/mL. The thrombin solution contains 500 IU/mL of thrombin and 40 μmol/mL of calcium chloride. These solutions are combined at the time of application by extrusion through a dual-ported syringe supplied by the manufacturer. The FDA-approved uses for Tisseel are for use as an adjunct to hemostasis in surgeries involving cardiopulmonary bypass and treatment of splenic injuries due to blunt or penetrating trauma to the abdomen, when control of bleeding by conventional surgical techniques, including suture, ligature, and cautery, is ineffective or impractical. Tisseel is not indicated for the treatment of massive and brisk arterial bleeding. The manufacturer indicates that 1.0 mL of Tisseel will cover no more than 8 cm^2. A heater/stirrer device can be obtained from the manufacturer for preheating and dissolution of the two components, or a water bath can be used for preheating the components and allowing another 10–15 min for complete dissolution. After application, the adhered tissue should be immobilized for at least 3–5 min while polymerization occurs. It reaches its ultimate strength in about 2 h, and achieves 70% of its ultimate strength after 10 min.

5. Fibrin Composites

Composites combine two or more materials in order to improve mechanical or other properties. Many possibilities exist for composites in the area of tissue adhesives. Fibrin glues are not very strong, and lose more strength when made porous to promote vascular ingrowth. Calcium phosphate salts could be used to form a mineralized composite similar to bone. Extracellular matrix components such as the glucosaminoglycans hyaluronic acid and chondroitin sulfate could be added to provide viscoelastic properties to a composite.

a. Hyaluronic Acid. Hyaluronic acid is a major component of the extracellular matrix, and high concentrations are associated with scarless wound healing. However, it is rapidly resorbed by the body from the site of application. When mixed with fibrin adhesive, hyaluronic acid dissolves much more slowly [227]. Sodium hyaluronate increases the viscosity of fibrin adhesive and reduces the amount of fibrin that enters blood vessels following re-anastomosis, resulting in higher patency rates than fibrin alone [211].

b. Fibrin-Collagen. A fibrin-collagen composite adhesive has been developed to overcome some of the limitations of fibrin sealants in certain applications [228]. These limitations include low viscosity prior to polymerization, low cohesive strength after polymerization, and short-term persistence (<2 weeks) in the site of application. (Note that for wound healing, these qualities are desirable.) In addition, there can be considerable variability in the mechanical properties of fibrin adhesives prepared from autologous or single-donor sources. The fibrin-collagen composite is prepared by adding fibrillar type I collagen to fibrinogen–Factor XIII solutions. Formulations using between 0.15 and 1.25 mg/mL fibrinogen together with 15 mg/mL collagen produce useful adhesive systems. Compared with fibrin sealant, the composites have greater storage, loss and complex moduli and a smaller tan δ, indicating a high binding strength between the collagen fibers and fibrin. The ability to alter the viscosity of fibrin adhesive with collagen permits use of a low viscosity fluid for skin graft attachment, for example, and a high viscosity putty for bone repair.

Collagen has been combined with fibroblast growth factor and heparin complexed to fibrin [229]. In vitro, FGF bound to this matrix is rapidly, but partially, released in the presence of heparin. Subcutaneous implantations of collagen sponges impregnated with the FGF/heparin/fibrin mixture promote tissue ingrowth within the sponges. The resulting fibroblast-infiltrated tissue resembles a normal dense connective tissue.

c. Other Composites. Orthopedic surgical uses include filling defects with bone fragments or calcium phosphate ceramic implants, reattachment of osteochondral fragments, bone fragment repositioning in compound or digital fractures, and tendon repair [230,231]. When bone grafting is performed adjacent to the spinal cord, stability of the grafted bone is improved by mixing fibrin adhesive with fine fragments of autologous bone [231]. In tendon repair, suturing is necessary for adequate mechanical strength, but the application of a FM improves the quality of healing scar tissue in both full and partial tears. Meniscal tears have been successfully repaired with arthroscopic application of fibrin matrix and have remained clinically successful during six years of clinical follow-up [232–234]. A mixture of Triosite and fibrin adhesive has been successfully used to narrow the nasal fossae in patients with atrophic rhinitis [235].

In dentistry, fibrin matrices have been used along with hydroxyapatite to fill extracted tooth sockets and to augment the alveolar ridge [236,237]. This composite did not induce bone formation or provoke bone resorption when placed subperiosteally adjacent to mandibular bone in rabbits [208].

6. *Future of Fibrin*

Many studies have documented the successful use of fibrin matrices for diverse clinical applications. The commercial availability of fibrin in the United States for hemostasis may herald subsequent approval for other applications already in use around the world. For many of these applications the systems are being optimized based on factors such as those listed in Table 1. For example, enhancing mechanical properties would be beneficial for many applications and efforts include fibrin composites as well as other chemical modifications.

Optimizing the scaffolding ability of the fibrin matrix is also beneficial for many applications. Although fibrin is the normal provisional matrix, providing a scaffold for granulation tissue, the system can still be enhanced. This would include structural changes that make the system more porous to enahance neocapillary ingrowth as well as the inclusion of biological response modifiers. Although growth factors can and have been incorporated into fibrin and released over time, drug release is primarily diffusion controlled. To take better advantage of biofeedback-controlled degradation, improved methods of incorporation need to be devised to permit biofeedback-controlled release.

B. Albumin

Serum albumin is the largest protein component of human blood, and is an important factor in the regulation of plasma volume and tissue fluid balance through its contribution to the colloid oncotic pressure of plasma. It normally constitutes 50–60% of plasma proteins and has a molecular weight of 66,300–69,000. The melting temperature of albumin is 63°C, obtained from differential scanning calorimetry [238]. Pooled source human serum albumin is used extensively in the United States for plasma volume expansion and for maintaining cardiac output in the treatment of hypovolemic shock. The albumin molecule possesses several drug-binding regions that bind a number of medications, such as aspirin, warfarin, and phenytoin.

1. *Types of Crosslinked Systems*

Albumin can be crosslinked to polymerize the protein and alter its mechanical properties. Specifically, adhesive properties can be imparted. Physiological mechanisms will degrade and remove albumin from its site of application during the healing process. A number of clinicians have developed “spot-welding” techniques for tissue repair using albumin. In addition, albumin is used to coat implants.

Albumin used for drug delivery has primarily been in the form of microspheres. Glutaraldehyde has been used to crosslink albumin microbeads for delivery of a number of proteins and polymers, including progesterone [239], insulin [240], heparin [241], and poly-(amidoamine)-PEG and poly(thioetheramido acid)-PEG [242]. Albumin microspheres with dexamethasone have been prepared using a spray-drying technique for joint injection [243].

Albumin has been crosslinked by heating, by mixing with glutaraldehyde, and by exposure to formaldehyde vapors [244]. Turbidity is linearly related to the amount of heat, glutaraldehyde or formaldehyde used, and inversely related with drug release rate, implying that turbidity could be used as a measure of crosslinking.

There are disadvantages associated with crosslinking with glutaraldehyde, including an inflammatory response [245,246], and an inhibitory effect on cells, due to residual glutaraldehyde [247]. Fibrous capsules form around the albumin microbeads at one and two months, retarding the release of insulin [240]. Another concern is that of infection. The envelope of the hepatitis B virus, for example, exhibits increased binding activity with human serum albumin following crosslinking with glutaraldehyde [248].

2. *Device Coating*

Devices rapidly become coated with a protein film on implantation, and albumin constitutes much of the film. Following protein coating, platelets adhere to implants, and then cells attach to the implant. Thus, albumin adhesive has been used in place of blood or fibrin to coat Dacron vascular grafts [245,246,249,250]. The rate of in vivo degradation of albumin-coated polyester knitted vascular prostheses depends on the site of implantation [249]. A carbodiimide crosslinking agent is as effective as glutaraldehyde in producing a prosthetic coating of albumin that is impervious to blood, and evokes less inflammatory response from the rat [245,250].

This crosslinked albumin coating also prevents prosthetic infection in rabbits [251]. The effects on healing of three materials for sealing Dacron grafts were evaluated in a 56-day implantation in a canine descending thoracic aorta [252]. Autogenic preclotting of the graft with blood was found to be superior to heat-coagulated autogenic blood and human albumin crosslinked by glutaraldehyde.

Crosslinking renders albumin less degradable and more resistant to tissue ingrowth. Thirty albumin-impregnated polyester vascular grafts were retrieved from patients following removal for complications or a reoperation, or following autopsy [246]. Mean duration of implantation was 8.4 months, with a range of 1 h to 26 months. After up to two years of implantation, the grafts are poorly infiltrated by healing tissues. An external capsule forms after two months of implantation with a chronic inflammatory reaction occurring between the capsule and the polyester yarns. Internal encapsulation and endothelialization does not occur. Large amounts of albumin sealant remain after two months, but it gradually degrades over time, and traces are still observed after two years implantation. Overall, healing was poor and the degradation rate of the albumin was delayed when compared to animal models.

Albumin has been used to coat titanium implants, both uncrosslinked [253,254] or crosslinked [251,255,256]. Coating titanium surfaces with carbodiimide-crosslinked albumin prevents adhesion of both *Staphylococcus aureus* and *Staphylococcus epidermis* [255]. Little degradation of the coating occurs up to 20 days and bacterial adherence inhibition is sustained at 85% [251]. Coating hydroxyapatite beads or human teeth surfaces with albumin has also been observed to significantly reduce adherence of oral bacteria [257].

Albumin has also been crosslinked using radiation [258]. Radiation requires special facilities, in addition to requiring exposure patients and workers to radiation.

3. *PEG Crosslinked Systems*

Hydrogels have been formed by crosslinking albumin with activated PEG (poly(ethylene glycol)) [259]. These hydrogels have been studied for diffusion delivery of drugs including acetominophen, theophylline, hydrocortisone, and gentamicin [260,261], and enzymes such as alkaline phosphatase and lysozyme [259,260].

When crosslinked with some activated PEG materials, albumin acquires strong adhesive properties [262–264]. When crosslinked with poly(ethylene glycol) disuccinimidylsuccinate, the peel force (a measure of adhesive strength) of albumin was similar to that of cyanoacrylate glue, and much greater than that of fibrin glue [262,263].

In a preliminary study of the mechanical shear strength of two adhesive albumin systems, high shear strengths superior to fibrin were achieved, between both skin-skin and metal-metal interfaces [264]. The higher shear strength, obtained with a 12.5% albumin–10% modified-PEG system, was approximately five times stronger than that of the fibrin adhesive. Shear strength for the 12.5% albumin–2.5% mod-PEG system was approximately 1.5 times that of the fibrin system.

Incisional wounds in rabbits have been used to compare the mechanical strength of crosslinked albumin with fibrin adhesive and sutures [265]. Ultimate tensile strength is highest in the specimens treated with crosslinked albumin (Table 2).

4. *Future of Albumin*

Because it appears that the degradation rate in vivo of crosslinked albumin can be controlled by regulating the extent of crosslinking, albumin adhesives could be designed for a number of tissue adhesive applications, including wound healing and drug delivery. If albumin can be degraded and replaced by regenerating tissue in an isomorphic manner without compromising its mechanical properties, it could be an ideal scaffold for wound healing. In addition, albumin adhesive could serve as a controlled drug delivery system, where medication or a biological response modifier is released from albumin-binding sites as the matrix dissolves, and thereby the medication is slowly released directly into the area of application. Albumin has demonstrated other potential advantages, including resistance to bacterial adhesion, an affinity for binding to certain surfaces, and biocompatibility.

Further work is needed to characterize an optimal crosslinking agent and procedure for modifying albumin so that adhesive properties are achieved and in vivo degradation rate can be optimized for the particular application. As with other adhesives, it appears that the greater the extent of crosslinking, the greater the mechanical properties are, and the less degradable the material is. Work in progress in our laboratories suggests that albumin adhesive is stronger than fibrin glue and does not interfere with wound healing. Because it is readily available in a sterile form and relatively inexpensive, albumin is a good choice for use as a tissue adhesive or drug delivery system for topical (external or internal) application.

C. Cyanoacrylates

The cyanoacrylates (CAs) were first synthesized in 1949 [266], and their adhesive properties were discovered a decade later [267]. The methyl-2-cyanoacrylates were the first systems used as surgical adhesives. Problems with wettability [268,269] and histotoxicity [270–275] led to the development of longer-chain homopolymers such as ethyl-2-cyanoacrylate (Krazy Glue), isobutyl-2 cyanoacrylate (Bucrylate) and butyl-2-cyanoacrylate (Hexacryl). Larger side chains increase the hardness and tensile strength of the adhesive, while flexibility and degradation rate decrease [276,277]. It was subsequently found that the isobutyl and butyl longer-chain derivatives were less toxic but still had good adhesive strength [278]. During the 1970s, reports on the histotoxicity of isobutyl-2-cyanoacrylate and others began to be published [279–288]. Since then, Hexacryl has been the system of choice, although it too has shown histotoxicity when placed subcutaneously [289]. Hexacryl is as effective an adhesive as Krazy Glue, more degradable, and produces less inflammation [290]. Nonetheless, some Hexacryl remains in tissues after one year. Systemic toxicity can occur with cyanoacrylate adhesives, including peripheral neuropathy [291]. Cyanoacrylates have a thrombotic effect which is normally undesirable, but useful for embolizing arteriovenous malformations [292,293].

Reports of histotoxicity as well as reports of carcinogenicity (methyl-2-cyanoacrylate) hindered FDA approval of cyanoacrylates for medical applications in the United States, although clinicians in Canada and Europe have used these products for more than 10 years. The methyl cyanoacrylates proved to be unsuitable for medical applications due to severe tissue necrosis [268,269], the lack of tissue wettability, and sarcoma formation (in rats) [285,294]. In addition, the local temperature increase resulting from curing is approximately 4°C [295]. Histotoxicity is primarily due to the release of relatively large amounts of formaldehyde as it is degraded.

Perhaps less toxic than other classes of cyanoacrylates, different formulations of n-butyl cyanoacrylates (Nexacryl, Avacryl, Occyldent, and TPS) have been developed for various applications: a tissue adhesive, a tissue sealant for periodontal applications, and a drug delivery system. The *n*-butyl cyanoacrylates, primarily Hexacryl, have been used in ophthalmological problems such as corneal perforation [296] and for embolizing genetic arteriovenous malformations [297–300]. There is an initial mild inflammatory response that resolves within a few weeks [289,301]. After degradation (about one year) only fibrous tissue remains. Local temperature rise is only about 1.5°C [295]. A total of 187 patients who underwent lung resection for cancer had *n*-butyl-2-cyanoacrylate glue applied to reinforce staples or sutures, and none had adverse effects or failures ascribed to the glue [302].

In general, the *n*-butyl cyanoacrylates show promise as a tissue adhesive because of their ease of application, reliability, and good bond strength. They provide a stronger attachment than sutures and are less expensive and easier to apply. There are concerns, however, when applied to well-vascularized tissue, as the cyanoacrylates are thrombogenic. In addition, granulation tissue cannot grow through the material, so true regenerative healing is hampered, unless the adhesive is made porous or only used to spot weld [126,290]. Spot welding causes a reduction in incision strength compared to sutures at 14 days even though cyanoacrylates have higher initial bond strength. However, spotting the adhesive is necessary to avoid patient discomfort and thermal injury from the exothermic polymerization reaction.

Isocyanates have been used as tissue adhesives, with the most successful formulation an isocyanate-terminated polyester. When mixed with pyridine, it rapidly bonds to tissue. However, moisture greatly weakens adhesive strength and thus limits its utility. Rubber lattices such as natural rubber/isoprene composites have also been used. Although good bond strengths have been achieved (400 g/cm^2) the time required for setting in vivo is relatively long. In addition, such composites do not degrade. Polyacrylates have shown promise because of the potential reactivity with acid chloride groups present in tissue. Polyacrylchloride does not bond well, while polyacrylamide and glyoxal bond rapidly, but have limited strength.

Cyanoacrylates have also been used in dental applications, for restoration of broken teeth, as a tooth sealant, and as a periodontal dressing [303].

Poly-*N*-alkyl cyanoacrylates are biodegradable, and much research has focused on using nanoparticles of this material as drug delivery systems [304–380]. Most studies involved cancer chemotherapeutic agents, but insulin, antibiotics, ocular medications, and gene therapy have also been investigated with poly-*N*-alkyl cyanoacrylate nanoparticle carrier systems.

A number of obstacles remain before cyanoacrylates achieve more widespread clinical use. Because of the exothermic polymerization process that cyanoacrylate adhesives undergo, adhesives must be carefully designed and applied to avoid thermal injury to tissues. An additional concern is that many therapeutic proteins and medications denature or degrade at higher temperatures. Although materials can be added to the cyanoacrylate to serve as heat sinks, this can adversely affect desired mechanical properties and the rate of bonding. It is also significant that healing tissue cannot grow through some cyanoacrylate adhesives.

In summary, cyanoacrylates possess great adhesive strength but can cause adverse tissue reactions. Degradation products have not been fully characterized with regard to potential toxicity. Cyanoacrylates remain in tissue for 12 months or longer. They form a barrier to vascular ingrowth in wounds and can be used to embolize arteries. Because of the heat of polymerization, however, added growth factors are likely to be inactivated.

D. Mucopolysaccharides

Because tissue adhesives must be cured in moist environments, researchers have looked at a substance produced by marine invertebrates. The blue mussel *Mytilus edulis* synthesizes a DOPA-rich polyphenolic protein, in a structure called the byssus, which mediates attachment along with other proteins found in the byssus through collagen threads or tethers [381–383]. It was found that this protein, designated mussel adhesive protein (MAP), serves to attach the mussel to rocks or other surfaces in turbulent tidal zones. This material cures on contact with water and has been patented.

After the constituents of MAP were identified, synthetic polymers were manufactured that were branch copolymers with a linear backbone onto which were grafted variants of the natural decapeptide [384]. These compounds have also been patented, but development has been hampered by a lack of funding.

Mussel adhesive protein appears to have affinity for both soluble and insoluble collagen. In vivo studies indicate that the MAP provides adequate strength within 1 min and increasing strength over several hours [385]. Although minimal toxicity has been reported in numerous cases [386,387], there has been a report of marked inflammation upon intraocular injection of undiluted MAP [388]. MAP with an enzyme polymerizer is less irritating than butyl-2-cyanoacrylate after instillation in the corneal stroma of rabbits [389].

As an artificial basement membrane, MAP permits diffusion of inulin and dextran, two large nonelectrolyte molecules [390]. This property may also permit its use as a wound dressing.

Because of its adhesiveness for cells MAP has been used for fixation of osteoblasts and chondrocytes [391]. Two commercial adhesive preparations—fibrin glue and MAP—were tested in vivo for their ability to fix an internal chondrocyte allograft. While results for the fibrin, including additional testing for chondroinductive/conductive properties, were at best inconclusive, the results for MAP were considered highly promising [392]. This natural adhesive may find clinical applications where permanent adhesion is required, such as implant fixation in hard tissue. In addition, if MAP's native strength under high shear conditions can be reproduced clinically, it may have applications in vascular procedures. Its properties under physiological conditions should be investigated more thoroughly, since its outstanding adhesion and durability in marine environments may be related to these conditions.

E. Other Adhesives

A *gelatin-resorcinol-formaldehyde* (GRF) system was developed as a tissue adhesive that would quickly set up and maintain its properties in a moist environment [393–395]. Gelatin is denatured collagen, the most common structural protein in the body. Formaldehyde increases the cohesive strength and decreases solubility by crosslinking the gelatin. Resorcinol reacts with the formaldehyde to further stabilize the system and also reduces the overall viscosity. By reducing the viscosity of the gelatin solution, the concentration of gelatin can be increased without adversely affecting handling characteristics.

Initially, formaldehyde was directly applied to the tissue prior to gelatin-resorcinol addition. Obvious cytotoxicity concerns led to the incorporation of the formaldehyde to create a GRF solution [396]. The product is not presently FDA approved and thus has had little clinical use in the United States. It is unlikely that the FDA will approve the use of formaldehyde for human use without substantial toxicity testing. Other less toxic means of crosslinking the gelatin need to be discovered.

Gelatin-resorcinol-formaldehyde has been used in animal studies [393,396,397]. The GRFG glue provides complete pneumostasis immediately after sealing of air leaks in a thermally injured rat lung in the presence of positive pressure ventilation, unlike fibrin glue or cyanoacrylate adhesive [398].

When aortic specimens from sheep are glued with warm GRF/glutaraldehyde under wet and dry conditions and submitted to defined degrees of compression, better adhesion is obtained under dry conditions [399]. At higher compression (20 N), GRF/glutaraldehyde tensile strength is significantly higher than cyanoacrylate gel or fibrin glue. Fibrin glue demonstrated only weak adhesive properties even under dry conditions.

Cryoprecipitate fibrin glue and a two-component fibrin glue are equally effective in controlling bleeding in dogs whose aortas and atria are incised and sutured, and whose ventricles are abraded until bleeding occurs, but gelatin-resorcinol-formaldehyde-glutaraldehyde (GRFG) glue is not as effective [400]. The two-component fibrin sealant is completely resorbed by 10 days. On histological examination, both fibrin glues cause minimal tissue reaction; in contrast, extensive fibroblastic proliferation is caused by the GRFG glue. The two-component and GRFG glues have suitable adhesive strength, while the cryoprecipitate glue does not. The GRFG glue has a significantly greater tensile strength than the two-component fibrin sealant.

Only a few clinical studies of GRF have been reported. In a series of 171 patients treated with GRF for acute aortic dissection, the authors concluded GRF was effective for emergency adjunctive use on aortic stumps prior to Dacron vascular repair [401,402]. Others have reported similar results for acute aortic dissection [403], although heart block has been reported as a potential complication of this use [404]. GRF may also be an effective sealant for lung air leaks [398]. It has been used to seal air leaks following excision of emphysematous lungs [405].

Similar to albumin, concerns of cytotoxicity of formaldehyde and glutaraldehyde must be overcome to bring GRF to clinical trials. If a less toxic crosslinker can be found or if residual unreacted aldehydes can be kept below a nontoxic level, this composite has promise because of its improved strength and other mechanical properties compared with fibrin glues. For use as a drug delivery system, the GRF mixture must be shown to be nonreactive with the drug.

To improve biocompatibility, investigators used two less toxic aldehydes, pentanedial and ethanedial, to replace the formaldehyde component from gelatin-resorcinol-formaldehyde glue [406]. To evaluate the adhesive strength of this new glue, lung incisions in rabbits were glued together. Each group ($n = 5$) was examined histologically after two days and one, two, and four weeks. The glue disintegrated gradually with good bioresorption when the incision was closed with a thin layer of glue. The healing process was favorable.

Anhydrides such as copolymers of maleic anhydride and methyl vinyl ether can be crosslinked with polyols, polyfunctional amines, and calcium salts [396]. These materials, however, have low cohesive strength. The dicarboxyanhydride of tartaric acid, which polymerizes with a small amount of base, does not bond well to tissue. Epoxy resins have been examined because of the reactivity of the epoxide group to amines, alcohols, and carboxylic acids. Formaldehyde condensation products of urea, phenol, melamine, casein, and guanidine have also been evaluated. These systems also have low cohesive strength. Vinyl containing polymers such as acrylated gelatin and acrylated polyvinylamine, which are crosslinked by free-radical catalysis, are slow to set up. Other systems such as crosslinked vinyl alcohol, polyvinylamine, polyvinylpyrrolidone, polyethyleneimine, nylon systems, and phosphonyl chloride-terminated polyols have also have been tried with little success.

Prolamine, a biodegradable protein based viscous gel, is another material that has been tried. It is an amino acid alcohol that polymerizes in the presence of water. It has been used clinically in Europe but is not presently approved in the United States [407].

II. GROWTH FACTORS

Normal wound healing requires the coordination of a number of cellular activities, including phagocytosis, chemotaxis, mitogenesis, collagen synthesis, and the synthesis of other extracellular matrix components [408]. A number of growth factors modulate these activities, including platelet-derived growth factor (PDGF), fibroblast growth factor (FGF), transforming growth factor-β (TGF-β), epidermal growth factor (EGF), interleukin-1 and -2 (IL-1 and IL-2), and tumor necrosis factor-α (TNF-α). Despite the names of these growth factors, most are multifunctional and stimulate more than one cell type. Most growth factors bind to multidomain transmembrane cell surface receptors, which may require dimerization and/or cofactors for activation, and these receptors in turn activate a cascade of intracellular kinases and phosphatases that stimulate or inhibit specific genes to alter cell function. Some have advocated a "cocktail" of these growth factors as an aid to wound healing, but it is likely that each factor's production and the expression of receptors for them is carefully regulated locally. Different portions of a wound may have different local conditions and requirements for growth factors, so it is difficult to predict how a given factor might influence the healing process when applied to all parts of the wound simultaneously.

Certain growth factors may serve to initiate the sequence of events in wound healing. Because platelets release PDGF, EGF and TGF-β, these factors are present in a wound shortly after injury. Macrophages arrive soon thereafter, and produce PDGF, FGF, TGF-α, TNF-α and TGF-β, so these growth factors also play a role in the early stages of wound healing. For example, they stimulate chemotaxis of keratinocytes [409]. FGF, insulin-like growth factor (IGF-1), and EGF stimulate DNA synthesis and keratinocyte outgrowth in human skin explants, with EGF having the greatest effect on epithelial growth [410]. The addition of one or more of these early phase growth factors to a tissue adhesive might stimulate more rapid healing. Early studies were performed on humans using an extract of autologous platelets in a collagen vehicle [411]. When this extract was applied topically on a daily basis to patients with nonhealing wounds, healing occurred in a mean time of about 10 weeks [412].

A. Fibroblast Growth Factor

Fibroblast growth factor is a mitogen for mesenchymal cells and stimulates angiogenesis and other wound-healing functions. There are a number of FGFs in this family, in which acidic FGF (FGF-1) and basic FGF (FGF-2) are the best characterized. Both FGF-1 and FGF-2 are potent angiogenic factors. In addition, they stimulate collagen synthesis, wound contraction, epithelialization, and fibronectin and proteoglycan synthesis. FGF-2 applied topically as a single dose after creating a rabbit dermal ulcer results in a twofold increase in reepithelialization

[413]. It also accelerates epithelialization of partial-thickness wounds in pigs [414,415]. FGF-2 injected into subcutaneous sponges in rats causes greater cellularity and collagen content [415]. Wound-breaking strength is greater in guinea pigs that are treated with FGF-2 in a collagen sponge than controls when dermal wounds are created [416]. Similarly, infiltration of an incisional wound in rats three days after injury with recombinant FGF-2 (400 ng) accelerates repair as measured by an increase in the tensile strength and wound-breaking energy [417]. In normal rats with full-thickness excisional wounds, FGF-1 significantly increases the rate of wound closure, and histological examination shows more granulation tissue and neovascularization, but with no affect on the rate of wound contraction [418]. FGF-2 applied in a single dose to rabbit ear dermal ulcer model accelerates healing by 30% and results in twice the volume of vascular granulation tissue as controls [419]. Topical FGF-2 prevents the impairments in wound contraction in rats caused by infection [420] or diabetes [421–423].

Sealed wound chambers have been created in porcine partial thickness excisional wounds for delivery of nanogram doses of FGF-2, PDGF-AB, insulin-like growth factor (IGF)-1, epidermal growth factor (EGF), and cholera toxin [424]. PDGF and EGF accelerate healing (by 1.1 days and 0.3 days, respectively), whereas FGF-2 and IGF-1 have no effect. Cholera toxin retards healing. EGF in the concentration range of 10 to 1000 ng/ml has the same stimulating effect on healing, but EGF at 10,000 ng/mL significantly delays healing.

The efficacy of topical acidic fibroblast growth factor (FGF-1) in healing of full-thickness wounds has been studied in a diabetic db+/db+ mouse model [425]. A viscous formulation of FGF-1 is as effective in wound healing as a nonviscous formulation. A formulation containing heparin (necessary for full biological and conformational stability of FGF-1) at a mass ratio of 3 : 1 to aFGF-1 is more efficacious than formulations with lower heparin: FGF-1 ratios. Wounds treated with three doses of 3.0 $\mu g/cm^2$ FGF-1 heal faster than those treated with a single dose of 3.0 $\mu g/cm^2$ FGF-1. Three applications of 3.0 or 0.6 $\mu g/cm^2$ FGF-1 are equally effective in accelerating wound healing.

These and other wound models demonstrate the efficacy of fibroblast growth factors on accelerating healing, by stimulating angiogenesis, granulation, contraction and reepithelialization. Overall, however, clinical trial results of topical growth factors have not been as successful as animal models of wound healing. FGF-2 was applied in a liquid vehicle in a randomized, prospective, double-blinded multicenter study of hospitalized patients with pressure ulcers [426]. When all subgroups of FGF-2 patients treated with different doses were combined, the change in wound volume was statistically significantly greater than controls. More FGF-2 patients achieved 70% or greater healing than control patients did. The treated wounds also contained more fibroblasts and capillaries when examined histologically.

Other potential uses of FGF include cardiovascular surgery. Intravenous administration of FGF-1 results in dose-dependent inhibition of intimal thickening with parallel promotion of endothelial regeneration in injured rat carotid arteries [427]. The prospect of causing new coronary arteries to develop in patients with atherosclerotic heart disease has great clinical implications because of the high death rate associated with myocardial infarctions.

In summary, FGF-1 and FGF-2 are potent angiogenic substances that have been successful in vivo in accelerating wound healing through neovascularization. Appropriate delivery vehicles should be developed to protect the growth factor from enzymatic breakdown and to deliver it at a controlled rate to enhance wound healing and promote tissue regeneration.

B. Other Growth Factors

1. *Platelet-Derived Growth Factor*

Platelet-derived growth factor is chemotactic for neutrophils and monocytes. PDGF stimulates both chemotaxis and mitogenesis in fibroblasts and smooth muscle cells. It also stimulates synthesis of collagen, fibronectin and hyaluronan. When subcutaneous sponges in rats are injected daily with PDGF-BB, they are more cellular and have a greater collagen content than uninjected controls [428]. Ten days after implantation, subcutaneous PTFE chambers containing PDGF have a greater DNA content than control chambers [429]. A single topical application of PDGF-BB to excisional wounds in rats significantly accelerates wound closure [430]. PDGF-BB in a collagen vehicle increases wound breaking strength when placed into incisional wounds in rats [431]. Incisional wounds in irradiated rats have greater wound-breaking strength when supplemented with PDGF-BB/collagen at the time of wounding [432]. Excisional wounds in diabetic mice heal more quickly when PDGF is applied [421].

Platelet-derived growth factor-BB has been applied topically in a liquid vehicle to chronic pressure ulcer patients in a randomized, prospective double-blind trial [426,433]. Wounds that were treated with a dose of 100 μg/mL experience a greater reduction in size than control wounds, although the difference is not statistically significant.

2. *Transforming Growth Factor-β*

Transforming growth factor-β is released by platelets by degranulation at the site of injury. TGF-β regulates many aspects of cellular function, including proliferation, differentiation and extracellular matrix metabolism. It belongs to a family that includes the bone morphogenetic proteins (BMPs). It stimulates monocytes to release other growth factors, including PDGF, FGF, TNF-α, and IL-1. It is chemotactic for macrophages and fibroblasts and stimulates fibroblast proliferation. TGF-β stimulates collagen synthesis and angiogenesis when injected subcutaneously into the necks of newborn mice [434]. However, it does not stimulate epithelialization in a rabbit ear model [413]. TGF-β-containing chambers planted subcutaneously in rats become more cellular and contain more collagen than control chambers [435]. Incisional wounds in guinea pigs that are left open and treated with TGF-β-containing sponges develop more granulation tissue than control wounds [436]. Breaking strength of healing wounds is increased by TGF-β when placed into rat incisional wounds in a collagen vehicle [437] or when placed into incisional wounds that are irradiated [438]. When applied in a collagen vehicle, TGF-β reverses steroid-induced healing deficits in rat incisional wounds [439]. A single dose of human recombinant TGF-β1 in a methylcellulose gel to skull defects in rabbits induces a dose-dependent increase in intramembranous bone formation [440].

Based on these and other observations, TGF-β may have clinical applications in soft and hard tissue healing. Due to its pleotropic nature, TGF-β could enhance or inhibit these processes, depending on the concentration or the timing of its presence at the desired site of action. High doses of TGF-β given intravenously for two weeks to rats results in significant fibrosis of the liver and kidneys and at the site of injection [441].

3. *Epidermal Growth Factor*

Epidermal growth factor is released by platelets when they degranulate. It is chemotactic for epithelial cells and fibroblasts, and is a potent mitogen for these cells as well as endothelial cells. It also stimulates angiogenesis and collagenase activity. In animal models of wound healing, EGF increases rates of epithelialization, collagen deposition, breaking strength of intestinal wounds, and reverses steroid-induced healing impairment. Increased rates of epithelialization have been observed in a rabbit ear wound model [442] and in split-thickness dermal wounds and partial-thickness burns [443] when EGF was applied topically every day. EGF-supplemented wound chambers in rats accumulate more collagen and are more cellular than untreated chambers [444]. Similar results occur when EGF-containing sponges are implanted subcutaneously in rats [445,446]. EGF accelerates collagen accumulation in a PTFE model in diabetic rats [447] and reverses the healing impairment produced by a systemic corticosteroid in rats [444].

Epidermal growth factor has been applied to skin graft donor sites using a Silvadene vehicle in a double-blinded clinical study [448]. The time for complete reepithelialization was reduced by a mean of 1.5 days in this study. In another study, nine patients with nonhealing ulcers were treated initially with Silvadene, then crossed over to EGF in a Silvadene vehicle [449]. Healing was achieved in eight of the nine patients in a mean time of 34 days. Application of topical EGF in a liquid vehicle results in greater healing of venous stasis ulcers compared with placebo [450]. Thus, EGF may have application in promoting reepithelialization, and may be most useful in repairing second degree burns and split-thickness skin graft donor sites.

4. *Interleukin-1*

Interleukin-1 is produced by macrophages and is chemotactic for epithelial cells, neutrophils, monocytes and lymphocytes. It stimulates fibroblast proliferation, collagen synthesis, and collagenase and hyaluronidase activity. Clinical studies in wound healing have not been performed, but because of its favorable actions on repair cells, it should be investigated.

5. *Tumor Necrosis Factor-α*

Tumor necrosis factor-α is produced by macrophages and is cytostatic or cytocidal for both tumor and normal cells. It is mitogenic for fibroblasts and stimulates collagen and collagenase synthesis. TNF-α has been implicated in the pathogenesis of autoimmune diseases such as rheumatoid arthritis. The implications of these actions for topical use in wound healing are unclear.

6. *Use of Growth Factors Clinically*

Although human clinical trials of growth factors have not been as successful as animal studies in wound healing, these factors have great potential. The mixed results for topical growth factors is most likely due to release of unprotected growth factor in a wound environment filled with proteolytic enzymes, as well as the lack of a scaffold to guide the tissue growth stimulated by the growth factor. Also in many cases, the optimal dose of growth factor has not been identified, nor whether single doses, multiple doses, or continuous application is best. In many studies, no evidence was presented to substantiate that the growth factor was biologically active at the time of use. To be clinically useful, the growth factors should be incorporated in a vehicle appropriate for the wound, and the effects of vehicles on growth factor performance have not been extensively investigated. Further, without a scaffold system the proliferation, chemotaxis, or angiogenesis stimulated by the growth factor will not be fully realized.

III. TISSUE ADHESIVES AS DRUG DELIVERY SYSTEMS

One of the most promising aspects of tissue adhesives is their potential as scaffolds and drug delivery systems. To enhance tissue regeneration, the nonmechanical characteristics of the adhesive system also need to be optimized. The addition of biological response modifiers such as cytokines, growth factors, or receptors can enhance tissue regeneration. Angiogenic substances are particularly promising additives for tissue adhesives, because wound healing is dependent on tissue perfusion or blood flow. Repair cells such as fibroblasts require oxygen for many functions, including proliferation, migration, attachment, and protein production. The diffusion of oxygen from vessels both causes an oxygen gradient as well as gives limits on the distance away cells can be from a blood supply and function or survive. Therefore, the ability of fibroblasts to migrate into a provisional repair matrix depends on the vascularity of the tissue bed and the formation of new capillaries in the provisional regenerative matrix. The blood vessels, however, need a scaffold to grow into. This can be the provisional matrix or the collagen laid down by the fibroblasts. The fibroblasts and the collagen framework they produce, therefore, grows in just ahead of the blood vessels. The oxygen gradient provides high oxygen to the fibroblasts at the wound edge but also creates a low oxygen region, in the wound center, to stimulate macrophages to release cytokines that stimulate chemotaxis, mitosis and other steps in the healing process.

In addition, substances such as growth factors or lactate, which are chemotactic for repair cells including macrophages and fibroblasts, could be incorporated within the implant to maintain the concentration gradient necessary to stimulate chemotactic migration into the implant. The expected result is more rapid tissue regeneration. Extracellular matrix components could also be investigated in tissue adhesives for their affect on wound healing.

For wound healing implants, fibrin seems to be a good choice for many applications because of its hemostatic and angiogenic effects. The addition of growth factors or other substances offers the potential to enhance healing even further. For substances that are able to diffuse out of the matrix, simple diffusion models can estimate the release rate of the medication. In other cases, such as growth factors or other therapeutic peptides, binding to fibrin may occur, and release of the growth factor would then be tied to the fibrin matrix degradation. Because tissue enzymes control fibrin matrix degradation, the release of bound drugs is biofeedback controlled. That is, drugs are released no faster than matrix degradation permits.

Albumin matrices are also promising vehicles for drug delivery in wound healing applications. Physiological processes degrade albumin, so, like fibrin, it can function as a provisional matrix. Continuing research suggests that its mechanical properties may be superior to fibrin. Because albumin contains drug-binding regions, it can also serve as a controlled-release system for therapeutic agents. Local drug concentrations would be maintained according to the binding affinity and extent of binding of the drug to albumin.

A. Drug Delivery Design

Drug delivery involves providing a sufficient amount of biologically active drug to a group of target cells to produce an intended effect for a period of time without unwanted effects. Besides tablets, injections and sprays, drugs can be administered by a variety of delivery systems. All surgical procedures result in wounds that need to heal. Enhancing surgical wound healing with therapeutic agents is a desirable goal that has substantial implications for reducing health care costs.

Besides wound-healing applications, tissue adhesives could be employed as drug delivery systems for providing medications to specific areas of the body and localizing the release of the therapeutic agent at the intended site. As a result, smaller, less toxic doses can be used than would be required of systemically administered medications, and adverse effects would largely be contained to the area of application. For example, cancer chemotherapeutic agents in a tissue adhesive could be applied in the area of a surgical tumor resection to destroy any remaining cancerous cells.

Drug delivery systems are designed to provide a predetermined dose of medication over a period of time. Systemically active medications are transferred to the vasculature, usually through absorption; they are then distributed throughout the body, metabolized and eliminated.

For such medications, bioavailability studies are performed on healthy human volunteers to determine clinical parameters related to the time course of drug concentrations in the blood after administration. These parameters allow characterization of blood levels achieved following a given dose, rate of elimination, apparent volume of distribution and half-life. Blood levels of topically applied medications should also be obtained, since some absorption and systemic distribution might be expected. The rate of administration of a medication should normally not exceed its rate of elimination, as toxic accumulation may occur.

Thus, drug delivery systems used in wounds should be characterized in regard to the rate of release of medication. This is a difficult task, as in vitro cell culture systems and other models do not duplicate the extensive biochemical environment of living tissue. Many enzymes and proteins are present in vivo that could either degrade or stabilize the therapeutic agent. In addition, concentrations in the blood following topical application should be determined for any medication that could potentially cause an adverse event. For many medications, the time course of serum drug concentrations in the blood has been characterized, so elimination rates are known. Release rates from the drug delivery system can be compared with elimination rates to estimate the system's pharmacokinetic properties. Serum concentrations achieved through the release of drug from the delivery system should be well below known toxic concentrations.

For biological agents or bioengineered polypeptides, such as human recombinant growth factors, an immunogenic response may occur, so the blood should be checked for the development of antibodies following application of the delivery system.

B. Diffusion Controlled Fibrin Systems

Fibrin matrices have been used to deliver drugs and antibiotics in preclinical studies, and clinically in trauma patients. Drug release appears to follow first order kinetics and be diffusion controlled. The in vitro release rates and antibacterial effects of fibrin matrix with carbenicillin, tetracycline, ciprofloxacin, gentamicin, teicoplanin, clindamycin, ampicillin, tobramycin, cefotaxime and mezlocillin have been evaluated [158,451–465].

Fibrin adhesive and cefotaxime mixtures have been utilized to treat osteitis in human subjects [459]. Serum concentrations of cefotaxime remained low for up to 36 h. Clinically effective concentrations were maintained at the wound site for 84 h. Clinically effective levels of cefoxitin, gentamicin, ciprofloxacin and teicoplanin can be released from a fibrin matrix in vitro [452].

In dogs, the use of fibrin-tobramycin mixtures for deliberately infected Dacron aortic grafts prevents significant infection compared with fibrin alone [460]. In split-thickness skin grafts, fibrin matrix alone has bacteriostatic effects [163]. Perhaps the large amount of fibrin overwhelms bacterial fibrinolytic capacity and thus prevents bacterial growth.

In an attempt to increase the duration of release of dibekacin sulfate from fibrin gels, glutaraldehyde was used as a covalent crosslinking agent to attach the antibiotic to the gel structure [463]. Gelation time increases as a function of increased glutaraldehyde concentration, although levels below 0.02% do not have a significant effect. The antibiotic remains at clinically effective levels in rat tibia for up to two weeks. High concentrations of glutaraldehyde in fibrin gels elicit a strong foreign body reaction, whereas lower concentrations have no noticeable effect on surrounding tissues.

Fibrin matrix–gelatin–Factor XIII and CA mixtures have also been used for intraperitoneal cancer therapy [466,467]. For fibrin matrix systems with doxorubicin and cisplatic, complete remission has been reported in 34 of 37 cases (92%); partial remission in 3 of 37 (8%). Gelatin was added to cause macrophage induction and to improve T-lymphocyte induction and activity. Drug delivery of 5-fluorouracil for up to 120 h via a fibrin matrix has been reported in an in vitro study [468,469].

C. Biofeedback Systems

Biofeedback systems are those in which degradation of the implant is controlled by the adjacent tissues. For example, in a healing wound the fibrin matrix is gradually replaced by regenerating tissue. In this case, degradation of the provisional fibrin matrix is accomplished by the physiological process of healing in which the temporary matrix is replaced by a new extracellular matrix created by fibroblasts. This occurs through controlled fibrinolysis. The fibrin degradation can be said to be under biofeedback control, because it is degraded at the same rate it is replaced by regenerating tissue.

1. System Design

With natural biomaterials, such as albumin or fibrin, delivery of a biological response modifier can be accomplished in a number of different ways. The biological response modifier can be impregnated within the matrix, attached to the polymer chain, or included through intrafibril entrapment. Incorporation during the polymerization process is similar to intrafibril entrapment. In this case, if the substance is larger than the intrafibril pores, it is released only when the natural material degrades. Therefore, the release is controlled by the rate of phagocytic cellular infiltration, and thus under biofeedback control. Additionally, the degradation is at the wound edge and thus gives the appropriate gradient to stimulate further angiogenesis and tissue healing.

The use of a degradable matrix to deliver a biological response modifier, such as a growth factor, can protect the growth factor until release, since growth factors seem to have a short half-life in vivo. The short half-life has potentially been a problem in clinical studies, leading to reduced efficacy, and/or increased expense associated with daily administration.

Thus, a natural polymeric matrix such as fibrin or albumin, can protect the biological response modifier until release, serve as a biofeedback controlled drug delivery system, and provide an adherent tissue scaffold during healing.

2. Fibrin Scaffold System Design

Based on previous work, FGF-1 was selected as the angiogenic agent for a wound scaffold. Both in vitro and in vivo studies have indicated that the tissue response is dose dependent and a maximal response is reached at an intermediate dose (Fig. 1) [470,471].

For the matrix, fibrin was selected for its adhesive properties [472,473] as well as its ability to serve as a degradable drug delivery system for wound healing [474–478]. It was found that the strength of the fibrin clot increased and degradation rate decreased with increasing concentration [264,265,472,479,480]. Additionally, in vivo and clinical studies have been done to test the use of autologous single and donor fibrin matrix for skin graft attachment in burn patients [480]. It was found that the fibrin hemostasis and early graft adherence led to an excellent graft take with reduced scarring. When compared to conventional treatments, this technique led to shorter hospital stays, minimal postoperative care and immobilization, no pressure dressings, and prompt start of ambulation and physical therapy with an early return to normal activities.

When designing a drug delivery system, the short biological half-life of FGF-1 is an important consideration. In the case of fibrin, FGF has been shown to be bound covalently [481,482], helping to provide protection prior to release. Additionally, immunolocalization studies have indicated that there is a uniform distribution of FGF-1 in the fibrin matrix [478].

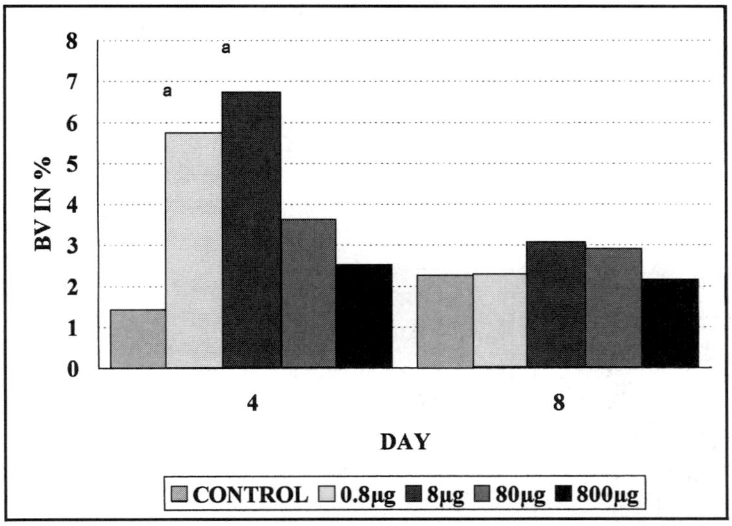

Figure 1 Angiogenic response to different doses of FGF-1 delivered through a fibrin matrix. The two doses significantly different than the controls are labeled with an "a."

Release studies using FGF and fibrin, formed in situ, have indicated that although there is an initial high release rate of about 30%, the subsequent release rate is relatively constant and proportional to the degradation rate (Fig. 2) [482–484]. To help show in vivo activity, a comparable enhancement of wound healing was seen with a topically applied dose, designed to mimic the fibrin/FGF-1 release kinetics, compared with the fibrin/FGF-1 system (Fig. 3) [477].

Although the fibrin matrix, due to its own biological activity [485], serves as a reasonable scaffold, better scaffolds can be made by optimizing the configuration as well as the bioactivity. In a rabbit ear ulcer model, full-thickness defects were treated with the fibrin matrices in two different pore configurations [475]. The more porous implant (modified fibrin) showed increased angiogenic response (Fig. 4). The levels of porosity and pore size are currently being optimized. Even in unoptimized systems, FGF-1 in a nonporous fibrin matrix was capable of complete epidermal regeneration with dermal filling of full-thickness defect and minimal contraction (20%) within two weeks, while controls took at least three weeks to heal and healed mostly by contraction [478].

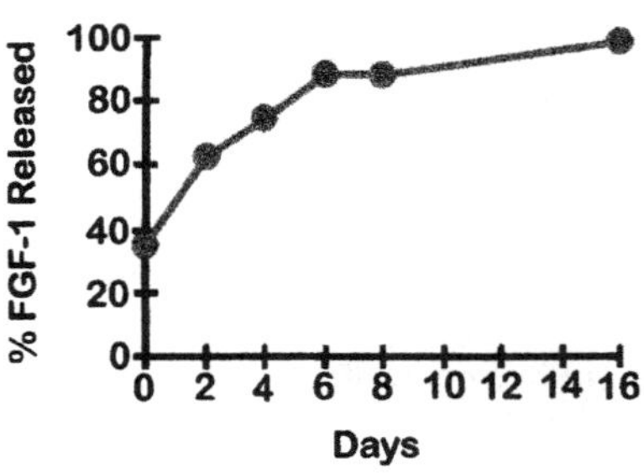

Figure 2 Release kinetics of FGF-1 delivered through a fibrin matrix.

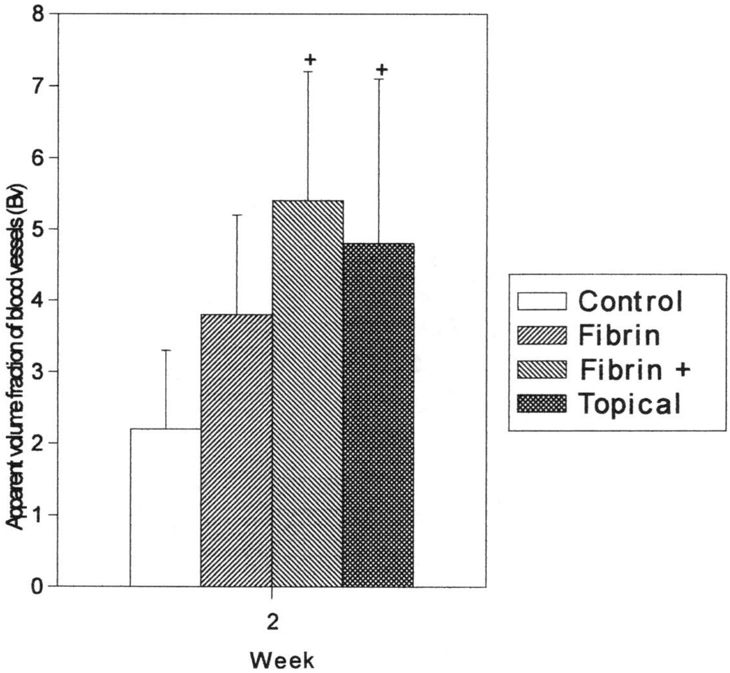

Figure 3 Angiogenic response to FGF-1 delivered topically versus through a fibrin matrix.

For skin grafts, it was anticipated that making the fibrin porous would allow for quicker graft take by providing a scaffold for the blood vessels to grow through, without the matrix having to be broken down first. In one study, however, as fibrin was made porous it appeared that the shear strength was inadequate to handle physiological loading [479]. Although the incorporation of FGF-1 reduced the incidence of graft loss, porous albumin, with its better adhesive strength, as well as other porous fibrin systems, are being investigated currently. It is also possible that using a thin adhesive layer (200 μm or less) would obviate the need for a porous structure.

Overall, the fibrin/FGF-1 system [471,475–478] when used in open wounds has led to the best overall healing, compared with all other treatments, with complete epithelialization and minimal contraction. This is likely due to the increased angiogenesis, mostly due to the FGF-1 as well as the increase in new tissue formation due to the fibrin scaffold. Similarly for meshed skin grafts, FGF-1 incorporated into a fibrin has led to the best graft healing. The FGF-1 concentration that has worked best overall is 10 μg/mL, which when spread in a 200 μm layer is 10 μg/50 cm^2.

3. *Fibrin System Applications*

The clinical efficacy of the fibrin/FGF-1 system is currently being evaluated in both a pressure ulcer and burn healing study at UAB [486–488]. For pressure ulcers, patients with Stage 2 and 3 ulcers are randomized into three groups: (1) control, (2) placebo control, and (3) FGF-1 in a fibrin matrix.

For burns, the fibrin is used for skin graft attachment and the wounds on the patients are randomized (double blinded) into two groups: (1) fibrin with FGF-1 and (2) fibrin with-

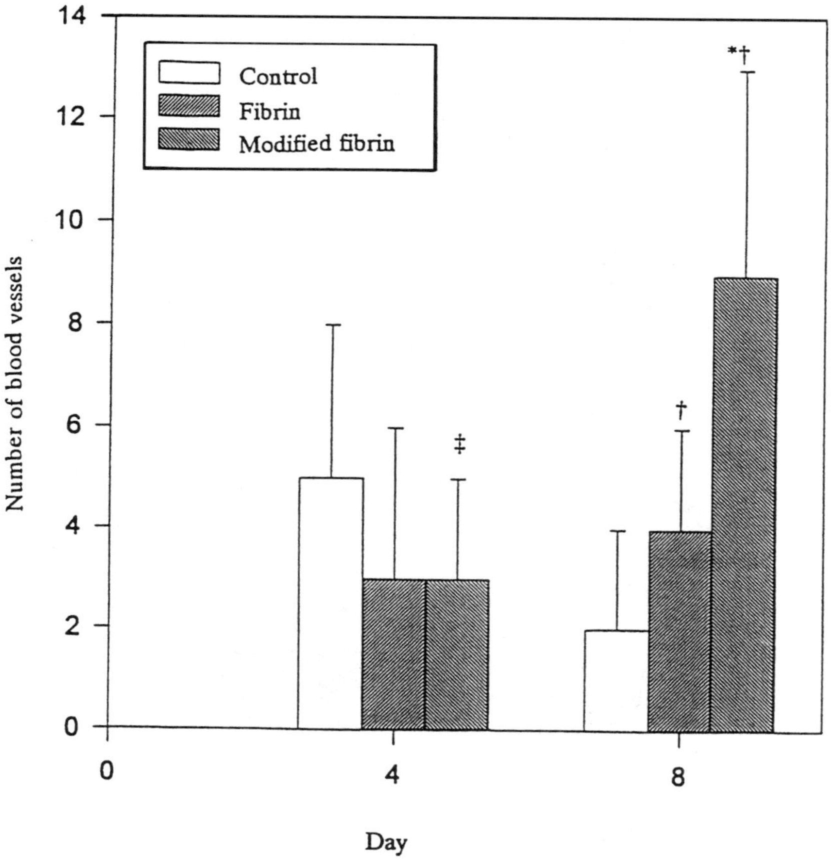

Figure 4 The effect of increasing fibrin porosity on the angiogenic response: an asterisk indicates different than the control; a superscript plus sign indicates difference between modified and nonmodified scaffold.

out FGF-1. Each patient also has a corresponding control (stapled or sutured graft) for comparison.

A critical part of each study has been the development of noninvasive clinical assessment tools to measure the effectiveness of the treatments. The three main variables for comparison are healing rate, angiogenesis, and tissue stiffness. These assessments are done weekly for the first few months and monthly thereafter. For healing rate, the epithelialization rate and contraction rate are determined in a manner that gives the rate independent of wound size. The angiogenesis is determined by a scanning laser doppler, which gives a 2-D map of the blood perfusion.

Preliminary results have indicated that the fibrin systems keep the blood perfusion level at a higher level longer than the controls. For meshed skin grafts, this means the vascularity is kept high until the mesh fills in. This is believed to be associated with increased new tissue formation and less scarring (Fig. 5).

FGF-1 mixed with fibrin adhesive was used to coat PTFE grafts that were placed into rabbit vessels [483]. At 28 days, all the FGF-1/FM grafts exhibited extensive capillary ingrowth and confluent endothelialized luminal surfaces, which were not observed in the un-

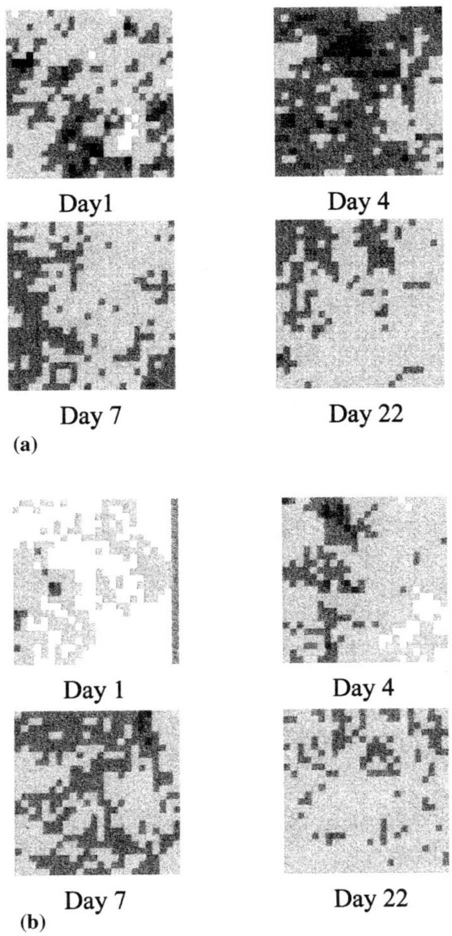

Figure 5 The clinical blood perfusion changes over time for meshed skin grafts adhered with (A) fibrin (1,4,7,22 days) and (B) sutures (1,4,7,22 days). The darker the area the higher is the blood perfusion. The fibrin system peaks at seven or later days while the sutured system peaks at four days.

treated or fibrin treated groups. In addition, the FGF-1/fibrin matrix was found to permit significantly less platelet deposition than uncoated PTFE [489].

A recent study found that FGF-1 in a fibrin adhesive carrier implanted with viable nerve into atrophic muscle stimulated the formation of functional, acetylcholine producing motor end plates [490]. Twelve rats underwent denervation of the hamstring muscle eight weeks before implantation of the nerve at a site distant from the original motor end plate. Robust axonal sprouting and formation of multiple motor end plates were found arborized in serial fashion equidistant around the implanted nerve ending. This study suggests that FGF-1 with fibrin adhesive can facilitate the reinnervation of atrophied muscle by enhancing the formation or revitalization of motor end plates.

The effect of fibrin sealant and endothelial cell growth factor on the healing of defects of the avascular portion of canine menisci was investigated in 30 menisci of 15 adult mongrel dogs [491]. The combination of fibrin sealant and endothelial cell growth factor enhanced the neovascularization and formation of granulation tissue, which resulted in increased healing in the avascular portion of the meniscus.

4. The Future

Overall, these studies show the feasibility of using natural biomaterials as not only tissue adhesives, but also angiogenic tissue scaffolds. These systems can protect the biological response modifier until release, serve as a biofeedback controlled drug delivery system, and provide an adherent tissue scaffold during healing. Although the characteristics of the tissue adhesives are being optimized for various applications, there still can be further optimization as well as development of additional applications.

Specifically, in the case of fibrin, adhesive properties are insufficient for a number of applications. Further, as the system is made porous; the adhesive properties decrease even more. Therfore improving the adhesive properties through chemical modifications or the use of composites is an important area to continue to explore.

Therefore, other stronger adhesive systems are under investigation. For example, preliminary data indicates that the albumin system performs as well or better than the fibrin system in terms of strength, in vitro and in vivo, and biocompatibility (Table 2). Studies are underway to develop it as a regenerative tissue scaffold.

Attempts are also underway to modify the release kinetics. Although growth factors such as FGF can bind to fibrin, the release still has a significant diffusion conrolled component. Different incorporation schemes are being investigated to provide release that is more completely biofeedback controlled.

IV. CONCLUSIONS

Because various applications (hemostatic agent, sealant, drug delivery system, tissue scaffold, etc.) have different requirements, the ideal tissue adhesive is defined in terms of its intended use. Characteristics include both biocompatiblity and handling issues. For handling characteristics, the adhesive should usually be moderately viscous so that once applied it remains in the field, yet readily spreads and conforms to the wound contours. It also needs to set rapidly in most applications. For biocompatiblity both the host reponse and the degradation of the implant need to be optimized for each application.

The tissue compatibility is a major concern with both cyanoacrylates and gelatin-resorcinol-formaldehyde, since both systems deliver formaldehyde and other toxic products, resulting in cytotoxicity and inflammation. Although these systems have the greatest initial strength, the type of host response limits their utility. On the other hand, natural tissue adhesives such as fibrin and albumin not only possess inherent bioactivity, but also the healing properties can be enhanced by the addition of growth factors or by making the matrices porous. Efforts, however, are underway to develop a clinically approved cyanoacrylate with less of a host response.

The mechanical properties at the wound interface, during healing, is important for many applications. Ideally the interface should have the same properties throughout: equivalent to the native tissue; with the tissue adhesive being replaced by regenerated tissue over time. This is not feasible, in most cases, with current technology. Although natural adhesives tend to be weaker; the sacrifice in initial strength can lead to systems that heal faster and more completely (Table 2). Even within these systems, however, optimization requires tradeoffs between, initial strength, degradation rate, and healing rate.

The ability to degrade at an appropriate rate is therefore a critical feature for tissue adhesives. The ultimate design being a system where the regenerative process controls the degradation—biofeedback control. In systems without biofeedback control the degradation rate has to be controlled to match the healing rate as best as possible by changes in structure or chemistry.

The degradation rate even for biofeedback systems can be controlled by chemistry and structure. Decreasing protein concentration or increasing porosity alters not only mechnical properties, but degradation rate. The biofeedback control, however, ensures that despiite modifications the healing rate is still tied to the degradation rate. This is because by increasing porosity, the matrix is degrading quicker due to it being a better scaffold and stimulating faster tissue ingrowth, not that it is less stable.

Therefore efforts to improve the scaffolding ability of these natural tissue adhesives is desirable. This can help not only in maintaining mechanical integrity at the interface, but also speed the process of returning to the normal tissue state.

Angiogenic factors in particular have demonstrated improved healing rates. When growth factors are combined with natural biodegradable tissue adhesives, enhanced tissue regeneration can occur. If the growth factor is bound to the tissue adhesive matrix, it will be released as the matrix degrades, thus exhibiting biofeedback-controlled delivery.

Although much progress has been made in the design of biofeedback controlled natural tissue adhesive systems, further improvements are needed. This includes both optimization for new applications and improvements in present uses. In general the mechanical properties, scaffolding ability, and handling properties of all systems can be improved. Optimization for each application, however, still requires trade-offs between these characteristics. For many applications, increasing the tissue adhesive strength without destroying the biofeedback control or significantly slowing the healing rate would be desirable. In addition, designs to speed regenerative healing without significant changes in mechanical or handling properties would be desirable.

ACKNOWLEDGMENTS

We wish to thank the UAB graduate students in the Biomedical Polymers Research Group responsible for the work presented in this chapter. In addition, we wish to thank J. Anthony Thompson for his help in providing FGF and collaboration in the design of degradable regenerative systems and the American Red Cross and S. Huang for supplying the fibrin and albumin matrices. Further, we acknowledge the funding agencies who helped support the research presented here: NIH, NSF, Department of Education, and the CDC.

REFERENCES

1. Hofer, S. O., Shrayer, D., Reichner, J. S., Hoekstra, H. J. and Wanebo, H. J., 1998. Wound-induced tumor progression: a probable role in recurrence after tumor resection. *Archives of Surgery*, 133: 383–389.
2. Bergel, S., 1909. Uber wirkungen des fibrins. *Dtsch Med Wochenschr*, 35:633–665.
3. Seddon, J. H. and Medawar, P. B., 1942. Fibrin suture of human nerves. *Lancet*, 2:87–92.
4. Cronkite, E. P., Lozner, E. L. and Deaver, J. M., 1944. Use of thrombin and fibrinogen in skin grafting. *Journal of the American Medical Association*, 124:976–978.
5. Spotnitz, W. D., 1995. Fibrin sealant in the United States: clinical use at the University of Virginia. *Thrombosis & Haemostasis*, 74:482–485.
6. Achauer, B. M., Miller, S. R. and Lee, T. E., 1994. The hemostatic effect of fibrin glue on graft donor sites. *Journal of Burn Care & Rehabilitation*, 15:24–28.

7. Sierra, D. H., 1993. Fibrin sealant adhesive systems: a review of their chemistry, material properties and clinical applications. *Journal of Biomaterials Applications*, 7:309–352.
8. Michel, D. and Harmand, M. F., 1990. Fibrin seal in wound healing: effect of thrombin and [Ca^{2+}] on human skin fibroblast growth and collagen production. *Journal of Dermatological Science*, 1: 325–333.
9. Francis, C. W. and Marder, V. J., 1982. A molecular model of plasmic degradation of crosslinked fibrin. *Seminars in Thrombosis & Hemostasis*, 8:25–35.
10. Clark, R. A., 1993. Regulation of fibroplasia in cutaneous wound repair. *American Journal of the Medical Sciences*, 306:42–48.
11. Liu, H. M., Wang, D. L. and Liu, C. Y., 1990. Interactions between fibrin, collagen and endothelial cells in angiogenesis. *Advances in Experimental Medicine & Biology*, 281:319–331.
12. Nowotony, R., Chalupka, A., Nowotony, C. and Bosch, P., 1982. Mechanical properties of fibrinogen adhesive material. In: Winter G. D., Gibbons G. F., Plenk H. (Editors). *Biomaterials* 1980. London: Wiley.
13. Anonymous, 1998. Package Insert for Tisseel. Baxter Corporation.
14. Okada, M., Blomback, B., Chang, M. D. and Horowitz, B., 1985. Fibronectin and fibrin gel structure. *Journal of Biological Chemistry*, 260:1811–1820.
15. Blomback, B. and Okada, M., 1983. Fibrin gels and their possible implication for surface hemorheology in health and disease. *Annals of the New York Academy of Sciences*, 416:397–409.
16. Kamykowski, G. W., Mosher, D. F., Lorand, L. and Ferry, J. D., 1981. Modification of shear modulus and creep compliance of fibrin clots by fibronectin. *Biophysical Chemistry*, 13:25–28.
17. Shen, L. L., Hermans, J., McDonagh, J., McDonagh, R. P. and Carr, M., 1975. Effects of calcium ion and covalent crosslinking on formation and elasticity of fibrin cells. *Thrombosis Research*, 6: 255–265.
18. Jorgensen, P. H., Jensen, K. H. and Andreassen, T. T., 1987. Mechanical strength in rat skin incisional wounds treated with fibrin sealant. *Journal of Surgical Research*, 42:237–241.
19. Burnouf-Radosevich, M., Burnouf, T. and Huart, J. J., 1990. Biochemical and physical properties of a solvent-detergent-treated fibrin glue. *Vox Sanguinis*, 58:77–84.
20. Byrne, D. J., Hardy, J., Wood, R. A., McIntosh, R. and Cuschieri, A., 1991. Effect of fibrin glues on the mechanical properties of healing wounds. *British Journal of Surgery*, 78:841–843.
21. Dahlstrom, K. K., Weis-Fogh, U. S., Medgyesi, S., Rostgaard, J. and Sorensen, H., 1992. The use of autologous fibrin adhesive in skin transplantation. *Plastic & Reconstructive Surgery*, 89:968–972; discussion 973–966.
22. Welch, M. T., Smith-Morse, B. R. and Brockbank, K. G. M., 1991. An assay for assessment of biomaterial properties of fibrin sealants. Transactions of the Society for *Biomaterials*, 14:136.
23. Wolner, E., 1982. Fibrin gluing in cardiovascular surgery. *Thoracic & Cardiovascular Surgeon*, 30:236–237.
24. Haverich, A., Walterbusch, G. and Borst, H. G., 1981. The use of fibrin glue for sealing vascular prostheses of high porosity. *Thoracic & Cardiovascular Surgeon*, 29:252–254.
25. Jonas, R. A., Schoen, F. J., Levy, R. J. and Castaneda, A. R., 1986. Biological sealants and knitted Dacron: porosity and histological comparisons of vascular graft materials with and without collagen and fibrin glue pretreatments. *Annals of Thoracic Surgery*, 41:657–663.
26. Leca, F., Karam, J., Vouhe, P. R., Khoury, W., Tamisier, D., Bical, O., Da Cruz, E. and Thibert, M., 1994. Surgical treatment of multiple ventricular septal defects using a biologic glue. *Journal of Thoracic & Cardiovascular Surgery*, 107:96–102.
27. Barbalinardo, R. J., Citrin, P., Franco, C. D. and Hobson, R. W. D., 1986. A comparison of isobutyl 2-cyanoacrylate glue, fibrin adhesive, and oxidized regenerated cellulose for control of needle hole bleeding from polytetrafluoroethylene vascular prostheses. *Journal of Vascular Surgery*, 4:220–223.
28. Mazzucotelli, J. P., Klein-Soyer, C., Beretz, A., Brisson, C., Archipoff, G. and Cazenave, J. P., 1991. Endothelial cell seeding: coating Dacron and expanded polytetrafluoroethylene vascular grafts with a biological glue allows adhesion and growth of human saphenous vein endothelial cells. *International Journal of Artificial Organs*, 14:482–490.

29. Cikrit, D. F., Dalsing, M. C., Weinstein, T. S., Palmer, K., Lalka, S. G. and Unthank, J. L., 1990. CO_2-welded venous anastomosis: enhancement of weld strength with heterologous fibrin glue. *Lasers in Surgery & Medicine*, 10:584–590.
30. Grubbs, P. E., Jr., Wang, S., Marini, C., Basu, S., Rose, D. M. and Cunningham, J. N., Jr., 1988. Enhancement of CO_2 laser microvascular anastomoses by fibrin glue. *Journal of Surgical Research*, 45:112–119.
31. Richling, B., 1982. Homologous controlled-viscosity fibrin for endovascular embolization. Part II: Catheterization technique, animal experiments. *Acta Neurochirurgica*, 64:109–124.
32. Berthelsen, B., Lofgren, J. and Svendsen, P., 1990. Embolization of cerebral arteriovenous malformations with bucrylate. Experience in a first series of 29 patients. *Acta Radiologica*, 31:13–21.
33. Dottori, V., Spagnolo, S., Passerone, G., Lijoi, A., Barberis, L., Agostini, M., De Gaetano, G., Parodi, E., Maccario, M. and Fumagalli, C., 1992. Ten years of surgery of aortic dissections and aneurysms. Clinical experience and original contributions. *Minerva Cardioangiologica*, 40: 431–436.
34. Shaffrey, C. I., Spotnitz, W. D., Shaffrey, M. E. and Jane, J. A., 1990. Neurosurgical applications of fibrin glue: augmentation of dural closure in 134 patients. *Neurosurgery*, 26:207–210.
35. Cain, J. E., Jr., Dryer, R. F. and Barton, B. R., 1988. Evaluation of dural closure techniques. Suture methods, fibrin adhesive sealant, and cyanoacrylate polymer. *Spine*, 13:720–725.
36. Cain, J. E., Jr., Rosenthal, H. G., Broom, M. J., Jauch, E. C., Borek, D. A. and Jacobs, R. R., 1990. Quantification of leakage pressures after durotomy repairs in the canine. *Spine*, 15:969–970.
37. Bento, R. F. and Miniti, A., 1989. Comparison between fibrin tissue adhesive, epineural suture and natural union in intratemporal facial nerve of cats. *Acta Oto-Laryngologica-Supplement*, 465: 1–36.
38. Bento, R. F. and Miniti, A., 1993. Anastomosis of the intratemporal facial nerve using fibrin tissue adhesive. *Ear, Nose, & Throat Journal*, 72:663.
39. Egloff, D. V. and Narakas, A., 1983. Nerve anastomoses with human fibrin. Preliminary clinical report (56 cases). *Annales de Chirurgie de la Main*, 2:101–115.
40. Maragh, H., Meyer, B. S., Davenport, D., Gould, J. D. and Terzis, J. K., 1990. Morphofunctional evaluation of fibrin glue versus microsuture nerve repairs. *Journal of Reconstructive Microsurgery*, 6:331–337.
41. Gilbert, C. E., Grierson, I. and McLeod, D., 1989. Retinal patching: a new approach to the management of selected retinal breaks. *Eye*, 3:19–26.
42. Gilbert, C. E., 1991. Adhesives in retinal detachment surgery. *British Journal of Ophthalmology*, 75:309–310.
43. Lagoutte, F. M., Gauthier, L. and Comte, P. R., 1989. A fibrin sealant for perforated and preperforated corneal ulcers. *British Journal of Ophthalmology*, 73:757–761.
44. Kim, M. S. and Kim, J. H., 1989. Effects of tissue adhesive (Tisseel) on corneal wound healing in lamellar keratoplasty in rabbits. *Korean Journal of Ophthalmology*, 3:14–21.
45. Chen, R. J., Fang, J. F., Lin, B. C., Hsu, Y. B., Kao, J. L., Kao, Y. C. and Chen, M. F., 1998. Selective application of laparoscopy and fibrin glue in the failure of nonoperative management of blunt hepatic trauma. *Journal of Trauma*, 44:691–695.
46. De la Garza, J. L. and Rumsey, E., Jr., 1990. Fibrin glue and hemostasis in liver trauma: a case report. *Journal of Trauma*, 30:512–513.
47. Kram, H. B., del Junco, T., Clark, S. R., Ocampo, H. P. and Shoemaker, W. C., 1990. Techniques of splenic preservation using fibrin glue. *Journal of Trauma*, 30:97–101.
48. Kram, H. B., Clark, S. R., Ocampo, H. P., Yamaguchi, M. A. and Shoemaker, W. C., 1991. Fibrin glue sealing of pancreatic injuries, resections, and anastomoses. *American Journal of Surgery*, 161: 479–481; discussion 482.
49. Kuzu, A., Aydintug, S., Karayalcin, K., Koksoy, C., Yerdel, M. A. and Eraslan, S., 1992. Use of autologous fibrin glue in the treatment of splenic trauma: an experimental study. *Journal of the Royal College of Surgeons of Edinburgh*, 37:162–164.
50. Ochsner, M. G., Maniscalco-Theberge, M. E. and Champion, H. R., 1990. Fibrin glue as a hemostatic agent in hepatic and splenic trauma. *Journal of Trauma*, 30:884–887.

51. Bardaxoglou, E., Campion, J. P., Landen, S., Manganas, D., Siriser, F., Chareton, B. and Launois, B., 1994. Oesophageal perforation: primary suture repair reinforced with absorbable mesh and fibrin glue. *British Journal of Surgery*, 81:399.
52. Bardaxoglou, E., Manganas, D., Meunier, B., Landen, S., Maddern, G. J., Campion, J. P. and Launois, B., 1997. New approach to surgical management of early esophageal thoracic perforation: primary suture repair reinforced with absorbable mesh and fibrin glue. *World Journal of Surgery*, 21:618–621.
53. Baumann, W. R., Ulmer, J. L., Ambrose, P. G., Garvey, M. J. and Jones, D. T., 1997. Closure of a bronchopleural fistula using decalcified human spongiosa and a fibrin sealant. *Annals of Thoracic Surgery*, 64:230–233.
54. Bayfield, M. S. and Spotnitz, W. D., 1996. Fibrin sealant in thoracic surgery. Pulmonary applications, including management of bronchopleural fistula. *Chest Surgery Clinics of North America*, 6:567–583.
55. Benko, I., Molnar, T. F. and Horvath, O. P., 1997. A case of fibrin sealant application for closing benign trachea-esophageal fistula (TEF). *Acta Chirurgica Hungarica*, 36:25–26.
56. Fleisher, A. G., Evans, K. G., Nelems, B. and Finley, R. J., 1990. Effect of routine fibrin glue use on the duration of air leaks after lobectomy. *Annals of Thoracic Surgery*, 49:133–134.
57. Hanawa, T., Ikeda, S., Funatsu, T., Matsubara, Y., Hatakenaka, R., Mitsuoka, A., Kosaba, S., Shiota, T., Ishida, H. and Konishi, T., 1990. Development of a new surgical procedure for repairing tracheobronchomalacia. *Journal of Thoracic & Cardiovascular Surgery*, 100:587–594.
58. Hansen, M. K., Kruse-Andersen, S., Watt-Boolsen, S. and Andersen, K., 1989. Spontaneous pneumothorax and fibrin glue sealant during thoracoscopy. *European Journal of Cardio-Thoracic Surgery*, 3:512–514.
59. Morgan, J. A., 1988. Closure of subarachnoid-pleural fistulae with fibrin sealant. *European Journal of Cardio-Thoracic Surgery*, 2:56–57.
60. Mouritzen, C., Dromer, M. and Keinecke, H. O., 1993. The effect of fibrin glueing to seal bronchial and alveolar leakages after pulmonary resections and decortications. *European Journal of Cardio-Thoracic Surgery*, 7:75–80.
61. Samuels, L. E., Shaw, P. M. and Blaum, L. C., 1996. Percutaneous technique for management of persistent airspace with prolonged air leak using fibrin glue. *Chest*, 109:1653–1655.
62. Wong, K. and Goldstraw, P., 1997. Effect of fibrin glue in the reduction of postthoracotomy alveolar air leak. *Annals of Thoracic Surgery*, 64:979–981.
63. Yasuda, Y., Mori, A., Kato, H., Fujino, S. and Asakura, S., 1991. Intrathoracic fibrin glue for postoperative pleuropulmonary fistula. *Annals of Thoracic Surgery*, 51:242–244.
64. Abel, M. E., Chiu, Y. S., Russell, T. R. and Volpe, P. A., 1993. Autologous fibrin glue in the treatment of rectovaginal and complex fistulas. *Diseases of the Colon & Rectum*, 36:447–449.
65. Antonelli, M., Cicconetti, F., Vivino, G. and Gasparetto, A., 1991. Closure of a tracheoesophageal fistula by bronchoscopic application of fibrin glue and decontamination of the oral cavity [see comments]. *Chest*, 100:578–579.
66. Brady, A. P., Malone, D. E., Tam, P. and McGrath, F. P., 1993. Closure of a duodenal fistula with fibrin sealant. *Journal of Vascular & Interventional Radiology*, 4:525–527; discussion 527–529.
67. Cellier, C., Landi, B., Faye, A., Wind, P., Frileux, P., Cugnenc, P. H. and Barbier, J. P., 1996. Upper gastrointestinal tract fistulae: endoscopic obliteration with fibrin sealant. *Gastrointestinal Endoscopy*, 44:731–733.
68. Eleftheriadis, E., Tzartinoglou, E., Kotzampassi, K. and Aletras, H., 1990. Early endoscopic fibrin sealing of high-output postoperative enterocutaneous fistulas. *Acta Chirurgica Scandinavica*, 156: 625–628.
69. Gutierrez, C., Barrios, J. E., Lluna, J., Vila, J. J., Garcia-Sala, C., Roca, A. and Ruiz Company, S., 1994. Recurrent tracheoesophageal fistula treated with fibrin glue. *Journal of Pediatric Surgery*, 29:1567–1569.
70. Hjörtrup, A., Moesgaard, F. and Kjaergard, J., 1991. Fibrin adhesive in the treatment of perineal fistulas. *Diseases of the Colon & Rectum*, 34:752–754.

71. Moesgaard, F., Hoffmann, S. and Nielsen, R., 1989. Successful fibrin seal closure of a contaminated fistula. Case report. *Acta Chirurgica Scandinavica*, 155:427–428.
72. Shand, A., Pendlebury, J., Reading, S., Papachrysostomou, M. and Ghosh, S., 1997. Endoscopic fibrin sealant injection: a novel method of closing a refractory gastrocutaneous fistula. *Gastrointestinal Endoscopy*, 46:357–358.
73. Shand, A., Reading, S., Ewing, J., Neil, B., Welsh, D., Parker, A. and Ghosh, S., 1997. Palliation of a malignant gastrocolic fistula by endoscopic human fibrin sealant injection. *European Journal of Gastroenterology & Hepatology*, 9:1009–1011.
74. Suzuki, Y., Kuroda, Y., Morita, A., Fujino, Y., Tanioka, Y., Kawamura, T. and Saitoh, Y., 1995. Fibrin glue sealing for the prevention of pancreatic fistulas following distal pancreatectomy. *Archives of Surgery*, 130:952–955.
75. York, E. L., Lewall, D. B., Hirji, M., Gelfand, E. T. and Modry, D. L., 1990. Endoscopic diagnosis and treatment of postoperative bronchopleural fistula. *Chest*, 97:1390–1392.
76. Akizuki, S., Yasukawa, Y. and Takizawa, T., 1997. A new method of hemostasis for cementless total knee arthroplasty. *Bulletin–Hospital for Joint Diseases*, 56:222–224.
77. Chang, H., Wu, G. J., Perng, W. L., Hwang, F. Y. and Hung, C. R., 1992. Effects of fibrin glue on hemostasis. *Journal of the Formosan Medical Association*, 91:601–607.
78. Green, D., Wong, C. A. and Twardowski, P., 1996. Efficacy of hemostatic agents in improving surgical hemostasis. *Transfusion Medicine Reviews*, 10:171–182.
79. Herold, G. and Stange, E. F., 1993. Sclerotherapy of duodenal varices using a fibrin tissue sealant. *Endoscopy*, 25:371–372.
80. Holcomb, J. B., Pusateri, A. E., Hess, J. R., Hetz, S. P., Harris, R. A., Tock, B. B., Drohan, W. N. and MacPhee, M. J., 1997. Implications of new dry fibrin sealant technology for trauma surgery. *Surgical Clinics of North America*, 77:943–952.
81. Holcomb, J., MacPhee, M., Hetz, S., Harris, R., Pusateri, A. and Hess, J., 1998. Efficacy of a dry fibrin sealant dressing for hemorrhage control after ballistic injury. *Archives of Surgery*, 133: 32–35.
82. Jackson, M. R., Danby, C. A. and Alving, B. M., 1997. Heparinoid anticoagulation and topical fibrin sealant in heparin-induced thrombocytopenia. *Annals of Thoracic Surgery*, 64:1815–1817.
83. Jackson, M. R., Friedman, S. A., Carter, A. J., Bayer, V., Burge, J. R., MacPhee, M. J., Drohan, W. N. and Alving, B. M., 1997. Hemostatic efficacy of a fibrin sealant-based topical agent in a femoral artery injury model: a randomized, blinded, placebo-controlled study. *Journal of Vascular Surgery*, 26:274–280.
84. Kram, H. B., Shoemaker, W. C., Clark, S. R., Macabee, J. R. and Yamaguchi, M. A., 1991. Spraying of aerosolized fibrin glue in the treatment of nonsuturable hemorrhage. *American Surgeon*, 57:381–384.
85. Liu, M. and Lui, W. Y., 1993. The use of fibrin adhesive for hemostasis after liver resection. *Chung Hua i Hsueh Tsa Chih—Chinese Medical Journal*, 51:19–22.
86. Milne, A. A., Murphy, W. G., Reading, S. J. and Ruckley, C. V., 1995. Fibrin sealant reduces suture line bleeding during carotid endarterectomy: a randomised trial. *European Journal of Vascular & Endovascular Surgery*, 10:91–94.
87. Song, S. Y., Chung, J. B., Moon, Y. M., Kang, J. K. and Park, I. S., 1997. Comparison of the hemostatic effect of endoscopic injection with fibrin glue and hypertonic saline-epinephrine for peptic ulcer bleeding: a prospective randomized trial. *Endoscopy*, 29:827–833.
88. Tawes, R. L., Jr., Sydorak, G. R. and DuVall, T. B., 1994. Autologous fibrin glue: the last step in operative hemostasis. *American Journal of Surgery*, 168:120–122.
89. Vatankhah, M., Moller, K. O., Lind, B. M. and Baretton, G., 1992. Experimental salvage of the spleen. A combined technique of cryosurgery and tissue sealant. *Acta Chirurgica Hungarica*, 33: 317–324.
90. Zusman, S. P., Lustig, J. P. and Baston, I., 1992. Postextraction hemostasis in patients on anticoagulant therapy: the use of a fibrin sealant. *Quintessence International*, 23:713–716.
91. Zusman, S. P., Lustig, J. P. and Bin Nun, G., 1993. Cost evaluation of two methods of post tooth extraction hemostasis in patients on anticoagulant therapy. *Community Dental Health*, 10:167–173.

92. Boris, W. J., Gu, J. and McGrath, L. B., 1996. Effectiveness of fibrin glue in the reduction of postoperative intrapericardial adhesions. *Journal of Investigative Surgery*, 9:327–333.
93. Caballero, J. and Tulandi, T., 1992. Effects of Ringer's lactate and fibrin glue on postsurgical adhesions. *Journal of Reproductive Medicine*, 37:141–143.
94. Frykman, E., Jacobsson, S. and Widenfalk, B., 1993. Fibrin sealant in prevention of flexor tendon adhesions: an experimental study in the rabbit. *Journal of Hand Surgery—American Volume*, 18: 68–75.
95. Gauwerky, J. F., Mann, J. and Bastert, G., 1990. The effect of fibrin glue and peritoneal grafts in the prevention of intraperitoneal adhesions. *Archives of Gynecology & Obstetrics*, 247:161–166.
96. Gauwerky, J. F., Klose, R. P., Vierneisel, P. and Bastert, G., 1992. Fibrin glue for reanastomosis of the fallopian tube in the rabbit: adhesions and fertility. *Human Reproduction*, 7:1274–1277.
97. Jack, D., 1998. Sticky situations: surgical adhesions and adhesives [news]. *Lancet*, 351:118.
98. Koltai, J. L. and Gerhard, A., 1990. Intraperitoneal application of fibrinogen gluing in the rat for adhesions prophylaxis. *Progress in Pediatric Surgery*, 25:71–80.
99. Schier, F., Srour, N. and Waldschmidt, J., 1991. Dura covered with fibrin glue reduces adhesions in abdominal wall defects. *European Journal of Pediatric Surgery*, 1:343–345.
100. Sheppard, B. B., De Virgilio, C., Bleiweis, M., Milliken, J. C. and Robertson, J. M., 1993. Inhibition of intra-abdominal adhesions: fibrin glue in a long term model. *American Surgeon*, 59:786–790.
101. Takeuchi, H., Awaji, M., Hashimoto, M., Nakano, Y., Mitsuhashi, N. and Kuwabara, Y., 1996. Reduction of adhesions with fibrin glue after laparoscopic excision of large ovarian endometriomas. *Journal of the American Association of Gynecologic Laparoscopists*, 3:575–579.
102. Takeuchi, H., Toyonari, Y., Mitsuhashi, N. and Kuwabara, Y., 1997. Effects of fibrin glue on postsurgical adhesions after uterine or ovarian surgery in rabbits. *Journal of Obstetrics and Gynaecology Research*, 23:479–484.
103. Tulandi, T., 1997. How can we avoid adhesions after laparoscopic surgery? *Current Opinion in Obstetrics & Gynecology*, 9:239–243.
104. Byrne, D. J., Hardy, J., Wood, R. A., McIntosh, R., Hopwood, D. and Cuschieri, A., 1992. Adverse influence of fibrin sealant on the healing of high-risk sutured colonic anastomoses. *Journal of the Royal College of Surgeons of Edinburgh*, 37:394–398.
105. Mandel, M. A., 1989. Autologous fibrin for blepharoplasty incisions [letter]. *JAMA*, 262:3271–3272.
106. Mandel, M. A., 1990. Closure of blepharoplasty incisions with autologous fibrin glue. *Archives of Ophthalmology*, 108:842–844.
107. Mandel, M. A., 1992. Minimal suture blepharoplasty: closure of incisions with autologous fibrin glue. *Aesthetic Plastic Surgery*, 16:269–272.
108. Mommaerts, M. Y., Beirne, J. C., Jacobs, W. I., Abeloos, J. S., De Clercq, C. A. and Neyt, L. F., 1996. Use of fibrin glue in lower blepharoplasties. *Journal of Cranio-Maxillo-Facial Surgery*, 24: 78–82.
109. Flemming, I., 1992. Fibrin glue in face lifts. *Facial Plastic Surgery*, 8:79–88.
110. Marchac, D. and Sandor, G., 1994. Face lifts and sprayed fibrin glue: an outcome analysis of 200 patients. *British Journal of Plastic Surgery*, 47:306–309.
111. Fontana, A., Muti, E., Cicerale, D. and Rizzotti, M., 1991. Cartilage chips synthesized with fibrin glue in rhinoplasty. *Aesthetic Plastic Surgery*, 15:237–240.
112. Marchac, D. and Renier, D., 1990. Fibrin glue in craniofacial surgery. *Journal of Craniofacial Surgery*, 1:32–34.
113. Vasconez, H., 1993. Craniofacial surgery/bone metabolism. *Proceedings of the Symposium on Surgical Tissue Adhesives*, 1:132.
114. Matras, H., 1982. The use of fibrin sealant in oral and maxillofacial surgery. *Journal of Oral & Maxillofacial Surgery*, 40:617–622.
115. Gregory, E. W. and Schaberg, S. J., 1986. Experimental use of fibrin sealant for skin graft fixation in mandibular vestibuloplasty. *Journal of Oral & Maxillofacial Surgery*, 44:171–176.

116. Staindl, O., 1979. Tissue adhesion with highly concentrated human fibrinogen in otolaryngology. *Annals of Otology, Rhinology & Laryngology*, 88:413–418.
117. AF, O. C. and Shea, J. J., 1982. A biologic adhesive for otologic practice. *Otolaryngology—Head & Neck Surgery*, 90:347–348.
118. Marquet, J., 1985. Fibrin glue in tympanoplasty. *American Journal of Otology*, 6:28–30.
119. Epstein, G. H., Weisman, R. A., Zwillenberg, S. and Schreiber, A. D., 1986. A new autologous fibrinogen-based adhesive for otologic surgery. *Annals of Otology, Rhinology & Laryngology*, 95: 40–45.
120. Filipo, R. and Barbara, M., 1986. Rehabilitation of radical mastoidectomy. *American Journal of Otology*, 7:248–252.
121. Palva, T. and Johnsson, L. G., 1986. Preservation of hearing after removal of the membranous canal with a cholesteatoma. *Archives of Otolaryngology—Head & Neck Surgery*, 112:982–985.
122. Harris, D. M., Siedentop, K. H., Ham, K. R. and Sanchez, B., 1987. Autologous fibrin tissue adhesive biodegration and systemic effects. *Laryngoscope*, 97:1141–1144.
123. Hayward, P. J. and Mackay, I. S., 1987. Fibrin glue in nasal septal surgery. *Journal of Laryngology & Otology*, 101:133–138.
124. Wood, A. P. and Harner, S. G., 1988. The effect of fibrin tissue adhesive on the middle and inner ears of chinchillas. *Otolaryngology—Head & Neck Surgery*, 98:104–110.
125. Takimoto, T., Ishikawa, S., Nishimura, T., Tanaka, S., Yoshizaki, T., Komori, T. and Umeda, R., 1989. Fibrin glue in the surgical treatment of ranulas. *Clinical Otolaryngology*, 14:429–431.
126. Ellis, D. A. and Shaikh, A., 1990. The ideal tissue adhesive in facial plastic and reconstructive surgery. *Journal of Otolaryngology*, 19:68–72.
127. Matthews, T. W. and Briant, T. D., 1991. The use of fibrin tissue glue in thyroid surgery: resource utilization implications. *Journal of Otolaryngology*, 20:276–278.
128. Siedentop, K. H. and Schobel, H., 1991. Stapedectomy modified by the application of fibrin tissue adhesive [see comments]. *American Journal of Otology*, 12:443–445.
129. Kojima, H., Omori, K., Shoji, K. and Kanaji, M., 1993. Reconstruction of the anterior commissure with a free mucosal flap. *American Journal of Otolaryngology*, 14:339–342.
130. Li, J. C., Leonetti, J. P. and Mokarry, V., 1993. The concept of transtympanic injection of fibrin caulk. *Archives of Otolaryngology—Head & Neck Surgery*, 119:854–857.
131. Nissen, A. J., Johnson, A. J., Perkins, R. C. and Welsh, J. E., 1993. Fibrin glue in otology and neurotology. *American Journal of Otology*, 14:147–150.
132. Moralee, S. J., Carney, A. S., Cash, M. P. and Murray, J. A., 1994. The effect of fibrin sealant haemostasis on post-operative pain in tonsillectomy. *Clinical Otolaryngology*, 19:526–528.
133. Murray, J. A., Willins, M. and Mountain, R. E., 1994. A comparison of glue and a tube as an anastomotic agent to repair the divided buccal branch of the rat facial nerve. *Clinical Otolaryngology*, 19:190–192.
134. Toriumi, D. M. and K, O. G., 1994. Surgical tissue adhesives in otolaryngology-head and neck surgery. *Otolaryngologic Clinics of North America*, 27:203–209.
135. Zeitouni, A. G., Frenkiel, S. and Mohr, G., 1994. Endoscopic repair of anterior skull base cerebrospinal fluid fistulas: an emphasis on postoperative nasal function maximization. *Journal of Otolaryngology*, 23:225–227.
136. Dornhoffer, J. L. and Milewski, C., 1995. Management of the open labyrinth. *Otolaryngology—Head & Neck Surgery*, 112:410–414.
137. Gillman, G. S. and Parnes, L. S., 1995. Acoustic neuroma management: a six-year review. *Journal of Otolaryngology*, 24:191–197.
138. Gleich, L. L., Rebeiz, E. E., Pankratov, M. M. and Shapshay, S. M., 1995. Autologous fibrin tissue adhesive in endoscopic sinus surgery. *Otolaryngology—Head & Neck Surgery*, 112:238–241.
139. Siedentop, K. H., Park, J. J. and Sanchez, B., 1995. An autologous fibrin tissue adhesive with greater bonding power. *Archives of Otolaryngology—Head & Neck Surgery*, 121:769–772.
140. Bold, E. L., Wanamaker, J. R., Zins, J. E. and Lavertu, P., 1996. The use of fibrin glue in the healing of skin flaps. *American Journal of Otolaryngology*, 17:27–30.

141. Isaacson, G. and Herman, J. H., 1996. Autologous plasma fibrin glue: rapid preparation and selective use. *American Journal of Otolaryngology*, 17:92–94.
142. Weber, R., Keerl, R., Draf, W., Schick, B., Mosler, P. and Saha, A., 1996. Management of dural lesions occurring during endonasal sinus surgery. *Archives of Otolaryngology—Head & Neck Surgery*, 122:732–736.
143. Siedentop, K. H., Chung, S. E., Park, J. J., Sanchez, B., Bhattacharya, T. and Marx, G., 1997. Evaluation of pooled fibrin sealant for ear surgery. *American Journal of Otology*, 18:660–664.
144. Selesnick, S. and al-Rawi, M., 1997. Adhesives in otology and neurotology. *American Journal of Otolaryngology*, 18:81–89.
145. Wax, M. K., Ramadan, H. H., Ortiz, O. and Wetmore, S. J., 1997. Contemporary management of cerebrospinal fluid rhinorrhea [published erratum appears in *Otolaryngol Head Neck Surg* 1997 Sep;117(3 Pt 1):197]. *Otolaryngology—Head & Neck Surgery*, 116:442–449.
146. Davis, B. R. and Sandor, G. K., 1998. Use of fibrin glue in maxillofacial surgery. *Journal of Otolaryngology*, 27:107–112.
147. Kollias, S. L. and Fox, J. M., 1996. Meniscal repair. Where do we go from here? *Clinics in Sports Medicine*, 15:621–630.
148. Homminga, G. N., Buma, P., Koot, H. W., van der Kraan, P. M. and van den Berg, W. B., 1993. Chondrocyte behavior in fibrin glue in vitro. *Acta Orthopaedica Scandinavica*, 64:441–445.
149. Peretti, G. M., Randolph, M. A., Caruso, E. M., Rossetti, F. and Zaleske, D. J., 1998. Bonding of cartilage matrices with cultured chondrocytes: an experimental model. *Journal of Orthopaedic Research*, 16:89–95.
150. Sims, C. D., Butler, P. E., Cao, Y. L., Casanova, R., Randolph, M. A., Black, A., Vacanti, C. A. and Yaremchuk, M. J., 1998. Tissue engineered neocartilage using plasma derived polymer substrates and chondrocytes. *Plastic & Reconstructive Surgery*, 101:1580–1585.
151. Ting, V., Sims, C. D., Brecht, L. E., McCarthy, J. G., Kasabian, A. K., Connelly, P. R., Elisseeff, J., Gittes, G. K. and Longaker, M. T., 1998. In vitro prefabrication of human cartilage shapes using fibrin glue and human chondrocytes. *Annals of Plastic Surgery*, 40:413–420; discussion 420–411.
152. Adamyan, L. V., Myinbayev, O. A. and Kulakov, V. I., 1991. Use of fibrin glue in obstetrics and gynecology: a review of the literature. *International Journal of Fertility*, 36:76–77, 81–78.
153. Papadopulos, N. A., Van Ballaer, P. P., Ordonez, J. L., Laermans, I. J., Vandenberghe, K., Lerut, T. E. and Deprest, J. A., 1998. Fetal membrane closure techniques after hysteroamniotomy in the midgestational rabbit model. *American Journal of Obstetrics & Gynecology*, 178:938–942.
154. Philippe, H. J., Perdu, M., Dompeyre, P., Wahid, A. and Dien, D. T., 1996. Transvaginal colpourethropexy with fibrin sealant: 4 years-follow up in 23 cases [published erratum appears in *Eur J Obstet Gynecol Reprod Biol* 1997 Apr;72(2):219–20]. *European Journal of Obstetrics, Gynecology, & Reproductive Biology*, 70:157–158.
155. Uchide, K., Terada, S., Hamasaki, H., Suzuki, N. and Akasofu, K., 1994. Intracervical fibrin instillation as an adjuvant to treatment for second trimester rupture of membranes. *Archives of Gynecology & Obstetrics*, 255:95–98.
156. Kjaergard, H. K. and Weis-Fogh, U. S., 1994. Important factors influencing the strength of autologous fibrin glue; the fibrin concentration and reaction time—comparison of strength with commercial fibrin glue. *European Surgical Research*, 26:273–276.
157. Boeckx, W., Vandevoort, M., Blondeel, P., Van Raemdonck, D. and Vandekerckhove, E., 1992. Fibrin glue in the treatment of dorsal hand burns. *Burns*, 18:395–400.
158. Boyce, S. T., Holder, I. A., Supp, A. P., Warden, G. D. and Greenhalgh, D. G., 1994. Delivery and activity of antimicrobial drugs released from human fibrin sealant. *Journal of Burn Care & Rehabilitation*, 15:251–255.
159. Brown, D. M., Barton, B. R., Young, V. L. and Pruitt, B. A., 1992. Decreased wound contraction with fibrin glue—treated skin grafts [published errata appear in *Arch Surg* 1992 Jul;127(7):822 and 1992 Aug;127(8):960]. *Archives of Surgery*, 127:404–406.
160. Cederholm-Williams, S. A., 1994. Benefits of adjuvant fibrin glue in skin grafting [letter; comment]. *Medical Journal of Australia*, 161:575.

161. Chakravorty, R. C. and Sosnowski, K. M., 1989. Autologous fibrin glue in full-thickness skin grafting. *Annals of Plastic Surgery*, 23:488–491.
162. Dean, M., Nicholls, M. and Wedderburn, C., 1994. Benefits of adjuvant fibrin glue in skin grafting [letter] [see comments]. *Medical Journal of Australia*, 160:526–527.
163. Jabs, A. D., Jr., Wider, T. M., DeBellis, J. and Hugo, N. E., 1992. The effect of fibrin glue on skin grafts in infected sites. *Plastic & Reconstructive Surgery*, 89:268–271.
164. Schumacher, J., Ford, T. S., Brumbaugh, G. W. and Honnas, C. M., 1996. Viability of split-thickness skin grafts attached with fibrin glue. *Canadian Journal of Veterinary Research*, 60: 158–160.
165. McGill, V., Kowal-Vern, A., Lee, M., Greenhalgh, D., Gomperts, E., Bray, G. and Gamelli, R., 1997. Use of fibrin sealant in thermal injury. *Journal of Burn Care & Rehabilitation*, 18:429–434.
166. Saltz, R., Dimick, A., Harris, C., Grotting, J. C., Psillakis, J. and Vasconez, L. O., 1989. Application of autologous fibrin glue in burn wounds. *Journal of Burn Care & Rehabilitation*, 10:504–507.
167. Stuart, J. D., Morgan, R. F. and Kenney, J. G., 1990. Single-donor fibrin glue for hand burns. *Annals of Plastic Surgery*, 24:524–527.
168. Vedung, S. and Hedlung, A., 1993. Fibrin glue: its use for skin grafting of contaminated burn wounds in areas difficult to immobilize. *Journal of Burn Care & Rehabilitation*, 14:356–358.
169. Vibe, P. and Pless, J., 1983. A new method of skin graft adhesion. *Scandinavian Journal of Plastic & Reconstructive Surgery*, 17:263–264.
170. Eroglu, E., Oral, S., Unal, E., Kalayci, M., Oksuz, O. and Tilmaz, M., 1996. Reducing seroma formation with fibrin glue in an animal mastectomy model. *European Journal of Surgical Oncology*, 22:137–139.
171. Hein, K. D., 1998. Use of fibrin sealant to prevent seromas in rats [letter]. *Plastic & Reconstructive Surgery*, 101:1744–1745.
172. Harada, R. N., Pressler, V. M. and McNamara, J. J., 1992. Fibrin glue reduces seroma formation in the rat after mastectomy. *Surgery, Gynecology & Obstetrics*, 175:450–454.
173. Kulber, D. A., Bacilious, N., Peters, E. D., Gayle, L. B. and Hoffman, L., 1997. The use of fibrin sealant in the prevention of seromas [see comments]. *Plastic & Reconstructive Surgery*, 99:842–849; discussion 850–841.
174. Lindsey, W. H., Masterson, T. M., Spotnitz, W. D., Wilhelm, M. C. and Morgan, R. F., 1990. Seroma prevention using fibrin glue in a rat mastectomy model. *Archives of Surgery*, 125:305–307.
175. Lindsey, W. H., Becker, D. G., Hoare, J. R., Cantrell, R. W. and Morgan, R. F., 1995. Comparison of topical fibrin glue, fibrinogen, and thrombin in preventing seroma formation in a rat model. *Laryngoscope*, 105:241–243.
176. Schwabegger, A. H., Ninkovic, M. M. and Anderl, H., 1998. Fibrin glue to prevent seroma formation [letter]. *Plastic & Reconstructive Surgery*, 101:1744.
177. Wang, J. Y., Goodman, N. C., Amiss, L. R., Jr., Nguyen, D. H., Rodeheaver, G. T., Moore, M. M., Morgan, R. F., Abbott, R. D. and Spotnitz, W. D., 1996. Seroma prevention in a rat mastectomy model: use of a light-activated fibrin sealant. *Annals of Plastic Surgery*, 37:400–405.
178. Lindsey, W. H., Masterson, T. M., Llaneras, M., Spotnitz, W. D., Wanebo, H. J. and Morgan, R. F., 1988. Seroma prevention using fibrin glue during modified radical neck dissection in a rat model. *American Journal of Surgery*, 156:310–313.
179. Sanders, R. P., Goodman, N. C., Amiss, L. R., Jr., Pierce, R. A., Moore, M. M., Marx, G., Morgan, R. F. and Spotnitz, W. D., 1996. Effect of fibrinogen and thrombin concentrations on mastectomy seroma prevention. *Journal of Surgical Research*, 61:65–70.
180. Jonk, A., van Dongen, J. A. and Kroon, B. B., 1987. Prevention of seroma following axillary lymph node dissection or radical mastectomy; ineffectiveness of fibrin glue sealing technique [letter]. *Netherlands Journal of Surgery*, 39:135.
181. Auger, F. A., Guignard, R., Lopez Valle, C. A. and Germain, L., 1993. Role and innocuity of Tisseel, a tissue glue, in the grafting process and in vivo evolution of human cultured epidermis. *British Journal of Plastic Surgery*, 46:136–142.
182. Ronfard, V., Broly, H., Mitchell, V., Galizia, J. P., Hochart, D., Chambon, E., Pellerin, P. and

Huart, J. J., 1991. Use of human keratinocytes cultured on fibrin glue in the treatment of burn wounds. *Burns*, 17:181–184.

183. Lasa, C. I., Jr., Kidd, R. R. D., Nunez, H. A. and Drohan, W. N., 1993. Effect of fibrin glue and opsite on open wounds in DB/DB mice. *Journal of Surgical Research*, 54:202–206.
184. Eden, C. G. and Coptcoat, M. J., 1996. Assessment of alternative tissue approximation techniques for laparoscopy. *British Journal of Urology*, 78:234–242.
185. Staindl, O., Galvan, G. and Macher, M., 1981. The influence of fibrin stabilization and fibrinolysis on the fibrin-adhesive system. A clinical study using radioactively marked fibrinogen as a tracer. *Archives of Oto-Rhino-Laryngology*, 233:105–116.
186. Milde, L. N., 1989. An anaphylactic reaction to fibrin glue. *Anesthesia & Analgesia*, 69:684–686.
187. Olsen, P. S. and Hjelms, E., 1989. Intravascular air after fibrin sealing by spray gun in cardiovascular surgery. *European Journal of Cardio-Thoracic Surgery*, 3:376–377.
188. Berguer, R., Staerkel, R. L., Moore, E. E., Moore, F. A., Galloway, W. B. and Mockus, M. B., 1991. Warning: fatal reaction to the use of fibrin glue in deep hepatic wounds. Case reports. *Journal of Trauma*, 31:408–411.
189. Hennis, H. L., Stewart, W. C. and Jeter, E. K., 1992. Infectious disease risks of fibrin glue [letter] [see comments]. *Ophthalmic Surgery*, 23:640.
190. Mitsuhata, H., Horiguchi, Y., Saitoh, J., Saitoh, K., Fukuda, H., Hirabayasi, Y., Togashi, H. and Shimizu, R., 1994. An anaphylactic reaction to topical fibrin glue. *Anesthesiology*, 81:1074–1077.
191. Muntean, W., Zenz, W., Finding, K., Zobel, G. and Beitzke, A., 1994. Inhibitor to factor V after exposure to fibrin sealant during cardiac surgery in a two-year-old child. *Acta Paediatrica*, 83: 84–87.
192. Ortel, T. L., Charles, L. A., Keller, F. G., Marcom, P. K., Oldham, H. N., Jr., Kane, W. H. and Macik, B. G., 1994. Topical thrombin and acquired coagulation factor inhibitors: clinical spectrum and laboratory diagnosis. *American Journal of Hematology*, 45:128–135.
193. Muntean, W., Zenz, W., Edlinger, G. and Beitzke, A., 1997. Severe bleeding due to factor V inhibitor after repeated operations using fibrin sealant containing bovine thrombin [letter; comment]. *Thrombosis & Haemostasis*, 77:1223.
194. Jackson, M. R., 1996. Tissue sealants: current status, future potential. *Nature Medicine*, 2:637–638.
195. Casali, B., Rodeghiero, F., Tosetto, A., Palmieri, B., Immovilli, R., Ghedini, C. and Rivasi, P., 1992. Fibrin glue from single-donation autologous plasmapheresis. *Transfusion*, 32:641–643.
196. Chisholm, R. A., Jones, S. N. and Lees, W. R., 1989. Fibrin sealant as a plug for the post liver biopsy needle track. *Clinical Radiology*, 40:627–628.
197. Raborn, G. W., Hohn, F. I., Grace, M. G. and Arora, B. K., 1990. Tisseel, a two component fibrin tissue sealant system: report of a trial involving anticoagulated dental patients. *Journal Canadian Dental Association. Journal de l Association Dentaire Canadienne*, 56:779–781.
198. Brennan, M., 1991. Fibrin glue. *Blood Reviews*, 5:240–244.
199. Henrick, A., Kalpakian, B., Gaster, R. N. and Vanley, C., 1991. Organic tissue glue in the closure of cataract incisions in rabbit eyes. *Journal of Cataract & Refractive Surgery*, 17:551–555.
200. Silverstein, J. I. and Mellinger, B. C., 1991. Fibrin glue vasal anastomosis compared to conventional sutured vasovasostomy in the rat. *Journal of Urology*, 145:1288–1291.
201. Ammar, A., 1992. Repair of skull base dural defects: the dura sandwich. Technical note. *Acta Neurochirurgica*, 119:174–175.
202. Kinahan, T. J. and Johnson, H. W., 1992. Tisseel in hypospadias repair. *Canadian Journal of Surgery*, 35:75–77.
203. Marczell, A. P. and Stierer, M., 1992. Partial pancreaticoduodenectomy (Whipple procedure) for pancreatic malignancy: occlusion of a non-anastomosed pancreatic stump with fibrin sealant. *HPB Surgery*, 5:251–259; discussion 259–260.
204. Warrer, K. and Karring, T., 1992. Guided tissue regeneration combined with osseous grafting in suprabony periodontal lesions. An experimental study in the dog. *Journal of Clinical Periodontology*, 19:373–380.
205. Warrer, K. and Karring, T., 1992. Effect of Tisseel on healing after periodontal flap surgery. *Journal of Clinical Periodontology*, 19:449–454.

206. Glimaker, H., Bjorck, C. G., Hallstensson, S., Ohlsen, L. and Westman, B., 1993. Avoiding blowout of the aortic stump by reinforcement with fibrin glue. A report of two cases. *European Journal of Vascular Surgery*, 7:346–348.
207. Lazovic, D. and Messner, K., 1993. Collagen repair not improved by fibrin adhesive. Cruciate ligament ruptures studied in dogs. *Acta Orthopaedica Scandinavica*, 64:583–586.
208. Oberg, S. and Kahnberg, K. E., 1993. Combined use of hydroxy-apatite and Tisseel in experimental bone defects in the rabbit. *Swedish Dental Journal*, 17:147–153.
209. Oosterlinck, W., Cheng, H., Hoebeke, P. and Verbeeck, R., 1993. Watertight sutures with fibrin glue: an experimental study. *European Urology*, 23:481–484.
210. Schwarz, N., Redl, H., Zeng, L., Schlag, G., Dinges, H. P. and Eschberger, J., 1993. Early osteoinduction in rats is not altered by fibrin sealant. *Clinical Orthopaedics & Related Research*, :353–359.
211. Wadstrom, J. and Wik, O., 1993. Fibrin glue (Tisseel) added with sodium hyaluronate in microvascular anastomosing. *Scandinavian Journal of Plastic & Reconstructive Surgery & Hand Surgery*, 27:257–261.
212. Canil, K., Fitzgerald, P. and Lau, G., 1994. Massive chylothorax associated with lymphangiomatosis of the bone. *Journal of Pediatric Surgery*, 29:1186–1188.
213. Oberg, S. and Rosenquist, J. B., 1994. Bone healing after implantation of hydroxyapatite granules and blocks (Interpore 200) combined with autolyzed antigen-extracted allogeneic bone and fibrin glue. Experimental studies on adult rabbits. *International Journal of Oral & Maxillofacial Surgery*, 23:110–114.
214. Povlsen, B., 1994. A new fibrin seal: functional evaluation of sensory regeneration following primary repair of peripheral nerves. *Journal of Hand Surgery—British Volume*, 19:250–254.
215. Povlsen, B., 1994. A new fibrin seal in primary repair of peripheral nerves. *Journal of Hand Surgery—British Volume*, 19:43–47.
216. Brady, A. P., Malone, D. E., Deignan, R. W., N, O. D. and McGrath, F. P., 1995. Fibrin sealant in interventional radiology: a preliminary evaluation. *Radiology*, 196:573–578.
217. Falconer, C. and Larsson, B., 1995. New and simplified vaginal approach for correction of urinary stress incontinence in women. *Neurourology & Urodynamics*, 14:365–370.
218. Kiilholma, P., Haarala, M., Polvi, H., Makinen, J. and Chancellor, M. B., 1995. Sutureless endoscopic colposuspension with fibrin sealant. *Techniques in Urology*, 1:81–83.
219. Kucukaydin, M., Okur, H., Kontas, O. and Patiroglu, T. E., 1995. Fibrin glue and conventional sutured vasal anastomosis in the rat. *Journal of Surgical Research*, 59:601–605.
220. Lee, K. H., Kim, M. S., Hahn, T. W. and Kim, J. H., 1995. Comparison of histologic findings in wound healing of rabbit scleral homografts with fibrin glue (Tisseel) and suture material. *Journal of Refractive Surgery*, 11:397–401.
221. Wiseman, N. E., 1995. Endoscopic closure of recurrent tracheoesophageal fistula using Tisseel. *Journal of Pediatric Surgery*, 30:1236–1237.
222. Bowen, C. V., Leach, D. H., Crosby, N. L. and Reynolds, R., 1996. Microvascular anastomoses. A comparative study of fibrinogen adhesive and interrupted suture techniques. *Plastic & Reconstructive Surgery*, 97:792–800.
223. Jarzem, P., Harvey, E. J., Shenker, R. and Hajipavlou, A., 1996. The effect of fibrin sealant on spinal fusions using allograft in dogs. *Spine*, 21:1307–1312.
224. Ochi, M., Mochizuki, Y., Deie, M. and Ikuta, Y., 1996. Augmented meniscal healing with free synovial autografts: an organ culture model. *Archives of Orthopaedic & Trauma Surgery*, 115:123–126.
225. Brittberg, M., Sjogren-Jansson, E., Lindahl, A. and Peterson, L., 1997. Influence of fibrin sealant (Tisseel) on osteochondral defect repair in the rabbit knee. *Biomaterials*, 18:235–242.
226. Ohsumi, N., 1998. Postoperative compressive penile dressing using fibrin seal (Tisseel) and tulle gauze for hypospadias repair [letter]. *Plastic & Reconstructive Surgery*, 101:1737–1738.
227. Wadstrom, J. and Tengblad, A., 1993. Fibrin glue reduces the dissolution rate of sodium hyaluronate. *Upsala Journal of Medical Sciences*, 98:159–167.
228. Sierra, D. H., 1996. Fibrin-collagen composite tissue adhesive. In: Sierra D. H., Saltz R. (Editors).

Surgical Adhesives and Sealants—Current Technology and Applications. Lancaster: Technomic, pp. 29–39.
229. DeBlois, C., Cote, M. F. and Doillon, C. J., 1994. Heparin-fibroblast growth factor-fibrin complex: in vitro and in vivo applications to collagen-based materials. *Biomaterials*, 15:665–672.
230. Schlag, G. and Redl, H., 1987. Fibrin sealant in orthopedic surgery. *Clinical Orthopedic Research*, 227:269–285.
231. Ono, K., Shikata, J., Shimizu, K. and Yamamuro, T., 1992. Bone-fibrin mixture in spinal surgery. *Clinical Orthopaedics & Related Research*, :133–139.
232. Ishimura, M., Tamai, S. and Fujisawa, Y., 1991. Arthroscopic meniscal repair with fibrin glue. *Arthroscopy*, 7:177–181.
233. Ishimura, M., Ohgushi, H., Habata, T., Tamai, S. and Fujisawa, Y., 1997. Arthroscopic meniscal repair using fibrin glue. Part I: Experimental study. *Arthroscopy*, 13:551–557.
234. Ishimura, M., Ohgushi, H., Habata, T., Tamai, S. and Fujisawa, Y., 1997. Arthroscopic meniscal repair using fibrin glue. Part II: Clinical applications. *Arthroscopy*, 13:558–563.
235. Bertrand, B., Doyen, A. and Eloy, P., 1996. Triosite implants and fibrin glue in the treatment of atrophic rhinitis: technique and results. *Laryngoscope*, 106:652–657.
236. Hotz, G., 1991. Alveolar ridge augmentation with hydroxylapatite using fibrin sealant for fixation. Part I: An experimental study. *International Journal of Oral & Maxillofacial Surgery*, 20:204–207.
237. Hotz, G., 1991. Alveolar ridge augmentation with hydroxylapatite using fibrin sealant for fixation. Part II: Clinical application. *International Journal of Oral & Maxillofacial Surgery*, 20:208–213.
238. Pico, G., 1995. Thermodynamic aspects of the thermal stability of human serum albumin. *Biochemistry & Molecular Biology International*, 36:1017–1023.
239. Lee, T. K., Sokolowski, T. D. and Royer, G. P., 1981. Serum albumin beads: an injectable, biodegradable system for the sustained release of drugs. *Science*, 213:233–235.
240. Goosen, M. F., Leung, Y. F., Chou, S. and Sun, A. M., 1982. Insulin-albumin microbeads: an implantable, biodegradable system. *Biomaterials, Medical Devices and Artificial Organs*, 10:205–218.
241. Cremers, H. F., Wolf, R. F., Blauww, E. H., Schakenraad, J. M., Lam, K. H., Nieuwehuis, P., Verrijk, R. and Kwon, G., 1994. Degradation and intrahepatic compatibility of albumin-heparin conjugate microspheres. *Biomaterials*, 15:577–585.
242. Lin, W., Garnett, M. C., Davies, M. C., Bignotti, F., Ferruti, P., Davis, S. S. and Illum, L., 1997. Preparation of surface-modified albumin nanospheres. *Biomaterials*, 18:559–565.
243. Pavanetto, F., Genta, I., Giunchedi, P., Conti, B. and Conte, U., 1994. Spray-dried albumin microspheres for the intra-articular delivery of dexamethasone. *Journal of Microencapsulation*, 11:445–454.
244. Rubino, O. P., Kowalsky, R. and Swarbrick, J., 1993. Albumin microspheres as a drug delivery system: relation among turbidity ratio, degree of cross-linking, and drug release. *Pharmaceutical Research*, 10:1059–1065.
245. Ben Slimane, S., Guidoin, R., Merhi, Y., King, M. W., Domurado, D. and Sigot-Luizard, M. F., 1988. In vivo evaluation of polyester arterial grafts coated with albumin: the role and importance of cross-linking agents. *European Surgical Research*, 20:66–74.
246. Chafke, N., Gasser, B., Lindner, V., Rouyer, N., Rooke, R., Kretz, J. G., Nicolini, P. and Eisenmann, B., 1996. Albumin as a sealant for a polyester vascular prosthesis: its impact on the healing sequence in humans. *Journal of Cardiovascular Surgery*, 37:431–440.
247. Desoize, B., Jardillier, J. C., Kanoun, K., Guerin, D. and Levy, M. C., 1986. *In-vitro* cytotoxic activity of cross-linked protein microcapsules. *Journal of Pharmaceutics and Pharmacology*, 38: 8–13.
248. Thung, S. N., Wang, D. F., Fasy, T. M., Hood, A. and Gerber, M. A., 1989. Hepatitis B surface antigen binds to human serum albumin cross-linked by transglutaminase. *Hepatology*, 9:726–730.
249. Ben Slimane, S., Guidoin, R., Mourad, W., Hebert, J., King, M. W. and Sigot-Luizard, M. F., 1988. Polyester arterial grafts impregnated with cross-linked albumin: the rate of degradation of the coating in vivo. *European Surgical Research*, 20:12–17.
250. Ben Slimane, S., Guidoin, R., Marceau, D., Merhi, Y., King, M. W. and Sigot-Luizard, M. F.,

1988. Characteristics of polyester arterial grafts coated with albumin: the role and importance of the cross-linking chemicals. *European Surgical Research*, 20:18–28.
251. An, Y. H., Stuart, G. W., McDowell, S. J., McDaniel, S. E., Kang, Q. and Friedman, R. J., 1996. Prevention of bacterial adherence to implant surfaces with a crosslinked albumin coating in vitro. *Journal of Orthopaedic Research*, 14:846–849.
252. Zammit, M., Wu, H. D., Mathisen, S. R. and Sauvage, L. R., 1986. Influence on healing in the canine thoracic aorta of three substances used to close the interstices of macroporous Dacron grafts. *American Surgeon*, 52:667–669.
253. Warkentin, P., Walivaara, B., Lundstrom, I. and Tengvall, P., 1994. Differential surface binding of albumin, immunoglobin G and fibrinogen. *Biomaterials*, 15:786–795.
254. Klinger, A., Steinberg, D., Kohavi, D. and Sela, M. N., 1997. Mechanism of adsorption of human albumin to titanium *in vitro*. *Journal of Biomedical Materials Research*, 38:387–392.
255. McDowell, S. G., An, Y. H., Draughn, R. A. and Friedman, R. J., 1995. Application of a fluorescent redox dye for enumeration of metabolically active bacteria on albumin-coated titanium surfaces. *Letters in Applied Microbiology*, 21:1–4.
256. An, Y. H., Bradley, J., Powers, D. L. and Friedman, R. J., 1997. The prevention of prosthetic infection using a cross-linked albumin coating in a rabbit model. *Journal of Bone and Joint Surgery*, 79B:816–819.
257. Yen, S. and Gibbons, R. J., 1987. The influence of albumin on adsorption of bacteria on hydroxyapatite beads *in vitro* and human tooth surfaces *in vivo*. *Archives of Oral Biology*, 32:531–533.
258. Luft, J. H., 1992. Fixation for biological ultrastructure. I. A viscometric analysis of the interaction between glutaraldehyde and bovine serum albumin. *Journal of Microscopy*, 167:247–258.
259. D'Urso, E. M., Jean-Francois, J., Doillon, C. J. and Fortier, G., 1995. Poly(ethylene glycol)-serum albumin hydrogel as matrix for enzyme immobilization: biomedical applications. *Art Cells Blood Subs and Immob Biotech*, 23:587–595.
260. Gayet, J.-C. and Fortier, G., 1995. Drug release from new artifical hydrogel. *Art Cells Blood Subs and Immob Biotech*, 23:605–611.
261. Weissleder, R., Poss, K., Wilkinson, R., Zhou, C. and Bogdanov Jr., A., 1995. Quantitation of slow drug release from an implantable and degradable gentamicin conjugate by *in vivo* magnetic resonance imaging. *Antimicrobial Agents and Chemotherapy*, 39:839–845.
262. Barrows, T. H., Truong, M. T., Lewis, T. W., Grussing, D. M., Kato, K. H., Gysbers, J. E. and Lamprecht, E. G. Evaluation of a new tissue sealant material: serum albumin crosslinked *in vivo* with poolyethylene glycol. In: Fifth World Biomaterials Conference; May 29–June 2, 1996; Toronto, p. 8.
263. Truong, M. T., Barrows, T. H. and Wilson, T. J. *In vitro* analysis of mechanical properties of a new tissue sealant material: polyethylene glycol crosslinked serum albumin. In: Fifth World Biomaterials Conference; May 29–June 2, 1996; Toronto, p. 73.
264. Huang, S. T., Kilpadi, D. V. and Feldman, D. S., 1997. A comparison of the shear strength of fibrin and albumin glues. *Transactions of the Wound Healing Society*, 7:63.
265. Huang, S. T., Blum, B. E., Barker, S. D., Barker, T. H., Kilpadi, D. V., Redden, R. A. and Feldman, D. S., 1999. *In vivo* evaluation of an adhesive albumin used for incision closure (submitted for publication). *Transactions of the Society for Biomaterials*.
266. Ardis, A. E., inventor, 1949. US patent 2467927.
267. Coover, H. W., Joyner, F. B., Shearer, N. H. and Wicker, T. H., 1959. Chemistry and performance of cyanoacrylate adhesives. *Journal of the Society of Plastic Engineering*, 15:413–417.
268. McKelvie, P., 1969. A trial of adhesives in reconstructive middle-ear surgery. *Journal of Laryngology & Otology*, 83:1105–1109.
269. Leonard, F., 1968. The *n*-alkylalphacyanoacrylate tissue adhesives. *Annals of the New York Academy of Sciences*, 146:203–213.
270. Arthaud, L. E., Lewellen, G. R. and Akers, W. A., 1972. The dermal toxicity of isoamyl-2-cyanoacrylate. *Journal of Biomedical Materials Research*, 6:201–214.
271. Lehman, R. A., Hayes, G. J. and Leonard, F., 1966. Toxicity of alkyl 2-cyanoacrylates. I. Peripheral nerve. *Arch Surg*, 93:441–446.

272. Lehman, R. A., West, R. L. and Leonard, F., 1966. Toxicity of alkyl 2-cyanoacrylates. II. Bacterial growth. *Arch Surg*, 93:447–450.
273. Aronson, S. B., McMaster, P. R., Moore, T. E., Jr. and Coon, M. A., 1970. Toxicity of the cyanoacrylates. *Arch Ophthalmol*, 84:342–349.
274. Stevenson, T. C. and Taylor, D. S., 1972. The effect of methyl cyanoacrylate tissue adhesive on the human Fallopian tube and endometrium. *Journal of Obstetrics & Gynaecology of the British Commonwealth*, 79:1028–1039.
275. Aleo, J. J. and DeRenzis, F. A., 1975. On the possible mechanism of cyanoacrylate histotoxicity. *Pharmacology & Therapeutics in Dentistry*, 2:21–24.
276. Coover, H. W. and McIntire, J. M., 1972. The chemistry of cyanoacrylate adhesives. In: Matsumoto T. (Editor). *Tissue Adhesives in Surgery*. New York: Medical Examination.
277. Smith, D. C., 1973. Lutes, glues, cements and adhesives in medicine and dentistry. *BioMedical Engineering*, 8:108–115 passim.
278. Bonutti, P. M., Weiker, G. G. and Andrish, J. T., 1988. Isobutyl cyanoacrylate as a soft tissue adhesive. An in vitro study in the rabbit Achilles tendon. *Clinical Orthopaedics & Related Research*, :241–248.
279. Kaufman, R. S., 1974. The use of tissue adhesive (isobutyl cyanoacrylate) and topical steroid (0.1 percent dexamethasone) in experimental tympanoplasty. *Laryngoscope*, 84:793–804.
280. Greer, R. O., Jr., 1975. Studies concerning the hsitotoxicity of isobutyl-2-cyanoacrylate tissue adhesive when employed as an oral hemostat. *Oral Surgery, Oral Medicine, Oral Pathology*, 40: 659–669.
281. Hunter, K. M., 1976. Cyanoacrylate tissue adhesive in osseous repair. *British Journal of Oral Surgery*, 14:80–86.
282. Diaz, F. G., Mastri, A. R. and Chou, S. N., 1978. Neural and vascular tissue reaction to aneurysm-coating adhesive (ethyl 2-cyanoacrylate). *Neurosurgery*, 3:45–49.
283. Hood, T. W., Mastri, A. R. and Chou, S. N., 1982. Neural and vascular tissue reaction of cyanoacrylate adhesives: a further report. *Neurosurgery*, 11:363–366.
284. Zumpano, B. J., Jacobs, L. R., Hall, J. B., Margolis, G. and Sachs, E., Jr., 1982. Bioadhesive and histotoxic properties of ethyl-2-cyanoacrylate. *Surgical Neurology*, 18:452–457.
285. Vinters, H. V., Galil, K. A., Lundie, M. J. and Kaufmann, J. C., 1985. The histotoxicity of cyanoacrylates. A selective review. *Neuroradiology*, 27:279–291.
286. Vinters, H. V., Lundie, M. J. and Kaufmann, J. C., 1986. Long-term pathological follow-up of cerebral arteriovenous malformations treated by embolization with bucrylate. *New England Journal of Medicine*, 314:477–483.
287. Kerr, A. G. and Smyth, G. D., 1971. Bucrylate (isobutyl cyanoacrylate) as an ossicular adhesive. *Archives of Otolaryngology*, 94:129–131.
288. Kerlan, R. K., Jr., Bank, W. O., Hoddick, W. K., Pogany, A. C. and Sollenberger, R. D., 1983. Occlusion of a hepatic artery to portal vein fistula with bucrylate. *Cardiovascular & Interventional Radiology*, 6:138–140.
289. Toriumi, D. M., Raslan, W. F., Friedman, M. and Tardy, M. E., Jr., 1991. Variable histotoxicity of histoacryl when used in a subcutaneous site: an experimental study. *Laryngoscope*, 101:339–343.
290. Toriumi, D. M., Raslan, W. F., Friedman, M. and Tardy, M. E., 1990. Histotoxicity of cyanoacrylate tissue adhesives. A comparative study. *Archives of Otolaryngology—Head & Neck Surgery*, 116:546–550.
291. Hanft, J. R., Kashuk, K. B., Toney, M. E. and McDonald, T. D., 1991. Peripheral neuropathy as a result of cyanoacrylate toxicity. *Journal of the American Podiatric Medical Association*, 81: 653–655.
292. Papatheofanis, F. J., 1989. Prothrombotic cytotoxicity of cyanoacrylate tissue adhesive. *Journal of Surgical Research*, 47:309–312.
293. Kazekawa, K., Iwata, H., Shimozuru, T., Sampei, K., Sakaki, N., Morikawa, N., Matsuda, S. and Ikada, Y., 1997. Nontoxic embolic liquids for treatment of arteriovenous malformations. *Journal of Biomedical Materials Research*, 38:79–86.

294. Samson, D. and Marshall, D., 1986. Carcinogenic potential of isobutyl-2-cyanoacrylate [letter]. *Journal of Neurosurgery*, 65:571–572.
295. Hida, T., Sheta, S. M., Proia, A. D. and McCuen, B. W. D., 1988. Retinal toxicity of cyanoacrylate tissue adhesive in the rabbit. *Retina*, 8:148–153.
296. Leahey, A. B., Gottsch, J. D. and Stark, W. J., 1993. Clinical experience with N-butyl cyanoacrylate (Nexacryl) tissue adhesive. *Ophthalmology*, 100:173–180.
297. Purdy, P. D., Batjer, H. H., Risser, R. C. and Samson, D., 1992. Arteriovenous malformations of the brain: choosing embolic materials to enhance safety and ease of excision [see comments]. *Journal of Neurosurgery*, 77:217–222.
298. Cognard, C., Miaux, Y., Pierot, L., Weill, A., Martin, N. and Chiras, J., 1996. The role of CT in evaluation of the effectiveness of embolisation of spinal dural arteriovenous fistulae with *N*-butyl cyanoacrylate. *Neuroradiology*, 38:603–608.
299. DeMeritt, J. S., Pile-Spellman, J., Mast, H., Moohan, N., Lu, D. C., Young, W. L., Hacein-Bey, L., Mohr, J. P. and Stein, B. M., 1995. Outcome analysis of preoperative embolization with *N*-butyl cyanoacrylate in cerebral arteriovenous malformations. *American Journal of Neuroradiology*, 16:1801–1807.
300. Jafar, J. J., Davis, A. J., Berenstein, A., Choi, I. S. and Kupersmith, M. J., 1993. The effect of embolization with *N*-butyl cyanoacrylate prior to surgical resection of cerebral arteriovenous malformations. *Journal of Neurosurgery*, 78:60–69.
301. Matsumoto, T., Hardaway, R. M., III Pani, K. C. and Margetis, P. M., 1968. Aron Alpha A "Sankyo," Japanese tissue adhesive, in surgery of internal organs. *American Surgeon*, 34:263–267.
302. Sabanathan, S., Eng, J. and Richardson, J., 1993. The use of tissue adhesive in pulmonary resections. *European Journal of Cardio-Thoracic Surgery*, 7:657–660.
303. McCabe, M. J., 1990. Use of histoacryl tissue adhesive to manage an avulsed tooth [see comments]. Bmj, 301:20–21.
304. Alonso, M. J., Sanchez, A., Torres, D., Seijo, B. and Vila-Jato, J. L., 1990. Joint effects of monomer and stabilizer concentrations on physico-chemical characteristics of poly(butyl 2-cyanoacrylate) nanoparticles. *Journal of Microencapsulation*, 7:517–526.
305. Alyautdin, R. N., Petrov, V. E., Langer, K., Berthold, A., Kharkevich, D. A. and Kreuter, J., 1997. Delivery of loperamide across the blood-brain barrier with polysorbate 80-coated polybutylcyanoacrylate nanoparticles. *Pharmaceutical Research*, 14:325–328.
306. Alyautdin, R. N., Tezikov, E. B., Ramge, P., Kharkevich, D. A., Begley, D. J. and Kreuter, J., 1998. Significant entry of tubocurarine into the brain of rats by adsorption to polysorbate 80-coated polybutylcyanoacrylate nanoparticles: an in situ brain perfusion study. *Journal of Microencapsulation*, 15:67–74.
307. Ammoury, N., Fessi, H., Devissaguet, J. P., Allix, M., Plotkine, M. and Boulu, R. G., 1990. Effect on cerebral blood flow of orally administered indomethacin-loaded poly(isobutylcyanoacrylate) and poly(DL-lactide) nanocapsules. *Journal of Pharmacy & Pharmacology*, 42:558–561.
308. Ammoury, N., Fessi, H., Devissaguet, J. P., Dubrasquet, M. and Benita, S., 1991. Jejunal absorption, pharmacological activity, and pharmacokinetic evaluation of indomethacin-loaded poly(D,L-lactide) and poly(isobutyl-cyanoacrylate) nanocapsules in rats. *Pharmaceutical Research*, 8:101–105.
309. Beck, P., Scherer, D. and Kreuter, J., 1990. Separation of drug-loaded nanoparticles from free drug by gel filtration. *Journal of Microencapsulation*, 7:491–496.
310. Beck, P., Kreuter, J., Reszka, R. and Fichtner, I., 1993. Influence of polybutylcyanoacrylate nanoparticles and liposomes on the efficacy and toxicity of the anticancer drug mitoxantrone in murine tumour models. *Journal of Microencapsulation*, 10:101–114.
311. Bender, A. R., von Briesen, H., Kreuter, J., Duncan, I. B. and Rubsamen-Waigmann, H., 1996. Efficiency of nanoparticles as a carrier system for antiviral agents in human immunodeficiency virus-infected human monocytes/macrophages in vitro. *Antimicrobial Agents & Chemotherapy*, 40:1467–1471.
312. Bennis, S., Chapey, C., Couvreur, P. and Robert, J., 1994. Enhanced cytotoxicity of doxorubicin

encapsulated in polyisohexylcyanoacrylate nanospheres against multidrug-resistant tumour cells in culture. *European Journal of Cancer*, 30A:89–93.

313. Blagoeva, P. M., Balansky, R. M., Mircheva, T. J. and Simeonova, M. I., 1992. Diminished genotoxicity of mitomycin C and farmorubicin included in polybutylcyanoacrylate nanoparticles. *Mutation Research*, 268:77–82.
314. Blum, G. N., Nolte, N. A. and Robertson, P., 1975. In vitro determination of the antimicrobial properties of two cyanoacrylate preparations. *Journal of Dental Research*, 54:500–503.
315. Bogush, T., Smirnova, G., Shubina, I., Syrkin, A. and Robert, J., 1995. Direct evaluation of intracellular accumulation of free and polymer-bound anthracyclines. *Cancer Chemotherapy & Pharmacology*, 35:501–505.
316. Bonduelle, S., Foucher, C., Leroux, J. C., Chouinard, F., Cadieux, C. and Lenaerts, V., 1992. Association of cyclosporin to isohexylcyanoacrylate nanospheres and subsequent release in human plasma in vitro. *Journal of Microencapsulation*, 9:173–182.
317. Brasseur, F., Couvreur, P., Kante, B., Deckers-Passau, L., Roland, M., Deckers, C. and Speiser, P., 1980. Actinomycin D absorbed on polymethylcyanoacrylate nanoparticles: increased efficiency against an experimental tumor. *European Journal of Cancer*, 16:1441–1445.
318. Cappel, M. J. and Kreuter, J., 1991. Effect of nanoparticles on transdermal drug delivery. *Journal of Microencapsulation*, 8:369–374.
319. Chavany, C., Le Doan, T., Couvreur, P., Puisieux, F. and Helene, C., 1992. Polyalkylcyanoacrylate nanoparticles as polymeric carriers for antisense oligonucleotides. *Pharmaceutical Research*, 9: 441–449.
320. Chavany, C., Saison-Behmoaras, T., Le Doan, T., Puisieux, F., Couvreur, P. and Helene, C., 1994. Adsorption of oligonucleotides onto polyisohexylcyanoacrylate nanoparticles protects them against nucleases and increases their cellular uptake. *Pharmaceutical Research*, 11:1370–1378.
321. Chiannilkulchai, N., Ammoury, N., Caillou, B., Devissaguet, J. P. and Couvreur, P., 1990. Hepatic tissue distribution of doxorubicin-loaded nanoparticles after i.v. administration in reticulosarcoma M 5076 metastasis-bearing mice. *Cancer Chemotherapy & Pharmacology*, 26:122–126.
322. Cicek, H., Tuncel, A., Tuncel, M. and Piskin, E., 1995. Degradation and drug release characteristics of monosize polyethylcyanoacrylate microspheres. *Journal of Biomaterials Science, Polymer Edition*, 6:845–856.
323. Couvreur, P., Kante, B., Roland, M., Guiot, P., Bauduin, P. and Speiser, P., 1979. Polycyanoacrylate nanocapsules as potential lysosomotropic carriers: preparation, morphological and sorptive properties. *Journal of Pharmacy & Pharmacology*, 31:331–332.
324. Couvreur, P., Kante, B., Roland, M. and Speiser, P., 1979. Adsorption of antineoplastic drugs to polyalkylcyanoacrylate nanoparticles and their release in calf serum. *Journal of Pharmaceutical Sciences*, 68:1521–1524.
325. Couvreur, P., Kante, B., Lenaerts, V., Scailteur, V., Roland, M. and Speiser, P., 1980. Tissue distribution of antitumor drugs associated with polyalkylcyanoacrylate nanoparticles. *Journal of Pharmaceutical Sciences*, 69:199–202.
326. Couvreur, P., Kante, B., Grislain, L., Roland, M. and Speiser, P., 1982. Toxicity of polyalkylcyanoacrylate nanoparticles. II: Doxorubicin-loaded nanoparticles. *Journal of Pharmaceutical Sciences*, 71:790–792.
327. Couvreur, P., 1988. Polyalkylcyanoacrylates as colloidal drug carriers. *Critical Reviews in Therapeutic Drug Carrier Systems*, 5:1–20.
328. Damge, C., Michel, C., Aprahamian, M. and Couvreur, P., 1988. New approach for oral administration of insulin with polyalkylcyanoacrylate nanocapsules as drug carrier. *Diabetes*, 37:246–251.
329. Damge, C., Vranckx, H., Balschmidt, P. and Couvreur, P., 1997. Poly(alkyl cyanoacrylate) nanospheres for oral administration of insulin. *Journal of Pharmaceutical Sciences*, 86:1403–1409.
330. Damge, C., Vonderscher, J., Marbach, P. and Pinget, M., 1997. Poly(alkyl cyanoacrylate) nanocapsules as a delivery system in the rat for octreotide, a long-acting somatostatin analogue. *Journal of Pharmacy & Pharmacology*, 49:949–954.
331. Das, S. K., Tucker, I. G., Hill, D. J. and Ganguly, N., 1995. Evaluation of poly(isobutylcyanoacry-

late) nanoparticles for mucoadhesive ocular drug delivery. I. Effect of formulation variables on physicochemical characteristics of nanoparticles. *Pharmaceutical Research*, 12:534–540.
332. de Verdiere, A. C., Dubernet, C., Nemati, F., Soma, E., Appel, M., Ferte, J., Bernard, S., Puisieux, F. and Couvreur, P., 1997. Reversion of multidrug resistance with polyalkylcyanoacrylate nanoparticles: towards a mechanism of action. *British Journal of Cancer*, 76:198–205.
333. Douglas, S. J., Davis, S. S. and Illum, L., 1987. Nanoparticles in drug delivery. *Critical Reviews in Therapeutic Drug Carrier Systems*, 3:233–261.
334. DT, O. H., Palin, K. J. and Davis, S. S., 1989. Poly(butyl-2-cyanoacrylate) particles as adjuvants for oral immunization. *Vaccine*, 7:213–216.
335. Egea, M. A., Gamisans, F., Valero, J., Garcia, M. E. and Garcia, M. L., 1994. Entrapment of cisplatin into biodegradable polyalkylcyanoacrylate nanoparticles. *Farmaco*, 49:211–217.
336. Fernandez-Urrusuno, R., Fattal, E., Porquet, D., Feger, J. and Couvreur, P., 1995. Evaluation of liver toxicological effects induced by polyalkylcyanoacrylate nanoparticles. *Toxicology & Applied Pharmacology*, 130:272–279.
337. Fernandez-Urrusuno, R., Fattal, E., Feger, J., Couvreur, P. and Therond, P., 1997. Evaluation of hepatic antioxidant systems after intravenous administration of polymeric nanoparticles. *Biomaterials*, 18:511–517.
338. Fresta, M., Cavallaro, G., Giammona, G., Wehrli, E. and Puglisi, G., 1996. Preparation and characterization of polyethyl-2-cyanoacrylate nanocapsules containing antiepileptic drugs. *Biomaterials*, 17:751–758.
339. Gasco, M. R., Morel, S., Trotta, M. and Viano, I., 1991. Doxorubicine englobed in polybutylcyanoacrylate nanocapsules: behaviour in vitro and in vivo. *Pharmaceutica Acta Helvetiae*, 66:47–49.
340. Gaspar, R., Preat, V., Opperdoes, F. R. and Roland, M., 1992. Macrophage activation by polymeric nanoparticles of polyalkylcyanoacrylates: activity against intracellular *Leishmania donovani* associated with hydrogen peroxide production. *Pharmaceutical Research*, 9:782–787.
341. Gaspar, R., Opperdoes, F. R., Preat, V. and Roland, M., 1992. Drug targeting with polyalkylcyanoacrylate nanoparticles: in vitro activity of primaquine-loaded nanoparticles against intracellular Leishmania donovani. *Annals of Tropical Medicine & Parasitology*, 86:41–49.
342. Gelderd, J. B., Matthews, M. A., St. Onge, M. F. and Faciane, C. L., 1980. Qualitative and quantitative effects of ACTH, piromen, cytoxan and isobutyl-2-cyanoacrylate treatments following spinal cord transection in rats. *Acta Neurobiologiae Experimentalis*, 40:489–500.
343. Gibaud, S., Demoy, M., Andreux, J. P., Weingarten, C., Gouritin, B. and Couvreur, P., 1996. Cells involved in the capture of nanoparticles in hematopoietic organs. *Journal of Pharmaceutical Sciences*, 85:944–950.
344. Gibaud, S., Andreux, J. P., Weingarten, C., Renard, M. and Couvreur, P., 1994. Increased bone marrow toxicity of doxorubicin bound to nanoparticles. *European Journal of Cancer*, 30A:820–826.
345. Ghanem, G. E., Joubran, C., Arnould, R., Lejeune, F. and Fruhling, J., 1993. Labelled polycyanoacrylate nanoparticles for human in vivo use. *Applied Radiation & Isotopes*, 44:1219–1224.
346. Guiot, P. and Couvreur, P., 1983. Quantitative study of the interaction between polybutylcyanoacrylate nanoparticles and mouse peritoneal macrophages in culture. *Journal de Pharmacie de Belgique*, 38:130–134.
347. Guzman, M., Molpeceres, J., Garcia, F., Aberturas, M. R. and Rodriguez, M., 1993. Formation and characterization of cyclosporine-loaded nanoparticles. *Journal of Pharmaceutical Sciences*, 82:498–502.
348. Harmia-Pulkkinen, T., Tuomi, A. and Kristoffersson, E., 1989. Manufacture of polyalkylcyanoacrylate nanoparticles with pilocarpine and timolol by micelle polymerization: factors influencing particle formation. *Journal of Microencapsulation*, 6:87–93.
349. Hu, Y. P., Jarillon, S., Dubernet, C., Couvreur, P. and Robert, J., 1996. On the mechanism of action of doxorubicin encapsulation in nanospheres for the reversal of multidrug resistance. *Cancer Chemotherapy & Pharmacology*, 37:556–560.
350. Kante, B., Couvreur, P., Dubois-Krack, G., De Meester, C., Guiot, P., Roland, M., Mercier, M.

and Speiser, P., 1982. Toxicity of polyalkylcyanoacrylate nanoparticles I: Free nanoparticles. *Journal of Pharmaceutical Sciences*, 71:786–790.

351. Kattan, J., Droz, J. P., Couvreur, P., Marino, J. P., Boutan-Laroze, A., Rougier, P., Brault, P., Vranckx, H., Grognet, J. M., Morge, X. et al., 1992. Phase I clinical trial and pharmacokinetic evaluation of doxorubicin carried by polyisohexylcyanoacrylate nanoparticles. *Investigational New Drugs*, 10:191–199.
352. Kreuter, J. and Hartmann, H. R., 1983. Comparative study on the cytostatic effects and the tissue distribution of 5-fluorouracil in a free form and bound to polybutylcyanoacrylate nanoparticles in sarcoma 180-bearing mice. *Oncology*, 40:363–366.
353. Kreuter, J., Wilson, C. G., Fry, J. R., Paterson, P. and Ratcliffe, J. H., 1984. Toxicity and association of polycyanoacrylate nanoparticles with hepatocytes. *Journal of Microencapsulation*, 1:253–257.
354. Kubiak, C., Couvreur, P., Manil, L. and Clausse, B., 1989. Increased cytotoxicity of nanoparticle-carried Adriamycin in vitro and potentiation by verapamil and amiodarone. *Biomaterials*, 10:553–556.
355. Labib, A., Lenaerts, V., Chouinard, F., Leroux, J. C., Ouellet, R. and van Lier, J. E., 1991. Biodegradable nanospheres containing phthalocyanines and naphthalocyanines for targeted photodynamic tumor therapy. *Pharmaceutical Research*, 8:1027–1031.
356. Lenaerts, V., Nagelkerke, J. F., Van Berkel, T. J., Couvreur, P., Grislain, L., Roland, M. and Speiser, P., 1984. In vivo uptake of polyisobutyl cyanoacrylate nanoparticles by rat liver Kupffer, endothelial, and parenchymal cells. *Journal of Pharmaceutical Sciences*, 73:980–982.
357. Lenaerts, V., Raymond, P., Juhasz, J., Simard, M. A. and Jolicoeur, C., 1989. New method for the preparation of cyanoacrylic nanoparticles with improved colloidal properties. *Journal of Pharmaceutical Sciences*, 78:1051–1052.
358. Lherm, C., Couvreur, P., Loiseau, P., Bories, C. and Gayral, P., 1987. Unloaded polyisobutylcyanoacrylate nanoparticles: efficiency against bloodstream trypanosomes. *Journal of Pharmacy & Pharmacology*, 39:650–652.
359. Liance, M., Nemati, F., Bories, C. and Couvreur, P., 1993. Experience with doxorubicin-bound polyisohexylcyanoacrylate nanoparticles on murine alveolar echinococcosis of the liver. *International Journal for Parasitology*, 23:427–429.
360. Losa, C., Calvo, P., Castro, E., Vila-Jato, J. L. and Alonso, M. J., 1991. Improvement of ocular penetration of amikacin sulphate by association to poly(butylcyanoacrylate) nanoparticles. *Journal of Pharmacy & Pharmacology*, 43:548–552.
361. Losa, C., Marchal-Heussler, L., Orallo, F., Vila Jato, J. L. and Alonso, M. J., 1993. Design of new formulations for topical ocular administration: polymeric nanocapsules containing metipranolol. *Pharmaceutical Research*, 10:80–87.
362. Lowe, P. J. and Temple, C. S., 1994. Calcitonin and insulin in isobutylcyanoacrylate nanocapsules: protection against proteases and effect on intestinal absorption in rats. *Journal of Pharmacy & Pharmacology*, 46:547–552.
363. Lux, G., Retterspitz, M., Stabenow-Lohbauer, U., Langer, M., Altendorf-Hofmann, A. and Bozkurt, T., 1997. Treatment of bleeding esophageal varices with cyanoacrylate and polidocanol, or polidocanol alone: results of a prospective study in an unselected group of patients with cirrhosis of the liver [see comments]. *Endoscopy*, 29:241–246.
364. Maincent, P., Devissaguet, J. P., LeVerge, R., Sado, P. A. and Couvreur, P., 1984. Preparation and in vivo studies of a new drug delivery system. Nanoparticles of alkylcyanoacrylate. *Applied Biochemistry & Biotechnology*, 10:263–265.
365. Maincent, P., Le Verge, R., Sado, P., Couvreur, P. and Devissaguet, J. P., 1986. Disposition kinetics and oral bioavailability of vincamine-loaded polyalkyl cyanoacrylate nanoparticles. *Journal of Pharmaceutical Sciences*, 75:955–958.
366. Manil, L., Couvreur, P. and Mahieu, P., 1995. Acute renal toxicity of doxorubicin (adriamycin)-loaded cyanoacrylate nanoparticles. *Pharmaceutical Research*, 12:85–87.
367. Nakada, Y., Fattal, E., Foulquier, M. and Couvreur, P., 1996. Pharmacokinetics and biodistribution

of oligonucleotide adsorbed onto poly(isobutylcyanoacrylate) nanoparticles after intravenous administration in mice. *Pharmaceutical Research*, 13:38–43.
368. Peracchia, M. T., Vauthier, C., Puisieux, F. and Couvreur, P., 1997. Development of sterically stabilized poly(isobutyl 2-cyanoacrylate) nanoparticles by chemical coupling of poly(ethylene glycol). *Journal of Biomedical Materials Research*, 34:317–326.
369. Pinto-Alphandary, H., Balland, O. and Couvreur, P., 1995. A new method to isolate polyalkylcyanoacrylate nanoparticle preparations. *Journal of Drug Targeting*, 3:167–169.
370. Polato, L., Benedetti, L. M., Callegaro, L. and Couvreur, P., 1994. In vitro evaluation of nanoparticle formulations containing gangliosides. *Journal of Drug Targeting*, 2:53–59.
371. Puglisi, G., Giammona, G., Fresta, M., Carlisi, B., Micali, N. and Villari, A., 1993. Evaluation of polyalkylcyanoacrylate nanoparticles as a potential drug carrier: preparation, morphological characterization and loading capacity. *Journal of Microencapsulation*, 10:353–366.
372. Sawant, K. K. and Murthy, R. S., 1993. In vivo evaluation of polyisobutylcyanoacrylate microparticles containing fluorouracil. *Pharmazie*, 48:783–784.
373. Schwab, G., Chavany, C., Duroux, I., Goubin, G., Lebeau, J., Helene, C. and Saison-Behmoaras, T., 1994. Antisense oligonucleotides adsorbed to polyalkylcyanoacrylate nanoparticles specifically inhibit mutated Ha-ras-mediated cell proliferation and tumorigenicity in nude mice. *Proceedings of the National Academy of Sciences of the United States of America*, 91:10460–10464.
374. Simeonova, M., Ilarionova, M., Ivanova, T., Konstantinov, C. and Todorov, D., 1991. Nanoparticles as drug carriers for vinblastine. Acute toxicity of vinblastine in a free form and associated to polybutylcyanoacrylate nanoparticles. *Acta Physiologica et Pharmacologica Bulgarica*, 17:43–49.
375. Simeonova, M. and Antcheva, M., 1994. In vitro study of cytotoxic activity of vinblastine in a free form and associated with nanoparticles. *Acta Physiologica et Pharmacologica Bulgarica*, 20: 31–35.
376. Tsai, T. R., Chuo, W. H., Lin, C. W. and Cham, T. M., 1998. Preparation and pharmacological evaluation of physostigmine-loaded polyisobutylcyanoacrylate nanoparticles in rats. *Kao-Hsiung i Hsueh Ko Hsueh Tsa Chih* [*Kaohsiung Journal of Medical Sciences*], 14:13–18.
377. Verdun, C., Brasseur, F., Vranckx, H., Couvreur, P. and Roland, M., 1990. Tissue distribution of doxorubicin associated with polyisohexylcyanoacrylate nanoparticles. *Cancer Chemotherapy & Pharmacology*, 26:13–18.
378. Vesin, W. R. and Florence, A. T., 1978. In vitro degradation rates of biodegradable poly-*N*-alkyl cyanoacrylates [proceedings]. *Journal of Pharmacy & Pharmacology*, 30:5P.
379. Youssef, M., Fattal, E., Alonso, M. J., Roblot-Treupel, L., Sauzieres, J., Tancrede, C., Omnes, A., Couvreur, P. and Andremont, A., 1988. Effectiveness of nanoparticle-bound ampicillin in the treatment of Listeria monocytogenes infection in athymic nude mice. *Antimicrobial Agents & Chemotherapy*, 32:1204–1207.
380. Zobel, H. P., Kreuter, J., Werner, D., Noe, C. R., Kumel, G. and Zimmer, A., 1997. Cationic polyhexylcyanoacrylate nanoparticles as carriers for antisense oligonucleotides. *Antisense and Nucleic Acid Drug Development*, 7:483–493.
381. Tamarin, A., Lewis, P. and Askey, J., 1976. The structure and formation of the byssus attachment plaque in Mytilus. *Journal of Morphology*, 149:199–221.
382. Strausberg, R. L. and Link, R. P., 1990. Protein-based medical adhesives. *Trends in Biotechnology*, 8:53–57.
383. Filpula, D. R., Lee, S. M., Link, R. P., Strausberg, S. L. and Strausberg, R. L., 1990. Structural and functional repetition in a marine mussel adhesive protein. *Biotechnology Progress*, 6:171–177.
384. Green, K., 1996. Mussel adhesive protein. In: Sierra D. H., Saltz R. (Editors). *Surgical Adhesives and Sealants—Current Technology and Applications*. Lancaster: Technomic, pp. 19–27.
385. Strausberg, R. L. and Link, R. P., 1990. *Protein-Based Medical Adhesives*. London: Elsevier Science.
386. Robin, J. B. and Salazar, J., 1987. The use of mussel adhesive protein in epikeratophakia. Proceedings of the Annual Spring Meeting of the Association for Research in Vision Ophthalmology.
387. Pittman, M. I., Menche, D., Song, E. K., Ben-Yishay, A., Gilbert, D. and Grande, D. A., 1989.

The use of adhesives in chondrocyte transplantation surgery: in vivo studies. *Bulletin of the Hospital for Joint Diseases Orthopedic Institute*, 49:213–221.
388. Liggett, P. E., Cano, M., Robin, J. B., Green, R. L. and Lean, J. S., 1990. Intravitreal biocompatibility of mussel adhesive protein. A preliminary study. *Retina*, 10:144–147.
389. Robin, J. B., Lee, C. F. and Riley, J. M., 1989. Preliminary evaluation of two experimental surgical adhesives in the rabbit cornea. *Refractive & Corneal Surgery*, 5:302–306.
390. Green, K., Berdecia, R. and Cheeks, L., 1987. Mussel adhesive protein: permeability characteristics when used as a basement membrane. *Current Eye Research*, 6:835–838.
391. Fulkerson, J. P., Norton, L. A., Gronowicz, G., Picciano, P., Massicotte, J. M. and Nissen, C. W., 1990. Attachment of epiphyseal cartilage cells and 17/28 rat osteosarcoma osteoblasts using mussel adhesive protein. *Journal of Orthopaedic Research*, 8:793–798.
392. Pitman, M. I., Menche, D., Song, E. K., Ben-Yishay, A., Gilbert, D. and Grande, D. A., 1989. The use of adhesives in chondrocyte transplantation surgery: in-vivo studies. *Bulletin of the Hospital for Joint Diseases Orthopaedic Institute*, 49:213–220.
393. Braunwald, N. S. and Tatooles, C. J., 1965. Use of a cross linked gelatin tissue adhesive to control hemorrhage from liver and kidney. *Surgical Forum*, 16:345–346.
394. Cooper, C. W. and Falb, R. D., 1968. Surgical adhesives. *Annals of the New York Academy of Sciences*, 146:214–224.
395. Bachet, J., Gigou, F., Laurian, C., Bical, O., Goudot, B. and Guilmet, D., 1982. Four-year clinical experience with the gelatin-resorcine-formol biological glue in acute aortic dissection. *Journal of Thoracic & Cardiovascular Surgery*, 83:212–217.
396. Braunwald, N. S., Gay, W. and Tatooles, C. J., 1966. Evaluation of crosslinked gelatin as a tissue adhesive and hemostatic agent: an experimental study. *Surgery*, 59:1024–1030.
397. Tatooles, C. J. and Braunwald, N. S., 1966. The use of crosslinked gelatin as a tissue adhesive to control hemorrhage from liver and kidney. *Surgery*, 60:857–861.
398. Kodama, K., Doi, O., Higashiyama, M. and Yokouchi, H., 1997. Pneumostatic effect of gelatin-resorcinol formaldehyde-glutaraldehyde glue on thermal injury of the lung: an experimental study on rats. *European Journal of Cardio-Thoracic Surgery*, 11:333–337.
399. Albes, J. M., Krettek, C., Hausen, B., Rohde, R., Haverich, A. and Borst, H. G., 1993. Biophysical properties of the gelatin-resorcin-formaldehyde/glutaraldehyde adhesive. *Annals of Thoracic Surgery*, 56:910–915.
400. Basu, S., Marini, C. P., Bauman, F. G., Shirazian, D., Damiani, P., Robertazzi, R., Jacobowitz, I. J., Acinapura, A. and Cunningham, J. N., Jr., 1995. Comparative study of biological glues: cryoprecipitate glue, two-component fibrin sealant, and "French" glue [see comments]. *Annals of Thoracic Surgery*, 60:1255–1262.
401. Bachet, J., Goudot, B., Teodori, G., Brodaty, D., Dubois, C., De Lentdecker, P. and Guilmet, D., 1990. Surgery of type A acute aortic dissection with Gelatine-Resorcine-Formol biological glue: a twelve-year experience. *Journal of Cardiovascular Surgery*, 31:263–273.
402. Bachet, J., Goudot, B., Dreyfus, G., Banfi, C., Ayle, N. A., Aota, M., Brodaty, D., Dubois, C., Delentdecker, P. and Guilmet, D., 1997. The proper use of glue: a 20-year experience with the GRF glue in acute aortic dissection. *Journal of Cardiac Surgery*, 12:243–253; discussion 253–245.
403. Goossens, D., Schepens, M., Hamerlijnck, R., Hartman, M., Suttorp, M. J., Koomen, E. and Vermeulen, F., 1998. Predictors of hospital mortality in type A aortic dissections: a retrospective analysis of 148 consecutive surgical patients. *Cardiovascular Surgery*, 6:76–80.
404. Von Oppell, U. O., Chimuka, D., Brink, J. G. and Zilla, P., 1995. Aortic dissection repair with GRF glue complicated by heart block. *Annals of Thoracic Surgery*, 59:761–763.
405. Nomori, H. and Horio, H., 1997. Gelatin-resorcinol-formaldehyde-glutaraldehyde glue-spread stapler prevents air leakage from the lung [published erratum appears in *Ann Thorac Surg* 1997 Sep; 64(3):892]. *Annals of Thoracic Surgery*, 63:352–355.
406. Ennker, I. C., Ennker, J., Schoon, D., Schoon, H. A., Rimpler, M. and Hetzer, R., 1994. Formaldehyde-free collagen glue in experimental lung gluing. *Annals of Thoracic Surgery*, 57:1622–1627.
407. Pigott, J. P., Donovan, D. L., Fink, J. A. and Sharp, W. V., 1989. Angioscope-assisted occlusion of

venous tributaries with prolamine in in situ femoropopliteal bypass: preliminary results of canine experiments. *Journal of Vascular Surgery*, 9:704–708; discussion 708–709.
408. Lawrence, W. T. and Diegelmann, R. F., 1994. Growth factors in wound healing. *Clinics in Dermatology*, 12:157–169.
409. Andresen, J. L. and Ehlers, N., 1998. Chemotaxis of human keratocytes is increased by platelet-derived growth factor-BB, epidermal growth factor, transforming growth factor-alpha, acidic fibroblast growth factor, insulin-like growth factor-I, and transforming growth factor-beta. *Current Eye Research*, 17:79–87.
410. Bhora, F. Y., Dunkin, B. J., Batzri, S., Aly, H. M., Bass, B. L., Sidawy, A. N. and Harmon, J. W., 1995. Effect of growth factors on cell proliferation and epithelialization in human skin. *Journal of Surgical Research*, 59:236–244.
411. Knighton, D. R., Ciresi, K. F., Fiegel, V. D., Austin, L. L. and Butler, E. L., 1986. Classification and treatment of chronic nonhealing wounds. Successful treatment with autologous platelet-derived wound healing factors (PDWHF). *Annals of Surgery*, 204:322–330.
412. Knighton, D. R., Ciresi, K., Fiegel, V. D., Schumerth, S., Butler, E. and Cerra, F., 1990. Stimulation of repair in chronic, nonhealing, cutaneous ulcers using platelet-derived wound healing formula. *Surgery, Gynecology & Obstetrics*, 170:56–60.
413. Mustoe, T. A., Pierce, G. F., Morishima, C. and Deuel, T. F., 1991. Growth factor-induced acceleration of tissue repair through direct and inductive activities in a rabbit dermal ulcer model. *Journal of Clinical Investigation*, 87:694–703.
414. Hebda, P. A., Klingbeil, C. K., Abraham, J. A. and Fiddes, J. C., 1990. Basic fibroblast growth factor stimulation of epidermal wound healing in pigs. *Journal of Investigative Dermatology*, 95: 626–631.
415. Fiddes, J. C., Hebda, P. A., Hayward, P., Robson, M. C., Abraham, J. A. and Klingbeil, C. K., 1991. Preclinical wound-healing studies with recombinant human basic fibroblast growth factor. *Annals of the New York Academy of Sciences*, 638:316–328.
416. Marks, M. G., Doillon, C. and Silver, F. H., 1991. Effects of fibroblasts and basic fibroblast growth factor on facilitation of dermal wound healing by type I collagen matrices. *Journal of Biomedical Materials Research*, 25:683–696.
417. McGee, G. S., Davidson, J. M., Buckley, A., Sommer, A., Woodward, S. C., Aquino, A. M., Barbour, R. and Demetriou, A. A., 1988. Recombinant basic fibroblast growth factor accelerates wound healing. *Journal of Surgical Research*, 45:145–153.
418. Mellin, T. N., Mennie, R. J., Cashen, D. E., Ronan, J. J., Capparella, J., James, M. L., Disalvo, J., Frank, J., Linemeyer, D., Gimenez-Gallego, G. et al., 1992. Acidic fibroblast growth factor accelerates dermal wound healing. *Growth Factors*, 7:1–14.
419. Pierce, G. F., Tarpley, J. E., Yanagihara, D., Mustoe, T. A., Fox, G. M. and Thomason, A., 1992. Platelet-derived growth factor (BB homodimer), transforming growth factor-beta 1, and basic fibroblast growth factor in dermal wound healing. Neovessel and matrix formation and cessation of repair. *American Journal of Pathology*, 140:1375–1388.
420. Hayward, P., Hokanson, J., Heggers, J., Fiddes, J., Klingbeil, C., Goeger, M. and Robson, M., 1992. Fibroblast growth factor reserves the bacterial retardation of wound contraction. *American Journal of Surgery*, 163:288–293.
421. Greenhalgh, D. G., Sprugel, K. H., Murray, M. J. and Ross, R., 1990. PDGF and FGF stimulate wound healing in the genetically diabetic mouse. *American Journal of Pathology*, 136:1235–1246.
422. Albertson, S., Hummel, R. P. D., Breeden, M. and Greenhalgh, D. G., 1993. PDGF and FGF reverse the healing impairment in protein-malnourished diabetic mice. *Surgery*, 114:368–372; discussion 372–363.
423. Tsuboi, R. and Rifkin, D. B., 1990. Recombinant basic fibroblast growth factor stimulates wound healing in healing-impaired db/db mice. *Journal of Experimental Medicine*, 172:245–251.
424. Breuing, K., Andree, C., Helo, G., Slama, J., Liu, P. Y. and Eriksson, E., 1997. Growth factors in the repair of partial thickness porcine skin wounds. *Plastic & Reconstructive Surgery*, 100: 657–664.
425. Matuszewska, B., Keogan, M., Fisher, D. M., Soper, K. A., Hoe, C. M., Huber, A. C. and Bondi,

J. V., 1994. Acidic fibroblast growth factor: evaluation of topical formulations in a diabetic mouse wound healing model. *Pharmaceutical Research*, 11:65–71.

426. Robson, M. C., Phillips, L. G., Lawrence, W. T., Bishop, J. B., Youngerman, J. S., Hayward, P. G., Broemeling, L. D. and Heggers, J. P., 1992. The safety and effect of topically applied recombinant basic fibroblast growth factor on the healing of chronic pressure sores. *Annals of Surgery*, 216:401–406; discussion 406–408.
427. Bjornsson, T. D., Dryjski, M., Tluczek, J., Mennie, R., Ronan, J., Mellin, T. N. and Thomas, K. A., 1991. Acidic fibroblast growth factor promotes vascular repair. *Proceedings of the National Academy of Sciences of the United States of America*, 88:8651–8655.
428. Lepisto, J., Laato, M., Niinikoski, J., Lundberg, C., Gerdin, B. and Heldin, C. H., 1992. Effects of homodimeric isoforms of platelet-derived growth factor (PDGF-AA and PDGF-BB) on wound healing in rat. *Journal of Surgical Research*, 53:596–601.
429. Sprugel, K. H., McPherson, J. M., Clowes, A. W. and Ross, R., 1987. Effects of growth factors in vivo. I. Cell ingrowth into porous subcutaneous chambers. *American Journal of Pathology*, 129: 601–613.
430. Pierce, G. F., Mustoe, T. A., Altrock, B. W., Deuel, T. F. and Thomason, A., 1991. Role of platelet-derived growth factor in wound healing. *Journal of Cellular Biochemistry*, 45:319–326.
431. Pierce, G. F., Mustoe, T. A., Senior, R. M., Reed, J., Griffin, G. L., Thomason, A. and Deuel, T. F., 1988. In vivo incisional wound healing augmented by platelet-derived growth factor and recombinant c-sis gene homodimeric proteins. *Journal of Experimental Medicine*, 167:974–987.
432. Mustoe, T. A., Purdy, J., Gramates, P., Deuel, T. F., Thomason, A. and Pierce, G. F., 1989. Reversal of impaired wound healing in irradiated rats by platelet-derived growth factor-BB. *American Journal of Surgery*, 158:345–350.
433. Robson, M. C., Phillips, L. G., Thomason, A., Robson, L. E. and Pierce, G. F., 1992. Platelet-derived growth factor BB for the treatment of chronic pressure ulcers. *Lancet*, 339:23–25.
434. Roberts, A. B., Sporn, M. B., Assoian, R. K., Smith, J. M., Roche, N. S., Wakefield, L. M., Heine, U. I., Liotta, L. A., Falanga, V., Kehrl, J. H. et al., 1986. Transforming growth factor type beta: rapid induction of fibrosis and angiogenesis in vivo and stimulation of collagen formation in vitro. *Proceedings of the National Academy of Sciences of the United States of America*, 83:4167–4171.
435. Sporn, M. B., Roberts, A. B., Shull, J. H., Smith, J. M., Ward, J. M. and Sodek, J., 1983. Polypeptide transforming growth factors isolated from bovine sources and used for wound healing in vivo. *Science*, 219:1329–1331.
436. Ksander, G. A., Ogawa, Y., Chu, G. H., McMullin, H., Rosenblatt, J. S. and McPherson, J. M., 1990. Exogenous transforming growth factor-beta 2 enhances connective tissue formation and wound strength in guinea pig dermal wounds healing by secondary intent. *Annals of Surgery*, 211: 288–294.
437. Mustoe, T. A., Pierce, G. F., Thomason, A., Gramates, P., Sporn, M. B. and Deuel, T. F., 1987. Accelerated healing of incisional wounds in rats induced by transforming growth factor-beta. *Science*, 237:1333–1336.
438. Bernstein, E. F., Harisiadis, L., Salomon, G., Norton, J., Sollberg, S., Uitto, J., Glatstein, E., Glass, J., Talbot, T., Russo, A. et al., 1991. Transforming growth factor-beta improves healing of radiation-impaired wounds. *Journal of Investigative Dermatology*, 97:430–434.
439. Pierce, G. F., Mustoe, T. A., Lingelbach, J., Masakowski, V. R., Gramates, P. and Deuel, T. F., 1989. Transforming growth factor beta reverses the glucocorticoid-induced wound-healing deficit in rats: possible regulation in macrophages by platelet-derived growth factor. *Proceedings of the National Academy of Sciences of the United States of America*, 86:2229–2233.
440. Beck, L. S., Amento, E. P., Xu, Y., Deguzman, L., Lee, W. P., Nguyen, T. and Gillett, N. A., 1993. TGF-beta 1 induces bone closure of skull defects: temporal dynamics of bone formation in defects exposed to rhTGF-beta 1. *Journal of Bone & Mineral Research*, 8:753–761.
441. Terrell, T. G., Working, P. K., Chow, C. P. and Green, J. D., 1993. Pathology of recombinant human transforming growth factor-beta 1 in rats and rabbits. *International Review of Experimental Pathology*, 34:43–67.
442. Franklin, J. D. and Lynch, J. B., 1979. Effects of topical applications of epidermal growth factor on

wound healing. Experimental study on rabbit ears. *Plastic & Reconstructive Surgery*, 64:766–770.
443. Brown, G. L., Curtsinger, L. D., Brightwell, J. R., Ackerman, D. M., Tobin, G. R., Polk, H. C., Jr., George-Nascimento, C., Valenzuela, P. and Schultz, G. S., 1986. Enhancement of epidermal regeneration by biosynthetic epidermal growth factor. *Journal of Experimental Medicine*, 163: 1319–1324.
444. Laato, M., Heino, J., Kahari, V. M., Niinikoski, J. and Gerdin, B., 1989. Epidermal growth factor (EGF) prevents methylprednisolone-induced inhibition of wound healing. *Journal of Surgical Research*, 47:354–359.
445. Buckley, A., Davidson, J. M., Kamerath, C. D., Wolt, T. B. and Woodward, S. C., 1985. Sustained release of epidermal growth factor accelerates wound repair. *Proceedings of the National Academy of Sciences of the United States of America*, 82:7340–7344.
446. Buckley, A., Davidson, J. M., Kamerath, C. D. and Woodward, S. C., 1987. Epidermal growth factor increases granulation tissue formation dose dependently. *Journal of Surgical Research*, 43: 322–328.
447. Hennessey, P. J., Black, C. T. and Andrassy, R. J., 1990. EGF increases short-term type I collagen accumulation during wound healing in diabetic rats. *Journal of Pediatric Surgery*, 25:893–897.
448. Brown, G. L., Nanney, L. B., Griffen, J., Cramer, A. B., Yancey, J. M., Curtsinger, L. J. D., Holtzin, L., Schultz, G. S., Jurkiewicz, M. J. and Lynch, J. B., 1989. Enhancement of wound healing by topical treatment with epidermal growth factor [see comments]. *New England Journal of Medicine*, 321:76–79.
449. Brown, G. L., Curtsinger, L., Jurkiewicz, M. J., Nahai, F. and Schultz, G., 1991. Stimulation of healing of chronic wounds by epidermal growth factor [see comments]. *Plastic & Reconstructive Surgery*, 88:189–194; discussion 195–186.
450. Falanga, V., Eaglstein, W. H., Bucalo, B., Katz, M. H., Harris, B. and Carson, P., 1992. Topical use of human recombinant epidermal growth factor (h-EGF) in venous ulcers. *Journal of Dermatologic Surgery & Oncology*, 18:604–606.
451. Itokazu, M., Yamamoto, K., Yang, W. Y., Aoki, T., Kato, N. and Watanabe, K., 1997. The sustained release of antibiotic from freeze-dried fibrin-antibiotic compound and efficacies in a rat model of osteomyelitis. *Infection*, 25:359–363.
452. Kram, H. B., Bansal, M., Timberlake, O. and Shoemaker, W. C., 1991. Antibacterial effects of fibrin glue-antibiotic mixtures. *Journal of Surgical Research*, 50:175–178.
453. Park, M. S. and Kim, Y. B., 1997. Sustained release of antibiotic from a fibrin-gelatin-antibiotic mixture. *Laryngoscope*, 107:1378–1381.
454. Thompson, D. F. and Davis, T. W., 1997. The addition of antibiotics to fibrin glue. *Southern Medical Journal*, 90:681–684.
455. van der Ham, A. C., Kort, W. J., Weijma, I. M., van den Ingh, H. F. and Jeekel, H., 1992. Effect of antibiotics in fibrin sealant on healing colonic anastomoses in the rat. *British Journal of Surgery*, 79:525–528.
456. Watanabe, G., Misaki, T. and Kotoh, K., 1997. Microfibrillar collagen (Avitene) and antibiotic-containing fibrin-glue after median sternotomy. *Journal of Cardiac Surgery*, 12:110–111.
457. Greco, F., de Palma, L., Spagnolo, N., Rossi, A., Specchia, N. and Gigante, A., 1991. Fibrin-antibiotic mixtures: an in vitro study assessing the possibility of using a biologic carrier for local drug delivery. *Journal of Biomedical Materials Research*, 25:39–51.
458. Redl, H., Schlag, G., Stanek, G., Hirschl, A. and Seelich, T., 1983. In vitro properties of mixtures of fibrin seal and antibiotics. *Biomaterials*, 4:29–32.
459. Zilch, H. and Lambiris, E., 1986. The sustained release of cefotaxim from a fibrin-cefotaxim compound in treatment of osteitis. Pharmacokinetic study and clinical results. *Archives of Orthopaedic & Traumatic Surgery*, 106:36–41.
460. Ney, A. L., Kelly, P. H., Tsukayama, D. T. and Bubrick, M. P., 1990. Fibrin glue-antibiotic suspension in the prevention of prosthetic graft infection. *Journal of Trauma*, 30:1000–1005; discussion 1005–1006.
461. Sugitachi, A., Shindo, T., Kido, T. and Kawahara, T., 1992. Bio-adhesiochemo-(BAC)-therapy. *Proceedings of the American Association of Cancer Research*, 33:216.

462. Kram, H. B., Bansul, M., Timberlake, O. et al., 1989. Antibacterial effects of fibrin glue and fibrin glue-antibiotic mixtures. *Transactions of the Society for Biomaterials*, 12:164.
463. Sato, H., Ono, K. and Oka, M., 1988. Antibiotic binding fibrinogen as a fibrin glue. Transactions of the Third World Biomaterials Congress, 11:531.
464. MacPhee, M. J., Singh, M. P., Brady, R., Akhyani, N., Liau, G., Lasa, C., Hue, C., Best, A. and Drohan, W., 1996. Fibrin sealant: a versatile delivery vehicle for drugs and biologics. In: Sierra D. H., Saltz R. (Editors). *Surgical Adhesives and Sealants—Current Technology and Applications*. Lancaster: Technomic, pp. 109–120.
465. Singh, M. P., Brady, R., Drohan, W. and MacPhee, M. J., 1996. Sustained release of antibiotics from fibrin sealant. In: Sierra D. H., Saltz R. (Editors). *Surgical Adhesives and Sealants—Current Technology and Applications*. Lancaster: Technomic, pp. 121–133.
466. Nunez, H., Hennigh, R., Campagna, A., Harding, S., Drohan, W. and MacPhee, M., 1993. Prolonged release of effective concentrations of antibiotics from fibrin sealant. *Proceedings of the Symposium on Surgical Tissue Adhesives*, 1:97.
467. MacPhee, M., 1993. Drug release. *Proceedings of the Symposium on Surgical Tissue Adhesives*, 1:97.
468. Sierra, D., Luck, E., Brown, D. and Yu, N., 1993. Fibrin-collagen adhesive drug delivery system for tumor therapy. *Transactions of the Society for Biomaterials*, 30:257.
469. MacPhee, M., Kidd, R., Campagna, A. and Drohan, W., 1993. Extended delivery of antiproliferative compounds from fibrin sealant. *Proceedings of the Symposium on Surgical Tissue Adhesives*, 1:24.
470. Feldman, D. and Sierra, D., 1995. Tissue adhesives in wound healing, *Encyclopedic Handbook of Biomaterials and Bioengineering*, Marcel Dekker, New York.
471. Pandit, A. S., Feldman, D. S., Caulfield, J., and Thompson, J. A., 1998. Stimulation of angiogenesis by FGF-1 delivered through a modified fibrin scaffold. *Growth Factors*, 15:113–123.
472. Sierra, D., Feldman, D. and Saltz, R., 1992. A method to determine the shear adhesive strength of fibrin sealants. *Journal of Applied Biomaterials*, 3:147–151.
473. Flahiff, C., Feldman, D., Saltz, R. and Huang, S., 1992. Mechanical testing of fibrin adhesives for blood vessel anastomosis. *Journal of Biomaterial Matter Research*, 26:481–491.
474. Pandit, A., 1998. Fibrin as a matrix for FGF-1 delivery in vivo. Ph.D. thesis, UAB.
475. Pandit, A. S., Feldman, D. S. and Caulfield, J., 1998. *In vivo* wound healing response to a modified degradable fibrin scaffold. *Journal of Biomaterials Applications*, 12:222–236.
476. Pandit, A., Feldman, D., and Thompson, J., 1995. Characterization of the delivery rate of FGF-1 through a porous fibrin scaffold. *Transactions of the Wound Healing Society*, 5:80.
477. Pandit, A., Feldman, D., Listinsky, C., and Thompson, J. A., 1997. The effect on wound healing by a modified fibrin scaffold delivering acidic fibroblast growth factor (FGF-1). *Journal of Bioactive and Compatible Polymers*, 12:99–111.
478. Wilson, D., Feldman, D., and Thompson, J., 1994. Acidic fibroblast growth factor delivery through a fibrin matrix without exogenous heparin. *Transactions of the Society for Biomaterials*, 30:255.
479. Feldman, D., Redden, R., Blum, B., and Osborne, S., 1997. Porous fibrin as a degradable adhesive and drug delivery system. *Transactions of the Society for Biomaterials*, 23:185.
480. Saltz, R., Feldman, D., Floyd, D. and Huang, S., 1991. Experimental and clinical applications of fibrin glue. *Plastic Reconstruction Surgery*, 88(6):1005–1015.
481. Sahni, A., Odrljin, T., and Francis, C., 1997. Binding of basic fibroblast growth factor to fibrinogen and fibrin. *Journal of Biological Chemistry*, 273(13):7554–7559.
482. Cheng, H., Fraidakis, M., Blomback, B., Lapchak, P., Hoffer, B. and Olson, L., 1998. Characterization of a fibrin glue-GDNF slow-release preparation. *Cell Transplantation*, 7:53–61.
483. Greisler, H. P., Cziperle, D. J., Kim, D. U., Garfield, J. D., Petsikas, D., Murchan, P. M., Applegren, E. O., Drohan, W. and Burgess, W. H., 1992. Enhanced endothelialization of expanded polytetrafluoroethylene grafts by FGF type 1 pretreatment. *Surgery*, 112(2):24–55.
484. Pandit, A., Feldman, D., and Thompson, J., 1994. In vivo dose response of acidic fibroblast growth factor delivered through a porous fibrin scaffold on wound healing. *Transactions of the Wound Healing Society*, 4:122.

485. France, R., DeBloise, C. and Doillon, C., 1993. Extracellular matrix analogs as carriers for growth factors: *in vitro* fibroblast behavior. *Journal of Biomedical Material Research*, 27:389–397.
486. Osborne, S., Blum, B., Feldman, Kelpke, S., Pandit, A., and Thompson, J., 1996. The effect of acidic fibroblast growth factor (FGF-1) delivered through a porous fibrin scaffold on meshed skin graft healing. *Trans. 5th World Biomaterial Congress*, 5:879.
487. Bowman, J., Blum B., Wehby, J., and Feldman, D., 1997. Obtaining investigator based clinical approval for FGF-1 in fibrin glue. *Transaction of the Society for Biomaterials*, 23:469.
488. Bowman, J., Blum, B., Feldman, D., Kilpadi, D., Roberts, M., and Wehby, J., 1997. Stability of FGF-1. *Transactions of the Wound Healing Society*, 7:93.
489. Gosselin, C., Ren, D., Ellinger, J. and Greisler, H. P., 1995. In vivo platelet deposition on polytetrafluoroethylene coated with fibrin glue containing fibroblast growth factor 1 and heparin in a canine model. *American Journal of Surgery*, 170:126–130.
490. Walter, M. A., Toriumi, D. M., Patt, B. S., Bhattacharyya, T. K., K, O. G., Caulfield, J. B. and Thompson, J. A., 1996. Fibroblast growth factor-induced motor end plate regeneration in atrophic muscle. *Archives of Otolaryngology—Head & Neck Surgery*, 122:425–430.
491. Hashimoto, J., Kurosaka, M., Yoshiya, S. and Hirohata, K., 1992. Meniscal repair using fibrin sealant and endothelial cell growth factor. An experimental study in dogs. *American Journal of Sports Medicine*, 20:537–541.

12

Drug Delivery Systems for the Treatment of Restenosis

Ilia Fishbein, Michael Chorny, Irit Gati, Laura Rabinovich, and Gershon Golomb
The Hebrew University of Jerusalem, Jerusalem, Israel

I. INTRODUCTION

Coronary atherosclerosis and its clinical sequela continue to be the leading cause of mortality in Western society. Over the past 15 years, mechanical means of achieving revascularization of obstructive atherosclerotic vessels have been greatly improved. Percutaneous transluminal coronary angioplasty (PTCA) procedures include balloon dilation, excisional atherectomy, endoluminal stenting and laser ablation [1]. However, revascularization induces thrombosis and neointimal hyperplasia, which in turn cause restenosis in 30% to 40% of coronary arteries within six months after successful balloon angioplasty and in over 60% of aorta-coronary saphenous vein bypass grafts within five years following surgical intervention [2]. Topol et al. estimated that the annual cost of restenosis in terms of human lives and dollars spent exceeded 2500 deaths and $3.5 billion in the United States alone in 1992 [3]. Despite extensive research on the incidence, timing, mechanisms, and pharmacological interventions in humans and animal models, no therapy consistently prevents this difficult problem [4,5].

II. PATHOPHYSIOLOGY OF RESTENOSIS

Many researchers claim that treatment failures probably result from a lack of understanding of the cellular mechanisms in restenotic neointimal formation. Accumulation of smooth muscle cells (SMC) in the intima of human arteries is a characteristic feature of atherosclerosis. These cells, which originate in the intima and media, undergo excessive and abnormally regulated proliferation that determines the size of the obstructive plaque [6].

The rat carotid artery injured by a balloon catheter has been widely used as a model of angioplasty. According to Clowes [7,8], a passage of embolectomy balloon catheter strips the endothelial cells away and stretches the underlying media. This form of injury causes immediate coagulation and thrombosis cascade in which platelets adhere, spread, and degranulate on the denuded surface of the artery. Approximately 24 h later SMC begin to proliferate. After four days, movement of SMC from the media to the intima can be detected, after which they continue to proliferate in the intima. The intimal thickening process is further augmented by the deposition of matrix around the intimal cells. From two weeks when the cellular proliferation is largely complete, and until three months, a steady state is reached in which the intima is approximately 20% cells and 80% extracellular matrix.

The underlying biochemical mechanisms controlling the proliferation and migration are described as follows. Injury to the artery causes endothelial and SMC disruption and release of intracellular mitogens such as basic fibroblast growth factor (bFGF) which stimulates SMC proliferation. Factors from platelets (such as platelet-derived growth factor {PDGF} and thrombin) regulate proliferation and migration of SMCs. In addition, the mural thrombus caused by the injury may contribute cytokines, various chemoattractants and growth factors, and angiotensin II, all of which affect the intimal thickening process.

Histopathological studies of human arteries obtained after PTCA [9] or of the restenotic lesions obtained by the use of transluminal atherectomy, revealed that SMC proliferation seems to play a pivotal role in the restenosis process. Furthermore, in a small but distinctive group, only intimal hyperplasia was observed without atherosclerotic plaque [10].

Schwartz, Holmes and Topol proposed an alternative model based on observations in the porcine coronary injury model [11]. In response to arterial injury by oversize stent, thrombus is formed. In the next stage, termed the cellular recruitment phase, the thrombus becomes covered by a layer of endothelium at roughly three to four days, followed by infiltration, from the lumen side first, of mononuclear cells consisting of macrophages and lymphocytes. In the third stage, the proliferative phase, cellular proliferation occurs in the final stage of arterial healing, approximately seven to nine days after injury. Beginning again at the lumen surface, cells that stain for alpha actin (SMC or myofibroblasts) colonize the degenerating thrombus mass as a thin "cap" just beneath the endothelium and progressively thicken the lesion. These cells do not arise from the media at the injury site. Extracellular matrix secretion and additional cell recruitment most likely add to the neointimal volume. The healing process is complete when all the residual thrombus is resorbed and replaced by mature neointima.

Recently it was realized that a final post-PTCA lumen diameter does not depend exclusively on neointimal mass formation but also on dynamic vessel constriction/dilatation, termed remodeling [12]. Intravascular ultrasound human studies have shown that up to 66% of ultimate lumen diameter might be determined by remodeling [13]. Pathological mechanisms and cell signaling processes underlying direction and degree of remodeling are still obscure.

An important insight in the pathophysiology of restenosis was recently made by studies in the porcine coronary injury model [14–17]. The authors focused on the role of adventitial myofibroblasts in neointimal formation. Using immunocytochemical methods they pulse-labeled proliferative cells in injured coronary arteries and showed migration of the labeled pool from adventitia (2–3 days postinjury) through the media to nascent neointima (10–14 days postinjury). Simultaneously with migration myofibroblasts lose the surface markers of fibroblast lineage and redifferentiate into SMC indistinguishable from medial SMC.

Another important recent contribution highlighted the role of macrophages and macrophage-derived cytokines in the pathophysiology of restenosis [18,19], especially in restenosis following stent deployment (Edelman, personal communication). An unusually high percentage of CD4-positive cells of macrophage origin in neointimal tissue of both rat and pig carotid arteries was demonstrated [18] by a highly sensitive modification of the immunocytochemical method.

In conclusion, in spite of many existing hypotheses about the mechanisms of restenosis, no single theory adequately reflects all of the components of its multifaceted pathogenesis. Moreover, different animal models accentuate different elements of integral pathogenesis. For example, the rat model might be more "proliferative," while the pig model more "thrombotic" than human restenosis [20]. Thus, all promising therapeutic approaches should be evaluated in several animal models before their proposal and consideration as potential treatment modalities for human restenosis.

III. SYSTEMIC PHARMACOLOGICAL INTERVENTIONS

From a simplistic point of view, restenosis appears to be an easy target for pharmacological prevention, since the complexity of its pathogenesis implies that there are many points where drugs might intervene and modify the process of vessel narrowing. On the other hand, such complexity in biological systems is often related to redundancy, i.e., the hypertrophied response of the uninhibited arms of pathogenesis that ultimately counterbalances the gains from the initial narrow-spectrum drug. Based on the straightforward logic of this view, the problem of restenosis was approached from different modes. A large number of drugs were investigated for restenosis prevention and could be categorized as antithrombotic, antiplatelet, antiproliferative, anti-inflammatory, calcium antagonists and lipid-lowering drugs [21,22]. Starting in the early 1980s more than 150 clinical studies were planned and conducted in a futile attempt to discover the "magic bullet" for prevention of postangioplasty restenosis and related disorders (for recent reviews see Refs. 4 and 5). An overwhelming majority of these studies evaluated effects of systemic therapeutics administered on an intermittent basis up to six months following intervention. Several studies utilized pretreatment protocols in order to cover the earliest events triggered by vascular trauma. All of them, with three important exceptions discussed below, failed to show any beneficial effect of therapy on incidence of restenosis. These disappointing results are even more confusing in light of the high antirestenotic efficacy the same drugs showed in animal models of restenosis.

The disappointing results obtained with the large number of drugs tested may be due in part to the use of suboptimal dosages in an effort to avoid systemic intolerance or difficulty in providing controlled administration of the drugs for an adequate period of time. A critical evaluation of the value of drug trials that have been performed in the past is extremely difficult because of differences in the selection of patients, methods of analysis, and definition of restenosis, as well as poor statistical designs [5].

Recently, several clinical trials have challenged the verdict of a total ineffectiveness of systemic therapy for human restenosis. Three unrelated drugs whose only common feature is their impact on the multiple arms of restenosis pathogenesis were shown to be at least partially effective blockers of restenosis.

Probucol was introduced in clinical practice as a hypolipidemic drug, but later was shown to be a very potent antioxidant in addition to its modest lipid-lowering properties [23,24]. Due to the ubiquitous involvement of lipid peroxidation in all aspects of vascular biology [25–28], probucol has been important in the context of restenosis antiadhesive [29,30], antiproliferative [31,32] and antimigration [32] activities. In a recent trial [33] probucol was shown to decrease significantly the incidence of restenosis (46.8% reduction of restenosis per segment), while an antioxidant regimen of vitamins A, C and E was ineffective.

The second promising drug, transilat, was developed as an antiallergic compound, but proved to be a potent antifibrotic [34,35] and antimigration [35] drug. In a preliminary small-scale trial [36] tranilast was highly effective (57.7% reduction of angiographic restenosis three months after directional coronary atherectomy), and a larger multicenter trial is currently underway.

The third drug, abciximab, is a monoclonal antibody to the glycoprotein IIb/IIIa receptor. This receptor is expressed on platelets and its engagement represents the final common step leading to platelets' activation and aggregation regardless of the nature of primary stimulus. In the EPIC study [37] bolus injection and infusion of abciximab resulted in a 26% reduction of clinical restenosis in a selected group of high-risk patients. Additionally, the rate of target vessel revascularization was reduced by 36%.

While statistically significant, the results of studies discussed above are still not good enough in clinical terms. Moreover, many experts in the field of vascular biology doubt that any other systemically administered drug could significantly decrease restenosis rates. In other words, it is not the drugs, but the route of their administration that has failed.

In summary, the failure to identify an effective therapy despite the numerous pharmacological trials could be due to one of the following: (1) interspecies differences in the process of neointimal formation render it impossible to directly extrapolate results obtained in animal models; (2) an effective agent has not yet been tested; (3) an effective agent has been overlooked due to incorrect dosage or timing; (4) more than a single agent is needed; (5) restenosis cannot be effectively treated by systemic therapy.

IV. LOCAL DRUG DELIVERY

There is a general consensus that an ultimate antirestenotic therapy will utilize the concept of local drug delivery. The unsuccessful attempts to control restenosis by systemic pharmacological interventions has prompted many researchers to look for more promising therapeutic approaches such as local drug delivery. Due to the focal nature of atherosclerotic disease, local delivery of the drug to the diseased artery appears to be a natural and advantageous way of treatment. Conceptually, local drug delivery might minimize toxic side effects of potent drugs by a decrease in two to three orders of magnitude the general drug dose which will lower the drug's exposure to circulation. Incorporation of the drug into a polymeric matrix allows sustained release of the drug, thus prolonging the period of treatment. This might be especially important in cases of chemically unstable drugs which would rapidly degrade, if unprotected by their vehicle. Moreover, by varying polymer class and composition different release patterns could be achieved, thus fitting the drug release to clinical demands. The compelling importance of proper release profile and timing was recently demonstrated by Edelman et al. [38]. The authors showed that rapid release of antisense c-myc oligonucleotides from periadventitially applied Pluronic gel did not inhibit neointimal formation in balloon-injured rat carotid artery, while more sustained release from ethylene-vinyl acetate copolymer (EVA) matrices efficiently blocked arterial narrowing. These results demonstrated a good correspondence with a prolonged two-peak pattern of c-myc expression following balloon injury.

There are two generic approaches in vascular local drug delivery [39,40]. Namely, a drug can be delivered to an injured arterial segment from an adventitial (external) aspect of vessel, or it can be applied intraluminally, i.e., from inside the artery. Recently, drug delivery from endoluminally deployed stents [41–43] has become a very intensive and developing area of research that might be viewed as an independent third route in vascular delivery, while being formally classified as a variant of intraluminal delivery. Both, adventitial and intraluminal, methods have benefits and limitations and have been implemented with a number of modifications. The remainder of this chapter will focus on a discussion of specific methods of vascular delivery.

A. Perivascular Delivery

The rationale behind perivascular delivery is to provide a local depot of a drug in immediate proximity to the adventitial (outer) aspect of the artery. The drug is leached from a repository by tissue fluid and passively diffuses into the vessel wall driven by a concentration gradient. This process is facilitated by the extensive vasa vasorum system in adventitia of nonrodent mammalian species. Vasa vasorum originate from a lumen of large arteries and run in adventitia along the main lumen, nourishing the components of the vessel wall. It may be noted in this context that vasa vasorum in atherosclerotic plaque is more extensive than in normal arteries and penetrates the vessel wall so deeply that it reaches the neointima [44,45]. When exposed to vasa vasorum the drug is distributed evenly in the whole arterial segment, including the innermost layers of media and intima.

A recent elucidation of the role that adventitial myofibroblasts play in initiation of neointimal thickening [14,17] revealed that an adventitial route of delivery is a very logical choice, since the drug is initially delivered to the principal site of its action. Similarly, therapeutic approaches aimed at the prevention of unfavorable remodeling should also target structures and processes localized in adventitia.

The pharmacokinetic aspects of the adventitial delivery were recently studied by Lovich et al. [46], who developed a methodology that discriminates between a truly local delivery of a drug from a perivascular depot and a secondary drug influx into the targeted arterial site through the general circulation. The authors showed that although 90% of the drug is cleared from the perivascular space through the extravascular capillaries, virtually all the deposited drug diffuses directly from the perivascular space.

B. Perivascular Delivery Systems

A number of vehicle systems were checked for the potential of steady drug delivery to the adventitial aspects of arteries (Table 1). The ultimate pharmaceutical challenge of these systems is to provide a release of the drug in question in a controlled fashion in order to fit the release to clinical needs. Other important factors to be considered are the capacity of a device, ease of manipulation and implantation, and the inflammatory response it causes. From a practical point of view periadventitial delivery systems are not directly applicable in a setting of PTCA. Nevertheless, due to their relative simplicity and reproducibility, they present an unique screening method for the evaluation of potentially useful antirestenotic drugs. Moreover, it is evident that an opportunity exists to utilize them in the treatment of accelerated atherosclerosis syndromes [47] that complicate venous or artificial conduit grafting procedures.

Edelman [48,49] pioneered the use of ethylene-vinyl acetate (EVA) copolymer for adventitial delivery. This nontoxic biocompatible FDA-approved biomaterial possesses excellent mechanical properties and enables several formulatory variables to modulate a drug release rate [50–54]. The molded EVA could be easily cut into the slabs of desired size and sutured around the artery. In a series of experiments from Edelman's group [48,51,55] both, unfractioned and low molecular weight heparin, were shown to reduce neointimal formation in the rat carotid model, while the systemic administration of heparin was ineffective. Our group has shown the antirestenotic efficacy of two tyrphostin drugs [52,54,56], colchicine [57] and verapamil [58], delivered periadventially from EVA matrices in the rat model. The antirestenotic potential of EVA-incorporated verapamil was also shown in a rabbit model of autologous venous grafts [58]. Recently, two studies demonstrated the successful incorporation of antisense oligonucleotides to c-myc [38] and to PDGFβ receptor [59] into EVA matrices. Notably, the formulation process used did not decrease the biological activity of the antisense molecules.

Another nonbiodegradable biocompatible polymer, silicone, was used for adventitial delivery of dexamethasone in rat carotid [60] and porcine stented carotid models [61], and for delivery of hirulog in the porcine model [53,62]. Sustained release local hirulog therapy decreased early thrombosis but not neointimal thickening after arterial stenting. Despite pro-

Table 1 Perivascular Drug Delivery Systems Studied in Animal Models

Animal model	Medication	Delivery system	Effect on restenosis [Ref.]
Rat carotid	antisense to NF-κB	osmotic pumps	+ 74
Dog femoral vascular graft	chimeric bFGF-saporin conjugate	osmotic pumps	+ 75
Rabbit carotid	peptide $\alpha_v\beta_3$ antagonist	osmotic pumps	+ 76
Rabbit carotid and femoral	dipyridamole	osmotic pumps	+ 77
Rat, inverted femoral anasthomosis	heparin	PVA into silastic shell	+ 130
Pig stented carotid	hirulog	silicone matrices	− 62
Pig stented carotid	dexamethasone	silicone matrices	− 61
Rat carotid	protamine	EVA matrices	+ 55
Rat carotid	heparin	EVA matrices	+ 50
Rat carotid	colchicine	EVA matrices	+ 57
Rat carotid	tyrphostin, AG-17	EVA matrices	+ 54
Rat carotid	tyrphostin, AG-1295	EVA matrices	+ 52
Rat carotid	antisense to PDGF receptor β	EVA matrices	+ 59
Rabbit stented and nonstented femoral arteries	heparin	EVA matrices	+/− 131
Rat carotid	antisense to c-myc	EVA matrices	+ 38
Rat carotid	antisense to c-myb	EVA matrices	+ 38
Rat carotid	plasmid bearing ras transdominant negative mutant	pluronic gel	+ 70
Rat carotid	peptide $\alpha_v\beta_3$ antagonist	pluronic gel	+ 69
Rat carotid	probucol	pluronic gel	+ 64
Rat carotid	antisense to c-myc	pluronic gel	−/+ 38; 65; 132
Rat carotid and aortic	antisense to c-myb	pluronic gel	+ 38; 66; 67
Rat carotid	heparin, inulin	pluronic gel	nonstudied 46
Rat carotid and aortic	nonspecific oligonucleotides	pluronic gel	+ 66; 71
Rat carotid	antisense to c-CDC2 and CDK2	pluronic gel	+ 68
Rat carotid	tyrphostin, AG-47	PLGA/pluronic matrices	− 72
Rat carotid	bFGF	alginate microcapsules incorporating heparine-sepharose beads	nonstudied 49
Rat carotid	heparin	alginate microcapsules incorporating heparine-sepharose beads	nonstudied 49
Pig coronary	IL-1β, tyrosine kinase inhibitor, ST 638	cotton mesh soaked with IL-1b-sepharose or ST 638-sepharose beads	+ 73; 133
Rat carotid	endothelial cells	Gelfoam biopolymeric matrices	+ 78; 79

tracted release kinetics of the drugs from silicone matrices and high focal drug levels found in stented arteries, neither of the two drugs was effective in the porcine model, while a high antirestenotic activity of dexamethasone was observed in a rat model [60]. This interspecies difference in response to vascular trauma is a well-known phenomenon and may be related to various sensitivities of SMC from different species to growth-promoting and growth-inhibiting factors [63].

Another widely used delivery vehicle is a poloxamer, Pluronic F-127. This substance is water-soluble at 4°C, but turns to gel immediately upon contact with body fluids at 37°C. An array of drug molecules, such as antioxidant probucol [64], antisense oligonucleotides to c-myc [65], c-myb [66,67], cdc2 [68], cdk2 [68], peptide inhibitor of integrin receptors [69] and plasmid DNA encoding for the ras transdominant negative mutant [70] formulated with Pluronic F-127 were used in rat and pig models of restenosis (Table 1). The main drawback of Pluronic F-127–based delivery systems is very rapid drug release, since a gel coating disappears from the arteries in 1–2 h [71]. Recently, Gottsauner-Wolf et al. [72] attempted to slow the release rate of a tyrphostin drug from Pluronic gel by coformulating F-127 with a biodegradable copolymer, polylactic/polyglycolic acid (PLGA). Despite the sustained in vitro release and a marked antiproliferative effect in cell culture, this delivery system of tyrphostin was ineffective in the rat carotid model of restenosis.

Ito et al. [73] developed a complex adventitial delivery system of protein tyrosine kinase inhibitor, ST 638, consisting of sterile cotton mesh impregnated with the drug preadsorbed on Sepharose microbeads. The in vitro release of ST 638 from this system was not determined. However, high and protracted arterial drug levels found in the treated segments suggest sustained release of the inhibitor.

In another study [49] the basic fibroblast growth factor (bFGF), ionically bound to heparin-Sepharose microbeads and encapsulated within calcium alginate microcapsules enabled controlled release of bFGF to balloon-injured rat carotid arteries. In these experiments microparticles were placed adjacent to injured arterial segments and were held in place by closing fascial planes over the artery.

Several groups [74–77] used osmotic pumps as a drug reservoir for perivascular delivery. The main advantage of this device is its relatively large volume of drug solution that might be delivered as well as its very accurate and predetermined delivery rates. On the other hand, the pumps deliver the drug with zero order kinetics and cannot be tailored to specific needs. In the cited studies the pumps were implanted in the back region of animals and connected to the arteries by polyethylene tubing. The tubing was firmly anchored in adjacent muscles and delivery was performed through side holes drilled in tubing [76], into a perivascular polyethylene cuff [74] or into a cuff-like minireservoir, glued in position on an ePTFE graft [75].

A novel and interesting approach for perivascular drug delivery was recently proposed by Nathan et al. [78]. The authors engrafted endothelial cells on Gelfoam biopolymeric matrices. Matrices were completely colonized by endothelial cells in 15 days. The three-dimensional scaffolds of endothelium on a Gelfoam were then implanted around the injured portion of rat carotid artery resulting in nearly total (92%) inhibition of neointimal formation. This marked antirestenotic effect was probably mediated by heparan sulfate glycosaminoglycans synthesized by endothelial cells, since local coadministration of protamine completely abolished the beneficial effects of Gelfoam-endothelium implants [79].

C. Perivascular Delivery of Tyrphostins

As stated above, the migration of medial SMC [80] and adventitial myofibroblasts [14] and their consequent intimal proliferation are governed by the combined effects of several growth

Table 2 Intraluminal Drug Delivery Systems Studied in Animal Models

Animal model	Medication	Delivery system	Effect on restenosis [Ref.]
Rabbit carotid	tissue factor pathway inhibitor	local irrigation, no specific delivery	+/– 92
Rat carotid	vascular endothelial growth factor	local irrigation, 30 min incubation, no specific delivery	+ 91
Rabbit femoral	protein adduct of nitric oxide	local irrigation, 15 min incubation, no specific delivery	+ 95
Rat iliofemoral	calreticulin	catheter-facilitated delivery	+ 134
Rabbit iliac and pig coronary	10% and 15% ethyl alcohol	catheter-facilitated delivery (Wolinsky porous balloon catheter)	+ 93; 94
Rabbit aortic	serine protease inhibitor, SERP-1	catheter-facilitated delivery (Wolinsky porous balloon catheter)	+ 135
Rabbit carotid	paxlitaxel (taxol)	catheter-facilitated delivery (microporous infusion catheter-Cordis)	+ 136
Pig coronary	antisense to c-myb	catheter-facilitated delivery (Transport catheter-SciMed)	+ 137
Pig carotid and iliac	antisense to c-myb	catheter-facilitated delivery (hydrogel-HydroPlus catheter)	nonstudied, proliferation reduced 138
Pig carotid and iliac	heparin	catheter-facilitated delivery (hydrogel-HydroPlus catheter)	nonstudied, proliferation reduced 139
Pig coronary	nonantisense oligonucleotides	catheter-facilitated delivery (iontophoretic catheter-CorTrak)	nonstudied 90
Dog graft	saporin	osmotic pump delivering chimeric bFGF-saporin	+ 114
Rat carotid	diphteria toxin	local irrigation +30 min incubation with chimeric EGF-diphteria toxin	+ 115
Rat carotid	antisense to cyclin G1	local irrigation +20 min incubation, retroviral vector	+ 140
Rabbit iliac	gax homeobox gene	catheter-facilitated delivery (channel balloon catheter, adenoviral vector)	+ 112
Rat carotid	C-type natriuretic peptide gene	local irrigation +15 min incubation, adenoviral vector	+ 109
Rat carotid and pig femoral	constitutively active form of Rb gene product	local irrigation +5 min incubation, adenoviral vector	+ 111
Rat carotid	full-length and truncated forms of Rb gene product	local irrigation +20 min incubation, adenoviral vector	+ 110

(continued)

Table 2 Continued

Animal model	Medication	Delivery system	Effect on restenosis [Ref.]
Rat carotid	dominant-negative H-ras	local irrigation +15 min incubation, adenoviral vector	+ 103
Rabbit iliac	herpes simplex virus thymidine kinase gene	catheter-facilitated delivery (hydrogel catheter), adenoviral vector	+ 106
Rat carotid	herpes simplex virus thymidine kinase gene	local irrigation +50 min incubation, adenoviral vector	+ 105
Rabbit femoral	cytosine deaminase gene	local irrigation +50 min incubation, adenoviral vector	+ 104
Rat carotid	p27	local irrigation +20 min incubation, adenoviral vector	+ 108
Rat carotid	p21	local irrigation +5 min incubation, adenoviral vector	+ 107
Rat carotid	hirudin gene	local irrigation + incubation, adenoviral vector	+ 113
Rat carotid	antisense to CDC2 kinase	local irrigation +15 min incubation, HVJ-liposome vector	+ 96
Rat carotid	antisense to CDC2 kinase and to cyclin B1	local irrigation +10 min incubation, HVJ-liposome vector	+ 97
Rat carotid	decoy sequence of E2F binding site	local irrigation +15 min incubation, HVJ-liposome vector	+ 99
Dog femoral vein graft	endothelial cell nitric oxide synthase gene	local irrigation +10 min incubation, HVJ-liposome vector	+ 141
Rat carotid	endothelial cell nitric oxide synthase gene	local irrigation +10 min incubation, HVJ-liposome vector	+ 101
Mice heart allograft	antisense to CDK2 kinase	ex vivo local irrigation +15 min incubation, HVJ-liposome vector	+ 100
Rat carotid	no drug	hydrogel endoluminal barrier	+ 116
Dog carotid	FGF/heparin/autologous endothelial cells	fibrin glue endoluminal barrier	+ 118
Rabbit femoral	colchicine and colchicine analogue	catheter-facilitated delivery, PLGA microparticles	− 120
Pig coronary	2-aminochromone	catheter-facilitated delivery (Dispatch dual lumen catheter), PLGA nanoparticles	+ 123
Rat carotid	dexamethasone	local irrigation and 3 min incubation, PLGA nanoparticles	+ 122
Rabbit aortic	hydrocortisone	catheter-facilitated delivery (Wolinsky porous balloon catheter), PLGA microparticles	+ 142
Pig femoral	tyrphostin, AG-1295	PLA nanoparticles	+ 125

(continued)

Table 2 Continued

Animal model	Medication	Delivery system	Effect on restenosis [Ref.]
Rabbit carotid	dexamethasone, colchicine	catheter-facilitated delivery (Wolinsky porous balloon catheter), PLA microparticles	+ 121
Pig coronary model	2-aminochromone	catheter-facilitated delivery (dispatch dual lumen catheter), surface-modified PLGA nanoparticles	+ 124
Rabbit stented iliac	vascular endothelial growth factor	catheter-facilitated delivery	+ 126
Pig stented coronary	dexamethasone	PLA-coated drug-eluting stent	− 127
Pig stented coronary	PTK inhibitor, ST638	PLA biodegradable stent stent	+ 128

factors released from aggregated platelets, injured endothelium and activated SMC [6,81]. Of all the growth factors and cytokines involved, platelet-derived growth factor (PDGF) is the one most intimately implicated in the process of intimal SMC cell accumulation.

Signaling cascades initiated by PDGF binding to its cognate receptors are meditated by the intrinsic protein tyrosine kinase (PTK) activity of receptors and several cytoplasmatic checkpoint proteins. Both, migration and proliferation of vascular SMC, critically depend on intact signal transduction through PTK. While it was recently shown that pathways for PDGF-driven SMC proliferation and migration bifurcate at an early point, they still share a common proximal section. Thus, inhibition of the autocrine/paracrine loop of PDGF could be achieved through the inhibition of PDGF receptor tyrosine kinase activity. It was shown that a series of low molecular weight PTK inhibitors, termed tyrphostins (for review see [82]), inhibit PDGF-dependent DNA synthesis and cell growth in cultured rabbit vascular SMC [83] and both proliferation and migration of cultured rat aortic SMC [84–86]. This inhibition correlates with the potency of tyrphostins to inhibit PDGF-dependent tyrosine autophosphorylation of intracellular substrates and activation of phospholipase Cγ [83]. More recently, it was shown that the PDGF-receptor PTK activity in a rat carotid artery is enhanced by cuff injury to the common carotid artery and is inhibited by a tyrphostin compound [87].

In a series of experiments [52,54] we evaluated the in vitro effects of a broad-spectrum tyrphostin, AG-17, and a PDGFRβ specific tyrphostin, AG-1295, on the proliferation of rat aortic SMC in culture. We also assessed the antirestenotic effects of this tyrphostin in the rat carotid model by a site-specific controlled release delivery system. Furthermore, we provided evidence that in vitro and in vivo effects of AG-1295 are mediated by its inhibitory effects on PDGFRβ-triggered tyrosine phosphorylation.

We formulated EVA-based delivery systems of both tyrphostins and assessed drug release from these systems in vitro and in vivo. The release of both drugs conformed to the t1/2 relationship and a good correlation was found between the amount of drug released and t1/2 for the various drug loads and matrix thicknesses. The release rate from the higher load AG-1295 matrices was somewhat slower than that obtained from the lower load matrices with the same thickness. Similarly, the release rate from the thicker AG-1295 matrices was slower than that obtained from the thinner matrices with the same drug content. The impact of thickness

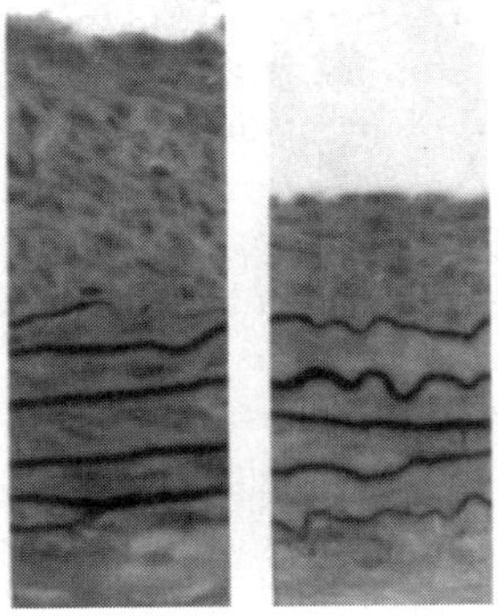

Figure 1 Microphotographs of representative histological sections: (a) section of rat carotid artery locally treated by blank EVA matrices; (b) section of rat carotid artery locally treated by 10% AG-1295 implants (Van Giesen elastic stain, magnification ×320) [52].

and drug load on release rate was not evident in the AG-17 matrices. The release of tyrphostins from EVA matrices in vivo was evaluated by analyzing the residual drug content in the matrices explanted at various time points after subcutaneous implantation in rats. In preliminary experiments the release rate from subcutaneously implanted devices was found to be the same as from perivascularly implanted matrices. The burst effect present in the release curves of AG-1295 in vitro was not evident with the subdermal implants. In general, the dependence of release rate on the matrices thickness and load found in vitro was replicated in the in vivo release pattern. Approximately 80–95% of both tyrphostins were released after 21–28 days from the matrices of different loads and thicknesses. This release rate corresponds to an average dose of 2 and 40 μg/kg/day of AG-1295 for 0.5% and 10% matrices, respectively and 75 μg/kg/day of AG-17 for 10% matrices. Local sustained delivery of the AG-1295 from perivascularly implanted polymeric matrices resulted in focal tyrphostin levels of 711 and 29.1 ng/mg of dry arterial tissue 1 and 14 days, respectively, after implantation in rats. Both tyrphostins delivered from polymeric matrices resulted in a significant (35–57%) reduction of neointimal formation on day 14 following balloon injury in the rat carotid model (Fig. 1). However, the focal treatment with 10% AG-17 matrices was associated with signs of local tissue toxicity, while the treatment with AG-1295 matrices was not. Tyrosine phosphorylation in arterial tissue extracts was significantly upregulated by balloon injury on day 3, but essentially returned to basal levels 14 days after injury. AG-1295 treatment decreased tyrosine phosphorylation at both time points below the basal levels. Moreover, PDGFRβ expression enhancement 3 and 14 days following arterial injury was strongly inhibited by AG-1295 treatment. Thus, it can be concluded that AG-1295 reduces neointimal formation by inhibiting PDGFβ-triggered tyrosine phosphorylation.

D. Intraluminal Delivery

The main advantage of intraluminal (intramural) delivery is that it is performed simultaneously with a vascular intervention, with no need for surgical procedures, and site-specific delivery. (The procedure of delivery is noninvasive, i.e., access to a diseased artery is through a percutaneous approach.) Since most of the delivered medication is distributed primarily in the intima and innermost media, the lumen-dependent processes, such as thrombosis, could be effectively treated by intraluminal delivery. The drug is then partially redistributed via vasa vasorum to outer layers of artery, reaching an adventitia [88,89]. Since drug redistribution occurs in a few hours [90], the adventitia-mediated processes, such as myofibroblast conversion and changes in a pattern of connective tissue proteins expression, could be targeted by intramural delivery. The concept of intraluminal delivery adds another requirement to a drug, vehicle and delivery system in comparison with perivascular applications. Since the drug is applied to an injured/diseased vessel wall that already has prothrombotic properties, it should not further increase local thrombogenicity. In addition, intraluminally delivered drug must be physically incorporated into the arterial wall. The long-term presence of a foreign material in immediate contact with the components of the artery emphasizes a problem of biocompatibility and biodegradability of the delivery system. The residence of a significant fraction of delivered dose in the intimal layer makes leaching of a drug more intensive because of the continuous blood flow through the treated arterial segment. Therefore, an intramural delivery requires, in general, a more pronounced sustained release than perivascular delivery. Finally, clinical embodiment of the concept of intraluminal delivery necessitates the development of dedicated equipment, such as special delivery catheters or endoluminal stents.

E. Intraluminal Delivery Vehicles

The delivery devices currently used allow for relatively small amounts of drug solution (usually less than 10 mL) to be introduced into the segment of artery. The drug driven by hydrostatic pressure and concentration gradient penetrates the vessel wall. Even with a high concentration of drug in solution, which is not achievable in hydrophobic molecules, the final amount entering the vessel wall is small. Moreover, most of the free drug is leached into general circulation by blood flow in the first minutes after restoration of circulation in the treated segment. The main directions in this field and specific embodiments of the research are summarized in Table 2.

In spite of several reports about the antirestenotic efficacy of intramurally delivered free drugs [91–95], the general trend is to bind drug to delivery vehicles that have an affinity to structural components of artery and/or have better residence properties than individual drug molecules.

Morishita and Dzau were the first to use liposomes for the introduction of a genetic material into injured arteries [96–99]. In order to increase the efficacy of liposome-mediated nucleic acid transfer, the authors developed a fusgenic liposome vector based on the principles of viral cell fusion. The fusion protein of hemagglutinating virus of Japan was complexed with liposomes encapsulating oligodeoxynucleotide or plasmid DNA. Subsequent fusion of HVJ-liposomes with plasma membranes directly introduced DNA into cytoplasm. Using this versatile delivery vector the authors showed high antirestenotic activity of antisense constructs to several cell cycle checkpoint proteins, such as CDC2 and PCNA [96], CDK2 [100], cyclin B1 [97] and a sense double-stranded decoy DNA for the transcription factor E2F [99]. Using the same liposomal vector, von der Leyen et al. [101] have shown that the transfer of complete gene sequence encoding for endothelial nitric oxide synthase restores NO synthesis in balloon-injured rat carotid arteries and effectively attenuates neointimal formation.

Several groups employed viral vectors to introduce genetic material into endothelial and smooth muscle cells of arterial wall. Historically, the first group of viruses used for engineering the delivery constructs were retroviruses. A potentially attractive feature of retroviral vectors is their ability to integrate into the DNA of target cells, thus conferring the long-lasting expression of transgene and a selective infection of proliferative cells. Taking into account the proliferative nature of restenotic lesions [9], this latter property is very important in the context of restenosis. However, the transduction efficacies of retroviral vectors are generally low due to a procedural failure to concentrate more than 10^6 transduction units per mL. Another unresolved issue is the safety profile of retroviral constructs. Since the site of viral integration into genome is accidental, the transgene may destroy a normal function of important "housekeeping" genes, including ones responsible for oncogenic transformation. The only reported study that explored antiproliferative properties of retroviral constructs demonstrated ~60% reduction of restenosis extent in the rat carotid artery locally treated with retroviral vehicle loaded with antisense sequence to cyclin G1 [102].

Recently adenovirus-based vectors became the central theme in virus-mediated arterial gene transfer. In comparison to retroviral counterparts, adeno- and adeno-associated viral vectors achieve 10–30% of transgene expression in vivo in targeted cell populations. The main disadvantages of adenoviral vectors are a marked local inflammatory and immune response and a relatively rapid (less than four weeks) disappearance of transgene from the transduced cells. The latter obstacle may be less important in the setting of postangioplasty restenosis, since the principal processes involved in the neointimal formation are essentially over by this time. In the past few years, marked antirestenotic properties of adenoviral constructs bearing genes for dominant-negative H-ras mutant [103], cytosine deaminase [104], thymidine kinase [105,106], p21 [107], p27 [108], C-type natriuretic peptide [109], truncated [110] or constitutively active [111] forms of Rb protein, growth-arresting gax homeobox protein [112] and recombinant hirudin [113] were shown in rat, rabbit and porcine models.

The viral vectors do not need high delivery pressures to achieve incorporation into arterial wall. This is because the entry of the vector into the blood vessel employs an intrinsic infective apparatus of a virus, conferring to vector affinity to the target tissue. Such affinity-facilitated delivery was approached from another angle in experiments assessing the antiproliferative and antirestenotic potential of chimeric toxins combining a potent cytotoxin molecule, such as saporin [114] or diphtheria toxin [115] with a targeting moiety that selectively binds activated SMC. This approach proved to be effective in the rat carotid and dog graft anasthomosis models, where the targeting molecules were basic FGF and epidermal growth factor, respectively [114,115].

Since platelets- and plasma-derived factors promote prorestenotic changes in injured arteries, some authors proposed to isolate a vessel wall from these factors immediately following vascular trauma. Pursuing this approach, West and Hubbell intraluminally applied hydrogel to the injured rat carotid artery, that following photopolymerization created a thin film separating lumen and vessel wall. This manipulation by itself drastically reduced the extent of restenosis in ballooned rat arteries [116]. A subsequent evolution of this endoluminal paving approach added an element of local drug delivery from intraluminally deposited hydrogel films [117].

Recently, a conceptually similar approach was proposed by Zarge et al., who reduced neointimal formation in balloon-injured dog carotid arteries by local intraluminal treatment with fibrin glue admixed with acidic FGF, heparin and autologous endothelial cells [118].

It was previously stated that the main drawback of intraluminal delivery is a rapid washout of the drug by blood flow. Most of the drug is leached from the infusion site within the initial period of 30 min with a very small residual amount lasting for a maximum period of 48 h [119]. Several groups attempted to slow the drug washout by incorporating the active drug into a biodegradable polymeric matrix formulated in the form of micro- or nanoparticles. From a theoretical point of view micro- and, especially, nanoparticulate systems based on biocompatible and biodegradable polymers are well suited for intraluminal delivery. Due to their extremely small size they should penetrate deeply into the vessel wall and create a depot relatively isolated from the arterial lumen. Nanoparticles that are 100–150 nm in size could be actively taken by several cell types present in atherosclerotic arteries. Additionally, several formulatory variables allow for a customized release rate of incorporated drug to fit the time course of pathological processes in the vessel wall. Finally, ingress of nanoparticles into the artery should be relatively nontraumatic, i.e., it should not add to the injury caused by vascular intervention. This area of research was pioneered by the work of Gradus-Pizlo et al., who showed that colchicine-loaded PLGA microparticles prolong drug residence in the wall of rabbit atherosclerotic arteries, but failed to demonstrate any beneficial effect of intraluminally infused colchicine nanoparticles on the extent of restenosis [120]. These results were recently confirmed by another group [121], using a different method for the preparation of colchicine- and dexamethasone-loaded microparticles. The authors achieved a marked sustained release in vitro, but neither the colchicine or the dexamethasone formulation was effective in the rabbit carotid model. Guzman et al. [122] used rhodamine B fluorescent-labeled PLGA nanoparticles in order to trace nanoparticle distribution and residence in balloon injured rat carotid arteries. Nanoparticles successfully penetrated the vessel wall and persisted up to 14 days following a single 3 min infusion. When loaded with dexamethasone, nanoparticles effectively reduced neointimal formation in the rat carotid model [122]. Recently, the same group demonstrated antirestenotic efficacy of intraluminally delivered 2-aminochromone-loaded PLGA nanoparticles in balloon-injured porcine coronary arteries [123]. Furthermore, in their latest work [124] these authors studied the impact of nanoparticle surface modification on arterial disposition and residence. They showed that nanoparticles surface-modified with a cationic compound, didodecyldimethylammonium bromide, demonstrated 7- to 10-fold greater arterial drug levels compared to the unmodified nanoparticles, both in ex vivo dog femoral and in vivo porcine coronary models [124].

Using a self-emulsification technique we prepared AG-1295 PLA-based nanoparticles. The mean diameter of nanoparticles formulated by this procedure was 110 nm with a narrow size distribution, and the incorporation efficiency of tyrphostin was about 40%. Fluorescent model compound (pyrene)-loaded nanoparticles formulated by the same technique were previously shown to effectively penetrate the injured rat carotid artery and were traced in the adventitial layer up to four days following delivery. Subsequent to balloon injury of pig femoral arteries about 1 mL of nanoparticles suspension was injected into a temporally isolated segment of artery. Following a 30 min delivery period the nanoparticle suspension was evacuated and the circulation in the treated segment was restored. HPLC analysis of the local drug content showed sustained tyrphostin levels in the injured artery up to 2 h postdelivery (the later timepoints were not checked), while AG-1295 blood levels were uniformly below the detection limits of the method. The extent of restenosis four weeks after injury and local delivery was reduced to 50% in the arteries treated with tyrphostin-loaded nanoparticles in comparison with blank PLA nanoparticles [125].

F. Endoluminal Stents and Vascular Drug Delivery

The last few years of restenosis research have highlighted a previously underestimated role that geometrical changes in angioplasticized vessels play in the final outcome of PTCA. A paradigm of restenosis as a solely hyperproliferative state was replaced by the concept of combined effects of neointimal formation and unfavorable remodeling, evenly contributing to lumen narrowing. These changes in our understanding of restenosis pathophysiology were reflected in an ever-growing use of endoluminal stents in cardiological practice. Stents represent rigid scaffolds introduced into treated arterial segments that due to their mechanical strength neutralize both early elastic recoil and late constrictive remodeling of the artery. While effective as preventers of arterial shrinkage, stents cause a neointimal proliferation that might be more robust than neointimal thickening after regular PTCA procedure. In comparison to "pure" postangioplasty restenosis this "in stent" restenosis has some unique histological properties. These properties are related to the admixtue of inflammatory cells, and of cytokines due to a local tissue reaction on the presence of foreign nonbiocompatible material. From the local delivery point of view stents represent a convenient platform to antirestenotic drug delivery. Three types of delivery systems were developed to meet a challenge of in-stent restenosis. In one type, a drug might be delivered intraluminally in the vicinity of a deployed stent [126]. In another type of system, the stent might be coated with drug-impregnated polymer [127]. Finally, the stent itself might be made from a biodegradable polymeric material loaded with a drug [128]. With the two latter approaches a marked sustained drug release can be achieved, which means these systems may be very promising in the context of in-stent restenosis. A detailed discussion of a stent-based vascular local drug delivery is not covered in this chapter. The reader is referred to three excellent recent reviews [41,42,129].

V. CONCLUSIONS

Analysis of recent literature on the therapy of postangioplasty restenosis and related disorders suggests that an ultimate solution of this multifaceted problem will arise from the field of local drug delivery. It will most probably be based on intraluminal approaches, and may utilize vectors facilitating targeting and ingress of the drug into the vessel wall, its long-term residence and local activity. Since pathological and pharmacological characteristics of human restenosis are only partially reflected in existing animal models, before any drug enters clinical study it should first be examined meticulously in an array of animal experiments that include a porcine or primate model as well as a rodent model.

REFERENCES

1. Baim, D. S. 1992. Interventional catheterization techniques: Percutaneous transluminal balloon angioplasty, valvuloplasty and related procedures. In Braunwald, E. (ed.), *Heart Disease: A Textbook of Cardiovascular Medicine*, Vol. 2, Saunders, Philadelphia, pp. 1365–1381.
2. Nobuyoshi, M., Kimura, T. and Ohishi, H., 1991. Restenosis after percutaneous transluminal coronary angioplasty: Pathologic observation in 20 patients. *J. Am. Coll. Cardiol.*, 17:433–439.
3. Topol, E. J., Ellis, S., Cosgrove, D., Bates, E., Muller, D., Schork, N. and Loop, F., 1993. Analysis of coronary angioplasty practice in the United States with an insurance-claims data base. *Circulation*, 87:1489–1497.
4. Landzberg, B. R., Frishman, W. H. and Lerrick, K., 1997. Pathophysiology and pharmacological approaches for prevention of coronary artery restenosis following coronary artery balloon angioplasty and related procedures. *Progr. Cardiovasc. Dis.*, 39:361–398.
5. Lefkovits, J. and Tolol, E. J., 1997. Pharmacological approaches for the prevention of restenosis after percutaneous coronary intervention. *Progr. Cardiovasc. Dis.*, 40:141–158.
6. Ross, R., 1993. The pathogenesis of atherosclerosis: a perspective for the 1990s. *Nature*, 362: 801–809.
7. Clowes, A. W., Reidy, M. A. and Clowes, M. M., 1983. Kinetics of cellular proliferation after arterial injury. I. Smooth muscle growth in the absense of endothelium. *Lab. Invest.*, 49:327–333.
8. Clowes, A. W., Reidy, M. A. and Clowes, M. M., 1983. Mechanisms of stenosis after arterial injury. *Lab. Invest.*, 49:208–215.
9. Gordon, D., Reidy, M. A., Benditt, E. P. and Schwartz, S. M., 1990. Cell proliferation in human coronary arteries. *Proc. Natl. Acad. Sci. USA*, 87:4600–4604.
10. Waller, B. F., Pinkerton, C. M., Orr, C. M., Slack, J. D., VanTassel, J. W. and Peters, T., 1991.

Restenosis 1 to 24 months after clinically successful coronary balloon angioplasty: a necroscopy study of 20 patients. *J. Am. Coll. Cardiol.*, 17:58B–70B.
11. Schwartz, R. S., Holmes, D. R. and Topol, E. J., 1992. The restenosis paradigm revisited: an alternative proposal for cellular mechanisms. *J. Am. Coll. Cardiol.*, 20:1284–1293.
12. Faxon, D. P., Coats, W. and Currier, J., 1997. Remodeling of the coronary artery after vascular injury. *Progr. Cardiovasc. Dis.*, 40:129–140.
13. Kovach, J. A., S, M. G. and M, K. K., 1993. Serial intravascular ultrasound studies indicate that chronic recoil is an important mechanism of restenosis following transcatheter therapy. *J. Am. Coll. Cardiol.*, 21:484A.
14. Shi, Y., O'Brien, J. E., Fard, A., Mannion, J. D., Wang, D. and Zalewski, A., 1996. Adventitial myofibroblasts contribute to neointimal formation in injured porcine coronary arteries. *Circulation*, 94:1655–1664.
15. Shi, Y., Pieniek, M., Fard, A., O'Brien, J., Mannion, J. D. and Zalewski, A., 1996. Adventitial remodeling after coronary arterial injury. *Circulation*, 93:340–348.
16. Wilcox, J. N. and Scott, N. A., 1996. Potential role of adventitia in arteritis and atherosclerosis. *Int. J. Cardiol.*, 54 Suppl.:S21–S35.
17. Wilcox, J. N., Cipolla, G. D., Martin, F. H., Simonet, L., Dunn, B., Ross, C. E. and Scott, N. A., 1997. Contribution of adventitial myofibroblasts to vascular remodeling and lesion formation after experimental angioplasty in pig coronary arteries. *Ann. N. Y. Acad. Sci.*, 811:437–447.
18. Hancock, W. W., Adams, D. H., Wyner, L. R., Sayegh, M. H. and Karnovsky, M. J., 1994. $CD4^+$ mononuclear cells induce cytokine expression, vascular smooth muscle cell proliferation, and arterial occlusion after endothelial injury. *Am. J. Pathol.*, 145:1008–1014.
19. Rubin, P., Williams, J. P., Riggs, P. N., Bartos, S., Sarac, T., Pomerantz, R., Castano, J., Schell, M. and Green, R. M., 1998. Cellular and molecular mechanisms of radiation inhibition of restenosis. Part 1: Role of the macrophage and platelet-derived growth factor. *Int. J. Radiation Oncol. Biol. Phys.*, 40:929–941.
20. Schwartz, R. S., 1994. Neointima and arterial injury: dogs, rats, pigs, and more. *Lab. Invest.*, 71: 789–791.
21. Herrman, J. P. R., Hermans, W. R. M., Vos, J. and Serruys, P. W., 1993. Pharmacological approaches to the prevention of restenosis following angioplasty: part I. *Drugs*, 46:18–52.
22. Herrman, J. P. R., Hermans, W. R. M., Vos, J. and Serruys, P. W., 1993. Pharmacological approaches to the prevention of restenosis following angioplasty: part II. *Drugs*, 46:249–262.
23. Jackson, R. L., Barnhart, R. L. and Mao, S. J. T. 1990. Probucol and its mechanism for reducing atherosclerosis. In Malmendier, C. L. (ed.), *Hypercholesterolemia, Hypocholesterolemia, Hypertriglyceridemia*, Plenum Press, New York, pp. 367–372.
24. Kuzuya, M. and Kuzuya, F., 1993. Probucol as an antioxidant and antiatherogenic drug. *Free Rad. Biol. Med.*, 14:67–77.
25. Holvoet, P. and Collen, D., 1994. Oxidized lipoproteins in atherosclerosis and thrombosis. *FASEB J.*, 8:1279–1284.
26. Penn, M. S. and Chisolm, G. M., 1994. Oxidized lipoproteins, altered cell function and atherosclerosis. *Atherosclerosis*, 108 (suppl):S21–S29.
27. Berliner, J. A., Navab, M., Fogelman, A. M., Frank, J. S., Demer, L. L., Edwards, P. A., Watson, A. D. and Lusis, A. J., 1995. Atherosclerosis: basic mechanisms: Oxidation, inflammation and genetics. *Circulation*, 91:2488–2496.
28. Steinberg, D., 1997. Oxidative modification of LDL and atherogenesis. *Circulation*, 95:1062–1067.
29. Barbeau, M. L., Whitman, S. C. and Rogers, K. A., 1995. Probucol but not maxEPA fish oil, inhibits mononuclear cell adhesion to the aortic intima in the rat model of atherosclerosis. *Biochem. Cell Biol.*, 73:283–288.
30. Ferns, G. A. A., Forster, L., Stewart-Lee, A., Nourooz-Zadeh, J. and Anggard, E. E., 1993. Probucol inhibits mononuclear cell adhesion to vascular endothelium in the cholesterol-fed rabbit. *Atherosclerosis*, 100:171–181.
31. Schneider, J. E., Berk, B. C., Gravanis, M. B., Santoian, E. C., Cipolla, G. D., Tarazona, N.,

Lassegue, B. and King, S. B., 1993. Probucol decreases neointimal formatioin in a swine model of coronary artery balloon injury. A possible role for antioxidants in restenosis. *Circulation*, 88: 628–637.
32. Shinomiya, M., Shirai, K., Saito, Y. and Yoshida, S., 1992. Inhibition of intimal thickening of the carotid artery of rabbits and of outgrowth of explants of aorta by probucol. *Atherosclerosis*, 97: 143–148.
33. Tardif, J. C., Cote, G., Lesperance, J., Bourassa, M., Lambert, J., Douset, S., Bilodeau, L., Nattel, S. and de Guise, P., 1997. Probucol and multivitamins in the prevention of restenosis after coronary angioplasty: Multivitamins and probucol study group. *N. Engl. J. Med.*, 337:365–372.
34. Miyazawa, K., Kikuchi, S., Fukuyama, J., Hamano, S. and Ujie, A., 1995. Inhibition of PDGF- and TGF-β1-induced collagen synthesis, migration and proliferation by tranilast in vascular smooth muscle cells from spontaneously hypertensive rats. *Atherosclerosis*, 118:213–221.
35. Tanaka, K., Honda, M., Kuramochi, T. and S, M., 1994. Prominent inhibitory effects of tranilast on migration and proliferation of and collagen synthesis by vascular smooth muscle cells. *Atherosclerosis*, 107:179–185.
36. Kosuga, K., Tamai, H., Ueda, K., Hsu, Y. S., Ono, S., Tanaka, S., Doi, T., Myou, U. W., Motohara, S. and Uehata, H., 1997. Effectiveness of tranilast on restenosis after directional coronary atheroctomy. *Am. Heart J.*, 134:712–718.
37. Topol, E. J., Califf, R. M. and Weismann, H. F., 1994. Randomised trial of coronary intervention with antibody against platelet IIb/IIIa integrin for reduction of clinical restenosis: results at 6 months. *Lancet*, 343:881–886.
38. Edelman, E. R., Simons, M., Sirois, M. G. and Rosenberg, R. D., 1995. C-myc in vasculoproliferative disease. *Circ. Res.*, 76:176–182.
39. Lincoff, A. M., Topol, E. J. and Ellis, S. G., 1994. Local drug delivery in the prevention of restenosis: Fact, fancy, and future. *Circulation*, 90:2070–2084.
40. Riessen, R. and Isner, J. M., 1994. Prospects for site-specific delivery of pharmacologic and molecular therapies. *J. Am. Coll. Cardiol.*, 23:1234–1244.
41. McKenna, C. J., Holmes, D. R. and Schwartz, R. S., 1997. Novel stents for the prevention of restenosis. *Trends Cardiovasc. Med.*, 7:245–249.
42. Belli, G., Ellis, S. G. and Topol, E. J., 1997. Stenting for ischemic heart disease. *Progr. Cardiovasc. Dis.*, 40:159–182.
43. Muller, D. W. M. and Topol, E. J., 1991. Implantable devices in the coronary artery from metal to genes. *Trends Cardiovasc. Med.*, 1:225–232.
44. Kumamoto, M., Nakasima, Y. and Sueishi, K., 1995. Intimal neovascularization in human coronary atherosclerosis: its origin and pathophysiological significance. *Hum. Pathol.*, 26:450–456.
45. Wolinsky, H. and Glagov, S., 1967. Nature of species differences in the medial distribution of aortic vasa vasorum in mammals. *Circ. Res.*, 20:409–421.
46. Lovich, M. A., Brown, L. and Edelman, E. R., 1997. Drug clearance and arterial uptake after local perivascular delivery to the rat carotid artery. *J. Am. Coll. Cardiol.*, 29:1645–1650.
47. Ip, J. H., Fuster, V., Badimon, L., Badimon, J., Taubman, M. B. and Chesebro, J. H., 1990. Syndromes of accelerated atherosclerosis: role of vascular injury and smooth muscle cell proliferation. *J. Am. Coll. Cardiol.*, 15:1667–1687.
48. Edelman, E. R., Adams, D. H. and Karnovsky, M. J., 1990. Effect of controlled adventitial heparin delivery on smooth muscle cell proliferation following endothelial injury. *Proc. Natl. Acad. Sci. USA*, 87:3773–3777.
49. Edelman, E. R., Nugent, M. A., Smith, L. T. and Karnovsky, M. J., 1992. Basic fibroblast growth factor enhances the coupling of intimal hyperplasia and proliferation of vasa vasorum in injured rat arteries. *J. Clin. Invest.*, 89:465–473.
50. Edelman, E. R., Nugent, M. A. and Karnovsky, M. J., 1993. Perivascular and intravenous administration of basic fibroblast growth factor: vascular and solid organ deposition. *Proc. Natl. Acad. Sci. USA*, 90:1513–1517.
51. Edelman, E. R. and Karnovsky, M. J., 1994. Contrasting effects of the intermittent and continuous administration of heparin in experimental restenosis. *Circulation*, 89:770–776.

52. Fishbein, I., Waltenberger, J., Banai, S., Rabinovich, L., Chorny, M., Levitzki, A., Gazit, A., Huber, R., Gertz, S. D. and Golomb, G., Local delivery of PDGF receptor-specific tyrphostin inhibits neointimal formation in rats. *Atheroscler. Thromb. Vasc. Bio.* (in press).
53. Golomb, G., Mayberg, M. and Domb, A. J. 1994. Polymeric perivascular delivery systems. In Domb, A. J. (ed.), *Polymeric Site-Specific Pharmacotherapy*, Wiley, pp. 205–219.
54. Golomb, G., Fishbein, I., Banai, S., Mishaly, D., Moscovitz, D., Gertz, S. D., Gazit, A., Poradosu, E. and Levitzki, A., 1996. Controlled delivery of a tyrphostin inhibits intimal hyperplasia in a rat carotid artery injury model. *Atherosclerosis*, 125:171–182.
55. Edelman, E. R., Pukac, L. A. and Karnovsky, M. J., 1993. Protamine and protamine-insulins exacerbate the vascular response to injury. *J. Clin. Invest.*, 91:2308–2313.
56. Golomb, G. and Fishbein, I., 1997. Tryphostins, inhibitors of protein tyrosine kinase, in restenosis. *Adv. Drug Deliv. Rev.*, 24:53–62.
57. Mishaly, D., Fishbein, I., Moscovitz, D. and Golomb, G., 1997. Site-specific delivery of colchicine in rat carotid artery model of restenosis. *J. Control. Rel.*, 45:65–73.
58. Brauner, R., Laks, H., Drinkwater, D. C. J., Chaudhuri, G., Shvarts, O., Drake, T., Bhuta, S., Mishaly, D., Fishbein, I. and Golomb, G., 1997. Controlled periadventitial administration of verapamil inhibits neointimal smooth muscle cell proliferation and ameliorates vasomotor abnormalities in experimental vein bypass grafts. *J. Thorac. Cardiovasc. Surg.*, 114:53–63.
59. Sirois, M. G., Simons, M. and Edelman, E. R., 1997. Antisense oligonucleotide inhibition of PDGF-β receptor subunit expression directs suppression of intimal thickening. *Circulation*, 95: 669–676.
60. Villa, A. E., Guzman, L. A., Golomb, G., Levy, R. J. and Topol, E. J., 1994. Local delivery of dexamethasone for prevention of neointimal proliferation in a rat model of balloon angioplasty. *J. Clin. Invest.*, 93:1243–1249.
61. Muller, D. W. M., Golomb, G., Gordon, D. and Levy, R. J., 1994. Site-specific dexamethasone delivery for the prevention of neointimal thickening after vascular stent implantation. *Coronary Artery Disease*, 5:435–442.
62. Muller, D. W. M., Gordon, D., Topol, E. J., Levy, R. J. and Golomb, G., 1996. Sustained release local hirulog therapy decreases early thrombosis but not neointimal thickening after arterial stenting. *Am. Heart J.*, 131:211–218.
63. Voisard, R., Seitzer, U., Baur, R., Dartsch, P. C., Osterhues, P. C., Hoher, M. and Hombach, V., 1995. A prescreening system for potential antiproliferative agents: implications for local treatment strategies of postangioplasty restenosis. *Int. J. Cardiol.*, 51:15–28.
64. Ishizaka, N., Kurokawa, K., Taguchi, J., Miki, K. and Ohno, M., 1995. Inhibitory effect of a single local probucol administration on neointimal formation in balloon-injured rat carotid artery. *Atherosclerosis*, 118:53–56.
65. Bennett, M. R., Anglin, S., McEwan, J. R., Jagoe, R., Newby, A. C. and Evan, G. I., 1994. Inhibition of vascular smooth muscle cell proliferation in vitro and in vivo by c-myc antisense oligodeoxynucleotides. *J. Clin. Invest.*, 93:820–828.
66. Castier, Y., Chemla, E., Nierat, J., Heudens, D., M A, Rainoch, C., Bruneval, P., Carpentier, A. and Fabiani, J. N., 1998. The activity of c-myb antisense oligonucleotide to prevent intimal hyperplasia is nonspecific. *J. Cardiovasc. Surg.*, 39:1–7.
67. Villa, A. E., Guzman, L. A., Poptic, E. J., Labhasetwar, V., D'Souza, S., Farrell, C. L., Plow, E. F., Levy, R. J., DiCorleto, P. E. and Topol, E. J., 1995. Effects of antisense c-myb oligonucleotides on vascular smooth muscle cell proliferation and response to vessel wall injury. *Circ. Res.*, 76:505–513.
68. Abe, J.-I., Zhou, W., Taguchi, J.-I., Takuwa, N., Miki, K., Okazaki, H., Kurokawa, K., Kumada, M. and Takuwa, Y., 1994. Suppression of neointimal smooth muscle cell accumulation in vivo by antisense CDC2 and CDK2 oligonucleotides in rat carotid artery. *Biochem. Biophys. Res. Commun.*, 198:16–24.
69. Slepian, M. J., Masia, S. P., Dehdashti, B., Fritz, A. and Whitesell, L., 1998. β3-integrins rather than β1-integrins dominate integrin-matrix interactions involved in postinjury smooth muscle cell migration. *Circulation*, 97:1818–1827.

70. Indolfi, C., Avvedimento, E. V., Rapacciuolo, A., Di Lorenzo, E., Esposito, G., Stabile, E., Feliciello, A., Mele, E., Giuliano, P., Condorelli, G. L. and Chiariello, M., 1995. Inhibition of cellular ras prevents smooth muscle cell proliferation after vascular injury in vivo. *Nature Medicine*, 1: 541–545.
71. Wang, W., Chen, H. J., Schwartz, A., Cannon, P. J., Stein, C. A. and Rabbani, L. E., 1996. Sequence independent inhibition of in vitro smooth muscle cell proliferation, migration, and in vivo neointimal formation by phosphorothioate oligodeoxynucleotides. *J. Clin. Invest.*, 98:443–450.
72. Gottsauner-Wolf, M., Jang, Y., Lincoff, A. M., Cohen, J. L., Labhasetwar, V., Poptic, E. J., Forudi, F., Guzman, L. A., DiCorleto, P. E., Levy, R. J., Topol, E. J. and Ellis, S. G., 1997. Influence of local delivery of the protein tyrosine kinase receptor inhibitor tyrphostin-47 on smooth-muscle cell proliferation in a rat carotid balloon-injury model. *Am. Heart J.*, 133:329–334.
73. Ito, A., Shimokawa, H., Kadokami, T., Fukumoto, Y., Owada, M. K., Shiraishi, T., Nakaike, R., Takayanagi, T., Egashira, K. and Takeshita, A., 1995. Tyrosine kinase inhibitor suppressess coronary arteriosclerotic changes and vasospastic responses induced by chronic treatment with interleukin-1β in pigs in vivo. *J. Clin. Invest.*, 96:1288–1294.
74. Autieri, M. V., Yue, T.-L., Feurstein, G. Z. and Ohlstein, E., 1995. Antisense oligonucleotides to the p65 subunit of NF-κB inhibit human vascular smooth muscle cell adherence and proliferation and prevent neointima formation in rat carotid arteries. *Biochem. Biophys. Res. Commun.*, 213: 827–836.
75. Chen, C., Mattar, S. G., Hughes, J. D., Pierce, G. F., Cook, J. E., Ku, D. N., Hanson, S. R. and Lumsden, A. B., 1996. Recombinant mitotoxin basic fibroblast growth factor-saporin reduces venous anastomotic intimal hyperplasia in the arteriovenous graft. *Circulation*, 94:1989–1995.
76. Choi, E. T., Engel, L., Callow, A. D., Sun, S., Trachtenberg, J., Santoro, S. and Ryan, U. S., 1994. Inhibition of neointimal hyperplasia by blocking αvβ3 integrin with a small peptide antagonist GpenGRGDSPCA. *J. Vasc. Surg.*, 19:125–134.
77. Singh, J. P., Rothfuss, K. J., Wiernicki, T. R., Lacefield, W. B., Kurtz, W. L., Brown, R. F., Brune, K. A., Bailey, D. and Dube, G. P., 1994. Dipyridamole directly inhibits vascular smooth muscle cell proliferation in vitro and in vivo: implication in the treatment of restenosis after angioplasty. *J. Am. Coll. Cardiol.*, 23:665–671.
78. Nathan, A., Nugent, M. A. and Edelman, E. R., 1995. Tissue engineered perivascular endothelial cell implants regulate vascular injury. *Proc. Natl. Acad. Sci. USA*, 92:8130–8134.
79. Han, R. O., Ettenson, D. S., Koo, E. W. Y. and Edelman, E. E., 1997. Heparin/heparan sulfate chelation inhibits control of vascular repair by tissue-engineered endothelial cells. *Am. J. Physiol.*, 273:H2586–H2595.
80. Casscells, W., 1992. Migration of smooth muscle and endothelial cells: critical events in restenosis. *Circulation*, 86:723–729.
81. Schwartz, S. M., deBlois, D. and O'Brien, E. R. M., 1995. The intima: Soil for atherosclerosis and restenosis. *Circ. Res.*, 77:445–465.
82. Levitzki, A. and Gazit, A., 1995. Tyrosine kinase inhibition: an approach to drug development. *Science*, 267:1782–1788.
83. Bilder, G. E., Krawiec, J. A., McVety, K., Gazit, A., Gilon, C., Lyall, R., Zilberstein, A., Levitzki, A., Perrone, M. H. and Schreiber, A. B., 1991. Tyrphostins inhibit PDGF-induced DNA synthesis and associated early events in smooth muscle cells. *Am. J. Physiol.*, 260:C721–C730.
84. Shimokado, K., Yokota, T., Umezawa, K., Sasaguri, T. and Ogata, J., 1994. Protein tyrosine kinase inhibitors inhibit chemotaxis of vascular smooth muscle cells. *Arterioscler. Thromb.*, 14:973–981.
85. Shimokado, K., Yokota, T., Kosaka, C., Zen, K., Sasaguri, T., Masuda, J. and Ogata, J., 1995. Protein tyrosine kinase inhibitors inhibit both proliferation and chemotaxis of vascular smooth muscle cells. *Ann. N. Y. Acad. Sci.*, 748:171–176.
86. Shimokado, K., Umezawa, K. and Ogata, J., 1995. Tyrosine kinase inhibitors inhibit multiple steps of the cell cycle of vascular smooth muscle cells. *Exp. Cell Res.*, 220:266–273.
87. Bilder, G. E., Kasiewski, C. J., Walczak, E. M. and Perrone, M. H., 1993. PDGF-receptor protein

tyrosine kinase activity in carotid artery is enhanced by injury and inhibited in vivo by tyrphostin RG 13291. *Drug Dev. Res.*, 29:158–166.

88. Rome, J. J., Shayani, V., Flugelman, M. Y., Newman, K. D., Farb, A., Virmani, R. and Dichek, D. A., 1994. Anatomic barriers influence the distribution of in vivo gene transfer into the arterial wall: Modeling with microscopic tracer particles and verification with a recombinant adenoviral vector. *Arterioscler. Thromb.*, 14:148–161.
89. Wilensky, R. L., March, K. L. and Hathaway, D. R., 1991. Direct intraarterial wall injection of microparticles via a catheter: A potential drug delivery strategy following angioplasty. *Am. Heart J.*, 22:1136–1140.
90. Robinson, K. A., Chronos, N. A. F., Schieffer, E., Palmer, S. J., Cipolla, G. D., Milner, P. G., Walsh, R. G. and King, S. B. I., 1997. Pharmacokinetics and tissue localization of antisense oligonucleotides in balloon-injured pig coronary arteries after local delivery with an iontophoretic balloon catheter. *Catheter. Cardiovasc. Diagn.*, 41:354–359.
91. Asahara, T., Bauters, C., Pastore, C., Kearney, M., Rossow, S., Bunting, S., Ferrara, N., Symes, J. F. and Isner, J. M., 1995. Local delivery of vascular endothelial growth factor accelerates reendothelialization and attenuates intimal hyperplasia in balloon-injured rat carotid artery. *Circulation*, 91:2793–2801.
92. Brown, D. M., Kania, N. M., Choi, E. T., Lantieri, L. A., Pasia, E. N., Wun, T.-C. and Khouri, R. K., 1996. Local irrigation with tissue factor pathway inhibitor inhibits intimal hyperplasia induced by arterial interventions. *Arch. Surg.*, 131:1086–1090.
93. Liu, M. W., Lin, S.-J. and Chen, Y.-L., 1996. Local alcohol delivery may reduce phenotype conversion of smooth muscle cells and neointimal formation in rabbit iliac arteries after balloon injury. *Atherosclerosis*, 127:221–227.
94. Liu, M. W., Anderson, P. G., Luo, J. F. and Roubin, G. S., 1997. Local delivery of ethanol inhibits intimal hyperplasia in pig coronary arteries after balloon injury. *Circulation*, 96:2295–2301.
95. Marks, D. S., Vita, J. A., Folts, J. D., Keaney, J. F., Welch, G. N. and Loscalzo, J., 1995. Inhibition of neointimal proliferation in rabbits after vascular injury by a single treatment with a protein adduct of nitric oxide. *J. Clin. Invest.*, 96:2630–2638.
96. Morishita, R., Gibbons, G. H., Ellison, K. E., Nakajima, M., Zhang, L., Kaneda, Y., Ogihara, T. and Dzau, V. J., 1993. Single intraluminal delivery of antisense cdc2 kinase and proliferating-cell nuclear antigen oligonucleotides results in chronic inhibition of neointimal hyperplasia. *Proc. Natl. Acad. Sci. USA*, 90:8474–8478.
97. Morishita, R., Gibbons, G. H., Kaneda, Y., Ogihara, T. and Dzau, V. J., 1994. Pharmacokinetics of antisense oligodeoxyribonucleotides (cyclin B_1 and CDC 2 kinase) in the vessel wall in vivo: enhanced therapeutic utility for restenosis by HVJ-liposome delivery. *Gene*, 149:13–19.
98. Morishita, R., Gibbons, G. H., Ellison, K. E., Lee, W., Zhang, L., Yu, H., Kaneda, Y., Ogihara, T. and Dzau, V. J., 1994. Evidence for direct local effect of angiotensin in vascular hypertrophy: In vivo gene transfer of angiotensin converting enzyme. *J. Clin. Invest.*, 94:978–984.
99. Morishita, R., Gibbons, G. H., Horiuchi, M., Ellison, K. E., Nakajima, M., Zhang, L., Kaneda, Y., Ogihara, T. and Dzau, V. J., 1995. A gene therapy strategy using a transcription factor decoy of the E2F binding site inhibits smooth muscle cell proliferation in vivo. *Proc. Natl. Acad. Sci. USA*, 92:5855–5859.
100. Suzuki, J.-I., Isobe, M., Morishita, R., Aoki, M., Horie, S., Okubo, Y., Kaneda, Y., Sawa, Y., Matsuda, H., Ogihara, T. and Sekiguchi, M., 1997. Prevention of graft coronary arteriosclerosis by antisense cdk2 kinase oligonucleotide. *Nature Med.*, 3:900–903.
101. von der Leyen, H. E., Gibbons, G. H., Morishita, R., Lewis, N. P., Zhang, L., Nakajima, M., Kaneda, Y., Cooke, J. P. and Dzau, V. J., 1995. Gene therapy inhibiting neointimal vascular lesion: in vivo transfer of endothelial cell nitric oxide synthase gene. *Proc. Natl. Acad. Sci. USA*, 92: 1137–1141.
102. Zhu, N. L., Wu, L., Liu, P. X., Gordon, E. M., Anderson, W. F., Starnes, W. A. and Hall, F. L., 1997. Downregulation of cyclin G1 expression by retrovirus-mediated antisense gene transfer inhibits vascular smooth muscle cell proliferation and neointima formation. *Circulation*, 96:628–635.
103. Ueno, H., Yamamoto, H., Ito, S.-I., Li, J.-J. and Takeshita, A., 1997. Adenovirus-mediated transfer

of a dominant-negative H-ras suppresses neointimal formation in balloon-injured arteries in vivo. *Arterioscler. Thromb. Vasc. Biol.*, 17:898–904.
104. Harrell, R. L., Rajanayagam, M. A. S., Doanes, A. M., Guzman, R. J., Hirschowitz, E. A., Crystal, R. G., Epstein, S. E. and Finkel, T., 1997. Inhibition of vascular smooth muscle cell proliferation and neointimal accumulation by adenovirus-mediated gene transfer of cytosine deaminase. *Circulation*, 96:621–627.
105. Guzman, R. J., Hirschowitz, E. A., Brody, S. L., Crystal, R. J., Epstein, S. E. and Finkel, T., 1994. In vivo suppression of injury-induced vascular smooth muscle cell accumulation using adenovirus-mediated transfer of the herpes simplex virus thymidine kinase gene. *Proc. Natl. Acad. Sci. USA*, 91:10732–10736.
106. Steg, P. G., Tahlil, Q., Aubailly, N., Caillaud, J.-M., Dedieu, J.-F., Berthelot, K., Le Roux, A., Feldman, L., Perricaudet, M., P and Branellec, D., 1997. Reduction of restenosis after angioplasty in an atheromatous rabbit model by suicide gene therapy. *Circulation*, 96:408–411.
107. Chang, M. W., Barr, E., Lu, M. M., Barton, K. and Leiden, J. M., 1995. Adenovirus-mediated over-expression of the cyclin/cyclin-dependent kinase inhibitor, p21 inhibits vascular smooth muscle cell proliferation and neointima formation in the rat carotid artery model of balloon angioplasty. *J. Clin. Invest.*, 96:2260–2268.
108. Chen, D., Krasinski, K., Chen, D., Sylvester, A., Chen, J., Nisen, P. D. and Andres, V., 1997. Downregulation of cyclin-dependent kinase 2 activity and cyclin A promoter activity in vascular smooth muscle cell by $p27^{kip1}$, an inhibitor of neointima formation in the rat carotid artery. *J. Clin. Invest.*, 99:2334–2341.
109. Ueno, H., Haruno, A., Morisaki, N., Furuya, M., Kangawa, K., Takeshita, A. and Sayto, Y., 1997. Local expression of C-type natriuretic peptide markedly suppresses neointimal formation in rat injured arteries through an autocrine/paracrine loop. *Circulation*, 96:2272–2279.
110. Smith, R. C., Wills, K. N., Antelman, D., Perlman, H., Truong, L. N., Krasinski, K. and Walsh, K., 1997. Adenoviral constructs encoding phosphorylation-competent full-length and truncated forms of the human retinoblastoma protein inhibit myocyte proliferation and neointima formation. *Circulation*, 96:1899–1905.
111. Chang, M. W., Barr, E., Seltzer, J., Jiang, Y.-Q., Nabel, G. J., Nabel, E. G., Parmacek, M. S. and Leiden, J. M., 1995. Cytostatic gene therapy for vascular proliferative disorders with a constitutively active form of the retinoblastoma gene product. *Science*, 267:518–522.
112. Maillard, L., Van Belle, E., Smith, R. C., Le Roux, A., Denefle, P., Steg, G., Barry, J. J., Branellec, D., Isner, J. M. and Walsh, K., 1997. Percutaneous delivery of the gax gene inhibits vessel stenosis in a rabbit model of balloon angioplasty. *Cardiovasc. Res.*, 35:536–546.
113. Rade, J. J., Schulick, A. H., Virmani, R. and Dichek, D. A., 1996. Local adenoviral-mediated expression of recombinant hirudin reduces neointima formation after arterial injury. *Nature Med.*, 2:293–298.
114. Mattar, S. G., Hanson, S. R., Pierce, G. F., Chen, C., Hughes, J. D., Cook, J. E., Shen, C., Noe, B. A., Suwyn, C. R., Scott, J. R. and Lumsden, A. B., 1996. Local infusion of FGF-saporin reduces intimal hyperplasia. *J. Surg. Res.*, 60:339–344.
115. Pastore, C. J., Isner, J. M., Bachc, P. A., Kearney, M. and Pickering, J. G., 1995. Epidermal growth factor receptor-targeted cytotoxin inhibits neointimal hyperplasia in vivo: Results of local versus systemic administration. *Circ. Res.*, 77:519–529.
116. West, J. L. and Hubbel, J. A., 1996. Separation of the arterial wall from blood contact using hydrogel barriers reduces intimal thickening after balloon injury in the rat: the roles of medial and luminal factors in arterial healing. *Proc. Natl. Acad. Sci. USA*, 93:13188–13193.
117. Slepian, M. J. and Hubbell, J. A., 1997. Polymeric endoluminal gel paving: hydrogel systems for local barrier creation and site-specific drug delivery. *Adv. Drug Deliv. Rev.*, 24:11–30.
118. Zarge, J. I., Huang, P., Husak, V., Kim, D. U., Haudenschild, C. C., Nord, R. M. and Greisler, H. P., 1997. Fibrin glue containing fibroblast growth factor type 1 and heparin with autologous endothelial cells reduces intimal hyperplasia in a canine carotid artery balloon injury model. *J. Vasc. Surg.*, 25:840–849.
119. Mitchel, J. F., Fram, D. B., Palme, D., Foster, R., Hirst, J. A., Azrin, M. A., Bow, L. M., Eldin,

A. M., Waters, D. D. and McKay, R. G., 1995. Enhanced intracoronary thrombolysis with urokinase using a novel, local drug delivery system, in vitro, in vivo, and in clinical studies. *Circulation*, 91:785–793.

120. Gradus-Pizlo, I., Wilenski, R. L., March, K. L., Fineberg, N., Michaels, M., Sandusky, G. E. and Hathaway, D. R., 1995. Local delivery of biodegradable microparticles containing colchicine or a colchicine analogue: effects on restenosis and implications for catheter-based drug delivery. *J. Am. Coll. Cardiol.*, 26:1549–1557.
121. Dev, V., Eigler, N., Fishbein, M. C., Tian, Y., Hickey, A., Rechavia, E., Forrester, J. S. and Litvack, F., 1997. Sustained local drug delivery to the arterial wall via biodegradable microspheres. *Catheter. Cardiovasc. Diagn.*, 41:324–332.
122. Guzman, L. A., Labhasetwar, V., Song, C., Jang, Y., Lincoff, A. M., Levy, R. and Topol, E. J., 1996. Local intraluminal infusion of biodegradable polymeric nanoparticles: A novel approach for prolonged drug delivery after balloon angioplasty. *Circulation*, 94:1441–1448.
123. Humphrey, W. R., Erickson, L. A., Simmons, C. A., Northrup, J. L., Wishka, D. G., Morris, J., Labhasetwar, V., Song, C., Levy, R. J. and Shebuski, R. J., 1997. The effect of intramural delivery of polymeric nanoparticles loaded with antiproliferative 2-aminochrome U-86983 on neointimal hyperplasia development in balloon-injured porcine coronary artery. *Adv. Drug Deliv. Rev.*, 24: 87–108.
124. Labhasetwar, V., Song, C., Humphrey, W., Shebuski, R. and Levy, R. J., 1998. Arterial uptake of biodegradable nanoparticles: effect of surface modifications. *J. Pharm. Sci.*, 87:1229–1234.
125. Banai, S., Wolf, Y., Golomb, G., Pearle, A., Waltenberger, J., Fishbein, I., Schneider, A., Gazit, A., Perez, L., Huber, R., Lazarovichi, G., Rabinovich, L., Levitzki, A. and Gertz, S. D., 1998. PDGF-receptor tyrosine kinase blocker AG1295 selectively attenuates smooth muscle cell growth in vitro and reduces neointimal formation after balloon angioplasty in swine. *Circulation*, 97: 1960–1969.
126. Van Belle, E., Maillard, L., Tio, F. O. and Isner, J. M., 1997. Accelerated endothelialization by local delivery of recombinant human vascular endothelial growth factor reduces in-stent intimal formation. Biochem. Biophys. Res. Commun., 235:311–316.
127. Lincoff, A. M., Furst, J. G., Ellis, S. G., Tuch, R. J. and Topol, E. J., 1997. Sustained local delivery of dexamethasone by a novel intravascular eluting stent to prevent restenosis in the porcine coronary injury model. *J. Am. Coll. Cardiol.*, 29:808–816.
128. Yamawaki, T., Shimokawa, H., Kozai, T., Miyata, K., Higo, T., Tanaka, E., Egashira, K., Shiraishi, T., Tamai, H., Igaki, K. and Takeshita, A., 1998. Intramural delivery of a specific tyrosine kinase inhibitor with biodegradable stent suppresses the restenotic changes of the coronary artery in pigs in vivo. *J. Am. Coll. Cardiol.*, 32:780–786.
129. Wilensky, R. L., March, K. L., Gradus-Pizlo, I., Spaedy, A. J. and Hathaway, D. R., 1993. Methods and devices for local drug delivery in coronary and peripheral arteries. *Trends Cardiovasc. Med.*, 3:163–170.
130. Mayberg, M. R., 1990. Localized release of perivascular heparin. *Persp. Neur. Surg.*, 1:77–95.
131. Rogers, C., Karnovsky, M. J. and Edelman, E. R., 1993. Inhibition of experimental neointimal hyperplasia and thrombosis depends on the type of vascular injury and the site of drug administration. *Circulation*, 88:1215–1221.
132. Bennett, M. R., Lindner, V., DeBlois, D., Reidy, M. A. and Schwartz, S. M., 1997. Effect of phosphorothioated oligonucleotides on neointima formation in the rat carotid artery: Dissecting the mechanism of action. *Arterioscler. Thromb. Vasc. Biol.*, 17:2326–2332.
133. Fukumoto, Y., Shimokawa, H., Kozai, T., Kadokami, T., Kuwata, K., Owada, M. K., Shiraishi, T., Kuga, T., Egashira, K. and Takeshita, A., 1996. Tyrosine kinase inhibitor suppresses the (re)-stenotic changes of the coronary artery after balloon injury in pigs. *Cardiovasc. Res.*, 32:1131–1140.
134. Dai, E., Stewart, M., Ritchie, B., Mesaeli, N., Raha, S., Kolodziejczyk, D., Hobman, M. L., Liu, L. Y., Etches, W., Nation, N., Michalak, M. and Lucas, A., 1997. Calreticulin, a potential vascular regulatory protein, reduces intimal hyperplasia after arterial injury. *Arterioscler. Thromb. Vasc. Biol.*, 17:2359–2368.

135. Lucas, A., Liu, L.-Y., Macen, J., Nash, P., Dai, E., Stewart, M., Graham, K., Etches, W., Boshkov, L., Nation, P. N., Humen, D., Hobman, M. L. and McFadden, G., 1996. Virus-encoded serine proteinase inhibitor SERP-1 inhibits atherosclerotic plague development after balloon angioplasty. *Circulation*, 94:2890–2900.
136. Axel, D. I., Kunert, W., Goggelmann, C., Oberhoff, M., Herdeg, C., Kuttner, A., Wild, D. H., Brehm, B. R., Riessen, R., Koveker, G. and Karsch, K. R., 1997. Paclitaxel inhibits arterial smooth muscle cell proliferation and migration in vitro and in vivo using local delivery system. *Circulation*, 96:636–645.
137. Gunn, J., Holt, C. M., Francis, S. E., Shepherd, L., Grohmann, M., Newman, C. M. H., Crossman, D. C. and Cumberland, D. C., 1997. The effects of oligonucleotides to c-myb on vascular smooth muscle cell proliferation and neointima formation after porcine coronary angioplasty. *Circ. Res.*, 80:520–531.
138. Azrin, M. A., Mitchell, J. F., Bow, L. M., Pedersen, C. A., Cartun, R. W., Aretz, T. H., Waters, D. D. and McKay, R. G., 1997. Local delivery of c-myb antisense oligonucleotides during balloon angioplasty. *Catheter. Cardiovasc. Diagn.*, 41:232–240.
139. Azrin, M. A., Mitchell, J. F., Fram, D. B., Pedersen, C. A., Cartun, R. W., Barry, J. J., Bow, L. M., Waters, D. D. and McKay, R. G., 1994. Decreased platelet deposition and smooth muscle cell proliferation after intramural heparin delivery with hydrogel-coated bballoons. *Circulation*, 90: 433–441.
140. Hanna, A. K., Fox, J. C., Neschis, D. G., Safford, S. D., Swain, J. L. and Golden, M. A., 1997. Antisense basic fibroblast growth factor gene transfer reduces neointimal thickening after arterial injury. *J. Vasc. Surg.*, 25:320–325.
141. Matsumoto, T., Komori, K., Yonemitsu, Y., Morishita, R., Sueishi, K., Kaneda, Y. and Sugimachi, K., 1998. Hemagglutinating virus of Japan-liposome-mediated gene transfer of endothelial cell nitric oxide synthase inhibits intimal hyperplasia of canine vein grafts under conditions of poor runoff. *J. Vasc. Surg.*, 27:135–144.
142. Valero, F., Hamon, M., Fournier, C., Meurice, T., Flautre, B., Van Belle, E., Lablanche, J.-M., Gosselin, B., Bauters, C. and Bertrand, M., 1998. Intramural injection of biodegradable microspheres as a local drug-delivery system to inhibit neointimal thickening in a rabbit model of balloon angioplasty. *J. Cardiovasc. Pharmacol.*, 31:513–519.

13

Liposomes Coated with Bioactive Polysaccharides for Use as Drug Delivery Systems

Didier Letourneur
CNRS, IFREMER, University Paris XIII, Villetaneuse, France

Maud Cansell
University Bordeaux I, Talence, France

I. INTRODUCTION

Polymers are commonly used in medicine as implants, polymeric drugs, or drug delivery systems. For drug delivery systems using natural components, numerous works have focused on polymers exhibiting biological activities associated with organized molecular systems consisting of lipid membranes, namely liposomes. They are lipid structures that delimit an aqueous internal volume. Liposomes have biocompatibility and biodegradability properties, and can be considered either as cell membrane models or as carriers for biologically active substances [1]. However, in vitro and in vivo studies have also pointed out that classical liposome preparations exhibit structural instability, rapid uptake by the mononuclear phagocyte system and low cell specificity. Indeed, liposomes in blood are rapidly intercepted by the reticuloendothelial system (RES) through endocytosis [2]. In particular, it was shown that as soon as liposomes are in the bloodstream, they interact with lipoproteins and opsonins [2,3]. These latter are specifically recognized by receptors on macrophage membranes, resulting in the rapid clearance of liposomes from the circulation [1–3]. Changes in liposome size and composition have proven to be effective in reducing the fast clearance by RES [2,4]. In parallel, to prolong circulation time in blood, liposome compositions have been formulated to provide new surface characteristics [2,4]. The sterically stabilized ("stealth") liposomes were first obtained by the incorporation of gangliosides [5–7]. Then, lipid derivatives of polyethylene glycol [8–10], vinyl polymers [11], and polysaccharides [4,12–15] were also used with promising results. In this context, hydrophilic polymers are now widely used to cover the liposome surface leading to a reduction of serum protein adsorption and an improvement of the pharmacokinetics profiles [3,5,16,17]. However, the presence on the liposome surface of specific ligands is required to achieve cell targetability. Interestingly, cell surface carbohydrate/membrane protein interactions are involved in numerous biological activities such as inflammation, fertilization and cell proliferation [18–21]. This gives the rationale for the design described below of liposomes bearing on their surface sugar moieties.

The achievements of liposome coating using monosaccharides or polysaccharides (including both natural and synthetic polymers) will be presented. The chemical modifications and characterizations of the coated liposomes as well as examples of biological applications will be discussed.

II. POLYSACCHARIDES AND THEIR BIOLOGICAL ACTIVITIES

The variability of carbohydrate sequences, prominent surface-exposed structures present on cell surfaces of all multicellular organisms, as well as on pathogens (viruses, bacteria) allows a large pattern of interactions. These carbohydrates are involved as recognition molecules in functionally important processes by interacting with other carbohydrate structures or with carbohydrate-recognizing ligands, like lectins [18,22–24]. Even if their affinities are rather low in general, multiple bindings are frequently reported [18,25].

Carbohydrate chains which offer simultaneously flexible, ordered, and easily modulable motives for specific recognition are of considerable value in nature. Numerous examples can be found among the glycosaminoglycans (GAGs) that are complex carbohydrates constituted of repeating disaccharide units. The GAG chains contain chemical groups such as sulfates defining binding sequences [26,27]. For instance, heparin produced in mast cells is an unbranched GAG of alternating residues of glucosamine and uronic acid, rich in N- and O-sulfate groups. A unique antithrombin pentasaccharide sequence in the heparin GAG chain has been clearly demonstrated [27,28], resulting in a potent anticoagulant activity. Heparin was thus coupled to liposomes to interfere with the coagulation pathway [29]. Beyond its anticoagulant activity, extensively used in clinical practice, heparin also binds and modulates the activity of proteins, including cell growth and angiogenic factors, complement components involved in the immunologic responses, and viral proteins [21,26,27,30,31].

Therefore, synthetic or vegetal polysaccharide analogs could have broad applications in biology and medicine [32–35]. In this context, different chemical procedures were developed for instance by our laboratory for the production of synthetic polysaccharides derived from dextran, a poly(α 1–6 glucose) (Fig. 1). The derivatized dextrans exhibit heparin-like properties, such as anticoagulation capacities [36], modulation of vascular cell growth [37,38], protection of heparin-binding growth factors [39], antiviral activities [40,41], and anticomplementary activity in vitro and in vivo [42–44]. Dextran was shown to increase the circulatory half-life of liposomes [45]. Coupled with liposomes, the steric barrier feature of the dextran macromolecule can be combined with a specific interaction capability due to the dextran functionalization [46].

On the other hand, biological properties were also demonstrated for sulfated polysaccharides of vegetal origin such as seaweeds [32,47]. Among them, fucans (Fig. 1) primarily composed of α(1,2) or α(1,3) 4-*O*-sulfated-L-fucose [48] show anticoagulant and antithrombotic properties [49–51], inhibit as well as heparin vascular smooth muscle cell growth [52,53], and modulate the immune system [54,55].

The wide-ranging relevance of carbohydrate structures explains the efforts to construct tailor-made tools to explore medical applications. Neoglycoproteins were already used for drug targeting, either directly [56–58] or coupled on liposomes [59,60]. Similarly, various carbohydrate chains of natural origin can directly target binding sites on proteins [21,61–63]. The synthesis and characterization of the liposome coating with carbohydrate moieties will be extensively described in this chapter. The use of fucans and derivatized dextrans as heparin-like polysaccharides [64] will also be described for the coating of liposomes.

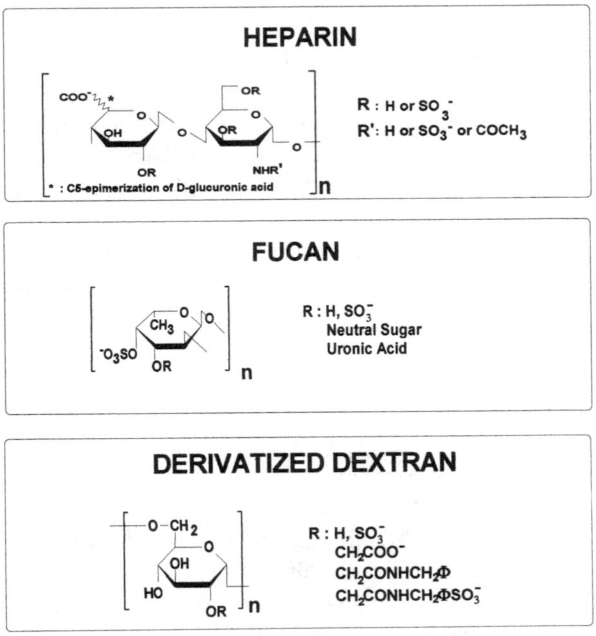

Figure 1 Chemical representations of biologically active polysaccharides. Heparin is a sulfated glycosaminoglycan extracted from mammalian tissues. Fucans are sulfated polysaccharides extracted from brown seaweeds and are primarily composed of 4-*O*-sulfated-L-fucose. Only α(1,2) linkages are shown. Derivatized dextrans are substituted from dextran (R=H) by a sequential reaction in three steps in order to obtain carboxymethyl groups ($R{=}CH_2COO^-$), coupled with benzylamine ($R{=}CH_2CONHCH_2\Phi$), and sulfonation ($R{=}CH_2CONHCH_2\Phi SO_3^-$) of some aromatic rings, or sulfation ($R{=}SO_3^-$) of hydroxyl groups. Only the main substitution in the C2 position is depicted. The compositions of the different derivatized dextrans are obtained by varying the chemical conditions.

III. POLYSACCHARIDE DERIVATIVES FOR LIPOSOME COATING

Saccharides have been added to liposome formulations from simple mono- or oligosaccharides (galactose, glucose, mannose, lactose, acetylglucosamine, maltopentose, mannopentose, etc.) to more complex structures (chitin, dextran, dextran sulfate, mannan, pullulan, amylopectin). Two main approaches have been undertaken in order to achieve a coating of the liposome surface by polysaccharides. The first consists of the spontaneous polysaccharide adsorption on the liposome outer bilayer. The second uses a hydrophobic lipid anchor to bind the saccharide moieties to the liposome membranes.

Spontaneous adsorption of polysaccharides on liposome surface occurs via hydrophobic interactions and/or electrostatic interactions. In the latter case, when the initial polysaccharide is nonionic, chemical groups are grafted to ensure a net ionic charge and allow the interaction with the charged membrane lipids. For instance, carboxymethyl groups have been grafted on chitin [65]. However, when a tight binding between the liposome surface and the polysaccharide is required, grafting procedures are carried out. This is obtained by a chemical reaction, either between the saccharide units and a lipid of the preformed membrane or by a preliminary grafting on the polysaccharide of an hydrophobic molecule that acts as an anchor in the liposome bilayer. We will describe both approaches, but the second will be discussed in more detail since it is most commonly used.

A. Direct Grafting of Polysaccharides

Few examples are found in the literature for the modification of saccharides with a direct grafting on the liposome membrane. Covalent coupling of the cyanogen bromide-activated dextran to the available amino groups of phosphatidylethanolamine (PE) has been reported [45]. Mono- and triantennary galactosyl ligands have also been grafted on preformed liposomes by reaction with a modified PE bearing maleimido groups [66]. More recently, phenylisothiocyanate monosaccharides based on β-galactose, α-glucose, α-mannose, β-lactose and β-*N*-acetylglucosamine have been covalently coupled to the polar head group of dimyristoylphosphatidylethanolamine (DMPE) [67]. However, liposome coating often requires the chemical synthesis of the phospholipid and/or glycoside derivatives [68]. Consequently, direct hydrophobization of the saccharides has also been developed.

B. Grafting of Hydrophobic Moieties

Classical procedures for the grafting of hydrophobic molecules usually require the presence of amino groups. Two main methods are thus performed. The first one consists of the use of amino groups present in a natural phospholipid (i.e., PE). Indeed, neoglycolipids were constructed from dipalmitoylphosphatidylethanolamine (DPPE) and oligosaccharides (pentose and mannose derivatives) [69]. When another type of hydrophobic anchor than PE is used, the lack, or low amount, of free amino groups both in the saccharide structure and in the lipid molecule implies a preliminary synthesis step with the grafting of the amino groups either on the lipids or on the polysaccharides. Such grafting has been performed on cholesterol [70,71], and on mono- or dialkyl chains [70–72]. On the other hand, several procedures have also been proposed with a saccharide modification consisting of carboxymethylation of polysaccharides such as dextran, pullulan, levan, mannan, amylopectine, inulin and amylose followed by a second stage with the introduction of ethylenediamine [4]. A simple one-step method with the reaction of bromopropylamine hydrobromide on dextran and biologically active polysaccharides (heparin, fucan) has also be proposed [64,73]. The quantification of amino groups can be obtained by acidic titration of the NH_2/NH_3^+ functions, spectrophotometric determination [74], and elemental analysis of nitrogen. Table 1 presents the chemical characterizations for aminated dextrans obtained by the bromopropylamine method and detected using phthaladehyde [74]. As can be seen in the table, an increase of bromopropylamine reagent increases amino groups grafted on the polymer. However, the grafting of amino groups increased logarithmically, probably due to a steric hindrance on the macromolecular structure.

Numerous polysaccharides were hydrophobized with a lipid molecule such as cholesterol or palmitoyl chain. Palmitoyl pullulan and amylopectin derivatives were prepared according to

Table 1 Amino Groups Grafted on Dextran

Polysaccharide	R^a	Spectroscopic assay[b]
Dextran	5	1.55
	1	0.85
	0.2	0.41
	0.05	0.14

[a]Molar ratio of bromopropylamine hydrobromide per dextran.
[b]Spectroscopic assay using phthaladehyde [74].

the procedures applied for palmitoyl-dextran [75,76]. Cholesterol grafting was described for pullulan, mannan, amylopectin and dextran by reaction of cholesteryl chloroformate on polysaccharides previously substituted by aminoethylcarbamoylmethyl groups [4]. A similar cholesterol grafting was applied to heparin, fucan, and biologically active dextran derivatives [29,77]. The main synthesis steps are presented in Fig. 2.

C. Chemical Characterizations

Effective chemical grafting of the lipid molecule on the polysaccharide can be assayed by infrared (IR) spectroscopy. The method was used for probing palmitoyl chain and cholesterol. In the former case, IR spectrum revealed the existence of an ester bond between pullulan and the palmitoyl group indicating that the two compounds were covalently bonded [78]. Effective cholesterol grafting was also assayed by Fourier transform infrared resonance spectroscopy (FTIR). The appearance of a vibration band $\nu(c{=}o)$ at 1720 cm^{-1} for cholesterol substituted dextran confirmed the presence of the carbonyl group resulting from the bond between the amino group of dextran and the carboxylic group of cholesteryl chloroformate. By UV spectroscopy, the presence of cholesterol was revealed by the appearance of an absorption peak at 257 nm. On the other hand, ^1H NMR and elemental analysis were used to determine the degree of substitution of palmitoyl chains [78] and of cholesterol groups on polysaccharides [4,71,79]. The amount of cholesterol grafted on dextran and bioactive polysaccharides was also determined using a cholesterol enzymatic assay [29]. Cholesterol grafting efficiencies on various polysaccharides are given in Table 2. As expected, we found that an increase of amino groups on the polysaccharides correlated with an increase of cholesterol grafting.

Integrity of the polysaccharides after the chemical modifications can be assayed by physicochemical or biological assays. As shown in Table 3, the apparent molecular weights of hydrophobized polysaccharides determined by high performance steric exclusion chromatography (HPSEC) are in the range of that of unmodified polysaccharides. The slight observed variations may be attributed to changes in hydrophobic and/or electrostatic interactions between the hydrophobized polysaccharides and the stationary phase. When cholesterol was grafted, for instance, on the heparin polysaccharide, the anticoagulant activity of hydrophobized heparin remained unchanged with 168 ± 5 UI mg^{-1} after grafting versus 173 ± 5 UI mg^{-1} for the native polysaccharide [29]. However, a higher amount of grafted cholesterol on the polysaccharide chain may decrease its specific anticoagulant activity [77].

The grafting of a hydrophobic moiety on the polysaccharide modifies its solubility in water. In particular, it was shown that polysaccharide derivatives bearing palmitoyl or cholesterol moieties form self-aggregates in aqueous solution [70,71]. The critical concentration at which nanoparticles are formed and the stability of the aggregates depend on various parameters such as (1) the nature of the polysaccharide; for instance, self-aggregation of cholesterol-

I) P—OH + $Br\text{-}(CH_2)_3\text{-}NH_2.HBr$ ⟶ $P\text{—}O\text{-}(CH_2)_3\text{-}NH_2$
(polysaccharide)

II) $P\text{—}O\text{-}(CH_2)_3\text{-}NH_2$ + Cl-C(O)-Chol ⟶ $P\text{—}O\text{-}(CH_2)_3\text{-}NH\text{-}C(O)\text{-}Chol$

Figure 2 Cholesterol grafting was applied to various polysaccharides (P-OH) such as heparin, fucan, and biologically active dextran derivatives. (I) Amino groups were attached to some hydroxyl groups of the polysaccharide. (II) Using cholesteryl chloroformate, a cholesteryl (Chol) moiety was introduced on the aminated polysaccharide.

Table 2 Number (N) of Cholesterol per 100 Monomer Units

Polysaccharide	Molecular Weight (g/mol)	N	Reference
Pullulan	50,000	1.3	4
Pullulan	50,000	1.8–5.5	70
Pullulan	55,000	1.7–2.5	72
Pullulan	108,000	1.3	72
Amylopectin	112,000	1.3	70
Amylopectin	112,000	0.7	12
Mannan	200,000	2.4	12
Mannan	85,000	2.3	72
Dextran	70,000	1.7	72
Dextran ($R^a = 5$)	27,000	1.2	this work
Dextran ($R^a = 1$)	27,000	0.3	this work
Dextran ($R^a = 0.2$)	27,000	0.2	this work
Fucan	15,000	0.7	this work
Heparin	21,000	0.4	29

[a]Number of bromopropylamine hydrobromide per saccharide unit.

pullulan occurs at a lower critical concentration than that of cholesterol-dextran and cholesterol-amylopectin [70,71]; (2) the nature of the hydrophobic anchor; for a given polysaccharide, the palmitoyl group is less effective than cholesterol in forming aggregates, and palmitoyl-bearing polysaccharides are less stable than cholesterol-bearing polysaccharides; (3) the substitution degree of the cholesterol moiety, i.e., the higher the substitution degree, the lower the critical concentration.

Table 3 Molecular Weights of Native and Hydrophobized Polysaccharides Determined by HPSEC [a]

Polysaccharide	Chromatographic molecular weight (kg/mol)	Mn (kg/mol)	Mp (kg/mol)
Dextran	23	22	33
Dextran-chol	34	24	36
Fucan	15	14	23
Fucan-chol	12	11	18
Heparin	21	19	24
Heparin-chol	21	18	24

[a]Determination of the molecular weights (Mc, Mn, and Mp) of native and hydrophobized polysaccharides was performed by high performance steric exclusion chromatography (HPSEC) in 0.15 M NaCl and 0.05 M NaH_2PO_4, solution buffered at pH 7, using a Licrospher Si 300 diol column (Merck) and Hema sec Bio40 column (Interchim). The two columns in series were calibrated with standard pullulans of narrow distribution in molecular weight.

IV. CHARACTERIZATION OF LIPOSOMES WITH SACCHARIDE DERIVATIVES

A. Adsorption of Saccharides on Preformed Liposomes

Polysaccharides such as dextran sulfate [80], chondroitin sulfate [81], or carboxymethyl-chitin (CM-chitin) [65,82] can spontaneously be adsorbed on liposome surface. Adsorption is performed by polysaccharide incubation with a suspension containing preformed liposomes often followed by a separation step (steric exclusion chromatography or centrifugation) to eliminate unbound polymers especially when coating quantification is needed. For instance, the quantity of CM-chitin adsorbed on liposome surface was determined by a colorimetric method, which assays a hydrolysis product of CM-chitin [65], or by the binding capacity of a specific dye to CM-chitin [82]. It was shown that the quantity of CM-chitin adsorbed on liposome surface increased as a function of initial CM-chitin concentration. In the case of dextran sulfate or chondroitin sulfate, binding only occurs on positively charged liposome surfaces, which indicated that adsorption mainly occurs via electrostatic interactions between sulfate moieties of the polysaccharides and the cationic liposome components. When neutral liposome formulations are considered, the binding requires the presence of a positively charged ion in the aqueous medium [80]. However, anionic CM-chitin is reported to adsorb both on positively and negatively charged membranes [65,82–84].

Absorption of CM-chitin is often studied in association with stearylamine based liposomes in order to reduce the cytotoxicity of this type of vesicles. It was shown that adsorption of CM-chitin varies as a function of the amount of stearylamine present in the liposome composition. Depending on the amount of lipid, the coating efficiency varies from 20% to 60% in the absence of stearylamine to 40% to 80% in the presence of 50% of charged lipids [84]. The mechanism by which CM-chitin adsorption on the liposome surfaces occurs is not fully elucidated. The amount of adsorbed CM-chitin could be the result of different phenomena, such as electrostatic interactions and a competition between adsorption and rearrangement kinetics [82]. Moreover, it was shown that the conformation of the adsorbed polyelectrolyte depends on its average molecular weight, the degree of carboxymethylation, the initial polymer concentration, the ionic strength and pH [65,82].

Spontaneous adsorption of uncharged polysaccharides on lipid surface occurs through hydrophobic interactions. For instance, coating efficiency of native pullulan on liposome surface is low (2%) and pullulan is detached from the liposome during gel chromatography [72]. It was shown that cholesteryl unsubstituted polysaccharide adsorption on the outer surface is enhanced when the fluidity of the phospholipid membrane increases and that the adsorbed polysaccharides could easily be desorbed from the liposome surface with dilution [85].

Negatively charged polysaccharides such as heparin and fucan can adsorb on neutral liposomes [29,77]. The presence of the polysaccharides on the surface was assayed by the measurements of plasma coagulation times since these polysaccharides exhibit anticoagulant properties. However, part of the adsorbed molecules can be desorbed by ultracentrifugation. Negatively charged heparin or fucan can also be found on the surface of dicetylphosphate (DCP)-containing liposomes. However, the adsorption is low and may be related to the small amount (5%) of DCP present in the liposome formulation, not high enough to completely prevent the polysaccharide adsorption. In contrast, a similar amount of stearylamine (5%) in the liposome membrane was sufficient to induce a rapid aggregation when these liposomes were in contact with heparin or fucan.

It is possible to study the behavior of the polysaccharides on the liposome membrane using, for instance, the fluorescence depolarization technique with fluoresceinylthiocarbamoyldextrans (FITC-dex) as probes. The adsorption of FITC-dex on liposomes significantly increased the fluorescence polarization of FITC fluorophore due to the restriction of the mobility of the dextrans [85]. However, the adsorption of dextrans did not cause a significant change in the fluidity of the liposomal membrane ascertained from the mobility of sodium-8-anilino-1-naphthalenesulfonate intercalated close to the surface of liposomes.

All these results suggest that adsorption is governed by both the nature of the lipid membrane (composition) and the structure of the polysaccharide (size, ionization, molecular weight). It appears that, even with a high coating efficiency, the adsorbed polysaccharides could easily desorb from the liposome surface either by a modification of the medium (dilution, pH variation) or by a mechanical stress (chromatography, ultracentrifugation). To overcome this problem, hydrophobized polysaccharides are synthesized, in which the grafted lipid acts as an anchor in the liposome membrane.

B. Coating of Liposomes with Hydrophobized Polysaccharides

Hydrophobized polysaccharides are added either on preformed liposomes, in order to coat the liposome surface, or prior to liposome formation, which implies their presence inserted in the membrane and encapsulated in the internal volume of the liposomes. The first method of coating is commonly used with hydrophobized saccharidic structures: neoglycolipids constructed from DPPE and oligosaccharides [69] or polysaccharides bearing either cholesteryl [4,13,72,86] or acyl chain moieties [13,15,72,78]. Incubation time and temperature of the liposome preparation greatly vary with authors from 30 min to 3 days for the incubation times, and 4°C to 50°C for the incubation temperatures. In our laboratory, we have chosen to perform simultaneously the coating of the liposome surface with the polysaccharide and its encapsulation in the internal volume during the liposome formation. Two methods of preparation were used depending on the final liposome size required [87]: (1) detergent dialysis from mixed lipids-octyl glucoside micellar solutions, which leads to large unilamellar vesicles, and (2) ultrasonic irradiation, which is known to give small unilamellar vesicles.

The efficiency of the liposome coating can be determined by direct analysis of the polysaccharide during or after binding to the membrane. The methods used are thin layer chromatography for quantitative analysis of the carbohydrate content, which gives, for instance, the number of DMPE bearing a sugar moiety relative to the total number of phospholipid molecules [67], spectrophotometric test based on the phenol-sulfuric acid method to assay free and bound polysaccharide concentrations [72], or fluorescence depolarization method to measure the binding velocity of the polysaccharide to the liposome membrane [72]. Coating efficiency can also be calculated by measurement of the biological activity of the liposomes resulting from the adsorption of the bioactive polymer, assuming that the specific activity detected is similar for free and bound polymer. This approach was used for liposomes coated with heparin and fucan using the anticoagulant properties of these polysaccharides [29,77].

Coating efficiencies are affected by various factors. For instance, the binding constants of hydrophobized polysaccharides increased with an increase in both the substitution degree of the cholesteryl moiety and the molecular weight of the polysaccharide, i.e., pullulan (MW 108,000 and 55,000), mannan (MW 85,000) and dextran (MW 70,000) [72]. The influence of the chemical polymer structure has also been investigated. For pullulan, mannan and dextran, the binding constants of these three polysaccharides appeared to be in the same order of magnitude [72]. On the other hand, different behaviors in their interactions with lipid monolayers have been shown for pullulan and amylopectin derivatives that could result from their orientation with respect to the interface, including the position of the cholesteryl moiety and the polysaccharide structure. Indeed, by modeling the liposome membrane by a phosphatidylcholine monolayer, it was shown that cholesteryl-pullulan was more efficient for covering the liposome surface than cholesteryl-amylopectin since (1) the former exhibits a relatively straight chain structure compared to the highly branched structure of amylopectin and (2) cholesterol moiety is completely embedded in the lipid environment for cholesteryl-pullulan, whereas, in the case of cholesteryl-amylopectin, only part of cholesterol enters the bilayer [13].

Cholesteryl derivatives of polysaccharides were shown to ensure a better coating of liposomes than simple monoalkyl or dialkyl chain derivatives [72]. By measurement of membrane fluidity of phosphatidylcholine-based large unilamellar liposomes after anchoring of palmitoyl-polysaccharide, it was shown that alkyl chains were rather located in the external region of the lipid bilayer and no perturbation of lipid organization for the hydrophobic region was observed [78]. No effect on the membrane fluidity could be detected up to concentrations of 4 mg of dextran/mg of lipid, which corresponds to two palmitoyl chain anchors per outer phospholipid molecule. At very high dextran concentrations (26 mg of dextran/mg of lipid, which corresponds to 13 palmitoyl chain anchors per outer phospholipid molecule), the movement of diphenyl hexatriene becomes restricted [15]. On the contrary, cholesteryl-pullulan insertion induces a remarkable change in the molecular order at low concentrations [86]. Pullulan derivatives disturb the molecular packing presumably due to the insertion of cholesterol moiety into the membranes.

It thus appears that the coating efficiency varies in large proportions depending on numerous parameters such as the nature of the polysaccharide, the type and substitution degree of the hydrophobic anchor present on the polymer, the lipid composition and the ratio lipid/modified polysaccharide, and the type of liposome according to the method of preparation.

V. PROPERTIES OF THE POLYSACCHARIDE-COATED LIPOSOMES

Numerous studies have showed that coated liposomes are of great interest for in vitro and in vivo applications. We will describe how polysaccharide-coated liposomes have gained stability and targetability in comparison to conventional noncoated liposomes.

A. Stability of the Coated Liposomes

In vitro stability is often assayed through permeability studies. In order to evaluate the bilayer integrity and the effect of polysaccharide coating, most techniques are related to the use of a soluble fluorescent probe and the study of its release as a function of time or other parameters, such as temperature, ionic strength, etc. However, some techniques based on fluorescent probes, surface tension measurements [13], or birefringence properties of liposome suspensions can also be used.

Permeability of liposome membrane covered or not by chondroitin sulfate adsorbed on the surface was measured by calcein leakage at 37°C [81]. Addition of chondroitin sulfate prevents calcein leakage from positively charged liposomes by a factor of 2.5. It is suggested that chondroitin sulfate adsorption modifies the molecular packing of the positive bilayer membrane, decreasing the membrane fluidity. In the case of this polymer, it seems that both the head groups and the acyl chain regions of the liposome membrane are affected by chondroitin sulfate adsorption, probably in relation to a modification of the hydration state of the lipids.

For the liposome coating with polysaccharides bearing acyl chain residues, it was shown that pullulan derivatives improved the liposome stability by decreasing membrane permeability. Indeed, carboxyfluorescein (CF) leakage from large unilamellar liposomes based on phosphatidylcholine (PC) and coated with palmitoyl-pullulan was measured as a function of time.

After 42 h of storage at 50°C, 30% of CF was released from coated liposomes instead of 67% for noncoated liposomes [78]. However, these results were dependent on the ratio palmitoyl-pullulan/PC, i.e., leakage of CF was reduced by a factor of 3 for liposomes prepared with a palmitoyl-pullulan/PC ratio of 3 instead of 1 [78]. Similar results were obtained when stability of the suspension was followed at room temperature over a longer period of time. Thus, coating the liposomes with modified dextran (30 mg of dextran/mg of lipid which corresponds to 15 palmitoyl chain anchors per outer phospholipid molecule) increased the initial leakage of CF. After an initial rapid loss of about 18% the first day, the liposomes stabilized and the loss was only of a few percent in the next 15 days. After 15 days of storage at room temperature, CF release was of 20% for coated liposomes instead of 62% for noncoated liposomes [15].

Coating of liposomes with neoglycolipids constructed from DPPE and oligosaccharides increases membrane impermeability, with more than 95% retention after three days of incubation at 4°C [69]. Similarly, coating of liposome surface with cholesteryl polysaccharides increases bilayer impermeability toward low molecular weight encapsulated compounds. For instance, leakage of CF from small unilamellar liposomes was decreased by a factor of 3 following coating by pullulan or amylopectin derivatives [13]. Stability of polysaccharide coated liposomes was also studied through the release of higher molecular weight polysaccharides, like heparin, fucan or dextran, at different temperatures [29,77]. Liposomes based on PC/PE/cholesterol/DCP retained in their aqueous volume 90% of the encapsulated heparin after one month of storage at 37°C.

Resistance of coated liposomes to biochemical lysis with enzymes (phospholipases and lypoxygenases) and detergent was also studied. Coating of liposomes with polysaccharides can protect the liposomal structures from enzymatic lysis. This was shown for two enzymes: (1) phospholipase D, which specifically destroys phosphatidylcholine by a hydrolytic cleavage of the bond between the phosphate and the choline moiety, and (2) pullulanase, which attacks the pullulan directed in the aqueous phase [12,79].

Addition of the detergent decyl-polyethylene glycol 300 caused a rapid release of CF, indicating that the integrity of the liposomes was lost. Coating of liposomes with modified dextran inhibited the rate of detergent-induced CF release. At 4 mg of dextran/mg of lipid, which corresponds to two palmitoyl chain anchors per outer phospholipid molecule, a significant protection against detergent degradation was obtained [15]. Solubilization by octyl glucoside of PC-based liposomes whether coated or not with dextran was followed by changes in turbidity, which reflect supramolecular rearrangements of detergent and lipid molecules in the aggregates. Micellization was characterized by a similar detergent to lipid ratio in the mixed aggregates ($[\text{det/lip}]_{\text{agg}}$) and concentration ($[\text{det}]_{\text{bulk}}$) of detergent molecules not associated with the lipids [87]. This suggested that either the hydrophilic core coating the vesicle surface did not inhibit detergent insertion in the bilayer, or the dextran was not in sufficient amount to prevent detergent incorporation.

B. Biological Properties of Polysaccharide-Coated Liposomes

Saccharides on the surface of cell membranes play an important role in cell-cell recognition, which is used to design a targetable drug carrier. Hence, natural and synthetic saccharide structures as recognition sites of the liposomal drug carrier were already used in numerous biological activities. Among all the approaches described in the literature, we will only present some applications for the coating of liposomes with saccharides and polysaccharides of natural and synthetic origins. Representative applications cover the treatment of various diseases, the targeting of organs or cells, and the attempts to avoid liposome uptake by the reticuloendothelial system.

When liposomes are used in vivo, one should thus distinguish between active and passive targeting. In the latter case, the majority of liposomes intravenously administered are retained in the liver and spleen, irrespective of liposome size, structural characteristics, or lipid composition. Active targeting attempts to modify this location and to target specific cells or tissues.

Targeting of glycolipid liposomes to specific liver cell types was shown in vitro and in vivo by simply varying the sugar residue on the surface of liposome [88]. The liver is an ideal organ for studying targeting strategies using a variety of liposomes inasmuch as its discontinuous capillaries have fenestrae through which liposomes less than 200 nm in diameter may escape into the extravascular space. Liver accumulation of liposomes depended on the sugar moieties such as galactosyl residues. Branched type galactosyl lipid derivatives for the targeting of asialoglycoprotein receptors on the surface of liver cells were thus designed [89]. For Das et al. [90], galactosylated and mannosylated liposomes were more efficient in transporting liposomes to liver compared to nonglycosylated liposomes. The glycoside-bearing liposomes were found to be cleared at a much faster rate than liposomes having no sugar on their surface. Mannosylated liposomes have much higher affinity for nonparenchymal cells, whereas the assimilation into hepatocytes is clearly favored for liposomes having galactose on their surface.

The biodistribution of liposomes modified by mannobiose residues was studied to target macrophages. After intravenous administration, mannobiose-modified liposomes were eliminated from the systemic circulation more rapidly than control liposomes without the modification. While the modification did not affect the distribution of liposomes to kidney, lung, or thymus, it increased the distribution to liver and spleen, suggesting that mannobiose mono fatty acid esters can be useful in the targeting of liposomes to Kupffer cells and other macrophages [91]. However, one major research goal of liposome pharmacology is the selective delivery of drugs to target cell populations while minimizing elimination by phagocytic macrophages and blood monocytes of the reticuloendothelial system. Different aspects of liposomal drug delivery to non-RES cells have been described. In one of the systems, by incorporating into liposomes neutral glycolipids with terminal beta-galactoside residue, it was possible to target liposomes to the liver parenchymal cells, partially bypassing the RES. Asialogangliosides were described as suitable for this purpose. In another approach, various factors that prolong the life span of circulating liposomes have been described with a number of saccharide-containing candidates, such as dextran and GM1 ganglioside [45,92]. Synthetic glycolipids prepared from synthetic cholesterol conjugates of D-mannose and 6-amino-6-deoxy-D-mannose also substantially alter the in vivo lifetime and distribution of liposomes outside the bloodstream [93].

For cancer chemotherapy, the overexpression of lectins by malignant cells was used [60]. It was applied for targeting of liposomes equipped with the saccharide vectors and loaded in the lipid phase with a lipid derivative of anticancer agents [56,58,94]. With the help of fluorescent lipid dye, it was shown that active saccharide ligands increased the level of the targeted liposome binding to malignant cells by 50–80% as compared to liposomes without vectors or with an inactive one [94]. Interaction between rat T-lymphocytes and a ganglioside (GM3, GD3, GT1b, or GQ1b)-containing liposome was also investigated by Kato et al. [6]. The GT1b- or GQ1b-containing liposome strongly stimulated the cells, while the GM3- or GD3-containing liposome showed much less effect. Free gangliosides did not stimulate the cells at all under the same conditions. The extent of the stimulation depended on the surface density of GT1b on the liposome accompanied by a clear threshold concentration. The efficiency of this direct T-cell stimulation with the ganglioside-containing liposome was closely related to the efficiency of the tumor growth suppression [6]. In another study, the effect of liposomal adriamycin (ADM) with tumor recognition molecule, 1-aminolactose (1-AL), on hepatoma transplanted into nude mice was studied [103]. 1-AL/Pullulan-coated liposomal ADM restrained tumor growth more when compared with pullulan-coated liposomal ADM. Thus, 1-

AL was assembled to the outer layer of liposomes as a tumor recognition molecule ADM, and pullulan acted to increase the stability in the bloodstream.

Different sugar-grafted liposomes were also prepared and tested for the treatment of diseases other than cancer, such as for the delivery of antimicrobial drugs and immunomodulating agents. When encapsulated in sugar-grafted liposomes, classical drugs were found to be more potent in vivo against experimental leishmaniasis in comparison to normal liposome-encapsulated drug or to the free drug. Moreover, the mannose-grafted liposomes were adjudged to be the best in lowering of spleen parasite load in comparison with those bearing glucose or galactose [95]. Medda et al. have tested asiaticoside, a plant glycoside with rhamnose as end sugar and having microbicidal properties against Mycobacterium leprae and Mycobacterium tuberculosis [96]. In the study of Van Bambeke et al., the binding of aminoglycoside antibiotics to acidic phospholipid membranes was shown to be an essential step in the development of their renal toxicities [97]. The authors conclude that aminoglycosides, like spermine, have the potential to prevent membrane fusion and that this capacity is determinant in aminoglycoside activity [98]. Appropriate glycosides, if used in liposomal form (incorporated or covalently grafted) have enhanced drug efficiencies and such glycosid-bearing liposomes could be used as targeted delivery systems for chemotherapeutic control of several diseases. For instance, liposomes bearing glycolipids greatly increase retention at the injection sites compared to control liposomes, which contain no saccharide determinants [99].

Among the natural polysaccharides, glycosaminoglycans can also be used to coat the surface of liposomes in order to target GAG receptors on cells. For instance, liposomes, in which hyaluronic acid is the surface-anchored bioadhesive ligand, have been tested in order to evaluate the retention of bound liposomes in the face of tissue-related events such as cell migration, proliferation, and death [100]. Synthetic GAG analogs were also used for cell targeting [46]. In this study, liposomes were coated with dextran (Dx) or functionalized dextran (FDx) both hydrophobized by a cholesterol anchor which penetrates the lipid bilayer during the vesicle formation. The coating of liposomes with these polysaccharides was performed because chemically modified Dx, but not native Dx, interacted with vascular cells. The liposome interaction by human endothelial cells was followed using uncoated and coated radiolabeled liposomes. The results indicated a preferential uptake by endothelial cells of FDx-coated liposomes as compared to uncoated or Dx-coated liposomes. Liposome membranes were also labeled with *N*-(lissamine rhodamine B sulfonyl)diacyl-PE, and liposome uptake by endothelial cells was observed by fluorescence microscopy. To avoid a potential disruption of the cell membrane when in contact with liposomes, the lipid composition of endothelial cells was determined [101], and liposome formulation was prepared accordingly. The punctate intracellular fluorescence of cells incubated at 37°C with fluorolabeled liposomes was indicative of the liposome localization within the endocytotic pathway of the cells. All together, the data demonstrate that coating of liposomes with FDx enable specific interactions with human endothelial cells in culture. Consequently, these liposomes coated with bioactive polymers represent an attractive approach for use as drug delivery systems targeting cells.

VI. CONCLUSION

Liposomes are widely recognized as a powerful system for delivery of active compounds. However, knowledge of their physical and chemical properties is prerequisite before use in any biological targeting. Integrity of phospholipid bilayer membrane as well as stability of the corresponding liposomes should be carefully analyzed, since liposomes will interact with the biological medium.

The use of saccharide or polysaccharides for the coating of liposomes is an interesting approach to improve the liposomal structure stability in blood. Moreover, coating the outermost surface of liposomes with polysaccharide derivatives allows targeting of specific cells or organs, and thus the use of coated liposomes as targetable drug carriers [4,6,12,14,46,102]. Such polysaccharide surface-modified liposomes may be useful for sustained release or selective delivery of therapeutic [103] or diagnostic agents [104]. It is also possible to simultaneously introduce hydrophobized polysaccharides and specific cell targetability using, for instance, a monoclonal antibody fragment [14]. Consequently, polysaccharide-coated liposomes appear to be very interesting for their wide applications.

ACKNOWLEDGMENTS

The authors want to thank the co-workers that have contributed to the research described in this manuscript, particularly C. Parisel, A. Sahli, J. Tapon-Bretaudière, A. M. Fischer, and J. Jozefonvicz. This work was supported by CNRS, IFREMER, and Université Paris 13.

REFERENCES

1. Allen, T. M., and Moase E. H. (1996) Therapeutic opportunities for targeted liposomal drug delivery. *Adv. Drug Del. Rev.*, **21**, 117–133.
2. Allen, T. M. (1994) The use of glycolipids and hydrophilic polymers in avoiding rapid uptake of liposomes by the mononuclear phacocyte system. *Adv. Drug Del. Rev.*, **13**, 285–309.
3. Allen, T. M., Hansen, C. B., and Lopes de Menezes, D. E. (1995). Pharmacokinetics of long-circulating liposomes. *Adv. Drug Del. Rev.*, **16**, 267–284.
4. Sato, T., and Sunamoto, J. (1992) Recent aspects in the use of liposomes in biotechnology and medicine. *Prog. Lipid Res.*, **31**, 345–372.
5. Gabizon, A., and Papahadjopoulos, D. (1988) Liposome formulations with prolonged circulation time in blood and enhanced uptake by tumors. *Proc. Natl. Acad. Sci. USA*, **85**, 6949–6953.
6. Kato, E., Akiyoshi, K., Furuno, T., Nakanishi, M., Kikuchi, A., Kataoka, K., and Sunamoto, J. (1994) Interactions between ganglioside-containing liposome and rat T-lymphocyte: confocal fluorescence microscopic study. *Biochim. Biophys. Res. Comm.*, **203**, 1750–1755.
7. Yamauchi, H., Yano, T., Kato, T., Tanaka, I., Nakabayashi, S., Higashi, K., Miyoshi, S., and Yamada, H. (1995) Effects of sialic acid derivative on long circulation time and tumor concentration of liposomes. *Int. J. Pharm.*, **113**, 141–148.
8. Parr, M. J., Ansell, S. M., Choi, L. S., and Cullis, P. R. (1994) Factors influencing the retention and chemical stability of poly(ethylene glycol)-lipid conjugates incorporated into large unilamellar vesicles. *Biochim. Biophys. Acta*, **1195**, 21–30.
9. Ishiwata, H., Vertut-Doï, A., Hirose, T., and Miyajima, K. (1995) Physical-chemistry characteristics and biodistribution of poly(ethylene glycol)-coated liposomes using poly(oxyetylene) cholestery ether. *Chem. Pharm. Bull.*, **43**, 1005–1011.
10. Torchilin, V. P., Omelyanenko, V. G., Papisov, M. I., Bogdanov, A. A., Trubetskoy, V. S., Herron, J. N., and Gentry, C. A. (1994) Poly(ethylene glycol) on the liposome surface: on the mechanism of polymer-coated liposome longevity. *Biochim. Biophys. Acta*, **1195**, 11–20.
11. Torchilin, V. P., Shtilman, M. I., Trubetskoy, V. S., Whiteman, K., and Milstein, A. M. (1994) Amphiphilic vinyl polymers effectively prolong liposome circulation time in vivo. *Biochim. Biophys. Acta*, **1195**, 181–184.
12. Sunamoto, J. (1986) Application of polysaccharide-coated liposomes in chemotherapy and immunotherapy. In: *Medical Applications of Liposomes*, Yapi Y. (Ed.), Japan Scientific Society Press, Tokyo, 121–129.

13. Rosilio, V., Baszkin, A., and Sunamoto, J. (1992) Polysaccharide-coated liposomes. In: *Liposome Technology: Liposome Preparation and Related Techniques*, Vol. 1, 2nd Ed., Gregoriadis G. (Ed.), 43–71.
14. Sunamoto, J., Sato, T., Hirota, M., Fukushima, K., Hiratani, K., and Hara, K. (1987) A newly developed immunoliposome. An egg phosphatidylcholine liposome coated with pullulan bearing both a cholesterol moiety and an IgMs fragment. *Biochim. Biophys. Acta*, **898**, 323–330.
15. Elferink, M. G. L., De Wit, J. G., In't Veld, G., Reichert, A., Driessen, A. J. M., Ringsdorf, H., and Konings, W. N. (1992) The stability and functional properties of proteoliposomes mixed with dextran derivatives bearing hydrophobic anchor groups. *Biochim. Biophys. Acta*, **1106**, 23–30.
16. Torchilin, V. P., and Papisov, M. I. (1994) Hypothesis: why do polyethylene glycol-coated liposomes circulate so long? *J. Liposome Res.*, **4**, 725–739.
17. Osterberg, E., Bergstrom, K., Holmberg, K., Schuman, T. P., Riggs, J. A., Burns, N. L., Van Alstine, J. M., and Harris, J. M. (1995) Protein-rejecting ability of surface-bound dextran in end-on and side-on configurations: comparison to PEG. *J. Biomed. Mater. Res.*, **29**, 741–747.
18. Varki, A. (1993) Biological role of oligosaccharides: All of the theories are correct. *Glycobiology*, **3**, 97–130.
19. Thomas, H., Maillet, F., Letourneur, D., Jozefonvicz, J., Kazatchkine, M. D., and Fischer, E. (1996) A substituted dextran derivative inhibits complement activation and complement-mediated cytotoxicity in an in vitro model of hyperacute xenograft rejection. *Molec. Immunol.*, **33**, 643–648.
20. Lasky, L. A. (1995) Selectin-carbohydrate interactions and the initiation of the inflammatory response. *Annu. Rev. Biochem.*, **64**, 113–139.
21. Letourneur, D., Caleb, B. L., and Castellot, J. J. (1995) Heparin binding, internalization and metabolism in vascular smooth muscle cells. I. Upregulation of heparin binding correlates with antiproliferative activity. *J. Cell. Physiol.*, **165**, 676–686.
22. Brandley, B. K., Kiso, M., Abbas, S., Nikrad, P., Srivasatava, O., Foxall, C., Oda, Y., and Hasegawa, A. (1993) Structure-function studies on selectin carbohydrate ligands. Modifications to fucose, sialic acid and sulphate as a sialic acid replacement. *Glycobiology*, **3**, 633–639.
23. Rosen, S. D., and Bertozzi, C. R. (1994) The selectins and their ligands. *Curr. Opin. Cell Biol.*, **6**, 663–673.
24. Imberty, A., and Pérez, S. (1994) Molecular modeling of protein-carbohydrate interactions. Understanding the specificities of two legume lectins towards oligosaccharides. *Glycobiology*, **4**, 351–366.
25. Spillmann, D., and Burger, M. M. (1996) Carbohydrate-carbohydrate interactions in adhesion. *J. Cell. Biochem.*, **61**, 562–568.
26. Hileman, R. E., Fromm, J. R., Weiler, J. M., and Linhardt, R. J. (1998) Glycosaminoglycan-protein interactions: definition of consensus sites in glycosaminoglycan binding proteins. *Bioessays*, **20**, 156–167.
27. Bourin, M. C., and Lindahl, U. (1993) Glycosaminoglycans and the regulation of blood coagulation. *Biochem. J.*, **289**, 313–330.
28. Petitou, M., Lormeau, J. C., and Choay, J. (1991) Chemical synthesis of glycosaminoglycans: new approaches to antithrombotic drugs. *Nature*, **350**, 30–33.
29. Sahli, A., Cansell, M., Tapon-Bretaudière, J., Letourneur, D., Jozefonvicz, J., and Fischer, A. M. (1998) The stability of heparin-coated liposomes in plasma and their effect on its coagulation. *Colloids Surfaces: B.*, **10**, 205–215.
30. Nelson, R. M., Cecconi, O., Roberts, W. G., Aruffo, A., Linhardt, R. J., and Bevilacqua, M. P. (1993) Heparin oligosaccharides bind L- and P-selectin and inhibit acute inflammation. *Blood*, **82**, 3253–3258.
31. Rider, C. C. (1997) The potential for heparin and its derivatives in the therapy and prevention of HIV-1 infection. *Glycoconj. J.*, **14**, 639–642.
32. Witvrouw, M., and De Clercq, E. (1997) Sulfated polysaccharides extracted from sea algae as potential antiviral drugs. *Gen. Pharmacol.*, **29**, 497–511.
33. Irie, T., and Uekama, K. (1997) Pharmaceutical applications of cyclodextrins. III. Toxicological issues and safety evaluation. *J. Pharm. Sci.*, **86**, 147–162.

34. Franz, G., and Alban, S. (1995) Structure-activity relationship of antithrombotic polysaccharide derivatives. *Int. J. Biol. Macromol.*, **17**, 311–314.
35. Jozefowicz, M., and Jozefonvicz, J. (1997) Randomness and biospecificity: random copolymers are capable of biospecific molecular recognition in living systems. *Biomaterials*, **18**, 1633–1644.
36. Chaubet, F., Champion, J., Maïaga, O., Mauray, S., and Jozefonvicz, J. (1995) Synthesis and structure-anticoagulant relationships of functionalized dextrans: CMDBS. *Carbohydr. Polym.*, **28**, 145–152.
37. Letourneur, D., Logeart, D., Avramoglou, T., and Jozefonvicz, J. (1993) Antiproliferative capacity of synthetic dextrans on smooth muscle cell growth—The model of derivatized dextrans as heparin-like polymers. *J. Biomater. Sci. Polym. Ed.*, **4**, 431–444.
38. Letourneur, D., Champion, J., Slaoui, F., and Jozefonvicz, J. (1993) In vitro stimulation of human endothelial cells by derivatized dextrans. *In Vitro Cell. Biol.*, **29**, 67–72.
39. Tardieu, M., Gamby, C., Avramoglou, T., Josefonvicz, J., and Barritault, D. (1992) Derivatized dextrans mimic heparin as stabilizers, potentiators, and protectors of acidic or basic FGF. *J. Cell. Physiol.*, **150**, 194–203.
40. Carré, V., Mbemba, E., Letourneur, D., Jozefonvicz, J., and Gattegno, L. (1995) Interactions of HIV-1 envelope glycoproteins with derivatized dextrans. *Biochim. Biophys. Acta*, **1243**, 175–180.
41. Neyts, J., Reymen, D., Letourneur, D., Jozefonvicz, J., Schols, D., Este, J., Andrei, G., Mc. Kenna P., Witvrouw, M., Ikeda, S., and De Clercq, E. (1995) Differential antiviral activity of derivatized dextrans. *Biochem. Pharmacol.*, **50**, 743–751.
42. Crepon, B., Maillet, F., Kazatchkine, M. D., and Jozefonvicz, J. (1987) Molecular weight dependence of the acquired anticomplementary and anticoagulant activities of specifically substituted dextrans. *Biomaterials*, **8** 248–253.
43. Thomas, H., Maillet, F., Letourneur, D., Jozefonvicz, J., and Kazatckine, M. D. (1995) Effect of substituted dextran derivative on complement activation in vivo. *Biomaterials*, **16**, 1163–1167.
44. Thomas, H., Maillet, F., Letourneur, D., Jozefonvicz, J., and Kazatchkine, M. D., and Fischer, E. (1996) A synthetic dextran derivative inhibits complement activation and complement-mediated cytotoxicity in an in vitro model of hyperacute xenograft rejection. *Trans. Proc.*, **28**, 593–594.
45. Pain, D., Das, P. K., Ghosh, P., and Bachhawat, B. K. (1984) Increased circulatory half-life of liposomes after conjunction with dextran. *J. Biosci.*, **6**, 811–816.
46. Cansell, M., Parisel, C., Jozefonvicz, J., and Letourneur, D. (1998) Dextran-coated liposomes for endothelial cell targeting: Effect of dextran functionalization. *J. Biomed. Mater. Res.*, in press.
47. Kolender, A. A., Pujol, C. A. Damonte, E. B., Matulewicz, M. C., and Cerezo, A. S. (1997) The system of sulfated alpha-(1 → 3)-linked D-mannans from the red seaweed *Nothogenia fastigiata*: structures, antiherpetic and anticoagulant properties. *Carbohydr. Res.*, **304**, 53– 60.
48. Patankar, M. S., Oehninger, S., Barnett, T., Williams, R. L., and Clark, G. F. (1993) A revised structure for fucoidan may explain some of its biological activities. *J. Biol. Chem.*, **268**, 21770–211776.
49. Grauffel, V., Kloareg, B., Mabeau, S., Durand, P., and Jozefonvicz, J. (1989) New natural polysaccharides with potent antithrombic activity: Fucans from brown algae. *Biomaterials*, **10**, 363–367.
50. Colliec, S., Boisson-Vidal, C., and Jozefonvicz, J. (1994) A low molecular weight fucoidan fraction from the brown seaweed pelvetia canaliculata. *Phytochem.*, **35**, 697–700.
51. Mauray, S., Sternberg, C., Theveniaux, J., Millet, J., Sinquin, C., Tapon-Bretaudière, J., and Fischer, A. M. (1995) Venous antithrombotic and anticoagulant activities of a fucoidan fraction. *Thromb. Haemost.*, **74**, 1280–1285.
52. Vischer, P., and Buddecke, E. (1991) Different action of heparin and fucoidan on arterial smooth muscle cell proliferation and thrombospondin and fibronectin metabolism. *Eur. J. Cell Biol.*, **56**, 407–414.
53. Logeart, D., Prigent-Richard, S., Jozefonvicz, J., and Letourneur, D. (1997) Fucan, a sulfated polysaccharide extracted from brown seaweeds, inhibits vascular SMC proliferation: Part I. Comparison with heparin for the antiproliferative activity, binding and internalization. *Eur. J. Cell. Biol.*, **74**, 376–384.
54. Blondin, C., Fischer, E., Boisson-Vidal, C., Kazatchkine, M. D., and Jozefonvicz, J. (1994) Inhibi-

tion of complement activation by natural sulfated polysaccharides (fucans) from brown seaweed. *Molec. Immunol.*, **31**, 247–253.
55. Blondin, C., Chaubet, F., Nardella, A., Sinquin, C., and Jozefonvicz, J. (1996) Relationships between chemical characteristics and anticomplementary activity of fucans. *Biomaterials*, **17**, 597–603.
56. Monsigny, M., Roche, A. C., and Midoux, P. (1984) Uptake of neoglycoproteins via membrane lectin(s) of L1210 cells evidenced by quantitative flow cytometry and drug targeting. *Biol. Cell*, **51**, 187–196.
57. André, S., Unverzagt, C., Kojima, S., Dong, X., Fink, C., Kayser, K., and Gabius, H. J. (1997) Neoglycoproteins with the synthetic complex biantennary nonasaccharide or its alpha 2,3/alpha 2,6-sialylated derivatives: their preparation, assessment of their ligand properties for purified lectins, for tumor cells in vitro, and in tissue sections, and their biodistribution in tumor-bearing mice. *Bioconjug. Chem.*, **8**, 845–855.
58. Roche, A. C., Barzilay, M., Midoux, P., Junqua, S., Sharon, N., and Monsigny, M. (1983) Sugar-specific endocytosis of glycoproteins by Lewis lung carcinoma cells. *J. Cell. Biochem.*, **22**, 131–140.
59. Eggens, I., Fenderson, B., Toyokuni, T., Dean, B., Stroud, M., and Hakomori, S. (1989) Specific interaction between Lex and Lex determinants. A possible basis for cell recognition in preimplantation embryos and in embryonal carcinoma cells. *J. Biol. Chem.*, **264**, 9476–9484.
60. Schäfer, H., Stahn, R., Schreiber, J., and Schmidt, W. (1996) Glycoprotein mediated cell binding of lectin coated liposomes. *J. Liposome Res.*, **6**, 479–496.
61. Lander, A. D. (1994) Targeting the glycosaminoglycan-binding sites on proteins. *Chem. Biol.*, **1**, 73–78.
62. Linhardt, R. J., and Hileman, R. E. (1995) Dermatan sulfate as a potential therapeutic agent. *Gen. Pharmacol.*, **26**, 443–451.
63. Mulloy, B., Ribeiro, A. C., Alves, A. P., Vieira, R. P., and Mourao, P. A. (1994) Sulfated fucans from echinoderms have a regular tetrasaccharide repeating unit defined by specific patterns of sulfation at the 0–2 and 0–4 positions. *J. Biol. Chem.*, **69**, 22113–22123.
64. Prigent-Richard, S., Cansell, M., Vassy, J., Viron, A., Puvion, E., Jozefonvicz, J., and Letourneur, D. (1998) Fluorescent and radiolabeling of polysaccharides: binding and internalization experiments on vascular cells. *J. Biomed. Mater. Res.*, **40**, 275–281.
65. Nishiya, T., and Robibo, D. (1992) Quantitative analysis of carboxymethyl chitin adsorbed on a liposome surface. *Carbohydr. Res.*, **235**, 239–245.
66. Haensler, J., and Schuber, F. (1991) Influence of the galactosyl ligand structure on the interaction of galactosylated liposomes with mouse peritoneal macrophages. *Glycoconjugate J.*, **8**, 116–124.
67. Stahn, R., Schäfer, H., Schreiber, J., and Brudel, M. (1995) Glycoliposomes—Simple preparation and specific binding to lectin. *J. Liposome Res.*, **5**, 61–73.
68. Focher, B., Savelli, G., and Torri, G. (1990) Neutral and ionic alkylglucopyranosides. Synthesis, characterization and properties. *Chem. Phys. Lipids.*, **53**, 141–155.
69. Sugimoto, M., Ohishi, K., Fukasawa, M., Shikata, K., Kawai, H., Itakura, H., Hatanaka, M., Sakakibara, R., Ishiguro, M., Nakata, M., and Mizuochi, T. (1995) Oligomannose-coated liposomes as an adjuvant for the induction of cell-mediated immunity. *FEBS Lett.*, **363**, 53–56.
70. Akiyoshi, K., Yamaguchi, S., and Sunamoto, J. (1991) Self-aggregates of hydrophobic polysaccharide derivatives. *Chem. Lett.* (*Japan Chem. Soc.*), 1263–1266.
71. Akiyoshi, K., Deguchi, S., Moriguchi, N., Yamaguchi, S., and Sunamoto, J. (1993) Self-aggregates of hydrophobic polysaccharides in water. Formation and characteristics of nanoparticles. *Macromolecules*, **26**, 3062–3068.
72. Kang, E. C., Akiyoshi, K., and Sunamoto, J. (1997) Surface coating of liposomes with hydrophobized polysaccharides. *J. Bioactive Compatible Polym.*, **12**, 14–26.
73. Stearns, N. A., Prigent-Richard, S., Letourneur, D., and Castellot, J. J. (1997) Synthesis and characterization of highly sensitive heparin probes for detection of heparin-binding proteins. *Anal. Biochem.*, **247**, 348–356.
74. Herman, S., Persijn, G., Vanderkerckhove, J., and Schacht, E. (1993) Synthesis of dextran deriva-

tives with thiol-specific reactive groups for the preparation of dextran-protein conjugate. *Bioconjugate Chem.*, **4**, 402–405.
75. Suzuki, M., Mikami, T., Matsumoto, T., and Suzuki, S. (1977) Preparation and antitumor activity of *O*-palmitoyldextran phosphates, *o*-palmitoyldextran, and dextran phosphates. *Carbohydr. Res.*, **53**, 223–229.
76. Takada, M., Yuzuriha, T., Katayama, K., Iwamoto, K., and Sunamoto, J. (1984) Increased lung uptake of liposomes coated with polysaccharides. *Biochim. Biophys. Acta*, **802**, 237–244.
77. Sahli, A. (1998) Interaction of liposomes and nanoparticles with the blood coagulation system. Ph. D. thesis, Paris North University, France.
78. Moreira, J. N., Almeida, L. M., Geraldes, C. F., and Costa, M. L. (1996) Evaluation of in vitro stability of large unilamellar liposomes coated with a modified polysaccharide (*o*-palmitoylpullulan). *J. Mater. Sci. Mat. Med.*, **7**, 301–304.
79. Sunamoto, J., Iwamoto, K., Takada, M., Yuzuriha, T., and Katayama, K. (1984) Improved drug delivery to target specific organs using liposomes as coated with polysaccharides. In: *Polymers in Medicine*, Chiellini E. and Giusti P (Eds.), Plenum Press, New York, 157–168.
80. Arnold, K., Ohki, S., and Krumbiegel, M. (1990) Interaction of dextran sulfate with phospholipid surfaces and liposome aggregation and fusion. *Chem. Phys. Lipids*, **55**, 301–307.
81. Yoshida, A., Yamauchi, H., Sakai, H., Kawashima, N., and Abe, M. (1997) Molecular interactions between lipid bilayers and a water-soluble polymer. *Colloids Surfaces B.*, **8**, 333–342.
82. Mobed, M., Eng, M., and Chang, T. M. S. (1996) The importance of standardization of carboxymethylchitin concentration by the dye-binding capacity of alcian blue before adsorption on liposomes. *Art. Cells Blood Subst. Immob. Biotech.*, **24**, 107–120.
83. Nishiya, T., and Lam, R. T. T. (1993) Mechanistic study on toxicity of positively charged liposomes containing stearylamine to blood: use of carboxymethyl chitin to reduce toxicity. *Colloids Surfaces*: *B.*, **1**, 213–219.
84. Nishiya, T., and Lam, R. T. T. (1995) Interaction of stearylamine-liposomes with erythrocyte ghosts: analysis of membrane lipid mixing and aqueous contents mixing, and the effect of carboxymethyl chitin on the interaction. *Colloids Surfaces*: *B.*, **4**, 55–63.
85. Iwamoto, K., and Sunamoto, J. (1982) Liposomal membranes. XII. Adsorption of polysaccharides on liposomal membranes as monitored by fluorescence depolarization. *J. Biochem.* (*Tokyo*), **91**, 975–979.
86. Mishima, K., Satoh, K., and Suzuki, K. (1997) Molecular order of lipid membranes modified with polysaccharide pullulan and a pullulan-derivative bearing a hydrophobic anchor. *Colloids Surfaces*: *B.*, **9**, 9–15.
87. Seras-Cansell, M., Sahli, H., Dubois, V., Letourneur, D., and Jozefonvicz, J. (1998) Characterization of vesicles for vascular cell targeting: Influence of dextran coating. In: *Frontiers in Biomedical Polymer Applications*, M. Ottenbrite (Ed.), Technomic, Lancaster, 95–108.
88. Ghosh, P. C., and Bachhawat, B. K. (1991) Targeting of liposomes to hepatocytes. *Targeted Diagn. Ther.*, **4**, 87–103.
89. Murahashi, N., Ishihara, H., Sasaki, A., Sakagami, M., and Hamana, H. (1997) Hepatic accumulation of glutamic acid branched neogalactosyllipid modified liposomes. *Biol. Pharm. Bull.*, **20**, 259–266.
90. Das, P. K., Murray, G. J., Zirzow, G. C., Brady, R. O., and Barranger, J. A. (1985) Lectin-specific targeting of beta-glucocerebrosidase to different liver cells via glycosylated liposomes. *Biochem. Med.*, **33**, 124–131.
91. Yachi, K., Kikuchi, H., Yamauchi, H., Hirota, S., and Tomikawa, M. (1995) Distribution of liposomes containing mannobiose esters of fatty acid in rats. *J. Microencapsul.*, **12**, 377–388.
92. Mumtaz, S., Ghosh, P. C., and Bachhawat, B. K. (1984) Design of liposomes for circumventing the reticuloendothelial cells. *Glycobiology*, **1**, 505–510.
93. Wu, M. S., Robbins, J. C., Bugianesi, R. L., Ponpipom, M. M., and Shen, T. Y. (1981) Modified in vivo behavior of liposomes containing synthetic glycolipids. *Biochim. Biophys. Acta*, **674**, 19–29.
94. Vodovozova, E. L., Gayenko, G. P., Razinkov, V. I., Korchagina, E. Y., Bovin, N. V., and Molot-

kovsky, J. G. (1998) Saccharide-assisted delivery of cytotoxic liposomes to human malignant cells. *Biochem. Mol. Biol. Int.*, **44**, 543–553.
95. Banerjee, G., Nandi, G., Mahato, S. B., Pakrashi, A., and Basu, M. K. (1996) Drug delivery system: targeting of pentamidines to specific sites using sugar grafted liposomes. *J. Antimicrob. Chemother.*, **38**, 145–150.
96. Medda, S., Das, N., Mahato, S. B., Mahadevan, P. R., and Basu, M. K. (1995) Glycoside-bearing liposomal delivery systems against macrophage-associated disorders involving Mycobacterium leprae and Mycobacterium tuberculosis. *Indian J. Biochem. Biophys.*, **32**, 147–151.
97. Van Bambeke, F., Tulkens, P. M., Brasseur, R., and Mingeot-Leclercq, M. P. (1995) Aminoglycoside antibiotics induce aggregation but not fusion of negatively-charged liposomes. *Eur. J. Pharmacol.*, **289**, 321–333.
98. Van Bambeke, F., Mingeot-Leclercq, M. P., Brasseur, R., Tulkens, P. M., and Schanck, A. (1996) Aminoglycoside antibiotics prevent the formation of non-bilayer structures in negatively-charged membranes. Comparative studies using fusogenic (bis(beta-diethylaminoethylether)hexestrol) and aggregating (spermine) agents. *Chem. Phys. Lipids*, **79**, 123–135.
99. Wu, P. S., Wu, H. M., Tin, G. W., Schuh, J. R., Croasmun, W. R., Baldeschwieler, J. D., Shen, T. Y., and Ponpipom, M. M. (1982) Stability of carbohydrate-modified vesicles in vivo: comparative effects of ceramide and cholesterol glycoconjugates. *Proc. Natl. Acad. Sci. USA*, **79**, 5490–5493.
100. Yerushalmi, N., and Margalit, R. (1998) Hyaluronic acid-modified bioadhesive liposomes as local drug depots: effects of cellular and fluid dynamics on liposome retention at target sites. *Arch. Biochem. Biophys.*, **349**, 21–26.
101. Cansell, M., Gouygou, J-P., Jozefonvicz, J., and Letourneur, D. (1997) Lipid composition of cultured endothelial cells in relation to their growth. *Lipids*, **32**, 39–44.
102. Sunamoto, J., and Iwamoto, K. (1986) Protein-coated and polysaccharide-coated liposomes as drug carriers. *Crit. Rev. Ther. Drug. Carrier, Syst.*, **2**, 117–136.
103. Ichinose, K., Yamamoto, M., Khoji, T., Ishii, N., Sunamoto, J., and Kanematsu, T. (1998) Antitumor effect of polysaccharide coated liposomal adriamycin on AH66 hepatoma in nude mice. *Anticancer Res.*, **18**, 401–404.
104. Gupta, H., and Weissleder, R. (1996) Targeted contrast agents in MR imaging. *Magn. Reson. Imaging Clin. N. Am.*, **4**, 171–184.

14

Polysaccharide Biomaterials As Drug Carriers

Chee-Youb Won
Hoffmann—La Roche, Inc., Nutley, New Jersey

Chih-Chang Chu
Cornell University, Ithaca, New York

I. INTRODUCTION

The attachment of biologically active compounds to synthetic and natural polymers has been investigated as a means of improving the efficacy of drug control/release devices through a constant but prolonged release of bioactive compounds with minimum side effects. The controlled release of the bioactive compounds, which are covalently coupled to the polymeric carrier via hydrolyzable linkages, can be achieved by hydrolytic or enzymatic cleavage of the linking bonds.

One of the main challenges in designing drug control/release devices is to construct biocompatible carriers that are be easily metabolized and eliminated, that do not adversely affect normal biological functions inside the human body, and that are inexpensive for reducing health care cost. A class of biomaterials that could be a potential candidate as the drug carrier is natural polymers, such as polysaccharides.

Naturally occurring polysaccharides are one of the most abundant and diverse families of biopolymers, and are widely used in various fields [1]. Recently, considerable attention has been focused on the chemical modification of polysaccharides to be used as drug carriers for controlled release system [2–4]. The repeating unit of polysaccharides may contain hydroxyl, carboxyl, amino, or sulfate functional groups, which provide the sites for the chemical modifications of polysaccharides.

Steroid hormones are endogenic biologically active compounds. They are composed of a rigid, stereochemically complex tetracyclic skeleton with a variety of substituents at the extremities. In steroid therapy, they have been used to treat various types of cancer and to correct disorders associated with fertility and menstrual cycle irregularity [5]. However, high immunogenicities, short circulatory half-life, and rapid excretion [6] limit the practical use of steroid drugs. In order to overcome these problems, a more precise mode of hormone release via drug carriers is essential. Only a few examples are available where steroids are covalently linked to a polymeric system [7–8].

In this chapter, we describe different chemical means to attach steroid hormones to several polysaccharides, including starch, dextran, cellulose, and chitosan.

II. STARCH

Starch, long-chain polymers of D-glucose, is one of the most frequently used polysaccharides in the drug delivery system because of its biocompatibility, biodegradability, nontoxicity, and abundant sources.

We first used degradable starch as a carrier for the delivery of estrone hormone [9]. The extensive inter- and intramolecular hydrogen bondings among hydroxy groups in starch make it difficult to dissolve, even in the LiCl/DMAc solvent system at high temperature. Therefore, the starch was pretreated with distilled water, methanol, and DMAc in order to facilitate dissolution with minimum degradation. The resulting solvent-exchanged swollen starch was readily dissolved in LiCl/DMAc solvent system at 40°C. In order to provide reactive sites on starch for subsequent coupling of estrone, as well as to introduce a suitable spacer between the carrier and the estrone, solvent-exchanged swollen starch was converted to a suitable reactive derivative with bromoacetyl bromide. The degree of substitution (D.S.) per anhydroglucose (AHG) unit was calculated from the bromine content and ranged from 0.11 to 2.29, depending on the ratio of bromoacetyl bromide to starch.

The structure of the bromoacetylated starches was characterized by FTIR, ^{1}H and ^{13}C NMR spectrometers. Figure 1b shows the FTIR spectrum of the bromoacetylated starch. The residual OH groups show its stretching vibration at around 3400 cm^{-1}. A new absorption peak

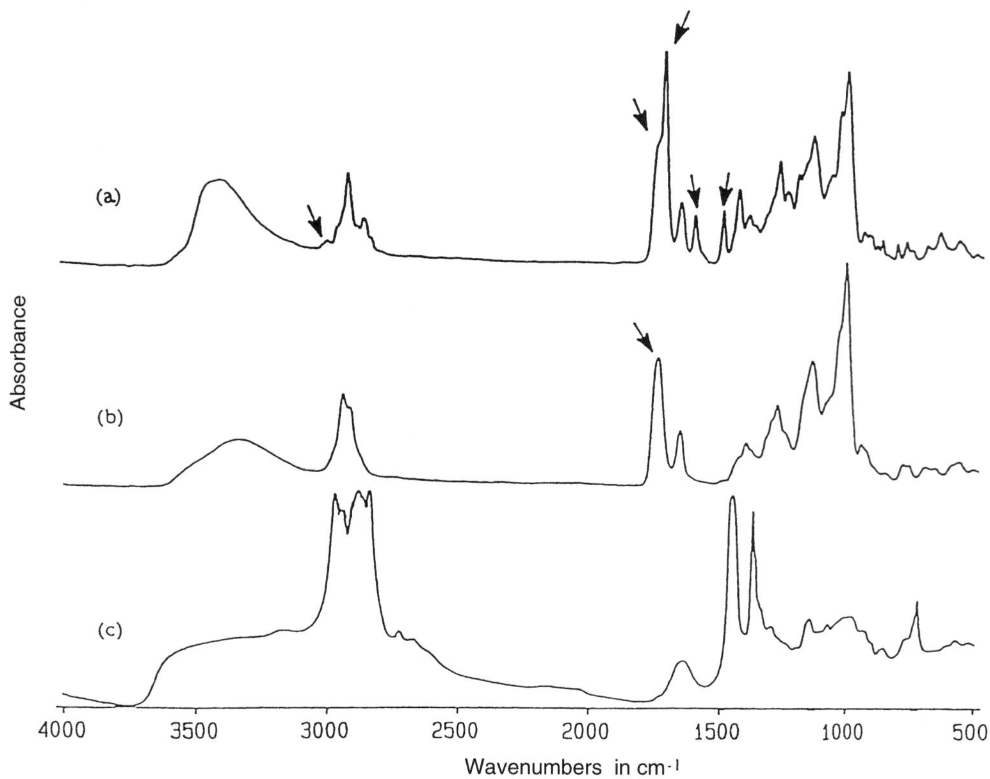

Figure 1 FTIR spectra of (a) unmodified starch, (b) bromoacetylated starch (D.S. = 2.3), and (c) starch-estrone conjugate. (A) Ester C=O stretch; (B) aromatic C—H stretch; (C) ketone C=O stretch in estrone; (D) ring C=C stretch. (From Ref. 9.)

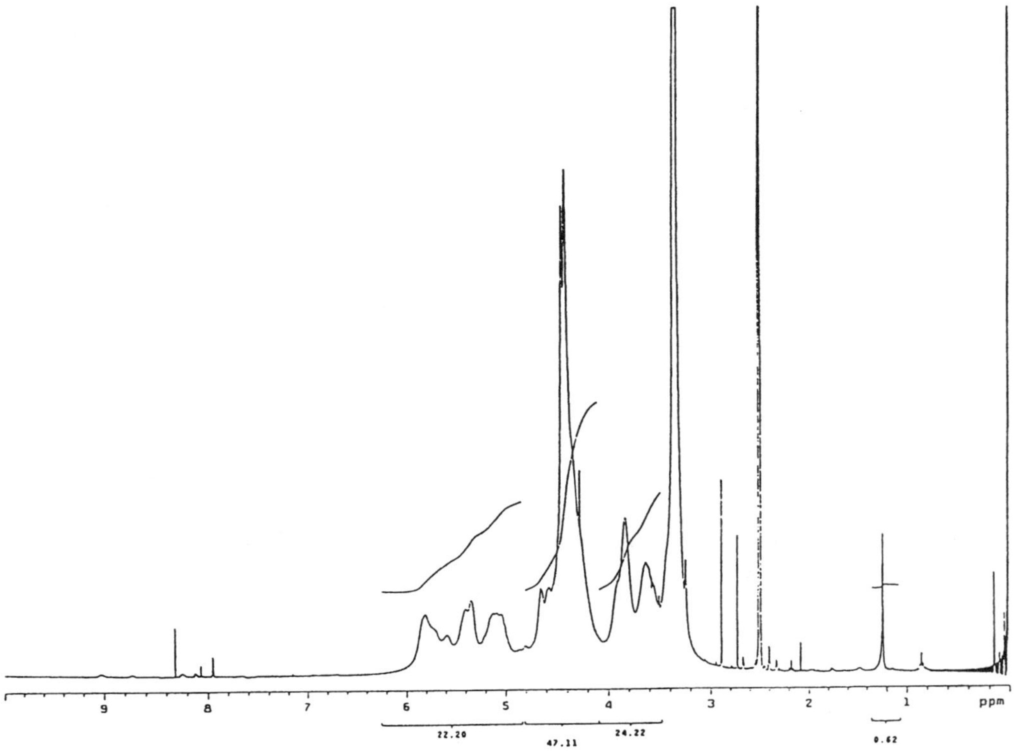

Figure 2 ^{1}H NMR spectrum of bromoacetylated starch (D.S. = 2.3) in DMSO-d_6. (From Ref. 9.)

at 1746 cm^{-1} can be assigned to the characteristic carbonyl groups in —COO—CH_2—Br. The relative intensities of these bands seem to depend on the modification extent in the biopolymer.

The ^{1}H NMR spectrum of the same modified starch (Fig. 2) shows a band at δ 4.27, which corresponds to the methylene protons of bromoacetate groups. There are also bands at δ 1.24 due to the methylene protons in the main backbone chain and a band at δ 5.09 which can be assigned to the resonance of the methine protons of modified starch [10].

The ^{13}C NMR spectrum shown in Fig. 3a exhibits the characteristic chemical shifts at δ 167.26 corresponding to the carbonyl carbon atom of ester groups. The peak δ 40.9 can be assigned to methylene carbon atom of bromoacetyl groups. The signals between δ 64.07 and 100.95 are due to sugar carbon atoms.

The coupling of pharmacologically active estrone hormone to the bromoacetylated starch was carried out by using the sodium salt of estrone via nucleophilic substitution as shown in Scheme 1. The estrone was readily converted to its sodium salt by treatment with sodium hydride at room temperature in THF. The ease of sodium salt formation depends on the relative acidity of alcohol. The aromatic alcohol of estrone (3-position) reacted immediately with sodium hydride under mild conditions.

The FTIR spectrum of the starch-estrone adducts is shown in Fig. 1c. The small peak at around 3005 cm^{-1} corresponds to the aromatic C—H bonds in estrone. The 17-carbonyl group of estrone is observed at 1726 cm^{-1}. It shows new absorption associated with the >C═C< stretching band of aromatic ring system in estrone at 1503 cm^{-1}. Three new bands at 1258, 1246 and 1218 cm^{-1} are observed, which were assigned to the characteristic C—O stretching in 3-benzoate ester of estrone.

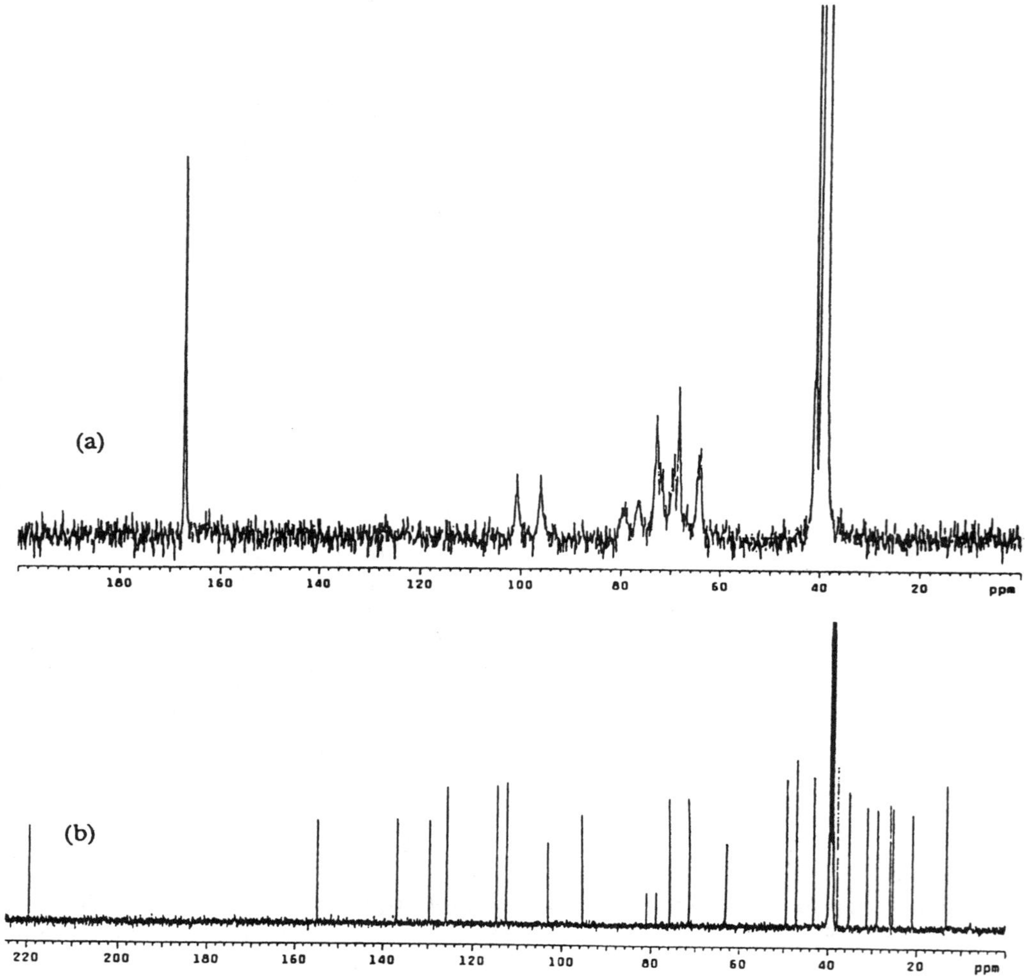

Figure 3 ^{13}C NMR spectra of (a) bromoacetylated starch (D.S. = 2.3); (b) starch-estrone conjugate in DMSO-d_6. (From Ref. 9.)

Figure 3b presents ^{13}C NMR spectrum of starch-estrone conjugate (DS = 2.3) in DMSO-d_6 solution, and shows new chemical shifts due to estrone moiety, as well as sugar carbon atoms (δ 63.67–110.15).

The bromoacetylated and estrone-conjugated starches were soluble at room temperature in polar aprotic solvents, such as DMF, DMSO, and NMP, whereas the unmodified starch did not dissolve in these solvents.

Sjoholm et al. [11] reported that low molecular weight antiparasitic drugs (primaquine and trimethoprim) have been covalently bound to the starch microparticles via tri-, tetra-, and pentapeptide spacer arms. The drug-starch particle conjugates were prepared by coupling different drug-peptide derivatives to the particles after activation of the starch with carbonyldiimidazole as an activating agent (Scheme 2).

The in vitro release of drug from the resulting drug-carrier conjugates showed that the conjugates were stable in serum, while free drugs were released in a lysosome fraction. Thus,

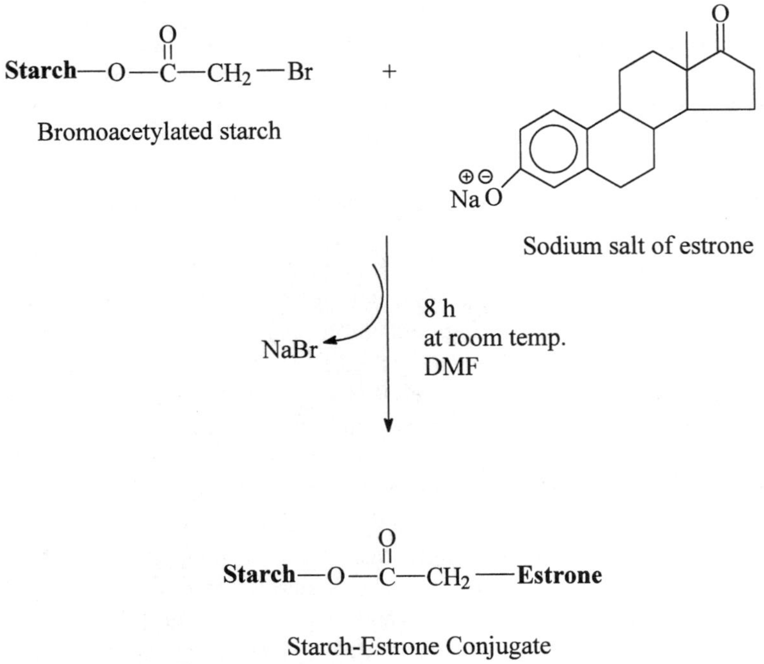

Scheme 1 Chemical synthesis of estrone-starch conjugate. (From Ref. 9.)

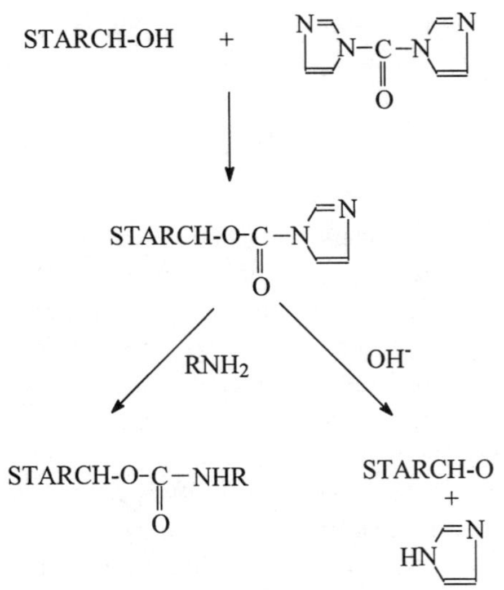

Scheme 2 Activation of the hydroxyl groups on starch with carbonyldiimidazole (CDI) and subsequent coupling of the amino-group-containing ligands, or hydrolysis of the active group. (From Ref. 11.)

$$\text{Starch-OH} + \text{ClCOCl} \longrightarrow \text{Starch-OCOCl} \xrightarrow{\text{NH}_2\text{CH}_2\text{CH}_2\text{C}_6\text{H}_5} \text{Starch-OCONHCH}_2\text{CH}_2\text{C}_6\text{H}$$

Scheme 3 Synthesis of starch-phenethylamine conjugate. (From Ref. 13.)

these conjugates fulfilled the basic requirements for a drug carrier for the treatment of lysosomal parasitic diseases.

Lenaerts et al. [12] demonstrated that crosslinked high amylose starches have been used as excipients for the formation of controlled release solid dosage forms for the oral delivery of drugs. Corn starch was highly crosslinked by using epichlorohydrin as a crosslinking agent. Water transport kinetics and dimensional changes were studied in matrices placed in water at 37°C by an image analysis technique. The studies showed that a gel was formed very quickly at the polymer surface and water continuously diffused into the core. This process went on for several hours until the core turned into a gel and equilibrium swelling was reached. In vivo results with different drugs (Efidac and Voltaren) showed that high amylose crosslinked starches were highly versatile, had a low intersubject variability and no food effect.

Zilkha et al. [13] reported the preparation of starch-phenethylamine conjugates and their preliminary pharmacological evaluation. As shown in Scheme 3, starch was first converted to the chlorocarbonate derivative by reaction with phosgene and then treated with the phenethylamines. The degree of substitution of amine per glucose ranged from 0.06 to 0.47. According to their pharmacological study in vivo, the duration of action was more than that of pure phenethylamine itself. Sedation was observed from 5 to 24 h after administrating starch-phenethylamine conjugate, and deaths occurred between two and six days.

Remon et al. [14] reported the production and characterization of thermally modified starch which might be used as hydrophilic matrices for oral drug delivery systems. Three native starches having a variety of amylose/amylopectin contents were pregelatinized by extrusion, drumdrying and a controlled pregelatinization-spray-drying technique in order to produce cold-water-swellable starches. Partial pregelatinized starch could only be produced by the controlled pregelatinization-spray-drying technique. As regards the application as hydrogel matrices for controlled release tablets, promising results were seen for the extruded and drum-dried starches containing medium to high amounts of amylopectin.

III. DEXTRAN

Dextran has been used as a plasma expander and has been frequently chosen as drug carrier due to its biocompatibility, biodegradability, hydrophilicity and water solubility [15].

We have recently extended the concept of polysaccharide-based drug carriers to dextran-estrone conjugate for delivery of estrogen [16]. In order to achieve this goal, we must first synthesize a suitable reactive derivative of estrone for its subsequent coupling with dextran. Estrone was succinylated by reaction with succinic anhydride. The carboxylic acid of succinoylated estrone then reacted with thionyl chloride to replace the hydroxy group of the carboxylic acid with chlorine to make a better leaving group. Since estrone receptor affinity is related to the 17-keto region, this modification of the 3-hydroxyl should not affect its biological activity. Because the functionally active center of estrone is intact, we thus expect that the attached estrone may be biologically active, even in the dextran conjugate form.

The dextran-estrone conjugate (**2**) was prepared by reacting the modified estrone with dextran (Scheme 4). The type of linkage between the drug and the polymeric carrier plays an

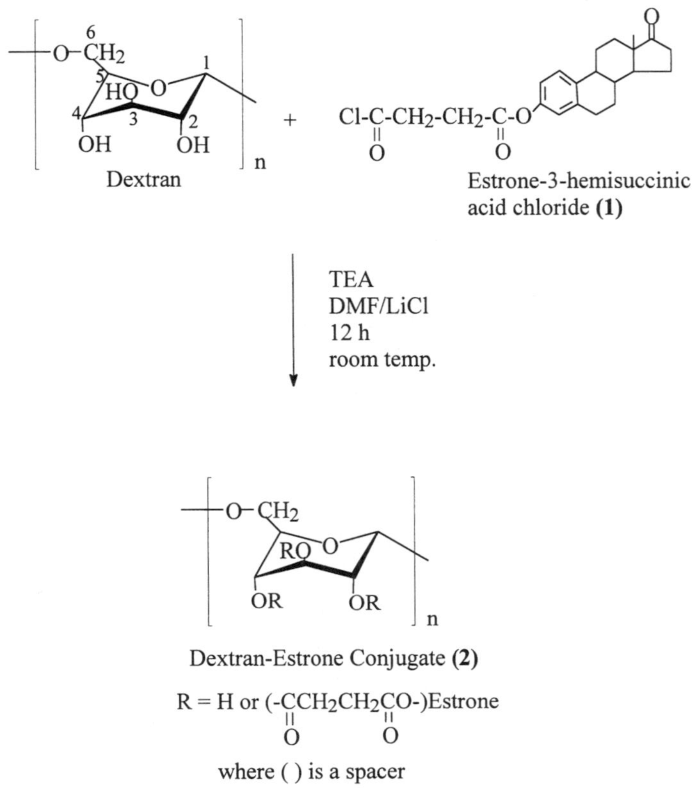

Scheme 4 Synthesis of dextran-estrone conjugate. (From Ref. 16.)

important role in the rate of drug release. The relative rates of hydrolysis of various labile bonds at pH 7.4 are known in the following order: carbonates (—OCOO—)> esters (—COO—)> urethanes (—NHCOO—)> amide (—CONH—). Carbonates [17] and amides [18] have been used for linking drugs to polymers. In this study, an ester linkage was designed to couple estrone to dextran.

The degree of substitution (D.S.) was estimated from the ^{1}H NMR spectrum by taking a ratio of the normalized, integrated intensities of the sum of the bands at δ 6.82, 7.30 and 7.95 ppm (corresponding to the aromatic ring protons of estrone) to the normalized, integrated intensities of the dextran backbone protons (the signals between δ 3.00 and 5.52 ppm). In our study, the D.S. of 0.33 (11.0 mol% of estrone moieties) was obtained when a 1 : 3 molar ratio of acid chloride of estrone-3-hemisuccinate to the hydroxyl groups of dextran was used.

The estrone-conjugated dextran (**2**) was soluble at room temperature in such polar, aprotic solvents as *N,N′*-dimethylformamide (DMF), dimethyl sulfoxide (DMSO) and *N*-methyl-2-pyrrolidone (NMP), whereas the unmodified dextran was dissolved in water, DMF/LiCl and DMSO/pyridine solvents systems at high temperature (80°C).

In order to evaluate estrone release from the dextran-estrone conjugate (**2**) through an in vitro hydrolytic process, the conjugate (11.0 mol% of estrone moieties) was hydrolyzed heterogeneously in aqueous phosphate buffer and ethanol mixtures of either pH 7.4 or 8.0 at 37°C. Figure 4 illustrates the release of estrone from the dextran-estrone conjugate at pH 8.0 and 7.4 and 37°C. The observed linear release of estrone versus time indicates that the release

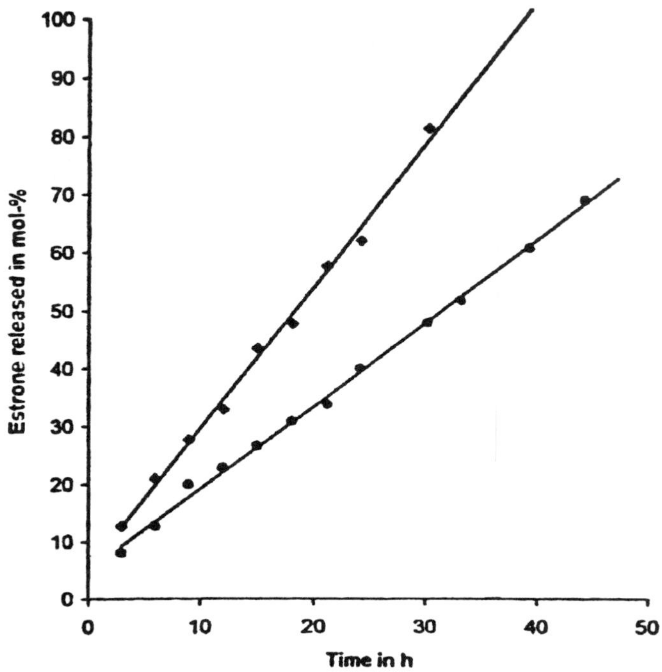

Figure 4 The release of estrone via heterogeneous hydrolysis of dextran-estrone conjugate (11.0 mol% of estrone) in buffer media at 37°C. (◆) -pH 8.0 and (●) -pH 7.4. (From Ref. 16.)

rate was zero order for up to 45 h. A faster release rate was observed at pH 8.0, and the total release of the estrone was achieved within a few days. The faster release rate at a higher pH is consistent with the well-known fact that the hydrolysis rate of simple organic esters increases with pH of the medium. The amount of ethanol (acting as a weak nucleophile for the cleavage of ester bonds), drug loading in the conjugate (i.e., degree of substitution), and hydrophilicity of the dextran may also control the rate of release. Because of the relative hydrophobic nature of estrone, a low amount of the conjugated estrone moieties (11.0 mol%) would make the dextran-estrone conjugate (**2**) more hydrophilic than a high loaded conjugate.

The hydrolyses of the two-ester linkages located in the spacer of the dextran-estrone conjugate were presumed to occur according to the reactions in Scheme 5. However, the hydrolysis of the enolic ester portion of the dextran-estrone conjugate (route B) may be faster than that of the aliphatic ester portion (route A) due to the higher sensitivity of the enolic ester toward water. If enough time is permitted for the two intermediate esters (**A-2** and **B-2**) to be continuously hydrolyzed, the eventual final products of hydrolysis would be estrone and dextran, regardless of A or B route of hydrolysis. The buffer medium obtained from 48 h hydrolysis of the dextran-estrone conjugate (**2**) was extracted with benzene and examined by FTIR. There was no characteristic FTIR ester peak in the medium, an indication of the possible lack of ester spacer group found in **A-2**. Therefore, the estrone-succinic acid ester derivative (**A-2**) has been further hydrolyzed and the parent estrone (**A-3**) or (**B-1**) was the end product. However, FTIR data are not sufficient to provide a definite answer to the question of whether **A-2** exists in the buffer media because the carbonyl IR band in the ester spacer could be masked by the ketone band of the estrone. Thus, other chemical means must be used to draw a definitive conclusion of the presence of **A-2** in the buffer medium.

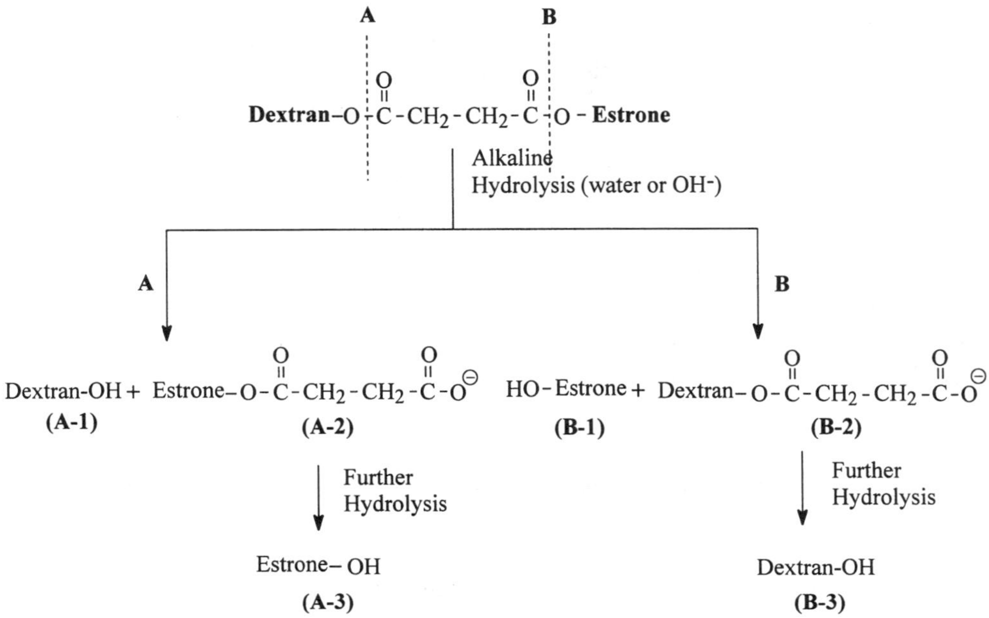

Scheme 5 The two possible routes of release of estrone from the hydrolytic degradation of dextran-estrone conjugate. (From Ref. 16.)

Mocanu et al. [19] reported the synthesis of dextran-theophylline conjugates and the in vitro release of the theophylline from these conjugates. Theophylline is a diuretic, cardiac stimulant, smooth muscle relaxant, vasodilator drug used in asthmatic and coronary disorders. Theophylline-dextran conjugates were prepared by two chemical methods. First, the theophylline was coupled onto chloroacetylated crosslinked dextran microparticles in organic basic aprotic solvents (Scheme 6).

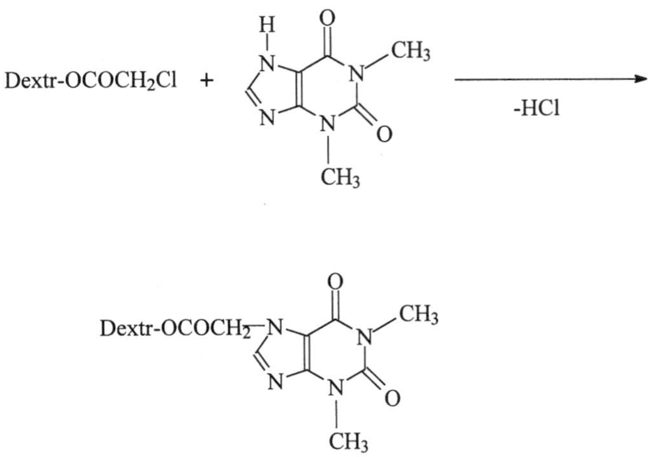

Scheme 6 The incorporation of theophylline into chloroacetylated dextran. (From Ref. 19.)

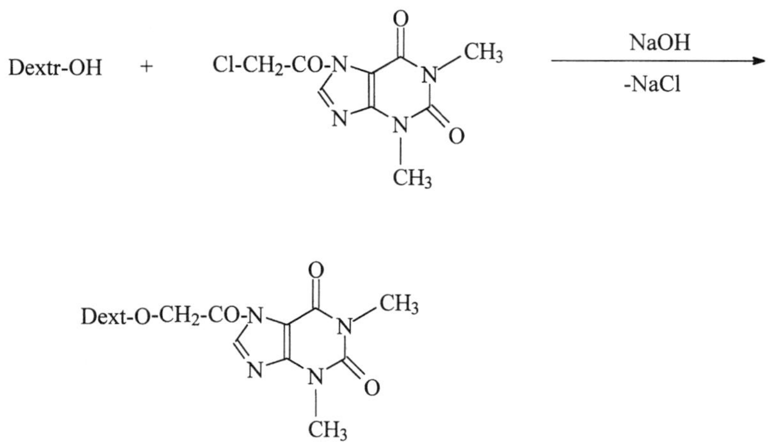

Scheme 7 The incorporation of 7-N-chloroacetyltheophylline into dextran. (From Ref. 19.)

Second, the conjugate was prepared by the reaction of 7-*N*-chloroacetyltheophylline with crosslinked dextran microparticles in basic media (Scheme 7). The in vitro behavior of the conjugates indicated that the drug was released gradually as function of pH and time. In acidic gastric medium, theophylline was released to a much smaller extent than in basic intestinal medium.

Schacht et al. [20] have reported the coupling of an antimicrobial drug, norfloxacin, having the tetrapeptides gly-phe-ala-leu or gly-phe-leu-gly substituted on the piperazine amino group onto chloroformate-activated dextran (Scheme 8). The conjugates were hydrolytically stable in buffers under physiological conditions (pH 5.5 and 7.4). In the presence of a lysosomal enzyme cathepsin B, stable amino acid derivatives of norfloxacin (leu-norfloxacin and gly-norfloxacin) were released. Aminopeptidases, however, can release the free drug from the amino-acid-based drug derivatives.

Hennink et al. [21] demonstrated the release of three model proteins (lysozyme, albumin, IgG) from nondegrading and enzymatically degrading dextran hydrogels. These hydrogels were obtained by radical polymerization of aqueous solutions of glycidyl methacrylate derivatized dextran shown in Fig. 5.

The release rate of the model proteins from dextran hydrogels depends on and can be manipulated by the initial water content and the crosslink density of the gel. In gels with a high hydration level, the release of proteins can be effectively described using the free volume theory. By incorporation of dextranase, degradable systems were obtained. The degradation rate of the gels was dependent on the concentration of dextranase in the gel and the crosslink

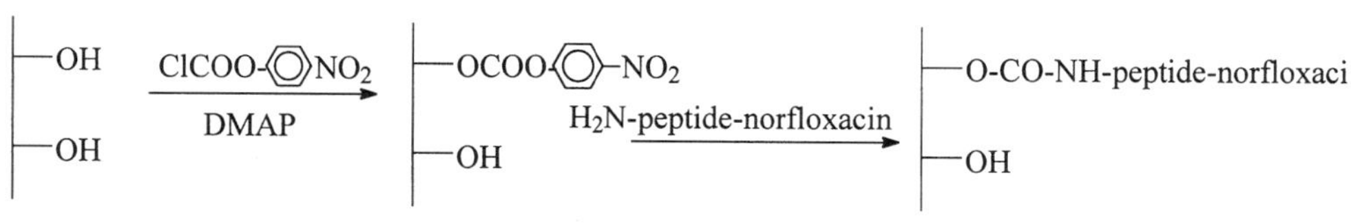

Scheme 8 Coupling of a peptide-norfloxacin onto dextran. (From Ref. 20.)

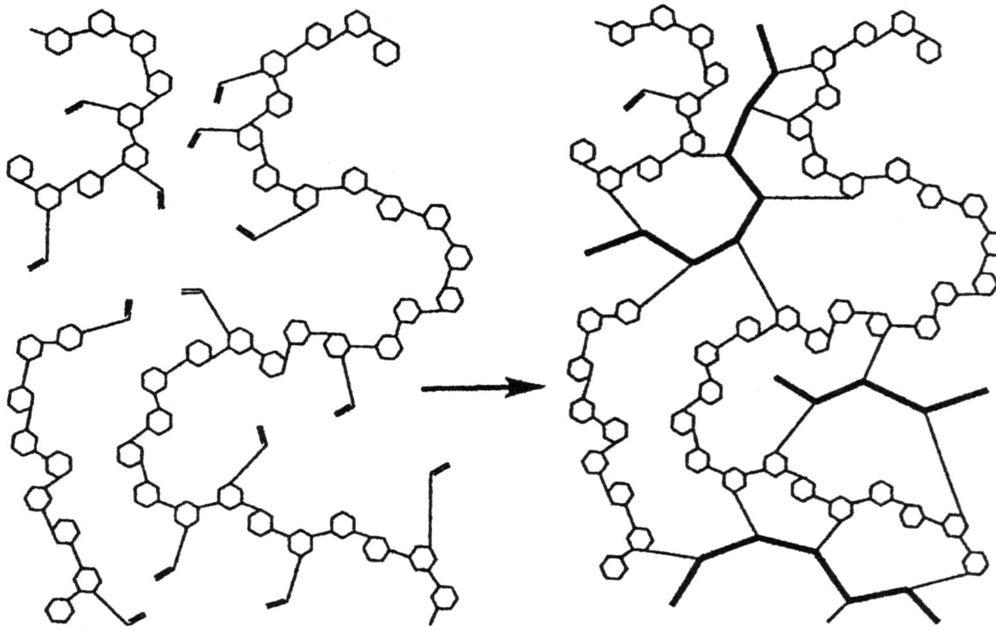

Figure 5 Schematic representation of the formation of dextran hydrogels. Hexagons denote monomer unit of dextran (α-1,6-glycopyranose residue), polymerizable methacrylate group are shown with =, polymerized methacrylate group are shown by bold lines. (From Ref. 21.)

density of the gels. The incorporation of dextranase in the gel could also be used to modulate the release of an encapsulated protein.

Very recently, Mestries et al., in France, developed a chemical method to modify dextran via its three —OH groups so that the resulting polysaccharide would have regenerating agents incorporated for in vivo tissue modulation, such as the control of the cellular proliferation and collagen synthesis [22]. The carboxymethylated or carboxymethyl sulfated dextran would reduce the proliferation of smooth muscle cells and the degree of suppression depended on the level of sulfation level. This new type of dextran derivative may have the potential for antifibrotic activity.

IV. CELLULOSE

Yolles et al. [23] reported the preparation of modified cellulose for attching steroids (Scheme 9). In their study, hydroxypropyl cellulose was first reacted with phosgene to form the corresponding chloroformate and then estrone and testosterone steroids were conjugated with these modified hydroxypropyl cellulose.

Khue and Galin [17] prepared polymer-steroid conjugates by the reaction of hydroxypropylcellulose with hydrocortisone and dexamethasone-^{21}C-chloroformates (Scheme 10). The steroids binding on the hydroxypropylcellulose through carbonate linkage was readily identified by their characteristic FTIR spectrum: C=O carbonate at 1760 cm^{-1}.

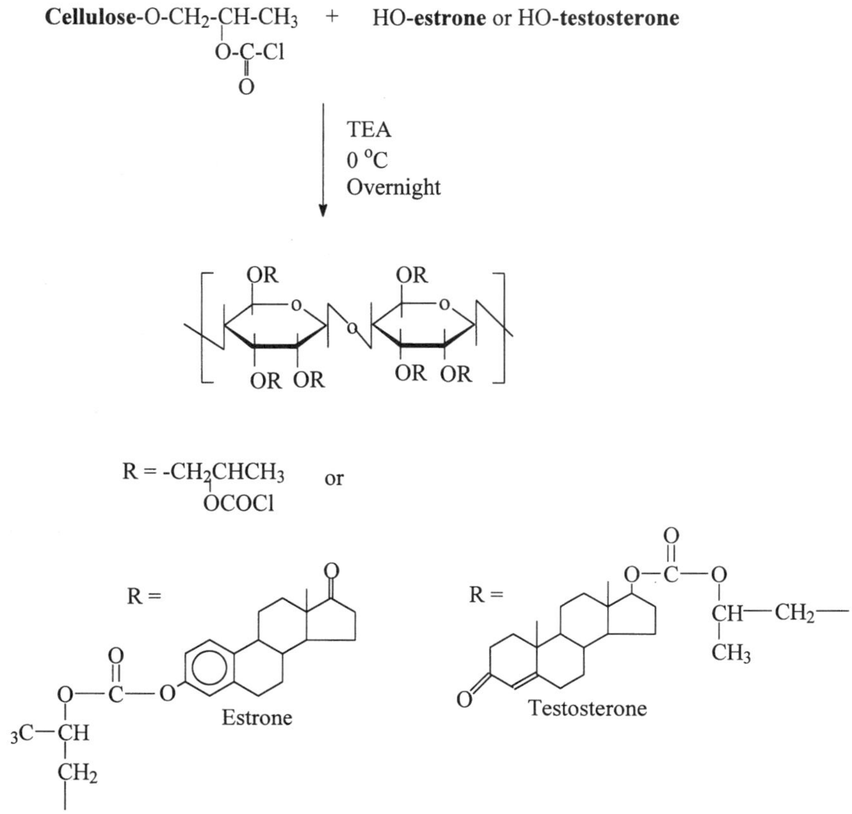

Scheme 9 Synthesis of estrone or testosterone conjugated cellulose. (From Ref. 23.)

Despite a high reactivity of the steroid-^{21}C-chloroformates toward the hydroxypropyl cellulose, the condensation yields were low (20–40%). Some attempts to improve the process were unsuccessful. This poor efficiency of the condensation process may be attributed partly to the very low concentration of reacting functional groups resulting from the poor solubility of the steroid derivatives in THF. Khue et al. had also studied the release of glucocorticoid from its conjugated hydroxypropyl cellulose via the hydrolysis of the carbonate linkages between the hydroxypropyl cellulose and steroid. The glucocorticoid released depended on pH and temperature and was highly sensitive to steric hindrance. In vivo testing of the anti-inflammatory activity of the cellulose bond dexamethasone in rats unfortunately showed negative results probably due to the low release of dexamethasone in the inflammatory region of acidic pH [24].

V. CHITOSAN

Chitosan is poly[β-(1,4)-2-amino-2-deoxy-D-glucose] and is obtained by deacetylation of chitin. Chitin is one of the most abundant natural polymers and exists largely in the cuticle of the crustacea, such as shrimp and crabs. Due to their biocompatibility, biodegradability, and nontoxicity, chitosan has potential uses as immobilization supports, drug carriers, and biomate-

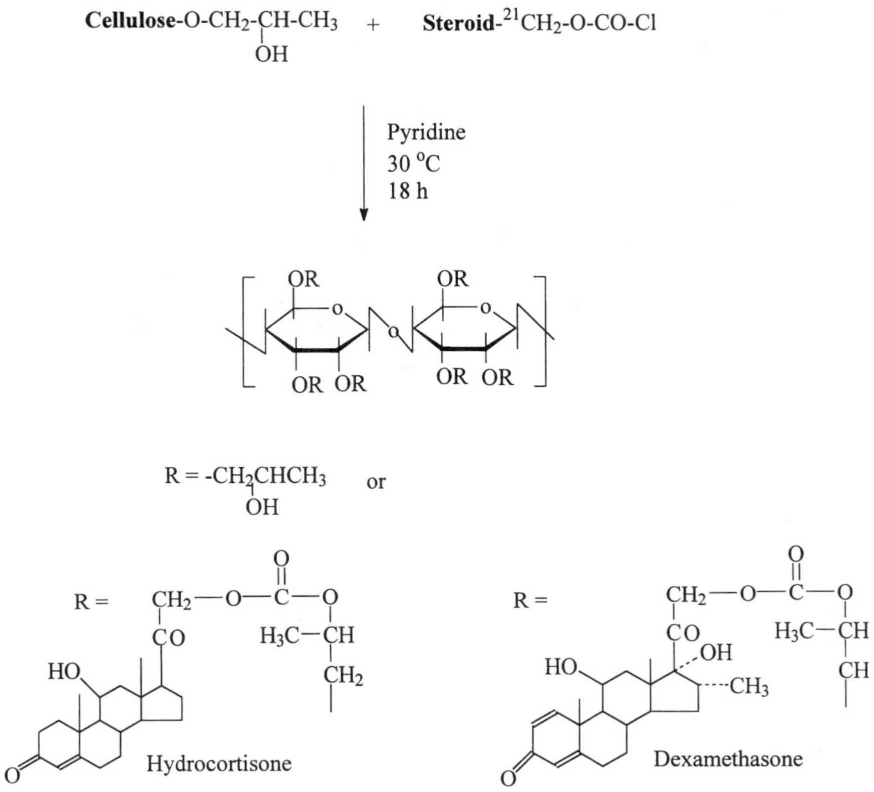

Scheme 10 Synthesis of cellulose-steroid conjugates via hydroxypropylcellulose. (From Ref. 17.)

rials. The chemical structure of chitosan is similar to that of cellulose, but chitosan has one amino group in addition to the hydroxyl groups in the repeating unit. This amino group could be easily accessed for modification with a variety of chemical reagents.

Ouchi et al. [25] prepared four types of chitosan-5-fluorouracil (5FU) conjugates having a variety of spacers via ether, amide, ester or carbonyl linkages. The contents of 5FU per glucosamine unit of the conjugates were relatively low (3.6–18%). This result might be due to their steric hindrances and low solubility of chitosan and chitosan derivatives. The release behaviors of 5FU and 5FU residues were studied in vitro at 37°C in various aqueous media. The antitumor activity was tested against p-388 *lymphocytic leukemia* in female CDF_1 mice. The chitosan-5FU conjugates did not display acute toxicity in the high dose range and mice treated with the conjugates lived longer than those treated by free 5FU.

Gallo et al. [26] reported the synthesis, characterization, and in vitro stability of chitosan-methotrexate conjugates (CMTX). 1-Ethyl-3-(3-dimethylaminopropyl) carbodiimide (ECDI) was used as the crosslinking agent in their study because it was considered suitable for bond formation between the carboxyl group of MTX and the amino group on chitosan. Drug content analysis showed that the percentage (w/w) MTX loading on the chitosan increased as the ratio of crosslinking agent, EDCI to MTX increased. The MTX-chitosan conjugates, which had drug loading ranging from 9% to 16%, had a maximum about 2.8% (w/w) of MTX released over a 24-h period in various testing media, an indication that the conjugate is stable. Gallo et al. also studied the lysosomal test medium in order to determine its effect on drug release from the

CMTX conjugates. The lack of significant differences in such a medium, compared to the others studied, suggested that the drug might be released by slow hydrolysis of the covalent bond rather than by enzymatic degradation.

Rao et al. [27] reported that acrylic and methacrylic acids could be graft copolymerized onto the chitosan backbone using a ceric ion initiation technique. Experimental parameters, such as monomer and initiator concentrations on grafting, were investigated. There was an increase in percent grafting and grafting efficiency up to a certain extent of increase in the monomer and initiator concentrations. These graft copolymers were subsequently utilized for the preparation of functionalized chitosan beads by a polymer dispersion technique. The prepared microspheres had a smooth surface and were found to be compact in nature. Particle size analysis showed a mean particle size of 27–36 μm. Sulfadiazine was incorporated in these microspheres and the release profiles were established in both simulated gastric and intestinal fluids.

VI. OTHER POLYSACCHARIDES

Nogusa et al. [28] reported the synthesis of the conjugates of doxorubicin (DXR) with carboxymethylpullulan (CMPul) through tetrapeptide spacers and their in vivo antitumor effect against Walker carcinosarcoma 256 (Walker 256) and Yoshida sarcoma in rats. CMPul-DXR conjugates in which DXR was attached through Gly-Dly-Phe-Gly and Gly-Phe-Gly-Gly spacers were found to be more potent than DXR after a single intravenous injection in rats bearing Walker 256 carcinosarcoma. These conjugates were also more effective than DXR in rats bearing Yoshida sarcoma. However, CMPul-DXR conjugate in which DXR was bound through Gly-Gly-Gly-Gly spacer was less effective against Walker 256-bearing rats than DXR.

Lin and Ayres [29] reported the utilization of calcium alginate beads as core carriers for delayed dissolution followed by burst release as a potential method of intestinal site specific drug delivery. 5-Aminosalicylic acid was spray-coated on dried calcium alginate beads and then coated with different percentages of enteric coating polymer and/or sustained-release polymer. Beads coated with more than 6% (w/w) methacrylic copolymer plasticized with dibutyl sebacate and triethylcitrate resisted release in 2-h acid fluid and showed immediate dissolution upon transferring to simulated intestinal fluid. With 6% (w/w) methacrylic copolymer on top 4% (w/w) ethylcellulose polymer, the major portion of drug did not release in 2 h of acid treatment or the next 3 h of simulated intestinal fluid treatment. This dosage form makes possible drug delivery to the lower intestinal tract with minimal early release, followed by sustained release in colon.

Kanke et al. [30] reported the use of curdlan, a natural β-1,3-glucan, as drug delivery device. They prepared curdlan tablets containing theophylline by compressing three different curdlan and theophylline mixtures (a physical mixture), spray-dried curdlan particles with theophylline powder, and spray-dried particles of curdlan/theophylline solution. The drug released from the tablets that were prepared from spray-dried particles of curdlan/theophylline was the lowest. The release rate was constant from 1 to 8 h, and 59% cumulative release was obtained at 8 h. Drug release from curdlan tablets was unaffected by pH or various ions and was diffusion-controlled.

Uryu et al. [31] prepared azidothymidine (AZT)-curdlan conjugate through biodegradable ester linkages shown in Scheme 11. The AZT-curdlan was subsequently sulfated with sulfur trioxide-pyridine complex to give sulfated AZT-curdlans (AZT-curdlan sulfates). AZT-curdlan sulfates exhibited anti-HIV activities in the EC_{50} range of 0.20 to 0.60 μg mL^{-1} in vitro, corre-

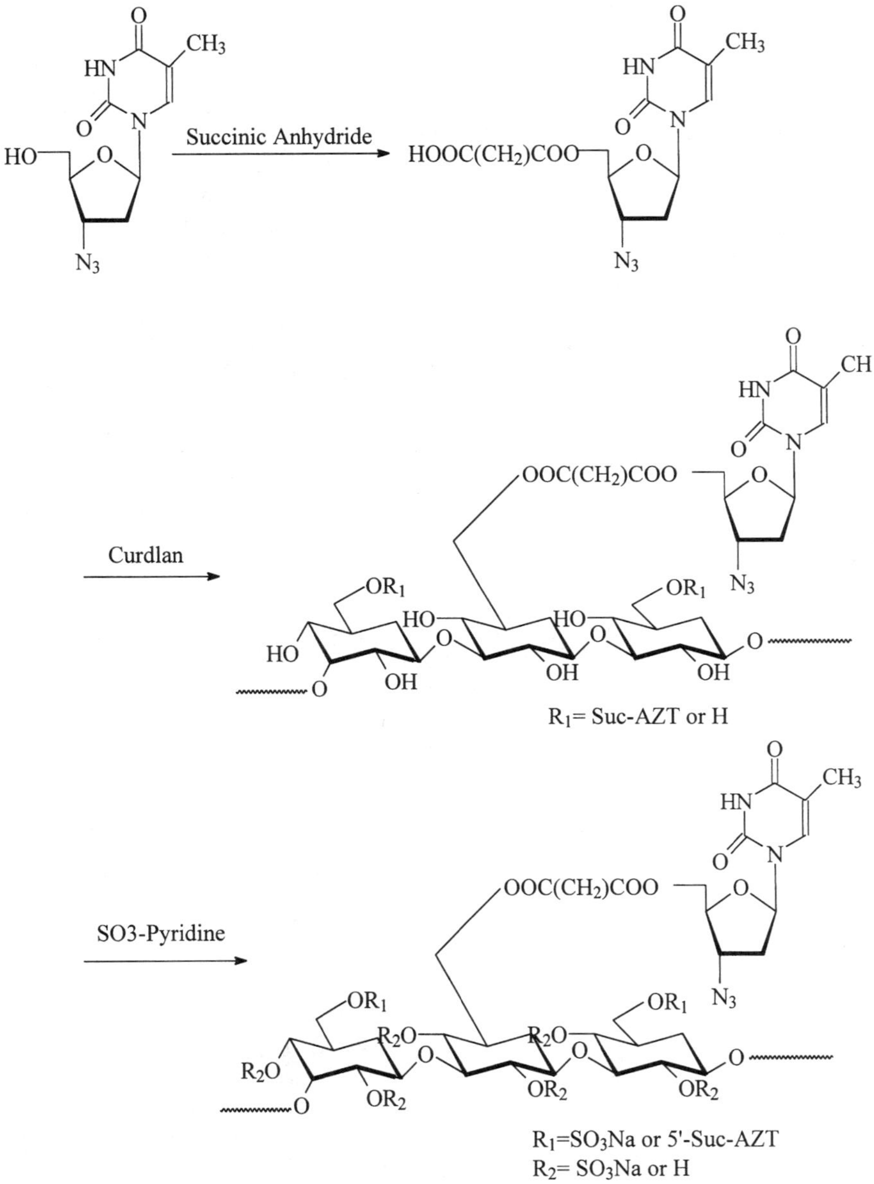

Scheme 11 Synthetic route of AZT-bound curdlan sulfate. (From Ref. 31.)

sponding to that ($EC_{50} = 0.50$ μg mL^{-1}) of highly active curdlan sulfate. The AZT-curdlan sulfate exhibited higher anti-HIV activity and cytotoxicity when the solution (in 90% buffer at pH 7.4 and 10% dimethyl sulfoxide mixture) was kept in a refrigerator for a few weeks. This finding indicated that anti-HIV active AZT was slowly released from the curdlan sulfate carrier. Furthermore, the AZT-curdlan sulfates exhibited low to moderate anticoagulant activities in vitro.

VII. CONCLUSION

This chapter describes the syntheses, characterization and biomedical applications of polysaccharide-based drug carriers. The major advantages of this class of natural biomaterials are abundant sources, low cost and variety in chemical, physical, and biological properties. The types of polysaccharides experimented for such a purpose included starch, dextran, cellulose, chitosan, calcium alginate and curdlan. In most cases, polysaccharides must be chemically modified for providing reactive sites for drug attachment. A wide variety of drugs had been incorporated into polysaccharides and included hormones/steroids (e.g., estrone, hydrocortisone, testosterone), antitumor agents (e.g., 5-fluorouracil, doxorubicin), anti-HIV agents (e.g., AZT), antiparasitic drugs (e.g., primaquine and trimethoprim), vasodilator agents (e.g., theophylline) and antibiotics (e.g., norfloxacin). This application is believed to have a great potential in the development of improved drug therapy for a variety of diseases at a reasonable cost. The slow release of bioactive agents from the polysaccharide carriers may provide an effective means to maintain drug concentrations within therapeutic ranges and to avoid undesirable side effects.

REFERENCES

1. Yalpani, M., 1988. *Polysaccharides in Studies in Organic Chemistry*, Elsevier, Amsterdam, pp. 36.
2. Chaves, M.S., Rodriguez, J.L., and Arranz, F., 1997. Synthesis and controlled release behaviour of dextran bioactive esters by direct reaction of dextran with the potassium salts of bioactive carboxylic acids. *Macromol. Chem. Phys.*, 198: 3465–3475.
3. Mocanu, G., Airinei, A., and Carpov, A., 1996. Macromolecular drug conjugates. V. Theophylline-dextran. *J. Control. Release.*, 40: 1–9.
4. Franz, G., Alban, S., and Kraus, J., 1995. Novel pharmaceutical applications of polysaccharides. *Macromol. Symp.*, 99: 187–200.
5. Goth, A., 1972. Gonadotropons and sex hormones. In: *Medical Pharmacology.*, C. V. Mosby, St. Louis, pp.485–489.
6. Bordor, N., 1981. Drug metabolism and drug design In: M. Briot, W. Cantreels, and R. Roncucci (Eds.), Center de recherches clinmidy, Montpellier, France, p. 217.
7. Panrin, E.F., Baikov, V.E., Timofeevsky, S.L., Nezhentsev, M.V., Vinogradov, O.V., and Kalinin, G.V., 1996. Polymer derivatives of glucocorticoid hormones. *Macromol. Symp.*, 103: 229–242.
8. Lescure, F., Gurny, R., and Doelker, E., 1994. *In vitro* and *in vivo* evaluation of progesterone implants based on new biodegradable poly(glutamic and glutamate esters) copolymers. *J. Biomater. Sci. Polym. Edn.*, 6: 797–807.
9. Won, C.Y., Chu, C.C., and Yu, T.J., 1997. Synthesis of starch-based drug carrier for the control/release of estrone hormone. *Carbohydrate Polym.*, 32: 239–244.
10. Moritani, T., and Fujiwara, Y., 1977. Solution studies of cellulose in lithium chloride. *Macromolecules*, 10: 532–538.
11. Laakso, T., Stjarnkvist, P., and Sjoholm, I., 1987. Biodegradable microspheres VI: Lysosomal release of covalently bound antiparasitic drugs from starch microparticles. *J. Pharm. Sci.*, 76: 134–140.
12. Lenaerts, V., Moussa, I., Dumoulin, Y., Mebsout, F. Chouinard, F., Szabo, P., Mateescu, M.A., Cartilier, L., and Marchessault, R., 1998. Crosslinked high amylose starch for controlled release of drugs: recent advances. *J. Control. Release*, 53: 225–234.
13. Weiner, B.Z., Tahan, M., and Zilkha, A., 1972. Polymers containing phenethylamines. *J. Med. Chem.*, 15: 410–413.
14. Herman, J., Remon, J. P., and De Vilder, J., 1989. Modified starches as hydrophilic matrices for controlled oral delivery. I. Production and characterization of thermally modified starches. *Int. J. Pharm.*, 56: 51–63.

15. Ohya, Y., Masunaga, T., Baba, T., and Ouchi, T., 1996. Synthesis and cytotoxic activity of dextran carrying *cis*-dichloro(cyclohexane-*trans*-diamine) platinum (II) complex. *J. Biomater. Sci. Polym. Ed.*, 7: 1085–1096.
16. Won, C.Y., and Chu, C.C., 1998, Dextran-estrone conjugate: Synthesis and *in vitro* release study. *Carbohyd. Polym.* 36(4): 327–334.
17. Khue, N.V., and Galin, J.C., 1985. Antiinflammatory polymer-bound steroids for topical applications. I. Synthesis and characterization. *J. Appl. Polym. Sci.*, 30: 2761–2778.
18. Negishi, N., Bennett, D.B., Cho, C.S., Jeong, S.Y., Van Heeswijk, W.A.R., Feijen, J., and Kim, S.W., 1987. Coupling of naltrexone to biodegradable poly(α-amino acids). *Pharm. Res.*, 4: 305–310.
19. Mocanu, G., Airinei, A., and Carpov, A., 1996. Macromolecular drug conjugates. V. Theophylline-dextran. *J. Control Release*, 40: 1–9.
20. Coessens, V., Schacht, E. H., and Domurado, D., 1997. Synthesis and in vitro stability of macromolecular prodrugs of norfloxacin. *J. Control Release*, 47: 283–291.
21. Hennink, W. E., Franssen, O., van Dijk-Wolthuis, W. N. E., and Talsma, H., 1997. Dextran hydrogels for the controlled release of proteins. *J. Control Release*, 48: 107–114.
22. Mestries, P., Borchiellini, C., Barbaud, C. et al., 1998. Chemically modified dextrans modulate expression of collagen phenotype by cultured smooth muscle cells in relation to the degree of carboxymethyl, benzylamide, and sulfation substitutions, *J. Biomed. Mater. Res.*, 42(2): 286–294.
23. Yolles, S., Morton, J.F., and Sartori, M.F., 1979. Preparation of steroid esters of hydroxypropyl cellulose. *J. Polym. Sci. Polym. Chem. Ed.*, 17: 4111–4113.
24. Khue, N.V., Jung, L., Galin, J.C., and Poindron, P., 1986. Antiinflammatory polymer-bound steroids for topical applications. II. Controlled release of the steroids. *J. Polym. Sci. Polym. Chem. Ed.*, 24: 359–373.
25. Ouchi, T., Banba, T., Fujimoto, M., and Hamamoto, S., 1989. Synthesis and antitumor activity of chitosan carrying 5-fluorouracils. *Makromol. Chem.*, 190: 1817–1825.
26. Sanzgiri, Y. D., Blanton, C. D., and Gallo, J. M., 1990. Synthesis, characterization, and *in vitro* stability of chitosan-methotrecate conjugates. *Pharm. Res.*, 7: 418–421.
27. Shantha, K., Bala, U., and Rao, K. P., 1995. Tailor-made chitosans for drug delivery. *Eur. Polym. J.*, 31: 377–382.
28. Nogusa, H., Yano, T., Kajiki, M., Gonsho, A., Hamana, H., and Okuno, S., 1997. Antitumor effects and toxicities of carboxymethylpullulan-peptide-doxorubicin conjugates. *Biol. Pharm. Bull.*, 20: 1061–1065.
29. Lin, S. Y. and Ayres, J. W., 1992. Calcium alginate beads as core carriers of 5-aminosalicylic acid. *Pharm. Res.*, 9: 1128–1131.
30. Kanke, M., Koda, K., Koda, Y., and Katayama, H., 1992. Application of curdlan to controlled drug delivery. I. The preparation and evaluation of theophylline-containing curdlan tablets. *Pharm. Res.*, 9: 414–418.
31. Gao, Y., Katsuraya, K., Kaneko, Y., Mimura, T., Nakashima, H., and Uryu, T., 1998. Synthesis of azidothymidine-bound curdlan sulfate with anti-human immunodeficiency virus activity *in vitro*. *Polym. J.*, 30: 31–36.

15

Bone Morphogenetic Proteins and Biomaterials as Carriers

Kazuhisa Bessho
Kyoto University, Kyoto, Japan

Joo L. Ong
University of Texas Health Science Center, San Antonio, Texas

Tadahiko Iizuka
Kyoto University, Kyoto, Japan

I. INTRODUCTION

Bone morphogenetic protein (BMP) is a cytokine which has been identified initially in the bone matrix. Messenger RNAs of some BMPs were later detected in early embryos and many tissues, suggesting that BMPs have important physiological activities other than bone induction [1]. Since 1965, when Urist first reported on BMP, many researchers have performed studies with the hope of finding clinical applications for this protein [2]. In experimental fracture sites, increased BMP activity was observed in and around the bone ends [3]. However, it is considered that the main role of BMPs in adults is bone formation [4].

Recently, BMPs have been used as a means to improve optimum bone-implant interactions. BMPs are known to induce bone tissue ectopically [5, 6]. These BMPs are present in the bone matrix and act on immature mesenchymal-type cells to initiate bone induction through endochondral bone differentiation [7, 8]. Other investigators have suggested osteoinduction to be accomplished by the diffusion of BMP from resorbing matrix and the broken bone ends, which then affect the initial proliferation and differentiation of mesenchymal cells into chondroblasts [9, 10].

Crude purified BMP (pBMP) does not always need a carrier for bone induction in vivo because some contaminated proteins coexisting with the BMP can serve as a natural carrier [11]. Excellent methods for the isolation and purification of BMP have been established [12, 13].

Unlike crude BMPs, highly pBMPs and recombinant human BMPs (rhBMP) are capable of dispersing themselves immediately after implantation and exert no effect on bone induction [14]. It has also been reported that when BMP was used alone, without any carrier, a large amount was required to achieve adequate bone induction [15]. Thus, an appropriate carrier acting as a slow delivery system for the pBMPs and rhBMPs is required. The use of guanidine-HCl as a useful carrier remains a problem because of its unknown macromolecules, which may cause undesirable immunological reactions [16]. In patients with periodontal disease, composites of allogenic demineralized bone matrix partially purified human BMP were observed to stimulated bone repair [17]. Enhanced bone regeneration was reported with implants containing bovine BMP and associated insoluble noncollagenous protein [18].

Numerous other delivery systems have been investigated, including the use of hydroxyapatite, biodegradable polymers, calcium phosphates, carbon and titanium [19]. However, an appropriate carrier is yet to be determined. The identification of a delivery system in the induction of bone formation and the evaluation of BMP activity using a pure carrier remains to be resolved prior to clinical use. Thus, it would be clinically useful to determine whether BMP, in a useful carrier system, will play any part in normal bone turnover or in the process of osseointegration. Due to these remaining problems, further studies have to be performed to correlate bone ingrowth to BMP stimulated implants. In this chapter, the bone inducing activities of BMPs and the different carriers currently used are discussed.

II. BMP

A. pBMP Derived from Bone Matrix

More than a decade ago, BMP derived from bone matrix could be purified using liquid chromatography [13]. Highly purified BMP was prepared using Bessho's BMP purification method [12]. The fresh cortical bone was crushed, defatted, demineralized, and treated with $CaCl_2$, EDTA, and LiCl. The specimen was then extracted chemically with 4 M guanidine-HCl (pH 5.2), and the extract was dissociated by ultrafiltration and ethanol partition. The dissociative fraction was dialyzed against distilled water. The water-insoluble fraction was fractionated by Sephacryl S-200 (Pharmacia, Sweden), hydroxyapatite (Pharmacia, Sweden), and TSK G2000SW (Tosoh, Japan) liquid chromatography. The molecular weight of the pBMPs derived from various animals thus obtained was 17.0–23.0 kDa as determined by SDS-PAGE and its pI value was 4.9 ± 0.3 [14]. This pBMP appeared as a single band. The rate of the pBMPs against cortical bone (starting material) were about 0.0003 wt% [14].

Implantation of pBMP with a carrier resulted in new bone formation not only on the outermost edge but also deep within the implant after three weeks. Immature mesenchymal-type cells were noted surrounding the new bone. The bone induced by pBMP had little bone marrow and rich bone matrix (Fig. 1A) [20]. There were only quantitative differences between these induced bones depending on the amount of pBMP implanted [21].

B. rhBMP Derived from Chinese Hamster Ovary Cell

Recently, BMP-1 through BMP-16 have been cloned by genetic engineering techniques [22], and rhBMP-2 has been particularly well studied. rhBMP-2 (Genetics Institute Inc., Cambridge, MA) was donated by Yamanouchi Pharmaceutical Co., Ltd. (Tokyo, Japan). The rhBMP-2 was synthesized in Chinese hamster ovary (CHO) cells [23]. The mature rhBMP-2 is a homodimer of two 114-residue subunits representing the C-terminal sequence of a long precursor protein of 396 amino acids. One of the four *N*-glycosylation sites present in the proprotein is retained and used in the mature rhBMP-2 polypeptide chains [24]. According to the amino acid sequence of the mature rhBMP-2 polypeptide chains, rhBMP-2 is a transforming growth factor-β (TGF-β) type protein (TGF-β superfamily).

The bone inducing activity and histological findings of the induced bone in the presence of rhBMP-2 with a carrier were compared with that of pBMP. The maturity level of rhBMP-2- and pBMP-induced bones at three weeks after implantation differed in histological findings.

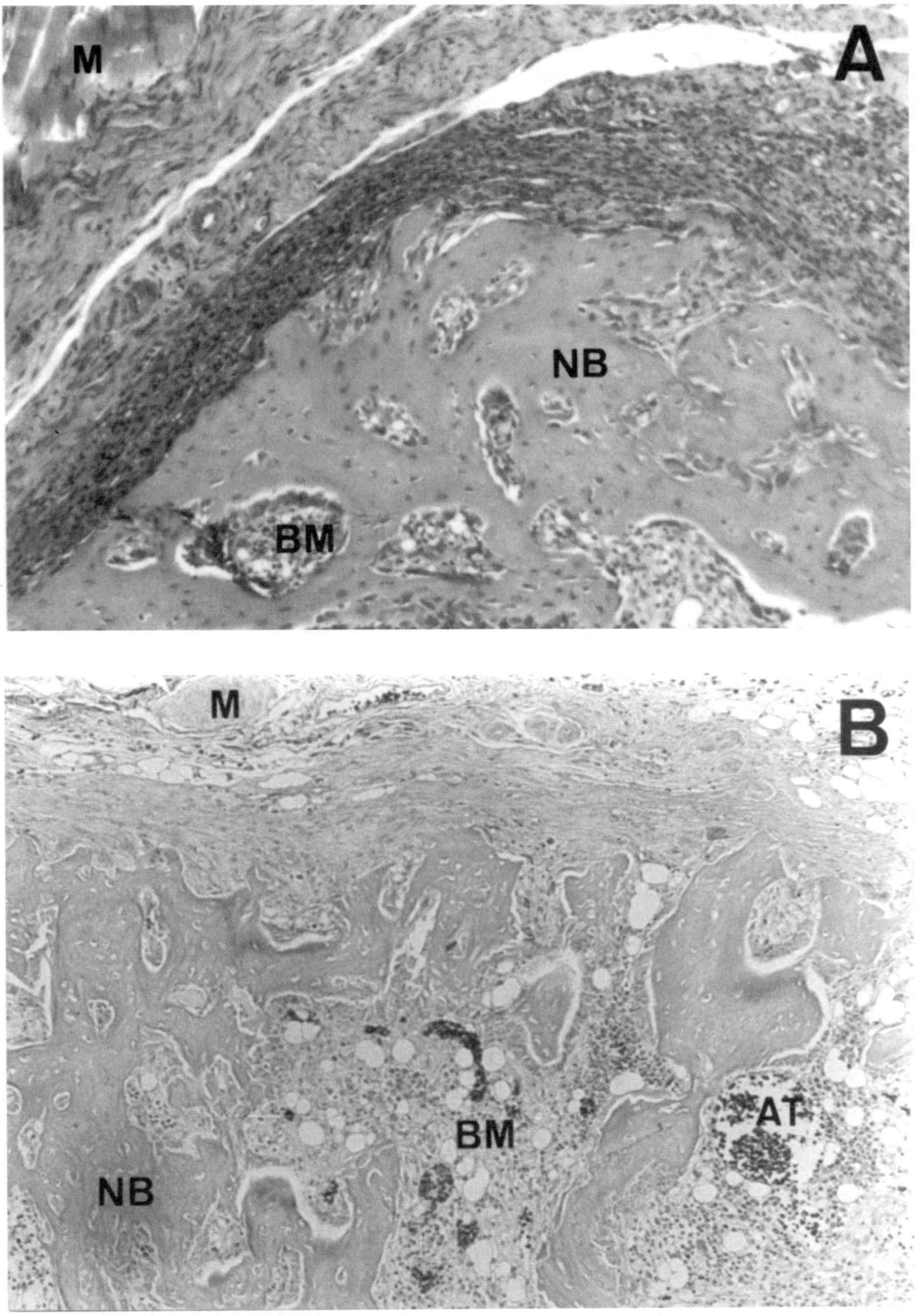

Figure 1 Photomicrographs of bones induced by 50 μg of pBMP (A), rhBMP-2 (B), or ErhBMP-2 (C) with collagen in the calf muscle pouch of Wistar rats at three weeks after implantation (hematoxylin-eosin staining; Original magnification × 100). MH: calf muscle of host, NB: newly formed bone, BM: bone marrow, AT: angioid tissue, FM: fatty marrow.

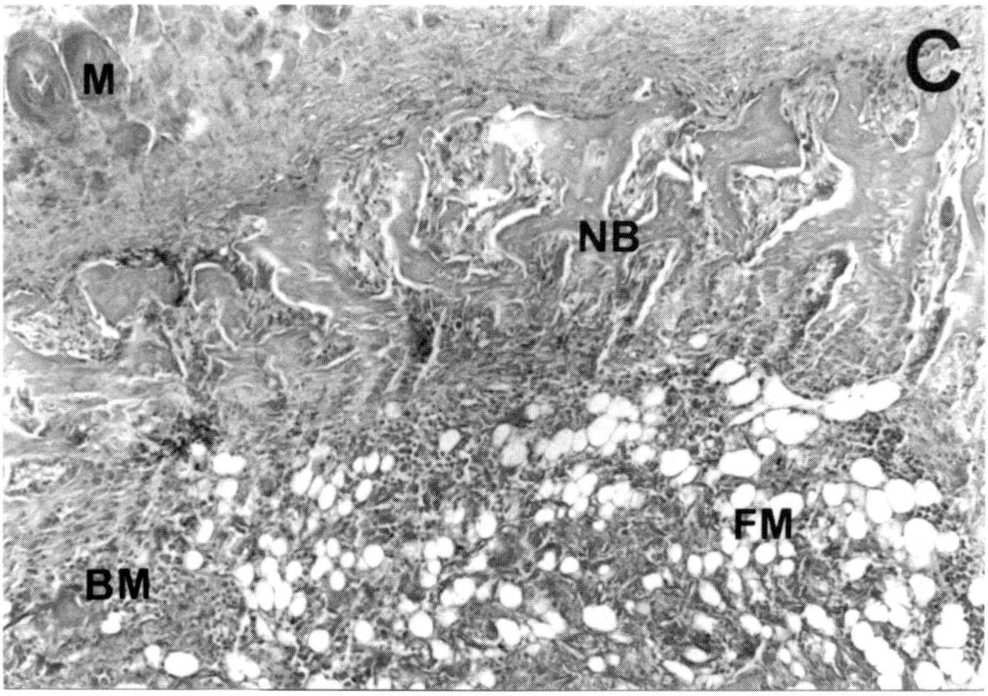

Figure 1 Continued.

Collectively, the bone induced by rhBMP-2 consisted of rich bone marrow containing angioid tissue and comparatively poor bone matrix (Fig. 1B) [20], while that induced by pBMP was that mentioned above. The bone-inducing activity of rhBMP-2 is lower than that of pBMP. By quantitative analysis of the results of bioassays performed under the same conditions in our previous study [21, 25], using ALP activity and Ca content as indices of bone formation, the bone-inducing activity of rhBMP-2 was below one-tenth that of pBMP. However, the clinical application of BMP would be more favorable if rhBMP could be easily synthesized and obtained. This is because pBMPs may have immunological problems, have prior risk of viral infections, and are limited in supply.

C. Variant rhBMP Derived from *Escherichia coli*

Recently, the rhBMP-2 derived from *Escherichia coli* (ErhBMP-2) could become recombinant. The variant rhBMP-2 was obtained from Professor Walter Sebald (Würzburg University, Germany). The ErhBMP-2 was generated by exchanging the sequence coding for the first 12 amino acids of BMP-2 by a sequence coding for the first 13 amino acids of human interleukin-2. The ErhBMP-2 variant with alternating *N*-terminal sequence was prepared with a purity greater than 98% and was lyophilized. A heparin-binding site was identified in the *N*-terminal segments of diametric rhBMP-2. Since the variant ErhBMP-2 was biological active, no obligatory receptor activation were reported on the heparin-binding sites [26].

The bone-inducing activity and histological findings of the induced bone in the presence of ErhBMP-2 with a carrier were compared with that of rhBMP-2. The histological findings of ErhBMP-2- and rhBMP-2-induced bones at three weeks after implantation were observed to be different. The bone induced by ErhBMP-2 consisted of much more expansive bone matrix toward the exterior and richer fatty marrow than that induced by rhBMP-2 (Fig. 1C). The bone-inducing activity of ErhBMP-2 is higher than that of rhBMP-2 [26]. Large quantity of ErhBMP-2 with high activity could be easily synthesized with low cost and thus could serve as a source for clinical application.

III. CARRIER

A. Human Fibrin

Human fibrin was not a slow enough delivery system for highly purified BMP and rhBMP. Highly purified BMP with fibrin could not induced new bone formation in our study. There is only one report in which crude pBMP with fibrin produced larger volumes of new bone than the crude pBMP alone. It was reported that the crude pBMP with fibrin induced approximately three times more bone matrix than the crude pBMP alone in the thigh muscle of a mouse [27].

B. Guanidine-HCl Extracted Residue of Demineralized Bone Matrix

In early studies of BMP, 4 M guanidine-HCl extracted residue of demineralized bone matrix (GER) was used as a carrier for pBMP and rhBMP [23, 28]. It was reported that rhBMP with GER induced bone and cartilage subcutaneously in rats. However, the composition of GER is not clear. Therefore, the possible presence of some additional factors, including harmful or bone-inducing factors, in GER could not be excluded. Furthermore, GER is limited in supply, and the effective use of GER as a delivery system for BMP is questionable because the quality of GER is unstable. Thus, GER is not a favorable carrier for use in the clinics.

C. Surface-Demineralized, Antigen-Extracted Autolyzed Allogeneic Bone

It has been reported that the use of surface-demineralized, antigen-extracted autolyzed allogeneic (AAA) bone with crude human pBMP induced bone formation, including bone marrow in an athymic rat [19]. A composite of human AAA bone and crude bovine pBMP was reported to regenerate 15 mm cranial bone defects in cynomolgus monkeys [29]. There is no report about AAA bone with rhBMP. Similar to the GER, the use of AAA bone as a carrier includes problems such as supply and its unstable effect. Therefore, AAA bone is difficult to use for a carrier of rhBMP in clinic.

D. Collagen

A purer collagen, atelopeptide type I collagen, was used as a carrier for pBMP and rhBMP in many studies [12, 30]. It was recognized that the collagen was a useful delivery system of highly purified BMP, rhBMP-2, and ErhBMP-2 from the findings of bone induction of pBMP, rhBMP-2, or ErhBMP-2 with the collagen (Fig. 1A–C). The collagen has little antigenicity because the telopeptide, which is a major factor of the collagen antigenicity, was removed (telopeptide-free), and does not inhibit bone induction. However, the use of collagen does present some problems with some unknown infection, such as prion, and strength. For these reasons, it is concluded that a synthetically stronger carrier or scaffold for BMP is required before the spread use of BMP in clinics.

E. Synthetic Hydroxyapatite

Through the purification of BMP, pBMP was found to be compatible with synthetic hydroxyapatite (HAP) [14]. The use of HAP as a carrier for BMP was considered. Porous HAP (pHAP), with porosity of 65% and pore sizes from 100 to 300 μm in diameter, was used in our study. Due to its rapid dissolution, pHAP was found unsuitable as a delivery system for highly purified BMP and rhBMP. The use of highly purified BMP and rhBMP with pHAP did not induce new bone formation in the pore of pHAP. However, the use of rhBMP with pHAP and collagen induced bone formation in rat muscle (Fig. 2) [31]. In this combination of rhBMP, pHAP, and collagen, the collagen acts as the carrier of rhBMP, and the pHAP acts as a scaffold (framing material).

F. Pure Titanium

Pure titanium (Ti) was used as a carrier for crude pBMP in some studies [32, 33]. It was recognized that Ti neither inhibited nor promoted pBMP activity. In addition, it was reported that Ti could not act as a slow delivery system for highly purified BMP and rhBMP. However, Ti strongly acted as a scaffold material. Highly purified BMP with collagen induced bone formation on the surface of Ti dental implant in dog mandible (Fig. 3). It was observed in our study that with pBMP-collagen stimulation, the new bone formation was in direct contact with the Ti implant interface after three weeks implantation. Histomorphometric evaluation of the

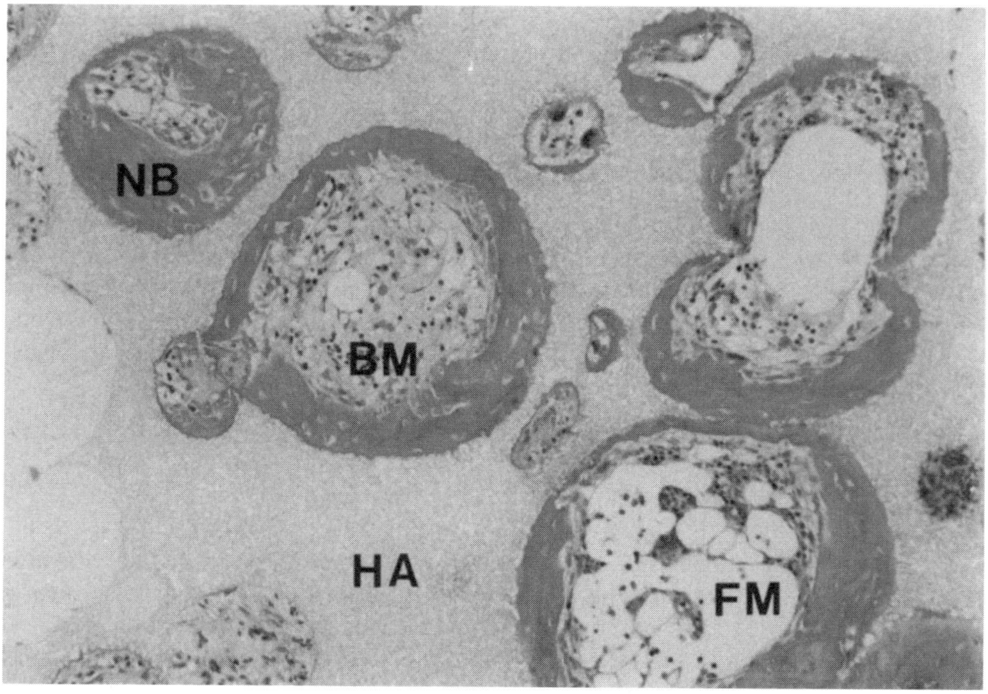

Figure 2 Photomicrographs of bones induced by 2 μg of rhBMP-2 with collagen and HAP in the calf muscle pouch of Wistar rats at three weeks after implantation (hematoxylin-eosin staining; original magnification × 100). MH: calf muscle of host, NB: newly formed bone, BM: bone marrow, FM: fatty marrow, HA: hydroxyapatite.

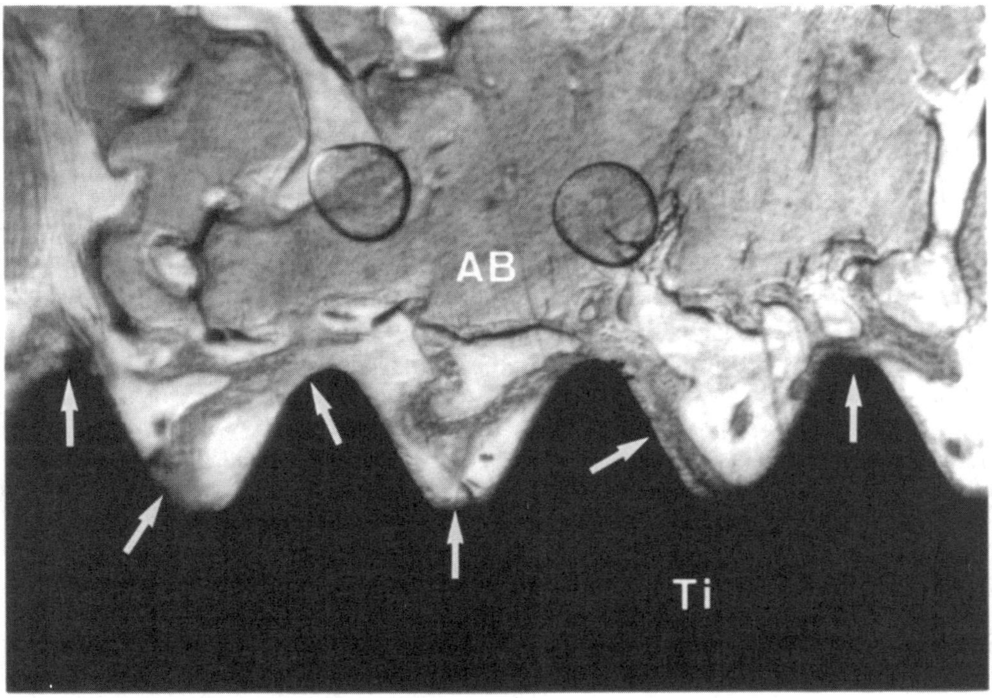

Figure 3 Photomicrographs of bones induced by 10 μg of rhBMP-2 with collagen on the surface of Ti dental implant in dog mandible at three weeks after implantation (Masson-Goldner staining; original magnification × 80). Ti: Ti dental implant, AB: alveolar bone of host, white arrow: newly formed bone.

bone-implant interface indicated a statistically higher bone contact length (81.6 ± 0.2%) for implants with BMP stimulation compared to implants without BMP stimulation (16.5 ± 2.0%) at three weeks implantation. In addition, reverse torque strength of the implants indicated a significantly higher bone-implant interfacial strength for implants stimulated with BMP (74.2 ± 5.2 N-cm) as compared to implants without BMP stimulation (32.8 ± 1.1 N-cm) at three weeks implantation.

G. Biodegradable Polymer

Various kinds of biodegradable polymers, for example, polylactic acid homopolymer (PLA) [19, 34], polylactic-polyglycolic acid coplymer (PLGA) [35, 36]. PLGA-gelatin sponge complex (PGS) [37], and polylactic acid-polyethylene glycol copolymer (PLA-PEG) [38], were used as carriers for BMPs. These biodegradable polymers have been considered for use as a delivery system as well as a scaffold for BMP. The PLA with high molecular weight is known to be the strongest among the biodegradable polymers described. However, despite being strongest the high molecular weight PLA was not strong enough for use as a scaffold in the repair of large bone defects such as jaw reconstruction. In addition, high strength biodegradable polymers are not appropriate for use as delivery systems for BMP due to their slow degradation rate, thereby resulting in very slow delivery of BMP. Hence, these polymers combined with BMP could not induce new bone formation. However, weak biodegradable polymers are suitable delivery systems for BMP. However, the occurrence of late noninfectious tissue response has been observed in cases using weak biodegradable polymers [39, 40].

IV. CONCLUSION

From the above discussion, it has been reported that pBMP possess unknown infection problems such as prion. In addition, pBMP are limited in supply. Since BMP-1 through BMP-16 can be cloned by genetic engineering techniques, rhBMP is more easily synthesized and obtained compared to pBMP. Similarly, the use of ErhBMP is more favorable because a large quantity of BMP with high activity could be easily synthesized with low cost.

Atelopeptide type I collagen was a useful delivery system of BMP. However, the use of collagen does present some problems with some unknown infection such as prion, and strength. For these reasons, it is concluded that a synthetically stronger carrier for BMP is required. The pHAP and Ti are strong for use as scaffolds, but because of their slow delivery rate they are not suitable for use as delivery systems for BMP. Weak biodegradable polymers are suitable for use as delivery systems for BMP, but are not strong enough for use as scaffolds. Therefore more than two kinds of synthetic biomaterials may be required for use with BMP in clinics. The two kinds of synthetic biomaterials must possess an optimum degradation rate for use as a delivery system and have sufficient strength for use as a scaffold.

ACKNOWLEDGMENTS

The authors thank Yamanouchi Pharmaceutical Co. Ltd. for providing rhBMP-2 and Nobel Bicare Co. Ltd., Japan, for providing the titanium implants. This work was supported in part by a Grant-in-Aid for Scientific Research (No. 09672044) of the Japanese Ministry of Education, Science and Culture, and by a grant from the Implant Dentistry Research and Education Foundation, a division of the International Congress of Oral Implantologists.

REFERENCES

1. Suzuki A, Nishimatsu S, Murakawa K, Ueno N. Differential expression of Xenopus BMPs in early embryos and tissues. *Zoo Sci* 10: 175–178, 1993.
2. Urist MR. Bone: Formation by autoinduction. *Science* 150: 893–899, 1965.
3. Hulth A, Johnell O, Henricson A. The implantation of demineralized fracture matrix yields more new bone formation than does intact matrix. *Clin Orthop* 235: 235–239, 1988.
4. Bessho K, Iizuka T. Clinical use of recombinant human bone morphogenetic protein-2 antibody in diagnosis of osteogenic lesions. In *Tissue Engineering Intelligence Unit. Bone Morphogenetic Proteins: Biology, Biochemistry and Reconstructive Surgery*. Lindholm TS (ed.), RG Landes, Austin, 1996, pp. 235–240.
5. Gerhart TN, Kirker-head CA, Vet MA, Kriz MJ, Holtrop ME, Henning GE, Hipp J, Schelling SH, Wang E. *J. Clin Orthop* 293: 317–326, 1993.
6. Yasko AW, Lane JM, Fellinger EJ, Rosen V, Wozney JM, Wang EA. The healing of segmental bone defects, induced by recombinant human bone morphogenetic protein (rhBMP-2). A radiographic, histological, and biomechanical study in rats. *J Bone Joint Surg* 74A: 659–670, 1992.
7. Urist MR, Leitze A, Mizutani H, Takagi K, Triffitt JT, Amstutz J, DeLange R, Termine J, Finerman GA. A bovine low molecular weight bone morphogenetic protein (BMP) fraction. *Clin Orthop* 162: 219–232, 1982.
8. Conover MA, Urist MR. Transmembrane bone morphogenesis by implants of dentin matrix. *J. Dent Res* 58: 1911, 1979.
9. Urist MR. Bone morphogenetic protein, bone regeneration, heterotopic ossification and the bone-bone marrow consortium. In *Bone and Mineral Research*, Volume 6. Peck WA (ed.), Elsevier, New York, 1989, pp. 57–112.

10. Simmons DJ. Fracture healing perspectives. *Clin Orthop* 200: 100–113, 1985.
11. Bessho K, Iizuka T. Activity and solubility of bone morphogenetic protein derived from porcine bone matrix. *Br J Oral Maxillofac Surg* 32: 86–90, 1994.
12. Bessho K, Tagawa T, Murata M. Purification of bone morphogenetic protein derived from bovine bone matrix. *Biochem Biophys Res Commun* 165: 595–601, 1989.
13. Urist MR, Huo YK, Brownell AG, Hohl WM, Buyske J, Lietze A, Tempst P, Hunkapiller M, DeLange RJ. Purification of bovine bone morphogenetic protein by hydroxyapatite chromatography. *Proc Natl Acad Sci USA* 81: 371–375, 1984.
14. Bessho K: Purification and characterization of bone morphogenetic protein. *Mie Med J* 40: 61–71, 1990.
15. Nilsson OS, Urist MR, Dawson EG, Schmalzried TP, Finerman GAM. Bone repair induced by bone morphogenetic protein in ulnar defects in dogs. *J Bone Joint Surg* 68B: 635–642, 1986.
16. Kuboki Y, Yamaguchi H, Yokoyama A, Murata M, Takita H, Tazaki M, Mizuno M, Hasegawa T, Iida S, Shigenobu K, Fujisawa R, Kawamura M, Atsuta T, Matsumoto A, Kata H, Zhou H-Y, Ono I, Taleshita N, Nagai N. Osteogenesis induced by BMP-coated biomaterials: Biochemical principles of bone reconstruction in dentistry. In *The Bone-Biomaterial Interface*. Davies JE (ed.), University of Toronto Press, Toronto, 1991, pp. 127–138.
17. Bowers G, Felton F, Middleton C, Glynn D, Sharp S, Mellonig J, Corio R, Emerson J, Park S, Suzuki J, Ma A, Romberg E, Reddi AH. Histologic comparison of regeneration in human intrabony defects when osteogenin is combined with demineralized freeze-dried bone allograft and with purified bovine collagen. *J Periodontal* 62: 690–702, 1991.
18. Nilsson O, Urist MR. Response of the rabbit metaphysis to implants of bovine bone morphogenetic protein (bBMP). *Clin Orthop* 1995: 275–281, 1985.
19. Urist MR. Experimental delivery systems for bone morphogenetic protein. In *Encyclopedic Handbook of Biomaterials and Bioengeering*. Part A: *Materials*. Volume 2. Wise DL, Trantolo DJ, Altobelli DE, Yaszemski MJ, Gresser JD, Schwartz ER (eds.), Marcel Dekker, New York, 1996, pp. 1093–1133.
20. Kusumoto K, Bessho K, Fujimura K, Konishi Y, Ogawa Y, Iizuka T. Comparative study of bone marrow induced by purified BMP and recombinant human BMP-2. *Biochem Biophys Res Commun* 215: 205–211, 1995.
21. Bessho K, Kusumoto K, Fujimura K, Konishi K, Ogawa Y, Tani Y, Iizuka T. Comparison of recombinant and purified human bone morphogenetic protein. *Br J Oral Maxillofac Surg 37*: 2–5, 1999.
22. Inada M, Katagiri T, Akiyama S, Namika M, Komaki M, Yamaguchi A, Kamoi K, Rosen V, Suda T. Bone Morphogenetic Protein-12 and -13 inhibit terminal differentiation into osteoblasts. *Biochem Biophys Res Commun* 222: 317–322, 1996.
23. Wang EA, Rosen V, D'Alessandro JS, Bauduy M, Cordes P, Harada T, Israel DI, Hewick RM, Kerns KM, LaPan P, Luxenberg DP, McQuaid D, Moutsatsos IK, Nove J, Wozney JM. Recombinant human bone morphogenetic protein induces bone formation. *Proc Natl Acad Sci USA* 87: 2220–2224, 1990.
24. Celeste AJ, Iannazzi JA, Taylor RC, Hewick RM, Rosen V, Wang EA, Wozney JM. Identification of transforming growth factor β family members present in bone-inductive protein purified from bovine bone. *Proc Natl Acad Sci USA* 87: 9843–9847, 1990.
25. Bessho K. Ectopic osteoinductive difference between purified bovine and recombinant human bone morphogenetic protein. In *Tissue Engineering Intelligence Unit. Bone Morphogenetic Proteins: Biology, Biochemistry and Reconstructive Surgery*. Lindholm TS (ed.), RG Landes, Austin, 1996, pp. 105–111.
26. Ruppert R, Hoffmann E, Sebald W. Human bone morphogenetic protein 2 contains a heparin-binding site which modifies its biological activity. *Eur J Biochem* 237: 295–302, 1996.
27. Kawamura M, Urist MR. Human fibrin is a physiologic derivery system for bone morphogenetic protein. *Clin Orthop* 235: 302–310, 1988.
28. Wozney JM, Rosen V, Celeste AJ, Mitsock LM, Whitters MJ, Kriz RW, Hewick RM, Wang EA. Novel regulators of bone formation: Molecular clones and activities. *Science* 242: 1528–1534, 1988.

29. Hollinger JO, Schmitz JP, Mark DE, Seyfer AE. Osseous wound healing with xenogeneic bone implants with a biodegradable carrier. *Surgery* 107: 50–54, 1990.
30. Fujimura K, Bessho K, Kusumoto K, Ogawa Y, Iizuka T. Experimental studies on bone inducing activity of composites of atelopeptide type I collagen as a carrier for ectopic osteoinducing by rhBMP-2. *Biochem Biophys Res Commun* 208: 316–322, 1995.
31. Kusumoto K, Bessho K, Fujimura K, Konishi Y, Ogawa Y, Iizuka T. Self-regenerating bone implant: ectopic osteoinduction following intramuscular implantation of a combination of rhBMP-2, atelopeptide Type I collagen and porous hydroxyapatite. *J Cranio-Maxillofac Surg* 24: 360–365, 1996.
32. Kawai T, Mieki A, Ohno Y, Umemura M, Kataoka H, Kurita S, Koie M, Jinde T, Hasegawa J, Urist MR. Osteoinductive activity of composites of bone morphogenetic protein and pure titanium. *Clin Orthop* 290: 296–305, 1993.
33. Herr G, Hartwig CH, Boll C, Kusswetter W. Ectopic bone formation by composites of BMP and metal implants in rats. *Acta Orthop Scand* 67: 606–610, 1996.
34. Miyamoto S, Takaoka K, Okada T, Yoshikawa H, Hashimoto J, Suzuki S, Ono K. Evaluation of polylactic acid homopolymers as carriers for bone morphogenetic protein. *Clin Orthop* 278: 274–285, 1992.
35. Kawamura M, Urist MR. Induction of callus formation by implants of bone morphogenetic protein and associated bone matrix noncollagenous proteins. *Clin Orthop* 236: 240–248, 1988.
36. Zellin G, Linde A. Importance of delivery systems for growth-stimulatory factors in combination with osteopromotive membranes. *J Biomed Mater Res* 35: 181–190, 1997.
37. Itoh T, Mochizuki M, Nishimura R, Matsunaga S, Kadosawa T, Kokubo S, Yokota S, Sasaki N. Repair of ulnar segmental defect by recombinant human bone morphogenetic protein-2 in dogs. *J Vet Med Sci 60*: 451–458, 1998.
38. Miyamoto S, Takaoka K, Okada T, Yoshikawa H, Hashimoto J, Suzuki S, Ono K. Polylactic acid-polyethylene glycol block copolymer. A new biodegradable synthetic carrier for bone morphogenetic protein. *Clin Orthop* 294: 333–343, 1993.
39. Böstman OM. Current concepts review, Absorbable implants for the fixation of fractures. *J Bone Joint Surg* 73-A: 148–153, 1991.
40. Bergsma EJ, Rozema FR, Bos RRM, Bruijn WC. Foreign body reactions to resorbable poly(l-lactide) bone plates and screws used for the fixation of unstable zygomatic fracture. *J Oral Maxillofac Surg* 51: 666–670, 1993.

16

Pyrolytic Carbon as an Implant Material for Small Joint Replacement

Stephen D. Cook, Laura Popich Patrón, Samantha L. Salkeld, and Jerome J. Klawitter
Tulane University School of Medicine, New Orleans, Louisiana

Robert D. Beckenbaugh
Mayo Clinic and Mayo Foundation, Rochester, Minnesota

Pyrolytic carbon was first used clinically in 1969 to make components of artificial heart valves. Over 2 million pyrolytic carbon/graphite heart valves have been implanted demonstrating its biocompatibility [1]. Pyrolytic carbon has an elastic modulus similar to cortical bone, making it an ideal material for implant-bone stress transfer. Animal models have shown pyrolytic carbon implants to be compatible with bone and capable of direct bone apposition [2–4]. Evaluation of articulating pyrolytic carbon metacarpophalangeal (MP) joints in primates revealed no evidence of wear and no inflammatory reaction, and excellent bone implant incorporation [2].

Long term clinical outcome of MP joint arthroplasties in humans has shown pyrolytic carbon implants to be biologically and biomechanically compatible [5]. The carbon implants improved range of motion as well as functional hand position, providing an acceptable option for MP joint arthroplasty. This chapter will discuss the potential applications and benefits of pyrolytic carbon as an implant material for replacement of small joints in the hand and foot.

I. PYROLYTIC CARBON

Carbon occurs naturally in many forms each having very different structures, material properties, and uses. Coal is an example of carbon in amorphous form with no crystalline structure. Graphite has an organized crystalline structure in which the carbon atoms are arranged in two-dimensional hexagonal sheets tightly joined by strong covalent bonds (Fig. 1A). Diamond is a third form of carbon having a three-dimensional cubic crystal structure with increased atomic bonds. Pyrolytic carbon is a man-made material formed by the pyrolysis or heating of a gaseous hydrocarbon. The resulting pyrolytic carbon has a turbostratic, or "wrinkled," two-dimensional crystal structure with a high concentration of three-dimensional diamond cross-link bonding (Fig. 1B). The small and highly dense crystalline structure is isotropic, meaning there is no preferred crystalline orientation. In general, the physical properties of pyrolytic carbon fall between those of graphite and diamond.

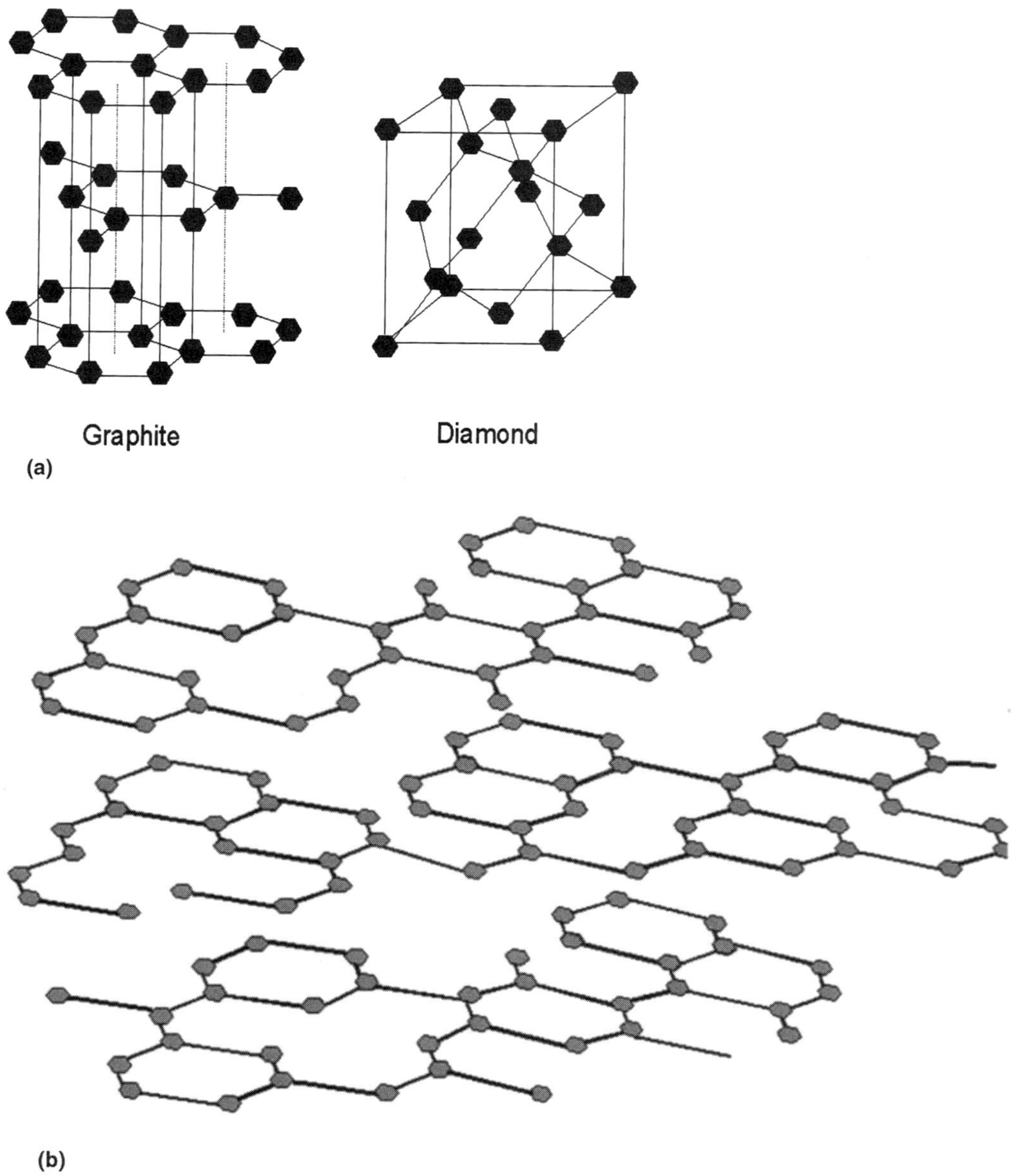

Figure 1 (A) Schematic of three-dimensional lattice crystalline structure of graphite and diamond. In graphite, carbon atoms are arranged in planes of ordered hexagonal arrays held together by covalent bonds. The layers of crystallites tend to be aligned in parallel with a regular sequence. Diamond has a three-dimensional cubic crystal structure with increased atomic bond strength.
(B) Schematic of a planar turbostratic crystalline structure similar to the structure of pyrolytic carbon. Some carbon atoms and covalent bonds are missing with several misaligned layers stacked in slight disarray. The imperfect stacking and random imperfections combined with a high concentration of cross link bonds between layers (not represented) contributes to the unique physical and mechanical properties of pyrolytic carbon.

Pyrolytic carbon was developed in the late 1950s as a material to coat uranium fuel particles used in high temperature, gas cooled, power generating, nuclear reactors. It is formed as a coating at high temperature by heating a hydrocarbon gas, such as propane or methane, to temperatures between 1000 and 1500°C. Pyrolytic carbons can exist in many different forms and have a range of structures and properties. A comprehensive review of the formation, structure, and material properties of pyrolytic carbon has been reported [6].

The form of pyrolytic carbon used as a surgical implant has a very fine grain structure (≈50Å) and isotropic physical and mechanical properties. Medical grade pyrolytic carbon is formed as a coating, usually 100 to 600μm thick, deposited onto a substrate such as graphite. The preformed or machined substrate, usually high strength graphite, is levitated in a fluidized bed reaction chamber (Fig. 2). The chemical vapor deposition process in which the carbon-hydrogen bonds of methane or propane gas are broken takes place under temperatures approaching 1400°C, which is low compared to the melting point of carbon. Hence, pyrolytic carbons are also referred to as low-temperature isotropic (LTI) carbons. Variations in the material properties of LTI carbons can be achieved by increasing the density, reducing the crystalline size, lowering the deposition temperature, and adding alloys such as silicon [7–8].

II. PROPERTIES OF PYROLYTIC CARBON

Beyond the various demanding industrial applications, there are many properties which make pyrolytic carbon an attractive material for surgical implant devices. Pyrolytic carbon is several

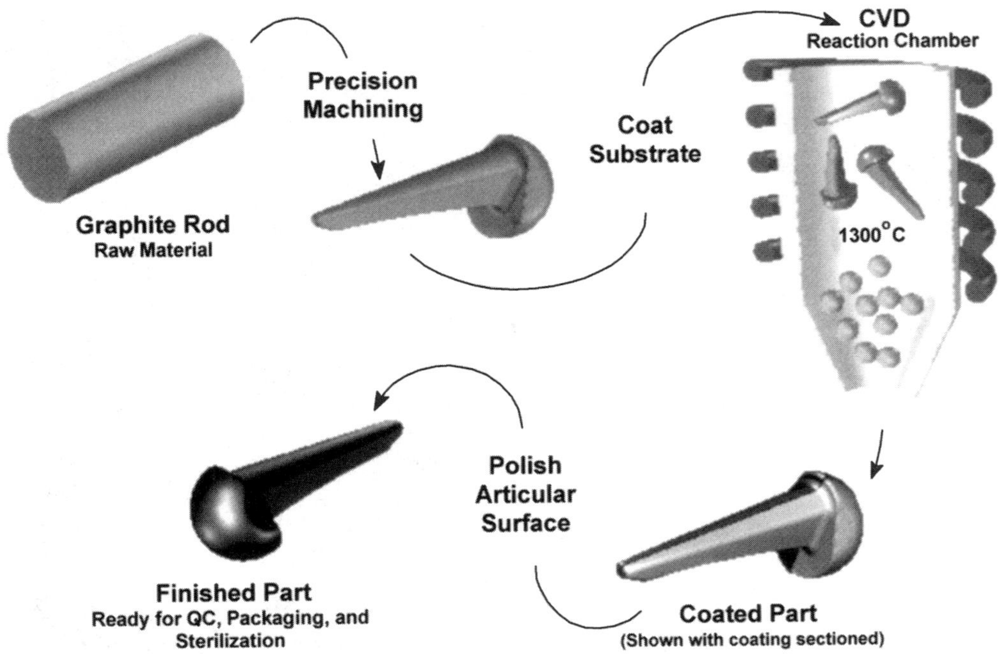

Figure 2 The manufacturing process for pyrolytic carbon implants starts with a substrate of high strength graphite which is machined to shape and suspended in a fluidized bed. Carbon is deposited by chemical vapor deposition (CVD) in which a hydrocarbon gas such as methane or propane is heated to temperatures over 1300°C. The implants are coated to final form with only polishing of intended articulating surfaces necessary. (QC = quality control)

times stronger than graphite, chemically inert, and resistant to wear and mechanical fatigue. It also has a high fracture strength and a low elastic modulus (21 to 26 GPa), which falls within the range of moduli reported for cortical bone [9–11].

The biocompatibility of LTI carbons is well documented and has been confirmed clinically in its extensive use in cardiovascular implants over the past 30 years [1,8,12]. Pyrolytic carbon has also proven to be biocompatible in soft tissues as well as in osseous and periarticular tissues. Microparticles of LTI carbon and carbon reinforced with carbon fibers were injected into the periarticular tissue and implanted in the knee joints of rats for up to six months without any local connective tissue reaction or systemic organ involvement [13]. Klasson and Scheman implanted graphite cloth and vitreous carbon in dogs with no evidence of inflammatory or foreign body reaction [14].

Histological evidence of LTI carbon's compatibility with bone has been demonstrated with endosseous blade type implants in the mandible of baboons [15–18] and with endosteal implants in dogs [19]. At two to three years postoperative, the carbon implants were shown to be capable of direct apposition with bone with no signs of resorption or inflammatory response as typically seen with metal alloy implants.

Durability and mechanical strength are well-known and attractive properties of pyrolytic carbon [7–8,12,20]. Due to its turbostratic nature, this material will not fail in fatigue when cyclic loads are applied at levels slightly below its fracture stress [7]. Pyrolytic carbon does

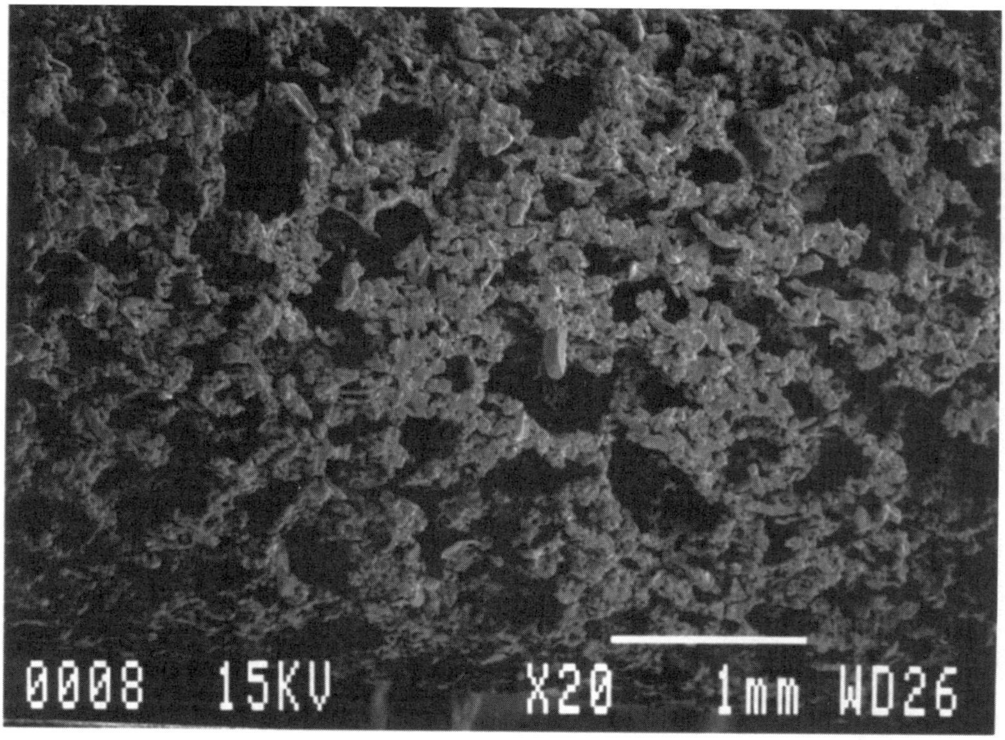

(a)

Figure 3 Scanning electron micrographs of the as-deposited surface of a pyrolytic carbon coated implant. Note the superficial porosity and roughness. (a) scale bar = 1 mm, 20× magnification, (b) scale bar = 100 μm, 200 × magnification.

not contain structural defects which can provide a mechanism for the initiation and propagation of cracks leading to mechanical failure as seen in other prosthetic materials, such as metal alloys, silicone elastomers, and polymers [8,12,21]. Progressive failure is not observed if the material is stressed below the single cycle fracture stress or endurance limit. Ely and Haubold also demonstrated that pyrolytic carbon is not subject to static fatigue or stress corrosion at loads within the tolerance of in situ conditions [22].

The wear resistance of pyrolytic carbon has been shown to be excellent [1,23–24]. The low frictional qualities of carbon on carbon and carbon in combination with metals are attributable to the toughness and hardness of the structural material. These properties are believed to contribute to pyrolytic carbon's ability to reduce and withstand high abrasive forces. The LTI carbons codeposited with silicon, which are widely used in heart valves, are considered the most wear resistant and hence are attractive for use in small joint applications [8].

The similarity of the elastic modulus of pyrolytic carbon to that of bone is advantageous for implants to function in contact with bone and distribute in vivo loads and stresses. Traditional prosthetic materials such as titanium, cobalt-chromium, and aluminum oxide have elastic moduli approximately one order of magnitude higher than that of bone and LTI carbons [25]. In contrast, silicone elastomers have an extremely low modulus of elasticity compared to LTI carbons. A mismatch in elastic properties would contribute to higher stresses being transferred to the implant-bone interface and eventually bone loss due to stress shielding, bone resorption, and possibly implant loosening [26].

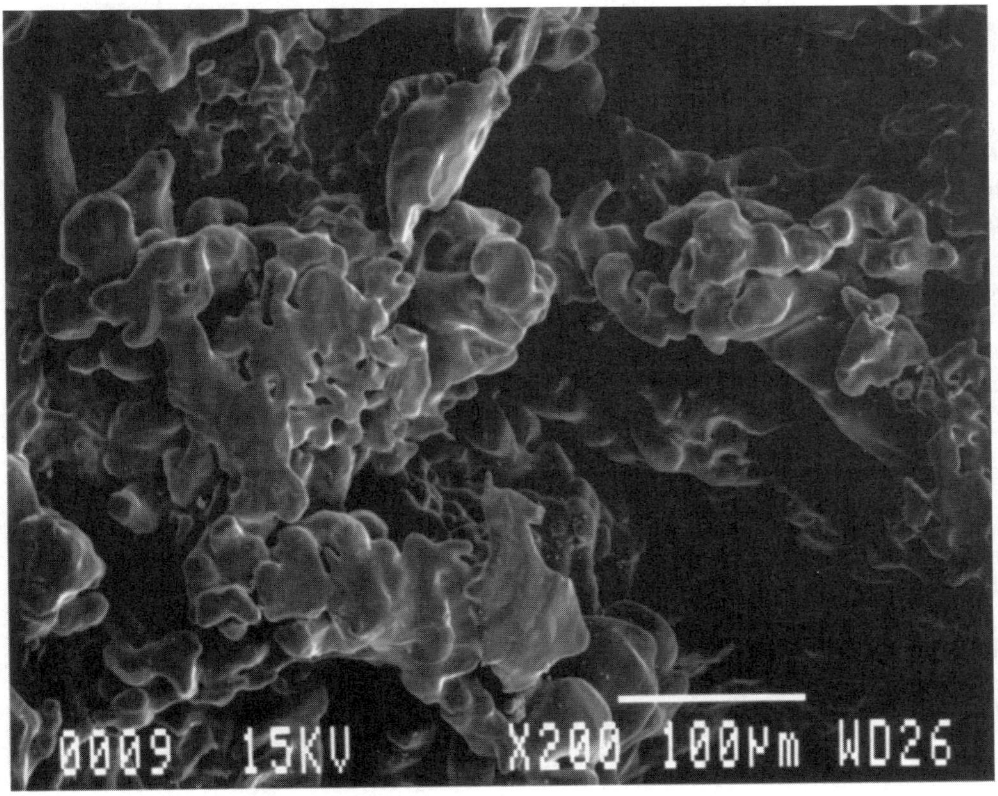

(b)

Figure 3 Continued.

The as-deposited surface of pyrolytic carbon is microporous and rough (Fig. 3). The pores are approximately 10 microns in size with a depth of about 20 microns [8]. The ability of new bone to form in apposition to this microporous surface has been demonstrated in a number of studies [4,27–31]. This allows for the biological fixation of bone directly to the surface of a carbon implant without the need for additional fixation with bone cement. The mechanical strength of the bone-implant interface to pyrolytic carbon may be enhanced with surface modifications or the presence of a hydroxylapatite coating [4,29–31].

The articulating surface of a pyrolytic carbon implant can also be polished to an extremely smooth finish [32]. This design characteristic is attractive for joint replacement and hemiarthroplasty considerations in which the minimization of wear and wear debris is desired. A highly polished LTI pyrolytic femoral prosthesis in articulation with cartilage in the canine acetabula exhibited significantly lower levels of gross wear, fibrillation, eburnation, glycosaminoglycan loss, and subchondral bone change than with metallic implant surfaces [33]. The increased potential for cartilage survival articulating with a pyrolytic carbon surface was attributed to the combination of a well-matched elastic modulus and improved biocompatibility over a metal surface.

III. CLINICAL APPLICATIONS

Pyrolytic carbon has been evaluated in a variety of forms as cardiovascular implants, dental implants, implant coatings as well as in soft tissue and osseous tissue applications.

A. Cardiovascular

Pyrolytic carbon is the material of choice for the construction of mechanical artificial heart valves (Fig. 4). The first pyrolytic carbon artificial heart valve component was implanted in 1969. Since that time, over 2 million pyrolytic carbon components have been implanted, resulting in the accumulation of over 15 million patient years of experience [1]. The human heart beats on average 100,000 times per day and 35 million times per year. The carbon on carbon pivot systems in pyrolytic carbon heart valves must resist stress and wear each time the heart beats. The history of successful function under the demanding mechanical conditions as a heart valve component demonstrates the outstanding wear resistance of the carbon on carbon articulation, biocompatibility, and structural durability of the pyrolytic carbon material.

B. Soft Tissue Evaluation

Carbon implant devices and coatings have been investigated for tendon and ligament repair and transcutaneous portals [8,34–35]. Pyrolytic carbon as a coating on carbon fibers was shown clinically and microscopically to reduce the sensitivity to shearing forces of a ligament replacement in the knee joint of rabbits [36]. Cook et al., in 1986, evaluated ultra low-temperature isotropic carbon-coated polyester suture histologically and mechanically in the repair of surgically incised medial collateral ligaments in dogs [37]. The response to the carbon coated sutures was better or equal to the uncoated sutures and showed a greater degree of tissue growth at most time periods.

Pyrolytic carbon is also being investigated as a possible inexpensive alternative to steroid-eluting pacing leads and as a protective coating around implanted silicone mammary expanders [38–39].

Figure 4 Pyrolytic carbon mechanical heart valve of the St. Jude design. A carbon to carbon articulation is present which shows no wear under physiologic loading of nearly 40 million cycles per year.

C. Dental and Endosseous Implants

The biocompatibility, high strength, and elastic modulus of pyrolytic carbon make it an attractive material for dental implants. Endosseous blade-type implants made of LTI pyrolytic carbon have been investigated for structural load-bearing applications in dentistry since the early 1970s primarily in primates [17–18,40–43]. The success of these implants mechanically and histologically in the mandible indicates the suitability of pyrolytic carbon as a material in joint replacement.

D. Orthopedic: Hemiarthroplasty

Pyrolytic carbon has shown promise in applications where an implant material must articulate with cartilage and interface with biological tissues. Hemijoint arthroplasty applications have included replacement in the hip joint, knee joint, great toe, and thumb [25–26,33,44]. The wear behavior and bone remodeling around LTI carbon, cobalt-chromium-molybdenum, and porous titanium femoral head prostheses were investigated in adult mongrel dogs [33,44]. Wear debris and inflammatory response were minimal in response to both the carbon and porous titanium devices [44]. However, radiographic and histological differences were noted. Trabecular hypertropy in the calcar area and dense cancellous bone conforming to the cap surface were noted with the LTI carbon implants without a loss of bone. Mineralized bone was present in pores of the load-bearing region of the titanium implants with a loss of bone and trabecular thinning in the calcar region and trabecular thickening at the distal stem.

Gross evaluation of the articulating cartilage in contact with different material surfaces revealed improved wear characteristics of LTI carbon compared to highly polished metal surfaces [33]. Progressive degeneration did occur over time with severe changes in the acetabula cartilage structure observed in the metallic devices. The survival rate of cartilage articulating with an LTI carbon surface was 92% at 4.5 to 18 months compared to titanium with a survival of 50% at 12 to 18 months and Co-Cr-Mo with a survival of 21% at 12 months [33].

Tian et al. reported the clinical results of a series of pyrolytic carbon hip joint replacements using a femoral head cup design and a pyrolytic carbon endoball-titanium stem prosthesis in China [25]. The early functional and clinical results were good, with one case of cup arthroplasty fracture occurring eight years postimplantation and one titanium stem implant fracture noted at five years postoperative. This author suggests the benefit of a pyrolytic carbon implant surface articulating with cartilage combined with an alternative porous material for bone ingrowth and fixation where micromotion may be a problem [25].

More recently the cartilage and bone response of a surgically created defect to pyrolytic carbon and cobalt-chromium alloy hemiarthroplasties was investigated in a canine knee joint model [45]. Full thickness cartilage defects created in the tibial plateau were in articulation with either a carbon or metal lateral femoral condyle implant for a period of one year. Fibrocartilage regeneration was seen in 86% of the carbon implants compared to 25% in the metal implants. Cracks on the surface of the tibial defect were observed in 100% of the metal and 14% of the carbon implants. These host responses were likely related to the differences in material and mechanical properties between pyrolytic carbon and cobalt-chromium. Cartilage controls showed increased cartilage regeneration and less subchondral bone thickening than the implanted knees. There was no evidence of residual adverse inflammatory response to either implant.

The short term results of hemijoint and total joint arthroplasties using a pyrolytic carbon implant in the basal joint of the thumb and great toe are reported by Kampner and Weinstein [26]. Four total joint thumb replacements and six basal joint hemiarthroplasties resulted in improved functional outcomes despite technical complications with fixation and loosening. Eighteen total joint toe prostheses showed a radiolucent zone surrounding the implant stems but no radiographic evidence of loosening or wear. Marked improvement from the preoperative clinical status was noted in all but two patients, who reported generalized moderate pain about the great toe.

E. Orthopedic: Total Joint Arthroplasty

The results of a nonconstrained pyrolytic carbon metacarpophalangeal (MP) joint replacement in baboons were first reported in 1983 [2]. Histological evidence of direct appositional bone fixation along the medullary stem was observed in one specimen, and a combination of bone fixation with an interposing fibrous tissue membrane was observed in another (Fig. 5). There was no evidence of bone resorption around the implant stems, and functional fixation was obtained with all of the uncemented carbon implants. No foreign body reaction was observed in the soft tissues, and no evidence of intracellular particles was present.

In human clinical trials, 151 pyrolytic carbon MP joint replacements were implanted as custom devices in 53 patients between 1979 and 1987 [5]. The presenting diagnoses included 45 rheumatoid arthritis, 5 posttraumatic, and 3 osteoarthritis cases. The implant was an articulating nonconstrained design with a hemispherical head and grooved, offset stems (Fig. 6). A retrospective study of the clinical experience revealed that 26 patients with 71 implants (45%) were available for an average long-term follow-up of 12.2 years (range, 10.1 to 16.0 years). Eighteen implants (12%) in 11 patients required revision. Thirteen of the revisions were due to persistent subluxation, dislocation, or soft tissue imbalance. Four implants were removed

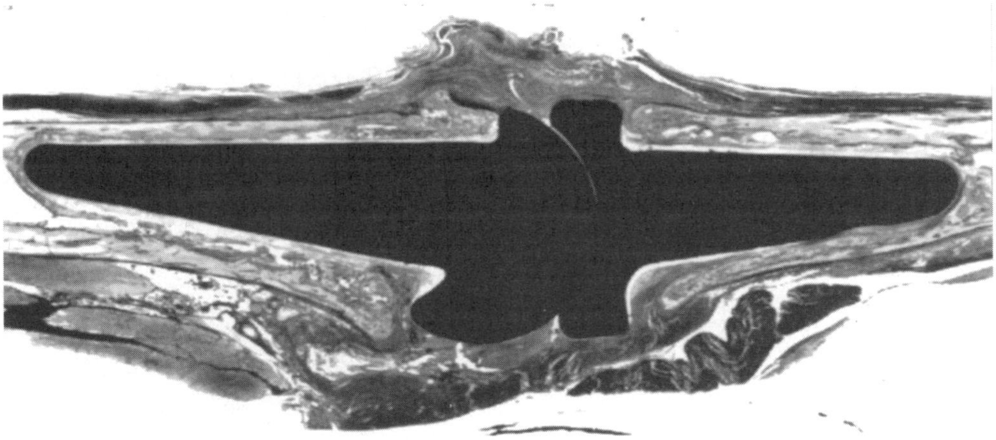

Figure 5 Histologic ground section of a nonconstrained pyrolytic carbon implant in the MP joint of a baboon [2] (1 × magnification, basic fuchsin and toluidine blue stain). In addition to a near normal clinical function provided by the polished articular surfaces, bone was observed in direct opposition along the implant stems and provided excellent fixation strength.

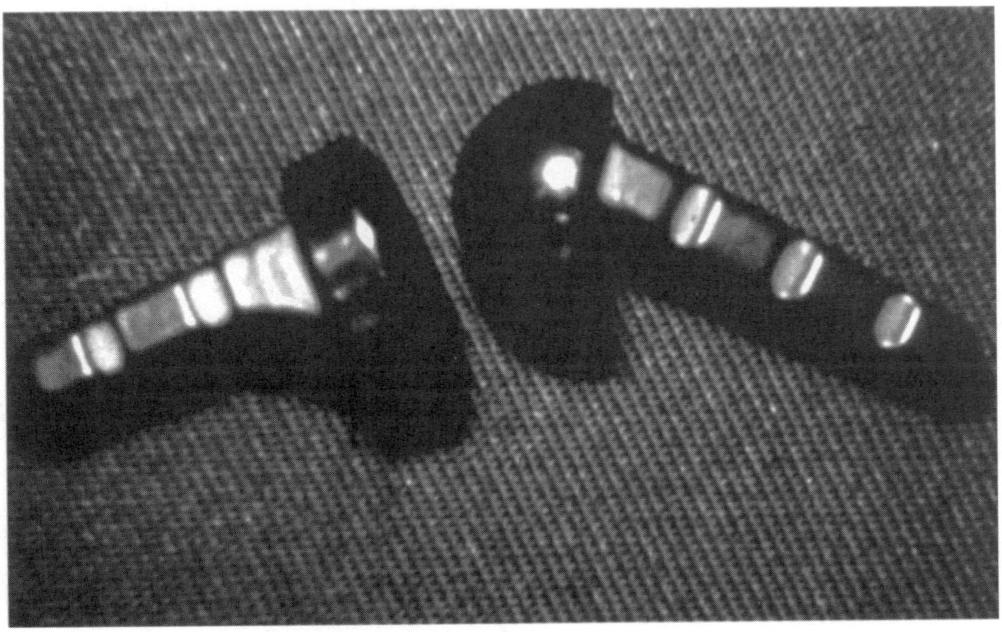

Figure 6 Nonconstrained metacarpophalangeal joint implant with hemispherical head and grooved, offset stems; metacarpal component (right) and proximal phalanx component (left). The implant device is formed as a machined graphite substrate which is then deposited with a coating of pyrolytic carbon. The stems remain in the as-deposited microporous finish for tissue attachment. The articulating wear surfaces are highly polished.

secondary to implant loosening. One implant fracture occurred after nine years in situ. All patients reported they were happy with the functional and cosmetic results of the surgery.

The pyrolytic MP joint implants improved the arc of motion by an average of 13° for all digits and elevated the arc by an average of 16° (Fig. 7). Ulnar deviation at long-term follow-up was 19° compared to 20° preoperatively and 12° in the early postoperative evaluation, indicating that the progression of ulnar deviation had been arrested. The majority of implants (94%) had radiographic evidence of periprosthetic sclerosis around the implant stems, indicating an osseous integration of the implants (Fig. 8). No adverse bone remodeling or resorption was noted in this series. Ten of the 13 revised hands had tissue submitted for pathological evaluation, and no evidence of intracellular particles or particulate synovitis was found. No signs of wear were visible on the carbon surfaces of the revised implants.

Using the survivorship method of analysis, implants that were revised, lost to follow-up or dislocated at late follow-up, were categorized as failures. The 5-, 10-, and 15-year survival rates of the pyrolytic carbon MP joints were 82%, 81%, and 70%, respectively. The annual failure rate over the 16-year period averaged approximately 2%. This retrospective clinical study demonstrated the biological and mechanical compatibility of pyrolytic carbon as a wear resistant and durable material for arthroplasty of the MP joint.

IV. CONCLUSIONS

There are many advantages of pyrolytic carbon as a material for small joint replacement applications. The outstanding biocompatibility and chemical stability of pyrolytic carbon allow for the intimate contact with living tissues without concern for biological reaction. New bone tissue is capable of direct apposition to the carbon surface which serves to stabilize an implant without additional fixation such as bone cement.

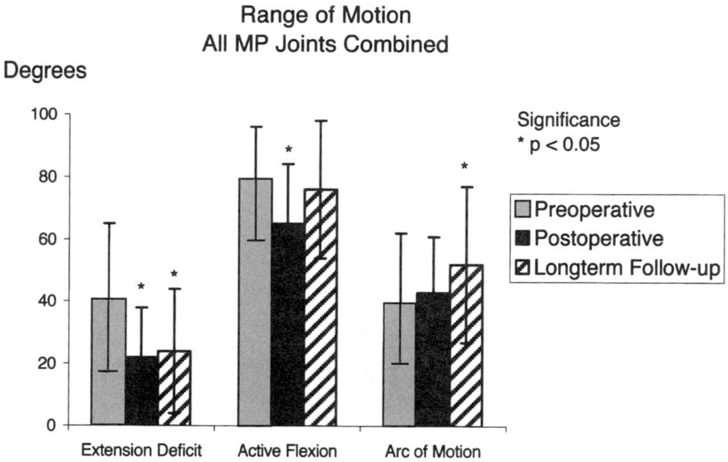

Figure 7 Average MP joint range of motion with standard deviations shown during the preoperative, early postoperative, and long term follow-up periods in the clinical pyrolytic carbon implant series. For all MP joint digits combined, the mean extension deficit was improved postoperatively and maintained in the long term. Although active flexion was reduced following surgical procedure, deterioration of flexion over the long term compared to preoperative motion was not observed. Overall, the arc of motion of the MP joints was improved at the long term evaluation. *ANOVA, significance level of $p < 0.05$.

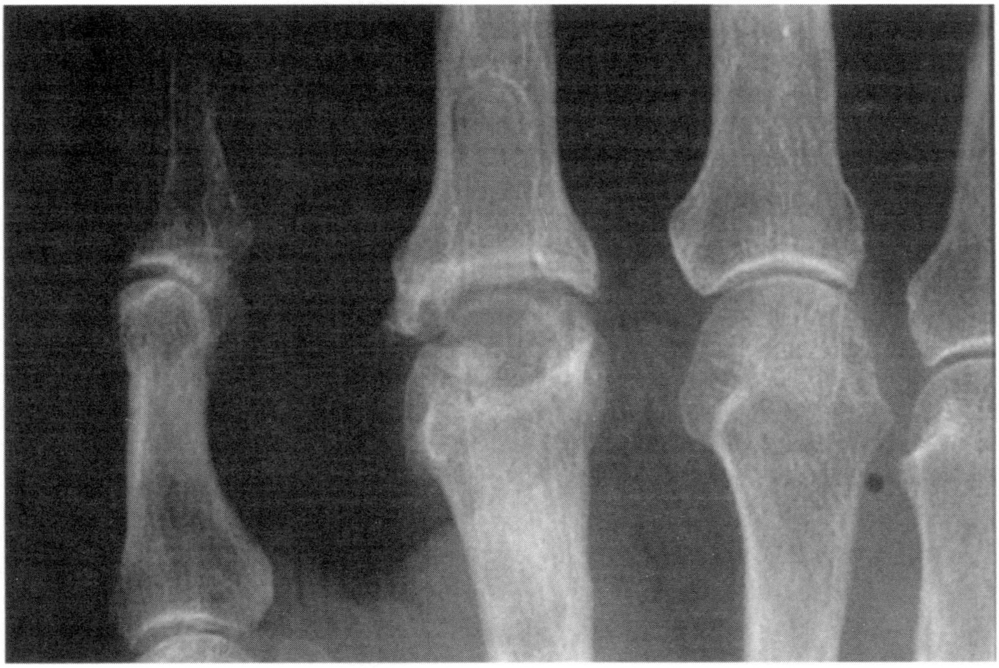

(a)

Figure 8 (a) Eleven-year follow-up radiograph of a posttraumatic arthroplasty of the right index MP joint in a 75 year old male. The pyrolytic carbon implant was stable and the joint reduced. The lucent line about the implant represents the thickness of the pyrolytic carbon coating which is radiolucent [5].

Pyrolytic carbon has been shown to provide a carbon on carbon articulation which is extremely resistant to wear. Complications and adverse reactions caused by wear debris resulting from metal on plastic implant designs are eliminated with the carbon on carbon wear surfaces in a small joint application. The use of pyrolytic carbon as a material in artificial heart valves and as a metacarpophalangeal joint replacement demonstrates its ability to withstand high levels of stress. Pyrolytic carbon has not been found to be subject to fatigue failure, which is ideal for artificial joint replacement.

Pyrolytic carbon has an elastic modulus which is essentially equal to that of bone. In contrast, metal alloys normally used in orthopedic implants have elastic moduli ranging from 5 to 10 times greater than bone. A joint replacement made of pyrolytic carbon has the potential to minimize bone loss due to stress shielding compared to a metal alloy.

The combined results of the animal and human clinical experience demonstrate the potential for use of pyrolytic carbon as a material for small joint replacements. Other areas of application for small joint replacement include the radial head, wrist, elbow, and ankle.

ACKNOWLEDGMENTS

This work was generously supported by the Joe W. and Dorothy Dorsett Brown Foundation, Metairie, Louisiana.

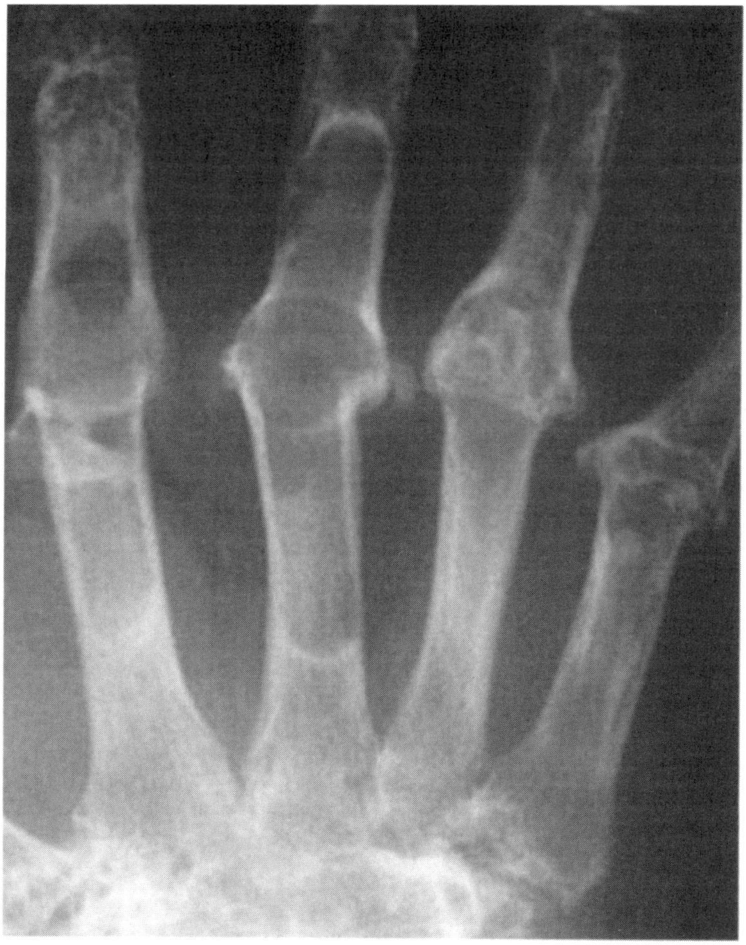

(b)

Figure 8 (b) Twelve-year follow-up radiograph of the index and long digits in a 48-year old female rheumatoid arthritis patient with pyrolytic carbon joint replacements. A thin sclerotic rim of bone is observed along the implant component stems with moderate implant subsidence [5].

REFERENCES

1. Haubold, A. D., 1994. On the durability of pyrolytic carbon in vivo. *Med. Progr. Technol.*, 20: 201–208.
2. Cook, S. D., Beckenbaugh, R., Weinstein, A. M., and Klawitter, J. J., 1983. Pyrolite carbon implants in the metacarpophalangeal joint of baboons. *Orthopedics* 6(8): 952–961.
3. Kent, J., Hulbert, S., Farrell, C., Klawitter, J., and Bokros, J., 1977. Clinical and histologic evaluation of functional LTI pyrolyte carbon endosseous blade implants in primates. In: *Transactions of the Society for Biomaterials*, 3rd Annual Meeting, New Orleans, LA, p.51.
4. Klawitter, J. J., Cook, S. D., Kay, J. F., Popich, L. S., and Rust, A. M., 1995. In vivo characterization of hydroxylapatite (HA) coated pyrolytic carbon implants. In: *Transactions of the Society for Biomaterials*, 21st Annual Meeting, San Francisco, CA, p.331.
5. Cook, S. D., Beckenbaugh, R. D., et al., 1999. Long-term follow-up of pyrolytic carbon metacarpophalangeal implants. *J. Bone Joint Surg. [Am] 81*: 635–648.

6. Bokros, J. C., 1969. Deposition, structure, and properties of pyrolytic carbon. In: P. L. Walker (Ed.), *Chemistry and Physics of Carbon*, Vol. 5, Marcel Decker, New York, pp.1–118.
7. Shim, H. S., 1974. The behavior of isotropic pyrolytic carbons under cyclic loading. *Biomat. Med. Dev. Art. Org.*, 2(1): 55–65.
8. Bokros, J. C., 1977. Carbon biomedical devices. *Carbon*, 15: 355–371.
9. Reilly, D. T., and Burstein, A. H., 1974. The mechanical properties of cortical bone. *J. Bone Joint Surg. [Am]*, 56(5): 1001–1022.
10. Reilly, D. T., Burstein, A. H., and Frankel, V. H., 1974. The elastic modulus for bone. *J. Biomech.*, 7: 271–275.
11. Carter, D. R., and Spengler, D. M., 1978. Mechanical properties and composition of cortical bone. *Clin. Orthop.*, 135: 192–217.
12. Bokros, J. C., LaGrange, L. D., and Schoen, F. J., 1973. Control of structure of carbon for use in bioengineering. In: P. L. Walker (Ed.), *Chemistry and Physics of Carbon*, Vol. 9, Marcel Decker, New York, pp.103–171.
13. Helbing, G., Wolter, D., Neugebauer, R., and Coldewey, J., 1977. The reaction of tissue to friction particles of LTI-carbon and carbon fibres reinforced carbon in the knee joint of the rat. In: *Transactions of the Society for Biomaterials*, 3rd Annual Meeting, New Orleans, LA, p.151.
14. Klasson, D. H., and Scheman, P., 1977. Purified carbon as a tissue replacement. *Int. Surg.* 62: 179–182.
15. Kent, J. N., and Bokros, J. C., 1980. Pyrolytic carbon and carbon-coated metallic dental implants. *Dental Clin. North Amer.*, 24(3): 465–485.
16. Kent, J. N., Farrell, S. D., Hulbert, S. D., Klawitter, J. J., and Bokros, J. C., 1977. Clinical evaluation of functional LTI pyrolite carbon blade implants in primates with design considerations. Extended Abstracts, 13th Biennial Conference on Carbon, Irvine, CA, Abstract No. B-2(3).
17. Hulbert, S. F., Kent, J. N., Bokros, J. C., Farrell, S. D., and Klawitter, J. J., 1975. Design and evaluation of LTI-Si carbon endosteal implants. *Oral Implant.*, 6: 79–86.
18. Cook, S. D., Weinstein, A. M., Klawitter, J. J., and Kent, J. N., 1983. Quantitative histologic evaluation of LTI carbon, carbon-coated aluminum oxide and uncoated aluminum oxide dental implants. *J. Biomed. Mater. Res.*, 17: 519–538.
19. Cranin, A. N., Rabkin, M., and Silverbrand, H., 1977. LTI carbon compared with chrome alloy implants in dogs—a clinical and histologic study. In: *Transactions of the Society for Biomaterials*, 3rd Annual Meeting, New Orleans, LA, p. 52.
20. Shim, H. S., 1976. The mechanical behavior of LTI carbon dental implants. *Biomat. Med. Dev. Art. Org.*, 4(2): 181–192.
21. Ely, J. L., Stupka, J., and Haubold, A. D., 1996. Pyrolytic carbon indentation crack morphology. *J. Heart Valve Dis.*, 5 (Suppl. 1): S65–71.
22. Ely, J. L., and Haubold, A. D., 1993. Static fatigue and stress corrosion in pyrolytic carbon. In: P. Ducheyne and D. Christiansen (Eds.), *Bioceramics*, Volume 6, Philadelphia, pp.199–204.
23. Shim, H. S., and Schoen, F. J., 1974. The wear resistance of pure and silicon-alloyed isotropic carbons. *Biomat. Med. Dev. Art. Org.*, 2(3): 103–118.
24. Vitale, E., and Giusti, P., 1995. Characterization of the resistance of pyrolytic carbon to abrasive wear. *Int. J. Artif. Organs*, 18(12): 777–785.
25. Tian, C. L., Hetherington, V. J., and Reed, S., 1993. A review of pyrolytic carbon: Application in bone and joint surgery. *J. Foot Ankle Surg.*, 32(5): 490–498.
26. Kampner, S. L., and Weinstein, A. M., 1984. Pyrolite carbon—an alternative implant material in orthopaedic surgery. *I. Mech. Eng.*, C202/84: 111–120.
27. Luedemann, R. E., and Cook, S. D., 1983. Radiographic and histologic evaluation of intramedullary implants intended for biological fixation. *Biomater. Med. Devices Artif. Organs*, 11(2–3): 197–210.
28. Anderson, R. C., Cook, S. D., Weinstein, A. M., and Haddad, R. J., 1984. An evaluation of skeletal attachment to LTI pyrolytic carbon, porous titanium, and carbon-coated porous titanium implants. *Clin. Orthop.*, 182: 242–257.
29. Thomas, K. A., and Cook, S. D., 1985. An evaluation of variables influencing implant fixation by direct bone apposition. *J. Biomed. Mater. Res.*, 19: 875–901.

30. Thomas, K. A., Cook, S. D., Renz, E. A., Anderson, R. A., Haddad, R. J., Haubold, A. D., and Yapp, R., 1985. The effect of surface treatments on the interface mechanics of LTI pyrolytic carbon implants. *J. Biomed. Mater. Res.*, 19: 145–159.
31. Hetherington, V. J., Lord, C. E., and Brown, S. A., 1995. Mechanical and histological fixation of hydroxylapatite-coated pyrolytic carbon and titanium alloy implants: A report of short-term results. *J. Appl. Biomat.*, 6: 243–248.
32. Pyrolite®, 1982. Intermedics Orthopedics Inc.™, Technical bulletin.
33. Cook, S. D., Thomas, K. A., and Kester, M. A., 1989. Wear characteristics of the canine acetabulum against different femoral prostheses. *J. Bone Joint Surg.*, 71-B(2): 189–197.
34. Jenkins, D. H. R., 1978. The repair of cruciate ligaments with flexible carbon fibre. A longer term study of the induction of new ligament and of the fate of the implanted carbon. *J. Bone Joint Surg.* [*Br*] 60: 520–522.
35. Jenkins, D. H. R., and McKibbin, B., 1980. The role of flexible carbon-fibre implants as tendon and ligament substitutes in clinical practice. *J. Bone Joint Surg.* [*Br*] 62:497–499.
36. Wolter, D., Fitzer, E., Helbing, G., and Coldewey, J., 1977. Ligament replacement in the knee joint with carbon fibres coated with pyrolytic carbon. In: *Transactions of the Society for Biomaterials*, 3rd Annual Meeting, New Orleans, LA, p.126.
37. Cook, S. D., Kester, M. A., Brunet, M. E., French, H. G., Haddad, R. J., and Haubold, A. D., 1986. A histologic and mechanical evaluation of carbon-coated polyester suture. *J. Biomed. Mater. Res.*, 20: 1347–1357.
38. Masini, M., Lazzari, M., Lorenzoni, R., Domicelli, A. M., Micheletti, A., Dianda, R., and Masini, G., 1996. Activated pyrolytic carbon tip pacing leads: an alternative to steroid-eluting pacing leads? *Pacing Clin. Electrophysiol.*, 19(11 Pt 2): 1832–1835.
39. Bosetti, M., Navone, R., Rizzo, E., and Cannas, M., 1998. Histochemical and morphometric observations on the new tissue formed around mammary expanders coated with pyrolytic carbon. *J. Biomed. Mater. Res.*, 40: 307–313.
40. James, R. A., Whittaker, J. M., and Schultz, R. L., 1977. An ultrastructural study of baboon tissue interfacing with pyrolytic carbon dental implants. In: *Transactions of the Society for Biomaterials*, 3rd Annual Meeting, New Orleans, LA, p.50.
41. Kent, J. N., Cook, S. D., Weinstein, A. M., and Klawitter, J. J., 1982. A clinical comparison of LTI carbon, alumina, and carbon-coated alumina blade-type implants in baboons. *J. Biomed. Mater. Res.*, 16: 887–899.
42. Cook, S. D., Weinstein, A. M., and Klawitter, J. J., 1983. The retention mechanics of LTI carbon, carbon-coated aluminum oxide, and uncoated aluminum oxide dental implants. *J. Biomed. Mater. Res.*, 17: 873–883.
43. Hetherington, V. J., Kavros, S. J., Conway, F., Mandracchia, V. J., Martin, W., and Haubold, A. D., 1982. Pyrolytic carbon as a joint replacement in the foot: a preliminary report. *J. Foot Surg.*, 21(3): 160–165.
44. Anderson, R. C., Cook, S. D., Skinner, H. B., Weinstein, A. M., and Haubold, A., 1983. An evaluation of LTI carbon and porous titanium load bearing prostheses. In: *Transactions of the Society for Biomaterials*, 9th Annual Meeting, Birmingham, AL, p.80.
45. Kawalec, J. S., Hetherington, V. J., Melillo, T. C., and Corbin, N., 1998. Evaluation of fibrocartilage regeneration and bone response at full-thickness cartilage defects in articulation with pyrolytic carbon or cobalt-chromium alloy hemiarthroplasties. *J. Biomed. Mater. Res.*, 41: 534–540.

17
Porous Ceramics for Intra-Articular Depression Fracture

Hajime Ohgushi, Masao Ishimura, Takashi Habata, and Susumu Tamai
Nara Medical University, Kashihara City, Japan

I. INTRODUCTION

Intra-articular depression fractures are difficult to treat and, occasionally, autogenous bone grafts have been used for the treatments. However, morbidity is clear when taking healthy bone from the patients and there is a limit to harvest the autogenous bone. To overcome the morbidity, allogeneic bone or synthetic graft has been used. Necessary amount of bones can easily be obtained in the case of allogeneic bone transplantation; however, there are risks in the allografts, such as known or unknown infectious disease and possible transplantation immunity. Calcium phosphate ceramics and certain types of glass ceramics have been proven to show bone bonding and excellent osteoconductive property [1–8]. Therefore, ceramics have recently been used as bone graft substitute [9,10]. Though the osteoconductive property resides in ceramics, they alone cannot show osteogenic ability and do not contribute to the healing potential in massive bone defects [11–16]. Bone morphogenetic protein (BMP) demonstrates osteogenesis when implanted at subcutaneous and intramuscular sites and thus the BMP has osteoinductive property [17–19]. Some reported the usefulness of a BMP/ceramic composite to enhance the inherent osteogenic ability of BMP [20–21]. In considering the clinical applications of BMP, purification of the protein is necessary to avoid complications such as immunological reaction caused by heterogeneous proteins, which are found in crude BMP preparations. Recently, many types of BMPs have been cloned and their recombinant BMP molecules are available [17,18]. However, purification of the recombinant molecules is difficult and the cost of the purification is high. Furthermore, a large amount of the recombinant protein is necessary to show the osteogenic response in primates [22] and the clinical outcome of recombinant BMP is not clear.

In contrast to the utilization of the osteogenic activity of BMP, we, along with other authors, reported that the composites of marrow cells and porous materials such as hydroxyapatite (HA) show an osteogenic response when implanted in vivo [11–15, 23–25]. The osteogenic capability of the marrow cells resides in not only fresh marrow cells but also in in vitro culture expanded mesenchymal stem cells derived from the marrow cells [23–25]. Though we can expand the number of cells having osteogenic property by using tissue culture technology, the cost of the culturing is very high and careful techniques, such as prevention of infection, are required. In this regard, fresh marrow from iliac crest of the patient during the operation can be easily obtained with no obvious morbidity; therefore, we utilized the fresh marrow/hydroxyapatite ceramic composite graft for the treatment of depression fractures of calcaneous and condyles of the tibia.

II. MATERIALS AND METHODS

A. Ceramics

Ceramics used were type F porous hydroxyapatite (mean pore size 200 μm and porosity 70%, Asahi Optics Co. Tokyo, Japan).

B. Animal Experiments

Femora and tibiae of five Fisher 344 rats (7-week-old males) were removed and place in saline. The marrow plugs from these bone shafts were then hydrostatically forced into a test tube containing 1 mL of heparinized phosphate buffered saline (PBS). The marrow in PBS was disaggregated by sequential passage through 18 G and then 20-G needles to obtain cell suspension. The cell suspension was centrifuged at 250 G for 5 min and the cell pellet was resuspended in 200 μL of the supernatant by vortex mixer. Following these procedures, the number of nucleated cells was about 5×10^8/mL [11–13]. To make the composite graft, the porous hydroxyapatite ceramics were soaked in this marrow cell suspension for about 5 min. Syngeneic Fischer rats were anesthetized by intraperitoneal injection of Nembutal (2.5 mg/100 g body weight). Subcutaneous pouches were created in back of the rat following small incisions. The ceramics combined with rat bone marrow cells were implanted subcutaneously in the back of rats and harvested one to four weeks postimplantation.

C. Surgical Procedure and Treatment

Bone marrow cells were obtained by needle aspiration from the patient's iliac bone. The ceramics soaked in the marrow cells were shaped by chisel to fit the cavity produced by the elevation of the depressed fragment as described below.

For the calcaneal fractures, a curved incision was made following the course of the peroneal tendons. The terminal branch of the sural nerve was protected. The sheath of the tendons was divided and talocalcaneal joint was opened. The depressed portion of the calcaneus was elevated by chisel and the marrow cells soaked ceramics were put beneath the chisel to maintain the normal position of the articular surface. The unstable fractures, Kirchner wire or staple between depressed bone fragment and postero-inferior part of the calcaneus was used to secure the reduction. Also, inferior fragments along the base of the bone which had spread laterally were compressed with the hand into a relatively normal position. We have routinely used plaster cast for four weeks and unrestricted weight bearing was started about two months after surgery.

For the tibial condylar fractures, an anterolateral or anteromedial incision (former was used for lateral plateau depression and the latter for medial depression) was made. The articular surface of the tibial condyle was exposed by incising the coronary ligament attachment of the meniscus and reflecting it proximally with the femoral condyles. Elevation of the depressed portion of the condyle and the ceramic implantation was done in the same way as the calcaneal fracture described above. Internal fixation with a tibia bolt, AO screw or buttress plate was used to maintain the normal position of the articular surface. Usually, the leg was placed in a plaster for four weeks with the knee flexed 30°. No weight bearing was allowed for about 10 weeks.

D. Classification of the Fractures

Essex-Loprestei classification [26] of calcaneal fractures (Type I: not involving subtalar joint; type II: involving subtalar joints; type IIa: without displacement; type IIb: tongue type without displacement; type IIc: centrolateral depression of joint; type IId: sustentaculum tali fracture along; type IIe: with gross comminution) and Hohl classification [27] of tibial condylar fractures (A: undisplaced, B: local compression, C: split compression, D: total condylar compression, E: split, F: comminuted) were adopted for the present cases.

E. Clinical Evaluation

The results of cases of calcaneal and tibial condylar fractures were judged according to Maxfield [28] and Hohl [27], respectively. They were judged as excellent, good, fair and poor.

III. RESULTS

A. Animal Experiments

The porous hydroxyapatite ceramics did not show bone formation after subcutaneous implantation; thus the ceramic is not osteogenic—in other words, not osteoinductive. When the ceram-

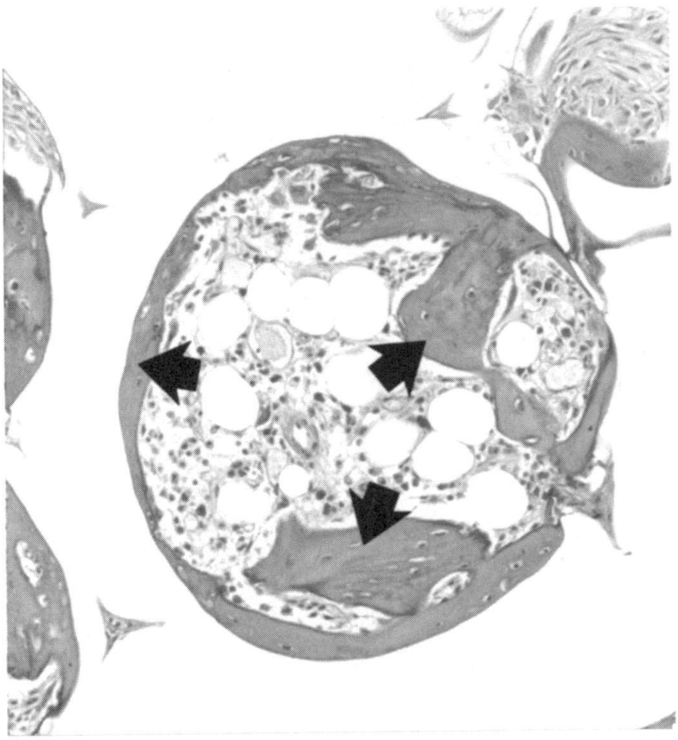

Figure 1 Composite of rat bone marrow cells and porous hydroxyapatite ceramic implanted in rat subcutaneous sites and harvested at four weeks after implantation. Bone formation (arrows) in the pore area of ceramic is clearly seen (Hematoxyline and Eosin stain, ×150).

ics were combined with fresh rat marrow cells, the composites started to show bone formation about three weeks postimplantation and all the composites showed extensive bone formation in the pore regions of the ceramics at four weeks postimplantation. As shown in Fig. 1, the marrow-cell-induced bone is directly apportioned on the pore surface and thus there was no intervening fibrous tissue between the bone and ceramic surface. The sequential and histological (Fig. 1) patterns of bone formation in pore areas of the ceramics were the same as our previous findings using other types of calcium phosphate ceramics [11–14]. These results confirmed the excellent osteogenic capability of the ceramics in the presence of fresh bone marrow cells.

B. Clinical Cases

For clinical application of the osteogenic capability of marrow cells, bone marrow was obtained by needle aspiration from the patient's iliac crest during the operation and the ceramics were soaked in the marrow cells. The marrow cell loaded ceramics were then used to hold the depressed articulate surface into a normal position.

Seventeen patients with tibial condylar fractures and calcaneal fractures were treated (Table 1). Only one case showed fair result, and other cases showed good or excellent results.

Table 1 Treatment of Intra-articular Fractures

Case	Fracture[a]	Sex	Age[b]	Classification[c]	Follow-up length (Year)	Clinical results[d]
1	Tibia	F	32	C	7	Good
2	Tibia	M	18	C	6.9	Good
3	Tibia	M	52	D	6.3	Excellent
4	Tibia	F	34	F	5	Good
5	Tibia	F	23	C	5.5	Fair
6	Tibia	F	25	C	3	Excellent
7	Calc.	M	45	IIc	6.8	Good
8	Calc.	M	58	IIc	6.8	Excellent
9	Calc.	M	47	IIc	6.6	Good
10	Calc.	M	59	IIe	6.5	Good
11	Calc.	F	42	IIe	6.4	Good
12	Calc.	M	33	IIc	6.3	Good
13	Calc.	F	35	IIc	6.3	Good
14	Calc.	M	45	IIe	4.1	Good
15	Calc.	M	25	IIe	4.3	Good
16	Calc.	M	40	IIe	4.3	Good
17	Calc.	F	48	IIe	3.2	Excellent

[a]Fractures of tibial condyle and calcaneus (calc.) were treated by the composites of patient's bone marrow and porous hydroxyapatite ceramics.
[b]Age at the time of operation.
[c]Hohl classification of tibial condylar fractures and Essex-Lopresti classification of calcaneal fractures were used.
[d]The results of cases of tibial condylar and calcaneal fractures were judged according to Maxfield and Hohl, respectively.

The fair case was a tibial condylar fracture which had anterior and posterior cruciate ligament ruptures, and at the time of surgery the ligaments were sutured. Thus, intra-articular ligaments were extensively injured resulting in decreased range of motion and instability of the joint.

Usually, homogeneous radiodense appearance around the ceramic implants began to appear about two months after the surgery, and eventually all ceramics showed the radiodense areas around the margin of the ceramics resulting in an obscure ceramic boundary (Figs. 2–4). Therefore, all cases showed that the implanted ceramics were well integrated with the surrounding bone areas. However, some cases having severely damaged joint showed the posttraumatic osteoarthritic change evidenced by osteophyte formation in the joints after several years. For the burst fracture in the joints, it is difficult to restore the normal joint architecture even when we used the porous ceramics; therefore the osteophyte formation was due to joint incongruence after the surgery. Osteoarthritic change in talocalcaneal joint is hard to detect because of the small joint. However, as shown in Fig. 4, CT scan demonstrated the osteophyte formation after three years postoperation in this case, although the patient had no complaint and continued playing sports, such as tennis.

None of the cases showed evidence of infection and obvious inflammatory response.

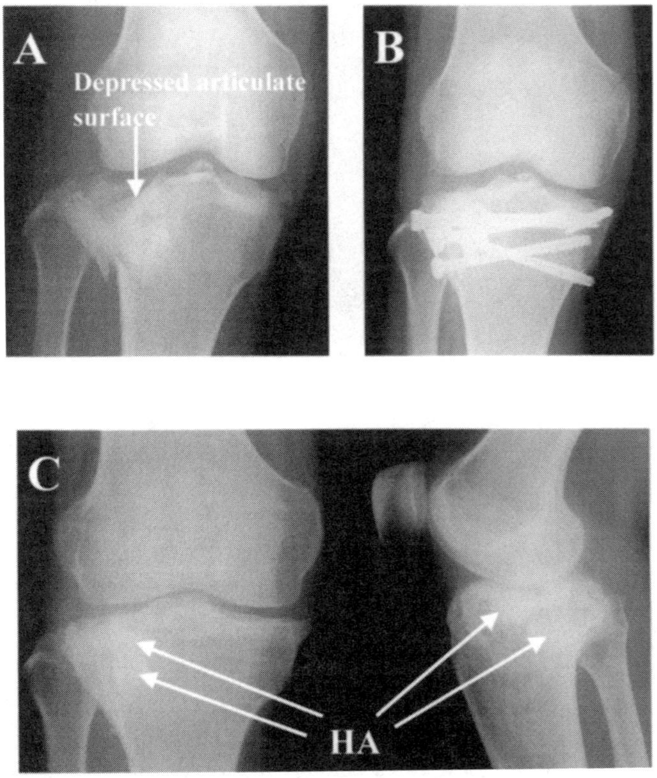

Figure 2 Radiograph of a depressed fracture of the lateral tibial condylar fracture. (A) Preoperative radiograph. (B) Postoperative radiograph taken one month after the operation. Depressed articulate surface was elevated and is well maintained to its normal position. (C) Postoperative radiograph taken three years after the operation. Homogeneous radiodense appearance around the hydroxyapatite (HA) ceramics is shown. The congruence of the lateral joint is the same as in (B) and thus reduction of the depressed tibial articulate surface is successful to show functional joint.

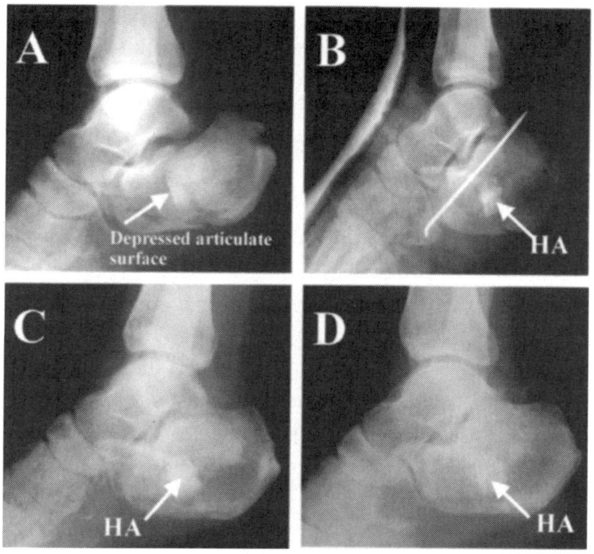

Figure 3 Radiographs of calcaneal fracture. (A) Preoperative radiograph. (B) Immediate postoperative radiograph. (C) Radiograph taken two months after the operation. (D) Radiograph taken one year after the operation. Boundary between the hydroxyapatite (HA) ceramic and host bone tissue is gradually obscured as time passed and therefore, regenerated bone tissue finally integrates to the entire ceramic surface.

Taking patient's bone marrow by aspiration through iliac bone is easy, and patients complained of little or no pain after the aspiration. These results indicate the usefulness of the composite of patient's bone marrow cells and porous hydroxyapatite for the treatment of depression fracture.

IV. DISCUSSION

We have used porous hydroxyapatite ceramics as a bone graft substitute and patient's bone marrow cells as an osteogenic cell source for the surgical treatments of 17 cases of intra-articular fractures. Our judgment of the operative indications are those with displaced fragments of type IIc, IIe for calcaneal fractures and type B,C,D,F for tibial condylar fractures, respectively. The indications about the depth of depressed articular surface of calcaneus and tibia are more than 3 and 5 mm, respectively. Most cases showed excellent or good functional results. We believe the satisfactory results were attributed to the osteogenic ceramics which restored the normal position of articular surface. Others also reported the usefulness of porous hydroxyapatite ceramics for the treatment of depression fractures [10].

Concerning bone graft substitutes and bone marrow cells, Burwell reported osteogenesis in the xenogeneic bovine cancellous bone (Kiel bone) after the impregnation of rat bone marrow cells [29]. And Salama reported good clinical results when using patient's bone marrow and Kiel bone for the treatment of bone tumor and fractures in patients [30]. However, Kiel bone is derived from bovine cancellous bone, and our results showed a superior osteogenic response of bone marrow cells in the porous hydroxyapatite ceramic when compared to that in Kiel bone. The superiority was based on these two findings [31]. When Kiel bone and the porous hydroxyapatite ceramics (marine coral derived hydroxyapatite; Interpore 200 produced

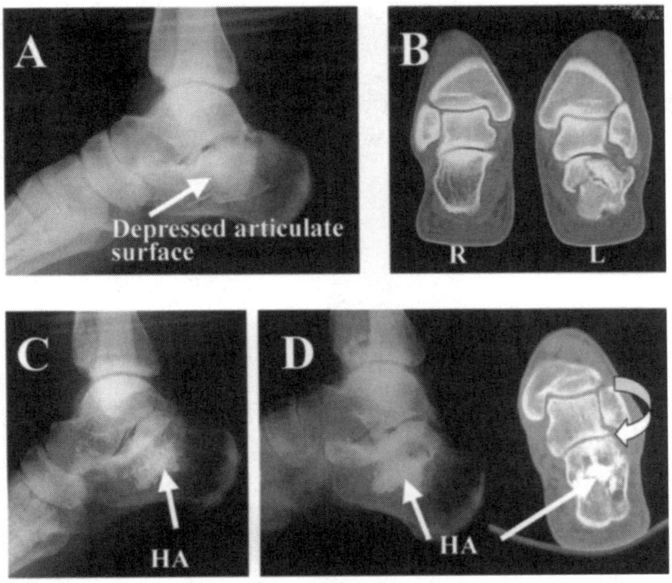

Figure 4 Radiographs and CT scans of calcaneal fracture. (A) Preoperative radiograph. (B) Preoperative CT scan of injured (L: left) and uninjured (R: right) calcaneus. (C) Radiograph taken two months after the operation. Elevated articulated surface is almost in normal position. (D) Radiograph (left) and CT scan (right) taken three years after the operation. Homogeneous radiodense appearance around the hydroxyapatite (HA) ceramic is seen in the radiograph. The CT scan well depicts integration of the regenerated bone to the HA ceramic and also shows that the depressed articulate surface is well maintained to its normal position. Although, small osteophytes are seen in lateral edge of the talocalcaneal joint (curved arrow).

by Interpore International, Irvine, CA) were combined with rat marrow cells and implanted subcutaneously, the amount and incidence of bone formation in porous ceramics were much more than that in Kiel bone. The bone formed in Kiel bone had intervening fibrous tissue between the de novo bone and Kiel bone, whereas the bone formed in the Interpore ceramics was always directly apportioned on the ceramic surface. These results indicate that Interpore hydroxyapatite ceramics can support good osteogenic response of marrow cells. In spite of the good osteogenic property of the Interpore ceramics, these are not approved for clinical application in our country. Therefore, we used synthetic porous hydroxyapatite (Asahi Optical Co., Japan) for the treatment of the fractures and evaluated whether the ceramics can support the osteogenic differentiation of marrow cells. As shown in Fig. 1, when the ceramics were combined with rat marrow cells and implanted subcutaneously, the ceramic showed excellent bone forming capability in the pore areas. The incidence of the bone formation was almost the same as Interpore ceramics, and thus the synthetic porous hydroxyapatite (Asahi Optical Co.) has comparable osteogenic capability to that of Interpore ceramics.

The animal experiment using rat marrow cells confirmed that porous ceramics can support the osteogenic phenotype expression of rat marrow cells, although an important question is whether the human bone marrow cells can show osteogenic expression. In this regard, our previous report showed that composite of the young and adult human bone marrow cells and porous hydroxyapatite ceramics showed new bone formation when implanted in an athymic nude mouse [23]. Interestingly, not only fresh but also cultured human marrow cells can show bone-forming capability when combined with porous ceramics [23,24]. Recently, a prospective and randomized study was conducted to compare the efficacy of autogenous bone graft and composite material composed of bovine collagen, calcium phosphate ceramic and autogenous marrow for the treatment of long bone fracture. The study showed no significant difference between two groups with respect to rates of union [32]. All results demonstrate the rationale to use porous ceramics loaded with patient's bone marrow cells for the treatment of intra-articular depression fractures.

REFERENCES

1. M. Jarcho, Calcium phosphate ceramics as hard tissue prosthetics, *Clin. Orthop.*, **157**, 259–278 (1981).
2. K. Kato, H. Aoki, T. Tabata, and M. Ogiso, Biocompatibility of apatite ceramics in mandibles, *Biomater. Med. Devices Artif. Organs*, **7**, 291–297 (1979).
3. K. DeGroot, Bioceramics consist of calcium phosphate salts, *Biomaterials*, **1**, 47–50 (1980).
4. L. L. Hench, R. J. Spliner, W. C. Allen, and T. K. Greenlee, Bonding mechanism at the interface of ceramic prosthetic materials, *J. Biomed. Mater. Res.*, Symp. **2**, 117–141 (1971).
5. L. L. Hench, Bioactive ceramics, *Ann. N.Y. Acad. Sci.*, **523**, 54–71 (1988).
6. U. M. Gross, H. J. Schmitz, and V. Strunz, Surface activities of bioactive glass, aluminum oxide and titanium in a living environment, *Ann. NY Acad. Sci.*, **523**, 211–226 (1988).
7. T. Kokubo, S. Ito, S. Sakka, and T. Yamamuro, Formation of a high-strength bioactive glass-ceramic in the system MgO-Ca-SiO_2-P_2O_5, *J. Mater. Sci.*, **21**, 536–540 (1986).
8. T. Nakamura, T. Yamamuro, S. Higashi, T. Kokubo, and S. Ito, A new glass-ceramic for bone replacement: Evaluation of its bonding to bone tissue, *J. Biomed. Mater. Res.*, **19**, 685–698 (1985).
9. T. Yamamuro, J. Shikata, H. Okumura, T. Kitsugi, Y. Kakutani, T. Matsui, and T. Kokubo, Replacement of the lumbar vertebrae of sheep with ceramic prosthesis, *J. Bone Joint Surg.*, **72-B**, 889–893 (1990).
10. M. Itokazu, T. Matsunaga, M. Ishii, H. Kusakabe, and Y. Wyni, Use of arthroscopy and interporus hydroxyapatite as a bone graft substitute in tibial plateau fractures, *Arch. Orthop. Surg.* **115**, 45–48 (1996).
11. H. Ohgushi, V. M. Goldberg, and A. I. Caplan, Heterotopic osteogenesis in porous ceramic induced by marrow cells, *J. Orthop. Res.*, **7**, 568–578 (1989).
12. H. Ohgushi, M. Okumura, S. Tamai, E. C. Shors, and A. I. Caplan, Marrow cell induced osteogenesis in porous hydroxyapatite and tricalcium phosphate, *J. Biomed. Mat. Res.*, **24**, 1563–1570 (1990).
13. M. Okumura, H. Ohgushi, and S. Tamai, Bonding osteogenesis in coralline hydroxyapatite combined with bone marrow cells, *Biomaterials*, **12**, 411–416 (1991).
14. H. Ohgushi, M. Okumura, T. Yoshikawa, K. Inoue, N. Senpuke, and S. Tamai, Bone formation process in porous calcium carbonate and hydroxyapatite, *J. Biomed. Mat. Res.*, **26**, 885–895 (1992).
15. H. Ohgushi, Y. Dohi, S. Tamai, and S. Tabata, Osteogenic differentiation of marrow stromal stem cells in porous hydroxyapatite ceramics, *J. Biomed. Mat. Res.*, **27**, 1401–1407 (1993).
16. H. Ohgushi, V. M. Goldberg, and A. I. Caplan, Repair of bone defects with marrow and porous ceramic. (Experiments in rats), *Acta Orthop. Scand.* **60**, 334–339 (1989).
17. J. M. Wozney, V. Rosen, A. J. Cheleste, L. M. Mitsock, M. J. Whitters, R. W. Kriz, R. M. Hewick, and E. A. Wang, Novel regulators of bone formation: molecular clones and activities, *Science* **242**, 1528–1533 (1988).
18. J. M. Wozney, The bone morphogenetic protein family and osteogenesis, *Molec. Rep. Dev.*, **32**, 160–167 (1992).
19. M. R. Urist, A. Lietze, and E. Dawson, Beta-tricalcium phosphate delivery system for bone morphogenetic protein, *Clin. Orthop.*, **187**, 277–280 (1984).
20. S. Ijiri, T. Nakamura, Y. Fujisawa, M. Hazama, and S. Komatsudani, Ectopic bone induction in

porous apatite-wollastonite-containing glass ceramic combined with bone morphogenetic protein, *J. Biomed. Mater. Res.*, **35**, 421–432 (1997).

21. Y. Kuboki, H. Takita, D. Kobayashi, E. Tsuruga, M. Inoue, M. Murata, N. Nagai, Y. Dohi, and H. Ohgushi, BMP-induced osteogenesis on the surface of hydroxyapatite with geometrically feasible and unfeasible structures: Topology of osteogenesis, *J. Biomed. Mat. Res.*, **39**, 190–199 (1998).
22. S. Miyamoto, K. Takaoka, and K. Ono, Bone induction in monkeys by bone morphogenetic protein. A trans-filter technique, *J. Bone Joint Surg.*, **75-B**, 107–110 (1993).
23. H. Ohgushi and M. Okumura, Osteogenic ability of rat and human marrow cells in porous ceramics, *Acta Orthop. Scand.*, **61**(5), 431–434 (1990).
24. S. E. Haynesworth, J. Goshima, V. M. Goldberg, and A. I. Caplan, Characterization of cells with osteogenic potential from human marrow, *Bone*, **13**, 81–88 (1992).
25. A. I. Caplan and S. P. Bruder, Cell and molecular engineering of bone regeneration, in *Principles of Tissue Engineering*, R. Lanza, R. Langer and W. Chick (eds.), R. G. Lands, 1997, pp. 603–618.
26. L. P. Essex-Lopresti, The mechanism, reduction technique and results in fractures of the os calcis, *Br. J. Surg.*, **39**, 395–419 (1952).
27. M. Hohl and J. V. Luch, Fracture of the tibial condyle, *J. Bone Joint. Surg.* **38A**, 1001–1018 (1956).
28. J. E. Maxfield, J. F. McDermott, and W. Falls, Experiences with palmer open reduction of fractures of the calcaneus, *J. Bone Joint. Surg.*, **37A**, 99–106 (1955).
29. R. G. Burwell, Studies in the transplantation of bone, *J. Bone Joint. Surg.*, **46B**, 110–140 (1964).
30. R. Salama, Xenogeneic bone grafting in humans, *Clin. Orthop.*, **174**, 103–112. (1983).
31. M. Okumura, C. A. van Blitterswijk, H. K. Koerten, and H. Ohgushi, Osteogenic response of rat bone marrow cells in porous alumina, hydrosyapatite and Kiel bone, in *Bioceramics*, W. Bonfield, W. Hastings and K. E. Tanner (eds.), Butterworth-Heinemann, Vol. 4, 1991, pp. 3–8.
32. M. W. Chapman, R. Bucholz, and C. Cornell, Treatment of acute fractures with a collagen-calcium phosphate graft material. A randomized clinical trial, *J. Bone Joint. Surg.*, **79A**, 495–502 (1997).

18

New Insights in the Biological Effects of Wear Debris from Total Joint Replacements

Michael C. D. Trindade, R. Lane Smith, Joshua Jacobs, and Stuart B. Goodman
Stanford University School of Medicine, Stanford, California

I. INTRODUCTION

Total joint replacement surgery continues as one of the most common procedures performed by orthopedic surgeons. The 10-year survival rate for current total joint implants is over 90% (1–4). Judet performed some of the first artificial joint replacements; however, these acrylic implants provided only temporary relief of pain before failing due to poor fixation. Initial total joint arthroplasties performed by Charnley in the 1950s had a high failure rate due to the metal on polytetrafluoroethylene (PTFE) articulation. PTFE became abraded, and a foreign body reaction to the particles ensued resulting in loosened prostheses. Subsequent implants by Charnley were composed of metal articulating on polyethylene. The majority of implants performed today use metal on ultrahigh molecular weight polyethylene. The implant materials include CoCr, Ti6Al4V, CP titanium, UHMWPE, and ceramics. All of these materials generate particulate debris in vivo that is thought to be a major contributing factor in osteolysis and aseptic loosening (5–9). Although ceramic implants appear to exhibit lower wear rates they continue to represent a small market share in North America compared to metal on polyethylene articulation due in part to concerns over femoral head fracture and cost. New (second) generation ceramic on ceramic and metal on metal implants fabricated with improved manufacturing methods and higher tolerances may prove to increase the life span of total joint replacements but long-term studies are necessary. Furthermore improved cross-linked polyethylenes and sterilization techniques show promise with respect to improving the life span of the typical metal on polyethylene implant.

The biology of the bone implant interface influences the outcome of total joint replacements. Longevity of total joint arthroplasty appears to be influenced by the generation and biological effects of wear debris that contributes to periprosthetic bone resorption (10–15). Granulomatous tissue that forms at sites of implant loosening is thought to play a role in particle mediated osteolysis (16–22). Analysis of retrieved granulomatous tissues reveals numerous macrophages, lymphocytes, fibroblasts, foreign body giant cells, and abundant particulate debris (11,15,20,23–26). Studies have documented the ability of macrophages to phagocytose orthopedic debris and release the pro-inflammatory mediators interleukin-1 (IL-1), interleukin-6 (IL-6), tumor necrosis factor-alpha (TNF-α), and prostaglandin E2 (PGE2) (10,12,27–32). The release of inflammatory mediators may then stimulate localized bone resorption (6,8,13,19,33–39).

In vitro studies have provided a vast amount of information about the responses of inflammatory cells to orthopedic biomaterials. Recent studies have elucidated new inflammatory mediators that are released by cells in response to particulate debris, and have described signal transduction mechanisms by which cells present in tissues surrounding failed total joint replacements are activated. Much of the early in vitro work focused on the macrophage and its role in periprosthetic osteolysis. It was previously thought that macrophages, osteoblasts, and osteoclasts were the only cells involved in particle-induced osteolysis. However over the past 10 years the involvement of other cells including fibroblasts, endothelial cells and T-cells have been studied. Recent studies have demonstrated that fibroblasts, lymphocytes, and osteoblasts respond to particle challenge as well as macrophages. Coculture systems have demonstrated important interactions between the different cell types following exposure to biomaterials. In addition, the process by which particulate debris effects bone and tissue remodeling has been examined using new animal models.

II. PRO-INFLAMMATORY MEDIATORS

Studies have documented the effects of particles on IL-6, TNF-α, interleukin-1 beta (IL-1β), transforming growth factor-beta (TGF-β), hexosaminidase and PGE2 release from macrophages along with their presence in tissues harvested from failed total joint replacements (5,10,12,18,19,28,30,36,40–43). Other factors including nitric oxide, matrix metalloproteinases, chemokines, and the colony-stimulating factors have been the subject of recent studies.

A. Nitric Oxide

Shanbhag et al. studied the effects of particulate debris on macrophage release of nitric oxide (44). Their findings demonstrated that release of nitric oxide was dependent on particle type, particle dose, and time of culture. Titanium-alloy particles stimulated the highest levels of nitric oxide, followed by commercially pure titanium and polymethylmethacrylate (PMMA), which is in agreement with the levels of cytokine expression observed in numerous other studies.

B. Matrix Metalloproteinases

Several studies on retrieved tissues harvested from failed total joint arthroplasties have demonstrated the presence of matrix metalloproteinases and their mRNAs (45–47). Two recent studies have examined the effect of particulate debris on matrix metalloproteinase (MMP) expression in vitro (46,48).

Yao et al. studied the effect of titanium particles on fibroblast expression of collagenase, stromelysin and the tissue inhibitor of metalloproteinases (TIMP) (48). The authors isolated fibroblasts from interfacial membranes of patients with failed total hip replacements and normal synovial tissue. The fibroblasts were subsequently challenged with titanium particles in vitro. The fibroblasts responded with increased expression and release of collagenase, stromelysin and TIMP. Addition of medium harvested from fibroblasts exposed to titanium particles to the cultures of the osteoblast-like MG-63 cell significantly suppressed the mRNA levels of procollagen alpha 1 (I) and alpha 3 (III). The authors concluded that the fibroblast response to particulate debris may enhance bone resorption through the induction of metalloproteinases and might inhibit osteoblast function through reduced collagen synthesis.

Studies by Nakashima et al. have demonstrated the involvement of matrix metalloproteinases in the macrophage response to orthopedic wear debris (46). Peripheral blood human macrophages were exposed to phagocytosable titanium and PMMA particles. Zymograms demonstrated that macrophages exposed to titanium alloy and PMMA particles released 92- and 72-kDa gelatinases in a particle dose- and time-dependent manner. Western blots confirmed that the 92- and 72-kDa gelatinolytic activities corresponded to matrix metalloproteinase-9 (MMP-9) and matrix metalloproteinase-2 (MMP-2), respectively. Western blots also indicated that titanium alloy and PMMA particles increased macrophage release of MMP-1. Northern blot analysis demonstrated elevated mRNA signal levels for MMP-1, MMP-2, and MMP-9 in response to titanium and PMMA particle challenge. Radioactive collagen substrate assays demonstrated increased collagenolytic activity in the macrophage culture medium after particle exposure.

C. Chemokines

Chemokines are a family of pro-inflammatory cytokines that act primarily as chemoattractants for cells of the immune system. In humans, four subfamilies of chemokines are recognized that exhibit homologous function and in some instances homologous protein structure. With the exception of lymphotactin and neurotactin, chemokines are subdivided based on arrangement of the first two of four highly conserved cysteines (49). The C-C or β subfamily consists of monocyte chemoattractant proteins 1, 2, 3, 4 (MCP-1, 2, 3, 4), macrophage inflammatory proteins 1α, 1β, 3α, 3β (matrix inhibitory protein, MIP-1α, 1β, 3α, 3β) and regulated upon activation normal T expressed and secreted protein (RANTES). Other recognized members of this subfamily include eotaxin, I309, HCC-1 and TARC. The members of the C-C chemokine subfamily are potent chemoattractants for monocytes, lymphocytes, basophils and eosinophils, while C-X-C chemokines act primarily on neutrophils (50–54). Lymphotactin, the sole member of the C subfamily, is a specific attractant for T-lymphocytes but not monocytes that lacks two of the cysteine residues that are characteristic of chemokines (55). Neurotactin (fractalkine), a C-X_3-C chemokine, exists as either an integral membrane protein or a shed 95K glycoprotein, exhibiting an *N*-terminal chemokine domain (56,57).

Nakashima et al. tested the effects of titanium alloy (Ti-alloy) and bone cement (polymethylmethacrylate, PMMA) particles on monocyte/macrophage expression of the C-C chemokines, MCP-1, MIP-1α, and RANTES (58). In addition, periprosthetic granulomatous tissues were analyzed for macrophage chemokine expression by immunohistochemistry. Polymerase chain reaction (RT-PCR) was used to examine mRNA for MCP-1 and MIP-1α in the in vitro arm of these studies. The effects of the chemokines MCP-1 and MIP-1α on monocyte migration were also determined. Immunohistochemical analysis of retrieved membranes demonstrated MCP-1 and MIP-1α expression in all specimens. Ti-alloy and PMMA particles increased MCP-1 and MIP-1α expression in macrophages in vitro in a dose- and time-dependent manner. RANTES expression was not observed. Following particle exposure MCP-1 and MIP-1α mRNA signal levels were observed. Monocyte migration was stimulated by culture medium collected from macrophages exposed to Ti-alloy and PMMA particles. Addition of antibodies to MCP-1 and MIP-1α inhibited monocyte migration. Recent unpublished results from our laboratory have confirmed that fibroblasts respond to PMMA and titanium particle challenge through a dose- and time-dependent release of the MCP-1.

Frokjaer et al. demonstrated the induction of MCP-1 in synovial mononuclear cells and

IL-8 in endothelial cells following injection of polyethylene particles into the knee joints of rabbits (59). Ishiguro et al. reported increased expression of MIP-1α, MCP-1, and IL-8 in the periprosthetic granulomatous tissue when compared to fibrous scar tissue (60). Chiba et al. detected MCP-1 gene expression in granulomatous tissues harvested from failed total joint arthroplasties. RANTES gene expression was not observed (61).

Onodera et al. demonstrated a possible role of macrophage migration inhibitory factor (MIF) in retrieved membranes harvested from failed total joint arthroplasties (62). Immunohistochemical analysis of human pseudosynovial tissues retrieved at revision total hip arthroplasty demonstrated mononuclear and multinuclear cells that were positively stained by both an anti-CD68 antibody and antihuman MIF antibody. In vitro challenge of the murine macrophage cell line RAW 264.7 with fluorescent-latex beads demonstrated that MIF was secreted in a particle dose-dependent manner. Northern blot analysis demonstrated elevation of MIF mRNA level. The authors also demonstrated that pretreatment of macrophages with recombinant MIF increased phagocytosis by 1.6-fold compared with untreated controls. A study by Lind et al. demonstrated the ability of MIF to inhibit migration of unchallenged, particle challenged, and LPS treated macrophages in vitro (25,63).

D. Colony-Stimulating Factors

There are several colony-stimulating factors (CSFs) including, IL-3, granulocyte-macrophage CSF, macrophage CSF, and granulocyte CSF. These are purported to have roles in the metabolism and recruitment of monocyte/macrophages, and in osteoclast development. The colony stimulating factors have been a topic of recent research by Trindade et al., Xu et al., AL-Saffar et al. and Takahashi et al. Most of the recent research has focused on granulocyte macrophage colony-stimulating factor (GM-CSF). GM-CSF is a pleiotrophic cytokine that stimulates maturation and proliferation of hematopoietic cells (64–66). GM-CSF is produced by a number of cell types including lymphocytes, macrophages, mast cells, endothelial cells and fibroblasts (67–69). GM-CSF enhances phagocytosis by macrophages and induces the synthesis and/or release of the pro-inflammatory cytokines TNF-α and IL-1β (64,70). Trindade et al. demonstrated that primary human monocytes release GM-CSF in a dose-dependent manner in response to PMMA particle challenge (42,71) (Fig. 1).

AL-Saffar et al. demonstrated the presence of GM-CSF in titanium particle challenged adherent cells (including stellate/dendritic cells, mononuclear phagocytes, fibroblasts, and endothelial cells) isolated from interfacial tissues retrieved from the bone-implant interface using intracellular staining (72). In contrast, significant expression of GM-CSF was not observed in adherent cells stimulated with polyethylene particles. GM-CSF was expressed by a distinct subset of phagocytic macrophages in the lining layer of membranes on the implant side using immunohistochemistry.

Granchi et al. examined the levels of bone-resorbing cytokines in the serum of both healthy volunteers and patients with aseptic loosening of total hip prostheses (73). The authors compared levels of IL-1β, TNF-α, IL-6 and GM-CSF for healthy volunteers and patients with either titanium alloy or cobalt-chrome alloy implants. The authors also measured the levels of soluble receptor interleukin-2 (sIL-2) as a measure of specific immune reaction against implants. The levels of sIL-2 and TNF-α did not change relative to control subjects. Levels of IL-6 were elevated (although not statistically significant) in the patients with titanium alloy implants compared to patients with cobalt-chrome alloy implants. Interleukin-1β levels were elevated in cemented-titanium groups relative to control subjects, uncemented titanium subjects, and cemented and uncemented cobalt-chrome alloy subjects. Levels of GM-CSF were elevated in all implant groups relative to control subjects with the exception of patients with

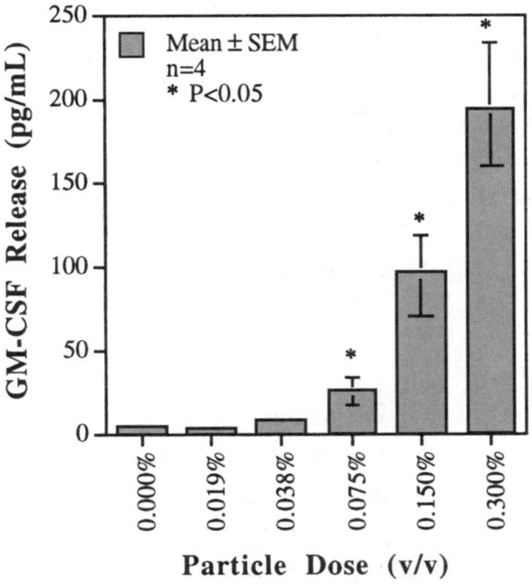

Figure 1 Dose-dependent effects of PMMA particles on macrophage granulocyte macrophage colony-stimulating factor (GM-CSF) release into culture medium. The levels of GM-CSF are expressed as pg/ml. The data represents the means for results obtained from four separate cell preparations in which each cell preparation was tested using triplicate samples for each particle dose (v/v). * denotes $p < 0.05$ vs. control cultures (0.000%).

cementless cobalt-chrome alloy implants. Moreover, the highest levels of GM-CSF levels were observed in patients taking nonsteroidal anti-inflammatory drugs (NSAIDs).

Kondo et al. described the effects of GM-CSF on osteoclast formation, bone resorption, and possible links to osteolysis (74). It has been proposed that GM-CSF can either stimulate bone resorption directly, or enhance the recruitment of osteoclasts (35,69). Xu et al. demonstrated that the number of GM-CSF positive cells per mm^2 tissue was higher in synovial-like membranes and pseudocapsules obtained from failed total joint arthroplasties than controls, using immunohistochemistry (75). The presence of M-CSF was also demonstrated (76).

Macrophage release of GM-CSF in response to particle challenge and presence in periprosthetic membranes may have diverse effects in the processes of prosthetic loosening. GM-CSF stimulates TNF-α and IL-1β release from neighboring macrophages. Furthermore GM-CSF can stimulate precursors of the macrophage lineage and increase the ability of macrophages to phagocytose orthopedic wear debris, thus increasing the input into the inflammatory cascade.

III. DIRECT EFFECTS OF PARTICLES ON OSTEOBLASTS

It was previously thought that the osteoblast response was primarily through interactions with soluble mediators released by inflammatory cells, primarily macrophages and fibroblasts. Recent studies have demonstrated that the osteoblast responds directly to particle challenge. Dean et al. demonstrated that UHMWPE particle challenge of the human osteoblast-like cell line MG-63 resulted in increased proliferation and $^{3-}$Hthymidine incorporation (77). Alkaline phosphatase activity, ^{35}S-sulfate incorporation, and osteocalcin production were inhibited in a particle-dose-dependent manner. In contrast, particle challenge increased the release of PGE2 in a dose-dependent manner.

Zambonin et al. studied the effects of PMMA powder on human osteoblast cell proliferation, collagen synthesis, osteocalcin production and IL-6 production (78). When the osteoblasts were exposed to particles, cell proliferation and collagen synthesis were inhibited. In contrast, IL-6 and osteocalcin production were increased following particle exposure. The authors hypothesized that effect of PMMA particles on bone resorption is twofold. Whereas particles act to decrease osteoblast proliferation, osteoblast release of IL-6 and osteocalcin may act to increase osteoclast recruitment and activation. Recent unpublished results from our laboratory have confirmed that osteoblasts respond to titanium particle challenge through a dose and time-dependent release of the inflammatory cytokine IL-6; release of IL-6 is known to directly activate osteoclasts to resorb bone.

Sun et al. have demonstrated the effects of hydroxyapatite and calcium phosphate particles on osteoblasts in vitro (79,80). The results demonstrated that hydroxyapatite particle challenge decreased osteoblast cell number. Furthermore, levels of TGF-β released into culture medium were suppressed relative to control cultures. Levels of the pro-inflammatory mediator PGE2 increased upon challenge with hydroxyapatite particles. In the second study, osteoblasts were challenged with four different types of calcium phosphate powders, including β-tricalcium phosphate (β-TCP), hydroxyapatite (HA), β-dicalcium pyrophosphate (β-DCP), and sintered β-dicalcium pyrophosphate (SDCP). All four treatments reduced cell counts after 1 h, 24 h and three day culture periods relative to unchallenged control cultures. With the exception of SDCP, cell counts were also reduced after seven days relative to controls. TGF-β release into culture medium varied over the time course analysis, however, elevated levels of TGF-β were observed after three days in all particle groups tested relative to control cultures. Alkaline phosphatase levels in culture medium were in general elevated over those of control cultures. PGE2 levels in culture medium were elevated relative to control cultures for all time periods tested. The authors concluded that the inhibiting effects observed on osteoblast cell proliferation were mediated by increased levels of PGE2.

IV. EFFECTS OF PARTICLES ON BONE RESORPTION

Particle containing mononuclear cells have been shown to differentiate into osteoclasts and resorb bone. Sabokbar et al. demonstrated the ability of murine macrophages cocultured with osteoblast-like cells in the presence of 1,25-dihydroxy-vitamin D3 and challenged with polymethylmethacrylate, ultrahigh molecular weight polyethylene, titanium and chromium-cobalt particles to differentiate into tartate-resistant acid phosphatase (TRAP)-positive multinucleated cells capable of lacunar bone resorption (osteoclasts) (81). Additional studies by Sabokbar et al. using a similar coculture system and PMMA particles on human cortical bone slices demonstrated osteoclast differentiation (TRAP-positive) and lacunar bone resorption (82). Interleukin-4 and leukemia inhibitory factor suppressed TRAP-positive expression and lacunar bone resorption.

Wang et al. examined direct effects of latex, PMMA and titanium particles on osteoclasts isolated from giant-cell tumors of bone and long bones in rats (83). This study demonstrated phagocytosis of particles by osteoclasts using light microscopy, energy-dispersive X-ray analysis and SEM. The osteoclasts observed were calcitonin-receptor-positive and showed an inhibitory response to calcitonin. These findings suggested that osteoclasts, having phagocytosed particles, remain functional, hormone responsive, bone resorbing cells. Further studies demonstrated the ability of osteoclasts cultured with similar particle treatments conditions on cortical bone slices were associated with the formation of resorption lacunae (84).

Anderson et al. studied the effects of hydroxyapatite, polymethylmethacrylate, and polyethylene particles on osteoclast activity using neonatal rabbit osteoclasts treated with 1,25-vitamin D3 and placed on top of bone slices (85). Bone slices were subsequently stained with antibovine collagen type I antibody and $3,3^1$-diaminobenzidine. The stained resorption pits were then quantified, and results were normalized to control cultures without particles. All three particle species increased bone resorption at lower concentrations. At high concentrations only polyethylene particles increased bone resorption relative to controls.

Bi et al. hypothesized that titanium-particle-induced osteolysis is due to differentiation of new osteoclasts rather than activation of existing osteoclasts (86). The authors challenged nucleated marrow cells from mice with varying concentrations of particles or culture alone. The filtered culture medium was added to cocultures of osteoclast precursors and mesenchymal support cells. The results demonstrated that conditioned medium increased the level of TRAP-positive activity by 30–40-fold.

V. NEW IN VITRO MODELS FOR STUDYING PARTICLE CELL INTERACTIONS

Kim et al. examined the effects of soluble mediators from high-density polyethylene, cobalt chrome alloy, titanium alloy, pure titanium, and hydroxyapatite particle challenged synovial cells on bone cells cultured on dentin slices (87). The conditioned synovial cell culture supernatants were added to unfractionated bone cells on dentin slices and resorbed areas were measured. Only supernatants from high-density polyethylene challenged synovial cells resulted in increased bone resorption.

Voronov et al. suspended UHMWPE (18–20 microns) and HDPE (4–10 microns) particles in soluble collagen type I and fixed the composite to glass coverslips (43). The study employed the use of the mouse macrophage cell line IC-21. Light and transmission electron microscopy (LM and TEM), as well as by scanning electron microscopy (SEM) were used to examine the particle-macrophage interactions. In addition, levels of inflammatory cytokines and lysosomal enzymes were measured. The results demonstrated that macrophages phagocytose small particles within 2 h, and surround larger particles within 24 h. Levels of IL-1α, IL-1β, PGE2, β-galactosidase, and hexosaminidase were significantly elevated when compared to unchallenged macrophages.

Gonzales et al. cultured murine macrophages on inverted tissue culture inserts and floated high-density polyethylene particles up to them to study the effects on macrophage metabolism. Macrophages responded to polyethylene challenge through the release of TNF-α and PGE2. Furthermore, this study added macrophage conditioned media to radiolabeled neonatal rat calvaria to assess the effects on bone resorption. Polyethylene particle challenge increased the level of bone resorption at 24, 72, and 96 h. These effects were inhibited with the addition of the bisphosphonate pamidronate (29).

Green et al. developed a model in which UHMWPE particles are imbedded into 1% agarose gel, and macrophages were placed on top (88,89). In this model, the macrophages migrate through the gel to ingest the polyethylene particles. The authors tested particles ranging from 0.24 to 88 μm as determined by scanning electron microscopy (SEM) and image analysis. The authors measured TNF-α, IL-6, and IL-1β in the cell culture supernatants and measured cell viability using a MTT (tetrazolium salt) assay following treatments. 0.24 μm particles stimulated the release of TNF-α, IL-6, and IL-1β only at the lower concentration. At the higher concentration the cytokine levels were similar to those observed in control cultures. Particles in the size range of 0.45–1.71 μm (means) stimulated production of IL-6 at lower particle concentrations and TNF-α, IL-6, and IL-1β at the higher particle concentration tested. Possible explanations for the lack of stimulation with the small particles are a feedback inhibition loop in the closed system, or the method by which small particles enter cells. The 0.24 μm particles may enter cells through pinocytosis; signal transduction cascades that activate pro-inflammatory cytokines may not be activated in the same manner as with phagocytosis of larger particles. The authors concluded that polyethylene particles of a critical size are required for the stimulation of cytokine release.

The use of cell lines for the study of particle cell interactions has focused primarily of mouse macrophage cell lines as they are well characterized. Bukata et al. has demonstrated that the human monocyte like cell line THP-1 responds to particle challenge and kinase inhibitors in a similar manner to that of peripheral blood monocytes (90). The use of cell lines is convenient for studies in which a large number of cells need to be employed, however it is not without its limitations. Glant et al. demonstrated that the response to particulate debris depend on the specific cell line chosen to study (28). In addition cell lines are known to transform spontaneously in culture.

Flow cytometry has been used to study the effects of particulate debris on macrophages. A study by Rhodes et al. demonstrated that a distinct population of highly granular cells are present in cells isolated from rats implanted with UHMWPE (91). These cells were identified as mature macrophages, and stained positive for MCP-1, The authors postulate that this subpopulation of cells may be responsible for the attraction of monocytes to the site of an implanted material. Catelas et al. used flow cytometry to establish a phagocytosis index based on the granularity and size of macrophage following exposure to ceramic (Al_2O_3 and ZrO_2) and high-density polyethylene particles (40). This study demonstrated that the phagocytosis index increased with the percentage of particles for particles <2.5 μm. When high-density polyethylene particles (4.5 μm) were added to the macrophages the phagocytosis index reached a maximum, independent of particle concentration.

VI. PARTICLES AND IMMUNE RESPONSE MECHANISMS

Activated lymphocytes are often observed in periprosthetic membranes retrieved from failed total joint arthroplasties. Patients with aseptic loosening of cemented total joint arthroplasties contain more activated T-lymphocytes than patients without aseptic loosening (92,93). Santavirta et al. examined interface membranes surrounding failed loose components and found that the membranes consisted of up to 25% CD2 positive T-cells (93,94). In addition, other immunological agents, such as interferon-gamma, are present in tissues harvested from the bone implant interface of failed orthopedic implants (16,24) One potential mechanism influencing macrophage activation and failure of total joint arthroplasties may involve lymphocyte derived factors.

Gil-Albarova et al. studied the effects of a dilute PMMA monomer solution on peripheral blood lymphocytes from patients undergoing revision surgery (92). After treatment the lymphocytes were stained for markers of activation including the IL-2 receptor, and subjected to lymphoblast transformation tests. Greater number of T- and B-lymphocytes, and IL-2 receptor positive lymphocytes were observed in patients with loose prostheses. In patients with a positive skin patch test for PMMA, a higher lymphoblast transformation rate was observed. Patients without loosening did not exhibit an increase in lymphoblast transformation upon PMMA particle challenge. These findings suggest a T-cell mediated immunologic response to PMMA in subsets of patients undergoing revision surgery.

In contrast to these findings, Kossovsky et al. studied the effect of in vivo administration of polytetrafluoroethylene (PTFE) particles, and demonstrated that lymphocyte activation in response to the polymer takes more than three months to develop (95). In vitro studies by Santavirta et al. on human peripheral blood mononuclear cells have demonstrated that exposure of blood monocytes and lymphocytes to nonphagocytosable PMMA particles (30–50 μm) led to the attachment of C3b positive monocytes, and active (CD2) positive lymphocytes following one to three days in culture (96). In addition major histocompatibility complex II molecules were seen after one day in culture with the PMMA particles. The authors found a slight increase in the IL-2 receptor (CD25) in lymphocytes cocultured with activated monocytes. Based on these data the authors suggested that the response to particle challenge is nonspecific and nonimmunological. Another study by Santavirta et al. using polyethylene particles did not demonstrate an increase in the lymphocyte IL2 receptor (97). In contrast, the marker of macrophage activation, CD11b, was upregulated upon particle exposure.

Gonzalez et al. studied the effects of PMMA particles on peripheral blood mononuclear cells (30). The cells were divided into three groups: monocytes alone, lymphocytes alone, and monocytes cocultured with lymphocytes at a ratio of 1 : 5, similar to that observed in peripheral blood. The results of the study demonstrated no significant change in IL-1β, IL-6, PGE2, or hexosaminidase in response to coculture of the cells with particles relative to particle challenge of the individual cell types. Further studies by Trindade et al., using a similar coculture system, examined the effects of particle challenge on monocytes and lymphocytes isolated from patients undergoing revision surgery (98). Macrophages were either isolated from peripheral blood, or from interfacial membranes harvested at the time of revision surgery. The cells were challenged with the particles of the same composition as the original failed prosthesis. In addition, monocytes were cocultured with peripheral blood lymphocytes at a ratio of 1 : 5, similar to that observed in peripheral blood. Cells isolated from interfacial membranes secreted increased basal levels of TNF-α and IL-6 relative to monocytes isolated from peripheral blood. No synergism was observed when the monocytes were cocultured with lymphocytes. This study was limited in specimen number, and the increased basal levels of cytokine release may have been a result of prior cell activation, particulate debris remaining in cells, or reactions to proteases used to isolate the cells.

Lee et al. studied the response of human peripheral blood monocytes of 10 normal volunteers and 14 patients with total hip replacements to particles of commercially pure titanium and chromium (99). In addition, cells isolated from the interfacial membranes were also studied. Particle challenge induced higher levels of IL-1 and PGE2 in cells isolated from patients with total hip replacements compared to cells isolated from control subjects. Bone resorption was also higher in cultures from subjects with total hip replacements. Interfacial membrane mononuclear cells secreted increased basal levels of PGE2 than that of peripheral blood monocytes stimulated with particulate wear debris. The authors suggested that the mononuclear cells from patients with total hip replacements were presensitized to metal particles.

Herman et al. tested the effects of PMMA particles on either unfractionated peripheral-blood mononuclear cells (consisting of lymphocytes and monocytes) and surface-adherent cells (monocyte-enriched) from control subjects who did and did not have clinical evidence of osteoarthrosis and from patients who had osteoarthrosis at the time of revision surgery (100). Cells were challenged with particles, or added to cell culture wells that were coated with polymerized cement. The cell culture supernatants from both methacrylate-stimulated cell populations had increased levels of IL-1, tumor necrosis factor, and PGE2 relative to control cultures. Interestingly, particle induced IL-1 release was decreased in macrophage/lymphocyte cocultures from control subjects compared to control subject macrophage cultures alone.

Haddad et al. used skin-patch test to assess the individual subjects reaction to metals, polyethylene, and the separated components of bone cement after hip replacement (101). Interestingly, of the 70 patients tested, 7 of the 15 who presented aseptic loosening in less than two years tested positive for the bone cement accelerator *N,N*-dimethylparatoluidine (DMT). Allergy to DMT is known in the dental profession as dental sore mouth syndrome. Patients could become presensitized to DMT through dental work or through exposure to industrial glues.

Trindade et al. examined the influence of interferon-gamma (IFN-gamma), a lymphocyte derived cytokine, on macrophage activation in response to orthopedic biomaterial challenge in vitro (102). Interferon-gamma was added to macrophages with and without PMMA (1–10 μm) particle challenge. Exposure of macrophages to IFN-gamma alone did not alter IL-6 release relative to controls. In contrast, when particles were added to the cultures, IFN-gamma increased levels of IL-6 release by macrophages in a dose-dependent manner (Fig. 2). IFN-gamma stimulated a dose-dependent release of TNF-α when added to the macrophages alone in contrast to the effects of IFN-gamma on IL-6 production. When particles were cocultured with IFN-gamma, TNF-α release by macrophages was increased by 17% and 171% with 1.0 and 10.0 ng/mL of IFN-gamma, respectively, when compared to particle challenged macrophages alone. Addition of neutralizing antibody to IFN-gamma inhibited the effect of IFN-gamma on particle induced IL-6 and TNF-α release by 75–85%. These data confirmed the hypothesis that the T-cell derived cytokine IFN-gamma modulates the monocyte/macrophage response to orthopedic biomaterial challenge in vitro. In vivo presence of IFN-gamma as evi-

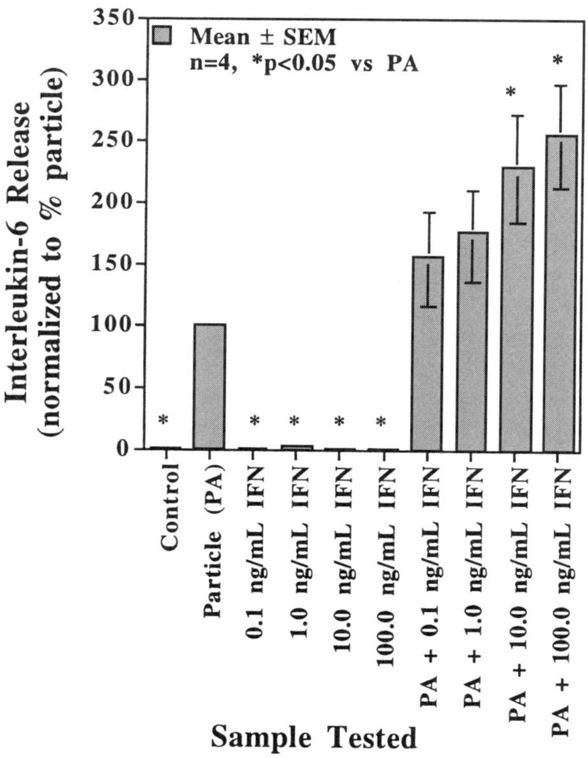

Figure 2 Effects of interferon-gamma on PMMA particle induced macrophage interleukin-6 release into culture medium. Each data plot represents the mean results for four separate cell preparations in which each experimental condition was tested in triplicate. All data are expressed as a function of the levels of interleukin-6 release observed with particle challenge (0.075% v/v). * denotes $p < 0.05$ vs. particle (PA).

denced in retrieved periprosthetic membranes is most likely to originate from T-cells, since the only other source is natural killer cells, which are rarely observed in tissues harvested from failed implants (103).

The influence of polyethylene particles on normal and T-cell deficient rats over a six week time period was examined (104). Based on histological characterization of retrieved tissues the authors concluded that T-cells are not necessary for the recruitment of macrophages to the site of particle implantation. Van Luyn et al. evaluated the effect of T-cells on macrophage recruitment in rats (nude versus control) in response to implanted cross-linked collagen (105). Their data suggests that T-cells play a crucial role in monocyte/macrophage attraction to the site of implant as evidenced by lower levels of cellular ingrowth in T-cell deficient rats at one, two, three, and four week periods. Since relatively few T-cells were detected at the implant sites, Van Luyn et al. hypothesized that either (1) only a small number of T-cells is needed at implant sites to induce macrophage attraction and/or phagocytosis, or (2) cytokines such as IFN-gamma are released further away from the implanted biomaterial which then act upon macrophages to induce infiltration and/or phagocytosis. Their data suggest that T-cells may play a role in particle mediated osteolysis.

The bone harvest chamber has provided insight into the process of bone ingrowth and remodeling in response to particle challenge, and provided a testing ground for growth factor intervention (106–112). A recent study by Trindade et al. demonstrated that tissues harvested from bone harvest chamber may closely mimic tissues harvested from failed total joint arthroplasties (113–115). This study implanted either polyethylene particles, polystyrene particles, the carrier solution (sodium hyaluronate, Healon), or nothing for periods of three, four, and six weeks with intervening rest periods. Following removal, the tissues were placed in culture medium, and levels of the pro-inflammatory mediators IL-6, TNF-α, and IL-1β release were measured using bioassays. The levels of cytokine release were elevated upon particle challenge at three and four weeks, but subsided at six weeks, possibly due in part to clearance mechanisms (Fig. 3).

Merkel et al. established a murine model of osteolysis in which PMMA particles are implanted subperiosially on the external calvaria of mice (116). Following one to three week treatments, the mice are sacrificed and the calvaria are excised, fixed and decalcified in EDTA. In this study 2–3 mm sections were taken and stained for either TNF-α or the osteoclast marker TRAP. Macrophages that were TNF-α positive and numerous cells that were TRAP-positive were predominant in the sections within one week of exposure. The tissue that was obtained had similar characteristics to that obtained from failed total joint arthroplasties. In addition, the authors demonstrated a crucial role of the p55 tumor necrosis factor receptor through the use of a mouse with the deletion of this gene. Another study by Merkel demonstrated that these effects are not due to possible endotoxin left on the particles with the use of endotoxin resistant HEJ mice (117).

VII. ANTI-INFLAMMATORY CYTOKINES

Several recent studies have focused on the effects of the anti-inflammatory cytokines IL-4 and IL-10. Interleukin-4 is a multifaceted anti-inflammatory cytokine that modulates the functions of numerous cell types including T-cells, B-cells, monocytes, macrophages, fibroblasts, neutrophils, hematopoietic progenitors, endothelial cells, and epithelial cells (118,119). IL-4 upregulates expression of class II MHC antigens, lymphocyte function associated-1 (LFA-1), and CD23 on macrophages. In addition, IL-4 inhibits macrophage expression of receptors for IgG,

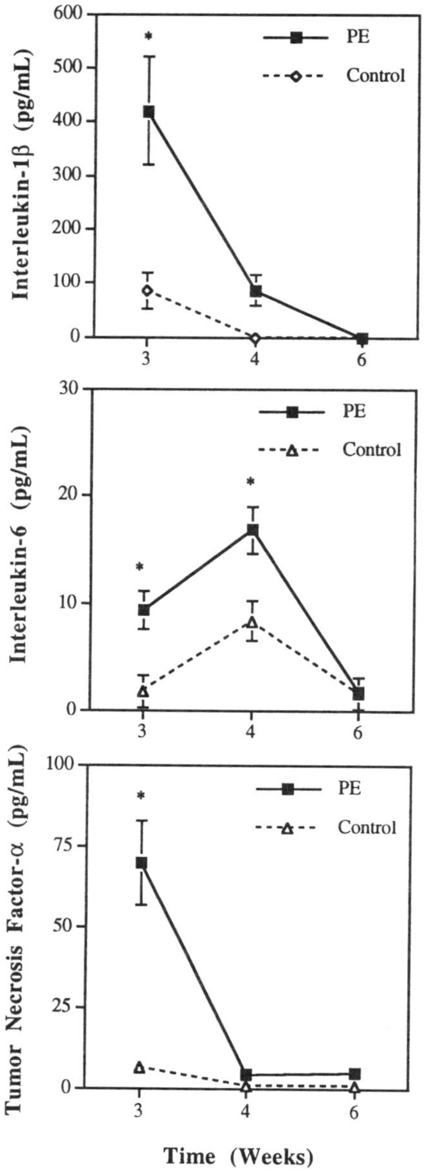

Figure 3 Interleukin-1β, interleukin-6, and tumor necrosis factor-α release by tissue explants from the bone harvest chamber at three, four, and six weeks. Data is presented as mean ± SEM. * denotes $p < 0.05$ vs. control. Control = no particles, PE = UHMWPE particles.

and expression of the pro-inflammatory cytokines IL-1, IL-6, IL-8, and TNF-α. Furthermore, IL-4 inhibits the release of superoxide anion and collagenase by macrophages (118–121).

A study by Pollice et al. demonstrated that both IL-4 and IL-10 inhibit titanium particle induced IL-6 and TNF-α expression (122). These studies also demonstrated the presence of these cytokines in membranes harvested from failed total joint replacements. These cytokines appeared to be adjacent to the bone/implant interface and were surrounded by populations of

T-lymphocytes. Trindade et al. demonstrated that IL-10 inhibits PMMA particle induced IL-6 and TNF-α release from particle challenged macrophages in vitro (102) (Figs. 4,5). Sabokbar et al. demonstrated that IL-4 reduced expression of TRAP by PMMA associated mouse monocytes cocultured with UMR 106 osteoblast-like cells (82). In other studies IL-4 stimulated osteoblasts, inhibited osteoclast-associated TRAP activity, and exerted a chemotactic effect on osteoblasts in vitro (123–126).

Hsu et al. studied the effects of titanium particles on macrophage and lymphocyte production of IL-10 (127). Medium from macrophages challenged with titanium particles induced the release of IL-10 from macrophages and lymphocytes in a dose-dependent manner. This effect was partially blocked by addition of antibodies to TNF-α.

Merkel et al. demonstrated that PMMA particle challenge of macrophages leads to the induction of TNF-α through transactivation of the TNF-promoter-CAT reporter (117). Interleukin-4 addition inhibited this effect. Studies by Trindade et al. tested the effects of IL-4 on PMMA particle challenged macrophage release of GM-CSF, IL-6, TNF-α, and MMPs. Interleukin-4 inhibited IL-6, TNF-α, GM-CSF, and MMP-9 expression by human monocytes in response to polymethylmethacrylate (PMMA) particle challenge in a dose-dependent manner (71,98) (Fig. 6).

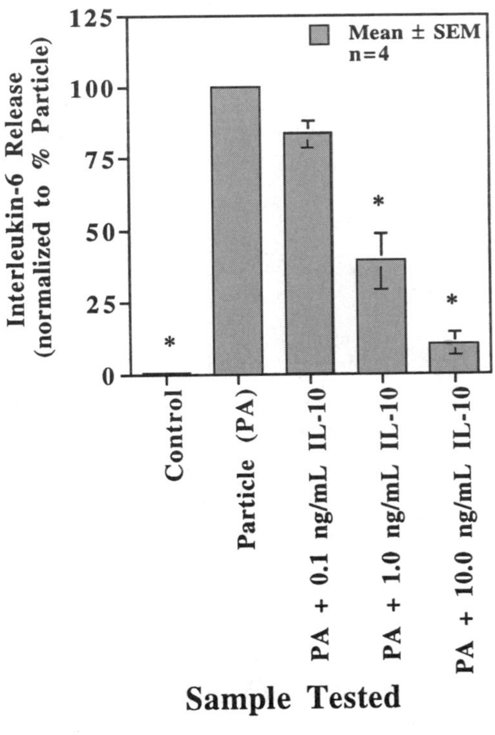

Figure 4 Effect of interleukin-10 on PMMA particle-induced macrophage interleukin-6 release into culture medium. The data represent the mean and standard error of the mean for results from four separate cell preparation in which the effects of the IL-10 were tested in triplicate. Statistical significance represents the difference between particle treated cells and either control cultures or cells exposed to particles in the presence of IL-10.

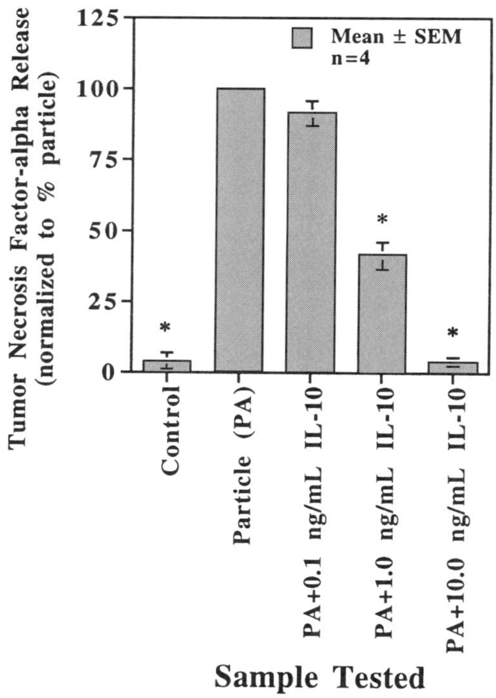

Figure 5 Effect of interleukin-10 on PMMA particle-induced macrophage tumor necrosis factor-α release into culture medium. The data represent the mean and standard error of the mean for results from four separate cell preparation in which the effects of the IL-10 were tested in triplicate. Statistical significance represents the difference between particle treated cells and either control cultures or cells exposed to particles in the presence of IL-10.

VIII. OPSONIZATION OF PARTICULATE DEBRIS

Work by Maloney et al. tested the effects of serum protein binding (opsonization) of titanium particles on macrophage metabolism in vitro (128). They assessed whether pretreatment of machined titanium-alloy particles with human serum would influence human macrophage activation and cytokine release in vitro. Their results demonstrated that opsonization of the titanium particles with human serum increased macrophage release of TNF-α, IL-1β, and IL-6 after 12 and 24 h exposure periods. Furthermore, the results of this study demonstrated that selected serum proteins bind titanium particles.

Sun et al. used one- and two-dimensional gel electrophoresis to characterize specific serum protein binding to titanium alloy, PMMA, hydroxyapatite and aluminum oxide particles (129–130) Their results demonstrated distinct profiles of serum protein binding for the different particle species. Five major bands were isolated from the PMMA particle extracts, and amino acid sequencing determined these corresponded to single proteins.

Recent work be Ragab et al. has demonstrated titanium particles used for particle challenge experiments have a high affinity to bind the bacterial endotoxin (LPS) (131). This discovery led to the development of new techniques to clean particles and improve the accuracy of results obtained from in vitro experiments. Whether titanium particles act as "endotoxin reservoirs" in vivo is yet to be determined. Hitchins and Merritt demonstrated that the standard

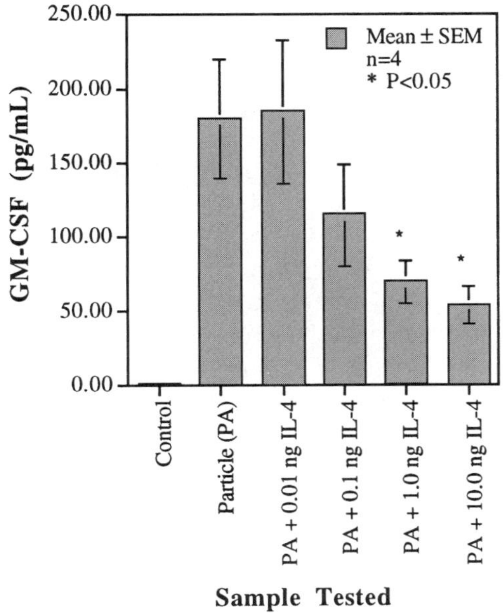

Figure 6 Effect of interleukin-4 on PMMA particle-induced macrophage granulocyte macrophage colony-stimulating factor (GM-CSF) release into culture medium. The data represent the mean and standard error of the mean for results from four separate cell preparation in which the effects of the IL-4 were tested in triplicate. Statistical significance represents the difference between particle treated cells and cell exposed to particles in the presence of IL-4.

ethanol and PBS washing method used in many laboratories is sufficient to remove endotoxin from PMMA particles (132).

IX. SIGNAL TRANSDUCTION/MECHANISM STUDIES

Studies on the signal transduction mechanisms by which macrophages and fibroblasts respond to particulate debris may provide insights into the process of osteolysis, and points for therapeutic intervention.

Nakashima et al. demonstrated the involvement of the transcription factors NF-κB and NF-IL-6 in the process of macrophage activation by orthopaedic biomaterials (41,133). Nakashima et al. also demonstrated the role of kinase A, C, and G in macrophage IL-6 and TNF-α expression. These studies also demonstrated a possible role of MAP kinase and differences in the activation of macrophages by LPS and particulate debris.

Using cyclohexamide, Nakashima et al demonstrated that phagocytosis of particulate debris is not required for the induction of macrophage activation.

Ritchie et al. demonstrated the role of cAMP in particle induced macrophage cytokine expression (134). In this study, cAMP, modulators, co-incubated with particles, inhibited TNF-α secretion by 57%, 89% and 97% with treatments of dibutyrl cyclic AMP, PGE2, and cholera toxin, respectively. The mechanisms of inhibition of these cAMP modulators differ. Dibutyrl cyclic AMP functions as a cAMP analog, whereas PGE2 and cholera toxin function to decrease intracellular levels of cAMP. The authors noted that the inability of PGE2 to inhibit TNF-α mRNA synthesis suggests that TNF-α regulation by particulate debris is posttranscriptionally regulated.This work is in agreement with that of Blaine et al., who also demonstrated the involvement of cAMP in particulate activation of macrophages in vitro (27).

Horowitz et al. has demonstrated the role of the macrophage-osteoblast interaction in stimulating increased release of pro-inflammatory mediators in response to particle challenge (5). Other coculture systems have been employed by Lind et al. (135). This group studied the interaction of the macrophage and fibroblast using a noncontact coculture system. In this system, macrophages were either exposed or not exposed to phagocytosable PMMA particles. In addition, the ratio of macrophages to fibroblasts was varied. IL-6 release in cocultures was increased 100-fold for both unchallenged and particle challenged cultures when compared with macrophage cultures alone. Addition of particles to the cocultures elevated IL-6 release threefold. TNF-α release was similar in cocultures and in macrophage cultures. IL-1β release in cocultures was independent of the macrophage-fibroblast ratio. Matrix metalloproteinase and hexosaminidase activity were elevated in cocultures relative to cultures of macrophages and fibroblasts alone. The increase IL-6 release in cocultures suggests that macrophage accumulation in fibrous tissue may lead to elevated IL-6 levels that are much higher than those caused by particle activation of macrophages alone. This is the first study that demonstrated the ability of the macrophage to modulate fibroblast expression of pro-inflammatory mediators.

Horowitz et al. studied mechanisms by which macrophages interact with osteoblasts and result in the secretion of cytokines (5,6). The authors confirmed the results of numerous studies showing that PMMA particle challenge of macrophages results in the release of tumor necrosis factor. The authors subsequently added macrophage conditioned medium to osteoblasts, and demonstrated an increase in the release of GM-CSF, IL-6, and PGE-2 relative to osteoblasts challenged with medium from unchallenged macrophages. The addition of antibodies to TNF to the conditioned medium prior to osteoblast exposure inhibited the release of GM-CSF, IL-6, and PGE-2 by the osteoblasts. The authors hypothesized from these results that TNF is a crucial mediator of osteoblast activity that may influence osteoclast activation and lead to prosthesis loosening.

X. INHIBITION OF DELETERIOUS EFFECTS OF PARTICLES ON BONE

Several recent studies have examined the effects of exogenous cytokines and growth factors on bone ingrowth and remodeling. These studies may lead to minimally invasive therapeutic treatments for osteolysis, possibly extending the life span of loosened implants in elderly patients in whom a revision surgery would pose to much of a challenge.

Studies by Goodman et al. using the bone harvest chamber model tested the effects of TGF-β on bone ingrowth in the presence of polyethylene particles (111). In this study the addition of TGF-β inhibited the negative effects of particles on bone ingrowth at four weeks. Interestingly, the opposite effect was observed at two weeks, demonstrating the pleiotrophic nature of TGF-β.

Canine models have provided great insights into the process of loosening, and a test bed for new implant designs. Moreover, there have been attempts to mimic the process by which aseptic loosening and osteolysis occur using canines. One study by Shanbhag et al. demonstrated that alendronate provides increased fixation in a canine osteolysis model (136). In this study, one group of dogs were treated with a mixture of UHMWPE, TiALV, and cobalt chrome alloy. The other groups were implanted without particles, or particles with oral drug therapy (5 mg alendronate sodium, once a day throughout the study starting seven days postoperatively). The animals were sacrificed at 24 week, and samples were taken for organ culture and histology. Radiographs were taken post operatively. Implants were considered loose in six of the seven animals exposed to particles without the addition of treatment drug. The addition of treatment drug improved fixation to the level of control animals (seven out of eight in both cases considered well-fixed). In addition, bioassays on the organ culture supernatants demonstrated that particulate debris exposure increased the levels of IL-1 and PGE2 in the particle challenged groups relative to controls. Alendronate had no effect on inflammatory mediator production. These results contrast those of Ritchie et al. (137). In this study, it was demonstrated that alendronate inhibits PMMA particle induced macrophage pro-inflammatory cytokine release in vitro. Alendronate is known to incorporate in the mineral matrix making it refractory to osteoclast resorption. The long-term effect of treatment with alendronate in terms of the mechanical properties of bone surrounding implants are unknown.

Studies by Jeppsson et al. tested the effects of OP-1 (BMP-7) and BMP-2 on bone formation using the bone harvest chamber model (138). In contrast to other studies, BMP-2 inhibited total tissue ingrowth but did not significantly affect bone percentage area. OP-1 inhibited total tissue ingrowth and bone percentage area, albeit only at one of the doses tested. These data suggest that BMP's are pleiotrophic, and their effects on bone formation are dose dependent. BMPs may inhibit the negative effects of particles on bone ingrowth; however, this hypothesis was not tested.

XI. DISCUSSION

Longevity of total joint arthroplasty appears to be influenced by the generation and biological effects of wear debris (139). Granulomatous tissue that forms at sites of implant loosening is thought to play a role in particle mediated osteolysis. Studies on retrieved periprosthetic membranes reveal numerous macrophages, lymphocytes, fibroblasts, foreign body giant cells, and abundant particulate debris. Macrophages phagocytose orthopedic debris and release the pro-inflammatory mediators IL-1, IL-6, TNF-α, and PGE2. The release of inflammatory mediators and cell differentiation may stimulate localized bone resorption (140).

Recruitment and accumulation of monocytes to inflammatory lesions remains a key factor that underlies granuloma formation. Recruitment of cells depends in part on chemotactic agonists generated at the site of inflammation. Several studies suggest that MCP-1 and MIP-1α expression following particle exposure in vivo may contribute to granuloma formation through the induction of macrophage recruitment. The regulation of cytokine and chemokine expression is complicated with overlapping effects and methods of stimuli. MCP-1 induces IL-1 and IL-6 release from monocytes (51). In addition, the pro-inflammatory cytokines IL-1 and TNF-α are know to induce macrophage and fibroblast MCP-1 gene expression (141). IL-1 and TNF-α also stimulate MIP-1α expression by macrophages, T-lymphocytes, and fibroblasts (141). Another C-C- chemokine, MIP-1α, stimulates production of IL-1, IL-6, and TNF-α by macrophages (49).

Remodeling of periprosthetic bone is a complex process that involves multiple cell types and signal transduction pathways. The expression of matrix metalloproteinases by both macrophages and fibroblasts and their effects on cellular activation and cytokine release may also play a crucial role in this process. The potentially devastating effects of wear particles can be inhibited by the use of anti-inflammatory cytokines, or other therapeutic drugs. Clearance mechanisms in vivo are usually not sufficient to overcome the amount of particles generated and thus, at the present time, the treatment for advanced particle induced osteolysis appears to be revision total joint arthroplasty. With the advent of new polyethylenes and other prosthetic materials, the life span of total joint arthroplasties should increase.

The role of lymphocytes in particle mediated osteolysis is debatable. Although no conclusive evidence has been demonstrated, both in vitro experiments using T-cell derived cytokines to modulate the macrophage response to particulate debris, and studies on retrieved membranes demonstrating activated lymphocytes add strength to the hypothesis that in some cases lymphocytes are involved in the modulation of particle induced osteolysis. The presence of lymphocytes in periprosthetic membranes retrieved from failed total joint arthroplasties has been demonstrated in several studies (24,28,93). Whether T-lymphocytes play a role in particle induced osteolysis is a subject of much debate. Many of the studies examined above add to this debate. The ability of T-cell derived cytokine to modulate the macrophage response to orthopedic biomaterial challenge was the primary goal of the study by Trindade et al. (102). In this study, we noted that the changes observed upon IFN-gamma addition may be due to changes in macrophage phagocytosis ability, cell-cell interactions, cytokine pathways, or a combination the above. We propose a mechanism in some cases of osteolysis in which T-cells are either activated, or stimulated by monocyte/macrophage release of IL-1β to secrete the cytokine IFN-gamma, which then acts on macrophages to increase their reactivity to surrounding particles. This may be in part through increased expression of class II MHC molecules. The expression of class II MHC molecules by macrophages is known to induce phagocytosis.

While in vitro models often lack complex cellular interactions that are inherent in vivo, they have provided insights into the biological response to wear debris. In vivo models such as the mouse, rabbit, and canine models described here have furthered our understanding of the biological response to wear debris. However, animal models are relatively short term compared to the time implants are in humans, and therefore may not accurately represent the process of osteolysis that occurs in humans. Treatments with therapeutic agents alters the mechanical properties of the bone surrounding loosened treatments. In addition, bones throughout the body may be affected.

In the future, total joint replacements will continue to be used for congenital conditions, posttrauma treatments, and degenerative inflammatory processes such as end-stage osteoarthritis and rheumatoid arthritis. Implants are being performed on younger more active patients; thus, the longevity must be improved. Osteolysis will continue to be a major problem, however, the life span of TJAs may be improved in part through the use of new materials, manufacturing techniques, therapeutic agents such as bisphosphonates and growth factors, hydroxyapatite coatings, and possibly implant prosthesis and material matching to specific individuals to mitigate possible adverse responses to specific materials.

REFERENCES

1. Mulroy, W. F. and Harris, W. H., 1997. Acetabular and femoral fixation 15 years after cemented total hip surgery. *Clin. Orthop.*, 337: 118–128.
2. Mulroy, W. F. and Harris, W. H., 1996. Revision total hip arthroplasty with use of so-called second-generation cementing techniques for aseptic loosening of the femoral component. A fifteen-year-average follow-up study. *J. Bone Joint Surg.*, 78: 325–30.
3. Mulroyk, W. F., Estok, D. M., and Harris, W. H., 1995. Total hip arthroplasty with use of so-called second-generation cementing techniques. A fifteen-year-average follow-up study. *J. Bone Joint Surg.*, 77: 1845–1852.
4. Scuderi, G. R., Insall, J. N., Windsor, R. E., and Moran, M. C., 1989. Survivorship of cemented knee replacements. *J. Bone Joint Surg.*, 71-B: 798–803.
5. Horowitz, S. M. and Gonzales, J. B., 1996. Inflammatory response to implant particulates in a macrophage/osteoblast coculture model. *Calcif. Tissue Int.*, 59: 392–396.
6. Horowitz, S. M. and Purdon, M. A., 1995. Mechanisms of cellular recruitment in aseptic loosening of prosthetic joint implants. *Calcif. Tissue Int.*, 57: 301–305.

7. Lewis, G., 1997. Polyethylene wear in total hip and knee arthroplasties. *J. Biomed. Mater. Res.*, 38: 55–75.
8. Maloney, W. J. and Smith, R. L., 1995. Periprosthetic osteolysis in total hip arthroplasty: the role of particulate wear debris. *J. Bone Joint Surg.*, 77: 1448–1461.
9. Willert, H. G., Buchhorn, G. and Semililiteritsch, M., 1981. Recognition and identification of wear products in surrounding tissues of artificial joint prostheses. In: J H Dumbleton (Editor), *Tribology of Natural and Artificial Joints*. Elsevier Scientific. Amsterdam, pp. 381–419.
10. Davis, R. G., Goodman, S. B., Smith, R. L., Lerman, J. A. and Williams III, R. J., 1993. The effects of bone cement powder on human adherent monocytes/macrophages in vitro. *J. Biomed. Mater. Res.*, 27: 1039–1046.
11. Freeman, M. A. R., Bradley, G. W., and Revell, P. A., 1982. Observations upon the interface between bone and polymethylmethacrylate cement. *J. Bone Joint Surg.*, 64-B: 489–493.
12. Glant, T. T., Jacobs, J. J., Molnar, G., Shanbhag, A. S., Valyon, M. and Galante, J. O., 1993. Bone resorption activity of particulate-stimulated macrophages. *J. Bone Miner. Res.*, 8: 1071–1079.
13. Goldring, S. R., Clark, C. R., and Wright, T. M., 1993. The problems in total joint arthroplasty. *J. Bone Joint Surg.*, 75: 799–801.
14. Harris, W. H., Schiller, A. L., Scholler, J. M., Freiberg, R. A., and Scott, R., 1976. Extensive localized bone resorption in the femur following total hip replacement. *J. Bone Joint Surg.*, 58: 612–618.
15. Maloney, W. J., Jasty, M., Harris, W. H., Galante, J. O., and Callaghan, J. J., 1990. Endosteal erosion in association with stable uncemented femoral components. *J. Bone Joint Surg.*, 72: 1025–1034.
16. Goodman, S. B., Knoblich, G., O'Connor, M., Song, Y., Huie, P., and Sibley, R., 1996. Heterogeneity in cellular and cytokine profiles from multiple samples of tissue surrounding revised hip prostheses. *J. Biomed. Mater. Res.*, 31: 421–428.
17. Greis, P. E., Geoegescu, H. I., Fu, F. H., and Evans C. H., 1994. Particle-induced synthesis of collagenase by synovial fibroblasts: an immunocytochemical study. *J. Orthop. Res.*, 12:286–293.
18. Horowitz, S. M., Frondoza, C. G., and Lennox, D. W., 1988. Effects of polymethylmethacrylate exposure upon macrophages: macrophage response to polymethylmethacrylate. *J. Orthop. Res.*, 6: 827–832.
19. Howie, D. W., 1990. Tissue response in relation to type of wear particles around failed hip arthroplasties. *J. Arthroplasty*, 5: 337–348.
20. Jasty, M., Floyd, W.E., Schiller, A. L., Goldring, S. R., and Harris, W. H., 1986. Localized osteolysis in stable non-septic total hip replacement. *J. Bone Joint Surg.*, 68: 912–919.
21. Johanson, N. A., Bullough, P. G., Wilson Jr, P. D., Salvati, E. A., and Ranawat, C. S., 1987. The microscopic anatomy of the bone-cement interface in failed total hip arthroplasties. *Clin. Orthop.*, 218: 123–135.
22. Mirra, J. M., Amstutz, H. C., Matos, M., and Gold, R., 1976. The pathology of the joint tissues and its clinical relevance in prosthesis failure. *Clin. Orthop.*, 117: 221–240.
23. Goldring, M. B. and Goldring, S. R., 1990. Skeletal tissue response to cytokines (review). *Clin. Orthop.*, 258: 245–278.
24. Goodman, S. B., Huie, P., Song, Y., Lee, K., Doshi, A., Rushdieh, B., Woolson, S., Maloney, W., Schurman, D., and Sibley, D., 1997. Loosening and osteolysis of cemented joint arthroplasties. *Clin. Orthop. Rel. Res.*, 337: 149–163.
25. Lind, M., Trindade, M. C. D., Nakashima, Y., Schurman, D. J., Goodman, S. B., and Smith, R. L. 1999. Chemotaxis and activation of particle challenged human monocytes in response to monocyte migration inhibitory factor and C-C chemokines. *J. Biomed. Mater. Res.*, *48*: 246–250.
26. Schmalzried, T. P., Jasty, M., and Harris, W. H., 1992. Periprosthetic bone loss in total hip arthroplasty. polyethylene wear debris and the concept of the effective joint space. *J. Bone Joint Surg.*, 74: 849–863.
27. Blaine, T. A., Pollice, P. F., Rosier, R. N., Reynolds, P. R., Puzas, J. E., and O'Keefe, R. J., 1997. Modulation of the production of cytokines in titanium-stimulated human peripheral blood

monocytes by pharmacological agents. The role of cAMP-mediated signaling mechanisms. *J. Bone Joint Surg.*, 79: 1519–1528.
28. Glant, T. T. and Jacobs, J. J. 1994. Response of three murine macrophage populations to particulate debris: Bone resorption in organ culture. *J. Orthop. Res.*, 12: 720–731.
29. Gonzales, J. B. and Horowitz, S. M., 1997. The cellular response to polyethylene particles. Transactions of the 43rd Annual Meeting of the Orthopaedic Research Society, 22: 731.
30. Gonzalez, O., Smith, R. L., and Goodman, S. B., 1997. The effects of PMMA particles on human macrophages and lymphocytes in vitro. Transactions of the 42nd Annual Meeting of the Orthopaedic Research Society, 21: 504.
31. Gonzalez, O., Smith, R. L., and Goodman, S. B., 1996. Effect of size, concentration, surface area and volume of polymethylmethacrylate particles on human macrophages in vitro. *J. Biomed. Mater. Res.*, 30: 463–473.
32. Shanbhag, A. S., Jacobs, J. J., Black, J., Galante, J. O., and Glant, T. T., 1995. Human monocyte response to particulate biomaterials generated in vivo and in vitro. *J. Orthop. Res.*, 13: 792–801.
33. Chambers, T. J., 1991. Regulation of osteoclastic bone resorption in vitro. In: B. K. Hall (Editor), *Bone*. Volume 2: *The osteoclast*. Boca Raton, CRC press, Inc., pp. 141–173.
34. Evert, V., Delaisse, J. M., Korper, W., Niehof, A., Vaes, G., and Beersten, W., 1992. Degradation of collagen in the bone-resorbing compartment underlying the osteoclast involves both cysteine-proteinases and matrix metalloproteinases. *J. Cell. Physio.*, 150: 221–231.
35. Mohan, S. and Baylink, D. J., 1991. Bone growth factors (review). *Clin. Orthop.*, 263: 30–48.
36. Maloney, W. J., James, R. E., and Smith, R. L., 1996. Human macrophage response to retrieved titanium alloy particles in vitro. *Clin. Orthop. Rel. Res.*, 322: 268–278.
37. Mundy, G. R., 1993. Cytokines and growth factors in the regulation of bone remodeling. *J. Bone Miner. Res.*, 8: 505–510.
38. Schmalzried, T. P., Maloney, W. J., Jasty, M., Kwong, L. M., and Harris, W. H., 1993. Autopsy studies of the bone-cement interface in well-fixed cemented total hip arthroplasties. *J. Arthroplasty*, 8: 179–188.
39. Vaes, G., 1988. Cellular biology and biochemical mechanism of bone resorption: a review of recent development on the formation, activation, and mode of action of osteoclasts. *Clin. Orthop.*, 231: 239–271.
40. Catelas, I., Petit, A., Zukor, D., Marchland, R., Yahia, L. H., and Huk, O. L., 1998. Ceramic and polyethylene particles induce TNF-a release after their phagocytosis by mouse macrophages in vitro. Transactions of the 24th Annual Meeting of the Society for Biomaterials, 21: 60.
41. Nakashima, Y., Trindade, M., Sun, D., Maloney, W., Goodman, S., Schurman, D., and Smith, R.L., 1997. The effects of signal transduction inhibitors on macrophage activation in vitro: requirement for tyrosine kinase activity. Orthopaedic Research Society Transactions of the 43rd Annual Meeting, 22: 323.
42. Trindade, M. C. D., Nakashima, Y., Lind, M., Goodman, S. B., Maloney, W. J., Schurman, D. J., and Smith, R. L., 1998. Inhibition of particle induced monocyte expression of GM-CSF by interleukin-4. Society For Biomaterials Transactions of the 24th Annual Meeting, 21: 59.
43. Voronov, I., Santerre, J. P., Hinek, A., Callahan, J. W., Sandhu, J., and Boynton, E. L., 1998. Macrophage phagocytosis of polyethylene particulate in vitro. *J. Biomed. Mater. Res.*, 39: 40–51.
44. Shanbhag, A. S., Macaulay, W., Stefanovic-Racic, M., and Rubash, H. E., 1998. Nitric oxide release by macrophages in response to particulate wear debris. *J. Biomed. Mater. Res.*, 41: 497–503.
45. Kim, K. J., Chiba, J., and Rubash, H., 1994. In vivo and in vitro analysis of membranes from hip prostheses inserted without cement. *J. Bone Joint Surg.*, 76: 172–180.
46. Nakashima, Y., Sun, D. H., Maloney, W. J., Goodman, S. B., Schurman, D. J., and Smith, R. L., 1998. Induction of matrix metalloproteinase expression in human macrophages by orthopaedic particulate debris in vitro. *J. Bone Joint Surg.*, 80-B: 694–700.
47. Yokohama, Y., Matsumoto, T., Hirakawa, M., Kuroki, Y., Fujimoto, N., Imai, K., and Okada, Y.,

1995. Production of matrix metalloproteinases at the bone-implant interface in loose total hip replacements. *Lab. Invest.*, 73: 899–911.
48. Yao, J., Glant, T. T., Lark, M. W., Mikecz, K., Jacobs, J. J., Hutchinson, N. I., Hoerrner, L. A., Kuettner, K. E., and Galante, J. O., 1995. The potential role of fibroblasts in periprosthetic osteolysis: fibroblast response to titanium particles. *J. Bone Miner. Res.*, 10: 1417–27.
49. Rollins, B. J., 1997. Chemokines. *Blood*, 90: 909–928.
50. Driscoll, K. E., 1994. Macrophage inflammatory proteins: biology and role in pulmonary inflammation. *Exp. Lung Res.*, 20: 473–490.
51. Jiang, Y., Beller, D. I., Frendl, G., Graves, D. T., 1992. Monocyte chemoattractant protein-1 regulates adhesion molecule expression and cytokine production in human monocytes. *J. Immunol.* 148: 2423–2428.
52. Miller, M. D., Krangel, M. S., 1992. Biology and biochemistry of the chemokines: a family of chemotactic and inflammatory cytokines. *Crit. Rev. Immunol.*, 12: 17–46.
53. Murphy, P. M., 1994. The molecular biology of leukocyte chemoattractant receptors. *Annu. Rev. Immunol.*, 12: 593–633.
54. Taub, D. D. and Oppenheim, J. J., 1993. Review of the chemokine meeting the Third International Symposium of Chemotactic Cytokines. Cytokine, 5: 175–179.
55. Kelner, G. S., Kennedy, J., Bacon, K. B., Kleyensteuber, S., Largaespada, D. A., Jenkins, N. A., Copeland, N. G., Bazan, J. F., Moore, K. W., Schall, T. J., and Zlotnik, A., 1994. Lymphotactin: A cytokine that represents a new class of chemokine. *Science*, 266: 1395–1399.
56. Bazan, J. F., Bacon, K. B., Hardlman, G., Wang, W., Soo, K., Rossi, D., Greaves, D. R., Zlotnik, A., and Schall, T. J., 1997. A new class of membrane-bound chemokine with a CX_3C motif. *Nature*, 385: 640–644.
57. Pan, Y., Lloyd, C., Zhou, H., Dolich, S., Deeds, J., Gonzalo, J. A., Vath, J., Gosselin, M., Ma, J., Dussault, B., Woolf, E., Alperin, G., Culpepper, J., Gutierrez-Ramos, J. C., and Gearling, D., 1997. Neurotactin, a membrane-anchored chemokine upregulated in brain inflammation. *Nature*, 387: 611–617.
58. Nakashima, Y., Sun, D. H., Trindade, M. C. D., Chun, L. E., Song, Y., Maloney, W. J., Goodman, S. B., Schurman, D. J., and Smith, R. L., 1999. Induction of Macrophage C-C Chemokine Expression by Titanium Alloy and Bone Cement Particles, *J. Bone Joint Surg.* (*Br*), 81: 155–162.
59. Frokjaer, J., Deleuran, B., Lind, M., Overgaard, S., Soballe, K., and Bunger C., 1995. Polyethylene particles stimulate monocyte chemotactic and activating factor production in synovial mononuclear cells in vivo. *Acta. Orthop. Scand.*, 66: 303–307.
60. Ishiguro, N., Kojima, T., Kurokouchi, K., Iwase, T., and Iwata, H., 1997. mRNA expression of chemokines in interface tissue around of loosening total hip arthroplasty components. *Trans. Orthop. Res. Soc.*, 22: 735.
61. Chiba, J., Oyama, M., Sugawara, S., Inoue, K., and Rubash, H. E., 1996. The role of chemokine, adhesion molecules, and cytokine receptor in femoral osteolysis after cementless total hip arthroplasty. *Trans. Orthop. Res. Soc.*, 21: 514.
62. Onodera, S., Suzuki, K., Matsuno, T., Kaneda, K., Takagi, M., and Nishihira, J. 1997., Macrophage migration inhibitory factor induces phagocytosis of foreign particles by macrophages in autocrine and paracrine fashion. *Immunology*, 92:131–137.
63. Lind, M., Trindade, M. C. D., Nakashima, Y., Schurman, D. J., Smith, R. L., and Goodman, S. B., 1998. Chemotaxis and activation of human monocytes in response to monocyte migration inhibitory factor and C-C chemokines, Orthopaedic Research Society Transactions of the 44th Annual Meeting, 23: 474.
64. Lopez, A. F., Eglinton, J. M., Lyons, A. B., Tapley, P. M., To, L. B., Park, L. S., Clark, S. C., and Vadas, M. A., 1990. Human interleukin-3 inhibits the binding of granulocyte-macrophage colony-stimulating factor and interleukin-5 to basophils and strongly enhances their functional activity. *Cell. Physiol.*, 145: 69–77.
65. Lopez, A. F., Elliot, M. J., Woodcock, J., and Vadas, M. A., 1992. GM-CSF, IL-3 and IL-5: cross-competition on human haemopoietic cells. *Immunol. Today.*, 13:495–500.
66. Nakata, K., Akagawa, K., Fukayama, M., Hayashi, Y., Kadokura, M., and Tokunaga, T., 1991.

Granulocyte-macrophage colony stimulating factor promotes the proliferation of alveolar macrophages in vitro. *J. Immunol.*, 147: 1266–1272.
67. Chen, B. D. M., Clark, C. R., and Chou, T. H., 1998. Granulocyte-macrophage colony stimulating factor stimulates monocyte and tissue macrophage proliferation and enhances their responsiveness to macrophage colony stimulating factor. *Blood*, 71: 997–1002.
68. Collins, H. L. and Bancroft, G. J., 1992. Cytokine enhancement of complement-dependent phagocytosis by macrophages: synergy of tumor necrosis factor-alpha and granulocyte-macrophage colony-stimulating factor for phagocytosis of Cryptococcus neoformans. *Eur. J. Immunol.*, 22: 1447–1454.
69. Takahashi, N., Udagawa, N., Akatsu, T., Tanaka, H., Shionome, M., and Suda, T., 1991. Role of colony stimulating factors in osteoclast development. *J. Bone Miner. Res.*, 6: 977–985.
70. Scheicher, C., Mehlig, M., Zecher, R., Seiler, F., Hintz-Obertreis, P., and Reske, K., 1993. Recombinant GM-CSF induces in vitro differentiation of dendritic cells from mouse bone marrow. In: E. W. A. Kamperdijk, P. Nieuwenhuis and E. C. M. Hoefsmit (Editors), *Dendritic Cells in Fundamental and Clinical Immunology*. Plenum Press, New York, V329, pp. 269–273.
71. Trindade, M. C. D., Nakashima, Y., Sun, D. H., Goodman, S. B., Schurman, D. J., and Smith, R. L., 1998. The effects of interleukin-4 on particle challenged macrophages in vitro, Orthopaedic Research Society Transactions of the 44th Annual Meeting, 23: 176.
72. AL-Saffar, N., Khwaja, H. A., Kadaoya, Y., and Revell, P. A., 1996. Assessment of the Role of GM-CSF in the Cellular Transformation and the Development of Erosive Lesions Around Orthopaedic Implants. *Am. J. Clin. Pathol.*, 105: 628–639.
73. Granchi, D., Verri, E., Ciapetti, G., Stea, S., Savarino, L., Sudanese, A., Mieti, M., Rotini, R., Dallari, D., Zinghi, G., and Montanaro, L., 1998. Bone-resorbing cytokines in serum of patients with aseptic loosening of hip prostheses. *J. Bone Joint Surg.*, 80-B: 912–917.
74. Kondo, Y., Sato, K., Ohkawa, H., Ueyama, Y., Okabe, T., Sato, N., Asano, S., Mori, M., Ohsawa, N., Kosaka, K., 1983. Association of hypercalcemia with tumors producing colony stimulating factor(s). *Cancer Res.*, 43: 2368–2374.
75. Xu, J. W., Konttinen, Y. T., Waris, V., Gomez-Barrena, E., Sorsa, T., and Santavirta, S., 1996. Granulocyte-macrophage colony stimulating factor in aseptic loosening of total hip arthroplasty. *J. Orthop. Rheumatol.*, 9: 128–134.
76. Xu, J. W., Konttinen, Y. T., Waris, V., Patiala, H., Sorsa, T., and Santavirta, S., 1997. Macrophage-colony stimulating factor (M-CSF) is increased in the synovial-like membrane of the periprosthetic tissues in the aseptic loosening of total hip replacement (THR). *Clin. Rheumatol.*, 16: 243–248.
77. Dean, D. D., Lohmann, C. H., Alderete, M., Sylvia, V. L., Agrawal, C. M., Blanchard, C., Mabrey, J. D., Schwartz, Z., and Boyan, B. D., 1998. Human hip wear debris particles affect proliferation, differentiation, and local factor production by MG63 osteoblast-like cells in vitro. Transactions of the 24th Annual Meeting of the Society for Biomaterials, 21: 57.
78. Zambonin, G., Colucci, S., Cantatore, F. and Grano, M., 1998. Response of human osteoblasts to polymethylmetacrylate in vitro. *Calcif. Tissue Int.*, 62: 362–365.
79. Sun, J. S., Liu, H. C., Chang, W. H., Li, J., Lin, F. H., and Tai, H. C., 1998. Influence of hydroxyapatite particle size on bone cell activities: an in vitro study. *J. Biomed. Mater. Res.*, 39: 390–397.
80. Sun, J. S., Tsuang, Y. H., Liao, C. J., Liu, H. C., Hang, Y. S., and Lin, F. H., 1997. The effects of calcium phosphate particles on the growth of osteoblasts. *J. Biomed. Mater. Res.*, 37: 324–334.
81. Sabokbar, A., Pandey, R., Quinn, J. M., and Athanasou, N. A., 1998. Osteoclastic differentiation by mononuclear phagocytes containing biomaterial particles. *Arch. Orthop. Trauma Surg.*, 117: 136–140.
82. Sabokbar, A., Fujikawa,Y., Brett, J., Murray, D. W., and Athanasou, N. A., 1996. Increased osteoclastic differentiation by PMMA particle-associated macrophages. Inhibitory effect by interleukin 4 and leukemia inhibitory factor. *Acta. Orthop. Scand.*, 67: 593–598.
83. Wang, W., Ferguson, D. J., Quinn, J. M., Simpson, A. H., and Athanasou, N. A., 1997. Biomaterial particle phagocytosis by bone-resorbing osteoclasts. *J. Bone Joint Surg.*, 79-B: 849–856.

84. Wang, W., Ferguson, D. J., Quinn, J. M., Simpson A. H., and Athanasou N. A., 1997. Osteoclasts are capable of particle phagocytosis and bone resorption. *J. Pathol.*, 182: 92–98.
85. Anderson, G. I., Tao, L., and Langman, M., 1998. Effect of polyethylene, hydroxyapatite and polymethylmethacrylate particles on osteoclast activity. Transactions of the 44th Annual Meeting of the Orthopaedic Research Society, 23: 371.
86. Bi, Y., Ragab, A. A., Goldberg, V. M., and Greenfield, E. M., 1998. Titanium wear particles stimulate osteoclast differentiation in vitro. Transactions of the 44th Annual Meeting of the Orthopaedic Research Society, 23: 372.
87. Kim, K. J., Itoh, T., Tanahashi, M., and Kumegawa, M., 1996. Activation of osteoclast-mediated bone resorption by the supernatant from a rabbit synovial cell line in response to polyethylene particles. *J. Biomed. Mater. Res.*, 32: 3–9
88. Green, T. R., Fisher, J., and Ingham, E., 1998. Polyethylene particles of a critical size are necessary for the induction of TNF-alpha by macrophages in vitro. Transactions of the 44th Annual Meeting of the Orthopaedic Research Society, 23: 124.
89. Green, T. R., Matthews, J. B., Fisher, J., and Ingham, E., 1998. Cytokine production by macrophages in response to clinically relevant UHMWPE debris of different sizes. Transactions of the 44th Annual Meeting of the Orthopaedic Research Society, 23: 374.
90. Bukata, S. V., Hsu, J. S., Puzas, J. E., Rosier, R. N., Reynolds, P. R., and O'Keefe, R. J., 1998. Characterization of a human monocyte cell line for the genetic analysis of cytokine expression in response to titanium wear debris. Transactions of the 44th Annual Meeting of the Orthopaedic Research Society, 23: 473.
91. Rhodes, N. P., Hunt, J. A., and Williams, D. F., 1997. Macrophage subpopulation differentiation by stimulation with biomaterials. *J. Biomed. Mater. Res.*, 37: 481–488.
92. Gil-Abarova, J., Lacleriga, A., Barrios, C., and Canadell, J., 1992. Lymphocyte response to polymethylmethacrylate in loose total hip prosthesis. *J. Bone. Joint. Surg.*, 74-B: 825–830.
93. Santavirta, S., Kontinnen, Y. T., Bergroth, V., Eskola, A., Tallroth, K., and Lindholm, T. S., 1990. Aggressive granulomatous lesions associated with hip arthroplasty. Immunopathological studies. *J. Bone Joint Surg.*, 72: 252–258.
94. Santavirta, S., Konttinen, Y. T., Hoikka, V., and Eskola, A., 1991. Cementless Acetabular Components. *J. Bone Joint Surg.*, 73-B, 38–42.
95. Kossovsky, N., Millet, D., Juma, S., Little, N., Briggs, P. C., Raz, S., and Berg, E., 1991. In vivo characterization of the inflammatory properties of poly(tetrafluoroethylene) particles. *J. Biomed. Mat. Res.*, 25: 1287–1301.
96. Santavirta, S., Kontinnen, Y. T., Bergroth, V., and Gronblad, M., 1991. Lack of immune response to methyl methacrylate in lymphocyte cultures. *Acta Orthop. Scan.*, 62: 29–32, 1991.
97. Santavirta, S., Nordstrom, D., Metsarinne, K., and Kontinnen, Y. T., 1993. Biocompatability of polyethylene and host response to loosening of cementless total hip replacement. *Clin. Orthop.*, 297: 100–110.
98. Trindade, M. C. D., Lind, M., Sun, D. H., Patel, R., Goodman, S. B., Schurman, D. J., and Smith, R. L., 1998. In vitro reactions to orthopaedic biomaterials by macrophages and lymphocytes isolated from patients undergoing revision surgery. Society For Biomaterials Transactions of the 24th Annual Meeting, 21: 315.
99. Lee, S. H., Brennan, F. R., Jacobs, J. J., Urban, R. M., Ragasa, D. R., and Glant, T. T., 1997. Human monocyte/macrophage response to cobalt-chromium corrosion products and titanium particles in patients with total joint replacements. *J. Orthop. Res.*, 15: 40–9.
100. Herman, J. H., Sowder, W. G., Anderson, D., Appel, A. M., and Hopson, C. N., 1989. Polymethylmethacrylate-induced release of bone-resorbing factors. *J. Bone Joint Surg.*, 71: 1530–1541.
101. Haddad, F. S., Cobb, A. G., Bentley, G., Levell, N. J., and Dowd, P. M., 1996. Hypersensitivity in aseptic loosening of total hip replacements. The role of constituents of bone cement. *J. Bone Joint Surg.*, 78-B: 546–549.
102. Trindade, M. C. D., Nakashima, Y., Sun, D. H., Goodman, S. B., Schurman, D. J., and Smith, R. L., 1998. The effects of lymphocyte generated cytokines on particle challenged macrophages in vitro. Orthopaedic Research Society Transactions of the 44th Annual Meeting, 23: 478.

103. Farrar, M. A., and Schreiber, R. D., 1993. The molecular biology of interferon-gamma and its receptor. *Annu. Rev. Immunol.*, 11: 571–611.
104. Goodman, S., Wang, J. S., Regula, D., and Aspenberg, P., 1994. T-lymphocytes are not necessary for particulate polyethylene-induced macrophage recruitment. Histological studies of the rat tibia. *Acta. Orthop. Scand.*, 65,: 157–160.
105. Van Luyn, M. J. A., Khouw, I. M. S. L., van Wachem, P. B., Blaauw, E. H., and Werkmeister, J. A., 1998. Modulation of the tissue reaction to biomaterials. II. The function of T cells in the inflammatory reaction to crosslinked collagen implanted in T-cell-deficient rats. *J. Biomed. Mater. Res.*, 39: 398–406.
106. Goodman, S. B., 1994. The effects of micromotion and particulate materials on tissue differentiation. Bone harvest chamber studies in rabbits. *Acta. Orthop. Scand.*, 65: 1–43.
107. Goodman, S. B., Aspenberg, P., Song, Y., Doshi, A., and Lidgren, L., 1995. Effects of particulate high-density polyethylene and titanium alloy on tissue ingrowth into the bone harvest chamber in rabbits. *J. Appl. Biomat.*, 6: 27–33.
108. Goodman, S. B., Aspenberg, P., Song, Y., Knoblich, G., Huie, P., Regula, D., and Lidgren, L., 1995. Tissue ingrowth and differentiation in the bone harvest chamber in the presence of cobalt chrome allow and high density polyethylene particles. *J. Bone Joint Surg.*, 77: 1025–1035.
109. Goodman, S. B., Aspenberg, P., Song, Y., Regula, D., and Lidgren, L., 1996. Polyethylene and titanium alloy particles reduce bone formation. *Acta. Orthop. Scand.*, 67: 599–605.
110. Goodman, S. B., Aspenberg, P., Song, Y., Regula, D., and Lidgren, L., 1996. Different effects of phagocytosable particles during bone formation versus remodeling. *J. Biomed. Mater. Res.*, 33: 153–158.
111. Goodman, S., Song, Y., Chun, L., Aspenberg, P., and Regula, D., 1998. Effects of TGF-β on bone ingrowth in the presence of polyethylene particles. *Trans. Soc. Biomater.*, 21: 61.
112. Goodman, S. B., Song, Y., Chun, L., Huie, P., Aspenberg, P., and Regula, D., 1998. Tissue differentiation in the prescence of polyethylene and polystyrene particles. *Trans. Soc. Biomater.*, 21: 58.
113. Trindade, M. C. D., Goodman, S. B., Song, Y., Nakashima, Y., Sun, D. H., and Smith, R. L., 1997. The effects of UHMWPE and polystyrene particles during bone ingrowth and remodeling: the bone harvest chamber model. Society for Biomaterials Transactions of the 23rd Annual Meeting, 20: 380.
114. Trindade, M. C. D., Song, Y., Smith, R. L., and Goodman, S. B., 1998. Pro-inflammatory mediator release in response to particle challenge: studies using the bone harvest chamber. Transactions of the third annual combined meeting of the Orthopaedic Research Societies of the United States, Canada, Europe and Japan, 57.
115. Trindade, M. C. D., Song, Y., Aspenberg, P., Smith, R. L., and Goodman, S. B., 1999. Pro-inflammatory mediator release in response to particle challenge: studies using the bone harvest chamber. *J. Biomed. Mater. Res.* 48: 434–439.
116. Merkel, K. D., Teitelbaum, S. L., Ross, F. P., and Erdmann, J. M., 1998. Polymethylmethacrylate induced osteolysis is mediated by the P55 TNF receptor in vivo. Transactions of the 44th Annual Meeting of the Orthopedic Research Society, 23: 126.
117. Merkel, K. D., Erdmann, J. M., Bercutt, L., McHugh, K., Abu-amer, Y., Ross, F. P., and Teitelbaum, S. L., 1997. Role of TNF in particle-induced osteolysis in-vitro and in-vivo. Transactions of the 43rd Annual Meeting of the Orthopaedic Research Society, 22,: 742.
118. Banchereau, J. and Rybak, M. E., 1994. Interleukin-4. In: A. Thomson (Editor), *The Cytokine Handbook*, 2nd ed, New York, Academic Press, pp. 99–126.
119. Paul, W. E., 1991. Interleukin-4: A prototypic immunoregulatory lymphokine. *Blood*, 77: 1859–1870.
120. Kikuchi, H., Hanazawa, S., Takeshita, A., Nakada, Y., Yamashita, Y., and Kitano, S., 1994. Interleukin-4 acts as a potent stimulator for expression of monocyte chemoattractant JE/MCP-1 in mouse peritoneal macrophages. *Biochem. Biophys. Res. Commun.*, 203: 562–569.
121. Orlofsky, A., Lin, E. Y., Prystowsky, M. B. J., 1994. Selective induction of the beta chemokine C10 by IL-4 in mouse macrophages. *J. Immunol.* 152: 5084–5091.
122. Pollice, P., Hicks, D. G., Rosier, R. N., Hsu, J., Brukata, S., Stroyer, B. F., Reynolds, P. R., Puzas,

J. E., and O'Keefe, R. J., 1997. Identification of an operative anti-inflammatory mechanism for the control of wear debris osteolysis. Transactions of the 43rd annual meeting of the Orthopaedic Research Society, 22–2: 728.

123. Lewis, D. B., Liggitt, H. D., Effmenn, E. L., Motley, S. T., Titelbaum, S. L., Jepsen, K. J., Goldstein, S. A., Bonadio, J., Carpenter, J., and Perlmutter, R. M., 1993. Osteoporosis induced in mice by overproduction on interleukin-4. *Proc. Natl. Acad. Sci.*, 90: 11618–11622.
124. Lind M, Deleuran B, Yssel H, Fink-Eriksen E., and Thestrup-Pedersen K: IL-4 and IL-13, but not IL-10, are chemotactic factors for human osteoblasts. *Cytokine*, 7: 78–82.
125. Mundy, G. R., Roodman, G. D., and Yoneda, T., 1995. Role of cytokines in regulation of bone remodeling and osteoporosis. In: B. B. Aaggarwal and R. K. Puri (Editors), *Human Cytokines: Their Role in Disease and Therapy*, Cambridge, Mass., Blackwell Science, pp. 285–292.
126. Ueno, K., Katayama, T., Miyamoto, T., and Koshihara, Y., 1992. Interleukin-4 enhances in vitro mineralization in human osteoblast-like cells. *Biochem. Biophys. Res. Commun.* 189: 1521–1526.
127. Hsu, J. C., Bukata, S. V., Lewis, G. D., Hicks, D. G., Puzas, J. E., and O'Keefe, R. J., 1998. Paracrine induction of IL-10 in periprosthetic osteolysis. Transactions of the 44th Annual Meeting of the Orthopaedic Research Society, 23: 125.
128. Maloney, W. J., Sun, D. H., Nakashima, Y., James, R., and Smith, R. L., 1998. Effects of serum protein opsonization on cytokine release by titanium-alloy particles. *J. Biomed. Mater. Res.*, 41: 497–503.
129. Sun, D. H., Nakashima, Y., Maloney, W. J., Goodman, S. B., Trindade, M., Schurman, D. J., and Smith, R. L., 1997. Opsonization of particulate debris: protein identification by 2-D gel analysis (PMMA and Ti-alloy particles). Transactions of the 23rd Annual Meeting of the Society for Biomaterials, 20: 259.
130. Sun, D. H., Nakashima, Y., Maloney, W. J., Goodman, S. B., Trindade, M., Schurman, D. J., and Smith, R. L., 1997. Opsonization of particulate debris: protein identification by 2-D gel analysis. Transactions of the 43rd Annual Meeting of the Orthopaedic Research Society, 22: 324.
131. Ragab, A. A., Bi, Y., Lavish, S. A., Van De Motter, R. R., Goldberg, V. M., and Greenfield, E. M., 1998. Stimulation of cytokine production by titanium wear particles is due to adherent endotoxin. *Trans. Orthop. Res. Soc.*, 23–1: 355.
132. Hitchins, V. M. and Merritt, K., 1998. Decontaminating particles exposed to bacterial endotoxin (LPS). *Trans. Soc. Biomater.*, 21, 311.
133. Nakashima, Y., Trindade, M., Sun, D. H., Maloney, W. J., Goodman, S. B., Schurman, D. J., and Smith, R. L., 1998. Signaling pathways of particulate debris-induced cytokine expression in human macrophages: Involvement of CD11b and CD18-mediated pathways. Transactions of the third annual combined meeting of the Orthopaedic Research Societies of United States, Canada, Europe and Japan, 75.
134. Ritchie, C. K, Paczkoskie, V. J., and Horowitz, S. M., 1997. Role of cyclic AMP in particulate-induced cytokine production in macrophages. *Trans. Orthop. Res. Soc.*, 22–1: 160.
135. Lind, M., Trindade, M. C. D., Yaszay, B., Goodman, S. B., Smith, R. L., 1998. Effects of particulate debris on macrophage-dependent fibroblast stimulation in coculture. *J. Bone Joint Surg.*, 80-B: 924–30.
136. Shanbhag, A. S., Hasselmann, C. T., and Rubash, H. E., 1997. Potential for inhibiting wear debris-mediated osteolysis in a canine total hip replacement (THR) model. Transactions of the 23rd annual meeting of the society for biomaterials, 20: 76.
137. Ritchie, C. K., Patel, M. A., Stamos, B. D., and Horowitz, S. M., 1998. Alendronate reduces particle-induced cytokine production in macrophages. *Trans. Orthop. Res. Soc.*, 23: 357.
138. Jeppsson, C. and Aspenberg, P., 1998. Negative effect on bone formation by BMP:S. chamber study in rabbits. Transactions of the 44th Annual Meeting of the Orthopaedic Research Society, 23: 472.
139. Maloney, W. J., Smith, R. L., Schmalzried, T. P., Chiba, J., Huene, D., and Rubash, H., 1995.

Isolation and characterization of wear particles generated in patients who have had failure of a hip arthroplasty without cement. *J. Bone Joint Surg.*, 77: 1301–1310.

140. Pandey, R., Quinn, J., Joyner, C., Murray, D. W., Triffitt, J. T., and Athanasou, N. A., 1996. Arthroplasty implant biomaterial particle associated macrophages differentiate into lacunar bone resorbing cells. *Ann. Rheum. Dis.*, 55: 388–395.

141. Brieland, J. K., Flory, C. M., Jones, M. L., Miller, G. R., Remick, D. G, Warren, J. S., and Fantone, J. C., 1995. Regulation of monocyte chemoattractant protein-1 gene expression and secretion in rat pulmonary alveolar macrophages by lipopolysaccharide, tumor necrosis factor-alpha, and interleukin-1 beta. *Am. J. Respir. Cell. Mol. Biol.*, 12: 104–109.

19

Clinical Application of Calcium Hydroxyapatite Ceramic in Bone Tumor Surgery

Hideki Yoshikawa
Osaka University Medical School, Osaka, Japan

Atsumasa Uchida
Mie University School of Medical, Tsu, Japan

I. INTRODUCTION

Curettage or resection with grafting of bone or some substitutes is one of the standard surgical procedures for benign bone tumors. Autogenous bone graft has been most commonly used for reconstruction of bone tumors, although a number of materials from human, animal, or nonbiological sources have been tried to fill defects [1,2]. However, it may be difficult to obtain the amount required, particularly in treating large bone tumors or in children. In addition, it may not be preferable to take autogenous bone for small bony defects such as in the phalanx. It would be ideal to have a bone substitute which is easily fabricated and available, biocompatible with host bone, and biodegradable. Calcium hydroxyapatite ceramics (CHA) are nontoxic substances which provoke little reaction from tissues and have many properties, both chemical and physical, that make them suitable alternatives to autogenous bone grafts. The biocompatibility of CHA to bone and bone marrow has already been demonstrated in animal experiments [3–9]. Based on those experimental results, we have used this ceramic for the treatment of bone tumors since 1983 [10]. We report here the clinical results for 120 benign bone tumors treated by curettage or resection followed by the implantation of CHA. Another application of CHA is as a delivery system for antibiotics and anticancer drugs. We also present the preliminary clinical results of local control for osteomyelitis and malignant bone tumors by CHA.

II. CALCIUM HYDROXYAPATITE CERAMIC

The CHA used in this study was in the forms of porous blocks (0.4–6.0 cm) with 55–70% porosity (pore size: 200–300 μm), looking like pumice, and of granules with 1–5 mm grain size (Figs. 1a, b) [10]. These ceramics were calcined at 1100°C to 1200°C and the compression strength of a porous block was about 15 MPa (1 MPa = 10 kg/cm^2). Blocks of various dimensions were used according to the needs of the case.

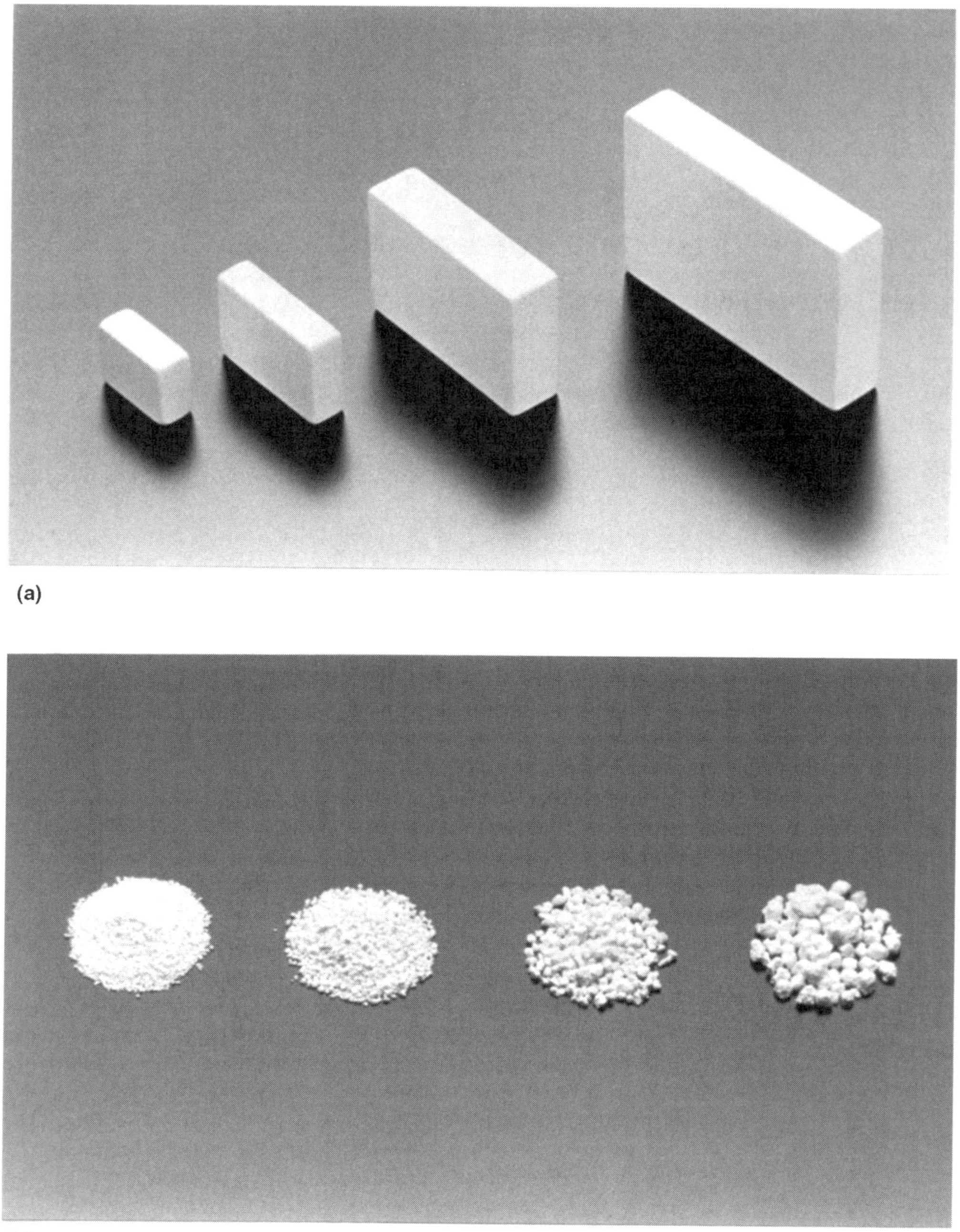

(a)

(b)

Figure 1 Photographs of CHA blocks (sizes: 0.4–6.0 cm) (a) and CHA granules (1 to 5 mm in diameter) (b) clinically available for orthopedic use.

Table 1 Histological Diagnosis and Location of Bone Tumors

Diagnosis	No.	Location	No.
Solitary bone cyst	27	Femur	36
Enchondroma	23	Humerus	27
Giant cell tumor	22	Tibia	18
Fibrous dysplasia	15	Finger bones	16
Chondroblastoma	8	Calcaneus	6
Aneurysmal bone cyst	8	Radius	5
Osteoid osteoma	6	Ilium	5
Osteofibrous dysplasia	5	Ulna	2
Nonossifying fibroma	3	Talus	2
Others	3	Scapula	2
		Fibula	1
Total	120	Total	120

A total of 120 patients were implanted with CHA for bony defects after the resection of benign bone tumors (Table 1). Only CHA blocks were implanted in 12 patients, only granules in 74, and both blocks and granules in 34. The average age of the patients was 28 years (range 2 to 68). In 38 patients, CHA implantation was performed before the closure of growth plates. The tumors included 27 solitary bone cysts, 23 enchondromas, 22 giant cell tumors, 15 fibrous dysplasias, 8 chondroblastomas, 8 aneurysmal bone cysts, and 17 others. The bone tumors were located in the femur (36 cases), in the humerus (27), in the tibia (18), in the finger bones (16), or in other bones (23). The defects varied in volume from 1 to 180 ml (mean 42 ml, median 36 ml). All blocks and granules of CHA were gas-sterilized before surgery. Follow-up ranged from 6 to 182 months (average 72).

III. OPERATIVE TECHNIQUES

After adequate resection or curettage of bone tumors, blocks and granules of CHA were heaped up to fill the bony defects. Blocks were contoured using rongeurs and an air-surgeotome to achieve a press fit with the surrounding cancellous bone. Variable-sized granules of CHA were inserted into the remaining small recesses or defects. Alternatively, only blocks or granules of CHA were applied to fill the bony defects. Fibrin glue as a "lid" was placed over the defects after packing CHA to avoid a scatter of the granules. The most important operative technique is to achieve a press fit with the surrounding cancellous and cortical bone. Satisfactory filling with CHA was achieved in almost all cases. Radiographs were taken before closure to confirm complete filling of the entire bony defect. Postoperative assessment was made from clinical, laboratory and radiological findings.

IV. BONE FORMATION

Radiological evaluation was by three criteria: (1) a change in the radiolucent line or halo around the CHA implant, (2) incorporation of each CHA, detected as an increase in density of the radiographic image, and (3) lack of displacement or dislocation of the implant. Any radiolucent line or halo around a CHA implant tended to decrease with time and eventually disappear (Figs. 2a–c). In patients with loosely packed implants, the loss of radiolucent lines was delayed compared to those with dense packing. With time after implantation, the radiographic density at the CHA implant sites appeared to increase, and the CHA blocks and granules seemed to become attached to each other (Figs. 3a–d). These findings were interpreted as bone regeneration around and within the implanted CHA. However, no quantitative measurements are available for the evaluation of bone ingrowth into porous materials, so further study is required of the histomorphometry of the implanted materials [11–12]. The CHA implant was biopsied in 10 patients at the time of the elective removal of metal or at other operations. Specimens taken in the first two to three months after implantation showed new-

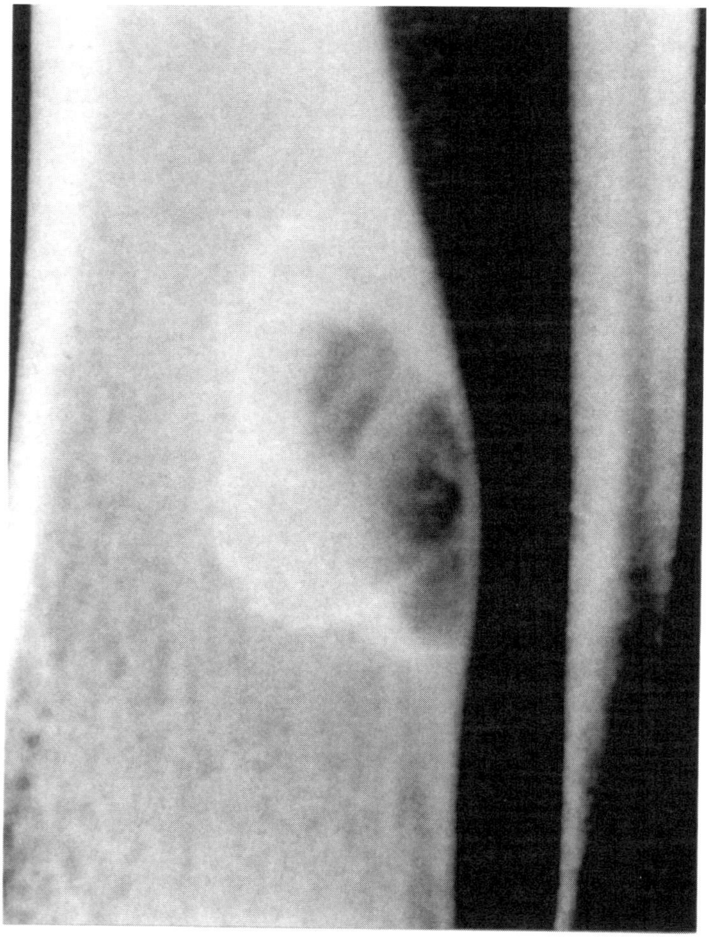

(a)

Figure 2 Preoperative (a), postoperative (b), six-months follow-up (c) radiographs of a nonossifying fibroma in a 10-year-old girl. The radiolucent line or halo (arrows) around the CHA implants in the postoperative radiograph (b) has vanished completely after six months (c).

(c)

(b)

Figure 2 Continued.

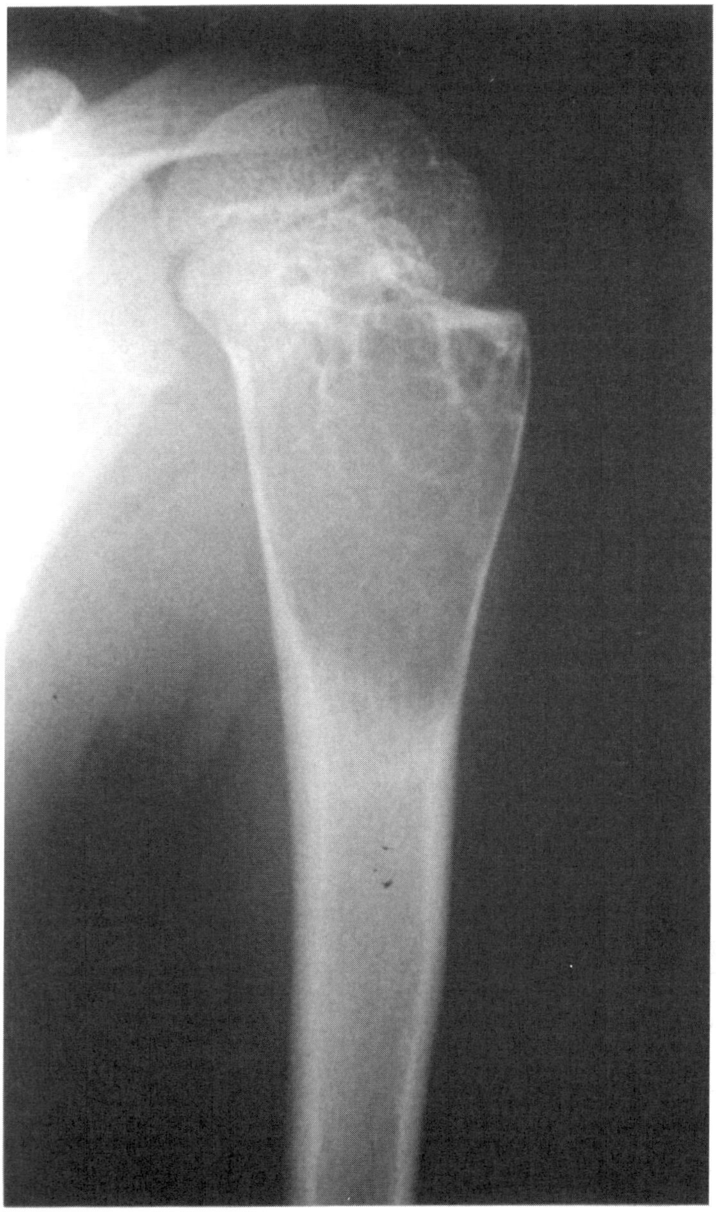

(a)

Figure 3 Preoperative (a), postoperative (b), 12-months follow-up (c) and eight-years follow-up (d) radiographs of a solitary bone cyst in a seven-year-old boy. The immediate radiograph shows the architecture of the CHA blocks and granules with their well-defined margins (b). One year later the implant margins have become indistinct, and there is little discrimination among the CHA granules (c). The CHA granules and blocks were implanted close to the growth plate (b), but a later view shows that the growth plate is functioning normally (d).

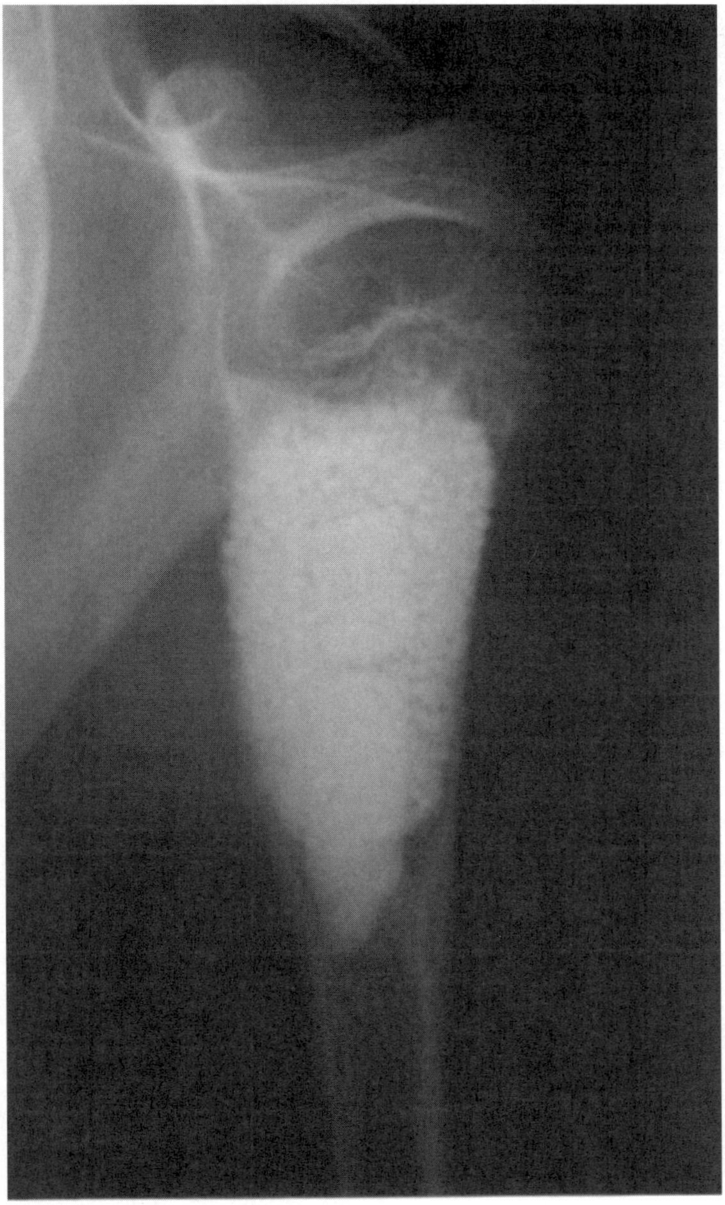

(b)

Figure 3 Continued.

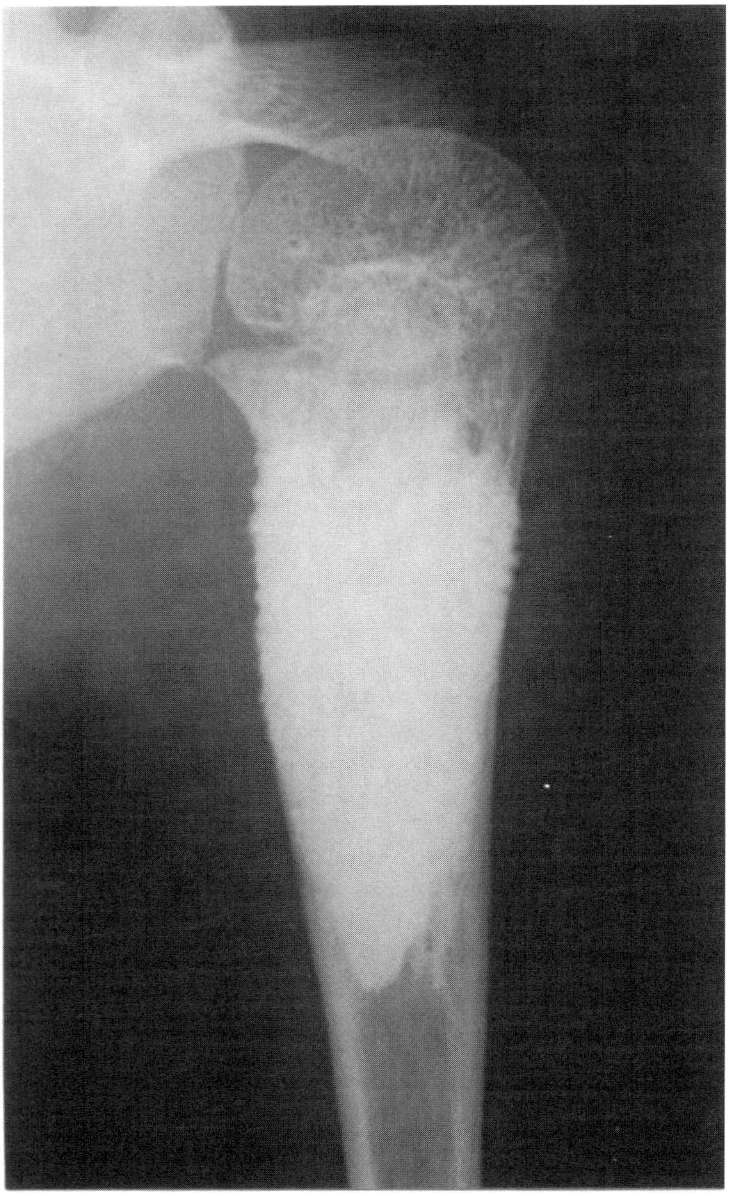

(c)

Figure 3 Continued.

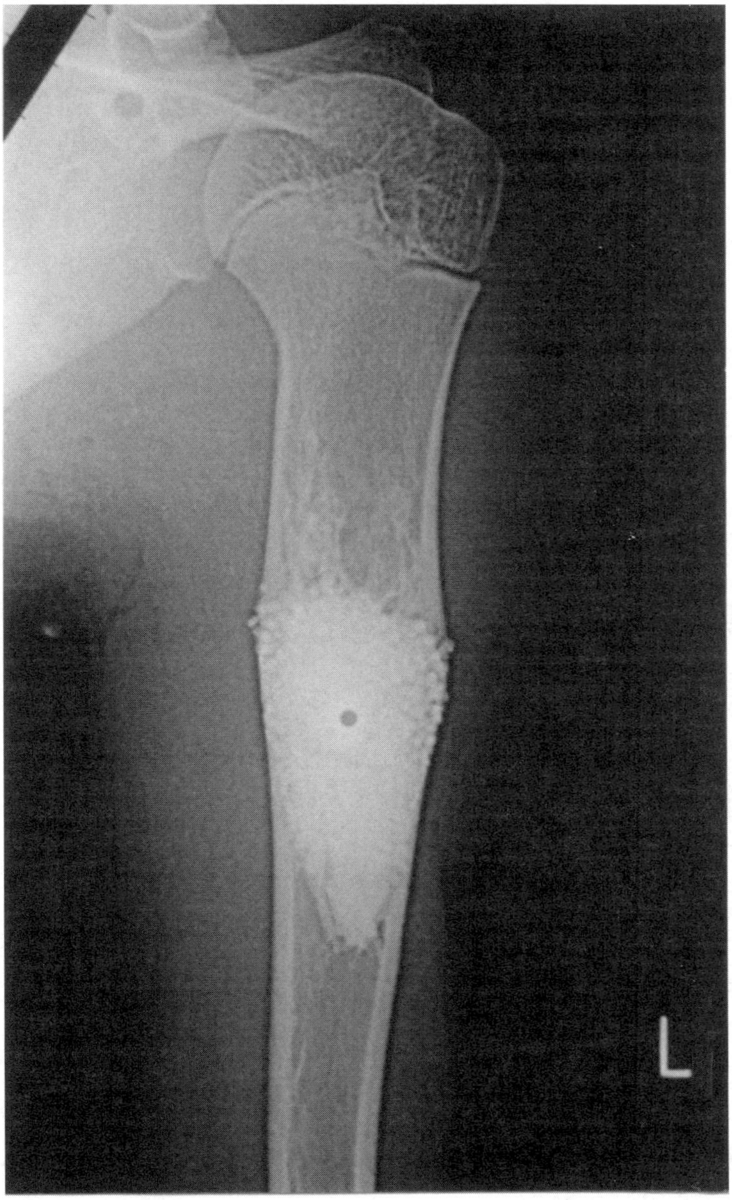

(d)

Figure 3 Continued.

ly formed bone only around the implanted CHA at the margins of a cavity. There was no evidence of inflammation or a foreign body response. Bone ingrowth into the pores increased with time and after about 12 months almost all pore structures were filled with newly formed bone. There was no fibrous tissue between the CHA and the newly formed bone, and the cortical portion of the specimens showed compact bone with regenerated osteons filling the CHA pores. In medullary specimens, newly formed bone in the pores had been partially absorbed and replaced by bone marrow tissue, giving the appearance of normal trabeculae in close apposition to the CHA (Fig. 4). These findings suggest that newly formed bone in the pores of the CHA has a normal turnover. Osteoblasts and osteoclasts were present around the newly formed bone in proportions suggesting normal bone remodeling. Continuing radiographic observations up to a maximum of 15 years after implantation have shown little evidence of biodegradation of the CHA. Biodegradation of the CHA is likely to be very slight, judging from our histological findings. These human histological findings confirm the results of animal experiments [7–9], though the rate of bone ingrowth into the pore structures was relatively delayed.

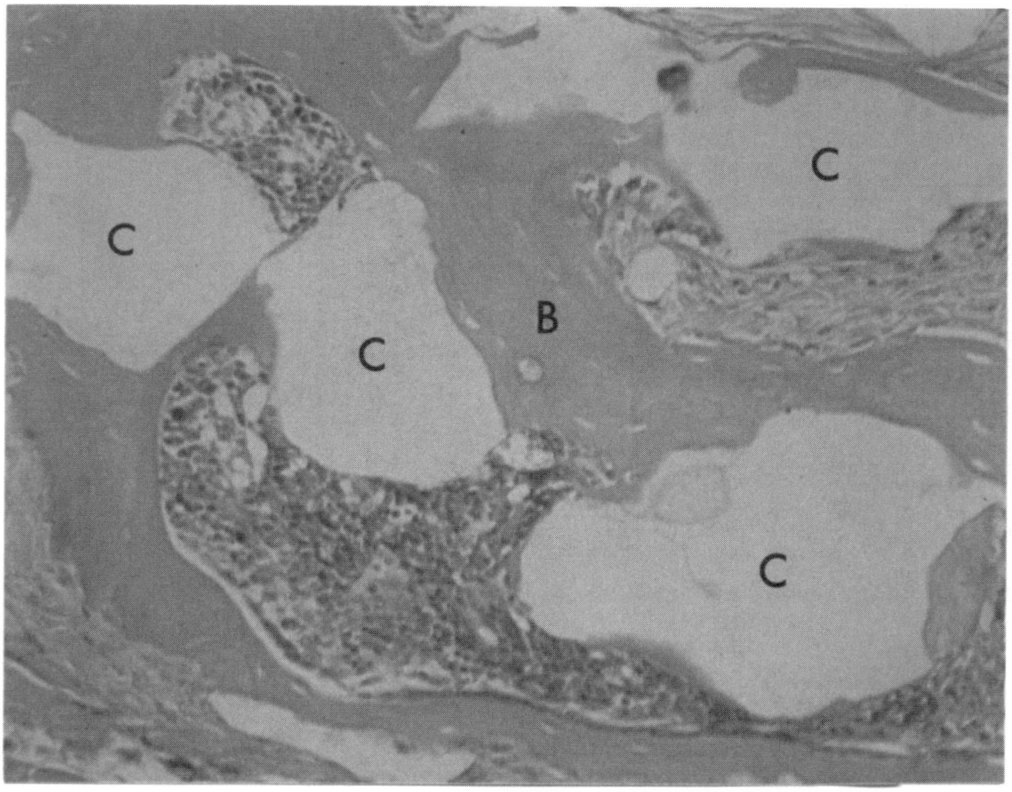

Figure 4 Histological appearance of a biopsy specimen taken from the diaphysis of the femur, one year after CHA (C) grafting for fibrous dysplasia of the femur. Bone ingrowth (B) into the pores and an intimate bonding between CHA and new bone are discernible (Hematoxylin and eosin, original magnification; ×100).

V. LOCAL RECURRENCE

Local recurrence of the original tumor developed around the implanted CHA in 11 (9.2%) of 120 cases. Five of 22 giant cell tumors, 4 of 27 solitary bone cysts, and 2 of 5 osteofibrous dysplasias showed local recurrence. Radiologically, a resorption and a scatter of CHA granules

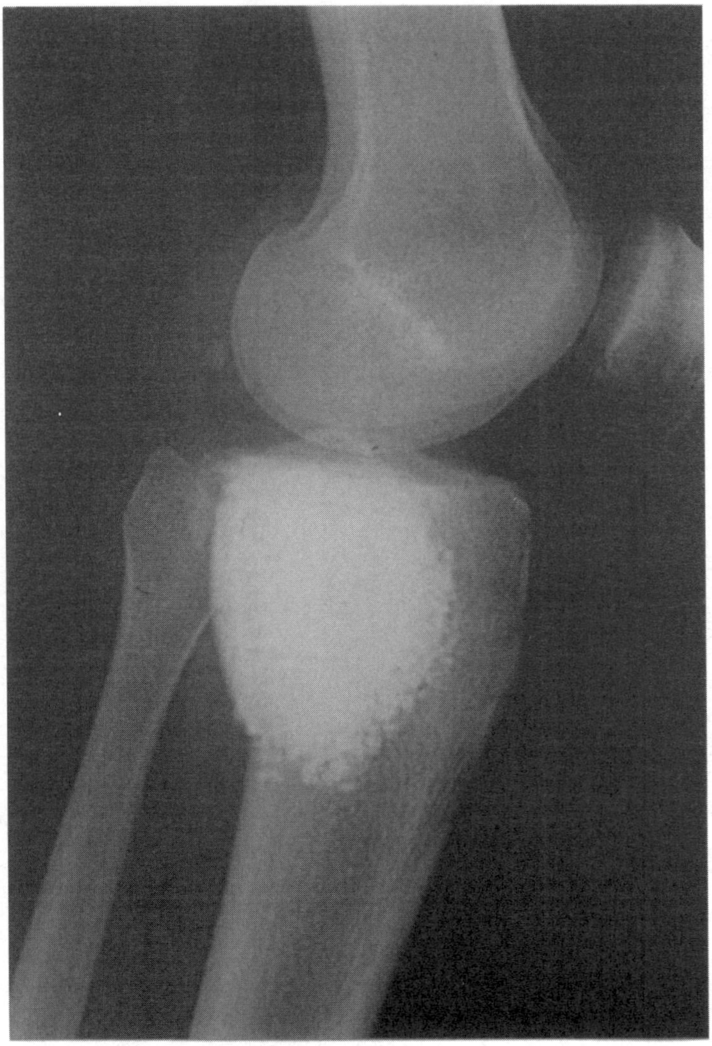

(a)

Figure 5 Radiographs of the proximal tibia in a 19-year-old girl with giant cell tumor. Six months after curettage and CHA implantation, the implant has been well incorporated and no local recurrence was observed (a). Ten months after surgery, an osteolytic area was detected and resorption of the implanted CHA granules was evaluated radiologically (b). Histologically, recurrence of the giant cell tumor and resorption of CHA granules (asterisk) by numerous osteoclasts (arrows) were observed (Hematoxylin and eosin, original magnification; ×100) (c).

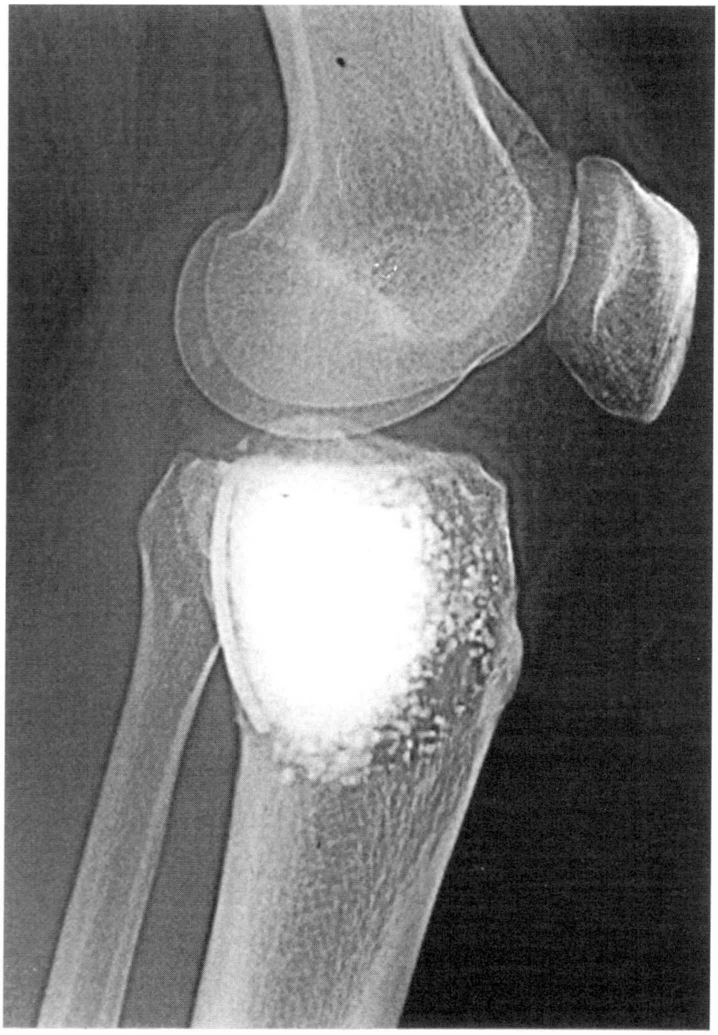

(b)

Figure 5 Continued.

were detected in the cases of giant cell tumor (Figs. 5a, b) and osteofibrous dysplasia, but not in the cases of solitary bone cyst. Histologically, in the cases of giant cell tumors or osteofibrous dysplasias, active osteoclasts absorbed CHA, and some CHA granules were obviously reduced in size (Fig. 5c). Recurrent lesions and the contaminated CHA were removed, and the adjacent intact bone was thoroughly curetted, and CHA was reimplanted in situ. One of the recurrent cases with giant cell tumor sustained recurettage twice to achieve a complete healing, but the others healed by a single recurettage. In the cases of solitary bone cysts, serous fluid was detected at the recurrent sites, but apparent absorption of CHA was not detected radiologically or macroscopically (Figs. 6a, b). Therefore, recurrent lesions were curetted and CHA was reimplanted without removal of the original CHA. One of the four recurrent cases with

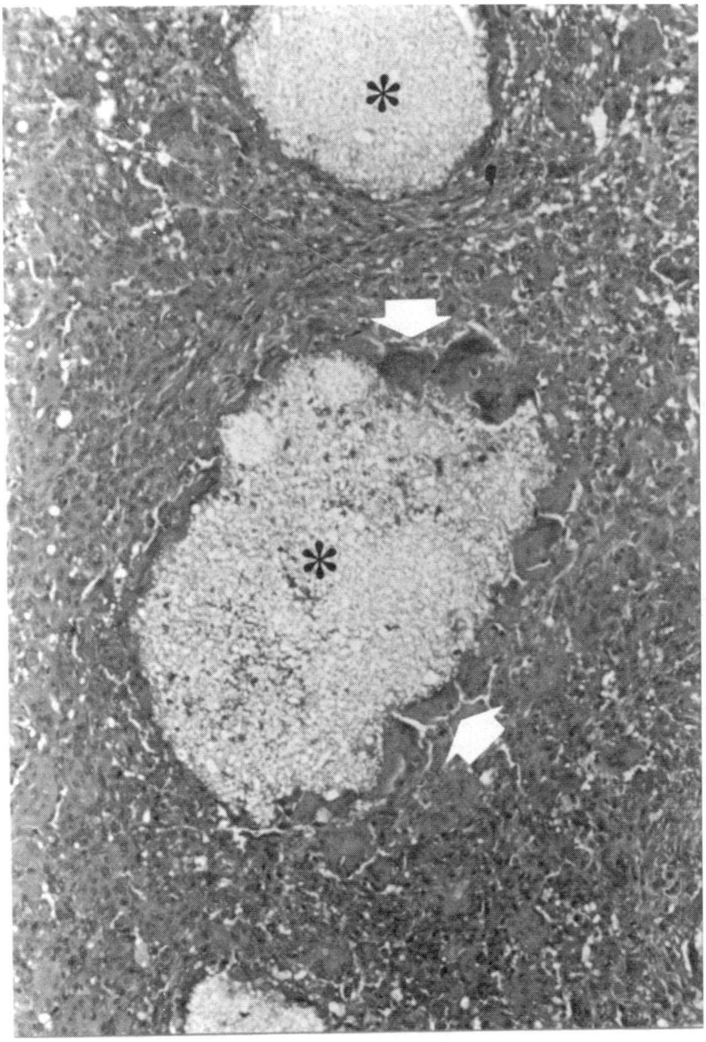

(c)

Figure 5 Continued.

solitary bone cyst sustained recurettage three times to achieve a complete healing, but the others healed after a single recurettage.

VI. COMPLICATIONS

There was no detachment of CHA from the implanted site, except in cases with postoperative pathological fractures. Longitudinal bone growth was not disturbed even when CHA was implanted in close proximity to the growth plate (Fig. 3d). There were no infections, toxic effects, or abnormal laboratory findings. Fever-up (37–38°C) was observed in 22 patients (18%)

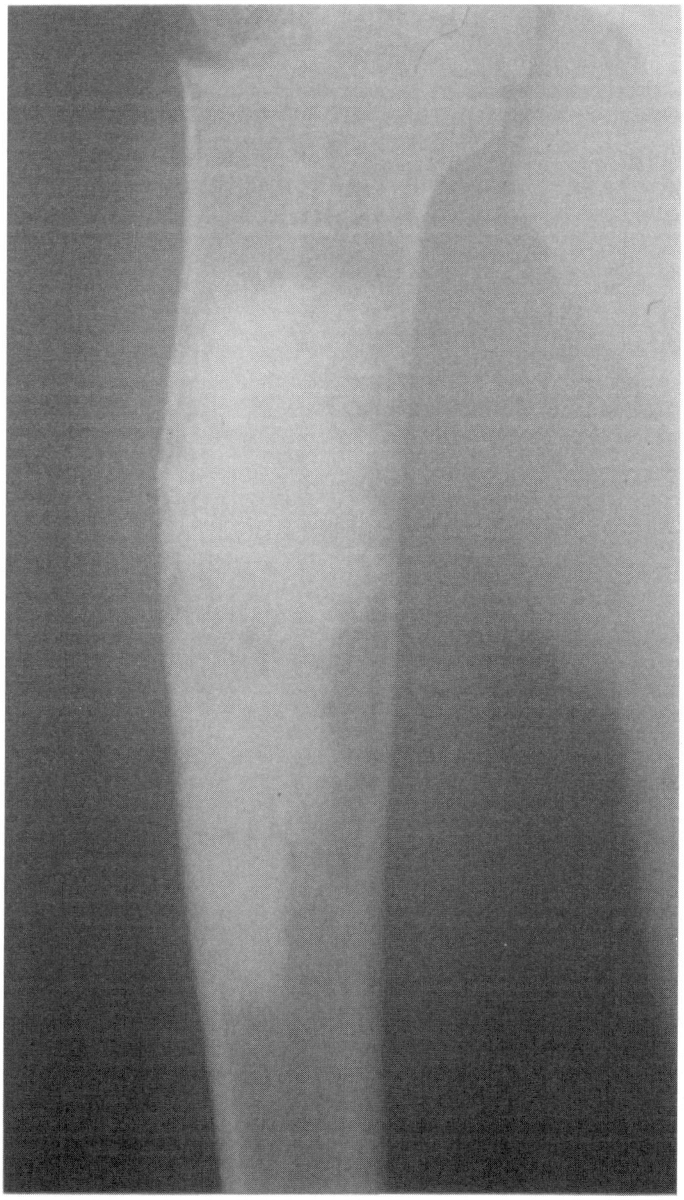

(a)

Figure 6 Postoperative radiograph of a solitary bone cyst in a 10-year-old boy (a). Nine months after curettage and CHA implantation, osteolytic lesions were detected superiorly and inferiorly to the CHA (arrows), but no apparent resorption of the CHA granules was detected (b).

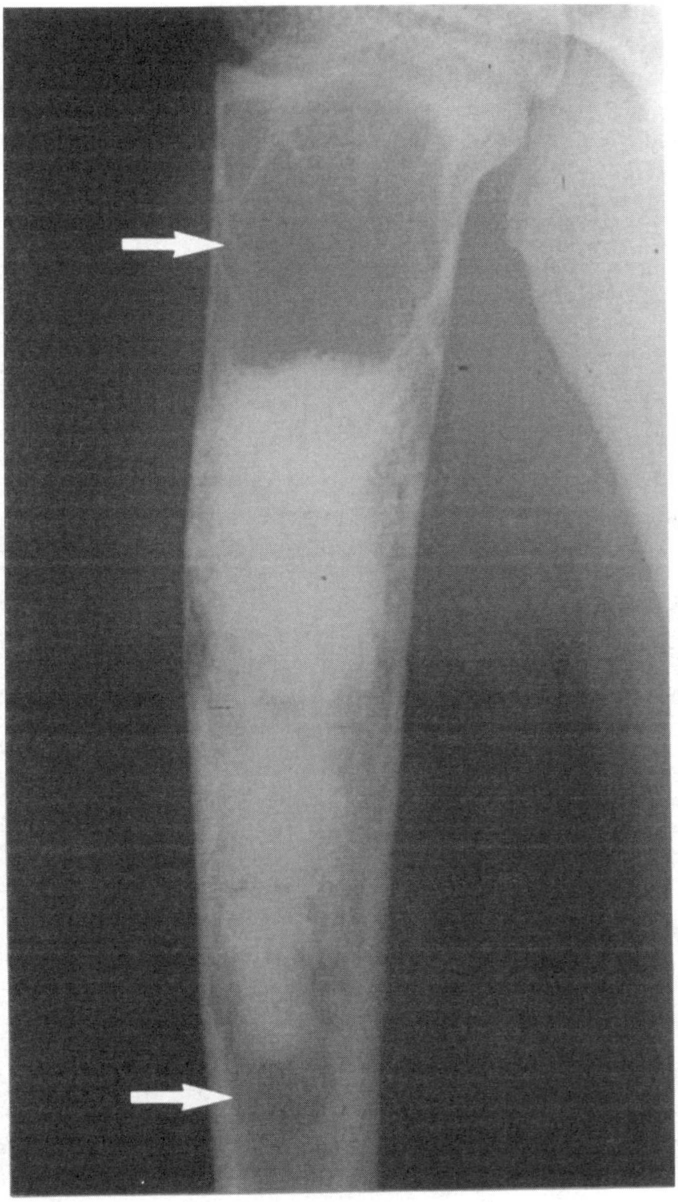

(b)

Figure 6 Continued.

during one week postoperatively. No adverse reactions to the CHA, such as excessive postoperative drainage, erythema, or other wound problems, were seen in any patient. In five cases, there were pathological fractures after CHA implantation. Four fractures were found within one or two months after operation, all in the femur: two in cases with fibrous dysplasia, one in a case of solitary bone cyst, and one in a case of enchondroma. These fractures united after internal fixation with retention of the CHA implant. In a case of osteofibrous dysplasia in the tibia, a late fracture at five years after surgery was observed (Figs. 7a–c). This fracture united by a conservative cast with retention of the CHA implant. From these experiences, CHA seemed not to inhibit the normal process of fracture healing.

However, the current available types of porous CHA lack substantial mechanical integrity because of their brittle nature, which makes them mechanically unsuitable for unstable segmental diaphyseal defect. Within three to six months, the composite of CHA and newly formed bone should have sufficient mechanical strength. The four pathological fractures in the femur appear to have been caused by faulty postoperative care rather than the fragility of CHA.

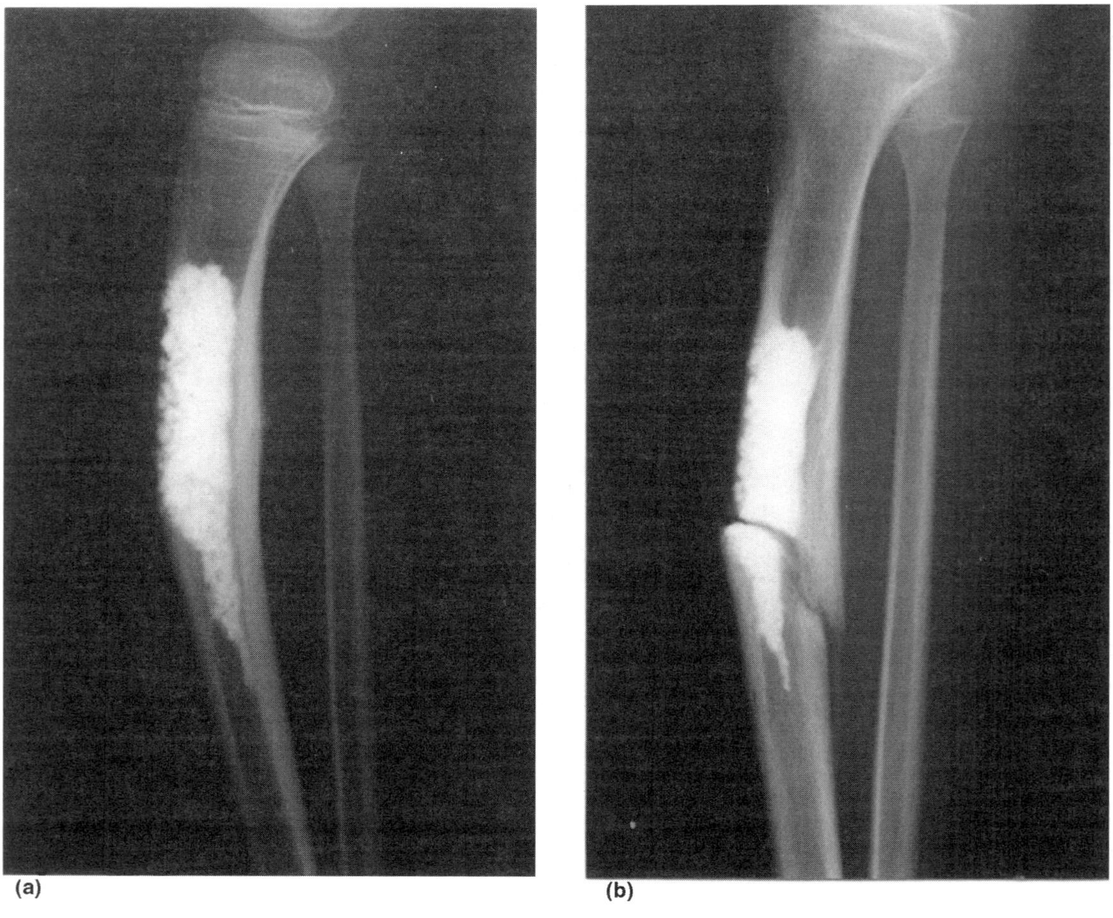

Figure 7 Postoperative radiograph of an osteofibrous dysplasia in a six-year-old boy (a). Five years after curettage and CHA implantation, he sustained a pathological fracture with a breakage of the CHA implant in the diaphysis of the tibia. The fracture healed by a conservative treatment with casting for three months, and the broken CHA was completely incorporated into the host bone after six months after fracture.

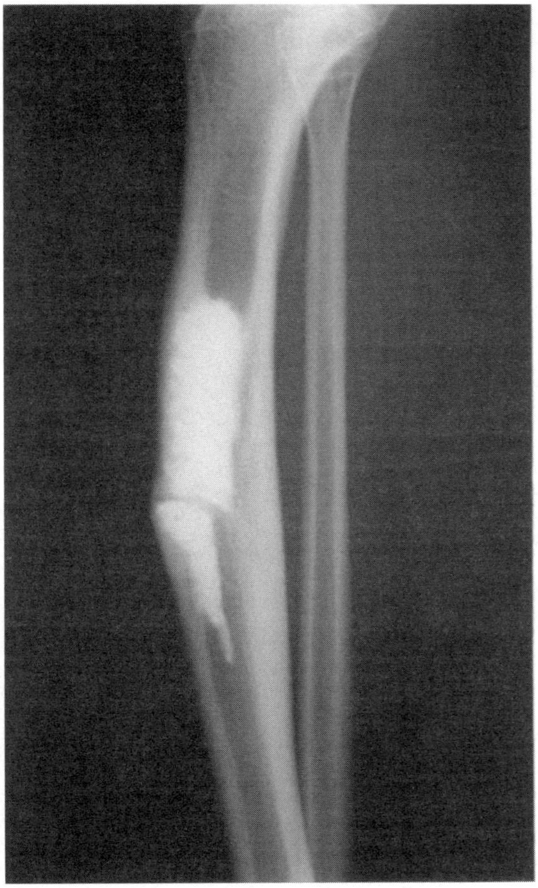

(c)

Figure 7 Continued.

Postoperative radiographs may pretend that the CHA-implanted bone has sufficient mechanical strength because of the dense radiographic image even just after operation. These radiographic findings may have led us an astray, so postoperative care is important, particularly after the resection of solid femoral tumors.

VII. APPLICATION OF CHA AS A DRUG DELIVERY SYSTEM

We have developed a new drug delivery system for sustained release of antibiotics or anticancer drugs by enclosure into blocks of porous CHA (shape: cube or column; size: $10 \times 10 \times 10$ mm or $6 \times 6 \times 6$ mm; sintering temperature: 1200°C; porosity: 30–40%; pore size: 40–150 μm) (Figs. 8a, b) [13,14]. We tested gentamicin sulfate, cefaperazone sodium, and flomoxef sodium in powder placed in a cylindrical cavity in the CHA blocks, and confirmed the continuous release of the antibiotics from the CHA at their effective doses, both in vitro and in vivo experiments [13]. Clinically, we implanted the CHA blocks with various antibiotics for 10 patients with osteomyelitis and obtained satisfactory results (Figs. 9a, b). This new system

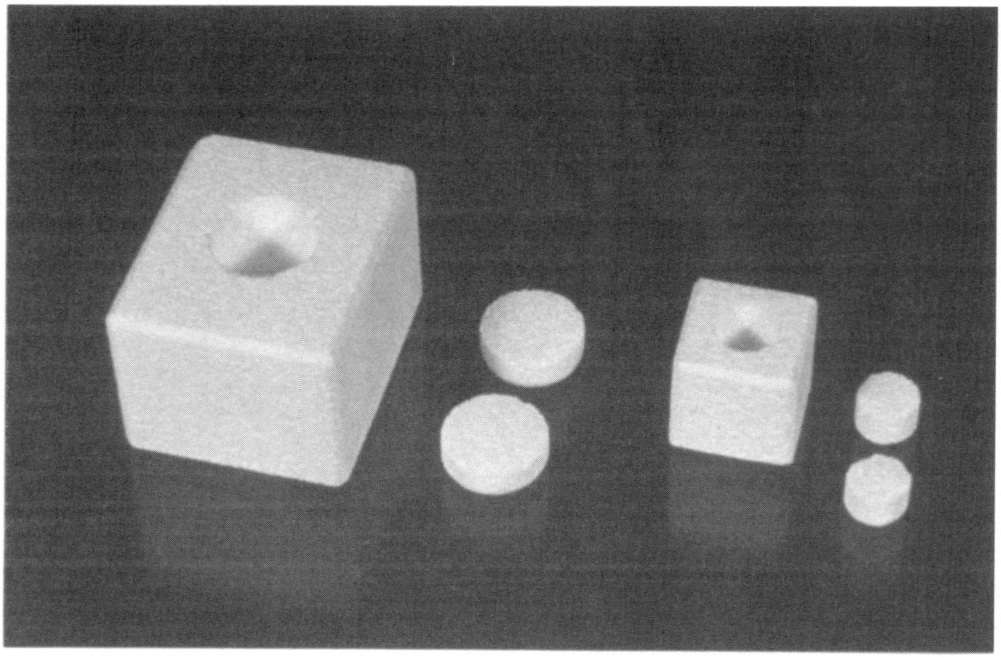

(a)

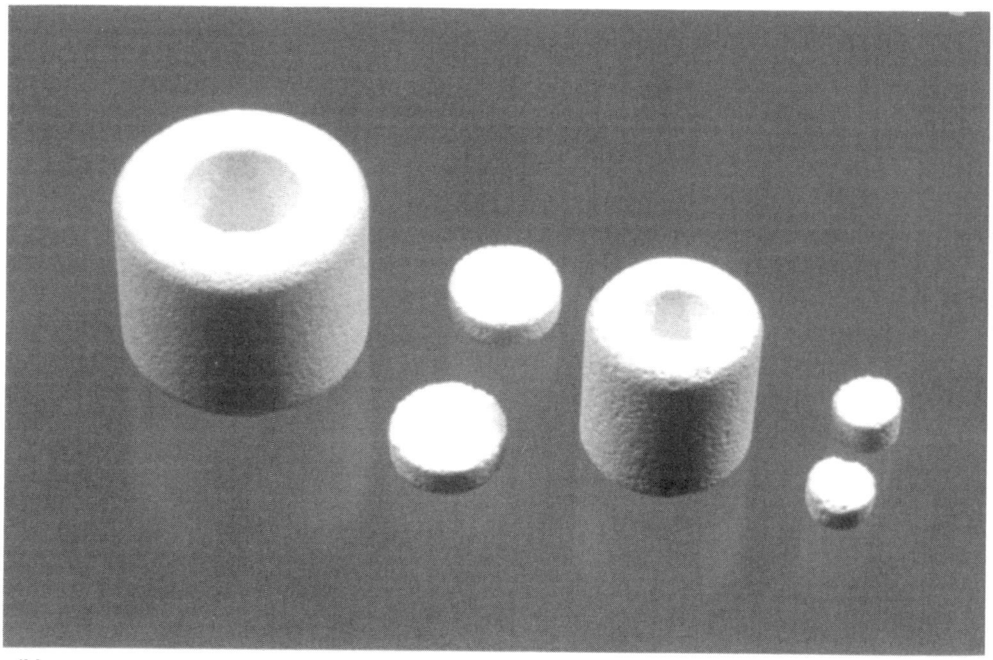

(b)

Figure 8 Photographs of CHA blocks as a drug delivery system. The small cylindrical hole is made by ultrasound at the center of a ceramic block. Powdered antibiotics or anticancer drugs are packed into the hole, which was sealed in with α-tricalcium phosphate (TCP) mixed in polylactic acid.

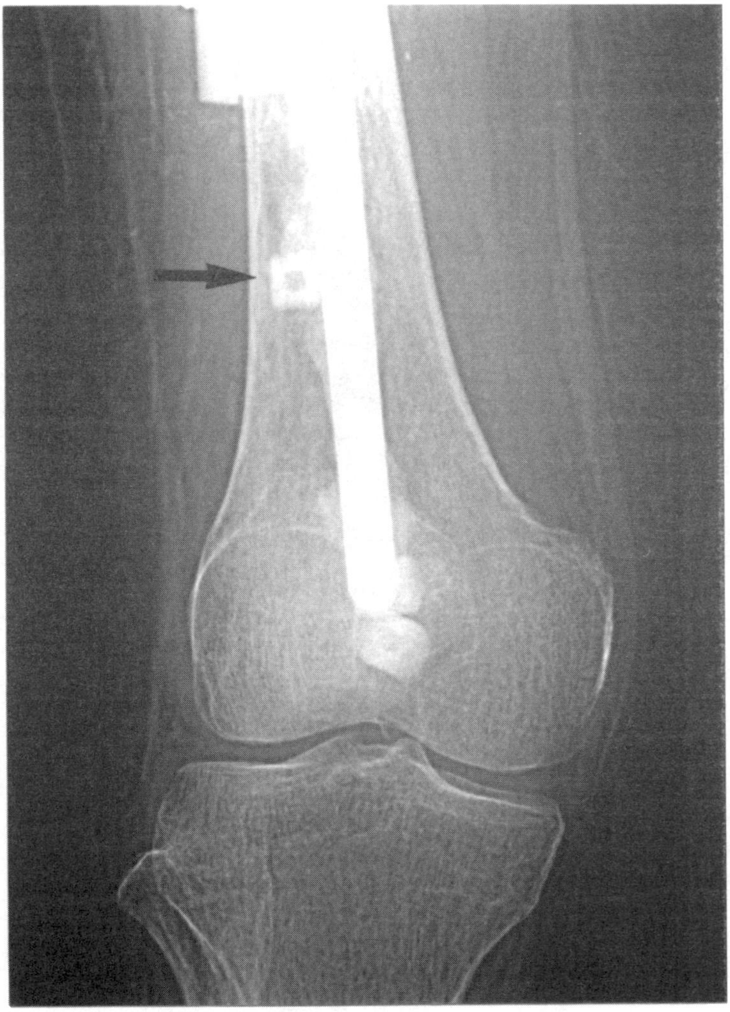

(a)

Figure 9 Anteroposterior (a) and lateral (b) radiographs at three years after implantation of the CHA blocks with gentamicin (arrows). The primary osteosarcoma at the proximal femur had been replaced by a stainless prosthesis six years before, and then infection around the stem of the prosthesis was detected three years ago. Removal of the prosthesis, curettage of the medullary cavity, implantation of the CHA blocks with gentamicin, and revision by the new prosthesis with a longer stem were performed. No infection sign has been detected for three years after the revision surgery and the CHA blocks has been well incorporated.

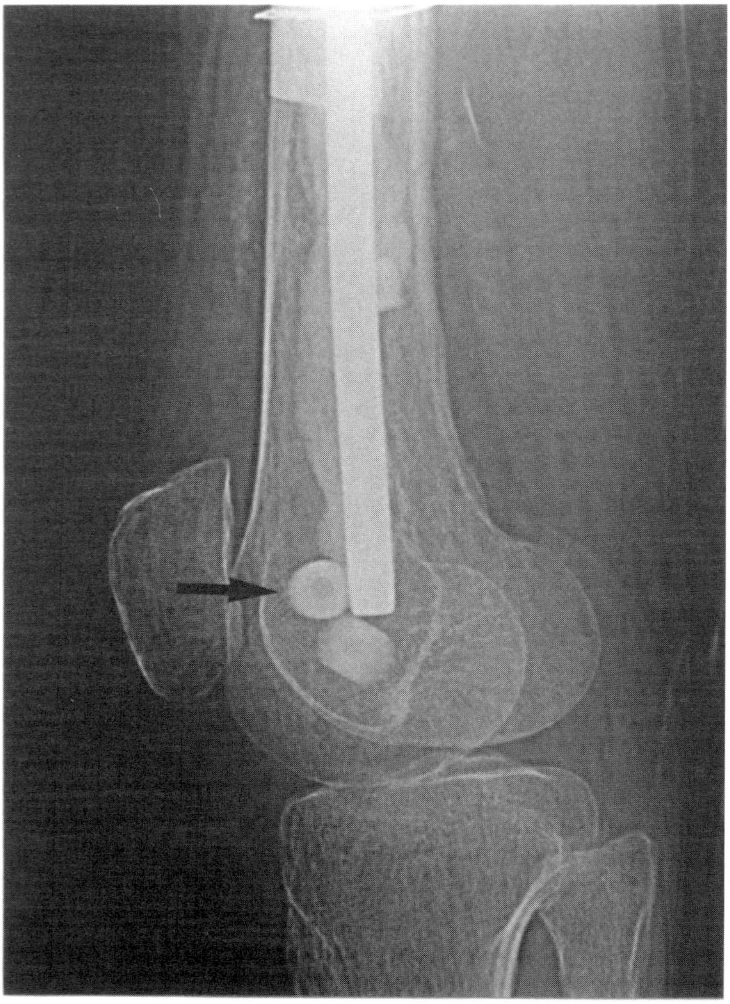

(b)

Figure 9 Continued.

overcomes the disadvantages of other drug delivery systems, avoiding thermal damage to the antibiotics and a second operation for the removal of the carrier. Some mechanical strength is provided by the ceramic, and healing may be accelerated by bone ingrowth into its micropores.

We tested an anticancer drug (cisplatinum) in powder placed in the cylindrical cavity in the same CHA blocks. The slow release of this drug from this system was confirmed in in vitro and in vivo experiments [14]. Clinically, we implanted the CHA blocks with cisplatinum for five malignant bone tumors, and obtained an evidence of tumor necrosis around the CHA block (Figs. 10a, b). These results indicate that this new approach to a drug delivery system may well have an important role in cancer chemotherapy. Especially in bone tumors, it is attractive because the mechanical strength of CHA permits partial surgical excision and replacement of the bony defect at the same time.

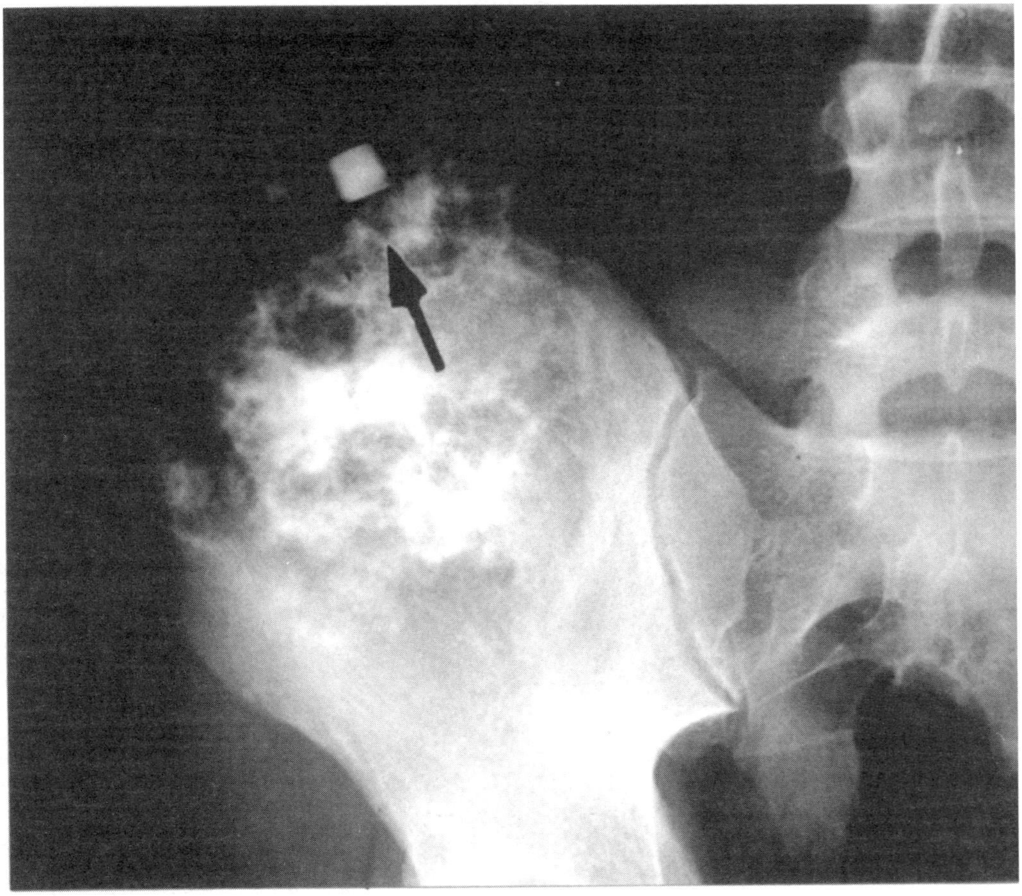

(a)

Figure 10 Postbiopsy radiograph of a chondrosarcoma (grade II) of the ilium in a 25-year-old man. A CHA block with cis-platinum (10 mg) was implanted into the site of biopsy (arrow) (a). Wide resection was performed at three weeks after biopsy. Histological specimen (b) showed apparent necrosis of the tumor only around the CHA implant (c) with cis-platinum.

VIII. CONCLUSIONS

Bone defects caused by resection and curettage for benign bone tumors should be filled with bone substitutes to obtain mechanical strength as soon as possible, by the stimulation of osteoconduction. Porous or granular CHA implants into bone defects can provide a suitable framework for osteogenesis and compares well with other bone substitutes such as allograft and xenografts. We believe that CHA has a favorable clinical result because of ease of sterilization, manufacture, preservation and shape adjustment. After follow-up of 6 months to 15 years (mean 72 months), no patients had infection or any serious adverse effects from the implants, although local recurrence was observed in giant cell tumor, solitary bone cyst, and osteofibrous dysplasia. In conclusions, CHA is a safe and convenient implant material which aids the regeneration of bone in defects produced by the surgical excision of benign bone

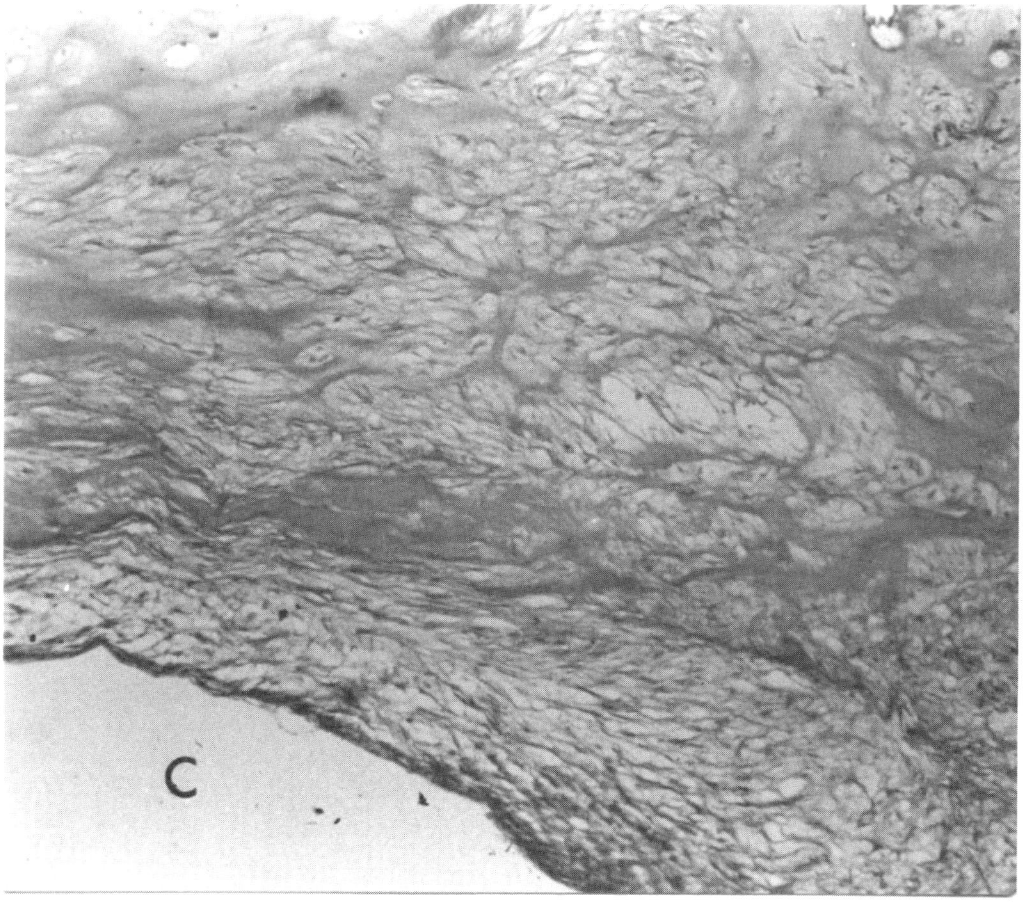

(b)

Figure 10 Continued.

tumors, and may be applicable as a drug delivery system to treat osteomyelitis or malignant bone tumors.

REFERENCES

1. Sartoris DJ, Gershuni DH, Akeson WH, Holmes RE, Resnick D. Coralline hydroxyapatite bone graft substitutes: preliminary report of radiographic evaluation. *Radiology* 1986;159:133–7.
2. Bucholz RW, Carlton A, Holmes RE. Hydroxyapatite and tricalcium phosphate bone graft substitutes. *Orthop Clin North Am* 1987:18:323–34.
3. Bhaskar SN, Brady JM, Getter L, Grower MF, Driskell T. Biodegradable ceramic implants in bone: electron and light microscopic analysis. *Oral Surg Oral Med Oral Path* 1971;32:336–46.
4. Cameron HU, Macnab I, Pilliar RM. Evaluation of biodegradable ceramic. *J Biomed Mater Res* 1977:11:179–86.
5. Hulbert SF, Klawitter JJ, Leonard RB. Compatibility of bioceramics with the physiological environment. *Mater Sci Res* 1971;5:417–34.

6. Jarcho M. Calcium phosphate ceramics as hard tissue prosthetics. *Clin Orthop* 1981;157:259–78.
7. Uchida A, Nade SML, McCartney ER, Ching W. The use of ceramics for bone replacement: a comparative study of three different porous ceramics. *J Bone Joint Surg (Br)* 1984;66:269–75.
8. Uchida A, Nade SML, McCartney ER, Ching W. Bone ingrowth into three different porous ceramics implanted into the tibia of rats and rabbits. *J Orthop Res* 1985;365–77.
9. Uchida A, Nade SML, McCartney ER, Ching W. Growth of bone marrow cells on porous ceramics in vitro. *J Biom Mater Res* 1987;21:1–10.
10. Uchida A, Araki N, Shinto Y, Yoshikawa H, Kurisaki E, Ono K. The use of calcium hydroxyapatite ceramic in bone tumour surgery. *J Bone Joint Surg (Br)* 1990;72:298–302.
11. Holmes RE, Bucholz RW, Mooney V. Porous hydroxyapatite as a bone graft substitute in diaphyseal defects: a histometric study. *J Orthop Res* 1987;5:114–21.
12. Shimizu T, Zerwekh JE, Videman T, Gill K, Mooney V, Holmes RE, Hagler HK. Bone ingrowth into calcium phosphate ceramics: influence of pulsing electromagnetic field. *J Orthop Res* 1988;6:248–58.
13. Shinto Y, Uchida A, Korkusuz F, Araki N, Ono K. Calcium hydroxyapatite ceramic used as a delivery system for antibiotics. *J Bone Joint Surg (Br)* 1992;74:600–4.
14. Uchida A, Shinto Y, Araki N, Ono K. Slow release of anticancer drugs from porous calcium hydroxyapatite ceramic. *J Orthop Res* 1992;10:440–5.

20
Bone Development and Bone Structure Depend on Surface Roughness and Structure of Metallic Implants

C. M. Müller-Mai, C. Voigt, and U. M. Gross
Universitätsklinikum Benjamin Franklin, Freie Universität Berlin, Berlin, Germany

G. Berger
Bundesanstalt für Materialforschung und -prüfung, Berlin, Germany

I. INTRODUCTION

Titanium and its alloys are frequently used as implant materials in dentistry and orthopedic surgery. Especially for load bearing applications the development of bone in the interface is of vital importance. Under ideal conditions the direct contact between titanium and bone will be achieved which was demonstrated in several papers [1–6]. Parameters which are important for direct bone-implant contact are the biocompatibility of the choosen implant material, macrostructure (shape) and microstructure (e.g., surface roughness, geometry of surface elevations and deepenings of the implant), surgical implantation procedure, direct bone-implant contact after insertion [2,4], time and mode of implant loading which may result in implant movement relative to the adjacent bone [5] and the implantation bed itself [6]. Recently it has become clearer that the microstructure of the implant surface significantly influences osteoblasts and, in turn, the amount of bone formed in the interface as well as the push-out or tensile strength of the interface [7–9]. Therefore, the biocompatibility of an implant is only one parameter influencing the tissue response to a metallic implant; additionally the morphology of microscopic surface structures becomes important, since one of the main problems limiting the application and function of metallic implants is the lack of reliable anchoring of the implant in the skeletal system. On this basis implants of different macrostructure and microstructure have been used to achieve a better anchoring and long-term stabilization of the implantation bed.

One way to solve the problem of bone anchorage on the implant surface is to use structured surfaces which allow for bone ingrowth, i.e., porous surfaces. Threaded implants or implants coated with spherical beads are known to allow bone contact and ingrowth within the implant interstices and were used clinically [10]. For many years it has been well established that a pore size of about 100 μm is required to achieve significant bone ingrowth [11]. The optimum pore size was determined for metal implants anchored in bone and coated with spherical beads using histology and push-out tests. Particles with diameters from 50 to 300 μm were used providing pores of about 100 to 400 μm [11–13]. Such structured surfaces allow bone ingrowth within implant interstices, but bone ingrowth does not necessarily provide appropriate direct bone to implant contact as well as sufficient implant fixation. It was shown in a former study using dental implants of Ti6Al4V alloy covered with beads producing pore sizes from 50 to 200 μm that a direct bone-implant contact corresponded with biomechanical strength of the interface (14]. In another study 3 different CoCrMo implants made of beads with interconnecting pores were used. Eighty-four days after implantation into the distal femur epiphysis of rabbits direct correlations were described between bone-metal contact in percent and the tensile strength of the interface (correlation coefficient 0.79) and between the area of bone in percent in a zone of 1 mm surrounding the implant and the tensile strength (correlation coefficient 0.9) [15]. But all implants mentioned in this paragraph were clinically of limited success. This is related to the smooth surface of single beads used to coat implants providing the porosity. Since the surface structure of the beads is smooth, micromovements were possible which resulted in pain and aseptic loosening in several implants. An example for a failed implant system with such a surface was the Lord prosthesis used for total hip replacement. Therefore the search for optimized surface structures is necessary.

A second way in improving the surface structure of metallic implants was to produce a certain microstructure, a surface roughness or rugosity in the range of several micrometers. This rugosity seems to be the pivotal point to reach the goal of load transmission in load-bearing metal implants. It was shown in former publications that the surface structure of a metallic implant should allow for interdigitation with newly deposited trabeculae to establish secondary implant fixation [9]. In tensile tests flame-sprayed surfaces provided excellent results. Viewing such implant surfaces in the scanning electron microscope (SCEM) they displayed a similarity to bone surfaces after resorption. It was shown in former publications that such resorbed bone surfaces served in many cases as anchoring points for the newly produced bone. This observation was explained as the ARF principle (activation, resorption, formation) by Frost [16]. Due to this knowledge one would expect, that implant surfaces mimicking bone surface structures after resorption should be more successful than others in being fixed by surrounding bone. Thus, implants having such surface structures would be useful in loaded applications, e.g., in total hip replacement. Therefore, it is of utmost importance to better describe qualitatively and quantitatively bone and implant surfaces. In the recent literature some papers described the surface morphology already in greater detail. This was performed according to 3D descriptions of the surfaces using parameters describing the height as well as the distances of single elevations using different quantitative methods [17–20].

The principle of direct bone to implant contact including mechanical interlocking becomes even more important, since ingrowing bone being separated from the implant by an intervening soft tissue layer, an interfacial membrane [21], enables micromovements in the interface. Micromovements are known to govern other misleading reactions such as destruction of the surface oxide layer which might be accompanied by corrosion and loss of the implant [22]. The higher surface area of structured implants in contact with the tissue promotes even more the risk of corrosive processes [23]. Additionally, stiff metal implants with a certain geometry fixed via bone contact might cause stress shielding resulting in bone resorption by disuse atrophy. These possible disadvantageous reactions point again to the importance of optimized surface structures, i.e., an optimized geometry, amplitude and arrangement of surface deepenings and elevations.

The tissue compatibility of titanium was shown to depend, among other parameters, on its surface chemistry and physics, i.e., especially on a stabile surface oxide layer. Therefore, the outermost atomic layers determine the primary interaction between implant and tissue. The chemical composition of the surface oxide layer might differ considerably from the bulk material. It can be influenced by various steps of implant production, e.g., machining, cleansing, and sterilization [24]. In the case of Ti many oxides are known, such as TiO_2, Ti_2O_3, TiO, of which TiO_2 is the most common [25]. In the case of Ti6Al4V alloy the surface consisted mainly of TiO_2, but included other constituents, such as the alloying element Al, but no V [26]. Impurities on the surface were detected using different methods of surface analysis, such as secondary ion mass spectroscopy (SIMS) and Auger electron spectroscopy (AES), etc. [24], and might influence the corrosion resistance of the implant and the host response which points to the importance of defining implant surfaces chemically prior to implantation.

Ultrastructural studies with titanium implants of different type demonstrated that the contact between titanium and tissue in vivo is mediated by a several-angstrom-thick layer which seems to contain proteoglycanes [27–28]. Some authors believe that the incorporation of Ti will be supported by chemical bonding forces due to the existence of several highly polarizable metal oxides on its surface [29]. Such a substrate seems to be a strongly attractive alternative to water in the implantation bed. The interaction between tissue and substrate was explained by the occurrence of strong, real chemical, short-range (1–10 Å) ionic and covalent bonds, and weak long-range (1–100 Å) van der Waals bonds [6,29–30]. In vitro, it was demonstrated, that amorphous Ca/P layers were deposited directly on the surface of titanium and Ti6Al4V and served as a bonding zone between metal substrate and collagen-rich extracellular matrix [31–32]. Such delicate "bonding" is probably influenced by impurities as described above.

In the light of the complexity of implant surfaces, e.g., due to different surface structures and surface roughnesses as well as different chemistries the purpose of this publication was to discuss the effect of titanium implants of different surface roughness and surface structure on mechanisms of implant incorporation, i.e., development and the shift of trabeculae due to bone remodeling after insertion and subsequent development of a tensile strength of the interface. Surface roughnesses of different titanium implants were measured and compared to bone surfaces after osteoclastic degradation, since these surfaces are known to be covered by newly deposited bone in many cases and therefore serve as anchoring points for the newly produced bone. Standardized titanium and Ti6Al4V implants of defined surface chemical composition and defined surface structure were used in a standardized animal model [33]. Ti6Al4V was chosen because of its superior mechanical properties compared to c.p. titanium. Implant surface structures were measured and related to tensile strength measurements and histomorphometric evaluations of the implant bone contact. From these observations and in comparison with the pertinent literature conclusions were drawn on how an optimized metallic implant surface structure designed for load-bearing applications could be.

II. MATERIALS AND METHODS

A. Implants

Cylinders of commercial pure titanium (ISO 5832/II, grade 2) and Ti6Al4V alloy (ISO 5832/III) of different surface roughness with a diameter of 4 mm and a length of 6 mm were used (composition in weight percent; see Table 1). Titanium cylinders had surface roughnesses of $R_t = 1$ μm ($n = 24$), 20 μm ($n = 25$), 50 μm ($n = 20$), and 100 μm ($n = 34$). Ti6Al4V ($n = 24$) was used with R_t of 0.5 μm. All R_t-values were given by the producers and were measured again during implant evaluation. The method of surface roughness measurement was not revealed by the producers. According to the Deutsches Institut für Normung [34] the R_t-value (also designated R_{max}) represents the distance between the highest and deepest points of the measurement. The surface roughnesses of 0.5 μm and 1 μm were produced by polishing implant surfaces which provided almost totally smooth surfaces with fine grooves (Fig. 1). Sur-

Table 1 Compositions of Implant Materials in Weight Percent (Ti balance, not listed)[a]

	Ti	Ti 100	Ti6Al4V
N_2	0.032	0.03	0.008
C	0.021	0.5	0.017
H_2	0.0011	0.015	0.0041
Fe	0.2	0.2	0.195
O_2	0.32	0.4	0.105
Zr	—	0.1	—
Al	—	—	6.07
V	—	—	4.06

[a]Ti = titanium 1 to 50 μm surface roughness, Ti 100 = 100 μm surface roughness (flame spray), Ti6Al4V = 0.5 μm surface roughness.

face roughnesses of $R_t = 20$ μm and 50 μm were produced by different size pellet-blasting (Zirshot) which provided comparable surface structures with deepenings and elevations which had no microroughness or rugosity. There was no possibility of mechanical interlocking (Fig. 2). The highly irregular microstructure of Ti $R_t = 100$ μm surface roughness was produced by flame-spraying of Ti powder onto Ti rods. This provided a surface with a porosity of 25–50 vol%, where 93% of pores had diameters lower than 50 μm, 5% between 50 and 100 μm, and 2% larger than 100 μm. Some interconnecting pores within the flame-spray coating were observed (Fig. 3).

After production, the implants with surface roughnesses of 0.5, 1, and 100 μm were cleaned by rinsing shortly in 5% HNO_3 and 1% HF, repeated rinsings in aqua destillata, and ultrasonication in methanol for at least 15 min. Implants with 20 and 50 μm surface roughness were subjected additionally to short rinsing with 50% HNO_3 prior to the procedure described above to remove surface contaminations. All implants were sterilized by gamma irradiation.

B. Surface Characterization

Randomly chosen implants were used to characterize the surface of the implants prior to implantation. The surface structure was analyzed morphologically by applying SCEM (see above) and quantitative data were provided by measuring the surface roughnesses by using an optical profilometer for light optical topometrical analysis (RM 600 Rodenstock, Munich, Germany). This was carried out by receiving reflected laser light with an infrared laser ($\lambda = 780$ nm). By using an autofocus the depth profile of the specimens could be measured with a resolution of about 50 nm. The speed of the specimens during the measurements was 8 mm/min and the examined distance between single measurements on the implant surfaces was 6.56 μm. The method provided data about the above-mentioned R_t-value as well as the R_a-value (arithmetic mean roughness of the absolute values of the surface departures from the mean plane within the measuring distance, according to Deutsches Institut für Normung, 34). The total length of a measurement on a single specimen was between 5 and 6 mm because of the limited specimen size. For comparative reasons, resorbed endosteal bone surfaces ($n = 8$, rabbit humerus) were measured applying the same procedure, except that additionally to the measured distance of 1.8 mm an area of 1.8 mm^2 was measured. The endosteal surfaces were prepared for SCEM and resorbed surfaces were selected for measurement. According to this procedure some draw-

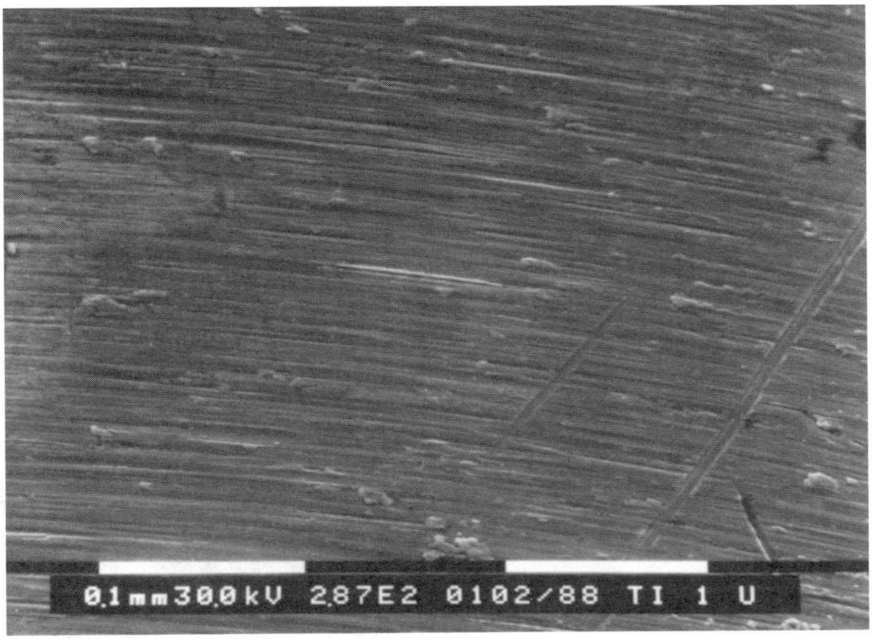

(a)

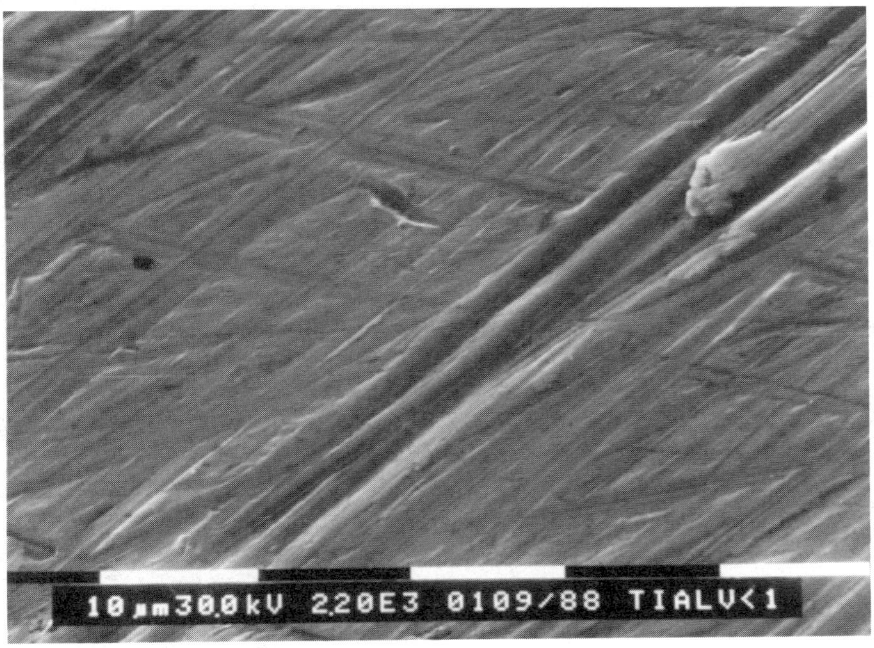

(b)

Figure 1 Surfaces of Ti 1 μm (a) and Ti6Al4V 0.5 μm (b) prior to implantation with fine grooves due to polishing. SEM, bar A: 100 μm, B: 10 μm.

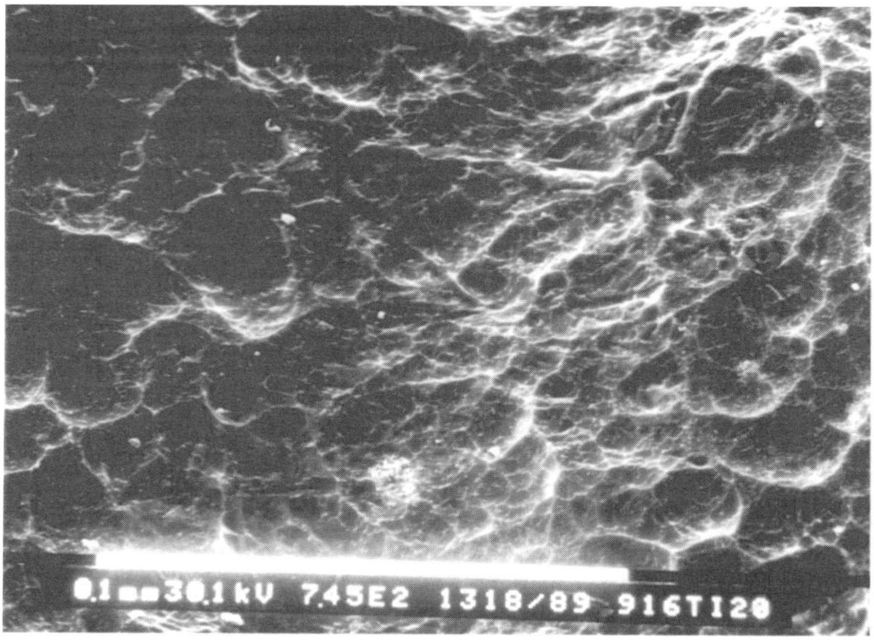

(a)

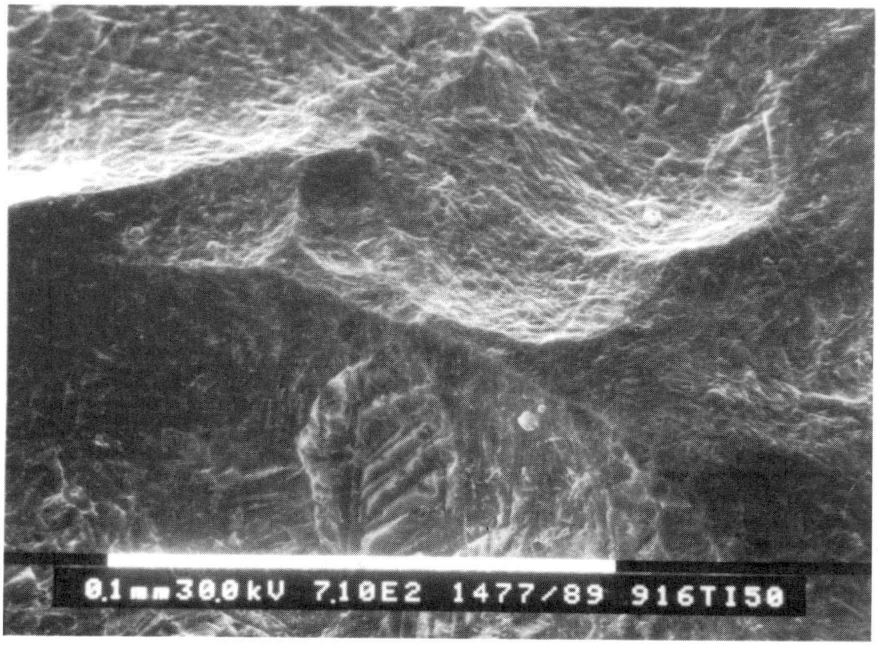

(b)

Figure 2 Ti cylinder prior to implantation with a surface roughness of 20 μm (a) and 50 μm (b) due to pellet blasting. SEM, bar 0.1 mm.

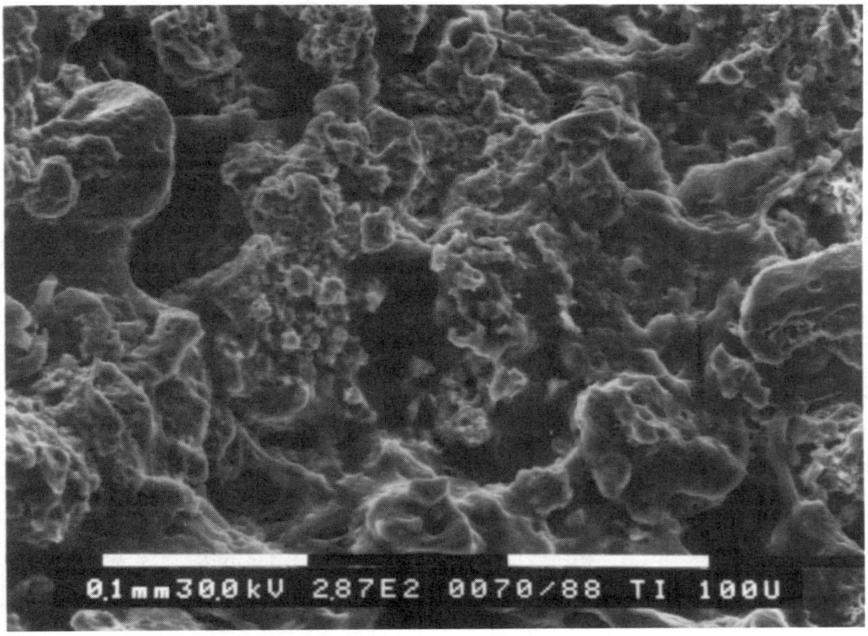

Figure 3 Surface of Ti 100 μm with highly irregular surface structure due to the flame-spray process. SEM, bar 100 μm.

backs and limitations were discovered on the bone surfaces. Due to the cavities on endosteal sites produced by openings of Volkmann vessel channels the laser light was scatterred in such areas and produced significant artefacts. Therefore, the surface roughness could only be described qualitatively.

For chemical surface analysis different methods have been applied. Static secondary ion mass spectroscopy (SIMS) was carried out using 3 keV argon sputtering ions (specimen angle of 60°) because it is extremely sensitive, observes the outermost surface layer, and does not significantly destroy the outermost layers of the surface. Auger electron spectroscopy (AES) was applied with an energy of 3 keV as described elsewhere [35] for analysis of the elemental surface composition and concentration depth profiles. Some implants (Ti 1, 20, 100 μm; Ti6-Al4V 0.5 μm, each $n = 3$) were subjected to SIMS surface analysis at 84 days of implantation. The implants were frozen immediately after explantation using liquid nitrogen, freeze-dried, and analyzed with static SIMS after fracture in the interface. Methods were described in detail elsewhere [35].

C. Implantation Procedure

Adult female Chinchilla rabbits with an average weight of 3500 g were used for implantation. They were fed with Altromin hard standard diet pellets (Altromin, Lage, Germany) and water ad libitum. A ketaminhydrochloride/Rompun (Bayer, Leverkusen, Germany) mixture (25 mg and 5 mg per kilogram body weight, respectively) was used for anesthesia. Techniques of operation and implant insertion into the trabecular bone of the distal femur below the patellar sliding plane perpendicular to the longitudinal axis of the femur were described in detail elsewhere [9,33]. This model was chosen, since large areas of joint arthroplasty implants contact cancellous bone. The cylinders were inserted press fit. All of the implants were partially loaded. The load seemed to act in a direction from proximal-ventral to distal-dorsal producing a typical healing sequence of bone in sequential fluorochrome labeling experiments [36]. The animals were subjected to gentamycin (20 mg) and 0.5 mL Bela-Pharm Antiphlogisticum 30% treatment perioperatively (Bela Pharm KG, Vechta, Germany). All animals were kept in larger cages for several hours every day for training of the musculoskeletal apparatus. The animals were sacrificed at 84 and 168 days in anesthesia with CO_2 inhalation.

D. Histology, Tensile Tests

For histological evaluation the implant containing distal femur was excised and processed as described previously [33]. A minimum of four sections, each section being about 50 μm thick, was prepared from each specimen with a sawing microtome (Leitz 1600, Leitz, Wetzlar, Germany). All specimens were polished after sectioning to avoid interface-covering artifacts. A von Kossa reaction/Fuchsin stain or Giemsa stain was performed and the sections mounted in a resin (Corbit-Balsam, I. Hecht, Kiel-Hassee, Germany). After histological descriptive evaluation the specimens were subjected to histomorphometry. At least four transverse sections of the bone-implant compound were measured. Histomorphometry comprised evaluation of the percentage of bone, osteoid, chondroid, and soft tissue in the bone-implant interface. The statistical evaluation was performed according to mean and standard error of the mean (SEM). Differences between the groups were calculated using the *U*-test with levels of significance of $p < 0.05$. Some specimens were ultrastructurally examined using scanning electron microscopy (SCEM) after tensile tests, and transmission electron microscopy (TEM). The TEM specimens were evaluated after fracture between implant and tissue, and subsequent processing as described in detail previously [37].

Tensile tests were carried out according to a well-established procedure [38]. For testing of the distal bone-implant-interface randomized implants were freed of adherent tissue on their proximal circumference. Only a small piece of tissue was left in contact to the distal interface because of the load transmitted here. The proximal circumference of the implants was freed from the tissue and attached to a clamp and the tissue in contact to the distal interface was fixed in a U-shaped retention bar. Implant and tissue were loaded in opposite directions with a speed of 1 mm/min in a testing machine (Schenk-Trebel, no. RM 100kn, Germany) and the load for rupture was recorded.

III. RESULTS

A. Surface Characterization

In SCEM the above-mentioned surface morphologies were observed (see Material and Methods, Figs. 1–3). It is important to note that only the Ti 100 allowed for interdigitation betweeen newly formed bone and the structures of the implant surface. Neither Ti 20 nor Ti 50 allowed for mechanical interdigitation in a microscopical scale, although there was a considerable surface structure, since the surface of the elevations and deepening was smooth. This points to the importance of a more exact description of the surfaces. A better understanding can be gained by using quantitative methods. The measurement of the surface roughness of the different implant types yielded different data as given by the producers. It is not clear how the roughness was determined. It has to be noted that the R_t-value was higher in every case as

Table 2 Surface Roughnesses of Ti and Ti6Al4V Specimens (μm) as Given by the Producers prior to Implantation in Comparison to Measured Data and to Resorbed Endosteal Bone (humerus, chinchilla rabbit)[a]

R_t (producer)	n	R_t	R_a
Ti6Al4V 0.5	1	3.36	0.45
Ti 1	1	3.56	0.52
Ti 20	1	31.96	3.00
Ti 50	1	60.63	5.87
Ti 100	2	158.13	14.96
		122.41	18.55
Bone, resorbed	8	n.m.[b]	n.m.[b]

[a] R_t maximal and R_a mean surface roughness of the specimens, n = number of specimens. n.m. = not measurable.
[b] The surface structure of resorbed bone surfaces was investigated as described in the text (Materials and Methods, Results). Due to the limitations mentioned the roughness was only estimated.

expected (Table 2). Since deposition of new bone starts usually on resorbed bone surfaces, such surfaces were also subjected to measurement of the surface roughness. Three-dimensional-profiles yielded surfaces with deepenings and elevations which were morphologically comparable to the pellet-blasted implants with roughnesses of 20 or 50 μm. The height of these structures on resorbed bone surfaces was about 20 μm. The 2D profiles yielded an additional rugosity on such surfaces of a few micrometers. This finding was the main difference in the pellet-blasted implants.

Chemical implant surface analysis prior to implantation using SIMS and AES demonstrated on Ti 1 μm and on Ti6Al4V 0.5 μm oxide layers which were a few monolayers thick and consisted mainly of TiO_2. Implants with higher surface roughness (Ti 20, 100 μm) had some defects in the oxide layer (Table 3). In these defects F and other elements up to a depth of 20 monolayers were detected. Other impurities consisted on Ti 1 μm mainly of Cl up to a

Table 3 Depth Profiles (monolayers) of the Amount of TiO_2 (calculated from measured concentrations of Ti, O) in Percent on Ti and Ti-Alloy Implants prior to Implantation as Determined by AES[a]

Depth	Ti 1	Ti 20	Ti 100	Ti6Al4V
Surface	96	68	67	99
2	76	67	65	87
5	67	65	64	72
20	60	60	62	73
100	38	38	60	66
400	36	43	52	61

[a] Ti = titanium, numbers refer to surface roughnesses. Ti 50 was not measured.
Source: From Ref. 35.

depth of 400 monolayers. In the outermost five monolayers Ca, Pb, and Sn impurities, and almost no F, were detected. On Ti 20 and 100 μm only 10% of Cl were detectable as compared to Ti 1 μm. Approximately 20% of negative secondary ions on the surface of Ti 20 μm consisted of F and TiF and around 70% of F and TiF were seen up to five monolayers in depth on Ti 100 μm. Almost no Ca was found on Ti 20 μm and up to 20 monolayers some Ca impurities were evident on Ti 100 μm. Ti6Al4V showed up to 10 monolayers mainly Cl in similar quantities as Ti 1 μm and Pb impurities up to 5 monolayers [35].

Surface analysis after implantation was carried out on three implants of each type (excluding Ti 50 μm) at 84 days of implantation to demonstrate possible changes in the composition and thickness of the oxide layer. Since after simple fracture between implant and tissue and rinsing with Ringer solution only positively charged aliphatic compounds were detectable, the specimens were gently rinsed in acetone for 3 min and afterwards in aqua destillata using ultrasonication. This procedure was shown to produce no artifacts within the surface oxide layer of test specimens [35]. No obvious changes of the oxide layer thickness were detected compared to the specimens which were analyzed prior to implantation. At 84 days the surface oxide was homogenous up to 20 monolayers. Impurities of N, K, and Cl increased on implants after implantation. A higher Ca/Ti-ratio on the surfaces of all the specimens as compared to nonimplanted specimens was demonstrated. Ti 100 μm showed higher values of Ca than Ti 20 μm, and this material in turn showed more than Ti 1 μm and Ti6Al4V 0.5 μm with just a few monolayers containing Ca. The Ca/P ratio was between 1 and 2.2. PO_2^-, and CN^-, which is considered to originate from proteinacious organic sources, showed similar distributions as Ca [35].

B. Histology and Morphometry

At 84 and 168 days postoperatively, light microscopy of the implants with surfaces which provided no possibility for mechanical interlocking (Ti 1, 20, 50 μm, and Ti6Al4V 0.5 μm) showed a bony frame surrounding the implants (Figs. 4–7). This bony frame was supported mainly at the distal pole by perpendicularly inserting trabeculae coming from old trabeculae surrounding the drill hole. From this bony frame, small foot-like projections of bone were developed toward the implant and fixed the implant to some extent inside this frame. The Ti implants with a surface roughness of 20 μm and 50 μm showed wider trabeculae between bony frame and implant than Ti 1 μm and Ti6Al4V. At the transition from old bone to newly formed trabeculae as well as at the transition from bony projections toward the implant surface typical surface profiles of trabeculae were developed (Fig. 5). The histomorphometric analysis yielded between 15% (Ti 1 μm) and 29% (Ti 20 μm) of bone in the interface at 84 days of implantation (Table 4). Ti6Al4V showed statistically significantly less bone in the interface than Ti 100 μm and Ti 1 μm developed significantly less than Ti 20 μm. At 168 days the amount of bone increased to 19% (Ti 1 μm), 36% (Ti 20 μm), and 48% in Ti 50 μm. Ti 1 μm and Ti6Al4V developed significantly less bone as compared to Ti 20 and 50 μm. The mineralization at the bone/metal interface was partially inhibited, demonstrated with the von Kossa reaction showing a considerable amount of osteoid in the interface (Fig. 6). This seam of osteoid had a thickness of approximately 15 μm. Using histomorphometry osteoid contacted between 3% (Ti 20 μm) and 8% to 9% of the implant surface (Ti 1 μm and Ti6Al4V, respectively) at 84 days of implantation. The amount increased to values of 15% (Ti6Al4V), 17% (Ti 20 μm), and 23% (Ti 1 μm), at 168 days. In the case of Ti 50 μm osteoid decreased from 21% at 84 days to 10% at 168 days. The increase in Ti 1 μm to 20 μm and Ti6Al4V was accompanied by a decrease of soft tissue in contact with the Ti (Table 4). One implant with 50 μm surface roughness was cut longitudinally. The amount of bone in the interface was

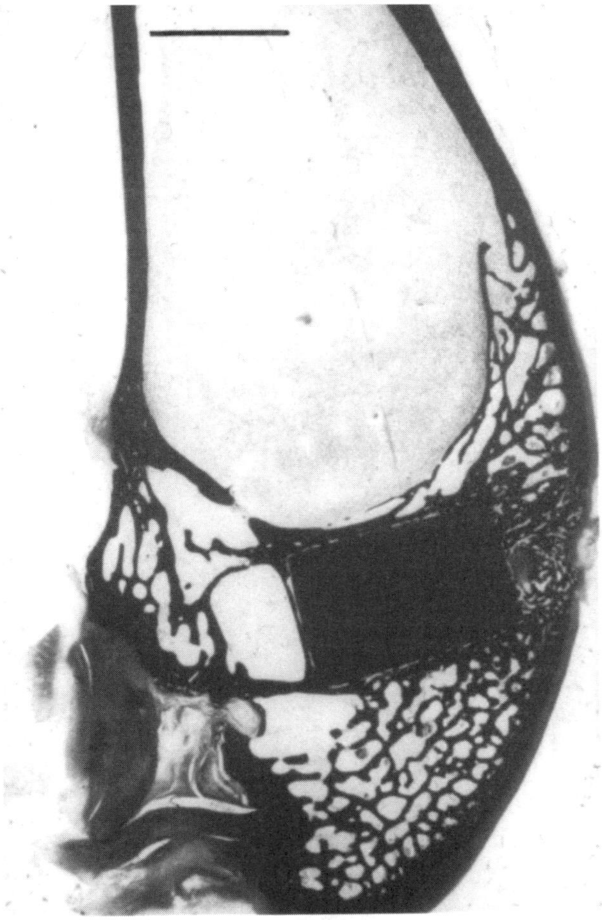

Figure 4 Sagittal section of a Ti cylinder with 50 μm surface roughness at 84 days of implantation into the spongy bone of the distal femur epiphysis of a rabbit. Dense bony shell encapsulates implant and corresponds with bony frame observed in cross sections. LM, von Kossa reaction/Fuchsin stain, bar 4 mm.

between 27% to 29% at proximal and distal aspects and 18% and 21% at dorsal and ventral aspects. Chondroid was only in contact with ventral aspects, which indicates a relation to the procedure of implant insertion (Table 5). Between the projections of bone there was soft tissue with fibers, vessels, and macrophages. Some macrophages seemed to contain tiny black particles. The source of these particles is not yet clear.

Around Ti implants with a surface roughness of 100 μm, there was no bony frame. Perpendicularly arranged abutment-like trabeculae inserted preferentially at the proximal and especially the distal circumferences (Fig. 8). Histomorphometry yielded 27% and 25% of bone in the interface at 84 and 168 days, respectively. There was almost no osteoid (Table 4). At proximal-dorsal aspects these trabeculae were sometimes replaced by bone marrow. At medial and lateral aspects only in some specimens similar almost perpendicularly inserting trabeculae were demonstrated. Soft tissue, occasionally with some macrophages, developed between the trabeculae. At 84 days and 168 days of implantation, there was no obvious difference.

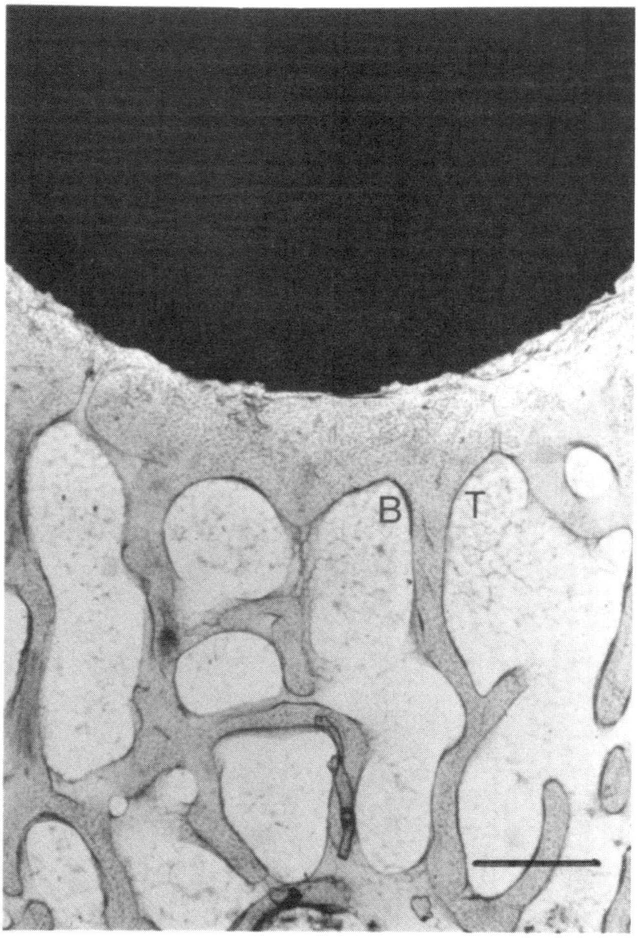

Figure 5 Distal interface of a Ti cylinder, 50 μm, with part of the bony frame and perpendicularly arranged trabeculae with respect to the implant surface between bony frame and spongy bone next to the cartilage of the knee joint. Trabeculae below bony frame showing surface profiles resembling Baud's curves (46). Trabecular surface profiles follow rules of avoiding notch stresses and having minimized weight. Note different profiles for tension (T) and bending (B). LM, Giemsa stain, bar 500 μm.

C. SCEM and TEM

After fractures in the interface between tissue and implant were performed, the observations gained with LM could be confirmed. All of the implants showed typical surface structures as demonstrated in Figs. 1–3. The tissue showed a surface structure reflecting the corresponding surface of the implants. Intact trabeculae were observed between soft tissue islands. Smooth implants had typical bone frames as described above (Fig. 7). On the other hand, in Ti 100 μm a fracture in the interface was not possible due to interlocking of bone trabeculae and rough surface. One implant (Ti 50 μm) was evaluated in the TEM at 168 days after performing a fracture in the interface and removal of the implant. Bone contact and soft tissue interfaces were detectable. Bone contact areas showed typical depressions and elevations produced by

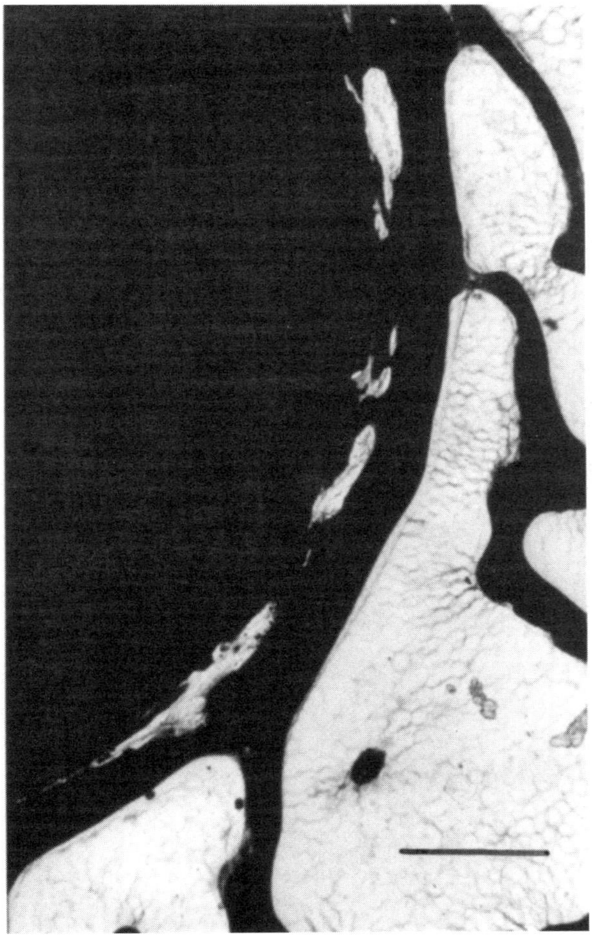

Figure 6 Part of the medial interface of a Ti 50 cylinder at 84 days of implantation. Disturbed mineralization with interposed osteoid and soft tissue. LM, von Kossa reaction/Fuchsin stain, bar 500 μm.

the surface roughness. There was no intervening interfacial membrane (Figs. 9–10). Occasionally, some resorbing cells and osteoblast-like cells were seen remodeling surface profiles of bone trabeculae-implant transitions (Figs. 9–10).

D. Tensile Tests

Tensile tests of the bone Ti interface yielded no tensile strength on smooth surfaces (Ti6Al4V, Ti 1 and 20 μm). Ti 50 μm yielded values close to zero. There was no statistically significant difference at 84 days as compared to 168 days after insertion. Ti 100 μm with a highly irregular surface which allowed mechanical interlocking yielded 5.53 and 4.66 N/mm^2 at 84 and 168 days, respectively, which was not statistically significantly different (Table 6). The fracture occurred in every case in the surrounding trabecular bone, thus actually yielding the tensile strength of the trabecular bone.

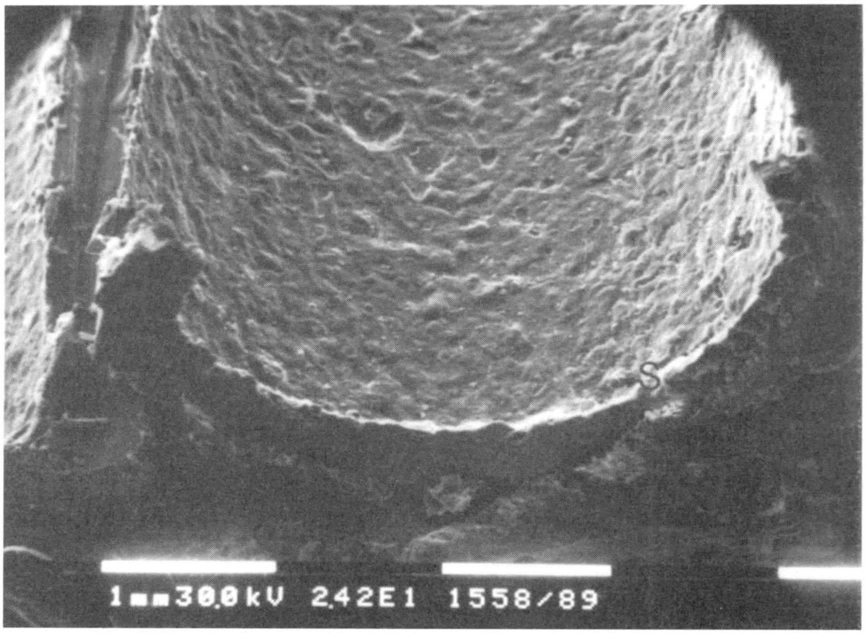

Figure 7 Tissue side of a Ti 20 cylinder at 168 days of implantation after fracture between tissue and implant. Note bony shell (S) and surface of the shell reflecting the implant surface roughness which did not allow mechanical interlocking. SEM, bar 1 mm.

Table 4 Bone, Osteoid (Os), Chondroid (Ch), and Soft Tissue (St) (% ± SEM) in the Interface of Ti and Ti6Al4V Cylinders of Different Surface Structure at 84 and 168 Days (d) of Implantation into the Trabecular Bone of the Distal Femur Epiphysis of Rabbits[a]

Material	R_t	*n*	*d*	Bone	Os/Ch	St
Ti6Al4V	0.5	4	84	17 ± 2	9 ± 4	74 ± 3
		6	168	21 ± 4	15 ± 2	64 ± 4
Ti	1	6	84	15 ± 4	8 ± 3	77 ± 4
		4	168	19 ± 2	23 ± 4	58 ± 1
	20	6	84	29 ± 4	3 ± 1	68 ± 4
		5	168	36 ± 5	17 ± 2	47 ± 3
	50	4	84	26 ± 2	21 ± 2	53 ± 4
	50L	1	84	26 ± 7	23 ± 7	51 ± 3
	50	5	168	48 ± 3	10 ± 2	42 ± 5
	100	6	84	27 ± 1	<1 ± <1	73 ± 1
		6	168	25 ± 3	<1 ± <1	75 ± 3

[a] R_t = surface roughness (μm), *n* = number of specimens, 50L = longitudinally cut implant.

Table 5 Mean Percentage ± SEM of Bone (B), Osteoid (Os), and Chondroid (Ch) (soft tissue not given in the table) at Proximal, Dorsal, Distal, and Ventral Aspects of a Longitudinally Cut Ti Cylinder with a Surface Roughness of 50 μm at 84 Days after Implantation into the Distal Femur Epiphysis of One Rabbit ($n = 4$ sections)

	Proximal	Dorsal	Distal	Ventral
B	27 ± 7	18 ± 6	29 ± 10	21 ± 7
Os	24 ± 4	26 ± 7	17 ± 8	17 ± 6
Ch	0 ± 0	0 ± 0	0 ± 0	15 ± 10

IV. DISCUSSION

A. Surface Characterization

It was demonstrated in this study that Ti and Ti alloy implants with different surface structures contain different kinds of impurities on the surface. Generally, the amount of impurities increased with decreasing surface roughness. The existence of impurities is well in accordance with former studies and can be explained by the production procedure [17,24,26,39]. To achieve different physical surface properties, processes such as polishing, lapping, pellet blasting, and flame spraying as well as many others were used. The content of Pb and Sn on polished implants seems to be related to Pb- and Sn-containing binders used on polishing devices. Fluoride and TiF defects on Ti 100 μm are probably related to 1% HF rinsing of the

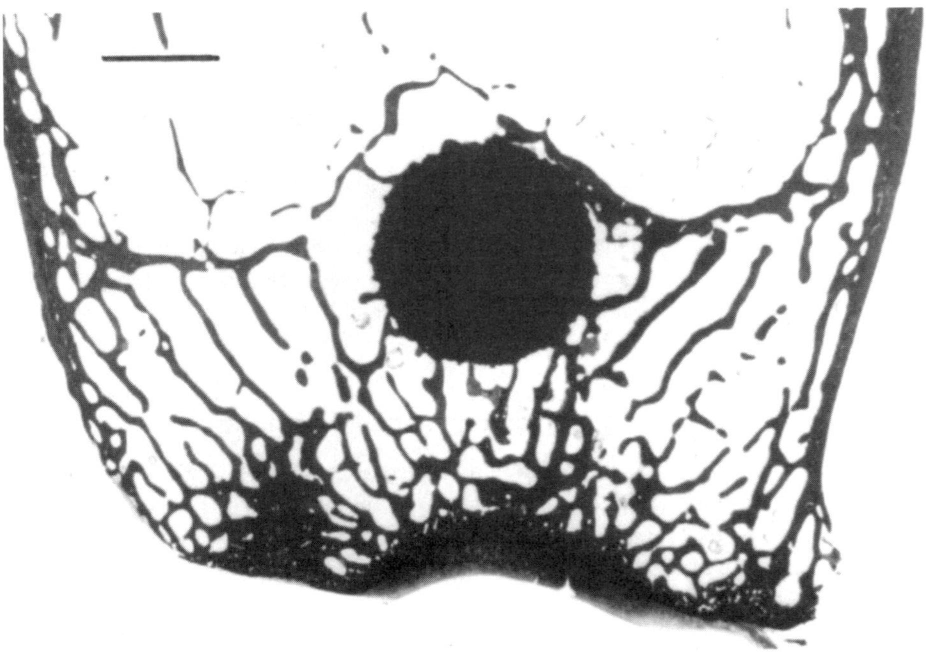

Figure 8 Ti cylinder with a surface roughness of 100 μm at 168 days of implantation with perpendicularly inserting reinforcing trabeculae at the distal interface. LM, von Kossa reaction, bar 2 mm.

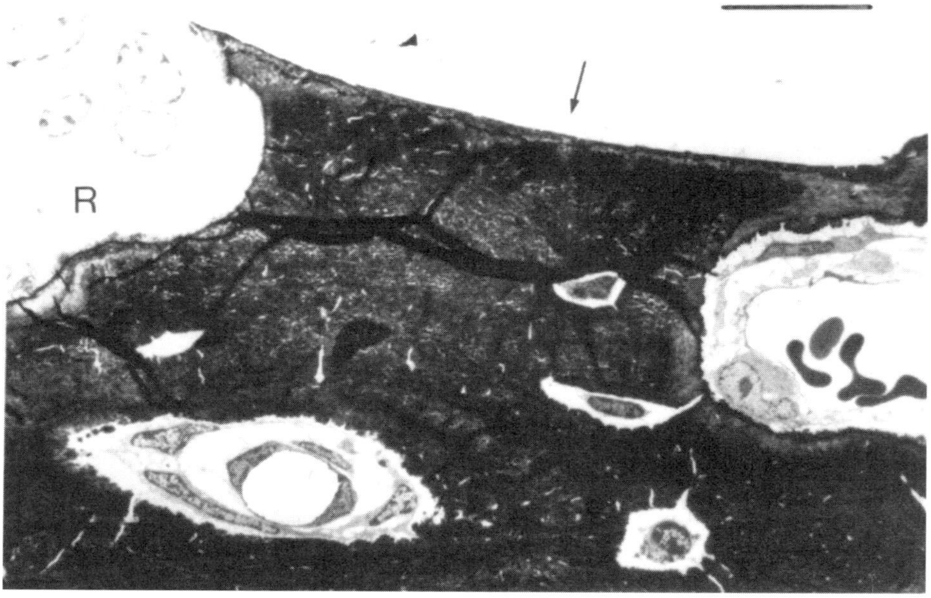

Figure 9 Distal interface of a Ti 50 μm cylinder at 168 days after fracture in the interface (implant removed, top, white). Bone-implant contact without intervening soft tissue (arrow). Still ongowing remodeling in the interface indicated by resorbing osteoclast-like cell (R) on a trabecular surface. Capillaries in the bone with erythrocytes (right). TEM, bar 20 μm.

implants prior to implantation to remove other impurities. Some defects or impurities of the surface oxide might be of minor importance in vivo, since it is well known that the oxide layer further grows in vivo [6]. This is well in accordance with the findings gained in this study. The rather sponge-like structure of the oxide layer with defects disappeared at 84 days suggesting a further growth without increasing thickness in the implantation bed [35]. All implants were incorporated in an acceptable manner, i.e., without remarkable disturbance of the healing sequence such as inflammation. But it should be considered whether higher percentages of osteoid are only related to smooth implant surfaces or may be associated with impurities on the implant's surface, since it is well known that corrosive processes increase with impurities and that especially metal oxides when leached from implants are able to disturb the mineralization of osteoid by delaying the calcification process [40–42].

B. Surface Characterization after Implantation and Mineral Deposition

Implant surface analyses at 84 days of implantation yielded after fracture in the interface level and rinsing with Ringer solution only aliphatic compounds on the surface. This was related to the rinsing process which led to a wide distribution of such compounds. Additionally, this process might account for the content of N, K, and Cl. After cleansing with acetone and aqua destillata using ultrasonication no changes of implant surface oxide layers of test specimens were detected before and after this procedure which corroborates former studies [43]. Therefore, this procedure was used for the 84 days specimens of this study. At 84 days Ca, PO_2^-, and CN^- impurities in different amounts and layers of different thickness were detected which correlated with different surface roughnesses. The thickness of Ca- and P-containing plaques

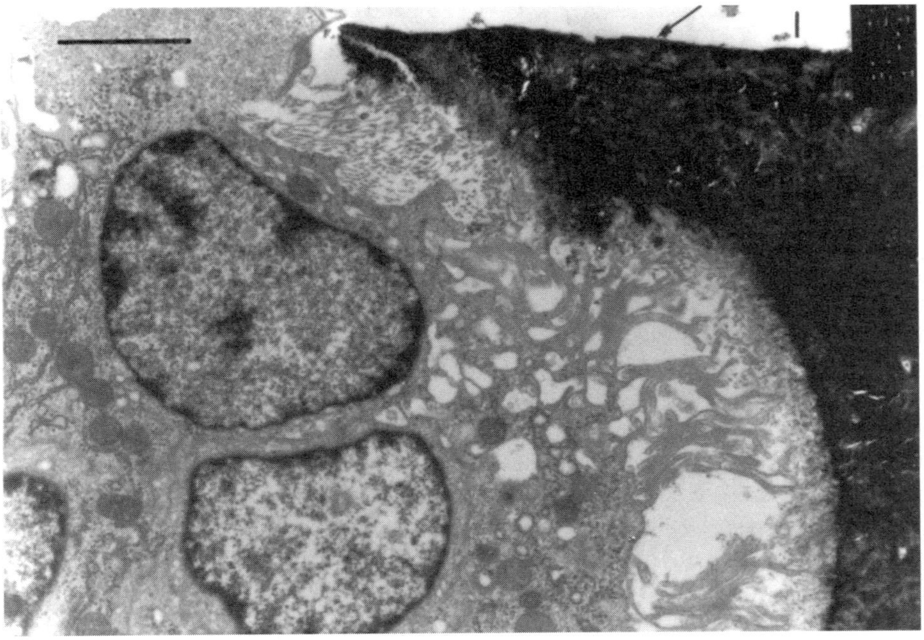

Figure 10 Higher magnification from Fig. 9. Mineralized bone at the interface (I). Implant removed (white, top). Bone-implant contact without intervening soft tissue. Bonding zone slightly darker (arrow) than collagen containing bone. TEM, bar 5 μm.

increased with increasing surface roughness. This suggests that with a higher surface roughness a better implant bone connection due to a better production of mineral by osteoblastic cells could be achieved. The occurrence of ions such as Ca^{2+} and PO_2^- might be explained by the highly polar oxide layer on the surface of Ti implants which attracts charged species. It was speculated that the particular configuration of the oxides on Ti provides a mosaic for matching with the cementing ground substance [22]. Comparable almost afibrillar Ca/P layers were observed in vivo [6,44] and in vitro [31–32] on Ti and Ti6Al4V surfaces. These layers acted as the initial bonding zone and enabled collagen fibers of the adjacent bone to anchor [31–32]. A thin afibrillar layer (at least 200 Å thick) consisting of partly calcified amorphous ground substances, such as proteoglycans and glycosaminoglycans, was related to good biocompatibil-

Table 6 Tensile Strength of the Interface between Bone and Ti and Ti6Al4V Cylinders ($N/mm^2 \pm SEM$) 84 and 168 Days (d) after Implantation into the Distal Femur Epiphysis of Rabbits[a]

Material	R_t	n	d	Tensile strength
Ti6Al4V; Ti	0.5–20	10	84,168	0
Ti	50	3	84	0.12 ± 0.08
		5	168	0.18 ± 0.06
Ti	100	10	84	5.53 ± 0.40
		7	168	4.66 ± 0.53

[a] R_t = surface roughness (μm), n = number of specimens.

ity, whereas Ti6Al4V showed coats of about 5000 Å thickness [44]. The afibrillar bonding zone seems to be inhomogeneous. In closest contact to the implant surface a mineralized layer containing osteopontin and bone sialoprotein provides the bonding. This layer in turn is covered by a zone containing single collagen fibers as well as higher levels of proteoglycans. This outer layer is the basis for collagen attachment, i.e., the bonding between the afibrillar layer and the "real" bone [45].

C. Trabecular Arrangement and Implant Surface Structure

Bony shell-like frames were observed encapsulating smooth implants, where mechanical interlocking is impossible. Small foot-like trabeculae fixed this frame in the old bone in the drilling hole and fixed the implant within the frame. At the implant surface inserting trabeculae showed often surface profiles at the transition from trabeculae to implant surfaces which were similar to Baud curves, which are mathematically calculated optimized surface profiles for T-joints [46]. Such surface profiles avoid notch stresses at T-joints of beams with different diameters and might function at the bone-implant transition in a similar manner. Similar surface profiles were observed on other biological structures, e.g., tree branches, deer antlers and bone trabeculae without implants and seem to represent a general biological principle [47]. Comparable bone collars around metal implants were also observed by other investigators, e.g., around bead covered implants with pore diameters of less than 50 μm, which prevented bony ingrowth, around implants with higher pore diameters allowing bone ingrowth, and surrounding screw-type mandibular implants [12–13,15,48–49]. On the other hand, the 100 μm implants yielded perpendicularly inserting trabeculae. These trabeculae also showed Baud curves at transitions to the implant surface or to other trabeculae without a bony shell and corresponded to a high resistance of the interface against tensile stresses. This points to the possibility that such surfaces with perpendicularly inserting trabeculae should be considered as the better implant in this model than a nonstructured surface or structured surfaces, which do not allow appropriate bone ingrowth or bone contact to fix the implant between trabeculae. This hypothesis is supported by experimental and clinical data. After injection of different fluorochromes using the same animal model and the same implant used in this study a higher turnover rate around smooth implants with surface roughnesses between 0.5 and 50 μm as compared to Ti 100 μm was demonstrated up to 84 days of implantation, indicating still ongoing remodeling in the interface region [36]. This might be explained by a slight instability of the interface in cases were mechanical interlocking or chemical bonding is not possible. On the other hand, the amount of bone on Ti 100 surfaces was similar at 84 and 168 days and there was almost no osteoid. This was coupled with a tensile strength of the interface, indicating optimized fixation of the implants and that load could be transmitted via the interface. Therefore, one can conclude, that the "optimal" amount of bone was already developed at 84 days in the case of Ti 100 μm since no more bone was needed to achieve appropriate fixation as evidenced by the tensile strength of the interface. This concept of faster bone development around structured metallic implants is supported by other experimental studies. In vitro, osteoblastic activity including alkaline phosphatase activity was higher on rougher Ti surfaces (R_t between 2.6 and 19.5 μm) [7]. Additionally, the responsibility of osteoblasts to 1α, 25-$(OH)_2D_3$ is higher if they settled on Ti implants with rougher as compared with smoother surfaces [8]. It is well established that osteoblasts grow but do not differentiate well on smooth surfaces [50].

The situation around Ti 100 implants is closer to the arrangement of trabeculae prior to implantation, which was shown earlier [15]. One could ask now whether an "ideal" implant should shift the trabecular arrangement as less as possible after implantation. A former study using different implants made of CoCrMo beads with bead diameters between 260 to 750 μm compared the amount and arrangement of bone trabeculae in the direct implant vicinity with control slices without implants by projection of implant containing specimens over controls. All implant containing specimens yielded a higher amount of bone related to the development of bony frames or partial frames. Fewer changes of the trabecular arrangement yielded implants with a bead diameter of 260 μm as compared with controls which correlated with the highest tensile strength of the interface. Therefore, these implants were considered to provide best results [15].

Ti plugs, smooth, sandblasted, and flame-sprayed prior to implantation yielded after insertion into the diaphysis of rats fastest bone-implant contact and most bone in the interface around flame-sprayed, i.e., rough implants [48]. On the other hand, fibrous tissue capsules developed around smooth titanium implants [51]. A clinical study using titanium-aluminum-niobium alloy total hip endoprostheses in humans totally covered with a surface roughness of 3 to 5 μm demonstrated that these implants did not yield bone collars, i.e., they were anchored mainly by trabecular structures. The implants did not cause cortical thinning and it was therefore concluded that the microroughness of the surface was sufficient in promoting secondary stabilization of the implant via ongrowing trabeculae and that there was functional load transmission avoiding stress shielding [52]. Better characterized implants, produced by a spark-eroding process with a surface roughness of S_a (equals R_a in three dimensions) of approximately 2.5 μm corroborated this result. There was a significant resistance against torque movements in case of the above-mentioned surface roughness [17]. A theoretical explanation for the incorporation of Ti cylinders with or without bone shells might provide calculations of the influence of different implant forms and different bone-implant connections on the arrangement of stress distributions around implants. From 3D finite element calculations and photoelastic experiments using identically shaped implants under axial compression as used in this study it was concluded that Young's modulus and the type of bone implant connection, i.e., bone-contact or bone bonding, lead to totally different stress distributions in the implant surrounding tissue. Relatively stiff implants with bone bonding showed a capability of load transfer over the whole interface which led to a less disturbed stress distribution around the implants [53]. Such a situation might be comparable to flame-sprayed Ti implants which simulate a bonding situation by mechanical interlocking which results in fixed implants. Smooth and stiff implants with bone contact generated certain stress concentrations because of incomplete stress transfer via the interface [53]. Therefore, the bony frames observed in smooth implants used here seem to indicate that forces are probably directed around the implant, resulting in higher stress concentrations in the surrounding tissue. This hypothesis was proven in an in vivo study using spherical implants (smooth and bead covered) in the spongy bone of pony patellas. Smooth implants yielded a greater remodeling response resulting in a bony shell. The histology correlated well with finite element models which predicted greater stress changes adjacent to smooth implants due to nonlinear boundary conditions [54].

Similar values of bone in the interface of Ti 100 μm at 84 and 168 days indicate that a stabile fixation of the implant has been already achieved which is supported by the resistance of the interface to tensile forces. The amount of bone in the interface (25% to 27%) was in the same range than in another study using loaded hip implants which were partially flame-spray coated (32.5% bone) [55]. Other experiments demonstrated higher values of bone in the interface. Porous surfaced tooth implants of Ti6Al4V covered with beads yielded 32% and 58% of bone in the interface at 4 weeks and 18 months, respectively [14,56]. This might be related to the other surface structure and to the different implant model. The trabecular arrangement and the amount of bone should be compared with specimens which do not contain implants.

From the other implant types of this study which do not allow mechanical interlocking only Ti 50 µm yielded a less intense delay of mineralization since at 168 days the amount of osteoid decreased and bone increased. A longer delay of the healing sequence was evidenced for Ti 1 and 20 µm and Ti6Al4V (Table 4). Therefore, an optimized implant design avoids a delay in calcification and undesirable reactions such as encapsulation with an outer layer of bone and an inner layer of soft tissue observed in contact to some implants of this study and in other investigations [57–58]. In cases of soft tissue implant interfaces a functional orientation indicated by obliquely and perpendicularly inserting soft tissue fibers was only observed in cases of structured surfaces such as flame-sprayed titanium [29,59] and bead-covered implants [57]. In cases of dental implants the fixing structures resembled the normal tooth root, indicating a functional orientation [14,33]. It was shown that similar oriented inserting fibers also resulted in a resistance against tensile forces [60].

D. Corrosion and Aseptic Loosening

In some cases, especially around smooth implants tiny black particles were observed in macrophages in the implant surrounding soft tissues. The source and kind of these particles remain unclear. It cannot be excluded that these particles are of implant origin, since it is well known that mechanically unstable implants show higher corrosion rates due to a breakdown of the surface oxide layer. Metallic debris was also observed in histiocytes surrounding Ti6Al4V hip implants [61]. Up to a certain amount corrosive reactions will be tolerated. But if a certain amount of particles is liberated from an implant factor released by macrophages, e.g., tumor necrosis factor [62], cytokines and interleukins lead to a stimulation of osteoblasts to release factors, e.g., interleukine 1 (IL 1), IL 6, tumor necrosis factor and transforming growth factor β as well as PG E_2 [62–64]. It is well established that these factors are able to stimulate bone degradation primarily through interaction with osteoclast precursors. At the end an aseptic loosing might result from these reactions, e.g., of a total hip implant system. The occurrence of such particles might be enhanced by disturbed mineralization of osteoid or increased resorption in the interface which was only observed here in significant amounts around smooth implants. The effect of resorption promoting factors, e.g., PG E_2, can be reduced by the addition of cyclooxygenase inhibiting drugs to implants. Such implants showed significantly higher values of bone to implant contact than implants without the drug [65]. Additionally, the elemental composition of the implant and possible impurities of the material's surface should be considered. It was reported that a Fe content higher than 0.05% leads to higher corrosion rates of Ti implants [6]. The normal level of Ti in human tissue is of about 50 ppm. This value arises when implant corrosion occurs [66]. Up to a certain amount of metal in the tissue macrophages and foreign-body giant cells might be able to remove corroded particles. If such a process does not stop, aseptical necrosis and fatigue failure might occur. Severe tissue reactions leading to implant loss were described around bead-covered CoCrMo human hip implants [67].

E. Bonding Strength

Biomechanical data showed that the concept of titanium being fixed via chemical forces by Ca/P-rich tissues seems to be of minor importance clinically. A bone titanium interface with a smooth surface yields no tensile strength at the interface. Therefore, chemical bonds are not suitable for implant fixation especially in load bearing applications. Many experiments using push-out tests are misleading to this result since structured surfaces without mechanical interlocking such as Ti 50 µm, yield positive results in push-out tests but fail tensile tests. Highly structured surfaces such as flame-sprayed Ti yielded tensile test values which were in the range of spongy bone. These values were composed mainly of biomechanical interlocking and maybe some additional chemical bonds. Chemical bonding forces might be represented by the tensile strength values of Ti 50 which were close to zero, but mechanical interlocking seemed to be excluded. This corresponds with other data. It was reported that the tear-off strength of sand-blasted surfaces reached 3.5 N/mm^2 at 210 days and that plasma-coated implants yielded 1.5 N/mm^2 at 120 days [68]. Additionally it was shown that flame-sprayed titanium tooth implants subjected to functional load in Maccaca Speciosa provide an "ancylotic" bond between implant and tissue. Fractures occurred only within the bone or between flame-spray coating and substrate [68] which stands for the tight bond between implant and bone. Therefore, the osteoid in the interface, observed only in implants which do not allow mechanical interlocking, indicates that the development depends, among other parameters, on the surface structure. It can be assumed that the healing sequence was delayed in Ti 50 μm as compared to Ti 100 μm and even more in Ti 1 μm and 20 μm as well as Ti6Al4V.

V. CONCLUSIONS

1. Metallic implant surfaces must be exactly characterized according to standardized procedures with regard to the height, width, and structure of the surface rugosity as well as the chemical composition, since impurities might be detected depending on the kind of implant production.
2. Some impurities and defects of the surface oxide layer are of minor importance in vivo, e.g., F.
3. The surface structure must be irregular, i.e., it must allow for mechanical interlocking between the surface rugosity and the bone. A morphological example is given with resorbed bone surfaces possessing a rugosity which provide an excellent substratum for bone deposition.
4. Depending on the surface roughness as well as surface structure, the amount of deposited amorphous layers containing Ca, P, and CN^- compounds, detected on implant surfaces at 84 days of implantation, increases with increasing surface roughness. Such plaques might adhere to the implant surface due to chemical forces, e.g., van der Waals bonds and covalent bonds especially in smooth implants.
5. Structured implant surfaces allowing mechanical interlocking provide a faster and biomechanically better incorporation of Ti implants which was evidenced by perpendicularly inserting trabeculae, lower remodeling rates in the interface, and a resistance of the interface against tensile forces. Although less bone contact of the interface is needed it allowed for appropriate implant fixation and load transmission.
6. Trabeculae seem to contact other trabeculae and implant surfaces, among other possibilities, according to rules of avoiding notch stresses on trabecular surfaces.

ACKNOWLEDGMENTS

This study was supported by the Ministry for Research and Technology, grant 01 VG 8601, Bonn, Germany. For implant production we are grateful to Battelle Institute, Frankfurt, Germany, and Aesculap AG, Tuttlingen, Germany. SIMS and AES analyses were by Dornier System, Friedrichshafen, Germany.

REFERENCES

1. Linder, L., Carlsson, Å., Marsal, L., Bjursten, L. M. and Brånemark, P. I., 1988. Clinical aspects of osseointegration in joint replacement. *J. Bone Joint Surg.*, 70-B: 550–555.
2. Carlsson, L., Röstlund, T., Albrektsson, B. and Albrektsson, T., 1988. Implant fixation improved by close fit. Cylindrical implant-bone interface studied in rabbits. *Acta Orthop. Scand.*, 59: 272–275.
3. Gross, U., Schmitz, H. J. and Strunz, V., 1988. Surface activities of bioactive glass, aluminum oxide, and titanium in a living environment. In: *Bioceramics: Material Characteristics versus in Vivo Behavior*, P. Ducheyne and J. E. Lemons (Editors), *Ann. N.Y. Acad. Sci.*, 523: 211–226.
4. Satomi, K., Akagawa, Y., Nikai, H. and Tsuru, H., 1988. Bone-implant interface structures after nontapping and tapping insertion of screw-type titanium alloy endosseous implants. *J. Prosthet. Dent.*, 59: 339–342.
5. Pilliar, R. M., Lee, J. M. and Maniatopoulos, C., 1986. Observations on the effect of movement on bone ingrowth into porous-surfaced implants. *Clin. Orthop. Rel. Res.*, 208: 108–113.
6. Albrektsson, T., 1984. The response of bone to titanium implants. In: *CRC Critical Reviews of Biocompatibility*, Vol. 1, CRC Press, Boca Raton, FL, pp. 53–84.
7. Nishimura, N. and Kawai, T., 1998. Effect of microstructure of titanium surface on the behaviour of osteogenic cell line MC3T3-E1. *J. Mater. Sci.: Mater. Med.*, 9: 99–102.
8. Boyan, B. D., Batzer, R., Kieswetter, K., Liu, Y., Cochran, D. L., Szmuckler-Moncler, S., Dean, D. D. and Schwartz, Z., 1998. Titanium surface roughness alters responsiveness of MG63 osteoblast-like cells to $1\alpha,25\text{-}(OH)_2D_3$. *J. Biomed. Mater. Res.*, 39: 77–85.
9. Gross, U., Müller-Mai, C., Fritz, T., Voigt, C., Knarse, W. and Schmitz, H. J., 1990. Implant surface roughness and mode of load transmission influence periimplant bone structure. In: *Clinical Implant Materials, Advances in Biomaterials* 9, G. Heimke, U. Soltész, A. J. C. Lee (Editors), Elsevier, Amsterdam, The Netherlands, pp. 303–308.
10. Engh, C. A., 1983. Hip arthroplasty with a moore prosthesis with porous coating. A five year study. *Clin. Orthop. Rel. Res.*, 176: 52–66.
11. Klawitter, J. J. and Hulbert, S. F., 1971. Application of porous ceramics for the attachment of load bearing internal orthopedic applications. *J. Biomed. Mater. Res. Symp.*, 2: 161–229.
12. Bobyn, J. D., Pilliar, R. M., Cameron, H. U. and Weatherly, G. C., 1980. The optimum pore size for the fixation of porous-surfaced metal implants by the ingrowth of bone. *Clin. Orthop. Rel. Res.*, 150: 263–270.
13. Pilliar, R. M., 1983. Powder metal-made orthopedic implants with porous surface for fixaton by tissue ingrowth. *Clin. Orthop. Rel. Res.*, 176: 42–51.
14. Deporter, D. A., Watson, P. A., Pilliar, R. M., Howley, T. P. and Winslow, J., 1988. A histological evaluation of a funtional endosseous, porous-surfaced, titanium alloy dental implant system in the dog. *J. Dent. Res.*, 67: 1190–1195.
15. Ziegler, S., 1991. Knöcherne Inkorporation von Metall-Implantaten mit unterschiedlicher Porengröße. Thesis, Freie Universität Berlin.
16. Frost, H. M., 1990. Skeletal structural adaptations to mechanical usage (SATMU): 2. Redefining Wolff's law: The remodeling problem. *Anat. Rec.*, 226: 414–422.
17. Wennerberg, A., Hallgren, C., Johhansson, C., Sawase, T., Lausmaa, J., 1997. Surface characterization and biological evaluation of spark-eroded surfaces. *J. Mater. Sci.: Mater. Med.*, 8: 757–763.
18. Wennerberg, A., Ohlsson, R., Rosen, B. G. and Andersson, B., 1996. Characterizing three-dimensional topography of engineering and biomaterial surfaces by confocal laser scanning and stylus techniques. *Med. Eng. Phys.*, 18: 548–556.
19. Wennerberg, A., Albrektsson, T. and Andersson, B., 1996. Bone tissue response to commercially pure titanium implants blasted with fine and coarse particles of aluminum oxide. *Int. J. Oral Maxillofac. Implants*, 11: 38–45.
20. Wennerberg, A. 1996. On surface roughness and implant incorporation. Thesis, University of Gothenburg.
21. Boss, J., 1994. The nature of the bone-implant interface. *Med. Prog. Tech.*, 20: 119–129.
22. Gross, U. M. and Müller-Mai, C., 1990. Hard materials–tissue interface: General considerations

and examples for bone-bonding and for epithelial attachment. In: *CRC Handbook of Bioactive Ceramics*, Vol. 1, *Bioactive Glasses and Glass-Ceramics*, T. Yamamuro, L. L. Hench, J. Wilson (Editors), CRC Press, Boca Raton, FL, pp. 25–39.

23. Ducheyne, P., Willems, G., Martens, M. and Helsen, J., 1984. In vitro metal-ion release from porous titanium-fiber material. *J. Biomed. Mater. Res.*, 18: 292–308.
24. Lausmaa, J., Kasemo, B. and Mattson, H., 1990. Surface spectroscopic characterization of titanium implant materials. *Appl. Surf. Sci.*, 44: 133–146.
25. Kasemo, B. and Lausmaa, J., 1985. Aspects of surface physics on titanium implants. *Swed. Dent. J., Suppl.*, 28: 19–36.
26. Ask, M., Lausmaa, J. and Kasemo, B., 1988–89. Preparation and surface spectroscopic characterization of oxide films on Ti6Al4V. *Appl. Surf. Sci.*, 35: 283–301.
27. Albrektsson, T. and Hansson, H. A., 1986. An ultrastructural characterization of the interface between bone and sputtered titanium or stainless steel surfaces. *Biomaterials*, 7: 201–205.
28. Linder, L., Albrektsson, T., Brånemark, P. I., Hansson, H. A., Ivarsson, B., Jönsson, U. and Lundström, I., 1983. Electron microscopic analysis of the bone-titanium interface. *Acta Orthop. Scand.*, 54: 45–52.
29. Schröder, A., 1984. Die Reaktion von Mukosa und Knochen auf Titanplasma bzw. Titanoxide (Rutil), *Fortschr. Zahnärztl. Implantol.*, I: 25–27.
30. Parsegian, V. A., 1983. Molecular forces governing tight contact between cellular surfaces and substrates. *J. Prosth. Dent.*, 49: 838–842.
31. Davies, J. E., Lowenberg, B. and Shiga, A., 1990. The bone-titanium interface in vitro. *J. Biomed. Mater. Res.*, 24: 1289–1306.
32. Lowenberg, B., Chernecky, R., Shiga, A. and Davies, J. E., 1991. Mineralized matrix production by osteoblasts on solid titanium in vitro. *Cells Mater.*, 1: 177–187.
33. Müller-Mai, C., Schmitz, H. J., Strunz, V., Fuhrmann, G., Fritz, T. and Gross, U. M., 1989. Tissues at the surface of the new composite material titanium/glass-ceramic for replacement of bone and teeth. *J. Biomed. Mater. Res.*, 23: 1149–1168.
34. DIN 4768, and DIN 4762/1E, Maximale Rauhtiefe, Begriffe, Mai 1978, Deutsches Institut für Normung e.V., Berlin, Germany.
35. Kerfin, W., Plog, C. and Karsch, R., 1990. Begleitende grenzflächenanalytische Untersuchungen zur Charakterisierung der Langzeitstabilität von Endoprothesen. report, Ministry for Research and Technology, no. 01 VG 8601, Bonn, Germany.
36. Draenert, K., unpublished data.
37. Müller-Mai, C. M., Voigt, C. and Gross, U., 1990. Incorporation and degradation of hydroxyapatite implants of different surface roughness and surface structure in bone. *Scanning Microsc.*, 4: 613–624.
38. Gross, U., Roggendorf, W., Schmitz, H. J. and Strunz, V., 1987. Biomechanical and morphological testing methods for porous and surface reactive biomaterials. In: *Quantitative Characterization and Performance of Porous Implants for Hard tissue Applications*, ASTM STP 953, J. E. Lemmons (Editor), American Society for Testing and Materials, Philadelphia, pp. 330–346.
39. Mäusli, P. A., Simpson, J. P., Burri, G. and Steinemann, S. G., 1987. Constitution of oxides on titanium alloys for surgical implants. In: Proc. 7th Europ. Conf. on Biomat., Amsterdam.
40. Blumenthal, N. C., Posner, A. S., Cosma, V. and Gross, U., 1988. The effect of glass-ceramic bone implant materials on the in vitro formation of hydroxyapatite. *J. Biomed. Mater. Res.*, 22: 1033–1041.
41. Blumenthal, N. C. and Cosma, V., 1989. Inhibition of apatite formation by titanium and vanadium ions. *J. Biomed. Mater. Res.*, 23: 13–22.
42. Müller-Mai, C. M., Amir, D., Schwartz, Z., Sela, J., Boyan, B. D., Wendland, H. and Gross, U. M., 1991. Ultrastructural histomorphometry of extracellular matrix vesicles in primary calcification around bone-bonding and non-bonding glass-ceramics. *Cells Mater.*, 1: 341–352.
43. Daculsi, G., LeGeros, R. Z. and Deudon, C., 1990. Scanning and transmission electron microscopy, and electron probe analysis of the interface between implants and host tissue. *Scanning Microsc.*, 4: 309–314.

44. Albrektsson, T. and Jacobsson, M., 1987. Bone-metal interface in osseointegration. *J. Prosth. Dent.*, 57: 597–607.
45. Davies, J. E., 1996. In vitro modeling of the bone/implant interface. *Anat. Rec.*, 245: 426–445.
46. Baud, R. V. 1934. Beiträge zur Kenntnis der Spannungs-verteilung in prismatischen und keilförmigen Konstruktionselementen mit Querschnittsübergängen. Bericht 29, Schweiz. Verband für die Materialprüfungen der Technik (Bericht 83 der Eidg. Materialprüfanstalt, Zürich).
47. Mattheck, C. 1990. Engineering components grow like trees. *Mat. wiss. Werkstofftech.*, 21: 143–168.
48. Kirsch, A. and Donath, K., 1984. Tierexperimentelle Untersuchungen zur Bedeutung der Mikromorphologie von Titanimplantatoberflächen. *Fortschr. Zahnärztl. Implantol.*, I: 35–40.
49. Hansson, H. A., Albrektsson, T. and Brånemark, P. I., 1983. Structural aspects of the interface between tissue and titanium implants. *J. Prosth. Dent.*, 50: 108–113.
50. Jones, D. B., Scholuebbers, J. G., Becker, M. and Matthiass, H. H., 1988. An osteoblast-like cell culture biocompatibility test of bone compatible materials. *Z. Zahnärztl. Implantol.*, IV: 290–294.
51. Thomas, K. A., Kay, J. F., Cook, S. D. and Jarcho, M., 1987. The effect of surface macrostructure and hydroxylapatite coating on the mechanical strengths and histologic profiles of titanium implant materials. *J. Biomed. Mater. Res.*, 21: 1395–1414.
52. Zweymüller, K. A., Lintner, F. K. and Semlitsch, M. F., 1988. Biologic fixation of a press-fit titanium hip joint endoprosthesis. *Clin. Orthop. Rel. Res.*, 235: 195–206.
53. Soltész, U., Siegele, D. and Baudendistel, E., 1990. Might biomechanical effects influence biocompatibility tests in bone? In: *Clinical Implant Materials, Advances in Biomaterials* 9, G. Heimke, U. Soltész, A. J. C. Lee (Editors), Elsevier, Amsterdam, pp. 657–662.
54. Cheal, E. J., Snyder, B. D., Nunamaker, D. M. and Hayes, W. C., 1987. Trabecular bone-remodeling around smooth and porous implants in an equine patellar model. *J. Biomech.*, 20: 1121–1134.
55. Turner, T. M., Sumner, D. R., Urban, R. M., Rivero, D. P. and Galante, J. O., 1986. A comparative study of porous coatings in a weight-bearing total hip-arthroplasty model. *J. Bone Joint Surg.*, 68-A: 1396–1409.
56. Deporter, D. A., Watson, P. A., Pilliar, R. M., Melcher, A. H., Winslow, J., Howley, T. P., Hansel, P., Maniatopoulos, C., Rodriguez, A., Abdulla, D., Parisien, K. and Smith, D. C., 1986. A histological assessment of the initial healing response adjacent to porous-surfaced, titanium alloy dental implants in dogs. *J. Dent. Res.*, 65: 1064–1070.
57. Pilliar, R. M., Cameron, H. U., Welsh, R. P. and Binnington, A. G., 1981. Radiographic and morphologic studies of load-bearing porous-surfaced structured implants. *Clin. Orthop. Rel. Res.*, 156: 249–257.
58. Bobyn, J. D., Engh, C. A. and Pilliar, R. M., 1987. Histological comparison of biological fixation and bone modeling with canine and human porous-coated hip prostheses. In: *Quantitative Characterization and Performance of Porous Implants for Hard Tissue Applications*, ASTM STP 953, J. E. Lemons (Editor), American Society for Testing and Materials, Philadelphia, pp. 185–206.
59. Schröder, A., van der Zypen, E., Stich, H. and Sutter, F., 1981. The reactions of bone, connective tissue, and epithelium to endosteal implants with titanium-sprayed surfaces. *J. Max.-Fac. Surg.*, 9: 15–25.
60. Bobyn, J. D., Wilson, G. J., MacGregor, D. C., Pilliar, R. M. and Weatherly, G. C., 1982. Effect of pore size on the peel strength of attachment of fibrous tissue to porous surfaced implants. *J. Biomed. Mater. Res.*, 16: 571–584.
61. Agins, H. J., Alcock, N. W., Bansal, M., Salvati, E. A., Wilson, P. D., Pellici, P. M. and Bullough, P. G., 1988. Metallic wear in failed titanium-alloy total hip replacements. A histological and quantitative analysis. *J. Bone Joint Surg.*, 70-A: 347–356.
62. Gonzales, J. B., Purdon, M. A., Horowitz, S. M., 1996. In vitro studies on the role of titanium in aseptic loosening. *Clin. Orthop. Rel. Res.*, 330: 244–250.
63. Greenfield, E. M., Alvarez, J. I., McLaurine, E. A., Oursler, M. J., Blair, H. C., Osdoby, P., Teitelbaum, S. L. and Ross, F. P., 1992. Avian osteoblast condidtioned media stimulate bone resorption by targeting multinucleating osteoclast precursors. *Calcif. Tissue Int.*, 51: 317–323.

64. Kim, K. J., Chiba, J. and Rubash, H. E., 1994. In vivo and in vitro analysis of membranes from hip prostheses inserted without cement. *J. Bone Joint Surg.*, 76-A: 172–180.
65. Müller-Mai, C., Rahmanzadeh, R., Lubnow, M., Voigt, C., Stupp, S. I. and Gross, U., 1998. Organo-apatite—neue degradierbare Apatite zur gerichteten Geweberegeneration. *Hefte zu "Der Unfallchirurg,"* 265: 295–302.
66. Parr, G. R., Gardner, L. K. and Toth, R. W., 1985. Titanium: The mystery metal of implant dentistry. Dental materials aspects. *J. Prosth. Dent.*, 54: 410–414.
67. Buchert, P. K., Vaughn, B. K., Mallory, T. H., Engh, C. A. and Bobyn, J. D., 1986. Excessive metal release due to loosening and fretting of sintered particles on porous-coated hip prostheses. *J. Bone Joint Surg.*, 68-A: 606–609.
68. Steinemann, S. G. and Straumann, F., 1984. Ankylotische Verankerung von Implantaten. *Schweiz. Mschr. Zahnmed.*, 94: 682–687.

21

Improved Bearing Systems for THP Using Zirconia Ceramic

B. Cales, L. Blaise, and F. Villermaux
Norton Desmarquest, Evreux, France

1. INTRODUCTION

Biomedical grade ceramics, alumina and zirconia, are highly wettable, hard and inert. As a consequence, they have low frictional torques, low wear rates and are extremely scratch resistant. This is the reason why these ceramic materials have been used in orthopedics for more than 20 years to solve the critical problem of polyethylene wear. Alumina ceramic was first introduced as hip joint heads by Boutin in France [1] and then by Griss in Germany [2]. Clinical results have clearly demonstrated the advantages of ceramic hip joint heads articulating with UHMWPE in total hip arthroplasty [3–5]. The polyethylene wear is significantly reduced when compared to metal/polyethylene systems.

Ceramic materials, in general, because of the absence of any plasticity, do not accommodate deformation over a few micrometers and exhibit a brittle type of fracture. However, biomedical grade ceramics can resist high stresses and cannot be considered as brittle materials. For instance, alumina ceramics are characterized by moderate fracture strengths, of 400 to 500 MPa, with a low fracture toughness. Thus, it is sensitive to microstructural flaws. Therefore, even when manufactured with limited designs, several alumina head fractures were reported over the past 20 years. To overcome the limitations of alumina, zirconia ceramics have been successfully introduced in orthopedics. Zirconia was first introduced in Europe in 1985 and then in the United States in 1989. The fracture strength of zirconia can reach four times that of alumina. The fracture toughness is at least twice as high [6]. Zirconia ceramic is considered the most fracture resistant commercially available monolithic ceramic material. The improved mechanical properties, combined with excellent biocompatibility and low wear behavior, have made zirconia the best candidate for the new generations of orthopedic prostheses. Zirconia ceramic is currently used in orthopedics replacing alumina ceramic and metals. More than 350,000 zirconia ceramic hip joint heads have been implanted throughout the world.

The physicochemical properties of surgical grade zirconia ceramic material, as well as the characteristics of zirconia hip joint heads, were reported in detail in the *Encyclopedic Handbook of Biomaterial and Bioengineering* [6]. H. McKellop, in the United States [7], W. Huber, in Germany [8], and P. Kumar, in Japan [9], have reported the outstanding wear behavior of zirconia femoral heads articulating against polyethylene sockets. Other issues, not described enough in detail in the literature, will be addressed in this chapter.

For example, one of the key issues regarding zirconia surgical grade ceramic is the long-term stability. This was recently questioned because of some recent publications on unexpected polyethylene wear with zirconia heads. There has been very little data and no detailed studies was reported, as of today, on this issue. Because long-term behavior is one of the main concerns of hip joint prostheses, this issue will be discussed in detail. The question of sterilization of zirconia heads will also be reviewed.

Another advantage of zirconia hip joint head is the opportunity to develop small 22.22 mm ceramic femoral heads. The 22.22 mm heads are recognized as one of the major solutions in reducing polyethylene wear. The use of small heads has been clearly shown to drastically decrease polyethylene wear [10]. Combining the use of zirconia ceramic material and small diameter heads should lead to extremely reduced polyethylene wear. The development of small 22.22 mm zirconia heads, even with 12/14 tapers, will also be presented in this chapter.

There is today a strong effort to completely remove polyethylene from hip joint prostheses. Even with drastically reduced wear by the use of ceramic heads and/or small diameter head, polyethylene debris could always be generated as third body wear. Therefore, hard-hard systems have been strongly promoted in the last years. The hard-hard hip systems have more than 20 years clinical experience. Development of ceramic-ceramic systems with zirconia heads have been claimed to be a poor combination [11]. Such an assertion was based on in vitro tests with ceramic materials that were not optimized and not representative of actual conditions. In opposition, hip joint simulator tests on ceramic-ceramic systems with commercial zirconia heads have shown extremely low wear. The wear was as low as the well-known clinically proven alumina-alumina combination. Developments of ceramic-ceramic systems with zirconia heads will be reported in the last part of this chapter.

2. LONG-TERM STABILITY OF SURGICAL GRADE ZIRCONIA CERAMICS

Surgical grade zirconia ceramic, which is usually made of 100% tetragonal grains, can transform into monoclinic grains under specific stress concentrations. Such a transformation is at the origin of the outstanding mechanical properties of this material [12]. The transformation may also occur at a low temperature in an acqueous environment. As a consequence a long-term strength and/or surface degradation may result. Several papers have reported contradictory results on the mechanical properties after short-term aging of zirconia ceramic either in vitro or in steam. However, the basic issue of the long-term (20–50 years) tetragonal to monoclinic ($T \rightarrow M$) phase transformation was never studied and reported in detail. The long-term behavior of zirconia ceramic is strongly dependent on the chemical composition and microstructure. No general behavior for all zirconia ceramics can be claimed. Specific compositions must be studied individually. The behavior of PROZYR zirconia ceramic, which is an yttria-stabilized tetragonal zirconia (Y-TZP) made of submicronic tetragonal grains, is described below.

A. Theory of the Tetragonal to Monoclinic Phase Transformation

The analysis of the phase diagram of the Y_2O_3-ZrO_2 system indicates the tetragonal phase is normally metastable at room temperature and may transform into a monoclinic phase under given thermodynamic conditions [13].

It has been shown in detail by Lange [14] that the $T \rightarrow M$ transformation rate in Y-TZP ceramic is controlled by the free energy change of the $T \rightarrow M$ transformation:

$$\Delta G_{T\rightarrow M} = \Delta G_c + \Delta G_{se} + \Delta G_s \qquad (1)$$

where ΔG_c, the *chemical free energy change*, is always highly negative and is controlled by the nature and content of the stabilizing oxide (here Y_2O_3 is used for the PROZYR zirconia ceramic); ΔG_s, the *surface free energy change*, is positive and strongly dependent on the grain size and ΔG_{se}, the *strain free energy change*, is slightly negative and controlled by the density, elastic modulus, compressive stresses, etc.

From a thermodynamic point of view, the tetragonal phase is stable only for $\Delta G_{T \to M} > 0$. This situation is only observed for given values of ΔG_c, ΔG_{se} and ΔG_s, which are controlled, as indicated above, by the physicochemical characteristics of the Y-TZP ceramic. Mainly the density, the mean grain size and the content of yttrium oxide are the determining characteristics.

For instance, if the grain size is small enough, the change in surface free energy is large enough to satisfy

$$\Delta G_s > | \Delta G_c + \Delta G_{se} | \tag{2}$$

As a consequence the free energy change $\Delta G_{T \to M}$ of the $T \to M$ transformation is positive and the tetragonal phase is perfectly stable. The stability of the tetragonal phase is also improved by an increase of the density, which affects ΔG_{se} and by an appropriate choice of the Y_2O_3 content, which controls the value of ΔG_c.

Therefore, there is a threshold value for the mean grain size, below which the tetragonal phase should be stable [14–15]. This value is dependent on the other physicochemical parameters of Y-TZP ceramics (density, Y_2O_3 content, chemical purity, etc.). This is the reason contradictory results have been reported on the existence and the value of the critical mean grain size. In practice, the critical mean grain size should be determined for each ceramic source.

If the mean grain size is larger than the critical value, then

$$\Delta G_s < | \Delta G_c + \Delta G_{se} | \tag{3}$$

and the tetragonal phase is no longer thermodynamically stable. It is metastable and the $T \to M$ transformation may occur.

Another reason for the $T \to M$ transformation is corrosion by water. This phenomenon is known as "low temperature degradation" (LTD). Water first reacts with the Zr-O-Zr bonds on the surface [16] (Fig. 1a). The corrosion of Y-TZP by water has been studied by Yoshimura et al. [17], who proposed a four-step mechanism (Fig. 1b):

Step 1 Chemical adsorption of water at the surface.
Step 2 Formation of Zr-OH and/or Y-OH bonds at the surface, where point stress is created.
Step 3 Accumulation of strain by migration of OH— ions at the surface and in the lattice, to prepare nucleating defects.
Step 4 Nucleation of the monoclinic phase in the tetragonal grains. Then the $T \to M$ transformation may yield microcracking.

This corrosion is thermally activated and was extensively studied [15] between 100 and 500°C.

It is therefore necessary to define the best compromise between transformation toughening and corrosion resistance. This is achieved by controlling the physicochemical parameters of Y-TZP ceramics, so the free energy change of the $T \to M$ transformation $\Delta G_{T \to M}$ is positive. It is admitted that outstanding mechanical properties and limited LTD is observed when [15]:

- The density is as close as possible to the theoretical density.
- The mean grain size is lower than 1 µm.
- The Y_2O_3 content is close to 3 mol%.

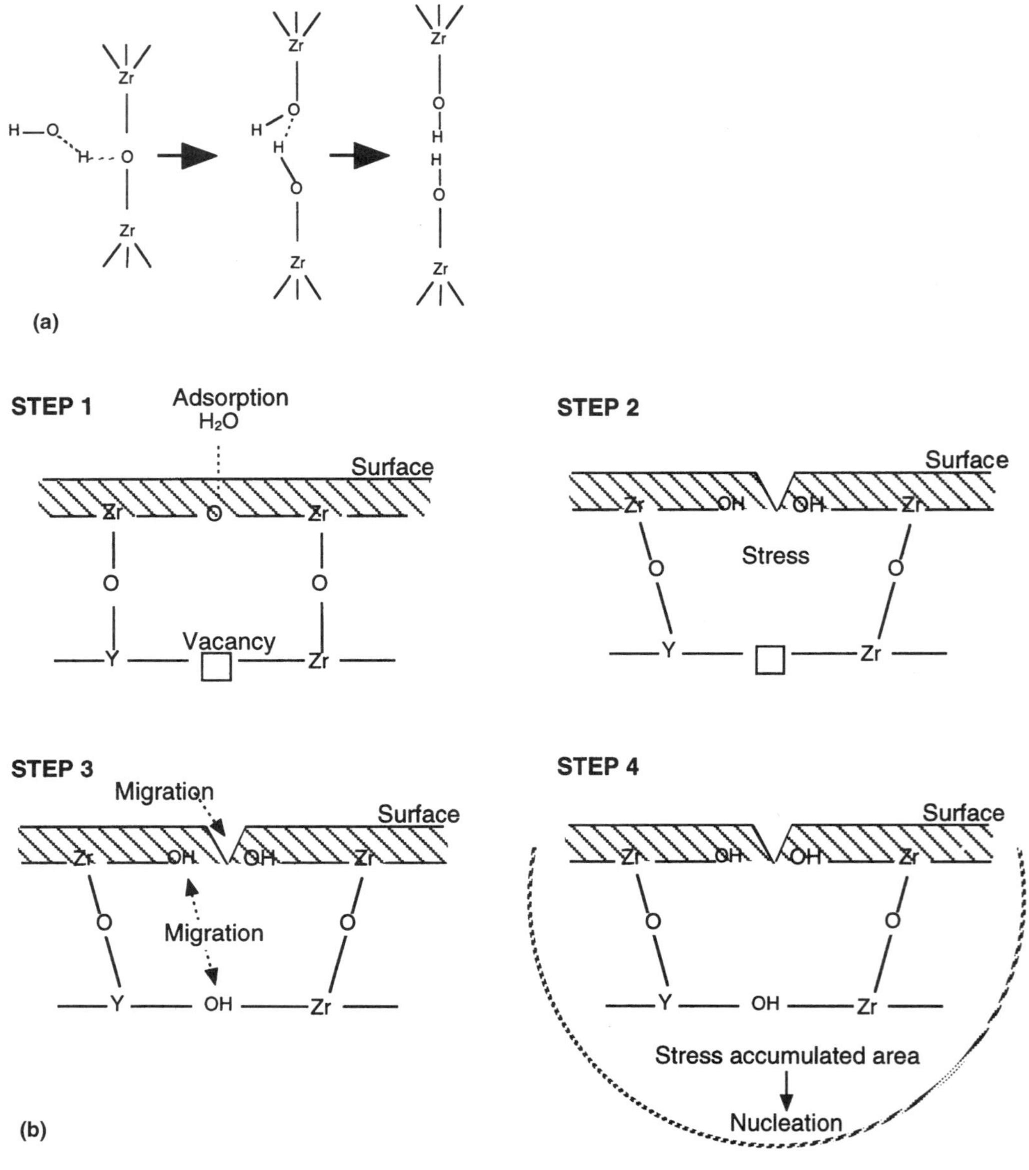

Figure 1 (a) Reaction scheme between water and Zr-O-Zr bonds (after Ref. 16). (b) T → M transformation process of Y-TZP by water (after Ref. 17).

However, even with such characteristics, low temperature degradation may occur in severe conditions, like autoclaving or very long aging.

B. Low Temperature Aging Behavior of Zirconia Hip Joint Heads

Experiments were conducted on commercially available zirconia hip joint heads (PROZYR), processed from 3Y-TZP powders, sintered and then hot isostatic pressed (HIPed) in order to reach theoretical density. The mean grain size, calculated from the linear intercept method, is in the range of 0.3 to 0.4 μm. Kinetic experiments were conducted using both steam sterilization at 134°C and 120°C and aging in Ringer's solution from 100°C down to 60°C, with femoral hip heads of different batches of zirconia heads in order to obtain a good insight into the process. At least three heads of each batch were tested for each aging temperature. Near surface monoclinic content was monitored by X-ray diffraction (CuKα, penetration depth ≈5 μm). The monoclinic content was calculated from the XRD diagram using the [111] lines of both the tetragonal and monoclinic phases and according to the following common relationship [18]:

$$M = \frac{I_M^{111} + I_M^{11\bar{1}}}{I_M^{111} + I_M^{11\bar{1}} + I_T^{111}} \tag{4}$$

Because the monoclinic phase growths from the tetragonal grains, there is a preferred orientation of the monoclinic grains (as in epitaxial growth) that makes the $11\bar{1}$ line disappear. However, the crystallographic $11\bar{1}$ planes still exist in the monoclinic phase and must be considered in the calculation. It was thus assumed that the 111 monoclinic line is not artificially increased by the preferred orientation of the monoclinic grains at the surface of the Y-TZP samples, and because the $I_M^{11\bar{1}}$ is equal to 68% of I_M^{111} [19], Eq. (4) was replaced by

$$M = \frac{1.68 I_M^{111}}{1.68 I_M^{111} + I_T^{111}} \tag{5}$$

Figure 2 shows the monoclinic content measured by X-ray diffraction on the surface of the zirconia heads for one given batch as a function of aging in steam and in Ringer's solution. It appears the variation of monoclinic content versus time is perfectly described with sigmoidal curves, showing a simple first order kinetic, which was sometimes used in literature [20,21], is unsuitable to fit low temperature degradation data. This also suggests the Mehl-Avrami-Jonhson (MAJ) law [22], already used in the literature for higher temperatures by Tsubakino

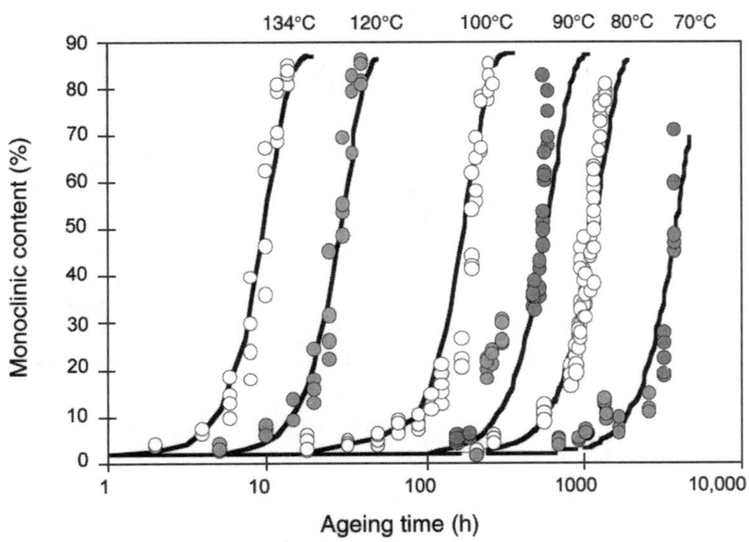

Figure 2 Kinetics of phase transformation between 60°C and 134°C.

et al. [23], is relevant. The MAJ law gives an expression of monoclinic content f versus time t in the form

$$f = 1 - \exp\,[-(bt)^n] \tag{6}$$

where n is a constant that must be determined for each zirconia ceramic and b gives the activation energy Q of the transformation mechanism by

$$b = A\,\exp\left(-\frac{Q}{RT}\right) \tag{7}$$

where T is the absolute temperature, A is a constant and R is the gas constant. The parameters n and b are deduced from a logarithmic plot of Eq. (6), which defines a straight line of slope n and ordinate origin $\ln(b)$. It has been observed that n is independent of temperature and equal to 3 ± 0.4. It should be noticed that, for n values between 3 and 4, the MAJ equation, which was often used to describe time-transformation isotherms in alloy systems, assumes an incubation period, then nucleation and growth [22].

The plot of $\ln(b)$, as a function of $1/RT$, gives a straight line of slope Q, found to be about 109 kJ/mol. for this zirconia. This activation energy is in fair agreement with information previously published in a smaller range of temperatures by Sato and Shimada [24] and reported by Chevalier for the so-called slow crack growth mechanism which involves the reaction of zirconia bonds with water at the crack tip [25]. The knowledge of this activation energy allows the prediction of the aging behavior at 37°C. Such a calculation is presented in Fig. 3, which takes into account the scattering from one batch to another. A low surface transformation occurs after an incubation period of about 10 years.

Another very important result of the aging study was the possibility of correlating aging in autoclave conditions at 134°C–2 bars and aging at 37°C in human body fluid. Previous studies indicated human fluid is similar to water for zirconia corrosion [26]. Based on the MAJ relationship 6 and 7, it can be shown, for the studied zirconia, 1 h in steam at 134°C–2 bars gives the same aging as four years at 37°C. Therefore, a 5 h autoclave test at 134°C–2 bars will perfectly simulate 20 years of aging in the human body. It is important to regularly check the aging behavior of zirconia orthopaedic components by a simple autoclave test.

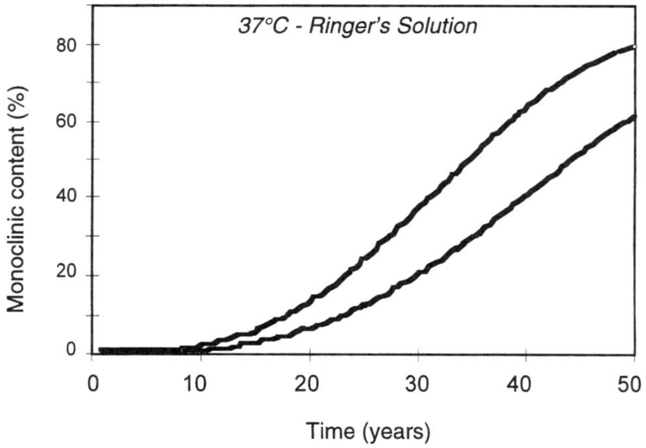

Figure 3 Predicted aging behavior at 37°C for PROZYR heads.

Table 1 Surface Roughness of PROZYR Zirconia Heads

Symbol	Surface roughness parameter	Specification (μm)	Observed value (μm)
R_a		<0.005	<0.004
R_p	Maximum peak	<0.030	<0.025
R_v	Maximum valley	>–0.035	>–0.030
PV	Peak to valley depth (R_{max})	<0.065	<0.055

Long-term stability is a key issue for orthopedic prostheses. The aging behavior is regularly checked by an autoclave test [134°C–2 bars–5 h] on each batch of PROZYR heads. All heads should exhibit a monoclinic content of 0% before the test and 2% to 3% after 5 h at 134°C–2 bars. This indicates the transformation rate, due to LTD, will be of only 2% to 3% after 20 years and will only affect the surface of the component and not the bulk. No strength degradation is predicted using this test method.

For a given source of zirconia ceramic, because the LTD proceeds from the surface, T $\rightarrow$ M transformation has been found to also be dependent on surface roughness: the lower the surface roughness, the lower the T $\rightarrow$ M transformation rate. In addition, the current surface roughness parameter of R_a is not sufficient to correctly describe the ceramic surface and understand the influence of surface roughness on LTD. Optical interferometer analysis is highly recommended. For example, PROZYR heads have a typical surface roughness as described in Table 1. In addition to a very low T $\rightarrow$ M transformation rate, a very low surface roughness also contributes to a decrease in polyethylene wear. Polyethylene wear has been shown to mainly be due to abrasive wear [54].

Wear of UHMWPE is one of the main concern in hip arthroplasty. One question is the effect of surface transformation on the wear behavior of zirconia femoral heads against polyethylene (PE) cups. Wear experiments were conducted on the same aged zirconia heads versus commercial PE acetabular cups. These couplings were tested on a MTS hip simulator. They were submitted to physiological loading (max. load 2500 N) in either synthetic serum (Plasmion, Roger Bellon) containing 25 g/l proteins or in bovine serum. The results in bovine serum are summarized in Table 2 for two different monoclinic contents, in terms of PE weight loss after 5 million cycles and monoclinic content after testing. The results in Table 2 show no significant effect of transformation rate on PE weight loss with even for 40% monoclinic content. Also, the wear tests had no effect on surface transformation. The monoclinic content was the same before and after the wear tests. Similar results were observed in synthetic serum (Plasmion).

Table 2 Wear Results after 5 Million Cycles in Bovine Serum

% Monoclinic before wear test	% Monoclinic after 5 M cycles	PE weight loss (±1 mg)	PE weight loss (mg/Mc)
0	0	86	17.2
40	40	94	18.8

C. Sterilization of Zirconia Ceramic Implants

Because the stability of the zirconia ceramic was found to be dependent on aging in water and the phenomenon to be thermally activated, the influence of autoclave sterilization on surface transformation has been questioned. The question was a result of a warning notice by the MDA. They pointed out some excessive polyethylene wear rates using zirconia hip joint heads. This issue was only observed on a very few non-HIPed zirconia femoral heads and sterilization was proposed as one of the possible origins of the issue.

Standard sterilization tests (134°C–2 bars–30 min) were done on HIPed PROZYR heads, to determine the effect of porosity on surface transformation. After just one sterilization cycle, the monoclinic content remained very low, equal or close to zero.

Non-HIPed and HIPed heads were investigated using three consecutive sterilization cycles at 134°C–2 bars–30 mn. The behavior of various HIPed heads manufactured from different raw material batches were compared. The results, summarized in Fig. 4, clearly indicate the monoclinic content at the surface of HIPed zirconia heads remains low (≤10%) even after three sterilization cycles. The monoclinic content after sterilization of HIPed heads does not vary from one batch to another and the scattering of the result is low. It may be concluded the T → M transformation at the surface of HIPed heads after steam sterilization is low and controlled.

For non-HIPed heads large variations in the monoclinic content were observed for heads manufactured from the same raw material batch. A monoclinic content of about 43% could be observed after three sterilization cycles (Fig. 4).

The origin of this difference may relate to the variations in density between HIPed and non-HIPed heads. The mean grain size and the Y_2O_3 content remain basically unchanged for HIPed and non-HIPed heads. In the case of HIPed heads, the density is equal to the theoretical density (6.1 g/cm^3). For non-HIPed heads, the density ranges from 99.5% to 99.8% of theoretical (6.07 to 6.09 g/cm^3). Even if small, this density difference may be significant. The porosity for the lower density is mainly made of micropores. These range from 1 to 10 μm. The number

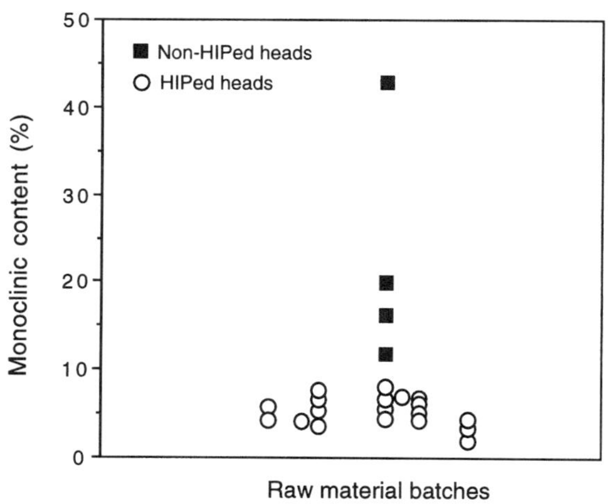

Figure 4 Variation of the monoclinic content at the surface of non-HIPed and HIPed zirconia balls from different raw material batches after autoclave at 134°C–2 bars–90 nm.

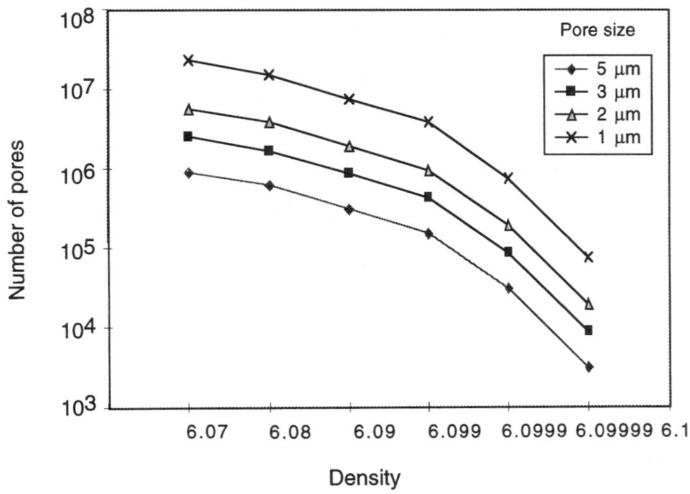

Figure 5 Variation of the number of micropores at the surface of non-HIPed heads versus density, for several sizes of pores.

of pores at the surface of non-HIPed heads have been estimated for micropores of different sizes and for different densities. The result of this calculation is shown in Fig. 5. It indicates, for a density as high as 6.099 g/cm^3 (99.98% of theoretical 6.1), the number of micropores could be high at the surface. Each of these pores plays a critical role in the T → M transformation rate. The grains at the surface of the pores are less stressed by the matrix as compared to the grains in the bulk (surrounded by other grains). The free energy change of the T → M transformation of these grains is lowered, leading to an easier T → M transformation. During steam sterilization, the grains around the micropores can transform more easily and lead to a significant monoclinic content at the surface. For a given density, depending on the micropore distribution (size and number) at the surface of the heads, the T → M transformation will be more or less important. This explains the strong scattering of monoclinic content for non-HIPed heads. Such a transformation has been suspected to generate some grain pullout when the head is sliding against the polyethylene. This pullout may lead to abrasive wear of the polyethylene cup.

In the case of HIPed heads, because the population of micropores is equal to or near zero, the opportunity for T → M transformation is extremely low. This is why no detectable monoclinic content results from steam sterilization of these heads. Thermodynamic studies show one steam sterilization cycle at 134°C–2 bars–30 mn induces an aging at the surface of the head equivalent to two years in the human body. Because of this, autoclaving is not recommended as a sterilization technique for zirconia heads.

Possible sterilization techniques are radiation (either gamma radiation or beta radiation), ethylene oxide or gas plasma. Each of these sterilization techniques should be validated on each zirconia composition of heads by measuring the monoclinic content with up to 10 sterilization cycles and after a poststerilization autoclave test. The monoclinic content should be compared to a control of nonsterilized heads. Because the sterilization conditions can vary from one apparatus to another, it is recommended that all sterilization procedures be validated before using them for implantable zirconia ceramic components.

III. SMALL 22.22 mm ZIRCONIA FEMORAL HEADS

Several studies have clearly demonstrated the advantage of using small diameter hip joint heads, 22.22 mm in diameter, as compared to 28 or 32 mm heads. As shown in Fig. 6a for alumina heads [27] and in figure 6b for both metallic and zirconia heads [28], small hip joint heads lead to significantly reduced UHMWPE wear rates. It appears the combination of a ceramic material and a small head diameter corresponds to an excellent solution for reducing polyethylene wear.

The present status of the art in this field is reported below and typical examples of 22.22 mm zirconia femoral head developments are presented.

A. Material Requirements

For small 22.22 mm ceramic femoral heads, zirconia ceramic, and especially yttria-stabilized zirconia (Y-TZP) ceramic, is the most appropriate material. Under loading on metallic trunnions, higher tensile stresses are observed in small femoral heads as compared to larger heads. A ceramic material with improved fracture strength and toughness is required. In addition, because of the higher Young's modulus of alumina ceramic, tensile stresses observed in alumina heads are significantly higher than in zirconia heads with the same design [29]. This is confirmed by the results reported by Fujiwara and Terui [30] which indicated a uniaxial burst strength for small alumina heads of about 45 kN, close to the FDA requirement for burst

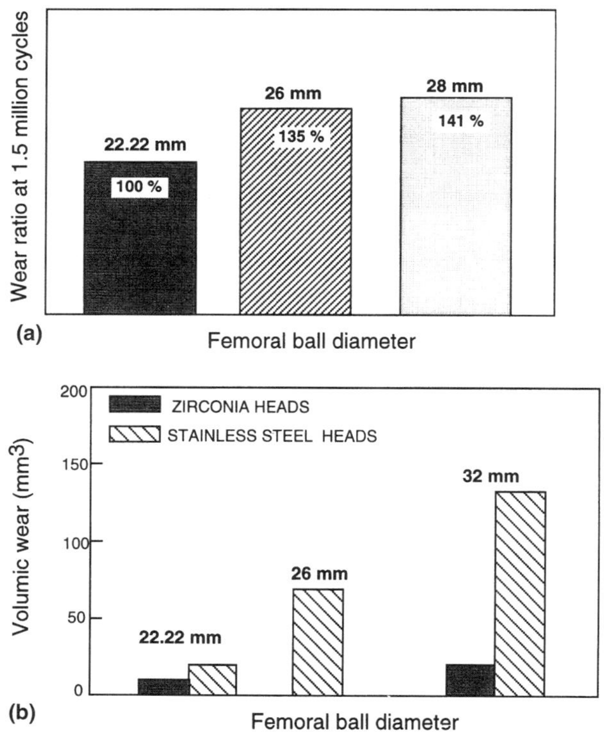

Figure 6 (a) In vitro wear of polyethylene versus the diameter of alumina femoral head. (b) In vitro wear of polyethylene versus the diameter of stainless steel and zirconia femoral head.

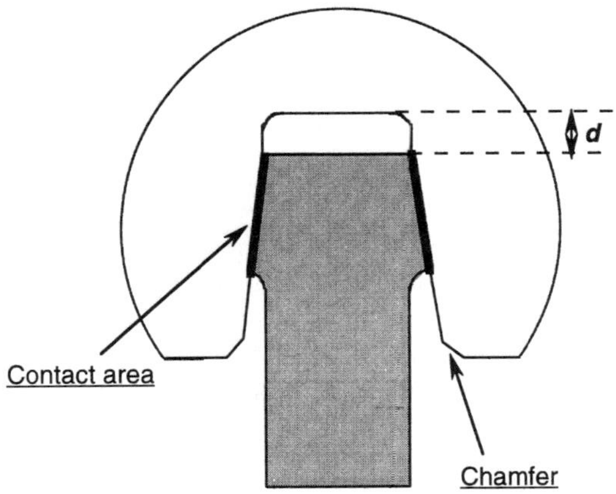

Figure 7 Main parameters for FEA design optimization of 22.22 mm femoral balls.

strength [31]. As a consequence, fatigue fracture of small alumina femoral heads may occur. This indicates 22.22 mm alumina femoral heads may not be considered safe enough to guarantee a high level of long-term reliability for orthopedic components.

The use of zirconia ceramics is highly recommended for small 22.22 mm femoral heads. Hot isostatic pressed (HIPed) Y-TZP ceramics with an appropriate microstructure, i.e., a mean grain size smaller than 1 μm, and a well-controlled chemical composition, i.e., an yttrium oxide content close to 3 mol%, are currently characterized by a bending strength higher than 1500 MPa and a fracture toughness, measured by indentation or the SENB technique, of about 10 Mpa $m^{1/2}$.

B. Optimization of Head Design and Head/Taper Fitting

Finite element analysis (FEA), with appropriate contact elements to simulate the metallic trunnion/ceramic head interface, is used to design highly reliable 22.22 mm zirconia heads for different types of metallic trunnions made of either stainless steel, Ti6Al4V or Co-Cr alloys.

As detailed in previous papers [32,33], such a design optimization is mainly concerned with the surface contact between head and trunnion, the distance *d* between the top of the metallic trunnion and the top of the ceramic bore and the chamfer at the bottom of the head (Fig. 7). The optimization of the chamfer size is especially critical in long neck designs. For long neck designs, the principal stresses are mainly located at the bottom of the head and an appropriate chamfer allows a reduction of the level of principal stresses in this area. As an example, the principal hoop stresses at the bottom of the head and the fracture strength for 22.22 mm zirconia heads using the same design and especially same distance *d*, but with two different chamfers, are reported in the Table 3. It clearly appears the simple optimization of the chamfer size leads to a 15% improvement of the fracture strength.

Table 3 Influence of Chamfer Size on Head Fracture Strength (22.22 mm Zirconia Heads, 10/12 Taper, 5°43′30″, Long Neck)

Chamfer size	0.5 mm	2 mm
Maximal hoop stress	719 MPa	626 MPa
Fracture strength	53 kN	61 kN

Figure 8 shows the principal stress σ1 distribution along the inner surface of the bore of 22.22 mm zirconia heads, for both Ti6Al4V and Co-Cr trunnions with the same geometrical characteristics and surface roughness ($R_{max} = 30$ μm). It clearly appears Co-Cr trunnions lead to higher tensile stresses in the femoral heads and to a lower fracture strength. A significant improvement can be obtained for Co-Cr trunnions after optimization of the trunnion surface roughness. As shown by FEA with appropriate contact elements between head and trunnion, for a given compressive loading, the surface and the location of contact area will depend on Young's modulus and the yield stress of the metal used for the trunnion. Therefore, the compressive fracture strength of the head can be managed by an optimization of trunnion surface roughness. The effect of the trunnion surface roughness is exemplified in Table 4. Up to 30% increase of the head fracture strength can be achieved with the adjustment of the trunnion surface roughness. The surface roughness of the trunnion must be carefully defined by the orthopedic prosthesis manufacturer. They must consider, not only the influence on the head fracture strength but also other consequences, like fretting and corrosion at head/trunnion interface.

C. Possible Designs for 22.22 mm Zirconia Hip Joint Heads

With the help of FEA and the appropriate trunnion surface roughness, it is possible to define several reliable Y-TZP 22.22 mm hip joint head designs. For instance, 22.22 mm zirconia heads with either 8/10, 9/11 or 10/12 trunnions can be used with two to three different neck lengths. The fracture strength of these head designs can reach outstanding values, close to or higher than 100 kN. (*Note the minimum FDA requirement is 46 kN* [31].) Figure 9 shows the variation of the uniaxial fracture strength (standard test method ISO 7206-5) for several 22.22 mm zirconia heads (PROZYR) with different taper diameters and different neck lengths. All the fracture strength values are higher than 60 kN. This value of 60 kN has been chosen as the minimum value during the design optimization by finite element analysis. It is easy to verify

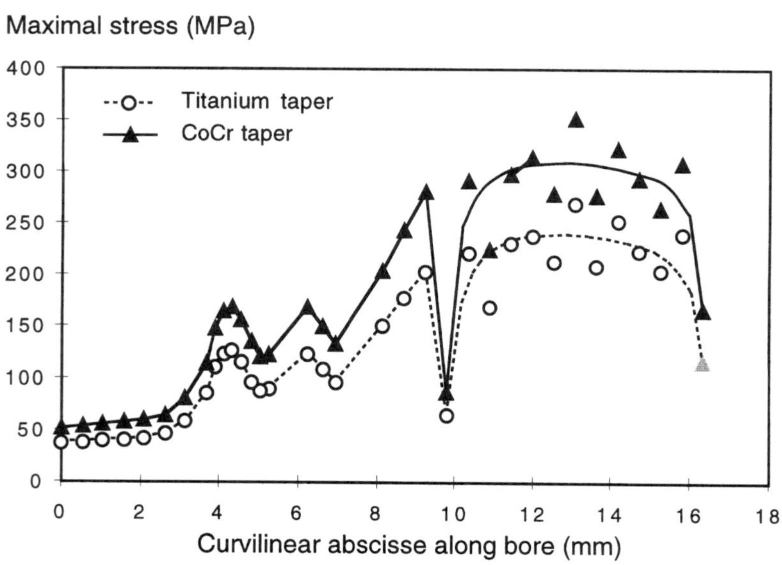

Figure 8 Principal stresses along the inner surface of the bore of 22.22 mm zirconia heads, for both $Ti6Al_4V$ and Co-Cr trunnions ($R_{max} = 30$ μm).

Table 4 Influence of Surface Roughness on Uniaxial Fracture Strength (22.22 mm PROZYR Heads, Neck Length: 0)

Metallic	Surface finish	Fracture (kN)
Ti6Al4V	 $R_{max} = 3$ μm	78
	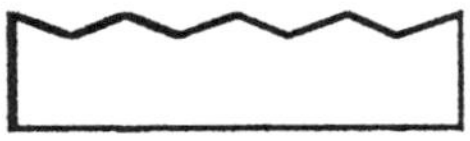 $R_{max} = 30$ μm	90
Co-Cr	 $R_{max} = 3$ μm	55
	 $R_{max} = 60$ μm	73

the fracture strength of 22.22 mm heads with an optimized design is directly related to the head/taper surface contact.

The maximum distance between short and long necks for 22.22 mm zirconia heads are listed in Table 5 for different bore sizes. It could reach about 5 mm for 8/10 tapers. This distance is directly dependent on trunnion characteristics (Young's modulus, hardness, surface roughness, etc.) and head design. Therefore, each fitting system must be optimized using finite element analysis.

Note it is possible to design reliable 22.22 mm zirconia heads with a 12/14 taper with a

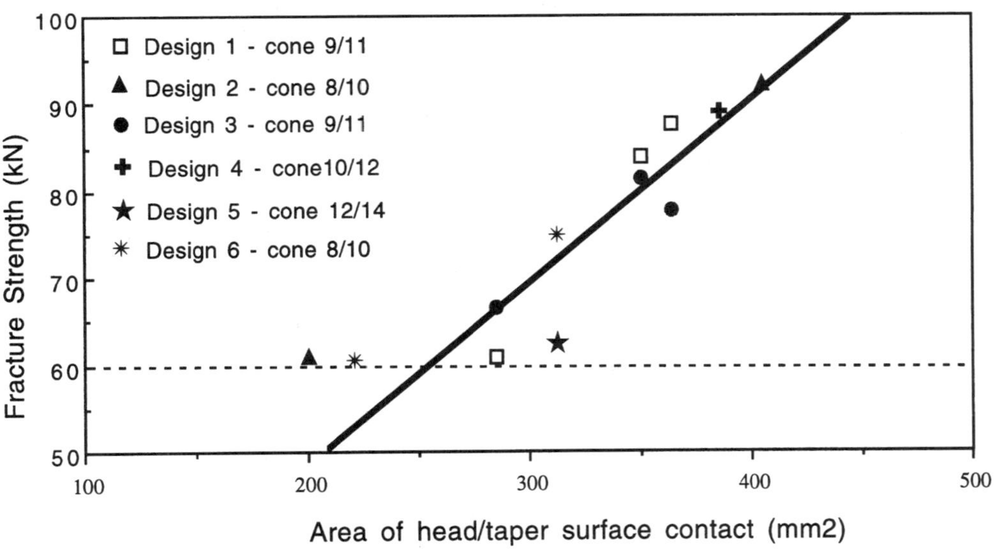

Figure 9 Fracture strength of 22.22 mm zirconia heads of different designs.

Table 5 Design Limit for 22.22 Zirconia Heads with Burst Strength ≥60 kN

Type of cone	8/10	10/12	12/14
Maximum length between short and long necks	5–6 mm	4–5 mm	3–3.5 mm

fracture strength higher than 60 kN (Fig. 9). This is of particular interest for 22.22 mm femoral heads since modular total hip prostheses (THP) with 12/14 taper represents about 50% of the THP implanted throughout the world.

The cyclic fatigue of 22.22 mm zirconia femoral heads was investigated at different maximum loadings between 25 kN to 40 kN, with a ratio R of the maximum to the minimum loading equal to 10 and a frequency of 10 Hz. This test is in accordance with FDA guidelines. The burst strength of ceramic heads before and after fatigue tests were measured using the standard uniaxial compressive test arrangement (ISO 7206-5). Most of the designs pass 10^7 cycles at 4–40 kN. For other 22.22 mm designs, the number of cycles before failure or the maximum cyclic loading for a survival after a minimum of 10^7 cycles was determined. The variation of the maximum loading versus number of cycles before failure, also called Wöhler's curve, was established.

An example of such a Wöhler's curve is shown in Fig. 10 for 22.22 mm PROZYR heads with a 10/12–6° bore and a +5 neck length. The number of cycles before failure rapidly increases when the maximum cyclic loading decreases from 40 to 15 kN. In addition, the lifetime tends to be quasi-infinite when the maximum loading becomes lower than a threshold value. This value is estimated to be close to 27.5 kN for the Ø 22.22 mm design tested. For cyclic fatigue, with a maximum loading of 27.5 kN, no failure was observed for 10^7 cycles and even 5×10^7 cycles. This corresponds to a period up to 50 years in humans. For a maximum loading of 28 kN the fracture occurred after only a few 100,000 cycles. The threshold value depends on the head design and on the geometrical and physical characteristics of the metallic trunnion (Young's modulus, surface roughness, etc.). The existence of such a threshold value in fatigue resistance is in good agreement with previous results on the fatigue behavior of Y-TZP zirconia ceramic [26]. These results displayed the existence of threshold values for

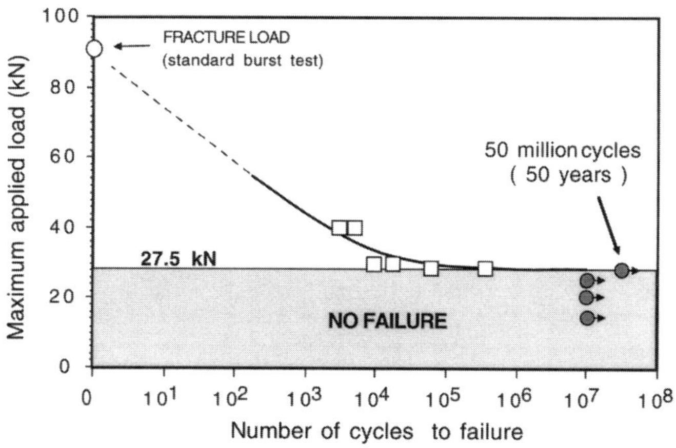

Figure 10 Cyclic fatigue of Ø22.22 mm PROZYR heads on Ti6Al4V trunnions.

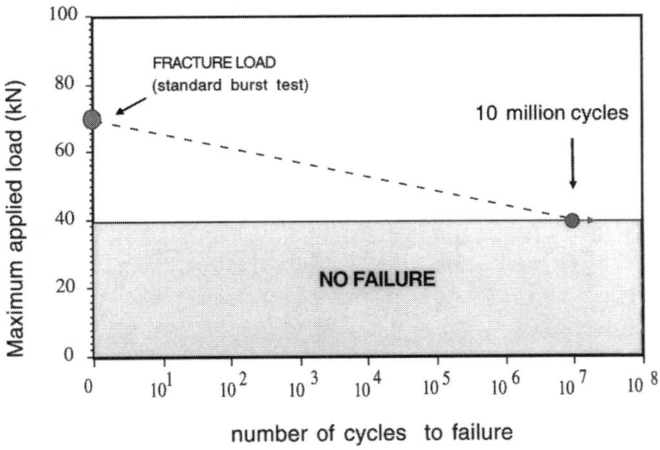

Figure 11 Cyclic fatigue of Ø22.22 mm PROZYR heads on CO-Cr (8/10) trunnions.

both static and cyclic fatigues, below which no crack growth occurs. As a consequence, a theoretical infinite lifetime may be expected below the threshold values.

Finally, it was observed the fatigue resistance of the same zirconia femoral heads was significantly higher on the Co-Cr tapers than on the Ti6Al4V tapers. They resist at 10^7 cycles at 4–40 kN (Fig. 11). This is due to the higher Young's modulus of Co-Cr alloys. This is very encouraging because of the importance of the Co-Cr taper in THP. For the static compression strength, as indicated previously, the classification was the opposite, Ti6Al4V tapers leading to higher burst strength than Co-Cr tapers.

The cyclic fatigue behavior of 22.22 mm zirconia head [12/14–5°43′30″] taper assemblies was also investigated. For Ti6Al4V tapers, the number of cycles before failure rapidly increases when the maximum cyclic loading decreases from 40 to 25 kN and is quasi-infinite ($\geq 5 \times 10^7$ cycles) when the maximum loading becomes lower than a threshold value. This value is estimated to be close 30 kN for the 22.22 mm/12–14 designs. These same designs resist 10^7 cycles at 4–40 kN on Co-Cr tapers.

Using FEA, several zirconia ceramic 22.22 mm femoral heads were designed with 12/14 tapers. They exhibits a high mechanical reliability on both Ti6AL4V and Co-Cr tapers. The maximum lengths between the short and long necks are between 3 and 3.5 mm depending on the head/trunnion combination. The main characteristics of these designs are indicated in Table 6. Today, it is estimated that about 25,000 of the 22.22 mm zirconia heads have been implanted throughout the world. About, 1,500 are of 12/14 designs.

Table 6 Mechanical Properties of 22.22 mm PROZYR Femoral Heads with a 12/14 Bore Size ($R_{max} = 30$ μm)

Neck length	Trunnion	Mean static fracture strength (kN)	Maximum loading to pass 10^7 cycles (kN)
0	Ti6Al4V	80	40
+3	Ti6Al4V	53	30
0	Co-Cr	55	40
+3	Co-Cr	46	30

IV. CERAMIC-CERAMIC BEARING SYSTEMS WITH ZIRCONIA HEADS

Polyethylene, and especially ultrahigh molecular weight polyethylene (UHMWPE), has been extensively used since the 1960s as bearing component face to metal or ceramic heads. Because of adverse tissue responses to polyethylene debris, the wear of polyethylene is one of the major concerns in the total hip arthroplasty community. As a result there is a growing interest in alternative "hard-on-hard" combinations such as metal-on-metal and ceramic-on-ceramic. The concept of the total hip prosthesis (THP) with hard materials for both the head and the cup was introduced in the 1960s with implants such as the Ring [34], McKee and Watson-Farrar [35] or Müller [36] metallic prostheses.

Metal-metal hip implants have shown mixed clinical results. On one hand, a number of the earlier McKee and Watson-Farrar implants were retained in the body even after more than 20 years with no evidence of surface degradation or tissue response [37]. However, discouraging results were also sometimes mentioned [38,39]. The issue of the metallic corrosion and ion release has been often questioned in the literature [38,39]. Some recent papers deal with the "second generation of metal-metal hip implants" exhibiting low in vitro wear rates [40]. Recent papers deal with cobalt (Co) and chromium (Cr) release in patients implanted with metal-metal THP [9,10].

Alumina-alumina hip prostheses have exhibited excellent clinical results. They have shown very low wear rates, typically less than 5 mm^3 per year as compared with the 100 mm^3 classically obtained for the polyethylene cups associated with metallic heads [41–43]. However, alumina is brittle and some fracture cases were reported [44,45]. The fracture rate seems to have been significantly higher for the alumina-alumina combination than for alumina-polyethylene. This is probably because of increased stresses in the head for a "hard-on-hard" system [45]. Fritsch and Gleitz [45] made a review of different case reports of THP using different ceramic sources. They found 12 fractures for a total of 5170 alumina-UHMWPE THPs with a fracture rate between 0 and 1.6% and a mean at 0.2%. Also for 5530 alumina-alumina THPs, they indicate 40 fractures with a fracture rate between 0 and 13.4% and a mean at 0.7%.

Because of zirconia high strength, its use against a ceramic acetabular cup is attractive. Zirconia heads will decrease the in vivo fracture probability of ceramic-ceramic THPs. However, little is known about the wear performance of a zirconia bearing against another ceramic. A recent publication has shown the wear behavior of a zirconia ceramic is highly dependent not only on microstructure parameters, but also on the test conditions (load, speed and lubrication) [46]. Some papers shown disastrous wear properties of "biomedical grade" zirconia ceramics in ceramic-ceramic combinations. However, little was known about the physicochemical parameters of the reported materials (processing, grain size, purity, density, etc.) [47]. In opposition, studies made using zirconia ceramic shown zirconia-zirconia or zirconia-alumina combinations could work as well as alumina-alumina [48]. Therefore, the wear properties of commercial zirconia heads (PROZYR) bearing against a biomedical grade alumina was investigated. Also, the stress analysis was evaluated using a zirconia-alumina combination.

A. Wear Behavior of the Zirconia-Alumina Combination

The wear behavior of zirconia-alumina combinations have been studied on an MTS machine at a frequency of 1 Hz using Paul's loading [49] with a max. of 2500 N with different lubricants containing the same protein content as synovial fluid (20–30 g/l); i.e., Plasmion, diluted bovine serum with addition of EDTA to complex calcium and a mixture of albumin plus globulin with EDTA. Co-Cr/UHMWPE and zirconia/UHMWPE combinations were used as

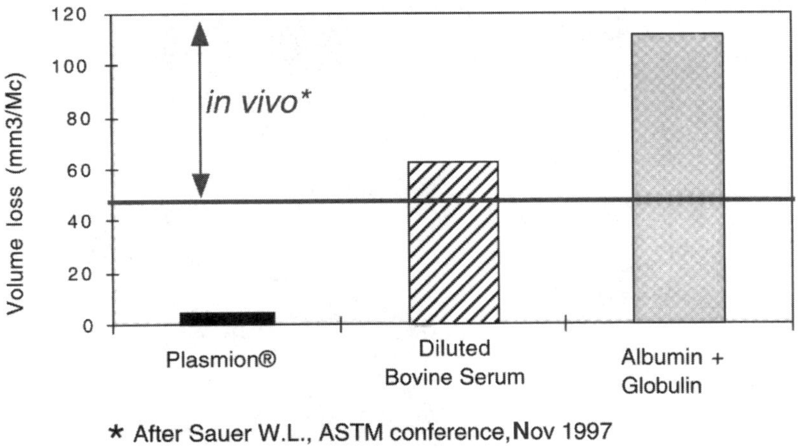

Figure 12 Polyethylene wear face to 28 mm Co-Cr heads in different lubricants.

controls. The heads were all 28 mm. The diametral clearance for the zirconia-alumina combination was 100 μm. The heads and cups were analyzed for roughness before and after wear testing using optical interferometer. Also using SEM for surface microstructure and X-ray diffraction for the monoclinic content, the surface of zirconia heads was analyzed.

The polyethylene volume loss for the Co-Cr/UHMWPE combination and the three lubricants is shown in Fig. 12. This was compared to the mean in vivo polyethylene volume loss [50], of about 50 mm^3/year, assuming 10^6 cycles is representative of about one year in vivo. It was observed that the dilute bovine serum and albumin plus globulin solution are the most representative lubricants. Plasmion lead to underestimated polyethylene wear. This could be due to the nature of the protein in Plasmion that is made from modified collagen.

The same influence was observed for the other combinations (Fig. 13). However, for the

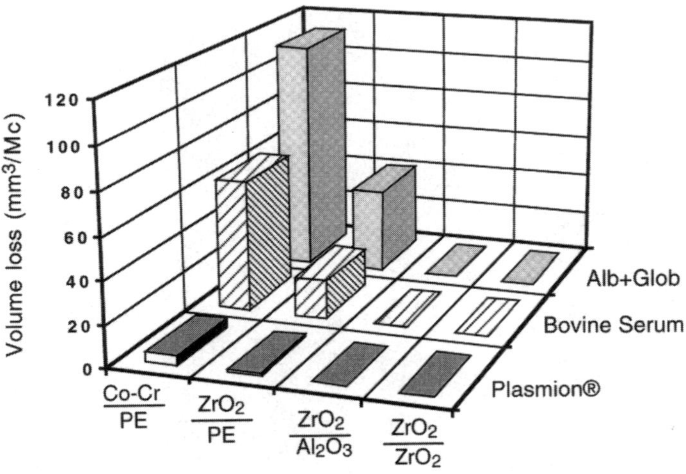

Figure 13 Wear rate of various bearing systems in different lubricants.

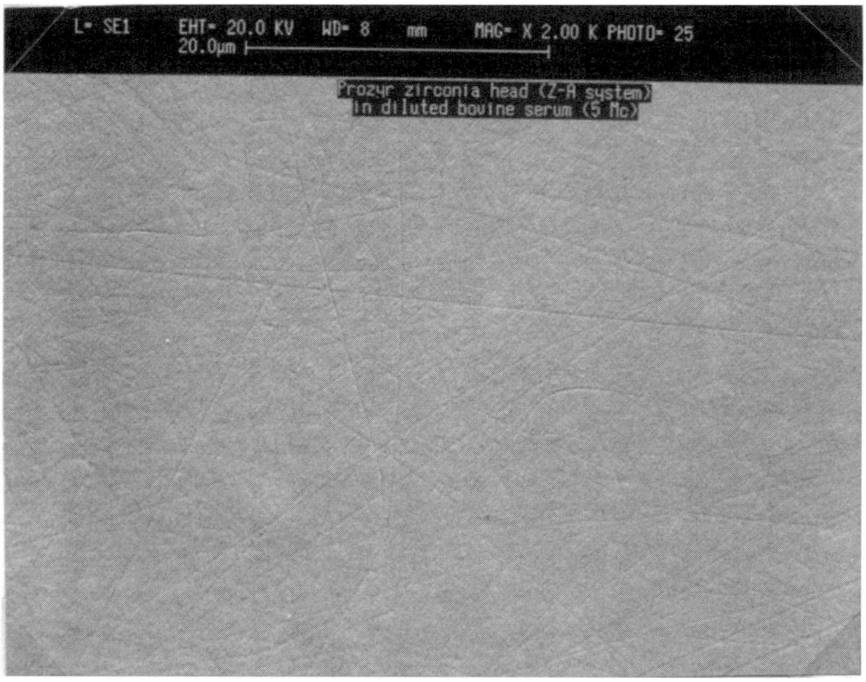

Figure 14 Microstructure of PROZYR zirconia head after sliding 5×10^6 cycles against alumina cup in diluted bovine serum (magnification $\times$ 2000).

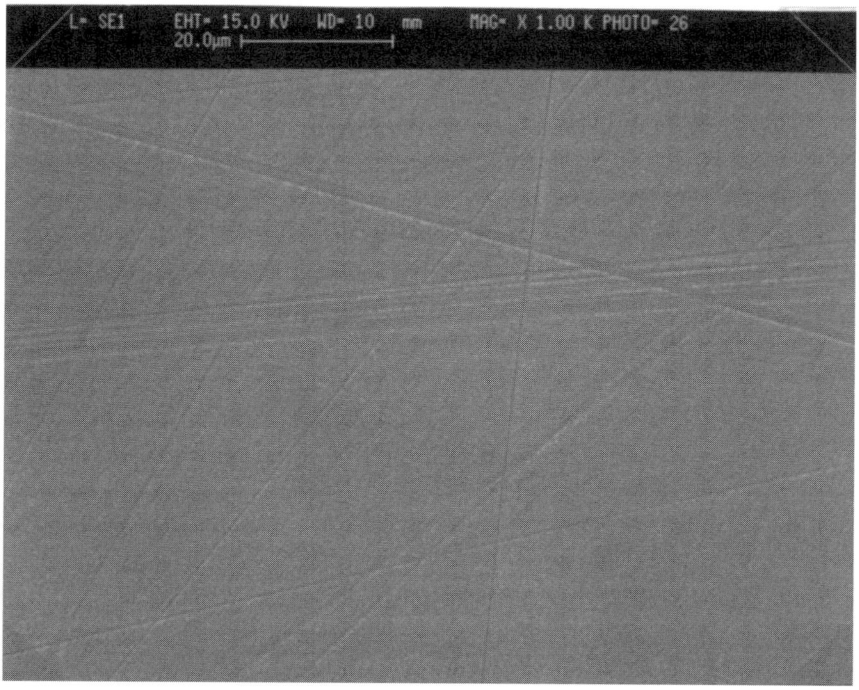

Figure 15 Microstructure of PROZYR zirconia head after sliding (5×10^6 cycles against alumina cup in albumin + globulin solution (magnification $\times$ 1000).

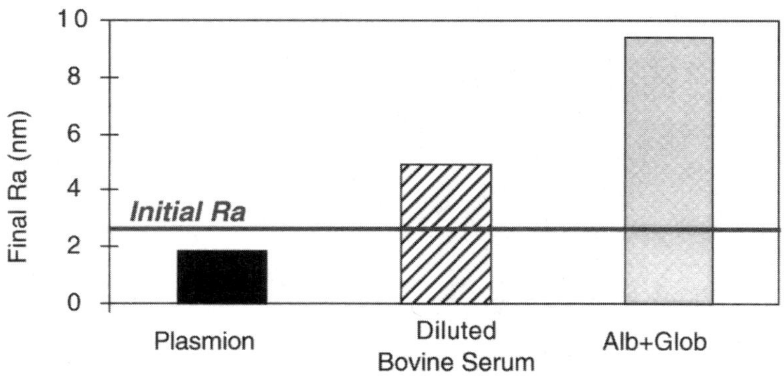

Figure 16 Surface roughness of zirconia heads after 5×10^6 cycles against alumina cup in different lubricants.

zirconia head-alumina cup combination, the wear was so low that no detectable weight loss was observed. No significant wear was detected by gravimetric method for the zirconia-alumina combination. The zirconia-alumina combination demonstrates to be as wear resistant as the common alumina-alumina combination. These results are in total disagreement with previously reported data by Willmann [11] and Oniishi [51], who observed, on another source of zirconia heads, a severe wear of zirconia ceramic with either the alumina-zirconia [11] or the zirconia-zirconia combination [51]. Examination by SEM of the surface of the zirconia heads after 5×10^6 cycles in diluted bovine serum only revealed few scratches (Fig. 14) which are the same as on as produced zirconia heads. In the case of the albumin plus globulin solution as lubricant, the scratches appear to be slightly more pronounced (Fig. 15), even though the wear volume is negligible. They were probably consecutive to the formation of the third body wear. The optical interferometer analysis confirmed this behavior. Figure 16 shows the modification of the surface roughness R_a after wear tests in the three different lubricants. The surface roughness increases from $R_a = 30$ Å on as manufactured heads to about 50 Å and 90 Å in dilute bovine serum and in albumin plus globulin solution after testing. In the case of Plasmion a self-polishing of the ceramic-ceramic pair is noted. This lubricant has been shown to be not representative of human body fluid.

The results of the wear tests up to 10^7 cycles on two similar zirconia (PROZYR) head/alumina cup pairs are summarized in Table 7. The data is again in disagreement with previously published papers [11,51].

Table 7 Results of Hip Simulator Test on Zirconia-Alumina after 10^7 Cycles in Plasmion

	Weight loss (g) [±.001]	Initial monoclinic content at surface of zirconia head	Final monoclinic content at surface of zirconia head
Zirconia head 1	+.0004	<2% (not detectable)	<2% (not detectable)
Alumina cup 1	+.0085 glue, metal transf.		
Zirconia head 2	+.0008	<2% (not detectable)	<2% (not detectable)
Alumina cup 2	ND glue, metal transf.		

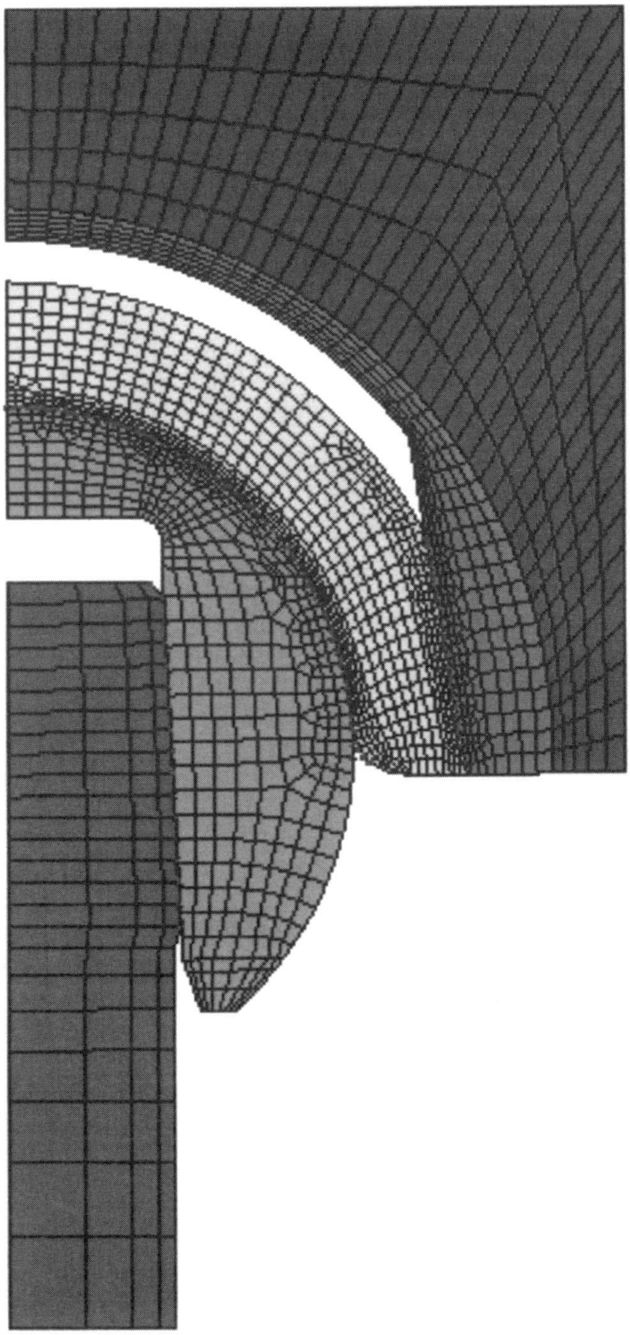

Figure 17 Mesh used for the FEA of the ceramic-ceramic prostheses.

Therefore, the zirconia-alumina combination appears to be as good as current combinations, if not the best candidate for a ceramic-ceramic THP. It combines an extremely low wear rate, as the current alumina-alumina coupling, and will exhibit improved mechanical properties, because of the higher fracture strength and toughness of the zirconia ceramic. A higher reliability of ceramic-ceramic THP can thus be reached. Nothing is known today on the clinical performance of the zirconia-alumina system.

B. Stress Analysis of Ceramic-Ceramic Pairing for THP

The optimization of the design of the ceramic heads and of the ceramic cup is more critical than for the current ceramic/polyethylene systems. It has been previously shown that the tensile stresses are significantly increased when a ceramic head is against a ceramic insert rather than a polyethylene cup [52]. The difference is mainly due to the lower surface contact between head and cup.

Ceramic-ceramic pairs must be designed to reduce the tensile stresses in the complete hip joint. FEA has been made on some complete hip systems, using a 12/14 Ti6A14V trunnion, a 28 mm zirconia head [neck length −3.5], an alumina insert, a Ti6A14V metal back and a cortical bone block (Fig. 17). This was done using Ansys 5.3 software, in nonlinear 2D analysis with an axisymmetrical model and an axial loading of 10 kN. The chosen characteristics of the different materials are indicated in Table 8. In addition, contact elements were used at each interface with the following friction coefficients: head/taper, 0.15; head/insert, 0; insert/metal back, 0.15.

Ceramic-ceramic THPs with a taper-locked insert in a metal back were investigated using FEA. Four areas are submitted to high tensile stresses (Fig. 18a,b) and must be followed during design optimization:

In the heads	Head/taper interface
	Top of the bore
In the insert	External top of the insert
	Inner surface of the insert

The two main parameters found to directly control the stresses in the ceramic head and insert were the head/insert radial clearance (defined as the difference between the radius of the insert and the radius of the head) and the insert thickness.

The influence of the clearance on the stresses in the head and the insert is shown in Fig. 19a,b. An increase of the clearance induces a drastic increase of the stress at the external top of the insert and at the top of the bore in the head. The stresses in the other areas are only slightly or not affected by the clearance change. It has been previously established, in order to

Table 8 FEA Conditions for Ceramic-Ceramic THP

	Young's modulus (GPa)	Poisson's ratio
Metallic taper	115	0.31
Zirconia head	220	0.29
Alumina insert	330	0.29
UHMWPE insert	0.34	0.42
Metal back	115	0.31
Bone block	20	0.42

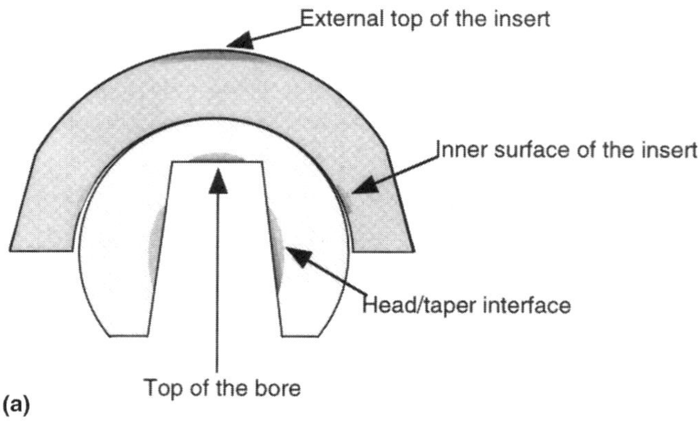
External top of the insert
Inner surface of the insert
Head/taper interface
Top of the bore

(a)

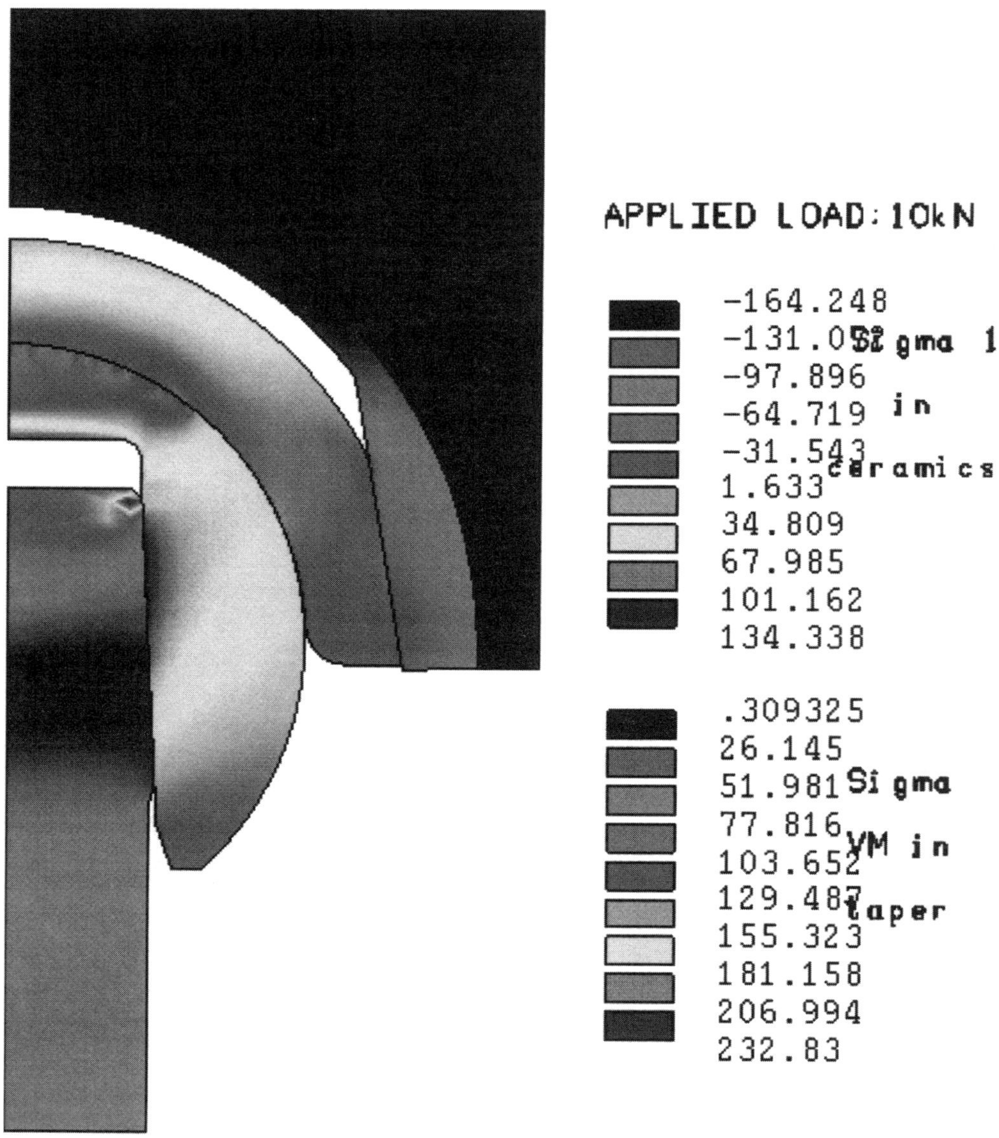
APPLIED LOAD: 10kN
-164.248
-131.072
-97.896
-64.719
-31.543
1.633
34.809
67.985
101.162
134.338
Sigma 1 in ceramics
.309325
26.145
51.981
77.816
103.652
129.487
155.323
181.158
206.994
232.83
Sigma VM in taper

(b)

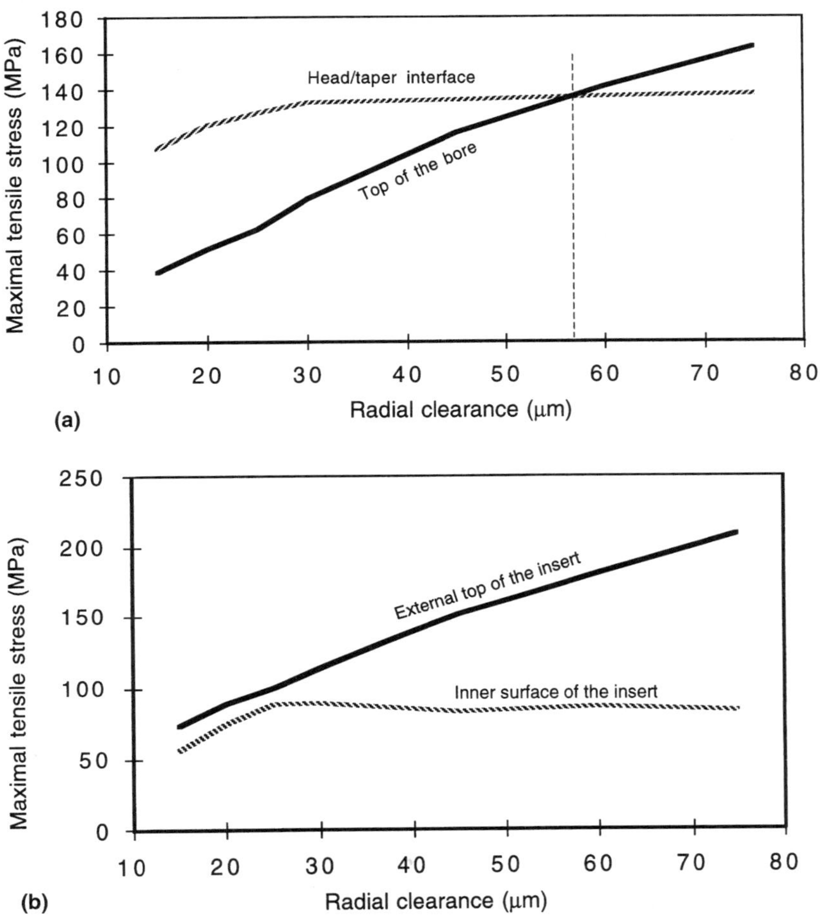

Figure 19a Stresses in zirconia head under 10 kN in the zirconia-alumina combination. **b.** Stresses in alumina insert under 10 kN in the zirconia-alumina combination.

guarantee a high reliability and avoid fatigue failure, the stress must remain lower than or equal to about 50% of the material fracture strength [53]. Because the fracture strength of alumina is typically 400–500 MPa and HIPed Y-TZP zirconia is higher than 1500 MPa, it appears the critical stress will be at the external top of the insert (Fig. 19b). This situation is less critical in the head because it is made of high strength zirconia ceramic. It would be a critical element if the head were alumina. The origin of the stress increase when the radial clearance increases is the change of the head/insert contact surface. The head/insert contact surface drastically decreases when the clearance increases (Fig. 20). This makes the tensile stresses higher in both the head and the insert.

Based on these results and on the calculated variation of the contact surface between

---

Figure 18 (a) Scheme of the ceramic-ceramic bearing system with the four principal areas of stresses. (b) Map of the principal stresses in the ceramic-ceramic prostheses calculated by FEA.

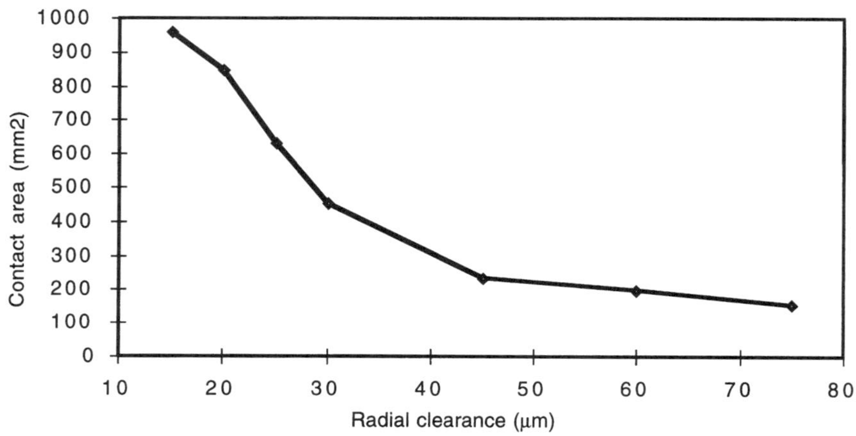

Figure 20 Influence of the head/insert clearance on the head/insert contact area at 10 kN zirconia-alumina combination.

head and insert versus the clearance (Fig. 20), it is recommended to maintain the radial clearance between the head and insert below 45 to 50 μm.

The variation of the insert thickness does not induce significant change of the stresses in the ceramic head. This is because the head/insert contact surface remains almost unchanged. It could however lead to a rapid increase of the stress at the inner surface of the insert (Fig. 21). Again, because of the fracture strength of the alumina ceramic, the insert thickness for taper-locked ceramic insert designs will be critical.

The influence of these two parameters on the stress location and intensity in ceramic-ceramic THPs with a sandwich design and a polyethylene liner between the ceramic insert and the metal back was also investigated by FEA. The stress location and intensity are similar in

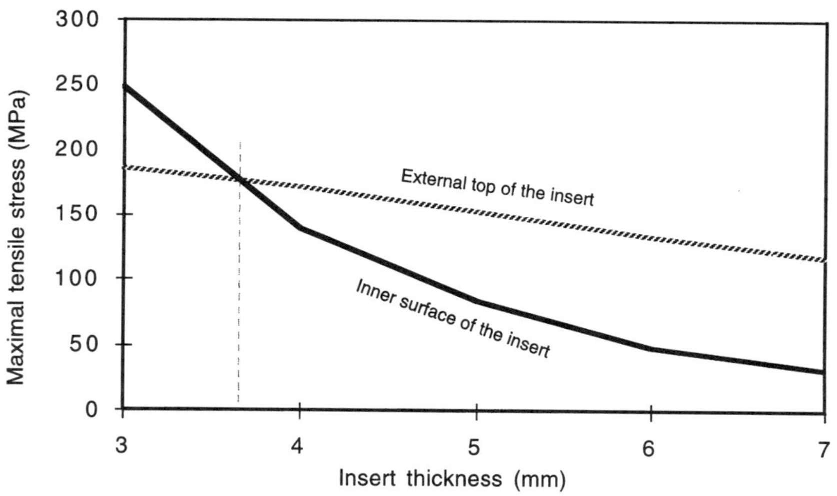

Figure 21 Influence of the alumina insert thickness on the maximal tensile stresses in the insert at 10 kN.

the ceramic head because the head/insert contact surface remains identical. Therefore, the same criteria would apply for the radial clearance (<45–50 μm). In the ceramic insert, the stresses are shielded by the polyethylene insert and are lower than in the taper-lock design. The stresses are about 100 to 130 MPa at the external top of the insert for a 3 mm thick insert instead of about 190 MPa for a taper-lock design (Fig. 21). Manufacturing capability will also need to be considered to determine the insert thickness.

An optimization of each design is necessary to find the best compromise between the stress in the ceramic head and insert and the size of the cup. In addition, it should be noted FEA results must be considered as only indicative. Real mechanical tests must be performed to validate all designs. In this respect, it is highly recommended to test the whole assembly (trunnion/head/insert/metal back). The contact conditions between the various components will be completely different from those in ceramic/polyethylene bearing systems. Standardization of mechanical tests, as well as a required fracture strength for ceramic-ceramic THPs do not exist, but need to be established.

Other developments in ceramic-ceramic THPs with zirconia heads concern the use of small head diameters. The use of zirconia ceramic for the head offers the possibility to develop small sized THPs. This would offer the advantage to drastically reduce the size of the insert and metal back. The 22.22 mm ceramic-ceramic THPs are presently under investigation.

REFERENCES

1. Boutin, P., *Rev. Chir. Ortop.*, **58**: 229–246 (1972).
2. Griss, P. and G. Heimke, *Arch. Orthop. Traumat. Surg.*, **98**: 157–164 (1981).
3. Zichner, L. P. and H-G. Willert, *Clin. Orthop. Rel. Res.*, **292**: 86–94 (1992).
4. Clarke, I. C., *Clin. Orthop. Rel. Res.*, **282**: 19–30 (1992).
5. Schuller, H. and R. Marti, *Acta Orthop. Scand.*, **61**: 240–243 (1990).
6. Calès, B. in *Encyclopedic Handbook of Biomaterials and Bioengineering*, Part B, Vol. 1: 415–452, Ed. D. L. Wise, D. J. Trantolo, D. E. Altobelli, M. J. Yaszemski, J. D. Gresser and E. R. Schwartz, Marcel Dekker, New York (1995).
7. McKellop, H. and B. Lu, *4th World Biomaterials Congress*, Berlin, Germany, April 24–28 (1992).
8. Huber, W. A., W. Plitz and H. J. Refior, *Proceedings 11th European Conference Biomaterials*, Pisa, Italy, Sept. 10–14 (1994).
9. Kumar, P. et al., *J. Biomed. Mat. Res.*, **25**, 813–828 (1991).
10. Clarke, I. C., A. Fujisawa and H. Jung, *19th Society for Biomaterials*, Birmingham, AL, April 28–May 2 (1993).
11. Willmann, G., Früth, H. J. and Pfaff, H. G., *Biomaterials*, Vol. 17, 1996, pp. 2157–2162.
12. Evans, A. G. and A. H. Heuer, *J. Am. Ceram. Soc.*, **63**: 241–248 (1981).
13. Stubican, V. S. and M. F. Goss, in *Advances in Ceramics*, **3**: 25, *Science and Technology of Zirconia*, The American Ceramic Society (1981).
14. Lange, F. F., *J. Mat. Sci.*, **17**, 235–262 (1982).
15. Lawson, S., *J. Europ. Ceram. Soc.*, **15**, 485–502 (1995).
16. Sato, T. and Shimada, M., *Am. Ceram. Soc. Bull.*, **64**, 1382–1384 (1985).
17. Yoshimura, M. et al., *J. Mat. Sci. Lett.*, **6**, 465–467 (1987).
18. Garvie, R. C. and Nicholson, P. S., *J. Am. Ceram. Soc.*, **55**, 6, 303 (1972).
19. Powder Diffraction Files, Inorganic Phases, n° 27–997, 37–1484, 42–1164, Ed. International Centre for Diffraction Data, Swarthmore, PA.
20. Drummond, J. L., *J. Am. Ceram. Soc.*, **72**(4): 675–676 (1989).
21. Thompson, I. and Rawlings, R. D., *Biomaterial*, **11**, 505–508 (1990).
22. W. Zhu, W., Lei, T. and Zhou, Y., *J. Mat. Sci.*, **28**: 6479 (1993).
23. H. Tzubakino, H., Sonoda, K. and Nozato, R., *J. Mat. Sci. Let.*, **12**: 196 (1993).

24. Sato, T. and Shimada, M., *J. Am. Ceram. Soc.*, **68**(6): 356 (1985).
25. Chevalier, J., Ph.D. thesis (in French), INSA de Lyon, France, 1996.
26. Calès, B., Stefani, Y., Olagnon, C. and Fantozzi, G., in Bioceramics 6, Ed. P. Ducheyne and D. Chritiansen, Butterworth Heinemann, Oxford (1993).
27. Clarke, I. C., Fujisawa, A. and Jung, H., *19th Annual meeting of the Society for Biomaterials*, Birmingham, AL, 1993.
28. Derbyshire, B., Fisher, J., Dowson, D., Hardaker, C. and Brummitt, K., *Med. Eng. Phys.*, **16**: 229–236 (1994).
29. Dingman, C. A. and Schwarts, G., *16th Annual meeting of the Society for Biomaterials*, Charleston, SC, 1990.
30. Fujisawa, A. and Terui, A., *Mat. Sci. Eng.*, **C1**, 149–151 (1994).
31. CERBAL 95.1, Guidance Document for the Preparation of Premarket Notifications for Ceramic Ball Hip Systems, FDA, Orthopedic Devices Branch, Jan. 1995.
32. Drouin, J. M. and Calès, B., Bioceramics, Vol. 7, Ed. Ö. H. Andersson, R. P. Happonen and A. Yli-Urpo, Butterworth-Heinemann, Oxford, UK (1994).
33. Drouin, J. M. and Calès, B., 10th European Conference on Biomaterials, Davos, Suitzerland, Sept. 1993.
34. Ring, P. A., *J. Bone Joint Surg.*, **50-B**: 720–731 (1968).
35. McKee, G.K. and Watson-Farrar, J., *J. Bone Joint Surg.*, **48-B**: 245–259 (1966).
36. Müller, M. E., *Clin. Orthop.*, **72**: 46–68 (1970).
37. Schmalzried, T. P., Peters, P. C., Maurer, B. T., Bragdon, C. R. and Harris, W. H., *J. Arthroplasty*, **11**: 322–331 (1996).
38. Jacobs, J. J., Skipor, A. K., Doorn, P. F., Campbell, P., Schmalzried, T. P., Black, J. and Amstutz, H. C., *Clin. Orthop. Rel. Res.*, **329S**: 256–263 (1996).
39. Black, J., *Clin. Orthop. Rel. Res.*, **329S**: 244–255 (1996).
40. Chan, F. W., Bobyn, J. D., Medley, J. B., Krygier, J. J., Yue, S. and Tanzer, M., *Clin. Orthop. Rel. Res.*, **333**: 96–107 (1996).
41. Sedel, L., Nizard, R. S., Kerboul, L. and Witvoet, J., *Clin. Orthop. Rel. Res.*, **298**: 175–183 (1994).
42. Boutin, P., Christel, P., Dorlot, J.-M., Meunier, A., de Roquancourt, A., Blanquaert, D., Herman, S., Sedel, L. and Witvoet, J., *J. Biomed. Mat. Res.*, **22**: 1203–1232 (1988).
43. Riska, E. B., *Clin. Orthop. Rel. Res.*, **297**: 87–94 (1993).
44. Michaud, R. J. and Rashad, S. Y., *J. Arthroplasty*, **10**: 863–867 (1995).
45. Fritsch, E. W. and Gleitz, M., *Clin. Orthop. Rel. Res.*, **328**: 129–136 (1996).
46. Lee, S. W., Hsu, S. M. and Shen, M. C., *J. Am. Ceram. Soc.*, **76**: 1937–1947 (1993).
47. Willmann, G., Früth, H. J. and Pfaff, H. G., *Biomaterials*, **17**: 2157–2162 (1996).
48. Chevalier, J., Calès, B., Drouin, J. M. and Stefani, Y., *Bioceramics*, Vol. 8, L. Sedel and C. Rey, Ed., Elsevier Science, Cambridge, 1997, pp. 271–274.
49. Paul, J., *Proc. Instn. Mech. Eng.*, **181**: 8–15 (1966–67).
50. Sauer, W. L., *Alternative Bearing Surfaces in Total Joint Replacement*, ASTM STP 1346, J. J. Jacobs and T. L. Craig, Eds., American Society for Testing and Materials, 1998.
51. Oniishi, H., Ueno, M., Okinatu, H. and Amino, H., Bioceramics 9, Ed. T. Kobuto, T. Nakamura, F. Miyaji, Kyoto University, Japan.
52. Cales, B. and Chevalier J., *Alternative Bearing Surfaces in Total Joint Replacement*, ASTM STP 1346, J. J. Jacobs and T. L. Craig, Eds., American Society for Testing and Materials, 1998.
53. Guim, F., Reere, J. and Vaugham, D. A. J., *J. Mater. Sci.*, **26**: 3275–3286 (1991).
54. Fisher, J. *Current Orthopaedics*, **8**: 166–169 (1996).

22
Biodegradable Fracture Fixation Devices

M. van der Elst
Reinier de Graaf Hospital, Delft, The Netherlands

C. P. A. T. Klein
Academic Center for Dentistry, Amsterdam, The Netherlands

P. Patka and H. J. Th. M. Haarman
Free University Hospital, Amsterdam, The Netherlands

I. INTRODUCTION

The main advantage of biodegradable materials in trauma surgery is the fact that biodegradable fracture fixation devices dissolve during implantation and obviate the need for removal of the implant during a second operation. Since the development of biocompatible polylactides, polyglycolides and copolymers, many efforts have been undertaken to introduce these materials into the field of trauma surgery and several polymer fracture fixation devices are currently in clinical use. Most implants are manufactured from hydroxy fatty acids such as polyglycolic acid (PGA), polylactic acid (PLA) and copolymers. Biodegradable sutures have been commercially available since 1970 and pins, rods and screws have been developed in the last decades. (Fig. 1) In view of the modest mechanical properties of the first generation implants, their use was restricted to low-stress applications in cancellous bone. Technical improvements have led to the manufacturing of stronger implants with optimal biomechanical characteristics for broad application in fracture treatment. This contribution provides an overview of the clinical applications reported in the literature and also deals with future developments in this field.

II. THE MATERIALS

PGA and PLA are aliphatic polyesters. They form long molecules resulting in fibers with interesting mechanical properties. Even high molecular mass subtypes prove to be hydrolyzable. Large polymer molecules are the result of a controlled polymerization of compounds from cheap and environment-compatible biomass materials like molasses or potato starch. Degradation of the polyesters occurs in aqueous media via hydrolysis and/or enzymatic reduction of the ester bonds in the polymer chain.

PLA and PGA degradation products enter the Krebs cycle and are ultimately excreted from the body by respiration and excretion in urine and faeces.

The mechanical properties and the degradation characteristics of the aliphatic polymers

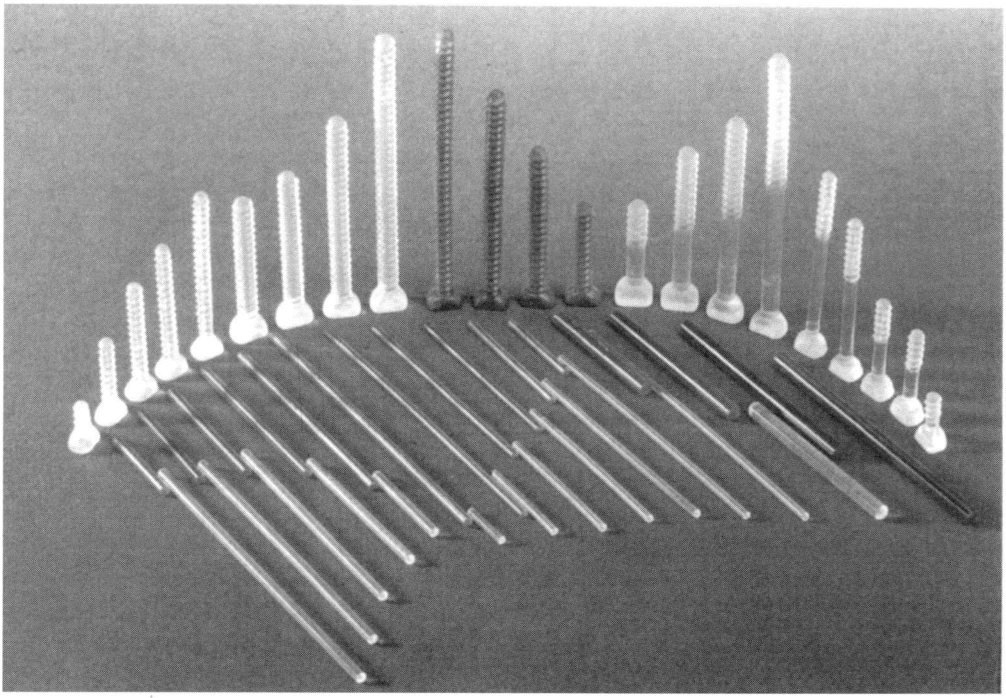

Figure 1 Biodegradable pins, rods and screws from self-reinforced composite material (Biofix).

show a considerable variation. These characteristics can be manipulated by varying the circumstances and ingredients of the polymerisation process. The composition of the implant also determines mechanical strength, strength retention and degradation behavior. Injection molding and drawing techniques are used to produce polymers in different sizes and shapes. The self-reinforcing (SR) technique in which oriented filaments are used as a scaffold for a matrix with the same chemical structure produces strong implants [1] (Fig. 2) Törmälä et al. introduced the sintering SR technique, by which implants can be built from strong fibrous polymer units [2]. The more advanced SR technique by partial fibrillation by orientational drawing has further strengthened the self-reinforced implants [3]. Rods, arrows, films and screws are now commercially available for clinical use (Fig. 3).

III. CLINICAL DATA

A. Ankle and Foot

PLA/PGA copolymer and SR-PGA implants have been in clinical use since 1984 [4]. Rokkanen and co-workers reported in 1985 on the use of PLA/PGA copolymer rods and threads that were used to fix simple malleolar fractures in humans [5]. In a prospective, randomized study, 44 adult patients were treated with a PLA/PGA implant or a standard steel implant. Similar favorable results were reported for both groups.

Böstman reported in 1989 on 102 patients with malleolar fractures treated with copolymer implants (Fig. 4). After a one year follow-up, fracture healing was excellent in 87% [6]. Six patients developed a nonbacterial wound disturbance which he called a "sterile sinus."

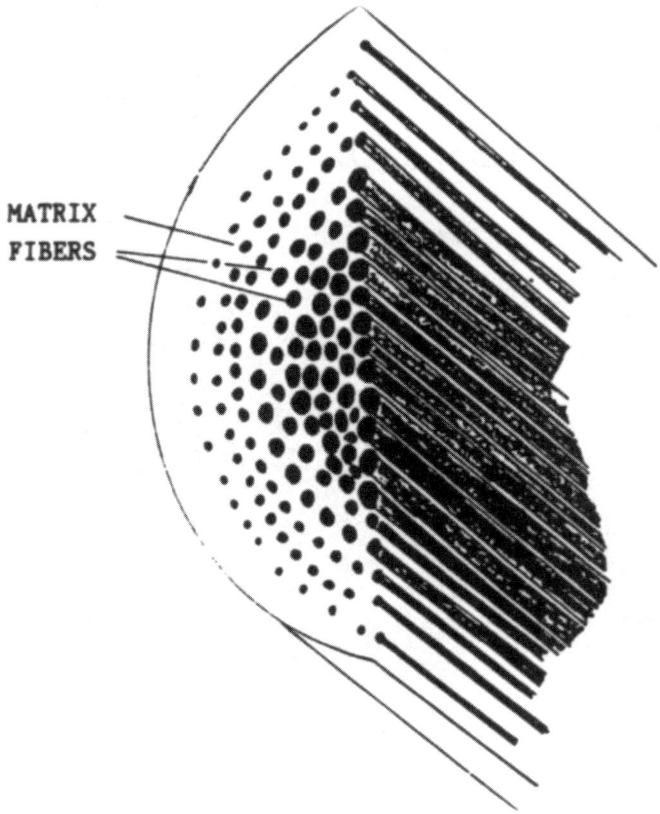

Figure 2 Schematic representing the self-reinforced composition technique by Törmälä and co-workers 1,2. The fibers function asa scaffold for the matrix substance of the same origin.

Survey of sinus material obtained between two and six months postoperatively demonstrated remnants of the degrading implants. All sinuses were cured by simple draining procedures. Swabs for bacteria were repeatedly negative in all cases. The authors concluded that the results were similar to those of patients treated with steel implants but they were hesitant about using biodegradable implants in osteopenic bone.

Partio treated uni-, bi- and trimalleolar ankle fractures with SR-PGA or SR-PLA screws (Figs. 5 and 6) [7]. The simple Weber type A and Weber type B fractures healed uneventfully after one year follow-up. In 39 out of 41 patients with more severe ankle damage as Weber type C or trimalleolar fractures, ankle function was good after one year. Ten patients out of the total 152 developed the sterile sinus foreign body reaction that was noticed before in other series [5,6].

Dijkema and co-workers performed a prospective randomized study in which 43 patients with ankle fractures were enrolled. The simple ankle fractures were treated with AO steel implants or SR-PGA rods. Results in both groups were favorable and no sinus formation was observed during the 5.5 months follow-up period [8]. Osteoarthrosis of the ankle joint caused by rheumatoid arthritis or trauma often leads to talocrural arthrodesis. In 12 patients, Partio used SR-PLA or SR-PGA screws to perform this arthrodesis [9]. The average follow-up time was 14 months. Solid fusion was achieved in 11 cases after nine weeks. The remaining patient needed additional surgery; one other patient developed a sterile sinus.

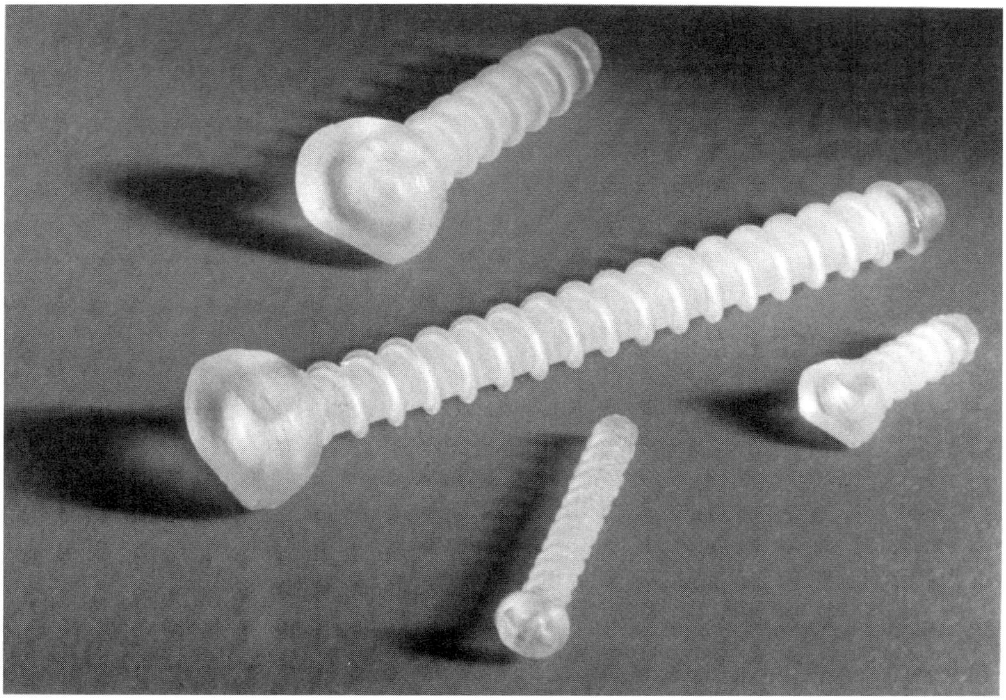

Figure 3 Commercially available self-reinforced PLA screws (Bionx Implants Inc.).

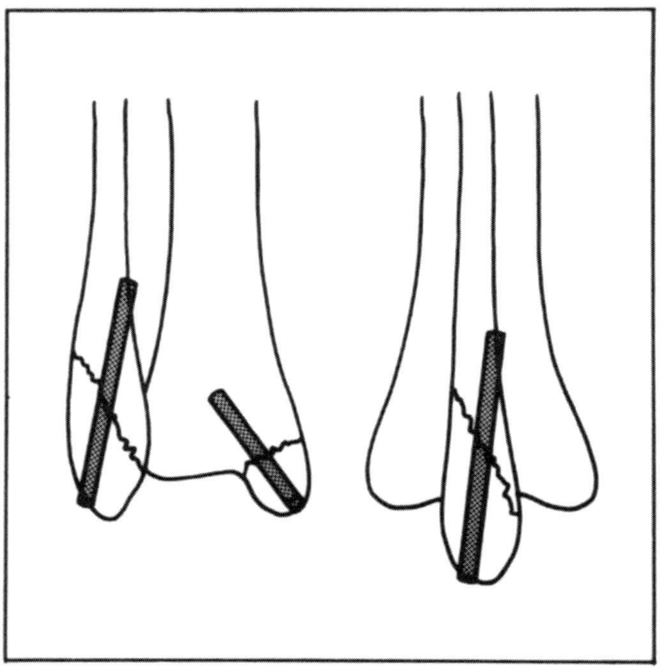

Figure 4 Schematic representation of the fixation of simple malleolar fractures using biodegradable pins.

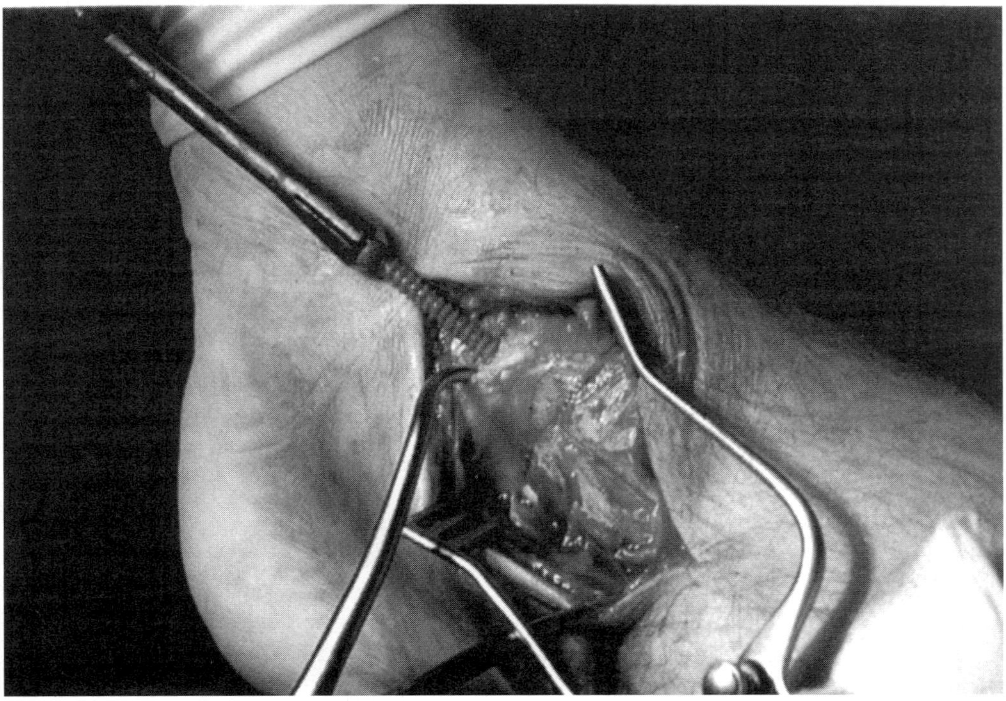

Figure 5 A biodegradable screw is inserted after open reposition of this malleolar fracture. A specific screwdriver is used to secure the screwhead (Bionx Implants Inc.).

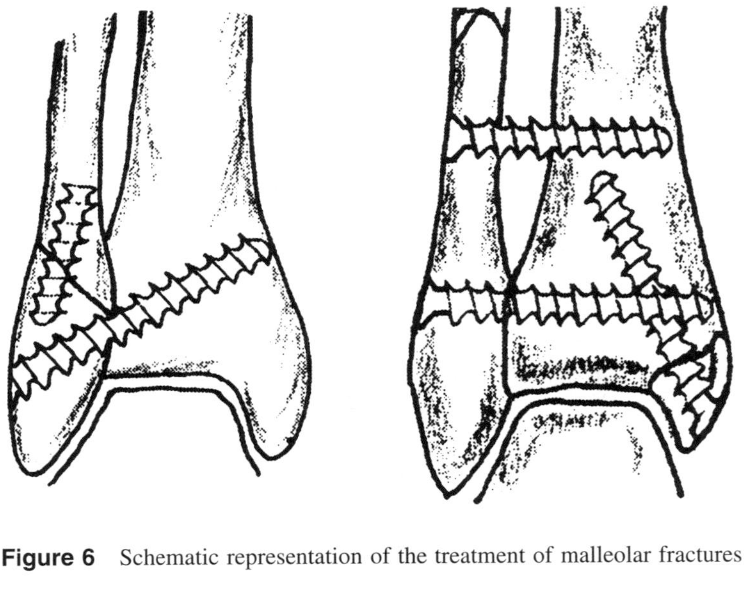

Figure 6 Schematic representation of the treatment of malleolar fractures using biodegradable screws.

Fourteen extra-articular arthrodeses were performed on seven children aged 5 to 15 years with neuromuscular disease and hindfoot valgus deformity [10]. One foot was implanted with a SR-PLA screw, the opposite foot was implanted with a stainless steel screw. Follow up after 17 months demonstrated radiological union in all ankles. One foreign body response was observed.

Arthrodesis of the first metatarsal phalangeal joint in rheumatoid arthritis was performed in 39 patients using either PLA rods or K-wires by Niskanen [11]. There were three nonunions in the PLA group and none in the K-wire group. The K-wires had to be removed in three cases. The union seemed superior in the K-wire group.

Biodegradable rods and screws have also been used in Chevron osteotomies for hallux valgus deformities. Hirvensalo and co-workers performed 78 chevron osteotomies using PGA pins [12]. Full weight bearing on the lateral side of the foot was allowed immediately. All osteotomies healed without redisplacement. One superficial bacterial wound infection and two aseptic sinus formations were reported during the follow-up.

B. Knee

Although the major contribution of biodegradables in trauma surgery is limited to small fractures in the foot and ankle regions, some devices have been used in the fixation of condylar fractures of the distal femur (Fig. 7). Osteochondritis dissecans has also been an indication for the use of biodegradables. Osteochondral flakes have been reattached to the corresponding defects in the femoral condyles with beneficial clinical results [13].

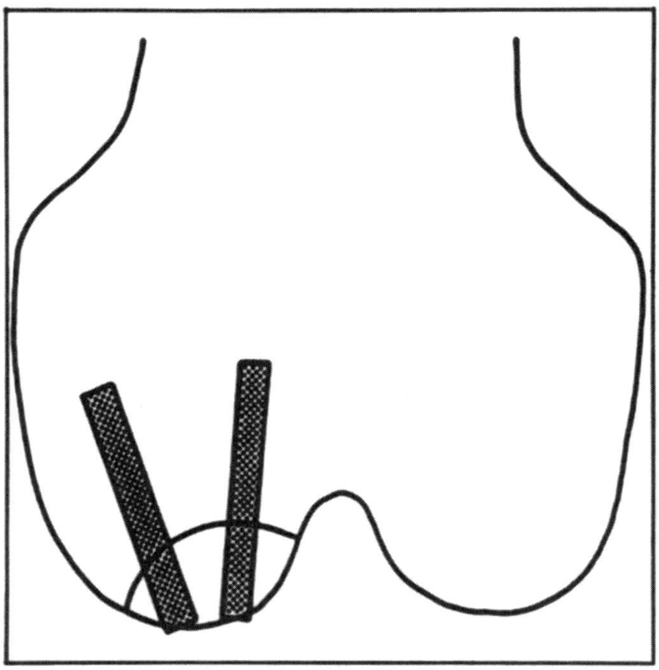

Figure 7 Small condylar fractures of the distal femur or osteochondritis dissecans lesions can be fixated with biodegradable pins or screws after open or arthroscopic reduction.

A new device is the meniscus arrow that is used in arthroscopy to fix meniscal tears (Fig. 8). (Meniscus Arrow by Bionx Implants Inc., Tampere, Finland, U.S. patent 4,873,976) Mechanical studies demonstrate a pull-out strength equivalent to horizontal suture techniques. Clinical studies demonstrate a healing rate equal to or greater than the other suturing techniques [14]. An aseptic synovitis of the knee after the use of intra-articular biodegradables has been described, and this reaction was reported to be probably related to the dose of the implant, the delay of operation or sensitization to PGA used previously [15].

C. Elbow

Hirvensalo reported in 1990 on the treatment of 24 displaced fractures of the radial head and radial neck using two diverging PGA pins [16] (Figs. 9, 10). The results were good or excellent in 91%. Two patients developed a transient sterile sinus and one bacterial infection was reported. There was redisplacement of 3 mm or less in four patients with comminuted fractures and minor incongruity in an additional four patients. The same author reported on the fixation of Type I humeral capitellum fractures using PGA rods in 1993 [17]. The pins were placed through a lateral incision and the ends of the rods were carefully buried below the bone surface. After immobilization with a cast for three weeks, full motion was allowed. Of eight patients, one developed a sterile sinus at the operated site which was treated by incision and draining twice.

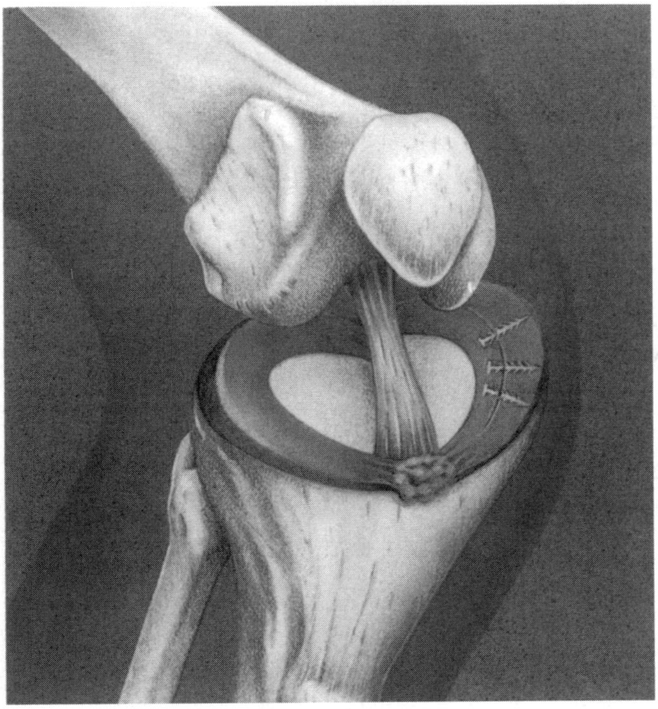

Figure 8 Drawing of the flexed right knee to illustrate the treatment of meniscal tears using the Meniscus Arrow from Bionx Implants Inc. The arrows can be placed during regular arthroscopy.

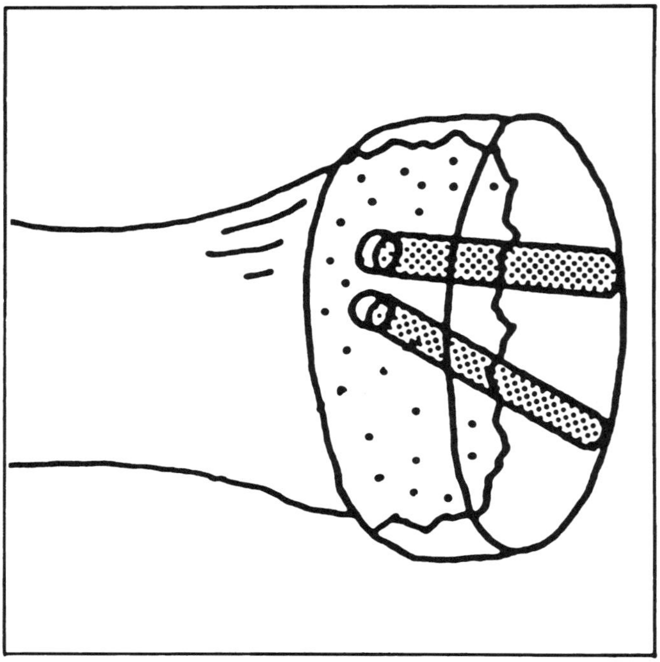

Figure 9 Radial head fracture fixed with biodegradable pins. A lateral approach will expose the radial head. The pins have been placed in a diverging direction to improve intrinsic stability.

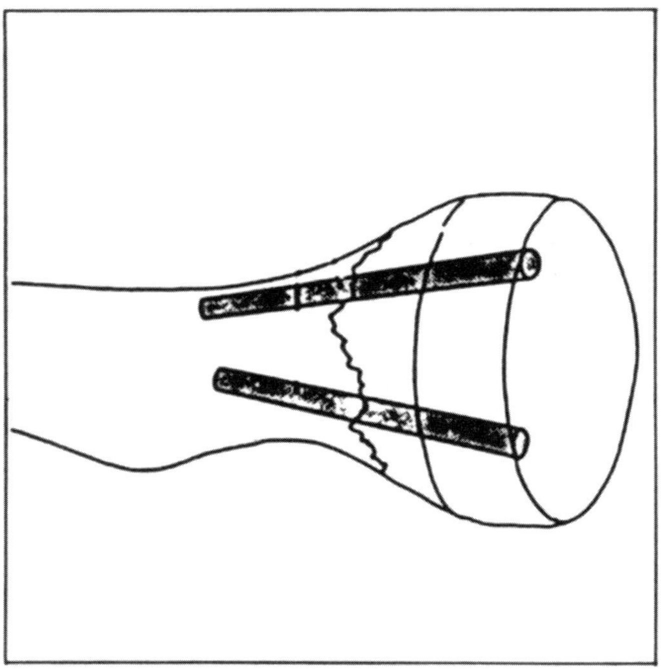

Figure 10 Radial neck fracture fixed with biodegradable pins. The ends of the pins are buried under the bone surface to limit damage to the joint and to avoid inflammatory tissue reactions.

Olecranon fractures have been treated by PGA pins and SR-PGA screws since 1986 [18,19] (Fig. 11). Partio treated 41 adult patients with good results. In two patients, secondary dislocation required a second operation. Sterile sinus formation was observed in three patients. The authors stress the fact that the combination of rods and screws together allowed treatment of more severe fractures of the olecranon compared to rods alone.

D. Wrist and Hand

The treatment of proximal ulnar collateral ligament avulsion was treated with PLA pins by Vihtonen and co-workers [20]. The avulsion was reattached to the intact bone and the patient received a circular plaster cast for five weeks. In all 20 patients, the ulnar collateral ligament was stable postoperatively and full range of motion was regained after three months.

The same authors reported in 1993 on the results of ulnar collateral ligament reconstruction using a SR-PLA tack [21]. Seventy patients were treated with a six months follow-up. Clinical normal pinch strength was observed in 66 patients. Two patients developed a postoperative infection which healed after removal of the implants.

Intramedullary SR-PGA rods were compared to K-wires for the treatment of metacarpal and phalangeal fractures [22] (Fig. 12). Kumta and co-authors randomized 15 patients in each group and results from both groups were similar with respect to the rate of union, the range of motion and the strength.

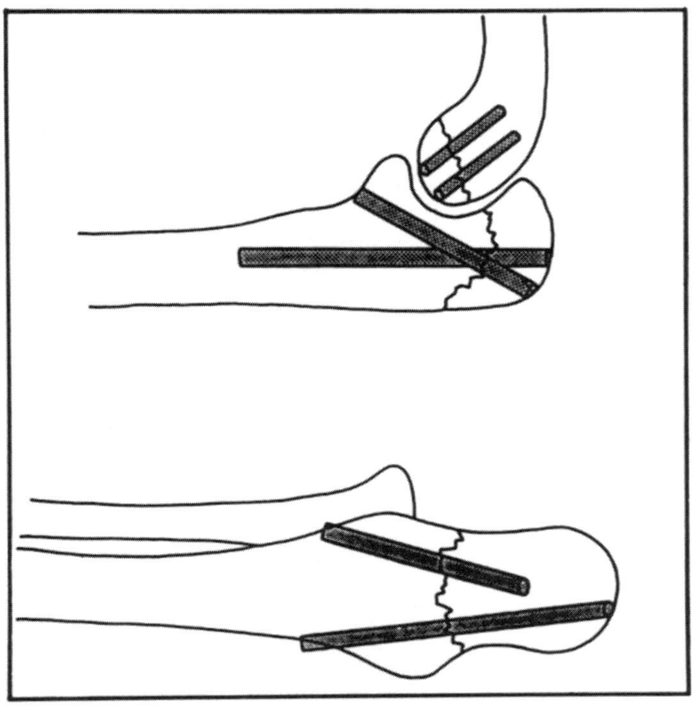

Figure 11 Schematic representation of the fixation of olecranon fractures using biodegradable pins or screws. A diverging position of the rod improves the stability of the construction. The end of the pins should be placed under the surface of the bone to avoid sinus formation.

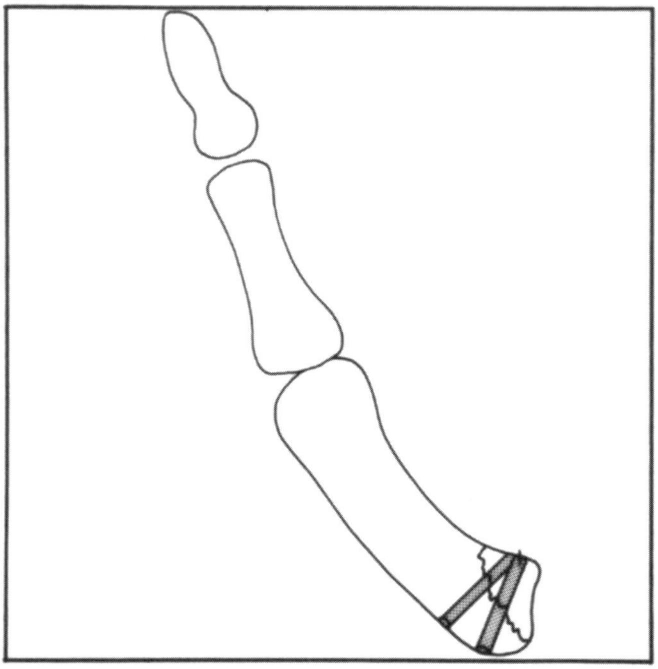

Figure 12 Bennett type fracture of the first metacarpal bone, fixed with a self-reinforced polymer osteosynthesis device (Biofix, Bionx Implants Inc.).

E. Physeal Fractures

Transphyseal fractures in children have been treated by Böstman and Mäkelä using PGA pins crossing the growth plate [23]. Three children (seven to nine years of age) with extension type supracondylar humerus fractures were treated with reduction and fixation with cylindrical PGA pins of 1.5 mm diameter, placed transphyseally crosswise through the fracture plane. Two out of three healed with perfect function and radiograph. The third patient was not reduced perfectly during the surgery and had a 10° varus deformity. Follow up was 10 months. In addition, three children with Salter-Harris type II and III fractures of the first metatarsal bone, the distal tibia and the medial epicondyle of the humerus were operated in a similar way. Recovery was uneventful. The same authors published in 1990 on a total of 19 children with physeal fractures of the humerus fixed with PGA pins [24]. Neither secondary displacement nor signs of growth disturbance were observed during the 26 month follow-up period. The authors claim good results but advise keeping the damage to the growth plate within 3% of its transversal surface to avoid growth disturbance.

IV. COMPLICATIONS

In trauma surgery, complications such as wound infections, bleeding or secondary displacement of the fracture parts are well-known effects of operative fracture treatment. With the introduction of the biodegradables, a new type of infectious complication has been observed, the sterile sinus formation. Santavirta reports an incidence of this aseptic inflammatory response that varies from 3% in Chevron osteotomies to 22% in distal radius fractures [25]. Partio, Böstman and co-workers reviewed the material from 216 patients with ankle fractures treated with biodegradables in order to determine the factors that possibly influence the incidence of this typical foreign body reaction [26]. Out of 216 patients with malleolar fractures, 24 developed a transient local inflammatory reaction with a painful erythematous fluctuant swelling. Average postoperative period was three months. All bacterial cultures were negative. Liquid obtained by aspiration was examined microscopically and showed remnants of the implanted screws. Biopsies showed fibroconnective tissue and polymer particles being fagocytized by macrophages and giant cells. All sinuses were cured by repeated needle aspiration or incision and draining. Fracture healing was not impaired by sinus formation. Although the colorant in the first generation of implants could play a role, the sinus did appear in patients with noncolored implants. Patients with noncolored implants, however, showed a significantly lower incidence of clinically manifest reactions than the colored implants, 4.5% versus 18%. An allergic reaction to the polymers is not likely to be an important factor since unilateral sinuses appeared in patients with bilateral implants. Partio and Böstman conclude that this sinus formation results from the hydrophilic nature of the polyglycolide being pressed out of the operation site. This causes fluid accumulation. A correct operation technique with burying of the noncolored implant's end below the bone surface will diminish sterile sinus incidence. Santavirta designed an immunological survey to ellicit the immune response on the alpha polyesters [25]. PGA seemed to be an inert material but it does induce inflammatory mononu-

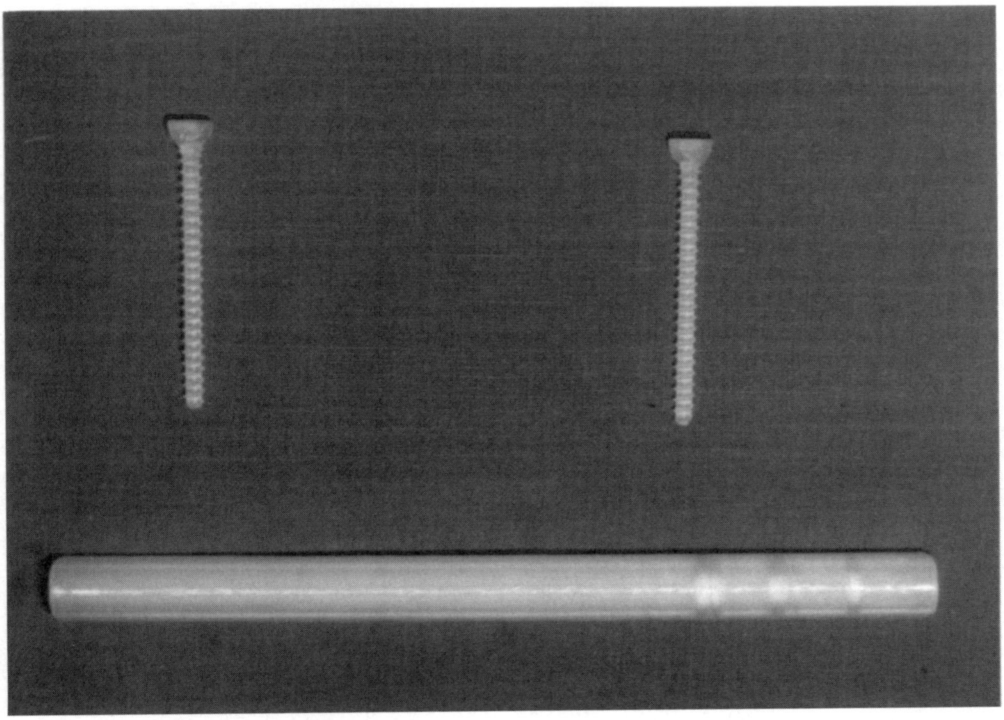

Figure 13 Biodegradable interlocking nail for fracture fixation. The nail is placed in the medullary canal after reaming through the trochanteric approach.The locking screws are placed proximally and distally from the fracture to secure rotational stability. This experimental rod has a diameter of 12 mm and a central canal to allow the use of a guidewire (Elst et al. [29–31]).

clear cell migration and adhesion, leading to a slight nonspecific lymphocyte activation. Cell type analysis of the clinical effusion samples indicated lymphocytes as the main cell type and some monocytes were found. The absence of neutrophils is seen as evidence for the sterile, noninfectious nature of the tissue response. Biopsies from patients with implants but without sinus formation demonstrated a similar image on histological survey.

The results from these histological data correspond to the observations described by Spector and co-workers in 1989 [27]. Spector describes the general tissue reaction to biomaterials as

- The presence of macrophages and multinucleated foreign body giant cells
- A fibrous capsule enveloping the implant
- A zone of reparative tissue bordering the tissue or organ into which the implant has been placed
- The infiltration of other cell types, e.g., lymphocytes, plasma cells, eosinophils etc.

V. FUTURE DEVELOPMENTS

Biodegradable materials from different origins are presently used as suture materials, fracture fixation devices, drug delivery systems, bloodvessels, bone replacement and other applications [28]. The mechanical properties of the currently available implants were the main reason for a limited clinical use; simple fractures in non-weight-bearing locations still are the common

Figure 14 Histological overview of medullary tissue around a stainless steel implant. In a circular fibrous tissue layer, some fibroblasts and macrophages are present. No signs of an intensive inflammatory response.

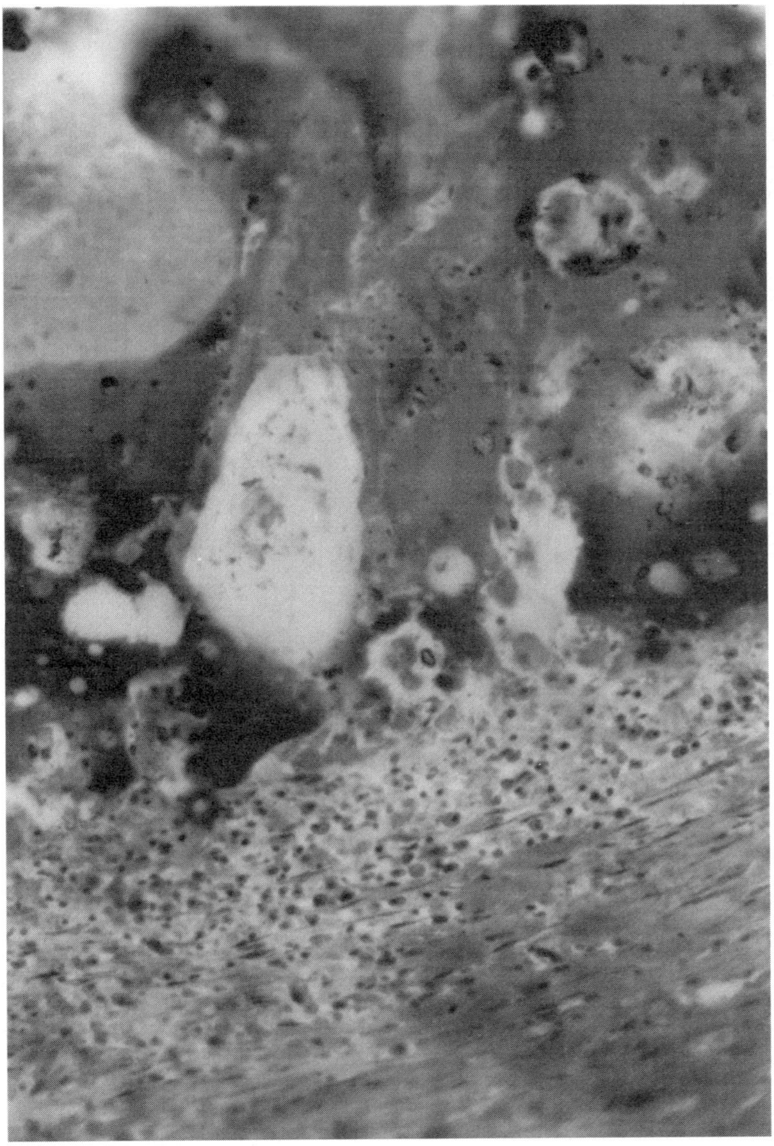

Figure 15 Histological overview of the inflammatory response on PLA interlocking nails. Remnants of the implant were still present after a 2 year implantation period. Multinucleated giant cells and lymphocytes were demonstrated in the presence of PLA particles.

applications in fracture surgery. As our knowledge about the composition of biodegradables and our technical skills improved, so did our ability to provide implants with better mechanical properties. In several experimental studies, strong biodegradable interlocking nails for the fixation of long bone fractures have been developed and tested [29–31] (Fig. 13). In an animal model using young pigs, a midshaft osteotomy was made to mimic a fracture [29]. A PLA interlocking nail was implanted intramedullary and proximal and distal interlocking with biodegradable pins of the same material were performed. Initial bending strength of the implants was comparable to that of human cortical long bone, so these implants allowed immediate full weight bearing. After termination of the animals after one and three months, mechanical and histological survey of the femurs was performed. The rods were not fully degraded and were tested separately. Fracture healing was fine although the rod had lost half of its bending strength in one month, which was considered unsafe for long bone fracture treatment in humans. A mild inflammatory response to the PLA was noticed without clinical signs of illness in the animals.

This response was simulated using predegraded PLA powder in a second experiment. Again an inflammatory response characterized by fibrous encapsulation of the PLA and lymphocytes and multinuclear giant cells activation was observed. The authors concluded that a long-term follow-up animal study after the histological response on these materials was necessary prior to human implantation [30].

After this simulation experiment, mechanically improved PLA/PGA copolymer and PLA interlocking nails were tested in a long-term, 2½ year follow-up study using adult sheep [31]. A unilateral midshaft femur osteotomy was performed to mimic a fracture, this osteotomy was fixated with an intramedullary nail of PLA, PLA/PGA copolymer or stainless steel origin. All three types of nails allowed immediate full loading. Two and one-half years after implantation, the animals were terminated and mechanical and histological analyses of the femurs were performed. The results of this experiment using two types of biodegradable interlocking nails showed an excellent fracture healing with no signs of stress atrophy. The rods had almost disappeared from the medullary cavity but remnants could still be found, even after this 2½ year period of implantation.

The histological response to the biodegradable rods was characterized as similar to the results of the former experiments: a prolonged inflammatory response with activation of multinucleated giant cells and lymphocytes (Figs. 14, 15). The exact clinical relevance of this response is not yet clarified. All animals with biodegradable implants performed equally to the control group animals implanted with a stainless steel interlocking nail in view of general condition, wound healing and gait patterns. Long-term follow-up animal studies beyond total degradation and immunological survey of the human immune response to the different polymers will have to be performed prior to a implantation of large amounts of polymer into the human body. In view of the technical developments, strong implants with reliable mechanical behavior can be produced [31]. Knowledge of the mechanical properties and degradation characteristics of the biodegradable materials as well as complete insight in the histological and systemic responses of the human body to these substances will guide us in making the decision whether to use a biodegradable fracture fixation device. Since the advantages of the use of these materials in fracture surgery are obvious and valuable, it seems worthwhile to further investigate the disadvantages.

REFERENCES

1. Ashammakhi N, Rokkanen P. Absorbable polyglycolide devices in trauma and bone surgery. *Biomaterials* 1997: 18; 3–9.
2. Törmälä P, Rokkanen P, Laiho J et al. Material for osteosynthesis devices. US Patent 4,743,257, 1988.
3. Törmälä P, Rokkanen P, Vainionpää S et al. Surgical materials and devices. US Patent 4,968,317, 1990.
4. Rokkanen P, Böstman O, Hirvensalo E et al. Absorbable implants in the fixation of fractures, osteotomies, arthrodeses and ligaments. *Acta Orthop Scand* 1994: 65(Suppl 260); 19–20.
5. Rokkanen P, Böstman O, Vainionpää S et al. Biodegradable implants in fracture fixation, early results of treatment of fractures of the ankle. *Lancet* 1985: 1; 1422–1424.

6. Böstman O, Hirvensalo E, Vainionpää S et al. Anklefractures treated using biodegradable internal fixation. *Clin Orth Rel Res* 1989: 72-B; 597–600.
7. Partio EK, Böstman O, Hirvensalo E et al. Self-reinforced absorbable screws in the fixation of displaced ankle fractures: a prospective clinical study of 152 patients. *J Orth Trauma* 1992: 6–2; 209–215.
8. Dijkema ARA, Elst, M van der, Verspui G et al. Surgical treatment of fracture dislocations of the ankle joint with biodegradable implants: a prospective randomized study. *J Trauma* 1993: 34; 82–85.
9. Partio EK, Hirvensalo E, Partio E et al. Talocrural arthrodesis with absorbable screws. *Acta Ortop Scand* 1992: 63–2; 170–172.
10. Partio EK, Merikanto J, Heikilä JT et al. Totally absorbable screws in fixation of subtalar extra articular arthrodesis in children with spastic neuromuscular disease. *J Pediatr Orth* 1992: 12–5; 646–65.
11. Niskanen RO, Lehtimaki MY, Hamalainen MMJ et al. Arthrodesis of the first metatarsalphalangeal joint in rheumatoid arthritis. *Acta Ortop Scand* 1993: 64; 100–102.
12. Hirvensalo E, Böstman O, Törmälä P et al. Chevron osteotomy fixed with absorbable polyglycolide pins. *Foot Ankle* 1991: 11; 212.
13. Rokkanen P, Böstman O, Hirvensalo E et al. Absorbable implants in the fixation of fractures and osteotomies. In: Leung KS, Hung LK, Leung PC, eds. *Biodegradable Implants in Fracture Fixation.* Chapter 4. Singapore, Hong Kong and World Scientific, 1994: 189–192.
14. Kristensen G et al. Biofix meniscus tacks versus inside-out suturing technique in the treatment of bucket-handle lesions, a randomized study. *Acta Ortop Scand* 1994: 65; 102–104.
15. Friden T, Rydholm U. Severe aseptic synovitis of the knee after biodegradable internal fixation. *Acta Ortop Scand* 1992: 63–1; 94–97.
16. Hirvensalo E, Böstman O, Partio EK et al. Absorbable polyglycolide pins in fixation of displaced fractures of the radial head. *Arch Ortop Trauma Surg* 1990: 109; 258–261.
17. Hirvensalo E, Böstman O, Partio E et al. Fracture of the humeral capitellum fixed with absorbable polyglycolide pins. *Acta Ortop Scand* 1993: 64; 85–86.
18. Partio EK, Hirvensalo E, Böstman O et al. Broches et vis bioresorbables: une nouvelle methode de fixation des fractures de l'olécrâne. *Int Orthop (SICOT)* 1992: 16; 250–254.
19. Hirvensalo E. Böstman O, Partio E et al. Biodegradable fixation in intra-articular fractures of the elbow joint. *Acta Ortop Scand* 1988: 227(suppl);78–79.
20. Vihtonen K, Patiala H, Rokkanen P et al. Preliminary results of reinsertion of ruptured ulnar collateral ligament of the first metacarpophalangeal joint with totally biodegradable polylactide (PLA) pin. *Acta Ortop Scand* 1990; 237(suppl): 44.
21. Vihtonen K, Juutilainen T, Pätiälä H et al. Reinsertion of the ruptured ulnar collateral ligament of the metacarpophalangeal joint with an absorbable self-reinforced polylactide tack. *J Hand Surg* 1993: 18B; 200–203.
22. Kumta SM, Spinner R, Leung PC. Absorbable intramedullary implants for hand fractures. *J Bone Joint Surg* 1992: 74b; 563–566.
23. Böstman O, Mäkelä EA, Törmälä P et al. Transphyseal fracture fixation using biodegradable pins. *J Bone Joint Surg* 1989: 71-B; 707–708.
24. Mäkelä EA, Böstman O, Kekomäki M et al. Biodegradable fixation of distal humeral physeal fractures. *Clin Orth Rel Res* 1990: 283; 237–243.
25. Santavirta S, Konttinen YT, Saito T et al. Immune response to polyglycolide acid implants. *J Bone Joint Surg* 1990: 72-B; 597–600.
26. Böstman O, Partio EK, Hirvensalo E et al. Foreign-body reactions to polyglycolide screws. *Acta Ortop Scand* 1992: 63; 173–176.
27. Spector M, Cease C, Tong-li X. The local tissue response to biomaterials. *Crit Rev Biocompatibility* 1989: 5: 269–295.

28. Nakamura T, Hitomi S, Watanabe S et al. Bioabsorption of polylactides with different molecular properties. *J Biomed Mat Res* 1989: 23;1115–1130.
29. Elst, M van der, Dijkema ARA, Klein CPAT et al. Tissue reaction on PLA versus stainless steel interlocking nails for fracture fixation; an animal study. *Biomaterials* 1995: 16; 103–106.
30. Elst, M van der, Kuiper I, Klein CPAT et al. The Burst phenomenon, an animal model simulating the long term tissue response on PLA interlocking nails. *J Biomed Mat Res* 1996: 30; 139–143.
31. Elst, M van der, Bramer JAM, Klein CPAT et al. Biodegradable interlocking nails for fracture fixation. *Clin Orth Rel Res* 1998: 357; 192–204.

23

Resorbable Buffered Internal Fixation Devices

Yu Wang, Donald L. Wise, and Yung-Yueh Hsu
Northeastern University, Boston, Massachusetts

I. BACKGROUND

Of the over 1 million fractures in the United States each year, more than 470,000 require some type of internal fixation device to stabilize the fracture during the healing process. Currently, the vast majority of these devices are made of metal. Following the healing of a fracture, a subsequent surgical procedure must be performed to remove the nonresorbable metal pins, plates and screws. While there is significant clinical demand for resorbable fixation devices to eliminate the added cost and patient discomfort associated with the removal process, currently available resorbable products have not been widely adopted because of their inability to replicate the structural strength of their metal counterparts and to their tendency to cause an inflammatory response due to their degradation products. When resorbable internal fixation devices fabricated from biodegradable polymers like polyglycolic acid (PGA), polylactic acid (PLA), and various copolymers such as polylactide-*co*-glycolide (PLGA) have been used in surgical repair of fractures, severe inflammatory responses (in approximately 8%) result in sterile abscess formation.

The biodegradation of the polymer only becomes evident after fairly long induction periods ranging from 5 to 20 weeks (average induction period of about 12 weeks). Degradation studies of solid PLGAs indicate that although a decrease of molecular weight begins rapidly, other processes including pH drop in the surrounding medium, loss of polymer weight, and the appearance of the monomeric hydrolysis products (lactic and glycolic acids) occur after a longer period of time. The length of this period depends on the comonomer ratio in the polymer. As an example, for PLGA with a lactic to glycolic ratio of 85 : 15, the induction period is about 10 weeks, with the rate of change of pH and weight loss reaching a maximum at about 16 weeks. This strongly suggests that the delayed inflammatory response is a direct consequence of the delayed appearance of acid hydrolysis products. This hypothesis is supported by the observation that PLGA degradation occurs at similar in vivo and in vitro rates.

II. BIODEGRADABLE MATERIALS

The trend in internal fixation devices is toward the use of resorbable, tissue compatible biopolymers such as PGA, PLA, and various copolymers of lactic and glycolic acids (PLGA).

Other polymers (such as polydioxanone) have also been considered. The PGA, PLA and PLGA polyesters have a long history of clinical use because they are biodegradable, biocompatible and exhibit moderate strength in tension, compression and bending. These properties make PLGA copolymers desirable for use in other medical applications. The physical advantages of PLGA copolymers include strength, hydrophobicity and pliability. The polymer is water insoluble but is degradable by hydrolysis to the monomers, lactic acid and glycolic acid, which are water soluble. PLGA is compatible with a wide variety of biologically active compounds. Solid formulations of polymer with a biologically active compound can therefore be prepared and designed to respond to an aqueous environment by slowly releasing both the compound and the degradation products—lactic and glycolic acids. The application of absorbable implants that negate the need to subsequent removal could offer major clinical advantages to the fixation of fractures. Implants for fixation of fractures have the common property of only being needed temporarily until the fracture has mended. Many trauma surgeons therefore recommend that all metallic implants used for the fixation of fractures be removed after the fracture has been mended and the strength is back in the bone [1]. Any financial benefits of the resorbable implant would be influenced both by the price of the absorbable implant in comparison with that of a conventional implant and by the price of removal for any particular fracture.

Biodegradable materials have been used for biomedical applications for many years. Surgical sutures are one of the earliest clinical implants. Sutures, obtained from ovine or bovine intestinal submucosa, data back to A.D. 150 in the time of Galen. Galen built his reputation by treating wounded gladiators. The enzymatic degradation of these sutures would result in the elimination of foreign materials that otherwise could serve as an Indus for infection.

Several naturally occurring polymers have found use as degradable implant materials. For example, partially oxidized cellulose has been used as absorbable suture material and as an absorbable homeostatic agent. The use of totally synthetic absorbable polymers began in the 1960s. Much pioneering work was done by suture companies in the United States of America including American Cyanamid Company and Johnson and Johnson, and at the U.S. Army Institute of Dental Research [2]. In 1962, polyglycolic acid was developed by the American Cyanamid Company as the first absorbable synthetic suture, Dexon. It has been commercially available since 1970. Commercial direction in the beginning stages was given to the homopolymer of lactic acid rather than the copolymers. This effort was primarily due to the limited availability of the glycolide monomer. The full range of monomers and polymers of PLGA has become rather easily accessible through major chemical companies since the 1970s. A copolymer of 92% polyglycolic acid and 8% polylactic acid came to the market in 1975 as a competitive resorbable suture. It was manufactured by the Ethicon Corporation and sold under the trade name Vicryl. The DuPont provided the lactide/glycolide polymers under the trade name Medisorb until it sold the business.

A. Chemical Synthesis

Polyglycolic acid, polylactic acid and various copolymers of polyglycolic acid and polylactic acid have been tested extensively. Chemically, these compounds are alpha polyesters. Because of the asymmetry of the lactic acid molecule, polylactic acid occurs in different stereoisomeric forms, including polylevolactic acid and a stereocopolymer poly(*d*, *l*-lactide-*co*-glycolide). Polyglycolic acid, polyglycolic acid: polylactic acid copolymer in a 10 : 90 ratio, and polydioxanone are used as absorbable sutures worldwide. The most efficient method to produce high-molecular-weight polyglycolic acid or polylactic acid is a ring opening polymerization of the corresponding cyclic diester, lactone glycolide or lactide by a catalyst [3]. Polyglycolic acid and polylactic acid are often called polyglycolide and polylactide, respectively. The polymerization is usually conducted over a period of 2–6 h at about 175°C in the melt. Organotin catalysts are normally utilized, stannous chloride and stannous octoate being the most common. Lauryl alcohol is often added to a control molecular weight during synthesis. PLGA polymers can be prepared in any molar ratio of lactic to glycolic acids. The proportion chosen is important in determining the in vivo degradation rate. Polymers prepared in a 50 : 50 proportion are hydrolyzed much faster than those which have a higher proportion of either monomer [4]. Bolus devices are often made as monolithic cylinders. Subcutaneous or intramuscular implantation can be constructed as rods, beads or suspended powders. The prepared materials can be sterilized by gamma radiation.

A broad spectrum of performance characteristics with the polylactides and polyglycolides can be obtained by careful manipulation of four key variables: monomer stereochemistry, comonomer ratio, polymer chain linearity and polymer molecular weight of the polymer. Because the mechanism of biodegradation is through the simple hydrolysis of the ester linkages, it is apparent how each of these factors plays an important role in the in vivo performance of the lactide and glycolide materials. Crystallinity and water uptake are key factors in determining the rates of in vivo degradation. An important characteristic that determines the behavior of these polymers is the glass transition temperature (T_g) at which the polymer becomes brittle and rigid [4]. Table 1 provides a summary of the glass transition temperatures of several lactide/glycolide polymers. T_g values range from about 40 to 65°C. Polylactide has the highest T_g at about 65°C.

The mechanical properties of biodegradable polymers proposed for use in internal fixation devices are crucial to the performance of the devices for support of healing. Because of PGA and PLA hydrolytic instability, they lose strength as they biodegrade. The rate at which they weaken is directly proportional to the amount they hydrolyc. Published reports on the biodegradation have shown that the hydrolytic degradation of ester bonds is apparently the means by which PGA and PLA are degraded.

Daniels [5] reviewed papers and abstracts from 1980 through 1988 and highlighted their findings. Compared to annealed stainless steel, unreinforced biodegradable polymers were initially up to 36% as strong in tension and 54% in bending, but only about 3% as stiff in either test mode. With fiber reinforcement, reported highest initial strengths exceeded that of stainless steel. Stiffness reached 62% of stainless steel with nondegradable carbon fibers, 15% with degradable inorganic fibers, but only 5% with degradable polymeric fibers. The slowest degrading unreinforced biodegradable polymers were polylactic acid and polydioxanone. Biodegradable composites with carbon or inorganic fibers would generally lose strength rapidly,

Table 1 Glass Transition Temperatures of Aliphatic Polymers

Polymer	Glass transition temperature, T_g (°C)	Melting temperature, T_m (°C)
Poly-L-lactide	60–67	172–174
Poly-DL-lactide	57–59	None
Polyglycolide	36	230
85 : 15 DL-lactide/glycolide copolymer	45	None
25 : 75 DL-lactide/glycolide copolymer	60	None
70 : 30 DL-lactide glycolide copolymer	58	None
Polycaprolactone	−65	63

with a slower loss of stiffness, suggesting the difficulty of fiber matrix coupling in these systems. The strength of composites was shown reinforced with a degradable polymeric fiber decreased more slowly. Low implant stiffness might be expected to allow too much bone motion for satisfactory healing. However, clinically unreinforced or degradable reinforced polymeric fiber materials have been used with initial clinically. Success has been careful selection of applications, plus use of designs and fixation methods distinctly different from those appropriate for stainless steel devices. Chu [6] studied hydrolytic degradation of PGA by examining the changes of tensile strength and the level of crystallinity of the suture material. It was found that the breaking stress decreased from 0.64 at day 0 to 0.00397 Pa at 49 days. The sigmodial shape of the stress strain curves gradually disappeared with an increase in the duration of the in vitro degradation. The degree of degradation was measured by the endpoint titration method. The percentages of PGA degraded were 42%, 56%, and 70% at 49, 60, and 90 days, respectively. The level of crystallinity of PGA at various duration of degradation exhibited an initial increase in the degree of crystallinity from 40% at day 0 to an upper limit of 52% at 21 days; this exhibited a gradual decrease to 23% at 90 days.

Tunc [7] established the process variables for fabricating polylactide with a high molecular weight. Both in vitro and in vivo testing showed that the modules of the material decreased linearly from 5.1 GPa initially to 3.6–3.9 GPa at three months. The initial and final yield strengths for the polymer were 57 MPa and 48 MPa, respectively. These had decreased to 22 and 18 MPa by two months. The history of absorbable implants in the repair of bone tissue began in the late 1960s. Initially, the implants to be studied were prepared by casting the polymers into sheets or films. This permitted basic investigations on the biological behavior of the compounds in bone tissue, but was not suitable for the fixation of fractures. Subsequently, the fabrication of implants was accomplished by melt-molding and extrusion into pins and rods. Most of the lactide/glycolide copolymers were injection-molded at temperature between 140 and 175°C. The polymer L-lactide requires 190–220°C for most molding operations. Implants of more complex design, such as screws and small plates, became possible in the late 1970s and 1980s. Promising results have been obtained using synthetic biodegradable polymers as a tendon replacement, in a resurfacing arthroplasty cup and as rods for the fixation of osteochondral fragments or osteotomies as well as in orthopedic surgeries [8]. The characteristics of the initial and retained strengths of rods and screws have been improved by the introduction of a fiber-reinforced composite texture in which the polymer matrix is reinforced with suture fibers of the same material. Table 2 shows the strength of different bones, polymers and steel [7].

Table 2 Strength of Different Bones, Polymers and Steel

Materials and bones	Strength (MPa)
Human femur bone	88.9–113.8
Human cancellous bone	10–20
Polylactide acid	59–72
Polyglycolide acid	57
Steel	480–1035
Titanium	240–550

B. Biodegradation Mechanism

Biodegradation of the aliphatic polyesters occurs by a process called bulk erosion. The lactide/glycolide polymer chains are cleaved by hydrolysis to the monomeric acids and are eliminated from the body primarily by exhaling as carbon dioxide and through the urine [6].

Because the rate of hydrolysis of the polymer chain is dependent only on significant changes in temperature and pH or the presence of catalyst, very little difference is observed in state of degradation at different body sites. The role of enzymatic involvement in the biodegradation of the lactide/glycolide polymer has been controversial. Most early literature concluded that bioerosion of these materials occurred strictly through hydrolysis with no enzymatic involvement. Other investigators [9,10] suggest that enzymes do play a significant role in the breakdown of the lactide/glycolide materials. Much of this speculation is based on the differences observed between in vivo and in vitro degradation rates. Holland [11] concluded that little enzyme involvement is expected in the early stages with polymers in the glassy state, whereas enzymes can play a significant role for polymers in the rubbery state. Ethoxylation of the carboxylic acid endgroups of aliphatic polyesters significantly changes the biodegradation rate as well as the crystallinity of these materials. Two degradation mechanism models have been proposed. Kenley [12] studied the hydrolysis of polylactide-*co*-glycolide and found the rate of cleavage per molecule was first order. Pitt [13] observed a first order loss of molecular weight of polycaprolactone and attributed this to an autocatalytic mechanism. In his earlier work, Pitt [14] determined that the polymer degrades to two states. The first stage in the biodegradation process is shown only by decrease in molecular weight caused by random hydrolytic cleavage of the ester linkage initial absorption of water. The second stage follows rapidly and exponentially with the onset of weight loss and a change in the rate of chain scission. These observations support a bulk erosion model. Most of literature on the degradation of PLGAs confirms the above observations of an initial slow phase of PLGA degradation followed by a stage of rapid hydrolysis. Table 3 shows that the 50 : 50 lactide/glycolide copolymer has the fastest degradation rate of the DL-lactide/glycolide materials [15]. The polymer degrades in about 50–60 days. The 65 : 35, 75 : 25, and 85 : 15 DL-lactide/glycolide have progressively longer in vivo lifetimes, with 85 : 15 copolymer lasting about 150 days. As the ratio of lactide is increased so is its lifetime. The lactide polymer therefore have the longest degradation rate of 12–16 months for the DL-lactide and 18–24 months for L-lactide polymer.

Kenley [12] compared the in vivo and in vitro decomposition of 10 lactide/glycolide copolymers, noted that the rate of hydrolysis was independent of pH in the range studied (pH

Table 3 Biodegradation of Lactide/Glycolide Polymers

Polymers	Approximate time for biodegradation (months)
Poly-L-lactide	18–24
Poly-DL-lactide	12–16
Polyglycolide	2–4
50 : 50 DL-lactide-*co*-glycolide	2
85 : 15 DL-lactide-*co*-glycolide	5
90 : 10 DL-lactide-*co*-caprolactone	2

4.5 to pH 7.4) and was virtually identical in vivo and in vitro. They suggested that the insensitivity of degradation rate to the pH of the medium was due to the formation of carboxylic acids (lactic, glycolic) subsequent to ester bond cleavage within the bulk of the polymer. This effectively lowers the pH in the microenvironment of the polymer, the acid content of the surrounding environment, due to the rapid hydrolysis—degradation of the polymer, supports the hypothesis that sterile abscesses are due to accumulation of these acids. They also studied polylactide-*co*-glycolide biodegradation kinetics in which they examined 10 lactide/glycolide copolymers representing a range of monomer ratios and molecular weights. The molecular distributions were characterized by intrinsic viscosity measurements and by size exclusion chromatography (SEC) using the "universal calibration" procedures. In this study, they prepared cylindrical samples and incubated them for timed intervals in pH 4.5–7.4 aqueous buffers at physiological 37°C. Parallel samples were also implanted subcutaneously in rats. They then monitored the time-independent changes in sample total weight (TW) and molecular weight (MW). TW and MW profiles were superimposed for all samples, demonstrating pH-dependent hydrolysis in vitro and equivalently in vivo versus in vitro copolymer degradation rates. TW loss lagged behind MW loss, indicating copolymer erosion via internal (versus surface) hydrolysis and dissolution. The copolymer MW loss rates adhered well to a pseudo-first-order kinetic model where the number of chain cleavages (X) per initial number average of the molecule is given by:

$$\ln[X] = -2.04 - 1.08t(\text{week}^{-1}) \qquad (1)$$

The most intensely studied of the 10 polymers was a poly-DL-lactide-*co*-glycolide with a 50 : 50 lactide: glycolide ratio (PLGA-50 : 50). They observed a reduction in molecular weight which followed first order kinetics and resulted in a decrease to about 13% of its initial value in three weeks. However, reduction in mass demonstrated a very different time profile: virtually no mass loss was noted for three weeks after which weight loss proceeded rapidly and exponentially. These observations support a bulk erosion mode in which hydrolysis is initiated by absorption of water and proceeds throughout the bulk of the polymer structure. In the phase of degradation corresponding to the onset of mass loss, the acid content of the interior can diffuse rapidly into the surrounding tissue and cause the delayed response noted by clinicians. Miller's early study [16] of PLGA degradation rates indicated degradation half-lives for 5–6 mg pellets of 5.0 months for polyglycolide (PGA), 0.65 month for PLGA-25 : 75, 1 week for PLGA-50 : 50, 0.6 month for PLGA-75 : 25, and 0.66 month for PLA. Based on these results, a PLGA-90 : 10 would be expected to have a half-life of three months and a PLGA-85 : 15 of two months.

However, in another study using cylinders measuring 1.60 × 1.75 mm, Cutright and Hunsuck [17] reported degradation rates for the same five polymers at 20 days intervals to 220 days. These results, summarized graphically in their paper, are reported in Table 4.

As reported, these data suggest a half-life of the PLGA-75 : 25 of between 113 and 140 days (3.8–4.7 months). The half-life of the PLGA 50 : 50 appears to be about 80 days. It is interesting to observe, however, that degradation appears to occur in two phases. In the first phase (to day 100) little degradation is observed after an initial loss of 18% in the first 20 days. A linear regression to day 100 (omitting the 0,0 point) gives a slope of –0.005, which can be taken as 0.5%/day degradation loss. A similar regression analysis from day 100 through day 180 gives a degradation loss of 0.875%/day with a regression coefficient of 0.912. In a study of degradation of C-14 labeled PLGA microspheres in the size range 63–125 microns, and ranging in composition from PLGA-74-26 to PLGA-96 : 4 (with DL-lactide as the comonomer), Tice [18] showed the following losses (see Table 5) of C-14 activity with time. It is

Table 4 Degradation Rate of Copolymers

Time (day)	Percent degradation of PLGA $X:Y$				
	100 : 0	75 : 25	50 : 50	25 : 75	0 : 100
20	15	18	18	1	2
40	18	20	25	3	4
60	14	26	53	31	11
80-	3	15	46	31	11
100	8	20	60	75	10
120	25	65	48	95	18
140	30	50	100	95	18
160	31	80	98	100	13
180	20	100	100	—	11
200	25	—	—	—	8
220	43	—	—	—	17

relevant to note that the PLGA-87 : 13 showed only 5% loss in the initial 5 weeks and only 20% by 10 weeks. Compared with the PLGA-96 : 4, the PLGA-87 : 13 showed less loss at 5 weeks and comparable loss at 10 weeks. Note also that based on these data the half-life of the PLGA-87 : 13 is about five months (Table 3 Pitt's study). Both Cutright's data for PLGA-75 : 25 and Tice's data for PLGA-87 : 13 show similar half-lives (3.8–4.7 months, Cutright; and 4.5 months, Tice). Indeed, Cutright's data suggest little degradation prior to three months, although Tice's data did not reflect this. The complex nature of PLGA degradation kinetics has been related to the extent of crystallinity. Miller [16], using different scanning calorimetry, rates the extent of crystallinity of various PLGAs given in Table 6.

Chu [6] pointed out that these polymers have both crystalline and amorphous regions and thus hydrolytic degradation proceeds through two main stages. First, water diffuses into the amorphous, but not the crystalline, regions. Hydrolytic degradation starts in the amorphous regions. As chain scission increases the mobility of new chain ends enabling them to reorganize into a more ordered state, thereby actually increasing the degree of crystallinity. Following

Table 5 Loss of C-14 Activity with Time

Time (weeks)	Percent of C-14 activity remaining			
	PLGA-96 : 4	PLGA-92 : 8	PLGA-87 : 13	PLGA-74 : 26
0	100	100	100	100
5	70	95	95	100
10	85	100	80	65
15	90	85	70	25
20	83	75	55	3
25	50	50	35	
25	50	50	35	
30	45	5		
35	75	20		
40	52			

Table 6 Crystallinity versus Copolymer Ratio

Glycolide lactide ratio	Crystallinity
100 : 0	high
75 : 25	moderate
50 : 50	low
25 : 75	low to moderate
0 : 100	moderate

hydrolysis of the amorphous regions, the second stage is characterized by hydrolysis of the new much more accessible crystalline regions. A recent study by Shih [19] suggests that the hydrolysis of poly-DL-lactide proceeds by chain end scission of monomer units rather than by random hydrolytic scission of internal ester bonds. The implication is that relatively little degradation is observed initially followed by progressively more rapid depolymerization. This is supported by the data presented above indicating relatively little degradation in the first several weeks.

C. Clinical Studies and the Inflammatory Response of the Internal Fixation Devices

Several investigators have demonstrated that implants of polyglycolic acid, polylactic acids are completely absorbable within bone tissue and that new bone is deposited on and within the implant as the degradation proceeds [20]. The degradation of these polymers occurs mainly by hydrolytic scission and through nonspecific enzymatic action. The differences in the final metabolism of the polymers are relatively slight, but the rates of degradation vary. The molecular weight, crystallinity, thermal history, and geometry of the implant influence the degradation considerably, a porous thin sheet depolymerizing much more rapidly than a dense block. However, the degradation process does not imply immediate absorption of an implant material. In an experimental study with pellets of carbon-14 and tritium-labeled polyglycolic acid implanted in the tibiae of rats, approximately 70% of the material from the implant remained at the site after three months. The principle route of ultimate elimination is respiration, with excretion in the urine and feces playing only a minor role [21]. The fixation properties of pins, rods, screws, and plates of alpha polyesters have been studied in experimental conditions simulating fractures of cancellous bone [22] and epiphysis separation in immature animals [23]. In a study on human cadavers, satisfactory fixation has been reported with 2.0 mm rods of polyglycolic acid in osteotomies of the distal part of radius and with 3.2 mm rods in osteotomies of the patella [24]. The experimental data on the response of tissue to absorbable polyesters implants have been contradictory. Several studies on animals have demonstrated signs of an inflammatory foreign body reaction, whereas in others there was no inflammation. The clinical use of absorbable polyester implants for fractures of the long bones had not been reported before 1985. The principal materials utilized have been polyglycolic acid, polyglycolic acid, polylactic acid copolymer in a 90 : 10 ratio and of polyparadioxanone [19]. In a 1984 randomized study [25], 56 open reduction and internal fixations of malleolar fractures of the ankle were done with metal ASIF screws and plates, or with rods of PLGA. All patients wore a plaster cast for six weeks postoperatively. Failure of fixation and redisplacement of the fracture was seen once in each group. There were two bacterial infections of the operative wound after metallic fixation, and two patients had a sterile inflammatory wound sinus three to four months after the operation in the injuries fixed with the polymer. The conclusion of this study was that there were no major differences, during the one year follow-up, in the results of treatment either in the control or study populations. 102 patients with unimalleolar or bimalleolar fracture were managed by open reduction and internal fixation with PLGA or PGA rods [26]. A minor postoperative displacement of the fracture was seen in nine patients and a late displacement of 1 to 2 mm, in an additional four. No reoperations were necessary. The appearance of a sterile sinus from the operative wound complicated the course in six patients (5.9%). In a study of rods of polyglycolic acid in 64 bimalleolar fractures [27,28] a minor redisplacement of the fracture occurred in nine patients (14.5%), and a late inflammatory sinus in five cases (8.1%).

In a series of 29 patients with intra-articular fractures about the elbow joint, internal fixation using PGA rods was followed by three weeks of immobilization in a plaster cast. The fixation failed in one patient who had a fracture of the olecranon, and slight redisplacement of a fracture occurred in four other patients. A late non-infectious inflammatory reaction occurred in four cases (13%) [29]. In a study, Hoffman [30] used PGA pins in the repair of 40 cases of displaced fractures of the distal part of the radius. They observed a late inflammatory response requiring debridement was observed in nine patients (22.5%). The inflammatory reaction to the polyglycolic acid was evident from 47 to 145 days after insertion. A study of 217 fracture cases, mostly of the ankle, involved repair with fiber reinforced PGA screws [31]. Reoperation for failure of the fixation was necessary in 7 patients (3.2%), and 10 (4.6%) had an inflammatory tissue reaction requiring operative drainage. Illi [32] used biodegradable fixation device in pediatric traumatology. They designed a driving system for screws consisting of an indwelling screw driver which transducers the torque force over the whole length of the thread. They performed a comparative study with either conservative treatment (reposition and casting) or osteosynthesis with biodegradable screws made from polylactic acid or metallic screws in 48 calves aged six weeks. Results from one to six weeks postoperatively were compared and good to excellent results obtained in equal amounts for PLA and steel screw osteosynthesis, whereas conservative treatment was only successful in one third of the cases. Clinical, radiological and histological follow-up proved the usefulness of biodegradable osteosynthesis in the growing skeleton. For clinical evaluation in children, the resorption time of biodegradable material was lowered to three to six months. Bos [33] performed a clinical application of resorbable polylactide plates and screws for the fixation of zygomatic fractures. Ten patients (mean age 40.3 years) with solitary displaced fractures of the malar complex were treated with PLA plates and screws. Fixation was performed in the front-zygomatic area. The results of the operative treatment were evaluated clinically starting immediately after the operation. Clinical follow-up was done during the period of hospitalization (two to three days) and at one, three, and six weeks and three months postoperatively. Based on the clinical evaluations, all fractures healed satisfactorily. No displacement occurred and no diplopia or restriction of eye ball movement could be detected.

Yamamuro [34] developed strong and bioabsorbable osteosynthetic screws, pins, and nails made of high strength polylactide and evaluated their use in 143 patients. The PLA loses about 10% initial strength within 8 weeks, about 40% within 4 weeks and nearly 100% within 20 weeks in vivo. The patients were aged from 9 to 78 years, and the follow up was from 2 to 6 years. The implants were used for fixation of bone grafts in 84 patients, peri- and intra-articular fractures in 49, after osteotomy in 8, and 2 in others. Bony union was not achieved in only one case. In the early states of the study, breakage of screws occurred in 12 cases (8%), mostly because of an inappropriate size of tap. These problems have not occurred after improvement of the tapping system. There were no cases with abnormal blood tests, infection or foreign body reaction. Bostman [8] point out that tissue response to resorbable implants fabricated from these polyesters has not been uniformly acceptable; investigators have reported a late sterile inflammatory foreign body response (sterile abscess). He estimates late sterile abscess formation in 8% of fractures repaired with these polymers. A review by Bostman of 516 patients who were treated with biodegradable rods at a single trauma center revealed rates of 1.2% for failure of fixation necessitating reoperation, 1.7% for bacterial infection of the operative wound, and 7.9% of the noninfectious inflammatory tissue response requiring operative drainage [35]. Internal fixation of the fractures was accomplished with devices fabricated from either PGA or a PLGA copolymer. He believed that the unique complication of these implants was the delayed inflammatory reaction. He observed that the clinical presentation of this complication was fairly consistent. The patients had no local or systemic sign of the problem with the wound in the immediate postoperative period, and then a painful erythematous, fluctuate swelling suddenly developed about the healed wound. The mean interval between fixation of the fracture and the clinical manifestation of the reaction was about 12 weeks (range, 7 to 20 weeks). A sinus draining the liquid remnants of the polyglycolide often formed. Bacterial cultures of the drainage were routinely negative. An inflammatory reaction may occur at any site of implantation, but there appear to be some differences in the frequency of occurrence of the reaction at various anatomical regions. The occurrence of the reaction was not influenced by the volume of the implanted polymer, by a subcutaneous location of the implant, or by bathing of the implant by synovial fluid. The histological characteristics of specimens obtained during debridement of the sinus tracts demonstrated a nonspecific foreign body reaction, with abundant giant cells phagocytosing the debris from the polymer. Similar histological findings were evident in specimens from patients who had had no clinically manifested reaction but had been reoperated on for mechanical failure of the fixation [27,28]. Bostman suggested that these observations appeared to indicate a biological response to the chemical irritation from the polymers. The factors that determined which patients would have a foreign body reaction and which would have a subclinical, asymptomatic reaction seemed to be the local tissue tolerance and the clearing capacity of the bone. This reaction is not limited to PGA as shown in a study by Eitenmuller [36], who reported on 19 patients with ankle fractures treated with absorbable plates and screws of polylactide. Of these 19, nine (47.7%) had an inflammatory reaction. In accordance with the slow degradation rate of polylactic acid, the reactions occurred as late as nearly three years after the operation.

III. MATERIALS

A. Poly-Lactide-*co*-Glycolide Acid 50 : 50

The polymer for this study was PLGA 50 : 50 (50% poly-*D*,*L*-lactic acid, 50% polyglycolic acid). This was purchased from Boehringer Ingelheim, lot number 954601. The repeating mer unit is

$$—[—O—CH(CH_3)—CO]—[O—CH_2—CO]$$

B. Calcium Carbonate

Calcium carbonate ($CaCo_3$) was used as buffer. It was purchased from Fisher Scientific (Pittsburgh, PA), lot number 34034.

C. Solvents

Acetone, glacial acetic acid and 2-propanol were obtained from Fisher Scientific (Pittsburgh, PA). Solvents were used without further purification.

IV. EXPERIMENTAL METHOD AND PROCEDURES

The salt used initially to buffer the acid formed on PLGA 50:50 hydrolysis was calcium carbonate. Calcium carbonate is an acceptable additive for human application and has been explored for use in other implants such as bone cements [15]. The rationale for its use is based on the following. An appropriate buffer should have a low aqueous solubility so that it will not be rapidly lost by dissolution. The basic component of the buffer (the anion) should react easily with the protons of the acid products of hydrolysis. Letting B^- represent the buffer anion and L^- the lactate (or glycolate) anion, the equilibrium can be expressed as

$$HL + B^- \Leftrightarrow L^- + HB \tag{2}$$

In other words, HB must be a weaker acid than HL (or B^- must be a stronger base than L^-). These relationships can be expressed quantitatively by ionization constants of the respective acids (K_a):

$$K_a^{HB} < K_a^{HL} \tag{3}$$

Thus a viable buffer would be calcium carbonate. The reaction of lactic acid with the carbonate anion is

$$HL + CO_3^{2-} \Leftrightarrow L^- + HCO_3^- \tag{4}$$

The product HCO_3^- anion (bicarbonate) has an acid dissociation constant $K_a^{HCO_3} = 6.31 \times 10^{-11}$, whereas the various racemes of lactic acid have dissociation constants in the range $K_a^{HL} = 1.38 - 1.62 \times 10^{-4}$. Taking 1.5×10^{-4} as a mean value of 50% glycolic acid and 50% D,L-lactic acid, the equilibrium constant for the above reaction may be calculated as

$$K_{eq} = \frac{K_a^{HL}}{K_a^{HCO3\_}} = 2.4 \times 10^6 \tag{5}$$

Thus the equilibrium lies to the right and protons produced by ionization of lactic or glycolic acids will be removed by the buffer. Another consideration is the extent to which the system should be buffered. Because slow generation of products at the end of the rapid phase may not cause clinical problems, it may not be necessary to provide total buffer capacity.

We can calculate the loading of calcium carbonate in the polymer matrix required to neutralize a given quantity of lactic and glycolic acid. We can calculate the quantity of polymer which can be neutralized by calculating the moles of monomeric acid produced at 100% hydrolysis. For PLGA of any composition (x, y = percents of lactide and glycolide, respectively):

$$\text{—[—O—CH(CH}_3\text{)—CO]}_{(x)}\text{—[O—CH}_2\text{—CO]}_{(1\text{-}x)}\text{—lactide MW} = 72$$

and glycolide MW = 58. Thus the average monomer residue molecular weight is $72x + 58(1 - x) = 14x + 58$. As an example, in 1 g of PLGA 50:50 ($x = 0.5$), a total of $1/[14(0.5) + 58] = 0.0154$ mole of monomeric acids will be generated. We can now calculate the grams of buffer needed to neutralize this. Taking $CaCo_3$ as an example:

$$2RCO_2H + CaCO_3 \rightarrow (RCO_2^-)_2Ca + H_2O + CO_2 \tag{6}$$

Thus we need $0.5 \times 0.0154 = 0.0077$ mole of $CaCO_3$ (molecular weight of $CaCO_3 = 100$), which weight 0.77 g. The loading will be $0.77 \times 100/(1.0 + 0.77) = 43.5\%$ by weight. We can also calculate what percent of the acid groups formed by hydrolysis can be neutralized by a given loading of buffer. The example given in Table 7 is for calcium carbonate in PLGA 50:50; if we choose to neutralize 25% of the acid formed, calcium carbonate loading can be reduced to 16.4%.

Table 7 Percent of Acid Groups Formed by Hydrolysis Which Are Neutralized by a Given Loading of Buffer

Percent loading (by weight)	Percent carboxyl groups neutralized
43.5	100
27.8	50
16.4	25
0	0

A. Experimental Outline

The goal of this study was to develop a resorbable polymeric fixation device which can control the pH of the microenvironment by incorporating some buffering (or neutralizing) capacity into the polymer. This work focused on determining the feasibility of a buffer-containing fixation device to (1) neutralize acid degradation products; (2) maintain a useful mechanical performance profile. The method was focused primarily on in vitro monitoring of degradation over time to include measurement of medium pH, device weight, polymer molecular weight, and mechanical properties in comparison with unbuffered controls. Thus this study provided a correlative basis for the design of a resorbable fixation device with improved compatibility and demonstrable mechanical performance.

B. Preparation of Buffered PLGA Samples

Considering degradation time and crystallinity, the polymer chosen for this study was the rapidly degrading PLGA 50 : 50. This polymer is commercially available. A flowchart of the buffered PLGA 50 : 50 samples preparation is shown in Fig. 1.

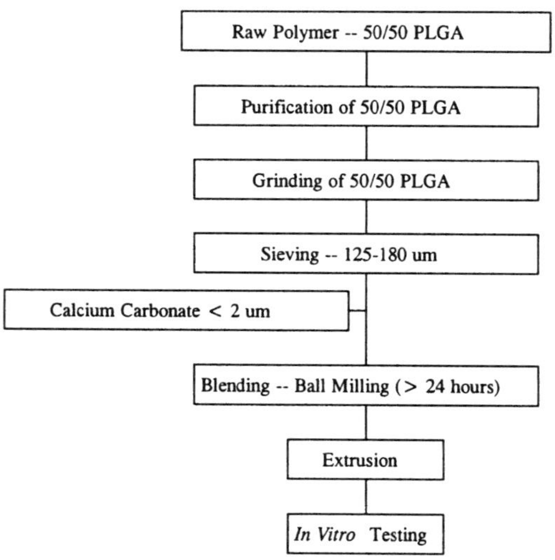

Figure 1 The flowchart of buffered PLGA 50 : 50 sample preparation purification of raw PLGA and calcium carbonate.

The raw PLGA 50 : 50 which was obtained commercially needed to be purified before use because of low molecular weight impurities. The system is shown in Fig. 2, in which 5 g of raw PLGA 50 : 50 was dissolved in 100 ml acetone. The purification was then done by adding the PLGA 50 : 50 solution dropwise into 500 ml of 2-propanol while stirring vigorously and collecting the precipitate on a metal rod. The precipitate was dried over vacuum at 1.0 mm Hg for two days, yielding the final product.

Calcium carbonate as precipitated powder was used. The average particle size of this product was 1.2 microns (μm) with 100% of the material less than 2 μm. The small particle size enabled intimate distribution of the calcium carbonate throughout the polymer matrix. After precipitation, the polymer was cryogenically grounded at liquid nitrogen temperature in a Tekmar A-10 analytical mill. The grounded polymer was sieved to retain particles between 125–180 μm. To ensure a homogeneous mixture, the calcium carbonate and PLGA 50 : 50 were blended in a ball mill without the use of grinding acids. Three blends were prepared with compositions indicated in Table 6. The full range of buffer capacity was explored in order to emphasize differences in the experimentally measured parameters as these relate to both acid neutralization and mechanical integrity.

The blends then were extruded as cylinders with a diameter of 0.116 inch (3.0 mm). Extrusion was carried out on a Compact Model MPC-40 hydraulic press equipped with a controller enabling maintenance of constant pressure. The extrusion die was heated with external heating taped to 57°C at which temperature the required pressure is about 13 tons/in^2. Figure 3 shows the schematic of the extrusion mold. Control consisted of PLGA 50 : 50 rods similarly prepared but without the calcium carbonate buffer.

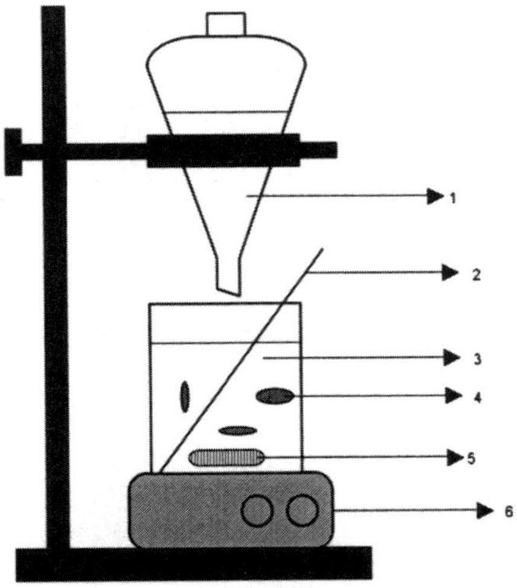

1. PLGA-50:50-acetone solution (50 mg/ml);
2. glass stick for catching precipitated PLGA-50:50;
3. biodegraded PLGA in alcohol solution;
4. precipitated PLGA-50:50;
5. stirring bar;
6. stirring machine.

Figure 2 Purification of PLGA (50 : 50).

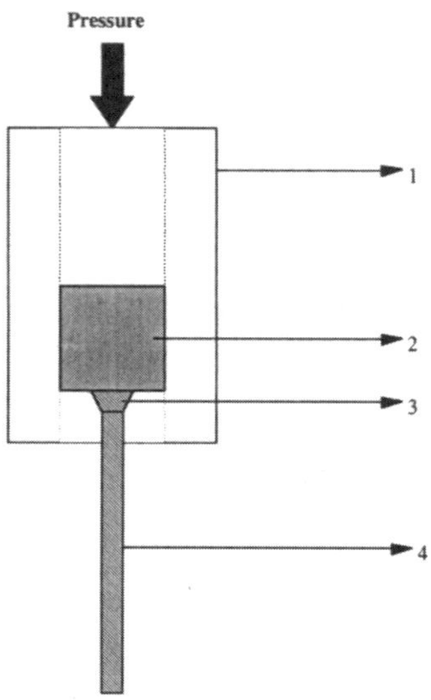

1. Extrusion Mold;
2. Mixture of $CaCO_3$ and PLGA 50:50;
3. Die;
4. Extruded Rod

Figure 3 Schematic of the extrusion mold.

C. In Vitro Studies

In vitro assays included the following measurements taken as a function of time: (a) pH change; (b) weight loss and molecular weight change and appearance of degradation products (low molecular weight oligomers and monomeric acids); and (c) mechanical properties testing. Each rod (3.0 mm diameter and 1.5 cm lengths) was incubated at 37°C in test tubes containing 50 ml of unbuffered water suitable for high performance liquid chromatography (HPLC grade, Fisher Scientific W-5) adjusted to pH 7.0. Tubes were purged with nitrogen gas to expel carbon dioxide prior to the experiment and after each measurement. Test tubes were tightly stoppered and held in a thermostatic shaking bath between measurements. Measurements were taken weekly, or more often during the period of rapid change. The use of unbuffered water in these experiments was preferred because it allowed us to follow the change in pH which would be masked in a buffered medium. The maximum drop in pH could be calculated. A 1.0 cm × 3.0 mm diameter rod weighed approximately 100 mg. With a 50% loading of calcium carbonate (greater than was used), the polymer content was 50 mg. For simplicity assume the polymer was polylactide with a residue molecular weight of 72 g/mol. At 100% hydrolysis the lactic acid concentration was calculated as $0.050/(72 \times 0.025) = 0.0278$ M. The K_a of lactic acid was 1.38×10^{-4}, and from this the expected pH was 2.72. A calculation for polyglycolide yielded similar results, and its $pK = 1.48 \times 10^{-4}$.

An experiment shown in Fig. 4 was designed to measure weight loss of both components, the calcium carbonate and the polymer. In addition polymer molecular weight and distribution were also determined. Sufficient samples were prepared so that each time point was taken in triplicate. As above, controls were pure (unbuffered) polymer rods. Samples were removed on schedule and dried at room temperature to constant weight in vacuum over phosphorous pentoxide. The weight loss reflected both polymer and calcium carbonate loss, as the latter was converted to soluble lactate and glycolate salts. Sample was dried in a vacuum desiccation

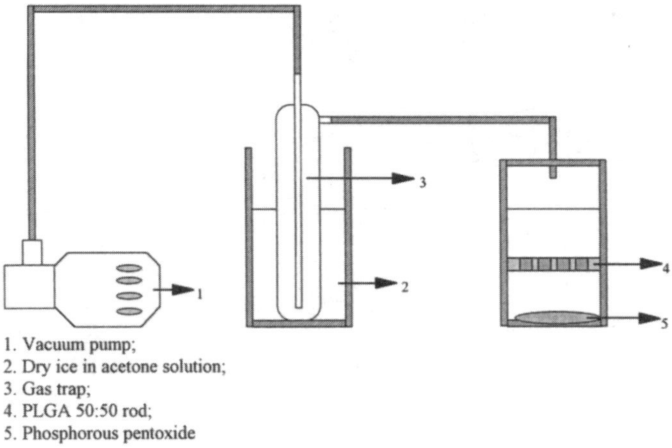

Figure 4 Experiment setup for drying the buffered PLGA 50 : 50 samples.

over 16 h with P_2O_5 present. Molecular weights were calculated by comparison with polystyrene standards. An aliquot sample of the remaining supernatant solution was analyzed by HPLC using a Vydac C-4 column, type 214TP54, and the 486 absorbance detector on the Waters analytical system previously described. Mobile phase was 0.1% H_3PO_4. Flow rate was 0.4 ml/min. It was to be noted that an initial concern in this feasibility survey was that the salt might release too quickly to offer any substantial buffering of the acidic hydrolysate. In this case, reasonable alternatives were considered in the contingency plans. These would induce modification of buffer particle size, encapsulation of the buffer within a portion of the PLGA 50 : 50 polymer (approximately a controlled release of the buffer), and use of less soluble buffer salt. Thus, there was a range of possible choices for introducing the buffer which should cover the continuation of the feasibility assessment.

D. Mechanical Properties Testing

Although knowledge of tensile and flexural properties was also important for design of internal fixation devices, this study limited measurement of mechanical properties to compressive failure strength and modules. Because of the small 3.0 mm diameter of the rods, specimens were supported vertically in soft foam cushions. Commercially available open celled polyurethane foams, available in sheets 0.25 to 2 inch thick, typically have densities 0.045 to 0.096 g/cm^3 and 25% deflection strength (compression) of 0.2 to 0.7 psi (1.4 to 4.8 kPa). Holes were drilled with a drill press so that mounted specimens remained vertical. The compressive deflection of the foam was eliminated by shaving the top surface as a pyramid or cone with its apex at the top of the specimen. Specimens were removed from the bath periodically, washed in distilled water and tested while wet. Compressive strength and modulus were measured on an Instron Model 1331 equipped with a 22,000 lb load cell coupled to a 500 lb load cell and operating with a compressive stroke of 0.5 mm/s.

V. EXPERIMENT RESULTS

pH changes over time were measured within a three-month period shown in Fig. 5. A pH of 7.0 was the initial value for all the polymer samples. HPLC grade water was used as the

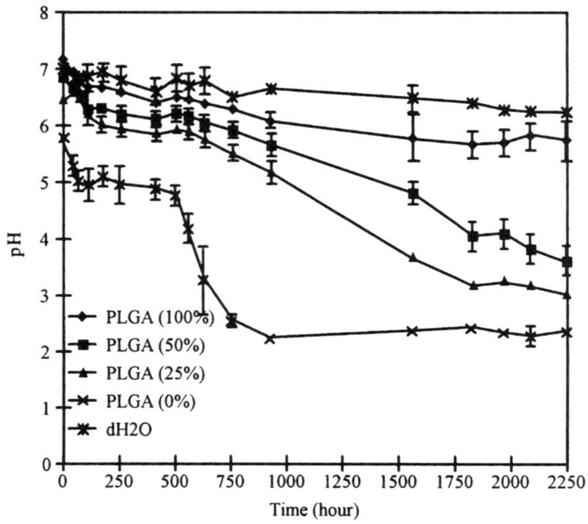

Figure 5 pH change over time on four different PLGA 50 : 50 rods.

standard. The pH in general showed a negative slope over time. The in vitro pH value of the 0% buffered PLGA 50 : 50 sample initially decreased sharply from pH 5.08 to pH 2.63 at 500–700 h. The pH value of the 0% buffered was lower than those of 25%, 50%, and 100% buffered PLGA 50 : 50 samples. The rate at which pH fell with time increased with decreasing content of the calcium carbonate buffer. These results would indicate that pH values of the PLGA 50 : 50 polymer were maintained at physiologically controlled ranges over a long period of time by adding buffers to the bioresorbable fixation devices.

The mechanical strengths of the four different buffered PLGA polymers were also measured with time as shown in Fig. 6. Not surprisingly, the strength of buffered rods were lower at all the time points than the unbuffered control rods. These strength measurements showed that mechanical strengths of the polymers decreased by incorporating buffers into the polymers. Also, the strengths steadily decreased over time.

It is a fact that PLGA 50 : 50 polymers will hydrolyze over time [10]. The weight of the polymers thus decreases with time, as shown in Fig. 7. The four PLGA 50 : 50 polymer samples lost their weight differently over a three month period. The weight loss of the polymers was faster with a lower percentage of buffer than those with a higher percentage. The four PLGA 50 : 50 polymers lost their weight over a three-month period.

A. The Dimensional Change of Volume/Diameter of the Rods with Time

The diameter change of these composites over time was very different when compared with the pH, strength, and weight loss change profiles. One would expect a decreasing diameter/volume with a concurrent weight loss as the data has shown. The diameter of the polymer rods shown an initially swelling and grew over time and then had a period of decreasing diameter. Figure 8 shows the change of the volume over the first month. The diameter profile was similar for these different buffer percentages of PLGA 50 : 50 polymers. The profiles occurred at different time points for the polymers, with that of the 25% buffered rod occurring first and being the most swollen, followed by the 50% buffer rod and then, least, the 100% buffered rods. At day 30, the diameter of the 25% buffered rods was 280% its original size; the diameter

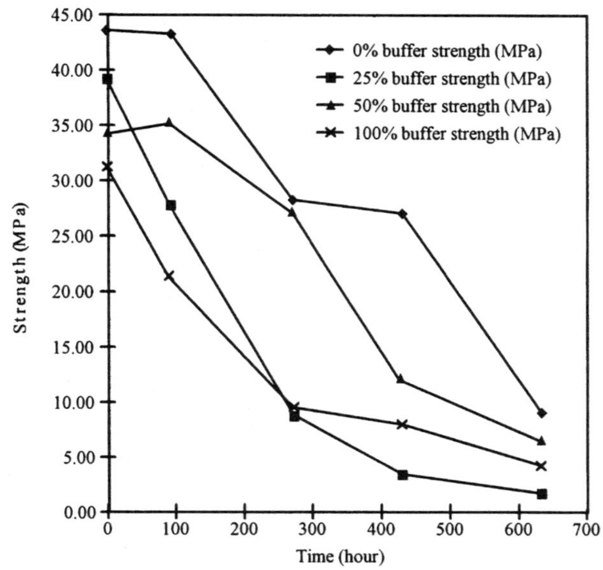

Figure 6 Strength test over time on four different PLGA 50 : 50 rods.

of the 50% buffered rods was 270% its original size; and the diameter of the 100% buffered rods was 260% its original size. The diameter of the unbuffered rods (control) did not change. And at day 90, most of the unbuffered, 25%, and 50% buffered rods were dissolved. At this time, the 100% buffered rods had shrunk/hydrolyzed to approximately their original size.

Electron micrographs were taken over the three-month study on an Amray-1000 scanning electron microscope (SEM). As shown in photos 1 to 4, the pores were bigger for those poly-

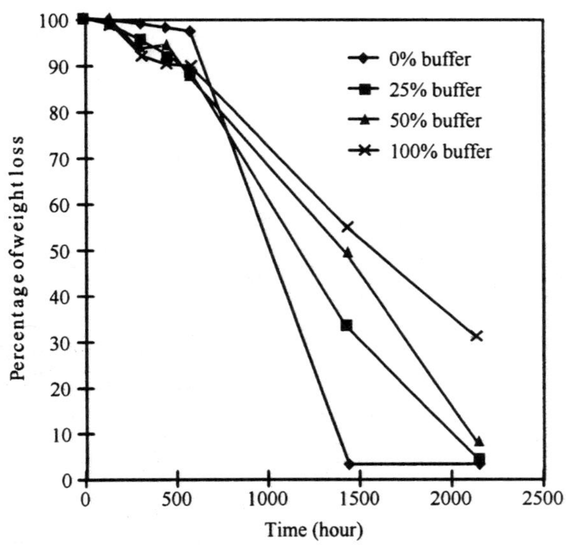

Figure 7 Weight loss over time on four different PLGA 50 : 50 rods.

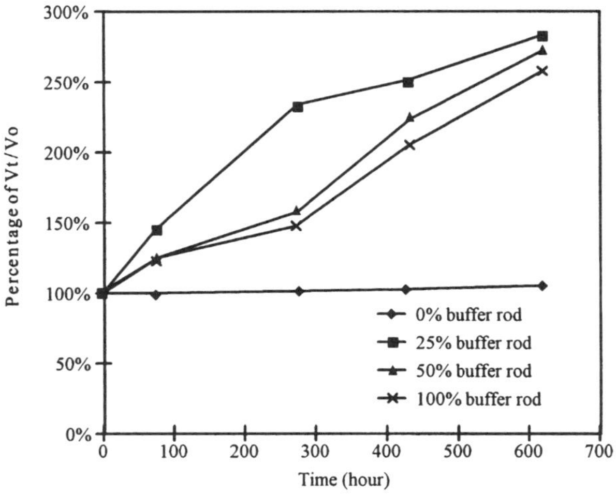

Figure 8 Volume change over time on four different PLGA 50 : 50 rods.

mer rods with a higher concentration of calcium carbonate. After three months, as the PLGA 50 : 50 was hydrolyzed, residuals were mostly calcium carbonate. This phenomenon was especially significant in the picture of the 100% buffered PLGA 50 : 50 rod.

Figure 9 is an SEM of the unbuffered rod at time point 90 days. Figure 10 is an SEM of the 25% buffered rod at time point 90 days. Figure 11 is an SEM of the 50% buffered rod at time point 90 days. Figure 12 is an SEM of the 100% buffered rod at time point 90 days.

B. HPLC Analysis Results

Over the three month time, monomer acid (glycolic acid and lactic acid) concentration increased with time for all of the polymer loadings, as shown in Fig. 13. This figure shows a

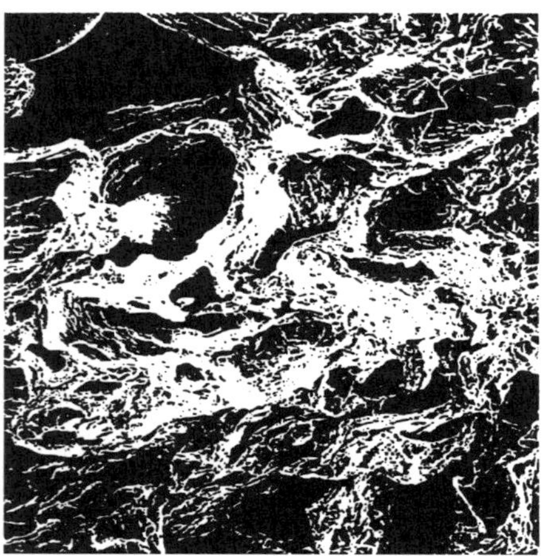

Figure 9 Scanning electron micrograph of the cross section of unbuffered rods, ×500, 5 kV at day 90.

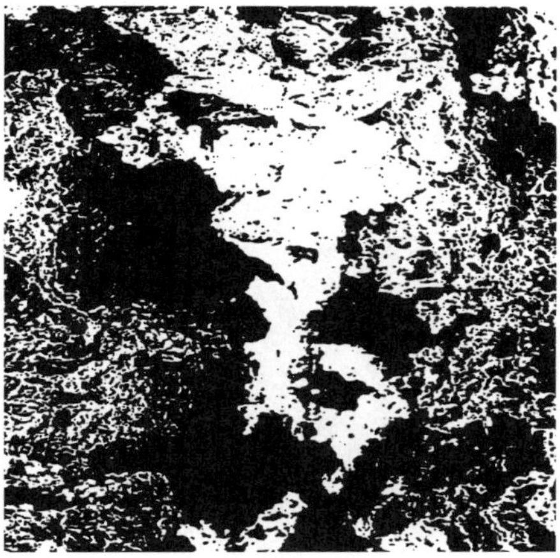

Figure 10 Scanning electron micrograph of the cross section of 25% unbuffered rods, ×500, 5 kV at day 90.

similar overall rate in release after hour 614 in all loadings. The more concentrated buffer correlates with a lower concentration of monomer acid present, due to the acidic neutralization.

VI. CONCLUSIONS AND RECOMMENDATIONS

Experimental results showed that pH drop due to the degradation of the biodegradable PLGA 50 : 50 polymer can be effectively controlled by incorporating a long-acting calcium carbonate buffer into the polymer. The buffer neutralizes the acid products as they are formed. Thus, the inflammatory response caused by the degradation of the fixation device should be reduced. As shown in Fig. 5, the pH value of the unbuffered PLGA 50 : 50 sample decreased sharply initially and at the 500–700 h range from pH 5.08 to pH 2.63. This drop in pH was caused by

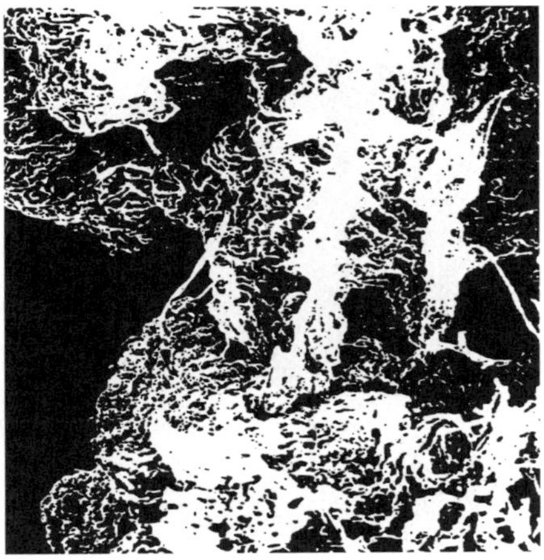

Figure 11 Scanning electron micrograph of the cross section of 50% unbuffered rods, ×500, 5 kV at day 90.

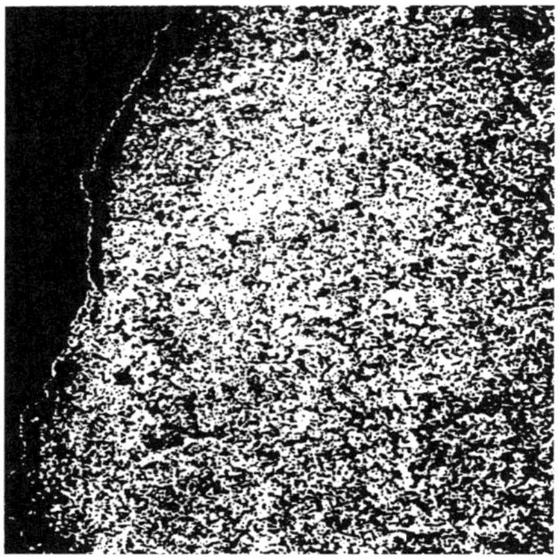

Figure 12 Scanning electron micrograph of the cross section of 100% unbuffered rods, ×500, 5 kV at day 90.

the acid production when PLGA 50 : 50 degraded and hydrolyzed, as shown by the following reaction:

$$—[CHR - CO_2]_n^- + H_2O \rightarrow nHO—CHR—CO_2H \tag{7}$$

Its pH value was significantly lower than that of 25%, 50%, and 100% buffered PLGA 50 : 50 samples. Distilled H_2O was run as a control; the distilled H_2O control shows a slight decrease in pH due to the absorption of CO_2 over the time span of the study.

The pH change over time was much smoother and slower for the buffered samples because of the buffering capacity of calcium carbonate. The neutralization reaction of calcium carbonate is

$$2HO—CHR—CO_2H + CaCO_3 \rightarrow Ca(HO—CHR—CO_2H)_2 + H_2CO_3 \tag{8}$$

and

$$H_2CO_3 \rightarrow H_2O + CO_2 \tag{9}$$

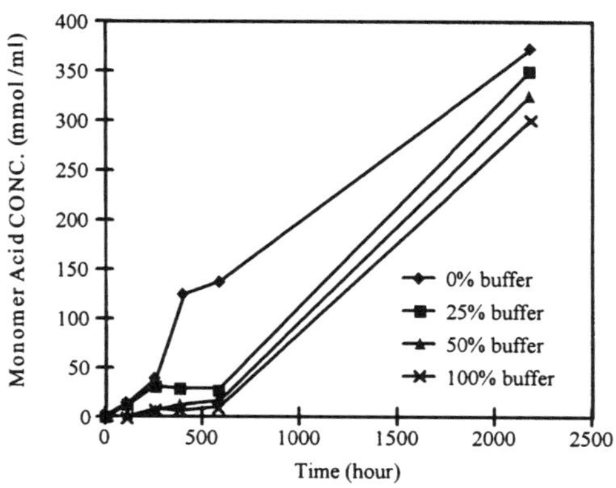

Figure 13 Monomer acid concentration over time on four different PLGA 50 : 50 rods.

The pH of the buffered PLGA 50 : 50 sample was controlled by hydrolysis and neutralization reactions, i.e., by the combination of the reaction given by Eqs. (7)–(9). These results indicated that the pH values of the PLGA 50 : 50 polymer samples were maintained within the controlled range over long periods of time by adding calcium carbonate buffers to the PLGA 50 : 50 fixation devices.

An increase in calcium carbonate buffer percentage improves the control of pH for PLGA 50 : 50 polymers. As shown in Fig. 5, the pH drop profile over the three month period became smoother as the buffer percentage increased. Thus, from the pH control standpoint, the 100% buffered PLGA 50 : 50 samples maintained the pH over a longer period of time than the 50% and 25% buffered PLGA samples. Higher calcium carbonate shifts both equations toward the right. More acid gets neutralized, so the pH can be maintained with in a well-controlled range.

The pH experiment results also confirmed that the pH change profile over time was that of a first order for unbuffered PLGA 50 : 50 polymers, as suggested by Kenley [12]. The pH change follows the equation $pH = 7.078 - 0.00208t$. The 100% buffered PLGA 50 : 50 polymer degradation rate was of a zero order, and it followed the equation: $-d[AC]/dt = 0.171$. The 100% buffered PLGA 50 : 50 pH change profile with time was also nearly a straight line. The excess calcium carbonate provides sufficient buffer capacity to prevent any pH drops beyond the controlled range for the experiment period. This demonstrated an extreme case for pH control.

The mechanical strength of both unbuffered and buffered PLGA 50 : 50 polymers decreased with time. The strength of buffered PLGA 50 : 50 samples was lower than that of the unbuffered samples. The decrease in mechanical strength of the samples was due to the degradation of the PLGA 50 : 50 polymers. With the disappearance of the PLGA polymer molecules, the support that these molecules once offered was gone. This was the primary reason for the decrease in strength of the unbuffered PLGA 50 : 50 samples. For the buffered PLGA 50 : 50 samples, the disappearance of calcium carbonate also contributed to the decrease in mechanical strength. One factor was that the calcium carbonate reacted with acid and the product dissolved into the solution. The strength provided by this calcium carbonate was hence lost. Another factor was that the reacted calcium carbonate left cavities inside the PLGA 50 : 50 sample rods as observed in the SEM photographs. The strength of the rods decreased because of the presence of these cavities.

The mechanical strength of the buffered rods seemed to decrease as the buffer percentage increased. We have to remember that the amount of PLGA 50 : 50 polymer in the sample rods decreases as buffer percentage increases. The original weight of all the samples was 3 mg. The amount of PLGA 50 : 50 polymer in the sample rods decreased as more calcium carbonate buffers were added to the sample rods, as shown in Table 8. This was demonstrated in Fig. 6. The strength of the original unbuffered PLGA 50 : 50 sample was the highest, followed by

Table 8 PLGA 50 : 50 and Calcium Carbonate Amount in the Sample Rods

Polymer rods	PLGA weight percentage	PLGA amount (mg)	Calcium carbonate weight percentage	Calcium carbonate amount (mg)
Unbuffered	100	3.0	0	0.0
25% buffered	83.6	2.5	16.4	0.5
50% buffered	72.2	2.2	27.8	0.8
25% buffered	56.5	1.7	43.5	1.3

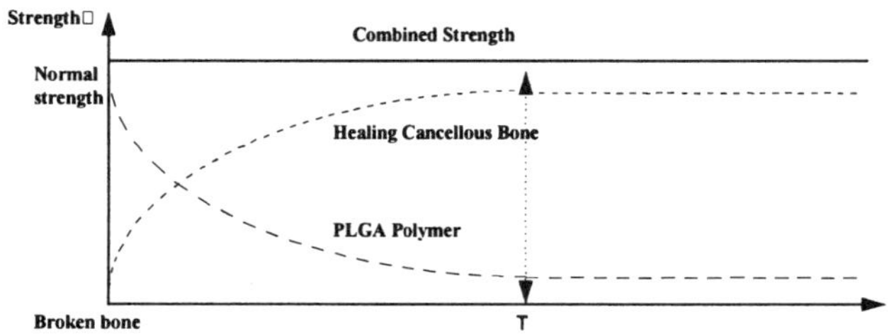

Figure 14 Combined strength of a PLGA fixation device and a cancellous bone.

25%, 50%, and 100% buffered PLGA 50 : 50 samples. As the samples degraded, the calcium carbonate reacted and disappeared. Cavities left by the calcium carbonate were formed: the higher the percentage of the calcium carbonate buffer, the more the cavities in the sample. This was another contributing factor. The cavities of lower buffer percentage rods seemed denser than those of the higher buffer percentage rods, as clearly observed in the SEM pictures. The different pore size distribution of the cavities might be another contributing factor.

Even though the mechanical strength of the PLGA 50 : 50 polymers degraded with time, the use of the PLGA 50 : 50 polymer biodegradable fixation devices is going to work well with lower strength bones such as cancellous bones. This is because when healing begins the strength of the bone starts to increase steadily. This is coupled with the gradual decrease in strength of the PLGA 50 : 50 fixation device; hence the combined strength of both the bone and the fixation device remains stable with time and is comparable to the strength of the healed bone (Fig. 14).

The swelling and shrinking of the diameters of the PLGA 50 : 50 polymer rods contributed to the irregular shapes of the strength versus time curve. As discussed in more detail in the next section, the diameters of the rods first swelled and then shrunk with time. Different percentage buffered PLGA rods swelled and shrunk at different time periods. The measured mechanical strength of the polymer rods was the average tensile strength across the rod cross section in units of MPa (pound/centimeter or lb/cm^2). Because of the changes in the rod cross section, the measured mechanical strength changed accordingly, thus contributing to the irregular shapes in the strength versus time profile curves.

An unexpected phenomenon occurred during the diameter measurement experiment. Instead of steady decrease in the strength and pH profile with time, the diameters of the rods first swelled and then shrunk. Different percentage buffered PLGA 50 : 50 rods swelled and shrunk at different time periods. Accurate data is not available; however, Fig. 15 represents a

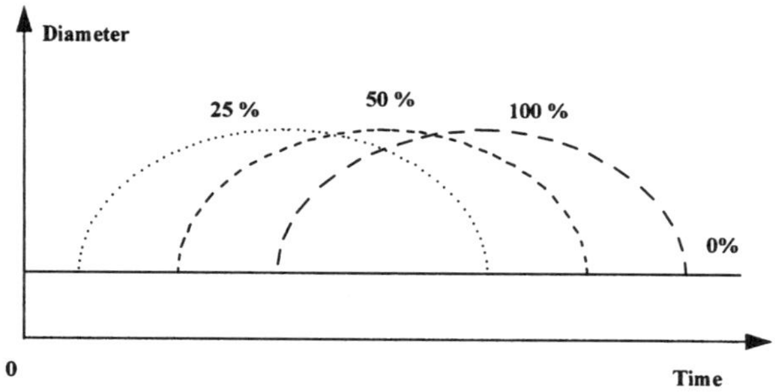

Figure 15 Diameter swell and shrink with time.

general profile of the phenomena and efficiently replaces the direct plots obtained from the actual data.

The cause of this unexpected phenomenon might be explained as follows. There are more small molecule polymers in the lower percentage buffered polymer rod. These smaller polymer molecules are easily degraded and hydrolyzed at first. The hydrolysis reaction produces carbon dioxide which evaporates and escapes from the polymers. This process is similar to the effect of opening a soda can or a champagne bottle. This process might enlarge the pores and cross sections of the polymer rods and make the diameter of the rods swell at first. Because the higher buffer percentage polymer samples had less small molecule polymer, the swelling step occurred later as larger molecules were hydrolyzed and released carbon dioxide gas. Because both PLGA 50 : 50 polymer and calcium carbonate reacted and disappeared, the total mass of the rods decreased. This resulted in the decrease of the diameter of the polymer rods. The change in the rod diameter was influenced by both carbon dioxide gas escape factor and total mass reduction factor. As smaller polymer molecules hydrolyzed, the release rate of the carbon dioxide slowed, which in turn slowed the swelling rate of the rods. In the meantime as more surface areas inside the polymer rods were exposed to the solution, polymer and calcium carbonate dissolving rates accelerated. At a certain time point, both factors contributed equally to the increase in the rod diameter, and the diameter reached its peak value. After that time, the total mass reduction force prevailed and resulted in a combined effect of diameter shrinkage. This theory explained the swelling and shrinking of the rod diameter, even though experimental data are not available.

The pH value of biodegradable PLGA 50 : 50 polymer rods was well controlled by incorporating a long-acting calcium carbonate buffer. The 100% buffered polymer worked better than the lower percentage buffered polymers. Unfortunately the strength of the polymer decreased with an increase in buffer percentage. Thus the 100% buffered polymer was found to be the weakest. Taking into account both the pH and the strength factors, it is recommended to use a 50% buffered PLGA 50 : 50 polymer as it may satisfy both the pH control and the strength requirements. Because the strength of bones such as cancellous bone increases after injuries, as demonstrated previously, the combined mechanical strength of both PLGA 50 : 50 polymer and the cancellous bone may be found to be stable during the healing process. It is recommended to use buffered PLGA 50 : 50 polymers in the treatment of low strength bones (such as cancellous bone).

NOMENCLATURE

B^-	Buffer anion
L^-	Lactate (or glycolate) anion
K_a	Ionization constant
K_{eq}	Equilibrium constant
MW	Molecular weight
w_0	Initial weight of rod (mg)
w_t	Weight of rod at time t (mg)
GPC	Gel permeation chromatography
HPLC	High performance liquid chromatography
PGA	Polyglycolic acid
PLA	Polylactic acid
PLGA	Polylactic-*co*-glycolic acid

T_g	Glass transition temperature (°C)
T_m	Glass melting temperature (°C)
THF	Tetrahydrofuran
TW	Total weight (mg)
TA	Total acid concentration of final supernatant (mg/ml)
t	Time (hour)
V_t	Volume of rod at time *t* (cm^3)
V_o	Volume of rod at initial time (cm^3)

APPENDIX

Sample Calculation

Percentage Buffer

The calculation of the loading of $CaCO_3$ in the polymer matrix required to neutralize a given quantity of PLGA 50 : 50 as following:

PLGA 50 : 50 repeating mer unit is

$$—[—O—CH(CH_3)CO]—[O—CH_2—CO]—$$

The monomer residue molecular weight for lactide is 72, and for glycolide is 58. Thus the average residue molecular weight is $72 \times 0.5 + 58 \times 0.5 = 65$. When PLGA 50 : 50 react with $CaCO_3$, it is following equation:

$$2RCO_2H + CaCO_3 \rightarrow (RCO_2^-)_2Ca + H_2O + CO_2$$

In 1 g of PLGA 50 : 50 there will be generated $1/65 = 0.0154$ moles of monomeric acids. Thus it need $0.5 \times 0.0154 = 0.0077$ moles of $CaCO_3$, and the $CaCO_3$ molecular weight is 100, which weighs is $0.0077 \times 100 = 0.77$ g. Thus the loading will be

$$0.77/(1.0 + 0.77) = 43.5\%$$

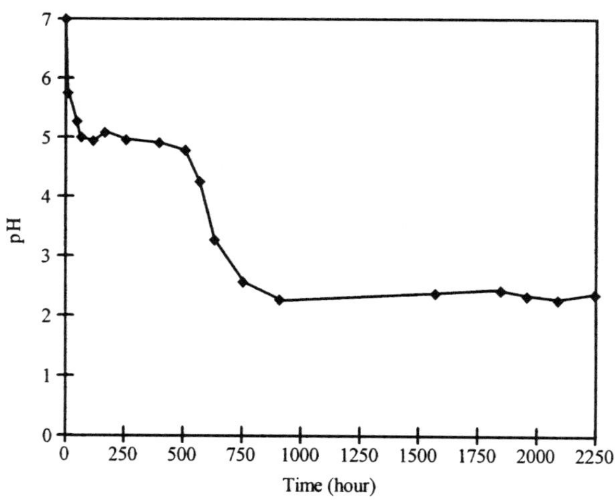

Figure A1 Unbuffer rods pH change vs. time.

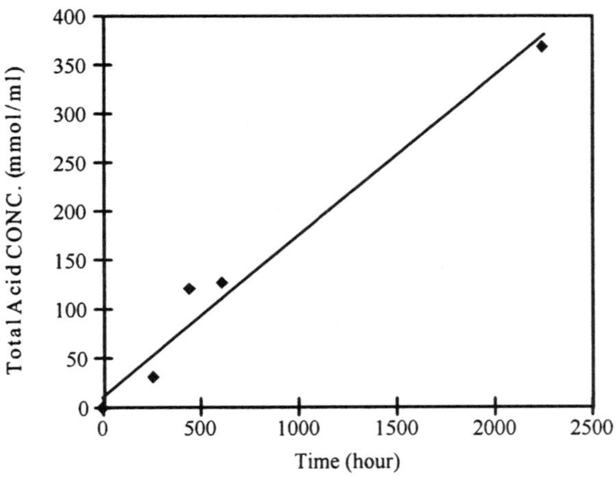

Figure A2 Total acid conc. of 100% buffer rods vs. time.

by weight. The same way calculate to neutralize the 25% of the acid formed and 50% of the acid formed. The $CaCO_3$ loading can be reduced to 16.4% and 27.8% by weight.

Kinetic Study

The unbuffer PLGA 50 : 50 rods were hydrolyzed in the water and it follows the equation:

$$-[-CHR-CO_2^-]_{n-} + H_2O \rightarrow n HO-CHR-COOH$$

The pH measurement directly shows us the change of sample concentration. From Fig. A1 the pH drop is linear at 500 h to 700 h, We know the $pH = -\log[H^+]$, so that tells us that the degradation rate is first order at this time and it follows the equation: $pH = 7.078 - 0.000208t$. This result is also confirmed by Kenley's (1987) study.

The 100% buffer PLGA 50 : 50 rod's were first hydrolyzed in the water, and then neutralized by $CaCO_3$ by following the equations:

$$-[CHR-CO_{2-}]_{n-} + H_2O \rightarrow n HO-CHR-COOH$$

and

$$2HO-CHR-COOH + CaCO_3 \rightarrow Ca(HOCHRCO_2)_2 + H_2O + CO_2$$

We assumed that the neutralization of RCOOH is much faster than hydrolysis of PLGA 50 : 50, so we can plot the concentration of total acid versus time to get degradation rate. From Fig. A2, it shown that the reaction is zero order, and it follows the equation: $-d[AC]/dt = 0.171$.

REFERENCES

1. ME Muller, M Allgower, and R Schneider. *Manual of Internal Fixation. Techniques Recommended by the AO Group*, Ed. 2, New York, Springer, 1979.
2. EJ Frazza and E Schmitt. A new absorbable suture, *J. Biomed. Mater. Res. Symp.*, **1**, p. 43, 1971.
3. DK Gilding and AM Reed. Biodegradable polymers for use in surgery-polyglycolic/poly(acetic acid) homo- and copolymers 1, *Polymer*, **20**, p. 1459–1464, 1979.

4. JO Hollinger and GC Battistone. Biodegradable bone repair materials, synthetic polymers and ceramics, *Clin. Orthop.*, **207**, p. 290–305, 1986.
5. AU Daniels, MK Chang, and KP Andrano. Mechanical properties of biodegradable polymers and composites proposed for internal fixation of bone, *J. Appl. Biomater.*, **1**, p. 57–78, 1990.
6. C Chu. Hydrolytic degradation of polyglycolic acid tensile strength and crystallinity study, *J. Appl. Poly. Sci.*, **26**, p. 1727–1734, 1981.
7. DC Tunc. A high strength absorbable polymer for internal bone fixation, *9th Annual Meeting, Society for Bioamaterials*, **6**, p. 47, 1983.
8. O Bostman, S Vainiopaa, and E Hirvensalo. Biodegradable internal fixation for malleolar fractures, *J. Bone Joint Surg.*, **69**-B:4, p. 615–620, 1987.
9. JB Herrman, RJ Kelly, and GA Higgins. Polyglycolic acid sutures, laboratory and clinical evaluation of a new absorbable structure material, *Arch. Surg.*, **100**, 1970.
10. DF Williams. Enzyme hydrolysis of polylactic acid, *Eng. Med.*, **10**, p. 5, 1981.
11. SJ Holland, BJ Tighe, and PJ Gould. Polymers for biodegradable medical devices, *J. Control. Rel.*, **4**, p. 155, 1986.
12. RA Kenley, MO Lee, and TR Mahoney. Poly-lactide-co-glycolide decomposition kinetics in vivo and in vitro, *Macromolecules*, **20**, p. 2398–2403, 1987.
13. CG Pitt. Polycaprolactone and its copolymers, M Chasin, R Langer (Eds.), *Biodegradable Polymer*, p. 97–103, Marcel Dekker, New York, 1990.
14. CG Pitt, M Gratzel, and GL Kimmel. Aliphatic polyesters, 2. the degradation of poly-*dl*-lactide, polycaprolactone and their copolymers in vivo, *Biomaterials*, **2**, p. 215, 1981.
15. AJ Domb, CT Laurencin, O Israeli, TN Gerhart, and R Langer. The formation of propylene fumarate poligomers for use in bioerodible bone cement composites, *J. Polym. Sci.*, **A(28)** 5, p. 973–985, 1990.
16. RA Miller, JM Brady, and DE Cutright. Degradation rates of oral resorbable implants: rate modification with changes in pla/pga copolymer ratios, *J. Biomed. Mater. Res.*, **11**, p. 711–719, 1977.
17. D Cutright, and E Hunsuck. The repair of fractures of the orbital floor using biodegradable polylactic acid, *Oral Surg.*, **33**, p. 28–34, 1972.
18. TR Tice, DH Lewis, RL Dunn, WE Meyers, RA Casper, and DR Cowsar. Biodegradation of microcapsules and biomedical devices prepared with resorbable polyesters, *Proc. Int. Symp. Control Rel. Bioact. Mater.*, **9**, p. 210, 1982.
19. C Shih. Chain-end scissionin acid catalyzed hydrolysis of poly(d,l-lactide) in solution, *J. Controlled Release*, **34**, p. 9–15, 1995.
20. P Christel, and F Chabot. Biodegradable composites for internal fixation. In *Biomaterials* 1980, p. 271–280, New York, Wiley, 1982.
21. J Brady, D Cutright, and R Miller. Resorption rate, route of elimination, and ultrastructure of the implant site of polylactic acid in the abdominal way of the rat, *J. Biomed. Mater. Res.*, **7**, p. 155–166, 1973.
22. S Vainiopaa, K Vihtonen, M Mero, and A Patiala. Fixation of experimental osteotomies of the distal femur of rabbits with biodegradable material, *Arch. Orthop. Traumat. Surg.*, **106**, p. 1–45, 1986.
23. EA Makela. Healing of epiphysis fracture fixed with a biodegradable polydioxanone implant or metallic pins. an experimental study on growing rabbits, *Clin. Mater.*, **3**, p. 61–71, 1988.
24. P Lyndrup, and S Andresen. Fixation of experimental osteotomies of the human patella with biodegradable material versus tension band, *Acta Orthop.*, *Scand.*, **Suppl. 231**, p. 11, 1989.
25. P Rokkanen, O Bostman, S Vainiopaa, and K Vihtonen. Biodegradable implants in fracture fixation: early results of treatment of fractures of the ankle, *Lancet*, **1**, p. 1422–1424, 1985.
26. O Bostman, E Makela, and P Tormala. Transphyseal fracture fixation using biodegradable pins, *J. Bone Joint Surg.*, **71**-B(4), p. 706–707, 1989.
27. O Bostman, E Hirvensalo, A Vainiopaa, A Makela, and K Vihtonen. Degradable polyglycolide rods for the internal fixation of displaced bimalleolar fractures, *Internal. Orthop.*, **14**, p. 1–8, 1990.
28. O Bostman, E Hirvensalo, and J Makinen. Foreign body reactions to fracture implants of biodegradable synthetic polymers, *J. Bone Joint Surg.*, **72**-B(4), p. 592–596, 1990.

29. E Hirvensalo, and O Bostman. Biodegradable fixation of intraarticular fractures of the elbow joint, *Acta Orthop. Scand.*, **Suppl. 227**, p. 78–79, 1988.
30. E Hoffmann, C Krettek, N Haas, and H Tscherne. Die distale radiusfraktur. frakturstabilisierung mit biodegradablen osteosynthes stifen(biofix), *Exp. Untersuch. klinische Erfahrungen, Unfallchirurgie*, **92**, p. 430–434, 1989.
31. EK Partio, O Bostman, E Hirvesalo, and H Patiala. The indication for the fixation of fractures with totally absorbable SR-PGA screws, *Acta Orthop. Scand.*, **Suppl. 237**, p. 43–44, 1990.
32. OE Illi, H Weigum, and F Misteli. Biodegradable implant materials in fracture fixation, *Clin. Mater.*, **10**, p. 69–73, 1992.
33. R Bos, G Boering, and F Rozema. Resorbable polylactide plates and screws for the fixation of zygomatic fractures, *J. Oral Maxillofac. Surg.*, **45**, p. 751–753, 1987.
34. T Yamamuro, A Matsusue, and A Uchida. Bioabsorbable osteosynthetic implants of ultra high strength polylactide: a clinical study, *Intern. Orthop.*, **18**, p. 332–340, 1994.
35. O Bostman, E Hirvensalo, A Vainiopaa, A Makela, and K Vihtonen. Ankle fractures treated using biodegradable internal fixation, *Clin. Orthop.*, **238**, p. 195–203, 1989.
36. J Eitenmuller, A David, and A Pommer. Die versorgung von sprungglenksfrakturn unter verwendung von platten und schraubeb aus resorbierbarem polymermaterial, *J. Deut. Gessel. Unfall.*, Berlin, Nov. 22, 1989.

24

In Vitro Study of Buffered Biodegradable Fracture Fixation Devices

Yankai Zhang, Donald L. Wise, and Yung-Yueh Hsu
Northeastern University, Boston, Massachusetts

I. BACKGROUND

Currently, the vast majority of the fracture fixation devices are made of metal composites. The use of metallic materials in the design of fracture fixation devices has traditionally been associated with early reduction and stabilization of the fracture site. With the advent of bioabsorbable materials, an obvious synergy exists between the use of such polymers in fracture fixation devices and the body's natural course of healing.

Despite their creditable performance in the repair of fractures, metallic implants have a significant disadvantage. Because of the irritation of the soft tissues following the healing of a fracture, a subsequent surgical procedure is usually performed to remove the nondegradable metal devices. Also because of their high strength and high stiffness, the load transfer from the metal devices to the healing bone is very limited, and thus may lead to refracture after the removal of the metal devices. Because of such disadvantages of metal devices, the application of absorbable polymeric implants, which would negate the need for subsequent removal of the metal devices and lead to better load transfer, could offer major clinical advantages for the fixation of fractures.

The resorbable, tissue compatible biopolymers such as polyglycolic acid (PGA), polylactic acid (PLA) and their copolymer (PLGA) have had a long history of clinical application. However, the materials of fixation devices like PGA, PLA and PLGA have not yet been widely accepted because of late sterile inflammation during the healing process and these biopolymers' inability to replicate the mechanical strength of their metal counterpart. Some reviews indicate that in more than 8% of the cases where PLA, PGA and PLGA were used, sterile abscess appeared; such abscess became evident only after fairly a long induction period ranging from 7–20 weeks (mean induction period of about 12 weeks). Degradation studies of PLGA indicate that although the decrease of molecular weight begins fairly early, other phenomena including a pH drop in the surrounding medium, loss of polymer weight, and the appearance of the monomeric hydrolysis products (lactic acid, glycolic acid) show an induction period, the length of which depends on the comonomer ratio in the polymer. The induction period and the point in time at which the inflammation appears closely coincide. This strongly suggests that delayed inflammatory response is caused by the accumulated hydrolysis products of the polymer. This hypothesis is also supported by the observation that there is no difference between the rate of in vitro hydrolysis and in vivo hydrolysis.

Early attempts to develop biodegradable fixation devices were based on the use of implants which were manufactured by traditional melted molding techniques (like extrusion and injection molding). These nonreinforced biodegradable polymeric implants cannot fulfill the demands of relatively high strength value and sufficient strength retention time. Therefore, improvements in these devices must be made. This study focused on the following: (1) Minimizing or eliminating inflammatory response caused by accumulation of degradation products (by use of acceptable buffers) and (2) improving the mechanical properties of biodegradable devices by fiber reinforcement.

There are more than 1 million bone fractures in the United States each year, of which, nearly half need some type of internal fixation device to stabilize the fracture during the healing process. These devices, which include plates, pins, rods and screws, are used for rigid internal fixation of bone which has been damaged by trauma, or reconfigured surgically to correct defects occurring congenitally, or reconfigured as a result of disease. Implants for the internal fixation of the fracture have a common property: there are needed only temporarily until the fracture has united. Currently the vast majority of these devices are made of metal composites. The use of metallic materials in the design of fracture fixation devices has traditionally been associated with early reduction and stabilization of the fracture site. Despite their creditable performance in the repair of fracture, metallic implants have significant disadvantages:

1. Mismatch of elastic modulus between metal and human bone. Although surgical grade metals and metal alloys possess high strength and excellent mechanical property retention, the mismatch of the elastic modulus between these materials and human bone is a major concern. Many studies have shown [1] that completion of healing is prevented by highly rigid fixation since much of the load that is normally carried by the bone is transferred across the fracture site by the implants. This is due to the elastic modulus mismatch between bone and metals:

   $$E_{\text{bone}} = 6\text{–}20\ \text{GPa},\ E_{\text{metal}} = 100\text{–}200\ \text{GPa} \qquad [2]$$

   This leads to a stress protection of the fractured bone while rigid material fixation prevents proliferation of periosteal callus and deprives the bone of its normal stress. This impairment of fracture healing may lead to delayed union of bone, nonunions and refractures after implant removal. The stiffer the material is, the more frequent is the development of delayed union and nonunions.
2. Metal implants have to be removed. Metal implants might cause chronic irritation of the surrounding soft tissues and need to be removed after complete healing. Thus, the patient's second hospitalization and operation are necessary with all the personal, medical and social consequences.

Apart from the above-mentioned disadvantages of the metal implants, research has led to a high standard of metal implant, and acceptable alternatives are still hard to find. Because of the disadvantages of metal implants, a great number of research groups in various countries have been working for more than two decades to develop recommendable biodegradable implants in orthopedic surgery.

The ideal situation for fracture fixation would be to develop an implant that can initially impose the necessary degree of fracture stabilization to promote rapid fracture union, and then gradually and biocompatibly absorb into the body, allowing for controlled stress transfer to the healed fracture and gradual transitioning of the bone toward its normal state following fracture healing. This would promote rapid fracture union while eliminating the occurrence of bone resorption due to postunion stress shielding [3,4] and, more importantly, would also eliminate the second operation for implant removal following complete healing at the fracture site.

Many organic macromolecular compounds are degradable and absorbable in the body, but very few possess the chemical and physical properties necessary for an internal fracture fixation device [5]. The best suited are α-polyesters such as polyglycolic acid (PGA), polylactic acid (PLA), and various copolymers of PGA and PLA. These polyesters have a long history of clinical use because they are biodegradable, tissue compatible and exhibit moderate strength in tension, compression and bending, and their elastic moduli are quite close to that of human bone ($E_{polymer}$ = 4–30 GPa, E_{bone} = 6–20 GPa) [1].

A. Characteristic of the Polyesters Used

Chemically, poly(lactic acid), poly(glycolic acid) and poly(lactic-*co*-glycolic acid) are all α-polyesters. Their chemical formula are

$—[CH_2—(CO)—O—]_{n}—$	poly(glycolic acid)
$—[CH(CH_3)—(CO)—O—]_{n}^{-}$	poly(lactic acid)
$—[CH_2—(CO)—O—]_{m}$	poly(lactic-*co*-glycolic acid)
$^{-}[—CH(CH_3)—(CO)—O—]_{n}—$	

Glycolic acid and lactic acid are the common names of the first two α-hydroxycarboxylic acids, α-hydroxyacetic and α-hydroxypropionic acid. The structures of these acids are

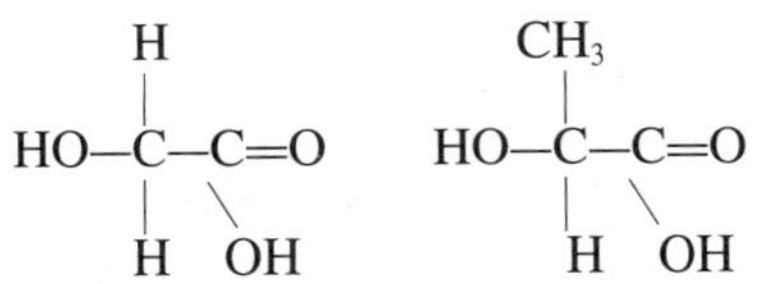

(a) glycolic acid (b) lactic acid

There are two routes to produce these polyesters: one is direct polymerization from linear monomer or oligomers, and the other is from cyclic dimers. The latter route can obtain the polyesters with higher molecular weight and therefore is used in most applications.

The polymerization via cyclic monomer is as follows:

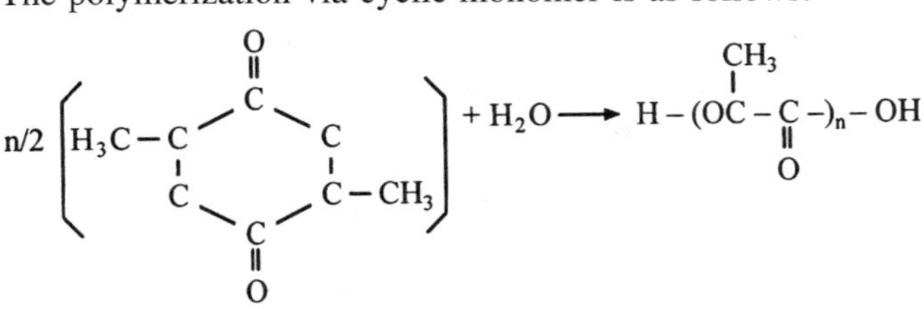

This polymerization is usually conducted over 200°C with the assistance of a catalyst such as stannous chloride and stannous octoate [6]. Various PLGA copolymers can be obtained by varying the molar ratio of lactic to glycolic acid monomers or cyclic dimers.

The α-polyesters are water insoluble but degradable by hydrolysis to the monomers, lactic acid and glycolic acid, which are water soluble. The degradation products (monomers) then are eliminated via metabolism.

PGA is the simplest linear α-polyester of the absorbable polymers and is very hydrophilic. It has a very high degree of crystallinity. PLA has more hydrophobic character than

PGA due to substitution of a methyl group in the place of a hydrogen atom. Because of the asymmetry, lactic acid exists in two forms: the optically active levo (L) enantiomer and the racemic mixture of dextro (D) and levo enantiomers. Polymerization of the two forms results in poly(L-lactic acid) and poly(D,L-lactic acid), referred to subsequently as PLLA and PDLLA (for copolymers, PLLGA and PDLLGA). Their mechanical properties and degradation rates differ since PLLA is predominantly crystalline and highly resistant to hydrolysis, whereas PDLLA is mainly amorphous and more susceptible to hydrolysis.

Biodegradable materials have long a history of clinical use. Surgical sutures are one of earliest clinical implants. The most widely used materials for absorbable fracture fixation devices are those that were initially developed for absorbable sutures and made from α-polyesters: PGA, PLA, and polyparadioxanone (PDS) [7]. Polyglycolic acid has been known since 1954 to be a potentially low cost, tough, fiber-forming polymer [6]. In 1962, PGA was developed as the first synthetic absorbable suture and given the commercial name Dexon by American Cyanamid Co. Du Pont considered the polymer of homologous alpha hydroxy acid, PLA, for the same application. Since 1970, PGA has been commercially available as the surgical suture Dexon. The copolymer of 92% GA, 8% LA has had a limited application as the competitive suture, Vicryl, since 1975 [7].

During the 1970s, PGA, PLA and their copolymers were widely investigated as biodegradable materials in dental, orthopedic, and drug delivery applications [5]. The first use of PGA polymer for fracture fixation devices was suggested by Schmitt and Polistina [7]. Some of the first reported uses of absorbable polymers for fracture fixation devices were conducted in 1971 by Kulkarni et al., who investigated the use of PLA rods, screws and plates to treat mandibular fractures in dogs. From these studies, it was quickly realized that one of the primary limitations/problems of these materials was their lack of strength. This led to the development of fiber reinforcement of the absorbable materials. Christel et al. are apparently the first to report the use of fully absorbable composites with PGA fiber-reinforced PLA plates for fracture fixation [7]. Numerous investigations followed, and by 1990 over 40 different formulations of absorbable polymers, copolymers, and their composites were developed for fracture fixation. Some of the formulations are now commercially available in the form of pins for cancellous bone fixation.

Although α-polyesters as the materials of absorbable fracture fixation devices have been investigated for more than two decades, and some of the devices have been commercially available, there are still some problems that hamper the widespread use of these devices in modern traumatology. First, they have relatively low initial strength and poor strength retention profile. These lead to the limitation of the use of these polymers to only lower load-bearing bone such as cancellous bone, but not high load-bearing bone such as cortical bone. Second, they may cause delayed noninfectious inflammatory response requiring operative drainage. These challenges will be discussed in detail in the next section.

B. Biodegradation Mechanism and the Inflammatory Response

Chemically, the degradation of the biodegradable polymers is simply depolymerization (reverse direction of polymerization of linear monomers):

$$—[—CH(CH_3)—CO—O—]— + H_2O \Rightarrow HO—CH(CH_3)—COOH$$

Several investigators have demonstrated that implants of PGA, PLA are completely absorbable within bone tissue and that new bone is deposited on and within the implant as degradation proceeds [8,9]. This process behaves as bulk erosion. First, water molecules diffuse into the polymer bulk. Then polymer chains are cleaved randomly through chain scission into relatively smaller molecules until the point where chain segments are small enough to diffuse out of the polymer bulk. While the polymers continues to degrade, they are eliminated by tissue metabolism. The differences in the final metabolism of the polymers are slight, but the rate of degradation varies. It is affected by many factors, the most important of which are

1. The chemical composition. Degradable synthetic α-polyesters of different molecular weight, or copolymers of different molar ratio and sequence length, exhibit markably different degradation rates. A general ranking in terms of decreasing degradation rates is [10]:

   DL.PLGA > PGA > DL.PLA > L.PLA

   The rate of degradation of PLGA copolymers is dependent on the molar ratio of lactide to glycolide. Increasing lactide content generally results in slowing the degradation of copolymers since the methyl group on the repeating lactide units lower the chemical reactivity of the ester group by steric hindrance [10]. PGA degrades more quickly than PLA due to a combination of (a) increased hydrophilicity; (b) a lower glass transition temperature T_g (36°C for PGA, 57–67°C for PLA) which results in increased chain flexibility for PGA at body temperature; and (c) the lactide ester group results in lower reactivity (increased hydrolysis resistance) [11–13].
2. The implant size and shape. Just as a porous thin sheet degrades much more rapidly than a dense block, the implant with greater surface area and porosity will degrade more rapidly than an implant with low surface area.
3. The morphology of the polymer, which is influenced by molecular weight, stereochemistry, and degree of crystallization. The polymers with higher molecular weight will have a slower degradation rate in a specified range, but at very high molecular weight (over 500,000) this will not be a decisive factor for the degradation rate.

The degradation rate of polyester is also related to the extent of crystallinity. Chu [14] points out that PLGA have both crystalline and amorphous regions, and thus hydrolytic degradation proceeds through two main stages. First, water diffuses into the amorphous, but not crystalline, regions. Hydrolysis starts in the amorphous regions. Chain scission increases the mobility of new chain ends, enabling them to reorganize into a more ordered state, thereby actually increasing the degree of crystallinity. Following hydrolysis of the amorphous regions, the second stage is characterized by hydrolysis of the now much more accessible crystalline regions.

Lewis [15] has summarized degradation times for PLGA as shown in Table 1.

Miller [16], using differential scanning calorimery, rated the crystallinity of various

Table 1 Degradation Time of PLGA

Polymer	Approximate degradation time (months)
Poly(L-lactic acid)	18–24
Poly(D, L-lactic acid)	12–16
Poly(glycolic acid)	2–4
Poly(D,L-lactide-*co*-glycolide)–50 : 50	2
Poly(D,L-lactide-*co*-glycolide)–85 : 15	5

Table 2 Crystallinity of PLGA

Glycolide/lactide ratio	Crystallinity
100 : 0	high
75 : 25	moderate
50 : 50	low
25 : 75	low to moderate
0 : 100	moderate

PLGA as follows: By comparing the information in Table 2 with that in Table 1, it is generally true that the higher degree of crystallinity will lead to a slower degradation rate.

4. The processing history. The processing condition such as the condition used to form the device, the storage history, processing temperature, purification process, etc., could also affect the rate of degradation.

The most important factor that affects mechanical strength of the polymers is the degree of orientation of the polymer chain (the degree of alignment of the crystalline and amorphous regions as the result of processing). The higher degree of molecular orientation leads to much better mechanical strength. Tormala [17] showed that significant improvement was achieved through self-reinforcement process to manufacture PLA and PGA samples (Table 3).

Many factors that affect the degradation rate of the implants also affect the mechanical property retention time. Such factors include chemical composition (monomer type, homopolymer, copolymer, or polymer blend), the isomerism or stereochemistry (the difference in structure arrangement), the molecular weight (a measure of chain length of the molecules that make up the polymer), the sequence distribution and arrangement (chain form, such as "random," "block," or "branch"), the crystallinity of the material, the device size and shape (different surface area of the device), and the processing history (storage, handling, packaging, sterilization, etc.). The effect of these factors is consistent with the effect of these to the degradation rate; that is, the faster the degradation rate of a polymer, the faster the declining of the mechanical strength of the polymer, and then the shorter the strength-retention profile. Besides, the degree of orientation also affects the mechanical retention profile. In general, a more oriented polymer retains its strength longer than one with a more amorphous structure [9].

The rate of degradation and absorption of the acidic breakdown products at the implant site constitutes a critical factor for tissue acceptance of an absorbable implant. According to Bostman [18], "inflammatory foreign body reactions, which include a discharging sinus without infection, have been encountered in nearly all clinical studies published on the use of absorbable fracture fixation implants made from polyglycolide (PGA)." A review of 516 pa-

Table 3 Comparison of Strength for SR- and IM-PLA/PGA

Material	Deformation mode	Bending strength (MPa)	Bending modulus (GPa)
SR-PGA	Free drawing	415	18
IM-PGA-		218	7
SR-PLA	Die drawing	300	10
IM-PLA-		119	3

tients by Bostman [19] with the implants of PGA and PLGA revealed that a late inflammatory response requiring operative drainage occurred in 7.9% of the cases. He comments: "the unique complication of these implants is the delayed inflammatory reaction. The clinical presentation of this complication is fairly consistent. The patient has no local or systematic sign of the problem with the wound in the immediate post operative period, and then a painful erythematous, fluctuant swelling suddenly develops about the healed wound. The mean interval between fixation of the fracture and the clinical manifestation of the reaction is 12 weeks (range from 7–20 weeks). This reaction is not limited to PGA or PLGA. Eitenmuller reported on 19 patients with ankle fractures treated with absorbable plates and screws of polylactide. Of these 19, nine (47.7%) had an inflammatory response.

The delayed appearance of an inflammatory response may be associated with the kinetics of the polymer degradation. Degradation studies of solid PLGA indicate that although decrease of molecular weight begins fairly early, other processes including pH drop in the surrounding medium, loss of polymer weight, and the appearance of monomeric hydrolysis products (lactic and glycolic acids) show an induction period, the length of which depends on various factors. Coombes [10] indicated that "biodegradation proceeds by a mechanism which involves random hydrolytic chain scission of ester linkages until the molecular weight has decreased to the point where chain segments are small enough to diffuse from the polymer bulk. . . . Chain cleavage is auto-catalyzed by carboxy end groups generated and is proportional to the ester and water concentration. . . . The presence of residual monomers, low molecular weight compounds and oligomers increases chain flexibility (plasticity offset), polymer hydrophilicity and the concentration of carboxylic acid groups." These can be confirmed by the factor that at the beginning period, little degradation can be observed, followed by a progressive increase of the degradation rate. These results of degradation studies and clinical observations strongly suggest that the delayed inflammatory response is a consequence of the delayed appearance of acid hydrolysis products. This hypothesis is supported by the observation [20] that PLGA degradation occurs at similar rates in both in vivo and in vitro degradation.

II. EXPERIMENTAL DESIGN AND PROCEDURE

The purpose of the current study was to develop and evaluate a biodegradable implant material which addresses the concerns and shortcomings of those devices which are currently used in osteosynthesis. Specifically, the project attempted to design a product that has an ideal strength retention curve while limiting the inflammatory response by altering the local environment within and around the implant.

Many devices made of PGA homopolymers demonstrate rapid loss of in vivo strength, leaving the fracture site unprotected, and ultimately leading to refracture [21–23]. Alternatively, remnants of some pure PLA (especially PLLA) implants have been identified up to five years after implantation [22,23], raising the question as to whether PLA is "too biostable" to be used as a biomaterial. Comparing with PGA and PLA, some copolymers of PGA and PLA have a moderate degradation rate and strength retention time (Table 2), and thus can be considered as ideal materials. PLGA(85 : 15) is one of such materials and is selected as the major material of the degradable fixation device in this project.

As discussed previously, the noninfectious inflammatory response is caused by the accumulation of degradation products, namely, lactic acid and/or glycolic acid. By incorporating a long-acting buffer into the fixation device matrix, the inflammatory response can be minimized or even eliminated.

An appropriate buffer should be retained within the polymer matrix and have no foreign body reaction with tissue, so it must be fairly insoluble as well as tissue compatible. The basic component of the buffer (the anion) should react easily with the protons of the acid products of hydrolysis. Because slow generation of the acid products at end of the rapid phase may not cause clinical problems, it may not be necessary to provide a total buffering capacity. Therefore the amount of the buffer utilized, the buffering capacity, should be optimized so that the pH change of the microenvironment of the implant site can be minimized while the mechanical properties of the device will not be compromised significantly.

Letting B^- represent the buffer anion and L^- the lactate/glycolate anion, the equilibrium can be expressed as

$$HL + B^- \Leftrightarrow L^- + HB$$

In order to keep reaction toward the right side, HB must have a much lower dissociation constant than HL:

$$K_a^{HB} << K_a^{HL}$$

Dibasic calcium phosphate ($CaHPO_4$, DCP) and hydroxylapatite ($Ca_{10}(OH)_2(PO_4)_6$, HAp) are acceptable additives. They have a long history of being used in dental clinical and bone cement, etc. [24], so they are tissue compatible. They both have low aqueous solubility ($K_{sp} = 2.6 \times 10^{-45}$ for HAp) and will not be dissolved in the water, so they meet the criteria of being long acting. Most importantly their dissociation constants are much lower [25]:

$$HPO_4^{2-} \Leftrightarrow H^+ + PO_4^{3-} \qquad K_{1a} = 4.2 \times 10^{-13}$$

$$H_2PO_{4-} \Leftrightarrow H^+ + HPO_4^{2-} \qquad K_{2a} = 6.2 \times 10^{-8}$$

$$H_3PO_4 \Leftrightarrow H^+ + H_2PO_4^- \qquad K_{3a} = 7.5 \times 10^{-3}$$

while the dissociation constants of lactic acid and glycolic acid are as follows:

$$HL \Leftrightarrow H^+ + L^- \qquad K_{La} = 1.62 \times 10^{-4}$$

$$HG \Leftrightarrow H^+ + G^- \qquad K_{Ga} = 1.48 \times 10^{-4}$$

Taking 1.6×10^{-4} as the mean value of PLGA(85 : 15), the equilibrium constant for the buffering reaction may be calculated as

$$HL + PO_4^{3-} \Leftrightarrow HPO_4^{2-} + L^- \qquad K_{1eq} = K_{La}/K_{1a} = 3.81 \times 10^8 \tag{1}$$

$$HL + HPO_4^{2-} \Leftrightarrow H_2PO^{4-} + L^- \qquad K_{2eq} = K_{La}/K_{2a} = 2.58 \times 10^3 \tag{2}$$

$$HL + H_2PO_4^- \Leftrightarrow H_3PO_4 + L^- \qquad K_{2eq} = K_{La}/K_{2a} = 0.0213 \tag{3}$$

The equilibrium of reaction (1), (2) lie to the right and HAp will be easily reacted with HL and/or HG. But the equilibrium of reaction (3) lies to the left, which means that it is very difficult for $H_2PO_4^-$ to react with HL.

According to the calculation above, the total amount of buffer required can be calculated as follows:

a. Molecular weight (MW)

$Ca_{10}(OH)_2(PO_4)_6$	$MW_{HAp} = 1004.64$ g/mol
$CaHP_4$	$MW_{HAp} = 136.06$ g/mol
Lactic acid	$MW_{La} = 72.06$ g/mol
Glycolic acid	$MW_{Ga} = 58.04$ g/mol
PLGA(85 : 15):	$MW_{PLGA} = 72.06 \times 0.85 + 58.04 \times 0.15 = 69.96$ g/mol

b. Total amount of buffer required to neutralize all degradation products

Table 4 Calculation of Different Buffering Capacity

Buffering capacity (%)	PLGA (85 : 15) (%)	DCP (%)	HAp (%)
0	100	0	0
25	79.59	32.72	20.41
50	66.10	49.30	33.90
100	49.36	66.04	50.64

From molecular formula, 1 molecule of HAp can be separated into two parts: 1 molecule of $Ca(OH)_2$ and 3 molecules of $Ca_3(PO_4)_2$. Every mole $Ca(OH)_2$ neutralizes 2 moles of HL, every mole $Ca_3(PO_4)_2$ can buffer 4 moles of HL. So 1 mole HAp can buffer total 14 moles of HL; that is, 1004.64 g of HAp can neutralize 979.44 g of PLGA(85 : 15). In another words, 1 g PLGA requires 1.0258 g HAp to neutralize it. For DCP, 1 mole can neutralize only 1 mole of HL; that is, 1 g of PLGA(85 : 15) requires 1.945g DCP to buffer it. It is notable that because a slow generation of products at the end rapid phase may not cause clinical problems, it may not be necessary to provide a total buffering capacity. Based on these, the different loading of buffers in the polymer matrix required to neutralize a given quantity of lactic and glycolic acids can be calculated as in Table 4.

For a biodegradable implant, the initial strength and strength retention profile are among the most critical requirements. Theoretically, after mixing with an inorganic buffer (which is almost strengthless), the strength of the PLGA matrix will decrease. Thus the extent of the effect of the buffers needs to be investigated. Different processing methods of PLGA implants will have significant differences in mechanical properties. Tormala [21] has investigated the different mechanical properties of PLA and PGA caused by different deformation methods. In order to maintain sufficient mechanical strength while incorporating inorganic buffers, reinforcement of the PLGA matrix must be conducted. The fiber reinforcement and self-reinforcement are studied.

III. EXPERIMENTAL DESIGN AND PROCEDURES

This study focuses on in vitro simulation of the degradation of PLGA(85 : 15). The effect of the different buffering capacity to the pH change of cylindrical rod samples was observed, and the effect of the buffer to the mechanical properties was studied. Finally, the reinforcement method was employed to improve the strength and strength durability of the samples.

The measurements in this study included the following: (1) pH change over time for the samples with different buffers and the different buffering capacity: The degradation of the samples with 25% and 50% buffering capacity, as well as PLGA(85 : 15) control, is carried out. From the degradation curve, the abilities of the buffers to minimize or eliminate the rapid pH drop are evaluated. (2) Weight loss and dimension change of the samples over time: Based on the total mass loss and dimension change of the samples during the degradation, the effect of the buffer to the rate of the degradation is analyzed and evaluated. (3) The test of mechanical properties: Three point bending tests for PLGA with buffer are carried out during degradation. The effect of the buffer to the initial strength and the elastic modulus, as well as the change of strength and elastic modulus during degradation, are studied. (4) Reinforcement of the PLGA matrix: The effect of the Al_2O_3 fiber to the mechanical properties of the samples was investigated.

A. Materials

Polymer: RESOMER	RG858 poly(lactic-*co*-glycolic acid) (85/15) Manufacturer: Boehringer Ingelheim of Germany Lot #: 25024 Molecular weight (MW): 150,000

The chemical structure of PLGA(85 : 15) is

$$—[—O—CH(CH_3)—CO]_{0.85}—[—O—CH_2—CO]_{0.15}—$$

Buffer 1: $CaHPO_4$	Aldrich Chemical Company, Inc. Lot #: 7757-93-9
Buffer 2: $Ca_{10}(OH)_2(PO_4)_6$	Aldrich Chemical Company, Inc. Lot #: 23093-6
Solvent: Acetone	Fisher Scientific Lot #: 963742
2-Propanol	Fisher Scientific Lot #: 966109
Fiber: Al_2O_3	Diameter = 10 micron

B. Purification of PLGA

The raw PLGA(85 : 15) purchased commercially contained certain amount of low molecular weight impurities and it will affect the mechanical properties of the samples significantly. It needs to be removed before use. Five grams of raw PLGA(85 : 15) were completely dissolved into 100 ml of acetone stirred by a magnetic bar. The solution then was poured into a 125 ml separatory funnel and dropped slowly into a 1000 ml beaker with 400 ml of 2-propanol. The mixture was stirred vigorously by a magnetic bar while the precipitate was collected by the metal stick. Precipitated PLGA was put on a dry paper towel to let the solvent be absorbed quickly. When the solution was unclear and precipitate became sticky, the solvent was changed. Every 5 g of PLGA needed 1000–1200 ml of 2-propanol. The PLGA then remained under the ventilated hood for more than 4 h for further drying. The predried PLGA then was dried under vacuum at 1.0 mmHg for 48 h, yielding the final purified PLGA(85 : 15) (Fig. 1).

C. Grinding Polymer

Purified PLGA polymer was ground in a Tekmar A-10 analytical mill. The Tekmar was cooled with liquid nitrogen to keep the grinding temperature below 0°C. The ground polymer was sieved to retain a particle size under 180 microns.

D. Mixing Buffer

The buffers were calcium phosphate powder which have a particle size under 10 microns. The calcium phosphate and ground PLGA were blended into a ball mill and rotated for 24 h to ensure a highly homogeneous mixture. The PLGA control rods (no buffer) were prepared in the same procedure.

E. Extrusion of Rod

The mixture of PLGA and calcium phosphate was put into an extrusion mold and extruded as a cylindrical rod through a die with diameter of 3.0 mm (Fig. 2). The extrusion process was

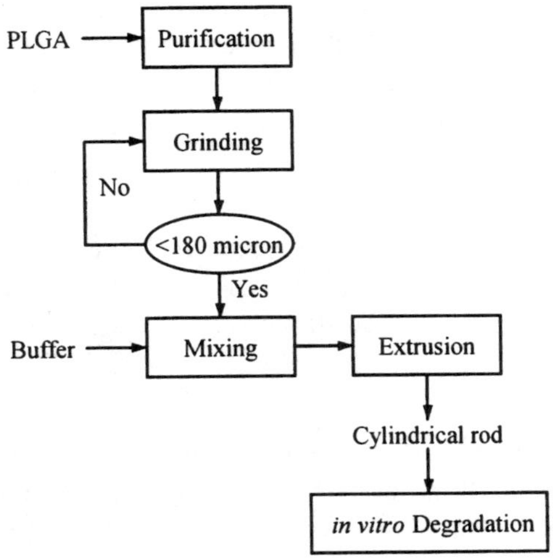

Figure 1 Diagram of preparation procedure of PLGA model fixture.

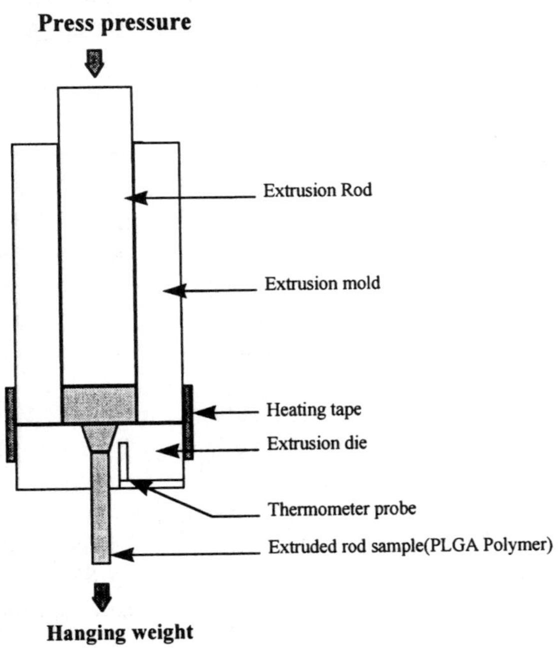

Figure 2 Diagram of extrusion process.

carried out on the Compact MPC-40 hydraulic press, and the pressure was maintained at 16 tons/in^2. An external heating tape was tied on the mold to maintain the temperature of the mold and die (thus the material) between 55–58°C. The probe of electronic thermal meter (OMEGA DP460) was placed into the mold to monitor the temperature.

F. Preparation of Fiber-Reinforced Samples

The Al_2O_3 fiber was cut to 5 mm and preliminarily mixed with the mixture of PLGA and buffer. These mixtures were mixed in Tekmar to ensure homogeneous mixing. Finally, the fiber-reinforced PLGA rods were extruded out on the COMPACT-40 hydraulic press.

G. In Vitro Degradation

The distilled water was first adjusted to pH = 7.0 with 0.1 N NaOH solution, and then distributed into a 50 ml glass tube (Pyrex, Lot # 5825) with 40 ml each. PLGA rods were put into the tubes, and the tubes were placed into a water bath (Precision Scientific, Inc., Lot #: 86146-0024). The incubation temperature was 37°C, and the shaking rate was 40 rpm.

H. In Vitro Measurements

1. pH measurement: The pH of the supernatant of the PLGA rods was measured at room temperature by pH meter (Fisher Accumet Model 520). The frequency of the measurement was ranged from 5 to 10 days depending on the rate of pH change.
2. Weight change over time: The initial weight of the samples was measured, then measured on every scheduled time point. First, samples were taken out of the water bath and then dried by paper towel. Then the samples were placed into a glass vacuum chamber with phosphorous pentoxide in it. The samples were dried under vacuum at 10 mmHg until there was no weight change. The samples were then weighed. The percentage of the weight change was calculated as follows: % wt change = $\{(W - W_i)W_i\} \times 100\%$ where W is the weight of the sample during the degradation (mg) and W_i is the initial weight of the sample (mg).
3. Dimension change over time: After vacuum drying to a constant weight, the diameter and length of each sample were measured four times by vernier caliper (Mitutoyo of Japan, 500-351) to get average value and standard deviation. The change of the dimension was expressed by the percentage of original dimension (diameter or length): % changed diameter = $D/D_i \times 100\%$; % changed length = $(L/L_i) \times 100\%$; here D is the diameter of the sample during the degradation (m), D_i is the initial diameter of the sample (m), L is the length of the sample during the degradation (m), L_i is the initial length of the sample (m).
4. Three-point bending test: Three-point bending tests were conducted at the Orthopedic Testing Laboratory of Beth Israel Deaconess Medical Center in Boston. The main testing equipment was INSTRON 8511. The PLGA rod had a diameter = 3 mm and a length = 30 mm; the testing temperature was room temperature; and the testing condition used wet samples. The specimen was tested shortly after being taken from the water bath to keep it wet. The strain rate of the INSTRON 8511 was set to 0.05 mm/s, and the total deflection was 5 mm. Figure 3 is the diagram of the three-point bending test.

The bending strength and bending modulus were calculated according to the following formulas [26]. Three-point bending strength = $8W_mL_s/\pi D^3$; three-point bending modulus = $(4/3)(P/s)(L_s^3/\pi D^4)$ where W_m is maximum load (N); L is span length (distance between two

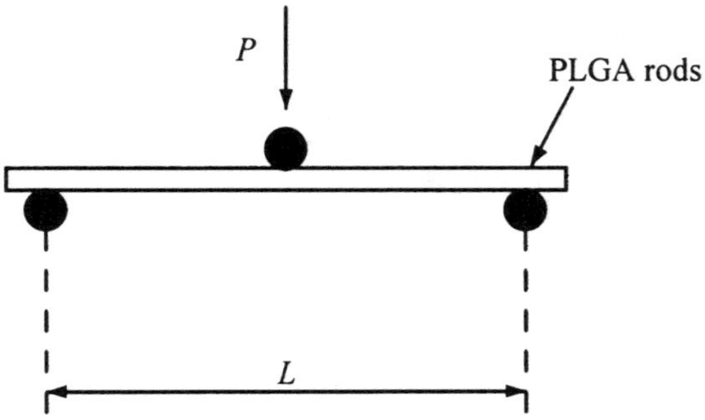

Figure 3 Diagram of three-point bending test.

support points) (m); P is load within elastic region (N); S is deflection under load P (m); D is specimen diameter (m).

IV. RESULTS AND DISCUSSION

The pH change versus time was measured over an eight-month period. The results are shown in Fig. 4. In the initial 13 weeks, the pH for all samples showed no difference. Then the PLGA control started to drop first and dropped to its lowest point (pH = 2.6) within five weeks. At week 17, all buffered samples started to drop, but the dropping rate varied. For samples with both buffers at 25% buffering capacity, the period of dropping to the lowest point was five to six weeks. On the other hand, for the samples with both buffers at 50% buffering capacity, the dropping period was 9–10 weeks, which was much longer time than former samples. Linear regression analysis for all samples over the dropping period also showed the same result and is presented in Table 5. From Table 5 it appears that the increase of the content of buffers (DCP, HAp) could lead to the decrease of the pH drop rate and higher final pH value.

A. The Weight Change versus Time

During degradation, the weight change of the samples was measured over four months; the results are shown in Fig. 5. In the initial 12 weeks, the weight of the samples dropped little

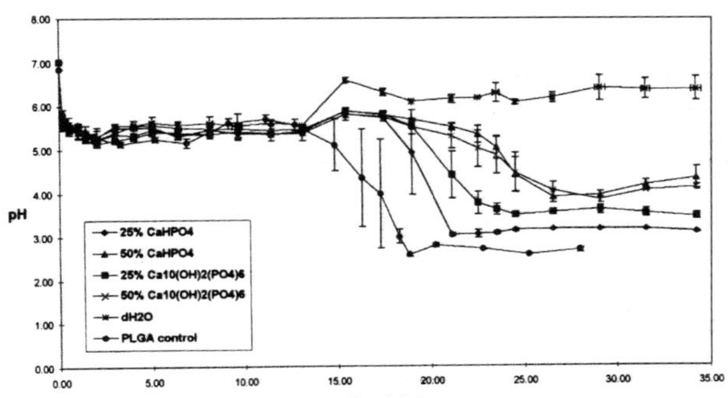

Figure 4 pH change over time for buffered PLGA (85 : 15).

Table 5 The Rate of pH Drop and Final pH Value

Buffering capacity	pH drop start-end (week)	pH drop rate (1/week)	Final pH (8 months)	Theoretical pH
PLGA control	13–18	0.49	2.68	2.60
25% DCP	17–22	0.46	3.08	2.63
50% DCP	17–27	0.16	4.17	2.88
25% HAp	17–23	0.39	3.44	2.66
50% HAp	17–27	0.18	4.10	2.81

(some were even increased). From week 13, the dropping rate of the weight of PLGA control went faster. But for buffered PLGA samples, the quick drop started at week 16. These phenomena seem to be consistent with the pH curve. It is notable that little change in pH and weight at the early period of degradation did not mean that little hydrolysis occurred. From analysis in the literature section, this was because at the beginning of the degradation, the polymer chains were very large. Even when they were cleaved, the cleaved parts were still large enough to stay in the polymer bulk. This cleavage continued until significant amounts of molecules were cleaved small enough to disperse out of the polymer bulk and into distilled water, causing the mass loss in the polymer bulk and the rapid drop of pH in surrounding area. In the polymer bulk, the porous structure was formed (greater surface area) due to the dispersion of the small molecules, producing a faster process until all of the polymer chains were degraded.

B. The Dimension Change versus Time

When the weight change was measured, the dimension change was also measured (Figs. 6, 7). Because of the degradation reaction, the geometry of the PLGA rods became irregularly shaped after about 16 weeks. At that point, the diameter and length were unreliable. Therefore the data was obtained through about the 16th week only. From Figs. 6 and 7, the trend of the change of the rod diameter during the degradation was expansion, while that of the length was shrinkage. One of the reasons might be the process of preparing the sample rods. Because the rods were made by extrusion method, during the process, the deformation of the PLGA rods was under high pressure (compression force) and tension (on axis direction, see Fig. 2), and

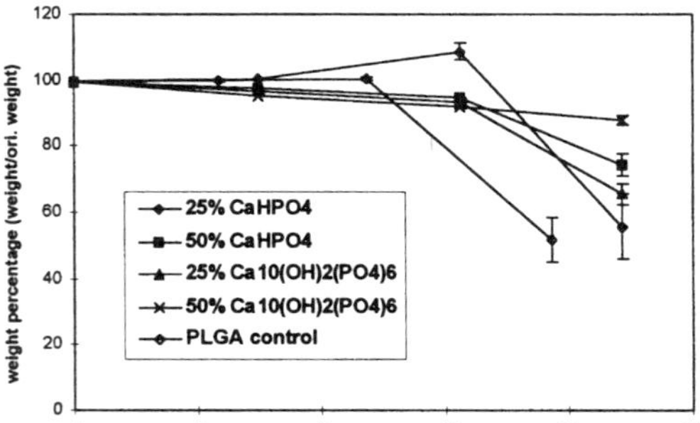

Figure 5 Weight change over time for buffered PLGA (85 : 15).

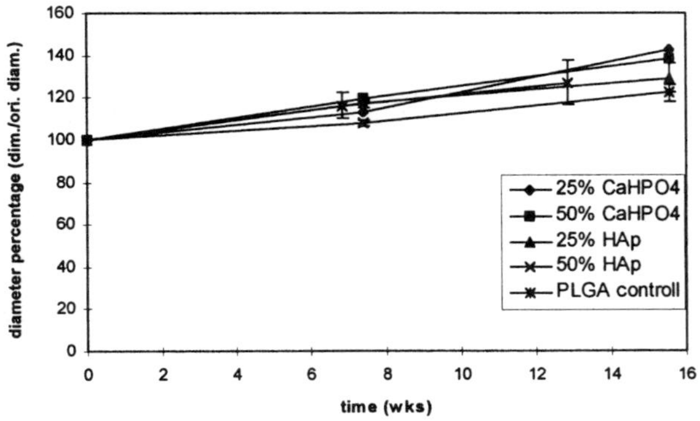

Figure 6 Change of the rod diameter over time for buffered PLGA (85 : 15).

the temperature was beyond the glass transition point, but much lower than the melting temperature. The internal stress was formed within the polymer structure when the rod was cooled to room temperature. During the degradation process, the polymer chains were cleaved and the internal stress was reduced or even disappeared; thus the dimension changed. The diameter expanded by 20% to 40%, and the length shrank by 5% to 25% over the study period.

C. Mechanical Properties

Because the mechanical properties are the most critical factors that affect the application, i.e., functionality, of the biodegradable fixation devices, the effect of the buffer additives to these properties and their retention profile must be investigated. Figures 8–11 show the results of the three-point bending test on the PLGA rod with different buffering capacity. HAp was selected as the buffer of the tested PLGA rods. The three-point bending strength was tested periodically over two months. Not surprisingly, the PLGA rods with different buffering capacity had different initial bending strength. For the samples with 0%, 25%, 50% buffering capacity, the mass percentage of of the PLGA(85 : 15) in the sample matrix was 100%, 79.59%, 66.10%, and the initial bending strength was 129 MPa, 120 MPa, 108 MPa, respectively (Fig. 8). The higher the PLGA content, the higher is the bending strength. Figure 9 shows the change

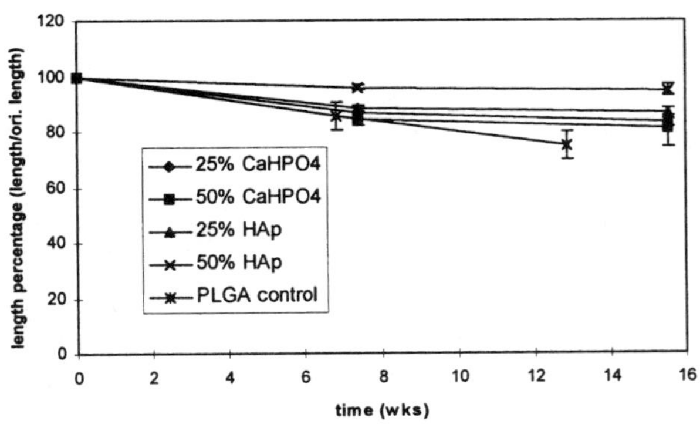

Figure 7 Change of the rod length over time for buffered PLGA (85 : 15).

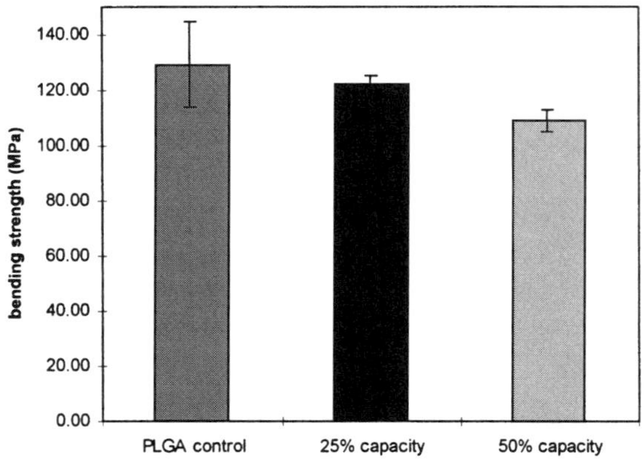

Figure 8 Initial bending strength of PLGA rods with different buffering capacity.

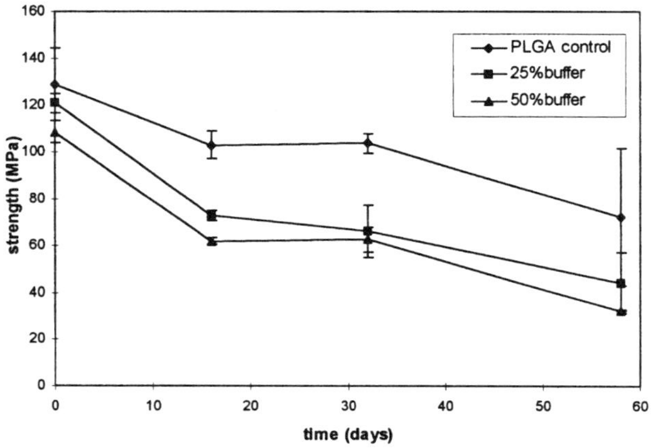

Figure 9 Change of the bending strength of different PLGA composite over time.

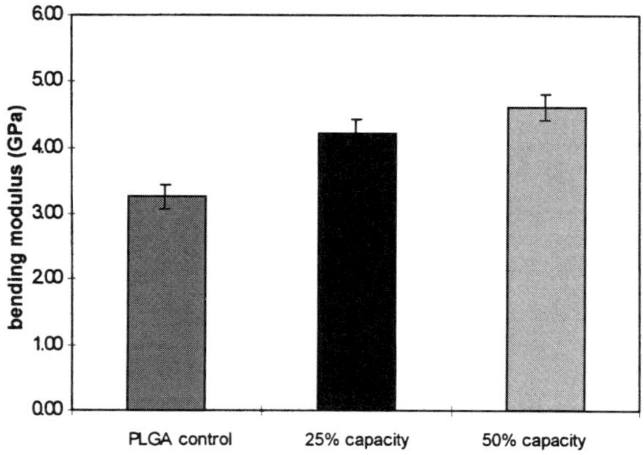

Figure 10 Initial bending modulus of PLGA rods with different buffering capacity.

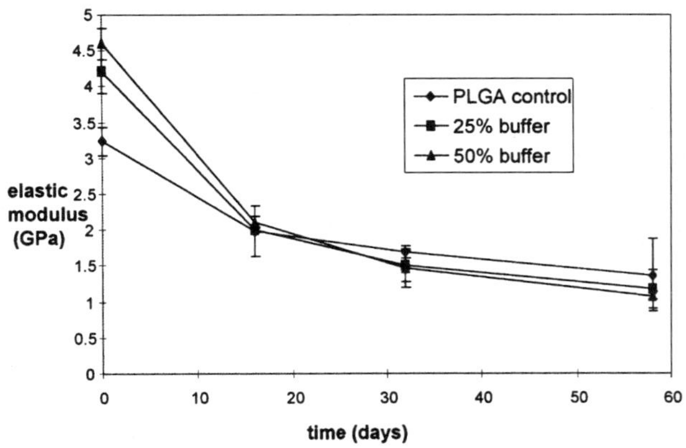

Figure 11 Change of the modulus over time on PLGA rods with different buffering capacity.

of the bending strength of the samples with different buffering capacity over the degradation time. Over the whole degradation period, the strength of all composites gradually dropped, and this drop still followed the pattern that a sample with higher content of the PLGA had higher strength. Even after almost two months, PLGA control still retained 56% of the initial strength while the strength of the samples with 25%, 50% buffering capacity retained only 37%, 30% of initial value respectively. It seems that the addition of the buffer increased the rate of degradation. This might be because the addition of the powder buffer created porosity inside the composite matrix, causing significant increase in the internal surface area. It is notable that the difference of the strength between buffered and unbuffered samples increased after two weeks of degradation. Table 6 shows the change of retained strength over time for the samples with different buffering capacity (initial value set to 100).

These phenomena indicated that although the effect of buffer on the bending strength of the samples was initially not very large, it tended to increase, and the higher content of buffer caused faster degradation. Thus new questions are raised: can the buffered sample retain sufficient strength to protect the healing bone from fracture? how can this problem be overcome? This is discussed later.

Elastic modulus is a very important factor that affects suitability of the biodegradable PLGA. As discussed previously, very high modulus will cause stress protection. Alternatively, low modulus allows too much motion of the fracture site and will cause poor union or have the risk of refracture. Figure 10 shows the effect of the buffer on elastic modulus.

It is interesting that although the buffer reduced the bending strength of the composites, it increased the bending modulus. For the sample with 25% buffering capacity (20.4% HAp by mass), its modulus increased by 30%, and for the sample with 50% buffering capacity (34% HAp by mass) the modulus increased by 42%. This may be because HAp is a very brittle

Table 6 Strength Retained over Time for Different Samples (%)

Buffering loading	0 day	16 days	32 days	58 days
PLGA control	100	80	80	56
25% loading	100	60	55	37
50% loading	100	57	57	30

Table 7 Elastic Modulus Retained over Degradation Period (% of initial modulus)

Buffer loading	0 day	16 days	32 days	58 days
PLGA control	100	61	52	43
25% buffer	100	47	36	28
50% buffer	100	46	32	23

material. Although the buffer can improve the initial bending modulus, the modulus of the buffered PLGA rods dropped faster than that of PLGA control over the degradation period (Fig. 11).

Initially, the sample with 25% buffering capacity had a bending modulus 30% higher than the PLGA control, and the sample with 50% buffering capacity had a modulus 42% higher than the PLGA control. But after 16 days, they all had almost same values, and after 58 days, samples with 25% and 50% buffering capacity dropped to 85% and 78% of the value of the PLGA control respectively. This, again, supported the hypothesis that the addition of the buffer increased the degradation rate.

It should be noticed that the drop rate of the elastic modulus was greater than that of the strength (see Table 6, initial modulus set to 100). Although the modulus of the samples with 50% buffering capacity had highest modulus initially, it had lowest value (23% of original value) after 58 days of degradation. This was consistent with the result of the bending strength that the higher buffer content would lead to faster degradation (Table 7).

D. Reinforcement

Because the addition of the buffer reduced the strength significantly over the degradation period, the reinforcement method was employed to retain or even improve it. One of the methods we studied was fiber reinforcement. The Al_2O_3 fiber was used as reinforcement fiber on the buffered PLGA matrix as well as unbuffered PLGA as a control. The three-point bending test of reinforced samples was conducted on INSTRON 8511, and the results are shown in Figs. 12–15.

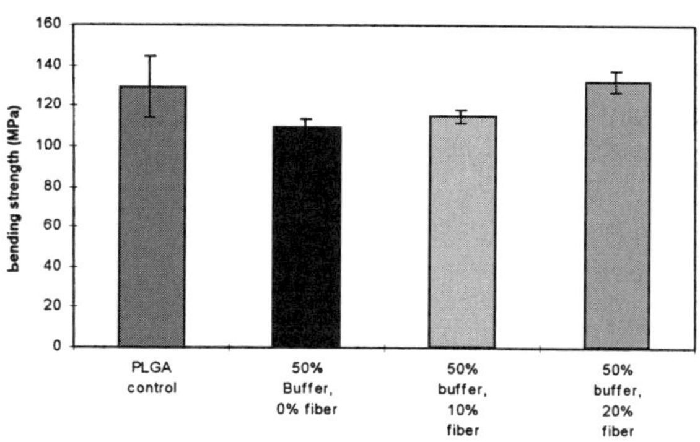

Figure 12 Effect of different fiber content on the strength of buffered PLGA rods.

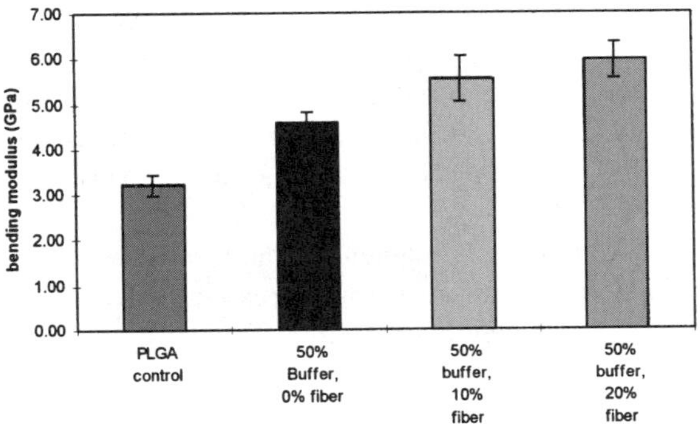

Figure 13 Effect of different fiber content on the modulus of buffered PLGA rods.

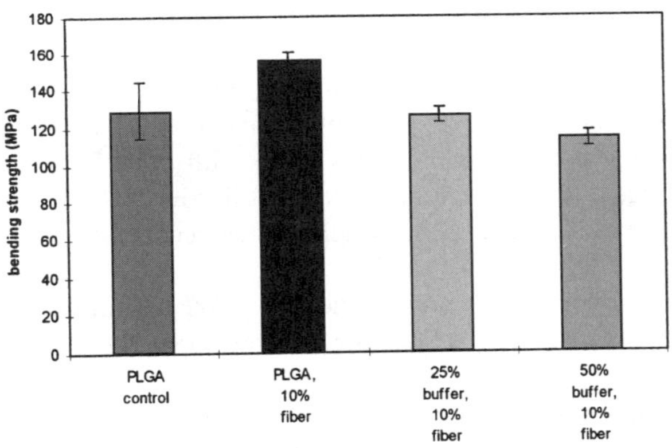

Figure 14 Effect of different buffer content on bending strength.

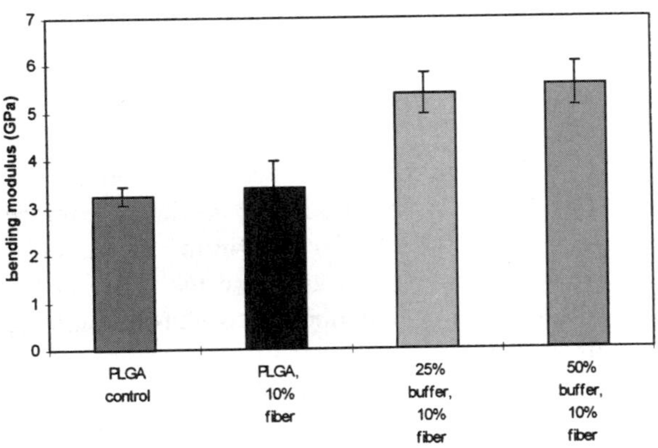

Figure 15 Effect of different buffer content on bending modulus.

Figures 12 and 13 showed the effect of the different fiber content on the mechanical properties of the buffered PLGA rods. The samples with 50% buffering capacity and 0%, 10%, 20% of three different fiber contents (by weight) were measured, and PLGA control also was measured as the comparison. Increase of fiber content led to increase in bending strength and modulus. This might be because the Al_2O_3 fiber was very brittle and had a relatively high density.

Figures 14 and 15 show the effect of the different buffer content on the mechanical properties of the fiber-reinforced PLGA rods. Under fiber reinforcement, increased buffer (HAp) content led to decreased strength yet also to increased bending modulus. This was consistent with nonreinforced PLGA rods.

Although fiber-reinforced PLGA rods showed increased strength, the increase was not very significant. This might be caused by the following factors:

1. The poor bonding force between the Al_2O_3 fiber and PLGA matrix. The fiber was inorganic material and possessed different characteristics when combined with PLGA. Also its surface was very smooth and could only provide relatively low frictional force between the fiber and the PLGA matrix. When a load was added on it, some of the fiber inside the matrix was pulled out, in another words, these fibers could not function as the reinforcing material.
2. The addition of the buffer reduced the advantage of the fiber reinforcement. From Figures 12 and 14, with 10% Al_2O_3 fiber added in both matrix, PLGA control showed 21% increase in bending strength, while the sample with 50% buffering capacity load had only 6% increase. This strongly indicated that this inorganic buffer functioned as a lubricant, leading to decrease of the bonding force between fiber and the PLGA matrix.
3. The processing method caused a low aspect ratio of the fiber. Aspect ratio is the ratio of the fiber length to its diameter. As a fiber for reinforcement purpose, its aspect ratio must be at least 40. During the process of the samples, the fiber was mixed with matrix materials in the Tekmar analytical mill. Because Al_2O_3 fiber was very brittle, part of the fiber was broken into very short pieces, and thus lost its reinforcement function.

In order to achieve high effective fiber reinforcement, the surface of the fiber needs to be roughened or chemically activated to increase the bonding force between the fiber and matrix materials. Also the mixing process also needs to be improved to avoid breaking the brittle fiber.

V. CONCLUSIONS

1. pH drop due to the accumulation of the biodegradation products can be effectively controlled by incorporating long-acting buffer. As shown in Fig. 4, the appearance of pH drop was delayed by DCP and HAp, and with 50% buffering capacity the drop rate of pH was significantly slowed down. This would give the tissue of the surgical site more time to eliminate the polymer degradation debris, thus minimizing or even eliminating the noninfectious inflammatory response.
2. Both DCP and HAp are effective long-acting buffers. Both DCP and HAp showed excellent buffering ability. They all could delay the appearance of pH drop and reduce the drop rate. At the same buffering capacity, two buffers showed no significant difference.

3. HAp has higher mass buffering ability than DCP. From Table 3, for 50% buffering capacity, 49.3% of DCP content (mass) is needed, while only 33.9% of HAp is needed. Because the higher buffer content will lead to lower mechanical properties, HAp is a better choice.
4. Degradation of the PLGA composite leads to expansion in diameter and shrinking in length for the cylindrical rods. In this study, these deformations were only observed without loading and in a free-standing position. For clinical purposes, the tension stress and compressive stress caused by these deformations on surgical site during fixation must be studied.
5. Bending strength and modulus of the PLGA composite show gradual decrease over two months of degradation. This pattern is ideal for a fixation device; the gradual decrease of the strength and modulus will lead to good transferring of the load to the healing bone and will prevent refracturing surgical site after completion of the healing. However, the drop rate in both bending strength and modulus of buffered PLGA rods is faster than that of unbuffered PLGA rods. This shows that the addition of the buffer (HAp) increases the degradation rate.
6. The drop rate of the bending modulus for all composites seems faster than that of the bending strength over the degradation period. After 58 days of degradation, the samples with 25% and 50% loading retained 37%, 30% of initial strength respectively, but only 28%, 23% of initial modulus.
7. The increase of the buffer content in the PLGA matrix leads to a decrease in bending strength and an increase in bending modulus. Although the initial bending modulus was improved by HAp buffer, this advantage quickly disappeared over the degradation period, and the modulus dropped lower than that of unbuffered PLGA.
8. Al_2O_3 fiber can increase the strength and modulus of PLGA matrix. This is true for buffered and unbuffered PLGA rods, but the fiber was much more effective on unbuffered PLGA than buffered PLGA. For instance, with 10% fiber the samples with 50% buffering capacity load improved only 6%, and the samples without buffer improved by 21%. This phenomenon shows that the existence of the buffer weakens the bonding force between the fiber and the PLGA matrix.
9. The process of the fiber-reinforced PLGA rods needs to be improved. Because of the problem mentioned above, the surface of the fiber needs to be roughened or chemically activated so that the fiber can bond well with matrix.

VI. RECOMMENDATIONS

1. The use of 50% buffering capacity loading of HAp to further study on various reinforcement and in vivo study is recommended. Although the sample with 25% buffer loading has better mechanical properties, the sample with 50% buffer loading has much better buffering performance. The mechanical properties can be improved through various reinforcement methods.
2. The effect of the dimension change as pertains to the ultimate clinical application must be investigated.
3. The process of the fiber-reinforced rods needs to be improved. The improvement includes treatment of the fiber surface and the method of mixing the PLGA/buffer with fiber.
4. The self-reinforcement method should be studied. This includes the use of PLGA fiber instead of inorganic fiber, and then using this PLGA fiber in the PLGA/buffer matrix.

NOMENCLATURE

B	Buffer anion
DCP	Calcium dibasic phosphate, $CaHPO_4$
IM	Injection molded
HAp	Hydroxylappatite, $Ca_{10}(OH)_2(PO_4)_6$
K_a	Dissociation constant, or ionization constant
K_{eq}	Equilibrium constant
K_{sp}	Solubility product constant
L^-	Lactate (or glycolate) anion
PGA	Polyglcolic acid
PLA	Polylactic acid
PDLLA or DLPLA	Poly(dextro, levo-lactic acid)
PLLA	Poly(levo-lactic acid)
PDLLGA	Poly(dextro, levo-lactic-*co*-glycolic acid)
SR	Self-reinforced
Buffering capacity	Amount of buffer loaded in the implant matrix which can neutralize certain portion of totally degraded polymer. For example, 50% buffering capacity means if all of the polymer loaded in the matrix is degraded totally into monomers, 50% of these monomers can be neutralized by the buffer loaded in the matrix. See Table 4.
D	Diameter of the rod, m
D_i	Initial diameter of the rod, m
E	Elastic modulus on three-point bending test, GPa
L	Length of the rod, m
L_i	Initial length of the rod, m
L_s	Span (distance between two support points) on three-point bending test, m
MW	Molecular weight, gram/mol
P	Load within elastic region on three-point bending test, N
S	Deflection under load P on three-point bending test, m
W	Weight of samples, mg
W_m	Maximum load on three-point bending test, N
W_i	Initial weight of samples, mg

REFERENCES

1. AU Daniels, MKO Chang, and KP Andriano. Mechanical properties of bio-degradable polymers and composites proposed for internal fixation of bone, *J. Appl. Biomater.*, **1**, p. 57–778, 1990.
2. GO Hofmann. Biodegradable implants in orthopedic surgery—a review on the state-of-the-art, *Arch. Orthop. Trauma Surg.*, **10**, p. 75–80, 1992.
3. N Gillett, SA Brown, RP Dumbleton, and RP Pool. The use of short carbon fiber reinforced thermoplastic plates for fracture fixation, *Biomaterials*, **6**, p. 113–121, 1985.
4. K Tayton, C Johnson-Nurse, B McKibbon, J Bradley, and G Hastings. The use of semi-rigid-carbon-fiber-reinforced plates for fixation of human fractures: result of preliminary trial, *J. Bone Joint Surg.*, **64-B**, p. 105–111, 1982.
5. OM Bostman. Current concepts review: absorbable implants for the fixation of fractures, *J. Bone Joint Surg.*, **73-A**, p. 148–152, 1991.

6. DK Gilding, and AM Reed. Bio-degradable polymers for use in surgery-polyglycolic/polylactic acid homo- and copolymers, **1**, *Polymer*, **20**, p. 1459–1463, December, 1979.
7. DL Wise. *Encyclopedic Handbook of Biomaterials and Bioengineering*, Pt. B, 2, p. 359–382, 1995.
8. DE Cutright, and EE Hunsuch. The repair of fractures of the orbital floor using biodegradable polylactic acid, *Oral Surg.*, **33**, p. 28–34, 1972.
9. H Greve, and J Holste. Refixation osteochondraler fragmente durch resorbierbare kunststoffstiffte, *Aktuel. Tramat*, **15**, p. 145–149, 1985.
10. AGA Coombes, and MC Meikle. Resorbable synthetic polymers as replacement for bone graft, *Clin. Mater.*, **17**, p. 35–67, 1994.
11. JO Hollinger, and GC Battistone. Biodegradable bone repair materials: synthetic polymers and ceramics, *Clin. Orthop. Rel. Res.*, **207**, p. 290–305, 1986.
12. S Li, H Carrean, and M Vert. Structure property relationships in the case of degradation of massive poly(α-hydroxyl acid) in a aqueous media. part 3. influence of the morphology of poly(lactic acid), *J. Mater. Med.*, **1**, p. 198–206, 1990.
13. KL Smith, ME Schlimpf, and KE Thompson. Bioerodible polymers for delivery of macromolecules, *Adv. Drug Del. Rev.*, **4**, p. 343–357, 1990.
14. CC Chu. Hydrolytic degradation of polyglycolic acid: tensile strength and crystallinity study, **26**, p. 1727–1734, 1981.
15. DH Lewis. Controlled release of bioactive agents from lactide/glycolide polymers, Chapter 1, in *Biodegradable Polymers as Drug Delivery Systems*, M Chasin and R Langer (Eds.), Marcel Dekker, NY, 1990.
16. RA Miller, JM Bradley, and DE Cutright. Degradation rate of oral resorbable implants (polylactates and polyglycolates): rte modification and changes in pla/pga copolymer ratios, *J. Biomed. Mater. Res.*, **11**, p. 711–719, 1977.
17. P Tormala. Ultra-high strength, self-reinforced absorbable polymeric composites for applications in different disciplines of surgery, *Clin. Mater.*, **13**, p. 35–40, 1993.
18. OM Bostman. Osteolytic change accompanying degradation of absorbable fracture fixation implants, *J. Bone Joint Surg.*, **73B** (4), p. 679–682, 1992.
19. OM Bostman, E Hirvensala, S Vainiopaa, K Vihtonnen, P Tormala, and P Rokkanen. Degradable polyglycolide rods for internal fixation of displaced bimalleolar fractures, *Int. Orthop.*, **14**, p. 1–8, 1990.
20. RA Kenley, MO Lee, TR Mahoney, and LM Sanders. Poly(lactide-co-glycolide) decomposition kinetics in vivo and in vitro, *Macromolecules*, **20**, p. 2398–2403, 1987.
21. P Tormala, J Vasenius, S Vainionpaa, J Laiho, T Pohjonen, and P Rokkanen. Ultra-high-strength absorbable self-reinforced polyglycolide (sr-pga) composite rods for internal fixation of bone fracture: in vitro and in vivo study, *J. Biomed. Mater. Res.*, **25**, p. 1–22, 1991.
22. J Vasenius, S Vainiopaa, K Vihtonnen, A Makela, P Rokkanen, M Mero, and P Tormala. Comparison of in vitro hydrolysis, subcutaneous and intramedully implantation to the strength retention of absorbable osteosynthesis implants, *Biomaterials*, **11**, p. 501–504, 1990.
23. J Vasenius, S Vainiopaa, K Vihtonnen, A Makela, P Tormala, and P Rokkanen. A histomorphological study on self-reinforced polyglycolide (SR-PGA) osteosynthesis implants coated with slowly absorbable polymers, *J. Biomed. Mater. Res.*, **24**, p. 1615–1635, 1990.
24. JC Elliot. Structure and Chemistry of the Apalites and Other Calcium Orthophosphates, p. 4, 1994.
25. *Merck Index*, 11th Ed. Merck and Co., Rahway, NJ, 1989.
26. CG Ives et al. *Handbook of Plastics Test Methods*, p. 146–148, 1971.

25

In Vitro and In Vivo Degradation Studies of Absorbable Poly(orthoester) Proposed for Internal Tissue Fixation Devices

Kirk P. Andriano
University of Utah School of Medicine, Salt Lake City, Utah

A.U. Daniels
University of Tennessee College of Medicine, Memphis, Tennessee

Jorge Heller
Advanced Polymer Systems, Inc., Redwood City, California

I. INTRODUCTION

In the field of medicine, absorbable polymers represent an important growing class of materials, especially in temporary applications for repair and regeneration of healing tissues. The interest in orthopedics in these materials for internal tissue fixation devices stem from their potential to stabilize damaged tissues during healing and obviate the need for second removal surgeries [1]. The absorbable polymers most widely investigated today for these fixation devices include poly(L-lactide) [2–4], poly(L/DL-lactide) [5], poly(glycolide) [6–8], poly(lactide-*co*-glycolide) [9,10], poly(ρ-dioxanone) [11,12], poly(ε-caprolactone) [13], and tyrosine-derived poly(carbonates) [14,15].

The initial mechanical properties, degradation kinetics, mechanism of degradation, and final breakdown products of these polymers are quite varied [16]. The design requirements for absorbable tissue fixation devices for specific applications are as yet unknown. These requirements are dependent upon many factors such as injury type, location, method of stabilization, implant geometry, age, sex, weight, and activity level of the patient [17]. Consequently, a wide variety of absorbable materials to choose from are needed for the design of effective and efficient tissue fixation devices.

Recently, a family of linear poly(orthoesters) has been proposed as an additional candidate material for internal tissue fixation devices. The incorporation of acidic or basic excipients into poly(orthoesters) has been used to bias surface hydrolysis for short- or long-term delivery of therapeutic agents, respectively. This work evaluates changes in mechanical and physical properties of excipient free poly(orthoesters) proposed for internal tissue fixation devices in a variety of in vitro studies and a long-term canine implantation study.

II. POLY(ORTHOESTER) POLYMERS

A. Polymer Background, Synthesis, and Degradation

Poly(orthoesters) are a family of synthetic absorbable polymers that have been under development for pharmaceutical applications for a number of years [18]. This family of highly hydrophobic absorbable polymers is completely amorphous [19]. Hydrolytic degradation of the bulk polymer is reported to occur predominately via surface hydrolysis [20]. POEs have been successfully used as erodible matrices for delivery of therapeutic agents [18]. Drug delivery devices exhibiting fairly constant, first-order, release kinetics and lifetimes ranging from hours up to a year have been produced.

The synthesis of POEs has been described [21] and is shown in Fig. 1. Briefly, the rigid diol, *trans*-cyclohexanedimethanol (t-CDM) and the flexible diol, 1,6-hexanediol (1,6-HD), in the desired ratio are added to tetrahydrofuran and the mixture is stirred until all solids dissolved. A solution of the diketene acetal 3,9-bis(ethylidene-2,4,8,10-tetraoxaspiro[5,5]undecane) is then added, and the polymerization is initiated by the addition of a catalytic amount of ρ-toluenesulfonic acid dissolved in tetrahydrofuran. After the initial exotherm subsides, the mixture is stirred for about 2 h, and polymer is isolated by precipitation into a large excess of methanol containing a small amount of triethylamine stabilizer, followed by filtration and vacuum drying at 60°C for 24 h. Polymers prepared by this procedure typically have molecular weights in the 80–100 kDa range. By giving special attention to exact stoichiometry and using very pure reagents (>99.5%), molecular weights in excess of 180 kDa have been achieved.

Polymers with a wide range in glass transition temperature (T_g) can be synthesized by varying the ratio of rigid to flexible diols (Fig. 2). However, the effect of T_g on elastic modulus is modest for POEs reported here since the T_g is well above room temperature and the service temperature for implants (i.e., 37°C). The designation $N_1 : N_2$ POE indicates that the polymer was synthesized with N_1 moles of t-CDM and N_2 moles of 1,6-HD.

When bulk POE of the type described above is exposed to an aqueous environment, the polymer undergoes an initial hydrolysis of the orthoester linkages to yield a mixture of the diols used in the synthesis and the mono- and dipropionates of pentaerythritol. These mono- and diesters eventually hydrolyze to pentaerythritol and propionic acid (Fig. 3) [22].

Even though orthoester linkages are much more labile than the ester linkages, the polymer is highly hydrophobic, so that initial hydrolysis is largely confined to the outer surface of a polymer specimen. Thus allowing the small water-soluble by-products to diffuse away at the solid-liquid interface and limit autocatalysis of the acid sensitive bulk polymer, which could be caused by the accumulation of propionic acid.

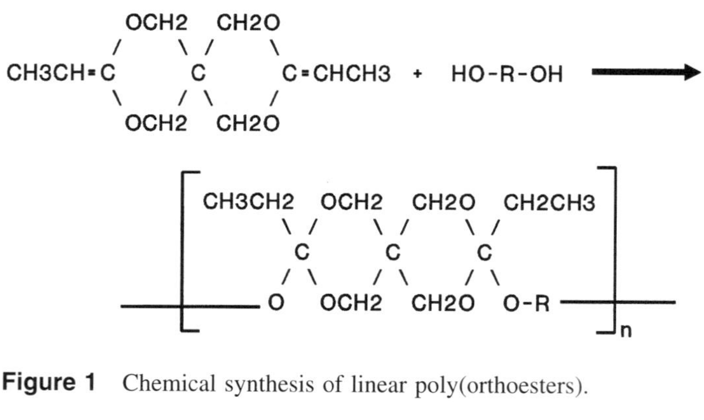

Figure 1 Chemical synthesis of linear poly(orthoesters).

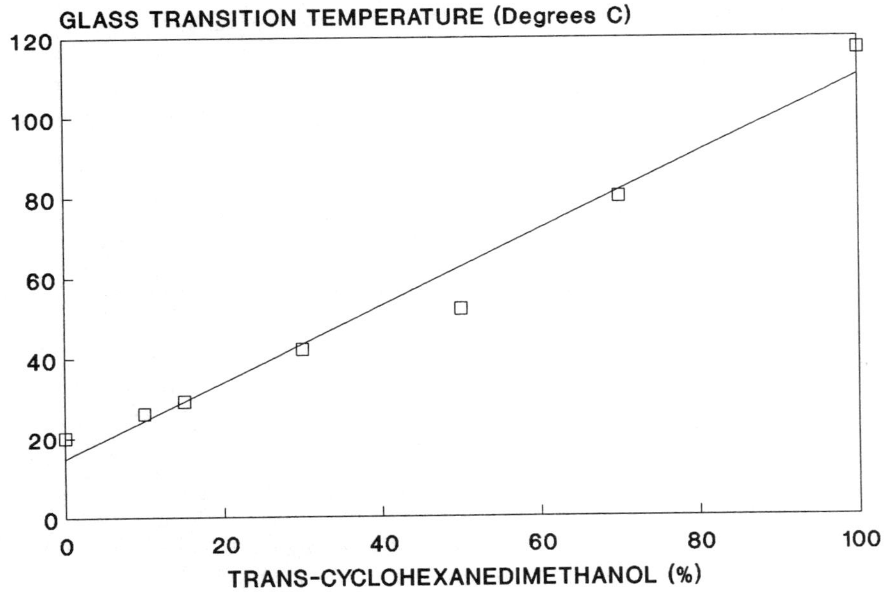

Figure 2 Effect of different molar ratios of trans-cyclohexanedimethanol and 1,6-hexanediol on the glass transition temperature of poly(orthoester).

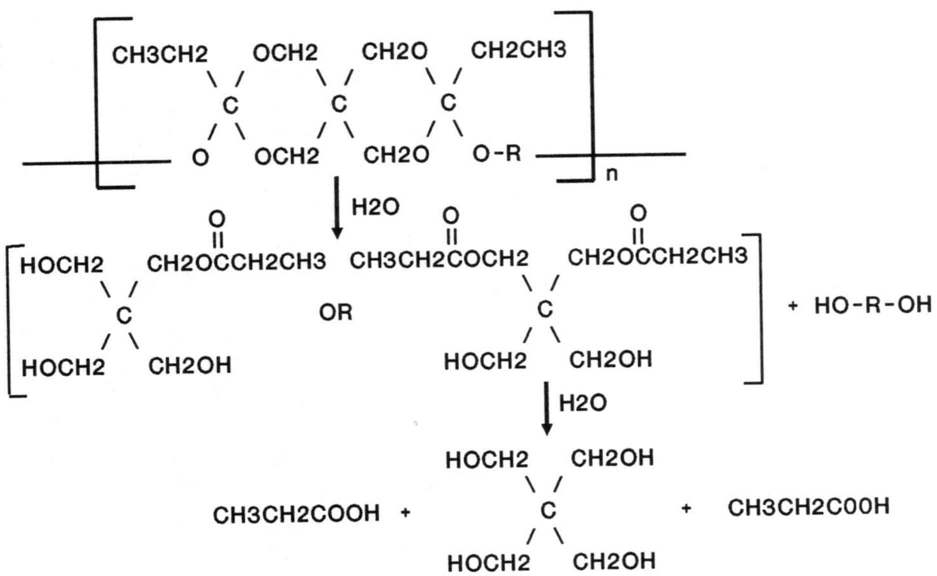

Figure 3 Two-step hydrolytic degradation of poly(orthoester).

B. Polymer in Vitro Degradation

1. *Effect of in Vitro Lipid Exposure on Solvent Cast Films*

The purpose of this study was to determine the mechanical and physical properties of 65 : 35 POE solvent cast film specimens after in vitro exposure to a physiological lipid and compare the results to those obtained from deionized water exposure.

Exposing a polymeric biomaterial to lipids in vitro would appear to be useful only if lipids are suspected to have an effect, but there are few guidelines for making such a prediction. Moacanin and Lawson [23] estimated the solubility parameter of physiological lipids and offered a rule of thumb that lipids may be soluble in polymers when their solubility parameters are similar ($\delta_{lipid} = \delta_{polymer} \pm 2$), where δ is given in Hildebrand units [$H = (cal/cm^3)^{1/2}$]. The solubility parameters of cholesterol, triglycerides, and fatty acids are in the range of 8–11 H [23], while POE has an estimated value of 9.9 H, thus suggesting certain physiological lipids may be soluble in poly(orthoesters).

A 65 : 35 POE polymer (SRI International) with a $T_g = 75°C$ and weight average molecular weight of 59 kDa was dissolved in spectrophotometry grade methylene chloride (Omnisolv) at concentration of 10.9% (w/w). The solvent was stabilized with trace amounts of triethylamine (Mallinckrodt) (less than 0.1% of solvent weight) before addition of the polymer. Solutions were left to stand covered at ambient temperature for 12 h, then gently stirred and covered for another 12 h for degassing. This was followed by cooling the polymer solutions to 5°C and storing for 24 h covered. The solutions were then hand-drawn on a polished glass plate with a variable height casting blade (Gardener AG-4303) set at 1.27 mm. Prior to drawing, both the glass plate and casting blade were cooled to 5°C by storing in a cold room for 30 min. After casting, films were stored under a glass cover with a desiccant ($CaCl_2$) at ambient temperature (20–25°C) for 48 h to allow initial evaporation of solvent. Films were removed from their glass plates and residual solvent was removed by vacuum drying at 0.05 torr and room temperature for 72 h. This procedure yielded consistently clear POE films.

Microdogbone tensile specimens were die cut from the solvent cast films in accordance with ASTM D 638-82 (Type V) [24]. Specimen thickness were measured with a tmi 549 μm (dead weight tester), and ranged from 130 to 229 μm.

Cholesterol emulsions were prepared by dissolving 0.1 g of cholesterol (Sigma Grade 99+%, Sigma Chemical Co.) in 5 mL of warm (30–40°C) ethanol (U.S.P. dehydrated 200 proof punctilious, U.S. Industrial Chemicals Co.). The cholesterol-ethanol solution was then sonically mixed with 100 mL of filtered deionized water using a Bransonic 220 bath for 10 min, producing cholesterol emulsions with approximately 1 g/L concentration. This is similar to cholesterol levels found in plasma [25]. Enzymatic lipids testing showed cholesterol levels in emulsions were stable for up to one week of storage at 37°C.

Fifteen specimens were tested prior to exposure to establish a baseline. The remaining specimens were sealed in 7-mL glass vials containing cholesterol emulsion or deionized water and placed in a water bath at 37°C. Both media were changed on a weekly basis. Specimens were tested after seven days, then weekly, for up to six weeks. After in vitro exposure, the specimens were rinsed with deionized water and mechanically tested immediately. Specimens were then dried for two days at 200 torr, 40°C before measuring dry mass change. Mechanical testing was performed at room temperature in ambient air on a free-standing Instron 1125 materials testing machine with hydraulic actuated rubber faced grips at a strain rate of 105%/min. Tensile yield strength, elongation to yield and to failure, and initial tensile modulus were calculated from initial specimen dimensions and load-elongation data, according the ASTM Standard D 638-82 [23].

The initial and final masses of each specimen were measured on a Mettler H51 analytical balance. The change in dry mass was calculated as percent dry mass change = [(initial mass − final mass)/(initial mass)] × 100.

The initial mechanical properties of 65 : 35 POE film baseline specimens are given in Table 1. Stress-strain curves indicated that all POE film specimens failed in a brittle fashion.

Tensile mechanical properties after in vitro exposure to deionized water or cholesterol emulsion at 37°C are shown in Fig. 4. There was little to no difference in tensile strength, elongation to yield, and tensile modulus between film specimens exposed to either deionized water of cholesterol emulsion up to six weeks. POE film specimens in deionized water retained 69% their tensile yield strength (30.5 MPa), 76% elongation to yield (10.4%), and 74% tensile modulus of elasticity (400 MPa) after six weeks exposure. While specimens exposed to the cholesterol emulsion after five weeks retained 91% of their tensile yield strength (40.5 MPa), 88% elongation to yield (12%), and 98% tensile modulus of elasticity (538 MPa).

Mass loss of 65 : 35 POE films after in vitro exposure to deionized water and cholesterol emulsion are shown in Fig. 4D. There was little to no change in dry mass in either water or cholesterol up to six weeks. Film specimens in deionized water showed a decrease in dry mass of 4%, while specimens in cholesterol emulsion showed a decrease of 2% after five weeks exposure.

The effect of many factors in the physiological environment, such as enzymes, lipids, and proteins, on the degradation of POE have not been simulated and studied in vitro. Enzymes are not expected to have much effect because of the rigidity of the polymer backbone and because an orthoester linkage has not been identified in nature yet.

POE precipitates in methanol, is soluble in solvents less polar than ether, and softens and swells in acetone. The solubility parameter of an amorphous polymer can be estimated as approximately equaling that of the solvent that causes maximum swelling [26], so an estimated of POEs solubility parameter is $\delta_{POE} = \delta_{Acetone} = 9.9$ H [9.9 $(cal/cm^3)^{1/2}$] [27].

Cholesterol was selected as the lipid exposure medium for several reasons: first, cholesterol is a major contributor of lipids found in human knee tissues (i.e., bone marrow, intra-articular fat pad, cartilage ligaments, meniscus, and synovium) [28,29] second, because cholesterol has been widely studied, it is readily available and inexpensive in pure form, and it can be handled relatively easily; and third, according to Moacanin and Lawson's rule of thumb [24], cholesterol may be soluble in POE.

Even though cholesterol was not soluble in POE, as indicated by the present data, increased exposure periods and/or increased cholesterol concentrations need to be evaluated to thoroughly answer this question.

2. *Effect of Molecular Weight on Solvent Cast Film Degradation*

The purpose of this study was to determine what effect molecular weight would have on POE polymer degradation rate of mechanical and physical properties.

Table 1. Initial Mechanical Properties of 65 : 35 POE Solvent Cast Films

Mechanical properties	Means ± SD
Tensile yield strength (MPa)	39.0 ± 7.5
Elongation to yield (%)	12.7 ± 1.6
Elongation to failure (%)	23.7 ± 10.8
Tensile modulus (MPa)	514.0 ± 50.0

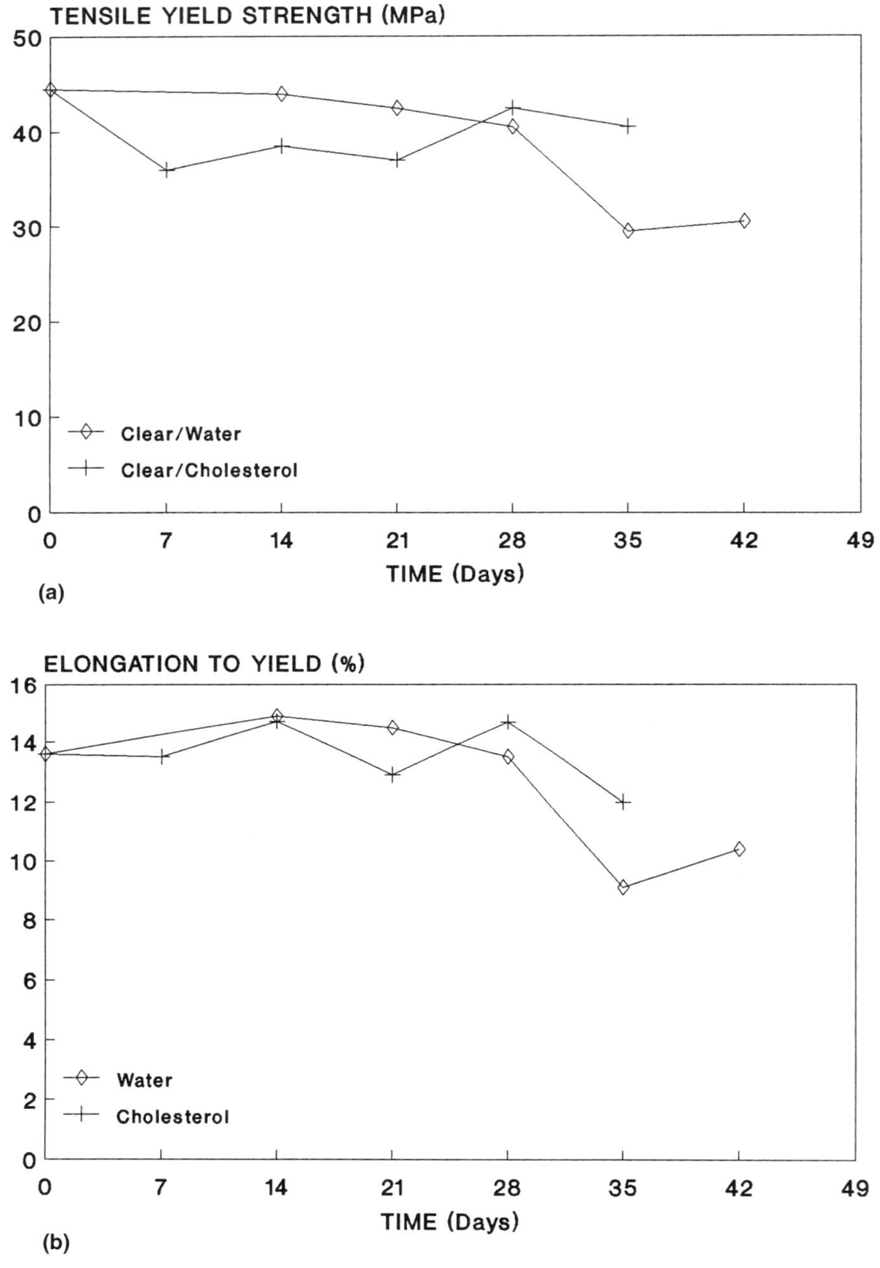

Figure 4 Effect of in vitro exposure on mechanical properties of 65 : 35 POE solvent cast films; deionized water versus cholesterol emulsions: (A) mean tensile yield strength (MPa); (B) mean elongation to yield (%); (C) mean modulus of elasticity (MPa); (D) mean dry mass loss (%).

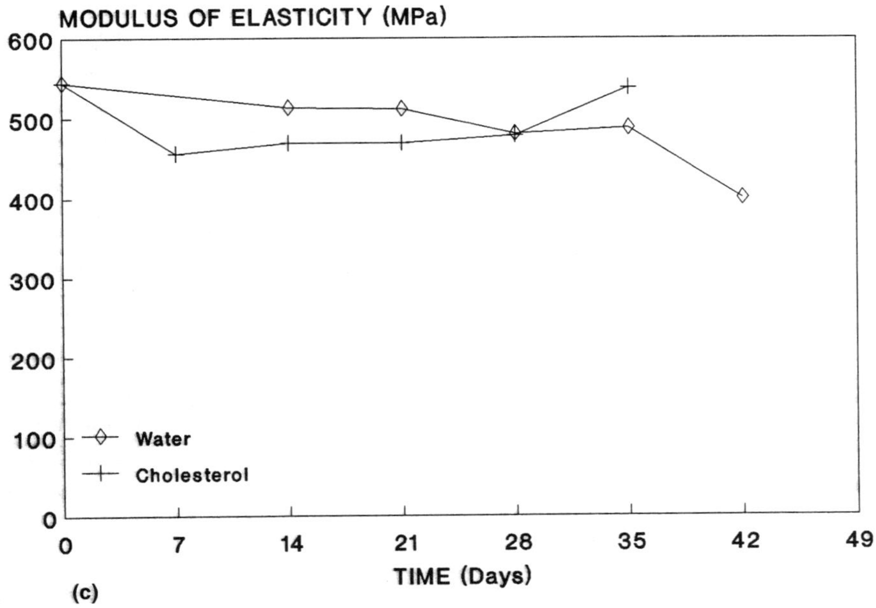

(c)

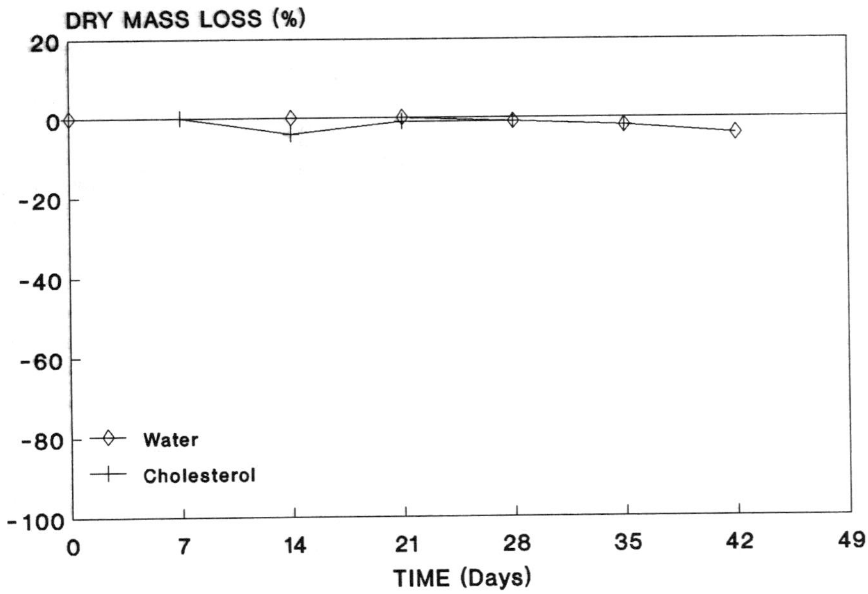

(d)

Figure 4 Continued

Additional POE polymer films, with a 60 : 40 ratio of rigid to flexible diols ($T_g = 70°C$) and a weight average molecular weights of 18, 27, 41, and 67 kDa were prepared using the above procedure at polymer to solvent concentration ratio of 20.0, 17.5, 15.0, and 9.1% (w/w), respectively. Individual film specimen thickness ranged from 90 to 210 μm.

POE film specimens were placed in individual 7-mL vials filled with Tris-buffered distilled water (0.05 M and pH 7.4), sealed and held at 37°C. Five specimens of each polymer molecular weight (18, 27, 41, and 67 kDa) were removed at 10, 18, and 31 days, and tensile stress-strain relationships were determined according to the ASTM Standard D 638–82 [23].

Absorbed wet mass was determined by rinsing each of the specimens with deionized water, blotting the surface to remove surface moisture, and then weighing immediately on an analytical balance. To determine dry mass change, specimens were then dried for two days at 200 torr and 40 to 50°C before having their final mass measured. The percent absorbed wet mass was calculated as follows: percent absorbed mass = (wet mass − final dry mass)/(final dry mass) × 100. Finally, inherent viscosity of a 0.5% solution of exposed, dried polymer in methylene chloride at 30°C was measured after each exposure interval with a Cannon-Ubbelohode Size 50 viscometer, with flow times averaged over five trials for each test.

Tensile mechanical properties after in vitro exposure to Tris-buffered distilled water are shown in Figs. 5–7. Films made from the lowest molecular weight 60 : 40 POE studied (18 kDa) were too brittle for mechanical testing; therefore no data are included. Films fabricated from other molecular weight POEs (27, 41, and 67 kDa) possessed similar initial tensile mechanical properties (Table 2). No change in tensile elastic modulus was observed for any of the three molecular weight POE films after 31 days in vitro exposure. However, the 27-kDa film showed a 21% decrease in tensile yield strength (25 ± 1.8 MPa) and a 34% decrease in elongation to yield (7.6 ± 0.2%) after 31 days exposure; the 41- and 67-kDa films showed no decrease in their mechanical properties.

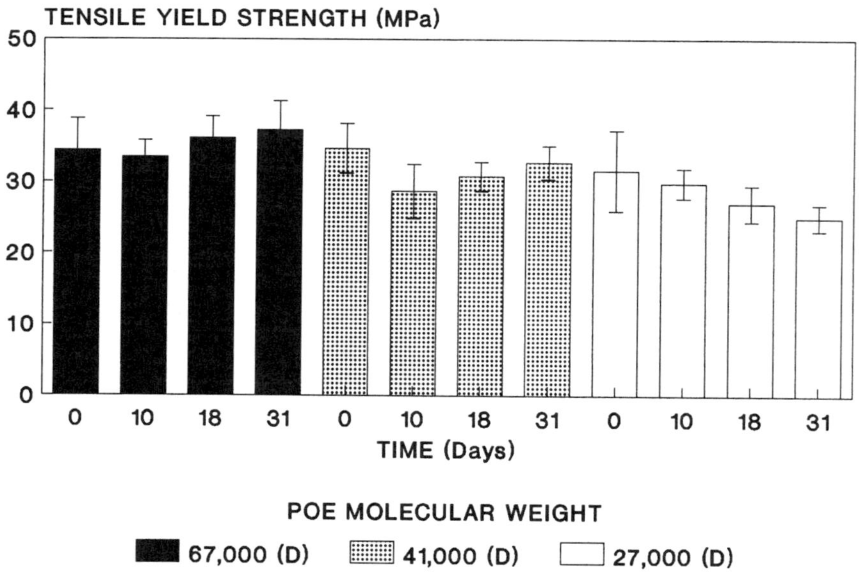

Figure 5 Effect of in vitro exposure to 7.4 pH Tris-buffered water at 37°C on mean tensile yield strength (MPa) of different molecular weight 60 : 40 POE solvent cast films (mean ± SD).

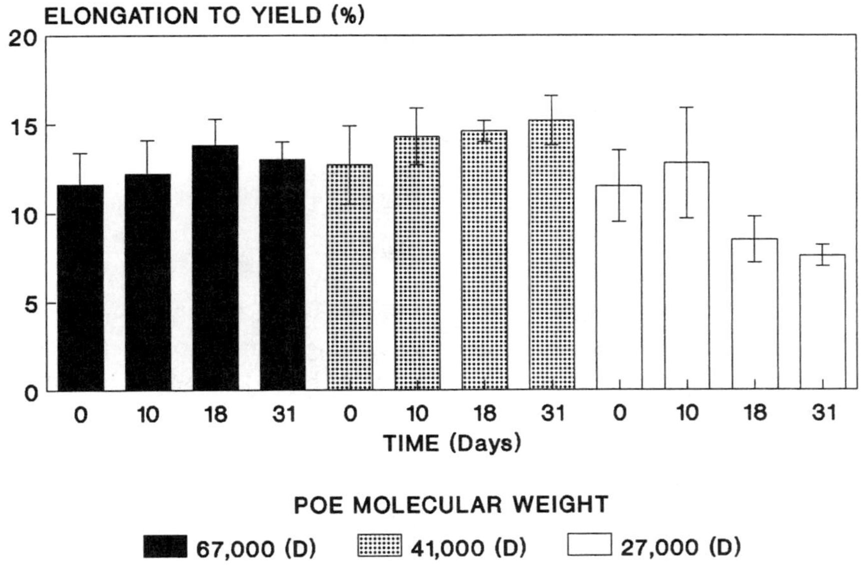

Figure 6 Effect of in vitro exposure to 7.4 pH Tris-buffered water at 37°C on mean elongation to yield (%) of different molecular weight 60 : 40 POE solvent cast films (mean ± SD).

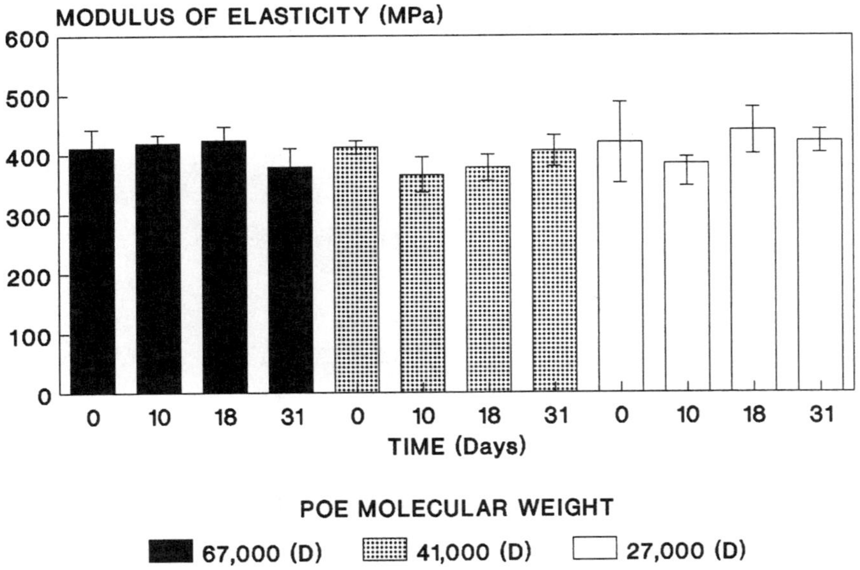

Figure 7 Effect of in vitro exposure to 7.4 pH Tris-buffered water at 37°C on mean modulus of elasticity (GPa) of different molecular weight 60 : 40 POE solvent cast films (mean ± SD).

Table 2 Initial Tensile Mechanical Properties of Different Weight Average Molecular Weight Solvent Cast 60 : 40 POE Films (mean ± SD)

Molecular weight (Da)	Strength (MPa)	Elongation to yield (%)	Modulus (MPa)
27000	34.4 ± 5.7	11.5 ± 2.0	420 ± 68
41000	34.7 ± 3.5	12.7 ± 2.0	412 ± 31
67000	34.7 ± 4.4	11.6 ± 1.8	411 ± 31

Specimen mean dry mass dropped for all three molecular weight films studied (Fig. 8). The 27-kDa film showed the largest decrease in dry mass ranging from 2.1 ± 1.8% (18 days) to 1.5 ± 0.2% (31 days), but inherent viscosity changed little (Fig. 9). Absorbed wet mass showed no increase with increasing exposure time for all three molecular weight POE films studied (Fig. 10).

The three higher molecular weight POE films (27, 41, and 67 kDa) were generally as strong or stronger than other absorbable films reported in the literature. The 27-kDa molecular weight film was much stronger and tougher than the 18-kDa film, indicating a sudden transition in microstructure (i.e., chain entanglement) occurs between 18 and 27 kDa. However, unlike the still higher molecular weight films, the 27-kDa film showed a decrease in yield strength and elongation to yield after 31 days exposure, suggesting that degradation is faster at lower molecular weights.

3. *Effect of pH Exposure on Hot-Molded Specimens*

The purpose of this study was to determine if exposure to different pHs would have any effect on the hydrolytic degradation rate of bulk POE, with respect to mechanical and physical properties.

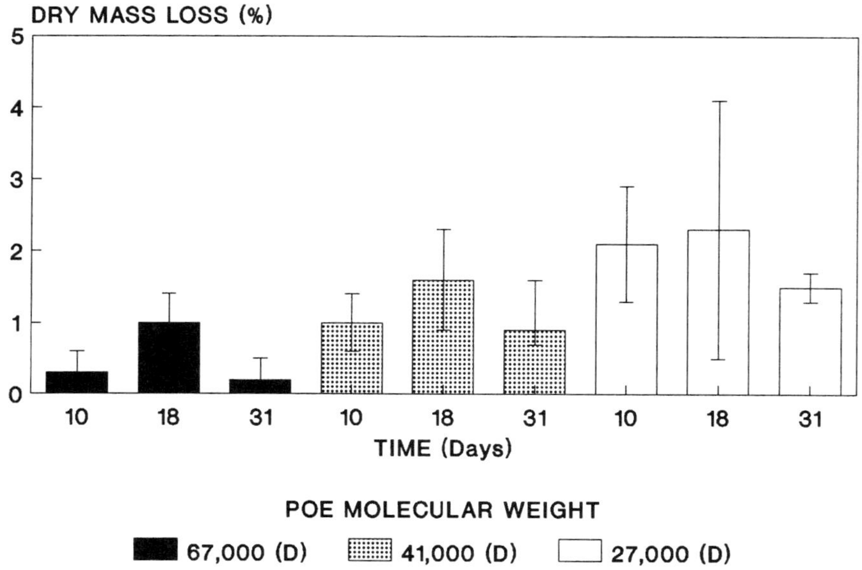

Figure 8 Effect of in vitro exposure to 7.4 pH Tris-buffered water at 37°C on mean dry mass loss (%) of different molecular weight 60 : 40 POE solvent cast films (mean ± SD).

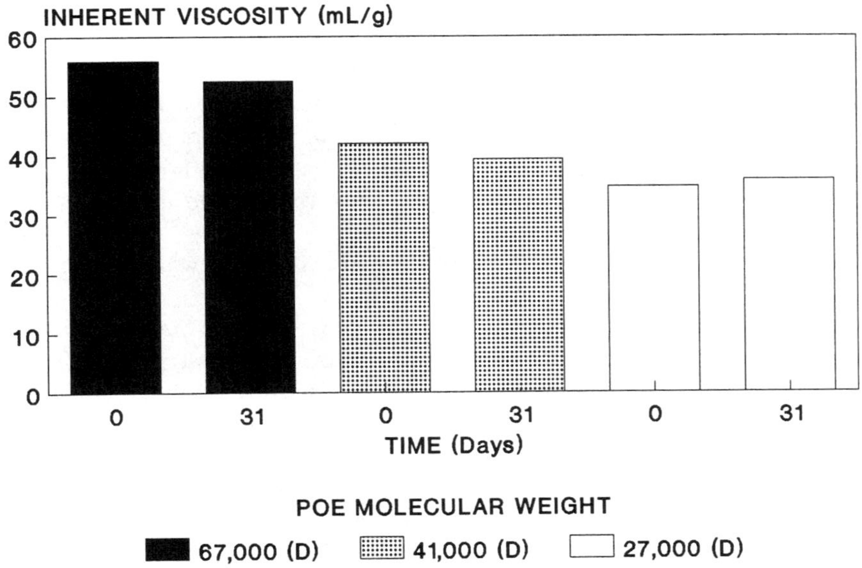

Figure 9 Effect of in vitro exposure to 7.4 pH Tris-buffered water at 37°C on inherent viscosity of different molecular weight 60 : 40 POE solvent cast films.

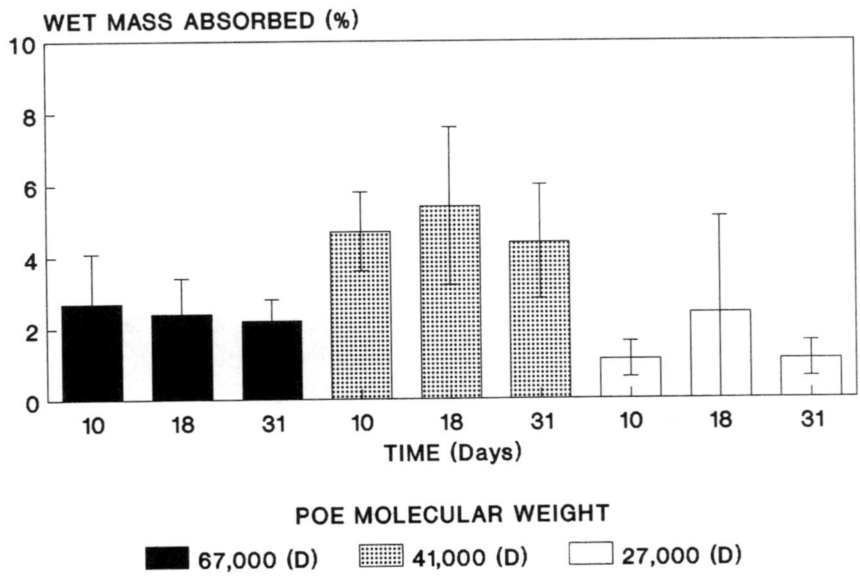

Figure 10 Effect of in vitro exposure to 7.4 pH Tris-buffered water at 37°C on wet mass absorbed (%) of different molecular weight 60 : 40 POE solvent cast films (mean ± SD).

A 60 : 40 POE polymer with a weight average molecular weight of 86 kDa was reduced to a coarse powder having a maximum particle diameter of less than 250 μm. This was accomplished by milling the polymer in a Thomas-Wiley Intermediate Mill at room temperature in open air and passing the materials through a 40-mesh screen. Test specimens of 60 : 40 POE (38.1 × 12.7 × 1.6 mm) were prepared by hot compression molding of 0.9 g of powdered polymer in a dual-plunger stainless die at 115°C and 4000 psi for 5 min using a hydraulic press with heated and water cooled platens. Polymer specimens were then cooled to ambient temperature while maintaining pressure at 4000 psi.

Twelve specimens of molded 60 : 40 POE were placed into vials filled with Tris-buffered saline (20 mM Tris/150 mM NaCl) at pH 7.4, sealed, and placed into a water bath at 37°C. Another twelve 60 : 40 POE specimens were placed in vials along with Tris-buffered saline at pH 5.0 and maintained at 37°C. Of these, three specimens were removed at 1, 3, 6, and 12 weeks from each pH solution and tested to failure in three-point bending, according to ASTM Standard D 790-81 Method 1 [30].

The effects of exposure to Tris-buffered saline pH 5.0 and pH 7.4 at 37°C are shown in Fig. 11. Initial flexural yield strength of 60 : 40 POE was 65.2 ± 0.8 MPa and the initial modulus of elasticity was 1.58 ± 0.02 GPa. POE retained approximately 80% of its initial flexural yield strength and 92% of its initial modulus of elasticity after 12 weeks in vitro exposure (pH 7.4). The final values for strength and stiffness were 50.9 ± 0.5 MPa and 1.45 ± 0.05 GPa, respectively.

Statistical analysis of variance and covariance for the two pHs showed no significant difference ($p < 0.05$) in the rate of degradation of mechanical properties:

Strength: $F(1,26) = 3.7270$, $p = 0.056$

Modulus: $F(1,26) = 0.9112$, $p = 0.394$

This seems extraordinary at first, since incorporation of as little as 0.1 wt% of an acidic excipient such as suberic acid into the polymer results in complete erosion of the polymer in a pH 7.4 buffer in as little as one day. Clearly, the insensitivity of the polymer to an external acidic environment is due to the high initial hydrophobicity of the polymer. This makes the acid-sensitive orthoester linkages virtually inaccessible to the external acid. The lack change in mechanical properties was further supported by weight loss data, showing virtually no weight loss at pH 7.4 and only minimal weight loss at pH 5.0 in 12 weeks (Fig. 12).

4. *Effect of Sterilization on Hot-Molded Specimens*

To determine the effect of radiation sterilization on both initial mechanical properties and the degradation rate of mechanical properties, nine molded 60 : 40 POE specimens, prepared as previously described, were sealed in 15-cc polystyrene vials with room air and irradiated with a cumulative dose of 2.52 mrad of gamma radiation at room temperature. Irradiation was accomplished using a cobalt 60 source, and a total exposure time was 171 min. Six of the irradiated polymer specimens were then placed in 15-cc polystyrene vials with Tris-buffered saline, at pH 7.4, sealed, and placed in a water bath at 37°C. Three specimens were removed at three and six weeks and tested to failure in three-point bending to determine flexural mechanical properties.

Additionally, three molded 60 : 40 POE specimens were sterilized with ethylene oxide (ETO). The prehumidification cycle was 1 h at 45% humidity, and the exposure cycle was 2.5 h with a 12% ETO concentration at 8 psig followed by degassing at 55°C for 48 h. Change in weight average molecular weight and polydispersity was determined before and after sterilization using gel permeation chromatography (GPC) calibrated with polystyrene

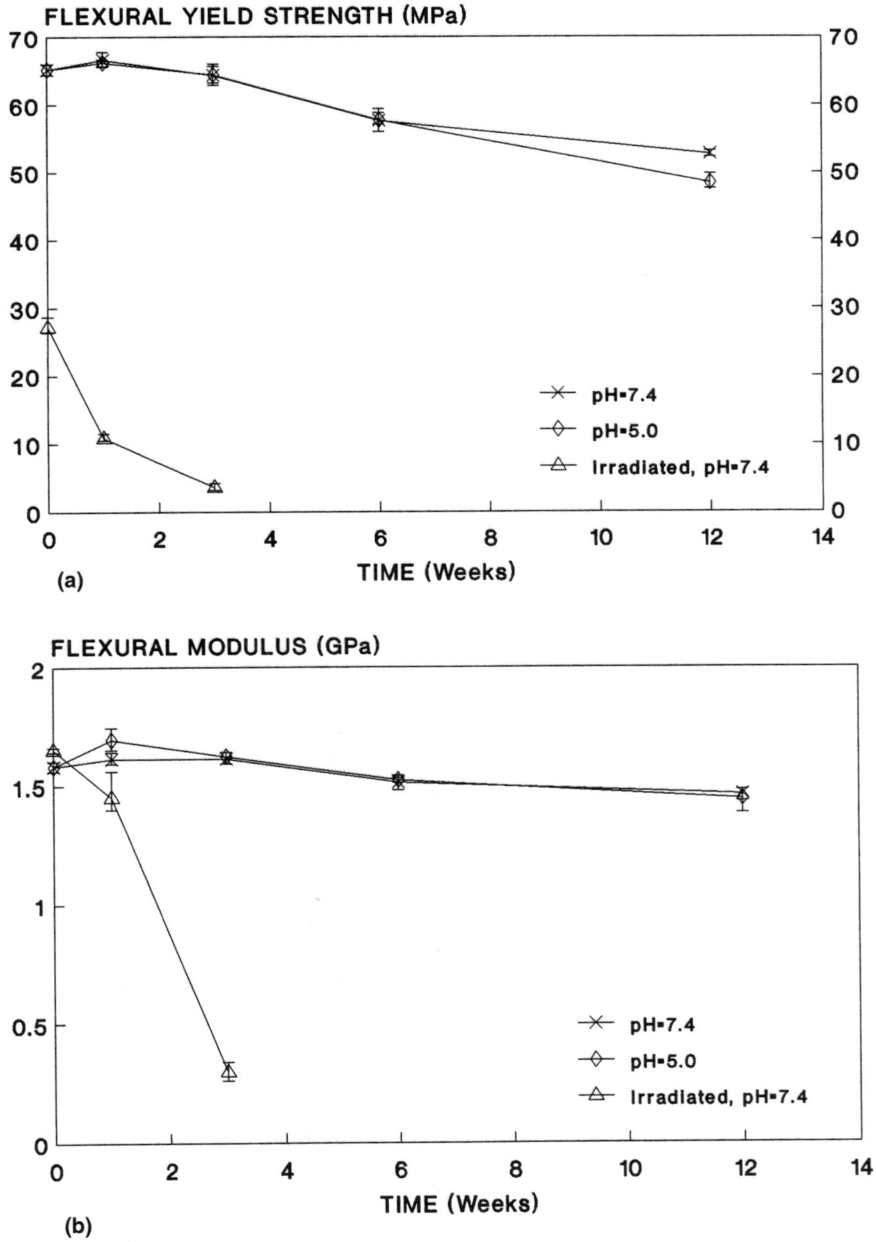

Figure 11 Effect of saline pH at 37°C in vitro and radiation sterilization on mechanical properties of hot-molded 60 : 40 POE: (A) mean flexural yield strength (MP) (mean ± SD); (B) mean flexural modulus (GPa) (mean ± SD).

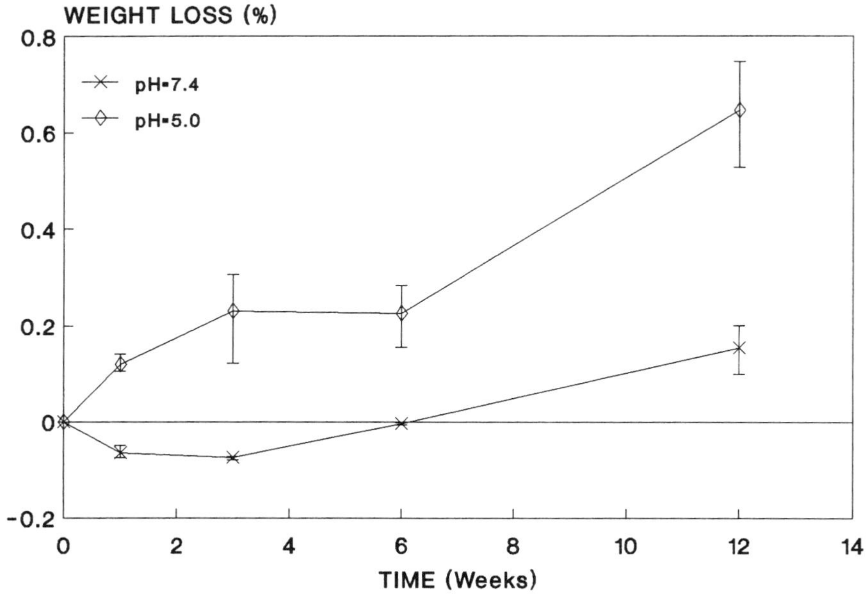

Figure 12 Percent weight loss of hot-molded 60 : 40 POE specimens after in vitro exposure to saline at pH 5.0 and 7.4 at 37°C (mean ± SD).

standards. Residual ETO was determined using a Hewlett-Packard 5890 with a flame ionization detector against ETO standards prepared weight to weight in parts per million (ppm).

Radiation sterilization (2.5 mrad) reduced the initial flexural strength by 60% (27.2 ± 1.5 MPa), and had a negligible effect on the initial modulus (1.65 ± 0.01 GPa) and markedly increased the degradation rate (Fig. 11). After three weeks exposure to Tris-buffered saline at pH 7.4 and 37°C, irradiated 60 : 40 POE retained 13% of its poststerilization flexural yield strength (3.5 ± 0.5 MPa) and 18% of its flexural modulus (0.3 ± 0.04 GPa). This latter effect is due presumably to reduction in molecular weight from radiation-induced chain scission. However, it may be possible to avoid this radiation effect. Preliminary experience suggest that there is little chain scission if radiation exposure is at dry ice temperature (−50°C) under an inert atmosphere such as argon (personal communication, J. Heller, Advanced Polymer Systems, Inc., Redwood City, CA).

Sterilization of the polymer with ETO had no measurable effect on the molecular weight or polydispersity of 60 : 40 POE. Residual ETO was measured at 3.8 ppm. The proposed limit for small (<10 g) implants is 25 ppm. No mechanical property tests or degradation studies were performed on ETO sterilized material. However, little effect is anticipated because no change in polymer molecular weight was noted.

5. *Exposure to in Vitro Mechanical Loading of Hot-Molded Specimens*

To determine the effect of in vitro exposure and mechanical loading on degradation of POEs mechanical properties, an in vitro mechanical loading study was performed.

Experimental details and fixture design have been described elsewhere [31]. Briefly, 18 molded 60 : 40 POE specimens, prepared as previously described, were exposed to intermittent cyclic loading in aerated Tris-buffered saline (pH 7.4) at 37°C. The specimens were immersed continually, but only loaded for 630 cycles at 1 Hz per day to simulate the activity of a sedentary postsurgical patient. This value was calculated from the time spent walking for a sedentary adult multiplied by an average cadence [32,33]. The load was 8.0 N. This corresponded to a stress in the outer regions of the specimen of 10% of the initial flexural yield strength of the POE polymer. Specimens were removed at 2, 4, 8, 16, 32, and 40 days and tested in three-point bending to detect changes in stress-strain behavior.

To determine to what extent loading augments the effect of the chemical environment on degradation rate, nine unloaded specimens were immersed in Tris-buffered saline at physiological pH and temperature. Three specimens were removed and tested at 7, 21, and 42 days.

Results of cyclic and unloaded exposure of 60 : 40 POE specimens to Tris-buffered saline at body temperature and pH are shown in Fig. 13. A combination of both exposure to Tris-buffered saline and intermittent cyclic loading decreased the initial flexural strength and stiffness by 75% after 40 days. The final flexural yield strength was 12.6 ± 9.3 MPa and the flexural modulus of elasticity was 0.37 ± 0.2 GPa.

For the unloaded exposures, both strength and stiffness decreased by less than 10% after 42 days. The final flexural yield strength was 57.6 ± 1.7 MPa and the flexural modulus of elasticity was 1.51 ± 0.03 GPa.

For data grouped according to type of load (no load and cyclic), strength and modulus for both loading conditions decreased in a linear fashion (Fig. 13). Flexural yield strength decreased significantly with exposure time for both no load and cyclic loading conditions. Modulus of elasticity decreased significantly with exposure time for both loading conditions (Table 3).

Analysis of variance indicated that the change in slope of flexural strength as a function of time differed significantly among the loading groups, $p < 0.000001$. In addition, the change in slope of the modulus of elasticity as a function of time also differed significantly among no loading and cyclic loading, $p < 0.000001$.

Data for the pH exposure study were obtained using specimens that were kept immersed in Tris-buffered saline in the absence of any mechanical load. Although no load degradation studies are typical for bioabsorbable polymers, they are not representative of actual use in mechanical stabilization of tissues. A real device is exposed to either a quasi-static load or, more likely, to a markedly dynamic load. The experiments reported here involved subjecting specimens to such loads during exposure. Loads were dynamic in nature and applied in four-point bending as reported elsewhere [30].

The data indicated that mechanical properties of specimens subjected to intermittent cyclic loading decreased at a significantly more rapid rate than specimens subjected no load conditions. The increase in degradation rate with mechanical loading is likely due to opening up of microscopic cracks when the surface of a specimen is in tension. These microscopic cracks then provide a path for water to penetrate into the device, which allows polymer hydrolysis to take place with consequent deterioration of mechanical properties.

This hypothesis is likely since cyclic loading of 60 : 40 POE specimens in air for 25,200 cycles at 1 Hz, the same number of cycles as for the 42-day intermittent cyclic loading exposure, showed no significant decrease in flexural strength or modulus (Fig. 14). This suggests the effects of hydrolytic exposure and mechanical loading are synergistic on the degradation of POE mechanical properties.

III. POLYMER IN VIVO DEGRADATION

A. Effect of Long-Term Canine Implantation

Granular, amorphous 60 : 40 POE polymer (Mw = 82 kDa) was hot-compression-molded as described previously, into pins 40 mm long by 4.7 mm in diameter, then sterilized with ethylene oxide.

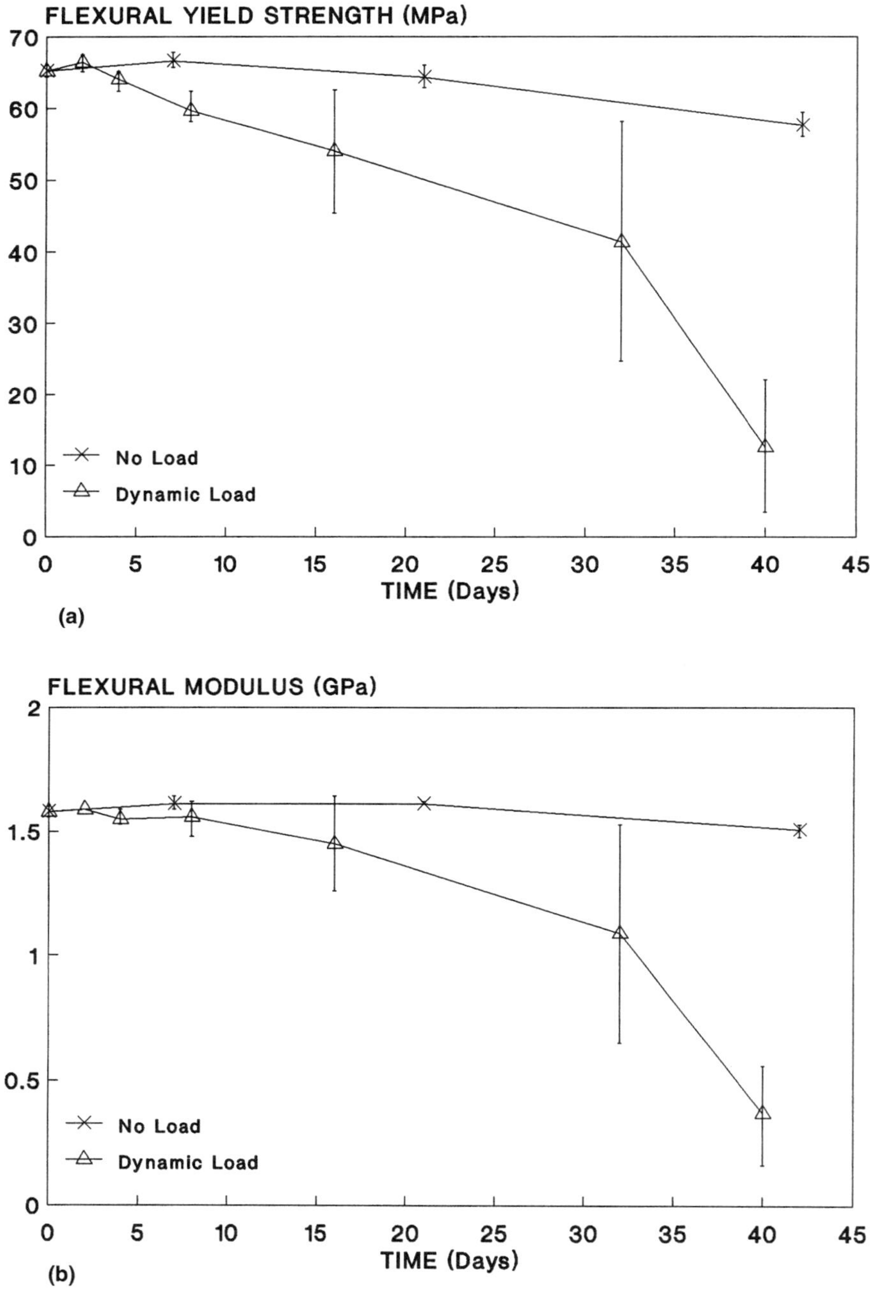

Figure 13 Comparison of in vitro mechanical loading conditions on mechanical properties of hot-molded 60 : 40 POE specimens at 37°C: (A) mean flexural yield strength (MPa) (mean ± SD); (B) mean flexural modulus (GPa) (mean ± SD).

Table 3 Linear Regression Equations for the in Vitro Mechanical Loading Study

	Flexural yield strength versus time		
No load:	$S_b = -0.2084\ (T) + 67.21$,	$r^2 = 0.7903$,	$p < 0.0001$
Cyclic load:	$S_b = -1.1727\ (T) + 69.91$,	$r^2 = 0.8131$,	$p < 0.00001$
	Flexural modulus of elasticity versus time		
No load:	$E_b = -0.0020\ (T) + 1.614$,	$r^2 = 0.4694$	$p < 0.0140$
Cyclic load:	$E_b = -0.0264\ (T) + 1.699$,	$r^2 = 0.7215$	$p < 0.00001$

S_b = flexural yield strength, E_b = flexural modulus of elasticity, T = time

A randomly chosen femur in each of 12 mongrel dogs was implanted with one pin under aseptic conditions by drilling a hole in the intertrochanteric fossa, then tapping the pins flush with the periosteum. Necropsy was performed on three dogs at 3, 6, 12, and 18 months. The femur-and-pin cross sections, 3 mm thick, were cut from each dog using a starting proximal landmark 5 mm below the lesser trochanter, then stored frozen until tested (Fig. 15).

Shear strength was recorded when pin specimens were pushed out of their host bone cross section. Pin specimens were then cleaned of soft tissue, measured for diameter and thickness to estimated volume (slight out-of-roundness resulted from implantation), then weighed. To provide a baseline state of hydration for all specimens prior to testing, and to better indicate weight loss, the specimens were vacuum dried over a few days until an asymptotic value could be inferred.

Pin and host bone cross section compressive strength and stiffness were obtained by compression testing of specimens in an Instron 1125 mechanical testing machine, using series IX software. Strength was given by the ratio of the maximum force to pin specimen initial cross-sectional area. Stiffness was the ratio of compressive displacement to initial specimen height measured between 20% and 90% of maximum force. The 20% point was convenient for eliminating initial data artifacts due to slight lack of parallelism in the specimens; the 90% point eliminated the region of plasticity from the modulus determination. Apparent pin density was estimated from measurements of dry mass and volume, which was calculated from pin dimensions assuming cylindrical pin geometry.

Figure 16 shows the compressive strength, modulus, and density of POE pins at the prescribed time periods. After 18 months of implantation, pin compressive strength and modulus decreased simultaneously in a linear fashion to less than 1% of their initial values (strength $r^2 = 0.9756$ and modulus $r^2 = 0.9603$). Pin density remained relatively constant through six months, then decreased to about 50% of its initial value at 18 months. Figure 17 shows bone mechanical properties and pin pushout shear strength at prescribed time periods. Despite the large interanimal variation in the data, both bone compressive strength and stiffness show a mean increase of 74% and 15%, respectively, after 18 months implantation, while the pin pushout shear strength decreased to 4.5% of its initial value.

Drug release studies using this family of POEs [18–22] have focused on the incorporation of acidic or basic excipients into hot-molded samples to catalyze hydrolysis at the surface or to stabilize the interior of the device against bulk hydrolysis, respectively. This is the first long-term implant study of this family of POEs that does not contain excipients. Similarity in the linear degradation rate between POE pin compressive strength and stiffness after 18 months suggests these excipient-free devices are undergoing initial surface hydrolysis. The lack of change in density in the first six months and the decrease in specimen diameter, which would

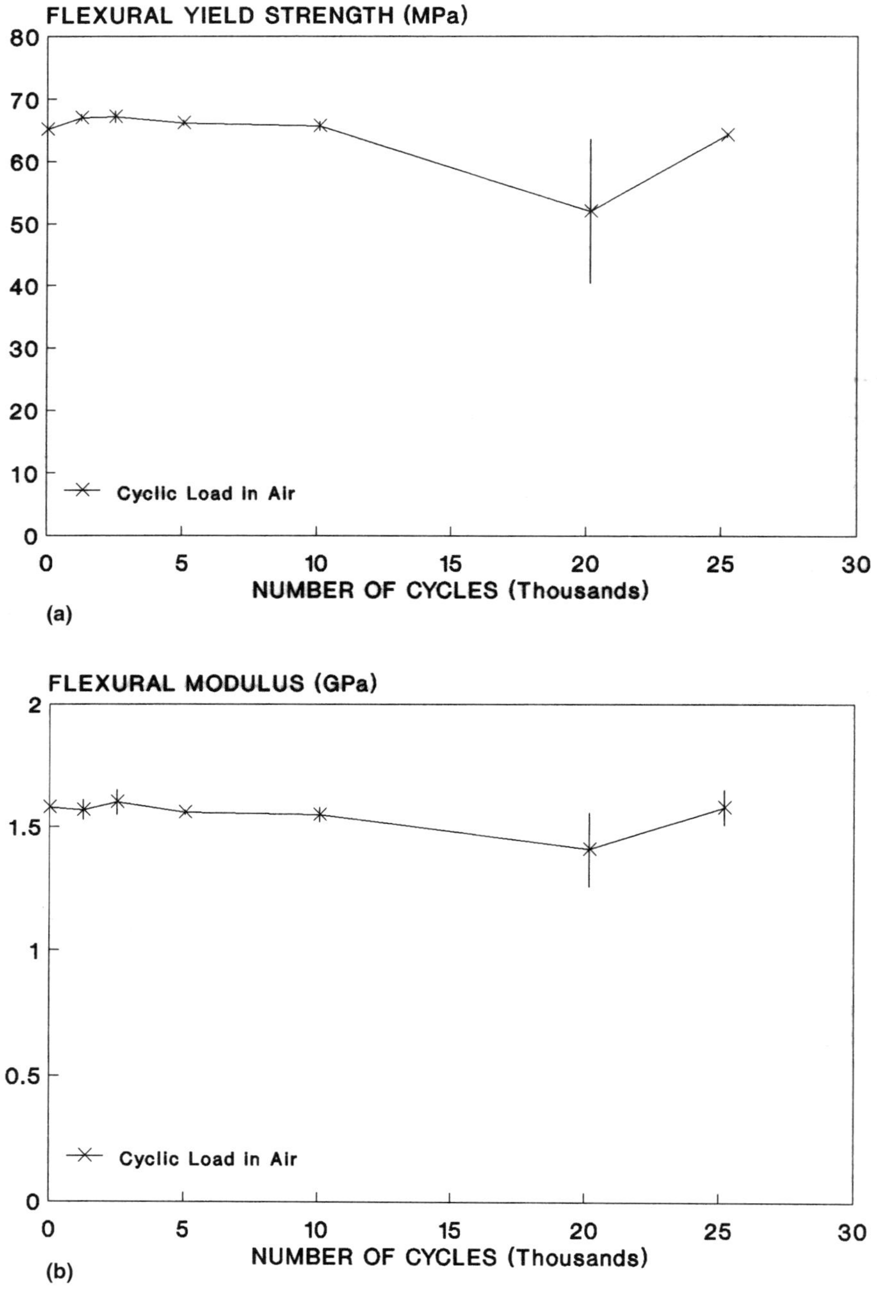

Figure 14 Effect of intermittent cyclic loading in air on the mechanical properties of hot-molded 60 : 40 POE specimens: (A) mean flexural yield strength (MPa) (mean ± SD); (B) mean flexural modulus (GPa) (mean ± SD).

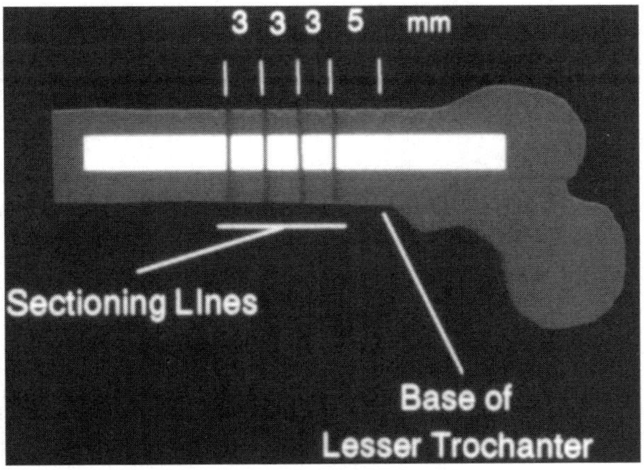

Figure 15 Schematic of the 60 : 40 POE pin implant location. Femur-and-pin cross sections, 3-mm-thick sections were cut from each femur using a starting proximal landmark 5 mm below the lesser trochanter.

also support this mechanism, were not apparent. The decrease in density at 12 months suggests eventually there is an increase in bulk hydrolysis for POE devices. This is demonstrated in POE cross sections after 18 months of implantation (Fig. 18), showing some internal resorption of the polymer. This effect is probably more pronounced in low-surface-area devices such as a rod or screw, rather than in high-surface-area devices such as films, fibers, or scaffolds. The increase in bone compressive strength and stiffness suggests a gradual transfer of load to the healing bone has occurred as the POE pin degrades. But the large data scatter from interanimal variation indicates that further study is needed to confirm this observation.

The appearance of granular particles as seen in Fig. 18 suggests that the internal resorption of POE polymer after 18 months of implantation may be associated with specimen processing. After 12 weeks of in vitro exposure, despite initial specimen transparency, particle grain boundaries appeared in hot-molded 60 : 40 POE specimens, suggesting specimen processing was a combination of melt molding and mold-sintering [34].

The appearance of grain boundaries over time would allow water to permeate the interior of the pin, and consequently lead to hydrolysis within the interior. As the hydrolysis rate increased, the buildup of degradation products in the interior, when compared to the surface layers would be less able to diffuse out of the device. The buildup of the propionic acid would cause autocatalysis within the specimen center, eventually resulting in resorption of polymer from the interior of the specimen. A similar diffusion-reaction model has been proposed for the autocatalysis noted in more hydrophilic absorbable polymers such as bulk hydrolyzing poly(α-hydroxy-acid) polymers [35–38].

IV. CONCLUSIONS

Bioabsorbable POEs based on two diols (trans-cyclohexanedimethanol and 1,6-hexanediol) and a diketene acetal (3,9-bis(ethylidene) 2,4,8,10-tetraoxaspiro[5,5]undecane), can be formed easily into solid shapes, both by solvent casting and hot compression molding. Forming, handling, and storage can be accomplished in dry air with little degradation.

Mechanical properties of solvent cast POE films are unaffected by physiological lipids

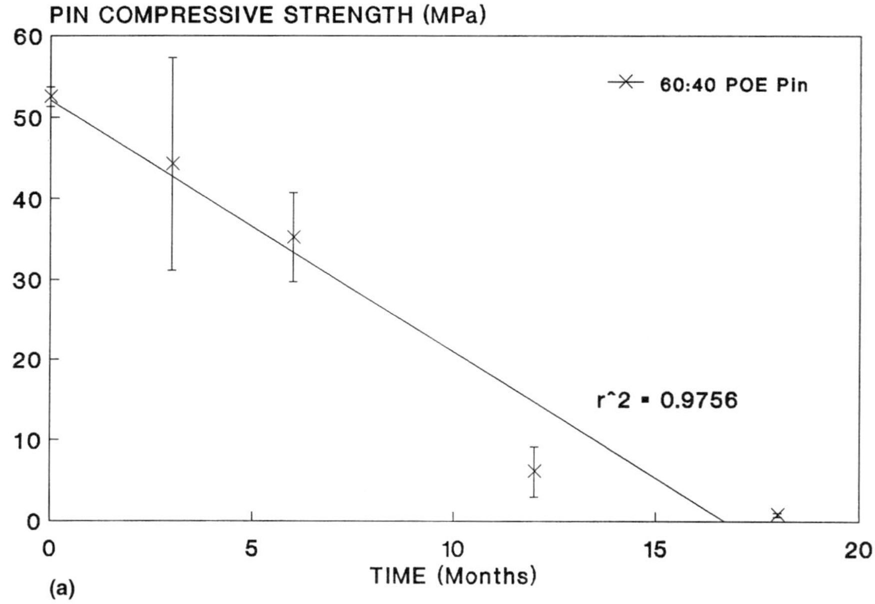

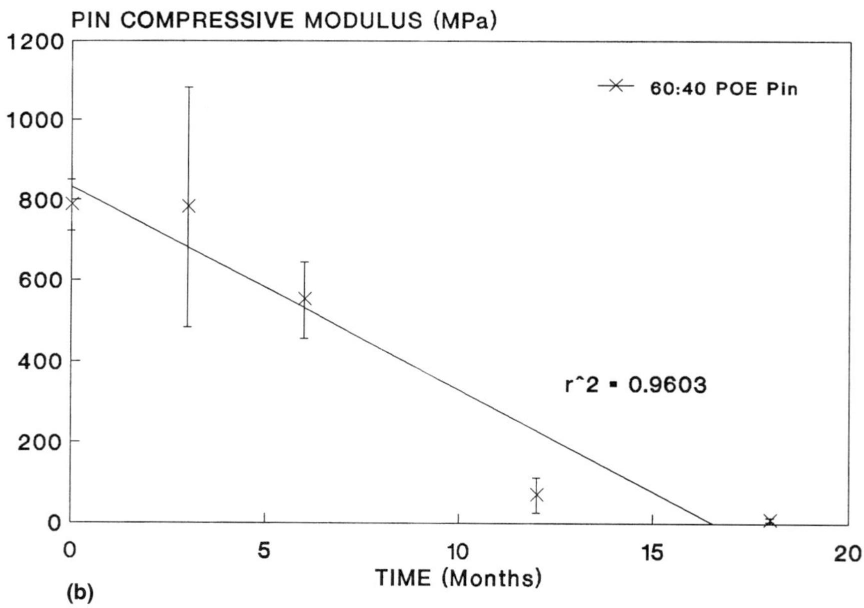

Figure 16 Effect of canine implantation time on the mechanical and physical properties of 60 : 40 POE polymer pins: (A) mean pin compressive strength (MPa) (mean ± SD); (B) mean pin compressive modulus (MPa) (mean ± SD); (C) mean pin density (g/cm^3) (mean ± SD).

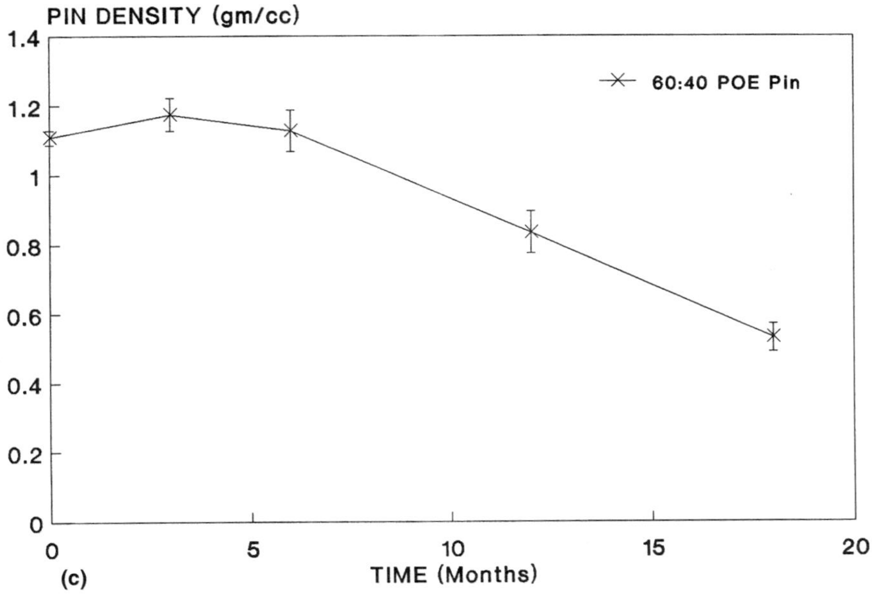

Figure 16 Continued

such as cholesterol through six weeks of in vitro exposure, beyond this time period the effect of lipid exposure is unknown. The films are strong in tension (30+ MPa tension yield) and reasonable tough (12–15% elongation to yield), except at low molecular weights. The lowest molecular weight film studied (18-kDa) was brittle. The higher molecular weight films (41 and 67 kDa) showed no decrease in mechanical properties after 31 days in physiological buffer. A film with a somewhat lower molecular weight (27 kDa) offered similar initial strength and toughness but showed a decrease in mechanical properties at 31 days. This faster degradation may be advantageous in some implant applications. The films showed a decrease in weight with exposure time but no change in either molecular weight or water absorption at 31 days. This combination supports the hypothesis that the relative hydrophobicity of POEs leads them to degrade initially by surface hydrolysis.

Hot-molded specimens degrade slowly in saline in the absence of applied loads, losing 10% of their stiffness and only about 20% of their strength in 12 weeks. The degradation is virtually unaffected by decreasing saline pH from 7.4 to 5.0. This demonstrates the relative hydrophobicity of POE because incorporating small amounts of acid within the polymer markedly increases the degradation rate. Degradation rates are increased substantially by adding dynamic mechanical loading to saline exposure. This may be the case for other absorbable polymers also, but no data of this type for other polymers were found in the open literature. Presumably, the mechanical loading opens microcracks, allowing water to enter POE specimens more easily.

This was the first long-term implant study of this family of POEs that does not contain excipients. The POE polymer pins showed a linear decrease in both compressive strength and stiffness that occurred simultaneously together over time. In contrast, many bulk hydrolyzing poly(α-hydroxy acid) polymers have been reported to show a decrease first in strength followed by a time delay before any significant decrease in stiffness [16]. As implantation time increases, there is an apparent increase in bulk hydrolysis for POE devices without excipients.

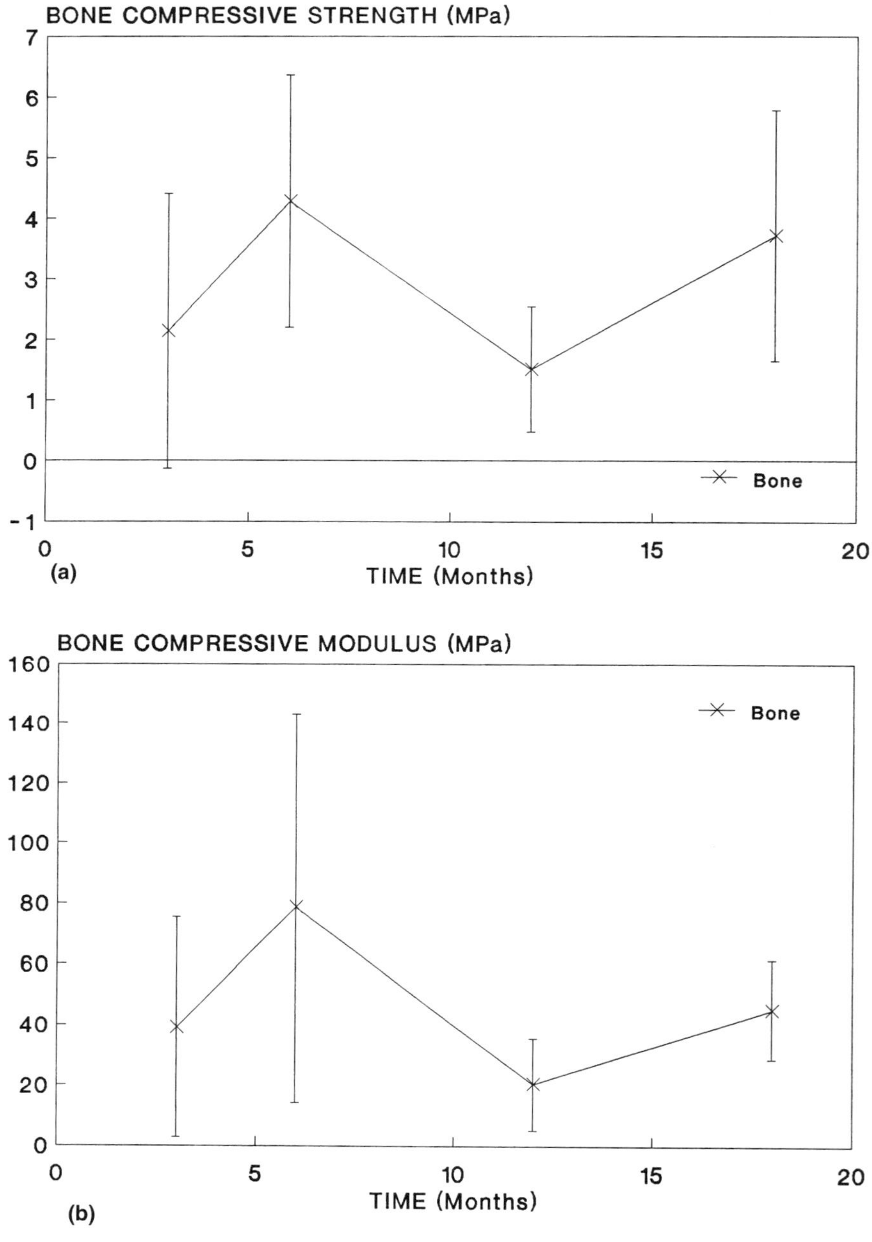

Figure 17 The 60 : 40 POE pin pushout shear strength and bone compressive mechanical properties versus implantation time: (A) mean bone compressive strength (MPa) (mean ± SD); (B) mean bone compressive modulus (MPa) (mean ± SD); (C) mean pushout shear strength (kPa) (mean ± SD).

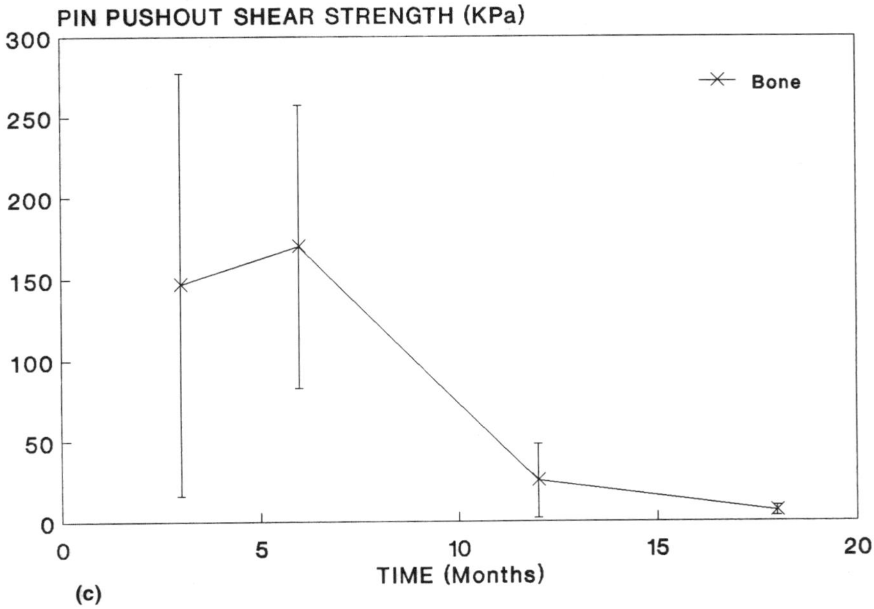

Figure 17 Continued

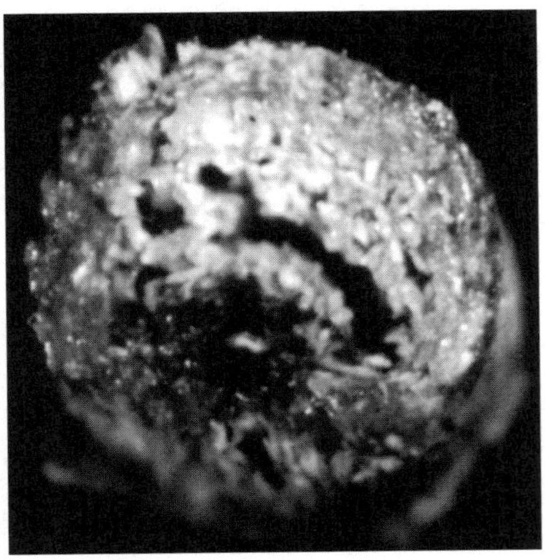

Figure 18 A representative POE cross section after 18 months of canine implantation time showing some internal resorption of polymer, suggesting bulk hydrolysis may be increasing with implantation time.

In summary, POEs offer a promising combination of fabricability, mechanical properties, and degradation rates for surgical implants intended for temporary tissue fixation devices. Initial strength and stiffness of POEs is less than that reported for some poly(α-hydroxy acid) polymers. However, the mechanical properties needed must be determined for each clinical application and POE mechanical properties appear to be sufficient for many cases. Further studies of this material system are warranted particularly to determine long-term degradation rates of actual fixation devices and biocompatibility of degradation products.

ACKNOWLEDGMENTS

Support was provided by Technology and Ventures Division, Baxter Health Care, Incorporated, Irvine, CA; Osteotech, Incorporated, Shrewsbury, NJ; Biomet, Incorporated, Warsaw, IN; SRI International, Menlo Park, CA; and the Orthopedic Bioengineering Laboratory at the University of Utah School of Medicine, Salt Lake City, UT.

REFERENCES

1. Andriano, K. P. and J. Heller, Bioerodible polymers: Use in fracture fixation. Salamone, J. C. ed. *The Polymeric Materials Encyclopedia: Synthesis, Properties, and Applications*. Boca Raton, FL: CRC Press, 1:610–615 (1996).
2. Bucholz, R. W., S. Henry, and M. B. Henley, Fixation with bioabsorbable screws for the treatment of fractures of the ankle. *J. Bone Joint Surg. Am.*, 76:319–324 (1994).
3. Pistner, H., R. Gutwald, R. Ordung, J. Reuther, and J. Muhling, Poly(L-lactide): A long-term degradation study in vivo. I. Biological results. *Biomaterials*, 14:671–677 (1993).
4. Matsusue, Y., S. Hanafusa, T. Yamamuro, Y. Shikinami, and Y. Ikada, Tissue reaction of bioabsorbable ultra high strength poly(L-lactide) rod: A long-term study in rabbits. *Clin. Orthop. Rel. Res.*, 317:246–253 (1995).
5. Claes, L. E., A. A. Ignatius, and C. Scholz, New bioresorbable pin for the reduction of small bony fragments: Design, mechanical properties, and in vitro degradation. *Biomaterials*, 17:1621–1626 (1996).
6. Bostman, O., E. Hirvensalo, S. Vainionpaa, A. Makela, K. Vihtonen, P. Tormala, and P. Rokkanen, Ankle fractures treated using biodegradable internal fixation. *Clin. Orthop. Rel. Res.*, 238:195–203 (1989).
7. Bostman, O. Absorbable implants for the fixation of fractures. *J. Bone Joint Surg.*, 73:148–153 (1991).
8. Tormala, P., J. Vasenius, S. Vainionpaa, and J. Laiho, Ultra-high-strength absorbable self-reinforced polyglycolide (SR-PGA) composite rods for internal fixation of bone fractures: in vitro and in vivo study. *J Biomed. Mater. Res.* 25:1–22 (1991).
9. Winet, H. and J. O. Hollinger, Incorporation of polylactide-polyglycolide in a cortical defect: Neo-osteogeneis in a bone chamber. *J. Biomed. Mater. Res.*, 27:667–676 (1993).
10. Gilding, D. K. and A. M. Reed, Biodegradable polymers for use in surgery—poly(glycolic)/poly(-lactic acid) homo- and copolymers. *Polymer*, 20:1459–1464 (1979).
11. Lin, H. L., C. C. Chu, and D. Grubb, Hydrolytic degradation and morphological study of poly-ρ-dioxanone. *J. Biomed. Mater. Res.*, 27:153–166 (1993).
12. Ray, J. A., N. Doddi, D. Regula, J. A. Williams, and A. Melveger, Polydioxanone (PDS), a novel monofilament synthetic absorbable suture. *Surg. Gynecol. Obstet.*, 153:497–507 (1981).
13. Plitt. C. G. Poly-ε-caprolactone and its copolymers. Chasin, M. and R. Langer, eds. *Biodegradable Polymers as Drug Delivery Systems*. New York: Marcel Dekker, 71–120 (1990).
14. Choueka, J., J. L. Charvet, K. J. Koval, H. Alexander, K. S. James, K. A. Hooper, and J. Kohn,

Canine bone response to tyrosine-derived polycarbonates and poly(L-lactic acid). *J. Biomed. Mater. Res.*, 31:35–41 (1996).
15. Ertel, S. I., J. Kohn, M. C. Zimmerman, and J. P. Parsons, Evaluation of poly(DTH carbonate), a tyrosine-derived degradable polymer, for orthopedic applications. *J. Biomed. Mater. Res.*, 29:1337–1348 (1995).
16. Daniels, A. U., M. K. O. Chang, K. P. Andriano, and J. Heller, Mechanical properties of biodegradable polymers and composites proposed for internal fixation of bone. *J. Appl. Biomater.*, 1:55–78 (1990).
17. Hofmann, G. O. Biodegradable implants in orthopedic surgery—A review on the state-of-the-art. *Clin. Mater.*, 10:75–80 (1992).
18. Heller, J., Poly(orthoesters). *Adv. Polym. Sci.*, 107:43–92 (1993).
19. Heller, J., D. W. H. Penhale, B. K. Fritzinger, and S. Y. Ng, Controlled release of contraceptive agents from poly(orthoesters). *Cont. Del. Sys.*, 4:43–53 (1983).
20. Heller, J. and K. Himmelstein, Poly(orthoester) biodegradable polymer systems. *Meth. Enzymol.*, 112:422–436 (1985).
21. Heller, J., Controlled drug release from poly(orthoesters). *Ann. N.Y. Acad. Sci.*, 466:51–66 (1985).
22. Heller, J., S. Y. Ng, D. W. H. Penhale, B. K. Fritzinger, L. M. Sanders, R. A. Burns, A. G. Gaymon, and S. S. Bhosale, The use of poly(orthoesters) for the controlled release of 5-fluroacil and a LHRH analogue. *J. Cont. Rel.*, 6:217–224 (1987).
23. American Society for Testing and Materials, ASTM D 638-82. Standard test method for tensile properties of plastics. Philadelphia: American Society for Testing and Materials (1982).
24. Moacanin, J. and D. D. Lawson, Prediction of lipid uptake by prosthetic heart value poppets from solubility parameters. *Biomater. Med. Dev. Artif. Organs*, 1(1):183–190 (1973).
25. Masoro, E. J., *Physiological Chemistry of Lipids in Mammals*. Philadelphia: W. B. Saunders (1968).
26. Young, R. J., *Introduction to Polymers*. New York: Chapman and Hall (1987).
27. Weast, R. C., ed., *CRC Handbook of Chemistry and Physics*, 57th ed., Ohio: CRC Press (1975).
28. Rabinowitz, J. L., J. R. Gregg, J. E. Nixon, H. R. Schumacher, Lipid composition of tissues human knee joints. I. Observations in normal joints (articular cartilage, meniscus, ligaments, synovial fluid, synovium intra-articular fat pad and bone marrow). *Clin. Orthop. Rel. Res.*, 143:260–265 (1979).
29. Rabinowitz, J. L., J. R. Gregg, J. E. Nixon, Lipid composition of the tissues of human knee joints. II. Synovial fluid in trauma. *Clin. Orthop. Rel. Res.*, 190:292–298 (1984).
30. American Society for Testing and Materials, ASTM D 790-81. Standard test methods for flexural properties of unreinforced and reinforced plastics and electrical insulating materials. Philadelphia: American Society for Testing and Materials (1981).
31. Smutz, W. P., A. U. Daniels, K. P. Andriano, E. P. France, and J. Heller, Mechanical test methodology for environmental exposure testing of biodegradable polymers. *J. Appl. Biomater.*, 2:13–22 (1991).
32. Cochran, G. V. B., *A Primer of Orthopedic Biomechanics*. New York: Church Livingstone (1982).
33. Schoenborn, C. A., Health habits of U.S. adults. *Public Health Reports*, 101(6):571–589 (1986).
34. Andriano, K. P., Development of Microfiber Reinforced Biodegradable Composites for Implant Use. University of Utah, Ph.D. thesis (1990).
35. Li, S., H. Garreau, and M. Vert, Structure-property relationships in the case of the degradation of massive aliphatic poly(alpha-hydroxy-acids) in aqueous media, Part 1: Poly(DL-lactic acid). *J. Mater. Sci.: Mater. Med.*, 1:123–130 (1990).
36. Li, S., J. Garreau, and M. Vert, Structure-property relationships in the case of the degradation of massive aliphatic poly(alpha-hydroxy-acids) in aqueous media, Part 2: Degradation of lactide-glycolide copolymers: PLA37.5GA25 and PLA75GA. *J. Mater. Sci.: Mater. Med.*, 1:131–139 (1990).
37. Li, S., J. Garreau, and M. Vert, Structure-property relationships in the case of the degradation of massive aliphatic poly(alpha-hydroxy-acids) in aqueous media, Part 3: Influence of morphology on poly(L-lactic acid). *J. Mater. Sci.: Mater. Med.*, 1:198–206 (1990).
38. Therin, M., P. Christel, S. Li, H. Garreau, and M. Vert, In vivo degradation of massive poly(alpha-hydroxy acids): validation of in vitro. *Biomaterials*, 13:594–600 (1992).

26

Buffered Biodegradable Internal Fixation Devices

Debra J. Trantolo, Joseph D. Gresser, and Donald L. Wise
Cambridge Scientific, Inc., Cambridge, Massachusetts

Kai-Uwe Lewandrowski
Massachusetts General Hospital, Boston, Massachusetts

I. INTRODUCTION

Of the more than 1.1 million fractures in the United States each year, greater than 470,000 require some type of internal fixation device to stabilize the fracture during the healing process. Currently, the vast majority of these devices are made of metal. Following the healing of a fracture, a subsequent surgical procedure must be performed to remove the nonresorbable metal pins, plates and/or screws. While there is significant clinical demand for resorbable fixation devices to eliminate the added cost and patient discomfort associated with the removal process, currently available resorbable products have not been widely adopted due to their inability replicate the structural strength of their metal counterparts and/or due to their tendency to cause an inflammatory response associated with acidic degradation products.

The reported work focuses on the issue of eliminating this inflammatory response, including the formation of sterile abscesses, while maintaining sufficient structural integrity of the device allowing bone healing to occur. It is our thesis that the inflammatory response can be eliminated by incorporating a long-acting buffer into a poly(lactic-*co*-glycolic acid), PLGA, fixation device which will neutralize the acid products as they form. The buffer must be retained within the polymer as degradation proceeds and must also be tissue compatible.

We have investigated the feasibility of incorporating a buffer into a resorbable PLGA-based internal fixation device (IFD). Hydroxyapatite (HA) and calcium carbonate (CC) were each investigated as buffering fillers. Initial studies demonstrated that the inclusion of CC or HA in a PLGA fixture can effectively moderate the rate of pH decline as the fixture degrades. Further, the rapid decline in pH can be offset without considering 100% neutralization of the lactic and glycolic components. Thus, even given that the polymeric fixture will be filled with an inorganic buffer, the mechanical characteristics of the fixture can be stabilized since the loading requirements for the buffer will not be nearly as compromising as expected at the outset.

While both CC and HA can ameliorate the decline in pH in the region of polymer hydrolysis, the use of hydroxyapatite as a filler also supports osteoconductivity [1–4]. Thus, HA not only acts as a buffer thereby preventing the formation of sterile abscesses that have been attributed to the acidic degradation products of PLGA implants [5–13], but also promotes bony ingrowth and obviates loosening of the fixture which additionally can contribute to implant failure. The resulting resorbable fixture should be capable of a buffered hydrolytic degradation and induction of bony ingrowth as resorption of the implant progresses. A resorbable buffered bone fixture with such properties could provide structural support to stabilize and support a fracture over the period of time required for natural healing to occur.

II. BACKGROUND AND SIGNIFICANCE

A. Clinical Indications for the Use of Resorbable Internal Fixation Devices

Major clinical applications of bioresorbable fixtures have included minimally displaced malleolar fractures and flake fractures around the ankle [14–16] and elbow [17]. Other areas of clinical application have included simple distal radial fractures [18] and hand fractures [19]. In most of these procedures removal of implants is not required and seldom performed unless there is discomfort at the implant site [20,21]. Typical complaints listed by patients treated for malleolar fractures that ultimately prompted implant removal included soreness over implant and cicatrix, reduced movement of the ankle joint, and strain-related pain. After implant removal, improvement was reported in about 75% of these patients [22]. Bioresorbable implants therefore offer an attractive alternative, ensuring fracture fixation yet eliminating the need for implant removal. This may be of particular advantage in the elderly in whom medical contraindications to anesthesia and the risk accompanied by surgical procedures are frequently present.

These advantages of resorbable fixtures are of increasing significance due to the expected raised incidence of osteoporotic ankle fractures. A European study has found evidence that the number of ankle fractures is increasing at a rate that cannot be explained simply by demographic changes because of osteoporosis; a condition that affects primarily the elderly patient population [23]. A similar study in the United States [24] has corroborated these findings and has identified other risk factors for ankle fractures such as one or more falls in the year prior to baseline, greater vigorous physical activity, weight gain since age 25, self-reported osteoarthritis, a sister's history of hip fracture after age 50, out of house ≤1 per week, and low distal radius BMD. Factors associated with foot fracture included insulin-dependent diabetes, use of seizure medications or of benzodiazepines, history of hyperthyroidism, poor far depth perception, and low distal radius BMD. Evidently, ankle and foot fractures have different profiles of risk factors that are not only related to low bone mass but appear also associated to decreased function in the elderly population. Reduced hospitalization and early return to home are therefore key components in the management of these patients and resorbable fixtures providing definite treatment without subsequent procedures may prove highly applicable for these patients.

Fractures suitable for fixation with the use of resorbable fixtures such as screws include those with minimal displacement commonly seen in fractures of the ankle [25–30] and foot [23,31]. As such, transverse fibular fractures at/or distal to the ankle joint (Danis-Weber Type A), low spiral fractures of the lateral malleolus and avulsion of the medial malleolus [Dyputren; Danis-Weber Type B, avulsion fractures of the posterior malleolus (Volkmann's Triangle)] may be treated with anatomical (less than 2 mm displacement or posterior malleolar fragment <25% of joint surface) [27,32] reconstruction by open reduction without casting and early mobilization both preventing complications of long-term immobilization such as thrombosis. Other fractures suitable for treatment with resorbable fixtures may include distal radial fractures (Type Colles), which present the most common type of fractures of the radius (85%). They typically occur at the metaphyseal portion of the bone proximal to the extensor tendon sheaths (loco typico sive classico). Of the various clinical types, isolated stable Colles fractures without major fragment dislocation and without (Malone Type I [33]) or with dorsal or volar deviation (Malone Type II [33]), frequently treated with K-wires, appear suitable for treatment with bioresorbable screws. This application has been reported in the literature [18]. Similarly, isolated dislocated fractures of the Olecranon at the elbow joint (Wadsworth Type I and II [34]), and metacarpal hand or metatarsal foot fractures, both commonly treated with wires, are suitable for treatment with bioresorbable fixtures.

B. Clinical Complications in the Use of Resorbable Internal Fixation Devices

The trend in internal fixation devices is toward the use of resorbable, tissue compatible biopolymers such as poly(glycolic acid) [6–8,11]/(PGA), poly(lactic acid) [13] (PLA), and copolymers of lactic and glycolic acids, the poly(lactide-*co*-glycolic acids) (PLGAs) [35]. Although other polymers have also been considered, such as poly(dioxanone) [36], the PGA, PLA and PLGA polyesters have a long history of clinical use (e.g., as sutures). However, the tissue response to resorbable implants fabricated from these polyesters has not been uniformly acceptable; investigators have reported a late sterile inflammatory foreign body response (sterile abscess) [8,11,15].

In-depth follow-up of several large clinical studies has substantiated the complications with the use of resorbable polymer fixtures. The resorbable systems now in use show promise, but problems such as sterile fluid collections and osteolytic changes around biodegradable implants and implant fatigue remain. Clinical results clearly indicate that problems still exist with these fixtures and given the great need for resorbable implants, there is room for more extensive examination and material improvement. In up to 22% of the cases where resorbable internal fixation devices fabricated from biodegradable lactic and glycolic polymers (polylactic-*co*-glycolic acids) are used in surgical repair of fractures, severe inflammatory responses reflected by sterile abscess formation are noted [11,13,14,19,20]. These processes become evident only after fairly long induction periods ranging from 7 to 20 weeks (mean induction period of about 12 weeks) [8].

In a randomized study of 56 open reduction and internal fixation of malleolar fractures of the ankle with metal ASIF screws and plates or with rods of PLGA, two cases of sterile inflammatory wound sinus were observed three to four months after the operation in the injuries fixed with the polymer rods [9]. In a study of 64 bimalleolar fractures fixed with PGA rods, Bostman [7] reported a late inflammatory sinus in five cases (8.1%). In a series of 29 patients with intra-articular fractures about the elbow joint, internal fixation using PGA rods resulted in a late noninfectious inflammatory reaction in four cases (13.8%) [17]. Hoffmann et al. used PGA pins in repair of 40 cases of displaced fractures of the distal radius [18]. They observed a late inflammatory response requiring debridement in nine patients (22.5%). A study of 217 fracture cases, mostly of the ankle, repaired with fiber-reinforced PGA screws revealed 10 cases (4.6%) of inflammatory tissue reaction requiring operative drainage [33]. Another study identified ovoid osteolytic foci in 34 patients of a total of 67 patients with displaced malleolar fractures treated by open reduction and internal fixation using absorbable polyglycolide rods [10]. Seventeen patients developed a discharging inflammatory foreign-body reaction, a complication unique to these fixation devices. These osteolytic foci, usually 5 to 10 mm in diameter, appeared within the implant channels 6 to 12 weeks after the operation. The same lesion occurred in 14 of the 17 patients who developed a foreign-body reaction, whereas only

20 out of the 50 patients with an uneventful course showed osteolytic areas. After one year the normal structure of the bone was restored. Finally, a review by Bostman of 516 patients revealed that a late inflammatory response requiring operative drainage occurred in 7.9% of these cases [5]. Internal fixation of the fractures was accomplished with devices fabricated from either PGA or a PLGA copolymer.

The clinical presentation of this complication is fairly consistent. Bacterial cultures of the drainage are routinely negative. It was therefore concluded that the biological response must be a consequence to the chemical irritation from the polymers [5]. The delayed appearance of the inflammatory response may be associated with the kinetics of the polymer degradation. Comparative in vitro and in vivo studies on the decomposition of 10 lactide/glycolide copolymers have shown that the rate of hydrolysis was independent of the pH range studied and was virtually identical in vivo and in vitro [37].

III. DESIGN OF BUFFERED PLGA IFDs

This project focused on the issue of eliminating the inflammatory response associated with the use of resorbable PLGA-based fixtures, while maintaining sufficient structural integrity of the device allowing bone healing to occur. It is our thesis that the inflammatory response can be eliminated by incorporating a long-acting buffer into a PGLA fixation device which will neutralize the acid products as they form. The buffer must be retained within the polymer as degradation proceeds and must also be tissue compatible.

We investigated the feasibility of incorporating a buffer into a resorbable PLGA-based internal fixation device (IFD). Two PLGA copolymers were investigated: (1) 50 : 50 PLGA and (2) 85 : 15 PLGA (where the ratios specify the molar relationship between the lactic, L, monomer and the glycolic, G, monomer). Calcium carbonate (CC) and the calcium phosphates, dibasic calcium phosphate (DBCP) and hydroxyapatite (HA), were investigated as buffering fillers. While all of these buffers can ameliorate the decline in pH in the region of polymer hydrolysis, the use of the hydroxyapatite as a filler also supports osteoconductivity [1–4]. Thus, HA not only promotes bony ingrowth and obviates loosening of the fixture, but also acts as a buffer thereby preventing the formation of sterile abscesses that have been attributed to the acidic degradative products of PLGA implants [5–12,38]. The resulting resorbable fixture should thus be capable of a buffered hydrolytic degradation and induction of bony ingrowth as resorption of the implant progresses. A resorbable buffered bone fixture with such properties could provide structural support to stabilize and support a fracture over the period of time required for natural healing to occur. This fixture appears particularly applicable to the minimally displaced malleolar fractures and flake fractures around the ankle [14–16] and elbow [17]. Other areas of clinical application include simple distal radial fractures [18] and hand fractures [19].

We demonstrated that the inclusion of CC, HA, or DBCP in a PLGA fixture can effectively moderate the rate of pH decline as the fixture degrades. Further, the rapid decline in pH can be offset without considering 100% neutralization of the lactic and glycolic components. Thus, even given that the polymeric fixture will be filled with an inorganic buffer, the mechanical characteristics of the fixture can be stabilized since the loading requirements for the buffer will not be nearly as compromising as expected at the outset.

Continued studies have resulted in further in vitro mechanical and in vivo histomorphological analyses both of which provide additional support to choice of an HA-filled fixture. It is the thesis that HA will ameliorate sterile abscess formation via control of both pH and implant loosening.

IV. RATIONALE FOR BUFFER SELECTION

The buffering, or neutralizing capacity, of the salts investigated in this study were based on the reactions of the acidic products of PLGA hydrolysis (lactic and glycolic acids) with these salts. The neutralization reactions are assumed to be stoichiometric because the neutralization products are the conjugate acids of strong bases and are consequently themselves very weakly ionized. Also, one of the agents, calcium carbonate, produces carbon dioxide as an end product, insuring the completion of the reaction. Relevant dissociation constants and reactions are given below, where HAn is either lactic or glycolic acid.

Dibasic calcium phosphate (DBCP)

$$CaHPO_4 + HAn \rightarrow Ca^{+2} + H_2PO_4^{-1} + An^{-1}$$

$$H_3PO_4 \rightarrow H^+ + H_2PO_4^{-1} \qquad K_1 = 7.1 \times 10^{-3}$$

$$H_2PO_4^{-1} \rightarrow H^+ + HPO_4^{-2} \qquad K_2 = 8.0 \times 10^{-8}$$

$$HPO_4^{-2} \rightarrow H^+ + PO_4^{-3} \qquad K_3 = 4.8 \times 10^{-13}$$

Hydroxy apatite (HA)

$$Ca_{10}(OH)_2(PO_4)_6 + 8HAn \rightarrow 10Ca^{+2} + 8An^{-1} + 2HOH + 6HPO_4^{-2}$$

$$HOH \rightarrow H^+ + OH^- \qquad Kw = 10^{-14}$$

Calcium carbonate (CC)

$$CaCO_3 + 2HAn \rightarrow Ca^{+2} + 2An^{-1} + H_2CO_3 \rightarrow HOH + CO_2$$

$$H_2CO_3 \rightarrow H^+ + HCO_3^{-1} \qquad K_1 = 4.2 \times 10^{-7}$$

$$HCO_3^{-1} \rightarrow H^+ + CO_3^{-2} \qquad K_2 = 5.6 \times 10^{-11}$$

For a poly(lactide-*co*-glycolide) of any lactide:glycolide ratio, the residue molecular weight (RMW) can be calculated from the polymer composition where x = the mole fraction of lactide:

$$\text{—(O—CH(CH}_3\text{)—CO)}_x\text{—(O—CH}_2\text{—CO)}_{(1-x)}$$

lactide (72.06) glycolide (58.04)

$$RMW = 72.06x + 58.04(1 - x) = 14.02x + 58.04$$

Thus, for PLGA-85 : 15 (where 85 : 15 is the lactide:glycolide ratio), RMW = 69.96 and 1.0 g comprises (69.96) – 1 = 0.01429 mole. The formula weight of calcium carbonate is 100.09. Thus for stoichiometric neutralization (100% buffering), sufficient CC must be present to neutralize 0.01429 mole of acids produced by polymer hydrolysis. This will be 0.5(0.01429)(100.09) = 0.715 g. The weight percent of CC needed will be 0.715/(1.0 + 0.715) = 41.7%. The loadings required to achieve any percent of neutralization can be similarly calculated for other lactide: glycolide ratios.

V. SAMPLE PREPARATION

PLGA copolymers (GMP quality, Boehringer Ingelheim, Resomer 858 for the 85 : 15 PLGA and Resomer 508 for the 50 : 50 PLGA) were purified by precipitation from an acetone solution into isopropanol. After vacuum drying, the polymer was cryogenically ground to <125 μm in a Tekmar A10 Analytical mill. The ground polymer was blended with the buffering agent milled to an average particle size of 1.2 μm. Compositions of the PLGA/buffer blends were chosen to achieve various percent neutralizations of the acidic degradation products of polymer hydrolysis. Compositions to achieve a desired percent neutralization were calculated as previously described. PLGA/buffer samples were blended overnight on a rotary blender, compressed at 10 tons force exerted on a 1 inch diameter ram and then extruded at 13 tons force (1 inch diameter ram) at 49–54°C.

VI. IN VITRO PH MEASUREMENTS

Extruded rods of approximately 3 cm length were incubated in stoppered vials at 37°C in 25 ml distilled water purged with nitrogen and adjusted to pH 7.0. (HPLC grade water was used, rather than a standard buffered medium, to allow for monitoring of pH.) Measurements of pH were taken periodically on replicates of five samples of each composition.

A. PLGA-85 : 15 Buffered with Calcium Phosphates

The buffering or neutralizing capacity of several calcium salts was tested in extruded rods of PLGA-85 : 15, prepared by extrusion as previously described. These included dibasic calcium phosphate (DBCP, $CaHPO_4$) and hydroxyapatite (HA, $Ca_{10}(OH)_2(PO_4)_6$) Both were included at loadings designed to achieve 25% and 50% neutralization based on the stoichiometry of the reaction with completely hydrolyzed polymer. Loadings on a weight percent basis are given in the table.

Neutralizing agent	Percent neutralization	Weight percent
$CaHPO_4$	25	32.7
$CaHPO_4$	50	49.3
$Ca_{10}(OH)_2(PO_4)_6$	25	30.5
$Ca_{10}(OH)_2(PO_4)_6$	50	46.8

Samples were incubated at 37°C in water, and the pH monitored periodically. The pH versus time results are presented in Fig. 1. During the period of rapid decline, the absolute value of the slopes of the pH versus time curves decrease with increasing buffer content, indicating the effectiveness of these agents in control of the rate of change of pH.

B. PLGA-85 : 15 and –50 : 50 Buffered with Calcium Carbonate

The pH of all samples decline with time, at rates decreasing with increasing $CaCO_3$ content. For the 50 : 50 PLGA, pH remains fairly steady before a decrease is observed at approximately 21 days; for the 85 : 15 PLGA, the decrease is observed at approximately 90 days. As reported upon in the literature [39], PLGA copolymers filled with calcium carbonate swell in vitro. Given the gross distortion of the samples, these samples were not attractive candidates for continued consideration.

C. System Comparison

The purpose of the buffer is twofold: (1) to neutralize at least a portion of the acids formed by polymer hydrolysis, and (2) to control the rate at which pH declines. It has been shown using

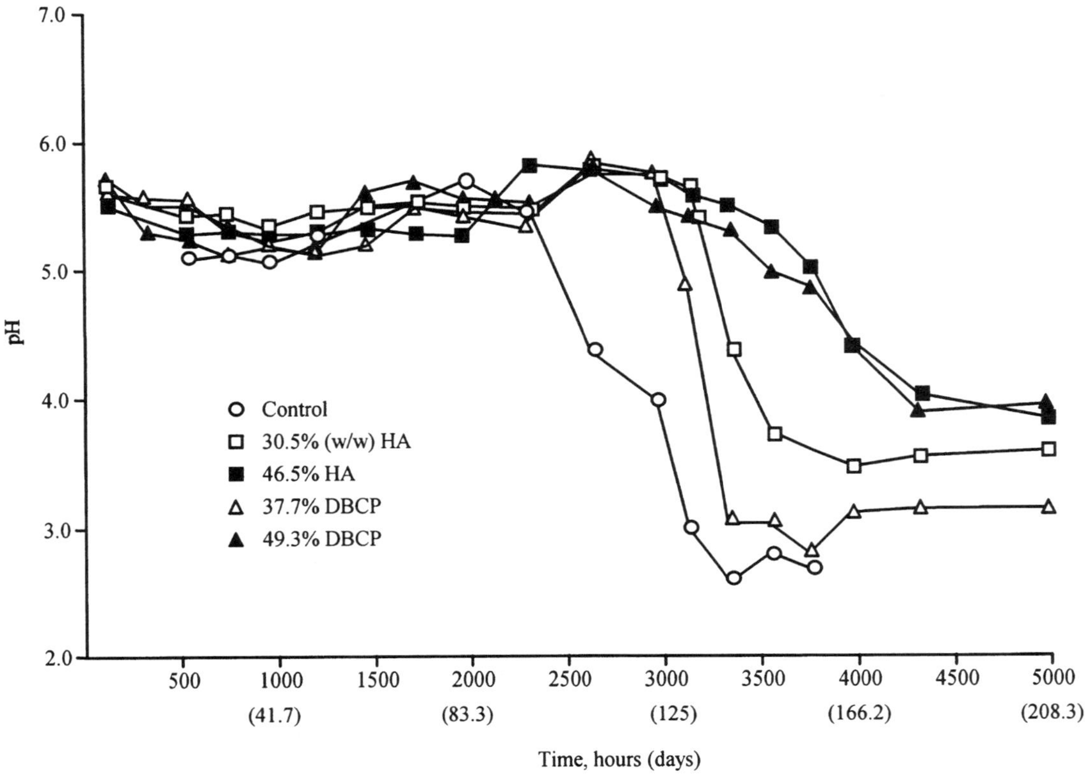

Figure 1 pH as a function of time for PLGA-85 : 15 buffered with DBCP or with HA (For clarity, error bars have been omitted. The standard deviation for each data point is within ± 3%.)

pH versus time curves for in vitro polymer hydrolysis that this may be accomplished with considerably less than the stoichiometric quantities of buffering agent. The pH versus time curves follow a pattern characterized by an initial period of relatively little change in pH. The duration of this period is about 500 h (21 days) for the rapidly degrading PLGA-50 : 50, and between 2000 and 3000 h (83–125 days) for the more slowly hydrolyzed PLGA-85 : 15. Following this quiescent period, the pH drops at rates which decline with increasing buffer loading. The pH again stabilizes at approximately 2.5.

The various polymer-buffer systems can be characterized by the values of the slopes, *b*, of the pH versus time curves in the period of rapid pH change. The slopes for the unbuffered PLGA-50 : 50 and PLGA-85 : 15 (indicated as b_0) are, respectively, -93.7×10^{-4} and -29.2×10^{-4} pH units/h, reflecting the more rapid hydrolysis of the former. (The ratio of the b_0 values for the 50 : 50 and 85 : 15 polymers is 3.2.)

The slopes of pH versus time in the period of rapid decline are shown in Fig. 2. The buffering ability of each agent appears reasonably similar in that the slopes at loadings greater than 0% are fairly close. Figure 3 compares the ratios of these slopes at a given loading to that for the unbuffered polymers at the time at which the rapid decline begins.

The above analysis indicates that even with significantly less than 100% buffering capacity the rate of pH decline can be effectively controlled.

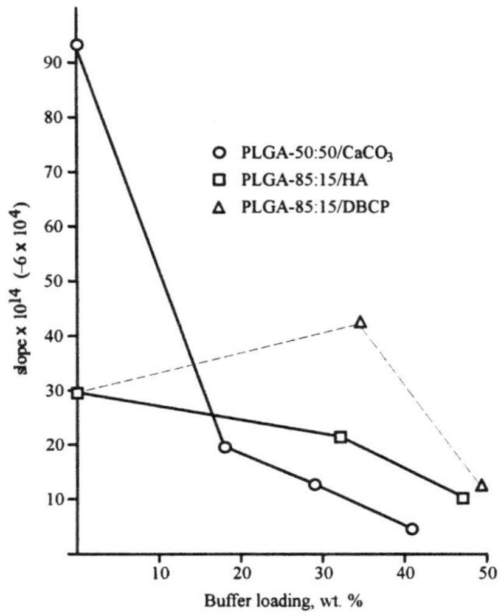

Figure 2 Slopes of pH vs. time in period of rapid decline.

VII. MECHANICAL TESTING: FLEXURAL AND TENSILE MODULUS AND STRENGTH

Mechanical tests were performed on an Instron Model 8511 equipped with a uniaxial 2500 lb capacity load cell and coupled to a 8500 computer program. Flexural modulus and strength were measured in a three-point bending test (displacement control) at a strain rate of 0.05 mm/s. Span length was 24 mm in most cases. Cylindrical supports were 1.57 mm diameter. Samples for modulus and strength testing were incubated at 37°C in phosphate buffered saline and

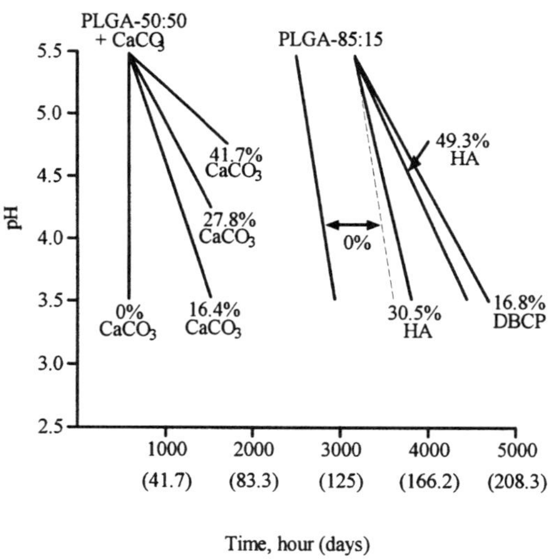

Figure 3 Slopes of pH vs. time curves in period of rapid decline. (The time of initial decent is also indicated.)

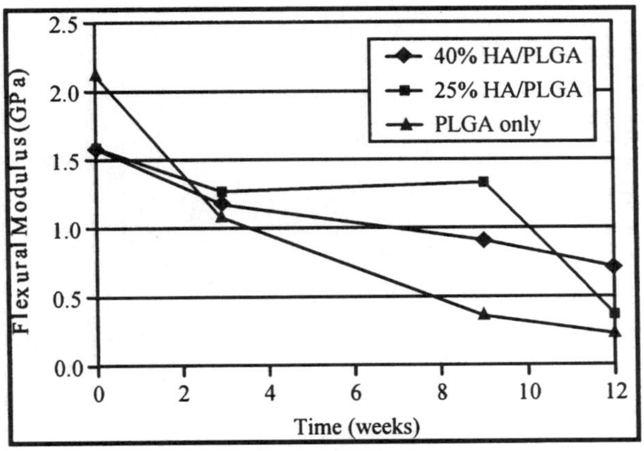

Figure 4 Flexural modulus of PLGA-85 : 15 buffered with hydroxy apatite. (The standard deviation for each data point is within ± 5%.)

removed at 4 days and at 1, 2, 3, 6, 9, and 12 weeks. Samples were tested wet; at t = 0, a measurement was also made on the dry samples before incubation.

Flexural strength and modulus were computed from the following relationships:

$$\text{Flexural modulus} = \frac{4}{3\pi}\left(\frac{P}{v}\right)\left(\frac{l^3}{d^4}\right)$$

where P = load at midpoint, N
v = vertical displacement, m
l = span between supports, m
d = diameter of rod, m

$$\text{Flexural strength} = \frac{8\text{Wl}}{\pi\text{d}^3}$$

where W = load at fracture, N

The results of the flexural strength and modulus tests for the candidate HA buffered 85 : 15 PLGA fixture are reported in Figs. 4 and 5. Although both modulus and strength decline with time for all fixtures in vitro, the trend is toward decreasing rates of mechanical losses with the HA filler. This suggests that the inclusion of the buffer can reinforce the composite, an observation consistent with others suggesting these effects on the basis of calcium ion bridging in carboxcylic acid polymer end groups [40].

VIII. IN VIVO BIOCOMPATIBILITY STUDIES

A. Tibial Defect Study

An in vivo biocompatibility study in rats was conducted using the tibia defect model of Gerhart et al. [41]. This limited study had as its objective the preliminary evaluation of local tissue responses to the buffered and unbuffered fixtures. In this regard sterile, 85 : 15 PLGA fixtures with and without a buffer (CC, 21.2% w/w) were used with animal sacrifice at 4 days, 1, 2, 3, 6, 9, and 12 weeks. Additionally, one set of animals received 50 : 50 PLGA-only fixtures with

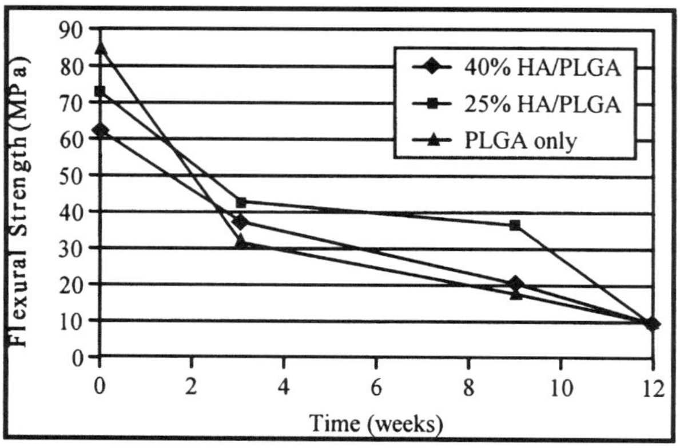

Figure 5 Flexural strength of PLGA-85 : 15 buffered with hydroxy apatite. (The standard deviation for each data point is within ± 5%.)

sacrifice at week 5 to investigate the tissue response to acid production estimated in the three- to four-week range (based on in vitro data for this polymer). However, given (1) sterile abscess formation has not been reported with these fixtures in rats [42] and (2) the limited statistical population, histological responses exclusive to the degradation of the fixtures was not expected.

Male Sprague-Dawley rats (Charles River Breeding Laboratories) weighing approximately 200 g were anesthetized with 0.3 cc of 10 mg/ml Ketamine (Fort Dodge NDA Labora-

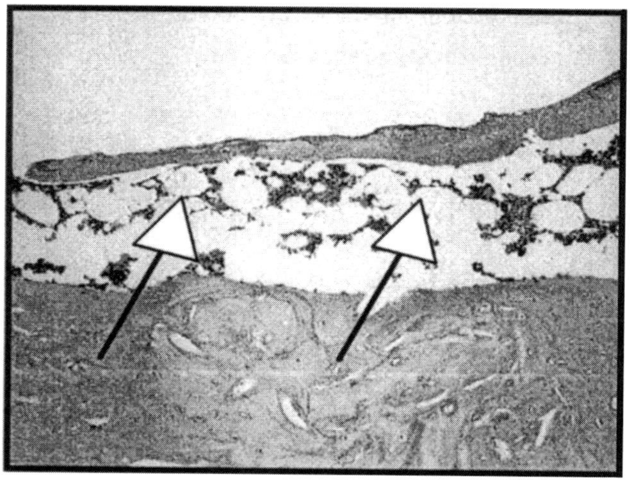

Figure 6 Photomicrograph of a longitudinal section (H&E, 20x) of a rat femur in which a 85 : 15-PLGA-system without HA was implanted. This section shows the rat femoral cortex on the bottom and the intramedullary canal at the top (implant site) 12 weeks postoperatively. The implant is fully resorbed and thus not present in this section. There is moderate new bone formation at the former interface with the implant (white arrowheads). Note, there is a single thin newly formed bone trabecule. Bone marrow is present in a relatively wide intertrabecular space. However, there is no fibrous tissue at the former interface to the implant. There are no PMNs or macrophages.

tory) and 0.15 cc of 20 mg/ml of Xylazine (Gemeni SA). The rear leg was shaved with an electric clipper and swabbed with betadine. An incision approximately 1/2 inch in length was made in the lower ventral portion of the left leg parallel to and exposing the tibia. A hole was drilled into the ventral cortex of the tibia with a dreml flex shaft drill equipped with a sterile 1.25 mm drill bit. Fixtures were inserted into the defect, the wound was closed in layers with Prolene 5.0. Groups of three animals were sacrificed at each time point.

Bone specimens were prepared for histological examination. Inflammation within the bone was graded on a scale of 0–5, with 5 being the most severe inflammation. This scale was defined by the number of inflamatory cells per high power field. Both control rods and buffered rods showed a high degree of biocompatibility. (All but one sample was graded 0/5; one was graded 1/5). At no time in any of the samples was there evidence of an inflammatory or foreign body response as would be suspected by the presence of a fibrous tissue surrounding the defect.

B. Intramedullary Implantation Study

An in vivo study in rats was performed using the intramedullary femoral model [43]. This study had initially two objectives: (1) the preliminary evaluation of histological and histomorphometric responses to the fixtures, and (2) the development of a mechanical versus degradative profiles for the fixture in a bony model. Thus, rather than the tibial defect model used exclusively for histological analysis in Part A, the intramedullary femoral model was used to

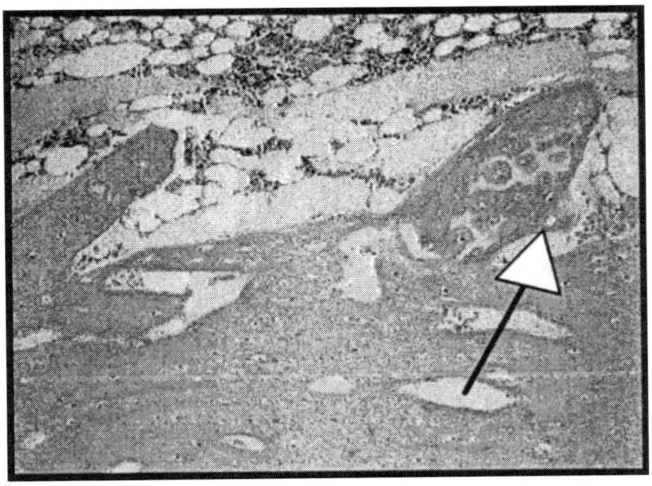

Figure 7 Photomicrograph of a longitudinal section (H&E, 20x) of a rat femur in which a 85 : 15-PLGA-system with 40% HA-filler was implanted. This section shows the rat femoral cortex on the bottom and the intramedullary canal at the top (implant site) 12 weeks postoperatively. The implant is fully resorbed and thus not present in this section. There are several bone trabeculae consisting of newly formed woven bone originating from the endosteal surface. At the interface with the former implant, there is evidence of active osteoclastic and osteoblastic activity suggesting that this woven bone layer is undergoing active remodeling. Endochondral bone formation is present in one area (arrow). Intertrabecular spaces are small and they contain bone marrow. Although there is no fibrous tissue at the former interface to the implant, the implant is undergoing active resorption as evidenced by the presence of macrophages.

allow also for the potential retrieval of fixtures of lengths sufficient for mechanical testing and supporting molecular weight analysis of the test debris.

Male, Sprague Dawley rats (300 g, Taconic Labs) were anesthetized with intraperitoneal ketamine and xylazine (8 : 1 ratio, 0.8–1.0 mg/kg). The left hind leg was shaved anteriorly and prepped with betadine. A median parapatellar capsulotomy was made through a midline incision. The patella was dislocated laterally with the leg extended. The knee was flexed, and the femoral notch exposed. The distal femur was reamed with a 0.042 Kirschner wire. A 20×1.2 mm fixture was placed into the femur in retrograde fashion. The patella was relocated, and the quadriceps tendon and medial retinaculum were reapproximated with 3.0 Vicryl suture. The skin was closed with a running 4.0 Vicryl suture.

At designated time intervals (3, 6, 9 and 12 weeks), the animals were sacrificed using carbon dioxide euthanasia. The left leg was removed and the soft tissues were stripped. At each time point, 3 specimens were reserved for histological examination and 11 were used for mechanical analysis. The histology specimens were fixed in 10% neutral buffered formalin for three to five days followed by decalcification in 0.50–0.75% nitric acid. The region of bone containing the implant was excised and the implant was left in place. The bone was dehydrated in graded alcohols, embedded in paraffin, cut into sections 5 microns thick, stained with hematoxylin and eosin, and examined microscopically.

The cutting of the bone samples could be done without difficulty. Thus, the interface between the endosteal surface of the femoral cortex and the implant could be analyzed in this histologic evaluation. For the mechanical samples, the femur was split with a small ronguer and the fixture was removed as intact as possible. The HA buffered fixtures were evaluated histologically. Hematoxylin and eosin sections of the buffered fixtures revealed new woven bone formation at the interface between the endosteal surface and the implant.

Histological evaluation of specimens showed that implantation of a 85 : 15 PLGA implant consistently resulted in formation of a thin trabecular bone cuff forming a tight interface between the implant and the rat femoral endosteal surface. The intertrabecular spaces were wide and filled with bone marrow. There was no fibrous tissue formation at the interface formed by implant and endosteal surface of the femoral cortex. In addition, there was little endosteal remodeling. In comparison, implants that had hydroxyapatite as a filler showed more woven bone formation at the interface with the implant. There were several newly formed bone trabeculae still undergoing active remodeling at 12 weeks postoperatively. Occasionally, there was enchondral bone formation extending from the endosteal surface to the implant. Intertrabecular spaces were small and they contained bone marrow. Although there is no fibrous tissue typically seen in foreign-body responses at the former interface to the implant, the implant is undergoing active resorption as evidenced by the presence of macrophages. In summary, the addition of HA to the PLGA formulation appeared to improve the osteoconductive properties of the implant.

In essence, the HA buffered fixtures showed osteoconductive properties and at no time in any of the samples studied was there evidence of an inflammatory or foreign-body response as one would have suspected by the presence of a fibrous tissue cuff surrounding the implant. While this model allowed us to view tissue/implant interfaces, it was not useful in the attempt to develop an initial degradative/mechanical corollary for the 1.2 mm fixture. At the earlier times of sacrifice (3 and 6 weeks), osteointegration had advanced to the point where fixtures were difficult to remove and at the latter times of sacrifice (9 and 12 weeks) degradation of the implant was such that removal for mechanical testing of the fixture was not feasible. These observations are important in that (1) the findings suggest good bone/implant interactions at the interface and resorption of the fixture, and (2) the deficiencies of this small animal model demonstrated the need of further preclinical evaluation to now focus on a more clinically relevant 3.2 mm (diameter) fixture in a fracture model.

IX. SUMMARY AND CONCLUSIONS

The results of pH studies can best be compared by examining the rate constants (slopes) of the pH versus time curves. Due to the more rapid degradation of PLGA-50 : 50, its drop in pH occurs earlier than for PLGA-85 : 15 (21 days as compared to 89 days.) The pH drop for PLGA-50 : 50 formulations containing calcium carbonate also begins at 21 days, but for PLGA-85 : 15 containing either DBCP or HA, the beginning of the decline is delayed to 119 days. Furthermore, we found that

- In the period of rapid decline the slope, d(pH)/*dt*, for unbuffered PLGA-50 : 50 is –00937 as compared with –0.00292 for the more slowly degrading PLGA-85 : 15. In this period the pH falls 3.2x more rapidly for the former.
- The effect of buffering on each polymer can be compared by the ratios b/b_0, where b_0 is the slope observed for the unbuffered polymer. The soluble buffer, CC, has a strong buffering effect on the rapidly degrading PLGA-50 : 50 The effect of HA is somewhat less, due, most likely, both to its very low solubility as well as to the slower rate of degradation of the PLGA-85 : 15 into which it was incorporated. The effect of DCP in PLGA-85 : 15 appears comparable to that of HA. The rate of decline of pH can be thus effectively reduced by addition of agents which neutralize the acidic products of polymer hydrolysis. Such agents include calcium carbonate, dibasic calcium phosphate, and hydroxyapatite.
- The flexural strength and modulus measured for PLGA-85 : 15 buffered with hydroxyapatite suggest that the inclusion of the buffer moderates the rate of mechanical losses.
- In preliminary biocompatibility studies, neither the PLGA-85 : 15 control rods nor buffered PLGA-85 : 15 rods produced an inflammatory response when implanted in rat tibial and intramedullary sites for up to 12 weeks. In vivo studies using an intramedullary site in rats suggested that a feature of HA additional to its buffering properties is its osteoconductivity.

In summary, these results demonstrate that the buffers can serve to neutralize the acids formed by polymer hydrolysis, to control the rate at which the pH falls, and to provide strength to the composite. Additionally, the use of HA encourages osteointegration of the implant and, thus, should prevent implant loosening. This work has resulted in identification of an HA buffered 85 : 15 PLGA fixture for further evaluation and development. The HA fixture buffers, shows osteoconductive properties, and does not provoke an inflammatory or foreign-body response. It induces various degrees of cortical remodeling as evidenced by the extent of new trabecular bone formation at the interface between the endosteal surface and the implant. Studies in larger animals which permit the use of implants of clinically relevant dimensions are now needed to develop a better understanding of the degradative/mechanical corollary to drive the concept of buffered resorbable implants aggressively toward clinical applicability. For this purpose, further research will now be directed to determining the functionality in vivo using fixtures optimized with respect to the loading of the filler. It is expected that these studies will reveal pertinent in vivo material properties with respect to implant degradation and osteointegration in relation to a healing osteotomy site. On the basis of these data, we anticipate determining which optimized formulation would be most suitable for clinical testing.

ACKNOWLEDGMENTS

The authors wish to thank Joseph Alroy, DVM, Associate Professor in Pathology, Tufts University Schools of Medicine and Veterinary Medicine, for his assistance in the histological analysis of this study. Furthermore, the authors are indebted to Shrikar Bondre and Eric Gusek for their assistance with animal care. This work was supported in part by NIH/NIAMS grant No. 1 R43 AR 44308-01 (to Debra J. Trantolo), and NIH/NIAMS grant AR 45062 (to Kai-Uwe Lewandrowski).

REFERENCES

1. T Antikainen, M Ruuskanen, R Taurio, M Kallioinen, W Serlo, P Tormala and T Waris. Polylactide and polyglycolic acid-reinforced coralline hydroxy-apatite for the reconstruction of cranial bone defects in the rabbit, *Acta Neurochirurg.*, **117**, 59–62, 1992.
2. S Higashi, T Yamamuro, T Nakamura, Y Ikada, SH Hyon and K Jamshidi. Polymer-hydroxyapatite composites for biodegradable bone fillers, *Biomaterials*, **7**, 183–187, 1986.
3. M Saito, A Maruoka, T Mori, N Sugano and K Hino. Experimental studies on a new bioactive bone cement: hydroxyapatite composite resin, *Biomaterials*, **15**, 156–160, 1994.
4. AF Tencer, V Mooney, KL Brown and PA Silva. Compressive properties of polymer coated synthetic hydroxyapatite for bone grafting, *J. Biomed. Mater. Res.*, **19**, 957–969, 1985.
5. O Bostman. Current concepts review: Absorbable implants for fixation of fractures, *J. Bone Joint Surg.—Am. Vol.*, **73**, 148–153, 1991.
6. O Bostman, E Hirvensalo, E Partio, P Tormala and P Rokkanen. [Resorbable rods and screws of polyglycolide in stabilizing malleolar fractures], *Unfallchirurg*, **95**, 109–112, 1992.
7. O Bostman, E Hirvensalo, S Vainionpaa, K Vihtonen, P Tormala and P Rokkanen. Degradable polyglycolide rods for the internal fixation of displaced bimalleolar fractures, *Int. Orthop.*, **14**, 1–8, 1990.
8. O Bostman, E Partio, E Hirvensalo and P Rokkanen. Foreign-body reactions to polyglycolide screws. Observations in 24/216 malleolar fracture cases, *Acta Orthop. Scand.*, **63**, 173–176, 1992.
9. O Bostman, S Vainionpaa, E Hirvensalo, A Makela, K Vihtonen, P Tormala and P Rokkanen. Biodegradable internal fixation for malleolar fractures. A prospective randomized trial, *J. Bone Joint Surg.—Br. Vol.*, **69**, 615–619, 1987.
10. OM Bostman. Osteolytic changes accompanying degradation of absorbable fracture fixation implants, *J. Bone Joint Surg.—Br. Vol.*, **73**, 679–682, 1991.
11. OM Bostman. Distal tibiofibular synostosis after malleolar fractures treated using absorbable implants, *Foot Ankle*, **14**, 38–43, 1993.
12. OM Bostman, U Paivarinta, E Partio, M Manninen, J Vesenius, A Majola and P Rokkanen. The tissue-implant interface during degradation of absorbable polyglycolide fracture fixation screws in the rabbit femur, *Clin. Orthop. Rel. Res.*, **285**, 263–272, 1992.
13. Y Matsusue, T Yamamuro, M Oka, Y Shikinami, S-H Hyon and Y Ikada. In vitro and in vivo studies on bioabsorbable ultra-high-strength poly(L-lactide) rods, *J. Biomed. Mater. Res.*, **26**, 1553–1567, 1992.
14. KA Athanasiou, GG Niederauer, CM Agrawal and AS Landsman. Applications of biodegradable lactides and glycolides in podiatry, [Review] *Clin. Podiatric Med. Surg.*, **12**, 475–495, 1995.
15. RW Bucholz, S Henry and MB Henley. Fixation with bioabsorbable screws for the treatment of fractures of the ankle, *J. Bone Joint Surg.—Am. Vol.*, **76**, 319–324, 1994.
16. E Partio, O Bostman, E Hirvensalo, H Patiala, S Vainionpaa, K Vihtonen, P Tormala and P Rokka-

nen. The indication of the fixation of fractures with totally absorbable SR-PGA screws, *Acta Orthop. Scand.*, Suppl. **237**, 43–44, 1990.

17. E Hirvensalo, O Bostman, S Vainionpaa, P Tormala and P Rokkanen. Biodegradable fixation of intraarticular fractures of the elbow joint, *Acta Orthop. Scand. Suppl.*, **227**, 78–79, 1988.
18. C Hoffmann, N Krettek, N Haas and H Tscherne. Die distale radiusfraktur. Frakturstabilisierung mit biodegradablen osteosynthesestiften (Biofix). Experimentelle unversuchungen und erste klinische erfahrungen, *Unfallchirurg*, **92**, 430–434, 1988.
19. SM Kumta, R Spinner and PC Leung. Absorbable intramedullary implants for hand fractures. Animal experiments and clinical trial, [see comments] *J. Bone Joint Surg.—Br. Vol.*, **74**, 563–566, 1992.
20. P Rokkanen, O Bostman, S Vainionpaa, EA Makela, E Hirvensalo, EK Partio, K Vihtonen, H Patiala and P Tormala. Absorbable devices in the fixation of fractures, *J. Trauma*, **40**, S123–127, 1996.
21. ML Strycker. Biodegradable internal fixation, [Review] *J. Foot Ankle Surg.*, **34**, 82–88, 1995.
22. S Jacobsen, M Honnens de Lichtenberg, CM Jensen and C Torholm. Removal of internal fixation–the effect on patients' complaints: A study of 66 cases of removal of internal fixation after malleolar fractures, *Foot Ankle Int.*, **15**, 170–171, 1994.
23. P Kannus, J Parkkari, S Niemi and M Palvanen. Epidemiology of osteoporotic ankle fractures in elderly persons in Finland, *Ann. Intern. Med.*, **125**, 975–978, 1996.
24. DG Seeley, J Kelsey, M Jergas and MC Nevitt. Predictors of ankle and foot fractures in older women. The study of osteoporotic fractures research group, *J. Bone Min. Res.*, **11**, 1347–1355, 1996.
25. T Ahl, N Dalen, S Holmberg and G Selvik. Early weight bearing of malleolar fractures, *Acta Orthop. Scand.*, **57**, 526–529, 1986.
26. T Ahl, N Dalen and G Selvik. Mobilization after operation of ankle fractures. Good results of early motion and weight bearing, *Acta Orthop. Scand.*, **59**, 302–306, 1988.
27. N Anand and L Klenerman. Ankle fractures in the elderly: MUA versus ORIF, *Injury*, **24**, 116–120, 1993.
28. M Bauer, B Bergstrom, A Hemborg and J Sandegard. Malleolar fractures: nonoperative versus operative treatment. A controlled study, *Clin. Orthop. Rel. Res.*, 17–27, 1985.
29. PL Broos and AP Bisschop. Operative treatment of ankle fractures in adults: correlation between types of fracture and final results, *Injury*, **22**, 403–406, 1991.
30. PL Broos and AP Bisschop. A new and easy classification system for ankle fractures, *Int. Surg.*, **77**, 309–312, 1992.
31. D Karasick. Fractures and dislocations of the foot, [Review] [31 refs]. *Sem. Roentgenol.*, **29**, 152–175, 1994.
32. MC Harper and G Hardin. Posterior malleolar fractures of the ankle associated with external rotation-abduction injuries. Results with and without internal fixation, *J. Bone Joint Surg.—Am. Vol.*, **70**, 1348–1356, 1988.
33. CP Malone. Articular fractures of the distal radius, *Orthop. Clin. N. Am.*, **15**, 217, 1984.
34. TG Wadsworth. *The Elbow*, Churchill Livingstone, Edingburgh, 1982.
35. DK Gilding and AM Reed. Biodegradable Polymers for use in surgery: Polyglycolic/polylactic homo- and copolymers: 1, *Polymer*, **20**, 1459–1464, 1979.
36. AM Donigian, BR Plaga and PM Caskey. Biodegradable fixation of physeal fractures in goat distal femur, *J. Pediatric Orthop.*, **13**, 349–354, 1993.
37. RA Kenley, MO Lee, TR Mahoney and LM Sanders. Poly(lactide-*co*-glycolide) decomposition kinetics in vivo and in vitro, *Macromolecules*, **20**, 2398–2403, 1987.
38. RB Heppenstall. Fracture healing, Chapter 2 in *Basic Orthopaedic Biomechanics*, VC Mow and WC Hayes, Eds., Raven Press, NY, 1991.
39. CM Agrawal and KA Athanasiou. Technique to control pH in the vicinity of biodegrading PLA-PGA implants, *J. Biomed. Mater. Res.: Appl. Biomat.*, **38**, 105–114, 1997.
40. AJ Domb, N Manor and O Elmalak. Biodegradable bone cement compositions based on acrylate

and epoxide terminated poly(propylene fumarate) oligomers and calcium salt compositions, *Biomaterials*, **17**, 411–417, 1996.

41. TN Gerhart, AA Renshaw, RL Miller, RJ Noecker and WC Hayes. In vivo histologic and biomechanical characterization of a biodegradable particulate composite bone cement, *J. Biomed. Mater. Res.*, **23**, 1–16, 1989.
42. Potential investigators for preclinical bone repair studies: A survey, Health Advances, Inc. (Wellesley, MA), March 1997.
43. A Majola, S Vainionpaa, K Vihtonen, M Mero, J Vasenius, P Tormala and P Rokkanen. Absorption biocompatibility, and fixation properties of polylactic acid in bone tissue: An experimental study in rats, *Clin. Orthop. Rel. Res.*, **268**, 260–269, 1991.

27

Biocompatibility of Self-Reinforced Poly(lactide-*co*-glycolide) Implants

Debra J. Trantolo, Joseph D. Gresser, and Donald L. Wise
Cambridge Scientific, Inc., Cambridge, Massachusetts

Kai-Uwe Lewandrowski
Massachusetts General Hospital, Boston, Massachusetts

Kevin J. Bozic
Beth Israel Deaconess Medical Center, Boston, Massachusetts

I. INTRODUCTION

Of the more than 1.1 million fractures in the United States each year, greater than 470,000 require some type of internal fixation device to stabilize the fracture during the healing process. Currently, the vast majority of these devices are made of metal. Following the healing of a fracture, a subsequent surgical procedure must be performed to remove the nonresorbable metal pins, plates and/or screws. While there is significant clinical demand for resorbable fixation devices to eliminate the added cost and patient discomfort associated with potentially necessary removal processes, currently available resorbable products have not been widely adopted due to their inability replicate the structural strength of their metal counterparts and/or due to their tendency to cause an inflammatory response associated with acidic degradation products.

This work focuses on the issue of eliminating this inflammatory response, including the formation of sterile abscesses, while maintaining sufficient structural integrity of the device allowing bone healing to occur. It is our thesis that the inflammatory response can be eliminated by incorporating a long acting buffer into a poly(lactic-*co*-glycolic acid), PGLA, fixation device which will neutralize the acid products as they form. In previous work, we demonstrated the feasibility of incorporating a buffer into a resorbable PLGA-based internal fixation device (IFD). The inclusion of the osteoconductive [1–4] calcium phosphate, hydroxyapatite (HA), as a buffer in a PLGA-based fixture effectively moderated the rate of pH decline as the fixture degraded. HA not only promotes bony ingrowth and obviates loosening of the fixture, but also acts as a buffer thereby preventing the formation of sterile abscesses that have been attributed to the acidic degradative products of PLGA implants [5, 6]. The resulting resorbable fixture should be capable of a buffered hydrolytic degradation and induction of bony ingrowth as resorption of the implant progresses. A resorbable buffered bone fixture with such properties could provide structural support to stabilize and support a fracture over the period of time required for natural healing to occur.

Given that this polymeric resorbable fixture is thus filled with an inorganic buffer, it was the objective of this project to demonstrate the feasibility of stabilizing the mechanical characteristics of the fixture via the use of self-reinforcing fibers. This approach has evolved from clinical and experimental observations indicating that *self*-reinforcement of resorbable implants may prove useful in preventing fatigue and precipitation of breakage [7]. In this project, the concept of *self*-reinforcement of the fixture implies the formation of a composite material comprised of reinforcing fibers of a composition similar in chemical structure to either the polymeric PLGA component of the fixture or the buffering calcium phosphate component of the fixture. Thus, various types of PLGA and calcium phosphate (CaP) fibers were first screened on the basis of their ability to strengthen the initial 85 : 15-PLGA/HA fixture (where the 85 : 15 designation specifies the lactide:glycolide mole ratio in the PLGA copolymer). Then, having identified suitable reinforcing candidates, temporal in vitro profiles of the mechanical properties of the selected fixtures were developed. Surface (plasma, CH_4) treated 95 : 5-PLGA fibers and surface treated CaP fibers each demonstrated promising contributions to the reinforcement of our 85 : 15-PLGA/HA bone repair fixture.

For the initial preclinical evaluation, these self-reinforced PLGA implants were placed into the intramedullary canal of the rat femur. At various time intervals, response of the endosteal bone to the presence of the implant was determined both by radiographic, histological and mechanical methods.

II. BACKGROUND AND SIGNIFICANCE

A. Clinical Indications for the Use of Resorbable Internal Fixation Devices

Major clinical applications of bioresorbable fixtures have included: minimally displaced malleolar fractures and flake fractures around the ankle [1–3] and elbow [8]. Other areas of clinical application have included simple distal radial fracture [9] and hand fractures [10]. In most of these procedures removal of implants is not required and seldom performed unless there is discomfort at the implant site. Typical complaints listed by patients treated for malleolar fractures that ultimately prompted implant removal included soreness over implant and cicatrix, reduced movement of the ankle joint, and strain-related pain. After implant removal, improvement was reported in about 75% of these patients [11]. Bioresorbable implants therefore offer an attractive alternative assuring fracture fixation yet eliminating the need for implant removal. This may be of particular advantage in the elderly in whom medical contraindications to anesthesia and the risk accompanied by surgical procedures are frequently present.

These advantages of resorbable fixtures are of increasing significance due to the expected raised incidence of osteoporotic ankle fractures. A European study has found evidence that the number of ankle fractures is increasing at a rate that cannot be explained simply by demographic changes because of osteoporosis, a condition that affects primarily the elderly patient population [12]. A similar study in the United States [13] has corroborated these findings and has identified other risk factors for ankle fractures such as one or more falls in the year prior to baseline, greater vigorous physical activity, weight gain since age 25, self-reported osteoarthritis, a sister's history of hip fracture after age 50, out of house less than once per week, and low distal radius BMD. Factors associated with foot fracture included insulin-dependent diabetes, use of seizure medications or of benzodiazepines, history of hyperthyroidism, poor far depth perception, and low distal radius BMD. Evidently, ankle and foot fractures have different profiles of risk factors that are not only related to low bone mass, but appear also associated to decreased function in the elderly population. Reduced hospitalization and early return to home are therefore key components in the management of these patients and resorbable fixtures providing definite treatment without subsequent procedures may prove highly applicable for these patients.

Fractures suitable for fixation with the use of resorbable fixtures such as screws include those with minimal displacement commonly seen in fractures of the ankle [14–19] and foot [20]. As such, transverse fibular fractures at/or distal to the ankle joint (Danis-Weber Type A), low spiral fractures of the lateral malleolus and avulsion of the medial malleolus (Dyputren; Danis-Weber Type B, avulsion fractures of the posterior malleolus (Volkmann's Triangle) may be treated with anatomical (less than 2 mm displacement or posterior malleolar fragment <25% of joint surface) [16, 21] reconstruction by open reduction without casting and early mobilization both preventing complications of long-term immobilization such as thrombosis. Other fractures suitable for treatment with resorbable fixtures may include distal radial fractures (Type Colles), which present the most common type of fractures of the radius (85%). They typically occur at the metaphyseal portion of the bone proximal to the extensor tendon sheaths (loco typico sive classico). Of the various clinical types, isolated stable Colles fractures without major fragment dislocation and without (Malone Type I [22]) or with dorsal or volar deviation (Malone Type II [22]), frequently treated with K-wires, appear suitable for treatment with bioresorbable screws. This application has been reported in the literature [23]. Similarly, isolated dislocated fractures of the Olecranon at the elbow joint (Wadsworth Type I and II [24]), and metacarpal hand or metatarsal foot fractures, both commonly treated with wires, are suitable for treatment with bioresorbable fixtures.

B. Clinical Complications in the Use of Resorbable Internal Fixation Devices

The trend in internal fixation devices is toward the use of resorbable, tissue compatible biopolymers such as poly(glycolic acid) [5, 25–27], (PGA), poly(lactic acid) [6], (PLA), and copolymers of lactic and glycolic acids, the poly(lactide-*co*-glycolic acids) (PLGA's). Although other polymers have also been considered, such as poly(dioxanone) [28], the PGA, PLA and PLGA polyesters have a long history of clinical use (e.g., as sutures). However, the tissue response to resorbable implants fabricated from these polyesters has not been uniformly acceptable; investigators have reported a late sterile inflammatory foreign body response (sterile abscess) [5, 27, 29].

In-depth follow-up of several large clinical studies has substantiated the complications with the use of resorbable polymer fixtures. The resorbable systems now in use show promise, but problems such as sterile fluid collections and osteolytic changes around biodegradable implants and implant fatigue remain. Clinical results clearly indicate that problems still exist with these fixtures and given the great need for resorbable implants, there is room for more extensive examination and material improvement. In up to 22% of the cases where resorbable internal fixation devices fabricated from biodegradable lactic and glycolic polymers PGLAs are used in surgical repair of fractures, severe inflammatory responses reflected by sterile abscess formation are noted [6, 27, 30–32]. These processes become evident only after fairly long induction periods ranging from 7 to 20 weeks (mean induction period of about 12 weeks) [5].

In a randomized study of 56 open reduction and internal fixation of malleolar fractures of the ankle with metal ASIF screws and plates or with rods of PLGA, two cases of sterile inflammatory wound sinus were observed three to four months after the operation in the injuries fixed with the polymer rods [33]. In a study of 64 bimalleolar fractures fixed with PGA rods, Bostman [26] reported a late inflammatory sinus in 5 cases (8.1%). In a series of 29 patients with intra-articular fractures about the elbow joint, internal fixation using PGA rods resulted in a late noninfectious inflammatory reaction in 4 cases (13.8%) [8]. Hoffman used PGA pins in repair of 40 cases of displaced fractures of the distal radius [23]. He observed a late inflammatory response requiring debridement in 9 patients (22.5%). A study of 217 fracture cases, mostly of the ankle, repaired with fiber reinforced PGA screws revealed 10 cases (4.6%) of inflammatory tissue reaction requiring operative drainage [22]. Another study identified ovoid osteolytic foci in 34 patients of a total of 67 patients with displaced malleolar fractures treated by open reduction and internal fixation using absorbable polyglycolide rods [34]. Seventeen patients developed a discharging inflammatory foreign-body reaction, a complication unique to these fixation devices. These osteolytic foci, usually 5 to 10 mm in diameter, appeared within the implant channels 6 to 12 weeks after the operation. The same lesion occurred in 14 of the 17 patients who developed a foreign-body reaction, whereas only 20 out of the 50 patients with an uneventful course showed osteolytic areas. After one year the normal structure of the bone was restored. Finally, a review by Bostman of 516 patients revealed that a late inflammatory response requiring operative drainage occurred in 7.9% of these cases [35]. Internal fixation of the fractures was accomplished with devices fabricated from either PGA or a PLGA copolymer.

The clinical presentation of this complication is fairly consistent. Bacterial cultures of the drainage are routinely negative. It was therefore concluded that the biological response must be a consequence to the chemical irritation from the polymers [35]. The delayed appearance of the inflammatory response may be associated with the kinetics of the polymer degradation. Comparative in vitro and in vivo studies on the decomposition of 10 lactide/glycolide copolymers have shown that the rate of hydrolysis was independent of the pH range studied and was virtually identical in vivo and in vitro [36].

C. Buffered PLGA Fixtures

In previous work, we substantiated the use of various calcium based salts as buffering fillers in a resorbable PLGA-based internal fixation device (IFD). Two PLGA copolymers were investigated: (1) 50 : 50-PLGA, and (2) 85 : 15-PLGA (where the ratios specify the molar relationship between the lactic, L, polymer and the glycolic, G, polymer). Calcium carbonate (CC) and the calcium phosphates, dibasic calcium phosphate (DBCP) and hydroxyapatite (HA), were investigated as buffering fillers. The buffering, or neutralizing capacity, of the salts investigated were based on the reactions of the acidic products of the PLGA hydrolysis (i.e., lactic and glycolic acids) with these salts. The neutralization reactions are assumed to be stoichiometric because the neutralization products are the conjugate acids of strong bases, and are consequently themselves very weakly ionized.

Our results demonstrated that the inclusion of CC, HA, or DBCP in a PLGA fixture can effectively moderate the rate of pH decline as the fixture degrades. Further, the rapid decline in pH can be offset without considering 100% neutralization of the lactic and glycolic components. Thus, even given that the polymeric fixture will be filled with an inorganic buffer, some of the mechanical characteristics of the fixture can be stabilized since the loading requirements for the buffer were not be nearly as grossly compromising as expected at the outset. While all of these buffers can ameliorate the decline in pH in the region of polymer hydrolysis, the use of the hydroxyapatite as a filler also supports osteoconductivity [1–4]. Thus, HA not only promotes bony ingrowth and obviates loosening of the fixture, but also acts as a buffer thereby preventing the formation of sterile abscesses that have been attributed to the acidic degradative products of PLGA implants [5, 6]. The pH versus time results for this HA buffered 85 : 15-PLGA system are presented in Fig. 1. During the period of rapid decline, the absolute value

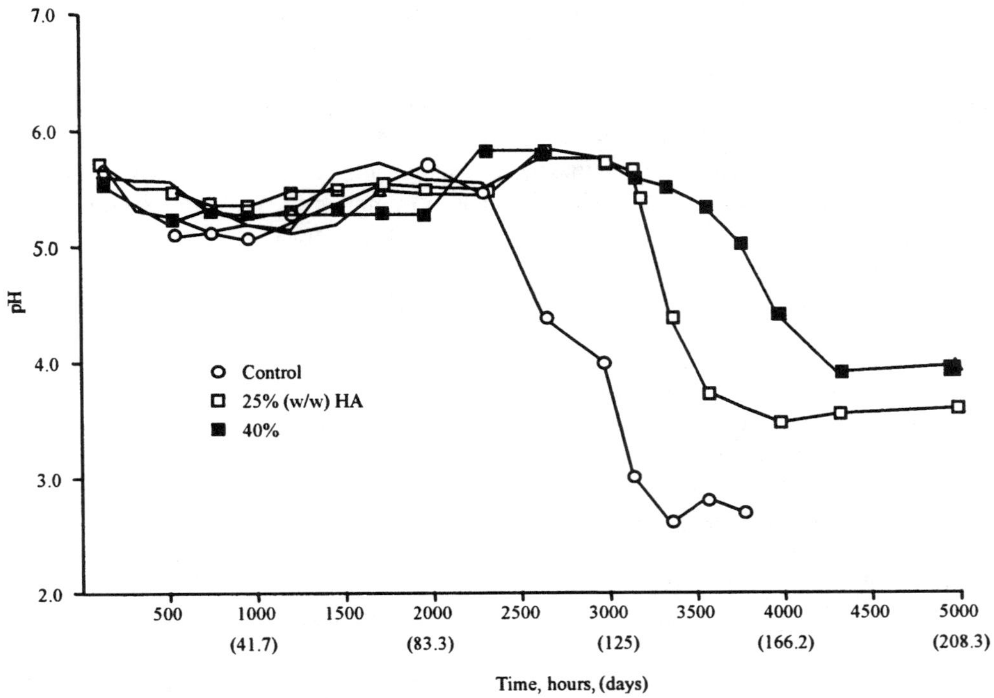

Figure 1 pH as a function of time for PLGA-85 : 15 buffered with HA. (For clarity, error bars have been omitted. The standard deviation for each data point is within ±3%.)

of the slopes of the pH versus time curves decrease with increasing buffer content, indicating the effectiveness of this agent in control of the rate of change of pH.

An in vivo biocompatibility study of these buffered fixtures was conducted using the rat tibia defect model of Gerhart [37]. This study had as its objective the preliminary evaluation of local tissue responses to buffered and unbuffered fixtures. In this regard, sterile 85 : 15 PLGA fixtures with and without a buffer were used with animal sacrifice at 4 days, 1, 2, 3, 6, 9, and 12 weeks. Additionally, one set of animals received 50 : 50 PLGA-only fixtures with sacrifice at week 5 to investigate the tissue response to acid production estimated in the 3–4 week range (based on in vitro data for this polymer). However, given (1) sterile abscess formation has not been reported with these fixtures in rats [38], and (2) the limited statistical, clinical population, histological responses exclusive to the degradation of the fixtures were not expected.

Fixtures were inserted into a tibial defect (measuring 1 mm in diameter) created in male Sprague-Dawley rats (Charles River Breeding Laboratories) weighing approximately 200 g. Groups of three animals were sacrificed at each time point. Hemoxylin and eosin sections were examined microscopically and inflammation within the bone was graded on a scale of 0–5, with 5 being the most severe inflammation.

Both control rods and buffered rods showed a high degree of biocompatibility. With the exception of one buffered specimen in the 12 week trial group there was no inflammation in the bone of any specimen at any time point. All specimens but one were graded 0/5. In the one bone at 12 weeks postimplantation, the inflammation was mild, consisted primarily of neutrophils, and was graded 1/5. This limited study suggested, then, that there were no acute responses to these types of implants.

III. RATIONALE FOR REINFORCEMENT OF BUFFERED PLGA FIXTURES

The work presented herein continues with our focus on eliminating the inflammatory response associated with the use of resorbable PLGA-based fixtures, while maintaining sufficient structural integrity of the device allowing bone healing to occur. It is our thesis that the inflammatory response can be eliminated by incorporating a long-acting buffer into a PGLA fixation device which will neutralize the acid products as they form. In our previous work (summarized above), we demonstrated the feasibility of incorporating a buffer into a resorbable PLGA-based internal fixation device (IFD).

Given that our polymeric resorbable fixture is thus filled in part with an inorganic buffer, it was the objective of previous experiments to demonstrate the feasibility of stabilizing the mechanical characteristics of the fixture via the use of self-reinforcing fibers. In this project, the concept of *self*-reinforcement of the fixture was derived from the use of fibers related to either the polymeric PLGA component of the fixture or the buffering calcium phosphate component of the fixture. Thus, various types of PLGA and calcium phosphate (CaP) fibers were first screened on the basis of their ability to strengthen the initial 85 : 15-PLGA/HA fixture (where the 85 : 15 designation specifies the lactide:glycolide mole ratio in the PLGA copolymer). Then, having identified suitable reinforcing candidates, temporal profiles of the mechanical properties of these fixtures in vitro were developed. Surface treated 95 : 5-PLGA fibers and surface treated CaP fibers each demonstrated promising contributions to the reinforcement of our 85 : 15-PLGA/HA fixture.

Ideally, the targeted reinforced IFD should degrade at rates commensurate with the course of bone healing The stages of healing have been described by Heppenstall [39]. On the basis of the clinical stages of fracture healing described by Heppenstall [39] and discussed earlier, we relied on the basic observation that soft callus is being formed between three to four weeks, and hard callus develops at one to four months. Remodeling will occur after several months to several years. In analogy to humans, we expected in rabbits that a complete hard callus will be formed in an average of about two months. After eight weeks, the fracture is sealed with hard immature bone which is capable of responding to stresses by remodeling like uninjured bone. Thus, a significant degradation of the implant at the one and two-month time point is acceptable and in fact desired because progressive healing of the osteotomy is expected.

With respect to mechanical properties of the osteotomized bone undergoing healing, we anticipated that half of the initial bending strength and modulus (prior to osteotomy) should be present at about three to four weeks. These values are typically achieved for cortical bone, and acceptable values may be lower for devices designed for repair of bones with a high proportion of cancellous (trabecular) structure such as ankle and wrist bones, where an initial flexural strength of about 40 MPa and an initial modulus of 4–6 GPA have been observed.

Mechanical properties of biodegradable polymers are well described by Daniels [40] for unreinforced polymers and copolymers of L-lactic, D,L-lactic, and glycolic acids and by Tormala [41] for self-reinforced homopolymers of glycolic acid and L-lactic acid. For the purpose of our studies, mechanical testing, done following in vitro degradation, showed both the strength and the modulus of the 3.2 mm diameter rods to decrease rapidly within the first week, and at more slower rate thereafter. In Fig. 2, these observations are summarized together with the expected kinetics of fracture healing in the rabbit animal model by expressing the strength of the implant and the healing osteotomy as percentage rates of their initial strength (prior to implantation or osteotomy). This comparative graphic evaluation shows the most significant decrease in mechanical strength of the fixture to be present during the first month when a significant increase of mechanical strength of the healing fracture is expected. Further-

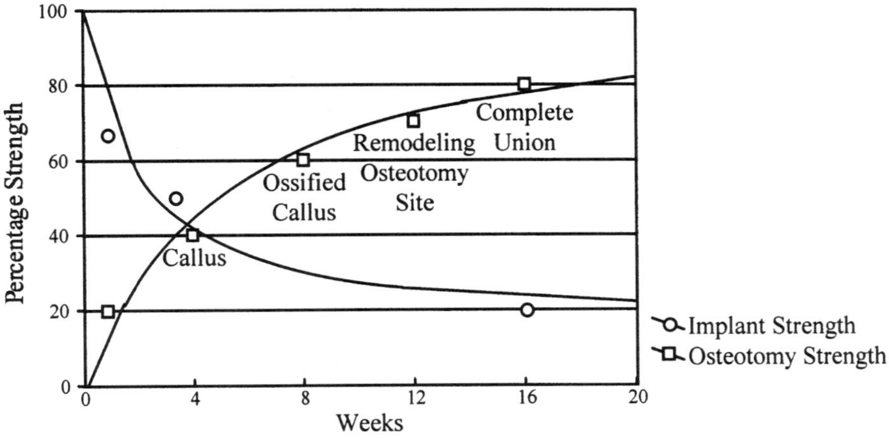

Figure 2 Model for implant and osteotomy strength over time.

more, another significant increase in rigidity of the healing fracture is expected during the second month when the soft callus is undergoing increasing ossification to form immature bone at the osteotomy site. Thereafter, a more gradual increase of the mechanical strength of the healing osteotomy site is anticipated because remodeling of the hard callus is known to take several weeks to months. In the rabbit model, it appears therefore appropriate to evaluate animals biomechanically and histologically at 4, 8, and 20 weeks because fundamental differences in the various stages of healing of the osteotomy in relation to the degrading fixture are expected at these time points.

When hard callus formation is complete, in an average of about two months, the fracture is considered to be clinically and radiographically healed. Thus, an IFD could be completely resorbed in three to four months (13 to 17 weeks). Actually after 25 to 50 days the fracture is sealed with hard immature bone which responds to stresses by remodeling like uninjured bone and thus we can accept significant degradation at about the two-month time point. The majority of the bone healing is virtually complete in about eight weeks, with about 50% recovery of mechanical properties of intact bone in about three weeks [42].

These reported mechanical properties of human bone and this brief description of bone healing allows the assignment of target values to the strength and modulus of a resorbable IFD (see Table 1). The initial bending strength and modulus for fixation of cortical bone should be about 170 MPa and 15 GPa, respectively, and about half these values at about three to four weeks. Acceptable values may be lower for devices designed for repair of bones with a high proportion of cancellous (trabecular) structure such as ankle and wrist bones. We are initially

Table 1 Properties of Human Bone

		Tensile	Compressive	Bending	Shear
Cortical	Strength, MPa	100–200	170	150–200	69
	Modulus, GPa	10–20		15	
Cancellous	Strength, MPa	95–156	106–201	40	
	Modulus, GPa	2.7	2.5		

Source: From Refs. 43–45.

targeting the clinical repair of cancellous bone for which an initial flexural strength of about 40 MPa and an initial modulus of 4–6 GPa is required.

Daniels [40] reviewed the literature on mechanical properties of biodegradable polymers. The range of values given in Table 2 are for unreinforced polymers and copolymers of L-lactic, D,L-lactic, and glycolic acids. The values are compared to those for self-reinforced homopolymers of glycolic acid and L-lactic acid as reported by Tormala [41].

IV. FIXTURE PREPARATION

A. Materials

1. Matrix polymer: PLGA-85 : 15 (PLGA, Boehringer Ingelheim, Resomer 858, Article No. 640671, Lot No. 261073)
2. Buffering filler: Hydroxy apatite (HA, calcium phosphate, tribasic, Aldrich Chemical Co., Cat. No. 23,093-6, Lot No. 01224EF)
3. Fibers
   a. Calcium metaphosphate (CaP), 15–20 μm diameter (as received from Prof. William LaCourse, Alfred University, Alfred, NY)
   b. Iron treated calcium phosphate (Fe-CaP) (as received from Prof. William LaCourse, Alfred University, Alfred, NY)
   c. PLGA-85 : 15, 123 ± 30 μm diameter (Southern Research Institute, Birmingham, AL, Lot No. J208-035-00)
   d. PLGA-95 : 5, 92 ± 23 μm diameter, (Southern Research Institute, Birmingham, AL, Lot No. J208-061-00)

The matrix polymer PLGA-85 : 15, was purified by precipitation from an acetone solution into isopropanol. After vacuum drying, the polymer was cryogenically ground to <125 μm a Tekmar A10 analytical mill. The ground polymer was blended with the buffering filler, hydroxy apatite (HA) which was previously ground in a Trost Gem-T air mill (particle size 80% <5 μm. Two compositions of the PLGA/HA blends (25% and 40%, by weight) were used to reflect differing percent neutralizations of the acidic degradation products of the PLGA hydrolysis (32% and 65% neutralization, respectively). The PLGA/HA blends were tumble mixed overnight on a rotary blender.

The PLGA/HA blends were then combined with the fibers at a loading of 10% fiber by weight of the PLGA/HA. Two types of CaP fibers and two types of PLGA fibers were used, each also used as received and after plasma (CH_4) treatment (Advanced Surface Technology, AST, Billerica, MA). The plasma treatment (denoted by "AST") was invoked to promote better long-term adhesion between the matrix material and the fibers [46]. Fibers were cut to

Table 2 Range of Mechanical Properties of Unreinforced and Self-Reinforced Polymers of Lactic and Glycolic Acids

	Unreinforced	Self-reinforced
Tensile strength, MPa	11.4–72	
Tensile modulus, GPa	2.8–5.1	
Flexural strength, MPa	40–150	215–400
Flexural modulus, GPa	2.4–10.3	10–18
Shear strength, MPa	53–68	

approximately 1 cm to yield aspect ratios of 80–110 for the PLGA fibers and 500 for the CaP fibers. Reinforced fixtures were then prepared by extrusion from the fiber loaded PLGA/HA blends.

Extrusions were performed on a Compac MPC 40-1 hydraulic press (Juelsmunde, Denmark). Smaller 1.4 mm diameter rods were extruded from a mold equipped with an 0.5 inch diameter ram and a 1.2 mm diameter die; a force of 3 tons was exerted on the ram. Larger 3.2 mm diameter rods were extruded through a 2.6 mm die from a mold fitted with a 1 inch diameter ram upon which a force of 10 tons was exerted. All extrusions were carried out at approximately 50°C.

V. *IN VITRO* TESTING

Extruded rods were cut to approximately 37 mm, weighed and diameters measured. Dental floss (Johnson and Johnson) was secured to within 5 mm of each with cyanoacrylate cement. The floss was secured through a hole in the screw cap lid of 50 ml test tubes so that the rod was suspended within the test tube. The tests tubes were filled with 45 ml of incubation medium and placed in a shaker bath (Polyscience Model SH28L, Cat. No. 040674) set at 37°C and programmed to shake at 60–62 cycles/min.

A. Mechanical Testing: Flexural Modulus And Strength

Three point bending tests were conducted on cylindrical extruded rods by one of two methods, depending on sample diameter and span.

Specimen diameter = 1.4 *mm, span* = 15 *mm:* Testing was performed on a TA-XT2 texture analyzer (Stable Micro Systems Texture Tech Corp., Scarsdale, NY) equipped with a 5 kg (12 lb) load cell. The cylindrical support diameter was 1.57 mm. The top loading point was displaced at a rate of 0.1 mm/s to failure. Load and displacement data were recorded using Stable Micro Systems XTRA-Dimensions software version 3.7J and a PC. A real time graph was immediately allows the maximum load and slope of the curve to be identified. From these data, the flexural modulus and strength were calculated.

Specimen diameter = 3.25 *mm, span* = 15–23 *mm:* Rods were tested to failure in three-point bending with a servohydraulic material test system Instron Model 8511 equipped with a 2500 lb load cell (Instron Corporation, Canton, MA). The upper load point was displaced at 0.1 mm/s until failure. Load and displacement data were collected using LabView software (Version 3.0.1, National Instruments, Austin, TX). Ultimate load and stiffness were determined from the load versus displacement curve, and flexural strength and modulus were calculated using the stiffness, aerial moment of inertia and the diameter.

Flexural strength and modulus were calculated from the following relationships:

$$\text{Flexural strength} = \frac{8Wl}{\pi d^3} \qquad \text{Flexural modulus} = \frac{4\pi}{3}\frac{P}{v}\frac{l^3}{d^4}$$

where W = load at fracture, N
l = span between supports, m
d = rod diameter, m
P = load at midpoint, N
v = vertical displacement, m

1. *Screening of Fiber Candidates*

The candidate systems were first screened at time 0 (prior to in vitro incubation). Fiber reinforced systems all included 10% (w/w) of the fiber and 90% (w/w) of a 60 : 40 blend of PLGA/HA. (Note: When reference is made to PLGA with no lactide:glycolide ratio indicated, it is understood to be the matrix polymer PLGA-85 : 15.) Three control systems contained no fiber reinforcement. The initial fixture compositions are

No fiber

0% HA:100% PLGA
25% HA:75% PLGA
40% HA:60% PLGA

PLGA fibers

90% (40% HA in PLGA):10% PLGA-85 : 15 fiber
90% (40% HA in PLGA):10% PLGA-95 : 5 fiber
90% (40% HA in PLGA):10% PLGA-95 : 5-AST fiber

CaP fibers

90% (40% HA in PLGA):10% FeCaP fiber
90% (40% HA in PLGA):10% FeCap-AST fiber
90% (40% HA in PLGA):10% CaP fiber
90% (40% HA in PLGA):10% CaP-AST fiber

Initial flexural moduli and strengths are shown in Figs. 3 and 4. These figures clearly show the high strength and modulus of fiber reinforced fixtures, particularly when contrasted to the filled HA fixture which exhibits a lower modulus and strength. HA is needed to control the rate of change of pH in the microenvironment of our fixture, thus demonstrating the rationale for reinforcement of the buffered device.

2. *In Vitro Analysis of Mechanical Strength*

All of the above reinforced candidates were investigated over a 12 week interval in vitro at 37°C in PBS. Given the initial screening results, though, the fiber reinforced formulations containing the CaP-AST and 95 : 5-PLGA-AST fibers were studied in more detail. (Even though the initial screening revealed no significant difference between the CaP and CaP-AST fiber reinforced systems, we chose to continue to work with the plasma treated fibers because the increased surface roughness should improve polymer fiber bonding over the term of fixture use [46].)

Three variables were included in this study of the AST-treated CaP and 95 : 5-PLGA fibers: (1) fiber composition, (2) HA loading, and (3) fixture diameter. These are summarized in Table 3.

The results of the flexural strength and modulus of the two systems are shown in Figs. 5–8 and are discussed as follows. The 3.2 mm diameter rods had moduli lower at all time points than the 1.4 mm rods. This trend was not so apparent for the strength. Both strength and modulus decreased rapidly within one week; further, the CaP-AST reinforced systems in general showed lower values of both modulus and strength at each time point beyond one week than did the PLGA-95 : 5 reinforced systems. Of most importance, however, is the fact that at time = 0, the CaP-AST reinforced materials have higher values of both parameters at both diameters, sufficient for fracture fixation.

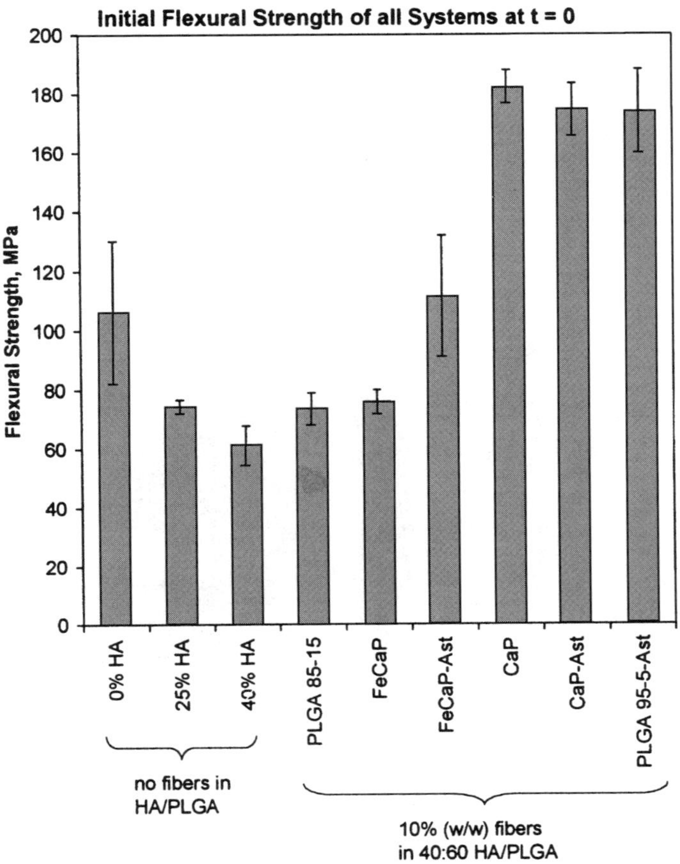

Figure 3 Initial flexural strength of all systems at $t = 0$.

Another trend is that both the modulus and strength of the CaP-AST systems with a 25% HA/PLGA ratio are higher at each time point than for the 40% HA/PLGA systems. This is a desirable result in keeping with the time 0 data for the unreinforced systems: strength increases with lower HA loadings, although the modulus was not particularly sensitive to the HA content. Our studies on pH versus degradation time in distilled water confirm that 25% loading of HA is sufficient to maintain the $d(\text{pH})/dt$ within targeted limits. This observation applies more rigorously to the 3.2 mm diameter rods, which are of clinical significance.

VI. PRELIMINARY *IN VIVO* TESTING

A preliminary in vivo study in rats was conducted using an intramedullary femoral model [47]. This study had initially two objectives: (1) the preliminary evaluation of histologic and histomorphometric responses to the fixtures, and (2) the development of a mechanical versus degradative profile for the fixture in a bony model. Thus, rather than the tibial defect model proposed for exclusively histological analysis, the intramedullary femoral model was used to allow also for the potential retrieval of fixtures of lengths sufficient for mechanical testing and supporting molecular weight analysis of the test debris.

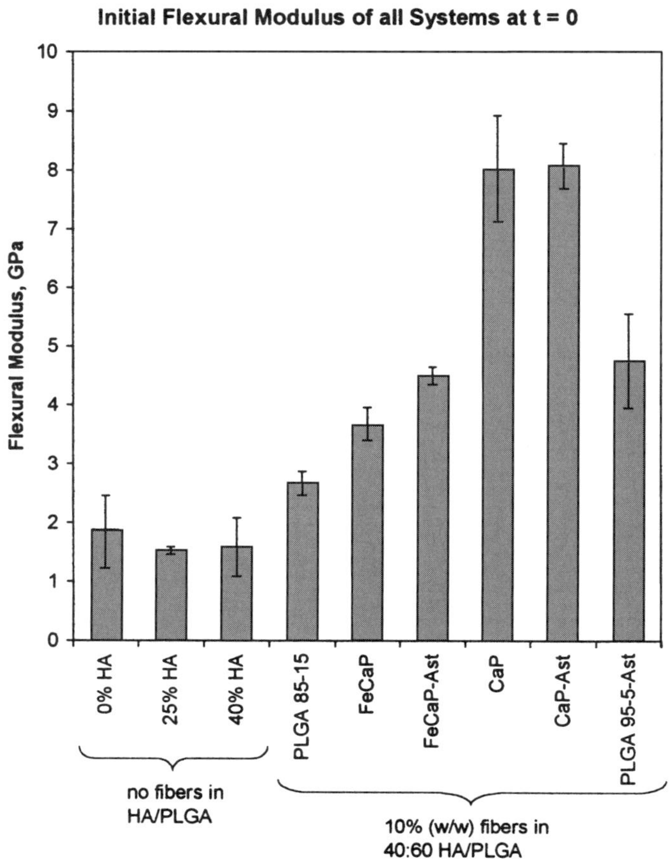

Figure 4 Initial flexural modulus of all systems at $t = 0$.

Table 3 In Vitro Reinforced Fixtures

	Fiber, 10% w/w	HA/PLGA, w/w	Diameter, mm	CSI ID
CaP	CaP-AST	40/60	1.4	84–43
	CaP-AST	25/75	1.4	84–47
	CaP-AST	40/60	3.2	96–13
	CaP-AST	25/75	3.2	96–15
95 : 5-PLGA	PLGA-95 : 5-AST	40/60	1.4	84–44
	PLGA-95 : 5-AST	25/75	1.4	84–48
	PLGA-95 : 5-AST	40/60	3.2	84–45
	PLGA-95 : 5-AST	25/75	3.2	84–50

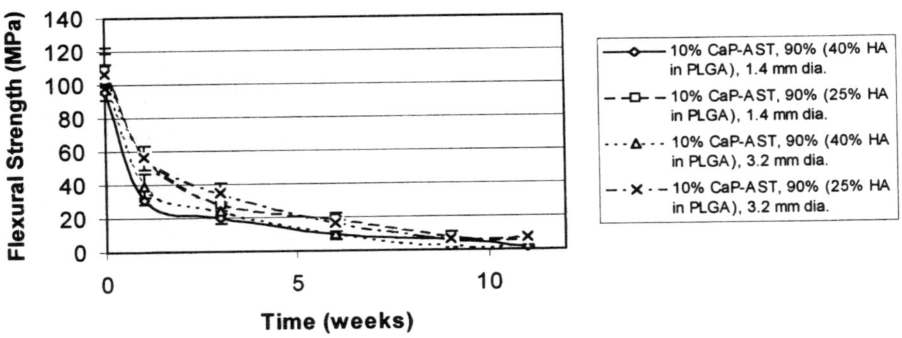

Figure 5 Flexural strength of CaP-AST fiber reinforced systems.

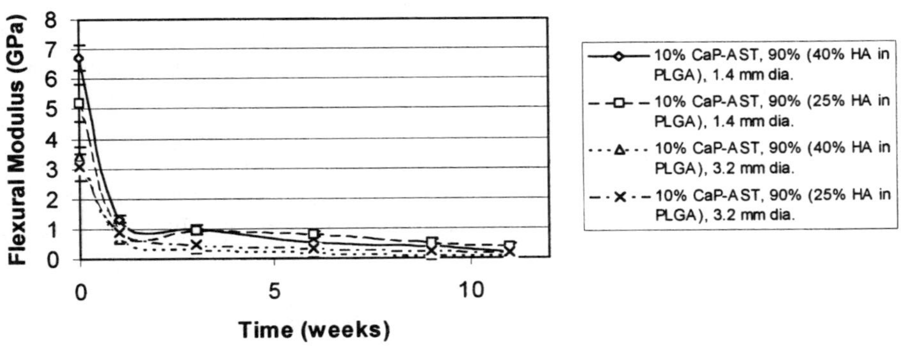

Figure 6 Flexural modulus of CaP-AST fiber reinforced systems.

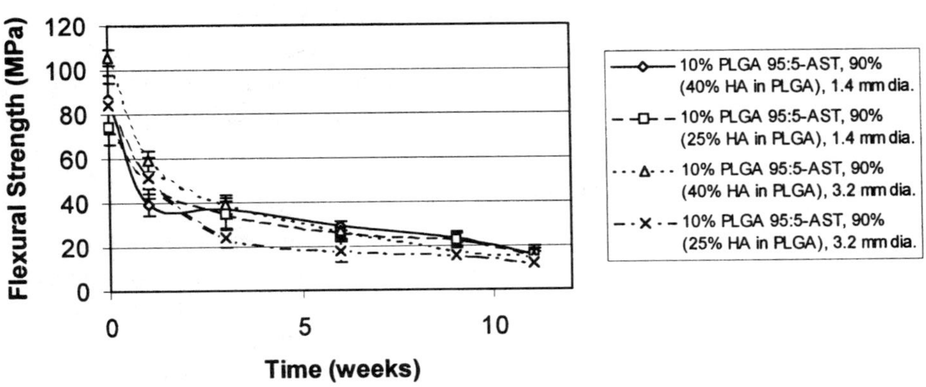

Figure 7 Flexural strength of PLGA 95 : 5-AST fiber reinforced systems.

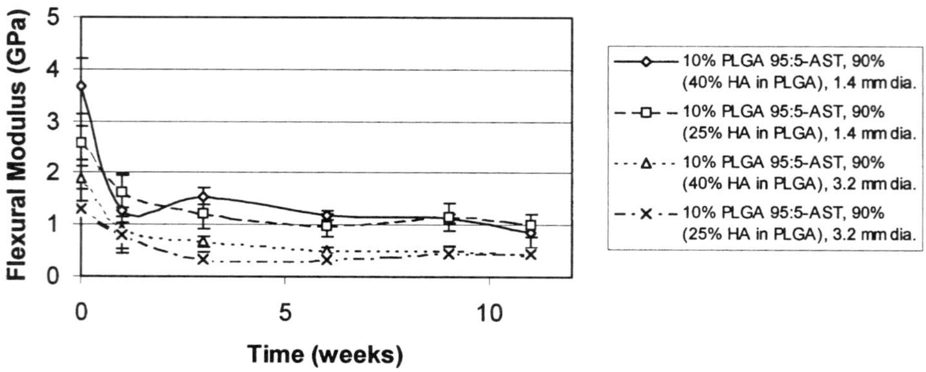

Figure 8 Flexural modulus of PLGA 95 : 5-AST fiber reinforced systems.

Male Sprague Dawley rats (300 g, Taconic Labs) were anesthetized with intraperitoneal ketamine and xylazine (8 : 1 ratio, 0.8–1.0 mg/kg). The left hind leg was shaved anteriorly and prepped with betadine. A median parapatellar capsulotomy was made through a midline incision. The patella was dislocated laterally with the leg extended. The knee was flexed, and the femoral notch exposed. The distal femur was reamed with a 0.042 Kirschner wire. A 20×1.2 mm fixture was placed into the femur in retrograde fashion. The patella was relocated, and the quadriceps tendon and medial retinaculum were reapproximated with 3.0 Vicryl suture. The wound was closed in layers with a running 4.0 Vicryl suture. Postoperatively, there were no restrictions or restraints.

At designated time intervals (3, 6, 9 and 12 weeks), the animals were sacrificed using carbon dioxide narcosis. The left leg was removed and the soft tissues were stripped. At each time point, 3 specimens were reserved for histological examination and 11 were used for mechanical analysis. The histology specimens were fixed in 10% neutral buffered formalin for three to five days followed by decalcification in 0.50–0.75% nitric acid. The region of bone containing the implant was excised and the implant was left in place. The bone was dehydrated in graded alcohols, embedded in paraffin, cut into sections 5 microns thick, stained with hematoxylin and eosin, and examined microscopically. Cutting of the three bone samples could be done without difficulty. Thus, the interface between the endosteal surface of the femoral cortex and the implant could be analyzed in this histological evaluation. For the mechanical samples, the femur was split with a small ronguer and the fixture was removed as intact as possible.

A. Results of In Vivo Biocompatibility Tests

Both the CaP-AST reinforced system and the 95 : 5-PLGA-AST reinforced system were evaluated histologically. Hematoxylin and eosin sections of the CaP-AST reinforced system revealed new trabecular woven bone formation at the interface between the endosteal surface and the implant. This was consistently observed at 6, 9, and, to a much greater extent, at 12 weeks (Fig. 9). In the PLGA reinforced system, similar observations were made but the endosteal trabecular bone cuff surrounding the implant appeared less extensive and cortical remodeling at the interface between the endosteal surface and the interface was more moderate (Fig. 10).

In essence, both the CaP-AST reinforced system and the 95 : 5-PLGA-AST reinforced system showed similar osteoconductive properties and at no time in any of the samples studied was there evidence of an inflammatory or foreign-body response as one would have suspected by the presence of a fibrous tissue cuff surrounding the implant.

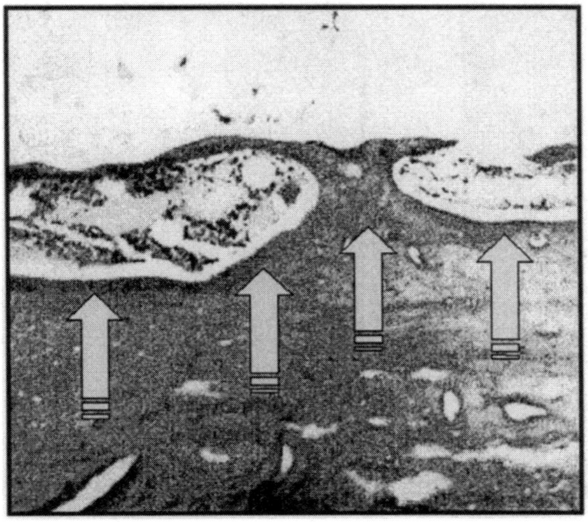

Figure 9 Photomicrograph of a longitudinal section (H&E, 10×) of a rat femur in which a 95:5-PLGA-AST fiber reinforced system was implanted. This section shows the rat femoral cortex on the bottom and the intramedullary canal at the top (implant site) nine weeks postoperatively. The implant is fully resorbed and thus not present in this section. There is moderate new trabecular bone formation at the endosteal surface (white arrow heads). Newly formed bone trabulae are thin and bone marrow is present in the wide intertrabecular spaces. Note, there is no fibrous tissue at the interface to the implant. There are no cells representative of macrophages or lymphocytes.

While this model allowed us to achieve the proposed goal, i.e., the preliminary evaluation of in vivo biocompatibility, it was not useful in the attempt to develop an initial degradative/mechanical corollary for the 1.2 mm fixture. At the earlier times of sacrifice (3 and 6 weeks), osteointegration had advanced to the point where fixtures were difficult to remove and at the latter times of sacrifice (9 and 12 weeks) degradation of the implant was such that removal for mechanical testing of the fixture was not feasible.

These observations are important in that (1) the findings suggest good bone/implant interactions at the interface and resorption of the fixture, and (2) the deficiencies of this small animal model demonstrated the need for the further preclinical evaluation to focus on the more clinically relevant 3.2 mm (diameter) fixture in a larger fracture model.

VII. CONCLUSIONS AND DISCUSSION

The major clinical application for a resorbable buffered fixture may include its use for fractures with little or no displacement that generally require open reduction and internal fixation (ORIF) with the use of screws, wires or plates (ankle, foot, forearm, hand, elbow fractures). The use of resorbable buffered fixtures would eliminate those implant-related complications usually necessitate their removal. The proposed resorbable fixture is buffered with a hydroxyapatite filler and self-reinforced with fibers. First, this results in prevention of formation of sterile abscesses due to collection of acidic degradation products, improvement of osteoconductive properties and stimulation of bony ingrowth during the resorption process of the implant. Second, it may reduce the overall risk for implant failure secondary to fatigue.

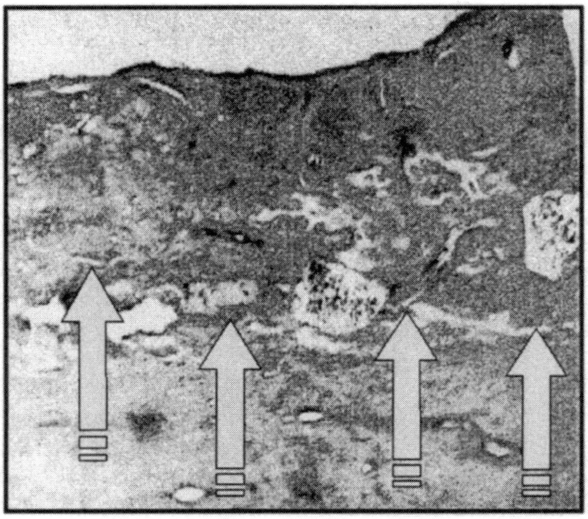

Figure 10 Photomicrograph of a longitudinal section (H&E, 10×) of a rat femur in which a CaP-AST fiber reinforced system was implanted. This section shows the rat femoral cortex on the bottom and the intramedullary canal at the top (implant site) nine weeks postoperatively. The implant is fully resorbed and thus not present in this section. There is a uniformly thick layer of newly formed woven bone at the endosteal surface (white arrow heads). This woven bone layer is undergoing active remodeling. Occasionally, there are small intertrabecular spaces in which bone marrow is present. Note: There is no fibrous tissue at the former interface to the implant. There are no cells representative of macrophages or lymphocytes.

This work has resulted in identification of two systems for further evaluation and development:

90% (25% HA in PLGA), 10% CaP-AST fiber; 3.2 mm diameter
90% (40% HA in PLGA), 10% PLGA-95 : 5 fiber; 3.2 mm diameter

Both the CaP-AST and the PLGA-AST reinforced systems have shown similar osteoconductive properties and did not provoke an inflammatory or foreign-body response. They have induced various degrees of cortical remodeling as evidenced by the extent of new trabecular bone formation at the interface between the endosteal surface and the implant. Studies in larger animals, which permit the use of actual size implants, are now needed to develop a better understanding of the degradative/mechanical corollary to drive the concept of self-reinforced resorbable implants aggressively toward clinical applicability. For this purpose, future work should be directed to determining the utility of one of the types of fixtures in vivo using fixtures optimized with respect to incorporation of the fiber. The choice of fiber, the fiber content in the rod, and alignment of the fibers should all be exploited to maximize reinforcement.

It is expected that these proposed studies could reveal pertinent in vivo material properties with respect to implant degradation and osteointegration in relation to a healing osteotomy site. This should allow us to determine which of the two candidate formulations will be most suitable for clinical testing.

While the basic premise is that some aspect of the inflammatory response is due the acidic degradation products of the fixture, it appears difficult to exclusively evaluate this aspect of the thesis as part of future study design. One approach might appear to be the histological evaluation of PLGA-only versus buffered PLGA fixtures in an animal model. However, our literature review, coupled with a survey of over 200 researchers in bone and related areas, has confirmed that it is unlikely that there would be an observable difference focusing only on the inflammatory response in animals [48]. This, develops the objective of future studies into one, which is strongly directed to substantiating the functionality of a reinforced buffered fixation device. The device could be approved given preclinical demonstration of functionality through large animal testing; further claim to a reduction in sterile abscesses would likely rely on data from clinical use of the functional device.

ACKNOWLEDGMENTS

The authors wish to thank Joseph Alroy, DVM, associate professor in pathology, Tufts University Schools of Medicine and Veterinary Medicine for his assistance in the histological analysis of this study. Furthermore, the authors are indebted to Shrikar Bondre and Eric Gusek for their assistance with animal care. This work was supported in part by NIH/NIAMS Grant No. 1 R43 AR 44600-01 (to Debra J. Trantolo), and NIH/NIAMS grant AR 45062 (to Kai-Uwe Lewandrowski).

REFERENCES

1. T Antikainen, M Ruuskanen, R Taurio, M Kallioinen, W Serlo, P Tormala and T Waris. "Polylactide and polyglycolic acid-reinforced coralline hydroxy-apatite for the reconstruction of cranial bone defects in the rabbit," *Acta Neurochirurgica*, **117**: 59–62, 1992.
2. S Higashi, T Yamamuro, T Nakamura, Y Ikada, SH Hyon and K Jamshidi. "Polymer-hydroxyapatite composites for biodegradable bone fillers," *Biomaterials*, **7**: 183–187, 1986.
3. E Partio, OM Bostman, E Hirvensalo, H Patiala, S Vainionpaa, K Vihtonen, P Tormala and P Rokkanen. "The indication of the fixation of fractures with totally absorbable SR-PGA screws," *Acta Orthopaedica Scandinavica Suppl.*, **237**: 43–44, 1990.
4. M Saito, A Maruoka, T Mori, N Sugano and K Hino. "Experimental studies on a new bioactive bone cement: hydroxyapatite composite resin," *Biomaterials*, **15**: 156–160, 1994.
5. OM Bostman, E Partio, E Hirvensalo and P Rokkanen. "Foreign-body reactions to polyglycolide screws. Observations in 24/216 malleolar fracture cases," *Acta Orthopaedica Scandinavica*, **63**: 173–176, 1992.
6. Y Matsusue, T Yamamuro, M Oka, Y Shikinami, S-H Hyon and Y Ikada. "*In vitro* and *in vivo* studies on bioabsorbable ultra-high-strength poly(L-lactide) rods," *Journal of Biomedical Materials Research*, **26**, 1553–1567, 1992.
7. P Tormala, J Vasenius, S Valnonpaa, J Laiho, T Pohjonen and P Rokkanen. "Ultra-high-strength absorbable self-reinforced polyglycolide (SR-PGA) composite rods for internal fixation of bone fractures," *In vitro* and *in vivo* study. *Journal of Biomedical Materials Research*, **25**: 1–22, 1991.
8. E Hirvensalo, OM Bostman, S Vainionpaa, P Tormala and P Rokkanen. "Biodegradable fixation of intraarticular fractures of the elbow joint," *Acta Orthopaedica Scandinavica Suppl.*, **227**: 78–79, 1988.
9. PP DeLuca, RC Mehata, AG Hausberger and BC Thanoo. "Biodegradable polyesters for drug and polypeptide delivery", Chapter 4 in Polymeric Delivery Systems: Properties and Applications, MA El-Nokaly, DM Piatt, and BA Charpentier, Eds., ACS Symposium Series 520, ACS, Washington, DC, 1993.
10. RA Kenley, MO Lee, TR Mahoney and LM Sanders. "Poly(lactide-co-glycolide) decomposition kinetics *in vivo* and *in vitro*," *Macromolecules*, **20**: 2398–2403, 1987.
11. Y-Y Hsu, JD Gresser, DJ Trantolo, CM Lyons, PRJ Gangadharam and DL Wise. *In vitro* controlled

release of isoniazid from poly(lactide-co-glycolide) matrices," *Journal of Controlled Release*, **31**, 223–228 1994.
12. S Jacobsen, M Honnens DeLichtenberg, CM Jensen and C Torholm. "Removal of internal fixation—the effect on patients' complaints: a study of 66 cases of removal of internal fixation after malleolar fractures," *Foot & Ankle International*, **15**: 170–171, 1994.
13. H Pistner, H Stallforth, R Gutwald, J Muhling, JF Ruether and C Michel. "Poly(L-lactide): a long-term degradation study in vivo. Part II. Physico-mechanical behavior of implants," *Biomaterials* **15(6)**: 439–450, 1993.
14. T Ahl, N Dalen, S Holmberg and G. Selvik. "Early weight bearing of malleolar fractures," *Acta Orthopaedica Scandinavica*, **57**: 526–529, 1986.
15. T Ahl, N Dalen and G Selvik. "Mobilization after operation of ankle fractures. Good results of early motion and weight bearing," *Acta Orthopaedica Scandinavica*, **59**: 302–306, 1988.
16. N Anand and L Klenerman. "Ankle fractures in the elderly: MUA versus ORIF," *Injury*, **24**: 116–120, 1993.
17. M Bauer, B Bergstrom, A Hemborg and J Sandegard. "Malleolar fractures: nonoperative versus operative treatment. A controlled study," *Clinical Orthopaedics & Related Research*, 17–27, 1985.
18. PL Broos and AP Bisschop. "Operative treatment of ankle fractures in adults: correlation between types of fracture and final results," *Injury*, **22**: 403–406, 1991.
19. PL Broos and AP Bisschop. "A new and easy classification system for ankle fractures," *International Surgery*, **77**: 309–312, 1992.
20. P Kannus, J Parkkari, S Niemi and M Palvanen. "Epidemiology of osteoporotic ankle fractures in elderly persons in Finland," *Annals of Internal Medicine*, **125**: 975–978, 1996.
21. MC Harper and G Hardin. "Posterior malleolar fractures of the ankle associated with external rotation-abduction injuries. Results with and without internal fixation," *Journal of Bone & Joint Surgery—American Volume*, **70**: 1348–1356, 1988.
22. CP Malone. "Articular fractures of the distal radius," *Orthopedic Clinics of North America*, **15**: 217, 1984.
23. C Hoffmann, N Krettek, N Haas and H. Tscherne. "H. Die distale radiusfraktur. Frakturstabilisierung mit biodegradablen osteosynthesestiften (Biofix)," *Experimentelle Unversuchungen und erste klinische Erfahrungen. Unfallchirurg*, **92**: 430–434, 1988.
24. DG Seeley, J Kelsey, M Jergas and MC Nevitt. "Predictors of ankle and foot fractures in older women. The study of osteoporotic fractures research group," *Journal of Bone & Mineral Research*, **11**: 1347–1355, 1996.
25. OM Bostman, E Hirvensalo, E Partio, P Tormala and P Rokkanen. "[Resorbable rods and screws of polyglycolide in stabilizing malleolar fractures]," *Unfallchirurg*, **95**: 109–112, 1992.
26. OM Bostman, E Hirvensalo, S Vainionpaa, K Vihtonen, P Tormala and P Rokkanen. "Degradable polyglycolide rods for the internal fixation of displaced bimalleolar fractures," *International Orthopaedics*, **14**: 1–8, 1990.
27. OM Bostman. "Distal tibiofibular synostosis after malleolar fractures treated using absorbable implants," *Foot & Ankle*, **14**: 38–43, 1993.
28. AM Donigian, BR Plaga and PM Caskey. "Biodegradable fixation of physeal fractures in goat distal femur," *Journal of Pediatric Orthopedics*, **13**: 349–354, 1993.
29. RW Busholz, S Henry and MB Henley. "Fixation with bioabsorbable screws for the treatment of fractures of the ankle," *Journal of Bone & Joint Surgery—American Volume*, **76**: 319–324, 1994.
30. K Athanasiou, GG Niederauer, CM Agrawal and AS Landsman. "Applications of biodegradable lactides and glycolides in podiatry. [Review]," *Clinics in Podiatric Medicine & Surgery*, **12**: 475–495, 1995.
31. SM Kumta, R Spinner and PC Leung. "Absorbable intramedullary implants for hand fractures. Animal experiments and clinical trial [see comments]," *Journal of Bone & Joint Surgery—British Volume*, **74**: 563–566, 1992.
32. P Rokkanen, OM Bostman, S Vainionpaa, EA Makela, E Hirvensalo, EK Partio, K Vihtonen, H Patiala and P Tormala. "Absorbable devices in the fixation of fractures," *Journal of Trauma*, **40**: S123–7, 1996.

33. OM Bostman, S Vainionpaa, E Hirvensalo, A Makela, K Vihtonen, P Tormala and P Rokkanen, "Biodegradable internal fixation for malleolar fractures. A prospective randomized trial," *Journal of Bone & Joint Surgery—British Volume*, **69**: 615–619, 1987.
34. OM Bostman. "Osteolytic changes accompanying degradation of absorbable fracture fixation implants," *Journal of Bone & Joint Surgery—British Volume*, **73**: 679–682, 1991.
35. OM Bostman. "Current concepts review: Absorbable implants for fixation of fractures," *Journal of Bone & Joint Surgery— American Volume*, **73**: 148–153, 1991.
36. D Karasick. "Fractures and dislocations of the foot. [Review] [31 refs]," *Seminars in Roentgenology*, **29**: 152–175, 1994.
37. TN Gerhart, AA Renshaw, RL Miller, RJ Noecker and WC Hayes. "*In vivo* histologic and biomechanical characterization of a biodegradable particulate composite bone cement," *Journal of Biomedical Materials Research*, **23**: 1–16, 1989.
38. Potential Investigators for Preclinical Bone Repair Studies: A Survey, Health Advances, Inc. (Wellesley, MA), March 1997.
39. RB Heppenstall. "Fracture Healing", Chapter 2 in Basic Orthopaedic Biomechanics, VC Mow and WC Hayes, Eds., Raven Press, NY, 1991.
40. AU Daniels, MKO Chang, KP Andriano and J Heller. "Mechanical properties of biodegradable polymers and composites proposed for internal fixation of bone", *J. Applied Biomaterials*, **1**, 57–78, 1990.
41. P Tormala. "Biodegradable self-reinforced composite materials; manufacturing, structure and mechanical properties," *Clinical Materials*, **10**, 129–134, 1992.
42. JC Adams. Outline of Fractures: Including Joint Injuries, Churchill Livingston Press, NY, 1978, p. 182 (BED-Vol 24, 1993).
43 DR Carter and WC Hayes. "Bone compressive strength: The influence of density and strain rate," *Science*, **194**, 1174–1176, 1996.
44. SC Cowin. Bone Mechanics, CRC Press, Boca Raton, 1989.
45. TM Keaveney, EF Wachtel, CM Ford and WC Hates. "Some compressive and tensile yield characteristics of trabecular bone," *Proceedings of the 1993 Bioengineering Conference*, LA Langrana, MH Friedman, and ES Groed, Eds. ASME, NY, 1993.
46. IH Loh, HL Lin and CC Chu. "Plasma surface modification of synthetic absorbable sutures," *Journal of Applied Biomaterials*, **3**, 131–146, 1992.
47. A Majola, S Vainionpaa, K Vihtonen, M Mero, J Vasenius, P Tormala and P Rokkanen. "Absorption biocompatibility, and fixation properties of polylactic acid in bone tissue: An experimental study in rats," *Clinical Orthopaedics and Related Research*, **268**: 260–269, 1991.
48. AJ Domb, N Manor and O Elmalak. "Biodegradable bone cement compositions based on acrylate and epoxide terminated poly(propylene fumarate) oligomers and calcium salt compositions," *Biomaterials*, **17**: 411–417, 1996.

28

Coating of Cortical Allografts with a Bioresorbable Polymer Foam for Surface Modification by Periosteal Cell Seeding

Kai-Uwe Lewandrowski and William W. Tomford
Massachusetts General Hospital, Boston, Massachusetts

Maurice V. Cattaneo, Joseph D. Gresser, Debra J. Trantolo, and Donald L. Wise
Cambridge Scientific, Inc., Cambridge, Massachusetts

Lawrence Bonassar
University of Massachusetts Medical Center, Worcester, Massachusetts

I. INTRODUCTION

Protection of the cortical allografts from the host by a tissue engineered osteoinductive surface modification appears to offer a reasonable method to avoid adverse effects of donor specific immune responses which could result in accelerated bone resorption and early mechanical failure of the graft. Coating of the graft with foam that is seeded with periosteal cells, possibly taken from the future recipient of the bone transplant, should result in direct bone formation on the graft and thus improve replacement of the allogeneic bone by the host's own bone. Second, coating with periosteal cells expressing the recipient's transplantation antigens should not only reduce immunogenicity but also prevent foreign body reactions such as its encapsulation by fibrous tissue. Third, new bone formation should by far exceed bone resorption which should improve the overall performance of these types of grafts. These conceivable advantages of structural cortical bone allografts having tissue engineered surface modification have stimulated the work presented herein. Production of a structural cortical bone allograft with decreased immunogenicity and dramatically improved osteoinductive properties is eminently worthwhile and could offer significant clinical advantages.

Although the bone transplants proposed here will remain true allografts, we hypothesized that these new types of grafts would resemble the biologic behavior of autografts because transplantation antigens expressed by cells on the surface of the graft are identical to the tissue type of the host as periosteal cells could be taken from the future recipient. Therefore, this new type of tissue engineered bone allograft should show significantly lower sensitization rates for antibody-mediated immunity. Although cellular-mediated cytotoxicity could be elicited by slow release of antigens from the allograft while it is actively undergoing remodeling, it should also be reduced simply by the formation of a dense autologous periosteal bone cuff surrounding the bone transplant. Furthermore, the additional use of a laser drilling and partial demineralization should result in enhanced incorporation for two reasons. First, it provides a highly osteoinductive matrix for the ingrowth of periosteal cells from the foam into the allograft. Second, it may release growth factors from the demineralized bone matrix stimulating the expansion of periosteal cells from the foam into the graft.

We have determined the feasibility of this approach not only by the capacity of the biocompatible polymer foam with an open cell structure to support cell attachment and growth but also by its capacity to form a construct with bone during the concurrent processes of bone healing and polymer degradation.

II. BACKGROUND AND SIGNIFICANCE

A. Clinical Management of Bone Defects Using Massive Allografts

The management of large skeletal defects continues to present a major challenge to orthopaedic surgeons, particularly when the problem arises in young patients in whom artificial devices and joint implants are likely to lead to early failure. Large frozen cortical bone allografts have been increasingly used in limb sparing procedures not only in the treatment of bone tumors [1–3]; after massive bone loss after traumatic injuries [4, 5], or in the treatment of avascular necrosis [6], but also in failed joint arthroplasties [7–11], where extensive bone loss due to osteolysis is commonly encountered.

Even though the overall success rate for massive cortical bone allografts, implying return to work and engagement in relatively normal activities without crutches or braces, is approximately 75–85%, only 50% of these patients have an entirely uncomplicated postoperative course. About a quarter of the total group require reoperations such as autologous grafting or replating for stress fractures delayed unions [12–14]. Some patients require excision of the graft because of infection [15–19], reimplantation, long-term bracing or in some cases amputation. These results clearly indicate that problems still exist with this procedure and that if the technique is to be more widely applied, it must be more extensively examined and materially improved.

An estimated 250,000 orthopedic allotransplants are now being used annually in the United States for a variety of disorders involving loss of skeletal structures, and there are currently no satisfactory alternatives for such grafts, particularly in young people, the population most affected by these conditions. Information gained by achievement of these aims is expected to result in a new type of structural bone allograft that potentially could significantly improve clinical results.

B. Biological Fate of Massive Allografts during Incorporation

The incorporation of a bone allograft is known to require cooperative interactions between the recipient site and the bone graft, each of which provide unique and indispensable contributions [20–22]. The graft provides a number of osteoinductive growth factors such as the TGF-β-like bone morphogenetic proteins (BMP) and other noncollagenous proteins present in the matrix [23–30], and an osteoconductive structure that supports new bone formation in the host. The host bed provides an inflammatory response which results in a fibrovascular stroma that will eventually revascularize the graft and provide a source for the recruitment and transformation of mesenchymal cells into osteogenic and chondrogenic cells [20]. The sequence of histological events in the incorporation of massive, nonvascularized, segmental bone allografts has been extensively described in animals [31–36]. A hematoma is initially formed, which is the origin of platelet derived growth factors, other growth factors and cytokines. A local inflammatory response is usually stimulated by the implantation procedures and by the presence of the graft [31, 32]. The formation of the fibrovascular stroma occurs within days. The connective tissue surrounding the graft conveys recipient-derived blood vessels and osteogenic precursors to the graft [21]. The invasion of vascular buds into cortical bone usually occurs through preexisting Haversian or Volkman channels. These channels are widened by osteoclastic activity that accompanies the neovascular response. Osteoclastic resorption of the bone allograft commonly occurs from the periosteal surface and at the junction with the host bone. In fresh frozen cortical bone allografts, this resorption may penetrate only a few millimeters into the graft [37, 38].

After the initiation of graft resorption with following revascularization, new bone formation begins. These two concurrent processes may result in an adaptive remodeling response of the graft to biomechanical loading [36, 39]. From a clinical standpoint, a graft has been successfully incorporated when the host-graft interface unites and the graft-host-bone construct tolerates physiological weight bearing without fracture or pain. From a basic science perspective, a bone allograft is successfully incorporated when the original graft bone has been completely substituted with new host bone. The incorporation of bone allografts can therefore be defined as the concurrent revascularization and substitution of nonviable allogeneic graft bone with viable autologous host bone without substantial loss of strength. The resulting composite of graft and host bone can bear physiological loads and can repair and remodel itself in response to changes in load or fatigue damage [40]. Experimental data suggest that the inflammatory response observed during the first one to three weeks is most likely to be the result of wound healing and a nonspecific foreign body response [31, 32, 34, 41]. While in autologous bone grafts these processes resolve in the direction of angiogenesis and osteogenesis with revascularization and new bone formation, the inflammatory response commonly persists in bone allografts resulting in continuous resorption. Since large segmental bone allografts present a depot of antigens, the ongoing resorption may provide a slow but steady release of antigens over a long period of time [22, 34, 42]. This could facilitate a persisting exposure of more antigens resulting in a chronic immune response. Interference with the initiation of the remodeling cycle due to impairment of osteoinductive processes and revascularization from the surrounding fibrovascular stroma is possible.

C. Immune Responses to Bone Allografts

The major complications of massive cortical bone allografts (fractures, nonunion, and infection) have been the subject, or included in, to several reviews. These studies have presented evidence that the causes of these complications are still poorly understood. However, many investigators hypothesized that at least some of these complications are immunologically mediated [43–52]. It appears likely that antigens carried by the bone allograft stimulate a specific immune response that activates immune competent cells in the host. In clinical studies, this response has been observed to develop as early as one month after surgery [43, 44, 53–56]. These cells secrete cytokines such as interleukin 1, tumor necrosis factor alpha and beta, which are also known as potent activators of the osteoclastic cell lineage and therefore may stimulate accelerated bone resorption [57–63]. Since bone resorption must occur before new bone can be formed, a delicate balance between the two concurrent processes must be maintained for the graft to be revascularized and substituted by host bone without appreciable loss of strength. A link between the presence of a chronic immune response and an accelerated resorption with premature mechanical failure of the bone allograft may exist, but no definite data is available indicating that these events are of any significance for the development of clinical complications. However, bone allografts are an excellent antigenic source and are capable of stimulating an immune response in the host.

Therefore, bone allografts generally appear to be subject to many of the same principles of transplantation as any other parenchymal organ transplant. However, bone allografts present with a number of tissue related characteristics which make the identification of a rejection more difficult. In comparison to parenchymal organs, such as in kidneys, in which there are readily identifiable markers of function, no systemic marker exists which would allow quantifying the rejection of an allogeneic bone transplant. Biopsies and techniques, such as bronchial alveolar lavage in lung transplants, allow a very specific diagnosis of rejection. However, biopsies are difficult to evaluate in allogeneic bone transplants because the cellular infiltrate is generally not uniform. In addition, it is often difficult to obtain biopsy specimens or the transplant is not expendable [64]. The definition of bone allograft rejection is therefore usually inferred from the resorption of the bone graft, the premature mechanical failure due to fatigue and the lack of incorporation into host bone.

The effect of histocompatibility matching on the incorporation of bone allografts has been investigated in controlled animal studies [39, 49, 65–69]. Humoral and antibody-dependent cell-mediated cytotoxicity were affected by matching with tissue antigens more than the cell-mediated immunity. The incorporation of canine nonvascularized bone allografts was enhanced by matching for tissue antigens. The magnitude of the immune response appeared to be modulated by the degree of tissue antigen matching [39, 69, 70]. The volume of fibrous connective tissue within the intratrabecular spaces was found to be directly proportional to the antigenicity of the graft. The percentage of osseous surface that was undergoing remodeling tended to be inversely proportional to the immunogenicity of the graft [39]. These data from an experimental animal model must be interpreted cautiously, and direct extrapolations to the clinical outcome maybe inappropriate. Cumulative observations suggest that MHC mismatches may have a deleterious influence particularly on the fate of massive osteochondral allografts.

In humans, knowledge of the influence of immune responses on these large grafts is still unclear. Small fresh, frozen [71, 72] and freeze-dried [53, 73] grafts have been shown to elicit cellular and humoral immune responses, but these did not appear to affect incorporation. In studies on large grafts, 85% of recipients of bone allografts were found to have an antibody response when tested against a cell panel [74], but an almost identical number was found in a control population, possibly as a result of receiving transfusions or having been pregnant [74]. Two other studies have failed to establish a clear correlation between the degree of donor/recipient histocompatibility and clinical complications [14, 54, 55]. Their incidence is therefore probably more related to a delay in incorporation, which at least in part may be mediated through the immune system.

It is, however, important to note that these complications are multifactorial in origin and hat the effect of factors such as fit of the graft, internal fixation device application, mechanics of the extremity, vascularity of the tissue bed, and the use of adjuvant radiation or chemotherapy must be included in any assessment of the response of the grafts to the immune reaction of the host. Studies to date have failed to provide satisfactory definition of the relationship of immune responses and the results of bone allograft transplantation in humans, in part due to their lack of consideration of these variables. Further immunological assessment of tissue engineered bone allografts is therefore indicated to determine how and indeed if the immune system affects their incorporation into host bone.

D. Polymer-Bone Biocompatibility

The trend in internal fixation devices is toward the use of resorbable, tissue compatible biopolymers such as poly(glycolic acid) [75–79], (PGA), poly(lactic acid) [80], (PLA), and copolymers of lactic and glycolic acids, the poly(lactide-*co*-glycolic acids) (PLGAs). Although other polymers have also been considered, such as poly(dioxanone) [81], the PGA, PLA and PLGA polyesters have a long history of clinical use (e.g., as sutures). In the new developing area of tissue engineering, scaffolds are needed as cell and tissue carriers emphasizing material superstructuring in the design of supports. These supports are increasingly engineered from bioabsorbable substrata made of biopolymers (collagen, laminin) or polymers (PLA, PLGA, PGA) to realize the three-dimensional nature of tissue equivalents. Biodegradable and biocompatible polymeric scaffolds are needed and have been recently introduced for orthopedic biomaterials and regeneration purposes [82–84]. In summary, there is a large body of literature supporting biocompatibility of biopolymers, such as PLGA. They have been demonstrated to facilitate bony healing and they have been demonstrated to represent excellent scaffolds for ingrowth of bony tissues. Implants made of PLGA have been shown to be capable of osteointegration [76–78]. Therefore, it is appropriate to use PLGA-based foam biopolymers as a scaffold for tissue-engineered surface modifications of cortical bone allografts.

E. Bone Regeneration by Periosteal Cells

The feasibility of such a tissue engineering approach depends on the ability to obtain a population of cells which have or will develop osteoblast-like function. For the purpose of this study, the source of cells was chosen to be long bone periosteum due to (1) its relatively easy surgical accessibility, and (2) the known osteogenic properties of its cells. The periosteum itself is known to play a critical role in fracture healing and callus formation [85, 86], which is presumably cell-mediated. Cells derived from periosteum are capable of forming bone in subcutaneous sites either alone [87], or when seeded onto ceramic calcium phosphate [88, 89] or polymeric PGA [90] substrates. Further, the healing of critical size defects in the cranium [91] and femur [92–94] have been demonstrated using periosteal cells seeded onto PGA scaffolds. Cells from the periosteum are known to have both osteogenic [87] and chondrogenic [95] potential, and the process of bone formation from periosteal cells involves transition through a cartilage-like phase [90]. Studies examining the use of chondrocytes on polymer scaffolds to fill bone defects have demonstrated the formation of stable cartilage that does not generate bone [96], while periosteal cells placed in the same material in the same site ultimately repair the defect with bone [91]. With this in mind, coating cryopreserved bone with PLGA foam, followed by seeding with periosteal cells appears to be an optimal process which takes advantage of the initial strength and geometry of the allograft, while accelerating the incorporation of the transplanted bone and minimizing the immune response to foreign tissue.

III. RATIONALE FOR SURFACE MODIFIED AND TISSUE ENGINEERED BONE ALLOGRAFTS

The approach of using hydrochloric acid demineralization for surface modification was chosen because it is known to result in the exposure of osteoinductive noncollagenous bone matrix factors [97] such as TGF-β bone morphogenetic proteins (BMP) [27, 28, 30, 98–100] and sialoproteins including osteopontin [99], bone Gla protein [101] osteocalcin [102–105] and others to the surrounding soft tissues. These factors contribute to the transformation of mesenchymal cells into osteogenic and chondrogenic cells [106] required for induction of bone resorption and formation of new bone [107]. Surface decalcification of cortical bone allografts has been shown to enhance their incorporation but problems such as bone resorption in excess of the rate of new bone formation and immune responses persisted [108–110].

Previous studies have attempted to improve host incorporation by altering the geometrical surface configuration of cortical bone [111–116]. The mechanism by which the presence of laser holes promoted osteogenesis and incorporation in partially demineralized grafts include the following possibilities: The greater surface area of partially demineralized bone or increased access to vascular tissue, or a combination of the two factors. Previous studies have pointed out that the geometry of an implant may influence the extent of bone formation. As such, the osteoconductive potential of cortical bone allografts treated in such a way has been attributed, at least in part, to its morphometric similarity to cancellous bone. In addition, demineralization alone would yield a geometrical surface configuration that is less advantageous for bony ingrowth when compared with cancellous bone because cortical bone is less porous and has a comparatively low surface area to volume ratio. The effect of drill holes and partial demineralization may therefore rely on the modification of cortical bone allografts into a more porous scaffold facilitating the development of focal centers of bone resorption and new bone formation [117].

There is sufficient evidence for both antibody-mediated and cellular-mediated cytotoxicity from in vivo and in vitro experiments with bone allografts [21, 43, 44, 65, 66, 71, 118–121]. These studies have indicated that the primary response of the host to the bone allograft is predominately a cellular-mediated response to the MHC encoded cell surface antigens that are carried by cells within the allograft and recognized by responding T-lymphocytes in the host [43, 44, 50, 51, 65, 66, 70, 72]. The use of demineralization for processing of bone allograft may have a twofold effect. It should improve incorporation by enhancing their osteoinductive potential. In addition, it should reduce immunogenicity of the graft because irrigation of the graft with solutions such as hydrochloric acid, Triton X or other defattening solutions has been shown to reduce the cellular antigen load within the graft.

The approach of using a bioresorbable foam having an open window structure as a carrier for tissue engineered surface modification stems from the fact that bone has a considerable potential for regeneration. In fact, it is considered by some the prototypic model for tissue engineering [82, 122]. Development of composites of extracellular matrix and stem cells responding to regulatory signals has been at the center of engineering skeletal tissues and is the main goal of this proposal. In fact, leading strategies on engineering and regeneration of skeletal tissues are largely based on observations on local bone induction after implantation of demineralized bone matrix into subcutaneous sites. These models have allowed study of sequential limb morphogenesis and has permitted the isolation of bone morphogens, such as bone morphogenetic proteins (BMPs), from demineralized adult bone matrix. BMPs initiate, promote, and maintain chondrogenesis and osteogenesis. Therefore, it appears reasonable to approach our biodegradable foam as a scaffold that will function as a cell and tissue carrier emphasizing material superstructuring in the design of surface modified and tissue engineered cortical bone allografts. Therefore, our work was aimed at engineering a biodegradable superstructure that will provide optimal spatial and nutritional conditions for cell maintenance by the arrangement of structural elements (e.g., pores or fibbers) so as to vary the order of cell to cell contact [123]. For example, three-dimensional cell-polymer matrices for tissue engineering are needed allowing osteoblast cells to maintain their phenotypic [83, 92].

The approach of using recipient periosteal cells stems from the observation that these cells can be readily cultured from periosteal tissue [87, 125]. In addition, they grow extremely well in synthetic polymeric, as well as in natural collagenous, matrixes and are therefore recognized as an extremely osteoinductive material [126]. Paralleling attempts to take dermal fibroblasts for the in vitro expansion and generation of skin grafting material for extensive burn wounds [124], or attempts to generate cartilaginous material for joint resurfacing by in vitro expansion of autologous articular cartilage cells [125], it is logical to take periosteal cells from the future recipient of bone allograft and to expand them in vitro for tissue engineering purposes. Breitbart et al. [93, 94] have already demonstrated the feasibility of the use of periosteal cells for tissue engineered bone repair of calvarial defects. In analogy, a new type of tissue engineered bone allograft using periosteal cells of the future recipient should show improved enhanced incorporation and significantly lower sensitization rates. Both antibody-mediated immunity and cellular-mediated cytotoxicity should be reduced simply by the formation of a dense autologous periosteal bone cuff surrounding the bone transplant.

In synopsis of the bone allograft processing techniques employed within the framework of this project, the additional use of laser drilling and partial demineralization in combination with tissue engineered surface modifications should further enhance incorporation for two reasons. The demineralized bone matrix surrounding the bone allograft should provide a highly osteoinductive matrix for the ingrowth of periosteal cells from the foam into the allograft. Second, it may release growth factors from the demineralized bone matrix stimulating the expansion of periosteal cells from the foam into the graft. Thus, further processing of cortical bone allografts by employing a tissue engineered surface modification is consistent with previous attempts at enhancing their incorporation into host bone.

IV. PRELIMINARY STUDIES

A. Bioresorbable Foam Suitable for Cell Culturing

Polymer foams are open-celled materials that appear to be ideal templates for cell growth. For cells requiring attachment, an open-celled porous scaffold can be prepared from a biodegradable (resorbable) polymer having a highly organized structure (Fig. 1) with a uniform pore size and characteristic density, which is dependent on the concentration of polymer in solution (Hsu, et al., 1997).

Scanning electron micrographs (SEMs) of our PLGA foams prepared as described show

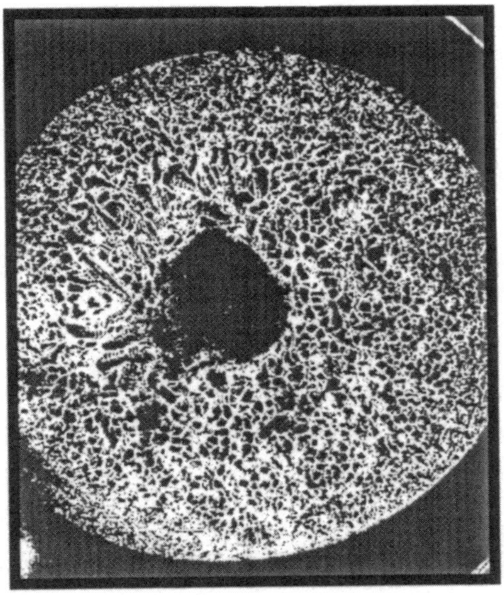

Figure 1 Scanning electron micrograph showing cross-sectional surface of a PLGA tubular foam.

a pore size distribution (median 15–20 μm) which depends on the location within the foam, with smaller pores near the surface. The surface of the foam, although still porous, is less so than the interior. This overall architecture is typical for all densities and offers high porosity for regenerative areas and low porosity for areas of foreign cell invasion.

The medical applications of open-celled polymeric foams includes use as guides for regeneration and healing of damaged nerves. In our related work, 85 : 15 PLGA polymer foam guides were inserted to join 10-mm gaps in sciatic nerves resected from outbreed Sprague-Dawley rats. This polymer of 15 to 20 micron median pore size presents a tremendous surface area for cell attachment, and nerve regeneration. Regenerated nerve cables were present across the entire 10-mm gap in all experimental animals. Marked vascularity was noted throughout the regenerated nerve cables only in the polymer scaffold guides. For tissue repair applications, control of pore size distribution is of prime importance. Not only must regenerating tissues be maintained within the polymer foam, but sufficient porosity must allow exchange of fluids with the environment outside of the guide. Porosity also determines the stiffness and elasticity of the device obviously affecting the ultimate utility of the device.

B. PLGA Foam Preparation

Our PLGA foam were prepared by lyophilization of polymer solutions in different solvents. Solvents were chosen on the basis of their ability to dissolve the polymer and their vapor pressure–temperature profiles. In addition to an easily accessible freezing point, a respectable vapor pressure is desirable. To date, studies have contributed preliminary evidence of the interconnected pore network of our foam as illustrated visually by SEM photographs (Fig. 2) supplemented by mercury intrusion porosimetry (MIP) and Brunauer-Emmett-Teller (BET) data measuring mean pore size and total surface area. Our pilot studies revealed the following additional information on the properties of the foam:

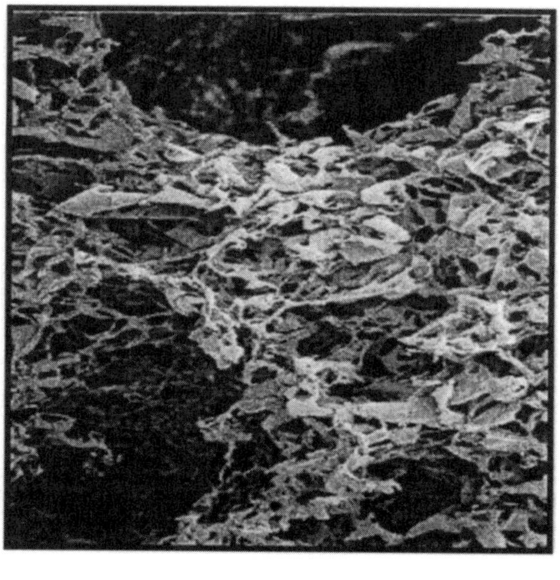

Figure 2 SEM photograph of 85 : 15 PLGA foam prepared by lyophilization of glacial acetic acid solution magnification 100×.

- *Density*: The concentrations of the PLGA used in our foam cuff preparations were in the range of 50–60 mg/ml in gl HAc, which yielded stable foams with densities of approximately 50 mg/cm^3 (Fig. 3).
- *Void volume*: Foams were prepared with theoretical void volumes ranging between 90% and 95.4%.
- *Percent open cell structure*: Our investigations have shown that foams produced by lyophilization of PLGA from glacial acetic acid (HAc) were essentially completely open celled, as confirmed by water imbibation. Open cell structure of 96% was measured with this procedure which confirmed the interconnection of the pores throughout the scaffold.
- *Morphology*: SEMs confirmed morphology of the cuff and appearance of the polymer/bone interface, as shown in Figs. 2, 3 and 4.
- *Pore size and distribution*: The MIP showed a median pore diameter of 60 microns.
- *Total surface area*: The BET data showed a total surface area of 242 m^2/g for a 33-mg/ml polymer concentration in acetic acid.

C. Design of PLGA Foams for Cell Seeding

A foam formulation should be tailored for periosteal cell seeding and expansion when applied to cortical bone allografts. Foam degradation by hydrolysis of the ester linkages results in chain scission and generation of lactic and glycolic acids. Degradation rates of PLGAs have been studied by many workers using a variety of polymer geometries subject to varying degradative conditions. A particularly useful study by Cutright et al. [126], the results of which are in accord with many subsequent observations, compared in vivo degradation of a series of

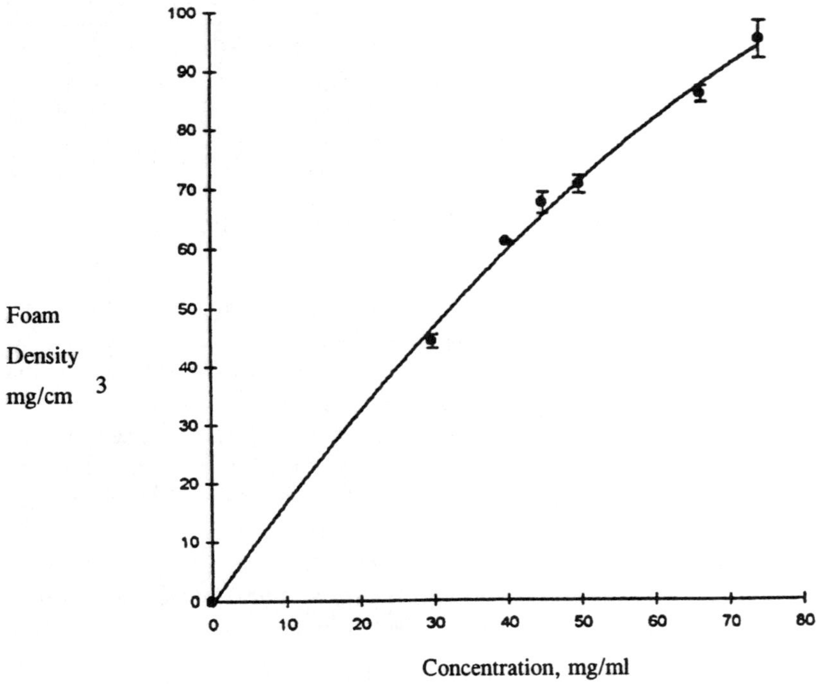

Figure 3 Foam density as a function of polymer solution concentration.

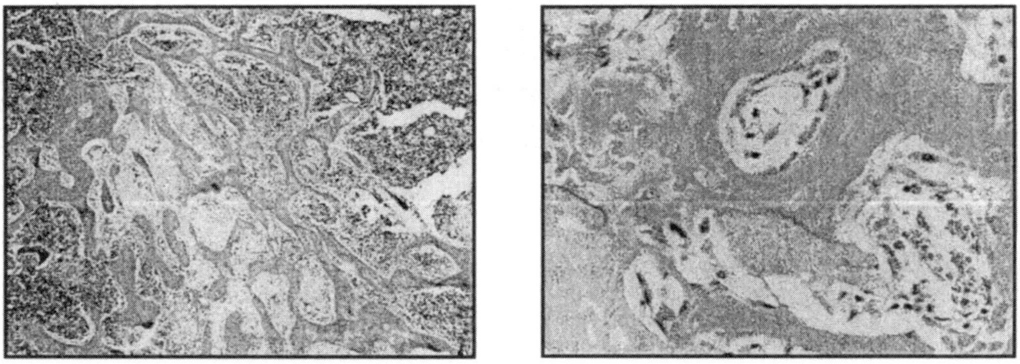

Figure 4 (left) Photomicrograph of a longitudinal section (H&E, 5×) of a rat into which bioresorbable foam was placed. This section shows the drill hole in the rat tibia us completely filled with newly formed woven bone. (right) Photomicrograph of a longitudinal section (H&E, 20×) of a rat. This section shows remnants of the foam between the trabeculae of the newly formed woven bone undergoing active resorption. In addition, there is osteoclastic and osteoblastic activity with neovascularization at the site of these foam remnants.

PLGAs including pure poly(lactic acid), PLA, PLGA-75 : 25, PLGA-50 : 50, PLGA-25 : 75, and pure poly(glycolic acid), PGA. PLA and PGA degraded significantly more slowly than any of the copolymers. At the completion of the study about 40% of the PLA and 20% of the PGA had been resorbed, while the three copolymers had been completely resorbed. Total resorption was observed between days 120 and 180. PLGA-50 : 50 degraded most rapidly with 50% resorption at 60 days. PLGA-75 : 25 and PLGA-25 : 75 showed 50% resorption at 110 and 90 days, respectively.

On the basis of these cumulative data, it appears reasonable to manufacture the foam cuff from PLGA-85 : 15. This formulation would be suitable from a polymer-engineering perspective because it will degrade somewhat more slowly, a desirable property because the high surface area presented by the foam will facilitate more rapid hydrolytic degradation.

D. Initial in Vivo Studies on Osteoconductivity and Biocompatibility of Bioresorbable PLGA Foams

To evaluate suitability of the bioresorbable PLGA-based foam for bony ingrowth prior to its use for periosteal cell seeding, an initial in vivo biocompatibility study in rats was conducted using the tibial defect model of Gerhart et al. [127]. The PLGA-based foam (PLGA 85 : 15) was implanted with use of an 18G needle into the into a tibial defect (measuring 1 mm in diameter) created in male Sprague-Dawley rats (Charles River Breeding Laboratories) weighing approximately 200 g. The foam implanted group was compared to a sham operated group having a drill hole but no implant. Groups of eight animals were sacrificed at one, three, five, and seven weeks postoperatively. Thus, a total of 64 rats was operated in eight groups. Hematoxylin and eosin sections of all specimens were examined microscopically for determination of early stage biocompatibility. Histological evaluation of these sections showed new woven bone formation in animals implanted with the PLGA-based foam as early as at postoperative week 1 (Fig. 4a, b).

Osteoclastic and osteoblastic activity and neovascularization was seen at the foam implantation site and it appeared that the foam served as a scaffold for new bone growth. The origin of this newly formed woven bone appeared to be the periosteum immediately adjacent to the drill hole. At postoperative week 3, the drill hole was completely healed in all animals injected with the foam. In comparison, this was not the case in control animals in which only a hole was drilled but no implant was injected. Although there was some periosteal bone formation at three and at four weeks postoperatively, complete healing of the hole was not observed in any of the sham operated animals.

This study has clearly demonstrated that the bioresorbable PLGA-based foam was capable of serving as a scaffold for bony ingrowth. In particular, this study suggested that the foam is suitable for seeding with periosteal cells because the bone that grew in the foam scaffold most likely stemmed from the periosteum adjacent to the drill hole and thus must have grown into the hole "per continuum" and appositional growth.

E. Preparation of Foam-Allograft Constructs

Out initial studies showed that polymer/bone constructs with adequate adherence of the foamed scaffold to a candidate bone can be prepared. Foam-bone constructs were prepared in the presence of the different types of grafts by impregnation and lyophilization—the net result being a polymer foam cuff around the bone. To make the foams, the dried PLGA was redissolved in glacial acetic acid. For a foam density of 0.04 g/ml, 2 g of PLGA were dissolved in 60 ml of glacial acetic acid. Three cm long, laser drilled bone grafts were immersed in the polymer/acetic acid solution. The graft-polymer mixture was then refrigerated overnight at 12°C. The frozen bone solution was lyophilized in 300 ml lyophilization flasks for 72 h at a vacuum of 5 microns and a temperature of –50°C. This will result in a uniform porous polymer cuff around the candidate bone allograft of 1 to 2 mm thickness. A range of foam thickness will be produced to determine its impact on cell loading and the ability of non-collagenous matrix factors from the demineralized bone matrix to affect the seeded foam cuff. In Figs. 5 and 6 are presented the SEMs of these constructs. Figure 5 is an SEM reproduction showing good, solid maintenance of cellular structure of the foam. The foam/bone interface is shown in Fig. 6. While mechanical demonstration of interface strength has yet to be verified, initial studies in vitro in periosteal cell culture visually demonstrate maintenance of construct integrity.

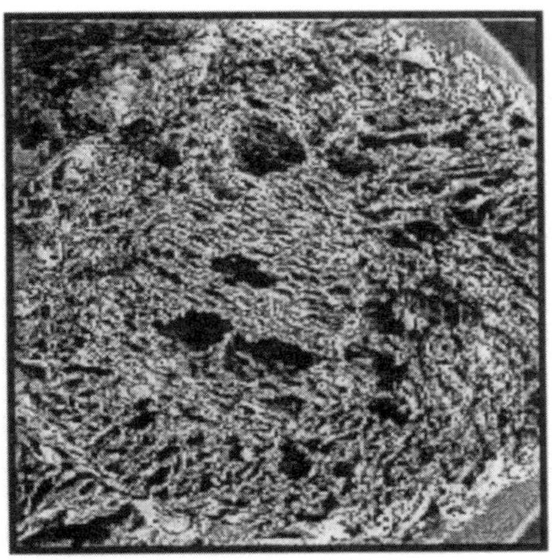

Figure 5 Polymer foam cuff.

Figure 6 Polymer foam/bone interface.

Figure 7 shows the biodegradable foam cuff deposited as a 10-mm-thick layer around a 4-cm cross section of human cortical bone.

F. Preliminary in Vitro Data on Foam Seeding with Periosteal Cells

Intact bovine tibia were obtained. Under sterile conditions, the periosteum was excised cut into 2 mm × 2 mm pieces. Individual pieces were placed in six well culture plates for 10–20 min to facilitate attachment to plastic, and washed twice with PBS containing antibiotics. Alternatively, pieces of periosteum were placed directly on top of PLGA scaffolds to induce attachment of the tissue to the polymer, and the tissue/polymer constructs were placed in six well culture dishes and washed twice with antibiotics. Subsequently, periosteal samples in dishes or tissue/polymer constructs were covered with 3 ml Medium 199 containing 10% FBS, 20

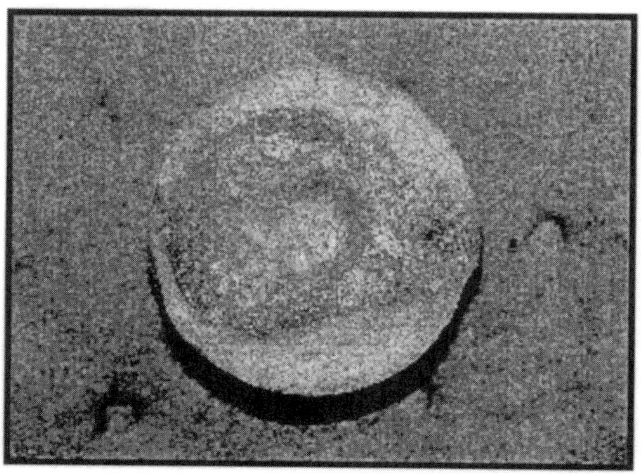

Figure 7 PLGA 85 : 15 foam cuff around human cortical bone.

mg/ml ascorbate, and antibiotics. Migration of cells from periosteal samples were monitored daily, with fresh media added every three days. Migration of cells out of periosteum samples was stopped upon reaching 60–80% confluence. Cells were removed from plates with 0.05% trypsin and 0.53 mM Na_2EDTA and either replated at a density of 2500 cells/cm^2 or seeded onto polymer scaffolds.

In these studies, the periosteal tissue adhered strongly to the 85 : 15 PLGA scaffold, however, very little cell migration or proliferation was observed. A small number of cells appeared to attach to the scaffold surface in the immediate vicinity of the periosteal tissue, but few penetrated the scaffold.

Our investigations have shown that the relative PLA/PGA content of the scaffolds is critical mediator of cell attachment. In particular, high lactic acid content was noted to adversely affect cell seeding. Specifically, fewer than half the number of cells attached to a 75% PLA/25% PGA scaffold than attached to a 0% PLA/100% PGA scaffold (Fig. 8). This suggested that minimizing the lactic acid content would optimize the cell seeding process and maximize the osteogenic potential of the constructs.

V. DISCUSSION

The major objective of this work was to initiate development of a surface modified cortical bone allograft for clinical application in the reconstruction of extensive skeletal defects. The development of an allogeneic cortical bone transplant with improved osteoinductive properties and reduced immunogenicity would potentially eliminate those transplant-related complications usually necessitating revision surgeries that may entail their removal. The proposed cortical bone allograft surface modified with a circumferential layer of a PLGA-based foam that is suitable for seeding with periosteal cells. Additional surface demineralization and perforation is expected to further enhance osteoinduction by improving adherence of the foam to the bone and by direct and indirect stimuli of the demineralized bone matrix on the expanding periosteal cells within the PLGA-based foam scaffold surrounding the graft.

On the basis of our initial in vivo testing of the foam in the rat tibia defect model, it is

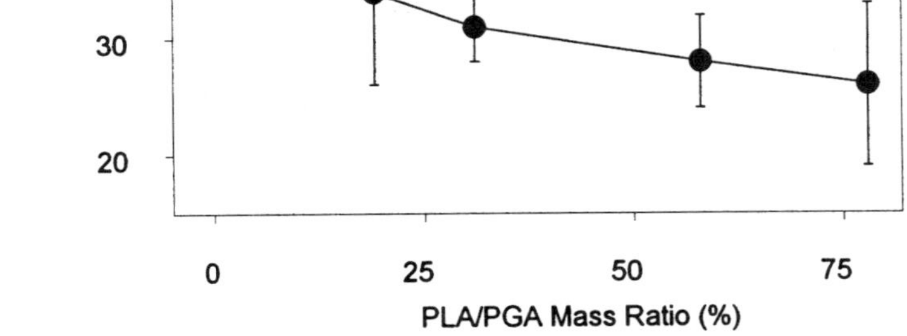

Figure 8 Effect of PLA content on cell attachment to PLA/PGA scaffolds.

anticipated that this highly osteoinductive coating could form new bone within the first postoperative week. Conceptually, this should initiate a variety of processes that may promote the incorporation process and yield superior biological outcomes than with conventional bone allografts. As such, the woven bone cuff that is expected to form around the graft should protect the graft from accelerated resorption, thus resulting in an improved balance between bone resorption and new bone formation throughout the incorporation process. Foreign body responses by encapsulation of the graft with fibrous tissue should be prevented and a new periosteal vascular supply should be rapidly established as evidenced by neovascularization observed in our preliminary in vivo studies using rodents. Incorporation should primarily proceed on the basis of direct bone formation with continuous replacement of nonviable allogeneic bone while periosteal cell expansion is taking place. Rather than relying on the transformation and recruitment of mesenchymal cells from the surrounding soft tissues to invoke the osteoinduction principal, the major source of the allografts osteoinductive potential should stem from direct bone formation that is initiated by the periosteal cells within the PLGA-based foam coating. The net result should be a new type of bone allograft capable of responding to physiological stresses appropriately because it is rapidly incorporated into the recipient's own bone without substantial loss of mechanical strength. These hypotheses developed the objective of this work.

This implies that the success of a tissue engineering approach to bone allografting depends on the ability of the allograft to act as a scaffold for autologous periosteal cells. Specifically, this depends on the growth of periosteal cells on the PLGA foam used to coat the allografts. Our investigations showed that seeding resulted in superficial migration of periosteal cells into the foam. Considering seeding the surface of an ideal porous medium, the drop in cell number was expected to be logarithmic with depth according to simple filtration theory [128]. However, the average pore size of the scaffolds used for initial studies was ~60 μm, which may have been small enough to inhibit penetration of the cells into the bulk of the material. Additionally, gas-water interfaces are known to severely limit cell transport into unsaturated porous media [129]. Therefore, future studies should focus on scaffolds with larger pores on the order of ~100–200 μm. In addition, several other factors should be considered in designing future studies, including scaffold composition and pore size, as well as method of delivery of cells to the scaffold. Given preliminary data presented above, it appears that the high lactic acid content (85%) of these scaffolds may have inhibited cell attachment and/or migration. Future studies should address this by directly examining the effect of the lactic acid content on cell migration from periosteum. The possibility also exists that the process of cell migration on PLGA is much different than that on standard tissue culture plasticware. In this case, alternative methods of delivery must be examined.

ACKNOWLEDGMENTS

The authors wish to thank Joseph Alroy, DVM, associate professor in pathology, Tufts University Schools of Medicine and Veterinary Medicine for his assistance in the histological analysis of this study. Furthermore, the authors are indebted to Shrikar Bondre and Eric Gusek for their assistance with animal care. This work was supported in part by an internal research project of Cambridge Scientific, Inc. and NIH/NIAMS Grant AR 45062 (to Kai-Uwe Lewandrowski).

REFERENCES

1. Clohisy and HJ Mankin. "Osteoarticular allografts for reconstruction after resection of a musculoskeletal tumor in the proximal end of the tibia," *J. Bone Joint Surg. Am.* **76**: 549–554, 1994.

2. HJ Mankin, MC Gebhardt, LC Jennings, DS Springfield and WW Tomford. "Long-term result of allograft replacement in the management of bone tumors," *Clin. Orthop.* **324**: 86–97, 1996.
3. BT Roughraff, MA Simon, JS Kneisl, DB Greenberg and HJ Mankin. "Limb salvage compared with amputation for osteosarcoma of the distal end of the femur. A long-term oncological, functional, and quality-of-life study," *J. Bone Joint Surg. Am.* **76**: 649–656, 1994.
4. KA Jaffe, SG Morris, RG Sorrell, MC Gebhardt and HJ Mankin. "Massive bone allografts for traumatic skeletal defects," *South. Med. J.* **84**, 975–982, 1991.
5. MN Mahomed, RJ Beaver and AE Gross. "The long-term success of fresh, small fragment osteochondral allografts used for intraarticular post-traumatic defects in the knee joint," *Orthopedics* **15**, 1191–1199, 1992.
6. JM Flynn, DS Springfield and HJ Mankin. "Osteoarticular allografts to treat distal femoral osteonecrosis," *Clin. Orthop.* **303**: 38–43, 1994.
7. FR Convery, C Minteer, M Convery, SD Devine and MH Meyers. "Acetabular augmentation in primary and revision total hip arthroplasty with cementless prostheses," *Clin. Orthop.* 167–175, 1990.
8. AE Gross, G Allen and G Lavoie. "Revision arthroplasty using allograft bone," *Instr. Course. Lect.* **42**, 363–380, 1993.
9. MH Huo, GE Friedlaender and EA Salvati. "Bone graft and total hip arthroplasty. A review," *J. Arthroplasty* **7**: 109–120, 1992.
10. MJ Kraay, VM, Goldberg, MP Figgie and HE Figgie. "Distal femoral replacement with allograft/prosthetic reconstruction for treatment of supracondylar fractures in patients with total knee arthroplasty," *J. Arthroplasty* **7**: 7–16, 1992.
11. JH Pak, WG Paprosky, WS Jablonsky and JM Lawrence. "Femoral strut allografts in cementless revision total hip arthroplasty," *Clin. Orthop.* 172–178, 1993.
12. BH Berrey, Jr., CF Lord, MC Gebhardt and HJ Mankin. "Fractures of allografts. Frequency, treatment, and end results," *J. Bone Joint Surg. Am.* **72**, 825–833, 1990.
13. HJ Mankin, DS Springfield, MC Gebhardt and WW Tomford. "Current status of allografting for bone tumors," *Orthopedics* **15**, 1147–1154, 1992.
14. WW Tomford, DS Springfield, HJ Mankin, HH Hung, KU Lewandrowski and TC Fuller. "The immunology of large frozen bone allograft transplantation in humans. Antibody and T-Lymphocyte response and their effects on results," *Trans. Orthop. Res. Soc.* **39**: 102, 1994.
15. HM Dick and RJ Strauch. "Infection of massive bone allografts," *Clin. Orthop.* 46–53, 1994.
16. P. Hernigou, G. Delepine and D Goutallier. "Infections after massive bone allografts in surgery of bone tumors of the limbs. Incidence, contributing factors, therapeutic problems," *Rev. Chir. Orthop.* **77**: 6–13, 1991.
17. CF Lord, MC Gebhardt, WW Tomford and HJ Mankin. "Infection in bone allografts. Incidence, nature, and treatment," *J. Bone Joint Surg. Am.* **70**, 369–376, 1988.
18. WW Tomford, NS Schachar, TC Fuller, WB Henry and HJ Mankin. "Immunogenicity of frozen osteoarticular allografts," *Transplant. Proc.* **13**, 888–890, 1981.
19. WW Tomford, J Thongphasuk, HJ Mankin and MJ Ferraro. "Frozen musculoskeletal allografts. A study of the clinical incidence and causes of infection associated with their use," *J. Bone Joint Surg. Am.* **72**, 1137–1143, 1990.
20. RG Burwell, GE Friedlaender and HJ Mankin. "Current perspectives and future directions: the 1983 Invitational Conference on Osteochondral Allografts," *Clin. Orthop.* 141–157, 1985.
21. GE Friedlaender. "Immune responses to osteochondral allografts. Current knowledge and future directions," *Clin. Orthop.* 58–68, 1983.
22. MC Horowitz and GE Friedlaender. "Immunologic aspects of bone transplantation. A rationale for future studies," *Orthop. Clin. North Am.* **18**, 227–233, 1987.
23. G Aldinger, G Herr, W Kusswetter, HJ Reis, FW Thielemann and U Holz. "Bone morphogenetic protein: a review," *Int. Orthop.* **15**, 169–177, 1991.
24. LJ Marden, AH Reddi and JO Hollinger. "Growth and differentiation factors: role in bone induction and potential application in craniofacial surgery," *J. Craniofac. Surg.* **1**, 154–160, 1990.

25. JB Mulliken, LB Kaban and J Glowacki. "Induced osteogenesis–the biological principle and clinical applications," *J. Surg. Res.* **37**, 487–496, 1984.
26. MR Urist, RJ Delange and GA Finerman. "Bone cell differentiation and growth factors," *Science* **220**, 680–686, 1983.
27. EA Wang, V Rosen, P Cordes, RM Hewick, MJ Kriz, DP Luxenberg, BS Sibley and JM Wozney. "Purification and characterization of other distinct bone-inducing factors," *Proc. Natl. Acad. Sci. U.S.A.* **85**, 9484–9488, 1988.
28. JM Wozney, V Rosen, AJ Celeste, LM Mitsock, MJ Whitters, RW Kriz, RM Hewick and EA Wang. "Novel regulators of bone formation: molecular clones and activities," *Science* **242**, 1528–1534, 1988.
29. JM Wozney. "Bone morphogenetic proteins," *Prog. Growth Factor Res.* **1**, 267–280, 1989.
30. JM Wozney. "The bone morphogenetic protein family and osteogenesis," *Mol. Reprod. Dev.* **32**, 160–167, 1992.
31. H Burchardt. "The biology of bone graft repair," *Clin. Orthop. 28–42, 1983.*
32. *H Burchardt. "Biology of bone transplantation," Orthop. Clin. North Am.* **18**, 187–196, 1987.
33. RG Burwell and G Gowland. "Studies in the transplantation of bone. III. The immune responses of lymph nodes draining components of fresh homologous cancellous bone and homologous bone treated by different methods," *J. Bone Joint Surg.* [Br] **41**, 160–171, 1959.
34. RG Burwell. "Studies in the transplantation of bone. V. The capacity of fresh and treated homografts of bone to evoke transplantation immunity," *J. Bone Joint Surg.* [Br] **45**, 386–401, 1963.
35. MR Urist "Bone: formation by autoinduction," *Science* **150**, 893–899, 1965.
36. MR Urist, BF Silverman, K Buring, FL Dubuc and JM Rosenberg. "The bone induction principle," *Clin. Orthop.* **53**, 243–283, 1967.
37. HJ Mankin, S Doppelt and W Tomford. "Clinical experience with allograft implantation. The first ten years," *Clin. Orthop.* 69–86, 1983.
38. HJ Mankin, Gebhardt, MC and WW Tomford. "The use of frozen cadaveric allografts in the management of patients with bone tumors of the extremities," *Orthop. Clin. North Am.* **18**, 275–289, 1987.
39. S Stevenson, XQ Li and B Martin. "The fate of cancellous and cortical bone after transplantation of fresh and frozen tissue-antigen-matched and mismatched osteochondral allografts in dogs," *J. Bone Joint Surg. Am.* **73**, 1143–1156, 1991.
40. S Stevenson and M Horowitz. "The response to bone allografts," *J. Bone Joint Surg. Am.* **74**, 939–950, 1992.
41. H Burchardt and WF Enneking. "Transplantation of bone," *Surg. Clin. North Am.* **58**, 403–427, 1978.
42. GE Friedlaender. "Bone grafts. The basic science rationale for clinical applications," *J. Bone Joint Surg. Am.* **69**, 786–790, 1987.
43. GE Friedlaender. "The antigenicity of preserved allografts," *Transplant. Proc.* **8**, 195–200, 1976.
44. GE Friedlaender, DM Strong and KW Sell. "Studies on the antigenicity of bone. I. Freeze-dried and deep-frozen bone allografts in rabbits," *J. Bone Joint Surg. Am.* **58**, 854–858, 1976.
45. GE Friedlaender, KW Sell and DM Strong. "Bone allograft antigenicity in an experimental model and in man," *Acta Med. Pol.* **19**, 197–305, 1978.
46. GE Friedlaender and MC Horowitz. "Immune responses to osteochondral allografts: Nature and significance," *Orthopedics* **15**, 1171–1175, 1992.
47. T Glant, C Hadhazy, L Bordan and S Harmati. "Antigenicity of bone tissue. I. Immunological and immunohistochemical study of non-collagenous proteins of the bovine cortical bone," *Acta Morphol. Acad. Sci. Hung* **23**, 111–122, 1975.
48. VM Goldberg and S Stevenson. "Natural history of autografts and allografts," *Clin. Orthop.* 7–16, 1987.
49. MC Horowitz and GE Friedlaender. "Induction of specific T-cell responsiveness to allogeneic bone," *J. Bone Joint Surg. Am.* **73**, 1157–1168, 1991.
50. DL Muscolo, S Kawai and RD Ray. "Cellular and humoral immune response analysis of bone-allografted rats," *J. Bone Joint Surg. Am.* **58**, 826–832, 1976.

51. DL Muscolo, S Kawai and RD Day. "*In vitro* studies of transplantation antigens present on bone cells in the rat," *J. Bone Joint Surg. Br.* **59**, 342–348, 1977.
52. FH Wezeman and RD Ray. "Effects of immune environments upon bone *in vitro*. I. Cytotoxicity of antibodies to bone homogenates," *Clin. Orthop.* **93**, 297–306, 1973.
53. GE Friedlaender, DM Strong and KW Sell. "Studies on the antigenicity of bone. II. Donor-specific anti-HLA antibodies in human recipients of freeze-dried allografts," *J. Bone Joint Surg. Am.* **66**, 107–112, 1984.
54. DL Muscolo, E Caletti, F Schajowicz, ES Araujo and A Makino. "Tissue-typing in human massive allografts of frozen bone," *J. Bone Joint Surg. Am.* **69**, 583–595, 1987.
55. WW Tomford, RJ Starkweather and MH Goldman. "A study of the clinical incidence of infection in the use of banked allograft bone," *J. Bone Joint Surg. Am.* **63**, 244–248, 1981.
56. EP Urovitz, F Langer and AE Gross. "Cell-mediated immunity in patients following joint allografting," *Trans. Orthop. Res. Soc.* **1**, 132, 1976.
57. DR Bertolini, GE Nedwin, TS Bringman, DD Smith and GR Mundy. "Stimulation of bone resorption and inhibition of bone formation *in vitro* by human tumor necrosis factors," *Nature* **319**, 516–518, 1986.
58. M Gowen, DD Wood, EJ Ihrie, MK McGuire and RG Russell. "An interleukin 1 like factor stimulates bone resorption *in vitro*," *Nature* **306**, 378–380, 1983.
59. M Gowen, BR MacDonald, DE Hughes, H Skjodt and RG Russell. "Immune cells and bone resorption," [Review]. *Adv. Exp. Med. Biol.* **208**, 261–273, 1986.
60. LG Raisz. "Mechanisms and regulation of bone resorption by osteoclastic cells," In Disorders of Bone and Mineral Metabolism, pp. 287–311. Edited by FL Coe and MJ Fauvus, New York, Raven Press, 1992.
61. DA Watrous and BS Andrews. "The metabolism and immunology of bone," *Semin. Arthritis Rheum.* **19**, 45–65, 1989.
62. H Yu and J Ferrier. "Interleukin-1 alpha induces a sustained increase in cytosolic free calcium in cultured rabbit osteoclasts," *Biochem. Biophys. Res. Commun.* **191**, 343–350, 1993.
63. MH Zheng, DJ Wood and JM Papadimitriou. "What's new in the role of cytokines on osteoblast proliferation and differentiation?" *Pathol. Res. Pract.* **188**, 1104–1121, 1992.
64. MW Elves and CH Ford. "A study of the humoral immune response to osteoarticular allografts in the sheep," *Clin. Exp. Immunol.* **17**, 497–508, 1974.
65. GD Bos, VM Goldberg, AE Powell, KG Heiple and JM Zika. "The effect of histocompatibility matching on canine frozen bone allografts," *J. Bone Joint Surg. Am.* **65**, 89–96, 1983.
66. GD Bos, VM Goldberg, JM Zika, KG Heiple and AE Powell. "Immune responses of rats to frozen bone allografts," *Bone Joint Surg. Am.* **65**, 239–246, 1983.
67. GD Bos, VM Goldberg, NH Gordon, BM Dollinger, JM Zika, AE Powell and KG Heiple. "The long-term fate of fresh and frozen orthotopic bone allografts in genetically defined rats," *Clin. Orthop.* 245–254, 1985.
68. VM Goldberg, A Powell, JW Shaffer, J Zika, GD Bos and KG Heiple. "Bone grafting: role of histocompatibility in transplantation," *J. Orthop. Res.* **3**, 389–404, 1985.
69. S Stevenson, RB Hohn and JW Templeton. "Effects of tissue antigen matching on the healing of fresh cancellous bone allografts in dogs," *Am. J. Vet. Res.* **44**, 201–206, 1983.
70. S Stevenson, GA Dannucci, NA Sharkey and RR Pool. "The fate of articular cartilage after transplantation of fresh and cryopreserved tissue-antigen-matched and mismatched osteochondral allografts in dogs," *J. Bone Joint Surg. Am.* **71**, 1297–1307, 1989.
71. F Langer, A Czitrom, KP Pritzker and AE Gross. "The immunogenicity of fresh and frozen allogeneic bone," *J. Bone Joint Surg. Am.* **57**, 216–220, 1975.
72. F Langer and A. Gross. "The clinical and immunological assessment of frozen bone allografts," *Acta Med. Pol.* **19**, 271–275, 1978.
73. WW Sepe, GM Bowers, JJ Lawrence, GE Friedlaender and RW Koch. "Clinical evaluation of freeze-dried bone allografts in periodontal osseous defects—part II," *J. Periodontol.* **49**, 9–14, 1978.

74. JJ Rodrigo, TC Fuller and HJ Mankin. "Cytotoxic HLA-anitbodies in patients with bone and cartilage allografts," *Trans. Orthop. Res. Soc.* **1**, 131, 1976.
75. OM Bostman, E Hirvensalo, S Vainionpaa, K Vihtonen, P Tormala and P. Rokkanen. "Degradable polyglycolide rods for the internal fixation of displaced bimalleolar fractures," *Int. Orthop.* **14**, 1–8, 1990.
76. OM Bostman, U Paivarinta, E Partio, M Manninen, J Vesenius and P Rokkanen. "Degradation and tissue replacement of an absorbable polyglycolide screw in the fixation of rabbit femoral osteotomies," *J. Bone Joint Surg.—Am. Vol.* **74**, 1021–1031, 1992.
77. OM Bostman, E Hirvensalo, E Partio, P Tormala and P Rokkanen. "Resorbable rods and screws of polyglycolide in stabilizing malleolar fractures," *Unfallchirurg* **95**, 109–112, 1992.
78. OM Bostman, E Partio, E Hirvensalo and P Rokkanen. "Foreign-body reactions to polyglycolide screws. Observations in 24/216 malleolar fracture cases," *Acta Orthop. Scand.* **63**, 173–176, 1992.
79. OM Bostman. "Distal tibiofibular synostosis after malleolar fractures treated using absorbable implants," *Foot & Ankle* **14**, 38–43, 1993.
80. Y Matsusue, T Yamamuro, M Oka, Y Shikinami, S-H Hyon and Y Ikada. "*In vitro* and *in vivo* studies on bioabsorbable ultra-high-strength poly(L-lactice) rods," *J. Biomed. Mater. Res.* **26**, 1553–1567, 1992.
81. AM Donigian, BR Plaga and PM Caskey. "Biodegradable fixation of physeal fractures in goat distal femur," *J. Pediatric Orthop.* **13**, 349–354, 1993.
82. CT Laurencin, MA Attawia, HE Elgendy and KM Herbert. "Tissue engineered bone-regeneration using degradable polymers: The formation of mineralized matrices," *Bone* **19**(1 Suppl), 93S–99S, 1996.
83. CH Rivard, C Chaput, S Rhalmi and A Selmani. "Bio-absorbable synthetic polyesters and tissue regeneration. A study of three-dimensional proliferation of ovine chondrocytes and osteoblasts," *Ann. Chirurgie* **50**(8), 651–658, 1996.
84. H Suh and C Lee. "Biodegradable ceramic-collagen composite implanted in rabbit tibiae," *ASAIO J.* **41**(3), M652–6, 1995 .
85. SE Utvag, O Grundes and O Reikeraos. "Effects of periosteal stripping on healing of segmental fractures in rats," *J. Orthop. Trauma* **10**, 279–284, 1996.
86. HT Aro, BW Wipperman, SF Hodgson and EF Chao. "Internal remodeling of periosteal new bone during fracture healing," *J. Orthop. Res.* **8**, 238–246, 1990.
87. H Nakahara, VM Goldberg and AI Caplan. "Culture-expanded human periosteal derived cells exhibit osteochondral potential *in vivo*," *J. Orthop. Res.* **9**, 465–476, 1991.
88. H Nakahara, VM Goldberg and AI Caplan. "Culture-expanded periosteal-derived cells exhibit osteochondrogenic potential in porous calcium phosphate ceramics *in vivo*," *Clin. Orthop.* **276**, 291–298, 1992.
89. H Nakahara, SP Bruder, VM Goldberg and AI Caplan. "*In vivo* osteochondrogenic potential of cultured cells derived from the periosteum," *Clin. Orthop.* **259**, 223–232, 1992.
90. TH Kim, C Janetta, JP Vacanti, J Upton and CA Vacanti. "Engineered bone from polyglycolic acid polymer scaffold and periosteum," *Mat. Res. Soc. Symp. Proc.* **394**, 91–97, 1995.
91. CA Vacanti, W Kim, J Upton, D Mooney and JP Vacanti. "The efficacy of periosteal cell compared to chondrocytes in the tissue engineered repair of bone defects," *Tiss. Eng.* **1**, 301–308, 1995.
92. WC Puelacher, JP Vacanti, NF Ferraro, B Schloo and CA Vacanti. "Femoral shaft reconstruction using tissue-engineered growth of bone," *Int. J. Oral Maxillofacial Surg.* **25**(3), 223–8, 1996
93. AS Breitbart, DA Grande, R Kessler, JT Ryaby and RJ Fitzsimmons. "Tissue engineered bone repair of calvarial defects using cultured periosteal cells," *Plast. Reconstruct. Surg.* **101**(3), 567–74; discussion 575–576, 1998.
94. AS Breitbart, DA Grande, R Kessler, JT Ryaby, RJ Fitzsimmons and RT Grant. "Tissue engineered bone repair of calvarial defects using cultured periosteal cells," *Plast. Reconstruct. Surg.* **101**, 567–574, 1998.
95. JM Rubak, M Poussa and V Ritisila. "Chondrogenesis in repair of articular cartilage defects by free periosteal grafts in rabbits," *Acta Orthop. Scand.* **53**, 181–186, 1982.

96. WS Kim, CA Vacanti, J Upton and JP Vacanti. "Bone defect repair with tissue-engineered cartilage," *Plast. Reconstruct. Surg.* **94**, 580–584, 1994.
97. MF Young, JM Kerr, K Ibaraki, AM Heegaard and PG Robey. "Structure, expression, and regulation of the major noncollagenous matrix proteins of bone," *Clin. Orthop.* **P 275-94**–94. 1992.
98. M Centrella, MC Horowitz, JM Wozney and TL McCarthy. "Transforming growth factor-beta gene family members and bone," [Review]. *End. Rev.* **15**, 27–39, 1994.
99. J Chen, K Singh, BB Mukherjee and J Sodek. "Developmental expression of osteopontin (OPN) mRNA in rat tissues: evidence for a role for OPN in bone formation and resorption," *Matrix* **13**, 113–123, 1993.
100. JM Wozney, V Rosen, M Byrne, AJ Celeste, I Moutsatsos, and EA Wang. "Growth factors influencing bone development," *J. Cell Sci. Suppl.* **13**, 149–156, 1990.
101. Y Dohi, H Ohgushi, S Tabata, T Yoshikawa, K Dohi and T Moriyama. "Osteogenesis associated with bone gla protein gene expression in diffusion chambers by bone marrow cells with demineralized bone matrix," *J. Bone Miner. Res.* **7**, 1173–1180, 1992.
102. JN Heersche, SM Reimers, JL Wrana, MM Waye and AK Gupta. "Changes in expression of alpha 1 type 1 collagen and osteocalcin mRNA in osteoblasts and odontoblasts at different stages of maturity as shown by in situ hybridization," *Proc. Finn. Dent. Soc.* **88 Suppl. 1**, 173–182, 1992.
103. B Hinrichs, T Dreyer, A Battmann and A Schulz. "Histomorphometry of active osteoblast surface labelled by antibodies against non-collagenous bone matrix proteins," *Bone* **14**, 469–472, 1993.
104. JB Lian, MD McKee, AM Todd and LC Gerstenfeld. "Induction of bone-related proteins, osteocalcin and osteopontin, and their matrix ultrastructural localization with development of chondrocyte hypertrophy *in vitro*," *J. Cell Biochem.* **52**, 206–219, 1993.
105. MD McKee, MJ Glimcher and A Nanci. "High-resolution immunolocalization of osteopontin and osteocalcin in bone and cartilage during endochondral ossification in the chicken tibia," *Anat. Rec.* **234**, 479–492, 1992.
106. J Glowacki. "Cellular reactions to bone-derived material," *Clin. Orthop. Rel. Res.* 47–54, 1996.
107. J Glowacki and JB Mulliken. "Demineralized bone implants," *Clin. Plast. Surg.* **12**, 233–241, 1985.
108. KU Lewandrowski, G Schollmeier, A Ekkernkamp, P Grosse-Wilde, V Rebmann and WW Tomford. "Immune response to laser-perforated and partially demineralized cortical bone allografts," submitted, 1998.
109. KU Lewandrowski, G Schollmeier, A Ekkernkamp, G Muhr, HK Uhthoff and WW Tomford. "Incorporation of laser-perforated and partially demineralized cortical bone allografts. Part II: A biomechanical study in sheep," submitted, 1998.
110. KU Lewandrowski, G Schollmeier, HK Uhthoff and WW Tomford. "Mechanical properties of laser-perforated and partially demineralized diaphyseal bone allografts," *Clin. Orthop.*, in press, 1998.
111. S Bernick, W Paule, D Ertl, SK Nishimoto and ME Nimni. "Cellular events associated with the induction of bone by demineralized bone," *J. Orthop. Res.* **7**, 1–11, 1989.
112. E Gendler. "Perforated demineralized bone matrix: A new form of osteoinductive biomaterial," *J. Biomed. Mater. Res.* **20**, 687–697, 1986.
113. E Gendler. Cartilage and bone induction by artificially perforated organic bone matrix. United States Patent 4,932,973, June 12, 1990.
114. RJ O'Donnell, TF Deutsch, TJ Flotte, CA Lorente, WW Tomford, HJ Mankin and KT Schomacker. "Effect of Er:YAG laser holes on osteoinduction in demineralized rat calvarial allografts," *J. Orthop. Res.* **14**, 108–113, 1996.
115. CE Scanlon. Analysis of laser-textured, demineralized bone allografts [Master's thesis]. Chicago, Northwestern University, Biomedical Engineering Department, March 27, 1991.
116. BS Sires. Bone allograft material and method. United States Patent, 5,112,354, May 12, 1992.
117. KU Lewandrowski, WW Tomford, KT Schomacker, TF Deutsch, and HJ Mankin. "Effect of Er: YAG laser-holes and partial demineralization on osteogenesis in rat diaphyseal bone allografts," *J. Orthop. Res.* **15**, 748–56, 1997.

118. H Daisaku. "Study on the immune response of mice receiving bone allografts," *Nippon. Seikei-geka. Gakkai. Zasshi* **62**, 71–83, 1988.
119. G Kaminska, M Kaminski and A Komender. "Immunogenicity of fresh and preserved cortical and cancellous allogeneic bone grafts as tested by modified migration inhibition test in mice," *Arch. Immunol. Ther. Exp. Warsz.* **26**, 1053–1057, 1978.
120. NW Nisbet. "Antigenicity of bone," *J. Bone Joint Surg. Br.* **59**, 263–266, 1977.
121. M Shigetomi, S Kawai and T Fukumoto. "Studies of allotransplantation of bone using immunohistochemistry and radioimmunoassay in rats," *Clin. Orthop.* 345–351, 1993.
122. AH Reddi. "Role of morphogenetic proteins in skeletal tissue engineering and regeneration," *Nature Biotechnol.* **16**(3), 247–52, 1998.
123. E Wintermantel, J Mayer, J Blum, KL Eckert, P Luscher and M Mathey. "Tissue engineering scaffolds using superstructures," *Biomaterials* **17**(2), 83–91, 1996 Jan.
124. Moll, P Houdek, H Schmidt and R Moll. "Characterization of epidermal wound healing in a human skin organ culture model: Acceleration by transplanted keratinocytes," *J. Invest. Derm.* **111-2**, 251–58, 1998.
125. M Brittberg, A Lindahl, A Nilsson, C Ohlsson, O Isaksson and L Peterson. "Treatment of deep cartilage defects in the knee with autologous chondrocyte transplantation," *N. Eng. J. Med.* **331**, 889–895, 1994.
126. DE Cutright, B Peez, JD Beasly III, WJ Larson and WR Posey. "Degradation rates of polymers and copolymers of polylactic acids," *Oral Surg.* **37**, 142–152, 1974.
127. TN Gerhart, AA Renshaw, RL Miller, RJ Noecker and WC Hayes. "*In vivo* histologic and biomechanical characterization of a biodegradable particulate composite bone cement," *J. Biomed. Mat. Res.* **23**, 1–16, 1989.
128. LM McDowell-Boyer, JR Hunt and N Sitar. "Particle transport through porous media," *Water Resources. Res.* **22**, 1901–1921, 1986.
129. MV Cattaneo, C Masson and CW Greer. "The influence of moisture on microbial transport, survival and 2.4-D biodegradation with a genetically marked *Burkholderia cepacia* in unsaturated columns," *Biodegradation* **8**, 87–96, 1997.

29

Repair of Articular Cartilage Defects Using Biomaterials and Tissue Engineering Methods

Todd D. Bell, Yuehuei H. An, and Richard J. Friedman
Medical University of South Carolina, Charleston, South Carolina

I. INTRODUCTION

Articular cartilage is an amazing tissue. It is capable of withstanding repeated load bearing activity and it evenly distributes these loads to the underlying subchondral bone. It has an extremely low coefficient of friction and a tremendous capacity to resist wear under normal conditions. However, in spite of its durability, damaged cartilage is not capable of effective repair or regeneration. Consequently, articular cartilage degeneration and injury is a major source of disability in the population.

Total joint arthroplasty has become one of the principal ways to manage terminal cartilage degeneration. Approximately half a million people undergo this procedure yearly in the United States at an estimated cost of $10 billion. Total joint arthroplasties do not produce a normal joint, however, and they have a limited life span as well as many potential complications, such as infection and loosening. Furthermore, this procedure is often not appropriate for use in younger patients or patients with less extensive disease. As a result many individuals are forced to live in discomfort or undergo more limited and unproven interventions.

Clearly, there is a tremendous need for methods of effectively regenerating damaged articular cartilage. One particularly exciting avenue of research is that of tissue engineering techniques aimed at producing this highly specialized tissue. Though the search for an effective mode of treatment is still in its infancy, much progress has already been made.

II. NORMAL ARTICULAR CARTILAGE STRUCTURE

Normal cartilage tissue is composed primarily of extracellular matrix. A single cell type, the chondrocyte, is located within the cartilage and is responsible for synthesizing and maintaining the extracellular matrix [1–3]. The cells make up less than 10% of the volume of articular cartilage in adult humans. Matrix components include water, collagens, proteoglycans and other minor constituents.

A. Structural Components

Type II collagen is the principal structural molecule found in articular cartilage. It makes up 50% of the dry weight of the tissue and comprises over 90% of the total collagen content [1,3].

Other types of collagen which are present include types V, VI, IX, X, and XI. Types IX and X appear to stabilize and strengthen the meshwork by acting as intrafibrillar and interfibrillar crosslinking agents for type II collagen fibrils. Type VI collagen probably helps to attach the chondrocytes to the surrounding matrix [4,5], while type X may have a role in cartilage calcification.

Water provides up to 80% of the wet weight of articular cartilage [1–3]. It has a dynamic role in the structure, function and metabolism of the tissue.

Proteoglycans are macromolecules which consist of numerous sulfated polysaccharide subunits (called glycosaminoglycans) bound to a protein core. The most important glycosaminoglycans in cartilage are chondroiton sulfate, dermatan sulfate, keratan sulfate and heparan sulfate. Multiple proteoglycan chains are, in turn, linked to a hyaluronic acid chain to form a proteoglycan aggregate (Fig. 1). These aggregates contain an abundance of electronegatively charged side chains in close proximity to each other. This has two important consequences. The negative charges attract cations, such as sodium, causing a great increase in osmolarity so that the substance is extremely hydrophilic. Secondly, the negative charges tend to repel each other and this keeps the molecule in a distended state [1]. The tensile strength of the collagen

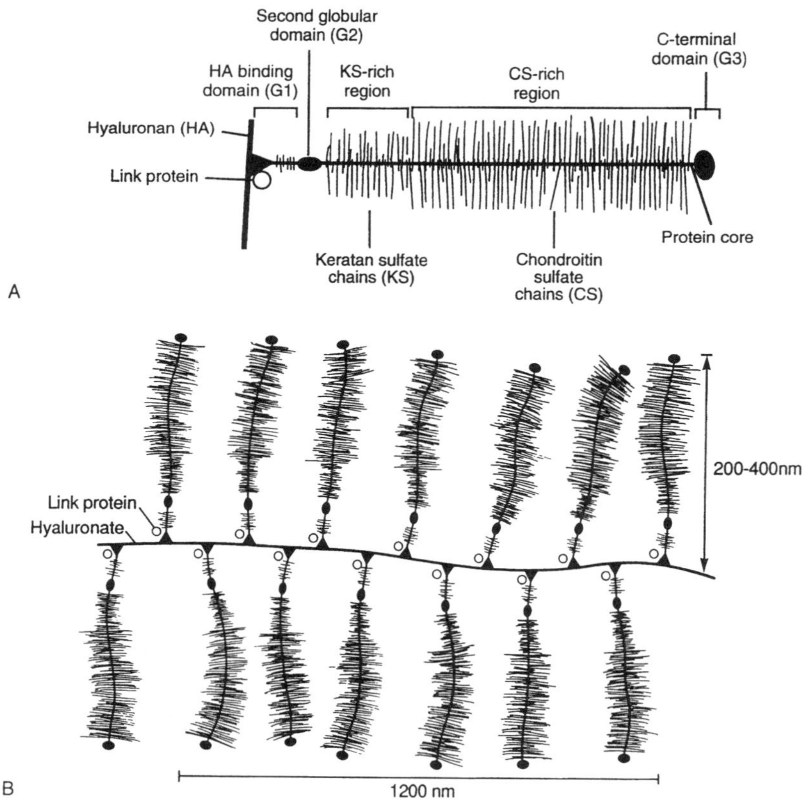

Figure 1 Numerous proteoglycans are bound to a hyaluronic acid chain to form a proteoglycan aggregate. These macromolecules are very hydrophilic due to the abundance of electronegative charges. (From Mow VC and Ratcliffe A, "Structure and function of articular cartilage and meniscus" in Mow VC and Hayes WC: Basic Orthopaedic Biomechanics, 2nd edition, Lippincott-Raven, Philadelphia, PA, 1997, with permission.)

fibrils counteracts this expansile force and helps to maintain the structural integrity of the matrix. This arrangement contributes to the compressive stiffness and durability of hyaline cartilage.

The minor components of articular cartilage include nonspecific lipids, noncollagenous proteins (link proteins, fibronectin and tenascin), a number of glycoproteins (such as anchorin CII and cartilage oligomeric protein) and trace amounts of various growth factors. In general these substances are not well characterized and their roles are poorly understood.

B. Zones of Cartilage

The morphology and physiology of the chondrocytes and the arrangement and distribution of the collagen matrix vary based on the depth from the articular surface. Articular cartilage can thus be divided into four morphological layers: the superficial zone, the middle or transitional zone, the deep or radial zone and the calcified cartilage zone [1,6,7]. Each layer has a unique structural arrangement and recent studies indicate that the different zones have functional significance as well (Fig. 2).

The superficial zone contains flattened chondrocytes, a relatively low concentration of proteoglycans and a high concentration of horizontally arranged collagen fibrils. The most superficial layer in this zone is completely acellular and corresponds to the lamina splendens which can be peeled away from the remaining cartilage in some areas. This layer forms the articulating surface of the cartilage and provides resistance to the sheer stresses which are present there.

The middle zone contains chondrocytes which are more spherical in shape. The collagen fibrils are larger and are arranged in a nonparallel fashion. There is a relatively high concentration of proteoglycans and less collagen and water in this layer.

The chondrocytes in the deep zone are spheroidal in shape and arranged in columns. The collagen fibrils have a parallel arrangement and are oriented perpendicular to the joint surface. Finally, the calcified cartilage zone corresponds to the tidemark line which can be seen on stained preparations of decalcified articular cartilage. The chondrocytes in this region are small, few in number and have a low level of metabolic activity. This area is rich in type II collagen and provides fixation between the articular cartilage and the underlying subchondral bone.

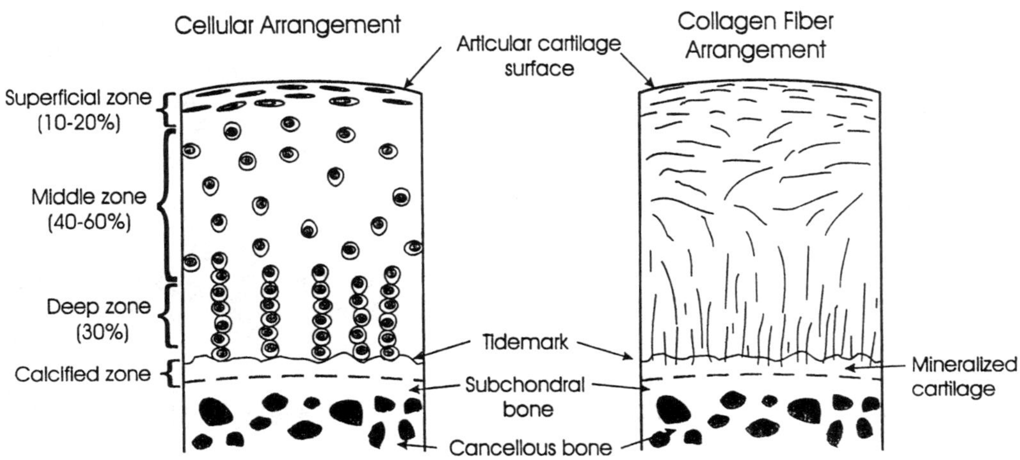

Figure 2 The zonal arrangement of articular cartilage. (From Jackson TW and Simon TM, 1996. Chondrocyte transplantation. Arthroscopy, 12:732–738, with permission.)

C. Physiology of Articular Cartilage

The superstructure of the cartilage creates a situation where the liquid phase can bear a large compressive load in a hydraulic fashion. Under compression the interstitial fluid within the cartilage flows out of the matrix and when the load is removed the fluid returns. The low permeability of the cartilage, particularly the superficial layer, prevents the fluid from being squeezed out too quickly [8,9]. This arrangement allows the fluid phase to bear most of the load during the first few seconds of compression. This helps to protect the solid phase from strong, rapidly applied compressive forces. If a constant compressive force is maintained long enough (for hundreds to thousands of seconds) the fluid phase will have time to flow out of the matrix until an equilibrium is reached. At this point the load is borne principally by the solid phase of the cartilage [9].

These fluid shifts also have a metabolic function. Since cartilage is avascular, it relies on synovial fluid to supply nutrients and carry away its metabolic by-products. The fluid movement associated with loading of the articular surface promotes the transfer of these substances between the intracartilaginous matrix and the synovial fluid within the joint [8]. This explains why immobilization of joints can lead to cartilage degeneration.

III. REPAIR PROCESS OF ARTICULAR CARTILAGE

Normal articular cartilage has an extremely limited capacity for repair. The response to injury depends largely upon whether it involves only the cartilage or extends to the subchondral bone. Since cartilage is an avascular tissue, superficial lesions do not have access to the blood supply. This absence of inflammatory and reparative mediators severely limits its ability to heal. Conversely, full thickness injuries which penetrate the subchondral plate are exposed to the rich blood supply of the underlying bone. This makes the response to these injuries a much more active process [10].

A. Partial-Thickness Injuries

Superficial lesions, which are restricted to the cartilage, have essentially no regenerative potential. After injury there is a brief period of chondrocyte proliferation and matrix synthesis, but the newly produced matrix does not fill the defect and there appears to be no progression of healing over time [10,16–22]. On the other hand, it has been demonstrated that these lesions do not tend to progress and rarely result in osteoarthritis [10,11,17,18,23–25].

B. Full-Thickness Injuries

Cartilaginous injuries that penetrate the subchondral bone are initially filled with a fibrin clot. Within three days undifferentiated mesenchymal cells from the bone marrow begin to migrate into the zone of injury. ^{3}H-Thymidine labeling studies have demonstrated that these mesenchymal cells, rather than the adjacent chondrocytes, mediate the repair process [26]. Within two weeks of the injury cancellous bone begins to appear at the base of the defect and by 12 weeks the subchondral bone has been completely regenerated [27]. Meanwhile the area of the cartilage defect is initially filled with irregular connective tissue. By two weeks the mesenchymal cells differentiate into cells which have the biochemical and morphological characteristics of chondrocytes and begin producing a matrix which contains collagen and proteoglycans [11,28,29]. However, the resulting "hyaline-like" cartilage fails to recreate the intricate laminated microstructure and biochemical composition of true articular cartilage, and the fibrillar framework does not appear to fully integrate into the surrounding uninjured tissue. Furthermore, type I collagen predominates early in the repair process and even at six months post-injury makes up an inordinately large portion (25–30%) of the total collagen content in the matrix [28].

Degenerative changes of this repair tissue can be seen as soon as 10–12 weeks after injury [10,11,26,28]. The earliest changes include superficial cartilage fibrillation, hypocellularity and a decrease in proteoglycan content. Worsening surface fibrillation and deeper fissures appear over the next 12–24 weeks and there is a gradual decrease in the amount of extracellular matrix. Extensive acellular areas may be seen as well. These changes are progressive over time [26], and extensive degeneration of the repair tissue usually becomes apparent by 6–12 months after injury [30].

IV. ANIMAL MODELS USED IN ARTICULAR CARTILAGE RESEARCH

In discussing the use of tissue engineering for cartilage repair it is relevant to examine the methods used in this type of research, particularly the animal models which can be employed and some of the practical aspects of their use.

A. Heterotopic Models of Chondrogenesis

Several of the major heterotopic models for testing the in vivo chondrogenic potential of various substances are the subcutaneous, intramuscular, and intraperitoneal implantation models. When in vitro studies such as cell culture or cell seeding experiments appear promising, subcutaneous implantation is often the initial step of in vivo studies. The animals used for heterotopic models include nude mice, syngenic mice, rats, and rabbits. Substances or constructs which have potential chondrogenic properties are implanted or injected into the above-mentioned sites. After two to four months the implants can then be explanted and studied histologically to identify new cartilage tissues. For example, chondrocytes from juvenile rabbits (2.2–3.2 kg) have been grafted into the anterior tibial compartment of mature rabbits (3.2–4.0 kg). Under these conditions cartilage tissue formation has been seen as early as 10 days after implantation [31]. Although Green [31] found similar amounts of cartilage formation using the intramuscular and subcutaneous models, the full effects of implantation sites and local environment are still not clear.

Another heterotopic model of chondrogenesis is the use of a diffusion chamber implanted intramuscularly into rats or nude mice. The diffusion chamber is made from a plastic ring (2 mm thick, 9 mm diameter) bounded on either side by microporous cellulose acetate and nitrate membranes of 100 micron thickness and 0.45 micron pore size. This produces a chamber volume of approximately 130 microliters (Millipore Corporation, MA) [32–35]. A diffusion chamber of this type, containing rabbit bone morphogenetic protein (BMP), was implanted into the abdominal muscle of the rat. Cartilage formation was seen within the tissue surrounding the chamber one to two weeks after implantation, and bone replaced the cartilage three to four weeks afterwards [35].

B. Cartilage Defect Models

There are two main types of cartilage defect models: focal full-thickness defects and partial-thickness defects. The most commonly used animals are rabbits and dogs.

1. Rabbit Distal Femoral Articular Surface Model

The basic steps in the testing of the regenerative potential of an implant or construct are the manufacturing of the implant, the creation of a cartilage defect, surgical implantation of the implant, specimen harvesting, and evaluation. The rabbit model, with cartilage defects created in the distal femoral surface, is ideal for this type of situation. The advantages of this model are (1) it has been widely used and is well studied, (2) the rabbit knee is a suitable size, making surgical procedures and specimen handling relatively simple, (3) it is consistently reproducible and (4) it is relatively economical. By comparison mice are too small for easy manipulation and specimen handling, and the use of dogs is more difficult to justify ethically and economically [36].

Cartilage defect models can be applied in various ways: these include partial-thickness versus full-thickness defects and isolated versus extensive defects. The isolated defect models are suitable for investigating the use of an engineered implant or construct, while extensive defect models are often used for examining other treatment methods such as abrasion burr arthroplasty, enzymatic treatment [37] or joint surface drilling.

The articulating surface of the femur in the rabbit knee can be divided into weight-bearing and partial weight-bearing regions. The femoral condyle, especially the posterior portion, is a weight-bearing region (rabbit knees stay in a flexed position most of the time). Partial weight-bearing areas are located in the intercondylar groove and the undersurface of the patella.

Typically, the rabbit is placed in a supine position and the knee joint is entered through a lateral parapatellar approach. The patella is dislocated medially to expose the articular cartilage of the patellar groove and the femoral condyles. In order to create an isolated, full-thickness chondral lesion a 1.5–3.0 mm deep defect is created using a 3.5 mm diameter drill bit in either the intercondylar groove or on the surface of the lateral or medial femoral condyles (Fig. 3). At least a 1.25 mm depth is needed to create a full thickness defect since the total thickness of the cartilage and the subchondral bone plate averages 1.0 mm for femoral condyle joint surfaces. Depths of 3.5–5.0 mm may be used in order to facilitate the anchoring of a cylindrical implant [38].

2. Alternative Models

In cases where a rabbit model is not sufficient, a dog model provides a suitable alternative. The dog knee joint is much larger than that of the rabbit, allowing a larger defect to be created.

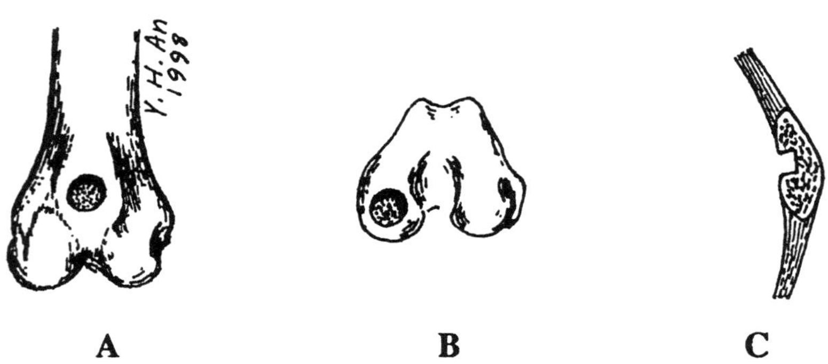

Figure 3 Isolated chondral lesions may be placed in partial weight-bearing regions, such as the interchondylar groove (A), or in full weight-bearing regions such as the posterior portion of the femoral condyle (B), or the articular surface of the patella (C). (From An YH and Friedman RJ: Animal Models in Orthopaedic Research, CRC Press, Boca Raton, FL, 1999, chapter 16, with permission.)

Also because of the larger volume of cartilage it is easier to harvest a sufficient amount of tissue for chondrocyte isolation used in autogenic cell seeding experiments. In our laboratory, we have experienced difficulty culturing rabbit chondrocytes from a small piece of cartilage tissue taken from a 3.5 mm diameter cartilage defect (at least a 50% failure rate) [36].

V. TRADITIONAL METHODS OF TREATING ISOLATED CHONDRAL DEFECTS

Though the true natural history of isolated articular cartilage lesions is unknown, they are undoubtedly a significant source of pain and disability, and in some cases they may lead to an increased risk of progression to degenerative arthritis. Clearly, effective surgical repair options are needed. Traditionally joint debridement with abrasion or drilling of the defect has been used with limited success. Newer treatment options include osteochondral transplantation, chondrocyte transplantation and transplantation of periostial or perichondrial flaps.

A. Pridie Drilling and Abrasion Arthroplasty

The drilling of articular chondral lesions was first described by Pridie in 1959 [39]. The procedure involves making multiple passes with a drill through the area of denuded cartilage. Abrasion arthroplasty was popularized during the 1980s [40] and consists of shaving the lesion back to bleeding subchondral bone with healthy cartilaginous edges. These techniques are similar in concept and outcome. Both are based on the idea that cartilaginous healing will be promoted by exposing the lesion to the rich vascular supply of the subchondral bone. However, numerous studies have indicated that the resulting repair tissue is fibrocartilaginous in nature rather than true hyaline cartilage [40,41].

Reports differ on the clinical outcome for these procedures [40,42–45]. Several reports have indicated that they provide little benefit. Although one recent study indicated that drilling may be superior to abrasion in treating full thickness lesions [46], neither technique has been proven to provide reliable long-term benefit. Further studies are needed to delineate the proper patient selection in terms of patient age, the nature of the underlying disorder, and the nature, location and extent of the lesion. It is possible that in some situations proper application of these techniques will result in better outcomes. However, at this point it appears that any improvement which occurs tends to deteriorate with time.

B. Chondrocyte Transplantation

The implantation of autologous cultured chondrocytes into isolated cartilage defects has shown great promise as a treatment for these lesions. Brittberg et al. [47] performed a prospective study in which 23 patients with localized defects of the femoral condyle or patella were treated in this manner. Chondrocytes were harvested from non-weight-bearing regions of the knee and amplified in culture for 14 to 21 days. The cultured chondrocytes were then injected into the area of the defect and covered with a periostial flap. Follow-up evaluations, two years after the procedure, have revealed that 14 of the 16 patients with condylar defects experienced good or excellent clinical results. Patients with patellar defects did not fare as well, with only two of seven having good or excellent clinical outcomes. Biopsies of the treated regions showed hyaline-like cartilage in many of the defects.

Chondrocyte transplantation has been used on a much larger scale in Sweden, with over 400 patients having been treated in this manner. Preliminary results of the first 100 patients who underwent this treatment are encouraging. Results with the use of this procedure in the United States have been similar [48]. Tissue engineering techniques are an attempt to improve on this early success.

VI. REPAIR OF CARTILAGE DEFECTS USING BIOMATERIALS AND TISSUE ENGINEERING TECHNIQUES

Tissue engineering is an interdisciplinary field that applies the principles of engineering and life sciences toward the development of biological substitutes that restore, maintain, or improve tissue function [49]. It encompasses a large field of research and may include the use of biological or synthetic matrices, growth factors and/or chondrocyte transplantation.

Much attention has been focused on the use of biocompatible and biodegradable synthetic matrices. Freed et al. [50] have set forth the following criteria for an ideal cell scaffold: the surface should permit cell adhesion and growth, neither the polymer nor its degradation products should provoke inflammation or toxicity when implanted in vivo, the material should be reproducibly processable into three-dimensional structures, the porosity should be at least 90% in order to provide a high surface area for cell-polymer interactions, sufficient space for extra cellular matrix regeneration and minimal diffusional constraints during in vitro culture, the scaffold should resorb once it has served its purpose of providing a template for the regenerating tissue. Finally, since foreign materials carry a permanent risk for inflammation, the scaffold degradation rate should be adjustable to match the rate of tissue regeneration by the cell type of interest.

A wide variety of different scaffold structures are under investigation, including collagen gels and sponges, agarose gels, thrombin, fibrin, polylactic acid, polyglycolic acid, polyurethane, polyester, and many others.

Itay [51] used a biologically resorbable immobilization vehicle (BRIV) as a carrier gel for cultured embryonal chick chondrocytes. The BRIV was composed of a mixture of chondrocyte extracellular matrix, trypsin inhibitor, fibrinogen and thrombin. Implantation of this aggregate into surgically created full-thickness cartilage defects was found to produce a hyaline-like repair cartilage with normal cellular content and collagen fiber organization. At six months the reparative cartilage tended to match the volume of the cartilage defect and integrated smoothly with the surrounding native cartilage. However, reformation of the subchondral bond plate remained incomplete. There appeared to be no signs of rejection of the allografted chondrocytes. No inflammatory reaction, synovitis, fibrosis, pannus formation or graft destruction were seen. In contrast, lesions which were left to heal by natural repair and those treated with BRIV alone (without chondrocytes) demonstrated incomplete healing with fibrous scar tissue containing horizontally arranged fibroblasts rather than cartilaginous tissue.

Similarly, Wakitani [52] used a collagen gel impregnated with cultured chondrocytes to treat full-thickness cartilage lesions in the patellofemoral groove of rabbits. The resulting repair tissue appeared histologically similar to normal hyaline cartilage at 24 weeks. It exhibited a smooth articular surface, a columnar cellular arrangement and a predominance of type II collagen. However, there appeared to be an insufficient union between the repair tissue and the normal surrounding tissue in some cases. On the other hand, control groups which were treated with the collagen gel alone or were allowed to heal naturally, with no treatment, tended to have incomplete fibrous repairs which consisted of predominantly type I collagen.

Speer et al. [53] also tested collagen as a carrier vehicle, but instead of a gel they produced a reconstituted, highly cross-linked collagen sponge (CS) and compared this to synthetic polyvinyl alcohol sponge (PVAS) implants. Neither of these matrices was seeded with chondrocytes. Instead they depended on the migration of repair cells into the area of the defect. The matrices were intended to serve as a mechanical support to allow for directed proliferation of natural repair tissue. These implants were placed within surgically created osteochondral defects in the patellofemoral groove of NZW rabbits. The CS treated defects were filled almost completely with a fibrous or fibrocartilaginous repair tissue, although a central dimple or crevice remained. The repair tissue seemed to be well integrated into the surrounding normal tissue and the junction was bridged by a continuity of collagen bundles. Furthermore, a reconstituted subchondral bone plate developed and appeared to be of similar thickness to the adjacent subchondral bone. The PVAS treated lesions healed poorly with incomplete filling of the central half of the defect and a tenuous junction with the surrounding tissue. The PVAS was felt to have acted as a barrier to ingrowth with repair tissue which could not be removed, bonded, covered over or replaced by the host tissue. The control group in this experiment, which received no implant at all, exhibited complete coverage of the defect, but with only a thin surface layer of fibrous and fibrocartilaginous tissue. Furthermore, it was accompanied by the formation of a thickened sclerotic subchondral plate which would presumably lead to accelerated degeneration.

One concern regarding the collagen sponge implant was the persistence of the cross-linking agent glutaraldehyde in the area of the repair. The investigators expressed concerns as to its possible chemical effects on cell ingrowth and proliferation; and they felt that a CS without the residual glutaraldehyde would serve as a better cell supporting matrix.

Kawabe and Yoshinao [54] isolated and cultured growth plate chondrocytes, allowing them to form a cohesive cartilaginous disk. These allografts were then attached to surgically created full thickness cartilage defects in the patellofemoral groove of NZW rabbits. In the early treatment period the cultured disks quickly formed a cartilaginous matrix and the chondrocytes reorganized into a columnar arrangement. However, an immunologic reaction to the transplanted cells ensued. Only 26% of the grafts had viable chondrocytes from the transplanted cartilage disk at three weeks and by four to six weeks 90% of the grafts contained mononuclear cell infiltrates. At six to nine weeks, most of the grafts had completely disappeared and were replaced with fibrous tissue. Histological and immunological analysis indicated that the graft rejection was primarily due to cell-mediated mechanisms rather than a humoral response.

Freed et al. [55] made an in vitro and in vivo comparison of two synthetic biodegradable polymer scaffolds: fibrous polyglycolic acid (PGA) and porous polylactic acid (PLA). The PGA that they utilized is composed of a filamentous mesh-like patch with a porosity of 97%. In vivo it undergoes a relatively rapid degradation by means of a hydrolytic deesterification to glycolic acid which is then enzymatically converted to glycine. PGA experiences approximately 99% degradation after four weeks of implantation [56]. It has been approved for clinical use by the FDA and is commercially available in the form of surgical sutures.

The PLA scaffold has a sponge-like arrangement with a porosity of 91% and degrades at a much slower rate than PGA. It exhibits essentially no change in scaffold geometry six months after implantation [57]. The monomeric breakdown product of PLA is lactic acid which is metabolized via the Krebs cycle.

In vitro comparisons of the PGA and PLA scaffolds were made. Each was seeded with chondrocytes and allowed to grow in culture. At two months the PGA scaffold exhibited rounded chondrocytes within lacunae surrounded by cartilaginous matrix which stained strongly positive for glycosaminoglycans (GAG) and types I and II collagen, whereas the PLA scaffold contained only spindle-shaped cells with minimal matrix staining for GAG or collagen. The polymer scaffolds also influenced chondrocyte proliferation. The final cellularity of the PGA constructs equaled that of the original donor cartilage specimen while the PLA construct final cellularity was less than half of that. Quantitation of sulfated glycosaminoglycan (S-GAG) content revealed that PGA supported samples reached a level equal to 25% of the amount of S-GAG in normal cartilage, while the PLA constructs only reached a level of 2.4% of normal.

In vivo comparison of the scaffolds involved subcutaneous implantation of the in vitro cultured constructs into athymic "nude" mice. At six months, both types of matrices exhibited the formation of rounded cells in lacunae with a surrounding matrix which stained strongly positive for GAG and types I and II collagen. The S-GAG content of the PGA based explants tended to increase rapidly and then either plateau or decrease over four to six months. On the other hand, the S-GAG content in the PLA explants increased steadily over four to six months. Importantly, all of the cartilage-laden implants retained their original shapes. Control implants of polymer without cultured chondrocytes all exhibited negative staining for GAG and types I and II collagen.

In summary, these investigators felt that while PGA proved to be a better scaffold in short-term in vitro tests, both PGA and PLA proved to be excellent scaffolds in longer term in vivo studies. They theorized that the early superiority of PGA was largely due to its more rapid degradation as well as the diffusional constraints imparted by the larger size and lower porosity of the PLA scaffolds.

Chu et al. [58] examined the suitability of porous PLA as a scaffold for cartilage repair. This compound differs from the PLA tested by the Freed group in that it does not require prewetting, so it is possible to seed the cores dry. The resulting capillary action then facilitates the initial entry of the seed chondrocytes into the scaffolds. In their study, Chu et al. seeded the polymer scaffolds with perichondrial chondrocytes and then implanted these constructs into surgically created drill holes in the posterior portion of the medial femoral condyle of NZW rabbits. At 16 weeks, 15 of 16 specimens exhibited biologically acceptable repair by gross examination. Microscopically, GAG matrix was seen surrounding clusters of chondrocyte-like cells. Cellular alignment was similar to the pore structure of the scaffold and cellular morphology differed between the superficial and deep portions of the cartilage. In the deep zones cells with round to polygonal nuclei were arranged in isogenous clusters. Superficially, more spindle-shaped cells with elongated nuclei were seen. Polarized light studies revealed an emerging zone of integration with the surrounding native cartilage and there was evidence of collagen fibril continuity. Apparently seamless peripheral attachment was seen in some specimens on gross examination. However, despite the gross and histological similarities, all biochemical parameters were significantly different than normal articular cartilage. Type II collagen composed only 19% of the engineered collagen content at six weeks and GAG content was only one-third that of normal articular cartilage.

Vacanti et al. [59] implanted chondrocyte-seeded PGA implants into surgically created full-thickness cartilage defects in the patellar groove of NZW rabbits. They then compared these to control lesions which received either no implant or an unseeded PGA polymer. At seven weeks, gross evidence of new cartilage filling the defect was seen in 11 of 12 lesions treated with seeded implants while no gross and scant histological evidence of articular repair was seen in the control groups. Microscopically, examination of the seeded implants revealed polymer remnants encased within the newly forming cartilage matrix. There was interdigitation of the new cartilage with the underlying bone forming a relatively normal bone-cartilage interface. Brd-U labeling of the cultured chondrocytes confirmed that the regenerative cartilage contained the seeded cells.

Similarly, von Schroeder et al. [60] compared the use of a PLA matrix with and without a periostial graft in the treatment of full-thickness lesions of the medial femoral condyle in NZW rabbits. At 12 weeks the grafted PLA more closely resembled hyaline cartilage histologically and contained 82% type II collagen versus 64% for PLA alone. The PLA-alone treated lesions tended to develop a thinner layer of repair tissue which was more fibrous in quality than the grafted PLA. However, the PLA-alone treated lesions showed better attachment to the surrounding native tissue. Biomechanically, there appeared to be no appreciable difference in compliance between the grafted and nongrafted lesions, but both were significantly more compliant than normal cartilage. Rare giant cells were the only evidence of an inflammatory reaction to the implanted PLA.

Messner et al. [61] compared the effectiveness of several different synthetic implants. They created full-thickness lesions in the posterior portion of the medial femoral condyle of rabbits and treated them with polyurethane-coated PTFE (Teflon), uncoated PTFE, polyester (Dacron) felt and periostial grafting. At three months the repair sites were examined. The polyurethane-coated implants had failed by separation within the implant. All uncoated implant repair sites demonstrated ingrowth of fibrous tissue and trabecular bone into the base and periphery of the implant, but migration of cells to the central region of the implant was rare. None of the lesions was noted to form hyaline cartilage.

Biomechanical testing of the repair sites was also performed. All repair sites had an approximately three times higher compliance than the normal cartilage-bone complex, although the synthetically treated lesions were closer to normal than the periostially grafted defects. This increased compliance is consistent with other studies using carbon fiber pads [62] and PLA matrices [60].

The Messner group attributed the elevated compliance to the presence of a poorly developed subchondral bone plate. In an effort to address this problem they subsequently tested hydroxylapatite (HA)-supported Dacron plugs. They found that the HA apparently enhanced growth of the subchondral plate but this did not translate into improved compliance of the repair site.

Another concern raised by the Messner group was that of an inflammatory reaction to the synthetic matrices. They noted that all repairs were associated with synovitis and approximately half of the knees treated with synthetic grafts contained particulate debris with formation of giant cells and foreign body reaction. This was worse with the coated implants.

VII. ROLE OF GROWTH FACTORS

With the increasing knowledge related to growth factors, it is becoming more clear that they play a very important role in the process of chondrogenesis. The major growth factors with chondrogenetic ability include TGF-β, FGF, EGF, IGF and BMP. These growth factors stimulate DNA synthesis (TGF-β [63,64], FGF [65,66], EGF [65,67], IGF-I [68], BMP [69,70]), chondrocyte proliferation (TGF-β [70,71], FGF [72,70], EGF [71], IGF-I [73,74], BMP [69]), chondrocyte differentiation (TGF-β [75,76], IGF-I [77], BMP [69]), and matrix production (TGF-β [63,70], FGF [70,78], EGF [78], IGF-I [79], BMP [80]). Several carriers have been reported to deliver growth factors for in vivo osteochondrogenesis such as devitalized collagen bone matrix (DVBM) [81], demineralized bone matrix (DBM) [82], bovine collagen [74], and polylactide-*co*-glycolide [82].

Another stimulus, continuous passive motion, has been used alone or in combination with osteoperiosteal grafts, and its positive effects demonstrated on the quality of the newly formed chondral tissue and the boundary between the graft and the defect [83]. Also, some evidence indicates that chondrocytes and progenitor cells respond to electrical or electromagnetic field stimulation [84,85].

VIII. CLINICAL STUDIES USING BIOMATERIALS FOR CARTILAGE REPAIR

While there has been a significant amount of research investigating the use of tissue engineering techniques in the treatment of articular cartilage lesions in animals, clinical trials in humans have been relatively limited. However, some experience with the use of carbon fiber implants does exist. The use of carbon fiber is advocated because of its purported strong fibrocartilage inducing effect but concerns remain regarding the resultant inflammatory reaction.

Muckle and Minns [86] used a loosely woven carbon fiber pad to repair osteochondral defects in the knees of 47 patients. These included 23 femorotibial lesions and 24 patellar lesions. The average age of the patients was 30 years. At a mean follow-up of three years the patients were assessed clinically and arthroscopically. Biopsies showed that the carbon pads had become filled with structurally strong fibrous tissue which was similar to that seen abrasion arthroplasty. It was predominantly type I collagen. Thirty-six patients (77%) had an excellent or good clinical outcome with no synovitis or significant particulate carbon fiber debris within the graft. Nine patients had mild generalized speckling within the joint but no gross or histological synovial reaction. In two patients the pad had failed.

Brittberg [87] tested carbon fiber scaffolds of two types: loosely woven pads were used for lesions on flat or concave joint surfaces and carbon fiber rods were used for convex weight-bearing surfaces. They used these implants to treat 37 patients with degenerative chondral lesions and exposed bone. At an average follow-up of 48 months, there was an overall improvement in the patients' functional and pain scores. Pain relief was the most dramatic benefit. Arthroscopic examination was performed on 20 of the 37 patients and all exhibited filling of the defect with a whitish fibrous tissue which was level with the surrounding cartilage. Biopsies showed a dense, carbon fiber anchored fibrocartilage. A few of the patients exhibited sparse carbon fiber speckling of the joint, but no gross signs of synovitis were seen. No differentiation in outcome was made between those patients treated with pads versus rods, and there was no control group in this study.

IX. CONCLUSION

Recent work in the field of tissue engineering has shown promise in the treatment of articular cartilage degeneration. However, many obstacles still remain and a breakthrough discovery remains elusive. No method of treatment has successfully replicated the intricate histological structure or the biomechanical properties of native cartilage. The issue of immunological rejection or inflammatory reaction to the engineered scaffold remains a concern and integration of the repair tissue with the normal surrounding cartilage has not been achieved. Despite these challenges, cartilage tissue engineering remains a steadily advancing field. It is possible that a combined approach using tissue engineering with cultured chondrocytes in conjunction with growth factors and other chondrogenic stimuli will provide the greatest opportunity for cartilage regeneration to occur.

REFERENCES

1. Buckwalter, JA, Hunziker, E, Rosenberg, L, et al, 1988. Articular cartilage: Composition and structure. In: SL-Y Woo and JA Buckwalter (Eds.), Injury and repair of the musculoskeletal soft tissues. AAOS, Park Ridge, IL, pp 405–425.

2. Buckwalter, JA, Mow, VC, Ratcliffe, A, 1994. Restoration of injured or degenerated cartilage. J Am Acad Orthop Surg, 2:192–201.
3. Buckwalter, JA, and Mankin, HJ, 1997. Articular cartilage part I: Tissue design and cartilage-matrix interaction. J Bone Joint Surg, 79 A:600–611.
4. Hagiwara, H, Schroter-Kermani, C, Merker, HJ, 1993 Localization of collagen type VI in articular cartilage of young and adult mice. Cell Tissue Res, 272:155–160.
5. Marcelino, J, McDevitt, CA, 1995. Attachment of articular cartilage to the tissue form of type VI collagen. Biomech Biophys Acta, 1249:180–188.
6. Buckwalter, JA, 1983. Articular cartilage. Instr Course Lect, 32:349–370.
7. Poole, AR, Pidoux, I, Reiner, A, et al, 1982. An immunoelectron microscope study of the organization of proteglycan monomer, link protein, and collagen in the matrix of articular cartilage. J Cell Bio, 93:921–937.
8. Mow, VC, Rosenwasser, M, 1988. Articular cartilage: Biomechanics. In: SL-Y Woo and JA Buckwalter (Eds.), Injury and repair of the musculoskeletal soft tissues. AAOS, Park Ridge, IL, pp 427–463.
9. Mow, VC, Ateshian, GA, Ratliffe, A, 1992. Anatomic form and biomechanical properties of articular cartilage of the knee joint. In: GAM Finerman and FR Noyes (Eds.), Biology and biomechanics of the traumatized synovial joint: The knee as a model. AAOS, Rosemont, IL, pp 55–81.
10. Mankin, HJ, 1982. The response of articular cartilage to mechanical injury. J Bone Joint Surg, 64A: 460–466.
11. Buckwalter, JA, Rosenburg, L. Coutts, R, et al, 1988. Articular cartilage: Injury and repair. In: SL-Y Woo and JA Buckwalter (Eds.), Injury and repair of the musculoskeletal soft tissues. AAOS, Park Ridge, IL, pp 465–482.
12. Campbell, CJ, 1969. The healing of cartilage defects. Clin Orthop, 64:45–63.
13. DePalma, AF, McKeever, CD, Subin, DK, 1966. Process of repair of articular cartilage demonstrated by histology and autoradiography with tritiated thymidine. Clin Orthop, 48:229–242.
14. Fuller, JA, Ghadially, FN, 1972. Ultrastructural observations on surgically produced partial-thickness defects in articular cartilage. Clin Orthop, 86:193–205.
15. Ghadially, FN, Thomas, I, Oryschk, AF, et al, 1977. Long term results of superficial defects in articular cartilage. A scanning electron-microscope study. J Pathol, 121:213–217.
16. Mankin, HJ, Boyle, CJ, 1967. The acute effects of lacerative injury on DNA and protein synthesis in articular cartilage. In: CAL Bassett (Ed.), Cartilage degradation and repair. National Research Council. National Academy of Sciences–National Academy of Engineering, Washington, DC, pp 185–199.
17. Mankin, HJ, 1974. The reaction of articular cartilage to injury and osteoarthritis: Part I. N Engl J Med, 291:1285–1292.
18. Mankin, HJ, 1974. The reaction of articular cartilage to injury and osteoarthritis: Part II. N Engl J Med, 291:1335–1340.
19. Bailey, RW, Habel, DW, 1960. The fate of surgically created osteochondral fractures in adult and young dogs. Surg Forum, 11:438–440.
20. Calandruccio, RA, Gilmer, WS, 1962. Proliferation, regeneration, and repair of articular cartilage of immature animals. J Bone Joint Surg, 44A:431–455.
21. Paget, J, 1969. Healing of cartilage. Clin Orthop, 64:7–8.
22. Shands, AR, Jr., 1931. The regeneration of hyaline cartilage in joints. An experimental study. Arch Surg, 22:137–178.
23. Meachim, G, 1963. The effect of scarification on articular cartilage in the rabbit. J Bone Joint Surg, 45B:150–161.
24. Meachim, G, Roberts, C, 1971. Repair of the joint surface from subarticular tissue in the rabbit knee. J Anat, 109:317–327.
25. Thompson, RC, 1975. An experimental study of surface injury to articular cartilage and enzyme responses within the joint. Clin Orthop, 107:239–248.
26. Shapiro, F, Koide, S, Glimcher, MJ, 1993. Cell origin and differentiation in the repair of full-thickness defects of articular cartilage. J Bone Joint Surg, 75A:532–553.

27. Wakitani, S, Goto, T, Pineda, SJ, et al, 1994. Mesenchymal cell-based repair of large, full-thickness defects of articular cartilage. J Bone Joint Surg, 76A:579–592.
28. Furukawa, T, Eyre, DR, Koide, S, et al, 1980. Biochemical studies on repair cartilage resurfacing experimental defects in the rabbit knee. J Bone Joint Surg, 62A:79–89.
29. Nakajima, H, Goto, T, Horikawa, D, et al, 1998. Characterization of the cells in the repair tissue of full-thickness articular cartilage defects. Histochem Cell Biol, 109:331–338.
30. Buckwalter, JA, Rosenburg, LC Hunziker, EB, 1990. Articular cartilage: Composition and structure, response to injury and methods of facilitating repair. In: JW Ewing (Ed.), Articular cartilage and knee joint function: Basic science and arthroscopy, Chap 2. Raven Press, New York.
31. Green, WT, Jr., 1977. Articular cartilage repair. Behavior of rabbit chondrocytes during tissue culture and subsequent allografting. Clin Orthop, 124:237–250.
32. Nakahara, H, Bruder, SP, Haynesworth, SE, et al, 1990. Bone and cartilage formation in diffusion chambers by subcultured cells derived from the periostium. Bone, 11:181–188.
33. Nakahara, H, Goldberg, VM, Caplan, AI, 1991, Culture-expanded human periostial-derived cells exhibit osteochondrogenic potential in vivo. J Orthop Res. 465–476.
34. Rooney, P, Walker, D, Grant, ME, et al., 1993. Cartilage and bone formation in repairing Achilles tendons within diffusion chambers: Evidence for tendon-cartilage and cartilage-bone conversion in vivo. J Pathol, 169:375–381.
35. Ono, Y, Kato, K, Oohira, A, et al, 1994. Cell function during chondrogenesis and osteogenesis induced by bone morphogenic protein enclosed in a diffusion chamber. Clin Orthop, 298:305–318.
36. An, YH, Friedman, RJ, 1999. Animal models of articular cartilage defect. In: YH An and RJ Friedman (Eds.), Animals in orthopaedic research, CRC Press, Boca Raton, FL, pp 309–325.
37. Mochizuki, Y, Goldberg, VM, Caplin, AI, 1993. Enzymatic digestion for the repair of superficial articular cartilage lesions. Trans Orthop Res Soc, 18:728.
38. Freed, LE, Grande, DA, Nohria, A, et al, 1994. Joint resurfacing using chondrocytes and synthetic biodegradable polymer scaffolds. J Biomed Mater Res, 28:891–899.
39. Pridie, KH, 1959. A method of resurfacing osteoarthritis knee joints. J Bone Joint Surg, 41B: 618–619.
40. Johnson, LL, 1986. Arthroscopic abrasion arthroplasty. Historical and pathological perspective: Present status. Arthroscopy, 2:54–69.
41. Mitchell, N, Shepard, N, 1976. The resurfacing of adult rabbit articular cartilage by multiple perforations through the subchondral bone. J Bone Joint Surg, 58A:230–233.
42. Rand, JA, 1991. Role of arthroscopy in osteoarthritis of the knee. Arthroscopy, 7:358–363.
43. Aichroth, PM, Patel, DV, Moyes, ST, 1991. A prospective review of arthroscopic debridement for degenerative joint disease of the knee. Int Orthop, 15:351–355.
44. Chandler, EJ, 1985. Abrasion Arthroplasty of the knee. Contemp Orthop, 11:21–29.
45. Friedman, MJ, Berasi, CC, Fox, JM, et al, 1984. Preliminary results with abrasion arthroplasty in the osteoarthritic knee. Clin Orthop, 182:200–205.
46. Menche, DS, Frenkel, SR, Blair, B, et al, 1996. A comparison of abrasion burr arthroplasty and subchondral drilling in the treatment of full-thickness cartilage lesions in the rabbit. Arthroscopy, 12:280–286.
47. Brittberg, M. Lindahl, A, Nilsson, A, et al, 1994. Treatment of deep cartilage defects in the knee with autologous chondrocyte transplantation. N Engl J Med, 331:889–895.
48. Cartilage Repair Registry: Genzyme tissue repair, 1997. ABT Associates, Cambridge, MA.
49. Skalak, R, Fox, CF (Eds.), 1988. Tissue Engineering. Liss, New York.
50. Freed, LE, Vunjak-Novakovic, G, Biron RJ, 1994. Biodegradable polymer scaffolds for tissue engineering. Bio Technology, 12:689–693.
51. Itay, S, Abramovic, A, Nevo, Z, 1987. Use of cultured embryonal chick epiphyseal chondrocytes as grafts for defects in chick articular cartilage. Clin Orthop, 222:284–303.
52. Wakitani, S, Kimura, T, Hirooka, A, 1989. Repair of rabbit articular surfaces with allograft chondrocytes embedded in collagen gel. J Bone Joint Surg, 71B:74–80.
53. Speer, DP, Chvapil, M, Volz, RG, 1979. Enhancement of healing in osteochondral defects by collagen sponge implants. Clin Orthop, 144:326–335.

54. Kawabe, N. Yoshinao, M, 1991. The repair of full-thickness articular cartilage defects: Immune responses to reparative tissue formed by allogenic growth plate chondrocyte implants. Clin Orthop, 268:279–293.
55. Freed, LE, Marquis, JC, Nohria, A, 1993. Neocartilage formation in vitro and in vivo using cells cultured on synthetic biodegradable polymers. J Biomed Mater Res, 27:11–23.
56. Chu, CC, Browning, A, 1978. The study of thermal and gross morphologic properties of polyglycolic acid upon annealing and degradation treatments. J Biomed Mater Res, 22:699–712.
57. Mikos, AF, Lai, H-L, Leite, SM, et al. Characterization of degradation of poly-L-lactic acid foams, unpublished results.
58. Chu, CR, Coutts, RD, Yoshioka, M, 1995. Articular cartilage repair using allogeneic perichondrocyte-seeded biodegradable porous polylactic acid (PLA): A tissue engineering study. J Biomed Mater Res, 29:1147–1154.
59. Vacanti, CA, Kim, W, Schloo, B, 1994. Joint Resurfacing with cartilage grown in situ from cell-polymer structures. Am J Sports Med, 22:485–488.
60. von Schroeder, HP, Kwan, M, Amiel, D, et al, 1991. The use of polylactic acid matrix and periosteal grafts for the reconstruction of rabbit knee articular defects. J Biomed Mater Res, 25:329–339.
61. Messner, K, Gillquist, J, 1993. Synthetic implants for the repair of osteochondral defects of the medial femoral condyle: A biomechanical and histological evaluation in the rabbit knee. Biomaterials, 14:513–521.
62. Minns, RJ, Muckle, DS, 1989. Mechanical and histological response of carbon fiber pads implanted in the rabbit patella. Biomechanicals, 10:273–276.
63. Livne, E, 1994. In vitro response of articular cartilage from mature mice to human transforming growth factor beta. ACTA Anat, 149:185–194.
64. Skantze, KA, Brinckerhoff, DE, Collier, JP, 1985. Use of agarose culture to measure the effect of transforming growth factor beta and epidermal growth factor on rabbit articular chondrocytes. Cancer Res, 45:4416–4421.
65. Corvol, M-T, Malemud, CJ, Sokoloff, L, 1972. A pituitary growth-promoting factor for articular chondrocytes in monolayer culture. Endocrinology, 90:262–271.
66. Jingushi, S, Heydemann, A, Kana, SK, et al, 1990. Acidic fibroblast growth factor injection stimulates cartilage enlargement and inhibits cartilage gene expression in rat fracture healing. J Orthop Res, 8:364–371.
67. Osborn, KD, Trippel, SB, Mankin, HJ, 1989. Growth factor stimulation of adult articular cartilage. J Orthop Res, 7:35–42.
68. Trippel, SB, Corvol, MT, Dumontier, MF, et al, 1989. Effect of somatomedin-C/insulin-like growth factor I and growth hormone on cultured growth plate and articular chondrocytes. Pediatr Res, 25: 76–82.
69. Sato, K, Urist, MR, 1984. Bone morphogenic protein induced cartilage development in tissue culture. Clin Orthop, 183:180–187.
70. Suzuki, F, 1992. Effects of various growth factors on chondrocyte differentiation model. Adv Exp Med Biol, 324:101–106.
71. Vivien, D, Galera, P, Lebrun, E, et al, 1990. Differential effects of transforming growth factor beta and epidermal growth factor on the cell cycle of cultured rabbit articular chondrocytes. J Cell Physiol, 143:534–545.
72. Kato, T, Iwamoto, M, Koike, T, 1987. Fibroblast growth factor stimulates colony formation of differentiated chondrocytes in soft agar. J Cell Phys Biol, 133:491–498.
73. Bohme, K, Conscience-Egli, M, Tchan, T, et al, 1992. Induction of proliferation or hypertrophy of chondrocytes in serum-free culture: The role of insulin-like growth factor I, insulin, or thyroxine. J Cell Biol, 116:1035–1042.
74. Frenkle, SR, Pachence, JM, Alexander, H, 1993. Optimization of a cell-seeded collagen implant for cartilage repair. Trans Soc Biomater, 16:730.
75. Rosen, DM, Stempien, SA, Thompson, AY, et al, 1988. Transforming growth factor beta modulates the expression of osteoblast phenotypes in vitro. J Cell Physiol, 134:337–346.
76. Seyedin, SM, Thompson, AY, Bentz, H, et al, 1986. J Biol Chem, 261:5693–5695.

77. Kemp, SF, Hintz, 1980. The action of somatomedian on glycosaminoglycan synthesis in cultured chick chondrocytes. Endocrinology, 106:744–749.
78. Hiraki, Y, Inoue, H, Asada, A, et al, 1990. Differential modulation of growth in phenotypic expression of chondrocytes in sparse and confluent culture by growth factors in cartilage. J Bone Miner Res, 5:1077–1085.
79. Curtis, AJ, Ng, CK, Handly, CJ, et al, 1992. Effect of insulin-like growth factor I on the synthesis and distribution of link protein and hyaluronan in explant cultures of articular cartilage. Biochem Biophys Acta, 1135:309–317.
80. Chen, P, Carrington, JL, Hammonds, RG, et al, 1991. Stimulation of chondrogenesis in limb bud mesoderm cells by recombinant human bone morphogenic protein 2B and modulation by transforming growth factor β1 and β2. Exp Cell Res, 195:509–515.
81. Damien, CJ, Parsons, JR, Benedict, JJ, et al, 1990. Investigation of a hydroxyapatite and calcium sulfate composite supplemented with an osteoinductive factor. J Biomed Mater Res, 24: 639–654.
82. Aspenberg, P, Lohmander, LS, 1989. Fibroblast growth factor stimulates bone formation. Bone induction studied in rats. Acta Orthop Scand, 60:473–476.
83. O'Driscoll, SW, Salter, RB, 1986. The repair of major osteochondral defects in joint surfaces by neochondrogenesis with autogenous osteoperiostial grafts stimulated by continuous passive motion. Clin Orthop, 208:131–140.
84. Baker, B, Spadaro, JA, Becker, RO, 1974. Electrical stimulation of articular cartilage. Ann NY Acad Sci, 238:491–499.
85. Brighton, CT, Unger, AS, Stambough, JL, 1984. In vitro growth of bovine articular cartilage chondrocytes in various capacitively coupled electrical fields. J Orthop Res, 2:15–22.
86. Muckle, DS, Minns, RJ, 1989. Biological response to woven carbon fiber pads in the knee. J Bone Joint Surg, 71B:60–62.
87. Brittberg, M, Faxen, E, Peterson, L, 1994. Carbon fiber scaffolds in the treatment of early knee osteoarthritis: A prospective 4 year follow up of 37 patients. Clin Orthop, 307:155–164.

30

Engineering of Resorbable Grafts for Craniofacial Reconstruction

Joseph D. Gresser, Debra J. Trantolo, and Donald L. Wise
Cambridge Scientific, Inc., Cambridge, Massachusetts

Kai-Uwe Lewandrowski
Massachusetts General Hospital, Boston, Massachusetts

I. INTRODUCTION

Bone is the second most implanted material in the body after blood. There are over 450,000 bone graft procedures annually in the United States (2.2 million worldwide) with a market potential of \$400 to \$600 million. Autografts and allografts are used in current bone graft procedures to repair defects caused by surgery, tumors, trauma, congenital deformities, implant revisions and infections and also for joint fusion. However, drawbacks such as the need for a second surgery to retrieve the graft (autograft) or the risk of viral infection, contamination and long-term complications (allografts) make bioresorbable bone graft substitutes a viable alternative to autografts and allografts. However, only 10% of these procedures use synthetic materials because the current approved synthetic grafts are considered to be inferior to the use of autograft or allograft. Significant drawbacks of these materials include lack of resorbability, inclusion of animal or marine derived components, poor handling characteristics, and lack of support for long-term dimensional stability resulting in deformation of the implant. It appears meaningful to develop cranial implants that do not contain foreign biological material, do not stay in the body after the healing process has occurred, and lend themselves to presurgical preparation and customization and rapid postsurgical cosmetic recovery.

Techniques such as CAD/CAM and stereolithography make precise craniofacial implant design and fabrication of cranial implants possible. Computer-aided implant manufacturing methods yield a better quality product. Implants which are customized to the individual patient decrease complications and the risks of surgery to the patient. There is a reduction in OR time and better implant fit and function. In addition, cost savings are likely to result from shortened hospital stays for patients requiring implant surgery. It is therefore reasonable to develop and apply the same risk-reducing techniques for manufacture of bioresorbable tissue engineered cranial implants.

We began the development of a resorbable bone graft substitute made from the unsaturated polyester, poly(propylene glycol-*co*-fumaric acid) ("PPF"), which can be crosslinked in the presence of CO_2 generators and osteoconductive filler salts and then cured before surgical implantation to yield a dimensionally stable scaffold that can be machined for best reconstructive fit. Rather than relying upon the inclusion of salts to yield porosity as others have done [1], the porosity of this scaffold is governed by both the rate of CO_2 generation (i.e., via the control of mixing) and the amount of CO_2 generated (i.e., via the control of the CO_2 generating reagents). The graft material provides an osteoconductive pathway by which bone will grow in faster without having to use auto- or allograft tissue, as well as an osteoinductive scaffold by which host cells are incorporated. In addition, the crosslinked network lends itself not only to bony ingrowth at the repair site, but also to the dimensional stability necessary for long-term cosmetic recovery of the repair site. The overall objective, though in any case, is to develop a scaffold which accommodates various defect configurations.

A dimensionally stable porous scaffold based on the crosslinking of the unsaturated polymer PPF with a vinyl monomer, vinyl pyrrolidone, (VP), in the presence of sodium bicarbonate and citric acid, and a filler, hydroxy apatite (HA). Upon mixing, the mixture cures via crosslinking of the PPF by the monomer and concomitant CO_2 generation resulting in a porous scaffold degradable by hydrolysis. The use of hydroxyapatite as part of the filler supports the osteoconductivity [2] of the scaffold, while the CO_2 generated pores provide regions for attachment and proliferation of cells and the hydrophilicity of the polymeric support encourages cellular migration.

A biodegradable scaffold that would provide immediate structural recovery and subsequently allow normal bone healing and remodeling would have immediate craniofacial applications. The utility of biodegradable, or resorbable, bone graft substitute for temporary void filling during bone reconstruction has been recognized [3–9]. Resorbability eliminates the need for implant removal and therefore reduces the potential for long-term implant complications associated with foreign materials [10]. A resorbable scaffold can in itself create an osteoconductive pathway and, when seeded with bony cells, serve also as an inductive path for bone cell ingrowth and void filling as bony juncture develops. This requires not only the optimization of the porosity of the scaffold to allow migration and proliferation of bone forming cells in a variety of geometrics, but also biodegradation of the scaffold at a rate commensurate with bony reconstruction.

The feasibility of the proposed approach was determined not only by the capacity to design the porous scaffold to optimally support cell attachment and proliferation in a variety of reconstructive scenarios but also by its capacity to support the reconstructive process of bone repair. Our experimental in vitro and in vivo data have been used to determine appropriate candidate scaffolds differing in their density and porosity and, thus, potential for bony ingrowth, by analyzing the efficiency of cell migration, histologic sequences of healing of bone repair sites, in support of establishing material functionality.

II. BACKGROUND AND SIGNIFICANCE

A. Clinical Management of Craniofacial Bone Defects

Craniofacial surgery encompasses a broad spectrum of reconstructive procedures of the cranium and face. The objective of these procedures is to correct deformities of the face and skull bones that result from birth defects, cranial deformities, trauma, or tumors. The highly specialized field of craniofacial surgery has made tremendous advancements in the treatment of major facial deformities over the past 25 years.

Craniofacial deformities, or alterations in the natural form of the face or skull, can be congenital or acquired. For those patients born with craniofacial anomalies, the obstetrician or pediatrician is the initial point of contact for appropriate medical treatment. Accurate diagnosis at an early age not only avoids unnecessary emotional distress for the parents and family, but it also minimizes potential future problems associated with the deformities by early correction. For those patients with acquired deformities as the result of trauma or tumor resection, referral to a craniofacial center may be desirable to help restore facial function and appearance.

While the pattern of embryonic craniofacial development has been well defined through extensive research, very little is known about the etiology of many craniofacial anomalies. Some are known to be primarily genetic in nature, while others are thought to be caused by environmental factors. A combination of both environment and genetics may play a role in the etiology; however, most of the time the cause is unknown.

When craniofacial surgery is recommended as treatment for a specific anomaly, there are two distinct goals of this surgery. The first goal is to attempt to restore the patient to as near normal function as possible and to prevent future dysfunction. Secondly, surgery may be necessary to correct structural disfigurement in order to achieve optimal appearance. Patients suffering from a facial deformity may experience problems in dealing with their disfigurement emotionally or socially. Often, improvements in appearance following craniofacial surgery can lead to increases in self-esteem, self-confidence and social acceptance. The psychological benefit of craniofacial surgery is an extremely important goal of the surgery.

Treatment of craniofacial problems does not end with surgical restoration, but continues for many years. As a child grows and develops, asymmetries may result if areas of the face fail to develop equally; therefore, a child's growth and development must be routinely followed. Sometimes it may be necessary to repair these asymmetries surgically. Often, major craniofacial deformities require multiple, staged procedures performed at different ages. Once treatment is initiated, it is important that follow-up care continues. This offers great development potential for biomaterials. Resorbable scaffolds will not only to shorten the initial surgical procedure, but also simplify or eliminate extended and repeated surgical treatment.

A number of advances have been made in surgical techniques and technologies as applied to craniofacial surgery. With the experimental evidence that membranous onlay bone grafts survive better than endochondral grafts, craniofacial surgeons, for the most part, have abandoned rib and iliac bone grafts and their associated morbidity for the longer surviving cranial bone grafts [11]. This bone is available in an assortment of sizes and shapes, and can be harvested as dust free grafts, or transposed as vascularized tissue. The techniques of harvesting and utilization of cranial bone have expanded from the field of congenital craniofacial surgery to correction of traumatic facial deformities or use in aesthetic procedures. As more experience is gained, the indications for cranial bone grafts will continue to expand rapidly.

Over the past 10 years, tremendous radiological advancements have been made that have improved our understanding of craniofacial deformities [12–20]. Improvements in diagnostic evaluation with the use of two- and three-dimensional CT scans have drastically enhanced the ability to analyze these complex deformities. In addition to plain radiographs, these scans are extremely helpful for evaluation of the upper face. The three-dimensional reconstructions have added a further dimension to facial bone analysis and provide valuable information for preoperative planning (Fig. 1).

Another major advancement has been the application of rigid skeletal fixation to craniofacial surgery [21–24]. The new techniques of rigid skeletal fixation combined with wide exposure have allowed the craniofacial surgeon to obtain much better stability and eliminate intermaxillary fixation in most cases. This technique offers significant advantages, particularly in children. It has improved the overall quality of results as well as decreasing morbidity.

New surgical techniques include the treatment of acute craniofacial trauma by complete exposure of the injury pattern followed by precise rigid fixation of the fractures (using bone grafts as needed) has improved both functional and aesthetic results [25–28]. The results achieved with this craniofacial approach have minimized revisional surgery and allowed better

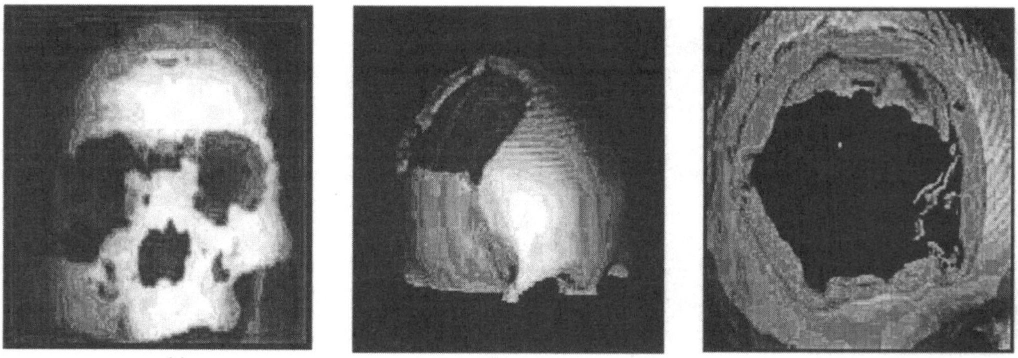

(a) (b) (c)

Figure 1 Example of a three-dimensional CT scan showing multiple facial fractures (a) and a cranial defect (b,c).

reproduction of exact preinjury facial bone architecture. These innovative techniques establish a higher standard of care for facial fracture treatment and call for development of new biocompatible materials.

B. Implant Design

The fabrication and surgical fitting of cranial implants by traditional methods is heavily dependent on subjective skills and procedures, producing, in most cases, a close, but imperfect fit that necessitates revisions during the surgical placement of the implant. Preliminary studies indicate that fabrication of cranial implants using CT data from the patient to design an implant in a CAD environment results in a "near-perfect" fit.

The design of an implant specific to a patient requires that data from that patient be translated into a three-dimensional representation of the form [29]. As shown in Fig. 2, this can be accomplished by building a computer model of the bony defect using CT scans of the patient.

A new state-of-the-art technique is stereolithography (Fig. 3). A skull model can be converted to a file suitable for stereolithography, the method used to subsequently generate a physical model of the data. The stereolithography approach is used to fabricate an implant and begins with constructing a three-dimensional model of the defect size. A computer-driven laser beam traces serial cross sections (defined as data points) on the surface of a liquid polymer

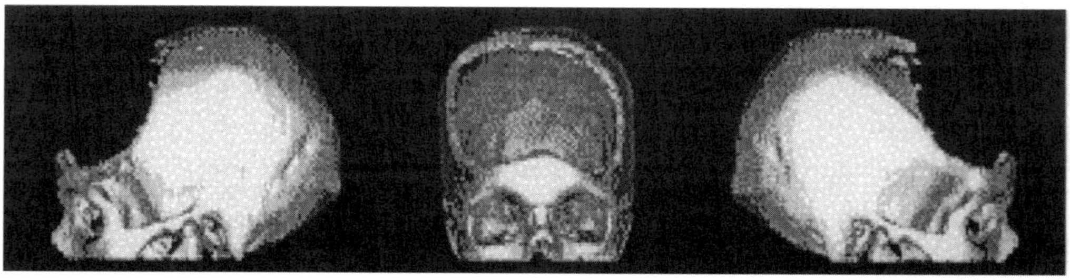

Figure 2 Three-dimensional model of a cranial bone defect.

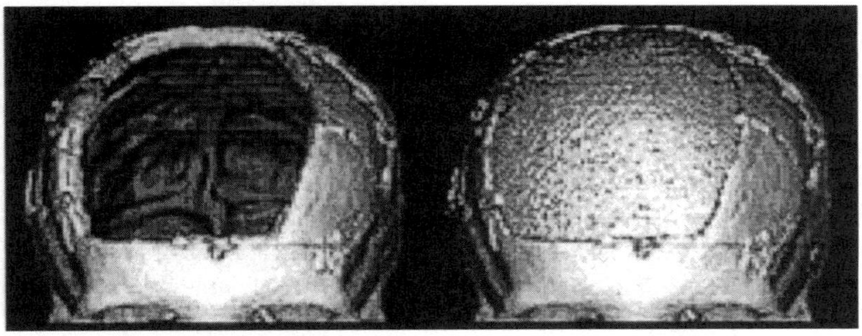

Figure 3 Model of a cranial defect (left) and of a cranial implant (right) designed with stereolithography.

(plastic). In this process, the energy of the laser beam cures and hardens the polymer as the individual cross sections are "drawn" on the liquid surface, and, upon completion of the process, the stacked cross sections combine to form a rigid, physical model of the bony defect [30].

C. Polymer-Bone Biocompatibility

In many of the reconstructive fields, there is an increasing recognition of the utility of resorbable, tissue compatible biopolymers for bone repair. In orthopedic applications, poly(glycolic acid) (PGA), poly(lactic acid) (PLA), and copolymers of lactic and glycolic acids, the poly(lactide-*co*-glycolic acids) (PLGAs) have been extensively used [31–36]. Although other polymers have also been considered, such as poly(dioxanone) [37], the PGA, PLA and PLGA polyesters have perhaps the longest history of clinical use (e.g., as sutures).

In the developing area of tissue engineering, scaffolds are needed as cell and tissue carriers that emphasize material superstructuring in the design of supports. These supports are increasingly being engineered from bioabsorbable substrata made of natural biopolymers or synthetic biopolymers to realize the three-dimensional nature of tissue equivalents.

Biodegradable and biocompatible polymeric scaffolds have been introduced for orthopedic biomaterials and tissue regeneration purposes [38–40]. There is a large body of literature supporting biocompatibility of degradable biopolymers. They have been demonstrated to facilitate bony healing [33] and they have additionally been shown to represent excellent substrates for cell and tissue culture [41, 42]. Therefore, it is certainly appropriate to use biopolymers as tissue-engineering scaffolds for maxillofacial reconstruction. The challenge is to develop a material that supports the reconstructive process by allowing (1) a high density of cells to be seeded into a device before transplantation, (2) diffusion of factors between the transplanted cells and the surrounding tissue following implantation, (3) vascularization, and (4) cosmetic recovery [43]. The PPF-based scaffold so proposed may meet this challenge by (1) presenting sufficient hydrophilicity for cell attachment and proliferation, (2) offering demonstrable porosity for cellular migration, (3) contributing a richness of surface area for neovascularization, and (4) providing sufficient dimensional stability for support of the reconstructive process.

D. Tissue Regeneration Approach to Polymer Scaffolds

The approach of using bone cells stems from the observation that these cells can be readily cultured from bone tissues [1, 44, 45]. In addition, these cells can grow extremely well in synthetic polymeric, as well as in natural collagenous, matrixes and are therefore recognized as an extremely bioactive material [46]. Paralleling attempts to take dermal fibroblasts for the in vitro expansion and generation of skin grafting material for extensive burn wounds [47], or attempts to generate cartilaginous material for joint resurfacing by in vitro expansion of autologous articular cartilage cells [48], it appears logical to take bone cells from the future recipient of the bone repair material and to expand them in vitro for engineering of repair tissues. Breitbart et al. [49] have already demonstrated the feasibility of using periosteal cells for tissue engineered bone repair of calvarial defects. By analogy, a dimensionally stable tissue engineered scaffold using cells obtained from the future recipient should show enhanced osteointegration, significantly lower immunological sensitization rates, and more rapid recovery with an improved cosmetic result.

The feasibility of using a tissue-engineered scaffold depends on the ability to obtain a population of cells which have or will develop osteoblast-like function. The periosteum itself is known to play a critical role in fracture healing and callus formation [50, 51]. Cells derived from periosteum are capable of forming bone in subcutaneous sites either alone [52], or when seeded into ceramic calcium phosphate [53] or polymeric PGA [54] substrates. Further, the healing of critical size defects in the cranium [55] and femur [49, 56] have been demonstrated using periosteal cells seeded onto PGA scaffolds. Cells from the periosteum are known to have both osteogenic [52] and chondrogenic [57] potential, and the process of bone formation from periosteal cells involves transition through a cartilage-like phase [54]. Studies examining the use of chondrocytes on polymer scaffolds to fill bone defects have demonstrated the formation of stable cartilage that does not generate bone [58], while periosteal cells placed in the same material in the same site ultimately repair the defect with bone [55].

Osteoblasts, directly involved in the activities of bone regeneration, have also been considered in building scaffolds. Osteoblast migration and proliferation is important to endochondral and intramembranous bone formation, remodeling and repair [9]. This is believed to be the first step in bone formation [59]. Ishaug et al. [8, 9] demonstrated that osteoblasts can grow and migrate throughout polymeric scaffolds in vitro. Ishaug-Riley et al. [1] then demonstrated that osteoblast function on synthetic materials was equal to that on nondegradable orthopedic materials. Seeded resorbable scaffolds are replaced by new bone when implanted into bony sites.

The approach of using a bioresorbable foam having an open window structure as a carrier for bony cells stems from the fact that bone has a considerable potential for regeneration. In fact, bone is considered by some the prototypic model for tissue engineering [38, 60]. Development of composites of cells responding to regulatory signals has been at the center of engineering skeletal tissues. In fact, leading strategies on engineering and regeneration of skeletal tissues are largely based on observations on local bone induction after implantation of bony matrices into subcutaneous sites. These models have allowed the study of sequential limb morphogenesis and have permitted the isolation of bone morphogens, such as bone morphogenetic proteins (BMPs), from demineralized adult bone matrix. BMPs initiate, promote, and maintain chondrogenesis and osteogenesis.

It appears reasonable to approach our biodegradable foam as a designable scaffold that will function as a cell and tissue carrier emphasizing material superstructuring. Therefore, this work was aimed at engineering a biodegradable superstructure that would provide optimal spatial and nutritional conditions for cell maintenance by the arrangement of structural elements (e.g., pores) so as to vary the order of cell to cell contact [61]. Predictable three-dimensional cell-polymer matrices for tissue engineering are needed so that osteoblast cells can maintain their phenotypic properties and form a mineralized matrix while seeded on the polymer surface [39, 56]. Therefore, optimization of a predictable and dimensionally stable scaffold which maintains the three-dimensional superstructure of the varying reconstructive sites was our main aim.

III. PRELIMINARY STUDIES

As outlined below, we have initially investigated a dimensionally stable porous scaffold based on the crosslinking of the unsaturated PPF polymer with a vinyl monomer, vinyl pyrrolidone (VP), in the presence of sodium bicarbonate and citric acid and a filler, hydroxy apatite (HA) [62]. Upon mixing, the mixture cures via crosslinking of the PPF by the monomer and concomitant CO_2 generation resulting in a porous scaffold degradable by hydrolysis. The use of hydroxyapatite as part of the filler supports the osteoconductivity [2] of the scaffold, while the CO_2 generated pores provide designed regions for attachment and proliferation of cells not only in situ, but ex situ, and the hydrophilicity of the polymeric support encourages cellular migration.

To demonstrate feasibility, the PPF scaffold was evaluated in vitro for morphological, mechanical and surface properties, and in vivo using a rat tibial defect, a model well established by Gerhart et al. [5]. Scaffolds, designed for controlled superstructural characteristics, were shown to be hydrophilic, mechanically comparable to trabecular bone, dimensionally stable, and porous. Histological and histomorphological examination of the implant region of rats suggested that the porosity of the scaffold supported bony ingrowth and the stability of the scaffold preserved the dimensional integrity of the defect site.

A. Preparation of PPF Foams

The foam is prepared as a two part system in order to separate the polymerizable entities (PPF, VP) from the initiator (BP). Because BP is, by weight, a minor ingredient, it is packaged with components inactive with respect to benzoyl peroxide to provide bulk and with the inhibitor. A typical formulation used in the pilot studies is given in Table 1. The foaming agent consists of granules composed of a stoichiometric ratio of citric acid (CA) and sodium bicarbonate (SB) (i.e., 1.0: 1.3 w/w).

The test specimens were circular cylinders prepared by placing the polymerizing composite material into a 10 mm diameter, 10 cm high cylindrical Teflon mold and letting the polymer foam crosslink. Samples used for the mechanical testing were 1.0 mm × 10 mm and 1.5 mm × 10 mm for the in vitro and in vivo testing. The example formulation yielded a foam with a density of 0.6484 ± 0.0834 g/ml.

Table 1 Foaming Cement Composition

Component	Wt. %
PPF	39.32
Hydroxyapatite	11.38
Vinyl pyrrolidone	12.10
Ethanol	12.70
Sodium bicarbonate	1.95
Benzoyl peroxide	1.71
Citric acid	1.47
Water	19.37

Table 2 Measurement of Contact Angles for Selected Biomaterials

Materials	Contact angle
PPF	47.2 ± 2.2
PLGA[a]	77.0 ± 6.3
PLLA	78.5
glass	34.5 ± 5.5
Teflon	77.9

[a]85:15 PLGA

B. In Vitro Foam Characterization

1. *Hydrophilicity and Water Uptake*

The PPF foam was characterized with respect to the surface hydrophilicity/hydrophobicity, an important indicator of eventual cell attachment and proliferation throughout a solid phase support. The hydrophilic/hydrophobic nature of a material can be semiquantitatively analyzed by measuring the contact angle formed at the interface between a droplet of water and a planar surface of the material. Glass and Teflon are hydrophilic and hydrophobic standard materials, respectively, as evidenced by the two extremes in their contact angles (see Table 2). PPF behaves like hydrophilic glass and poly-L-lactic acid like hydrophobic Teflon. PPF is a material which wets easily and hence should facilitate migration of osteoblasts into a porous interior.

Greater surface hydrophility translates into higher water uptake for the PPF biopolymer compared to PLGA. As shown in Fig. 4, the water uptake for the PPF foam was around 0.5 g water/g of dry polymer. On the other hand, PLGA absorbed 0.3 g water/g of dry polymer, reaching equilibrium in 120 hs. Greater water imbibition translates into higher diffusion and convection fluxes in and out of the polymer, thus aiding cell proliferation.

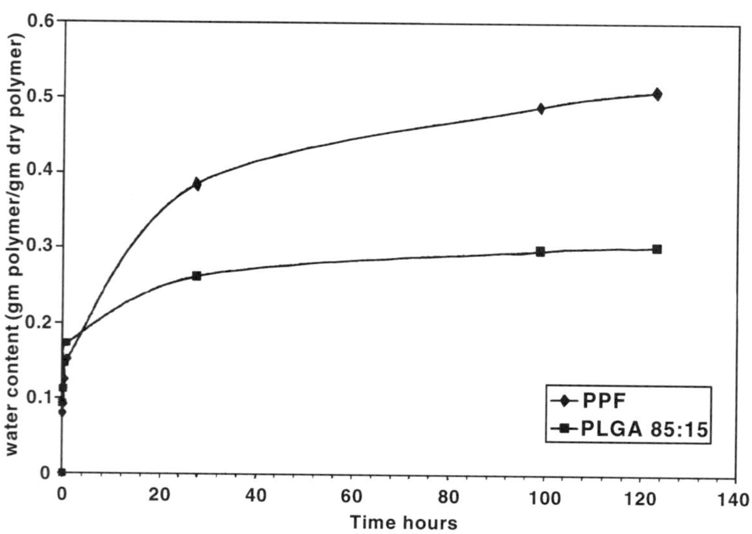

Figure 4 Water absorption by PPF and PLGA.

2. Mechanical Characterization

The mechanical testing portion of the pilot study analyzed the initial compressive strengths and moduli of the composite material. The testing was done in accordance with ASTM F451-86 for bone cements. The samples were tested in compression on an Instron Model 8511 materials testing machine. The Instron was fitted with a 500-lb load cell, and the load measuring apparatus was zeroed, balanced and calibrated. The samples were compressed to failure at a crosshead speed of 1 cm/min and the load deformation curve was recorded. From these data the ultimate compressive stress (σ) and Young's modulus (γ) could be calculated. The ultimate compressive stress was calculated as the applied load at failure divided by the original cross-sectional area of the test specimen, whereas Young's modulus was calculated as the slope of the load-deformation curve in its linear portion.

The compression testing yielded values of 6.4 MPa for the compressive strength and 130 MPa for the compressive modulus, values which are appropriate for a human trabecular bone replacement material [63]. The compressive strength may be compared with values reported by Carter and Hayes [64], who measured this property of bone as a function of strain rate and density. The relationship between these parameters is $S = 68\varepsilon^{0.06}\rho^2$, where S is the compressive strength (MPA), ε is the strain rate (s^{-1}), and ρ is the density (g/cm^3). At a strain rate of 0.01 s^{-1}, the compressive strength of trabecular bone ($\rho = 0.31$ g/cm^3) is 5.0 MPa, and for cortical bone ($\rho = 2.0$ g/cm^3) it is 200 MPa.

3. Morphological Characterization

The pore size of a cellular solid has no influence on the strength or stiffness but it can have considerable importance on the biological outcome. A porous material is normally used in bone-compatible implants to encourage bony ingrowth. In general, for bone implants, the minimum pore size is on the order of the diameter of the osteons in normal Haversian bone, or about 75 microns. However, in in vitro studies it is difficult to maintain viable osteons in materials with pore sizes less than 150 microns [65].

Given osteoblast migration into the porous scaffold as the primary goal, pore sizes in the range 100–300 microns are desirable. The design of the porosity is attainable by control of the sodium bicarbonate/citric acid content and by the size of the SB particles. The reaction of SB with CA produces carbon dioxide, the blowing agent responsible for porous expansion and foam production.

Our PPF foams are characterized by small and large pores. A scanning electron micrograph showing the presence of both types of pores is shown in Fig. 5. The foam has a wide pore size distribution (median pore size 70 microns) with at least 30% of pores with an average diameter greater than 200 microns (as confirmed via mercury intrusion porosimetry). After 24 days in vitro (PBS, pH 7.2, 37°C), there is no gross change in the dimensions of the implants and scanning electron microscopy confirms maintenance of trabecular structure (Fig. 6).

C. Initial in Vivo Studies on Osteoconductivity and Biocompatibility of a PPF Foam

To evaluate the suitability of a bioresorbable PPF-based foam for bony ingrowth prior to its use for cell seeding, an initial in vivo study in rats was conducted using the tibial defect model of Gerhart et al. [5]. This model was also recently utilized by Yaszemski et al. for in vivo evaluation of PPF-cements [66]. Adult, male Sprague Dawley rats weighing approximately 400 g were used as the animal model (Zivic Miller, Zelienople, PA, USA). The formulations were evaluated by implanting foam plugs of appropriate sizes into 3-mm holes that were made

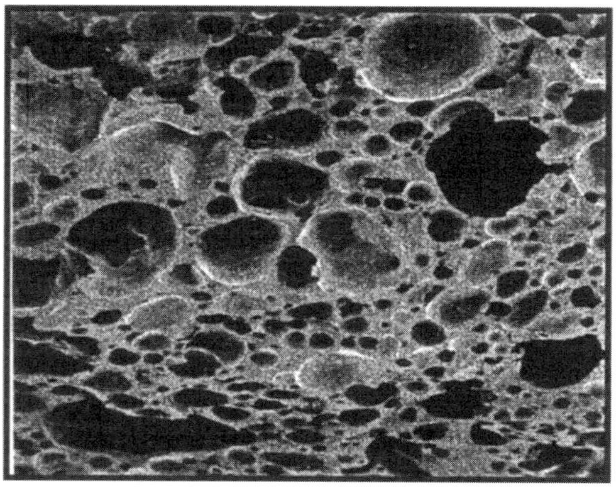

Figure 5 Scanning electron micrograph of PPF Foam at $t = 0$.

into the anteromedial tibial metaphysis of rats. In this transitional portion of the rat tibia, the cortex consists of dense lamellar cortical bone and the intramedullary canal is comprised of trabecular bone forming the metaphysis. A drill hole placed at this site will therefore allow evaluation of the repair response of both dense cortical and loose metaphyseal bone to the implanation of the PPF-based foaming cement.

Animals were anesthetized using an intramuscular injection of ketamine HCl (100 mg/kg) and xylazine (5 mg/kg). The rats were also given an intramuscular prophylactic dose of penicillin G (25,000 U/kg), and the surgical site was shaved and prepared with a solution of Betadine (povidone-iodine) and alcohol (Dura-Prep; 3M Health Care, St. Paul, MN, USA). A percutaneous incision was made in the anterior left hind leg, and the tibial metaphysis exposed. The periosteum from the adjacent bone and the bone marrow from the ends of the tibial bone were left intact. This allowed us to solely evaluate the osteoconductive properties of the resorb-

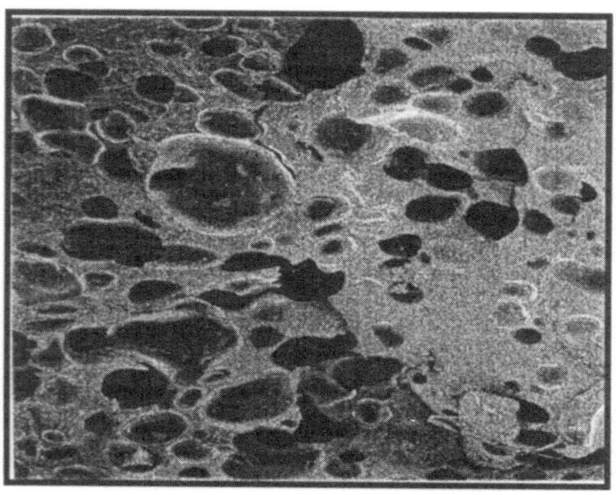

Figure 6 Scanning electron micrograph of PPF foam after 24 days *in vitro* (PBS, pH 7.23,37°).

able foam material by ingrowth through metaphyseal and periosteal bone formation. The soft tissues and skin were closed in layers with interrupted absorbable sutures.

Two groups were included in this study, allowing us to compare the foam in the experimental group against sham operated animals only having a drill hole but no implant in the control group. Eight animals were included in each group. Animals were sacrificed at one, three and seven weeks postoperatively. Thus, a total of 48 animals was included in this study.

Following sacrifice, 10-mm-long segments of the tibial bone, including the section in which the foam material was implanted, were harvested and processed for histological analysis by fixation in 10% buffered formalin. Specimens which included residual foam material were decalcified in EDTA and paraffin embedded. Longitudinal sections (5 μm thick) of the total specimen were then cut and stained with hematoxyline and eosin. Slides were examined for resorptive activity and new bone formation and inflammatory responses at the site where residual foam material was present.

Histomorphometry was done by acquiring images of serial longitudinal hematoxyline and eosin stained sections of the specimen using a CCD video camera system (TM-745; PULNiX, Sunnyvale, CA, USA) that was mounted on a Zeiss microscope. Images were digitized and analyzed using Image Pro Plus software. For each bone specimen, the total cross-sectional area of trabecular bone at both the metaphyseal and cortical drill hole sites was determined on sequential longitudinal sections. In addition, the intertrabecular space was approximated by measuring the area occupied between the bone trabeculae. A minimum of 10 sections obtained from different levels of the specimen was included for this analysis. The spacing between sections of adjacent levels was typically 300 μm. This allowed us to obtain an approximate absolute bone volume for both the metaphyseal and the cortical drill hole defect, which is given as an average percentage rate (mean ± standard deviation) of these volume measures for each bone specimen. To compare the extent of bone remodeling at the metaphyseal and the cortical drill hole defect between the experimental and the control group, the metaphyseal and cortical remodeling indices were determined. They were defined as the volume ratio of new bone and the volume of the whole defect based on eight animals per study group. They are thus given as average percentage rates. Differences in the remodeling index and the metaphyseal trabecular bone volume at sacrifice were analyzed for statistical significance by employing an ANOVA test. A p-level of 0.05 was considered statistically significant.

At *one week* postoperatively, there was extensive new bone formation in animals in which the foam material was implanted (Fig. 7a). The drill hole was filled with newly formed woven bone that appeared to have originated from the adjacent metaphyseal bone. The foam material was almost entirely replaced by new bone. However, residuals of the foam material were present. These foam remnants were sites of active bone formation, suggesting that new bone growth was facilitated by the foam scaffold. As shown in Fig. 8, there was osteoblastic activity at the interface between foam remnants and newly formed woven bone trabeculae. Moreover, neovascularization was noted in close proximity to these foam remnants. Cells suggestive of inflammatory responses were not seen. In addition to healing of the defect made in the tibial metaphysis, there was notable bone healing at the cortical drill hole site. Loosely packed woven bone with wide intertrabecular spaces was present in the drill hole (Fig. 9).

In comparison to the experimental group, there was no evidence of either metaphyseal or cortical bone healing in the control group (Fig. 7b). At one week postoperatively, both the metaphyseal defect and the cortical drill hole showed virtually no new bone formation. Occasionally, there was formation of a postoperative hematoma undergoing organization with replacement by fibrous tissue and invasion of bone marrow.

At *three weeks* postoperatively, there was less new woven bone within the metaphyseal defect; however, the cortical drill hole was entirely filled with more densely packed woven

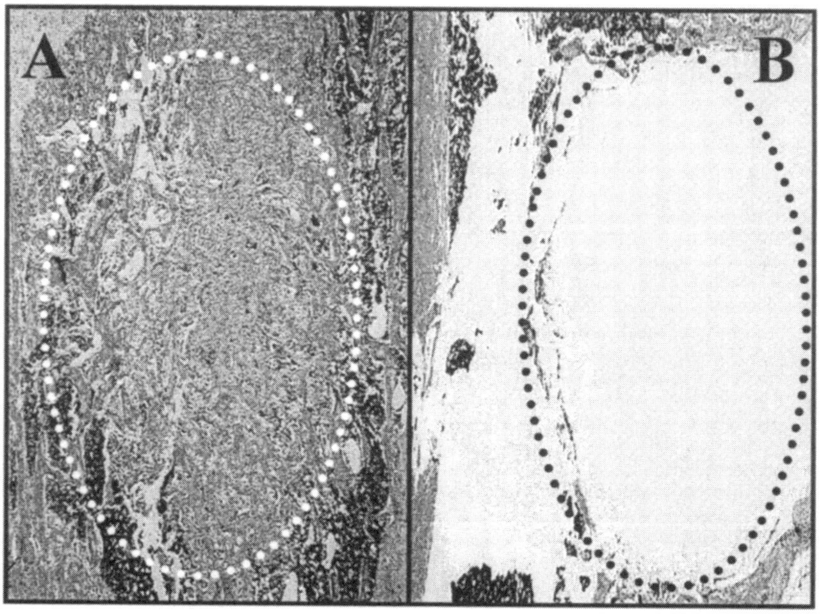

Figure 7 Photomicrograph of a longitudinal section (H&E) of rat tibias in which a drill hole defect was placed. Sections A and B show (A) a foam implant at *one week postoperatively* (5×) which had stimulated vigorous new bone formation in the rat tibial metaphysis. The drill hole defect is outlined with a white dotted line; (B) a sham operated rat tibia at *three weeks postoperatively* (5×). The drill hole defect made in the rat tibia metaphysis, which is outlined with a black dotted line, is empty.

bone. Almost no foam residuals were found. In addition, there was intimate contact between tibial cortical bone and the newly formed bone (Fig. 10). In the control group, both the metaphyseal and the cortical drill hole defect were still present at three weeks postoperatively. However, new bone formation originating from the adjacent tibial periosteum was noted. In comparison, there was little new bone formation at the site of the metaphyseal defect.

At *seven weeks* postoperatively, the entire metaphyseal area had remodeled in the experimental group resembling a normal tibial rat metaphysis (Fig. 11). In fact, most of the new bone occupying the defect was absent. No foam remnants were identifiable. In comparison, the cortical drill hole defect had completely healed. Although the proximal and distal rims of the tibial drill hole were still appreciable on longitudinal sectioning, the new bone filling the defect had remodeled so that intertrabecular gaps were absent. Instead only vascular channels and voids filled with chondrocytes were found. In the control group, there was no appreciable difference to findings noted at three weeks postoperatively.

Results of the histomorphometric analysis corroborated findings seen on histological examination. The metaphyseal and cortical remodeling indices were determined as approximated average percentage rates based on eight animals per study group. This analysis showed that there was significantly more new bone formation in the experimental group when compared to the control group ($p < 0.05$). Interestingly, new metaphyseal bone formation peaked at one week postoperatively and declined thereafter in the experimental group. In the control group, metaphyseal new bone formation was minimal and did not change over the postoperative follow-up period. In comparison, bone formation at the cortical drill hole defect increased over time and led to complete healing of the defect in the experimental group, an observation that

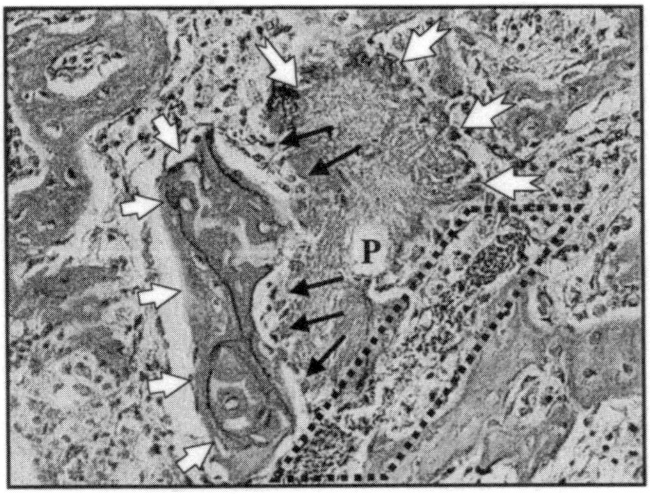

Figure 8 Photomicrograph of a longitudinal section (H&E) of a rat tibia in which a drill hole defect was placed and filled with the foam material. The specimen was retrieved at *one week postoperatively* (40×). Porous foam remnants (outlined by white notched arrows; **P** indicates a pore) were sites of active bone formation. There is a newly formed woven bone trabecule (white arrowheads) in close proximity to the foam remnant with osteoblasts (black arrows) at the interface between the foam remnant and the new bone. In addition, new bone formation was accompanied by neovascularization as evidenced by the presence of capillaries (outlined by black dotted line).

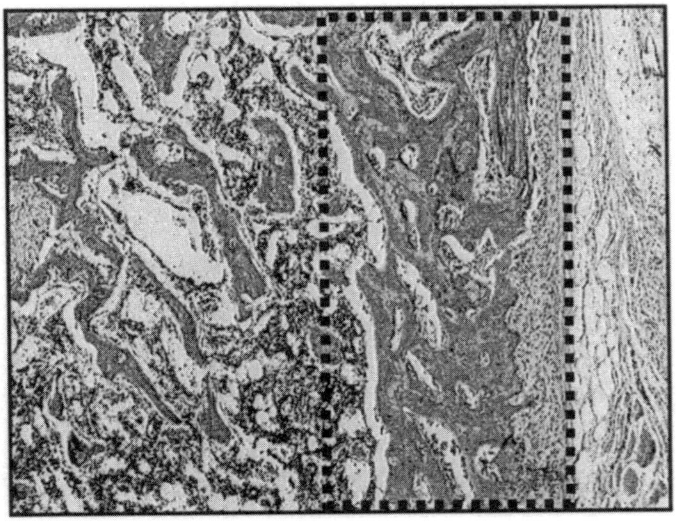

Figure 9 Photomicrograph of a longitudinal section (H&E) of a rat tibia in which a drill hole defect was placed and filled with the foam material. The specimen was retrieved at *one week postoperatively* (10×). The cortical drill hole defect, which is outlined with a black dotted line, is filled with loosely packed new woven bone.

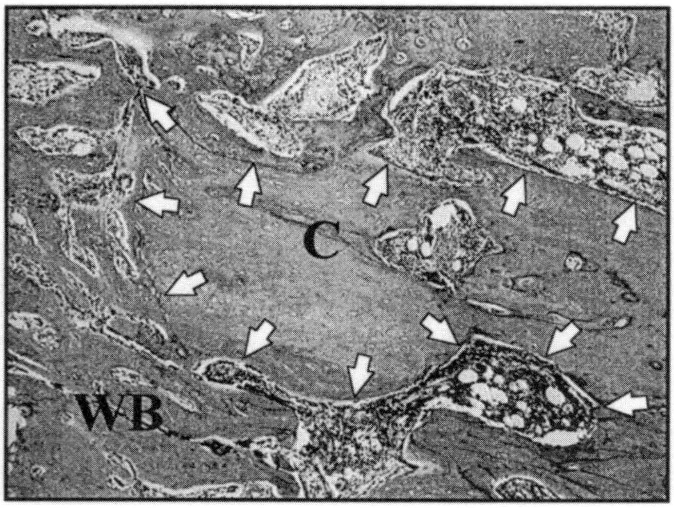

Figure 10 Photomicrograph of a longitudinal section (H&E) of a rat tibia in which a drill hole defect was placed and filled with the foam material. The specimen was retrieved at *three weeks postoperatively* (20×). On the right, the distal rim of the drill hole is shown. A portion of the drill hole is shown on the left. The intact tibial cortex (C) is outlined with white arrows. There is tightly packed new woven bone (WB) in the drill hole forming an intimate interface with the tibial cortex.

could not be made in the control group. The respective data for the experimental and control group are shown in Tables 3 and 4.

Results of this study show that defects did not heal in sham operated animals, which had a drill hole placed but did not receive a tissue regenerate implant. A similar critical size defect without bony healing was employed in another animal study, suggesting that spontaneous healing did not occur in this animal model [5]. In fact, both metaphyseal and cortical bone healing was absent in control animals of our study throughout the postoperative course.

In the experimental group, metaphyseal drill hole defects were essentially healed within the first postoperative week as evidenced by the presence of immature woven bone. Similar observations were made at the site of the cortical drill hole defect, where healing was then noted to progress to a complete closure by newly formed bone. Histomorphometry corroborated these findings by a statistical analysis, which showed that there was significantly more newly formed bone in the experimental than in the control group. Interestingly, metaphyseal bone remodeling peaked at one week postoperatively and then decreased at healing of the cortical defect progressed. This suggest that healing of the cortical drill hole defect was primarily fueled by the new bone formed at the metaphyseal defect site and that this new metaphyseal bone then underwent remodeling upon completion of the healing at the cortical defect. In other words, near complete restoration of the original state of the tibial bone occurred in this animal model at seven weeks postoperatively, which is supportive of the concept of in situ bone regeneration by application of engineered biodegradable porous scaffolds.

IV. SEEDING STUDIES

The feasibility of the proposed approach was determined by the capacity of the resorbable polymer foam to support cell attachment and growth. In a preliminary investigation of a resorb-

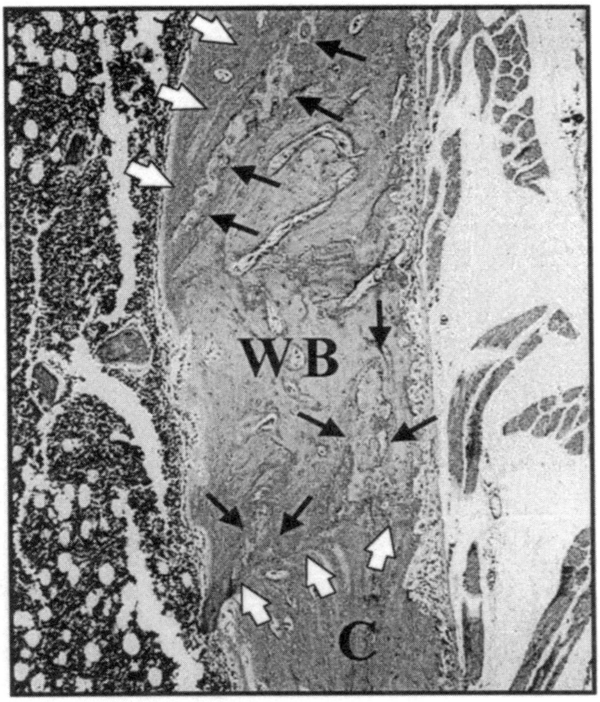

Figure 11 Photomicrograph of a longitudinal section (H&E) of a rat tibia in which a drill hole defect was placed and filled with the foam material. The specimen was retrieved at *seven weeks postoperatively* (10×). Shown is the entire cortical drill hole site (outlined by white arrows) filled with new woven bone (WB). There is intimate contact with complete healing at both the proximal and distal rim of the drill hole through the tibial cortex (C). At these junction sites, there is endochondral bone formation as evidenced by chondrocytes (black arrows), which were noted to fill remaining intertrabecular spaces. There is no new bone present at the tibial metaphysis, which is only filled with bone marrow.

able bone graft substitute, a seeding technique, which should ultimately allow reliable seeding of polymeric cranial implants, was developed. In analogy to previous studies, we used fetal bovine articular cartilage cells for these seeding experiments. PPF foams were evaluated in vitro as cellular supports using a cell culture model.

A. Chondrocyte Isolation and Culture

Articular cartilage was harvested from newborn calf wrist joints under sterile conditions. For digestion, tissue was placed in a sterile spinner culture bottle with 0.08% collagenase in Dul-

Table 3 Histomorphometric Analysis of New Bone Formation at the Metaphyseal Drill Hole Defect

Postoperative week	Metaphyseal remodeling index [%] Foam injected group	Metaphyseal remodeling Index [%] sham Operated group
1	68.5 ± 14.5	12.3 ± 6.9
3	45.8 ± 11.2	18.4 ± 7.2
7	16.7 ± 7.9	20.1 ± 8.1

Table 4 Histomorphometric Analysis of New Bone Formation at the Cortical Drill Hole Defect

Postoperative week	Cortical remodeling index [%] Foam injected group	Cortical remodeling Index [%] sham Operated group
1	75.2 ± 10.4	3.1 ± 3.9
3	91.6 ± 6.2	8.5 ± 6.7
7	95.3 ± 4.6	8.6 ± 7.1

becco's Modified Eagle Medium (DMEM), supplemented with 10% fetal calf serum and incubated overnight at 37°C in an atmosphere of 92% air and 8% CO_2. Undigested connective tissue and debris were removed by filtering the cell suspension using sterile gauze. Cells were subsequently washed with PBS (phosphate buffered saline). Viability was assessed with trypan blue exclusion. A cell count was established using a hemocytometer. The cell suspension was then adjusted to a concentration of 1.5×10^{-6} cells/ml.

Multilayer culturing was performed as previously described by placing 5 ml of cell suspension in a 15-ml standard test tube [67, 68]. Centrifugation at 1300 rpm for 5 min resulted in the formation of an approximately 1.5-mm-tall cell pellet on the bottom of the test tube. All chondrocyte pellets were incubated in an atmosphere of 92% air and 8% CO_2 and cultured in DMEM supplemented with 50 U/ml penicillin, 50 g/ml streptomycin, 10% of fetal calf serum, and 50 g/ml ascorbic acid and cultured for two days.

B. Coculturing of Chrondrocytes with PPF Foam Specimens

Foam samples were immersed in culture medium for three days to ensure filling of the void volume. Then the foam samples were immersed in the chondrocytes cell solution using 24 well culture plates. PPF specimens were cocultured with the calf chondrocytes for a minimum period of 7 days (group 1), 14 days (group 2), or 21 days (group 3), and a maximum period of 28 days (group 4). Six specimens were included per time point and the culture medium was changed every three days. The foam/chondrocyte cultures were evaluated on the basis of cell expansion, migration, and penetration depth using confocal and light microscopy.

Results of this initial seeding study showed that PPF foams were suitable to stimulate chondrocyte ingrowth. As early as one week of coculturing, chondrocytes were present in the pores of the foam (Fig. 12). Chondrocytes were noted to advance further into the foam specimens as evidenced by an increasing penetration depth (Fig. 13). At four weeks of coculturing, individual chondrocytes were noted within the foam scaffold with the maximum penetration reaching 350 μm. However, chondrocytes appeared well rounded and loosely packed, suggestive of minimum pain and production and poor adherence to each other and the surrounding polymer scaffold. Specimens were then processed for regular light microscopy, which demonstrated a cellular approach in similar pores increasing 60–120 μm in diameter (Fig. 14). Cells could not be demonstrated in large pores.

V. DISCUSSION AND CONCLUSION

The bioresorbable cranial implant as proposed could serve immediately as an osteoconductive path to bone reconstruction. The goals of a biodegradable material for void filling address

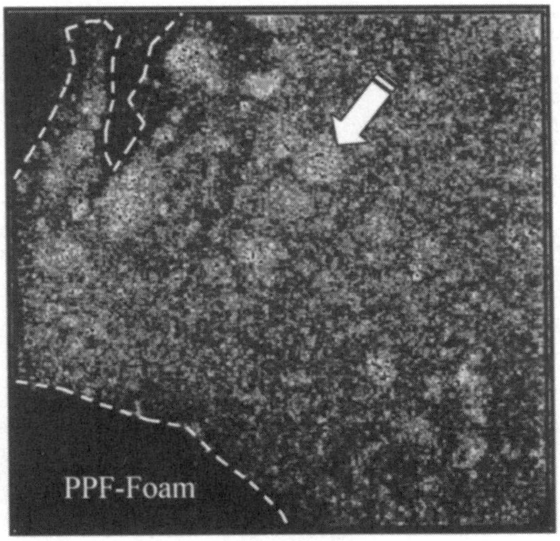

Figure 12 Confocal microscopic image of a PPF foam sample cocultured with fetal bovine articular cartilage cells for one week (magnification ×30) showing a pore filled with chondrocytes at a penetration depth of 126 μm. The pore is outlined by the white dashed line. A chondrocyte is indicated by a white arrow.

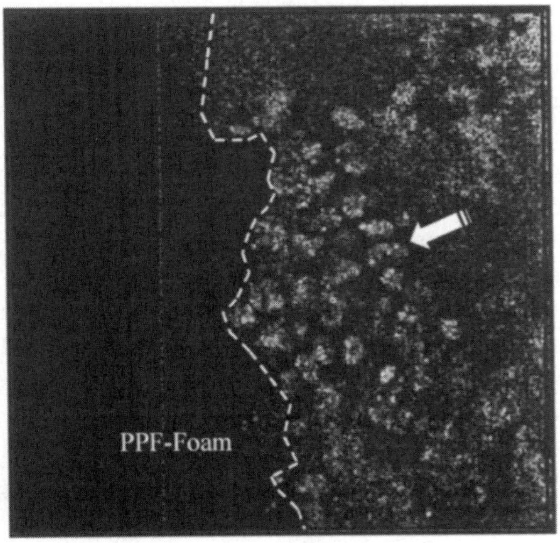

Figure 13 Confocal microscopic image of a PPF foam sample cocultured with fetal bovine articular cartilage cells for four weeks (magnification ×30) showing a pore filled with chondrocytes at a penetration depth of 351 μm. The pore is outlined by the white dashed line. Individual chondrocytes (white arrow) are present in close proximity to the PPF foam pore.

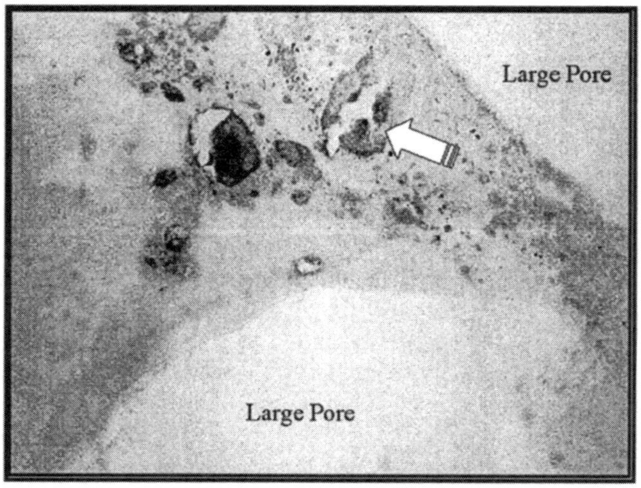

Figure 14 Photomicrograph of a cross section through the PPF foam at four weeks of cocultured (H&E × 10) showing chondrocytes but not large pores.

degradation [69], porosity [70], ex situ scaffold morphologies, and the ultimate flexibility in geometry to permit use in reconstructive situations calling for varying rates of degradation and bony recovery.

The biodegradable porous scaffold used in this study was engineered to promote osteoconductivity both by the development of extreme porosity and the presence of hydroxyapatite. Although the foaming cement did not have an interconnected pore structure, it appeared to have favorable osteoconductive properties in the rat tibial implantation model as a result of its rapid resorption, which allowed vascular invasion to occur. Since interconnectivity of the pores appears to be the key for other combinatorial tissue engineering applications, such as PPF foaming cement with living cells, it appears of interest that the polymer composite system described in the present study offers considerable flexibility for optimizing the osteoconductive properties for any given application. The relative contents of PPF, HA, and VP allow for control of the mechanical properties of the implant, which may be necessary for treatments of different skeletal defects. Furthermore, the adjustment of the HA content or other less soluble calcium salts may allow for control of the migration rate of osteoblasts into defects sites. Additionally, changing the content of the blowing agents (SB and CA) may allow for control of both the overall porosity and the pore size, which may, in turn, regulate the degradation rate of the construct and influence cellular migration rates as well. The in vivo study provided sufficient data to conclude that the PPF foam would be suitable for seeding with bone forming cells.

Initial in vitro seeding studies with use of bovine fetal chondrocytes demonstrated that ingrowth occurs as demonstrated on fresh unprocessed samples with use of confocal microscopy. The absence of cells in large pores on light microscopy cells could only be demonstrated in surface pores, indicating poor adherence and better in small pores. Most likely this was due to a processing artifact which resulted in removal of the cellular components from the polymer foam scaffold. However, this would corroborate the hypothesis that adherence of the chondrocytes to the PPF foam scaffold was poor in large pores. This is consistent with findings by other investigators, who have used other polymeric scaffolds, such as PLGA, and algenate. Paralleling work by Tong [71, 72], Patel [73], Ranieri [74], and Shakesheff [75] on peptide surface modification with adhesion molecules such as laminin, integrin, and RGD, future developmental efforts would have to focus on improvement of cellular attachment so that cells which migrate into the polymeric scaffold would ultimately remain there. Although the in vivo study demonstrated that open-celled polymeric scaffolds can serve as three-dimensional support systems for ingrowth of bone cells, our in vitro seeding study clearly showed that engineering of functional polymer-cell tissue composites appears more complex and that factors such as hydrophobicity or hydrophobility, porosity, pore size, and pore size distribution are essential. As such, small-diameter pores are preferable to yield high surface area to volume ratios as long as the pore size is greater than the diameter of a cell in suspension. In the case of bone regeneration, there is an optimal pore size for maximum tissue ingrowth ranging from 200 to 400 μm [76]. Knowledge of both growth factor and cell transport into the foam matrix is necessary for the optimization of the geometry of the foam scaffold. When seeding the surface of an ideal porous medium, the drop in cell number is expected to be logarithmic with depth, suggesting that the bulk of the cells may be confined within the first few hundred micrometers depth of the foam [77]. Porosity also determines the stiffness and elasticity of the device, obviously affecting the ultimate utility of the device. This hypothesis should form the basis for continuing research focusing on preparing a dimensionally stable polymeric scaffold as a resorbable support to the developing bony structure.

ACKNOWLEDGMENTS

The authors wish to thank Joseph Alroy, DVM, associate professor in pathology, Tufts University Schools of Medicine and Veterinary Medicine for his assistance in the histological analysis of this study. Furthermore, the authors are indebted to Shrikar Bondre and Eric Gusek for their assistance with animal care. This work was supported in part by an internal research project of Cambridge Scientific, Inc., and NIH/NIAMS Grant AR 45062 (to Kai-Uwe Lewandrowski).

REFERENCES

1. SL Ishaug-Riley, GM Crane-Kruger, MJ Yaszemski and AG Mikos. "Three-dimensional culture of rat calvarial osteoblasts in porous biodegradable polymers," *Biomaterials*, **19**, 1405–1412, 1998.
2. M Saito, A Maruoka, T Mori, N Sugano and K Hino. "Experimental studies on a new bioactive bone cement: hydroxyapatite composite resin," *Biomaterials*, **15**, 156–160, 1994.
3. SC Cowin, Bone Mechanics, CRC Press, Boca Raton, 1989.
4. AJ Domb, N Manor and O Elmalak. "Biodegradable bone cement compositions based on acrylate and epoxide terminated poly(propylene fumarate) oligomers and calcium salt compositions," *Biomaterials*, **17**, 411–417, 1996.
5. TN Gerhart, AA Renshaw, RL Miller, RJ Noecker and WC Hayes. "*In vivo* histologic and biomechanical characterization of a biodegradable particulate composite bone cement," *Journal of Biomedical Materials Research*, **23**, 1–16, 1989.
6. DK Gilding and AM Reed. "Biodegradable polymers for use in surgery: polyglycolic/polylactic homo- and copolymers: 1," *Polymer*, **20**, 1459–1464, 1979.
7. MC Harper and G Hardin. "Posterior malleolar fractures of the ankle associated with external rotation-abduction injuries. Results with and without internal fixation," *Journal of Bone and Joint Surgery—American Volume*, **70**, 1348–1356, 1988.
8. SL Ishaug, MJ Yaszemski, R Bizios and AG Mikos. "Osteoblast function on synthetic biodegradable polymers," *Journal of Biomedical Materials Research*, **28**, 1445–1453, 1994.

9. SL Ishaug, RG Payne, MJ Yaszemski, TB Aufdemorte, R Bizios and AG Mikos. "Oesteoblast migration on poly(α-hydroxy esters)," *Biotechnology and Bioengineering*, **50**, 443–451, 1996.
10. P Rokkanen, O Bostman, S. Vainionpaa, EA Makela, E Hirvensalo, EK Partio, K Vihtonen, H Patiala and P Tormala, "Absorbable devices in the fixation of fractures," *Journal of Trauma*, **40**, S123-7, 1996.
11. LA Sargent Tennessee Craniofacial Center, 975 East Third Street, Chattanooga, TN 37403, 1998.
12. T Nakajima, Y Yoshimura, Y Nakanishi, S Koga and K Katada. "Integrated life-sized solid model of bone and soft tissue: application for cleft lipand palate infants," *Plastic and Reconstructive Surgery*, **95**(5), 1020–1025, 1995.
13. JL Marsh and MW Vannier. "Cartographic mapping of the skull from computed tomography scans," *Journals of Craniofacial Surgery*, **5**(3), 188–194, 1994.
14. A Furst, S Reinert, C Passelk, P Kuhn, J Lentrodt and U Modder, "The value of 2- and 3-dimensional computed tomography in the diagnosis and classification of midfacial and orbital fractures," *Deutsche Zahn-, Mund-, und Kieferheilkunde Mit Zentrablatt*, **80**(4), 199–207, 1992.
15. W Lill, P Solar, C Ulm, G Watzek, R Blahout, and M Matejka, "Reproducibility of three-dimensional CT-assisted model production in the maxillofacial area," *British Journal of Oral and Maxillofacial Surgery*, **30**(4), 233–236, 1992.
16. AA Waitzman, JC Posnick, DC Armstrong and GE Pron. "Craniofacial skeletal measurements based on computed tomography: Part I. Accuracy and reproducibility," *Clef Palate-Craniofacial Journal*, **29**(2), 112–117, 1992.
17. MJ Yaremchuk. "Changing concepts in the management of secondary orbital deformities," *Clinics in Plastic Surgery*, **19**(1), 113–124, 1992.
18. S Chapman, JH Goldin, RG Hendel, AD Hockley, MC Wake and P Weale. "The median cleft face syndrome with associated cleft mandible, bifid odontoid pog and agenesis of the anterior arch of atlas," *British Journal of Oral and Maxillofacial Surgery*, **29**(4), 279–281, 1991.
19. JT Richtsmeier, HM Grausz, GR Morris, JL Marsh and MW Vannier. "Growth of the cranial base in craniosynostosis," *Cleft Palate-Craniofacial Journal*, **28**(1), 55–67, 1991.
20. JC Carr, WR Fright and RK Beatson. "Surface interpolation with radial basis functions for medical imaging," *IEEE Transactions on Medical Imaging*, **16**(1), 96–107, 1997.
21. JS Orringer, V Barcelona and SR Buchman. "Reasons for removal of rigid internal fixation devices in cranofacial surgery," *Journal of Craniofacial Surgery*, **9**(1), 40–44, 1998.
22. KW Gurstein, AH Sather, KN An and BE Larson. "Stability after inferior or anterior maxillary repositioning by Le Fort I osteotomy: a biplanar stereocephalometric study," *International Journal of Adult Orthodontics and Orthognathic Surgery*, **13**(2), 131–143, 1998.
23. W Tharanon. "Comparison between the rigidity of bicortical screws and a miniplate for fixation of a mandibular setback after a simulated bilateral sagittal split osteotomy," *Journal of Oral and Maxillofacial Surgery*, **56**(9), 1055–1058, 1998.
24. JW Polley, KF Hung, A Figueroa, T Lakars and H Heiberger. "Alternative techniques in rigid fixation of the craniomaxillofacial skeleton," *Journal of Craniofacial Surgery*, **9**(3), 249–253, 1998.
25. RD Marciani and AA Gonty, "Principles of management of complex craniofacial trauma," *Journal of Oral and Maxillofacial Surgery*, **51**(5), 535–542, 1993.
26. E Robotti, T Forcht Dagi, M Ravegnani and G Bocchiotti. "A new prospect on the approach to open, complex, craniofacial trauma," *Journal of Neurosurgical Sciences*, **36**(2), 89–99, 1992.
27. DL Benzil, E Robotti, TF Dagi, P Sullivan, JR Bevivino and NW Knuckey. "Early single-stage repair of complex craniofacial trauma," *Neurosurgery*, **30**(2), 166–171, discussion 171–172, 1992.
28. C Lauritzen, J Lilja and B Vallfors. "The craniofacial approach to trauma," *Annals of Plastic Surgery*, **17**(6), 503–512, 1986.
29. JC Carr, AH Gee, RW Prager and KJ Dalton. "Quantitative visualization of surfaces from volumetric data," *Proceedings of WSGC '98—The Sixth International Conference in Central Europe on Computer Graphics and Visualization*, Plzen, Czech Republic, February 10–14, 1998.
30. RK Beatson and G Newsam. "Fast evaluation of radial basis functions: I," *Computers and Mathematics with Applications*, **24**(12), 7–19, 1992.
31. O Bostman, E Hirvensalo, S Vainionpaa, K Vihtonen, P Tormala and P Rokkanen. "Degradable

polyglycolide rods for the internal fixation of displaced bimalleolar fractures," *International Orthopaedics*, **14**, 1–8, 1990.

32. OM Bostman, U Paivarinta, E Partio, M Manninen, J Vesenius and P Rokkanen. "Degradation and tissue replacement of an absorbable polyglycolide screw in the fixation of rabbit femoral osteotomies," *Journal of Bone and Joint Surgery*—American Volume, **74**, 1021–1031, 1992a.
33. O Bostman, E Hirvensalo, E Partio, P Tormala and P Rokkanen, "Resorbable rods and screws of polyglycolide in stabilizing malleolar fractures," *Unfallchirurg*, **95**, 109–112, 1992b.
34. OM Bostman. "Distal tibiofibular synostosis after malleolar fractures treated using absorbable implants," *Foot and Ankle*, **14**, 38–43, 1993.
35. M Kellman, SC Huckins, J King, D Humphrey, L Marentette and DC Osborn. "Bioresorbable screws for facial bone reconstruction: a pilot study in rabbits," *Laryngoscope*, **104**(5 pt 1), 556–561, 1994.
36. Y Matsusue, T Yamamuro, M Oka, Y Shikinami, SH Hyon and Y Ikada, "*In vitro* and *in vivo* studies on bioabsorbable ultra-high-strength poly(L-lactide) rods," *Journal of Biomedical Materials Research*, **26**, 1553–1567, 1992.
37. AM Donigian, BR Laga and PM Caskey. "Biodegradable fixation of physeal fractures in goal distal femur," *Journal of Pediatric Orthopedics*, **13**, 349–354, 1993.
38. CT Laurencin, MA Attawia, HE Elgendy and KM Herbert. "Tissue engineered bone regeneration using degradable polymers: the formation of mineralized matrices," *Bone*, **19**(1 Supp), 93S–99S, 1996.
39. CH Rivard, C Chaput, S Rhalmi and A Selmani. "Bio-absorbable synthetic polyesters and tissue regeneration. A study of three-dimensional proliferation of ovine chondrocytes and osteoblasts," *Annales de Chirurgie*, **50**(8), 651–658, 1996.
40. H Suh and C Lee. "Biodegradable ceramics-collagen composite implanted in rabbit tibiae," *ASAIO Journal*, **41**(3), M652-6, 1995 July–Sept.
41. AG Mikos, MD Lyman, LE Freed and R Langer. "Wetting of poly(L-lactic acid) and poly(DL-lactic-co-glycolic acid) foams for tissue culture," *Biomaterials*, **15**(1), 1994.
42. DJ Mooney, S Park, PM Kaufmann et al., "Biodegradable sponges for hepatocyte transplantation," *Journal of Biomedical Materials Research*, **29**, 959–966, 1995.
43. DJ Mooney, DF Baldwin, NP Suh, JP Vacanti and Robert Langer, "Novel approach to fabricate porous sponges of poly(D,L-lactic-co-glycolic acid) without the use of organic solvents," *Biomaterials*, **17**(14), 1996.
44. LE Freed and G Vunjak Novakovic. "Tissue culture bioreactors: Chondrogensis as a model system," Principles of Tissue Engineering, edited by R Lanza, R Langer and W Chick, RG Landes Company, 17.
45. Y Koshihara, M Kawamura, S Endo, C Tsutsumi, H Kodama, H Oda and S Higaki. "Establishment of human osteoblastic cells derived from periosteum in culture," *In Vitro Cellular and Developmental Biology*, **25**, 37–43, 1989.
46. A Uchida, T Kikuchi and Y Shimomura. "Osteogenic capacity of cultured human periosteal cells," *Acta Orthopaedica Scandinavica*, **59**, 29–33, 1988.
47. I Moll, P Houdek, H Schmidt and R Moll. "Characterization of epidermal wound healing in a human skin organ culture model: Acceleration by transplanted keratinocytes," *J. Invest. Derm*, **111–112**, 251–258, 1998.
48. M Brittberg, A Lindahl, A Nilsson, C Ohlsson, O Isaksson and L Peterson. "Treatment of deep cartilage defects in the knee with autologous chondrocyte transplantation," *N. Eng. J. Med.*, **331**, 889–895, 1994.
49. AS Breitbart, DA Grande, R Kessler, JT Ryaby and RJ Fitzsimmons. "Tissue engineered bone repair of calvarial defects using cultured periosteal cells," *Plastic and Reconstructive Surgery*, **101**(3), 567–574, discussion 575–576, 1998.
50. SE Utvag, O Grundnes and O Reikeraos. "Effects of periosteal stripping on healing of segmental fractures in rat," *J. Orthop. Trauma*, **10**, 279–284, 1996.
51. HT Aro, BW Wipperman, SF Hodgson and EF Chao. "Internal remodeling of periosteal new bone during fracture healing," *J. Orthop. Res.*, **8**, 238–246, 1990.

52. H Nakahara, VM Goldberg and AI Caplan. "Culture-expanded human periosteal derived cells exhibit osteochondral potential *in vivo*," *J. Orthop. Res.*, **9**, 465–476, 1991.
53. H Nakahara, SP Bruder, VM Goldberg and AI Caplan. "*In vivo* osteochondrogenic potential of cultured cells derived from the periosteum," *Clin. Orthop.*, **259**, 223–232, 1992.
54. TH Kim, C Janetta, JP Vacanti, J Upton and CA Vacanti. "Engineered bone from polyglycolic acid polymer scaffold and periosteum," *Mat. Res. Soc. Symp. Proc.*, **394**, 91–97, 1995.
55. CA Vacanti, W Kim, J Upton, D Mooney and JP Vacanti. "The efficacy of periosteal cell compared to chondrocytes in the tissue engineered repair of bone defects," *Tiss. Eng.*, **1**, 301–308, 1995.
56. WC Puelacher, JP Vacanti, NF Ferraro, B Schloo and CA Vacanti. "Femoral shaft reconstruction using tissue-engineered growth of bone," *International Journal of Oral and Maxillofacial Surgery*, **25**(3), 223–228, 1996.
57. JM Rubak, M Poussa and V Ritislia. Chondrogenesis in repair of articular cartilage defects by free periosteal grafts in rabbits. *Acta Orthop. Scand.*, **53**, 181–186, 1982.
58. WS Kim, CA Vacanti, J Upton and JP Vacanti. "Bone defect repair with tissue-engineered cartilage," *Plast. Reconstruct. Surg.*, **94**, 580–584, 1994.
59. LS Beck, L Deguzman, WP Lee, Y Xu, LA McFatridge, NA Gillett and EP Amento. "TGF-β_1 induces bone closure of skull defects," *J. Bone Miner. Res.*, **6**, 1257–1265.
60. AH Reddi. Role of morphogenetic proteins in skeletal tissue engineering and regeneration. *Nature Biotechnology*, **16**(3), 247–52, 1998.
61. E Wintermantel, J Mayer, J Blum, KL Eckert, P Luscher and M Mathey. "Tissue engineering scaffolds using superstructures," *Biomaterials*, **17**(2), 83–91, 1996.
62. KU Lewandrowski, MV Cattaneo, JD Gresser, DL Wise, RL White, L Bonassar and DJ Trantolo. "Effect of a poly(propylene fumarate) foam of healing of critical size bone defects," submitted for publication in *Tissue Engineering*, 1999.
63. SA Goldstein, DL Wilson, DA Sonstegard and LS Matthews. "The mechanical properties of human tibial trabecular bone as a function of metaphyseal location," *J. Biomech.*, **16**, 965, 1983.
64. DR Carter and WC Hayes. "Bone compressive strength: The influence of density and strain rate," *Science*, **194**:1774–1776, (1976).
65. JB Park and RS Lakes. "Biomaterials: An Introduction," Plenum Press, New York, 1992.
66. MJ Yaszemski, RG Payne, WS Hayes, RS Langer, TB Aufdemorte and AG Mikos. "The ingrowth of new bone tissue and initial mechanical properties of a degrading polymeric composite scaffold. Tissue Engineering," **1**, 41, 1995.
67. RT Ballock and AH Reddi. "Thyroxine is the serum factor that regulates morphogenesis of columnar cartilage from isolated chondrocytes in chemically defined medium," *J. Cell Biol.*, **126**:1311–1318, 1994.
68. Y Kato, M Iwamoto, T Koike, F Suzuki, Y Takano. "Terminal differentiation and calcification in rabbit chondrocyte cultures grown in centrifuge tubes: regulation by transforming growth factor beta and serum factors," *Proc. Natl. Acad. Sci.*, **85**:9552–9556, 1988.
69. RA Kenley, K Yim, J Abrams, E Ron, T Turek, LJ Marden and JO Hollinger. "Biotechnology and bone graft substitutes," *Pharmaceutical Research*, **10**:1393–1401 (1993).
70. AR Gazdag, JM Lane, D Galser and RA Forster. "Alternatives to autogenous bone graft: efficacy and indications," *J. Amer. Acad. Orthopaedic Surgeons*, **3**, 1–8, 1995.
71. YW Tong and MS Shoichet. "Enhancing the interaction of central nervous system neurons with poly(tetrafluoroethylene-co-hexafluoropropylene) via a novel surface amine-functionalization reaction followed by peptide modification," *J. Biomater. Sci. Polym. Ed.*, **9**(7), 713–29, 1998.
72. YW Tong and MS Shoichet. "Peptide surface modification of poly(tetrafluorethylene-co-hexafluoropropylene) enhances its interaction with central nervous system neurons," *J. Biomed. Mater. Res.*, **42**(1), 85–95, 1998.
73. N Patel, R Padera, GH Sanders, SM Cannizzaro, MC Davies, R Langer, CJ Roberts, SJ Tendler, PM Williams and KM Shakesheff. "Spatially controlled cell engineering on biodegradable polymer surfaces," FASEB J. **12**(14), 1447–54, 1998.

74. P Ranieri, R Bellamkonda, EJ Bekos, JA Gardella Jr., HJ Mathieu, L Ruiz and P Aebischer. "Spatial control of neuronal cell attachment and differentiation on covalently patterned laminin oligopeptide substrates," *Int. J. Dev. Neurosci.*, **12**(8), 725–35, 1994.
75. Shakesheff, S. Cannizzaro, R Langer. "Creating biomimetic micro-environments with synthetic polymer-peptide hybrid molecules," *J. Biomater. Sci. Polym. Ed.*, **9**(5), 507–18, 1998.
76. BD Boyan, TW Hummert, DD Dean and Z Schwartz. *Biomaterials*, **17**, 137, 1996.
77. MV Cattaneo, C Masson and CW Greer. "The influence of moisture on microbial transport, survival and 2.4-D biodegradation with a genetically marked Burkholderia cepacia in unsaturated columns," *Biodegradation*, **8**, 87–96, 1997.

31
Preclinical Evaluation of Bone Graft Substitutes

Christopher V. Bensen, Yuehuei H. An, and Richard J. Friedman
Medical University of South Carolina, Charleston, South Carolina

I. INTRODUCTION

The past several decades have witnessed a remarkable revolution in the technology of bone grafting. Since the very first bone grafting techniques were introduced over a century ago, both the technology and clinical applications have expanded to such a degree that over 1 million procedures involving bone grafting are performed annually in the United States alone. Bone banks are now found in every major medical center to accommodate the routine use of allograft materials in orthopedic, maxillofacial, and trauma surgery. Limb salvage and reconstructive procedures which were impossible only a very short time ago are now routine as a result of this emerging technology.

Along with this unprecedented growth in the clinical arena has been a concomitant rise in the basic science research of bone graft materials and the methods to evaluate their effectiveness. New materials and procedures are constantly being introduced in pursuit of the ideal graft substitute.

In this chapter we will review the types of grafts currently used in the reconstruction of bone defects as well as substitute materials that are currently under investigation. The major animal models used in the evaluation of these materials will also be introduced as well as methods of analysis. Finally, we will explore the future of bone graft substitutes and some promising new materials which will be implemented in the 21st century.

II. ANIMAL MODELS FOR BONE DEFECT REPAIR AND EVALUATION

A. Ectopic Models of Osteogenesis

Prior to evaluating a bone graft substitute in a bone defect model, materials with osteogenic potential may be evaluated first in an ectopic model. Several major models have been described, including the subcutaneous, intramuscular, and intraperitoneal models. These models are relatively inexpensive and are often able to yield rapid results in the evaluation of new materials, allowing efficient progression to more clinically relevant models. Constructs, or substances containing osteogenic materials, are implanted in one of the three aforementioned anatomical sites. After 3–24 weeks the implants are explanted and examined radiographically and histologically to identify the existence and evaluate the quality of new bone formation.

Substances which could rapidly dissipate into surrounding tissues in vivo may be implanted by use of a diffusion chamber. These chambers have been described by several authors and usually include porous membranes contained in a plastic housing with a volume of approximately 130 μl [1–4]. In one study, a diffusion chamber containing rabbit bone morphogenetic protein was implanted in the abdominal muscle of the rat. After implantation, cartilage differentiated around the chamber in one to two weeks and bone replaced the cartilage in three to four weeks [3].

B. Bone Defect Models

Although numerous animal models have been described and used for the evaluation of bone graft substitutes, four major types of defects have been most frequently utilized. These include the calvarial defect, long bone or segmental defect, partial cortical defect (cortical window, wedge defect, or transcortical drill hole), and cancellous bone defect (drill holes). Of these, the calvarial and segmental defects have been the most widely used.

Whenever a bone defect model is chosen, consideration must be given to the size of the defect and the material to be tested. Schmitz and Hollinger have defined the smallest defect that would not heal spontaneously over a given period of time as the "critical size defect" or CSD [5]. Each animal model has a specific CSD and the size can vary with age, weight, and even strain in a given species [6]. For example, the CSD of adult NZW rabbit calvarial defects is approximately 15 mm [7], whereas the CSD of rat calvarial defects ranges from 4 to 8 mm [8,9]. Figure 1 shows incomplete healing of a 12 mm rabbit radius defect at eight weeks postoperatively.

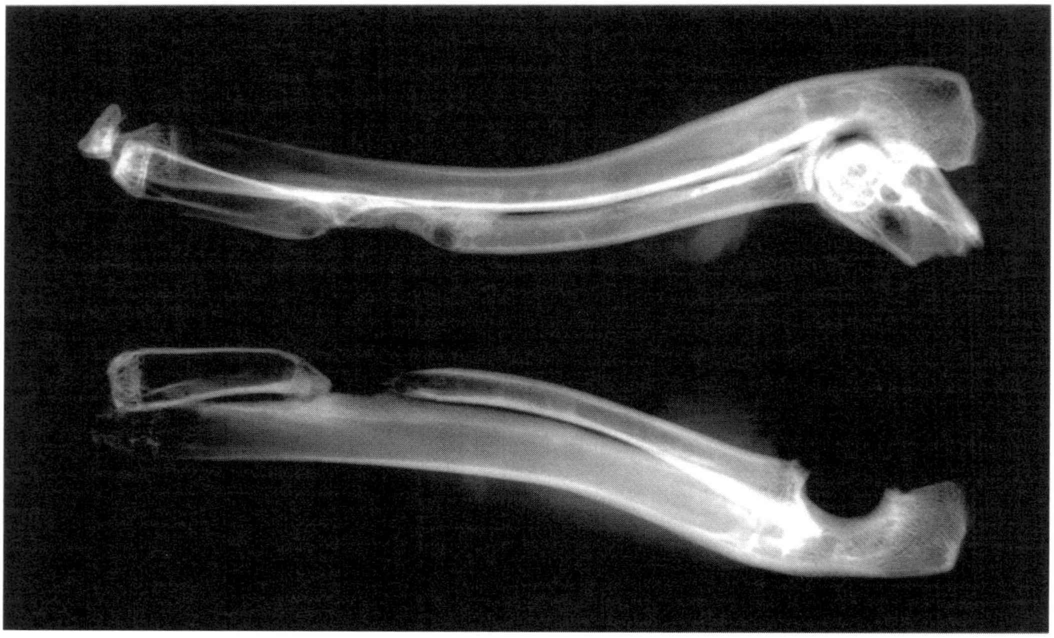

Figure 1 Radiograph comparing bridging callous formation in a 12 mm rabbit radius segmental defect (above) vs. fibrous nonunion seen in a 15 mm defect (below).

Another important consideration is to match the type of material being tested with an appropriate model. For example, graft substitutes which are particulate or which will be used for cavitary defects may be better evaluated in a calvarial model. Conversely, a segmental long bone model is more appropriate for grafts which have structural properties.

Rabbits and rats are usually the first choice for calvarial models, and rabbits and dogs are best for segmental defects. Dogs and sheep are also most often used for experiments using heavy internal or external fixations. A summary of some popular animal models used for the evaluation of bone graft substitutes is given in Table 1.

Table 1 Selected Bone Defect Models in the Literature

Animal	Bones	Type of defect	Author, year
Rat	Cranial bone	Circular	Ray 1957 [10], Takagi 1982 [9], Kenley 1994 [11], Miki 1994 [12], Sweeney 1995 [13], McKinney 1996 [14]
	Radius	Segmental	Herold 1971 [15], Gepstein 1987 [16], Alper 1989 [17], Solheim 1992 [18], Nyman 1995 [19]
	Femur	Segmental	Melcher 1962 [20], Einhorn 1984 [21], Pelker 1989 [22], Feighan 1995 [23], Hunt 1996 [24], Stevenson 1997 [25]
	Fibula	Segmental	Narang 1971 [26], Chakkalakal 1994 [27], Bluhm 1995 [28]
Rabbit	Cranial bone	Circular	Frame 1980 [7], Schmitz 1988 [29], Damien 1990 [30], Arnaud 1994 [31], Robinson 1995 [32], Ashby 1996 [33]
	Radius or ulna	Segmental	Herold 1971 [15], Tuli 1981 [34] Bolander 1986 [35], Hopp 1989 [36], Iyoda 1993 [37], Yang 1994 [38]
	Tibia	Circumscribed	Shimazaki 1985 [39], Suh 1995 [40], Uchida 1985 [41]
	Fibula	Segmental	Yang 1994 [38], Taguchi 1995 [42]
Dog	Cranial bone	Circular	Oklund 1986 [43]
	Radius or ulna	Segmental	Key 1934 [44], Nilsson 1986 [45], Delloye 1990 [46], Grundel 1991 [47], Cook 1994 [48], Johnson 1996 [49]
	Ulna	Segmental	Frayssinet 1991 [50]
	Femur	Segmental	de Pablos 1994 [51]
	Tibia	Segmental	Tiedeman 1989 [52], Markel 1991 [53]
	Fibula	Segmental	Enneking 1975 [54], Welter 1990 [55]
	Mandible	Segmental	Holmes 1979 [56], Toriumi 1991 [57]
Goat	Femur	Transcortical hole	Radder 1996 [58]
Primate	Cranial bone	Circular	Hollinger 1989 [59]
	Mandible	Grooves	Drury 1991 [60]
Sheep	Cranial bone	Circular	Lindholm 1988 [61], Viljanen 1996 [62]
	Femur	Segmental	Ehrnberg 1993 [63], Brunner 1994 [64]
	Tibia	Drill hole	Hallfeldt 1995 [65]
		Segmental	Gao 1995 [66]

1. Calvarial Defect Models

The calvarial defect model is a well-studied and appropriate model for the following reasons: (1) the calvarial bone is a platelike structure which allows creation of a uniform circular defect that enables convenient radiographic and histological analysis; (2) the calvarial bone has a good size for easier surgical procedures and specimen handling; (3) no fixation is required because of the support provided by the dura and the overlying skin; (4) the model has been well studied, is easily reproducible, and permits precise comparison of a variety of graft substances; and (5) it is a relatively economical model.

The rabbit and rat calvarial defects are created using very simple operative procedures. In the rabbit model, an anteroposterior midline skin incision (4–5 cm long) is made over the cranial vault and the periosteum is elevated and retracted to expose the cranial bone. A 15 mm diameter defect is created using a trephine. The drill should be centered at a point on the midline 5 mm posterior to the transverse bone suture between the frontal and parietal bones (Fig. 2a). Copious saline irrigation is required during the drilling. Care must be taken to avoid perforating the dura, which is usually achieved by periodic examination of the osteotomy depth. Occasionally the bone disk needs to be taken out of the defect by breaking the leftover connection points between the bone disk and the edge of the defect. A neural elevator is very helpful for checking the depth, breaking bone connections within the osteotomy, and separating the dura from the inner surface of the bone disk. The key is to avoid complete cutting of the bone plate. If this does occur, it is likely the dura has already been compromised, which could cause an intracranial hemotoma. After implantation of the bone substitute (Fig. 2b), the periosteum is closed and strengthened by closing the adjacent muscles and subcutaneous tissue. Occasionally, a hemotoma may occur postoperatively, which may require drainage by cutting several of the skin sutures. Wound infection is a rare complication in this model.

2. Segmental Long Bone Defect Models

For evaluation of segmental graft materials, the rabbit radius offers the following advantages: (1) the radius is tubular, which allows creation of a segmental defect that enables convenient radiographic and histological evaluation; (2) the radius has a good size for easier surgical procedures and specimen handling; (3) no fixation is required because of the support of ulna; (4) the model has been well studied and is reproducible, which permits precise comparison of a variety of graft substances; and (5) it is relatively economical. The rabbit radial defect model

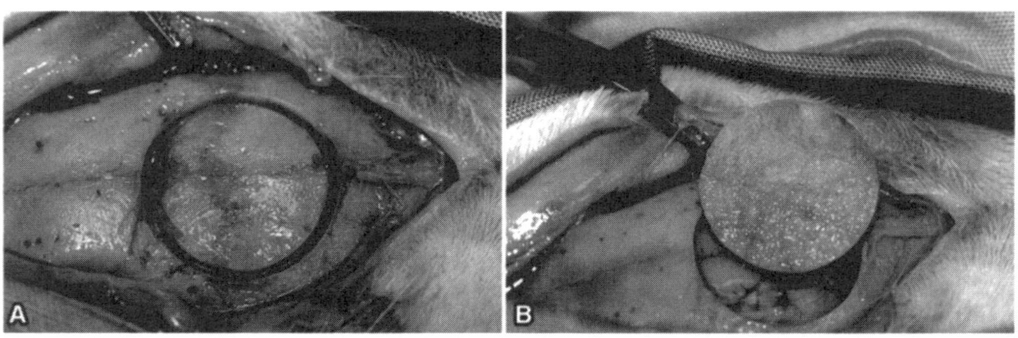

Figure 2 (a) Creation of a calvarial defect in a rabbit model showing the bone disk still in situ. (b) Replacement of the native bone disk with a hydroxyapatite-collagen complex composite graft. (From An, YH & Friedman, RJ, Eds. *Animal Models in Orthopaedic Research*, Chapter 13, with permission)

was first described by Herold in 1971 to test the effect of growth hormone on the healing of bone defect [15]. It is well accepted that the CSD of long bones is approximately two times that of the bone diameter. Because the diameter of the radius of adult NZW rabbits is about 5–6 mm, a radial defect should be no less than 12 mm long. In the authors' laboratory, three out of twenty 12 mm defects healed spontaneously in 12 weeks. Therefore, it is the authors' opinion that a 15 mm defect should be created.

Surgically, a longitudinal skin incision is made over the radius at the middle one third of the front leg. The periosteum is carefully separated from the surrounding muscles. A 15 mm defect beginning about 2.0 to 2.5 cm proximal to the radiocarpal joint and extending proximally is created by using a circular saw attached to a minidriver or dental handpiece. The defect should be checked carefully and any residual periosteum removed. Thorough irrigation with saline is carried out and the defect is then grafted with a bone substitute or spacer or left empty. Postoperatively, unrestricted weight bearing is normally allowed. Radiographs should be taken immediately after surgery and at any designated time periods postoperatively. Figure 3 is a radiographic image at eight weeks postoperatively showing a healed 15 mm radial defect by implantation of a porous form of demineralized bone matrix.

The rabbit ulnar defect has also been described and utilized, but it has several disadvantages. Most importantly, the ulna does not have the round shape of the radius, which creates difficulties and errors in the processes of implant preparation, implant positioning, and sample evaluation. Also, because of its shape, mechanical testing, such as bending or torsional testing,

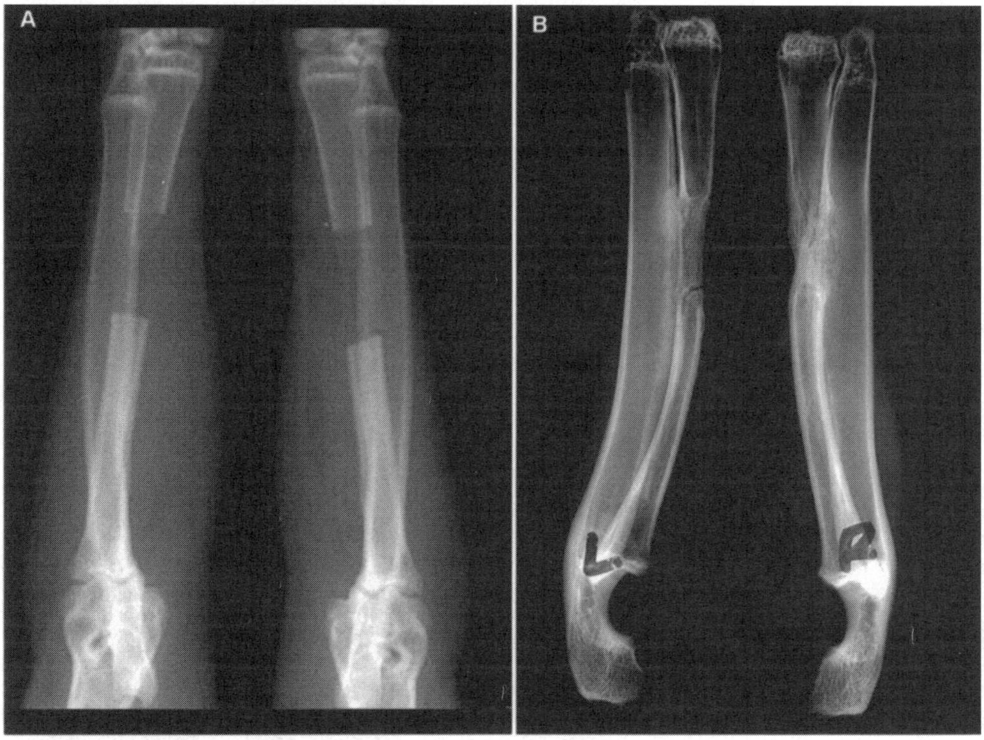

Figure 3 (a) Intraoperative radiograph demonstrating bilateral 15 mm rabbit radius defects. (b) Eight weeks after grafting with porous DBM, the defects are healed. (From An, YH & Friedman, RJ, Eds. *Animal Models in Orthopaedic Research*, Chapter 13, with permission)

is more difficult. For these reasons, the authors' preferred segmental defect model is the rabbit radius.

C. Methods of Evaluation

1. Radiographic Analysis

Periodic radiographic evaluation at regular intervals during the postoperative period provides an easily reproducible, inexpensive, and noninvasive method of examining the graft site. The amount and structure of new bone as well as the degree of continuity of the graft with the recipient bone can be evaluated. In the case of segmental defects, several scoring systems have been described, including those of Bos et al. and Lane and Sandhu [67,68]. These allow quantitation of parameters such as peristeal reaction, quality of union, and degree of remodeling. Figure 4 shows radiographic evidence of early healing of a rabbit calvarial defect using a hydroxyapatite-collagen complex composite graft substitute.

2. Biomechanical Testing

A variety of means of mechanical testing have been described by numerous authors to evaluate bone substitutes. Torsional testing [22, 36, 38, 45, 47, 69–71], three-point bending [72], and tensile testing [73] have all been described and used effectively to evaluate the incorporation of a graft. In the case of segmental defects, torsional testing appears to be the most popular and perhaps most appropriate method of mechanical testing. Mechanical testing systems are commercially available with PC linkage to facilitate data aquisition and processing. Computer-generated load-deformation curves are obtained, and biomechanical parameters, including ulti-

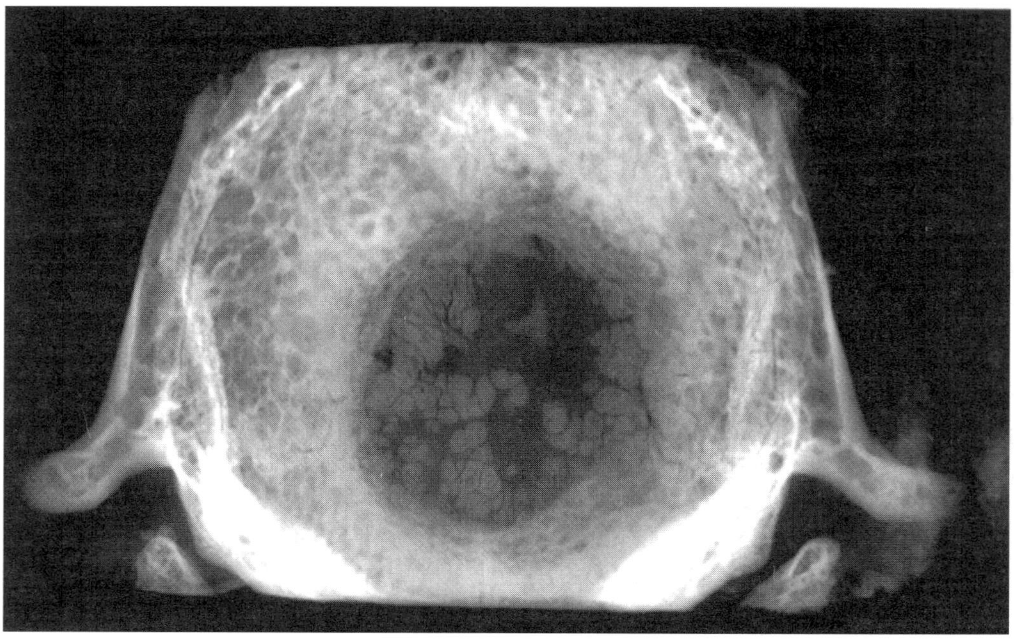

Figure 4 Radiograph taken eight weeks postoperatively showing early healing of a rabbit calvarial defect treated with a hydroxyapatite-collagen complex composite graft.

mate load, stiffness, displacement to failure, and ultimate strength, can subsequently be calculated.

3. Histological Analysis

Although radiographic and mechanical data are important in the overall evaluation of bone graft substitutes, histology is the most powerful method to examine the healing of bone defects. Common histological parameters include the following categories: bone union at the two osteotomies, callus formation, new bone formation in the defect, resorption of the bone graft, marrow changes, and cortex remodeling. These parameters can be quantitated based on a variety of scoring systems described in the literature [40, 45, 67, 68, 74]. Based on the nature of the graft to be used, the scoring system may be customized to suit the individual situation.

Computerized image analysis has made histomorphometry more efficient, especially for calculation of the percentage of defect filling by the repair tissues and the fractions of different tissues which are present. For histomorphometrical analysis of ectopic bone formation, the following elements can be quantified: area and penetration depth of mineralized "bonelike" tissue, area and thickness of nonmineralized bonelike tissue, area of osteoblast-covered surfaces, thickness of trabeculae, area and thickness of cartilage tissue, area of fibrovascular tissue, and void space [75,76]. For analysis of repair tissues of bone defect, the areas of mineralized bone, nonmineralized bone, lamellar bone, woven bone, chondral tissue, fibrocartilage, or fibrous vascular tissue can be used as parameters for quantification [77, 125].

III. TRADITIONAL BONE GRAFTS

A. Mechanisms of Bone Repair by Grafting

There are three primary mechanisms by which bone defects are repaired by a graft material. The first is osteoconduction, or the provision of a scaffold or substrate which will allow rapid vascularization and distribution of osteogenic cells. Materials with osteoconductive properties, including hydroxyapatite, demineralized bone matrix (DBM), ceramics, and many others, facilitate the host response to replace the defect with new bone. Without this medium, a good defect repair will be less likely. The second mechanism is osteoinduction. This can be defined as both the recruitment and induction of circulating osteoprogenitor cells to the graft site as well as the differentiation of local mesenchymal stem cells and other undifferentiated cells into osteoblasts. This is usually accomplished in the presence of growth factors which are added to the graft substitute or are present in the graft itself. An example of the latter case is DBM. Growth factors are discussed later in this chapter. Finally, direct osteogenesis involves the production of bone by osteoblasts at the margins of the graft site. Alternatively, this can be accomplished by implanting autogenic osteoblasts with the graft, providing immediate osteogenic potential from within the graft itself.

Implants which can provide all three of these factors for osteogenesis which are absent or insufficient in a bone defect or bone nonunion should be considered ideal bone grafts. Various substances investigated for bone defect repair are briefly described below.

B. Autograft and Allograft

Since Mowlem reported 75 bone graft cases using cancellous bone from the iliac crest in 1944, autogenous grafts have become the gold standard for skeletal reconstruction [78]. These grafts are by far the most widely used clinically and have a very high rate of success for several important reasons: the osteoconductive ability for the ingrowth of blood vessels and new bone, the presence of preexisting differentiated osteogeneic cells for immediate osteogenesis, the presence of preexisting growth promoting factors for osteoinduction, and, uniquely, the lack of immunological rejection from host defense system. However, an autologous graft has several disadvantages in skeletal reconstruction. First, it is available in limited quantity, thus making it inappropriate for the repair of large defects. Second, a separate operative procedure is required to harvest the graft and there is the risk of morbidity of the donor site. One author reported major complications in 8% of a large series of patients [79]. Finally, the difficulty of fabricating the graft into a desired shape for filling the defect often precludes its use in larger defects.

If a bone defect is not amenable to reconstruction with autograft, allograft bone is a viable alternative. These cadaveric grafts are usually freeze-dried or frozen to reduce their immunogenic potential and are available in numerous forms which range from powder to large osteoarticular grafts. Allogenic grafts are well tolerated, nontoxic, incorporated by host bone, and last a long time [80]. Since it is readily available from bone banks, allograft bone is one of the major methods for bone grafting in numerous orthopedic procedures, including revision total joint arthroplasty, tumor reconstruction, and limb salvage surgery.

C. Demineralized Bone Matrix

DBM may be considered a subtype of allograft. Using chemical or physical methods, allogenic, xenogenic, or even autogenous bone can be demineralized and made into DBM either in a porous or powder form or molded into any solid shape. It is also available commercially in paste, gel, and putty forms. DBM has been extensively investigated since the 1960s [17, 18, 34, 35, 38, 52, 61, 81, 82]. The osteogenetic mechanisms of DBM include osteoconduction [83, 84] and osteoinduction [38, 85]. Because of its availability, diverse uses, and limited immunological response, DBM will continue to be widely used in the management of skeletal defects and, perhaps, cartilage defect repair [86].

D. Bone Marrow

The addition of autologous bone marrow to bone graft substitutes (allograft, xenograft, demineralized bone matrix, or biomaterials) has been investigated since the early 1960s [20, 87–90]. The purpose is to increase the osteogenic ability of the bone grafts with autologous osteogenic precursor cells which differentiate into bone forming cells at the recipient site [4]. Autologous marrow is now being used clinically as an osteogenic material for treating nonunions and for repairing skeletal defects [91–93].

IV. BONE DEFECT REPAIR USING BIOMATERIALS AND TISSUE ENGINEERING TECHNIQUES

A. Biomaterials

There is an ever-expanding list of synthetic materials which have been used alone or in combination with organic material to repair bone defects. Ceramics, polymers, and even metals have been extensively investigated as bone graft substitutes with varying degrees of success. Hydroxyapatite (HA) [17, 94–96], calcium phosphate, tricalcium phosphate (TCP) [97, 99] tetracalcium phosphate cement, or composite forms such as HA/tricalcium phosphate ceramic [100] and calcium hydroxyapatite/calcium triphosphate ceramic [98], have been used to heal bone defects.

Since the implanted materials are inorganic, they act as osteoconductive substrates with little or no osteoinductive capacity. The rate of bone regeneration into the material pores varied depending upon the implant location (metaphyseal versus diaphyseal), cortical versus cancellous, the availability of local osteogenic cells, and the stability of the implant. Clinical results reported by Bucholz et al. showed that filling traumatic and tumor defects with HA and TCP resulted in healing in most cases [98]. Basically, these materials act as osteoconductive scaffolds and have no osteoinductive ability. Consequently, the addition of DBM [18], bone marrow [88, 102] and growth factors [101] to the implants has been investigated.

B. Growth Factors

Bone tissue harbors many growth promoting factors, such as bone morphogenetic protein (BMP), transforming growth factor (TGF), platelet-derived growth factor (PDGF), epidermal growth factor (EGF), fibroblast growth factor (FGF), and insulin-like growth factor (IGF). It is known that normal bone structure is maintained through a continuous balancing process of bone formation and resorption which occurs at the cellular level and is regulated by systemic hormones and local growth factors. Significant evidence supports the potential use of growth factors for bone repair [103, 104].

Perhaps the most promising of the osteoinductive substances are the bone morphogenetic proteins. BMPs are low molecular weight proteins that have been purified from bovine and human bones. BMPs have many isoforms and the well-known ones are BMP-1, BMP-2A, and BMP-3 or osteogenin [105]. Since the pioneering work of Urist in the 1960s [106, 107], it has been well known that bone morphogenetic proteins are very important local regulators of osteogenesis, and extensive work has been done on their function of stimulating bone formation [108–112]. BMPs have been used alone, as well as in combination with autogenic or allogenic bone graft, demineralized bone matrix, and some biomaterials. Even nonabsorbable materials such as porous titanium have been tried for repairing bone defects with impregnated bone morphogenetic protein and the result showed the BMP-Ti composite had interesting properties as an osteoinductive implant [101]. In the last several years, BMPs have been used clinically with several successful applications reported [113, 114]. The efficacy of human recombinant BMP-2 was first studied by Wang et al. [124], who demonstrated that it could induce heterotopic bone formation in a rat model. Alpaslan et al. reported that when combined with blood clot and polylactic acid–polyglycolic acid microspheres, rhBMP-2 "induces rapid proliferation of mesenchymal cells that leads to formation of cartilage and bone by 7 days which had matured by 9 weeks" [115]. These studies suggest that rhBMP-2 could potentially have a significant clinical application.

C. Tissue Engineered Composite Grafts

Cell-seeding a substrate to make an implantable graft is not a new concept [116, 117], but it has not been well developed, especially in the field of bone and cartilage repair. Recently, a few groups reported some preliminary and very promising data on the use of cell-seeded implants for repairing osseous or chondral defects [50, 118–121]. Different osteogenetic cells have been used. In 1991, Frayssinet et al. reported that cancellous bone cells from canine humeri were grown on HA granules and the cell-HA composite was placed in a bioreactor and implanted into a canine ulna defect. Osteogenesis was seen in active bioreactors three weeks after implantation but not in the controls [50]. Cultured rabbit chondrocytes bound to an HA block were implanted to repair a rabbit ulna defect by Iyoda [37]. More recently, periosteum-derived osteoblasts from newborn calf scapulae were seeded to polyglycolic acid mesh. The cell-polymer products were implanted subcutaneously in mice and all of the specimens demonstrated organized bone formation at 18 weeks [120]. The same product was also used to repair rat cranial defect [122]. Osteoblast-like cells (MC3T3-E1) have also been used to study the potential of bioabsorbable polymers and ceramics to support osteoblast growth for a proposed

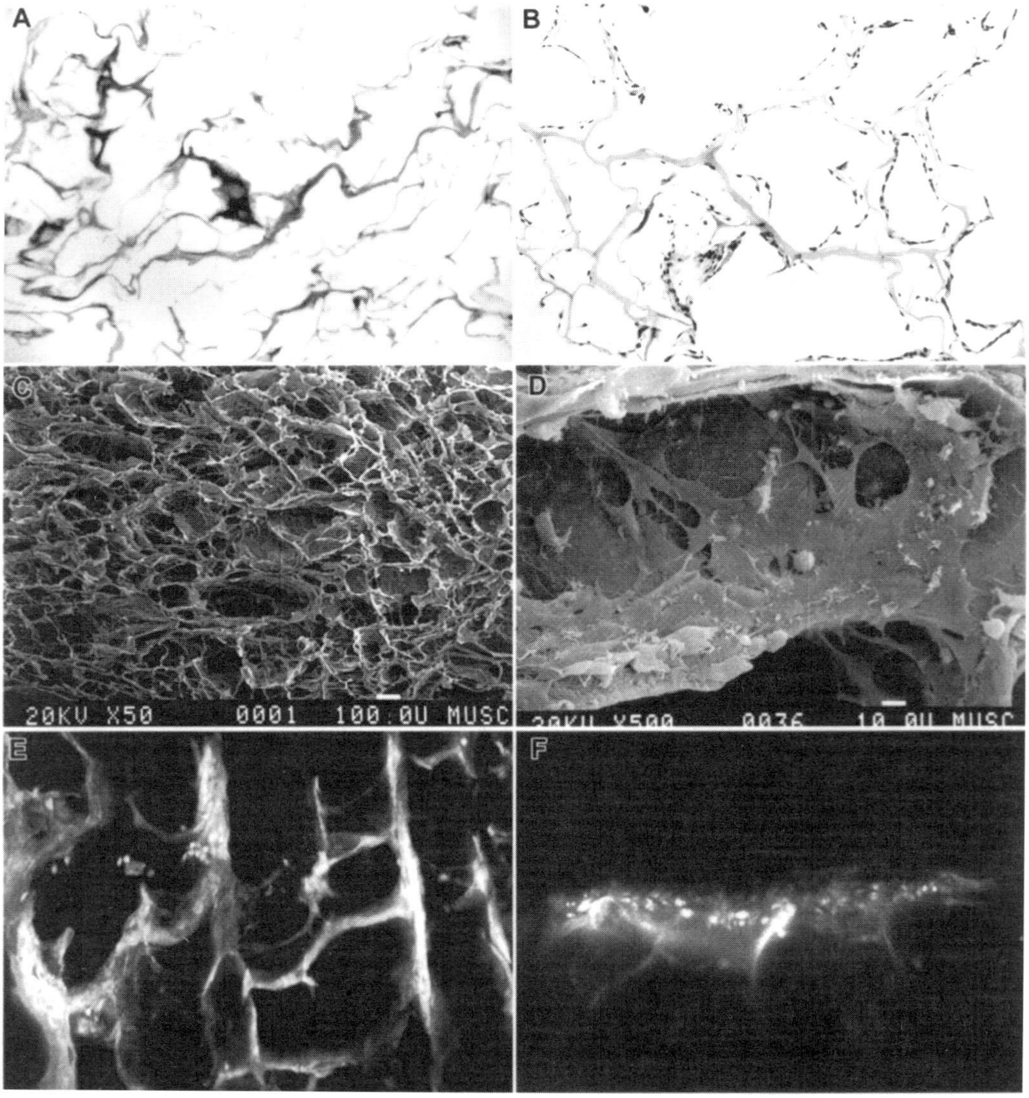

Figure 5 Histological section (Fig. 5a) and SEM (Fig. 5c, dry sample, mag. 50× bar at lower edge = 100 μm) of the HA-CC graft material investigated in the authors' laboratory. Confocal microscopy (wet sample, Fig. 5e) showing that the surfaces of the HA-CC showing porous structure. Histological section (Fig. 5b), SEM (dry sample, bar at lower edge = 10 μm, Fig. 5d), and confocal microscopy (wet sample, Fig. 5f) showing many osteoblasts adhered to collagen fibers on the implant surface and inside the pores.

bone-polymer composite for bone repair [119]. A novel hydroxyapatite-collagen complex (HA-CC) composite graft material has been investigated in the authors' laboratory. This material, when seeded with autologous osteoblasts has demonstrated promising results when used to heal a rabbit calvarial defect. The ultrastructure of this composite material is shown in Fig. 5.

These studies demonstrate that cell-seeded composite implants can induce bone tissue formation, leading to defect repair. In general, osteogenic cell-seeded composite implants act similarly to bone marrow transplantation. This technique is referred to by Langer and Vacanti as "tissue engineering" [123]. It has certain advantages over current methods, such as the autogeneic characteristics (if cells are taken from the same individual), minor morbidity of the donor site, and the unique property of permitting the production of a large amount of bone in vitro. Finally, it does not have the risk of disease transmission. For these reasons, the clinical future of these composite bone graft substitutes is quite promising.

V. THE FUTURE

Despite the improved technology and astounding number of bone graft procedures performed annually, the demand for more and better bone graft materials will continue to increase. The reasons for this are many. The number of trauma cases requiring bone grafts is sharply rising. The improved surgical instrumentation is now allowing reconstructive procedures which will undoubtedly require new graft technology. Finally, as discussed previously, there are numerous bone graft materials used clinically with excellent track records. However, these materials not only have a limited availability, but they are expensive, antigenic and, in some cases, carry the risk of infectious disease transmission. Consequently, the search continues to find a superior graft substitute.

The ideal bone graft substitute should have the following properties, which most of the current methods do not have: (1) osteoconductive ability for the ingrowth of blood vessels and new bone, (2) the presence of preexisting differentiated osteogenic cells for immediate osteogenesis, (3) the presence of preexisting growth promoting factors for osteoinduction, which can transform some mesenchymal or undifferentiated cells into bone-forming cells, (4) no immunological rejection by the host defense system, (5) minor (or no) donor site morbidity and no risk of infectious or transmitted diseases, and (6) the ability to be produced in large volume in vitro or in vivo and easily fabricated into any desired shape. It is the authors' opinion that this material will be a composite of resorbable synthetic material and autogenous cells which could be impregnated with growth factors. Research on several such composite materials is presently under way in our laboratories.

Considering the intense investigation currently under way to find such an ideal material, the future of this technology is bright. If this goal can be achieved, the potential clinical use of this method will be widespread. Patients with bone defects or fracture nonunions secondary to a number of different pathological conditions may have a new treatment option.

REFERENCES

1. Gundle, R., Joyner, C. J., Triffitt, J. T., Human bone tissue formation in diffusion chamber culture in vivo by bone-derived cells and marrow stromal fibroblastic cells. *Bone* 16: 597, 1995.
2. Rooney, P., Walker, D., Grant, M. E., McClure, J., Cartilage and bone formation in repairing Achilles tendons within diffusion chambers: evidence for tendon-cartilage and cartilage-bone conversion in vivo. *J Pathol* 169: 375, 1993.

3. Ono, Y., Kato, K., Oohira, A., Katoh, R., Nogami, H., Cell function during chondrogenesis and osteogenesis induced by bone morphogenetic protein enclosed indiffusion chamber. *Clin Orthop* 298: 305, 1994.
4. Ashton, B. A., Allen, T. D., Howlet, C. R., Eaglesom, C. C., Hattori, A., Owen M., Formation of bone and cartilage by marrow stromal cells in diffusion chambers in vivo. *Clin Orthop* 151: 294, 1980.
5. Schmitz, J. P., Hollinger, J. O., The critical size defect as an experimental model for craniofacial nonunions. *Clin Orthop* 205: 299, 1986.
6. Kleinschmidt, J. C., Hollinger, J. O., Animal models in bone research, in *Bone Grafts and Bone Substitutes*, Habal, A., and Reddi, M., Eds., W. B. Saunders, Philadelphia, 1992.
7. Frame, J. W., A convenient animal model for testing bone substitute materials. *J Oral Surg* 38: 176, 1980.
8. Mulliken, J. B., Glowacki, J., Induced osteogenesis for repair and construction in the craniofacial region. *Plast Reconstr Surg* 65: 553, 1980.
9. Takagi, K., Urist, M. R., The reaction of the dura to bone morphogenetic protein (BMP) in repair of skull defects. *Ann Surg* 196: 100, 1982.
10. Ray, B., Holloway, J. A., Bone implants. *J Bone Joint Surg* 39A: 1119, 1957.
11. Kenley, R., Marden, L., Turek, T., Jin, L., Ron, E., Hollinger, J. O., Osseous regeneration in the rat calvarium using novel delivery systems for recombinant human bone morphogenetic protein-2 (rhBMP-2). *J Biomed Mater Res* 28: 1139, 1994.
12. Miki, T., Harada, K., Imai, Y., Enomoto, S., Effect of freeze-dried poly-L-lactic acid discs mixed with bone morphogenetic protein on the healing of rat skull defects. *J Oral Maxillofac Surg* 52: 387, 1994.
13. Sweeney, T. M., Opperman, L. A., Persing, J. A., Ogle, R. C., Repair of critical size rat calvarial defects using extracellular matrix protein gels. *J Neurosurg* 83: 710, 1995.
14. McKinney, L., Hollinger, J. O., A bone regeneration study: transforming growth factor-β1 and its delivery. *J Craniofac Surg* 7: 36, 1996.
15. Herold, H. Z., Hurvitz, A., Tadmor, A., The effect of growth hormone on the healing of experimental bone defects. *Acta Orthop Scand* 42: 377, 1971.
16. Gepstein, R., Weiss, R. E., Saba, K., Hallel, T., Israel, R. A., Bridging large defects in bone by demineralized bone matrix in the form of a powder. *J Bone Joint Surg* 69A: 984, 1987.
17. Alper, G., Bernick, S., Yazdi, M., Nimni, M. E., Osteogenesis in bone defects in rats: The effects of hydroxyapatite and demineralized bone matrix. *Am J Med Sci* 298: 371, 1989.
18. Solheim, E., Pinholt, E. M., Andersen, R., Bang, G., Sudmann, B., Sudmann, E., The effect of a composite of polyorthoester and demineralized bone on the healing of large segmental defects of the radius in rats. *J Bone Joint Surg* 74A: 1456, 1992.
19. Nyman, R., Magnusson, M., Sennerby, L., Nyman, S., Lundgren, D., Membrane-guided bone regeneration. Segmental radius defects studied in the rabbit. *Acta Orthop Scand* 66: 169, 1995.
20. Melcher, A. H., Irving, J. T., The healing mechanism in artificially created circumscribed defects in the femora of albino rats. *J Bone Joint Surg* 44B: 928, 1962.
21. Einhorn, T. A., Lane, J. M., Burstein, A. H., Kopman, C. R., Vigorita, V. J., The healing of segmental bone defects induced by demineralized bone matrix. *J Bone Joint Surg* 66A: 274, 1984.
22. Pelker, R. R., Mckay, J., Jr., Troiano, N., Panjabi, M. M., Friedlaender, G. E., Allograft incorporation: A biomechanical evaluation in a rat model. *J Orthop Res* 7: 585, 1989.
23. Feighan, J. E., Davy, D., Prewett, A. B., Stevenson, S., Induction of bone by a demineralized bone matrix gel: a study in a rat femoral defect model. *J Orthop Res* 13: 881, 1995.
24. Hunt, T. R., Schwappach, J. R., Anderson, H. C., Healing of a segmental defect in the rat femur with use of an extract from a cultured human osteosarcoma cell-line (Saos-2). A preliminary report. *J Bone Joint Surg* 78A: 41, 1996.
25. Stevenson, S., Li, X. Q., Davy, D. T., Klein, L., Goldberg, V. M., Critical biological determinants of incorporation of non-vascularized cortical bone grafts. Quantification of a complex process and structure. *J Bone Joint Surg* 79A: 1, 1997.

26. Narang, R., Lloyd, W., Wells, H., Grafts of decalcified allogeneic bone matrix promote the healing of fibular fracture gaps in rats. *Clin Orthop* 80: 174, 1971.
27. Chakkalakal, D. A., Mashoof, A. A., Novak, J., Strates, B. S., McGuire, M. H., Mineralization and pH relationships in healing skeletal defects grafted with demineralized bone matrix. *J Biomed Mater Res* 28: 1439, 1994.
28. Bluhm, A. E., Laskin, D. M., The effect of polytetrafluoroethylene cylinders on osteogenesis in rat fibular defects: a preliminary study. *J Oral Maxillofac Surg* 53: 163, 1995.
29. Schmitz, J. P., Hollinger, J. O., A preliminary study of the osteogenic potential of a biodegradable alloplastic-osteoinductive alloimplant. *Clin Orthop* 237: 245, 1988.
30. Damien, C. J., Parsons, J. R., Benedict, J. J., Weisman, D. S., Investigation of a hydroxyapatite and calcium sulfate composite supplemented with an osteoinductive factor. *J Biomed Mater Res* 24: 639, 1990.
31. Arnaud, E., Morieux, C., Wybier, M., de Vernejoul, M. C., Potentiation of transforming growth factor (TGF-β1) by natural coral and fibrin in a rabbit cranioplasty model. *Calcif Tissue Int* 54: 493, 1994.
32. Robinson, B. P., Hollinger, J. O., Szachowicz, E. H., Brekke, J., Calvarial bone repair with porous D,L-polylactide. *Otolaryngol Head Neck Surg* 112: 707, 1995.
33. Ashby, E. R., Rudkin, G. H., Ishida, K., Miller, T. A., Evaluation of a novel osteogenic factor, bone cell stimulating substance, in a rabbit cranial defect model. *Plast Reconstr Surg* 98: 420, 1996.
34. Tuli, S. M., Gupta, K. B., Bridging of large chronic osteoperiosteal gaps by allogeneic decalcified bone matrix implants in rabbits. *J Trauma* 21: 894, 1981.
35. Bolander, M. E., Balian, G., The use of demineralized bone matrix in the repair of segmental defects. *J Bone Joint Surg* 68A: 1264, 1986.
36. Hopp, G., Dahners, L. E., Gilbert, J. A., A study of the mechanical strength of long bone defects treated with various bone autograft substitutes; an experimental investigation in the rabbit. *J Orthop Res* 7: 579, 1989.
37. Iyoda, K., Miura, T., Nogami, H., Repair of bone defect with cultured chondrocytes bound to hydroxyapatite. *Clin Orthop* 288: 287, 1993.
38. Yang, C. Y., Simmons, D. J., Lozano, R., The healing of grafts combining freeze-dried and demineralized allogeneic bone in rabbits. *Clin Orthop* 298: 286, 1994.
39. Shimazaki, K., Mooney, V., Comparative study of porous hydroxyapatite and tricalcium phosphate as bone substitute. *J Orthop Res* 3: 301, 1985.
40. Suh, H., Lee, C., Biodegradable ceramic-collagen composite implanted in rabbit tibiae. *ASAIO J* 41: M652, 1995.
41. Uchida, A., Nade, S., McCartney, E., Ching, W., Bone ingrowth into three different porous ceramics implanted into the tibia of rats and rabbits. *J Orthop Res* 3: 65, 1985.
42. Taguchi, Y., Pereira, B. P., Kour, A. K., Pho, R. W., Lee, Y. S., Autoclaved autograft bone combined with vascularized bone and bone marrow. *Clin Orthop* 320: 220, 1995.
43. Oklund, S. A., Prolo, D. J., Gutierrez, R. V., King, S. E., Quantitative comparisons of healing in cranial fresh autografts, frozen autografts and processed autografts, and allografts in canine skull defects. *Clin Orthop* 205: 269, 1986.
44. Key, J. A., The effect of a local calcium depot on osteogenesis and healing of fractures. *J Bone Joint Surg* 16: 176, 1934.
45. Nilsson, O. S., Urist, M. R., Dawson, E. G., Schmalzried, T. P., Finerman, G. A., Bone repair induced by bone morphogenetic protein in ulnar defects in dogs. *J Bone Joint Surg* 68B: 635, 1986.
46. Delloye, C., Verhelpen, M., d'Hemricourt, J., Govaerts, B., Morphometric and physical investigations of segmental cortical bone autografts and allografts in canine ulnar defects. *Clin Orthop* 282: 273, 1990.
47. Grundel, R. E., Chapman, M. W., Yee, T., Moore, D. C., Autogeneic bone marrow and porous biphasic calcium phosphate ceramic for segmental bone defects in the canine ulna. *Clin Orthop* 266: 244, 1991.

48. Cook, S. D., Baffes, G. C., Wolfe, M. W., Sampath, T. K., Rueger, D. C., Recombinant human bone morphogenetic protein-7 induces healing in a canine long-bone segmental defect model. *Clin Orthop* 301: 302, 1994.
49. Johnson, K. D., August, A., Sciadini, M. F., Smith, C., Evaluation of ground cortical autograft as a bone graft material in a new canine bilateral segmental long bone defect model. *J Orthop Trauma* 10: 28, 1996.
50. Frayssinet, P., Primout, I., Rouquet, N., Autefage, A., Guilhem, A., Bonnevialle, P., Bone cell grafts in bioreactor: a study of feasibility of bone cell autograft in large defects. *J Mater Sci Mater Med* 2: 217, 1991.
51. de Pablos, J., Barrios, C., Alfaro, C., Canadell, J., Large experimental segmental bone defects treated by bone transportation with monolateral external distractors. *Clin Orthop* 298: 259, 1994.
52. Tiedeman, J. J., Connolly, J. F., Etrates, B. S., Lippiello, L., Treatment of nonunion by percutaneous injection of bone marrow and demineralized bone matrix. An experimental study in dogs. *Clin Orthop* 268: 294, 1991.
53. Markel, M. D., Wikenheiser, M. A., Chao, E. Y., Formation of bone in tibial defects in a canine model. Histomorphometric and biomechanical studies. *J Bone Joint Surg* 73A: 914, 1991.
54. Enneking, W. F., Burchardt, H., Puhl, J. J., Piotrowski, G., Physical and biological aspects of repair in dog cortical-bone transplants. *J Bone Joint Surg* 57A: 237, 1975.
55. Welter, J. F., Shaffer, J. W., Stevenson, S. et al., Cyclosporin A and tissue antigen matching in bone transplantation. *Acta Orthop Scand* 61: 517, 1990.
56. Holmes, R., Bone regeneration within a coralline hydroxyapatite implant. *Plast Reconstr Surg* 63: 626, 1979.
57. Gantous, A., Phillips, J. H., Catton, P., Holmberg, D., Distraction osteogenesis in the irradiated canine mandible. *Plast Reconstr Surg* 93: 164, 1994.
58. Radder, A. M., Leenders, H., van Blitterswijk, C. A., Application of porous PEO/PBT copolymers for bone replacement. *J Biomed Mater Res* 30: 341, 1996.
59. Hollinger, J., Mark, D. E., Bach, D. E., Reddi, A. H., Seyfer, A. E., Calvarial bone regeneration using osteogenin. *J Oral Maxillofac Surg* 47: 1182, 1989.
60. Drury, G. I., Yukna, R. A., Histologic evaluation of combining tetracycline and allogeneic freeze-dried bone on bone regeneration in experimental defects in baboons. *J Periodontol* 62: 652, 1991.
61. Lindholm, T. C., Lindholm, T. S., Alitalo, I., Urist, M. R., Bovine bone morphogenetic protein (bBMP) induced repair of skull trephine defects in sheep. *Clin Orthop* 227: 265, 1988.
62. Viljanen, V. V., Gao, T. J., Lindholm, T. C., Lindholm, T. S., Kommonen, B., Xenogeneic moose (*Alces alces*) bone morphogenetic protein (mBMP)-induced repair of critical-size skull defects in sheep. *Int J Oral Maxillofac Surg* 25: 217, 1996.
63. Ehrnberg, A., De Pablos, J., Martinez-Lotti, G., Kreicbergs, A., Nilsson, O., Comparison of demineralized allogeneic bone matrix grafting (the Urist procedure) and the Ilizarov procedure in large diaphyseal defects in sheep. *J Orthop Res* 113: 438, 1993.
64. Brunner, U. H., Cordey, J., Schweiberer, L., Perren, S. M., Force required for bone segment transport in the treatment of large bone defects using medullary nail fixation. *Clin Orthop* 301: 147, 1994.
65. Hallfeldt, K. K., Stutzle, H., Puhlmann, M., Kessler, S., Schweiberer, L., Sterilization of partially demineralized bone matrix: the effects of different sterilization techniques on osteogenetic properties. *J Surg Res* 59: 614, 1995.
66. Gao, T. J., Lindholm, T. S., Kommonen, B., Ragni, P., Paronzini, A., Lindholm, T. C., Microscopic evaluation of bone-implant contact between hydroxyapatite, bioactive glass and tricalcium phosphate implanted in sheep diaphyseal defects. *Biomaterials* 16: 1175, 1995.
67. Bos, G. D., Goldberg, V. M., Powell, A. E., Heiple, K. G., Zika, J. M., The effect of histocompatibility matching on canine frozen bone allografts. *J Bone Joint Surg* 65A: 89, 1983.
68. Lane, J. M., Sandhu, H. S., Current approaches to experimental bone grafting. *Orthop Clin North Am* 18: 213, 1987.
69. Johnson, K. D., Frierson, K. E., Keller, T. S. et al., Porous ceramics as bone graft substitutes in

long bone defects: a biomechanical, histological, and radiographic analysis. *J Orthop Res* 14: 351, 1996.
70. Burstein, A. H., Frankel, V. H., A standard test for laboratory animal bone. *J Biomech* 4: 155, 1971.
71. Black, R. J., Zardiackas, L. D., Teasdall, R., Hughes, J. L., Jr., The mechanical integrity of healed diaphyseal bone defects grafted with calcium hydroxyapatite/calcium triphosphate ceramic in a new animal model. *Clin Mater* 6: 251, 1990.
72. Toriumi, D. M., Kotler, H. S., Luxenberg, D. P., Holtrop, M. E., Wang, E. A., Mandibular reconstruction with a recombinant bone-inducing factor. *Arch Otolaryngol Head Neck Surg* 117: 1101, 1991.
73. Paley, D., Young, M. C., Wiley, A. M., Fornasier, V. L., Jackson, R. W., Percutaneous bone marrow grafting of fractures and bony defects. An experimental study in rabbits. *Clin Orthop* 208: 300, 1986.
74. Heiple, K. G., Goldberg, V. M., Powell, A. E., Bos, G. D., Zika, J. M., Biology of cancellous bone grafts. *Orthop Clin North Am* 18: 179, 1987.
75. Mohr, H., Kragstrup, J., Morphostereometry of heterotopic ossicles in the rat. *Acta Orthop Scand* 62: 257, 1991.
76. Ishaug-Riley, S. L., Crane, G. M., Gurlek, A., Miller, M. J., Yasko, A. W. et al., Ectopic bone formation by marrow stromal osteoblast transplantation using poly(DL-lactic-co-glycolic acid) foams implanted into the rat mesentery. *J Biomed Mater Res* 36: 1, 1997.
77. DeVries, W. J., Runyon, C. L., Martinez, S. A., Ireland, W. P., Effect of volume variations on osteogenic capabilities of autogenous cancellous bone graft in dogs. *Am J Vet Res* 57: 1501, 1996.
78. Mowlem, R., Cancellous chip bone-grafts. Report on 75 cases. *Lancet* 1: 746–748, 1944.
79. Younger, E. M., Chapman, M. W., Morbidity at bone graft donor sites. *J Orthop Trauma* 3: 192, 1989.
80. Mankin, H. J., Gebhardt, M. C., Tomford, W. W., The use of frozen cadaveric allografts in the management of patients with bone tumors of the extremities. *Orthop Clin North Am* 18: 275, 1987.
81. Sciadini, M. F., Dawson, J. M., Johnson, K. D., Bovine-derived bone protein as a bone graft substitute in a canine segmental defect model. *J Orthop Trauma* 11: 496, 1997.
82. Harakas, N. K., Demineralized bone matrix-induced osteogenesis. *Clin Orthop* 188: 239, 1984.
83. Kleinschmidt, J. C., Marden, L. J., Kent, D., Quigley, N., Hollinger, J. O., A multiphase system bone implant for regenerating the calvaria. *Plast Reconstr Surg* 91: 581, 1993.
84. Nade, S., Burwell, R. G., Decalcified bone as a substrate for osteogenesis. An appraisal of the interrelation of bone and marrow in combined grafts. *J Bone Joint Surg* 59B: 189, 1977.
85. Strates, B. S., Connolly, J. F., Osteogenesis in cranial defects and diffusion chambers. Comparison in rabbits of bone matrix, marrow, and collagen implants. *Acta Orthop Scand* 60: 200, 1989.
86. Glowacki, J., Mulliken, J. B., Demineralized bone implants. *Clin Plast Surg* 12: 233, 1985.
87. Aspenberg, P., Wittbjer, J., Thorngren, K.-G., Bone matrix and marrow versus cancellous bone in rabbit radius defects. *Arch Orthop Trauma Surg* 106: 335, 1987.
88. Lane, J. M., Yasko, A. W., Tomin, E., Cole, B. J., Waller, S., Browne, M., Turek, T., Gross, J., Bone marrow and recombinant human bone morphogenetic protein-2 in osseous repair. *Clin Orthop* 361: 216, 1999.
89. Ohgushi, H., Goldberg, V. M., Caplan, A. I., Heterotopic osteogenesis in porous ceramic induced by bone marrow cells. *J Orthop Res* 7: 568, 1989.
90. Takagi, K., Urist, M. R., The role of bone marrow induced repair of femoral massive diaphyseal defects. *Clin Orthop* 171: 224–31, 1982.
91. Burwell, R. C., Studies in the transplantation of bone. VII. The fresh composite homograft-autograft of cancellous bone: An analysis of factors leading to osteogenesis in marrow transplants and in marrow-containing bone grafts. *J Bone Joint Surg* 46B: 110, 1964.
92. Salama, R., Weissman, S. L., The clinical use of combined xenografts of bone and autologous red marrow. *J Bone Joint Surg* 60B: 111, 1985.
93. Healey, J. H., Zimmerman, P. A., McDonnell, J. M. et al. Percutaneous bone marrow grafting of delayed union and nonunion in cancer patients. *Clin Orthop* 256: 280, 1990.

94. Bay, B. K., Martin, R. B., Sharkey, N. A., Chapman, M. W., Repair of large cortical defects with block coralline hydroxyapatite. *Bone* 14: 225, 1993.
95. Levine, J. P., Bradley, J., Turk, A. E., Ricci, J. L., Benedict, J. J., Steiner, G., Longaker, M. T., McCarthy, J. G., Bone morphogenetic protein promotes vascularization and osteoinduction in preformed hydroxyapatite in the rabbit. *Ann Plast Surg* 39: 158, 1997.
96. Holmes, R. E., Bucholz, R. W., Mooney, V., Porous hydroxyapatite as a bone graft substitute in diaphyseal defects: A histometric study. *J Orthop Res* 5: 114, 1987.
97. St John, K. R., Zardiackas, L. D., Terry, R. C., Teasdall, R. D., Cooke, S. E., Mitias, H. M., Histological and electron microscopic analysis of tissue response to synthetic composite bone graft in the canine. *J Appl Biomater* 6: 89, 1995.
98. Bucholz, R. W., Carlton, A., Holmes, R. E., Hydroxyapatite and tricalcium phosphate bone graft substitute. *Orthop Clin North Am* 18: 323, 1987.
99. Daculsi, G., Passuti, N., Martin, S., Deudon, C., Legeros, R. Z., Raher, S., Macroporous calcium phosphate ceramic for long bone surgery in humans and dogs. Clinical and histological study. *J Biomed Mater Res* 24: 379, 1990.
100. Moore, D. C., Chapman, M. W., Manske, D., The evaluation of a biphasic calcium phosphate ceramic for use in grafting long-bone diaphyseal defects. *J Orthop Res* 5: 356, 1987.
101. Hosokawa, R., Kubo, T., Wadamoto, M., Sato, Y., Kimoto, T., Direct bone induction in the subperiosteal space of rat calvaria with demineralized bone allografts. *J. Oral Implantol* 25: 30, 1999.
102. Goshima, J., Goldberg, V. M., Caplan, A. I., The origin of bone formed in composite grafts of porous calcium phosphate ceramic loaded with marrow cells. *Clin Orthop* 269: 274, 1991.
103. Lind, M., Growth factors: possible new clinical tools. A review. *Acta Orthop Scand* 67: 407, 1996.
104. Linkhart, T. A., Mohan, S., Baylink, D. J., Growth factors for bone growth and repair: IGF, TGF beta and BMP. *Bone* 19: 1S, 1996.
105. Wozney, J. M., Rosen, V., Celeste, A. J., Mitsock, L. M., Whitters, M. J., Kriz, R. W., Hewick, Wang, E. A., Novel regulators of bone formation: molecular clones and activities. *Science* 242: 1528–1534, 1988.
106. Urist, M. R., Bone: formation by autoinduction. *Science* 150: 893, 1965.
107. Urist, M. R., Delange, R. J., Finnerman, G. A. M., Bone cell differentiation and growth factors. *Science* 220: 680, 1983.
108. Lieberman, J. R., Daluiski, A., Stevenson, S., Wu, L., McAllister, P., Lee, Y. P., Kabo, J. M., Finerman, G. A., Berk, A. J., Witte, O. N., The effect of regional gene therapy with bone morphogenetic protein-2 producing bone-marrow cells on the repair of segmental femoral defects in rats. *J Bone Joint Surg* 81A: 905, 1999.
109. Mark, D. E., Hollinger, J. O., Hastings, C., Chen, G., Marden, J. M., Reddi, A. H., Repair of calvarial nonunions by osteogenin, a bone-inductive protein. *Plast Reconstr Surg* 86: 623, 1990.
110. Laffargue, P., Hildebrand, H. F., Rtaimate, M., Frayssinet, P., Amourex, J. P., Marchandise, X., Evaluation of human recombinant bone morphogenetic protein-2-loaded tricalcium phosphate implants in rabbits' bone defects. *Bone* 25 (Suppl 2): 55S, 1999.
111. Sato, K., Urist, M. R., Induced regeneration of calvaria by bone morphogenetic protein (BMP) in dogs. *Clin Orthop* 197: 301–311, 1985.
112. Kawai, T., Mieki, A., Ohno, Y., Umemura, M., Kataoka, H., Kurita, S., Koie, M., Jinde, T., Hasegawa, J., Urist, M. R., Osteoinductive activity of composites of bone morphogenetic protein and pure titanium. *Clin Orthop* 290: 296–305, 1993.
113. Sailer, H. F., Kolb, E., Application of purified bone morphogenetic protein (BMP) in cranio-maxillo-facial surgery. BMP in compressed surgical reconstructions using titanium implants. *J Crainomaxillofac Surg* 22: 2–11, 1994.
114. Sailer, H. F., Kolb, E., Application of purified bone morphogenetic protein (BMP) preparations in cranio-maxillo-facial surgery. Reconstruction in craniofacial malformations and post-traumatic or operative defects of the skull with lyophilized cartilage and BMP. *J Craniomaxillofac Surg* 22: 191–99, 1994.
115. Alpaslan, C., Irie, K., Takahashi, K., Ohashi, N., Sakai, T., Nakajima, T., Ozawa, H., Long-term

evaluation of recombinant human bone morphogenetic protein-2 induced bone formation with a biologic and synthetic delivery system. *Br J Oral Max Surg* 34: 414, 1996.
116. Green, W. T., Jr., Articular cartilage repair. Behavior of rabbit chondrocytes duringtissue culture and subsequent allografting. *Clin Orthop* 124: 237, 1977.
117. Herring, M., Gardner, A., Glover, J., A single-staged technique for seeding vascular grafts with autogenous endothelium. *Surgery* 84: 498, 1978.
118. Vacanti, C. A., Langer, R., Schloo, B., Vacanti, J. P., Synthetic polymers seeded with chondrocytes provide a template for new cartilage formation. *Plast Reconstr Surg* 88: 753, 1991.
119. Elgendy, H. M., Norman, M. E., Keaton, A. R., Laurencin, C. T., Osteoblast-like cell (MC3T3-E1) proliferation on bioerodable polymers: an approach towards the development of a bone-bioerodable polymer composite material. *Biomaterials* 14: 263, 1993.
120. Vacanti, C. A., Kim, W., Mooney, D., Schloo, B., Vacanti, J. P., Tissue engineered growth of organized bone containing marrow, utilizing synthetic polymer templates. *Trans Orthop Res Soc* 18: 479, 1993.
121. Frenkel, S. R., Toolan, B., Menche, D., Pitman, M. I., Pachence, J. M., Chondrocyte transplantation using a collagen bilayer matrix for cartilage repair. *J Bone Joint Surg* 79B: 831, 1997.
122. Kim, W., Vacanti, J. P., Upton, J., Vacanti, M. P., Vacanti, C. A., The repair of cranial defects in rats with cartilage grown from chondrocytes on synthetic polymer templates. *Trans Orthop Res Soc* 18: 487, 1993.
123. Langer, R., Vacanti, J. P., Tissue engineering. *Science* 260: 920, 1993.
124. Wang, E. A., Rosen, V. et al., Recombinant human bone morphogenetic protein induces bone formation. *Proc Natl Acad USA* 87: 2220, 1990.

32

Stabilization of Collagen in Medical Devices

John A. M. Ramshaw, Veronica Glattauer, and Jerome A. Werkmeister
CSIRO Molecular Science, Parkville, Victoria, Australia

I. INTRODUCTION

The collagen family is one of the most abundant groups of proteins as they are the principal protein component of all connective tissues, including skin, tendon, ligament and cartilage [1]. Because of their vital role in tissue, collagens have long been seen as having the potential to provide the basis of biomaterials, either as native, unmodified tissue grafts or as manufactured products for use in various medical areas [2]. When used in manufactured medical devices, stabilization of the collagen by either a chemical or physical method is a prerequisite for effective performance of the device in most cases [2].

A broad range of potential manufactured products based on collagen has been suggested, covering many medical disciplines (Table 1) [3], and a range of different product types is now available commercially. These products have generally proved very successful and include injectable collagen for soft tissue augmentation [4], hemostatic powder or fleece [5] and replacement components for the cardiovascular system, such as bioprosthetic heart valves [6] and vascular replacements [7,8]. Tissue-based heart valves are a good example of a collagen-based biomaterial product that has provided enhanced quality of life to thousands of people. The extent of stabilization is normally greater for devices that are intended for long implant times, often ideally whole of life, where the durability of the device is critical. In devices that are intended for short-term application, for example as scaffolds to assist in wound repair, the extent of stabilization required is generally low, and is intended only to control the resorption rate of the device. Stabilization also provides further benefits, including minimizing the immunogenicity of the tissue or collagen [9].

Although collagen-based biomaterials are normally discussed as a single group, the different products form a wide range of materials with distinct formats and characteristics [2,3]. Most notably, there are two distinct groups of collagen-based biomaterials. These are the tissue-based biomaterials, where intact tissue is adapted so that the product meets the medical needs, and purified collagen, either as a comminuted product or as a soluble product, which may then be re-formed into a particular shape or form. The advantages of the collagen that are being exploited vary between these different products (Table 2).

Various strategies are possible for the stabilization of tissue or purified collagen. Developments in this field were initially driven by the development of bioprosthetic heart valves [10], and this demanding application has continued to be a principle driving force for further improvements. Currently, glutaraldehyde treatment is the predominant method used for stabili-

Table 1 Examples of Applications of Collagen-Based Medical Devices

Medical area	Application
Cardiovascular surgery	Vessel replacement, heart valves
Dentistry	Periodontal attachment, alveolar ridge augmentation
Dermatology	Tissue augmentation
General surgery	Hernia repair, adhesion barriers, tissue adhesives
Neurosurgery	Nerve conduits, nerve repair
Ophthalmology	Corneal graft, vitreous replacement, retinal reattachment
Orthopedics	Bone repair, cartilage reconstruction, ligament reconstruction
Otology	Tympanic membrane replacement
Urology	Ureter replacement, renal repair, urinary incontinence
Wound management	Dressings

zation [11,12]. However, it is well recognized that there are potential limitations with this reagent, most notably in its propensity to lead to unwanted calcification of devices in certain applications such as in bioprosthetic heart valves [13], and in its potential to cause cytotoxic reactions [14]. These problems have led to many variations in the use of glutaraldehyde, mainly through use of auxiliary treatments to the stabilized materials [12]. Also, many other nonaldehyde methods have been examined, some of which show promise, for example dye-mediated photooxidation [15,16], although none have yet replaced glutaraldehyde as the predominant reagent used in the industry. Some have proved ineffective for devices intended for long-term implant, for example dialdehyde starch [17], but these approaches should not be discounted for short-term, resorbable devices where a lower potential for cytotoxicity may be of benefit. For example, dialdehyde starch has recently been used in the development of a collagen-based adhesive [18].

II. CHEMISTRY OF COLLAGEN

The name collagen is frequently used as a generic term for a large family of proteins that are key structural components of the extracellular matrix. The characteristic feature of all members of the collagen family is the presence of the specific structural motif, the triple helix. The

Table 2 Examples of Advantages and Applications of Collagen in Devices of Different Formats

Collagen format	
Tissue-based	Purified
Advantages	
Low immunogenicity	Low immunogenicity
Durability and persistence	Controlled turnover
Cell-matrix interactions	Cell-matrix interactions
Strength	Fibril reformation
Examples of applications	
Heart valves	Injectable collagen
Blood vessels	Hemostats
Ligaments and tendons	Wound and burn dressings

collagen family now comprises at least 19 distinct types and includes the 5 homologous interstitial collagens (types I, II, III, V and XI) that are found in the fibril network of most tissues and which are characterized by a 67 nm axial period [19]. Type I is the major collagen in skin, bone and tendon, while type II is found in cartilage, and type III is found, together with type I, in skin, blood vessels and other extensible tissues. Thus, type I collagen is the predominant collagen type found in all biomaterials applications. In addition, there are at least 14 nonfibrillar collagens, including basement membrane collagens and "fibril-associated collagens with interrupted triple helices" (FACIT) collagens, which have triple helical domains of varying lengths and are found in a variety of supramolecular forms in tissues [20,21].

The triple helix comprises three polypeptide chains, each in an extended polyproline II–like helix, which are supercoiled about each other to form a rod-like structure [22,23]. The close packing of the three chains requires that every third residue is Gly as this amino acid is small enough to fit where the three chains pack in close proximity. This steric constraint generates the repeating sequence pattern (Gly-X-Y)$_n$. The triple helix is stabilized by the close packing and interchain hydrogen bonding of the three chains, by an extensive, highly ordered hydration network, confirmed recently by crystal structure analysis [24,25], and by a high imino acid content. Thus the imino acid proline (Pro) is frequently found in the X position while hydroxyproline (Hyp), formed by posttranslational hydroxylation of proline during collagen synthesis, is frequently found in the Y position [26].

The three residues in the Gly-X-Y repeating unit occupy distinct positions within the supercoiled helix. The Gly residues are buried in the center and are solvent inaccessible. All the residues in the X and Y positions of collagens lie on the surface of the molecule and are solvent accessible. However, the residues in the X position are more exposed than those in the Y position [27]. The residues in the X and Y positions provide the functional groups that are available for chemical stabilization strategies.

In tissues, the highly abundant interstitial types I, II and III collagens are found in fibre bundles built up from collagen fibrils [1]. During biosynthesis, these collagens form initially as more complex procollagen molecules with additional globular domains at both amino and carboxy terminals of the triple helical domain. As part of their secretion from the cell and incorporation into functional fibrils these globular domains are removed by specific enzymatic cleavage. This process leaves short non-triple-helical segments at each end of the triple helix, the telopeptides, that are critical in the subsequent formation of covalent cross-links that provide the mechanical and structural stability to the fibrils [1].

The natural cross-links involving the telopeptides are retained in tissue-based devices. The telopeptide regions are generally regarded as the most immunogenic domains of the collagen found in tissue. This was initially demonstrated by the significant reduction of antibody responses in rabbits injected with calf collagens that had been treated with pepsin [28], and later confirmed by various other studies [29–31]. Stabilization of tissue collagen that takes place during manufacture has the potential to mask the collagen and minimize its immunogenicity. This also applies to other tissue proteins that would be present, particularly if washing steps prior to stabilization are limited. For example, comparisons between stabilized and native bovine collagens injected into rabbits showed that only the unmodified collagen had high levels of immunogenicity [32]. However, while glutaraldehyde treatment is often considered to render the tissue implant nonimmunogenic, this cannot always be assumed [33]. There is evidence of an apparent antigen depressant effect of glutaraldehyde [34,35], although other reports note that an immunogenic response is still found [36–39]; these studies have been reviewed by Bajpai [33]. A detailed discussion of the immunogenicity of purified and tissue collagens, and the effects of modification treatments, has been presented elsewhere [9].

In the production of soluble collagen the telopeptides are cleaved from the helical domain, usually using proteolysis by pepsin, breaking the natural cross-link network and allowing isolation of individual molecules in a soluble form [3]. Thus, soluble collagens lack the natural strength of native tissues and need stabilization if strength is a requirement of the final product. An advantage of removal of the telopeptides is that the immunogenicity is lowered in the soluble, purified collagen-based devices.

In fibrils the individual collagen molecules are packed in a staggered overlap arrangement where individual molecules are related by a 67 nm stagger, called the D-period [1]. As each molecule is about 4.4 D long, each fibril has zones where there are gaps between the ends of molecules. These hole zones lead to the characteristic banding pattern of collagen fibrils that is observed in transmission electron microscopy due to the differential uptake of stain into these hole zones compared to other parts of the fibril where full overlap of collagen molecules occurs. This differential accessibility to chemicals may also impact on the accessibility of stabilizing agents to these different regions of the collagen fibril. In tissue-based devices the collagen is also in close association with a wide range of other macromolecular components, including structural proteins such as elastin and a range of proteoglycan components. The nature and extent of these associations and interactions are dependent on the specific architecture of individual tissues, and could also lead to differences in the penetration and reaction of stabilizing reagents with various tissues.

III. STRATEGIES FOR CROSS-LINKING

The strategies that are used for cross-linking are based on either chemical methods or physical methods, and generally utilize specific amino acid side chains in the X and Y positions of the collagen helix. Most chemical approaches target cross-linking of Lys residues using bifunctional reagents or formation of interchain amide bonds utilizing Lys and either an Asp or Glu residue from the adjacent chain. Because these amide bond cross-links do not introduce any additional chemical entity they are sometimes referred to as "zero length" cross-links. A physical method, dry heat [40] also seeks to form these zero-length, amide bond cross-links. Another physical method, photooxidation, believed to involve His residues, also possibly forms zero-length cross-links. Zero-length cross-link approaches are seen as potentially advantageous as the methods do not introduce any additional compounds that could lead to cytotoxicity. On the other hand, the degradation products of the new cross-linked amino acids could possibly have longer-term detrimental effects [12], although there is no evidence that this is the case.

The frequency and distribution of amino acids between X and Y positions and their availability for cross-link formation is characteristic of a particular amino acid [26]. The interstitial collagens contain about 3%–4% of each of the amino containing Lys and carboxyl containing Asp and Glu [26]. Lys is found more frequently in the Y position [26], which could potentially limit its accessibility to reagents, depending on the residue at the adjacent X position. Asp residues are found equally in the X and Y positions, whereas Glu is found predominantly in the X position. Studies on triple helical peptide models have indicated that charged side chains are generally not involved in intramolecular hydrogen bonding within the triple helix, with the exception of Asp in the Y position, but are involved in hydrogen bonding with the extensive hydration network that surrounds the triple helix [41]. In collagens that have been given treatment at high pH, deamidation of Asn and Gln residues to yield Asp and Glu would increase the availability of these sites, particularly of Glu in the Y position.

If one examines other potential functional groups, opportunities for stabilization using sulfydryl group are essentially absent as Cys residues are extremely rare or absent in all collagens. Histidine, which is believed to play an important role in cross-links introduced by photooxidation, occurs at low frequency in all collagens. On the other hand, although Arg is frequently found in collagens, particularly in the Y position where it contributes to triple helical stability [42], its guanidino functional group has yet to be exploited in collagen cross-linking strategies.

Overall, the key outcomes that are sought from cross-linking include strength, durability, control of turnover and minimization of immunogenicity. However, in achieving these aims it is important that appropriate physical properties for the material are retained. Since the addition of cross-links can affect physical properties, for example by increasing the stiffness, the extent of cross-linking must necessarily reflect a balance between desirable and undesirable effects.

IV. EFFECTIVENESS OF CROSS-LINKING

Various methods have been used to establish the extent and effectiveness of cross-linking, based on thermal or chemical stability. For some stabilization strategies, for example glutaraldehyde treatment, the bonds formed are stable to acid hydrolysis [43], whereas for others, for example amide bonds formed by carbodiimide coupling, the bonds are readily hydrolyzed. In other cases, for example photooxidation stabilization, the role of particular amino acids such as histidine has been inferred rather than directly demonstrated [15,16]. When stable bonds are present, amino acid analysis allows identification of the loss of functional amino acids. However, this approach is limited since it indicates only the extent of modification of amino acids, but not how many are then involved in cross-link formation. On the other hand, if the nature of the cross-link is known, then it may be possible to identify it on amino acid analysis, as has been shown for naturally occurring mature cross-links [44]. Alternatively, an analytical system designed specifically for cross-link identification could be appropriate [45].

Both biochemical and physical methods can be used to examine the extent of stabilization. The most common biochemical approach is to examine its susceptibility to enzyme attack, for example by pepsin to give soluble collagen, or by bacterial collagenase, which degrades the collagen to soluble peptide material. However, interpreting the significance of these data can present difficulties. The major physical approach used for characterization is thermal denaturation, generally using either an isometric tension method [46] or differential scanning calorimetry. These approaches allow, for example, the denaturation temperature, otherwise called the shrinkage temperature (T_s), of the modified material to be determined. Addition of cross-links increases the T_s of the material. Further, mechanical testing of materials that have been fully heat denatured can be used to estimate the average molecular weight of the collagen chain between cross-links [47] but these data should be treated with caution. When examining T_s data, consideration should be given to the chemical nature of the cross-links present in the sample, particularly if they may be thermally labile, as this may make comparisons between the extent of cross-linking in different materials difficult to interpret. In all cases, the key indicator of success is the biological performance of the material.

A. Chemical Approaches

1. *Aldehydes*

a. Glutaraldehyde. The use of glutaraldehyde, more than any other reagent, has been the cross-linking strategy of choice for collagen-based biomaterials, particularly for heart valve bioprostheses, and as such has been the subject of many reviews [11,12,48]. However, as discussed below, this reagent is not without problems and research on alternative cross-linking approaches is continually in progress.

The use of glutaraldehyde for tanning collagen started in the 1950s when the reagent first became readily available in commercial quantities [49], and was successfully adapted to bioprosthesis manufacture about a decade later [10]. The mechanism of stabilization by glutaraldehyde, although probably based initially on aldimine formation between the reagent and lysine side chain amino groups, is considerably more complex as a result of the complexity of components which make up the reagent [48,50]. Glutaraldehyde exists in solution as a mixture of chemical species of which only a small proportion is the dialdehyde itself. Various workers [51,52] established that the composition of glutaraldehyde solutions at 25°C was approximately 8% linear monohydrate $CHO—CH_2—CH_2—CH_2—CH(OH)_2$, 54% of the linear dihydrate, 34% of a hydrated cyclic hemiacetal (tetrahydropyran) and only about 4% of the dialdehyde. All four components interchange readily and therefore 100% analysis for aldehyde is obtained. The proportions of the various components do not vary greatly with concentration or pH, although high pH induces rapid conversion of the glutaraldehyde into addition products. However, the concentration of free aldehyde increases rapidly with temperature [52,53]. Glutaraldehyde also forms a polymer based upon the cyclic hemiacetal [51].

The most common application for glutaraldehyde fixation is in the manufacture of porcine heart valve bioprostheses [12]. The clinical performance of these devices, while providing significant benefits to patients, has continued to bring to the fore certain deficiencies in this reagent, particularly in relation to the extent of tissue penetration, cytotoxicity, mechanical properties and most importantly the propensity of glutaraldehyde fixed valves to show calcification. These issues have led to a considerable body of research aimed at minimizing or eliminating them.

The rapid rate of reaction at neutral pH and the tendency to form polymers have suggested that tissue penetration could be a problem and that inner tissues would not be adequately fixed [54]. The polymer may become trapped in the structure or bound through single attachment points to the collagen and may act to hinder penetration into the full thickness of the material [48,50]. Therefore, it has been suggested that low reagent concentrations are best suited for bulk tissue fixation [55].

The main problem, however, that can potentially arise from polymer formation appears to be cytotoxicity of the stabilized tissue. The amount of polymer contained in the stabilized material may be reduced by washing procedures, and the immediate level of free aldehyde eliminated by a glycine wash [56]. However, after the tissue is implanted the residual polymer degrades over an extended period of time releasing monomers or small polymers that are cytotoxic [57]. This is also the case for stabilized collagen sponges where leaching of glutaraldehyde is sufficient to produce potentially adverse cellular effects [14,56,58]. This may have short-term benefits as a result of bactericidal effects, but in the longer term the lack of cellular infiltration appears potentially detrimental to the performance of the product. Nevertheless, it appears that the rate of release is sufficiently slow and the clearance rapid so that the leached chemical may be tolerable [12]. More important than the long-term release may be the high initial release from devices that are shipped in a glutaraldehyde sterilizing solution.

A wide range of studies have been aimed at optimizing the mechanical properties of the stabilized tissue, and obtaining the best match to natural tissue. A particularly challenging issue for heart valve bioprostheses, compared to other glutaraldehyde stabilized materials, is achieving durability despite the constant flexing of the tissue. Examination of failed devices shows that the damage occurs at locations of high stress [59]. Thus for heart valves, stabilization of the tissue under pressure has been used, although it has now been shown that a low-pressure approach best retains the mechanical characteristics of the original heart valve leaflet tissue [60]. Alternatively, it has been suggested that dynamic fixation may provide materials that are better suited to resist long-term mechanical fatigue and degradation [61]. For pericardial tissue, and potentially other tissues, it has been shown that tethering during fixation can control the mechanical behavior of xenografts and produce a material with extensibility nearly identical to that of fresh tissue [62].

Despite these mechanical issues, the predominant failure mode for heart valve bioprostheses is calcification [12], and this has been attributed specifically to the glutaraldehyde used for stabilization [13]. The mechanism of this process is not understood, although as initiation may be cell-based, the cytotoxic effects of the depolymerized reagent may be involved. A variety of suggestions have been made for additives to the processing steps, or pretreatments of tissue, whereas other approaches have examined the role of noncollagenous components. For example, the negative charges of phospholipids have been proposed as participating in the calcification, and treatment to remove these lipids has indicated some beneficial effects [63]. Beneficial effects from treatment with sodium dodecyl sulfate (SDS) and other surfactants have been suggested as useful in lowering calcification, with effects varying between reagents [64,65]; it has been suggested that the benefits result from changes to lipid composition [65]. In other cases, the loss of proteoglycans has been suggested as providing vacant sites for calcification [63]. Other additives, supported from animal data, include trivalent metal ions [66], coupling of sulfonated polyethyleneoxide to the tissue [67] and inclusion of aminooleic acid [68]. A class of compounds that has attracted attention has been diphosphonates, these being synthetic analogs of natural inhibitors of calcification [69]. In some cases devices have been modified to allow continual slow release of these compounds. Most recently, the use of organic solvents has shown some promise. Thus, treatment with high concentrations of dimethyl sulfoxide (DMSO) has been shown to provide benefits by lowering calcification, while apparently also enhancing the extent of cross-linking [70]. Pretreatment with aqueous ethanol has also proved of benefit, probably through multicomponent effects involving tissue material and modification of macromolecular characteristics [71].

Other approaches have involved changes to the method of application of the reagent. For porcine valves, various fixation strategies have been used, although prolonged time at moderate (0.5%) glutaraldehyde concentration has been frequently employed [12]. It has been suggested, however, that high glutaraldehyde concentrations reduce rather than increase the calcification that occurs in certain tissues [72]. Another approach, particularly aimed at reducing the amount of polymeric forms of glutaraldehyde incorporated into the material, is to react at lower pH [14]. At lower pH the amount of polymeric material formed is low. However, particularly low pH can lead to swelling, although this can be controlled by addition of NaCl. This approach leads to lower calcification, as well as minimizing the likelihood of cytotoxic effects [73]. Alternatively, changing the solution conditions by inclusion of organic solvents leads to significant improvements to the properties of the stabilized collagen. Thus, for example, cross-linking using a glutaraldehyde/*n*-butanol solution produced fully stabilized pericardial tissue with tensile mechanical behavior that was very close to that of fresh tissue, although of interest, the flexural properties of this treated tissue were different from those of fresh tissue [74].

b. Other Bisaldehydes. A number of other aliphatic dialdehydes, and also dialdehyde starch, have been investigated for their ability to cross-link collagen. It is apparent that the cross-linking resulting from these reagents is different from that of glutaraldehyde. The differences arise from solution behavior where glutaraldehyde, compared with lower and higher homologues, shows a greater tendency to form hydrates and polymers [75]. The other aldehydes mostly react with protein much more slowly than glutaraldehyde [43,76], and generally with much reduced efficiencies and with lower bond stabilities [43,77], so that they would be unsuitable for products requiring long-term durability. Dialdehyde starch, for example, had been proposed for use in vascular prostheses, but did not provide materials with sufficient durability in the long term [17].

Further study, nevertheless, may lead to certain of these alternative dialdehydes being useful in replacing glutaraldehyde in some specific product areas. In some cases, however, the reagents are not as readily available and may need to be made immediately prior to application from precursors. Thus, dialdehyde starch has been proposed recently as a component for making a collagen/gelatine based surgical adhesive [18]. Also, there has been renewed interest in the biochemistry of malonaldehyde [78]. This dialdehyde is formed in vivo during oxidation of certain polyunsaturated lipids and reacts with a wide range of proteins [79] and has been shown to be effective in cross-linking collagen [80]. It has disadvantages in low stability and low solubility in water [77]. Malonaldehyde prepared by hydrolysis of tetraethoxypropane, was used to show that significant increases in shrinkage temperature (>20°C) can be achieved within 3 h. These data also suggest that optimal stabilization may take place at around pH 4, much lower than for glutaraldehyde [80]. Model studies have suggested that malonaldehyde reacts with lysine to form a dihydropyridine derivative rather than the unstable imidopropene Schiff base [78].

c. Formaldehyde. Although it contains only a single functional aldehyde group, formaldehyde has been known for a long time to be an effective cross-linking agent for proteins, and the chemistry and reactivity of the reagent have been long established [81]. Thus, although lysine side chain amino groups are key targets for the initial reactivity of this reagent the complex cross-linking that occurs also includes the side chains of a wide range of other amino acids [82].

Initially for the stabilization of heart valve tissue bioprostheses, formaldehyde was seen as having good characteristics, and initial in vivo performance data seemed promising [83]. However, this optimism was not matched by the performance over more extended periods and devices were failing after only a year or two [84]. Concerns were also raised over the mechanical properties of the fixed materials, where stretching and deformations were observed, and of the cytotoxicity of the reagents and the leachables during implant. Although the use of this reagent for tissue fixation for bioprostheses has been discontinued, it is still used in sterilizing solutions, although efficient washing is needed to minimize cytotoxic effects [12]. On the other hand, the reagent may yet be valuable for fixation of devices intended for short-term applications, such as wound dressings, where lack of long-term durability is not a problem [85].

d. Sugars. An important natural cross-linking system is that caused by the nonenzymatic modification of proteins with reducing sugars, the Maillard reaction. This process is seen as playing an important role in protein aging and particularly in the complications of diabetes.

The complete mechanism by which cross-links are formed and the nature of the final cross-links is still under discussion. Initially, however, a reducing sugar (glucose) forms a reversible Schiff base with a lysine side chain. This relatively unstable Schiff base leads to a more stable Amadori product through a rearrangement process. This product, over extended periods of time, undergoes further reactions, probably through dicarbonyl intermediates, to form "advanced glycation end products" (AGEs), also called advanced Maillard products [86]. The AGEs comprise a heterogeneous mixture of chemical structures that, in some cases, form cross-links between adjacent amino acids. Clear identification of AGEs cross-links is so far limited, although pentosidine is an example of a compound that has been identified [87], in this case apparently forming a lysine to arginine cross-link. Thus in collagens, which have a very long half-life and thus turn over very slowly in the body, the AGE compounds accumulate and affect the physical and biochemical properties of extracellular matrix. For example, the increase of cross-linking with age can lead to arterial wall stiffening and decline in vessel function [88], as well as a wide variety of other tissue changes; these changes are accelerated in diabetes.

As well as glucose, other reducing sugars, for example ribose [89], or metabolites such as methyl glyoxal [86] can also lead to AGEs. These alternatives can show much greater rates of reaction compared with glucose. Thus, ribose has been proposed as a reagent for cross-linking collagen in medical devices [90]. For example, collagen sponges treated with 0.2 M ribose showed significantly increased stiffness after a few days of treatment [90], and were not immunogenic or toxic. Ribose cross-linked collagen shows a large increase in tensile stress, with electron microscopy showing increased fibril packing density [91]. The ribose cross-linked collagen is highly insoluble [91,92] and resistant to pepsin solubilization at 4°C, although extensively cross-linked material is susceptible to pepsin at 30°C [92].

A range of modifications may be possible to enhance further the rate of cross-linking and therefore improve the utility of this approach. These include use of more reactive sugars, addition of free lysine [93] and inclusion of hydrogen peroxide [94].

2. *Other Approaches Introducing Cross-Link Chemicals*

a. Epoxides. A wide range of poly-glycidyl ether (polyepoxy) compounds (Denacols) have been described that may prove useful cross-linking agents for collagens [95,96]. The cross-linked collagen remains pliable, hydrophilic and does not discolor. These polyepoxy compounds are available for cross-linking with two, three or more reactive epoxy groups on a variety of frameworks, such as sorbitol and glycerol [97]. The reactions leading to cross-links primarily involve (about 80%) amine groups of lysine residues, although reactions with other amino acids including histidine also occur [98].

The compounds are water soluble and do not readily self-polymerize. The water solubility means that excess reagent can be readily removed after reaction by washing. The rate of reaction is slower than for glutaraldehyde. The rate of fixation is particularly slow compared with the reagent diffusion rate [99], even though the larger size of these compounds compared to many other reagents would potentially slow the diffusion rate. This leads to a uniform distribution of cross-linking throughout the depth of the samples [100]. The extent of reaction was favored, as expected, by increasing temperature and concentration of reagent and by using a high pH for the reaction [99,101]. The rate of stabilization is also affected by the strain on the collagen during the reaction, with an increase in strain leading to materials with greater tensile strength [102]. The mechanical performance of polyepoxy stabilized tissues may be better than for glutaraldehyde treated material, perhaps reflecting the lower cross-link density that may be found [101].

The biocompatibility of these agents seems satisfactory. Thus, cell culture methods have demonstrated that the polyepoxy compounds exhibit a lower cytotoxicity than, for example, glutaraldehyde, formaldehyde or carbodiimide [97,103]. Also, in vivo examination of stabilized collagen in the rat subcutaneous model showed that there was no calcification, in contrast to the problems associated with glutaraldehyde [95]. Other studies have demonstrated the blood compatibility of vessel surfaces [98].

This method for collagen stabilization has now been successfully applied to a range of tissues, including pericardium [102], heart valves [95] including the aortic wall [100], tendons [101] and blood vessels [99]. Initial biological performance studies have proved promising, although long-term data is not yet readily available.

b. Isocyanates. Isocyanates are homobifunctional reagents that allow cross-links between the amine groups of lysine residues. The most commonly used reagent has been hexamethylene diisocyanate (HMDI). An early clinical product used this reagent to modify gelatine as a plasma expander [104,105]. Subsequently, this reagent was examined for stabilization of collagen for biomaterial applications [14,106]. These data showed that HMDI is effective and does not leave residues in the product since it can be extracted effectively, and it does not form polymeric compounds which release slowly. Subcutaneous implantation in rats of HMDI stabilized collagen sponges showed that early tissue reaction was minimal and that cellular ingrowth gradually occurred, leading to thorough infiltration of the device [14]. Thus the HMDI cross-linked material provided a better environment for in-growing cells than glutaraldehyde treated material, and has been reported as leading to lower levels of calcification [12]. The most effective cross-linking by HMDI uses nonaqueous solvent. This may provide additional benefits as removal of lipid components may allow better access of reagent [12], as well as minimizing the subsequent calcification.

Other studies have demonstrated the effectiveness of collagen-based products for wound dressing that use HMDI cross-linking [107], and have shown that these materials are less cytotoxic than glutaraldehyde treated materials, particularly after careful washing, although a low level of cytotoxicity could still be detected [108,109]. For devices where load bearing ability is important, HMDI cross-linked materials have also been shown to provide material with suitable mechanical properties [110].

c. Bis imidates. Bis imidates have been used for a long time in protein chemistry studies that require a rapid, mild and specific cross-linking technique.They allow cross-linking between amino groups, generally from lysine side chains. Thus, for example, dimethylsuberimidate (DMS) has been used extensively in studies aimed at determining the number of subunits in oligomeric protein assemblies [111]. Although DMS is the most commonly used reagent, a variety of bis imidates are available, including dimethyladipimidate to dimethylsebacimidate [112], giving a range of cross-link lengths that potentially allow the involvement of different numbers of lysine residues. The use of DMS has been successfully applied to the preservation of reconstituted collagen fibrils [113]. Examination of the collagen by transmission electron microscopy suggested that the fibril structural preservation was better than that achieved by using glutaraldehyde. The durability of the collagen for medical applications was not addressed; however, it was suggested that DMS could be used as a prefixative prior to glutaraldehyde stabilization [113], which could allow structural preservation while providing durability. Studies on collagenous membranes for dental application have indicated that cytotoxicity of the reagent can be controlled by postfixation washing [114]. Examination of collagen films, prepared by drying a slurry of bovine collagen fibrils, [115] showed that a high pH, greater than pH 9.0, was necessary for effective collagen cross-linking, in accord with studies on globular proteins [111]; at lower pH less effective cross-linking is obtained [115]. At high pH, partial resistance to proteolysis could be achieved and an increase in T_s of about 13°C could be achieved, compared with an increase of about 18°C using glutaraldehyde [115]. Chemical analysis and mechanical measurements suggested that up to seven cross-links per molecule were formed, about half the number formed using glutaraldehyde. In vivo studies, using a rat subcutaneous model, indicated that, unlike glutaraldehyde treated samples, there was no marked inflammatory cell reaction to the material, and no calcification, although the persistence of the implant was less that for glutaraldehyde treated controls [115].

d. Other Approaches. A variety of other cross-linking or stabilizing agents have also been suggested as suitable for preparation of collagen-based medical devices. These include, for example, the use of tannic acid [116,117], 3,4-dihydroxyphenylalanine [118], and glycerol [119]. The use of glycerol was initially proposed as an approach for sterilization and preservation of homografts [120]. The mechanism of its stabilization has not been fully established, but at high concentrations (98%) it renders tissue nonviable while providing good mechanical properties [119]. Glycerol stabilized pericardial tissue was shown to exhibit a lower extent of calcification in a rat model compared with glutaraldehyde cross-linked tissue. Also, a glycerol stabilized, pericardium-based mitral valve was shown to be durable in sheep and to provide a nonthrombogenic surface that allowed endothelial cell growth over 6 to 12 months implant time [119]. Other chemical cross-linking approaches that have been proposed for cross-linking of collagens include the use of chloro-*s*-triazines [121], bis(nitrophenyl) esters of dicarboxylic acids [122], and a variety of classical protein chemistry reagents [123]. In general, these have not been examined in detail or have not been shown to have utility for biomedical applications. More recently the concept of introducing sulfydryl groups into gelatine or collagen and forming cross-links by oxidation [124] has been examined [125]. Control of the amount of introduced thiol allows control of the mechanical properties, and can provide materials that are more resistant to collagenase digestion than glutaraldehyde treated material [125]. This approach has also been used for development of collagen-based adhesives [126]. Other approaches, such as chrome tanning, have been proposed for wound dressings [127]. Although still used in collagen-based suture manufacture, the potential toxic, mutagenic and teratogenic risks mean that this approach has not been pursued.

3. Approaches Introducing Zero-Length Cross-Links

Zero-length cross-link approaches, where the stabilization strategy leads to direct cross-linking of adjacent amino acids from different collagen molecules, are seen as potentially advantageous as these methods do not introduce any additional compounds that could lead to cytotoxicity or other detrimental effects.

a. Carbodiimides. Water-soluble carbodiimides have been long established as effective reagents for peptide bond formation [128], finding extensive use in peptide synthesis applications. It was also recognized that these reagents could be useful in protein cross-linking, and it was shown that aqueous gelatine reacted rapidly to form a cross-linked gel [128]. The most commonly used reagent now is 1-ethyl-3(3-dimethylaminopropyl)carbodiimide (EDC). Initially the reagent reacts with a carboxyl group, from either an aspartic or glutamic acid side chain, to form an *o*-acylurea. This then reacts with an adjacent amino group, from a lysine side chain, to form a peptide bond cross-link, ideally between the chains, while releasing the urea derivative of the reagent.This approach has also been applied to cross-linking of collagen tissues [129]. It was shown that effective cross-linking of dermal collagen could be obtained, with the thermal stability increasing from 56°C to 73°C. It was further found that inclusion of *N*-hydroxysuccinimide (NHS) in the reaction both increased the rate of the cross-linking and the effectiveness as shown by a higher thermal stability, 86°C [129]. This modification or, alternatively, use of *N*-hydroxysulfosuccinimide [130], is well established in peptide chemistry, and enhances the process by reacting through the *o*-acylurea to form an activated ester that is more stable in solution. This can then react with an amine to form the cross-link or can hydrolyse to restore the free carboxyl group. When NHS is not used, the rate of hydrolysis of the acylurea is very rapid and a lower efficiency of cross-linking is found [129]. Examination of the in vitro degradation properties of EDC-NHS cross-linked dermal collagen showed that it was particularly stable to proteolysis, with little collagen loss over 24 h. This is in contrast to unmodified dermal collagen that degraded at about 18% per hour [131,132]. Stabilized pericardium also showed a similar resistance to CNBr digestion as glutaraldehyde treated tissue [132]. Extensive mechanical testing [132] has shown that carbodiimide cross-linked pericardium was more extensible and more elastic than glutaraldehyde treated material. However, Lee and colleagues [132] cautioned that a by-product, behaving like histidine on amino acid analysis, could be present and would need consideration prior to clinical application of this method.

Cyanamide, which can be regarded as the smallest carbodiimide [132] has also been used for collagen cross-linking, both in solution and in the vapor phase. Initially, it was used as a second step after dehydrothermal cross-linking [133,134] for treatment of reconstituted collagen materials. However, Lee and colleagues [132,135] have shown that the effectiveness of cyanamide alone is low for intact pericardial tissue, particularly in aqueous solutions, and have suggested that in the combination approaches, the dehydrothermal cross-linking may be the dominant effect.

b. Acyl Azides. The acyl azide approach also provides a zero length cross-linking approach. Initially it was applied to collagen for covalent attachment of enzymes to preformed collagen membranes [136]. Subsequently, Petite and colleagues [137] proposed this approach for stabilization of tissue collagen-based devices, using pericardial tissue. In their initial approach, carboxyl groups from aspartyl or glutaminyl residues were transformed in a three-step procedure utilising hydrazine to give acyl azides that are then able to react with adjacent lysine derived amino groups. Measurement of the thermal stability indicated that under optimal conditions, where 1 M NaCl was included to inhibit swelling, stability (83°C) comparable to glutaraldehyde treatment (85°C) could be achieved. Further, both resistance to CNBr solubilization and collagenase digestion also indicated the effectiveness of the procedure. A key advantage of this method was that the products did not calcify readily, as shown in a rat subcutaneous model [138]. In both untreated and glutaraldehyde treated sponges extensive calcification occurred within 15 days, whereas for the acyl azide stabilized sponge calcification was absent at 90 days. Further studies [139] confirmed the effectiveness of the method for treatment of dermis and aorta, with better fixation than a range of other methods, and no cytotoxicity. This hydrazine method, however, is not convenient as it involves many reactions and washing steps over several days. Consequently, Petite and colleagues [140] introduced the use of diphenylphorylazide. This reagent, which was developed for use in peptide synthesis [141], allows direct formation of the acyl azide without the need to isolate intermediates, although it requires the use of an appropriate solvent, such as dimethyl formamide. This reagent also led to good thermal stability, 81°C, and resistance to CNBr and collagenase, while chemical analyses showed that phosphorus residues were absent [140]. Pericardial tissue stabilized by this method was shown to be noncytotoxic, performing better than the hydrazine treated material [142]. Clinically, collagen membranes stabilized using diphenylphorylazide performed well when used in the treatment of periodontal disease and showed no adverse reactions [143].

c. Transglutaminase. The enzyme transglutaminase (protein-glutamine:amine γ-glutamyl transferase) enables the cross-linking of proteins through formation of *N*-ε-(γ-L-glutamyl)-L-lysine bonds. Evidence suggests that members of this enzyme family play an important role in stabilizing tissue at the dermal-epidermal junction in wound healing, where type VII collagen is the probable key substrate [144]. The enzyme has been shown to be useful in preparation of protein-collagen [145] and protein-gelatine [146] conjugates and for cross-linking gelatine to form an insoluble matrix that can be used to entrap enzymes [147]. While the use of this enzyme has been described for cross-linking gelatine to make wound dressings [148], it has been suggested that its utility in forming cross-linked collagen materials may be limited due to steric considerations [146]. However, it is possible that this enzyme may be an important cell-derived component in the formation of tissue-engineered skin substitutes [149,150], so that in certain circumstances enzyme augmentation may be of benefit.

B. Physical Methods

1. Dehydrothermal

Collagen is normally surrounded by an extensive hydration network that is important in fibril organization and in stabilization of the triple helical structure [25]. Gelatine is similarly hydrated. It was discovered that if the water could be removed from gelatine, by high vacuum and through application of heat, then it was possible to cross-link the gelatine [40]. It was suggested that from mass action considerations alone, the drastic removal of the aqueous product of an amide condensation reaction could cause the formation of intermolecular amide (peptide bond) linkages involving lysine and aspartic acid or glutamic acid residues. Subsequently, it was shown that this dehydrothermal cross-linking was also possible for native collagen [151], and that in the absence of water the denaturation of the collagen triple helix did not occur, although a low level of fragmentation of the collagen has been reported [152]. The earlier studies [151] also demonstrated that the cross-linking led to a decrease in free acidic and basic residues, confirming the likely formation of interchain peptide bonds. A range of studies have reported on conditions for optimal cross-linking [133,151,153,154] defining the vacuum, temperature and time required.

The dehydrothermal cross-linking method has been used to generate reconstituted collagen fibrils that have acceptable strength [152,154]. However, the principle use of this approach has been in the development of wound and burns dressings [155]. In these products [156] tensile strength is less critical, while the lack of any additional chemical entities in cross-links, that could provide cytotoxic products either directly or on degradation of the material, is seen as a clear advantage.

2. *Irradiation*

A wide variety of studies have shown that either UV or gamma-ray irradiation are effective for introducing cross-linking to collagen-based biomaterials, while having the advantage of not introducing any toxic components.

UV irradiation of collagen solutions leads to an increase in viscosity, with gels forming in the absence of oxygen [157]. The effectiveness of this process is greater for native collagen compared to atelocollagen where tyrosine residues have been lost [157]. However, the stabilization produced must be balanced against the degradation that occurs, illustrated, for example by the decrease in fibril reforming capacity [97]. For reconstituted collagen fibril materials, UV irradiation leads to an increase in T_s and in tensile strength after relatively short exposure times, but can be accompanied by partial fragmentation, as demonstrated by increased susceptibility to trypsin digestion [152,158]. UV irradiation has also proved to be a successful method for stabilizing collagen-based sponges for wound repair [159]. In all applications, UV irradiation has the advantage of being rapid and easily controlled.

Gamma-ray irradiation of collagen solutions leads to a lowering of solubility, and changes to the collagen chain composition that can be shown on SDS-polyacrylamide gel electrophoresis and by an increase in susceptibility to trypsin digestion [160]. These changes are accompanied by an increase in fibril diameters, although the axial staining pattern in electron microscopy remains essentially unchanged [160]. Gamma-ray irradiation has been used, for example, to cross-link soluble collagen to form clear insoluble gel for contact lenses [97]. However, the principle impact for gamma-ray irradiation arises from its use in sterilizing collagen products [161]. The doses of irradiation commonly used for sterilization lead to significant damage to the collagen, with significant numbers of peptide bonds being cleaved [162], although if the collagen is stabilized by other methods, adequate functional performance after irradiation sterilization can still be achieved [163]. The nature and extent of the damage depends on whether the collagen is in a wet or dry format.

3. *Photo-oxidation*

Dye mediated photooxidation was suggested in the early 1960s as an approach for collagen biomaterial stabilization [164], but it is only more recently that renewed interest in this approach has emerged [15,16,165]. Although many potential dyes could be used [164], methylene blue has been the dye of choice, since it is readily removed from stabilized tissue by washing, whereas other dyes can remain bound to the collagen.

Dye mediated photooxidation has been shown to modify a range of amino acids, including histidine, tyrosine, tryptophan and methionine, with the extent of modification dependent on pH [166]. In collagen treated at neutral pH, the major modification is to histidine, which is rapidly lost in a time-dependent manner, with minor changes to methionine and tyrosine [15]; tryptophan is absent from interstitial collagens. Accompanying this loss of histidine there is evidence of new, higher molecular weight CNBr fragments [15], loss of collagen extraction [167,168] and increased resistance to tissue solubilization by pepsin treatment [167,168], indicative of cross-link formation. Compared with aldehyde stabilization methods, little change is observed in T_s [15,167], although increases in tensile strength [15] and thermoelastic properties [169] have been reported that support the presence of new cross-link formation. The nature of the cross-links that are formed have not yet been defined, although it is assumed that they involve the modified histidine residues. However, separation of distinct, newly formed cross-links [170] after derivatization by diphenylboronic acid [45] suggests a route that will allow identification and characterization.

The effectiveness of this stabilization method has been evaluated in vivo. Initially it was shown in rats that photooxidized pericardium was stable compared to untreated controls [167]. Subsequently, this cross-linked tissue was shown to be noncytotoxic and passed a wide range of general tests for biocompatibility [171]. Also, it was shown that the material elicited only low antibody levels, and that these were lower than similar tissue samples treated with glutaraldehyde [171]. Histopathology of implants in rabbits showed no significant macroscopic reaction at 90 days. The use of photooxidation has now been applied successfully to stabilization of pericardium for use in bioprosthetic heart valves [171,172]. This led to the development of an iodine-based sterilization approach [173], since photooxidation avoids the use of reactive chemicals such as aldehydes that would modify the tissue, and physical methods of sterilization such as irradiation would degrade the tissue significantly. In a juvenile sheep model, after two years of implant, the valves were fully functional, retaining excellent flexibility, and showed no evidence of calcification despite this being a very aggressive model that leads to rapid failure of glutaraldehyde stabilized devices [16]. These experiments also showed evidence for integration by the host with a new layer of host endothelial cells having formed on part of both inflow and outflow leaflet surfaces [171], consistent with previous in vitro findings [172].

V. CONCLUSIONS

Glutaraldehyde still remains the method of choice for stabilization of collagen devices that require long-term durability, for example heart valves, despite the problems associated with calcification of the devices. After many attempts, this calcification problem has not yet been fully overcome. On the other hand, some new, nonglutaraldehyde strategies such as photooxidation are being reevaluated that may yet prove to be suitable alternatives. For devices that do not require long-term durability, such as scaffolds for wound management, the use of glutaraldehyde is not ideal, as there is the potential for cytotoxicity. Other strategies, including those that are not suitable for long-term durability may be better. For all devices, the use of a combination of strategies is an approach that may deserve further examination. A brief prestabilization with one reagent may, for example, optimize tissue mechanics, while the second reagent could provide, for example, durability or control of immunogenicity.

REFERENCES

1. Bateman, J. F., Lamande, S. and Ramshaw, J. A. M., 1995. Collagen superfamily. In: Comper, W. D. (Ed.), *Extracellular Matrix*: Vol. 2, *Molecular Components and Interactions*, Harwood Academic Publishers, Amsterdam, pp. 22–67.

2. Ramshaw, J. A. M., Werkmeister, J. A. and Peters, D. E., 1990. Collagen as a biomaterial. In: Williams, D. F. (Ed.), *Current Perspectives on Implantable Devices*, Vol. 2, JAI Press, London, pp. 151–220.
3. Ramshaw, J. A. M., Werkmeister, J. A. and Glattauer, V., 1996. Collagen-based biomaterials. *Biotech. Genetic Eng. Rev.* 13: 335–382.
4. Keefe, J., Wauk, L., Chu, S. and DeLustro, F., 1992. Clinical use of injectable bovine collagen: a decade of experience. *Clin. Mater.* 6: 155–162.
5. Hait, M. R., 1970. Microcrystalline collagen. A new hemostatic agent. *Am. J. Surg.* 120: 330.
6. Ferrans, V. J., Hilbert, S. L., Tomita, Y., Jones, M. and Roberts, W. C., 1988. Morphology of collagen in bioprosthetic heart valves. In: Nimni, M. E. (Ed.), *Collagen*, Vol. 3, CRC Press, Boca Raton, FL, pp. 145–190.
7. Ramshaw, J. A. M., Edwards, G. A. and Werkmeister, J. A., 1995. Tissue-polymer composite vascular prostheses. In: Wise, D. L., Trantolo, D. J., Altobelli, D. E., Yaszemski, M. J., Gresser, J. D. and Swartz, E. R. (Eds.), *Encyclopedic Handbook of Biomaterials and Engineering*, Part B, Vol. 2, Marcel Dekker, New York, pp. 953–978.
8. Werkmeister, J. A., Edwards, G. A. and Ramshaw, J. A. M., 1999. Collagen-based vascular prostheses. In: Wise, D. L. (Ed.), *Biomaterials Engineering and Devices: Human Applications*, Humana Press, in press.
9. Werkmeister, J. A. and Ramshaw, J. A. M., 1999. Immunology of collagen-based biomaterials. In: Wise, D. L. (Ed.), *Handbook of Biomaterials Engineering*, Marcel Dekker, in press.
10. Carpentier, A., Lemaigre, G., Robert, L., Carpentier, S. and Dubost, C., 1969. Biological factors affecting long-term results of valvular heterografts. *J. Thorac. Cardiovasc. Surg.* 58: 467–483.
11. Jayakrishnan, A. and Jameela, S. R., 1996. Glutaraldehyde as a fixative in bioprostheses and drug delivery matrices. *Biomaterials* 17: 471–484.
12. Khor, E., 1997. Methods for the treatment of collagenous tissues for bioprostheses. *Biomaterials* 18: 95–105.
13. Golomb, G., Schoen, F. J., Smith, M. S., Linden, J., Dixon, M. and Levy, R. J., 1987. The role of glutaraldehyde-induced crosslinks in calcification of bovine pericardium used in cardiac valve bioprostheses. *Am. J. Pathol.* 127: 122–130.
14. Chvapil, M., Speer, D., Mora, W. and Eskelson, C., 1983. Effect of tanning agent on tissue reaction to tissue implanted collagen sponge. *J. Surg. Res.* 35: 402–409.
15. Ramshaw, J. A. M., Stephens, L. J. and Tulloch, P. A., 1994. Methylene blue sensitized photooxidation of collagen fibrils. *Biochim. Biophys. Acta* 1206: 225–230.
16. Moore, M. A., 1997. Pericardial tissue stabilized by dye-mediated photooxidation: a review article. *J. Heart Valve Dis.* 6: 521–526.
17. Rosenberg, N., Martinez, A., Sawyer, P. N., Wesolowski, S. A., Postlethwait, R. W. and Dillon, M. L., 1966. Tanned collagen arterial prosthesis of bovine carotid origin in man. *Ann. Surg.* 164: 247–256.
18. Tammishetti, S., Wallace, D. G. and Reich, C. J., 1997. Dialdehyde starch crosslinked gelatine: a new tissue sealant. In, Proceedings, 23rd Annual Meeting of the Society for Biomaterials, 1997, New Orleans, p. 275.
19. Kadler, K., 1994. Extracellular matrix 1: Fibril-forming collagens. *Protein Profile* 1: 519–638.
20. Kielty, C. M., Hopkinson, I. and Grant, M. E., 1993. Collagen: The collagen family: Structure, assembly, and organisation in the extracellular matrix. In: Royce, P. M. and Steinmann, B. (Eds.), *Connective Tissue and Its Heritable Disorders*, Wiley, New York, pp. 103–148.
21. Prockop, D. J. and Kivirikko, K. I., 1995. Collagens: Molecular biology, diseases, and potentials for therapy. *Annu. Rev. Biochem.* 64: 403–434.
22. Rich, A. and Crick, F. H. C., 1961. The molecular structure of collagen. *J. Mol. Biol.* 3: 483–506.
23. Fraser, R. D. B., MacRae, T. P. and Suzuki, E., 1979. Chain conformation in the collagen molecule. *J. Mol. Biol.* 129: 463–481.
24. Bella, J., Eaton, M., Brodsky, B. and Berman, H. M., 1994. Crystal and molecular structure of a collagen-like peptide at 1.9 A resolution. *Science* 266: 75–81.

25. Bella, J., Brodsky, B. and Berman, H. M., 1995. Hydration structure of a collagen peptide. *Structure* 3: 893–906.
26. Ramshaw, J. A. M., Shah, N. K. and Brodsky, B., 1998. A host-guest peptide approach to triple-helix stability. *J. Struct. Biol.* 122: 86–91.
27. Jones, E. Y. and Miller, A., 1991. Analysis of structural design features in collagen. *J. Mol. Biol.* 218: 209–219.
28. Schmitt, F. O., Levine, L., Drake, M. P., Rubin, A. L., Pfahl, D. and Davison, P. F., 1964. The antigenicity of tropocollagen. *Proc. Natl. Acad. Sci. USA.* 51: 493–497.
29. Timpl, R., 1984. Immunology of the collagens. In: Piez, K. A. and Reddi, A. H. (Eds.), *Extracellular Matrix Biochemistry*, Elsevier, New York, pp. 159–190.
30. Furthmayr, H. and Timpl, R., 1976. Immunochemistry of collagens and procollagens. *Int. Rev. Connective Tissue Res.* 7: 61–99.
31. Furthmayr, H., Stoltz, M., Becker, U., Beil, W. and Timpl, R., 1972. Chicken antibodies to soluble rat collagen. II. Specificity of the reactions with individual polypeptide chains and cyanogen bromide peptides of collagen. *Immunochem.* 9: 789–798.
32. Meade, K. R. and Silver, F. H., 1990. Immunogenicity of collagenous implants. *Biomaterials* 11: 176–180.
33. Bajpai, P. K., 1985. Immunological aspects of treated natural tissue prostheses. In: Williams, D. F. (Ed.), *Biocompatibility of Tissue Analogs*, Vol. 1, CRC Press, Boca Raton, FL, pp. 5–25.
34. Axthelm, S. C., Porter, J. M., Strickland, S. and Baur, G. M., 1979. Antigenicity of venous allografts. *Ann. Surg.* 189: 290–293.
35. Okamura, K., Chiba, C., Iriyama, T., Itoh, T., Maeta, H., Ijima, H., Mitsui, T. and Hori, M., 1980. Antigen depressant effect of glutaraldehyde for aortic heterografts with a valve, with special reference to a concentration right fit for the preservation of grafts. *Surgery* 87: 170–176.
36. Tauro, J. C., Parsons, J. R., Ricci, J. and Alexander, H., 1991. Comparison of bovine collagen xenografts to autografts in the rabbit. *Clin. Orthop.* 266: 271–284.
37. Dahm, M., Lyman, W. D., Schwell, A. B., Factor, S. M. and Frater, R. W. M., 1990. Immunogenicity of glutaraldehyde-tanned bovine pericardium. *J. Thorac. Cardiovasc. Surg.* 99: 1082–1090.
38. van Gulik, T. M. and Klopper, P. J., 1987. The processing of sheepskin for use as a dermal collagen graft—an experimental study. *Netherlands J. Surg.* 39: 90–94.
39. Dahm, M., Husmann, M., Mayer, E., Prufer, D., Groh, E. and Oelert, H., 1995. Relevance of immunologic reactions for tissue failure of bioprosthetic heart valves. *Ann. Thorac. Surg.* 60: S348–S352.
40. Yannas, I. V. and Tobolsky, A. V., 1967. Cross-linking of gelatine by dehydration. *Nature* 215: 509–510.
41. Chan, V. C., Ramshaw, J. A. M., Kirkpatrick, A., Beck, K. and Brodsky, B., 1997. Positional preferences of ionizable residues in Gly-X-Y triplets of the collagen triple-helix. *J. Biol. Chem.* 272: 31441–31446.
42. Yang, W., Chan, V. C., Kirkpatrick, A., Ramshaw, J. A. M. and Brodsky, B., 1997. Gly-Pro-Arg confers stability similar to Gly-Pro-Hyp in the collagen triple-helix of host-guest peptides. *J. Biol. Chem.* 272: 28837–28840.
43. Bowes, J. H. and Cater, C. W., 1968. The interaction of aldehydes with collagen. *Biochim. Biophys. Acta* 168: 341–352.
44. Yamaguchi, M. and Mechanic, G. L., 1988. Cross-linking of collagen. In: Nimni, M. E. (Ed.), *Collagen*, Vol. 1, CRC Press, Boca Raton, FL, pp. 157–172.
45. Graham, L. and Gallop, P. M., 1994. Covalent protein cross-links: general detection, quantitation, and characterisation via modification with diphenylboronic acid. *Anal. Biochem.* 217: 298–305.
46. Rigby, B. J., 1967. Increases with time in the thermal stability of rat-tail tendon stored in 0.9% NaCl. *Biochim. Biophys. Acta* 140: 548–551.
47. Flory, P. J. and Rehner, J., 1943. Statistical mechanics of cross-linked polymer networks. *J. Chem. Phys.* 11: 521–520.
48. Nimni, M. E., Cheung, D. T., Strates, B., Kodama, M. and Sheikh, K., 1988. Bioprosthesis derived

from cross-linked and chemically modified collagenous tissues. In: Nimni, M. E. (Ed.), *Collagen*, Vol. 3, CRC Press, Boca Raton, FL, pp. 1–38.
49. Fein, M. L. and Filachione, E. M., 1957. Tanning studies with aldehydes. *J. Am. Leather Chem. Assoc.* 52: 17–23.
50. Nimni, M. E., Cheung, D., Strates, B., Kodama, M. and Sheikh, K., 1987. Chemically modified collagen: a natural biomaterials for tissue replacement. *J. Biomed. Mater. Res.* 21: 741–771.
51. Hardy, P. M., Nicholls, A. C. and Rydon, H. W., 1969. The nature of glutaraldehyde in solution. *J. Chem. Soc. Chem. Commun.* 1969: 565–566.
52. Holloway, C. E. and Dean, F. H., 1975. Carbon-13 NMR study of aqueous glutaraldehyde equilibria. *J. Pharm. Sci.* 64: 1078–1079.
53. Korn, A. H., Feairheller, S. H. and Filachione, E. M., 1972. Glutaraldehyde: the nature of the reagent. *J. Mol. Biol.* 65: 525–529.
54. Fisher, J., Gorham, S. D., Howie, A. M. and Wheatley, D. J., 1987. Examination of fixative penetration in glutaraldehyde-treated bovine pericardium by stratigraphic analysis of shrinkage temperature measurements using differential scanning calorimetry. *Life Supp. Syst.* 5: 189–193.
55. Cheung, D. T., Perelman, N., Ko, E. C. and Nimni, M. E., 1985. Mechanism of crosslinking of proteins by glutaraldehyde. III: Reactions with collagen in tissues. *Connect. Tissue Res.* 13: 109–115.
56. Gendler, E., Gendler, S. and Nimni, M. E., 1984. Toxic reactions evoked by glutaraldehyde-fixed pericardium and cardiac valve tissue bioprosthesis. *J. Biomed. Mater. Res.* 18: 727–736.
57. Speer, D. P., Chvapil, M., Volz, R. G. and Holmes, M. D., 1979. Enhancement of healing in osteochondral defects by collagen sponge implants. *Clin. Orthop.* 144: 326–335.
58. Speer, D. P., Chvapil, M., Eskelson, C. D. and Ulreich, J., 1980. Biological effects of residual glutaraldehyde in glutaraldehyde-tanned collagen biomaterials. *J. Biomed. Mater. Res.* 14: 753–764.
59. Broom, N. D., 1977. The stress/strain and fatigue behaviour of glutaraldehyde preserved heart valve tissue. *J. Biomech.* 10: 707–727.
60. Broom, N. D. and Thompson, F. J., 1979. Influence of fixation conditions on the performance of glutaraldehyde-treated porcine aortic valves: towards a more scientific basis. *Thorax* 34: 166–176.
61. Duncan, A. C. and Boughner, D., 1998. Effect of dynamic glutaraldehyde fixation on the viscoelastic properties of bovine pericardial tissue. *Biomaterials* 19: 777–783.
62. Lee, J. M., Corrente, R. and Haberer, S. A., 1989. The bovine pericardial xenograft: II. Effect of tethering or pressurisation during fixation on the tensile viscoelastic properties of bovine pericardium. *J. Biomed. Mater. Res.* 23: 477–489.
63. Jorge-Herrero, E., Fernandez, P., Gutierrez, M. and Castillo-Olivares, J. L., 1991. Study of the calcification of bovine pericardium: analysis of the implication of lipids and proteoglycans. *Biomaterials* 12: 683–689.
64. Carpentier, A., Nashef, A., Carpentier, S., Ahmed, A. and Gouseff, N., 1984. Techniques for prevention of calcification of valvular bioprostheses. *Circulation* 70 (Suppl. I): I165-I168.
65. David, D. E., Pollick, C. and Bos, J., 1990. Aortic valve replacement with stentless porcine aortic bioprosthesis. *J. Thorac. Cardiovasc. Surg.* 99: 113–118.
66. Tan, W. M., Loke, W. K., Tan, B. L., Wee, A., Khor, E. and Goh, K. S., 1993. Trivalent metal ions in the prevention of calcification in glutaraldehyde treated biological tissue. Is there a chemical correlation? *Biomaterials* 14: 1003–1007.
67. Park, K. D., Yun, J. Y., Han, D. K., Jeong, S. Y., Kim, Y. H., Choi, K. S., Kim, H. M., Kim, H. J. and Kim, K. T., 1994. Chemical modification of implantable biologic tissue for anti-calcification. *ASAIO J.* 40: M377–M382.
68. Gott, J. P., Chih, P., Dorsey, L. M. A., Jay, J. L., Jett, G. K., Schoen, F. J., Girardot, J.-M. and Guyton, R. A., 1992. Calcification of porcine valves: a successful new method of antimineralisation. *Ann. Thorac. Surg.* 53: 207–216.
69. Levy, R. J., Wolfrum, J., Schoen, F. J., Hawley, M. A., Lund, S. A. and Langer, R., 1985. Inhibition of calcification of bioprosthetic heart valves by local controlled-released diphosphonate. *Science* 228: 190–192.

70. Khor, E., Wee, A., Feng, T. C. and Goh, D. C. L., 1998. Glutaraldehyde-fixed biological tissue calcification: effectiveness of mitigation by dimethylsulphoxide. *J. Mat. Sci.: Mat. Med.* 9 : 39–45.
71. Vyavahare, N. R., Hirsch, D., Lerner, E., Baskin, J. Z., Zand, R., Schoen, F. J. and Levy, R. J., 1998. Prevention of calcification of glutaraldehyde-crosslinked porcine aortic cusps by ethanol preincubation—mechanistic studies of protein structure and water-biomaterial relationships. *J. Biomed. Mater. Res.* 40: 577–585.
72. Zilla, P., Weissenstein, C., Bracher, M., Zhang, Y., Koen, W., Human, P. and von Oppell, U., 1997. High glutaraldehyde concentrations reduce rather than increase the calcification of aortic wall tissue. *J. Heart Valve Dis.* 6: 502–509.
73. Casagranda, F., Werkmeister, J. A., Ramshaw, J. A. M. and Ellender, G., 1994. Evaluation of alternative glutaraldehyde stabilization strategies for collagenous biomaterials. *J. Mat. Sci.: Mat. Med.* 5: 332–337.
74. Gratzer, P. F., Periera, C. A. and Lee, J. M., 1996. Solvent environment modulates effects of glutaraldehyde crosslinking on tissue derived biomaterials. *J. Biomed. Mater. Res.* 31: 533–543.
75. Hardy, P. M., Nichols, A. C. and Rydon, H. N., 1972. The hydration and polymerisation of succinaldehyde, glutaraldehyde, and adipaldehyde. *J. Chem. Soc. Perkin II* 1972: 2270–2278.
76. Branner-Jorgenson, S., 1978. On the mechanism of protein cross-linking with glutaraldehyde. *Enzyme. Eng.* 4: 393–394.
77. Cater, C. W., 1965. Investigations into the efficiency of dialdehydes and other compounds as cross-linking agents for collagen. *J. Soc. Leather Tech. Chem.* 49: 455–469.
78. Slatter, D. A., Murray, M. and Bailey, A. J., 1998. Formation of a dihydropyridine derivative as a potential cross-link derived from malonaldehyde in physiological systems. *FEBS Lett.* 421: 180–184.
79. Chio, K. S. and Tappel, A. L., 1969. Inactivation of ribonuclease and other enzymes by peroxidizing lipids and by malonaldehyde. *Biochemistry* 8: 2827–2832.
80. Davidkova, E., Svadlenka, I. and Rosmus, J., 1973. Interaction of malonaldehyde with collagen. *Z. Lebensm. Unters.-Forsch.* 153: 13–16.
81. Fraenkel-Conrat, H., Cooper, M. and Olcott, H. S., 1945. The reaction of formaldehyde with proteins. *J. Am. Chem. Soc.* 67: 950–954.
82. Kunkel, G. R., Mehrabian, M. and Martinson, H. G., 1981. Contact-site cross-linking agents. *Mol. Cell Biochem.* 34: 3–13.
83. O'Brien, M. F., 1967. Heterograft aortic valves for human use: valve bank, techniques of measurement, and implantation. *J. Thorac. Cardiovasc. Surg.* 53: 392–397.
84. Buch, W. S., Kosek, J. C. and Angell, W. W., 1970. Deterioration of formalin-treated aortic valve heterografts. *J. Thorac. Cardiovasc. Surg.* 60: 673–677.
85. Barker, H., Oliver, R., Grant, R. and Stephen, L., 1980. Formaldehyde as a pre-treatment for dermal collagen heterografts. *Biochim. Biophys. Acta* 632: 589–597.
86. Nagaraj, R. H., Shipanova, I. N. and Faust, F. M., 1996. Isolation, characterization, and in vivo detection of a lysine-lysine cross-link derived from methylglyoxal. *J. Biol. Chem.* 271: 19338–19345.
87. Sell, D. R. and Monier, V. M., 1989. Structure elucidation of a senescence crosslink from human extracellular matrix: Implication of pentoses in ageing process. *J. Biol. Chem.* 264: 21597–21602.
88. Sims, T. J., Rasmussen, L. M., Oxlund, H. and Bailey, A. J., 1996. The role of glycation cross-links in diabetic vascular stiffening. *Diabetologia* 39: 946–951.
89. Tanaka, S., Avigad, G., Brodsky, B. and Eikenberry, E. F., 1988. Glycation induces expansion of the molecular packing of collagen. *J. Mol. Biol.* 203: 495–505.
90. Brodsky, B., Berg, R. A., Avigad, G., Eikenberry, E., Jain, M. and Tanaka, S., 1990. Collagen-based matrices ribose cross-linked. U.S. Patent 4,971,954.
91. Bai, P., Phua, K., Hardt, T., Cernadas, M. and Brodsky, B., 1992. Glycation alters collagen fibril organization. *Connective Tissue Res.* 28: 1–12.
92. Tian, S. F., Toda, S., Higashino, H. and Matsumara, S., 1996. Glycation decreases the stability of the triple-helical strands of fibrous collagen against proteolytic degradation by pepsin. *J. Biochem. (Tokyo)* 120: 1153–1162.

93. Simionescu, A., Simionescu, D. and Deac, R., 1991. Lysine enhanced glutaraldehyde crosslinking of collagenous biomaterials. *J. Biomed. Mater. Res.* 25: 1495–1505.
94. Elgawish, A., Glomb, M., Friedlander, M. and Monnier, V. M., 1996. Involvement of hydrogen peroxide in collagen cross-linking by high glucose in vitro and in vivo. *J. Biol. Chem.* 271: 12964–12971.
95. Imamura, E., Sawatani, O., Koyanagi, H., Noishiki, Y. and Miyata, T., 1989. Epoxy compounds as a new cross-linking agent for porcine aortic leaflets: subcutaneous implant studies in rats. *J. Cardiac Surg.* 4: 50–57.
96. Imamura, E., Noishiki, Y., Koyanagi, H., Miyata, T. and Furuse, M., 1992. Bioprosthetic valve. U.S. Patent 5,080,670.
97. Miyata, T., 1992. Collagen engineering for biomaterial use. *Clin. Mater.* 9: 139–148.
98. Wang, E., Thyagarajan, K., Tu, R., Lin, D., Hata, C., Shen, S. H. and Quijano, R. C., 1993. Evaluation of collagen modification and surface properties of a bovine artery via polyepoxy compound fixation. *Int. J. Artif. Organs* 16: 530–536.
99. Tu, R., Lu, C.-L., Thyagarajan, K., Wang, E., Nguyen, H., Shen, S., Hata, C. and Quijano, R. C., 1993. Kinetic study of collagen fixation with polyepoxy derivatives. *J. Biomed. Mater. Res.* 27: 3–9.
100. Shen, S.-H., Sung, H. W., Tu, R., Hata, C., Lin, D., Noishiki, Y. and Quijano, R. C., 1994. Characterisation of a polyepoxy compound fixed porcine heart valve bioprosthesis. *J. Appl. Biomater.* 5: 159–162.
101. Sung, H.-W., Cheng, W.-H., Chiu, I.-S., Hsu, H.-L. and Liu, S.-A., 1996. Studies on epoxy compound fixation. *J. Biomed. Mater. Res.* (*Appl. Biomater.*) 33: 177–186.
102. Chachra, D., Gratzer, P. F., Periera, C. A. and Lee, J. M., 1996. Effect of applied uniaxial stress on the rate and mechanical effects of cross-linking in tissue-derived biomaterials. *Biomaterials* 17: 1865–1875.
103. Nishi, C., Nakajima, N. and Ikada, Y., 1995. *In vitro* evaluation of cytotoxicity of diepoxy compounds used for biomaterial modification. *J. Biomed. Mater. Res.* 29: 829–834.
104. Schwick, G. and Freund, U., 1962. Immunoligische unterfuchungen mit haemacell. *Dtch. Med. Wschr.* 87: 737–741.
105. Moeller, J. and Sykudes, A., 1962. Die vertraglichkeit von Haema-cell. *Dtch. Med. Wschr.* 87: 726–729.
106. Chvapil, M., 1982. Considerations on manufacturing principles of a synthetic burn dressing: a review. *J. Biomed. Mater. Res.* 16: 245–263.
107. Chvapil, M., Chvapil, T. A. and Owen, J. A., 1986. Reaction of various skin wounds in the rat to collagen sponge dressing. *J. Surg. Res.* 41: 410–418.
108. van Luyn, M. J., van Wachem, P. B., Damink, L. O., Dijkstra, P. J., Feijen, J. and Nieuwenhuis, P., 1992. Relations between in vitro cytotoxicity and crosslinked dermal sheep collagens. *J. Biomed. Mater. Res.* 26: 1091–1110.
109. van Luyn, M. J., van Wachem, P. B., Olde Damink, L. H., Dijkstra, P. J., Feijen, J. and Neiwenhuis, P., 1992. Secondary cytotoxicity of cross-linked dermal sheep collagens during repeated exposure to human fibroblasts. *Biomaterials* 13: 1017–1024.
110. Olde Damink, L. H., Dijkstra, P. J., van Luyn, M. J., van Wachem, P. B., Nieuwenhuis, P. and Feijen, J., 1995. Crosslinking of dermal sheep collagen using hexamethylene diisocyanate. *J. Mater. Sci.: Mat. Med.* 6: 429–434.
111. Davis, G. E. and Stark, G. R., 1970. Use of dimethyl suberimidate, a cross-linking reagent, in studying the subunit structure of oligomeric proteins. *Proc. Natl. Acad. Sci. USA* 66: 651–656.
112. Watanabe, K., 1997. Comparison of properties of the collagen cross-linked with aliphatic or aromatic reagent. *Polym. J.* 29: 286–289.
113. Tzaphlidou, M., 1983. The effects of fixation by combination of glutaraldehyde/dimethyl suberimidate. Use of collagen as a model system. *J. Histochem. Cytochem.* 31: 1274–1278.
114. Hey, K. B., Lachs, C. M., Raxworthy, M. J. and Wood, E. J., 1990. Crosslinked fibrous collagen for use as a dermal implant: control of the cytotoxic effects of glutaraldehyde and dimethylsuberimidate. *Biotechnol. Appl. Biochem.* 12: 85–93.

115. Charulatha, V. and Rajaram, A., 1997. Crosslinking density and resorption of dimethyl suberimidate-treated collagen. *J. Biomed. Mater. Res.* 36: 478–486.
116. Heijmen, F. H., du Pont, J. S., Middelkoop, E., Kreis, R. W. and Hoekstra, M. J., 1997. Cross-linking of dermal sheep collagen with tannic acid. *Biomaterials* 18: 749–754.
117. Koide, T. and Daito, M., 1997. Effects of various collagen crosslinking techniques on mechanical properties of collagen film. *Dental Mater. J.* 16: 1–9.
118. Gade, J. N., Fellman, J. H. and Bentley, J. P., 1991. The stabilization of fibrillar collagen matrices with 3,4-dihydroxyphenylalanine. *J. Biomed. Mater. Res.* 25: 799–811.
119. Liao, K., Seifter, E., Gong, G., Yellin, E. L. and Frater, R. W. M., 1991. Improved biocompatibility of bovine pericardium using a new method of cross linking. *ASAIO Trans.* 37: M175–M176.
120. Rose, A. G., 1987. *Pathology of Heart Valve Replacement*. MTP Press, Boston, MA.
121. Zerlotti, E., 1967. Cross-linking of rat tail tendons with chloro-*s*-triazines. *Nature* 214: 1304–1306.
122. Zahn, H., Schade, F. and Siepmann, E., 1963. Collagen. VII. Reactions of nitrophenyl esters of mono and dicarboxylic acids with collagen. *Leder.* 14: 299–304.
123. Wong, S. S., 1991. *Chemistry of Protein Conjugation and Cross-linking*. CRC Press, Boca Raton, FL.
124. Benesch, R. and Benesch, R. E., 1958. Thiolation of proteins. *Proc. Natl. Acad. Sci. USA* 44: 848–853.
125. Nicholas, F. L. and Gagnieu, C. H., 1997. Denatured thiolated collagen. II. Cross-linking by oxidation. *Biomaterials* 18: 815–821.
126. Tiollier, J., Constancis, A., Gagnieu, C., Tardy, M., Bryson, N., Dupont, D., Meyrueix, R., Gravagna, P., Grosselin, J. M., Pouradier, X., Ulrich, S., Soula, G. and Tayot, J. L., 1995. Novel developments of collagen/gelatine surgical adhesives for surgical soft tissue applications. In Proceedings 21st Annual Meeting, Society for Biomaterials, 1995, San Francisco, p. 257.
127. Sinha, R. N., Verma, P. K. and Maden, P., 1972. Collagen sheets as a biological dressing in burns. *Ind. J. Plastic Surg.* 5: 56–60.
128. Sheehan, J. C. and Hlavka, J. J., 1957. The cross-linking of gelatin using a water-soluble carbodiimide. *J. Am. Chem. Soc.* 79: 4528–4529.
129. Olde Damink, L. H., Dijkstra, P. J., van Luyn, M. J., van Wachem, P. B., Nieuwenhuis, P. and Feijen, J., 1996. Cross-linking of dermal sheep collagen using a water-soluble carbodiimide. *Biomaterials* 17: 765–773.
130. Staros, J. V., Wright, R. W. and Swingle, D. M., 1986. Enhancement by *N*-hydroxysulfosuccinimide of water-soluble carbodiimide-mediated coupling reactions. *Anal. Biochem.* 156: 220–222.
131. Olde Damink, L. H., Dijkstra, P. J., van Luyn, M. J., van Wachem, P. B., Nieuwenhuis, P. and Feijen, J., 1996. In vitro degradation of dermal sheep collagen cross-linked using a water-soluble carbodiimide. *Biomaterials* 17: 679–684.
132. Lee, J. M., Edwards, H. H. L., Pereira, C. A. and Samii, S. I., 1996. Crosslinking of tissue-derived biomaterials in 1-ethyl-3-(3-dimethylaminopropyl)-carbodiimide. *J. Mater. Sci.: Mat. Med.* 7: 531–541.
133. Weadock, K., Olson, R. M. and Silver, F. H., 1983. Evaluation of collagen crosslinking techniques. *Biomat. Med. Device Artif. Organs* 11: 293–318.
134. Kato, Y. P., Dunn, M. G., Zawadsky, J. P., Tria, A. J. and Silver, F. H., 1991. Regeneration of achilles tendon with a collagen tendon prosthesis. Results of a one-year implantation study. *J. Bone Joint Surg.* 73: 561–574.
135. Pereira, C. A., Lee, J. M. and Haberer, S. A., 1990. Effect of alternative crosslinking methods on the low strain rate viscoelastic properties of bovine pericardial bioprosthetic material. *J. Biomed. Mater. Res.* 24: 345–361.
136. Coulet, P. R. and Gautheron, D. C., 1981. Enzymes immobilized on collagen membranes: a tool for fundamental research and enzyme engineering. *J. Chromatogr.* 215: 65–72.
137. Petite, H., Rault, I., Huc, A., Menasche, P. and Herbage, D., 1990. Use of the acyl azide method for cross-linking collagen-rich tissues such as pericardium. *J. Biomed. Mater. Res.* 24: 179–187.
138. Anselme, K., Petite, H. and Herbage, D., 1992. Inhibition of calcification in vivo by acyl azide cross-linking of a collagen-glycosaminoglycan sponge. *Matrix* 12: 264–273.

139. Simmons, D. M. and Kearney, J. N., 1993. Evaluation of collagen cross-linking techniques for the stabilization of tissue matrices. *Biotechnol. Appl. Biochem.* 17: 23–29.
140. Petite, H., Frei, V., Huc, A. and Herbage, D., 1994. Use of diphenylphosphorylazide for cross-linking collagen-based biomaterials. *J. Biomed. Mater. Res.* 28: 159–165.
141. Yokoyama, Y., Shiori, T. and Yamada, S., 1977. Phosphorus in organic synthesis. XVI. Diphenylphosphorylazide and diethylphosphocyanidate. Two reagents for the preparation of thiol esters from carboxylic acids and thiols. *Chem. Pharm. Bull.* 25: 2423–2429.
142. Petite, H., Duval, J. L., Frei, V., Abdul-Malak, N., Sigot-Luizard, M. F. and Herbage, D., 1995. Cytocompatibility of calf pericardium treated by glutaraldehyde and by the acyl azide methods in an organotypic culture model. *Biomaterials* 16: 1003–1008.
143. Benque, E., Zahedi, S., Brocard, D., Oscaby, F., Justumus, P. and Brunel, G., 1997. Guided tissue regeneration using a collagen membrane in chronic adult and rapidly progressive periodontitis patients in the treatment of 3-wall intrabony defects. *J. Clin. Periodontol.* 24: 544–549.
144. Raghunath, M., Hopfner, B., Aeschlimann, D., Luthi, U., Meuli, M., Altermatt, S., Gobet, R., Bruckner-Tuderman, L. and Steinmann, B., 1996. Cross-linking of the dermo-epidermal junction of skin regeneration from keratinocyte autografts. Anchoring fibrils are a target for tissue transglutaminase. *J. Clin. Invest.* 98: 1174–1184.
145. Mosher, D. F., 1984. Crosslinking of fibronectin to collagenous proteins. *Mol. Cell. Biochem.* 58: 63–68.
146. Jelenska, M. M., Fesus, L. and Kopec, M., 1980. The comparative ability of plasma and tissue transglutaminases to use collagen as a substrate. *Biochim. Biophys. Acta.* 616: 167–178.
147. Fuchsbauer, H. L., Gerber, U., Engelmann, J., Seeger, T., Sinks, C. and Hecht, T., 1996. Influence of gelatin matrices cross-linked with transglutaminase on the properties of an enclosed bioactive material using β-galactosidase as model system. *Biomaterials* 17: 1481–1488.
148. Motoda, T., Takehiro, T. and Masao, M., 1989. Gelatin-based cataplasm. Japanese Patent 89-316647.
149. Boyce, S., Michel, S., Reichert, U., Shroot, B. and Schmidt, R., 1990. Reconstructed skin from cultured human keratinocytes and fibroblasts on a collagen-glycosaminoglycan biopolymer substrate. *Skin Pharmacol.* 3: 136–143.
150. Auger, F. A., Lopez Valle, C. A., Guignard, R., Tremblay, N., Noel, B., Goulet, F. and Germain, L., 1995. Skin equivalent produced with human collagen. *In Vitro Cell. Dev. Biol. Anim.* 31: 432–439.
151. Silver, F. H., Yannas, I. V. and Salzman, E. W., 1979. In vitro blood compatibility of glycosaminoglycan-precipitated collagens. *J. Biomed. Mater. Res.* 13: 701–716.
152. Weadock, K. S., Miller, E. J., Bellincampi, L. D., Zawadsky, J. P. and Dunn, M. G., 1995. Physical crosslinking of collagen fibers: comparison of ultraviolet irradiation and dehydrothermal treatment. *J. Biomed. Mater. Res.* 29: 1373–1379.
153. Gorham, S. D., Light, N. D., Diamond, A. M., Willins, M. J., Bailey, A. J., Wess, T. J. and Leslie, N. J., 1992. Effect of chemical modifications on the susceptibility of collagen to proteolysis. II. Dehydrothermal crosslinking. *Int. J. Biol. Macromol.* 14: 129–138.
154. Wang, M. C., Pins, G. D. and Silver, F. H., 1993. Collagen fibres with improved strength for the repair of soft tissue injuries. *Biomaterials* 15: 507–512.
155. Yannas, I. V., Burke, J. F., Orgill, D. P. and Skrabut, E. M., 1982. Wound tissue can utilise a polymeric template to synthesize a functional extension of skin. *Science* 215: 174–176.
156. Geesin, J. C., Brown, L. J., Liu, Z. and Berg, R. A., 1996. Development of a skin model based on insoluble fibrillar collagen. *J. Biomed. Mater. Res.* 33: 1–8.
157. Miyata, T., Sohde, T., Rubin, A. L., and Stenzel, K. H., 1971. Effects of ultraviolet irradiation on native and telopeptide-poor collagen. *Biochim. Biophys. Acta.* 229: 672–680.
158. Weadock, K. S., Miller, E. J., Keuffel, E. L. and Dunn, M. G., 1996. Effect of physical cross-linking methods on collagen-fiber durability in proteolytic solutions. *J. Biomed. Mater. Res.* 32: 221–226.
159. Shibata, H., Shioya, N. and Kuroyanagi, Y., 1997. Development of new wound dressing composed

of spongy collagen sheet containing dibutyryl cyclic AMP. *J. Biomed. Sci. Polym, Edn.* 8: 601–621.

160. Liu, B., Harrel, R., Davis, R. H., Dresden, M. H. and Spira, M., 1989. The effect of gamma irradiation on injectable human amnion collagen. *J. Biomed. Mater. Res.* 23: 833–844.
161. Bechtol, L. D. and Artandi, C., 1969. Radiating tanned and untanned collagen prosthesis with 5 to 25 megarads of ionizing radiation. U.S. Patent 3,451,394.
162. Cheung, D. T., Perelman, N., Tong, D. and Nimni, M. E., 1990. The effect of γ-irradiation on collagen molecules, isolated α-chains, and crosslinked native fibers. *J. Biomed. Mater. Res.* 24: 581–589.
163. Roe, S. C., Milthorpe, B. K., True, K., Rogers, G. J. and Schindhelm, K., 1992. The effect of gamma irradiation on a xenograft tendon bioprosthesis. *Clin. Mater.* 9: 149–154.
164. Kuntz, E., 1964. Preparation of collagenous materials. U.S. Patent 3,152,976.
165. Mechanic, G. L., 1992. Process for cross-linking collagenous material and resulting product. U.S. Patent 5,147,514.
166. Verweij, H., Dubbleman, T. M. A. R. and van Steveninck, J., 1981. Photodynamic protein cross-linking. *Biochim. Biophys. Acta.* 647: 87–94.
167. Moore, M. A., Chen, W. M., Phillips, R. E., Bohachevsky, I .K. and McIlroy, B. K., 1996. Shrinkage temperature versus protein extraction as a measure of stabilization of photooxidized tissue. *J. Biomed. Mater. Res.* 32: 209–214.
168. Moore, M. A., Bohachevsky, I. K., Cheung, D. T., Boyan, B. D., Chen, W. M., Bickers, R. R. and McIlroy, B. K., 1994. Stabilization of pericardial tissue by dye-mediated photooxidation. *J. Biomed. Mater. Res.* 28: 611–618.
169. Lee, J. M., Kujath, M. F., Doherty, P. A., McIlroy, B. K., Grear, D. A. and Moore, M. A., 1997. Hydrothermal isometric tension (HIT) testing demonstrates collagen cross-linking by dye mediated photooxidation. In Proceedings 23rd Annual Meeting, Society for Biomaterials, 1997, New Orleans, p. 177.
170. McIlroy, B. K., Robinson, M. D., Chen, W. M. and Moore, M. A., 1997. Chemical modification of bovine tissues by dye-mediated photooxidation. *J. Heart Valve Dis.* 6: 416–423.
171. Moore, M. A. and Phillips, R. E., 1997. Biocompatibility and immunologic properties of pericardial tissue stabilized by dye-mediated photooxidation. *J. Heart Valve Dis.* 6: 307–315.
172. Bianco, R. W., Phillips, R., Mrachek, J. and Witson, J., 1996. Feasibility evaluation of a new pericardial bioprosthesis with dye mediated photo-oxidized bovine pericardial tissue. *J. Heart Valve Dis.* 5: 317–322.
173. Moore, M. A., McIlroy, B. K. and Phillips, R. E., 1997. Nonaldehyde sterilization of biologic tissue for use in implantable medical devices. *ASAIO J.* 43: 23–30.

33
Immunology of Collagen-Based Biomaterials

Jerome A. Werkmeister and John A. M. Ramshaw
CSIRO, Parkville, Victoria, Australia

I. INTRODUCTION

A key issue for any biomaterial used in a medical implant is that it has to be immunologically inert. For collagen-based medical devices, the possibility of immunological reactions remains a theme of continued interest. Prior to the mid-1950s, collagen was thought to be nonimmunogenic, but there is now a large amount of evidence that clearly demonstrates that antibodies can be generated against collagens [1,2]. Indeed, with the advent of monoclonal antibody technology, antibodies can now be generated that are highly specific in relation to both the collagen type and its species of origin [2–6], and it is possible to map the epitopes recognized by these antibodies [7].

If any adverse immunological properties of collagens can be minimized or eliminated, then collagen has many advantages that make it an ideal choice of biomaterial in a wide range of clinical applications. These advantages arise from its being the main protein component of all connective tissues, where it provides strength and where it fulfils a wide range of biological functions, principally through its wide range of interactions with other biological macromolecules [8]. Also, because it is the natural protein of connective tissues, medical products made from collagen are seen as being able to integrate fully with the host.

Thus, a wide range of commercially produced devices are now available that are successful and have achieved consumer acceptance [9]. These successful products include, for example, hemostatic fleece [10], burn and wound dressings [11, 12], injectable collagen for soft tissue augmentation [13], and various replacement components for the cardiovascular system [14, 15].

Nevertheless, despite the clear functional success of these varied products, questions continue to be asked regarding the immunological safety of collagen products.

II. COLLAGEN STRUCTURE

The name collagen is frequently used as a generic term to cover a wide range of molecules. When a range of tissues is examined, a variety of different, genetically distinct, collagens can be found, each with characteristic tissue distributions [16]. All these different collagens share a distinct and characteristic structural motif, the triple helix.

The conformation of the triple helix consists of three polypeptide chains, each in an extended polyproline II–like helix, that are supercoiled around a common axis in a coiled-coil ropelike structure [17]. A consequence of this close packing of three staggered chains is that every third residue must be glycine, since every third residue in each polypeptide chain has its side chain internal in the helix and only glycine is small enough to fit in this sterically constrained position. Thus collagens all share a readily identifiable primary sequence pattern of a $(\text{Gly-X-Y})_n$ repeat, where the number of repeats, n, may vary between collagen types [18]. In this repeating structure, X and Y can be any amino acid, but are most frequently proline (Pro) (X-position) or hydroxyproline (Hyp) (Y-position) as a high content of these imino acids is required to stabilize the conformation. In the triple-helical structure, Gly residues are buried and solvent inaccessible. Residues in the X-position are highly exposed to solvent, while residues in the Y-position are somewhat less accessible to solvent than those in the X-position because of their proximity to the neighboring chain [19]. Thus, in the triple-helical conformation the amino acids in the X- and Y-positions of the triple helix are all available for interactions with other protein molecules, including antibodies.

Because type I collagen is the most abundant, being the predominant component of skin, tendon, and most collagenous tissues and accounting for up to 90% of the collagen found in the body, it is the most studied. It is also, therefore, the collagen normally encountered in biomaterial applications. The other collagen type that may be present in significant quantities in biomaterials is type III collagen, as it is a significant component of young skin which is often the source of collagen for biomaterials.

The various collagen types fall into distinct structural groups [18]. Type I collagen and the other interstitial collagens (types II, III, V and XI) all share the same basic structural arrangement (Fig. 1). When the interstitial collagens are synthesized within the cell they are present as a more complex structure with additional globular propeptide domains at both the N- and C-terminals (Fig. 1) [18]. Prior to folding into the trimeric molecule, these individual chains undergo a range of important secondary modification reactions. These include the hydroxylation in the Y-position of most Pro and some lysine (Lys) residues; certain of the resulting hydroxylysine (Hyl) residues are then further glycosylated [16]. Subsequent folding into the trimer, procollagen, terminates these hydroxylation and glycosylation reactions. The procollagen is then secreted from the cell and further processed, with specific enzymes normally removing the globular propeptide domains. These cleavage reactions leave short, non-triple-helical domains, called telopeptides, at each end of the collagen molecule. Thus, in the extracellular matrix, the triple-helical domain of collagen forms the predominant structural unit. During the accumulation of the processed collagen into the fibrils of the extracellular matrix, the telopeptides play an essential role in the maturation of the collagen matrix through involvement in intermolecular cross-link formation. Specific Lys residues in the telopeptides are converted to aldehydes by the enzyme lysyl oxidase, and these aldehydes then undergo a range of condensation reactions with adjacent Lys or Hyl residues to form cross-links that stabilize the collagen fibrils. The other amino acids involved in these reactions are also highly specific to particular collagen chains and sequence positions [16].

III. ANTIGENIC DOMAINS IN COLLAGEN

Antigenic determinants in proteins can be classified as either being "sequential" or "conformational" [20]. Sequential determinants are dependent solely on the amino acid sequence of the protein and are not dependent on any secondary or tertiary structure. Thus, they comprise consecutive amino acids in the primary structure, although the importance of some residues along the determinant sequence may be considerably less than for others. Conformational deter-

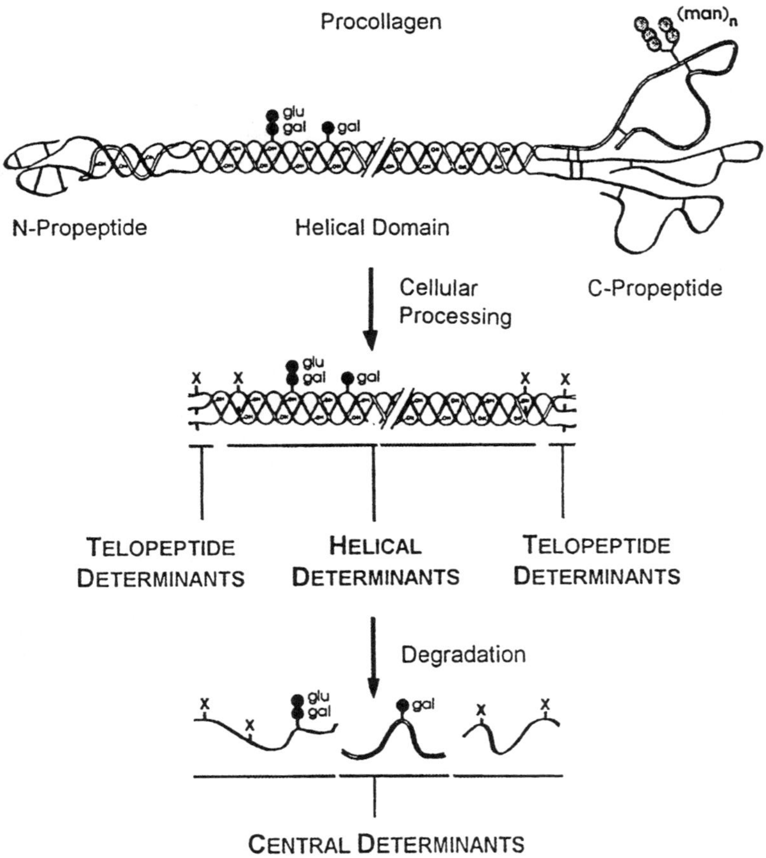

Figure 1 Structure and antigenic domains of interstitial collagens. X represents the locations of potential crosslink sites, and gal (galactose) and glu (glucose) represent potential glycosylation of hydroxylysine residues.

minants, on the other hand, are dependent on the tertiary structure of the proteins, and so, in addition to requiring a defined conformation, these determinants may comprise amino acid segments from distinct, noncontinuous regions of the protein molecule.

For collagens, both these types of antigenic determinant have been observed. For processed collagen, lacking the propeptide domains, conformational determinants depend on the intact triple helix (Fig. 1). These determinants are lost when the collagen is denatured and the triple helix is unfolded. These determinants are referred to as "helical determinants." Because of the rigid, rodlike structure of the triple helix, it is not possible for different regions on individual collagen molecules to contribute to a single antigenic determinant, and so these helical determinants also comprise short continuous regions in the amino acid sequence of the protein. In tissue it may be possible that adjacent collagen molecules could contribute to a single conformational determinant, but evidence for such quaternary structure determinants has not yet been observed.

In collagens, sequential determinants are provided by the non-triple-helical telopeptide regions at the ends of the processed collagen molecule. These regions may have defined conformations in native tissue [21] and so could also provide conformational determinants. Also, when collagen is denatured, the individual unfolded chains or chain fragments provide sequential determinants, called "central determinants" (Fig. 1).

In addition, it is possible that the various secondary modification processes that affect the collagen molecule, and any chemical modifications during manufacturing of collagen-based devices, may lead to specific determinants. For example, the glycosylation of Hyl residues and the cross-links between the telopeptides and an adjacent triple-helical domain could lead to specific determinants.

For collagen-based biomaterials, the importance of these different potential determinants will depend on the form and methods used in the manufacture of the collagen device.

IV. FORMS OF COLLAGEN-BASED BIOMATERIALS

Biomaterials made from collagen come in two distinct forms. In one, the collagen is prepared in a purified form, generally as a soluble form, and then this collagen is used to make a reconstituted collagen biomaterial. In the other, the native structure of a collagenous tissue is used and the material adapted for use as a biomaterial. Because of the different approaches to manufacture, there are differences in the immunological properties of these two classes of material.

Purified, soluble collagen can be prepared easily and purified to provide medical grade material. Generally, an enzyme treatment of minced tissue, such as young skin, is used in collagen preparation. The enzyme normally used is the acid protease pepsin, acting by cleaving the telopeptide cross-linking regions. This process is effective as the triple-helical domain is particularly resistant to proteolysis. The solubilized collagen can then be purified, for example, by a combination of various precipitation methods [22]. For the production of commercial materials, a variety of additional proprietary approaches exist.

An advantage of this purification process is that the enzyme treatment removes, or minimizes, the contamination by intact serum components that are present in the starting tissue. Also, during the enzyme digestion the telopeptides are cleaved in part from the triple-helical domain, thus reducing the amount of a potential immunogenic domain. On the other hand, this cleavage may leave peptide fragments cross-linked to the triple-helical domain, so introducing a new, potential antigenic determinant.

Chemical approaches have also been described for making soluble collagen [23]. These lack the specificity of the enzyme process and can lead to cleavage in the collagen helix domain while not removing the telopeptide domains with the same efficiency. This type of approach would, therefore, be less effective from an immunological perspective, since it could introduce new determinants based on denatured fragments of collagen and more of the potential immunogenic determinants from the telopeptide domains are retained.

Another approach to providing a purified collagen material is to mince the tissue until it is a fine dispersion, and then to wash the resulting fibrillar particles [10]. This washing, however, may not always be fully effective and potentially immunogenic serum and tissue proteins may remain with the collagen product and lead to immunological reactions [24]. Mild enzyme treatments, for example using ficin, have been used to minimize the level of impurities in this type of product.

Tissue-based devices are made using various approaches. Thus, in some devices, for example bioprosthetic heart valves, the native organisation of the tissue is used and this structure is preserved by chemical stabilization along with mounting the tissue on a suitable frame to allow clinical insertion. In others, a natural tissue, such as pericardium, is stabilized and then the device, for example a heart valve, is manufactured by cutting the material to the desired shape and mounting it in a frame. In other devices, for example arterial replacements, tissue can be grown in a heterologous host using a mandrel [15]. The explanted tissue with the desired shape and form is then stabilized prior to use. Since mechanical stability is a key feature of these devices, the collagen is retained in an intact form, with the telopeptides and the natural crosslinks still present.

In all tissue-based devices the tissue is stabilized prior to use. This is essential if the device, such as a cardiovascular replacement, is to have long-term durability. The most frequently used stabilizing agent has been glutaraldehyde, which cross-links proteins via the lysine residues [25]. A disadvantage attributed to glutaraldehyde is that it may increase the propensity for the treated tissue to calcify [26]. For this reason alternative approaches for use of glutaraldehyde [27] and other stabilization reagents have also been examined [28]. These alternatives include, for example, epoxides, which also react with lysine residues and carbodimides that form bonds between carboxyl side chains of aspartic acid or glutamic acid and lysine amino groups [28, 29].

Specific stabilization processes are used by different companies, but are generally not preceded by a significant wash procedure. Therefore, it is likely that these devices contain a significant amount of other noncollagenous proteins, and that these other proteins would be potential immunogens. These proteins would also react with the stabilizing agent and so would become an integral part of the device. However, the chemical modifications that take place have the potential to mask the collagen and other proteins and render them inaccessible as immunogens. On the other hand, it is possible that the modified proteins could, in certain cases, introduce new antigenic determinants not present originally. In all cases, the nature and extent of the specific processing used will determine the immunological properties of the particular product.

V. IMMUNOLOGICAL STUDIES ON PURIFIED COLLAGEN

The majority of collagens are poorly immunogenic, certainly when compared to many other globular proteins [1, 30]. The reason for this is poorly understood but the interspecies similarity in sequence homology as well as the overall similarity in structure are some key features which might contribute to its poor immunogenicity. Indeed the many studies that have been undertaken to generate polyclonal and monoclonal antibodies against collagens have required the use of adjuvants, particularly the use of complete Freund's adjuvant, which enhances the persistence and delivery of the immunogen, and which generally augments the immune response. For production of high titre antibodies or monoclonal antibodies with good affinity, high dosages, repetitive injections and often techniques to allow for selection of low frequency of antibody-producing cells are necessary [3–6].

Proteins in general can possess a number of immunogenic determinants which can induce an immune response, and a number of antigenic determinants or epitopes which will react with one or more antibodies. The natures of these determinants are not necessarily identical and each will independently depend on a number of factors including the size, sequence and structure of the presenting molecular determinant, and how each interacts with the immune system. In the case of immunogenicity, the determinants have to interact with a complex system of different types of T cells, B cells and antigen presenting cells, while antigenic determinants are largely defined as parts of the protein which react with either the B cell (or the differentiated antibody producing plasma cell), or more simply with the antibodies produced by these cells. It is this last aspect of monitoring the reactivity of sera from immunized animals or humans, which is predominantly measured and is used as an index of immune reactivity.

As discussed in the previous section, there are a number of well-defined regions within the procollagen molecule which can act as potential immunogenic or antigenic determinants. The part of the protein molecule that is recognized by antibodies or involved in eliciting an immune response is dependent on the species of investigation [1, 30]. Like most immune responses, the nature of this response is classically T cell dependent [31], although an early study had suggested a T cell independence for rat collagen [32], which might have been due to an early IgM response. More typically, Nowack and colleagues [31] elegantly demonstrated the necessity for T cells by observing an absence of an antibody response in athymic nude mice and mice whose immune systems were totally ablated by thymectomy and irradiation. Restoration of an antibody response could only be achieved after adoptive transfer of T cells to the nude mice or by transferring both T and B cells to the immuno-incompetent mouse.

Strong supporting evidence for a T-cell-dependent immune response to collagens has been the identification of immune response (Ir) gene(s) located in the IA or IB subregion of the major histocompatibility (H-2) locus [33]. The clarification of Ir gene control for collagen challenge and response in humans is less defined. Some work has suggested that the DR4 antigen [34] and either the DR2 alone or in combination with the DR4 [35] was positively associated with an individual's capacity to mount an immune response. Other studies have found a lack of association with DR4 and immunoreactivity, and have even suggested that DR4 inhibits anticollagen reactivity [36]. In humans injected with bovine collagen, no single immunodominant region on the collagen was detected [37]. Similar studies in mice [4], chicken [38] and rabbits [30] show that there appear to be immunoreactive sites along the entire length of the collagen.

The antibody response to collagens is not only genetically regulated, but also varies between species. For example, chickens, mice and rats will more often produce antibodies to native helical conformational determinants, while rabbits generally produce antibodies toward nonconformational terminal globular and telopeptide determinants, sequential central determinants and to some extent helical determinants. The most immunogenic domains of the collagen are perhaps the globular procollagen and the non-triple-helical telopeptide regions, although the globular domains are not present in processed collagens in tissues. The first indication of these immunodominant domains came from studies which showed a severe reduction of antibody responses in rabbits with calf collagens that had been treated with pepsin [39]. This has now been confirmed by a large number of subsequent investigations, largely in rabbits, and predominantly with collagen type I [1, 30, 38]. These studies have indicated the sequential nature of the epitope on individual chains and have defined a number of antigenic determinants for bovine, human and rat type I collagen [30]. The telopeptide domains of collagen are regions where there are large degrees of species variability [40], and are therefore, not surprisingly, immunogenic in nature [30]. Tyrosine residues are known to augment the immunogenicity of collagen [41], and are prominently involved in almost all terminal telopeptide determinants [30].

Antibodies directed to the conformational helical determinants of collagen are of greatest importance with respect to the use of reconstituted and purified collagens for medical use. This type of antibody response to helical determinants is the most commonly observed pattern in chickens, rats, and mice [3–6, 42]. These antibodies are generally not cross-reactive to other collagen types and can have varying degrees of species specificity [3–6]. Given the importance of these triple-helical domains in collagens, it is a little surprising that there are only limited reports on the nature of the epitopes being recognised. In polyclonal human sera, multiple bands have been detected after electrophoresis [43], although there appears to be distinct immunodominant "hotspots" particularly in $\alpha 1$(I)-CB6 and $\alpha 2$(I)-CB4 fragments [37]. In mice, using a library of monoclonal antibodies to collagen type III, a clearer picture has emerged [4]. Again there is an apparent spread of immunoreactive conformational determinants along the length of the helix [4]. Interestingly, one of these sites has recently been mapped to a region close to a biologically important domain of the collagen helix, one which is associated with the $\alpha_2\beta_1$ integrin receptor that is involved in platelet interactions [7]. Antibody responses directed to the nonhelical central determinants (after immunization with native collagen) are rare and have been found in chickens and some rabbits [30] and as expected are highly cross-reactive between species.

Deliberate immunization with denatured protease-treated collagens can induce antibodies preferentially active against the single denatured α chains and not against conformational helical determinants [42]. It is evident here that the native collagen, upon denaturation, exposes new antigenic sites. This type of response can also occur when immunizing with fish collagens where the melting temperature of the collagen is below the body temperature of the host animal [Werkmeister et al., unpublished]. For biomaterial applications, it is important to keep in mind that the resorbing or degraded collagen implant could lead to this type of immune response. It is also important to keep this type of response in perspective since it is unlikely to lead to antibodies that will cross-react with native host collagens and lead to an autoimmune response.

The mechanism of immune activation when collagen is used as a biomedical implant has not been well defined. As mentioned above, for the collagen to generate an immune response, typically manifested by antibody generation, it requires cooperation of a number of distinct immune cells. Antigen presenting cells are clearly important and these are normally macrophages or a related cell. Dendritic cells such as epidermal Langerhans cells could be a likely candidate for priming of collagen-reactive T cells since it is the vicinity of the collagen implant. It is perhaps worth noting that it is fortuitous that these cells cannot present native collagen type I or II, but only small sequential peptide fragments [44] similar to what one might expect for a denatured collagen.

In all the above discussion, only one arm of the immune response has been addressed, namely the humoral or antibody response. A complete understanding on the immunological aspects of collagens is incomplete without mention of cell-mediated immunity (CMI) or delayed type hypersensitivity (DTH) responses. Characterization of CMI against collagen has come from early studies primarily using guinea pig skin tests and purified forms of collagens [45–48]. It is known that injection of collagen in complete Freund's adjuvant will produce a long-lasting DTH response independent of antibody generation [45]. Both native and denatured collagen can induce cellular responses in guinea pigs presensitized with only native collagen type I [49]. Conversely, sensitization with denatured collagen also led to positive skin tests after challenge with native collagen [46, 47]. Clearly, the antigenic determinants involved in the cellular type responses were different to those described for the humoral response to collagens. The lack of differences found in skin tests using pepsin-treated collagen in either the sensitization or the challenge phase of CMI provides further evidence [48]. When collagen is used as a medical implant for soft tissue augmentation, it is this type of "adverse" response which is detected in patients, and which is sometimes associated with antibody responses and extrapolated to autoimmune type responses.

VI. ANIMAL STUDIES OF COLLAGEN BIOMATERIAL PRODUCTS

Collagen, prepared as a soluble protein and not in any fabricated biomaterial form, when administered with adjuvant and injected repetitively will induce an antibody response, as discussed in the previous section. Similarly, it can elicit a cell-mediated response, which again, highlights the use of adjuvant during the multiple injections during the sensitization phase.

While these tests for humoral and cellular immunity are standard for assessing immunological potential, they may not relate directly to the immunogenic status of a prototype or fabricated collagen product, intended to be used without adjuvant and used mostly under less stringent "immunizing" schedules. One of the commonest applications of collagen as a biomaterial has been as injectable bovine collagen for soft tissue augmentation. Reactions to the material have been reported in animals and humans (see later), albeit of low incidence and intensity. Certainly in the short term, there seems to be no correlation with the presence of antibodies and the clinical performance.

Certainly the method of preparation of the collagen implant is important, namely whether it has been prepared as an acid-soluble collagen or as a pepsin-treated collagen product. It has been known for a long time that the former preparation with the telopeptides still present is considerably more immunogenic [39]. Also for tissue-based bioprostheses, some form of stabilization, generally glutaraldehyde treatment, is normally used to manufacture a durable device [9]. This type of chemical modification has also been used in implants of purified collagen in an attempt to enhance the persistence of the injected material. While glutaraldehyde treatment is often said to render the biological material inert, nonviable and nonimmunogenic, this cannot always be assumed [50].

A. Bovine Collagens for Soft Tissue Augmentation

Currently, bovine collagen is the major source of collagen for soft tissue augmentation and other biomedical applications, and is in widespread use for correction of dermal contour deficiencies, which might include wrinkles, scars, maxillofacial abnormalities and other cosmetic conditions. The largest studies on assessment of immunogenicity of collagen products has come from animal models using Zyderm in comparison with a number of other collagen products [24, 51]. In one hemostatic model using rabbit splenic laceration, Avitene (Avicon Inc., Humacao, Puerto Rico) implantation stimulated a significant antibody response, whereas Zyderm did not. In the more conventional guinea pig model, Zyderm was again found to induce a much lower antibody response than the range of common collagen hemostatic materials, namely Avitene, Collastat (Collastat, Kendall Co., Boston, MA) and Gelfoam (Gelfoam, Upjohn Co., Kalamazoo, MI). Avitene induced a marked immune response to a number of components in the implant [24, 51], including elevated titres against denatured collagen type III. Collastat collagen also elicited a strong antibody response, while Gelfoam, atelocollagen and even glutaraldehyde cross-linked collagen induced a lower level of humoral response. In a discussion paper in response to one of these studies [51], Fisher [52] questioned the accelerating disappearance of collagen after multiple injections. The point being made was that the disappearance of the collagen implant was due to an induced immune response. It is possible that an absence of a perceived response might simply mean that the kinetics of the immune response was such that it was not detected, or a more sensitive assay is needed to detect the response, either humoral or cellular. Certainly, assays like immunodiffusion and hemagglutination, that have been used and shown not to detect anticollagen antibodies [53], are not particularly sensitive. The enzyme-linked immunosorbent assay (ELISA) is more sensitive and now more commonly used. Nonetheless, even this assay might not be sensitive enough to detect levels of allergenic IgE antibodies that could arguably still elicit an immediate type hypersensitivity. In this case it may be wiser to include assays like the passive cutaneous anaphylactic (PCA) response for assessment of IgE functional activity.

B. Other Studies with Bovine Collagens

There are a number of other studies using bovine collagen distinct from the processed Zyderm I or II commercial collagens intended for soft tissue contour correction. Immunological responses, performed in a wide variety of models using bovine collagen materials intended for other applications, primarily as hemostatic or wound dressings, demonstrate good tolerance and wound healing with no apparent immunological activity [54]. Haemostagen, a collagen sponge manufactured by Bioetica (Lyon, France) and a collagen fabricated dermal wound dressing were totally nonimmunogenic [55, 56]. In the latter case, the source materials were also nonimmunogenic, whereas the tropocollagen material was immunogenic, consistent with previous experimental data [39]. Importantly, this study also performed a very sensitive assay for allergenic IgE antibody response, namely the PCA response, and found no responses with either the source material or the final product.

C. Studies with Nonbovine Collagens

While bovine collagens are relatively nonimmunogenic in most animal studies, they are still foreign proteins with the potential to elicit immune responses which can be detected in a proportion of individuals. Other sources of collagens have also been investigated as possible substitutes or alternatives. The equine collagen hemostatic and wound dressing agent (Tachotop) was totally nonimmunogenic in simulated model applications in the guinea pig with respect to both humoral and cellular response [57]. In this study, the immune response could be elicited in the presence of adjuvant. Porcine collagen, either as a collagen membrane (Bio-Gide) [58] for use in periodontal repair, or as a dermal xenograft for burns [56, 59] is nonimmunogenic in rabbits and rats.

For ultimate applications in humans, an autogenous collagenous substitute is considered to be the most ideal. Animal assessments, particularly aimed at immunogenicity, have been thus performed on human collagens. Liu and colleagues [60] compared the immunogenicity of human collagen from either skin or amnion in rats. The skin-derived collagen was more inflammatory, less persistent and more immunogenic with higher levels of antibodies, compared with the amnion collagen. The composition of the two preparations was different with respect to the ratios of collagen type I to type III. The authors have suggested that the type III collagen, which is enriched in the amnion extract, is the cause for the difference. Alternatively, the reactivity may be caused by impurities within the skin extract that are not present in the amnion preparation. Experience with attempts to raise monoclonal antibodies to human collagen types I and III indicate that type III is far more immunogenic than type I collagen [3, 4]. A more recent study using human amnion collagen in rats has found a significant increase in the level of circulating antibodies to human collagen, as well as an elevated lymphoblastic cellular response [61]. What was critical in this report was the timing of the response which peaked at 14 to 28 days and then subsided abruptly. In these animal studies with human collagens, the animal would naturally see the human collagen as foreign; the response in humans with these materials is likely to be negligible if anything.

D. Effect of Collagen Stabilization on Immunogenicity

By far the most widely used example is the injectable collagen product, Zyplast (Collagen Corp., Palo Alto, CA), and in this instance the concentration of glutaraldehyde used is around 0.01%. Apart from the cross-linking, Zyplast is similar to the original Zyderm I collagen [62]. The results are not clear on whether glutaraldehyde is able to remove immunogenic determinants in the bovine collagen implants. The treatment of the collagen may reduce some immunogenicity [63], although the typical hypersensitivity associated with a proportion of individuals still occurs [62]. In another study, the treated collagen was more immunogenic compared with the original Zyderm implant when antibodies were tested against their respective implant material [24].

In rabbits injected with untreated bovine collagen or glutaraldehyde- or cyanamide-treated collagen, only the group with untreated collagen showed high levels of immunogenicity and the level of activity increased with reexposure to immunogen [64]. Extensive cross-linking reduced but did not eliminate the humoral response to collagen. In other studies with glutaraldehyde cross-linked human collagens that had been γ-irradiated, the antibody response to collagen was not only significantly elevated but also more lasting [61]. Of greater concern, the specificity of the antibodies generated changed compared with antibodies produced in rats injected with untreated collagen. After cross-linking, the antibodies also reacted with bovine collagen and the host rat collagen.

E. Noncollagenous Immunogenic Components within the Collagen Implant

While collagens may have the potential to elicit humoral or cellular responses under the right conditions, these are generally low. When used as an implant, however, either as a natural tissue or as a reconstituted product, collagen implants might still be immunogenic for other reasons.

Allograft rejection of skin substitutes is a well-known immunological phenomenon. While the components of skin are multifactorial, it appears that the major noncellular component of collagen is not responsible for eliciting graft rejection [65]. In 84 dermal collagen allografts in rats, no rejection was observed and all implants were nonantigenic compared with typical, total allograft, skin substitutes. This process of removing "antigenic" cells as well as proteoglycans and glycoproteins has been used in biomaterial applications including vascular and tendon replacements. For instance, untreated bovine tendon xenografts, similar to the dermal allografts, stimulate a severe inflammatory response and extraction of cells with chloroform/methanol removes immunogenicity, although there was still an apparent transient humoral response [66]. The cells do not have to be viable to induce an immune response and, glutaraldehyde treatment did not remove immunogenicity at all time points investigated. However, there was no antibody response to the bovine collagen component.

Contamination of the final collagen biomaterial product, in the above case by cells or cell debris, can present as major immunodominant entities within the poorly immunogenic collagenous milieu. The contamination can sometimes arise from the reagents used in the preparative process or from noncollagenous proteins or other components within the source tissue being extracted. Trace amounts of pepsin, which is commonly used to prepare nonimmunogenic atelocollagen, may itself lead to the induction of unwanted antibodies [67]. Avitene, a commercial microfibrillar collagen hemostat, is highly immunogenic due to the presence of contaminating bovine serum albumin from the source material, not collagen [24]. Preclinical and clinical studies with various forms of Bio-Gide (Geistlich Biomaterials, Switzerland), a porcine collagenous membrane, also demonstrated a lack of immunogenicity of the source collagens types I and III, but possible reactivity against lipoproteins [58].

F. Enhanced Anticollagen Immune Responses due to Factors Not Related to the Implant

Depending on the application of the collagen biomaterial and the underlying status of the recipient host, immune responses may already exist to collagen for unknown reasons or may develop merely as a result of disease state or due to the surgery performed for implantation.

Certainly for patients with autoimmune type disorders, particularly rheumatoid arthritis, there is already an elevated immunity and it may be wise for stringent testing prior to further deliberate collagen priming. While the anticollagen antibodies are primarily against collagen type II, there may be some degree of cross-reactivity against collagens type I and III. In other vascular diseases like thromboanglitis obliterans (Buerger's disease) more generalized cellular and humoral immune responses including those directed to collagen types I and III are apparent [68]. Cellular immunity to collagen types I, II and III, assessed by blastogenesis, occurred in rabbits after partial medial meniscectomy [69]. In this study, the reactivity was likely due to the immune response to denatured or damaged collagens during the surgery, and may not be a problem for further implantation of a native collagenous implant. The converse has been proposed where, in hosts with preexisting antibodies directed to native bovine collagen type I, no enhanced activity is detected after treatment with a gelatin matrix implant [70]. In other applications like periodontal repair, the story may be different where there is both endogenous cellular responses to denatured homologous collagen [71] and humoral responses to native homologous collagens [72].

VII. IMMUNOLOGICAL RESPONSES TO COLLAGENOUS TISSUE IMPLANTS

Generation of an immune response against tissue prostheses, particularly the collagenous content, is generally regarded as undesirable, although, in some cases, an immune response can sometimes be desirable to augment collagenous resorption of the implant in order to enhance host tissue integration. Most tissue-based biomaterials rely on their durability for performance, as discussed earlier in this chapter. Depending on the intended application of the collagen biomaterial, it is usually necessary to use a chemical or radiation treatment to stabilize the proteins in the material to prevent or partially delay the breakdown or resorption of the implant.

There are only limited studies on immunogenicity against the natural untreated tissue intended for replacement or repair of damaged tissue. Venous allografts are mildly immunogenic [73], although this did not accelerate rejection [74]. A human acellular dermal allograft (Alloderm) has been used for the treatment of full-thickness burn injuries [75]. Similar to the animal studies of Oliver et al. [65], there was no immune response to the collagenous components in the skin and no rejection of the grafts. Like the animal allografts it was important to remove the immunogenic foreign cellular material. Use of the patient's (autologous) cells is naturally nonimmunogenic. In addition to burns applications, this procedure has been used in the development of a tissue equivalent for capsuloligamentary reconstruction of the knee [76]. Bovine collagen was used as the matrix for the cell seeding. No cellular reactions and only transient humoral responses were found.

In other applications, collagen has been used as a coating around or within the tissue material. The potential here for the collagen coating the device to mount an adverse immune response has been investigated. Two early studies involving collagen-impregnated vascular Dacron prostheses have shown positive results [77, 78]. Norgren and colleagues found an exceptionally high rate of patients (4 out of 11) with the collagen-Dacron graft elicited elevated antibovine antibodies; the reactivity was surprisingly against denatured, not native collagen [77]. In a larger study of 128 patients [78], no cell-mediated immune responses were found, although five patients did show seroconversion to bovine type I collagen. In this study, one patient who had received a control Dacron implant without collagen also apparently developed antibodies to collagen, which throws some questions onto the interpretation of the data. In both studies, the immune status did not effect the clinical performance of the graft. In a more recent study of 37 patients with gelatin-impregnated Dacron grafts (Unigraft) and 33 patients with the collagen-impregnated (Hemashield) graft, no antibodies were detected against native bovine collagen types I, II or III in the perioperative period or up to three months postoperatively [79].

A. Effect of Tissue Stabilization on Immunogenicity

Perhaps the best-known examples of biomedical stabilized devices that contain high amounts of collagen within the natural harvested tissue, are those used for cardiovascular applications, either as replacement blood vessels or as heart valves [14, 15]. There is enough evidence to suggest that stabilization of collagen-based devices does not necessarily make them exempt from immune recognition. Certainly, there is some evidence where there is an apparent antigen depressant effect of glutaraldehyde, the most commonly used chemical cross-linking agent [80, 81]. On the other hand, there are a number of convincing reports showing significant antibody responses and sometimes cell-mediated response to chemical or other cross-linked material [66, 82–84]. The degree of response can vary depending on the nature of the tissue, the amount of collagen in the material, and the cross-linking agent used. What is important is the specificity of the antibody (or cellular) reactivity and the relationship of the reactivity, if any, to the biological performance of the implant. In many cases studied, immunoreactivity was simply assessed against the native collagen molecule, where there is likely to be a reduction in reactivity. When the antigen assayed is the modified treated collagen, then immunoreactivity can sometimes be detected. With respect to the relevance of the reactivity, this is either unproved or is irrelevant. In the case of biological heart valves, where calcification of glutaraldehyde-preserved bioprostheses is a frequent long-term complication, this unwanted immunoreactivity may be an issue if proven to be related to the calcific degeneration. A large number of studies have shown that glutaraldehyde-treated porcine heart valves [85–87] and glutaraldehyde-treated pericardial tissue [88] are immunogenic; these studies have been reviewed extensively by Bajpai [50]. In some of the studies both the untreated and treated tissue has been compared as immunogen and as antigen in both cell-mediated and antibody assessment. It appears that there was immunoreactivity in both untreated and treated tissues [50]. These responses can be both humoral and cellular and vary in intensity and cross-reactivity. Both bovine and porcine pericardial tissues are immunogenic [50, 89], although bovine tissue, in some reports, was more immunogenic [84, 90]. Generation of new epitopes is not unique to glutaraldehyde cross-linking and can also occur with dye-mediated photo-oxidation [89].

Overall then, any form of stabilization has the potential to mask antigenic determinants, but can under some conditions, reveal other determinants not normally present in the native tissue. Whether this induced reactivity is involved in the pathogenesis of implant failure, or whether it can lead to reactivity against native tissue is unknown. At the very least, generation of these immunoreactive determinants might have some bearing on the longevity of the implanted foreign material and each type of tissue-derived bioprosthesis should be judged on it own merits.

VIII. CLINICAL IMMUNOLOGICAL RESPONSES TO PURIFIED COLLAGENS

Clinical evaluation of collagen immunogenicity is usually by a skin hypersensitivity test (an index for cellular immunity, delayed type IV) or by the presence of reactive antibodies (an index for humoral immunity). Experimentally it is reasonable to assess both types in patients

[51, 63, 91], although in the dermatologist's clinic it is more than likely that only the skin testing will be performed. The determinants on the collagen that elicit each type of response are not necessarily the same [46].

The majority of information has arisen directly from clinical analyses of patients receiving injectable bovine collagen implants, namely Zyderm I or II or Zyplast [51,62]. While there are studies from animal implantation of collagenous materials, it is the vast studies on humans, which have assessed both clinical reactions and antibody responses, that are more extensive and indeed more pertinent [43, 62, 92, 93]. Like most collagens used for medical applications, the Zyderm or Zyplast products are prepared from bovine dermis as pepsin-soluble collagens. The patients receiving injectable collagens are predominantly Caucasian (>90%) and female (>80%), with a median age range of 36–40 years [94, 95].

Prior to treatment, patients are skin tested with a sensitizing dose of implant collagen. The most common response reported is one of hypersensitivity, and around 3% of the population will have an underlying reaction [63, 94]. While this pretreatment skin test is routinely performed to assess the sensitivity and suitability of prospective patients, it does not screen out some patients who will develop an immune response after repeated injections. Around a further 1% to 2% of patients will develop clinical symptoms of delayed type hypersensitivity [62, 94]. Typical symptoms include a local edematous and erythematous reaction with induration and sometimes pruritis which usually lasts from four to six months but can persist up to periods in excess of one year.

There have been several reports demonstrating the presence of serum anticollagen antibodies in these patients with clinical hypersensitivity symptoms [43, 92, 93]. The class of antibody was usually IgG, although in a proportion of sera IgA was also detected; IgM was rare and IgE has not been detected [43, 63]. In patients where there was no visible clinical signs of reactivity, there have been reports of high percentage of sera with anticollagen antibodies [43, 96]. The reason for this phenomenon is unclear but may well be related to a response to beef in the normal diet, similar to what is seen with the detection of anti-milk-protein antibodies [97]. While there is no reason to associate the presence of these anticollagen antibodies to any subsequent clinical sequelae, screening for these may be beneficial to further select patients more likely to have an adverse reaction.

Although the immunoreactivity of collagen is a problem in only a small number of patients, the total population pool is large with significant numbers of patients likely to be affected. It is perhaps surprising that little information is available on the nature of the epitopes being recognized in sera or the determinants which trigger such severe hypersensitivity reactions. Multiple bands after electrophoresis were detected with a variety of sera [43]. Analyses of cyanogen bromide (CB) fragments of isolated collagen chains with reactive sera have highlighted certain hot spots particularly in α1(I)-CB6 and α2(I)-CB4 fragments [37]. No data exist on putative cellular antigenic determinants.

A. Other Studies with Bovine Collagen Excluding Zyderm and Zyplast

Collagen has been used extensively in the treatment of patients with severe burns and with chronic ulcers [98, 99] with no reports of apparent adverse immunological effects. In addition, as a fibrillar collagen which can aggregate platelets and promote coagulation, it has been used as a hemostat in a variety of surgical procedures with minimal or no immune interference [100]. An alternate bovine collagen implant, Atelocollagen (Koken Co., Ltd., Japan) has also been investigated in a large follow-up study of 705 patients [101]. Both skin testing and antibody assessments were assayed. Twenty-seven patients (3.8%) were positive in the preliminary skin tests, and a further 2.3% developed positive reactions after implant. This incidence of immunoreactivity against the Koken collagen was reasonably similar to that found with Zyderm or Zyplast collagens. A strong association (92%) was found between the presence of the antibovine collagen antibodies and positive skin testing. Furthermore, in a prospective study of 420 patients, 6% presented with some degree of hypersensitive skin test while a further 8.3% with a negative skin reaction had antibodies to collagen in their sera [102]. Collagen therapy proceeded only with patients who were negative in both cellular and humoral immunological testing. Under these stringent guidelines, no major clinical adverse reactions were detected in the study, although a small percentage (0.5%) did develop indications of some form of minor reactions. While some authors have rightly said that there is no reason to associate a positive anticollagen antibody response with the cascade of clinical reactions, perhaps this simple assay night be used to screen out a broader population of potentially reactive individuals.

Injectable bovine collagen has also been used as a bulking agent for the treatment of urinary stress incontinence. The commercial collagen, Contigen Bard Collagen Implant (Collagen Corp., Palo Alto, CA) is a glutaraldehyde cross-linked bovine collagen implant. Around 5% of patients treated with the collagen had preexisting antibodies to native bovine collagen type I, and an additional 23% developed antibodies after treatment [103]. IgG was again the major antibody class detected in all positive cases, IgA and IgM were found in 40% and 0.6% of positive sera, respectively, and IgE was not detected. This rate of antibody response is much higher than those reported in studies of correction of skin defects discussed above. The antibodies did not cross-react with human collagens and were not associated with clinical performance. Like the skin augmentation studies, IgE, the predominant immunoglobulin involved in allergic immediate hypersensitivity reactions, was negative. The lowest detectable concentrations mentioned for all other antibodies is around 1 to 10 mg/L. It is quite reasonable that levels of IgE well below this detection level could still be present and capable of binding to mast cells and basophils. Recently, two cases of allergic (IgE mediated) reaction to bovine collagen corneal shields have been described [104]. Both patients presented with subconjunctival edema, and reacted positively in skin tests using extracts of bovine collagen from the corneal shield as well as from catgut sutures and food gelatin. While the nature and the specificity of the response is puzzling, it does draw further caution particularly to patients who might have a previous history of allergic reactions to bovine collagens or even dietary gelatin products.

Serological data also exist on two bone-grafting substitutes comprising collagen, Alveoform (Collagen Corporation, Palo Alto) and Collagraft (Collagen Corporation, Palo Alto; Zimmer, Warsaw, IN) [105]. Both Alveoform and Collagraft comprise bovine fibrillar collagen; in the former, the product also contains hydroxylapatite (HA) particles, in the latter the product contains a mixture of HA and tricalcium phosphate. Out of 77 patients receiving Alveoform, 5 had preexisting antibovine antibodies and 5 additional patients developed antibodies after surgery. In the 134 patients receiving Collagraft, 10 had antibodies directed to the native bovine collagen implant. In both studies, the elevated immune response was not associated with the onset of any adverse effect, particularly the ability to interfere with fracture healing.

XI. THE QUESTION OF AUTOIMMUNE DISEASE

How safe is collagen and can one be guaranteed of no adverse side effects? With more than 750,000 people treated with injectable collagens since 1976, and only some 1% to 2% showing hypersensitive responses, the product would seem reasonably good. All medical treatments are subject to some form of risk. The question is the nature and intensity of the risk. The most important question regarding immunogenicity of collagen-based biomaterials is indeed whether there is cross-reactivity in the immune response to the implanted collagen to allow the development of autoimmune diseases. Autoimmune diseases develop basically when the allergic-type responses can cause destruction of the body's own tissues. The normal individual will not produce antibodies or elicit cellular responses to any part of its own body; immunologically, one does not produce immunity to "self" antigens which are made tolerant during development.

Indeed there are many connective tissue autoimmune and heritable diseases where anticollagen antibodies have been detected and may well be associated with the disease. These include, for example, rheumatoid arthritis (anticollagen II) [106], Goodpasture syndrome (anticollagen IV) [107], X-linked Alport syndrome (anticollagen IV) [108], epidermolysis bullosa acquisita (anticollagen VII) [109], and some forms of vasculitis (anticollagen IV) [110]. The question as to whether the anticollagen type antibodies are responsible for the pathogenesis of the disease or a consequence of the progression of the disease remains contentious. In the many studies discussed earlier on collagen biomaterial implants, and where antibovine collagen antibodies have been detected, no cross-reactivity has been demonstrated in the vast majority of patients studied [37, 92, 93]. There is one isolated report of a 45-year-old female patient who had received eight injections of Zyderm and two injections of Zyplast over a three-year period [111]. High levels of antibodies were detected against bovine collagen, and these antibodies also cross-reacted strongly with human collagen type III, C1q and weakly with human collagen type I. A similar finding in rats on the response to human placental type I collagen, mentioned earlier [61], reported cross-reactivity against the host collagens which may be a sign of an autoimmune response.

The primary autoimmune diseases where there has been some concern are polymyositis and dermatomyositis (PM/DM), two very rare conditions which involve the destruction of the body's own muscle tissue. Questions have been raised regarding the onset of PM/DM after treatment with bovine collagen. Mostly these have been anecdotal estimates fueled by media and litigation in the United States. The absolute numbers of patients receiving bovine collagen implants is very high, and because of this, there will be a "normal" expected percentage of these patients who will develop PM/DM for reasons unrelated to the collagen. The key scientific issue is whether the incidence of observed cases of PM/DM in collagen-treated patients exceeds this background level. This kind of assessment has now been performed by epidemiological analyses of large numbers of patients receiving collagen implants. The majority of investigations have found no scientific data to support the alleged link between injected bovine collagen and PM/DM [112–114]. In all these studies the consistent finding was that there were far fewer cases than expected among the collagen users. Cukier and colleagues [115] also found no elevated frequency in the incidence of the disease, but make an important point that this and all other studies rely heavily on the assumption that all patients have been monitored scrupulously for PM/DM, the symptoms of which are rather vague and not easy to diagnose. In this study, an average interval of 6.4 months after collagen treatment was found for onset of PM/DM in the positive patients, and the authors point out that this rate of onset is an extremely rare and significantly unlikely event as assessed by a Monte Carlo simulation statistical test.

It appears from the epidemiological data that it is very unlikely that bovine collagen implants have anything to do with the onset of these autoimmune diseases. All these reports are based on interpretation of statistics. No causal relationship between the collagen itself and the onset of disease has been established.

X. CONCLUSION

Extensive experimentation and clinical use has shown that heterologous collagens can be made into safe and effective devices if due care is taken to reduce or eliminate potential immunological components during manufacture. In the future, more attention may be given to homologous collagens, since materials without risk of disease can be made using recombinant techniques [116]. However, even with these new materials, as well as with variations on existing approaches, constant vigilance on potential immunological effects will still be needed.

REFERENCES

1. Timpl, R., 1984. Immunology of the collagens. In: K. A. Piez and A. H. Reddi (Editors), *Extracellular Matrix Biochemistry*, *Elsevier*, New York, pp. 159–190.
2. Glattauer, V., Ramshaw, J. A. M., Tebb, T. A. and Werkmeister, J. A., 1991. Conformational epitopes on interstitial collagens. *Int. J. Biol. Macromol.*, 13: 140–146.
3. Werkmeister, J. A., Ramshaw, J. A. M. and Ellender, G., 1990. Characterisation of a monoclonal antibody against native human type I collagen. *Eur. J. Biochem.*, 187: 439–443.
4. Werkmeister, J. A. and Ramshaw, J. A. M., 1991. Multiple antigenic determinants on type III collagen. *Biochem. J.*, 274: 895–898.
5. Werkmeister, J. A. and Ramshaw, J. A. M., 1991. Monoclonal antibodies to type V collagen as markers for new tissue deposition associated with biomaterial implants. *J. Histochem. Cytochem.*, 39: 1215–1220.
6. Werkmeister, J. A., Tebb, T. A., White, J. F. and Ramshaw, J. A. M., 1993. Monoclonal antibodies to type VI collagen demonstrate new tissue augmentation of a collagen-based biomaterial implant. *J. Histochem. Cytochem.*, 41: 1701–1706.
7. Glattauer, V., Werkmeister, J. A., Kirkpatrick, A. and Ramshaw, J. A. M., 1997. Identification of the epitope for a monoclonal antibody that blocks platelet aggregation by type III collagen. *Biochem. J.*, 323: 45–49.
8. Kadler, K., 1994. Extracellular matrix 1: Fibril-forming collagens. *Protein Profile*, 5: 519–638.
9. Ramshaw, J. A. M., Werkmeister, J. A. and Glattauer, V., 1995. Collagen-based biomaterials. *Biotechnol. Genetic Eng. Rev.*, 13: 335–382.
10. Hait, M. R., 1970. Microcrystalline collagen. A new hemostatic agent. *Am. J. Surg.*, 120: 330.
11. Yannas, I. V. and Burke, J. F., 1980. Design of an artificial skin. I. Basic design principles. *J. Biomed. Mater. Res.*, 14: 65–81.
12. Miyata, T., Taira, T. and Noishiki, Y., 1992. Collagen engineering for biomaterial use. *Clin. Mat.*, 9: 139–148.
13. Knapp, T. R., Kaplan, E. N. and Daniels, J. R., 1977. Injectable collagen for soft tissue augmentation. *Plast. Reconst. Surg.*, 60: 398–405.
14. Ferrans, V. J., Hilbert, S. L., Tomita, Y., Jones, M. and Roberts, W. C., 1988. Morphology of collagen in bioprosthetic heart valves. In: M. E. Nimni (Editor), *Collagen*, Vol. 3, CRC Press, Boca Raton, pp. 145–190.
15. Ramshaw, J. A. M., Edwards, G. A. and Werkmeister, J. A., 1995. Tissue-polymer composite vascular prostheses. In: D. L. Wise, D. J. Trantolo, D. E. Altobelli, M. J. Yaszemski, J. D. Gresser and E. R. Schwartz (Editors), *Encyclopedic Handbook of Biomaterials and Bioengineering*, Part B, Vol. 2, Marcel Dekker, New York, pp. 953–978.
16. Bateman, J. F., Lamande, S. and Ramshaw, J. A. M., 1995. Collagen superfamily. In: W. D. Comper (Editor), *Extracellular Matrix*: Vol. 2, *Molecular Components and Interactions*, Harwood Academic, Amsterdam, pp. 22–67.
17. Fraser, R. D. B., MacRae, T. P. and Suzuki, E., 1979. Chain conformation in the collagen molecule. *J. Mol. Biol.*, 129: 463–481.
18. Ayad, S., Boot-Handford, R., Humphries, M. J., Kadler, K. and Shuttleworth, A., 1994. *The Extracellular Matrix Facts Book*. Academic Press Harcourt Brace, London.
19. Jones, E. Y. and Miller, A., 1991. Analysis of structural design features in collagen. *J. Mol. Biol.*, 218: 209–219.
20. Sela, M., Schechter, B., Schetcher, M. and Borek, F., 1967. Antibodies to sequential and conformational antigenic determinants. *Cold Spring Harbor Symp. Quant. Biol.*, 32: 537–545.

21. Jones, E. Y. and Miller, A., 1987. Structural models for the N- and C-telopeptide regions of interstitial collagens. *Biopolymers*, 26: 463–480.
22. Trelstad, R. L., 1982. Native collagen fractionation. In: H. Furthmayr (Editor), *Immunochemistry of the Extracellular Matrix*, Vol. 1, CRC Press, Boca Raton, pp. 31–41.
23. Chvapil, M., Kronenthal, R. L. and Van Winkle, W., Jr., 1973. Medical and surgical applications of collagen. *Int. Rev. Conn. Tissue Res.*, 6: 1–61.
24. DeLustro, F., Condell, R. A., Nguyen, M. A. and McPherson, J. M., 1986. A comparative study of the biologic and immunologic response to medical devices derived from dermal collagen. *J. Biomed. Mater. Res.*, 20: 109–120.
25. Nimni, M. E., Cheung, D., Strates, B., Kodama, M. and Sheikh, K., 1987. Chemically modified collagen: A natural biomaterial for tissue replacement. *J. Biomed. Mater. Res.*, 21: 741–771.
26. Levy, R. J., Schoen, F. J., Levy, J. T., Nelson, A. C., Howard, S. L. and Oshry, L. J., 1983. Biologic determinants of dystrophic calcification and osteocalcin deposition in glutaraldehyde-preserved porcine aortic valve leaflets implanted subcutaneously in rats. *Am. J. Pathol.*, 113: 143–155.
27. Casagranda, F., Ellender, G., Werkmeister, J. A. and Ramshaw, J. A. M., 1994. Evaluation of alternative glutaraldehyde stabilisation strategies for collagenous biomaterials. *J. Mat. Sci. Mat. Med.*, 5: 332–337.
28. Khor, E., 1997. Methods for the treatment of collagenous tissues for bioprostheses. *Biomaterials*, 18: 95–105.
29. Lee, J. M., Pereira, C. A. and Kan, W. K., 1994. Effect of molecular structure of poly(glycidyl ether) reagents on crosslinking and mechanical properties of bovine pericardial xenograft materials. *J. Biomed. Mater. Res.*, 28: 981–992.
30. Furthmayr, H. and Timpl, R., 1976. Immunochemistry of collagens and procollagens. *Int. Rev. Connect. Tissue Res.*, 7: 61–99.
31. Nowack, H., Hahn, E. and Timpl, R., 1976. Requirement for T cells in the antibody response of mice to calf skin collagen. *Immunology*, 30: 29–32.
32. Fuchs, S., Mozes, E., Maoz, A. and Sela, M., 1974. Thymus independence of a collagen-like synthetic polypeptide and of collagen and the need for thymus and bone marrow-cell cooperation in the immune response to gelatin. *J. Exp. Med.*, 139: 148–158.
33. Nowack, H., Hahn, E., David, C. S., Timpl, R. and Gotze, D., 1975. Immune response control to calf collagen type I in mice: A combined control of Ir-1A and non H-2 linked genes. *Immunogenetics*, 2: 331–335.
34. Stuart, J. M. and Kang, A. H., 1986. Monkeying around with collagen autoimmunity and arthritis. *Lab. Invest.*, 54: 1–3.
35. Cooperman, L. S., Garovoy, M. R. and Sondel, P. M., 1986. Association of the HLA-DR2/DR4 phenotype with skin test responses to bovine dermal collagen: A potential interaction of two MHC alleles in regulating an immune response. *Human Immunol.*, 17: 471–479.
36. Vanderveen, E. E., McCoy, J. P., Schade, W., Kapur, J. J., Hamilton, T., Ragsdale, C., Grekin, R. C. and Swanson, N. A., 1986. The association of HLA and immune responses to bovine collagen implants. *Arch. Dermatol.*, 122: 650–654.
37. Ellingsworth, L. R., DeLustro, F., Brennan, J. E., Sawamura, S. and McPherson, J. M., 1986. The human immune response to reconstituted bovine collagen. *J. Immunol.*, 136: 877–882.
38. Furthmayr, H., Stoltz, M., Becker, U., Beil, W. and Timpl, R., 1972. Chicken antibodies to soluble rat collagen. II. Specificity of the reactions with individual polypeptide chains and cyanogen bromide peptides of collagen. *Immunochemistry*, 9: 789–798.
39. Schmitt, F. O., Levine, L., Drake, M. P., Rubin, A. L., Pfahl, D. and Davison, P. F., 1964. The antigenicity of tropocollagen. *Proc. Natl. Acad. Sci. USA*, 51: 493–497.
40. Michaeli, D., Martin, G. R., Kettman, J., Benjamini, E., Leung, D. Y. K. and Blatt, B. A., 1969. Localization of antigenic determinants in the polypeptide chains of collagen. *Science*, 166: 1522–1524.
41. Kirrane, J. A. and Robertson, W. B., 1968. The antigenicity of native and tyrosylated neutral-salt-soluble rat collagen. *Immunology*, 14: 139–148.

42. Beil, W., Timpl, R. and Furthmayr, H., 1973. Conformation dependence of antigenic determinants on the collagen molecule. *Immunology*, 24: 13–24.
43. McCoy, J. P., Jr., Schade, W., Siegle, R. J., Waldinger, T. P., Vanderveen, E. E. and Swanson, N. A., 1985. Characterization of the humoral immune response to bovine collagen implants. *Arch. Dermatol.*, 121: 990–994.
44. Michaelsson, E., Holmdahl, M., Engstrom, A., Burkhardt, H., Scheynius, A. and Holmdahl, R., 1995. Macrophages, but not dendritic cells, present collagen to T cells. *Eur. J. Immunol.*, 25: 2234–2241.
45. Adelmann, B. C., Kirrane, J. and Glynn, L. E., 1972. The structural basis of cell-mediated immunological reactions of collagen. Characteristics of cutaneous delayed hypersensitivity reactions in specifically sensitised guinea-pigs. *Immunology*, 23: 723–737.
46. Adelmann, B. C., 1972. The structural basis of cell-mediated immunological reactions of collagen. Reactivity of separated α-chains of calf and rat collagen in cutaneous delayed hypersensitivity reactions. *Immunology*, 23: 739–748.
47. Adelmann, B. C., 1973. The structural basis of cell-mediated immunological reactions of collagen. Recognition by the cutaneous delayed hypersensitivity reaction in Guinea-pigs of conformational alterations of rat and calf skin collagen. *Immunology*, 24: 871–877.
48. Adelmann, B. C. and Kirrane, J., 1973. The structural basis of cell-mediated immunological reactions of collagen. The species specificity of the cutaneous delayed hypersensitivity reaction. *Immunology*, 25: 123–130.
49. Senyk, G. and Michaeli, D., 1973. Introduction of cell-mediated immunity and tolerance to homologous collagen in Guinea pigs: Demonstration of antigen-reactive cells for a self-antigen. *J. Immunol.*, 111: 1381–1388.
50. Bajpai, P. K., 1985. Immunological aspects of treated natural tissue prostheses. In: D. F. Williams (Editor) *Biocompatibility of tissue analogs*, Vol. 1. CRC Press, Boca Raton, pp. 5–25.
51. DeLustro, F., Smith, S. T., Sundsmo, J., Salem, G., Kincaid, S. and Ellingsworth L., 1986. Reaction to injectable collagen: Results in animal models and clinical use. *Plastic Reconst. Surg.*, 79: 581–594.
52. Fisher, J. C., 1986. Discussion on "Reaction to injectable collagen: Results in animal models and clinical use" by Delustro et al. *Plastic Reconst. Surg.*, 79: 593–594.
53. Streckbein, R. G., 1979. Tierexperimentelle versuche zur induktion humoraler antikorper durch wiederholte implantation von dentalen gelatine-praparaten und von trockenfibrinschaum. *Dtsch. Zahnarztl. Z.*, 34: 843–845.
54. Chvapil, M., Chvapil, T. A. and Owen, J. A., 1986. Reaction of various skin wound in the rat to collagen sponge dressings. *J. Surg. Res.*, 41: 410–418.
55. Anselme, K., Bacques, C., Charriere, G., Hartmann, D. J., Herbage, D. and Garrone, R., 1990. Tissue reaction to subcutaneous implantation of a collagen sponge. A histological, ultrastructural, and immunological study. *J. Biomed. Mater. Res.*, 24: 689–703.
56. Takeda, U., Izawa, M., Koeda, T. and Shibata, U., 1983. Laboratory study of collagen wound dressings (CAS). II. An immunological study (I) Immunogenicity in rabbits and mice. *J. Dermatol.*, 10: 593–601.
57. Adelmann-Grill, B. C. and Otto, K., 1987. Immunological safety evaluation of a haemostatic agent and wound dressing made from horse collagen fibrils. *Arzneimittel-Forsch.*, 37: 802–805.
58. Schlegal, A. K., Möhler, H., Busch, F. and Mehl, A., 1997. Preclinical and clinical studies of a collagen membrane (Bio-Gide®). *Biomaterials*, 18: 535–538.
59. Wang, H. J., Chen, T. M. and Cheng, T. Y., 1997. Use of a porcine dermis template to enhance widely expanded mesh autologous split-thickness skin graft growth: preliminary report. *J. Trauma*, 42: 177–182.
60. Liu, B., Harrell, R., Xu, Z., Dresden, M. H. and Spira, M., 1989. Immune response to γ-irradiated injectable human amnion and human skin collagens in the rat. *Arch. Dermatol.*, 125: 1084–1089.
61. Quteish, D. and Dolby, A. E., 1991. Immune responses to implanted human collagen graft in rats. *J. Periodont. Res.*, 26: 114–121.

62. Keefe, J., Wauk, L., Chu, S. and DeLustro, F., 1992. Clinical use of injectable bovine collagen: A decade of experience. *Clin. Mat.*, 6: 155–162.
63. DeLustro, F., Mackinnon, V. and Swanson, N. A., 1988. Immunology of injectable collagen in human subjects. *J. Dermatol. Surg. Oncol.*, 14: 49–56.
64. Meade, K. R. and Silver, F. H., 1990. Immunogenicity of collagenous implants. *Biomaterials*, 11: 176–180.
65. Oliver, R. F., Hulme, M. J., Mudie, A. and Grant, R. A., 1975. Skin collagen allografts in rats. *Nature (London)*, 258: 537–538.
66. Tauro, J. C., Parsons, J. R., Ricci, J. and Alexander, H., 1991. Comparison of bovine collagen xenografts to autografts in the rabbit. *Clin. Orthop. Rel. Res.*, 266: 271–284.
67. Zlabinger, G. J., Menzel, E. J. and Steffen, C., 1989. Induction of anti-pepsin antibodies after immunization with pepsin-extracted collagen. *Matrix*, 9: 135–139.
68. Hada, M., Sakihama, T., Kamiya, K., Tasaka, K. and Ueno, A., 1993. Cellular and humoral immune responses to vascular components in thromboangiitis obliterans. *Angiology*, 44: 533–540.
69. Champion, B. R. and Poole, A. R., 1982. Immunity to homologous type II collagen after partial meniscectomy and sham surgery in rabbits. *Arth. Rheum.*, 25: 274–287.
70. Rosen, T. and Watkins, H. C., 1990. Use of gelatine matrix implant in patients hypersensitive to bovine collagen. *J. Am. Acad. Dermatol.*, 22: 848–849.
71. Mammo, W., Singh, G. and Dolby, A. E., 1982. Enhanced cellular immune response to type I collagen in patients with periodontal disease. *Int. Arch. Allergy Appl. Immunol.*, 67: 149–154.
72. Hyder, P., Singh, G. and Adam, S., 1992. Humoral responses to type I collagen after surgical curettage procedures employing bovine collagen implants. *Biomaterials*, 13: 693–696.
73. Perloff, J. L., Reckhard, C. R., Rowlands, D. T., Jr. and Barker, C. F., 1972. The venous homograft: An immunological question. *Surgery*, 72: 961–970.
74. Schwartz, S. I., Kutner, F. R., Neistadt, A., Barner, H., Resnicoff, S. and Vaughan, J., 1967. Antigenicity of homografted veins. *Surgery*, 61: 471–477.
75. Wainwright, D. J., 1995. Use of an acellular allograft dermal matrix (Alloderm) in the management of full-thickness burns. *Burns*, 21: 243–248.
76. Chamson, A., Bousquet, G., Le Petit, J. C., Perier, C., Pomier, G. and Frey, J., 1986. Acceptance of a connective tissue equivalent for grafting and capsuloligamentary reconstruction. *Clin. Physiol. Biochem.*, 4: 281–284.
77. Norgren, L., Holtås, S., Persson, G., Ribbe, E., Saxne, T. and Thörne, J., 1990. Immune response to collagen-impregnated Dacron double-velour grafts for aortic and aorto-femoral reconstructions. *Eur. J. Vasc. Surg.*, 4: 379–384.
78. The Canadian Multicenter Hemashield Study Group, 1990. Immunologic response to collagen-impregnated vascular grafts: a randomized prospective study. *J. Vasc. Surg.*, 12: 741–746.
79. Schmiedt, W., Neufang, A., Scholl, E., Schmid, F. X. and Oelert, H., 1996. Immune response to gelatin- and collagen-impregnated aortic Dacron grafts. A randomized study. *Vasc. Surg.*, 30: 513–518.
80. Axthelm, S. C., Porter, J. M., Strickland, S. and Baur, G. M., 1979. Antigenicity of venous allografts. *Ann. Surg.*, 189: 290–293.
81. Okamura, K., Chiba, C., Iriyame, T., Itoh, T., Maeta, H., Ijima, H., Mitsui, T. and Hori, M., 1990. Antigen depressant effect of glutaraldehyde for aortic heterografts with a valve, with special reference to a concentration right fit for the preservation of grafts. *Surgery*, 87: 170–176.
82. Dahm, M., Lyman, W. D., Schwell, A. B., Factor, S. M. and Frater, R. W., 1990. Immunogenicity of glutaraldehyde-tanned bovine pericardium. *J. Thorac. Cardiovasc. Surg.*, 99: 1082–1090.
83. van Gulik, T. M. and Klopper, P. J., 1987. The processing of sheepskin for use as a dermal collagen graft: an experimental study. *Neth. J. Surg.*, 39: 90–94.
84. Dahm, M., Husmann, M., Mayer, E., Prufer, D., Groh, E. and Oelert, H., 1995. Relevance of immunologic reactions for tissue failure of bioprosthetic heart valves. *Ann. Thoracic Surg.*, 60(Supp 2): S348–352.
85. Frovola, M., Barbarash, L., Gudkova, R. and Karpinskaya, J., 1973. Effect of methods of preserva-

tion on the immunogenicity and antigenic composition of xenograft heart valve tissues. *Byull. Eksp. Biol. Med.*, 75: 83–86.
86. Spray, L. T. and Roberts, W. C., 1977. Structural changes in porcine xenografts used as substitute cardiac valves. Gross and histologic observations in 51 glutaraldehyde-preserved Hancock valves in 41 patients. *Am. J. Cardiol.*, 40: 319–330.
87. Bajpai, P. K. and Stull, P. A., 1980. Glutaraldehyde cross-linked porcine heart valve xenografts and cell-mediated immune response. *Int. Res. Commun. Syst. Med. Sci.*, 8: 642–644.
88. Salgaller, M. L. and Bajpai, P. K., 1985. Immunogenicity of glutaraldehyde-treated bovine pericardial tissue xenografts in rabbits. *J. Biomed. Mat. Res.*, 19: 1–12.
89. Moore, M. A. and Phillips, R. E., 1997. Biocompatibility and immunologic properties of pericardial tissue stabilized by dye-mediated photooxidation. *J. Heart Valve Disease*, 6: 307–315.
90. Gong, G., Seifter, E., Lyman, W. D., Factor, S. M., Blau, S. and Frater, R. W., 1993. Bioprosthetic cardiac valve degeneration; role of inflammatory and immune reactions. *J. Heart Valve Disease*, 2: 684–693.
91. Oliver, R. F., Grant, R. A., Cox, R. W. and Cooke, A., 1980. Effect of aldehyde cross-linking of human dermal collagen implants in the rat. *Br. J. Exp. Pathol.*, 61: 544–549.
92. Cooperman, L. and Michaeli, D., 1984. The immunogenicity of injectable collagen. I. A 1-year prospective study. *J. Am. Acad. Dermat.*, 10: 638–646.
93. Cooperman, L. and Michaeli, D., 1984. The immunogenicity of injectable collagen. II. A retrospective review of seventy-two tested and treated patients. *J. Am. Acad. Dermat.*, 10: 647–651.
94. Cooperman, L., Mackinnon, V., Bechler, G. and Pharris, B. B., 1985. Injectable collagen: a six-year clinical investigation. *Aesth. Plastic Surg.*, 9: 145–152.
95. Wallace, D. G., McPherson, J. M., Ellingsworth, L., Cooperman, L., Armstrong, R. and Piez, K. A., 1988. Injectable collagen for tissue augmentation. In: M. E. Nimni (Editor), *Collagen*, Vol. 3, CRC Press, Boca Raton, pp. 117–144.
96. Sellem, P. H., Caranzan, F. R., Bene, M. C. and Faure, G. C., 1987. Immunogenicity of injectable collagen implants. *J. Dermatol. Surg. Oncol.*, 13: 1199–1202.
97. André, C., Heremans, J. F., Vaerman, J. P. and Cambiaso, C. L., 1975. A mechanism for the induction of immunological tolerance by antigen feeding: Antigen-antibody complexes. *J. Exp. Med.*, 142: 1509–1519.
98. Burton, J. L., Etherington, D. J. and Peachy, R. D., 1978. Collagen sponge for leg ulcers. *Br. J. Dermatol.*, 99: 681–685.
99. Michaeli, D. and McPherson, M., 1990. Immunologic study of artificial skin used in the treatment of thermal injuries. *J. Burn Care Rehabil.*, 11: 21–26.
100. Browder, I. W. and Litwin, M. S., 1986. Use of absorbable collagen for hemostasis in general surgical patients. *Am. Surg.*, 52: 492–494.
101. Charriere, G., Bejot, M., Schnitzler, L., Ville, G. and Hartmann, D. J., 1989. Reactions to a bovine collagen implant. Clinical and immunologic study in 705 patients. *J. Am. Acad. Dermatol.*, 21: 1203–1208.
102. Hartmann, D. J., Charriere, G. and Ville, G., 1992. Immunoassay techniques for the detection of circulating antibodies to collagen following the use of collagen medical devices. *Clin. Mat.*, 9: 163–167.
103. McClelland, M. and DeLustro, F., 1996. Evaluation of antibody class in response to bovine collagen treatment in patients with urinary incontinence. *J. Urol.*, 155: 2068–2073.
104. Mullins, R. J., Richards, C. and Walker, T., 1996. Allergic reactions to oral, surgical and topical bovine collagen. Anaphylactic risk for surgeons. *Aust. N.Z. J. Ophthalmol.*, 24: 257–260.
105. DeLustro, F., Dasch, J., Keefe, J. and Ellingsworth, L., 1990. Immune responses to allogeneic and xenogeneic implants of collagen and collagen derivatives. *Clin. Orthopaedics*, 260: 263–279.
106. Myers, L. K., Cooper, S. W., Terato, K., Seyer, J. M., Stuart, J. M. and Kang, A. H., 1995. Identification and characterization of a tolerogenic T cell determinant within residues 181-209 of chick type II collagen. *Clin. Immunol. Immunopathol.*, 75: 33–38.
107. Dehan, P., Weber, M., Zhang, X., Reeders, S. T., Foidart, J. M. and Tryggvason, K., 1996. Sera from patients with anti-GBM nephritis including goodpasture syndrome show heterologous reac-

tivity to recombinant NC1 domain of type IV collagen α-chains. *Nephrol. Dialysis Transplant.*, 11: 2215–2222.

108. Dehan, P., Van den Heuval, L. P., Smeets, H. J., Tryggvason, K. and Foidart, J. M., 1996. Identification of post-transplant anti α5(IV) collagen alloantibodies in X-linked Alport syndrome. *Nephrol. Dialysis Transplant.*, 11: 1983–1988.
109. Chen, M., Chan, L. S., O'Toole, E. A., Sample, J. C. and Woodley, D. T., 1997. Development of an ELISA for rapid detection of anti-type VII collagen autoantibodies in epidermolysis bullosa acquisita. *J. Invest. Dermatol.*, 108: 68–72.
110. Direskeneli, H., D'Cruz, D., Khamashta, M. A. and Hughes, G. R., 1994. Autoantibodies against endothelial cells, extracellular matrix, and human collagen type IV in patients with systemic vasculitis. *Clin. Immunol. Immunopathol.*, 70: 206–210.
111. Trautinger, F., Kokoschka, E. M. and Menzel, E. J., 1991. Antibody formation against human collagen and C1q in response to a bovine collagen implant. *Arch. Dermatol. Res.*, 283: 395–399.
112. Elson, M. L., 1993. Injectable collagen and autoimmune disease. *J. Dermatol. Surg. Oncol.*, 19: 165–168.
113. Rosenberg, M. J. and Reichlin, M., 1994. Is there an association between injectable collagen and polymyositis/dermatomyositis? *Arth. Rheum.*, 37: 747–753.
114. Hanke, C. W., Thomas, J. A., Lee, W. T., Jolivette, D. M. and Rosenberg, M. J., 1996. Risk assessment of polymyositis/dermatomyositis after treatment with injectable bovine collagen implants. *J. Am. Acad. Dermatol.*, 34: 450–454.
115. Cukier, J., Beauchamp, R. A., Spindler, J. S., Spindler, S., Lorenzo, C. and Trentham, D. E., 1993. Association between bovine collagen dermal implants and a dermatomyositis or a polymyositis-like syndrome. *Ann. Internal Med.*, 118: 920–928.
116. Vaughan, P. R., Galanis, M., Werkmeister, J. A. and Ramshaw, J. A. M., 1997. Production of recombinant hydroxylated collagen in yeast. In: Transactions of the 23rd Annual Meeting of the Society for Biomaterials, May 1997, New Orleans, Society for Biomaterials, Minneapolis, p. 253.

34

Collagen Scaffolds for Tissue Regeneration

Frederick H. Silver and David L. Christiansen
U.M.D.N.J.-Robert Wood Johnson Medical School, Piscataway, New Jersey

I. INTRODUCTION

The ability of humans to maintain their physiological functions involve many complex interwoven biological pathways. Part of this homeostasis involves normal turnover of tissues found throughout the body as well as repair and regeneration responses that occur after chemical, mechanical or electrical trauma. Healing and regeneration are multistep processes that involve biological components found in blood and elements that make up extracellular matrix. These elements are components of systems that prevent excessive bleeding, remove exogenous debris, promote new tissue deposition and allow resumption of normal physiological processes [1].

One of the most challenging questions facing medical scientists in the twentieth century has been why do some tissues heal by regeneration and others heal by the formation of scar tissue consisting of aligned collagen fibers. In mammals, injury to most tissues, whether external or internal, results in blood clotting, fibrosis and functional repair; however, most tissues do not regain their original structure and function. For instance when an injury to the skin occurs that results in loss of only the epidermal or superficial layer, the epidermis will regenerate by cell migration from the wound edges. However, if both the epidermis and the underlying layer of dermis is injured, then dermis that is lost will be replaced with scar tissue and the epidermis will migrate over scar tissue [1]. On a superficial level we understand that the presence of an intact basement membrane over the dermis supports epidermal regeneration; we do not understand the exact information contained in the basement membrane that signals basal epithelium to divide and migrate. Therefore much of our understanding of wound healing is descriptive.

Although isolated cells have been used in an attempt to repair and regenerate tissues and organs, the majority of tissue engineering applications requires a combination of cells and scaffolding materials that can be "molded" into a variety of shapes and sizes. In addition, most cells do not function as isolated elements but interact with other cell types and the extracellular matrices that they sit on. Therefore, interaction between cells and matrices that they associate with are important aspects of repair and regeneration.

Much of the information about optimizing cell attachment to substrates has come from designing surfaces and coatings for cell culture plates. The role of surface morphology has been studied extensively; cells become oriented in response to underlying surface topography, commonly known as contact guidance. A number of physicochemical surface properties such as surface composition, surface charge, surface energy, surface oxidation, curvature and morphology affect cell attachment and behavior [2,3].

Many studies report the use of collagen and other extracellular matrix macromolecules (ECM) as substrates for attachment and growth of cells. The interaction between ECM and cells is an important aspect of tissue engineering since cell shape, cytoskeleton, cell migration, control of cell growth and differentiation all involve the extracellular matrix found around cells and in connective tissue.

ECM is a structural material that interacts with the cell cytoskeleton as well as binds growth factors. These interactions allow the transduction of chemical and mechanical signals across the cell membrane and result in changes in cell shape, protein synthesis and other cellular functions (Fig. 1). While collagen provides tensile strength and creep resistance to ECM, elastic fibers and proteoglycans help dissipate loads elastically and viscously. Attachment between the cell membrane and the ECM occurs via cell surface PGs, integrins and cell attachment glycoproteins. These factors influence adhesion and migration of cells to substrates.

In this chapter we review our progress in engineering scaffolding materials that mimic the structure of normal extracellular matrix which in conjunction with either differentiated cells or pluripotential stem cells can be used to repair and regenerate tissue function.

II. STRUCTURE OF EXTRACELLULAR MATRIX

A. Background

The extracellular matrix is composed of fiber and non-fiber-forming collagens, elastic fibers, fibronectin, laminin, dermatan sulfate and chondroitin sulfate proteoglycans, laminins, integrins and hyaluronan. The primary components include the fibrous collagens, types I, II and III [2]. These collagens form the networks that prevent premature mechanical failure of most tissues and contain molecular sequences of approximately 1000 amino acids in the form of Gly-X-Y with small nonhelical ends. All of these collagen types form continuous triple-helices that pack laterally into a quarter-stagger structure in tissues to form characteristic D-periodic fibrils as shown in Fig. 2. These fibrils range in diameter from about 20 nm in cornea to over 100 nm in tendon. In tendon, collagen fibrils are packed into fibril bundles that are aligned along the tendon axis. In skin, type I and III collagen fibrils form a nonwoven network of collagen fibrils that when stretched aligns with the direction of force. In cartilage, type II collagen fibrils form oriented networks that vary from the surface where they are parallel, to the deep layer where they are perpendicular to the surface.

The other fibrous component found in large amounts in ECM is elastic tissue which forms the networks of skin and cardiovascular tissue (elastic arteries) and is associated with elastic recovery. Elastic fibers are composed of a core of elastin surrounded by microfibrils 10 to 15 nm in diameter composed of a family of glycoproteins recently termed fibrillins [4].

Laminins [2] are a family of extracellular matrix proteins that are found in basement membranes and have binding sites for cell surface integrins and other extracellular matrix components. They consist of α, β and γ chains with molecular weights between 140,000 and 400,000. Eight different laminin chains have been identified: $\alpha1$, $\alpha2$, $\alpha3$, $\beta1$, $\beta2$, $\beta3$, $\gamma1$ and $\gamma2$. The most extensively characterized of the seven forms of laminin is laminin-1 ($\alpha1$ $\beta1$ $\gamma1$) which assembles in the presence of calcium to form higher ordered structures in basement membranes with type IV collagen, a non-fiber-forming collagen.

The fibronectins [2] are a class of high molecular weight multifunctional glycoproteins which are present in soluble form in plasma (0.3 g/l), other bodily fluids, and in fibrillar form

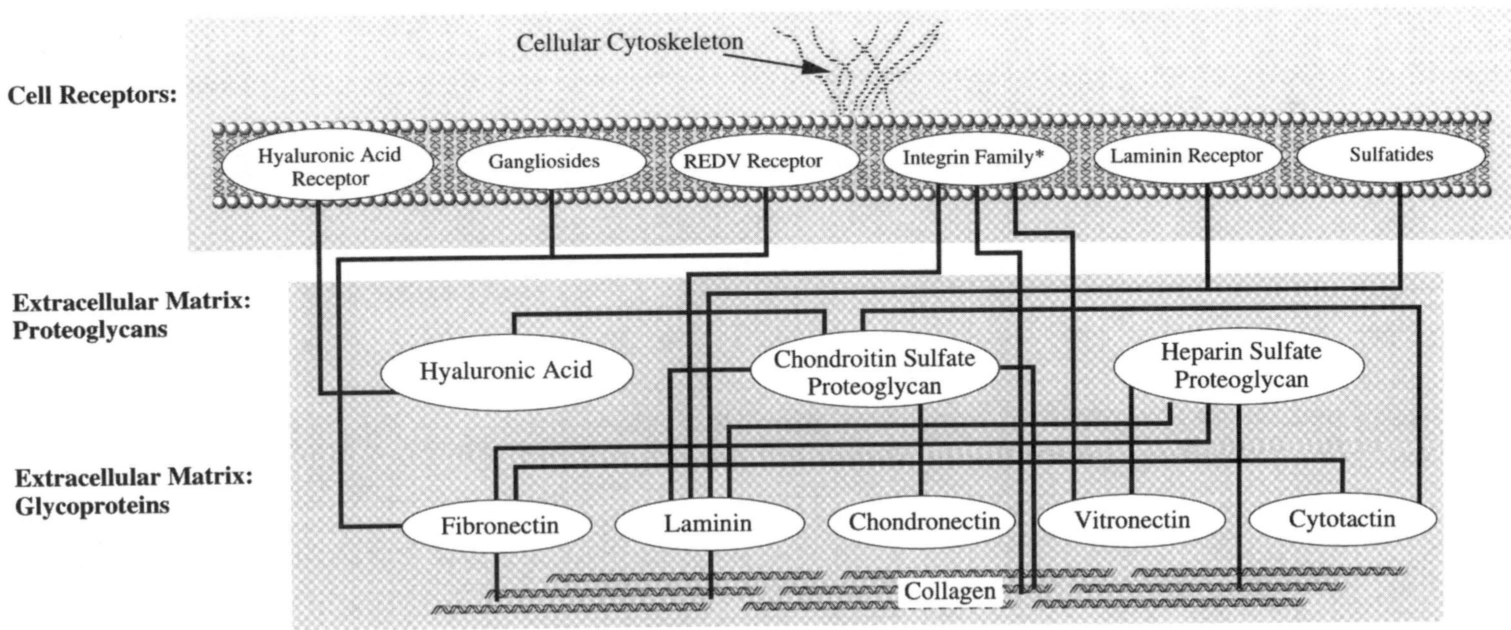

Figure 1 Relationship between cell membrane components and the extracellular matrix. Diagram showing the relationship between cell surface receptors and extracellular matrix components that are involved in adhesion and upregulation of cellular pathways.

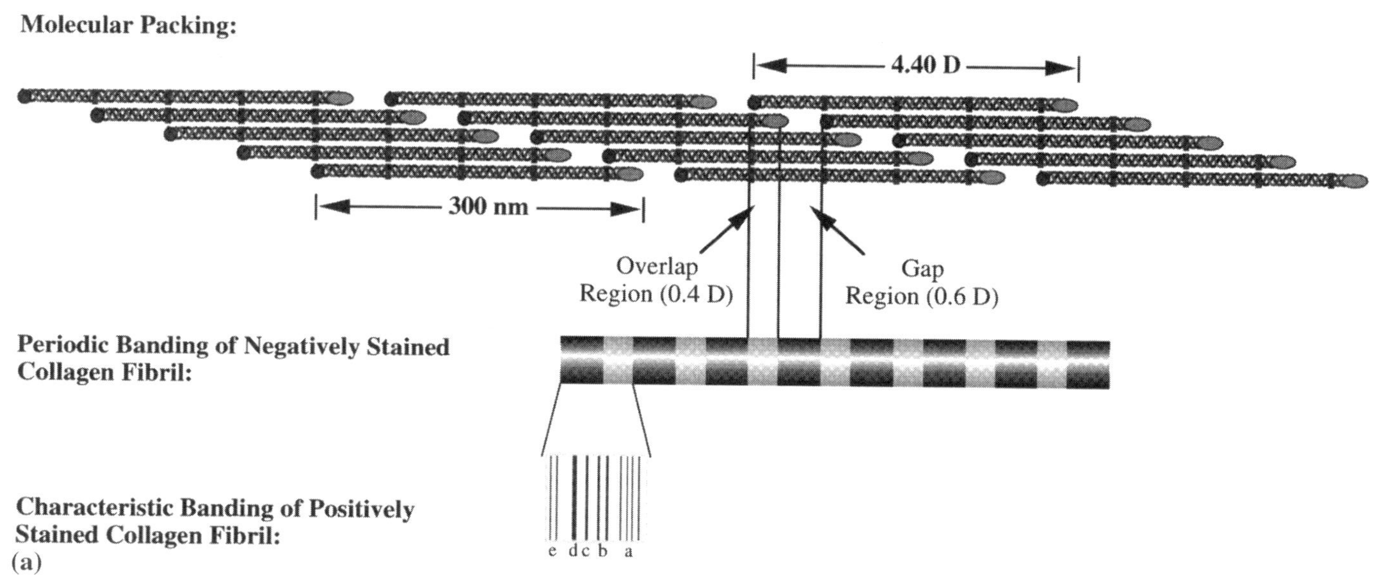

Figure 2 D-periodic structure in type I collagen. (a) Schematic showing molecular packing and origin of the D-period in collagen fibrils.

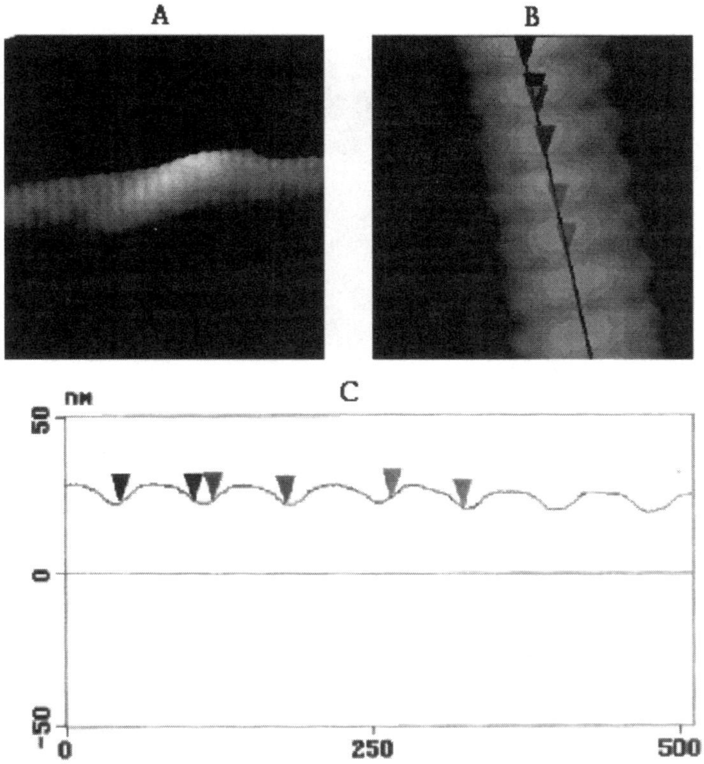

(b)

Figure 2 (continued) (b) Surface topology of a reconstituted collagen fibril obtained using an atomic force microscope at low (A) and high resolution (B) as well as a scan along the axis (C) showing 67 nm repeat.

in extracellular matrix. They bind to cell surfaces and other macromolecules including collagen and gelatin, the unfolded form of collagen, as well as fibrinogen and DNA.

Proteoglycans (PGs) [2] are a diverse family of glycosylated proteins which contain sulfated polysaccharides as a principle constituent. Diverse structures are found for these molecules that depend on the type of glycosaminoglycan side chains, length and net charge. Aggrecan found in large amounts in cartilage (50 mg/g of tissue) is highly glycosylated with 200 chains containing chondroitin sulfate and keratan sulfate. In contrast, decorin and biglycan have relatively small core proteins that have a leucine rich repeat and have one (decorin) and two (biglycan) chondroitin/dermatan sulfate side chains. Decorin binds to collagen fibrils while biglycan does not (Fig. 3).

Hyaluronan (HA) [2] is a component of every tissue or tissue fluid in higher animals. The highest concentrations are found in cartilage, vitreous humor and umbilical cord while even blood contains some HA [2]. At physiological pH, the molecule can adopt a helical structure in solution and therefore the molecule can be modeled as a series of helical segments that are connected with flexible segments.

A unique heparan sulfate proteoglycan, perlecan, is the major PG found in basement membranes. It has a large protein core containing about 3500 amino acids, consisting of multiple domains. Perlecan is able to self-associate or interact with several other basement membrane macromolecules including laminin and type IV collagen. In the kidney, heparan sulfate

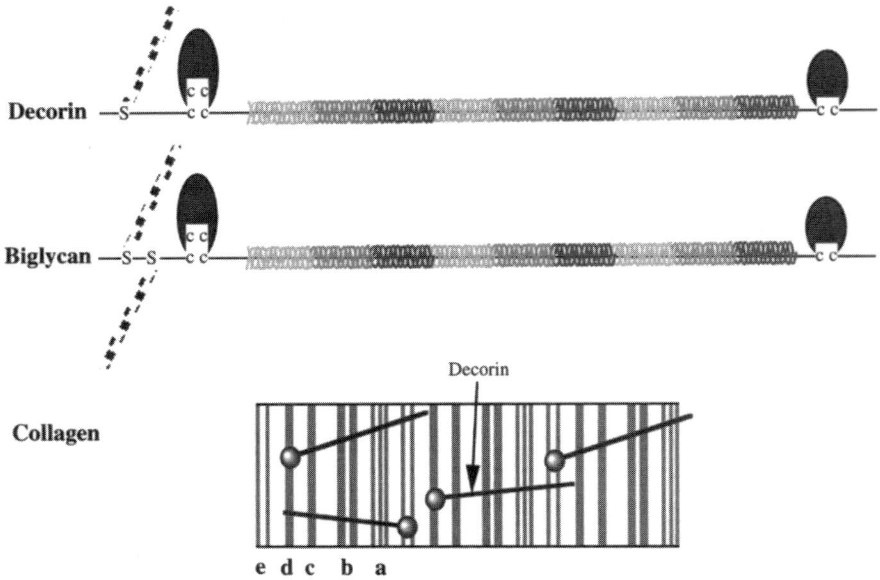

Figure 3 Structure of Decorin. Diagram illustrating the structure of decorin and biglycan (top), small proteoglycans. Decorin is shown attaching to collagen fibrils (bottom).

PG contributes a negative charge to the basement membrane and is thought to exclude serum proteins from being filtered out of the blood.

Virtually every mammalian cell has heparan sulfate proteoglycans as a plasma membrane component [2]. They are inserted into the cell membrane either through a transmembrane domain in their core protein (syndecans), or via a modified glycosaminoglycan region that is linked to the cell membrane (glypican). Heparan sulfate PGs interact with numerous molecules such a growth factors, cytokines, extracellular matrix proteins, enzymes, and protease inhibitors. They are believed to act to transduce signals that emanate from the interplay between components in the extracellular matrix.

Integrins [2] are a family of membrane glycoproteins consisting of α and β subunits. The binding site appears to contain sequences from both subunits, and their cytoplasmic domains form connections with the cytoskeleton. In this manner integrins form a connection between the cytoskeleton and extracellular matrix.

All these components are found in ECM; we are only beginning to understand the relationship between these components and the normal functioning of tissues. However, to engineer replacement tissues and organs it is essential to understand how to duplicate the properties of the critical components of the original tissue structures. Below we report on our progress in assembling analogs of ECM to be used as scaffolds for tissue engineered structures.

B. Model of Self-Assembly of Collagen Molecules

It was first observed about 40 years ago that solubilized collagen in acidic solutions when neutralized would sponstaneously self-assemble into fibers that resembled those found in extracellular matrix [5]. The fibers had the characteristic 67 nm macroperiod when the pH was close to neutral and the temperature was between 22 and 40°C. Subsequent studies showed that ions, alcohol, and other substances that affect both electrostatic, hydrophobic and covalent bonds modified self-assembly [6–8].

The mechanism by which collagen molecules assemble into fibrillar aggregates is still a subject of debate. Although the original kinetic studies of self-assembly indicated that it involved a phase transition with no change in molecular conformation [6–8] and could be modeled by nucleation and growth kinetics; later studies suggested that a range of subfibrillar components were observed [9].

We now believe that the mechanism involves conversion of single molecules to linear aggregates two and three molecules long that laterally aggregate into nanofibrillar subunits containing about 15 molecules [10]. The nanofibrillar subunits are probably twisted (that is why linear growth occurs before lateral growth) and are held together by lysine derived cross-links at the ends of the molecules. Fibril growth proceeds by linear and lateral fusion of 4 nm subunits yielding fibrils with cross sections that are not intitially circular (Fig. 4).

III. EXPERIMENTAL PROCEDURE FOR MAKING SELF-ASSEMBLED COLLAGEN FIBERS WITH FIBRILLAR SUBSTRUCTURE

To properly self-assemble collagen fibrils into fibers it is necessary to first start with solutiuons of intact collagen molecules. Preparation of soluble type I collagen dates back to the 1960s when Piez and co-workers reported extraction of collagen from rat skin using either a Tris buffer at pH 7.5 containing 0.5 M NaCl or 0.5 M acetic acid [11]. Solublilized collagen is precipitated using either 2.2 to 2.5 M NaCl or 0.7 M NaCl for neutral salt soluble or acid

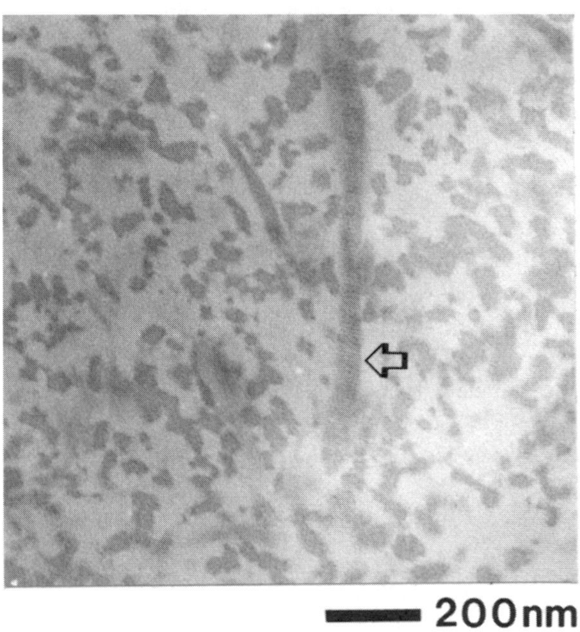

Figure 4 Cross-section of self-assembled type I collagen fibrils. Transmission electron micrograph of self-assembled collagen fibrils showing irregular and circular cross sections. It is proposed that fibrils form by fusion of a 4 nm nanofibrillar unit giving these fibrils their irregular cross sections [10].

extracted collagen, respectively and resolubilized in the starting buffer. The precipitation is repeated until the product is free of noncollagenous materials based on electrophoresis, chromatography and amino acid analysis. State of collagen aggregation is analyzed by measurement of the intrinsic viscosity which is usually about 75% monomers and 25% dimers [12].

Fibers are formed by first dispersing soluble collagen at a concentration of 1% (w/v) in distilled water containing 10 mM HCl at pH 2.0. Purified acid soluble type I collagen is coextruded with fiber formation buffer (FFB-30 mM Trizma Base (Tris) and 5 mM sodium phosphate dibasic pH 7.4), into a 37°C bath of FFB. The collagen is allowed to self-assemble in the FFB for a period of 24 h. The pH and temperature of FFB can be varied to control the fibril diameter [10]. Fibers immersed in FFB are rinsed in distilled water for 60 min and then air-dried under tension. Self-assembled collagen fibers can be be further aligned by stretching from 0 to 50% of their original length on a rack immersed in FFB as previously described [13].

Other molecules that are soluble in neutral buffers are incorporated into FFB and coextruded with collagen during the self-assembly process. We have reported coextrusion of decorin, a small proteoglycan found in connective tissue, with type I collagen [12]. In this manner the effects of a variety of molecules on collagen self-assembly can be studied and scaffolds containing these molecules can be made.

IV. MORPHOLOGICAL CHARACTERISTICS OF SELF-ASSEMBLED COLLAGEN FIBERS

Collagen fibers extruded using the self-assembly process are smooth when observed by scanning and light microscopy and appear to be composed of subfibrillar components. Transmission electron microscopy of longitudinal and transverse stained sections show that the fibrils in unstretched collagen fibers have D periodic banding patterns characteristic of those observed in native collagen fibrils, but have only partial orientation with respect to the fiber axis [12]. Cross sections of unstretched fibers indicate that fibrils are oriented obliquely to the fiber axis. Fibers stretched 30% were primarily composed of axially oriented collagen fibrils with only a few collagen fibrils oriented obliquely. At 50% stretch, fibers were composed of densely packed collagen fibrils that lie parallel to the fiber axis (Fig. 5). Fibril diameters were unchanged after stretching or decorin incorporation [12].

Analysis of fibrils formed at neutral pH show that they have an average diameter of 21.5 ± 9.9 nm. Results of correlation analyses suggest that fibrils have subfibrillar substructure and that they are composed of 4 nm "nanofibrils" [10]; fibrils grow by lateral and longitudinal fusion of these units. Large fibril diameters are obtained by promoting collagen self-assembly at pHs between 5.0 and 7.5 while inhibiting collagen crosslinking. In the presence of glycine, which inhibits collagen crosslinking, fibril diameters as large as 40.2 ± 18.2 are formed. These fibrils are about the size seen in developing tissues and are smaller than those seen in fully mature tissue.

V. CROSSLINKING OF COLLAGEN FIBERS

In vivo collagen fibers have about two crosslinks per α chain or a total of at least six crosslinks per molecule. These crosslinks mechanically stabilize connective tissue as well as block potential sites that induce immunity when collagen is transplanted across species. The crosslinks also play a role in self-assembly since the presence of lysine derived aldehydes that crosslink

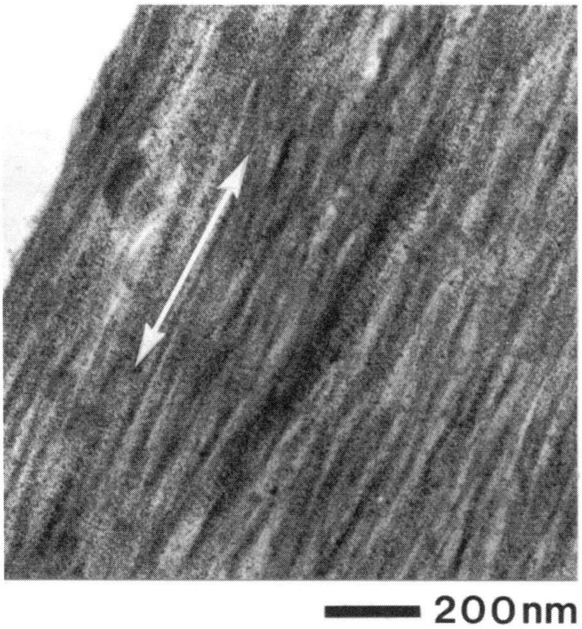

Figure 5 Longitudinal section of self-assembled collagen fibrils. Transmission electron micrograph of self-assembled collagen fibrils showing alignment of the fibrils parallel to the fiber axis. Note periodicity of the fibrils.

the ends of the molecules inhibits lateral fibril growth [10]. In the absence of crosslinks or when crosslinks are deleted in patients with genetic connective tissue disorders, collagen fibrils are disorganized and do not prevent premature mechanical failure.

Crosslinking can be achieved by exposure to chemicals such as aldehydes or ethylene oxide, to ulraviolet light or to γ-rays, to severe dehydration at elevated temperatures or through a process of self-crosslinking at standard temperature and pressure (STP) [14, 15]. Self-crosslinking at STP occurs in collagens that have intact nonhelical ends and that have lysine derived aldehydes that form crosslinks during self-assembly. The benefit of the later process is that the triple helix is unchanged during the crosslinking process and therefore scaffolds made out of collagen fibers crosslinked in this manner will be identical to collagen fibers found in normal tissues.

VI. MECHANICAL PROPERTIES OF COLLAGEN FIBERS

A. Type I Collagen

The static and viscoelastic mechanical properties of collagen fibers are dependent on a number of design parameters including the collagen fibril diameter, fibril length, fibril alignment, the extent of crosslinking and the presence of other macromolecules added during self-assembly. Static mechanical behavior is determined from measurement of stress-strain behavior while viscoelastic properties are obtained from incremental stress-relaxation curves. Typical stress-strain and incremental stress-strain curves are shown in Figs. 6 and 7. Parameters analyzed from these data include low strain modulus, high strain modulus, ultimate tensile strength (UTS), ultimate strain, elastic fraction (EF) and load at failure.

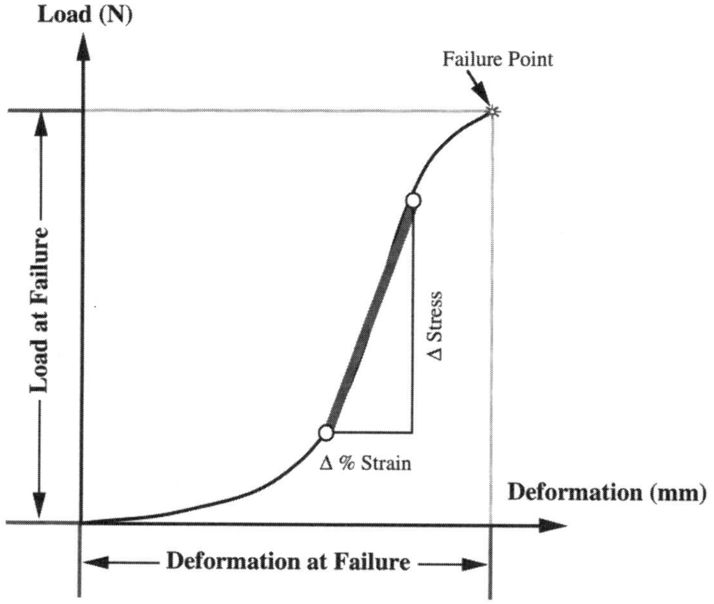

Definition of Mechanical Properties

- **Stress** (MPa) = Load (N) / Cross-sectional area (mm^2)
- **Ultimate Tensile Strength** (UTS) = Stress at Failure
- **% Strain at Failure** = Deformation at Failure (mm) / Gage Length (mm)
- **Modulus** = Δ Stress / Δ % Strain (values extrapolated from linear region of the load - deformation curve)

Figure 6 Typical stress-strain curve for a collagen fiber. Diagram showing definitions used to determine the mechanical properties of collagen fibers including stress, ultimate tensile strength, modulus and strain at failure.

The mechanical behavior of uncrosslinked collagen fibers has been studied extesively in our lab [10, 12, 13]. The stress-strain behavior of uncrosslinked collagen fibers is unusual in that the curve approximates a straight line with a modulus of about 2.2 MPa, UTS of between 0.5 and 1.5 MPa, strain at failure between 35% and 45% and elastic fractions between 0.2 and 0.3. When the low strain modulus is plotted versus fibril diameter a straight line results. High strain modulus and UTS are correlated with the viscosity of the fiber obtained from the relaxation part of incremental stress-strain curves. The viscosity of the fiber is related to the number of crosslinks within fibrils and therefore UTS and high strain modulus are related to the extent of crosslinking. The modulus and UTS of collagen fibers aligned by stretching increase to about 40 and 13 MPa, respectively, while the strain at failure decreases to about 20% when the fibers are prestretched up to 50%.

In contrast, the mechanical behavior of crosslinked collagen fibers are quite different. Collagen fibers after crosslinking by severe dehydration have moduli up to 450 MPa (unstretched) and 750 MPa stretched (50%) and the UTS increases from 45 (unstretched) to 70

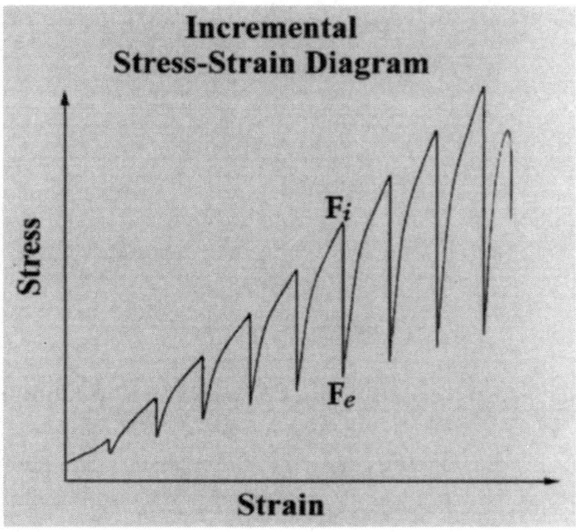

Figure 7 Incremental stress-strain curve. Diagram illustrating how the elastic fraction (Fe/Fi) is determined by stretching a fiber in tension through a strain increment of 2% to 5% and then allowing the fiber to relax to equilibrium before an additional strain is imposed.

(stretched 50%) MPa. Strain at failure decreases from about 14% (unstretched) to 12% (stretched 50%). Fibers self-crosslinked at STP have similar mechanical properties and in addition have elastic fractions of about 0.8. Thus crosslinking increases fibril viscosity and in this manner the UTS and high strain moduli are also increased.

B. Influence of Decorin

Decorin is small interstitial dermatan sulfate proteoglycan that is regularly and specifically associated with the surface of fibrillar type I collagen in tissues. It has been hypothesized that decorin may help limit fibril diameter growth, inhibit mineralization of fibrillar collagen, and may align collagen fibrils during fibrillogenesis. Addition of small amounts of decorin during type I collagen fibrillogenesis does not change collagen fibril diameters, however; it does increase collagen UTS [12], suggesting that decorin may form linkages between collagen fibrils that inhibit viscous sliding during deformation. It is possible that covalent crosslinks increase collagen fibril lengths within a fibril and that interactions with decorin may increase "effective" collagen fibril lengths between fibrils.

VII. SUMMARY

Now that we understand self-assembly of collagen fibers and the relationship between structure and mechanical properties, it is possible to design and assemble collagen scaffolds that mimic the structure of different types of connective tissue. The easiest design is an aligned collection of collagen fibers to be used as a replacement for tendon/ligament. Other applications include biaxially oriented networks for replacement of skin and dura mater. The ability to make composite structures out of collagen that resemble normal tissues structures and to crosslink the collagen fibers without exogenous chemicals leads to the next generation of scaffolding materials. The fabrication of viable tissue scaffolds must take into consideration the viscoelastic properties of the host tissue to be replaced. Creation of this new generation of scaffolding materials will ultimately lead to progress in tissue regeneration as opposed to limiting ourselves to only tissue repair.

REFERENCES

1. Silver, F. H. and Doillon, C. J., *Biocompatibility: Interaction of Biological and Implantable Materials*. VCH, New York, chapter 4 (1989).
2. Silver, F. H. and Christiansen, D. L., *Biomaterials Science and Biocompatibility*, Springer-Verlag, New York, in press, 1999.
3. Singhvi, R., Stephanopoulos, G. and Wang, D. I. C., Review: Effects of Substratum Morphology on Cell Physiology, *Biotechnol. Bioeng*. 43, 64 (1994).
4. Reinhardt, D. P. Keene, D. R., Corson, G. M., Poschl, E., Bachinger, H. P., Gambee, J. E. and Sakai, L. Y., Fibrillin-1: Organization in Microfibrils and Structural Properties, *J. Mol. Biol*. 258, 104 (1996).
5. Gross, J., Highberger, J. H. and Schmitt, F. O., Some Factors Involved in the Fibrogenesis of Collagen in vitro, *Proc. Soc. Exp. Biol. Med*. 80, 462 (1952).
6. Bensusan, H. B. and Scanu, A., Fiber Formation From Solutions of Collagen. II. The Role of Tyrosine Residues, *J. Am. Chem. Soc*., 80, 2990 (1960).
7. Hayashi, T. and Nagai, Y. J., Factors Affecting The Interactions of Collagen Molecules, as Observed by in vitro Fibril Formation. II. Effects of Species and Concentration of Anions, *J. Biochem*., 74, 253 (1973).
8. Cassel, J. M., Mandelkern, L. and Roberts, D. E., The Kinetics of The Heat Precipitation of Collagen, *J. Am. Leather Chem. Assoc*., 57, 556 (1962).
9. Trelstad, R. L., Hyashi, K. and Gross, J., Collagen Fibrillogenesis: Intermediate Aggregates and Suprafibrillar Order, *Proc. Natl. Acad. Sci. USA*, 73, 4027 (1976).
10. Christiansen, D. L., Huang, E. K. and Silver, F. H., Assembly of Type I Collagen: Fusion of Fibril Subunits and the Influence of Fibril Diameter on Mechanical Properties, submitted (1999).
11. Piez, K. A., Lewis, M. S., Martin, G. R. and Gross, J., Subunits of the Collagen Molecule, *Biochem. Biophys. Acta*, 53, 596 (1961).
12. Pins, G. D., Christiansen, D. L., Patel, R. and Silver, F. H., Self-Assembly of Collagen Fibers. Influence of Fibrillar Alignment and Decorin on Mechanical Properties, *Biophys. J*., 73, 2164 (1997).
13. Pins, G. D., Huang, E. K., Christiansen, D. L., and Silver, F. H., Effects of Axial Strain on the Tensile Properties and Failure Mechanism of Self-Assembled Collagen Fibers, *J. Appl. Polym. Sci*., 63, 1429 (1997).
14. Silver, F. H. and Garg, A., Collagen: Characterization, Processing and Medical Applications, In: *Handbook of Biodegradable Polymers*, edited by A. J. Domb, J. Kost and D. M. Wiseman, Chapter 17, Hardwood Academic, Australia (1997).
15. Chen, Y. H., Viscoelastic Properties of Native Rat Tail Tendon And Self-Assembled Type I Soluble Collagen Fibers: Effect of Aging, M.S. thesis in Biomedical Engineering, Rutgers University (1998).

35

Tissue Assessment for Skin Substitutes

Dale Feldman, Tom Barker, Barbara Blum, John Bowman, Deepak Kilpadi, and Robert Redden
University of Alabama at Birmingham, Birmingham, Alabama

I. INTRODUCTION

To accurately assess effectiveness of skin regenerative biomaterials, quantitative assessments are essential. These measures should not only be quantitative, but easy to use, noninvasive, and repeatable. The lack of these clinically used tools poses problems in comparing new treatments or skin substitutes to traditional treatments. Traditional clinical measures of wound healing include estimates of skin graft take, area healed, time to subjective healing, and length of hospital stay. Even the more quantitative measure of change in surface area healed is subject to variability, due to the influence of wound size on healing rate. Other more quantitative measures of wound healing, such as oximetry and single point laser Doppler flowmetry, require contact with the wound. Further, either they provide a single measure to represent an entire area (oximetry) or they measure at only one point (flowmetry). Subsequently, it is difficult to identify any problem regions.

A. Tissue State

Another goal of assessments should be to measure tissue state as a comparison to the normal state as well as track the tissue state during treatment. There, however, needs to be consensus on what are the normal properties of tissue in order to determine the difference in properties of the selected tissue. The goal of the treatment can then be defined as moving the present state closer to the normal state. Therefore, interventions can be assessed on how well they push the wound toward the normal state. Once standardized quantitative clinical assessment tools are in place, sufficient tissue state data could be collected to allow predictions or prognosis to be made based on the current state. Eventually this could also be used to predict how a specific intervention would alter the tissue state and thus aid in development of treatment protocols.

This would then be useful to clinicians as well as researchers trying to assess the efficacy of new interventions. The problem is both not having a consensus on appropriate parameters to characterize normal tissue or a good healing response as well as the lack of good proven quantitative measures for assessing these parameters. In addition, the demonstration of correlation of consensus quantitative tissue state measures to clinical outcome would help prove the utility of these assessments for evaluation of new treatments in clinical trials. In addition to being helpful in clinical studies, any technique which could help determine the rate, or stage of healing, on a continuing basis would also be useful in animal studies by reducing the need to sacrifice animals at numerous time points.

For skin wounds, such as skin ulcers or burns, assessments need to be done in three groups: healing rate, tissue health, and tissue function. Healing rate is the speed in which the wound is closed and should be determined independent of wound size in order to accurately compare treatments. For healing rate, wound volume and surface area are needed to obtain tissue fill rate. Also the wound margin and epithelial margin need to be measured to determine epithelialization rate and contraction rate. Tissue health would be an assessment of the wound microenvironment in an effort to determine the ability of the wound to heal. An assessment of blood flow or vascularity, which in fact helps determine tissue oxygen level, is a measure of tissue health. Tissue function would be an assessment of the skin's ability to perform its normal functions. Assessments of mechanical properties and epidermal barrier functions can be used to determine tissue function.

Ultimately standards should be developed to determine what is the minimum number of tests needed to assess a skin wound and how can these measures be standardized from one clinic to another. Critical would be to select systems that are easy to use, portable, low cost, and readily available in all clinical settings.

B. Healing Rate

The larger the wound, the more tissue that can be laid down in each time period, and the larger the change in volume or surface area. The optimal parameter to assess rate of healing should be independent of wound size [1]. It has been shown that the epidermal migration rate is relatively constant at approximately 1–2 mm/week [2]. In shallow wounds, the epidermal migration rate determines the change in average wound diameter. This rate, therefore, should not be influenced by the original wound size. In shallow wounds, such as first and second stage pressure ulcers, the average change in wound diameter can be calculated by taking the change in surface area, per unit time, divided by the average of the original perimeter and the final perimeter. Total healing rate can be divided into epithelialization rate and contraction rate [1].

For deep wounds, such as stage 3 or higher pressure ulcers, the average change in wound diameter can be calculated by taking the change in wound volume per unit time, divided by the average of the original wound surface area and the final wound surface area. This measure can be used to obtain an overall healing rate in these wounds. In deep wounds, total healing is due to tissue fill, in addition to epithelialization and contraction. Epithelialization rate and contracture rate, for these wounds, would be calculated by measuring the two-dimensional changes, as for shallow wounds [1].

Traditionally, wound area and volume measurements have been obtained using contact methods. The easiest method of obtaining area is to trace the wound onto transparent plastic. Area can be measured by cutting out the shape and weighing it or by digitizing the image and using image analysis. The most common method of measuring wound volume is to use calcium alginate dental molding material. Over the years a number of noncontact methods have been developed to measure wound volume and surface area. Standardization of assessment will require consensus of what needs to be measured to determine wound healing rate as well as the best systems to obtain these measures.

C. Tissue Health

Since oxygenation and, thus, angiogenesis, are critical for wound healing, methods to assess them would provide insight into the wound healing process and overall tissue health. For open wounds, techniques which provide a 2-D map of the wound provide many advantages, such as determining quick spatial comparisons between areas within the wound or adjacent to it and ease in making correlations to histological results.

Recently both thermography and scanning laser Doppler have been used to assess vascular perfusion and flow, giving a measure of tissue oxygenation [3–6]. Precision and ease of use should be considered in determination of the best tools to be used for clinical wound assessment. Since these techniques measure tissue oxygenation indirectly, they should be helpful in assessing tissue health, which will indicate the stage of healing.

D. Tissue Function

One of the biggest causes of impaired tissue function of healed wounds is that of excessive scarring, which can result in increased tissue stiffness. Thus, a method of measuring tissue stiffness would provide a way to evaluate tissue function. Additionally, a technique to nondestructively assess wound stiffness could also give an indication of progression of healing, since stiffness is a function of collagen orientation and density. Tissue stiffness measurements could also potentially be used to detect underlying tissue death or other changes, which would require therapeutic interventions.

Tissue stiffness can be evaluated histologically by quantifying orientations and cross-sectional area of collagen fibers. Tissue strength of excised tissue samples, such as sections of treated wounds, can be measured via tensile testing, using an MTS (mechanical testing system) or Instron mechanical testing machine. Both methods, however, require excision of tissue, and thus are most practical in animal studies after sacrifice. While histological assessment may be performed on tissue excised from small biopsies, mechanical testing would require a piece of tissue too large to be practically excised from a patient.

The biomechanical tissue characterization system (BTC1000, SRLI Technologies, Nashville, TN) has been devised to measure breaking strength or stiffness of soft tissue, in vivo. With the system, two reflective markers are placed on the wound or skin and are imaged as a suction pressure is applied. The pressure can be limited to avoid tissue damage. A stiffness value can be obtained from the displacement of the markers.

II. TECHNIQUES

A. Healing Rate

The assessment of healing rate is an important clinical tool both for comparison of treatments as well as for prognostic value. Clinicians use a variety of methods to assess healing rate and determine clinical endpoints. Traditional measures of healing can generally be divided into methods that determine healing rates over time and those that use clinical endpoints. The former method includes changes in surface area or volume healed, percent epithelialization or closure, and changes in surface area epithelialized or contracted. The latter includes length of hospitalization, time for complete healing or closure, and time for complete epithelialization.

When comparing wounds of different sizes, these methods tend to be inaccurate or subjective. For example, a large, shallow wound has an increased wound edge perimeter and can lay down more tissue in a given time than a smaller wound. Similarly, a deep wound has more surface area than a shallow wound and can create more new tissue in a given time in order to fill the wound. This causes a deceptively faster healing in larger or deeper wounds compared to smaller or more shallow wounds, even if the two wounds are healing linearly from the wound periphery at the same speed. These limitations cause deceptive, inaccurate results and prevent quantitative comparison of the healing of different wounds or even the same wound at different times. The ideal model of healing rate then should be independent of the area and/or volume of the wound.

1. *Justification*

Traditional models often combine all mechanisms of healing into a single measurement. Healing of full-thickness or deeper wounds is, however, a combination of three mechanisms: epithelialization, contraction, and tissue fill [1,7,8]. The amount of healing via each mechanism depends on the type of wound, depth of wound, and many other variables [9]. A better measurement would isolate these mechanisms. In particular, contraction rate, associated with scarring and thus desirable to minimize, should be determined separately. Clinical trials that use endpoints such as total closure can often take months or years to show efficacy, especially if dealing with chronic wounds like diabetic ulcers. The proper model should be able to give clinically relevant results at all points during the course of healing. Finally, the model should be able to accurately compare the healing of different wounds by being independent of size and shape.

2. *Healing Rate Model*

A quantitative model for calculating healing rate proposed by Gilman [10] and later illustrated by Gorin et al. [11] fits the necessary requirements of independence of wound size and shape. The method allows for the calculation of a linear healing rate perpendicular to [12,13] the wound edge as a function of wound area and perimeter: healing rate (MR) $= \Delta SA/(P_{avg} \times \Delta t)$, where ΔSA is the change in surface area over the time period and P_{avg} is the average wound perimeter between initial and final time points (Δt). This model provides an accurate, quantitative way to assess healing rates or closure rates at various times during the course of healing. Researchers have used this model of a linear healing rate to assess the healing of various skin wounds such as diabetic ulcers [14,15], pressure ulcers [16], and skin grafts [17].

For shallow wounds like skin grafts or partial thickness wounds, this model is adequate for calculating healing rates. For deeper wounds, such as a stage 3 or 4 pressure ulcer, the 3-D healing rate of the wound should be determined along with the 2-D measures of epithelialization and contraction rates. This 3-D healing rate, or tissue fill rate, can also be determined independent of wound volume [16]: tissue fill (TF) $= \Delta V/(SA_{avg} \times \Delta t)$, where ΔV is the change in volume over the time period, and SA_{avg} is the average surface area of the wound (in 3-D) between the two time points (time). The equation parallels the 2-D closure rate with a change in dimension of the variables. These two calculations work together to quantitatively assess the healing of a wound independent of size, depth, or shape.

Using these calculations, healing rate (HR) can be separated into its three basic components: HR = ER + CR in 2-D and HR = CR + TF in 3-D, where ER is the epithelialization rate, CR is the contraction rate, and TF is tissue fill. Epithelialization rate is defined as the growth of new epithelium either solely from the wound periphery, as in deep wounds like third-degree burns, or also radially from epithelial islands present around hair follicles and sweat glands, as in partial-thickness wounds:

$$ER = \frac{|\Delta SA_{epithelium}|}{P_{avg} \times \Delta t}$$

Contraction rate represents the change in wound size, using the original wound margin, due to centripetal contraction:

$$\text{CR} = \frac{\Delta \text{SA}}{P_{\text{avg}} \times \Delta t} \qquad \text{margin}$$

Tissue fill represents the change in wound volume centripetally "inward" from the 3-D volumetric surface area of the wound. The equation has been described earlier:

$$\text{TF} = \frac{\Delta V}{\text{SA}_{\text{avg}} \times \Delta t}$$

Combining these equations allows for comparisons of wounds of all dimensions.

An even more accurate measure can be obtained by adjusting ER and TF. Because significant contraction can give artificially high ER and TF values, a better value is obtained by adjusting ER and TF to the values that would have occurred with no contraction.

In addition to the independence of size and volume considerations, different types of wounds heal differently and one component of healing may be more important in a particular application than others. By using these equations, the contribution of each healing mechanism can be investigated. The ability to isolate healing rate into components allows for unique analysis of different types of wounds. For example, while it may be perceived that skin grafted regions in humans heal primarily by epithelialization, in reality there is a combination of epithelialization and contraction involved in graft healing. Pressure ulcers, on the other hand, heal mainly through contraction and tissue fill. In each case, the individual mechanisms may be either isolated or combined for specific types of analyses. With this analytical model of healing rate, wounds of different shapes and sizes can be accurately and consistently compared to assess the effectiveness of different treatments on wound healing.

3. *Model Justification*

Different healing rates are obtained with different types of assessments. The importance of developing measures independent of wound size is illustrated for shallow wounds in Fig. 1 and full-thickness wounds in Fig. 2. In both analytical models, a constant healing rate was selected (1 mm/wk). For shallow wounds (Fig. 1) this corresponds to a constant inward linear growth (or closure) rate. Only the measure of HR [$\text{HR} = \Delta\text{SA}/(P_{\text{avg}} \times \Delta t)$] as described above demonstrates a measure of healing rate independent of wound size. In a clinical situation wounds of many different sizes and many different healing rates are encountered. This independence of wound size and shape is critical for accurate quantitative comparison of healing.

In Fig. 1a,b, change in wound surface area (i.e., a traditional measure of healing rate) of shallow wounds of different sizes is plotted against time, assuming the constant change in radius of 1 mm/wk. In both cases three different wound sizes were selected (2–6 cm). In 1a the original surface area normalizes the change in surface area, and in 1b the healing rate is the change in surface area per unit time. In 1c the percent epithelialization is shown. Using the actual healing rate $\Delta\text{SA}/(P_{\text{avg}} \times \Delta t)$, a straight line would result (1 mm/wk) and be the same line for each wound size. Figs. 2a,b show the same phenomenom, but for deep wounds. Again the $\Delta V/V_0$ and $\Delta V/\Delta t$ are not independent of wound size unlike $\Delta V/(\text{SA}_{\text{avg}} \times \Delta t)$.

The derivation for $\Delta V/(\text{SA}_{\text{avg}} \times \Delta t)$ is similar to the one for 2-D [10] and is used in stereology, equations to obtain 3-D measures from 2-D images such as microscope slides, and thus can be found in any stereology textbook. Based on stereological principles $V_f = A_f = P_f$ (volume fraction equals area fraction equals point count fraction). For a hemispherical wound, the change in radius in a 2-D histology slide would be the same as the 2-D measure $\Delta A/(P_{\text{avg}} \times \Delta t)$, which is equivalent to $\Delta V/(\text{SA}_{\text{avg}} \times \Delta t)$.

Specifically for a hemispheric wound that changes radius from r_0 to r_1 (ΔR) in a given time:

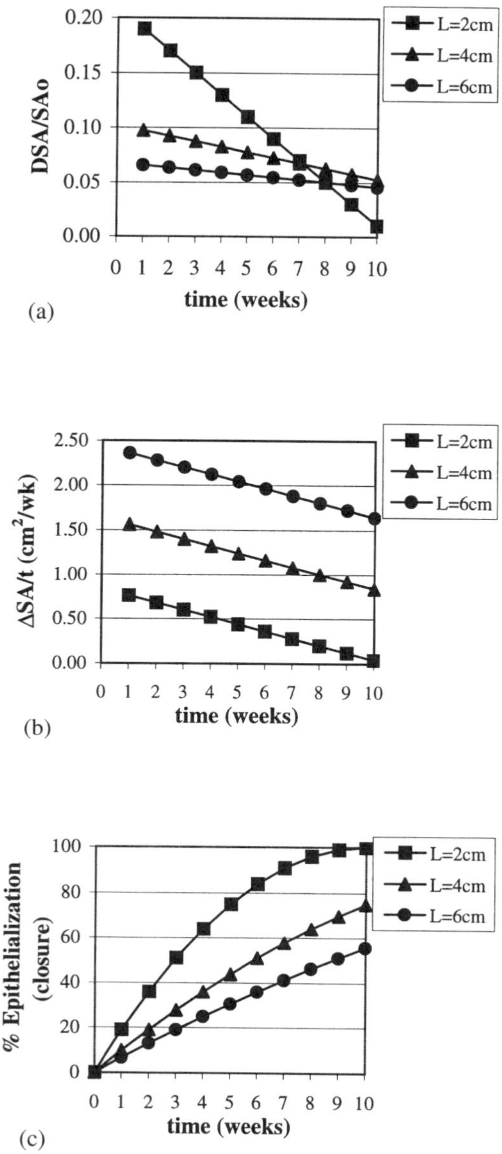

Figure 1 Model justification (2-D). Mathematical model of wound healing shows that traditional measurements of healing change over time. Models assume square shaped wound (side of L) with constant linear inward healing of 1 mm/week. Traditional measurements symbolized are as follows: (a) change in surface area of wound (ΔSA) divided by original wound area (SA_o), (b) change in wound surface area over time, and (c) percent epithelialization (or closure) over time.

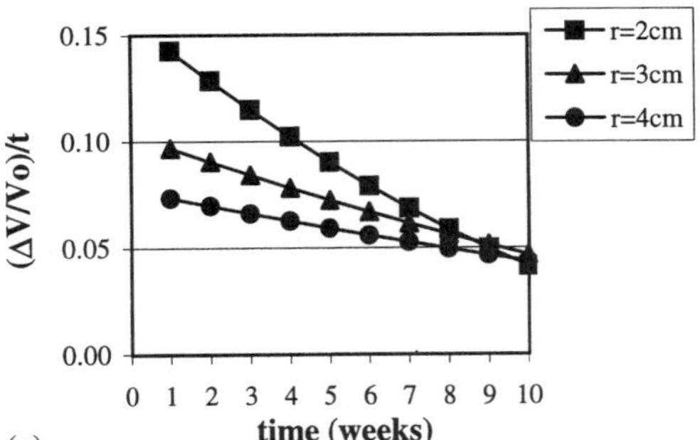

(a)

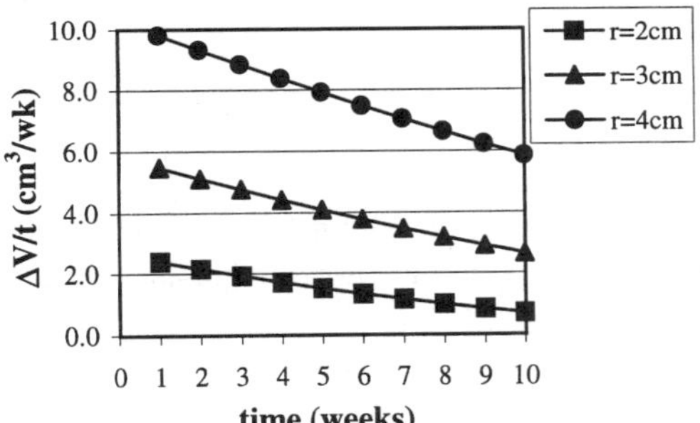

(b)

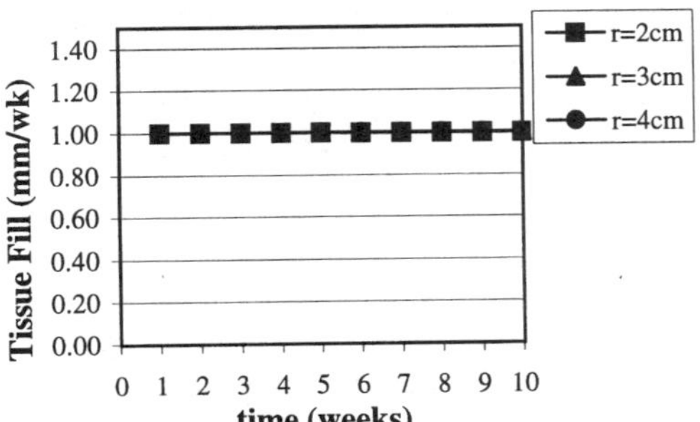

(c)

Figure 2 Model justification (3-D). Mathematical model of wound healing shows that traditional measurements of three-dimensional healing change over time. Model assumes hemispherical wound (radius R) with constant linear inward healing of 1 mm/week. Traditional measurements symbolized are as follows: (a) change in volume of wound (ΔV) divided by original volume (V_0), (b) change in volume over time, and (c) tissue fill over time.

$$\begin{aligned}
\frac{\Delta V}{\mathrm{SA}_{avg}} &= \frac{[(4/3)\pi r_1^3 - (4/3)\pi r_0^3]/2}{[4\pi r_1^2 + 4\pi r_0^2]/2/2} \\
&= \frac{4}{6}\frac{r_1^3 - r_0^3}{r_1^2 + r_0^2} \\
&= \frac{2}{3}\frac{[(r_0 - r_1)\,(r_0^2 + r_0 r_1 + r_1^2)}{r_1^2 + r_0^2} \\
&= \frac{2}{3}\frac{\Delta R[(\Delta R + r_1)^2 + (\Delta R + r_1)\,r_1 + r_1^2]}{r_1^2 + (\Delta R + r_1)^2} \qquad \text{since} \Delta R = r_0 - r_1 \\
&= \frac{2}{3}\frac{\Delta R[\Delta R^2 + 3\Delta R r_1 + 3r_1^2]}{\Delta R^2 + 2\Delta R r_1 + 2r_1^2} \\
&= \frac{2}{3}\frac{\Delta R[3r_1(\Delta R + r_1)]}{2r_1(\Delta R + r_1)} \qquad \text{since } \Delta R^2 \ll r_1^2, \quad \Delta R^2 \to 0 \\
&= \Delta R
\end{aligned}$$

In deep open wounds, any epithelialization will occur at the wound edge. In shallow wounds, epithelial growth may also occur from epithelial islands within the wound. In these wounds, epithelialization rate is a sum of epithelial growth from the wound edge and from the islands, or $\mathrm{ER} = \mathrm{ER}_{\mathrm{edge}} + \mathrm{ER}_{\mathrm{islands}}$. Growth from the wound edge is inward, while epithelial growth from islands is outward.

4. *Two-Dimensional Healing Rate Techniques*

The techniques used to measure 2-D wound areas have gotten more sophisticated, but the basic principles are the same. The simplest area measure is done by tracing the wound margins on a material of known density, cutting away the excess along the perimeter, and weighing the "wound area." A simple calculation will give the size of the tracing, and thus the size of the wound.

With the widespread availability of image analysis equipment and software packages, the most common method today is to either trace or photograph the wound, scan the image into a computer, and use software to calculate relevant distances or areas. At UAB, highly magnified pictures of the wounds are taken, scanned into an image file, and analyzed with SigmaScan Pro Image software (Jandel Scientific, Mountain View, CA). It is crucial, however, that a reference distance marker, such as a transparent ruler, be included in the picture. From the image, the software can be used to calculate surface areas and perimeters of any region of the wound. The identification of neoepithelium allows for the calculation of ER and the identification of the wound periphery allows for the calculation of CR. For skin grafts, ER can be calculated from the average neoepithelialization from a group of individual units of the mesh graft as seen in Fig. 3 [17]. The contraction rate can be calculated by examining the change in area of a given number of individual units of the mesh as seen in Fig. 4 [17]. The true ER (ER_t), can then be determined by normalizing for the CR. Thus for shallow wounds such as skin grafts, the ER and CR independent of wound size can be isolated for accurate analysis of healing.

Combining this analytic model of wound healing and image analysis software allows for quantitative assessments of healing rate independent of wound size. The model also allows for the accurate comparison of various underlying disease states, treatments, drugs, or therapies on the rate of healing of a variety of wounds. In addition, the contributions of contraction and epithelialization can be investigated in shallow wounds such as partial-thickness injuries or

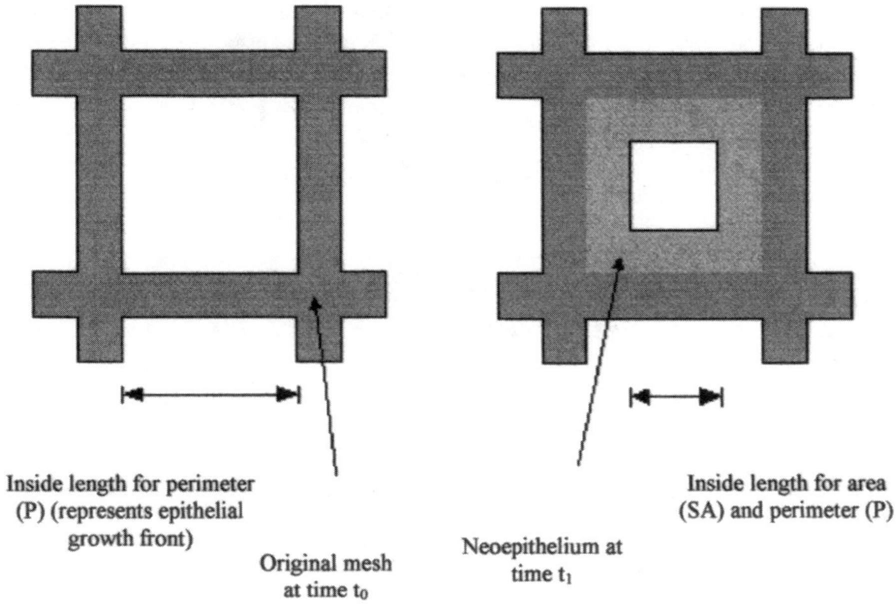

Figure 3 Calculation of epithelialization rate (ER) in a single mesh of a skin graft. For the calculation, the following formula is used: $ER = (SA_0 - SA_1)/[(P_0 + P_1)/2] \times (t_1 - t_0)$.

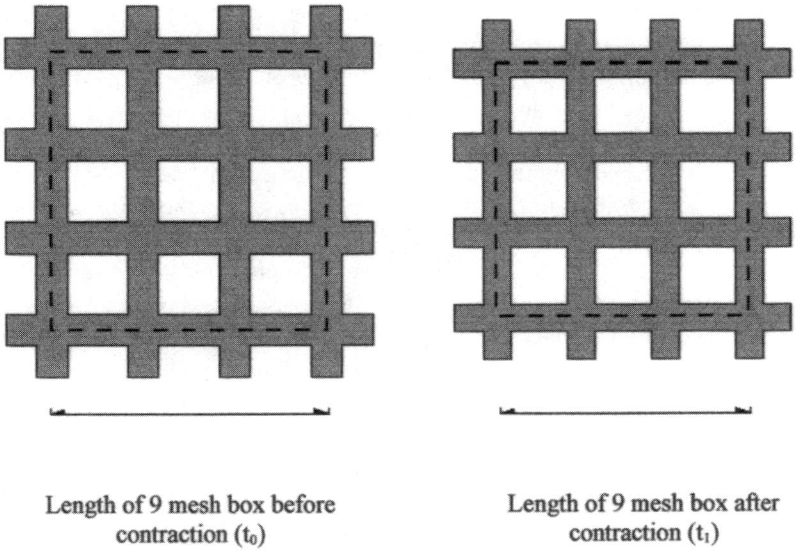

Figure 4 Calculation of contraction rate (CR) from a "box" of meshes of a skin graft. For the calculation, the following formula is used: $CR = (SA_0 - SA_1)/[(P_0 + P_1)/2] \times \text{time}$.

skin grafts; and the effect of treatments designed to limit contraction can be quantitatively analyzed. This technique finally offers advantages in that the equipment is sufficiently portable, noninvasive, and inexpensive to warrant its use in any clinical wound healing center. The consistency gained by the widespread use of this technique could contribute greatly to the understanding of wound healing and the ability to quantitatively and statistically evaluate the benefits of various treatments and therapeutic regimes.

5. Three-Dimensional Healing Rate Techniques

Traditionally, a number of methods have been employed to measure wound volume. While these methods are relatively inexpensive to perform, most require wound contact, leading to possible wound disruption, contamination, and increased chance of infection. Mathematical estimations have typically been based on one-dimensional wound measures, also requiring some degree of contact [18, 19].

a. Traditional Measures.

Saline Fill. Probably the oldest method of measuring wound volume is that of filling the wound with saline or another inert fluid, and measuring the volume used. This method is simple, inexpensive, and safe, if sterile saline is used. The measure, however, gives only volume, not surface area. More importantly, it cannot be used for all wounds, such as shallow wounds and those in contoured regions, e.g., the heel [20]. In addition, accuracy is dependent on positioning the patient so that the wound surface is parallel to the ground. The technique, however, does have the potential to measure volume associated with undermining, which is difficult for most other techniques.

Casting. Another common method of measuring wound volume is that of casting, or creating a mold of, the wound with a foreign material. Commonly used casting materials include calcium alginate (Jeltrate, Dentsply, Milford, DE) and vinyl polysiloxane (Reprosil, Dentsply, Milford, DE), both dental molding materials that are inert and set within a few minutes. Casting is inexpensive and simple, and casts can be saved for a permanent record of the wound.

The primary disadvantage of casting, like saline fill, is the requirement of wound contact. Filling the wound with a foreign material increases the chance of wound contamination, irritation or an allergic reaction [21]. Like saline fill, casting is not feasible in all patients, due to insufficient wound depth, or the inability to position the patient appropriately [20]. Also, calcium alginate molds must be wrapped or sealed in an airtight wrapping immediately, to prevent loss of size and weight associated with drying.

A number of methods have been employed to obtain volume from the casting. Some groups have also measured surface area from the molds. The simplest methods of assessing volume include measuring water displacement, or calculating the volume based on weight and the density of the casting material [22]. Some groups have taken a more sophisticated approach by digitizing the shape and performing calculations based on those images. Hayward and associates used NMR (nuclear magnetic resonance) imaging for digitization [23], while others have used CT (computerized tomography). Although these methods provide an accurate method for obtaining surface area [23], they still require wound contact and use of equipment that is not available to everyone.

Mathematical Estimations. While calculations based on single dimensional values, such as maximum and minimum wound diameter, are simple and require only minimal contact, they are actually only an estimate [18, 19]. Most estimative calculations are only for area, but Johnson proposed finding volume by multiplying surface area by maximum depth. For deep wounds, however, which are not typically uniformly deep, this technique would not provide an accurate measurement of volume [19].

Kundin Wound Gauge. Kundin developed an instrument to measure wound volume [24–26]. The plastic, disposable device consists of four crossarms and a vertical component, which is placed over and into the wound. The graduated arms are used to measure four radii, and the vertical component is used to measure depth. Area and volume are calculated using simple equations. Like the previously discussed methods, the Kundin wound gauge requires contact. Further, like the other mathematical methods, values obtained are estimates, not considering shape irregularity. Despite this, r values of 0.99 for area and volume have been observed for pressure ulcers. Error rates were 19% and 9% for volume and area, respectively [24].

Thomas and Wysocki compared area measurements of decubitus and venous stasis ulcers made from photographs, wound tracings and the Kundin wound gauge [27]. They found that while all three methods were highly correlated ($r = 0.93$, or above $p \leq 0.001$), they were all significantly different from each other ($p \leq 0.001$). The most accurate measure of actual wound area was obtained from acetate wound tracings.

b. Noncontact Methods of Wound Volume Measurement. More sophisticated methods of wound volume measurement exist. They offer the advantages of being noncontact, and they directly provide a permanent, digitized model. These methods include stereophotogrammetry, three-dimensional laser scanning and structured light. All of these methods are similar in that they involve obtaining digitized images using photographic techniques, and determining coordinates, dimensions and measures either from the produced images or from directly produced coordinate data. Using computer software, volume, surface area, and other dimensions can be obtained from the digitized images or coordinates with a high degree of accuracy and precision [28].

Stereophotogrammetry. Stereophotogrammetry has been used for two decades to measure contours on the body. The most common application in the medical field has been for digitization and reconstruction of facial contours. More recently, stereophotogrammetry has been employed to measure wound volume of pressure ulcers and deep burn wounds.

Stereophotogrammetry is based on the principle that two 2-D images taken from two different angles can be combined to create a 3-D image. Three-dimensional images of wounds have been constructed by manually selecting points with a metrograph or by using a stereocomparator [21, 28]. A setup involving two cameras, placed on both sides of the surface, captures images to be used for image reconstruction. Specialized software is used to combine these images to calculate wound volume, area, and perimeter.

Bulstrode et al. found high precision and accuracy of area and volume measures of leg ulcers using stereophotogrammetry [28]. They also observed high precision and accuracy associated with detection of the epithelialization edge in full-thickness excisional wounds in pigs [28]. Primary advantages of steroeophotogrammetry are the high degree of accuracy and precision and the acquisition of a permanent, digitized image. The main disadvantages include high equipment cost, the time needed to collect the data, low portability, and the elaborate training required.

Laser Scanning for 3-D Reconstruction. Two types of laser scanners have been employed in the measurement of wound volume. These systems have been used to digitize facial contours for burn masks, design other burn dressings or pressure garments, plan plastic surgery and to calculate volume of foot ulcers.

The first system consists of a laser beam projected onto a wound, positioned horizontally on an x-y table. As the table moves, distance to the wound surface is recorded and used to reconstruct the wound surface. The primary disadvantage of this system is the expense and the lack of portability. Also, the patient must be lying with the wound horizontal. Systems have also shown relatively low accuracy with some measures. While Patete et al. observed a volume error rate of less than 1% with wounds of 0.5 and 1 mL, they only measured very small wounds, and the error rate jumped to 4% for a 2 mL wound [29]. Ibbett et al. observed precision values between 0.6% and 8.3% and accuracy values between 0.1% and 25% for volume values up to 60 mL [30].

A second, larger system employs a projected laser line and a camera that is rotated around the study surface (e.g., a face or wound), taking images every few degrees. Images are based on the concept that a straight line projected onto a curved surface will appear as a distorted line when viewed at an angle. Coordinates are identified by measuring the distance from the axis of rotation to a line representing the study surface. Custom software is used to construct a model of the surfaces scanned and to make calculations. Systems like this used for wound volume and surface area measurement are made by Cyberware [31, 32] and Cencit [33]. The main advantages of these laser scanning systems include direct data acquisition and processing, the ability to record color, and very fast scanning time. The main disadvantages are the high cost, the lack of portability, and the necessity of positioning wounds vertically, which may be difficult with some patients.

c. Structured Light. Like large 3-D laser scanning systems, structured light is based on the concept that a vertical line projected onto a curved surface appears curved when viewed at an angle. Unlike 3-D laser scanning, however, a structured light system does not necessarily require costly equipment. Structured light systems consist of a light projector system, a camera, a method of positioning the components appropriately, and software to help calculate wound and surface area from the obtained image. Costs can be further reduced by making the necessary calculations directly off the structured light photograph, instead of using software. Like stereophotogrammetry, structured light uses images obtained at an angle from the surface. For structured light used with wounds, stripes or dots have been projected directly onto the area to be examined [34–37]. By spacing the lines or dots apart a known distance, depth on the image can be easily calibrated to obtain volume and area values.

Plassman and Jones have developed a structured light system for wound measurement that employs color-coded lines [38]. Using the system, they take two pictures of each wound: one, with a homogeneous beam of light shone on the wound, to identify the wound boundary and to map reflectance, and a second picture, with 60 color-coded lines shone on the surface, to obtain the structured light image for analysis. The structured light images are then analyzed using custom software, which includes algorithms to adjust for reflectance, camera no-linearity, and lens error. With this system, they obtained depth measure accuracy of ±0.4 mm in real wounds and ±0.02 mm in an idealized situation. Advantages to their particular system include portability and easy use due to a handheld device and direct transmission of the images to a computer.

In a comparative study of volume measurements of mock ulcers, Plassman et al. obtained the smallest range of standard deviation values (per wound type) with structured light, when compared to saline fill and casting [39]. The standard deviation values were 3–15%, 5–16% and 9–18% of actual volume, for structured light, casting and saline fill, respectively.

At UAB, an inexpensive, portable structured light system is being developed. The system consists of a 35 mm camera on a tripod, a light situated on a heavy-duty stand with a projection attachment to project a slide of vertical lines, and commercially available image analysis software (SigmaScan Pro, Jandel Scientific, Mountain View, CA). The light source and projection attachment (Dedolight, Visual Departures, Riverside, CT) can be positioned at any angle, to ensure perpendicularity to the wound surface.

The projection device is used to project a slide of evenly spaced vertical lines (Fig. 5). It was found that yellow lines within a black background provide the best contrast for the photograph. A different slide could easily be employed, to provide different colored lines. A

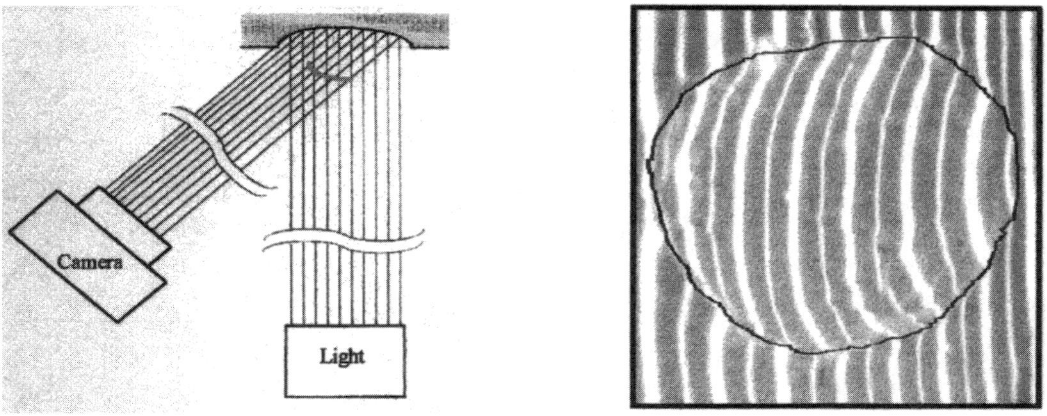

A B

Figure 5 (A) Parallel lines are projected from a light, onto the wound. A picture is taken with a camera located 45° from the line of projection. (B) Photos are digitized, and the volume and surface area of each slice is calculated based on depth, which, at any vertical point, is equal to horizontal deflection multiplied by $\sqrt{2}$.

photograph is taken using the camera, which is positioned 45° from the wound using a custom attachment on a standard tripod. The resultant photo is scanned for digitization. Wound area and volume are calculated from digitized photographs using SigmaScan Pro software.

At 45°, such as viewed in the resultant photograph, horizontal displacement of the parallel lines is equivalent to wound depth or surface topography. Total volume and wound surface area are calculated by summing up the volume and surface area of each "slice." The volume and surface area of each slice of a known thickness are calculated by multiplying the average of the cross-sectional areas at two adjacent projected lines by the thickness of the slice between lines. Wound surface area is computed by multiplying the average of length of the lines along the wound surface, at two adjacent projected lines, by the thickness of the slice. The volume and surface area values obtained at given time periods can be used to calculate tissue fill.

Because the camera is at 45° from the image, depth is easily calculated from the image:

$$\text{Depth} = \text{deflection distance} \times \text{square root} \tag{2}$$

To measure the deflection distance, first a line is extended across the wound to represent an image of the intact surface by connecting the lines on each edge of the wound. Deflection distance is the distance between the theoretical surface line and the actual line obtained from the contoured surface (Fig. 5).

Advantages of the system at UAB include minimal equipment costs and portability. Further, both the wound volume and wound surface area can easily be obtained using this method to allow calculation of tissue fill independent of wound size. Also, because the light source and camera can be rotated, a wound positioned at any angle can be accommodated.

B. Tissue Health

Tissue health may be defined as the potential of a selected tissue to undergo the healing process, and combat infection. This variable should describe the microenvironment within a wound. The assessment of tissue health should identify not only the ability of the wound to carry out natural healing but will also act as an indicator of where the wound is in the healing process, and help determine how far the wound is from normal.

One assessment of tissue health in skin is level of oxygenation. Research has shown that hypoxia reduces fibroblast proliferation, collagen synthesis, angiogenesis, and epithelialization [40–43]. Oxygen is also vital in the prevention of infection [44]. Investigators have determined that a tissue pO_2 of at least 30 mm Hg is necessary for efficient bacterial killing [42]. In addition, it has been shown that wounds with decreased oxygen display characteristics of delayed granulation and epithelialization [45]. This tissue health can be determined by directly measuring oxygen tension or indirectly by skin temperature or blood flow.

1. Traditional Techniques

a. Oxygen Tension. Oxygen tension, a measure for oxygenation, may be determined directly by either transcutaneous or subcutaneous oxygen tension measurements. Transcutaneous pO_2/CO_2 has been used in previous studies to assess wound healing [46]. It is used to assess both the current status of the wound and its ability to heal as well as its resistance to infection, since it is a measure of vascularity and perfusion [7,16]. Transcutaneous oximetry, however, has been criticized, in part, due to the requirement of heating the tissue. This step, vital to the measurement, actually skews the results to yield higher levels of tissue oxygenation than truly exist. The technique is also only partially sensitive to changes in the perfusion, and yields only one value for the entire wound. It is also a contact technique that requires a minimum of 15 min to obtain a measurement. Therefore, it appears not to be sensitive enough to measure the dynamics of the wound or model the entire wound microenvironment.

Subcutaneous pO_2 can be measured by two techniques: polarographic tonometry or optical fluorescence [47,48]. Both techniques make use of a tissue tonometer. The tonometer is a 1 mm diameter Silastic tube that is implanted into the subcutaneous tissue such that each end protrudes from the skin. Although the tonometer is inserted using sterile techniques, there are concerns about the level of invasiveness and that a percutaneous device, with its inherent infection risks, remains implanted during the healing process.

b. Oxyscope. Oxygenation may also be measured by an alternative technique which is claimed to be both noninvasive and quantitative. This method determines the tissue oxygen concentration via a light excitation and emission. *Porphyrine* dye is injected into the tissue of interest. A detection device measures the phosphorescence of the dye as it is being quenched by the oxygen within the tissue. Two devices presently on the market (both from Medical Systems Corp., Greenvale, NY) include the OXYSPOT, which measures a single point, and the OXYMAP, which is able to image an entire area. Both techniques provide a quick assessment or the local tissue oxygen levels.

c. Thermography. Thermography is the measurement of tissue heat, and in the case of skin wounds this would correspond to skin temperature. Although many variables influence skin temperature, among them ambient temperature, radiation and evaporation, if it is measured with relatively constant environmental factors, local changes in skin temperature can be determined [49]. These changes are a mainly a result of heat conduction followed by local heat production. Heat conduction is a result of blood and lymph flow, while local heat production is due to energy producing processes. Therefore, thermography acts as an indirect measure of blood flow and thus perfusion and oxygenation. This technique is a noncontact method that is able to measure and model an entire wound with surrounding tissue, assessing both the spatial and temporal dynamics of a wound [50,51]. This technique, however, is limited in that, at best, it is only an indirect measure of blood flow and difficulty in maintaining constant environmental factors hampers the ability to accurately assess changes in the wound over time.

d. Xe Wash. ^{133}Xe-washout method was one of the first "noninvasive" techniques established to determine blood flow directly. The technique begins with an injection of the radioactive labeled tracer ^{133}Xe. The tracer is deposited by means of the atraumatic epicutaneous technique [52,53]. Portable CdTe(Cl) semiconductor detectors are then fixed directly to the skin surface above the tracer deposits, with a single layer of Mylar membrane between the skin and the detector. Counts are taken at various intervals and stored for blood flow analysis.

e. Laser Doppler Flowmetry. Laser Doppler flowmetry (LDF) utilizes an optical probe which emits laser light of a known frequency. The laser light penetrates into the tissue of interest. The light is shifted due the motion of the light impacted particles, assumed to be blood cells [54]. The shifted light is either backscattered or absorbed. That light which is backscattered and emitted from the illuminated tissue is detected and analyzed. A series of algorithms determines the velocity of the particles based on the wavelength shift, giving a direct measure of blood flow.

This technique, although a direct measure of tissue perfusion, requires wound contact and is only able to measure blood flow over a single point (or up to seven points in some models). Although the probe may be sterilized, the close contact of the probe and/or adhesive ring to a healing wound is both difficult to achieve and difficult to justify. These limitations severely hinder its ability to model an entire wound or to be used in certain types of wounds.

f. Laser Doppler Imaging. Laser Doppler imaging (LDI) was introduced in the late eighties in an effort to overcome the limitations of LDF [55–58]. LDI is based on similar principles, yet is able to measure the Doppler shift, and thus blood flow, over a large area within a short time and with no tissue contact. This measurement is done with a laser head positioned 7 to 30 cm from the tissue. The laser head consists of a raster, or scanning, laser and a detection device, which is able to pick up the Doppler shifted light emitted from the tissue (Fig. 6). A data acquisition board is used to input this data to an analysis program, which then produces a map of the wound perfusion. These characteristics allow the LDI to provide a portable, noncontact blood perfusion assessment of the spatial variability within the entire wound [59,60].

LDI has been used in numerous applications both in the clinic and for research to measure blood perfusion in tissues such as skin, brain, muscle, and liver [60–68]. The device is capable of penetrating the tissue, to various depths, depending on the wavelength and tissue

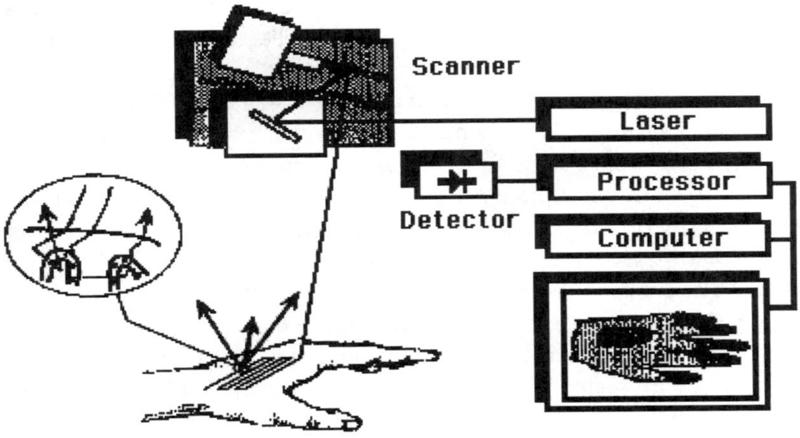

Figure 6 Diagram of the principle behind LDI.

density [69]. For skin a He-Ne laser is capable of penetrating up to 200–300 μm, which is into the dermis but above the underlying muscle and away from the major vessels. This allows assessment of the tissue health of skin from a vascular standpoint. These devices have been used in previous studies to help determine healing and healing rate in a number of different wounds [5,6].

Laser Doppler imaging is quickly emerging as a valuable assessment technique in clinics. For accurate quantification of the clinical assessment, it is important, however, to know the limitations of the technique and to quantify the physical variables that affect the perfusion values so that improper results are not obtained. The effect on perfusion level of physical variables, such as ambient light, distance from the tissue and presence of wound dressings should be known and quantified in order to develop standardized clinical assessment protocols. In addition, it has been shown that the amount of skin pigmentation has a direct effect on the perfusion values obtained. Several researchers are attempting to solve this limitation by the use of lasers with differing wavelengths [70]. Another concern has been the ability to match perfusion scans with wound landmarks. This is critical in evaluating temporal changes of many types of wounds.

2. *Tissue Health Measurement Development*

At UAB, a laser Doppler perfusion scanner (LISCA, Hewitt, NJ), with both high- and low-resolution scanning heads, is currently used in order to determine the effect of various treatment types on grafted burn victims and pressure ulcer patients.

a. Burn Patients. LDF has been used in the past for the early assessment of wound state and potential to heal [71,72]. In these studies, burns were classified, upon patient admission, by medical examiners as either partial-thickness or full-thickness wounds and clinical outcome was classified as healed, completely epithelialized by postburn day 21, or unhealed. Perfusion of healed wounds was found to be significantly greater than those of unhealed wounds at all time points. This and other studies have shown the usefulness of laser Doppler flowmetry and imaging in the determination of wound state [73,74].

In an ongoing clinical study, the tissue health is being evaluated for grafted wounds over a three-month period (Fig. 7). Laser Doppler images are taken, with the scanning head position parallel to and 23 cm away from the wound surface. At this distance, each pixel is approximately 1 mm^2. Scans of treatment sites and control sites are taken three times during the first week, postgraft, beginning with day 1. Images are then taken weekly for a period of three months. Both the average perfusion and the spatial variability are determined for each site. To account for patient-to-patient variability and measurement errors, the perfusion values used in the statistical analysis are the differences between the treatment sites and their respective internal controls. This allows more accurate comparison of temporal changes in blood perfusion.

Others have suggested that change in perfusion level due to a provocation (temperature, pressure, occlusion) can be used to get a more accurate assessment than the absolute levels. At this time, both the concerns of wound contact and wound manipulation, as well as the satisfactory results using controls, have obviated the need to do this.

b. Pressure Ulcer Patients. In an ongoing clinical study the tissue health of II-IV pressure ulcers is being evaluated over a two-month period for outpatient spinal cord injured (SCI) patients (Fig. 8). Scans are taken at the time of treatment and at each monthly visit thereafter. From the scans, the averaging blood flow values, above the control, of both the healed and unhealed portions of the pressure ulcer are determined. Three metallic markers are placed at fixed distances around the wound to give landmarks that show up in the scan and in photographs taken. This allows better determination of the wound regions in the scan and

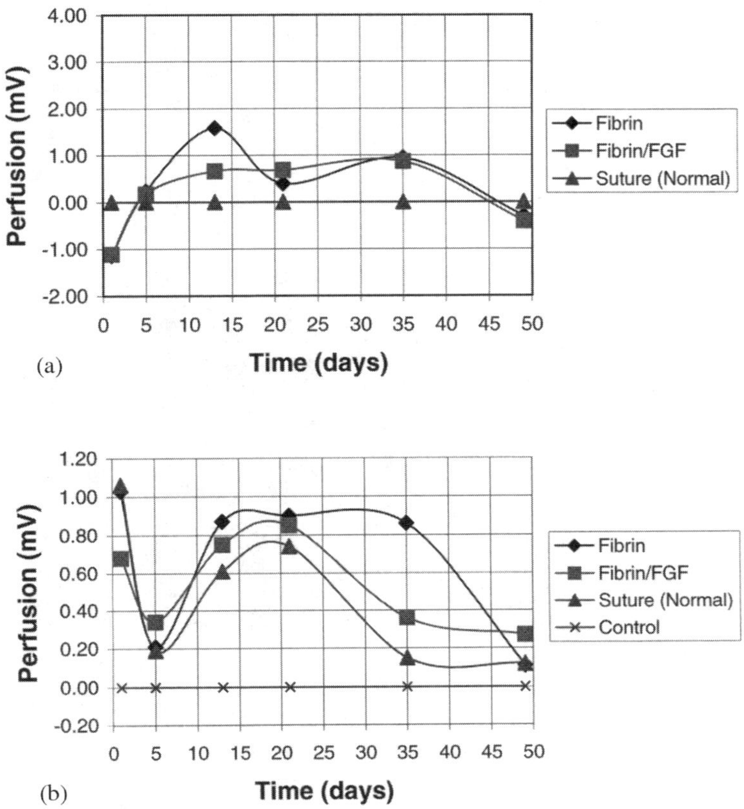

Figure 7 LDI results for a burn patient. (a) Mean perfusion of alternate treatment types minus the mean perfusion of the standard treatment type. LDI appears to distinguish the temporal changes of each treatment type. (b) Standard deviation of perfusion within a scanned region (alternate treatment types minus the standard treatment type). The deviation within the scanned area decreases in conjunction with the healing of the wound.

distances within the scan. Again at this point in time, provocations have not been necessary to obtain satisfactory results.

c. Quantitative Associations. In order to help obtain better quantitative data from these measurements, studies have been done to determine the effect of measurement variables on perfusion level. Specifically the effects of ambient light, distance of the scanning head from the sample, and the angle of the scanner head have been evaluated.

Preliminary studies have shown that scans taken, by the LISCA laser Doppler imager, with ambient light levels between 2 and 20 lux give equivalent readings (Figs. 9, 10). Additionally, the low-resolution scanner head showed consistent readings for laser head-tissue distances between 18 and 26 cm. The high-resolution head, however, yielded greater perfusion values as the head got closer to the tissue.

d. Future. Additional studies are necessary to determine the maximum light levels that produce consistent readings. Furthermore, studies are needed to determine the quantification of physical variables which may affect perfusion readings, such as the ability to penetrate wound dressings, the effect of scanning head angle, room temperature, and other variables to help develop clinical protocols which provide the most accurate results. These studies must be continued to help determine the exact limits of this system for use in a clinical setting.

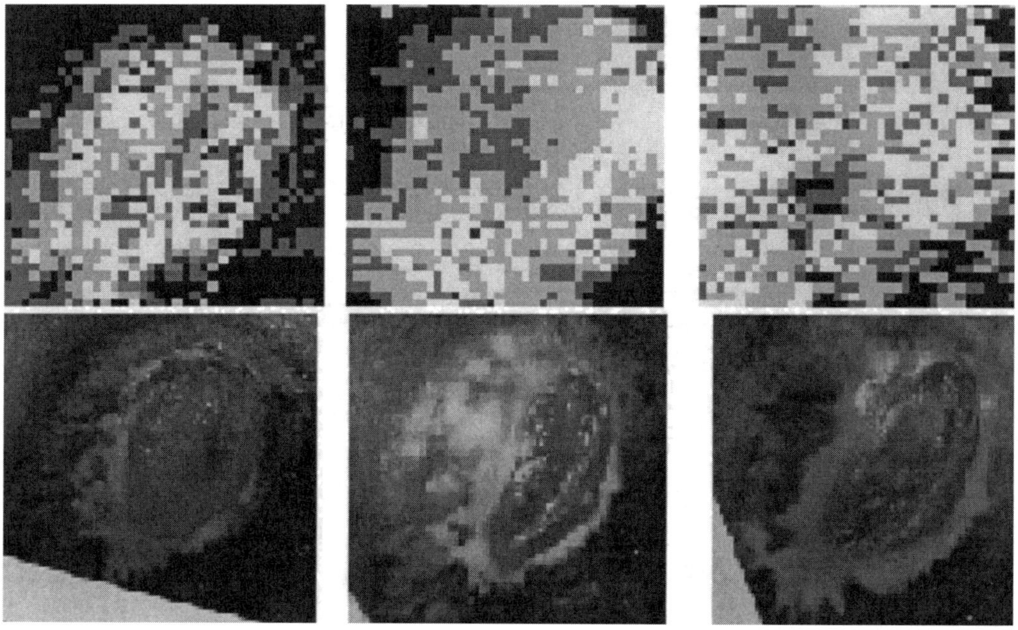

Figure 8 Pressure ulcer and corresponding LDI at one month intervals.

C. Tissue Function

1. *Structure/Function of Skin*

The skin is the largest organ in the body, and its normal functions are essential to survival. Skin provides coverage and protection against water loss, thermal loss, and chemical, infectious, and other environmental agents. It provides mechanical integrity and protects against external forces. The skin is a multilayer composite structure with varying thicknesses over the body. Anatomically, the skin is divided into two structurally distinct layers, the epidermis and the dermis, which as a unit provide the functions of the skin (Fig. 11). The epidermis is the cellular

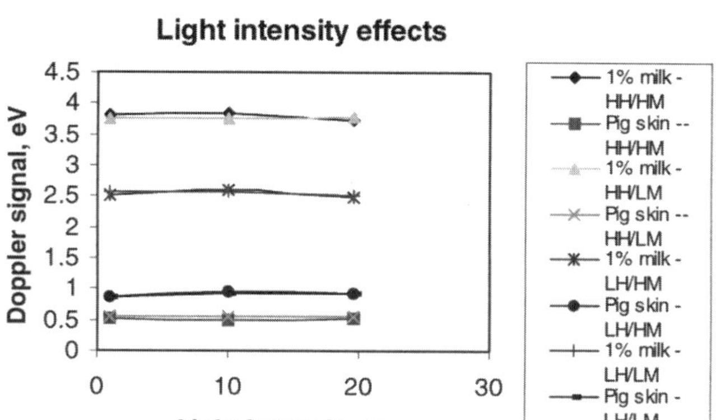

Figure 9 The effect of light levels on perfusion measurements.

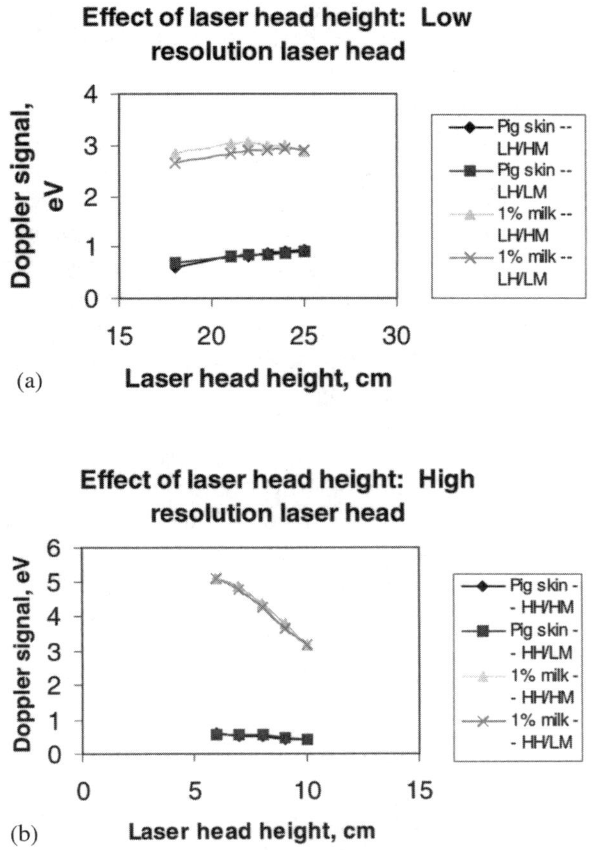

Figure 10 The effect of scanning head distance from sample on perfusion measurements: (a) low resolution scanning head; (b) high resolution scanning head.

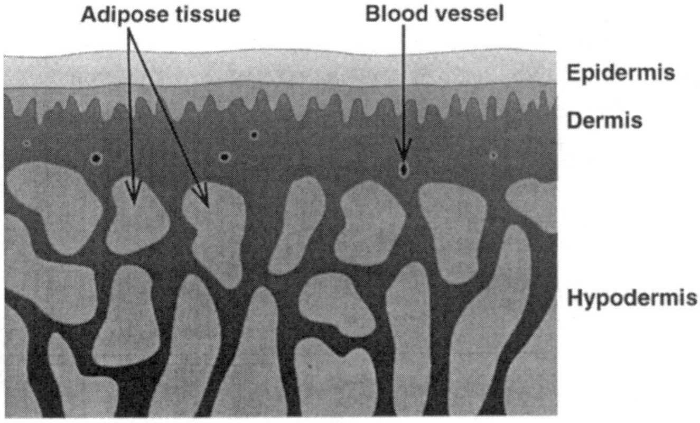

Figure 11 Schematic of skin.

outermost layer that is primarily responsible for barrier functions. The dermis is a thicker, mostly acellular layer that gives the skin its mechanical properties and the epidermis its blood supply.

Healing of skin injuries proceeds through a combination of epithelialization, new dermal tissue formation, and contraction to achieve wound closure. The thickness of the injury is an important factor in determining the extent of each component in the healing process. Injuries to the epidermis and superficial dermis primarily heal by epithelialization from the wound perimeter as well as from epithelial islands formed by hair follicles and sweat glands. Injuries to the deeper dermis, the reticular dermis, primarily heal by contraction. These deeper wounds, in some cases, exhibit excessive healing and the formation of hypertrophic scar tissue. Contracture, the pathological shortening of scar tissue, can also occur and is both deforming and debilitating. In both cases, this results in an unacceptable clinical result due to the increased stiffness, or reduced elasticity, of the skin and poor cosmetic appearance [75–78].

Although the goal of skin wound healing is a complete regeneration of original structure and function, it is difficult, if not impossible, to currently achieve this clinically. In reality, the best result is probably a functional repair that closely approximates the original tissue and provides virtually all the functions. Skin barrier functions are important in the early stages of healing to prevent desiccation and infection. The use of skin grafts for burns and specific wound dressing, however, can provide adequate coverage and protection until the epidermis can regenerate. The restoration of acceptable stiffness or pliability long term is still a primary challenge to clinicians.

2. *Justification of Mechanical Testing*

Researchers and clinicians have devised techniques and instruments to measure the various functions of the skin. Techniques to measure the barrier function have been used by several researchers, especially techniques to measure transepidermal water loss (TEWL) [79,80]. These instruments measure the amount of water vapor passing through the skin by passive diffusion, and have been used as an indirect measure of reepithelialization of partial-thickness wounds [81]. The devices have been shown to be good indicators of the barrier function of epidermis, but numerous environmental and external factors such as temperature, humidity, air turbulence, and probe pressure, must be carefully controlled.

Restoration of epidermal barrier function is crucial in the early stages of healing of large wounds, but wound dressings have been developed that may serve adequate barrier function while the epidermis regenerates underneath. The regenerative ability of the epidermis, then, makes measurements of barrier function less important than assessments of ultimate skin functionality. A main cause of impaired tissue function in wound healing is the increased stiffness, or decreased pliability, due to hypertrophic scarring. A nondestructive measure of tissue stiffness would be best able to assess ultimate tissue function in order to accurately compare treatment and therapy modalities for long-term wound healing.

3. *Clinical Assessment of Skin Stiffness*

The assessment of scar characteristics is an important clinical tool to determine tissue function and there have been several attempts at developing clinical rating scales to better quantify scar healing [82–84]. These rating scales are often based primarily on visual assessment of scar color and thickness, rather than stiffness, and are rather subjective and inconsistent. In addition, the sheer number of different scales used by different institutions makes comparisons of treatments among institutions difficult, if not impossible, with any accuracy. More often, the clinician or therapist makes a highly subjective estimation of the stiffness and functionality of scar tissue based upon his or her experience. The obvious drawback is that the effect of different therapies on scar stiffness cannot be accurately and statistically compared in order to identify better treatments. There is also a significant problem with inconsistencies between individual clinicians, so-called interrater reliability, which limits the validity of multicenter or even multi-investigator studies. More objective, quantitative measures of the mechanical properties of scars are obviously needed to accurately assess tissue function.

4. *In Vivo Measurement of Tissue Stiffness*

Many of the mechanical characteristics of skin and other soft tissues have been measured in vitro. These techniques generally use excised tissue tested with a materials testing system such as an Instron. There are some advantages to in vitro systems. Uniaxial testing can give simple stress-strain relationships that can be modeled and quantified. In addition to the mechanical analysis, the skin can be analyzed histologically to determine collagen or elastin content and orientation as well as the thickness of scar tissue.

There are, however, key disadvantages to the use of in vitro systems for determining or even predicting skin function clinically. Besides the obvious impossibility of using in vitro tests to determine the state of tissue function in a clinical situation, the mechanical properties of skin in vivo are different than those of excised tissue. Skin is an anisotropic material and in vivo is under a constant state of tension in varying degrees from all directions. It is difficult to duplicate that in vitro. In vitro systems are also generally uniaxial or biaxial and have difficulty mimicking the natural state of skin tensions. Underlying tissues, which are removed in the excision, may also have significant effects on the mechanical properties of skin in vivo. In addition, it is difficult to excise skin without damaging it, especially in wound healing applications. There are also numerous problems with actual testing of soft tissues, not the least of which is gripping the tissue without compromising and weakening the areas near the grips.

The ideal clinical system should be minimally invasive and nondestructive while providing accurate, quantitative measurements of stiffness and function. There have been a variety of methods employed to quantitatively determine the mechanical properties of skin in vivo in order to ascertain tissue function. In fact, a primary problem has been the sheer number of different methods and the different parameters they measure. It is often difficult to compare the various methods and the results, and the clinical significance is unclear. It is vitally important with in vivo tests to clearly identify all testing parameters, body locations, orientations, and so on, so that proper comparisons can accurately be made. Because of the need for nondestructive evaluation, in vivo tests cannot determine ultimate strengths of tissue and must determine mechanical parameters by measuring the response to small loads or strains. There have generally been six in vivo approaches to determine the mechanical properties of skin: horizontal traction (most similar to in vitro tensile testing), horizontal torque, vibration (by acoustic shear wave propagation and vertical mechanical vibrations), indention, elevation, and suction [85,86]. In addition, several researchers have concurrently used ultrasound to measure skin thickness and to localize and delineate scar tissue [87].

a. Horizontal Traction. Horizontal traction systems, such as the quasi-static extensometer and the bio-skin tension meter, most nearly mimic in vitro testing and have been used by many researchers to measure the mechanical properties of skin [84, 88–104]. Tabs are attached to the skin by cyanoacrylate glue or adhesive and a tensile force or displacement is applied to separate the tabs. The resulting strain or stress, respectively, can be measured by strain gauges and a stress-strain curve may be plotted. The sigmoidal curve features an initial "toe region" followed by a linear region, the slope of which represents a modulus of elasticity.

In addition to the stress-strain curve, in vivo tensile tests can also demonstrate the viscoelastic nature of skin by measuring stress relaxation, creep, hysteresis, and preconditioning. Care must be taken with these types of devices to carefully standardize tab separation, tab geometry, and applied vertical pressure on the tabs by the device, as well as direction of load application, due to the anisotropy of skin.

b. Horizontal Torque (Twistometry). In vivo torsion devices involve the application of a rotational force, or torque, to the skin in a plane parallel to the surface [105–108]. Like the traction tests, a disk or ring is attached to the skin and a force or displacement is applied. Again, a stress-strain curve may be generated from these devices to obtain elasticity, though by definition the value obtained is not technically Young's modulus of elasticity. Agache et al. and others have applied a step torque to measure instantaneous angular deformation, creep during a constant torque, and recovery following the release of the load [106]. The anisotropy of skin gives an advantage to torsional tests over uniaxial or biaxial torsion, since the torsional tests twist the skin a small angle through all orientations and uniaxial tests fail to take into account the anisotropy of skin and scar tissue. The radius of the attached ring has significant influence on the depth of tissue affected; a wider ring records the properties of deeper layers of skin, fat, or even muscle. In addition, the attached torque ring may be fitted with an immobilized outside guard ring so that only the area inside the guard ring is subjected to the applied torque. This serves to limit the amount of sliding of the skin over the subcutis and to enhance the importance of the mechanical properties of the stratum corneum alone [105].

c. Vibration. Another technique to determine the mechanical properties of skin has involved the measurement of the skin's response to physical and acoustical vibrations [109–113]. The ballistometer and gas-bearing electrodynamometer induce a vibration physically and measure the induced mechanical response. It has been argued that the former measures elastic parameters below the surface, while the latter primarily measures stiffness in the surface plane of the skin [111,112]. Some researchers have measured the response to acoustical shear waves in the skin, such as speed of transmission and amount of dampening, to determine skin properties [113].

d. Indention. Indention techniques measure skin mechanical properties in response to the application of a vertical pressure [114–121]. As such, these devices show a high dependence on underlying structures such as fat, muscle, and bone except at very low pressures. Indentometers have been developed with both flat and spherical application heads with varying success [114,121]. Tonometers, originally used by ophthalmologists to measure endocular pressure to diagnose glaucoma, have been modified by clinicians to measure skin stiffness. These devices use a spring to measure resistance of the test site against indention. Esposito et al. reported on the use of a modified Schiotz tonometer to measure skin tone in burn scars and showed good correlation between the tonometer measurements and clinical evaluation. One problem they noted, however, was the occasional discrepancy over time between the clinical rating of the scar and the tonometer measurement [117]. Spann et al. showed similar results, but noted that an instrument with a larger diameter for testing (versus the less than 7 mm diameter plunger of the tonometer) should better evaluate stiffness and reduce localized variations [118]. Durometers, originally designed for hardness measurements of nonmetallic polymers, have also been used to evaluate normal and sclerodermatous skin [119]. Finally, indention systems have been used in conjunction with ultrasound to determine mechanical properties and corresponding skin thickness [120].

e. Elevation. Elevation techniques involve the application of a vertical pressure to the skin through an attached tab with or without a guard ring around the test area to limit the involvement of a larger region [121–123]. It is comparable to a clinician's pinch to determine suppleness. Like most methods, the measured properties are greatly affected by the test region, with larger test area corresponding to deeper tissue affects. Levarometry is one such technique that has been used to measure skin "slackness" and exerts a vertical pull on the skin without a guard ring [123].

f. Vacuum Methods. One of the more common mechanical testing systems involves the application of suction and the measurement of the resulting deformation. Several systems are commercially available and relatively inexpensive and portable. The primary difference between the systems lies in the diameter of the vacuum chamber and consequent test area. The diameter of the aperture affects the depth of tissue affected by the vacuum. For example, Gniadecka and Serup argued that the standard Cutometer SEM 474 (2 mm diameter) primarily measures the mechanical properties of the epidermis and superficial dermis, while the Dermaflex A (10 mm diameter) measurements are primarily influenced by the mechanical properties of the dermis [124]. Vacuum-driven systems have been used to derive mechanical properties of skin for a variety of applications and can obtain many potentially useful mechanical parameters.

Cutometer. The Cutometer SEM 474 is probably the most widely used commercially available vacuum-driven device for measuring the in vivo mechanical properties of skin [125–134]. It features a standard 2 mm diameter vacuum chamber (approximately 3 mm^2 test area), but larger probes are available. The system measures the vertical deformation of the skin by an optical system that detects the reduction of the intensity of an infrared laser beam.

Researchers using the cutometer have primarily focused on creep and stress relaxation parameters. The system applies a load as a square wave, in which the load is applied instantaneously, held for a time, and instantaneously released. Since the skin is a viscoelastic material, the applied load causes an instantaneous deformation with the step and a slow increase in deformation during constant load (creep). When the vacuum is released, the skin shows a partially instantaneous recovery followed by a slower return to baseline. Using the notation described by Agache [105, 106], Barel [126] and others the following parameters are isolated: instantaneous deformation (U_e), deformation due to creep (U_v), instantaneous relaxation (U_r), and total deformation ($U_f = U_e + U_v$). Obviously, these parameters are dependent on the strength and duration of the load, so these parameters must be standardized in order to obtain accurate results. The measurements are also a function of skin thickness, which varies according to gender, age, and a variety of other variables. Thus, the raw deformation numbers are inadequate for comparisons. Instead, most researchers use the ratio of recovery to immediate deformation (U_r/U_e) as a parameter of elasticity independent of thickness. For test sites with controlled contralateral areas, the total deformation may also be used to quantify differences in elasticity. These parameters have been used by researchers to investigate normal skin, wound healing, scar mechanics, topical agents, pathological states, age, and so on.

Dermaflex. The Dermaflex system is vacuum driven, similar to the cutometer, but has a significantly larger test area (10 mm diameter) and as such is claimed to be more useful in determining mechanical properties of the dermis [124, 134–137]. Three parameters are isolated from the Dermaflex: tensile distensibility, resilient distension, and hysteresis. Tensile distensibility represents the maximum deformation at the end of the load application period, a parameter comparable to U_f. The resilient distension represents the deformation remaining immediately after the removal of the load (comparable to $U_f - U_t$). The hysteresis is the difference between the maximum deformation after the first load and the final load of a given number of loading cycles [134]. The phenomenon of hysteresis reflects the viscous nature of skin. The larger test area of the Dermaflex affects deeper tissues and may be a better assessment tool than the cutometer for determining functionality of deeper dermal scar tissue.

5. Tissue Function Measurement

The in vivo mechanical properties of skin can also be measured using the biomechanical tissue characterization system (BTC1000, SRLI Technologies, Nashville, TN) [138–142]. The system consists of a vacuum chamber and a video camera for application of load and measurement of tissue response, respectively. Basically, a 2.5 cm diameter ring is adhered to the site using double-sided adhesive tape. Two infrared reflective stickers (diameter 2 mm) are attached at consistent locations within the ring, 5 mm apart. The vacuum chamber, consisting of a glass tube connected to a vacuum pump (Precision Scientific), is then fitted securely to the ring. A negative pressure is applied to the site at a constant rate to a given negative pressure and released. The tissue deformation defined by the separation of the reflective stickers is monitored in real time at 10 Hz by a video camera mounted at the top of the vacuum chamber (Fig. 12) The system calculates deformation with corresponding pressure value to generate a pressure-deformation, or stress-strain, curve. From the resultant pressure-displacement curves, an elastic modulus and total energy under the curve can be calculated. Both measures may be representative of distinct mechanical characteristics of a healing wound. In addition, the area under the "toe" region of the stress-strain curve and the slope of this early region may reveal specific elastic contributions. This reflects the distinctly different histomorphometry of hypertrophic scars primarily in terms of collagen and elastin fiber concentration, structure, and orientation.

The BTC system has been used to determine the strength of incisional wounds in animal models. More recently, the BTC system has been used by our lab in a nondestructive manner

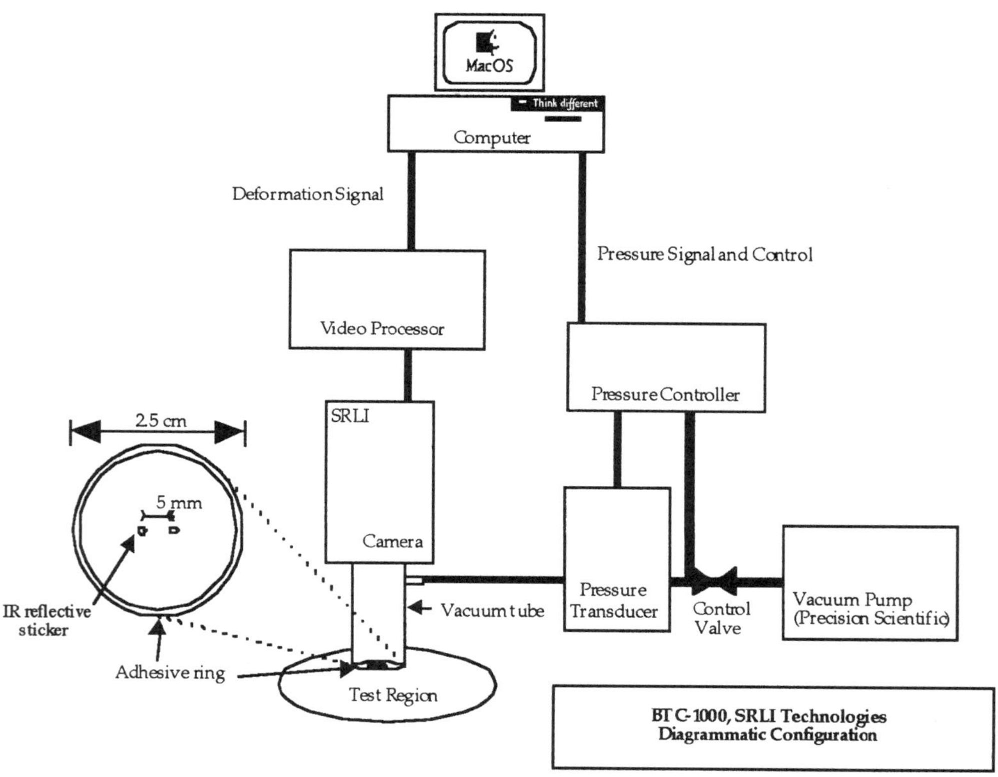

Figure 12 Schematic of the BTC system.

to quantitatively assess mechanical parameters of normal skin and burn scars following skin grafts [138]. The BTC system tests a circular area of approximately 5 cm^2 and as such is considerably larger than other vacuum systems. The BTC system, then, should measure the mechanical properties of deeper layers of the skin or scar tissue over a larger surface area. It should better quantify the mechanical properties of the dermis or hypertrophic scar tissue and reduce the error introduced by the influence of localized surface variance. It is important to be consistent with the orientation of the reflective stickers if comparing a given area over time. For this reason, it is useful to photograph the test area and use the pictures to insure consistency over time. The use of this quasi-biaxial system can be used, however, to compare the stiffness of a given region in different directions. In that regard, it is similar to the horizontal traction systems.

The results of a typical stiffness test are shown in Fig. 13. A two-month postgraft site and a contralateral control area were each tested twice [138]. Figure 13a shows the deformation versus time curves for the two regions. The control site had an approximately 40% increased total deformation. Figure 13b shows the pressure versus strain (percent deformation) curves for the regions. The total area under the curve for the control site was 50% higher than that of the graft site. As illustrated by the regression lines, there was a significant difference between the slopes (elastic modulus) of the two curves subjected to small pressures. Specifically, from zero applied load to 30 mm Hg applied load, the elastic modulus of the graft site was 201 kPa compared to 92 kPa for the control site. At higher applied loads, up to the final 150 mm Hg, the moduli of the two sites are similar. Results similar to this have been reported for in vitro mechanical testing of skin as well [143]. In ongoing clinical trials, in order to determine clinical relevance, the BTC measurements are compared to a clinical rating scale, the Vancouver scar scale [82,83]. The difference between skin graft scars and normal skin are being evaluated at various time points of healing as well as being followed longitudinally. The comparison of the BTC to clinical evaluation tools is important in order to ensure that a clinically relevant parameter can be detected with the system.

6. *Clinical Relevance and Future*

Though it is clear that in vitro and in vivo measurements have greatly contributed to our knowledge of the skin, it is important that the parameters measured have clinical relevance. For example, it is difficult to prove at this point whether a given mechanical parameter such as elasticity in tensile tests or the ratio of immediate deformation to immediate recovery in cutometer measurements really indicates valid clinical parameters. The mechanical measurements of healing skin should also be correlated with both normal skin and the evaluations of experienced clinicians and therapists. The correlation should help both parties to better understand wound healing and tissue assessment.

The use of pressure dressings to reduce the formation of hypertrophic scarring has been done for years, but there have been no multicenter, quantitative studies to prove actual improvement in scar elasticity. Similarly, many researchers have investigated the use of topical and systemic agents in wound healing with little quantitative data to support their findings. A quantitative measurement could enhance statistical capacity to accurately compare treatment and therapy regimes. An understanding of normal skin and the "normal" progression of wound healing by objective, quantitative measurements could also assist in the early detection of abnormalities. A big problem lies in the wide array of mechanical tests with questionable ability for cross-comparison. There are so many variables involved in the different tests and such a wide range of possible parameters that multicenter studies are difficult to obtain. In addition, many of the mechanical tests are complicated and time consuming. These types of

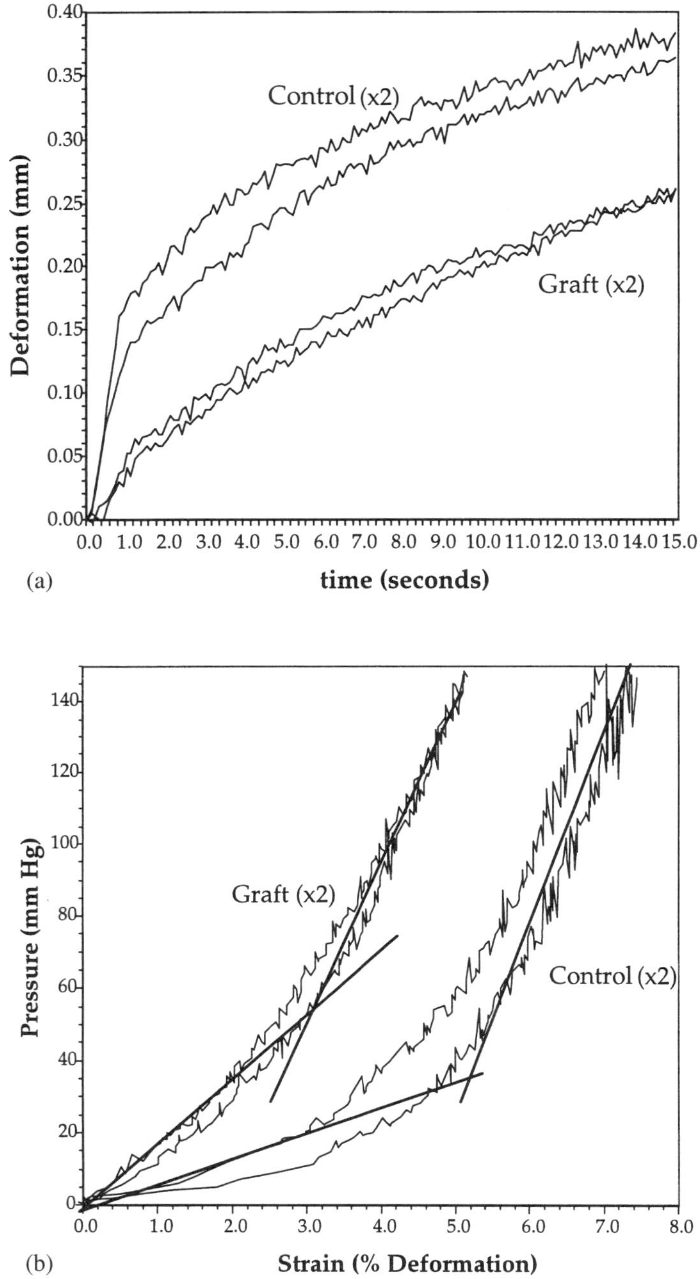

Figure 13 Results from a typical BTC1000 test of a grafted site and a contralateral control: (a) deformation vs. time; (b) pressure vs. strain.

problems reduce the chance of their incorporation into routine outpatient examinations, and decrease the availability of hard data that could only enhance our knowledge of skin structure and function.

III. CONCLUSIONS

To accurately assess effectiveness of skin regenerative biomaterials, quantitative assessments are essential. The lack of these clinically used tools poses problems in comparing new treatments or skin substitutes to traditional treatments. Another goal of assessments should be to measure tissue state as a comparison to the normal state as well as track the tissue state during treatment. Therefore, interventions can be assessed on how well they push the wound toward the normal state.

For skin wounds, such as skin ulcers or burns, assessments need to be done in three groups: healing rate, tissue health, and tissue function. In this chapter noninvasive, portable, quantitative techniques were described to obtain each of these three measures: healing rate independent of wound size, blood perfusion as a measure of tissue health, and tissue stiffness as a measure of tissue function.

Now there needs to be consensus among researchers and clinicians to determine what is the minimum number of tests needed to assess a skin wound and how can these measures be standardized from one clinic to another. This will permit sufficient tissue state data to be collected to allow predictions or prognosis to be made based on the current state. Eventually this could also be used to predict how a specific intervention would alter the tissue state and thus aid in development of treatment protocols. This would then be useful to clinicians as well as researchers trying to assess the efficacy of new interventions. The problem is not having a consensus on appropriate parameters to characterize normal tissue or a good healing response as well as the lack of good proven quantitative measures for assessing these parameters. In addition, the demonstration of correlation of consensus quantitative tissue state measures to clinical outcome would help prove the utility of these assessments for evaluation of new treatments in clinical trials.

REFERENCES

1. Feldman, D. and Osborne, S., 1994. Use of non-invasive imaging techniques to accurately quantify wound healing. *Trans. Wound Healing Soc.*, 4:21.
2. Winter, G. W., 1972. Epidermal regeneration studied in the domestic pig. In: H. I. Maibach and D. T. Rovee (Editors), *Epidermal Wound Healing*, Chicago, Year Book Publishers, pp. 71–112.
3. Mian, E., Martini, P., Beconcini, D. and Mian, M., 1992. Healing of open skin surfaces with collagen foils. *Int. J. Tiss. Reac.*, 14 (suppl): 27–34.
4. Cravalho, E., 1994. Thermography. Technologies to Assess Wound Healing. *Workshop at Wound Healing Soc.*, 4: 72–75.
5. Pan, F., 1994. Laser Doppler. Technologies to Assess Wound Healing. *Workshop at Wound Healing Soc.*, 4: 30–31.
6. Shannon, R., 1994. Evaluation of neovascularization in a porcine full-thickness wound using a laser doppler imaging technique. Technologies to Assess Wound Healing. *Workshop at Wound Healing Soc.*, 4: 93.
7. Blum, B., Feldman, D., Kilpadi, D. and Redden, R. A., 1997. Non-invasive assessment of regenerative skin systems. *Trans. Soc. Biomater.* 23: 470.
8. Snowden, J. M., 1984. Wound closure: an analysis of the relative contributions of contraction and epithelialization. *J. Surg. Res.*, 37:453–463.

9. Carrel, A. and Hartmann, A., 1916. Cicatrization of wounds. I. The relationship between the size of the wound and the rate of cicatrization. *J. Exp. Med.*, 24:429.
10. Gilman, T. H. 1990. Parameter for measurement of wound closure. *Wounds*, 3:95–101.
11. Gorin, D. R., Cordts, P. R., LaMorte, W. W. and Manzoian, J. O., 1996. The influence of wound geometry on the measurement of wound healing rates in clinical trials. *J. Vasc. Surg.*, 23(3): 524–528.
12. Van den Brenk, H. S. A., 1956. Studies in restorative growth processes in mammalian wound healing. *Br. J. Surg.*, 43:525.
13. Catty, R. H. C., 1965. Healing and contraction of experimental full-thickness wounds in the human. *Br. J. Surg.*, 52:542.
14. Pecoraro, R. E., Ahroni, J. H., Boyko, E. J. and Stensel, V. L., 1991. Chronology and determinants of tissue repair in diabetic lower extremity ulcers. *Diabetes*, 40:1305–1313.
15. Margolis, D. J., Gross, E. A., Wood, C. R. and Lazarus, G. S., 1993. Planimetric rate of healing in venous ulcers of the leg treated with pressure bandage and hydrocolloid dressing. *J. Am. Acad. Dermatol.*, 28: 418–421.
16. Blum, B., Feldman, D., Kilpadi, D. and Redden, R. A., 1997. Quantifying wound healing in pressure ulcers. *Wound Rep. Reg.* 5(1): A99.
17. Redden, R. A., Blum, B., Kilpadi, D. and Feldman, D., 1998. Quantitative assessment of wound healing rate. *Wound Rep. Reg.*, 6(3): A246.
18. Stacey, M. C., Burnand, K. G., Layer, G. T., Pattison, M. and Browse, N. L., 1991. Measurement of the healing of venous ulcers. *Aust. N. Z. J. Surg.*, 61: 844–848.
19. Johnson, J. D., 1995. Using ulcer surface area and volume to document wound size. *J. Am. Podiatr. Med. Assoc.*, 85(2): 91–95.
20. Covington, J. S., Griffin, J. W., Mendius, R. K., Tooms, R. E. and Cliff, J. K., 1989. Measurement of pressure ulcer volume using dental impression materials: suggestion from the field. *Phys. Ther.*, 69(8): 690–694.
21. Eriksson, G., Eklund, A.-E., Torlegard, K. and Dauphin, E., 1979. Evaluation of leg ulcer treatment with stereophotogrammetry. *Br. J. Dermatol.*, 101: 123–131.
22. Resch, C. S., Kerner, E., Robson, M. C., Heggers, J. P., Scherer, M., Boertman, J. A. and Schileru, R., 1988. Pressure sore colume measurement: a technique to document and record wound healing. *J. Am. Geriatr. Soc.*, 36(5): 444–446.
23. Hayward, P. G., Hillman, G. R., Quast, M. J. and Robson, M. C., 1993. Surface area measurement of pressure sores using wound molds and computerized imaging. *J. Am. Geriatr. Soc.*, 41(3): 238–240.
24. Johnson, M. and Miller, R., 1996. Measuring healing in leg ulcers: practice considerations. *Appl. Nurs. Res.*, 9(4): 204–208.
25. Kundin, J. I., 1989. A new way to size up a wound. *Am. J. Nurs.*, 1: 206–207.
26. Kundin, J. I., 1985. Designing and developing a new measuring instrument. *Periop. Nurs. Quart.*, 1(4): 40–45.
27. Thomas, A. C. and Wysocki, A. B., 1990. The healing wound: a comparison of three clinically useful methods of measurement. *Decubitus*, 3(1): 18–20.
28. Bulstrode, C. J. K., Goode, A. W. and Scott, P. J., 1986. Stereophotogrammetry for measuring rates of cutaneous healing: a comparison with conventional techniques. *Clin. Sci.*, 71:437–443.
29. Patete, P. V., Bulgrin, J. P., Shabani, M. M. and Smith, D. J., 1996. A non-invasive, three-dimensional, diagnostic laser imaging system for accurate wound analysis. *Physiol. Meas.*, 17: 71–79.
30. Ibbett, D. A., Dugdale, R. E., Hart, G. C., Vowden, K. R. and Vowden, P., 1994. Measuring leg ulcers using a laser displacement sensor. *Physiol. Meas.*, 15: 325–332.
31. Cutting, C., McCarthy, J. G. and Karron, D., 1988. Three dimensional input of body surface data using a laser light camera. *Ann. Plast. Surg.*, 21: 38–45.
32. McQuiston, B. and Whitestone, 1995. The aplication of laser surface scanning for quantifying human wound progression. *Trans. Wound Healing Soc.*, 5:33.
33. Bhatia, G., Vannier, M. W., Smith, K. E., Commean, P. K., Riolo, J. and Young, L. V., 1994.

Quantification of facial surface change using a structured light scanner. *Plast. Reconstr. Surg.*, 94(6): 768–774.
34. Gregory, A. and Lipcznyski, R. T., 1994. The three dimensional reconstruction and monitoring of facial surfaces. *Med. Eng. Phys.*, 16: 249–252.
35. Wood, F. M., Currie, K., Backman, B. and Cena, B., 1996. Current difficulties and the possible future directions in scar assessment. *Burns*, 22(6): 455–458.
36. Blum, B. E., Feldman, D. S. and Osborne, S. W., 1996. Measuring wound healing using structured light and an analytical model. *Ann. Biomed. Eng.*, 24 Suppl. 1: S54.
37. Ozturk, C., Nissanov, J., Dubin, S., Shi, W.-Y., Nichols, J. and Mark, R., 1995. Measurement of wound healing by image analysis. *Biomed. Sci. Instrum.*, 189–193.
38. Plassman, P. and Jones, B. F., 1992. Measuring area and volume of human leg ulcers by colour coded structured light. *J. Wound Care*, 1: 35–38.
39. Plassman, P., Melhuish, J. M. and Harding, K. G., 1994. Methods of measuring wound size: a comparative study. *Ostomy Wound Manage.*, 40(7): 50–60.
40. Hunt, T. K. and Pai, M. P., 1972. The effect of varying oxygen tensions on wound metabolism and collagen synthesis. *Surg. Gynecol. Obstet.*, 135: 561.
41. Jonsson, K., Jensen, J. A., Goodson, W. H., et al., 1986. Wound healing in subcutaneous tissue of surgical patients in relation to oxygen availability. *Surg. Forum*, 37: 86.
42. Knighton, D. R., Halliday, B. and Hunt T. K., 1984. Oxygen as an antibiotic: The effect of inspired oxygen on infection. *Arch. Surg.*, 119: 199.
43. Niinikoski, J., 1980. Cellular and nutritional interactions in healing wounds. *Med. Biol.*, 58: 303.
44. Wipke-Tevis, D., 1995. Subcutaneous tissue oximetry; implications for wound healing and monitoring critically ill patients. *New Technol. Criti. Care*, 7(2): 275–283.
45. Ahn, S. T. and Mustoe, T. A., 1990. Effects of ischemia on ulcer wound healing: A new model in the rabbit ear. *Ann. Plast. Surg.*, 24: 17.
46. Pandit, A., Feldman, D. and Estridge, T., 1991. Effect of oxygen and oxygen permeability on wound healing using polyurethane and polyacrylonitrile membranes. *Trans. Soc. Biomat.*, 17:138.
47. Gottrup, F., Firmin R., Chang, N., et al., 1983. Continuous direct tissue oxygen tension measurement by a new method using an implantable silastic tonometer and oxygen polarography. *Am. J. Surg.*, 146: 399.
48. Hopt, H. W. and Hunt T. K., 1994. Comparison of clark electrode and optode for measurement of tissue oxygen tension. *Oyxgen Transport to Tissue* (Advances in Experimental Medicine and Biology), eds. Vaupel, P., Zander, R., Bruley, D. F., vol. 345: p 841.
49. Viitanen, S. M. and Viljanto, J., 1972. Wound healing: a thermographic study. *Ann. Chir. Gynaecol. Fenn.*, 60: 101–6.
50. Cole, R. P., Jones, S. G. and Shakespeare, P. G., 1990. Thermographic assessment of hand burns. *Burns*, 16(1): 60–3.
51. Cole, R. P., Shakespeare, P. G., Chissell, H. G. and Jones, S. G., 1991. Thermographic assessment of burns using a non-permeable membrane as wound covering. *Burns*, 17(2): 117–22.
52. Sindrup, J. H., Kastrup, J. and Jorgensen, B., 1991. Regional variations in nocturnal fluctuations in subcutaneous blood flow rate in the lower leg of man. *Clin. Physiol.*, 11(6): 491–9.
53. Sindrup, J. H., Kastrup, J., Madsen, P. L., Christensen, H., Jorgensen, B. and Wildschiodtz G., 1992. Nocturnal variations in human lower leg subcutaneous blood flow related to sleep stages. *J. Appl. Physiol.*, 73(4):1246–52.
54. Bircher, A., 1995. Laser Doppler measurement of skin blood flux: Variation and validation. In: J. Serup and G. B. E. Jemec (Editors), *Handbook of Non-invasive Methods and the Skin*, CRC Press, Boca Raton, Florida, pp. 399–403.
55. Quinn, A., McLelland, J., Essex, T. and Farr, P., 1991. Measurement of cutaneous inflammatory reactions using a scanning laser Doppler velocimeter. *Br. J. Dermatol.*, 125: 30–37.
56. Yamamoto, Y., Ohura, T., Nohira, K., Sugihara, T., Minakawa, H., Igawa, H., Shintomi, Y. and Fujii, H., 1993. Laserflowgraphy: A visual blood flow meter utilizing a dynamic laser speckle effect. *Plast. Reconstr. Surg.*, 91(5): 884–894.

57. Wardell, K., 1992. *Laser Doppler Perfusion Imaging*. Biomedical Engineering. Linkoping, Sweden, Linkoping University: 40.
58. Wardell, K., Jakobsson, A. and Nilsson, G., 1993. Laser Doppler perfusion imaging by dynamic light scattering. *IEEE Trans. Biomed. Eng.*, 40(4): 309–316.
59. Wardell, K., Naver, H. K., Nilsson, G. and Wallin, B. G., 1993. The cutaneous vascular axon reflex in humans characterized by laser Doppler perfusion imaging. *J. Physiol.*, 460: 185–199.
60. Wardell, K. and Nilsson, G., 1995. Laser Doppler imaging of skin. In: J. Serup and G. B. E. Jemec (Editors), *Handbook of Non-invasive Methods and the Skin*, CRC Press, Boca Raton, Florida, pp. 421–428.
61. Linden, M., Sirsjo, A., Lindbom, L., Nilsson, G. and Gidlof, A., 1995. Laser Doppler perfusion imaging of microvascular blood flow in rabbit tenuissimus muscle. *Am. J. Physiol.*, 269: H1496–H1500, 1995.
62. Eriksson, E., Miles, R. and Le, H., 1986. Changes in blood flow and metabolism in a neurovascular island skin flap after a burn. *Ann. Plast. Surg.*, 17(1): 79–81.
63. Sindrup, J. H., Petersen, L. J., Madsen, S. M., Kristensen, J. K. and Kastrup, J., 1995. Nocturnal temperature and subcutaneous blood flow in humans. *Clin. Physiol.*, 15: 611–622.
64. Bornmyr, S., Arner, M. and Svensson, H., 1994. Laser Doppler imaging of finger skin blood flow in patients after microvascular repair of the ulnar artery at the wrist. *J. Hand Surg.* [*Br*]. 19(3): 295–300.
65. Lauritzen, M. and Fabricius, M., 1995. Real time laser Doppler paerfusion imaging of cortical spreading decompression in rat neocortex. *NeuroReport*, 6: 1271–1273.
66. Anderson, C., Andersson, T. and Wardell, K., 1994. Changes in skin circulation after insertion of a microdialysis probe visualized by laser Doppler perfusion imaging. *J. Invest. Dermatol.* 102(5): 807–811.
67. Wang, I., Andersson-Engels, S., Nilsson, G., Wardell, K. and Svanberg, K., 1997. Superficial blood flow following photodynamic therapy of malignant non-melanoma skin tumors measured by laser Doppler perfusion imaging. *Br. J. Dermatol.*, 136: 184–189.
68. Barker, T. H., Kilpadi, D. V., Blum, B. S., and Feldman, D. S., 1998. Laser Doppler imaging as a noninvasive technique of wound state. *Wound Rep. Reg.*, 6(3): A241.
69. Jakobsson, A. and Nilsson, G., 1993. Prediction of sampling depth and pathlength in laser Doppler flowmetry. *Med. Biol. Eng. Comput.*, 31: 301–307.
70. Abbot, N. C., Ferrel, W. R., Lockhart, J. C. and Lowe, J. G., 1996. Laser Doppler perfusion imaging of skin blood flow using red and near-infrared sources. *J. Invest. Dermatol.*, 107(6): 882–886.
71. Atiles, L., Mileski, W., Purdue, G., Hunt, J. and Baxter, C., 1995. Laser Doppler flowmetry in burn wounds. *J. Burn Care Rehab.*, 16(4): 388–393.
72. Atiles, L., Mileski, W., Spann, K., Purdue, G., Hunt, J. and Baxter, C., 1995. Early assessment of pediatric burn wounds by laser Doppler flowmetry. *J. Burn Care Rehab.*, 16(6):596–601.
73. Heimbach, D., Engrav, L., Grube, B. and Marvin, J., 1992. Burn depth: a review. *World J. Surg.*, 16: 10–15.
74. Shakespeare, P. G., 1992. Looking at burn wounds: The A.B. Wallace Memorial Lecture 1991. *Burns*, 18(4): 287–295.
75. Deitch, E. A., Wheelman, T. M., Rose, M. P., Clothier, J. and Cotter, J. 1983. Hypertrophic burn scars: analysis of variables. *J. Trauma*, 23(10):895–898.
76. Rudolph, R., Vande Berg, J. and Ehrlich, H. P. 1992. Wound contraction and scar contracture. In: I. K. Cohen, R. F. Diegelmann, and W. J. Lindblad (Editors), *Wound Healing: Biochemical and Clinical Aspects*, W. B. Saunders, Philadelphia, pp. 96–114.
77. Linares, H. A., 1996. Pathophysiology of the burn scar. In: D. N. Herndon (Editor), *Total Burn Care*, W. B. Saunders, London, pp. 383–397.
78. McCauley, R. L. and Hollyoak, M. 1996. Medical therapy and surgical approach to the burn scar. In: D. N. Herndon (Editor), *Total Burn Care*, W. B. Saunders, London, pp. 473–478.
79. Distante, F. and Berardesca, E., 1995. Transepidermal water loss. In: E. Berardesca et al. (Editors),

Bioengineering of the Skin: Methods and Instrumentation, CRC Press, Boca Raton, Florida, pp. 1–4.

80. Barel, A. O. and Clarys, P., 1995. Comparison of methods for measurement of transepidermal water loss. In: J. Serup and G. B. E. Jemec (Editors), *Handbook of Non-invasive Methods and the Skin*, CRC Press, Boca Raton, Florida, pp. 179–184.
81. Jonkman, M. F., Molenaar, I., Nieuwenhuis, P. and Klasen, H. J., 1989. Evaporative water loss and epidermis regeneration in partial-thickness wounds dresses with by a fluid-retaining versus a clot-inducing wound covering in guinea pigs. *Scand. J. Plast. Reconstr. Surg.*, 23:29.
82. Sullivan, T., Smith, J., Kermode, J., McIver, E. and Courtemanche, D. J., 1990. Rating the burn scar. *J. Burn Care Rehab.*, 11(3): 256–260.
83. Baryza, M. J. and Baryza, G. A. 1995. The Vancouver Scar Scale: an administration tool and its interrater reliability. *J. Burn Care Rehab.*, 16(5): 535–538.
84. Leung, K. S., Cheng, J. C. Y., Ma, G. F. Y., Clark, J. A. and Leung, P. C., 1984. In vivo study of the mechanical property of post-burn hypertrophic scar tissues. *J. Burn Care Rehab.*, 5: 458–462.
85. Pierard, G. E., 1989. A critical approach to in vivo mechanical testing of the skin. In: J-L. Leveque (Editor), *Cutaneous Investigation in Health and Disease: Noninvasive Methods and Instrumentation*, Marcel Dekker. New York, pp. 215–240.
86. Vogel, H. G., 1994. Mechanical measurements of skin. *Acta. Derm. Venereol.*, Suppl. 185: 39–43.
87. Hambleton, J., Shakespeare, P. G. and Pratt, B. J., 1992. The progress of hypertrophic scars monitored by ultrasound measurements of thickness. *Burns*, 18(4): 301–307.
88. Clark, J. A., Cheng, J. C. Y., Leung, K. S. and Leung, P. C., 1987. Mechanical characterization of human postburn hypertrophic skin during pressure therapy. *J. Biomech.*, 20(4): 397–406.
89. Clark, J. A., Cheng, J. C. Y. and Leung, K. S., 1996. Mechanical properties of normal skin and hypertrophic scars. *Burns*, 22(6): 443–446.
90. Bartell, T. H., Monafo, W. W. and Mustoe, T. A., 1988. A new instrument for serial measurements of elasticity in hypertrophic scar. *J. Burn Care Rehab.*, 9(6): 657–660.
91. Chu, B. M. and Brody, G., 1975. Nondestructive measurements of the properties of healing burn scars. *Med. Instrum.*, 9(3): 139–142.
92. Jagtman, B. A., Wijn, P. F. F., Brakee, A. J. M. and Kuiper, J. P., 1982. Clinical investigation of the mechanical properties of human skin *in vivo* for small deformations. *Bioeng. Skin.*, 4: 32–40.
93. Berardesca, E., Borroni, G., Gabba, P., Borlone, R. and Rabbiosi, G., 1986. Evidence for elastic changes in aged skin revealed in an *in vivo* extensometric study at low loads. *Bioeng. Skin*, 2: 261–270.
94. Quan, M. B., Edwards, C. and Marks, R., 1997. Non-invasive *in vivo* techniques to differentiate photodamage and ageing in human skin. *Acta. Derm. Venereol.*, 77(6): 416–419.
95. Gunner, C. W., Hutton, W. C. and Burlin, T. E., 1979. The mechanical properties of skin *in vivo*—a portable hand-held extensometer. *Br. J. Dermatol.*, 100(2): 161–163.
96. Thacker, J. G., Lachetta, F. A. and Allaire, P. E., 1977. *In vivo* extensometer for measurement of the biomechanical properties of human skin. *Rev. Sci. Instrum.*, 48(2): 181–185.
97. Ridge, M. D. and Wright, V., 1966. An extensometer for skin—its construction and application. *Med. Biol. Eng.*, 4(6): 533–542.
98. Fung, Y. C. B., 1967. Elasticity of simple tissues in simple elongation. *Am. J. Physiol.*, 213: 1532.
99. Lanir, Y. and Fung, Y. C. B., 1974. Two-dimensional mechanical properties of rabbit skin. I. Experimental model. *J. Biomech.*, 7: 29.
100. Sodema, W. A. and Burch, G. E., 1938. A direct method for the estimation of skin distensibility with its application to the study of vascular states. *J. Clin. Invest.*, 17: 785.
101. Barbanel, J. C. and Evans, J. H., 1977. Time-dependent mechanical properties of skin. *J. Invest. Dermatol.*, 69(3): 318–320.
102. Ballou, S. P., Mackiewicz, A., Lysikiewicz, A. and Neuman, M. R., 1990. Direct quantitation of skin elasticity in systemic sclerosis. *J. Rheumatol.*, 17(6): 790–794.
103. Sugihara, T., Ohura, T., Homma, K. and Igawadic, H. H., 1991. The extensibility in human skin: variation according to age and site. *Br. J. Plast. Surg.*, 44(6): 418–422.

104. Ohura, T., Siguhara, T. and Honda, K., 1980. Postoperative evaluation in plastic surgery using the Bio-Skin Tension Meter. *Ann. Plast. Surg.*, 5: 74.
105. Agache, P. G., 1995. Twistometry measurement of skin elasticity. In: J. Serup and G. B. E. Jemec (Editors), *Handbook of Non-invasive Methods and the Skin*, CRC Press, Boca Raton, Florida, pp. 319–328.
106. Agache, P. G., Monneur, C., Leveque, J. L. and de Rigal, J., 1980. Mechanical properties and Young's modulus of human skin *in vivo*. *Arch. Dermatol. Res.*, 269: 221.
107. Finlay, B., 1971. The torsional characteristics of human skin *in vivo*. *J. Biomed. Eng.*, 6: 567–573.
108. Kallis, B., De Rigal, J., Leonard, F., Leveque, J-L., Riche, O., Le Corre, Y. and De Lacharriere, O., 1990. *In vivo* study of sclerodoma by noninvasive techniques. *Br. J. Dermatol.*, 122: 785–791.
109. McHugh, A. A., Fowlkes, B. J., Maevsky, E. I., Smith, D. J., Rodriguez, J. L. and Garner, W. L., 1997. Biomechanical alterations in normal skin and hypertrophic scar after thermal injury. *J. Burn Care Rehab.*, 18(2): 104–108.
110. Pereira, J. M., Mansour, J. M. and Davis, B. R., 1990. Analysis of shear wave propagation in skin; application to an experimental procedure. *J. Biomech.*, 23(8): 745–751.
111. Hargens, C. W., 1995. The gas-bearing electrodynamometer. In: J. Serup and G. B. E. Jemec (Editors), *Handbook of Non-invasive Methods and the Skin*, CRC Press, Boca Raton, Florida, pp. 353–357.
112. Hargens, C. W., 1995. Ballistometry. In: J. Serup and G. B. E. Jemec (Editors), *Handbook of Non-invasive Methods and the Skin*, CRC Press, Boca Raton, Florida, pp. 359–364.
113. Bjerring, P., 1986. Skin elasticity measured by dynamic admittance. A new technique for mechanical measurements in patients with sclerodoma. *Acta Derm. Venereol.* Suppl., 120: 84.
114. Manny-Aframian, V. and Dikstein, S., 1995. Indentometry. In: J. Serup and G. B. E. Jemec (Editors), *Handbook of Non-invasive Methods and the Skin*, CRC Press, Boca Raton, Florida, pp. 349–352.
115. Lanir, Y., Dikstein, S., Hartzshtark, A. and Manny V. 1990. In-vivo indentation of human skin. *J. Biomech. Eng.* 112(1): 63–69.
116. Katz, S. M., Frank, D. H., Leopold, G. R. and Wachtel, T. L., 1985. Objective measurement of hypertrophic burn scar: a preliminary study of tonometry and ultrasonography. *Ann. Plast. Surg.*, 14(2): 121–127.
117. Esposito, G., Ziccardi, P., Scioli, M., Pappone, N. and Scuderi, N., 1990. The use of a modified tonometer in burn scar therapy. *J. Burn Care Rehab.*, 11(1): 86–90.
118. Spann, K., Mileski, W. J., Atiles, L., Purdue, G. and Hunt, J., 1996. Use of a pneumatonometer in burn scar assessment. *J. Burn Care Rehab.*, 17(6, pt 1): 515–517.
119. Falanga, V. and Bucalo, B., 1993. Use of a durometer to assess skin hardness. *J. Am. Acad. Dermatol.*, 29(1): 47–51.
120. Zheng, Y. P. and Mak, A. F., 1996. An ultrasound indentation system for biomechanical properties assessment of soft tissues in-vivo. *IEEE Trans. Biomed. Eng.*, 43(9): 912–918.
121. Dikstein, S., 1979. *In vivo* mechanical properties of the skin measured by indentometry and levarometry. *Bioeng. Skin*, 2: 23–24.
122. Pierard, G. E. and Lapiere, C. M., 1977. Physiopathological variations in the mechanical properties of the skin. *Arch. Dermatol. Res.*, 260: 231–239.
123. Manny-Aframian, V. and Dikstein, S., 1995. Levarometry. In: J. Serup and G. B. E. Jemec (Editors), *Handbook of Non-invasive Methods and the Skin*, CRC Press, Boca Raton, Florida, pp. 345–347.
124. Gniadecka, M. and Serup, J., 1995. Suction chamber method for the measurement of skin mechanical properties: the dermaflex. In: J. Serup and G. B. E. Jemec (Editors), *Handbook of Non-invasive Methods and the Skin*, CRC Press, Boca Raton, Florida, pp. 329–334.
125. Matsuzaki, K., Kumagai, N., Fukushi, S., Ohshima, H., Tanabe, M. and Ishida, H., 1995. Cultured epithelial autografting on meshed skin graft scars: Evaluation of skin elasticity. *J. Burn Care Rehab.*, 16(5): 497–502.
126. Barel, A. O., Courage, W. and Clarys, P., 1995. Suction method for measurement of skin mechani-

cal properties: the cutometer. In: J. Serup and G. B. E. Jemec (Editors), *Handbook of Non-invasive Methods and the Skin*, CRC Press, Boca Raton, Florida, pp. 335–340.

127. Fong, S. S., Hung, L. K. and Cheng, J. C., 1997. The cutometer and ultrasonography in the assessment of postburn hypertrophic scar—a preliminary study. *Burns*, 23 Suppl. 1:S12–8.

128. Enomoto, D. N., Mekkes, J. R., Bossuyt, P. M., Hoekzema, R. and Bos, J. D., 1996. Quantification of cutaneous sclerosis with a skin elasticity meter in patients with generalized scleroderma. *J. Am. Acad. Dermatol.*, 35(3 pt 1): 381–387.

129. Krusche, T. and Worret, W. I., 1995. Mechanical properties of keloids *in vivo* during treatment with intralesional triamcinolone acetonide. *Arch. Dermatol. Res.*, 287(3–4): 289–293.

130. Kumagai, N., Oshima, H., Tanabe, M., Ishida, H. and Uchikoshi, T., 1997. Favorable donor site for epidermal cultivation for the treatment of burn scars with autologous cultured epithelium. *Ann. Plast. Surg.*, 38(5): 506–513.

131. Henry, F., Van Look, R., Goffin, V., Fissette, J. and Pierard, G. E., 1996. Mechanical properties of skin and liposuction. *Dermatol. Surg.*, 22(6): 566–568.

132. Pierard, G. E., Nikkels-Tassoudji, N. and Pierard-Franchimont, C., 1995. Influence of the test area on the mechanical properties of skin. *Dermatology*, 191(1): 9–15.

133. Cua, A. B., Wilhelm, K. P. and Maibach, H. I., 1990. Elastic properties of human skin: relation to age, sex, and anatomical region. *Arch. Dermatol. Res.*, 282(5): 283–288.

134. Elsner, P., 1995. Skin elasticity. In: E. Berardesca et al. (Editors), *Bioengineering of the Skin: Methods and Instrumentation*. CRC Press, Boca Raton, Florida, pp. 53–64. (cutometer, dermaflex)

135. Serup, J. and Northeved, A., 1985. Skin elasticity in localized scleroderma (morpheoa). Introduction of a biaxial *in vivo* method for measurement of tensile distensibility, hysteresis, and resilient distension of diseased and normal skin. *J. Dermatol.*, 12: 52.

136. Serup, J. and Northeved, A., 1985. Skin elasticity in psoriasis. *In vivo* measurement of tensile distensibility, hysteresis, and resilient distension with a new method. Comparison with skin thickness as measured with high frequency ultrasound. *J. Dermatol.*, 12: 318.

137. Jemec, G. B. and Serup, J., 1990. Epidermal hydration and skin mechanics. The relationship between electrical capacitance and the mechanical properties of human skin *in vivo*. *Acta. Derm. Venereol.*, 70(3): 245–247.

138. Redden, R. A., Eberhardt, A. W. and Feldman, D. S., 1998. Quantitative in vivo mechanical testing of burn scar stiffness. *Ann. Biomed. Eng.*, 26 Suppl. 1: S-80.

139. Cahill, B. P., An, K-N., Fisher, J. and Chao, E. Y. S., 1987. In vivo test system for characterization of skin wound healing. *Biomed. Sci. Instrum.*, 23: 29–33.

140. Charles, D., Williams, K., Perry, L. C., Fisher, J. and Rees, R. S., 1992. An improved method of *in vivo* wound disruption and measurement. *J. Surg. Res.*, 52: 214–218.

141. Perry, L. C., Connors, A. W., Matrisian, L. M., Nanney, L. B., Charles, P. D., Reyes, D. P., Kerr, L. D. and Fisher, J., 1993. Role of transforming growth factor-1 and epidermal growth factor in the wound-healing process: an *in vivo* biomechanical evaluation. *Wound Rep. Reg.*, 1(1): 41–46.

142. Clugston, P. A., Vistnes, M. D., Perry, L. C., Maxwell, G. P. and Fisher, J., 1995. Evaluation of silicone-gel sheeting on early wound healing of linear incisions. *Ann. Plast. Surg.*, 34: 12–15.

143 Dunn, M. G., Silver, F. H. and Swann, D. A., 1985. Mechanical analysis of hypertrophic scar tissue: structural basis for apparent increased rigidity. *J. Invest. Dermatol.*, 84(1): 9–13.

36
Biomaterial-Enhanced Regeneration for Skin Wounds

Dale Feldman, Tom Barker, Barbara Blum, John Bowman, Deepak Kilpadi, and Robert Redden
University of Alabama at Birmingham, Birmingham, Alabama

I. INTRODUCTION

A. Biomaterial Enhanced Regeneration

Biomaterial enhanced regeneration is the use of biomaterials to stimulate regeneration. This requires an understanding of methods to enhance regeneration as well as ways in which biomaterials can be used to enhance the regenerative process.

Biomaterial enhanced regeneration falls under the broad heading of tissue engineering: the application of scientific principles to the design, construction, modification, growth, and maintenance of living tissues [1]. More specifically, tissue engineering is the use of materials (synthetic and natural) usually in conjunction with cells (both native and gentically engineered) and/or biological response modifiers (growth factors, cytokines and other recombinant products). The goal is to use these systems to replace tissue and organ functions (biochemical and/or structural).

Biomaterial enhanced regeneration is the branch of tissue engineering related to biomaterials. This is the designing of materials to better deliver and protect the cells and biological response modifiers and, in many cases, to better serve as degradable scaffolds to help promote the healing and regenerative process. Historically, however, the focus of tissue engineering has been on cells and cell seeding, whereas the focus of biomaterial enhanced regeneration is on the use of biomaterials with biological response modifiers.

The focus of this chapter will therefore be on the use of biomaterials with biological response modifiers: biomaterial enhanced regeneration. Although this concept has been used to regenerate tissue in many applications, the focus of this chapter will be on skin wounds.

B. Biocompatibility

In order to use the biomaterials to their fullest, it is critical to determine the optimal implant design for each application. A biocompatibility hierarchy can be developed and used to help determine the optimal implant design in each case.

Biocompatibility is the study of how the host affects the implant (e.g., corrosion and degradation) and how the implant affects the host (e.g., inflammation and allergic response).

By developing a ranking system of host and implant responses, the most biocompatible implant system can be determined and then designed for each application. To develop a biocompatibility hierarchy it is important to understand how biocompatibility is determined. When implants were first developed, a biocompatible material was defined as one that was capable of being implanted, caused no systemic toxic reaction, had no carcinogenic qualities, and one for which the local tissue response neither compromised function nor caused pain, swelling, or necrosis. This definition applies only for an inert biomaterial. It, however, only tells us what an inert biomaterial cannot do, not what it can. Over the years as biomaterials research has moved from designing inert implants to biointegrated implants to bioactive implants, this definition has become less and less meaningful.

Biocompatibility has been viewed in many different ways. In the 1970s biocompatibility was described as an interfacial problem [2]. The body interacts with the surface of the biomaterial, which can differ from the bulk of the material. This emphasis led to the critical attention to surface analysis and characterization that is still a major part of biomaterials research today.

Others have described biocompatibility as what one wants the response to be in a particular situation [3]. For example, a degradable collagen implant can be used subcutaneously or as a wound dressing to stimulate the ingrowth of blood vessels, angiogenesis, and thus help in regeneration. The same implant, however, may be used as a corneal shield to protect the eye after surgery. Angiogenesis in this case would be undesirable and lead to loss of vision. Therefore, biocompatibility is dependent on the application.

Biocompatibility is also dependent on location. Different tissues in the body respond differently to the same implant. The eye of a rabbit, for example, is used for toxicity testing because of the sensitivity to any type of irritation.

Further, biocompatibility is dependent on time. An implant can be biocompatible for short-term applications, but not long-term ones. In addition, an implant may trigger a "bad" response in the short term in order to elicit a "good" response in the long term.

Biocompatibility is also dependent on the animal model used. The ultimate goal is use in humans, but no animal can perfectly duplicate the human clinical response. In addition, each animal model elicits different responses. The approach, however, has been to optimize implant design in one or more animal models and it should be at least close to the optimal design clinically.

II. BIOCOMPATIBILITY HIERARCHY

In order to develop a biocompatibility hierarchy it is necessary to look at the possible responses. For biocompatibility, both implant and host responses need to be examined (Table 1). Biocompatibility is both a study of these responses and an assessment of whether these responses are "good" (biocompatible) or "bad" (poor biocompatibility).

Table 1 Biocompatible Responses

Implant	Host
Inert	Inert
Surface active	Biointegration
Drug delivery	Severe
Degradable	Regeneration

The implant can be inert or modified in the biological environment. In many cases, these modifications can be detrimental and lead to poor biocompatibility. These responses, however, can also be selected as part of the implant design and thus be biocompatible.

Inert implants are the most common type of implants currently on the market. Most implants like artificial joints are designed to serve a function without being altered in any way. The implant, however, can be designed to be modified in vivo. For example, it can be surface active having a bioglass, calcium phosphate, or biochemically active surface that can stimulate an in vivo response. One example of this surface activity is the use of hydroxyapatite coatings on orthopedic and dental implants. The implant can also serve as a drug delivery system for biochemical agents as in wound dressings that release antibiotics. Alternatively, the implant can serve its function and dissolve, like degradable sutures. Similarly, the implant can stimulate an inert host response or an active response. These responses would be detrimental or beneficial depending on the application.

Again, most implants presently on the market are designed to be inert in the host environment, as in the original definition. They perform a function with as little modification of the host as possible. In some cases, however, it is beneficial to have the implant integrated with the host. A porous implant can be used to stimulate tissue ingrowth. A bioactive calcium phosphate surface has been used to get direct bone attachment. Both of these responses have been used to achieve better long-term stability.

Sometimes, however, a severe initial response is needed or desired to get a desirable long-term response. For example, some implants are used to stimulate a fibrous capsule that can structurally regenerate a blood vessel or tendon sheath. Finally, some implants are designed to trigger a regenerative response. Since bone is regenerative, fracture-fixation systems such as bone plates are designed to heal fractures by regeneration.

Although the field of biomaterials has come a long way, and many responses can be stimulated, they are still not capable of completely duplicating the structure and function of the replaced part. In addition, man-made materials will lose part of their function and properties over time and cannot "heal" like biological tissue. Therefore, the best biocompatible response is for the implant to stimulate tissue regeneration. For regeneration to work, however, the implant must resorb or degrade as it stimulates regeneration—the process of biomaterial enhanced regeneration.

In addition, the best design is where the degradation and regeneration occur at the same rate. This has been called isomorphous tissue replacement [4]. If degradation proceeds too quickly, the matrix loses its ability to serve as a tissue scaffold. If the regeneration rate is faster than the degradation rate, the matrix will slow the regenerative process and become the rate-limiting step. To achieve isomorphous tissue replacement of synthetic degradable materials as well as artificially crosslinked natural materials the degradation rate must be adjusted to match the healing rate [5]. The ultimate design, however, would be a system where the regenerative process controls the degradation. This would allow the system to automatically take into account patient-to-patient variability in healing rate, adjust healing rate continually rather than just approximate the overall rate, and could be easily modified for use in different tissues with different healing rates. This biofeedback control can occur in a number of ways. One example would be how the body normally does this: as the cells come in to repair the tissue they break down the provisional matrix.

If a biocompatibility hierarchy is established (Table 2) this response would be at the top. Sometimes, however, the tissue cannot regenerate adequately, using present technology, so different options down the hierarchy should be selected first.

For example, artificial hips are used in cases of severe arthritis or avascular necrosis. Present capabilities for regeneration of bone and cartilage are not sufficient to have a degrad-

Table 2 Biocompatibility Hierarchy

Host	Implant
Regeneration	Degradable
Integration	Bioactive
Minimal inflammation	Inert
Inert	

able regenerative implant. Therefore, a prosthesis is used to replace the joint surface and part of the femur as well as the acetabulum. Most of these implants were designed to be as inert as possible; just perform a function and lead to as few problems as possible. Research, however, has continued in moving up the hierarchy. For example, techniques to reduce implant breakdown and inflammatory response are still being pursued. More recently, however, porous and bioactive (calcium-phosphate-coated) implants have been used to increase stabilization and prevent long-term failures. In addition, research is ongoing to regenerate cartilage and large sections of bone. Although progress has been made, many technical problems, such as restoring the macro- and microstructure of the joint as well as long-term functionality, still need to be solved before joint regeneration will be a clinical reality; but it still should be the ultimate goal.

In addition, sometimes the implant function is needed and the implant cannot be degraded away. For example, catheters need to provide access from outside to inside the body (a percutaneous device), and therefore are not designed to degrade. Similarly, however, biomaterials research has moved the design up the hierarchy. Original implants were designed to transport blood, fluids, or drugs across the skin and to cause minimal inflammation. In an effort to reduce the inflammation and long-term infection problems, porous catheter cuffs (Fig. 1) have been used to allow the implant to become more integrated with the surrounding tissue [6,7]. This integration along with the incorporation of antimicrobial drug delivery systems has increased the life of these devices. Infection, however, still remains a problem with these devices partly because the tissue that grows into the cuffs is not similar enough to the healthy noninflamed surrounding tissue [8,9]. Therefore, research continues toward design of optimal porous

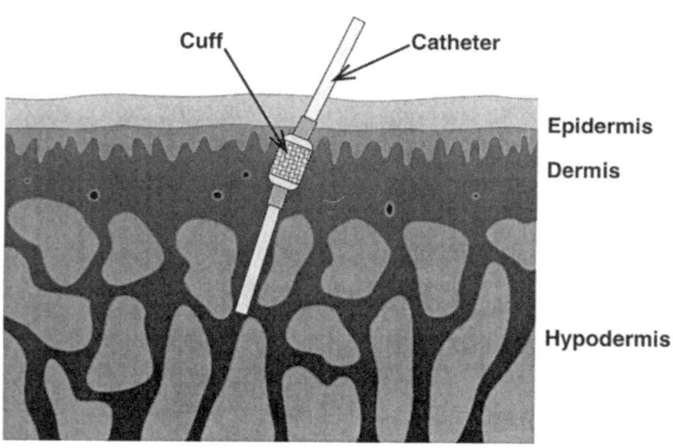

Figure 1 Schematic of a percutaneous device.

implants to help regenerate the normal tissue seal found around natural percutaneous devices such as found in teeth.

Therefore, a hierarchy can be established to assist in implant design. The degradable regenerative response should be the first choice to investigate and the inert response should be the last. The goal of a degradable regenerative system can be applied in virtually every case, although some systems are further along than others are. Although each may start at a different place on the hierarchy, they all should continually move upwards.

An important issue then becomes how good we are at regeneration (Table 3). Tissue can regenerate, grow larger to fill the space (hypertrophy) or use scar tissue to fill the space. Of the four tissue types (epithelium, muscle, nerve, and connective tissue) only epithelium is totally regenerative. The regenerative ability, however, is not unlimited and requires assistance in some cases. For example, skin grafting is required in burn patients to enhance the healing process and reduce secondary complications [10]. For muscle, the tissue usually scars or hypertrophies [11]. There is evidence of smooth muscle regeneration in artificial blood vessels and the artificial bladder as well as in cell culture [12–15]. Although there is some evidence of skeletal muscle mitosis, this tissue heals mostly by scarring [11].

For nerves, healing occurs mostly by scarring. Recently, however, nerves have been shown to regenerate for distances close to 1 centimeter as well as being sutured together to regain some function [16].

For connective tissue, three tissues of interest are cartilage, bone, and dense connective tissue like that found in the dermis of skin. Bone is regenerative, but only up to a certain size [17]. Presently dermis usually scars but work with artificial skin has been able to restore much of the structure and function [18]. Cartilage is relatively avascular and has a limited regenerative capacity. Recent work, however, with degradable scaffolds and cell seeding has shown promise [17].

Therefore, research in many disciplines has concentrated on understanding how to make these types of tissue heal by regeneration rather than by scarring. In biomaterials, this research has centered on how implants can be used to enhance the regeneration process.

This chapter covers the design of implants for skin. This organ is composed of regenerative tissue (epidermis) and mostly reparative tissue (dermis).

III. RAMIFICATIONS IN IMPLANT DESIGN

Another important issue then becomes how degradable regenerative implants are designed. The process of biomaterial enhanced regeneration involves development of bioactive materials by merging an optimal biomaterial with an optimal bioactivity. These systems are tested and optimized in vitro and in vivo prior to clinical testing (Fig. 2).

Table 3 Tissue Healing

		Muscle			Connective tissue		
	Epithelium	Smooth	Skeletal	Nerve	Cartilage	Bone	Dermis
Regeneration	X	Some	?	Some	Some	X	Some
Hypertrophy		X	X	X			
Scar		X	X	X	X	X	X

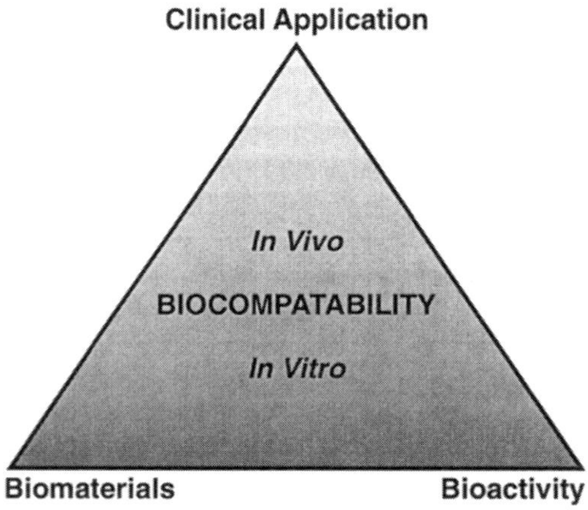

Figure 2 Biomaterial enhanced regeneration design.

A. General Wound Healing

Optimal bioactivity is the enhancement and optimization of regenerative healing—in other words, controlling the speed and completeness of tissue repair by modifying the biological activity of an implant. It is essential for this optimal biological activity to be coupled with a biomaterial scaffold to achieve this regenerative response. The optimal scaffolds are ones that allow regenerative tissue ingrowth as the scaffold degrades. Therefore, optimization of the scaffold requires optimization of the implant porosity as well as the implant degradation rate. In addition, if the biological activity is incorporated into the implant, optimization of the delivery rate is also important.

Many investigators have examined and continue to examine ways to enhance and control healing. Healing can be altered by changing the wound environment or the wound biological activity. Each of these can significantly affect the progression of healing, the rate of healing, and the type of tissue formed. In addition, if a wound dressing or implant is used the healing can be altered by changing the implant configuration, the implant surface, the implant biological activity, or the implant degradation rate. The goal in virtually all cases is tissue regeneration at the fastest possible rate.

Of course, the fastest rate does not always lead to regeneration. In skin, the process of contraction leads to a faster healing response because less new tissue needs to be formed. This process, however, is accompanied by scarring. Therefore, regenerative healing in skin requires minimal contraction and speed is sacrificed for control [5]. Once control of the process is achieved, the process can be accelerated.

Altering physical parameters such as temperature, humidity, air composition or pressure, and electric or magnetic fields can change the wound macroenvironment. These alterations plus biological response modifiers can also alter the wound microenvironment.

It has been demonstrated that both the tissue oxygen level and oxygen gradient are critical factors in the healing process [19–23]. Oxygen level is often the rate-limiting step in tissue reconstruction [1,24]. Oxygen is required for various cell functions such as attachment, spreading, and protein production. The gradient is critical in a wound to allow fibroblasts a high oxygen level, at the wound margin, to produce collagen and form the structure for vessels to grow into. The gradient also creates a low oxygen region, in the wound center, to stimulate macrophages to release cytokines that stimulate chemotaxis, mitosis and other steps in the healing process. The diffusion of oxygen from vessels at the wound margin creates the gradient and limits the thickness of tissue that can survive. The maximum thickness of tissue or cell masses [5] is about 100–200 μm, which is the maximum distance cells are away from the blood supply in vivo [5,24,25].

A number of studies have attempted to determine the optimum oxygen gradient along with a clinical strategy to create it [20–23]. Similarly, a number of studies have attempted to determine the optimum electric and magnetic fields for wound healing [26–31].

The biological activity can be altered by addition of biological response modifiers such as growth factors or cytokines, integrins and extracellular matrix (ECM) molecules. These substances are used to influence cell-cell, cell-matrix, and cell-growth factor interactions in an effort to control cell proliferation, adhesion, migration, uptake, secretion, and differentiation [32,33]. For example, growth factors have been shown to affect migration, proliferation and differentiation of all the cells involved in tissue repair. Several in vitro and in vivo studies have demonstrated the effect of growth factors on wound healing. Transforming growth factor-β (TGF-β), platelet derived growth factor (PDGF), fibroblast growth factor (FGF), and epidermal growth factor (EGF) are some of the growth factors that have been investigated for their role in wound healing [14,16,34–45].

Specifically, fibroblast growth factor (FGF) has been shown to induce angiogenesis [36,37,46]. In lower vertebrates, FGF promotes limb regeneration [47]. The mitogenic, chemotactic and differentiation properties of this growth factor suggest that it is involved in embryonic development and is probably a major trophic factor operating at all stages of embryogenesis [48]. Unlike other growth factors, FGF can stimulate in vivo as well as in vitro proliferation of all cell types involved in would healing [49,50]. In vitro studies have demonstrated that FGF inhibits contraction while enhancing wound healing [51–52]. The inhibition of contraction may have therapeutic implications in the prevention of contracture scars. FGF has also been shown to increase graft survival by stimulating epithelialization by cultured keratinocytes and vascularization in the wound bed in athymic mice [53].

Both basic FGF (FGF-2) and acidic FGF (FGF-1) are extremely angiogenic in vivo [54,55] and mitogenic for fibroblasts in vitro [56]. A preferential response by keratinocytes to FGF-1 in either the presence or absence of heparin, compared to FGF-2, has been reported [56]. Stimulation of angiogenesis, granulation tissue formation and neoepithelialization have been demonstrated in response to FGF-1, in vivo, in dermal wounds [39]. Increased wound strength, with increased cellularity and collagen deposition, has also been observed, without increased contraction. Dose-dependent inhibition of intimal thickening with parallel promotion of endothelial regeneration over the injured area has also been reported for FGF-1 [40].

Over the past 10 years, numerous clinical studies with topically applied growth factors (PDGF, FGF, EGF, and platelet derived wound factor) for skin wounds have been performed [57–63], but results have been mixed. The mixed results are most likely due to release of unprotected growth factor in a wound environment filled with proteolytic enzymes, as well as the lack of a scaffold to guide the tissue growth stimulated by the growth factor. It has been shown, however, that continuous dosing of growth factors like FGF-1, at a lower total dose, is more effective for skin wounds than just one application. In addition, since growth factors have been shown to increase the rate of healing, there has been concern that increasing cell proliferation might lead to uncontrolled (malignant) growth or hypertrophic scar formation. However, there is not evidence that growth factor induced proliferation is caused by cellular transformation as in cancerous cell lines. The effect of growth factors, particularly FGF-1, seems to be self-limiting, by promoting an increase in healing rate rather than uncontrolled healing [44–63].

Various other biological response modifiers have been incorporated into implants to enhance healing or ingrowth [64–67]. This includes polymerizable ECM molecules (fibrinogen, albumin, collagen, and hyaluronate) used not only to stimulate cell migration and activity but also to serve as scaffolds and drug delivery systems [64,65]. In addition, other ECM molecules such as glycosaminoglycans (GAGs) (e.g., heparan sulfate, thrombin) have been used [5]. Attachment of both heavy metal ions (silver and copper) and biochemical agents (heparin) to Dacron and collagen have been examined both subcutaneously and percutaneously. Silver has been used for its antimicrobial properties [8,9,68–70], while angiogenic heparin/copper complex has been used to enhance tissue ingrowth [64,65,71].

B. Scaffold Design

When a wound dressing or implant is used, wound healing can be altered by changing the implant configuration (pore size, porosity, fiber diameter, etc.), the implant surface (composition, charge, surface energy, etc.), the implant biochemical activity (incorporation of growth factors or other biological response modifiers), or the implant physical activity (degradation rate and drug delivery rate).

1. Configuration

Earlier work showed the importance of implant configuration on the implant's effectiveness as a tissue scaffold for skin wounds [4,5] as well as other soft tissue defects [72–80]. Optimization of implant configuration leads to significant enhancement of regenerative skin healing [4,5]. Variables such as pore size, percent porosity, fiber diameter, and surface roughness of the matrix determine the speed and completion of ingrowth [68,72,73,81]. Pore size must be large enough to allow blood vessel ingrowth (at least 40 μm) [81]. At larger pore sizes, however, the scaffolding effect decreases. The optimal pore size for collagen-based wound dressings is approximately 100 μm [4,5]. Many others have examined the effect of pore size and porosity on soft tissue healing in a number of different natural and synthetic polymeric systems [75,76,81–82].

Since some of these porous structures are fibrous it has also been necessary to examine the effect of fiber diameter and fiber spacing (akin to pore size) on soft tissue ingrowth. In vivo and in vitro studies have shown that increasing fiber diameters (≥50 μm) leads to significant reduction in the chronic inflammatory response [72,73] (Fig. 3). Further, the average fiber spacing requirements are similar to the pore size requirements (40 μm) for vascular ingrowth [72,73,81].

When the system is degradable, the porosity and average pore size change over time. A number of researchers have tried to optimize the scaffolding effect of these systems [8,9,79,80]. These systems have been made from fibers (e.g., collagen, polyurethanes, or PLA) or in the form of sponges, (e.g., collagen/GAGS or PLA) [80–90].

2. Drug Delivery Systems

A number of degradable systems have been studied for use as potential drug delivery/wound dressing systems, including poly(lactic acid) (PLA), PEO/PBT, collagen and hyaluronic acid and fibrin. Synthetic polymers, such as PLA, are typically made into a performed matrix, which is then placed on the wound and sutured into place. Of these materials, fibrin is the only one with notable adhesive properties, allowing for adhesion to the wound bed, as well as the potential to adhere skin grafts. In open skin wounds, adhesiveness can also decrease the loss of the drug delivery matrix due to abrasion of wound coverings and clothes, and provide a uniform contact between the tissue and the drug delivery system.

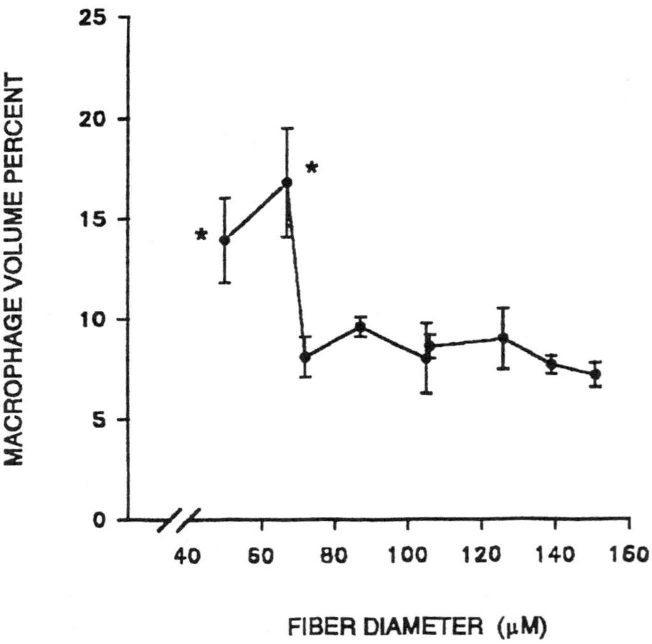

Figure 3 Macrophage response to fibers of different diameters. Macrophage volume percent around dacron fibers of different fiber diameters. The two points significantly different from the rest are denoted with an asterisk.

An ideal drug delivery system should be biocompatible, easily delivered, degradable, allow for control of the drug delivery rate, and should not alter the activity of the drug to be delivered. The natural materials, including collagen, hyaluronic acid, chitosan and fibrin, have all been applied topically to wounds. Topical applications that set up in situ eliminates the need for preshaping and measuring an implant prior to application, and allows for easy delivery. Implants used in the skin, as tissue scaffolds, should be degradable to allow for complete skin regeneration. Natural materials are advantageous over synthetic materials because degradation is primarily cell-controlled, allowing for biofeedback control. Further, the degradation by-products are biocompatible.

As previously mentioned, the mixed results for topical growth factors are most likely due to release of unprotected growth factor in a wound environment filled with proteolytic enzymes, as well as the lack of a scaffold to guide the tissue growth stimulated by the growth factor. Also since healing has been enhanced by continuous dosing, a controlled release system would be more convenient than daily dosing, and more effective in controlling the healing process. A biodegradable controlled release system could serve as a protective reservoir of the growth factor prior to release, provide a constant rate of medication and serve as a tissue regeneration scaffold. In addition, regenerative tissue adhesive scaffolds that degrade primarily due to cellular invasion, such as natural materials, provide biofeedback control of both the degradation and release rate of the growth factor. This means the patient's own healing rate controls the scaffold degradation, which also controls the introduction of the growth factor.

IV. SKIN APPLICATIONS

Although the use of degradable regenerative systems has been used in many applications such as nerves, blood vessels, bone, and cartilage; this chapter will concentrate on the skin applications. A review of the use of biomaterial enhanced regeneration in these applications is given in Ref. 91.

A. Skin

One example of how the biocompatibility hierarchy can be applied to implant design is in the area of skin research. Pioneers have been developing skin implant systems for over two decades that are degradable and regenerative [4,5,39,92–96].

1. *Structure*

Skin is the largest organ in the body. It is responsible for helping to maintain body temperature and to prevent fluid loss [97]. In addition, the hydrophilicity and mechanical integrity of skin help prevent damage due to external substances or forces.

The skin is divided into two layers: the epidermis and the dermis (Fig. 1). The epidermis is the top protective layer, while the dermis provides the mechanical properties as well as supplies the epidermis with its blood supply. The epidermis is divided into five layers (stratum corneum, stratum lucidum, stratum granulosum, stratum spinosum, and stratum basale). The bottom two layers, also called the stratum germinativum, are responsible for the normal turnover of the epidermis. The process, which takes about a month, gradually turns a squamous epithelial cell into keratin as it migrates upward and is eventually flaked off [10].

The dermis is divided into two layers. The upper layer (papillary dermis) has a rich blood supply, necessary for epidermal survival, as well as pain and temperature sensors. The next layer, the reticular dermis, is mostly a connective tissue matrix with collagen, elastin, and proteoglycans that give structural support to the skin. This layer also contains hair follicles and sweat glands. Underneath the skin is the subcutaneous tissue, which is mostly adipose tissue [10,97].

2. *Clinical Problems*

Of the two skin layers, only the epidermis is innately regenerative. The underlying dermis normally heals by scarring [10]. In addition, superficial wounds allow epidermal regeneration from around the hair follicles that invaginate into the dermis [10]. Often a wound may heal by contraction due to scarring which reduces or obviates the need for epidermal regeneration. The results of this process are usually unacceptable, however, due to the ultimate loss of skin elasticity, which generally causes loss of function and poor cosmetic results [96]. One challenge to researchers is to restrict this contraction and stimulate epithelial coverage of the wound.

a. Burns. Clinically, the two major problems are skin ulcers and burns. Burns are classified by depth of tissue damage. Partial-thickness burns (first and second degree) do not go all the way through the skin, as do full-thickness burns (third degree) [98–99]. It has been estimated that more than 500,000 persons are treated in a hospital emergency department each year due to thermal injury. Additionally, as many as 70,000 to 100,000 are hospitalized, and of these, 10,000 to 12,000 will die [100–101]. Death is mainly due to loss of body fluids and bacterial infection, both normally prevented by an intact epidermis.

b. Ulcers. Two of the most prevalent skin ulcers are pressure ulcers and diabetic ulcers. Pressure ulcers are localized areas of tissue necrosis that develop when soft tissue is compressed between a bony prominence and an external surface for a prolonged period of time. Pressure ulcers are found in 20–30% of spinal cord injury patients, 3–35% percent of nursing home residents, and in 3–11% of acute injuries [102–104]. This corresponds to about 11% of the hospitalized patients. In addition, nursing home residents with pressure ulcers have a fourfold increase in mortality [102,103]. Also about 1.5 million of the 10 million people in the United States with diabetes have ulcers on their feet or ankles [105–106].

c. Clinical Impact. Slow or impaired healing has been the most prominent factor in these lesions. Skin ulcers generally tend to be chronic due to continual irritation or pathologies associated with this population (paralysis or diabetes). These wounds can ultimately result in the loss of epidermis, dermis, subcutaneous fat, muscle, and bone. Amputations for both diabetic and paralyzed patients have been linked to this lack of healing [106]. Burns present not only an initial danger in terms of fluid loss and infection, but also a long-term potential for loss of function and poor cosmetic appearance due to contraction and scarring. Therefore, for both skin ulcers and burns effectively stimulating the rate of regenerative healing will reduce the likelihood and effect of secondary complications.

Current nonsurgical treatment for skin wounds is usually limited to passive modalities such as debridement and cleansing. The problem is the additional cost to the patient (both medical and lost wages) due to the long time required for these skin wounds to heal and the bed rest that is usually required during the healing process. Even with surgical skin grafting, bed rest can be up to six weeks.

For burns, the problem is also the speed of healing for second-degree burns and donor sites. For third-degree burns, which are usually grafted, the problem is graft "take" and supply of grafts in patients with a high percentage of body surface area burned. Therefore, systems are needed to stimulate regeneration is open skin wounds as well as to help in skin graft viability.

Specifically, approaches that would enhance burn wound healing would significantly reduce health care costs. Cost savings could result from quicker graft healing, a shorter hospital stay, and, in some cases, obviation of surgery if the patient's wounds heal quickly enough on their own as a result of the proposed treatments.

Hospital costs for burn patients requiring admission have been estimated at approximately $90,000 per admission if surgery is required and about $30,000 for nonsurgical patients. This comes to about $5.5 billion dollars per year in the United States. For example, if the amount of surgery and the subsequent increased hospital stay could be reduced by at least 25%, the cost savings could be over $2 billion. Additionally, the mortality rate and days of lost work could be reduced for this population of burned patients as well.

Due to pressure ulcers, it is estimated that patients incur at least an additional $15,000 per year in health-care costs with the average cost of treatment for a pressure ulcer at about $120,000 [104,107]. Total costs for treating pressure ulcers are estimated at $3–$5 billion per year [108].

The mental and emotional costs of a pressure ulcer are hard to quantify. Other aspects of this problem derive from delays in achieving rehabilitation goals, reduced educational opportunities and the resulting long-term impact on vocational potential, separation from the family unit with its impact on psychological and social development, and loss of general personal independence and productivity that contribute to self-esteem. The lack of effective conservative treatment interferes with mobility, increases the burden for caregivers and frequently leads to lost employment opportunities and concomitant economic impact on the individual and family regardless of whether that person had been very successful in his or her rehabilitation program.

A pressure ulcer patient typically has two options: nonsurgical bedrest for three to six months or skin flap surgery that still requires six weeks bedrest. A reasonable goal would be to increase the healing rate by 100% (reduce healing time by half). This makes the time of bedrest for the nonsurgical option more comparable to the surgical option and therefore a more viable alternative. For these patients, cost savings would result from a reduction in the amount of surgery and subsequent hospitalization. In studies [109], the cost savings for conventional treatment over myocutaneous flap surgery was about $17,000, per patient, for nursing home residents, with over $30,000 for outpatient SCI patients [110]. If even half of these patients are treated without surgery, savings could be more than $3 billion. Further, decreases in the mortality rate and the amount of bedrest for nonsurgical patients would also result. Therefore lost wages for patients, both surgical and nonsurgical, could also be reduced.

It has been reported that successful treatment of diabetic foot ulcers would result in an annual cost saving of about $200 million [105]. Diabetics, however, have a 15-fold increase in the risk of amputation. The 50,000 diabetic amputations, in 1985, accounted for about half of the nontraumatic amputations, at a cost of about $500 million [105,106]. This corresponds to about 0.5% per year of diabetics (almost 10% per year if over age 65) requiring amputation [106]. Amputation leads to a 3–7% mortality rate and a 36% major complication rate in this population [106]. Although foot ulcers are the primary reason for hospitalization in only 20% of diabetic admissions, they account for more hospital days than all other complications combined [105].

d. Treatment Options. In order to reduce the economic, morbidity, and mortality effects of these types of wounds effectively, it is important to develop a systematic approach toward intervention. Repair of these wounds requires both short- and long-term implant systems based on the severity of the wound (partial- or full-thickness wounds). For both systems, the goal for the implant is to increase the rate and quality of healing as compared with passively treated wounds by simulating and/or stimulating epidermal and dermal regeneration.

For purposes of discussion, it is important to distinguish between an implant and a wound dressing. Wound dressings have an advantage over internal devices, since they can fall off or be peeled off and thus do not have to be degradable if their only function is to simulate the epidermis. An implant would be the material under this simulated epidermis. A degradable regenerative system, in this case, however, implies that the implant actively stimulates regeneration by serving as a scaffold and thus should be degradable. Although the implant can serve as its own wound dressing, the epidermal simulation needs to be considered separately from the dermal regenerative component since each layer may be composed of different and separate materials.

Partial-Thickness Wounds. In most cases partial-thickness wounds are treated with wound dressings rather than degradable regenerative systems, since the lost epithelium can regenerate on its own with little or no dermal contraction. Immediate concerns do, however, exist about blood loss, bacterial invasion, and fluid loss in partial-thickness wounds [80]. Intervention will therefore be done to stabilize the wound site and increase the normal rate of epithelial coverage. Present clinical options include synthetic wound dressing and topical ointments.

The wound dressings, composed of a variety of synthetic polymers and some organic materials, provide varying degrees of occlusion (wound sterility, hydrophilicity, and gas permeability) in an attempt to replace the functions of the epidermis temporarily [111,112]. These dressings usually adhere to the wound site and in some cases are used primarily to help in the debridement process. Some of these wound dressings even have incorporated bioactivity to help in the healing process. Typically, however, this has been mostly limited to antibiotics

[113–116]. For, example silver has been used in topical agents as well as releasable systems, such as the Vitacuff [6–9] and the Acticoat dressings [117].

Over the past few years, a number of modified allograft and heterograft autograft substitutes have been introduced. These include systems made entirely of natural polymers like collagen, from human or animal dermis (typically acellular like Alloderm), and from human and animal cells (typically in a natural or synthetic polymer mesh like Dermagraft). Most are marketed for full-thickness defects. One, Dermagraft, is designed for partial thickness wounds. This system is made from neonatal fibroblast cells cultured in a nylon mesh. The implanted sheet does not have live cells, but claims to have the components necessary for partial-thickness wound healing [118].

Further, a number of topical ointments have also been used in an effort to speed healing. Again, antimicrobial systems are the most common. This treatment is controversial because these skin wounds are contaminated and reducing the contamination does not necessarily increase the healing unless it is a gross infection. In fact, some have suggested that a moderate amount of infection actually enhances healing [119]. Other topical treatments include Preparation H, hyaluronic acid, and growth factors [120–123]. For virtually all of these topical treatments the clinical efficacy is unknown, however, and randomized controlled clinical trials for a number of these substances are ongoing at many burn and pressure ulcer treatment centers.

Although growth factors have been used mostly for full-thickness defects they have also been used in partial-thickness wounds and administered topically in a liquid or in a cream base. Even though numerous clinical studies with topically applied growth factors (PDGF, FGF, EGF, and platelet derived wound factor) for skin wounds have been performed [58–62], the results have been mixed. Even in reports where growth factors have been shown to speed healing rate [62], it has not been at the rate (100%) necessary to be comparable to surgical intervention, particularly in the case of pressure ulcers. The mixed results are most likely due to release of unprotected growth factor in a wound environment filled with proteolytic enzymes, as well as the lack of a scaffold to guide the tissue growth stimulated by the growth factor.

These clinical trials use weekly doses of topically applied growth factors. It has been shown that continuous dosing of FGF-1, at a lower total dose, is more effective for skin wounds than just one application. A controlled release system would be preferable, due to the short biological half-life [63] of growth factors, especially in chronic wounds with omnipresent degradative enzymes [32,124,125]. Such a system could provide for easy application, as well as the possibility of biofeedback controlled release of the agent. Currently, there is one approved growth factor system, Regranex with others in various stages of clinical evaluation. This system uses PDGF in a cream base and is currently approved for use on foot ulcers [126].

Although these partial-thickness wounds typically heal naturally, many of the skin ulcers can progress to deeper wounds due to pathology or continual irritation [10,104]. In addition, in all cases speeding up the regenerative healing would be beneficial. Therefore, there is a place for degradable regenerative systems even for these shallow wounds.

Full-Thickness Wounds. Unlike partial-thickness wounds, full-thickness (loss of epithelium and dermis) wounds usually necessitate more active treatments than just wound dressings. Many of the same treatments, however, are still used clinically. Since the dermis does not normally regenerate itself, healing in full-thickness defects occurs primarily through the development of granulation tissue and a scar that causes the wound area to contract and lose its elasticity [10,18]. Therefore, the optimal implant must provide a scaffold that promotes the development of a new dermis over which the epidermis can grow with minimal contraction.

Clinical approaches toward healing these types of wounds can again range from environmental control through dressing applications to surgery in the form of skin grafting and skin flaps. In addition, cultured epidermal autografts (CEAs) have been used to cover burn wounds when the supply of donor sites is limited [18]. Dressing change regimens for deep (Grades 3 and 4) skin ulcers can take from six weeks to six months of bedrest to heal, depending on the size and patient compliance [10,104,123].

Surgical intervention in the form of pedicle flaps and skin grafts is not an ideal solution either. Skin grafting, for burns or skin ulcers, requires creation of a full- or partial-thickness donor site depending on whether the surgeon prefers partial-thickness grafts, split-thickness mesh grafts or full-thickness grafts [10,18]. Although contraction and scarring are reduced when a large portion of the dermis is transferred with the graft, it is not practical to create another large full-thickness defect routinely to heal another unless it is for cosmetic purposes, as for the hand or face. In addition, a higher graft take is obtained with thinner grafts, and mesh grafts can cover a larger area, which is critical for burn patients [18,123,127]. Scarring, however, occurs in the mesh interstices, reducing the cosmetic effect [4,5].

For skin ulcers, skin flaps do not always "take"; there are a limited number of sites available for these wounds, and bedrest typically is still one to two months [10,18]. Nonautologous grafts (cadaver and pig skin) have also been used, but because of the immune rejection these are typically only used for temporary coverage [4,5].

Clinically, there are both dermal and epidermal systems (CEAs) used to replace skin grafting. The dermal systems, such as for partial-thickness wounds, are made entirely of natural polymers like collagen, from human or animal dermis (typically acellular like Alloderm) or from human and animal cells (typically in a natural or synthetic polymer mesh like Dermagraft). Integra, a porous collagen/glycosaminoglycan composite (collagen/GAG), is the current clinical system, designed originally by Yannas in the 1970s. Integra is currently recommending a combination approach: a later (two to three week) thin meshed autograft or CEA application. Both require a two to three week healing time, and, if they heal, result in 30% contraction.

Work is ongoing to enhance the bioactivity of these systems by incorporation of antibiotics or cytokines. Although the optimal release kinetics or biological half-lives of the biochemical factors used are not yet completely known, the sustained release would provide many advantages over topical or systemic dosing. Presently polyurethane/poly(L-lactide) microspheres are being added to the collagen matrix of Stage I membranes to deliver antibiotics as the spheres degrade [128]. Also TGF-β has been incorporated into collagen sponges [44].

The CEAs alone have had mixed results. Although some have reported graft take in the 60–80% range, others have reported much lower. The strength and durability of the CEAs as well as the limited neodermis formation are concerns. Even in the best cases it takes five years to obtain dermal-like tissue [18]. It appears, therefore, that formation of a neodermis is a critical part of a successful treatment.

3. *Design Based on the Hierarchy*

At UAB, research has focused on evaluating optimal bioactivity and the optimal implant scaffolding. For optimal bioactivity, both environmental changes (oxygen and PEMFs) and biochemical modifications (growth factors) have been assessed in vitro and in vivo in order to optimize the regenerative response. For optimizing the scaffold, different materials with different configurations, degradation rates and drug delivery kinetics have been assessed in vitro and in vivo. The ultimate goal has been to design systems suitable for the treatment of pressure ulcers and burns that could be used in open wounds as well as in conjunction with skin grafts [129].

a. Bioactivity. In order to determine the optimal level of oxygen needed for healing, a series of in vitro and in vivo studies were done. In vitro oxygen at 20% O_2 (160 mmHg) [20,21] was found to stimulate the greatest increase in fibroblast activity with a concomitant decrease in macrophage activity. In an in vivo study, both the oxygen level and oxygen gradient were modified, based on preceding in vitro studies, to help determine the optimal clinical oxygen treatment protocol [22,23]. Oxygen treatment, corresponding to the 160 mmHg in vitro level (70%), significantly accelerated the healing response with a more occlusive (oxygen impermeable) wound dressing in the early healing stages. A more oxygen permeable wound dressing, however, provided the better cellular and tissue response at later healing stages [22,23]. A further study examined a lower oxygen dosage (40%), which is closer to the more clinically acceptable 6 L/min, and found a similar acceleration in the healing response [130]. It appears, therefore, that the oxygen gradient is only helpful in the early inflammatory stages when macrophages are present, until the granulation tissue is formed and optimum oxygen levels can be achieved without hyperbaric oxygen.

The use of low-frequency pulsating electromagnetic fields (PEMFs) was also examined to more fully understand its effects in the treatment of full-thickness defects in a rabbit model [28–31]. It was found that a magnetic field of 2–2.8 mT at a frequency of 75 Hz applied for 240 min daily for one week significantly accelerated the healing response [28,29]. An additional in vivo study was done to determine the optimum parameters for the PEMFs to be implemented for soft tissue regeneration and overall wound healing [30,31]. This study demonstrated that although PEMF accelerates the healing response in all cases, specific combinations of frequency and intensity levels produce a specific cellular response. It is possible that the optimal PEMF system may involve a series of different frequencies and intensities at various stages of the healing process [30,31].

In vitro growth factor studies have been done with PDGF, FGF-1, and TGF-β with or without collagen, PLA or fibrin substrates. Although maximum values were found in the nanogram/mL range (optimal fibroblast proliferation with and without PLA and collagen implants) [20,21], in vivo, these levels showed no significant effect [67].

Therefore, for in vivo studies increased growth factor concentrations were evaluated. An in vivo study was done to compare the effectiveness of TGF-β and FGF-1 for the treatment of full-thickness wounds created on the dorsum of New Zealand white rabbits [44,45]. Each animal had a control (no treatment), TGF-β (2 $\mu g/cm^2$) incorporated in a collagen matrix, plain collagen, and collagen with added FGF-1 (100 $\mu g/cm^2$). Even though growth factors were incorporated into the matrices, the total release was so quick (a few days) it was actually more similar to a topical dose than a controlled release system. Although the TGF-β incorporated matrix showed enhanced angiogenesis, it was concluded the wounds treated with the topical FGF-1 with the collagen matrix healed slightly better overall [66].

b. Scaffold. To optimize the scaffold, different materials have been evaluated, including collagen, PLA, and fibrin [66,67,129,130]. At this point, the fibrin systems have worked the best overall. Fibrin matrices are not only adhesives that can set up in situ, filling voids and irregular shapes, but can also have the growth factor incorporated at the time of polymerization. In addition, the ability to tie the drug delivery and degradation to cellular infiltration establishes a biofeedback system that is tailored to the individual patient's healing rate [41,129].

Fibrin. Fibrin is derived from fibrinogen in blood. As a tissue adhesive, it is generally supplied as a two-component kit consisting of human fibrinogen/Factor XIII and bovine thrombin/$CaCl_2$. These fibrin sealants have been used since 1972 in Europe, where a commercial version is presently available, and have been recently approved in the United States. Therefore, until recently, studies in the United States have used only autologous or single donor preparations [131–133]. Clinically, the fibrin matrix has been used as a hemostatic agent, for tissue anastomosis, as a fluid barrier, as a drug delivery vehicles, as a tissue scaffold, and as a matrix for cultured keratinocytes [134–136]. Fibrin sealant used for skin grafting has been shown to increase strength of attachment to the wound bed compared with staples [137], leading to less seroma formation [132], and wound contraction [138]. Making the fibrin porous allows for quicker graft take by providing a scaffold for the blood vessels to grow through, without the matrix having to be broken down first. In one study, however, as fibrin was made porous it appeared that the shear strength was inadequate to handle physiological loading (Fig. 4) [139].

In full-thickness defects, a degradable fibrin scaffold has been shown to increase the angiogenic and tissue response over controls [79]. Apart from the scaffolding effect, these degradable systems demonstrate other desirable characteristics. Studies have also shown the utility of fibrin as a degradable adhesive for both blood vessel anastomosis [140,141] as well as skin graft attachment [131–133]. The concentration of fibrinogen has been varied between 15 and 70 mg/mL in an effort to determine minimum setting time and maximum adhesive strength. It was found that the strength of the fibrin clot increased linearly [131,132,140,141] while degradation rate decreased [137,138] with increasing concentration (Fig. 5). In addition, near maximum strength was achieved in the first few minutes [131,132,140,141]. Interestingly the commercial fibrin glue (Tissel or Tissucol) made by a modified cryoprecipitation technique is not as strong as autologous cryoprecipitate preparations made at UAB (due to its slightly lower concentration) and attains full strength at a slower rate (Fig. 5) [131–133].

Albumin. While fibrinogen can be treated with detergent and organic solvents to kill enveloped viruses, such as HIV-1 and hepatitis B, the process does not inactivate viruses that lack lipid envelopes, such as hepatitis A and human parvovirus B19 [142]. This was one of the major reasons that the use of pooled fibrinogen has only recently been approved by the FDA, but only for hemostasis applications. In other clinical applications, therefore, fibrin must be extracted from the blood or plasma of the patient. This requires a lag time between blood or plasma extraction and availability of the fibrinogen for surgical use, as well as the need to take blood from a patient whose health is already compromised.

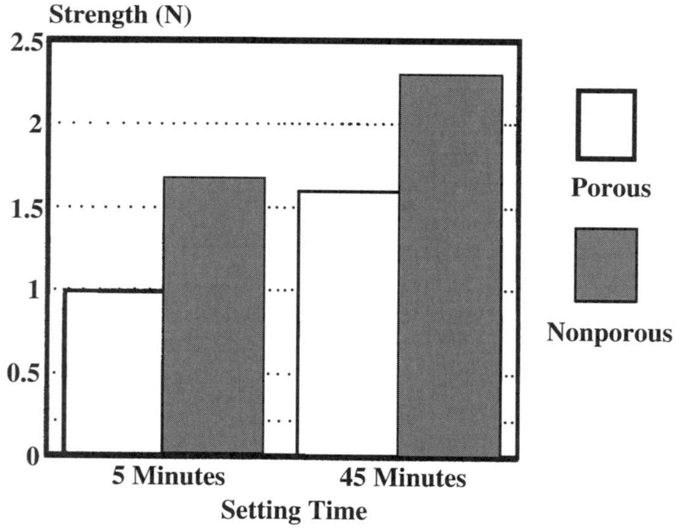

Figure 4 The effect of porosity on the shear adhesive strength of fibrin glue. Strength (*N*) is compared at two setting times.

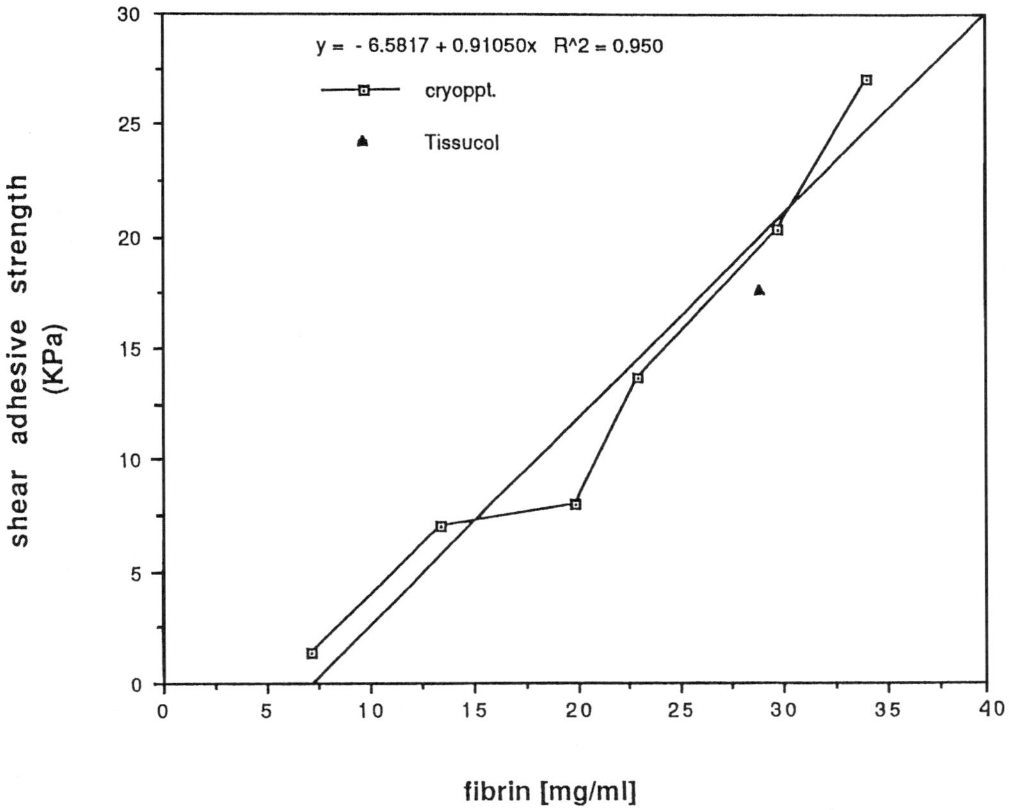

Figure 5 The relationship between fibrinogen concentration and the shear adhesive strength of fibrin glue.

The concerns and logistical difficulties with fibrin glue have provided impetus to look at other tissue adhesives. Currently, an albumin system is being evaluated as regenerative tissue adhesive scaffold. Albumin, while also a blood product, can be processed at high enough temperatures to inactivate viruses that are potential problems in fibrin systems. Albumin is widely used and accepted by the medical community, and has been approved by the FDA for clinical use. In addition, because albumin is derived from pooled human plasma, it is a more consistent product than autologous fibrin. It is also more convenient for the same reason, with no advance blood donation and processing prior to clinical use as in the case of fibrin. Further, adhesive albumin, such as that produced by crosslinking with poly(ethylene glycol), has been shown to possess mechanical properties superior to both autologous and commercially available fibrin glues [143].

Human serum albumin is the most prevalent soluble protein in blood. The linear pattern of loops of the 66 kDa polypeptide, with short-range coupling between half-cystines, provides for both flexibility of the albumin molecule and resistance to harsh conditions. The loops can associate, forming a globular structure, or can separate reversibly.

Since World War II, albumin has been widely used in circulatory therapy, primarily for circulatory support during shock. Polymerized microaggregates of albumin can serve as agents for radiocontrast and ultrasonic imaging of the circulatory system. Albumin is also used to coat prosthetic blood vessel grafts and vascular catheters to passivate that surfaces so as to minimize platelet aggregation and thrombotic consequences [144–148].

The use of albumin as a medical adhesive is relatively new. It must be crosslinked to achieve the necessary strength, but could be used in the same applications as fibrin: hemostasis and wound closure. In addition, its added strength of adhesion would allow additional applications.

Disuccinate crosslinked albumin has also been shown to inhibit bacterial growth on titanium (cp-Ti) surfaces [148,149] as well as effectively deliver gentamicin and increase the half-life of the antibiotic [150]. This is a significant advantage over fibrin, which is claimed to accelerate infection. Because it is a natural human protein, tissue proteinases will degrade albumin. Thus, in a healing wound, crosslinked albumin adhesive should degrade as the healing tissue advances.

A preliminary study of the mechanical shear strength of two adhesive albumin systems has been conducted, and further studies are in progress [143]. A solution of 25% (w/v) human albumin was mixed with modified-PEG (mod-PEG), to form solutions with either 20% or 5% (w/v) final PEG concentration. Both adhesive albumin systems were tested on samples of skin (skin-skin) by imposing a shear force until failure, using an Instron mechanical testing machine. Results were compared to simultaneous tests using fibrin adhesive. The 25% albumin–20% mod–PEG adhesive albumin system was also tested between aluminum samples (Al-Al; no skin).

The highest shear strength, obtained with the 25% albumin–20% mod–PEG system, was approximately five times that of the fibrin glue. The 25% albumin–5% mod–PEG system shear strength was approximately 1.5 times stronger than that of the fibrin system. Shear strength of the 25% albumin–5% mod–PEG system was comparable for both the skin-skin and Al-Al tests.

In pilot studies, albumin glue, fibrin glue, cyanoacrylate, or sutures were used to close incisional wounds on the dorsa of rats and rabbits [151]. Wounds closed with albumin and fibrin had excellent healing, with no inflammatory reaction observed histologically. There was a minimal lymphocytic reaction in the sites treated with albumin. In the wounds closed with silk sutures, there was complete healing, but with a more prominent scar than in the albumin and fibrin treated wounds, and some foreign body reaction around the sutures. In the wounds closed with cyanoacrylate, there were ulcerations, necrosis and a severe inflammatory reaction observed at the wound site.

Additionally it was found that the incisional strength for an albumin adhesive could be as strong or stronger than sutured wounds after one and two weeks in vivo as well as significantly stronger than the fibrin adhesive both initially and after one week. Interestingly at high albumin concentration the adhesive does not show the twofold increase in incisional strength, between the first and second weeks, seen in sutures and other albumin glues, possibly due to too slow a degradation rate.

c. Design of Degradable Regenerative Systems.

Current Systems. The difficulties with the CEAs due mostly to lack of dermal has increased interest in dermal substitutes. Attempts to provide a dermal replacement, however, began in 1962 when Chardack et al. developed a formalinized porous polyvinyl alcohol sponge covered with a silicone rubber sheet [123]. Later in the 1970s, Yannas and co-workers developed Stage I membranes, which are a composite of a biodegradable collagenous/chondroitin sulfate dermal substitute in conjunction with a silicone rubber wound dressing [4,5,92–96,127]. This implant is not completely regenerative, however, without epidermal autografting. The autografting can be in the form of cell seeding or partial thickness skin grafts. Cell-seeded implants are called Stage II implants [127].

Several variations on the bilayer membrane have been developed in an attempt to provide a more bioactive dermal substrate capable of fibroblast recruitment and neovascularization [152–155]. Much investigation has been done concerning the bioactive nature of the extracellular matrix glycosaminoglycans and proteoglycans such as chondroitin-6-sulfate and fibronectin, respectively. In addition, the type, structure, and porosity of the collagen substrate have been investigated to optimize the rate and quality of cellular ingrowth better [155–160]. In the best cases, these implants have been successful in retarding the scarring process since the dermal scaffold provides structural integrity until epidermal grafting or cell mitosis occurs [127]. There are currently two different approaches. One in which the dermal substitute is placed on quickly, at the time of a normal graft procedure and the epidermal part either CEAs or thin autografts two to three weeks later (time delay either to produce CEAs or let the dermal substitute heal in) [4,5,90]. The other approach is to wait the two to three weeks and implant the dermal substitute with the epidermal covering [53]. Both procedures require two separate surgical procedures plus a biopsy for systems with CEAs.

An additional concern, however, is the "take" and infection rate of these implants, especially when compared with the high percentage take of autografts [10,18,115]. Angiogenesis and adhesion to the graft bed are keys to this successful take, suggesting the utility of an adhesive regenerative scaffold. Other critical parameters are angiogenesis to keep the graft viable and not prone to infection, adhesive to keep the graft attached to the graft bed to allow angiogenesis and nutrient exchange with the host, and a scaffold to allow and stimulate fibroblast and blood vessel ingrowth. Further, if the scaffold provides angiogenesis quickly enough, the epidermal layer can be placed immediately, eliminating one surgical procedure.

Design Considerations. *Bioactivity.* Critical to determining the appropriate system are the rates of migration of the key cells, since fibroblasts need to be within 100 μm of the blood supply to have sufficient oxygen and nutrients to survive and produce collagen. Fibroblasts migrate up to 200 μm/day [4,5], but will only do so if the blood vessels and nutrients are within 100 μm. Angiogenesis for a 0.5 mm collagen/GAG system takes seven to nine days [4,5] (50–70 μm/day). This slows the fibroblast ingrowth across 500 μm to 5–8 days (60–100 μm/day) versus 2.5 days (200 μm/day). Therefore angiogenesis becomes the rate-limiting step. Angiogenesis is also needed to support the epidermal layer. Epidermis migrates at about 0.33 mm week over viable vascularized dermis, but only half that if it has to burrow through tissue to find viable tissue [161]. Therefore the speed in which vessels permeate the scaffold determines the earliest that the matrix can support epidermal cells—at least a week for the 500 μm collagen/GAG template.

Angiogenesis is also the key in reducing the risk of infection in the graft because it can bring cells for an acute inflammatory response to kill bacteria [161,175]. Researchers have tried to circumvent the slow angiogenesis by providing a cell culture like medium for the cells (with antibiotics) [53]. This may not be logistically possible in all cases, and could be avoided with a different design.

Therefore, strategies to speed up angiogenesis are needed. This would include use of angiogenic growth factors such as FGF. Additionally modification of the matrix configuration, such as optimizing the porosity or making the matrix less than 200 μm thick, could ensure that angiogenesis was not the rate-limiting step.

Drug delivery. With natural biomaterials, such as albumin or fibrin, delivery of a biological response modifier can be accomplished in different ways. The biological response modifier can be impregnated within the matrix, attached to the polymer chain, or included through intrafibril entrapment [135]. Incorporation during the polymerization process is similar to intrafibril entrapment. In this case, if the substance is larger than the intrafibril pores, it is released only when the natural material degrades [134,135,74]. Therefore, the release is controlled by the rate of phagocytic cellular infiltration, and thus under biofeedback control. Additionally, the degradation is at the wound edge and thus gives the appropriate gradient to stimulate further angiogenesis and tissue healing.

The use of a degradable matrix to deliver a biological response modifier, such as a growth factor, can protect the growth factor until release, since growth factors seem to have a short half-life in vivo. The short half-life has potentially been a problem in clinical studies, leading to reduced efficacy, and/or increased expense associated with daily administration.

Thus, a natural polymeric matrix such as fibrin or albumin can protect the biological response modifier until release, serve as a biofeedback controlled drug delivery system, and provide an adherent tissue scaffold during healing.

System Design. Based on previous work, FGF-1 was selected as the angiogenic agent for skin applications. Both in vitro and in vivo studies have indicated that the tissue response is dose dependent and a maximal response is reached at an intermediate dose (Fig. 6) [162].

For the matrix, fibrin was selected for its adhesive properties [131–135,140,141] as well as its ability to serve as a degradable drug delivery system for wound healing [163–167]. It was found that the strength of the fibrin clot increased and degradation rate decreased with increasing concentration [131–135,139–143]. Additionally, in vivo and clinical studies have been done to test the use of autologous single and donor fibrin matrices for skin graft attachment in burn patients [132]. It was found that the fibrin hemostasis and early graft adherence led to an excellent graft take with reduced scarring. When compared to conventional treatments, this technique led to shorter hospital stays, minimal postoperative care and immobilization, no pressure dressings, and prompt start of ambulation and physical therapy with an early return to normal activities.

For a drug delivery system, there was a concern due to the short biological half-life of FGF-1. This would require that the FGF-1 be protected within the matrix. In the case of fibrin, it has been shown to covalently bind with FGF-1 [168]. Additionally, immunolocalization studies have indicated that there is a uniform distribution of FGF-1 in the fibrin matrix [166,167]. Release studies using FGF and fibrin have indicated that although there is an initial

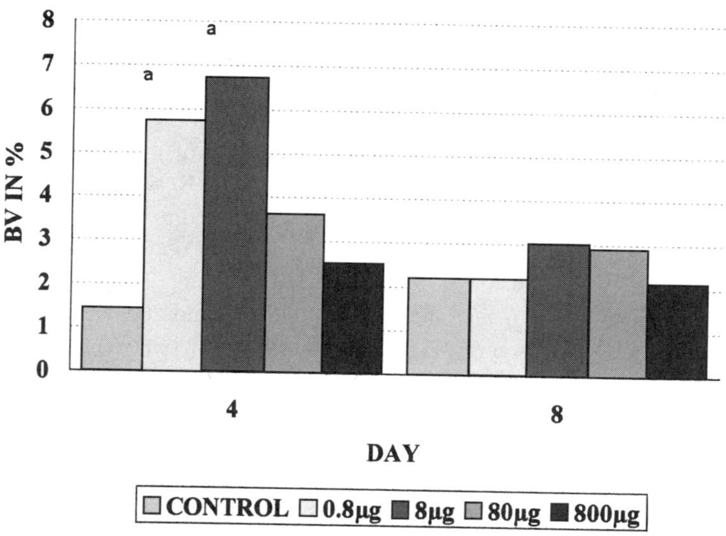

Figure 6 Angiogenic response to different doses of FGF-1 delivered through a fibrin matrix. The two doses significantly different than the controls are labeled with an asterisk.

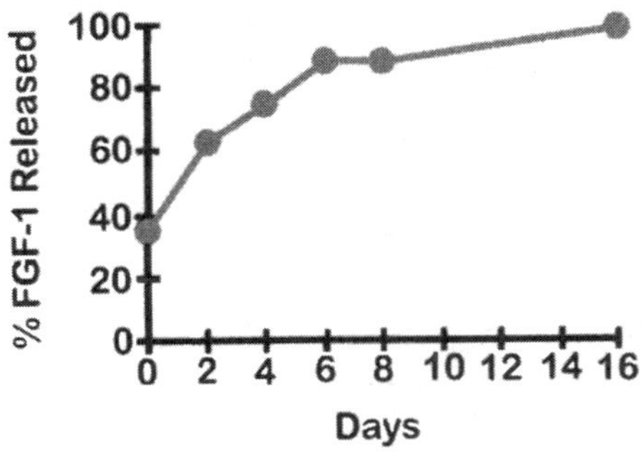

Figure 7 Release kinetics of FGF-1 delivered through a fibrin matrix.

high release rate of about 30%, the subsequent release rate is relatively constant and proportional to the degradation rate (Fig. 7) [74,168,169,170]. To help show in vivo activity, a comparable enhancement of wound healing was seen with a topically applied dose, designed to mimic the fibrin /FGF-1 release kinetics, compared with the fibrin/FGF-1 system (Fig. 8) [165,169,170].

Although the fibrin matrix, due to its own biological activity [171], serves as a reasonable scaffold, better scaffolds can be made by optimizing the configuration as well as the bioactiv-

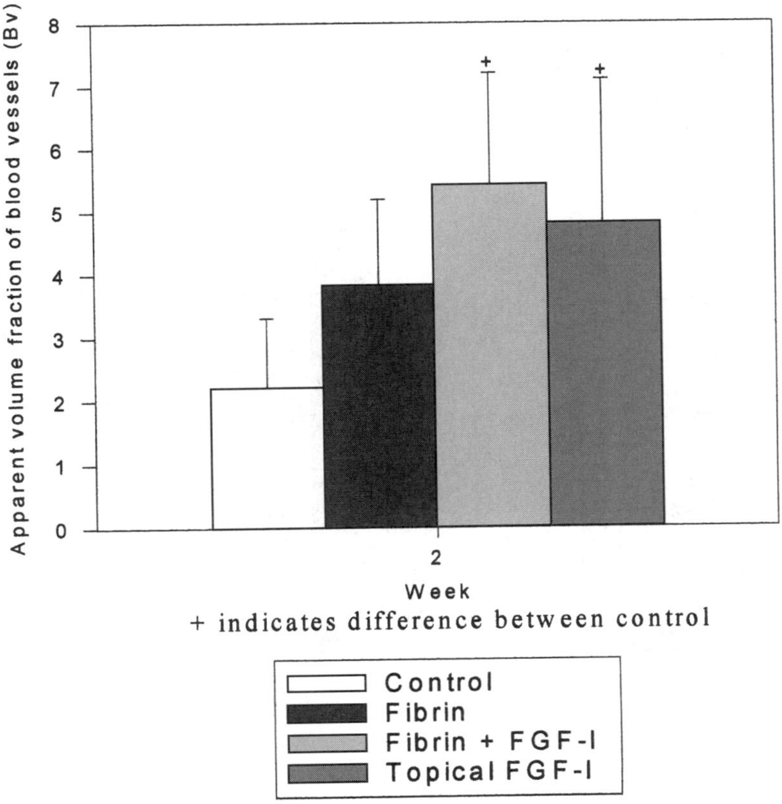

Figure 8 Angiogenic response of FGF-1 delivered topically versus through a fibrin matrix.

ity. In a rabbit ear ulcer model, full-thickness defects were treated with the fibrin matrices in two different pore configurations [79,163]. The more porous implant (modified fibrin) showed increased angiogenic response (Fig. 9). The levels of porosity and pore size are currently being optimized. Even in unoptimized systems, FGF-1 in a nonporous fibrin matrix was capable of complete epidermal regeneration with dermal filling of full-thickness defect and minimal contraction (20%) within two weeks, while controls took at least three weeks to heal and healed mostly by contraction [166,167].

For skin grafts, it was anticipated that making the fibrin porous would allow for quicker graft take by providing a scaffold for the blood vessels to grow through, without the matrix having to be broken down first. In one study, however, as fibrin was made porous it appeared that the shear strength was inadequate to handle physiological loading (Fig. 4) [139]. Although the incorporation of FGF-1 reduced the incidence of graft loss, porous albumin, with its better adhesive strength, as well as other porous fibrin systems is being investigated. It is also possible that using a thin adhesive layer (200 μm or less) would obviate the need for a porous structure.

Overall, the fibrin/FGF-1 system [162–167,169,170] when used in open wounds has led to the best overall healing, compared with all other treatments, with complete epithelialization and minimal contraction. This is likely due to the increased angiogenesis, mostly due to the FGF-1 as well as the increase in new tissue formation due to the fibrin scaffold. Similarly for meshed skin grafts, FGF-1 incorporated into a fibrin has led to the best graft healing. The

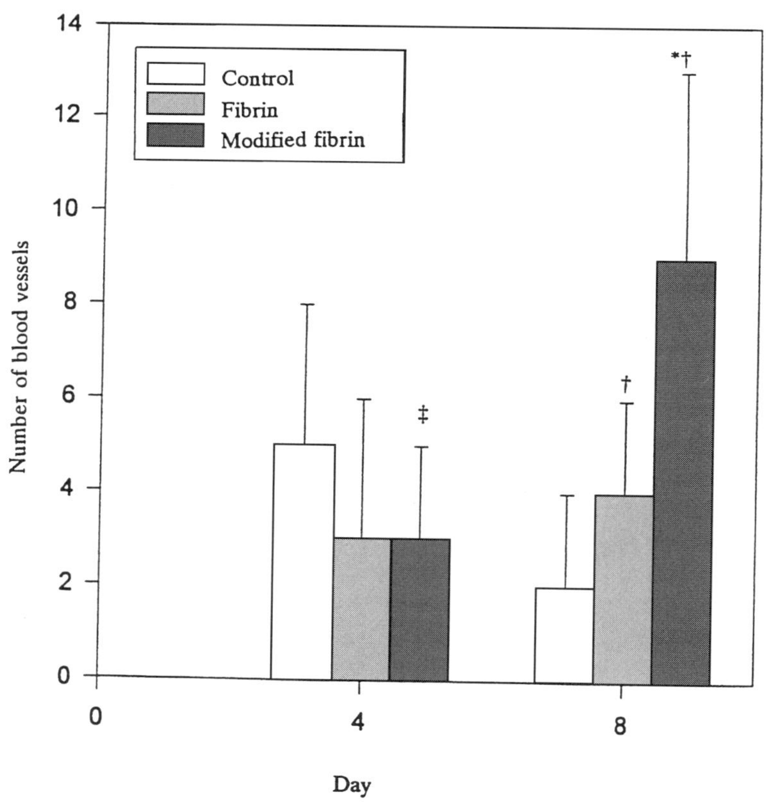

Figure 9 The effect of increasing fibrin porosity on the angiogenic response: (*) indicates different than the control; (†) indicates difference between modified and nonmodified scaffold.

FGF-1 concentration that has worked best overall is 10 μg/mL, which when spread in a 200 μm layer is 10 μg/50 cm^2.

Preliminary data indicates that the albumin system performs as well or better than the fibrin system in terms of strength, in vitro and in vivo, and biocompatibility. Studies are underway to develop it as a regenerative tissue scaffold.

Clinical Studies. The clinical efficacy of the fibrin/FGF-1 system is currently being evaluated in both a pressure ulcer and burn healing study at UAB [172–174]. For pressure ulcers, patients with Stage II and III ulcers are randomized into three groups: (1) control, (2) placebo control, and (3) FGF-1 in a fibrin matrix.

For burns, the fibrin is used for skin graft attachment and the wounds on the patients are randomized (double blinded) into two groups: (1) fibrin with FGF-1 and (2) fibrin without FGF-1. Each patient also has a corresponding control (stapled or sutured graft) for comparison.

A critical part of each study has been the development of noninvasive clinical assessment tools to measure the effectiveness of the treatments. The three main variables for comparison are healing rate, angiogenesis, and tissue stiffness. These assessments are done weekly for the first few months and monthly thereafter. For healing rate, the epithelialization rate and contraction rate are determined in a manner that gives the rate independent of wound size. The angiogenesis is determined by a scanning laser Doppler, which gives a 2-D map of the blood perfusion.

Preliminary results have indicated that the fibrin systems keep the blood perfusion level at a higher level longer than the controls. For meshed skin grafts, this means the vascularity is kept high until the mesh fills in. This is believed to be associated with increased new tissue formation and less scarring (Fig. 10).

Overall, these studies serve as an example of several variables that can be evaluated and optimized for the ultimate development of a degradable scaffold for the regeneration of cells and tissue damaged by pressure ulcers and burns.

B. Percutaneous Devices

1. *Clinical Problem*

Percutaneous devices (PDs) are implants that permanently penetrate the skin through a surgically created cutaneous opening (Fig. 1). Presently, there are implants designed for use as dental prostheses, catheters for perfusion or dialysis, for electrical and pneumatic connections for artificial internal organs, skeletally attached artificial limbs, and external fracture-fixation devices [175–178].

Despite systematic studies of the pathophysiology of percutaneous interface healing by Lee et al. [176], Hall et al. [177], Winter [179], Freed et al. [178], Daly et al. [180], von Recum and Park [175], and Feldman [77], consistent long-term viability of the PD tissue interface has not been achieved. The consensus from a percutaneous device workshop held in San Antonio, Texas (in 1982), still holds today: that much work needs to be done before a successful PD will be developed. Different variables need to be examined separately and continued emphasis should be placed on the study of clinical failure and the basic pathophysiological phenomena responsible for PD failure. For example, a polyurethane peritoneal dialysis access port developed by Daly et al. showed a 19% failure rate by 137 days, which extrapolates to approximately a 50% failure rate by one year [180].

2. *Failure Modes*

The most successful PDs are usually small and well anchored. In many cases, however, this design is not feasible. All PDs are susceptible to failure modes, caused by epidermal down-

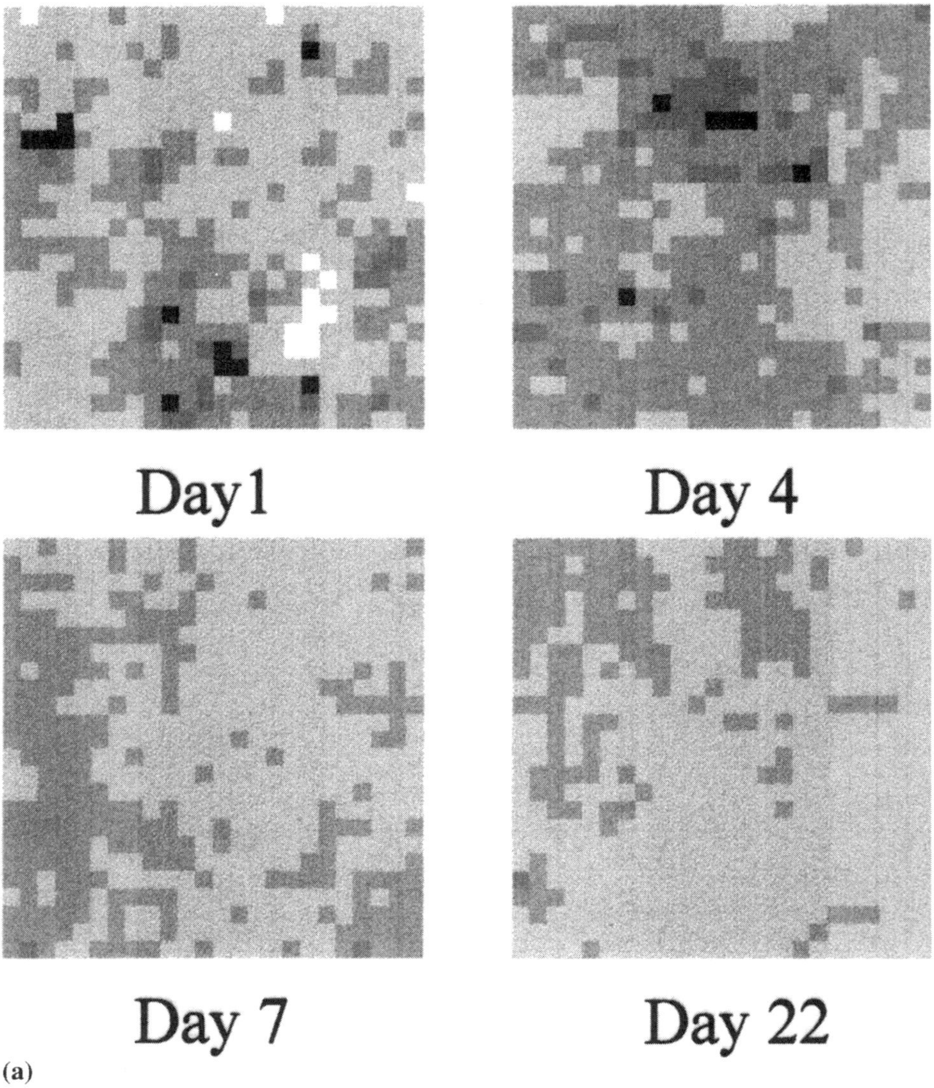

Figure 10 The clinical blood perfusion changes over time for meshed skin grafts adhered with (a) fibrin (1,4,7,22 days), (b) sutures (1,4,7,22 days). The darker the area the higher the blood perfusion. The fibrin system peaks at seven days or later, whereas the sutured system peaks at four days.

growth, infection, and mechanical trauma [175,181–183]. Both the epidermally induced failure modes (sinus tract formation, marsupialization and extrusion due to basal cell maturation) and the nonepidermally induced failure modes (failure due to interconnecting porosity and failure due to avulsion) have been described in the literature [8,9,77,175,182].

An implant needs to be designed to eliminate both epidermally and nonepidermally induced failure modes. Prevention of epidermal downgrowth by tissue growth has been the route taken by most investigators to stop epidermal failure. Fabric materials are the materials of choice because they allow rapid connective tissue ingrowth [77,183]. Since epidermal basal cells do not migrate through intact tissue or collagen layers, "good" ingrowth should stop sinus

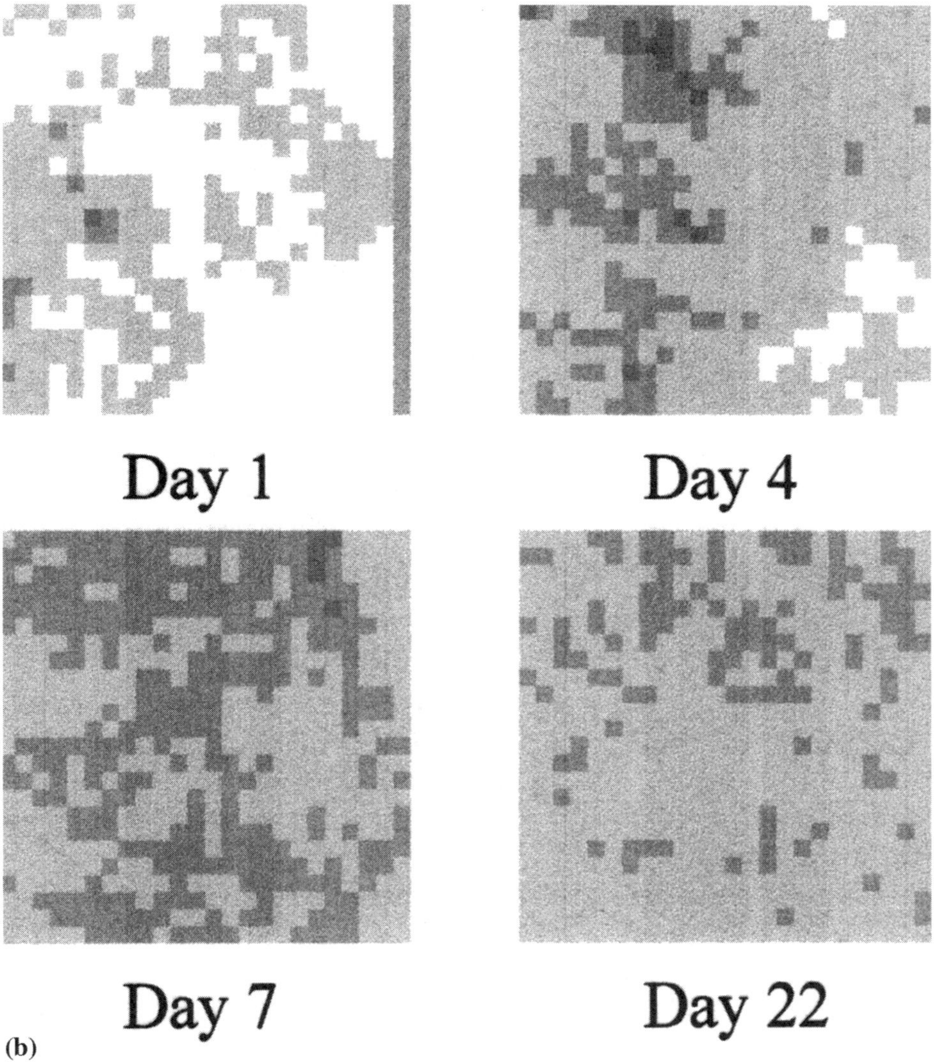

Figure 10 Continued.

tract formation and marsupialization [175]. Extrusion due to basal cell maturation can occur, however, if the epidermis juxtaposes a porous material [175–177]. This, however, does not occur if the tissue ingrowth is mature and a noninterconnecting porous interface is used [77,175].

The major clinical sign of percutaneous device failure is infection. Infections around heart valves and vascular grafts occur in 1–6% of the cases while patients with ventricular assist devices (VADs) or total artificial hearts (TAHs) in situ longer than three months have an infection rate of 25–100%. The percutaneous interface is partly responsible for this high incidence. The failure at the percutaneous interface due to infection is historically the reason for terminating all long-term animal experiments with implanted blood pumps as well as for patients with all types of percutaneous interfaces (blood pumps, dental implants, catheters, kidney dialysis, etc.) [6,8,175–178].

3. Design Based On the Hierarchy

a. General Design. Using the hierarchy, percutaneous implants need to be designed to regenerate the natural tissue interface seen in naturally occurring PDs. Although many different devices and techniques have been investigated over the past 15 years, little progress has been made toward the goal of a permanent percutaneous implant. Based on previous studies [8,9,77,81,175–180], all presently known failure modes can be prevented by decreasing the chronic inflammatory response to the implant and increasing the rate of tissue regeneration. In addition, initial failure by infection must also be prevented, until the percutaneous interface heals. Based on the hierarchy, both the regenerative response around the PD and the scaffolding effect of the implant need to be optimized. The goal is to design a material that can be used as a skin interface for a PD and alleviate all PD failure modes.

As for all implants, many factors affect the chronic inflammatory response or healing at the percutaneous interface. These factors would include implant configuration, implant stress, implant material, implant surface properties, implant surface texture, implant location, implant biochemical environment, and the implant model. Improving implant fixation [175–177], which could be achieved by material configuration alteration, can lessen the effect of stress. Investigations on PDs have indicated that configuration is more important than location and stress in determining the chronic inflammatory response [77,81,180].

Alterations in surface texture and implant material have been investigated in a number of studies. Dacron/collagen/silver composites have been implanted subcutaneously and percutaneously for tissue ingrowth [70], porous textured implants have been placed subcutaneously for increased ingrowth and mechanical attachment [76,184], and silver/collagen and collagen cuffs have been studied for infection prevention [8,9,68,69]. In addition, in some cases, implant preparation (cleaning and sterilization) protocol has been found to have a dramatic effect on tissue response by alteration of surface chemistry and surface energy [185,186].

The use of biochemically active surfaces can help induce increased tissue response [12,13,61–64]. Both growth factors (e.g., FGF) [62–66] and extracellular matrix proteins [153–160] have been shown to enhance cellular and tissue attachment as well as tissue ingrowth.

This general information has been used in the design of three different types of PDs: artificial skin, long-term intravenous (IV) catheters, and dental implants. Artificial skin has been described above and IV catheter systems are stressed next.

b. Intravenous Catheters.

Clinical Problem. Approximately 35,000 medical device infections per year are catheter related [178,187]. About 8% of the patients with intravenous catheters become septic because of catheter-associated infections, with a mortality rate of 20–40% for this group [188]. Percutaneous implant-associated infection can be the result of nonsterile surgical technique, bacterial contamination, excessive inflammatory response, and lowered patient defense mechanisms [6,189]. Most catheter-related sepsis, however, starts with the infection at the catheter entry site. The patient's own cutaneous flora typically provide the invading microorganisms that cause the infection [7]. The intravenous catheter provides a pathway for the microorganisms to traverse the epithelial barriers to the underlying tissue. These microorganisms may eventually end up in the bloodstream. In addition, the catheter itself has been reported to provide support for the growth of the microorganisms [190].

Research on IV Catheter Infection. Research into the prevention of infection has included the incorporation of antimicrobial or antibiotic substances into the implant body or surface. Trooskin et al. studied catheters that had antibiotics bonded to their surface [190]. Silver compounds have been used because of their antimicrobial qualities [191,192,193].

A multicenter study of a catheter having a collagen-silver cuff (Vitacuff) has been performed. The Vitacuff material was made of silver chelated onto a Type I collagen cylindrical cuff. The study consisted of 234 patients; 135 were controls and 99 had the silver-collagen cuff catheter. The catheters, with the cuff, were left in place by the patient's personal physicians an average of two days longer than the control catheters ($p \leq 0.05$). In addition, the silver-collagen catheter cuffs were well tolerated and the cuffs reduced bacterial colonization by 68% and septicemia by 73% [6].

Design Based on the Hierarchy. To maintain a long-term percutaneous device, prevention of device-related infections and a stabilizing connective tissue ingrowth are needed. The optimization of percutaneous devices as an antimicrobial barrier system as well as for stimulating tissue ingrowth into the device to promote a good seal could prevent infection and provide implant stability.

Optimization of tissue ingrowth and attachment to provide a good epidermal and dermal seal can be done by altering the implant configuration and implant biochemical activity. Building on previous studies the optimal configuration ranges can be selected [72,73,76–78]. Biochemical activity can be used to increase cellular attachment and spreading (fibronectin, laminin, Type IV collagen, and extracellular matrix proteins) [193] as well as cellular proliferation and activity (growth factors) [130]. Based on previous studies, these alterations lead to increased mature tissue ingrowth at a faster rate, although there is some concern that tissue response would be greater than desired. Therefore, basic systematic studies to determine the effect of each of these alterations individually as well as in combination is required.

Prevention of bacterial colonization can be achieved by using antimicrobial agents or antifouling coatings. Antimicrobial coatings work better than systemic antibiotics because the amount of agent is reduced, and, thus, toxicity is reduced. Further the act of bacterial attachment itself can be prevented. A current clinically approved delivery system (Vitacuff using silver/collagen) [6] as well as other systems have been explored. These systems include attachment of the antimicrobial agent by chemical means, radio frequency discharge activation, or photoimmobilization.

In addition, studies have shown that bacterial colonization can be prevented by using hydrophilic polymeric coatings. Increasing the wettability of the relatively hydrophobic silicone rubber has been shown to reduce the affinity for protein and bacteria [194]. These hydrophilic coatings can be applied in ways similar to the antimicrobial coatings. Further, attachment of extracellular matrix proteins can also reduce hydrophobicity and thus decreases bacterial adhesion [193].

In determining the optimal systems for a long-term IV catheter, optimization of the tissue response to each of the critical layers (epidermal, dermal, subcutaneous, and blood vessel) needs to be investigated. This will probably lead to a device with different material systems at each interface that need to be evaluated in vitro and in vivo.

c. Dental Percutaneous Devices.

Clinical Problem. The percutaneous interface around dental implants or teeth is critical. In the case of these periodontal defects, poor healing of the degenerated ligament space, bone defects, and associated gingival defects are a problem. These problems limit the types and success of clinical treatments. Repair of the periodontal ligament spaces often leads to a fibrous capsule-like tissue, which has limited functional value.

Enhancing and controlling the healing for the soft-tissue defects could reduce the incidence of continuing disease and improve the prognosis for periodontal therapies. In addition, healing and reattachment of a regenerated tissue would be beneficial in many other dental applications (dental implant interfaces, tooth reimplantation, ridge augmentation, maxillofacial reconstruction, etc.) as well as orthopedic applications (bone defects, ligament tears, etc.).

Most treatments of soft-tissue periodontal defects have been limited to the use of barrier substances, the goal being to prevent epithelial downgrowth into the lesion spaces prior to connective tissue regeneration and attachment. This is similar to the goals for other percutaneous device systems. The main biomaterial used in this application has been expanded polytetrafluoroethylene (Gore-Tex). Although somewhat successful as a treatment modality a functional periodontal ligament is not restored. Similarly, for dental implants, if implant-to-bone attachment is not achieved, what results is an oriented fibrous tissue capsule rather than a functional ligament. A major advantage for implant interfaces would be to orient the collagen across the lesion space rather than parallel to the implant surface.

The regeneration of bone lesion sites around teeth and dental implants would be similar to that required for other bone applications. The only difference would be the oriented attachment to the tooth or implant.

Design Based on the Hierarchy. One goal, therefore, has been to regenerate functional tissue for the treatment of periodontal defects, dental implants, or tooth reimplantation. In the most favorable circumstance, this will require regeneration of oriented collagen fibers (perpendicular to the tooth or implant surface) that are anchored to the bone and tooth or implant with fibers like Sharpey's. This will require design of degradable regenerative templates with optimum configuration, mechanical properties, and regenerative abilities (both structural and attachment regeneration). Therefore, initial steps have been in the fabrications and characterization (mechanical, configuration, and surface composition) of the degradable and nondegradable systems.

Optimization of regeneration will require control of both implant configuration and biochemical activity. This research will build on the results from previous work to choose the materials, configurations, biochemical activity, and factor release systems. This research has direct significance and ramifications for the treatment of periodontal defects (periodontal ligament, gingiva, and alveolar bone) and oriented tissue regeneration around teeth and dental implants. This then has direct implications for periodontal treatment, dental implant anchoring, tooth reimplantation, extraction socket defects and bone defect filling.

ACKNOWLEDGMENTS

We wish to thank the UAB graduate students in the Biomedical Polymers Research Group responsible for the work presented in this chapter. In addition, we wish to thank J. Anthony Thompson for his help in providing FGF and collaboration in the design of degradable regenerative systems and the American Red Cross and S. Huang for supplying the fibrin and albumin matrices. Further, we acknowledge the funding agencies who helped support the research presented here: NIH, NSF, Department of Education, and the CDC.

REFERENCES

1. Berthiaume, F. and Yarmush, M., 1995. Tissue engineering, in *The Biomedical Engineering Handbook*, ed. Bronzino, J., CRC Press, Hartfod, CT, pp. 1556–1565.
2. Hench, L. and Ethridge, C., 1974. Biomaterials—the interfacial problem. *Adv. Biomed. Eng.*, 5: 35–150.
3. Black, J., 1981. *Biological Performance of Materials: Fundamentals of Biocompatibility*, Marcel Dekker, New York.
4. Yannas, I. V., Burke, J. F., Gordon, P. L., Huang, C. and Rubenstein, R., 1981. Design of an artificial skin. II. Control of chemical composition. *J. Biomed. Mater. Res.*, 14:107–131.
5. Yannas, I. V., Lee, E., Orgill, D. P., Skrabut, E. M. and Murphy, S. F., 1989. Synthesis and

characterization of a model extracellular matrix that induces partial regeneration of adult mammalian skin. *Proc. Natl. Acad. Sci. U.S.A.*, 86:933–937.
6. Maki, D. G., Cobb, L., Garman, J. K., Shapiro, J. M., Ringer, M. and Helgerson, R. B., 1988. An attachable silver-impregnated cuff for prevention of infection with central venous catheters: A prospective randomized multicenter trial. *Am. J. Med.*, 85:307–314.
7. Maki, D. G. and Ringer, M., 1987. Evaluation of dresing regimens for prevention of infection with peripheral intravenous catheters. *JAMA*, 258, No. 17:2396–2403.
8. Solovay, D., 1991. Histological investigation of dacron and collagen percutaneous catheter cuffs, M.S. thesis, University of Alabama at Birmingham.
9. Solovay, K., Bond, S., Estridge, T., Feldman, D. and Schoumacher, R., 1991. Evaluation of porous cuff materials for IV catheters. *Trans. Soc. Biomater.*, 17:57.
10. Mast, B., 1992. The skin, in *Wound Healing: Biochemical & Clinical Aspects*, ed. Cohen, I., Diegelman, R., and Lindblad, A., W. B. Saunders, Philadelphia, pp. 344–355.
11. Goss, R., 1992. Regeneration versus repair, in *Wound Healing: Biochemical & Clinical Aspects*, ed. Cohen, I., Diegelman, R. and Lindblad, A., W. B. Saunders, Philadelphia, pp. 20–39.
12. Bowald, S., Busch, C. and Erikkson, I., 1979. Arterial regeneration following polyglactin 910 suture mesh grafting. *Surgery*, 86:722–729.
13. Kambic, H., Novick, A. and Nose, Y., 1982. Observations on experimental bladder substitution: review of the literature and reports of studies using natural tissue biodegradable graft material, in *Biocompatibility in Clinical Practice*, ed. Williams, D., CRC Press, pp. 73–106.
14. Torrente, J., 1991. The effects of ECGF and PDGF on the co-cultures of vascular cell lines, M.S. thesis, University of Alabama at Birmingham.
15. Torrente, J. and Feldman, D., 1991. The effect of ECGF and PDGF on proliferation and interaction of co-cultured vascular cell lines on collagen membranes. *Trans. Soc. Biomater.*, 17:75.
16. Madison, R., Archibald, S. and Krarup, M., 1992. Peripheral nerve injury, in *Wound Healing: Biochemical & Clinical Aspects*, ed. Cohen, I., Diegelman, R. and Lindblad, A., W. B. Saunders, Philadelphia, pp. 450–489.
17. Wormom, M. and Buchman, S., 1992. Bone and cartilaginous tissue, in *Wound Healing: Biochemical & Clinical Aspects*, ed. Cohen, I., Diegelman, R. and Lindblad, A., W. B. Saunders, Philadelphia, pp. 450–489.
18. Boykin, J. and Molnar, J., 1992. Burn scar and skin equivalents, in *Wound Healing: Biochemical & Clinical Aspects*, ed. Cohen, I., Diegelman, R. and Lindblad, A., W. B. Saunders, Philadelphia, pp. 450–489.
19. Hunt, T. K. and Pai, M. P., 1972. The effect of varying ambient oxygen tensions on wound metabolism and collagen synthesis. *Surg. Gynecol. Obstet.*, 135:561–567.
20. Estridge, T. D., 1991. The use of oxygen, growth factors and implants to effect cellular activity *in vitro*, Ph.D. dissertation, University of Alabama at Birmingham.
21. Estridge, T. and Feldman, D., 1991. The use of oxygen for optimal fibroblast activation. *Trans. FASEB*, 75:7230.
22. Pandit, A., 1991. The effect of oxygen treatment and dressing oxygen permeability on wound healing, M.S. thesis, University of Alabama at Birmingham.
23. Pandit, A. and Feldman, D., 1994. The effect of oxygen permeability on full-thickness skin defects. *Wound Healing & Regeneration*, 2:130–137.
24. Fleichaker, R. and Sinskey, A., 1981. Oxygen demand and supply in culture. *Eur. J. Appl. Microbiol. Biotech.*, 12:93.
25. Lightfoot, E., 1995. The roles of mass transfer in tissue function, in *The Biomedical Engineering Handbook*, ed. Bronzino, J., CRC Press, Hartford, CT, pp. 1656–1670.
26. Rodemann, H. P., Bayreuther, K. and Pfleiderer, G., 1989. The differentiation of normal and transformed human fibroblast *in vitro* is influenced by electromagnetic fields. *Exp. Cell Res.*, 182: 610–621.
27. Greenough, C. G., 1989. The effects of pulsed electromagnetic fields on blood vessel growth in the rabbit ear chamber. *J. Orth. Res.*, 10:256–262.
28. Andino, R., 1991. The use of a low frequency PEMF in the treatment of full thickness skin defects in the rabbit model, M.S. thesis, University of Alabama at Birmingham.

29. Andino, R. and Feldman, D., 1991. Pulsating electromagnetic fields used to treat full thickness defects in the rabbit model. *Trans. FASEB*, 75:4378, 1991.
30. Kelpke, S. and Feldman, D., 1994. Pulsed electromagnetic field to accelerate wound healing, M.S. thesis, University of Alabama at Birmingham.
31. Kelpke, A. and Feldman, D., 1993. Polyurethane dressing in combination with pulsed electromagnetic fields to accelerate wound healing. *Trans. Soc. Biomater.*, 19:56.
32. Lauffenburger, D., 1995. Cell engineering, in *The Biomedical Engineering Handbook*, ed. Bronzino, J., CRC Press, Hartford, CT, pp. 1536–1543.
33. Long, M., 1995. Tissue microenvironments, in *The Biomedical Engineering Handbook*, ed. Bronzino, J., CRC Press, Hartford, CT, pp. 1692–1709.
34. Dijke, P. T. and Iwata, K., 1989. Growth factors for wound healing. *Biotechnology*, 7:793–798.
35. Ksander, G. A., 1989. Exogenous growth factors in dermal wound healing. *Annual Reports in Medicinal Chemistry*, pp. 223–230.
36. Shing, Y., Folkman, J., Haudenschild, C., Lund, ?., Crum, R. and Klagsburn, M., 1985. Angiogenesis is stimulated by a tumor-derived endothelial cell growth factor. *J. Cell Biochem.*, 29:275.
37. Davidson, J., Klagsburn, M. and Hill, K., et al., 1985. Accelerated wound repair, cell proliferation, and collagen accumulation are produced by a cartilage derived growth factor. *J. Cell Biol.*, 100:1219.
38. Dubert, L., Brunner-Ferber, F., Misiti, J., Thomas, K. and Dubertet, M., 1991. Activities of human a-FGF in an in vitro dermal equivalent model. *J. Invest. Derm.*, 97(5):793–798.
39. Mellen, T., Mennie, R. and Cashen, D., et al., 1992. Acidic fibroblast growth factor accelerates dermal wound healing. *Growth Factors*, 7:1–4.
40. Bjornsson, T., Dryjski, M. and Tluczek, J., et al., 1991. Acidic FGF promotes vascular repair. *Proc. Natl. Acad. Sci. USA*, 88:8651–8655.
41. Griesler, H., Cziperie, D. and Kim, D., 1992. Enhanced endothelialization of expanded polytetrafluoroethylene grafts by FGF type 1 pretreatment. *Surgery*, 112(2):24–55.
42. Pandit, A. And Feldman, D., 1993. In vitro optimization of acidic fibroblast growth factor (a-FGF) for wound healing applications. *Ann. Meeting of the Wound Healing Soc. and the European Tissue Repair Soc.*
43. Estridge, T. and Feldman, D., 1991. The design and evaluation of PDGF releasable implants for optimal fibroblast activation. *Trans. Soc. Biomater.*, 17:76.
44. Pandit, A., Ashar, R., and Feldman, D., 1997. The effect of TGF-β delivered through a collagen scaffold in stimulation of healing in full-thickness defects, accepted in *J. Investigative Surgery*, 12:89–100, 1999.
45. Pandit, A., Ashar, R., Feldman, D. and Thompson, J., 1996. The effect of acidic fibroblast growth factor delivered through a collagen scaffold in stimulation of healing in full thickness skin defects, accepted in *Plastic and Reconstructive Surgery*, 101(3):766–775, 1997.
46. Gospodarowicz, D., 1981. Growth control of mammalian cells. Growth factors and extracellular matrix, in *The Biology of Normal Human Growth*, ed. Ritzer, M., Apeia, A., Hall, K., Larson, A., Zeterberg, A., and Zestterstrom, R., Raven Press, New York, pp. 109–134.
47. Gospodarowicz, D., Neufeld, G., Schweigerer, L., 1987. Fibroblast growth factor; structural and biological properties. *J. Cell Physiol.*, 5:15.
48. Gospodarowicz, D., Neufeld, G., Schweigerer, L., 1986. Fibroblast growth factor. *Mol. Cell Endon.*, 46:187.
49. Gospodarowicz, D., Neufeld, G., Schweigerer, L., 1986. Molecular and biological characterization of fibroblast growth factor, an angiogenic factor which also controls the proliferation and differentiation of mesoderm and neuroectoderm derived cells. *Cell Differ.*, 19:1.
50. Gospodarowicz, D. and Greenburg, G., 1985. Biological activity in vivo and in vitro of pituitary and brain fibroblast growth factor. In G. Guroff (Ed.) *Growth and maturation factors* (Vol. 2) (pp. 1–36), New York: John Wiley, 1983.
51. Finesmith, T. H., Broadley, K. N., and Davidson, J. M., 1990. Fibroblasts from wounds of different stages of repair vary in their ability to contract a collagen gel in response to growth factors. *J. Cell Phys.*, 144(1):99–107.

52. Stenberg, B., Phillips, L., Hokanson, J., Heggers, J., and Robson, M., 1991. Effect of b-FGF on the inhibition of contraction caused by bacteria. *J. Surg. Res.*, 50:47.
53. Boyce, S., Foreman, T., and English, K., et al., 1991. Skin wound closure in athymic mice with cultured human cells, biopolymers, and growth factors. *Surgery*, 110:5:866.
54. Pessa, M., Bland, K., and Copeland, E., 1987. Growth factors and determinants of wound repair. *J. Surg. Res.*, 42:207.
55. Folkman, J. and Klagsburn , M., 1987. Angiogenic factors. *Science*, 235:445.
56. Shipley, G., Keeble, W., Hendrickson, J., Coffey, R., and Pittelkow, M., 1989. Growth of normal human keratinocytes and fibroblasts in serum-free medium is stimulated by acidic and basic fibroblast growth factor. *J. Cell Physiol.*, 138:511–519.
57. Brown, G. L., Nanney, L. B., Griffen, J., Cramer, A. B., Yancey, J. M., Curtsinger, L. J. III, Holtzin, L., Schultz, G. S., Jurkiewicz, M. J., and Lynch, J. B., 1989. Enhancement of wound healing by topical treatment with epidermal growth factor. *N. Engl. J. Med.*, 321(2):76–79.
58. Schultz, G., Rotatori, D. S., and Clark, W., 1991. EGF and TGF-alpha in wound healing and repair. *J. Cell. Biochem.*, 45(4): 346–352.
59. Cohen, I. K., Crossland, M. C., Garrett, A., and Diegelmann, R. F., 1995. Topical application of epidermal growth factor onto partial-thickness wounds in human volunteers does not enhance reepithelialization. *Plastic Reconstr. Surg.*, 96(2): 251–254.
60. Mustoe, T. A., Cutler, N. R., Allman, R. M., Goode, P. S. Deuel, T. F., Prause, J. A., Bear, M., Serdar, C. M., and Pierce, G. F., 1994. A phase II study to evaluate recombinant platelet-derived growth factor-BB in the treatment of stage 3 and 4 pressure ulcers. *Arch. Surg.*, 129(2): 213–219.
61. Falanga, V., Eagelstein, W. H., Bucalo, B., Katz, M. H., Harris, B., and Carson, P., 1992. Topical use of human recombinant epidermal growth factor (h-EGF) in venous ulcers. *J. Derm. Surg. Onc.*, 18(7): 604–606.
62. Robson, M., Phillips, L., Heggers, J., McPherson, M., and Heiner, L., 1991. Clinical studies on growth factors in pressure ulcers, in *Clinical and Experimental Approaches to Dermal and Epidermal Repair* (Preliminary report), Wiley Liss, New York, p. 95.
63. Damon, D. H., Lobb, R. R., D'Amore, P. A., and Wagner, J. A., 1989. Heparin potentiates the action of acidic fibroblast growth factor by prolonging its biological half life. *J. Cell. Physiol.*, 138:221–226.
64. Feldman, D. and McConnell, G., 1991. The effect of angiogenic heparin-copper complex on tissue ingrowth strength and maturation. *Trans. Soc. Biomater.*, 17:56.
65. McConnell, K. and Feldman, D., 1988. The effects of a heparin-copper complex on the stability of tissue ingrowth into dacron velour, *Digest of Papers of the Seventh Southern Biomedical Engineering Conference*, ed. D. Moyle, pp. 194–196.
66. Pandit, A., Ashar, R. and Feldman, D., 1993. Acidic fibroblast growth factor and transforming growth factor beta in stimulation of healing in full thickness skin defects. *Ann. Meeting of the Wound Healing Soc. and the European Tissue Repair Soc.*
67. Estridge, T., Feldman, D., Pandit, A. and Andino, R., 1993. The effect of wound matrices on the healing of full thickness defects in the rabbit model. *Trans. 19th Ann. Meeting Soc. Biomaterials*, 19:44.
68. Schoumacher, B., 1990. Antimicrobial vascular catheters, M.S. thesis, University of Alabama at Birmingham.
69. Schoumacher, B., Estridge, T. and Feldman, D., 1988. Antimicrobial vascular catheters. *Proc. Seventh Southern Biomedical Engineering Conf.*, 3.
70. Harris, K., Estridge, T. and Feldman, D., 1988. The effects of chelated silver on the success of percutaneous devices, *Digest of Papers of the Seventh Southern Biomedical Engineering Conference*, ed. D. Moyle, pp. 192–193.
71. Feldman, D., Estridge, T., McConnell, K. and Schoumacher, R., 1988. Stimulating tissue ingrowth into dacron velour with heparin-copper complex. *Trans. Third World Biomaterials Congress.*
72. Ferguson, D., 1991. The effect of fiber diameter and fiber spacing on soft tissue ingrowth. M.S. thesis, University of Alabama at Birmingham.

73. Ferguson, D. and Feldman, D., 1991. The effect of fiber diameter and fiber spacing on the tissue response to dacron. *Trans. Soc. Biomater.*, 17:124.
74. Greisler, H. and Kim D., 1986. Aspects of biodegradable vascularprosthesis, in *Vascular Graft Update*, ed. Kahalic, H., Kantrowitz, A., and Sung, P., ASTM, Fairfield, PA, pp. 197–218.
75. Annis, D., Burnat, A., Edwards, R., Highan, A., Loveday, K. B. and Wilson, J., 1978. An elastomeric vascular prosthesis. *Trans. Am. Soc. Artificial Internal Organs*, 24:209–214.
76. Feldman, D. and Solovay, K., 1989. Development of an antimicrobial percutaneous cuff for long-term vascular catheterization procedure, *Report to Vitaphore Corp.*, NIH PHS/HCFA, 88-1.
77. Feldman, D., 1982. Biological spaces around a percutaneous implant; histological interface study. Ph.D. dissertation, Clemson University.
78. Hong, J., Feldman, D. and Engles, B., 1991. The effect of fiber diameter and carbon coating treatment on the in vitro and in vivo cellular response to dacron fabric materials. *Trans. 17th Annual Meeting Soc. Biomaterials.*
79. Pandit, A. and Feldman, D., 1994. The effect of a porous degradable fibrin scaffold on wound healing. *Trans. Soc. Biomater.*, 20:34.
80. Mooney, D. and Langer, S., 1995. Engineering biomaterials for tissue engineering: the 10–100 micron size scale, in *The Biomedical Engineering Handbook*, ed. Bronzino, J., CRC Press, Hartford, CT, pp. 1609–1617.
81. Feldman, D., Hultman, S., Colaizzo, R. and von Recum, A., 1983. Electron microscopic investigation of soft tissue ingrowth into dacron velour with dogs. *Biomaterials*, 4(2):105–111.
82. Mikos, A., Sarakinos, G., and Lyman, L., 1993. Laminated three-dimensional biodegradable foams for use in tissue engineering. *Biomaterials*, 14(5):323.
83. Mooney, D., Organ, G., Vacanti, J., 1994. Design and fabrication of biodegradable polymer devices to engineer tubular tissues. *Cell Trans.*, 3(2):203.
84. Vacanti, C., Langer, R., and Schloo, B., 1991. Synthetic polymers seeded with chondrocytes provide a template for new cartilage formation. *Plast. Reconstr. Surg.*, 88(5):753.
85. Wang, M., Pins, G., and Silver, F., 1994. Collagen Fibers with improved strength for the repair of soft tissue injuries. *Biomaterials*, 15:507.
86. Cavallaro, J., Kemp, P., and Kraus, K., 1994. Collagen fabrics as biomaterials. *Biotech. Bioeng.*, 43:781.
87. Bell, E., Ehrlich, H., Buttle, D., Nakatsuji, T., 1981. Living tissue formed in vitro and accepted as skin-equivalent tissue of full thickness. *Science*, 211:1052.
88. Chvapil, M., 1979. Industrial uses of collagen, in *Fibrous Proteins: Scientific, Industrial, and Medical Aspects*, ed. Parry, D. and Creamer, L., Academic Press, London.
89. Stenzel, K., Miyata, T., and Rubin, A., 1974. Collagen as a biomaterial. *Ann. Rev. Biophys. Bioeng.*, 3:231.
90. Hansbrough, J., Cooper, M., and Cohen, R., 1992. Evaluation of a biodegradable matrix containing cultured human fibroblasts as a dermal replacement beneath meshed skin grafts on athymic mice. *Surgery*, 111(4):438.
91. Feldman, D., Czuwala, P., Kelpke, S., Pandit, A., and Wilson, D., 1995. A biocompatibility hierarchy: justification for biomaterial enhanced regeneration, in *Encyclopedic Handbook of Biomaterials and Bioengineering*, Marcel Dekker, New York, pp. 223–268.
92. Yannas, I. V. and Burke, J. F., 1980. Design of an artificial skin. I. Basic design priniciples. *J. Biomed. Mater. Res.*, 14:65–81.
93. Yannas, I. V., Burke, J. F., Warpehoski, M., Stasikelis, P., Skrabut, E. M., Orgill, D. and Giard, D. J., 1981. Pompt, Long-term functional replacement of skin. *Trans. Am. Soc. Artificial Internal Organs*, 27: 19–23.
94. Burke, J. F., Yannas, I. V., Quinby, W. C., Bondoc, C. C. and Jung, W. K., 1981. Successful use of physiologically acceptable artificial skin in the treatment of extensive burn injury. *Ann. Surg.*, 194: 413–428.
95. Yannas, I. V., Burke, J. F., Orgill, D. P. and Skrabut, E. M., 1982. Wound tissue can utilize a polymeric template to synthesize a functional extension to skin. *Science*, 215: 174–176.

96. Burke, J. F., 1983. Observations on the development of an artificial skin. Presidential address, 1982 American Burn Association Meeting, *J. Trauma*, 23: 543–551.
97. Guyton, A. C., 1986. Circulation in the skin, in *Textbook of Medical Physiology*, W. B. Saunders, Philadelphia, pp. 343–345.
98. Rees, J. M., and Dimick, A. R., 1992. The cost of burn care and the federal governments response in the 1990s. *Clin. Plast. Surg.*, 19(3): 561–568.
99. Salisbury, R., 1983. Synthetic skin substitutes–1980. *Biomaterials in Reconstructive Surgery*, C. V. Mosby, St. Louis, pp. 1–26.
100. American Burn Association, 1990. Hospital and prehospital resources for optimal care of patients with burn injury: guidelines for development and operation of burn centers. *J. Burn Care Rehabil.*, 11:97–104.
101. Renz, B. M. and Sherman, R., 1992. The burn unit experience at Grady Memorial Hospital: 844 cases. *J. Burn Care Rehabil.*, 13:426–436.
102. Sanders, S., 1992. Pressure ulcer, part I: prevention strategies. *J. Am. Acad. Nurse Prac.*, 4(2):63–70.
103. Allman, R., Laprade, C., and Noel, L., 1986. Pressure sores among hospitalized patients. *Ann. Intern. Med.*, 105(3):337–342.
104. Curtin, L., 1985. Wound management: care and cost—an overview. *Nur. Manage.*, 15: 22–25.
105. Grunfeld, C., 1991. Diabetic foot ulcers: etiology, treatment, and prevention. *Adv. Internal Med.*, 37: 103–132.
106. Keyser, J. E., 1992. Foot wounds in diabetic patients. A comprehensive approach incorporating the use of topical growth factors. *Diabetic Food Wounds*, 19(4): 98–109.
107. Wharton, G., Milani, J., and Dean, L., 1987. Pressure sore profile: cost of management. *Proc. Am. Spinal Inj. Assoc.*, 11:115–119.
108. Maklebust, J., Mondoux, L., and Sieggreen, S., 1986. Pressure relief characteristics of various support surfaces used in prevention and treatment of pressure ulcers. *J. Enterostom. Ther.*, 13:85–89.
109. Siegler, E. and Lavizzo-Mourey, R., 1991. Management of stage III pressure ulcers in moderately demented nursing home residents. *J. General Internal Med.*, 6(6):507–513.
110. Sanders, J., Goldstein, B., and Leotta, D., 1995. Skin response to mechanical stress: Adaption rather than breakdown—A review of the literature. *J. Rehab. Res. Dev.*, 32(3):214–226.
111. Knauth, A., Gordin, M., McNelis, W. and Baumgart, S., 1989. Semipermeable polyurethane membrane as an artificial skin for the premature neonate. *Pediatrics*, 83: 945–949.
112. Feldman, D. L., 1991. Which dressing for split-thickness skin graft donor sites? *Ann. Plastic Surg.*, 27: 288–291.
113. Feldman, D. L., Rogers, A. and Karpinski, R., 1991. A prospective trial comparing Biobrane^{R}, Duoderm^{R}, and Xerororm for skin graft donor sites. *Surg. Gynecol. Obstet..*, 173: 1–5.
114. Quinn, K. J., Courtney, J. M., Evans, J. H., Gaylor, J. D. S. and Reid, W. H., 1985. Priniciples of burn dressings. *Biomaterials*, 6: 369–377.
115. Nangia, A., Ganbhir, R. and Maibach, H., 1991. Factors influencing the performance of temporary skin substitutes. *Clin. Mater.*, 7: 3–13.
116. Sternberg, I., Sternberg, N., Seelenfreund, M. H. and Levine, M. R., 1987. The use of artificial skin in the prevention of early wound healing. *Ann. Opthalmol*, 19: 127–128.
117. Tredget, E., Shankowsky, H., Groeneveld, A., and Burrell, R., 1998. A matched-pair, randomized study evaluating the efficacy and safety of Acticoat^{R} silver coated dressing for the treatment of burn wounds. *J. Burn Care Rehab*, 19(6):531–537.
118. Gentzkow, G., Iwaski, S., Hershon, K., et al., 1996. Use of dermagraft, a cultured human dermis, to treat diabetic foot ulcers. *Diabetes Care*, 19(4):350–354.
119. Tenorio, A., Jindrak, K. and Weiner, M., 1976. Accelerated healing in infected wounds. *Surg. Gynecol. Obstet.* 142:537–543.
120. Kaplan, J., 1984. Acceleration of wound healing by a live yeast cell derivative. *Arch. Surg.*, 119: 1005–1008.
121. Mertz, P., Davis, S., Polarek, J., and Franzens, L., 1993. Effects of a RGD peptide coupled to hyaluronic acid on second degree burn wound healing. *Proc. Joint Meeeting of the Wound Healing Society and the European Tissue Repair Soc.*, p. 27.

122. Himmel, H., 1993. Plastic and reconstructive surgery. *Proc. Symp. Surgical Tissue Adhesives*, 1: 129.
123. DeVries, H., Mekkes, J., Middelkoop, E., Hinrichs, W., Wildevuur, C. and Westerhof, W., 1993. Dermal substitutes for full-thickness wounds in a one-stage grafting model. *Wound Rep. Reg.*, 1: 244–252.
124. Buckley, A., Davidson, J., and Kamerath, C., 1985. Sustained release of epidermal growth factor accelerates wound repair. *Proc. Natl. Acad. Sci. USA*, 82:7340.
125. Reddy, C., Wells, A., and Lauffenburger, D., 1994. Proliferative response of fibroblasts expressing internalization-deficient EGF receptors is altered via differential EGF depletion effects. *Biotech. Prog.*, 10:377.
126. Robson, M., 1997. The role of growth factors in the healing of chronic wounds. *Wound Rep. Reg.*, 5a:12–17.
127. Yannas, I. V., Burke, J. F., Orgill, D. and Skrabut, E. M., 1982. Stage 2 artificial skin: A polymeric template for regeneration of wounded skin. *Trans. Soc. Biomater.*, 8:67.
128. Matsuda, K., Suzuki, S., Issiki, N., Yoshioka, K., Okada, T., Hyon, S. H. and Ikada, Y., 1991. A bilayer "artificial skin" capable of sustained release of an antiobiotic. *Br. J. Plastic Surg.*, 44: 142–146.
129. Feldman, D., Pandit, A., Kelpke, S. and Wilson, D., 1992. *Final Report, Treatment of Pressure Sores, Spain Rehab. Res. Center*, NIDRR/H13B80012.
130. Wilson, D., Feldman, D. and Thompson, T., 1993. Fibrin glue as a matrix for a-FGF delivery in vivo. *Trans. 19th Annual Meeting Soc. Biomaterials.*
131. Sierra, D., 1991. The evaluation of fibrin glue adhesive strength, M.S. thesis, University of Alabama at Birmingham.
132. Saltz, R., Feldman, D., Floyd, D. and Huang, S., 1991. Experimental and clinical applications of fibrin glue. *Plast. Reconstr. Surg.*, 88(6):1005–1015.
133. Sierra, D., Feldman, D. and Saltz, R., 1992. A method to determine the shear adhesive strength of fibrin sealants. *J. Appl. Biomater.*, 3:147–151.
134. Sierra, D., 1993. Fibrin sealant adhesive systems: a review of their chemistry, material properties and clinical applications. *J. Biomater. Appl.*, 7:309–352.
135. Feldman, D. and Sierra, D., 1995. Tissue adhesives in wound healing, in *Encyclopedic Handbook of Biomaterials and Bioengineering*, Marcel Dekker, New York.
136. Ronfard, V., Broly, H., Mitchell, V., et al., 1991. Use of human keratinocytes cultured on fibrin glue in the treatment of burn wounds. *Burns*, 17:181–184.
137. Dahlstrom, K. K., Weis-Fogh, U. S., Medgyesi, S., Rostgaard, J., and Sorenson, H., 19??. The use of autologous fibrin adhesive in skin transplantation. *Plast. Reconstr.*, 89:968–972.
138. Brown, D. M., Barton, B. R., Young, V. L., and Pruitt, B. A., 1992. Decreased wound contraction with fibrin glue-treated skin grafts. *Arch. Surg.*, 125:404–406.
139. Feldman, D., Redden, R., Blum, B., and Osborne, S., 1997. Porous fibrin as a degradable adhesive and drug delivery system. *Trans. Soc. Biomater.*, 23:185.
140. Flahiff, C., 1990. Mechanical testing of fibrin adhesives for blood vessel anastomosis. M.S. thesis, University of Alabama at Birmingham.
141. Flahiff, C., Feldman, D., Saltz, R. and Huang, S., 1992. Mechanical testing of fibrin adhesives for blood vessel anastomosis. *J. Biomater. Mat. Res.*, 26:481–491.
142. Featherstone, C., 1997. Fibrin sealants for haemostasis and drug delivery. *Lancet*, 349: 334.
143. Huang, S., Kilpadi, D., and Feldman, D., 1997. A comparison of the shear strength of fibrin and albumin glues. *Trans. Wound Healing Soc.*, 7: 63.
144. Slimane, S. B., Guidoin, R., Mourad, W., Hebert, J., King, M. W., and Sigot-Luizard, M.-F., 1988. Polyester arterial grafts impregnated with cross-linked albumin: The rate of degradation of the coating *in vivo*. *Eur. J. Surg. Res.*, 20:12–17.
145. Slimane, S. B., Guidoin, R. Marceau, D., Merhi, Y., King, M. W., and Sigot-Luizard, M.-F., 1988. Characteristics of polyester arterial grafts coated with albumin: The role and importance of the cross-linking chemicals. *Eur. J. Surg. Res.*, 20:18–28.
146. Slimane, S. B., Guidoin, R., Merhi, Y., King, M. W., Domurado, D., and Sigot-Luizard, M.-F.,

1988. *In vivo* evaluation of polyester arterial grafts coated with albumin: The role and importance of cross-linking agents. *Eur. J. Surg. Res.*, 20:66–74.

147. Chafke, N, Gasser, B., Lindner, V., Rouyer, N., Rooke, R., Kretz, J. G., Nicolini, P., and Eisenmann, B., 1996. Albumin as a sealant for a polyester vascular prosthesis: Its impact on the healing sequence in humans. *J. Cardiovasc. Surg.*, 37(5):431–440.
148. An, Y. H., Stuart, G. W., McDowell, S. J., McDaniel, S. E., Kang, Q., and Friedman, R. J., 1996. Prevention of bacterial adherence to implant surfaces with a crosslinked albumin coating *in vivo*. *J. Ortho. Res.*, 14(5):846–849.
149. An, H., Bradley, J., Powers, D. L., and Friedman, R. J., 1996. Preventing prosthetic infection using a crosslinked albumin coating *in vivo*. *Proc. 5th World Biomaterials Conf.*, 256, Toronto, Canada.
150. Weissleder, R., Poss, K., Wilkinson, R., Zhou, C., and Bogdanov, A., Jr, 1995. Quantitation of slow drug release from an implantable and degradable gentamicin conjugate by *in vivo* magnetic resonance imaging. *Antimicrob. Agents Chemother.*, 29: 839–845.
151. Unpublished data.
152. Suzuki, S., Matsuda, K., Isshiki, N., Tamada, Y., Koshioka, K. and Ikada, Y., 1990. Clinical evaluation of a new bilayer "artificial skin" composed of collagen sponge and silicone layer. *Br. J. Plastic Surg.*, 43: 47–54.
153. Michaeli, D. and McPherson, M., 1990. Immunologic study of artificial skin used in the treatment of thermal injuries. *J. Burn Care Rehabil.*, 11: 21–26.
154. Stern, R., McPherson, M. and Longaker, M. T., 1990. Histologic study of artificial skin used in the treatment of full-thickness thermal injury. *J. Burn Care Rehabil.*, 11: 7–13.
155. Tompkins, R. G. and Burke, J. F., 1990. Progress in burn treatment and the use of artificial skin. *World J. Surg.*, 14: 819–824.
156. Matsuda, K., Suzuki, S., Isshiki, N. and Ikada, Y., 1993. Refreeze dried bilayer artificial skin. *Biomaterials*, 14:1030–1035.
157. Matsuda, K., Suzuki, S., Isshiki, N., Yoshioka, K., Okada, T. and Ikada, Y., 1990. Influence of glycosaminoglycans on the collagen sponge component of a bilayer artificial skin. *Biomaterials*, 11: 351–355.
158. Murphy, G., Orgill, D. and Yannas, I. V., 1990. Partial dermal regeneration is induced by biodegradable collagen-glycosaminoglycan grafts. *Lab. Inves.*, 63: 305–313.
159. Doillon, C. J., Whyne, C. F., Brandwein, S. and Silver, F. A., 1986. Collagen-based wound dressings: Control of the pore structure and morphology. *J. Biomed. Mater. Res.*, 20: 121–128.
160. Koide, M., Osaki, J. K., Oyamada, T. K. and Takahashi, A., 1993. A new type of biomaterial for artificial skin: Dehydrothermally cross-linked composites of fibrillar and denatured collagens. *J. Biomed. Mater. Res.*, 27: 79–87.
161. Winter, G. W., 1972. Epidermal regeneration studied in the domestic pig, in *Epidermal Wound Healing*, ed. Maibach, H. I., and Rovee, D. T., Year Book Publishers, Chicago, pp. 71–112.
162. Pandit, A. S., Feldman, D. S., Caulfield, J., and Thompson, J. A., 1998. Stimulation of angiogenesis by FGF-1 delivered through a modified fibrin scaffold. *Growth Factors*, 15:113–123.
163. Pandit, A. S., Feldman, D. S., and Caulfield, J., in press. *In vivo* wound healing response to a modified degradable fibrin scaffold. *J. Biomater. Appl.*, 12:222–236.
164. Pandit, A., Feldman, D., and Thompson, J., 1995. Characterization of the delivery rate of FGF-1 through a porous fibrin scaffold. *Trans. Wound Healing Soc.*, 5:80.
165. Pandit, A., Feldman, D., Listinsky, C., and Thompson, J. A., 1997. The effect on wound healing by a modified fibrin scaffold delivering acidic fibroblast growth factor (FGF-1). *J. Bioactive Compatible Polym.*, 12:99–111.
166. Wilson, D., 1994. Fibrin as a matrix for FGF-1 delivery *in vivo*. M.S. thesis, UAB.
167. Wilson, D., Feldman, D., and Thompson, J., 1994. Acidic fibroblast growth factor delivery through a fibrin matrix without exogenous heparin. *Trans. Soc. Biomater.*, 30:255.
168. Sahni, A., Odrljin, T., and Francis, C., 1997. Binding of basic fibroblast growth factor to fibrinogen and fibrin. *J. Biol. Chem.*, 273(13):7554–7559.
169. Pandit, A., 1998. Fibrin as a matrix for FGF-1 delivery *in vivo*. Ph.D. thesis, UAB.
170. Pandit, A., Feldman, D., and Thompson, J., 1994. *In vivo* dose response of acidic fibroblast growth

factor delivered through a porous fibrin scaffold on wound healing. *Trans. Wound Healing Soc.*, 4:122.
171. France, R., DeBloise, C. and Doillon, C., 1993. Extracellular matrix analogs as carriers for growth factors: *in vitro* fibroblast behavior. *J. Biomed. Mater. Res.*, 27:389–397.
172. Osborne, S., Blum, B., Feldman, D., Kelpke, S., Pandit, A., and Thompson, J., 1996. The effect of acidic fibroblast growth factor (FGF-1) delivered through a porous fibrin scaffold on meshed skin graft healing. *Trans. 5th World Biomaterial Congress*, 5:879.
173. Bowman, J., Blum, B., Wehby, J., and Feldman, D., 1997. Obtaining investigator based clinical approval for FGF-1 in fibrin glue. *Trans. Soc. Biomater.*, 23: 469.
174. Bowman, J., Blum, B., Feldman, D., Kilpadi, D., Roberts, M., and Wehby, J., 1997. Stability of FGF-1. *Trans. Wound Healing Soc.*, 7: 93.
175. von Recum, A. and Park, J., 1981. Permanent percutaneous devices. *CRC Crit. Rev. Bioeng.*, 5(1): 37–77.
176. Lee, H., Ocumpaugh, D., Cupples, L., and Culp, G., 1970. Development of satisfactory long-term percutaneous leads, Final Report, NIH Contract PH43-67-1108.
177. Hall, C., Adams, L. and Ghidoni, J., 1975. Development of skin interfacing cannula. *Trans. Am. Artif. Intern. Organs*, 21:281.
178. Freed, P. S., Wasfie, T., Bar-Lev, A., Higiwara, K., Vemuri, D., Vaughan, D., Bernstam, L., Gray, R., Bernstein, I. and Kantrowitz, A., 1985. Long-term percutaneous access device. *Trans. Am. Soc. Artif. Intern. Organs*, 31:230–234.
179. Winter, B., 1974. Transcutaneous implants: Reactions of the skin implant interface. *J. Biomed. Mater. Res. (Symp. 5)*, 99.
180. Daly, B. D. T., Dasse, K. A., Gould, K. E., Smith, T. J., Bousquet, G. G., Poirier, V. L. and Cleveland, R. J., 1987. A new percutaneous access device for peritoneal dialysis. *Trans. Am. Soc. Artif. Intern. Organs*, 33:664–671.
181. Feldman, D. and Hultman, S., 1985. Characterization of tissue response to subcutaneous and percutaneous dacron velour implants. *Trans. 11th Ann. Meeting Soc. Biomaterials.*
182. Feldman, D. and von Recum, A., 1985. Non-epidermally induced failure modes of percutaneous devices. *Biomaterials*, 6(5): 352–356.
183. Feldman, D., Hultman, S., Colaizzo, R., and von Recum, A., 1983. Electron microscopic investigation of soft tissue ingrowth into dacron velour with dogs. *Biomaterials*, 4(2): 105–111.
184. Bence, J. and Picha, G., 1984. Effect of surface microtexture and implant orientation on percutaneous wound healing in the pig. *Trans. 2nd World Congress on Biomaterials*, 40.
185. Baier, R., Natiella, J., Meyer, A., Carter, A. and Turnbull, T., 1984. Preferred sterilization techniques for biomaterials with diverse surface properties. *Trans. 2nd World Congress on Biomaterials*, 173.
186. Feldman, D., Estridge, T. and Johnston, S., 1987. Surface modifications to enhance ingrowth into dacron velour implants. *AL. Mater. Res. Conf.*
187. Stamm, W. E., 1978. Infections related to medical devices. *Ann. Intern. Med.*, 89 (part 2), 764–769.
188. Collignon, P., Chan, R., and Munro, R., 1987. Rapid diagnosis of intravascular catheter-related sepsis. *Arch. Intern. Med.*, 147, 1609–1612.
189. Dasse, K. A., 1984. Infection of Percutaneous Devices: Prevention, monitoring, and treatment. *J. Biomed. Mater. Res.*, 18: 403–411.
190. Trooskin, S. Z., Donetz, A. P., Harvey, R. A. and Greco, R. S., 1985. Prevention of catheter sepsis by antibiotic bonding. *Surgery*, 97, No. 5: 547–551.
191. Spadaro, J. A., Chase, S. E. and Webster, D. A., 1986. Bacterial inhibition by electrical activation of percutaneous silver implants. *J. Biomed. Mater. Res.*, 20:565–577.
192. Stemberger, A., Glas, A., Ascherl, R., Sorg, K., Geibdorfer, K., Scherer, M. and Blumel, G., 1989. Experimental studies on covering burn injuries by silver impregnated collagen films. *Trans. Soc. Biomater.*, 15:165. (Abstract)
193. Clapper, D. L. and Guire, P. E., 1990. Covalent immobilization of cell adhesion proteins and peptides to promote cell attachment and growth on biomaterials. *Trans. Soc. Biomater.*, 13: 159.
194. Dunkirk, S. G. and Guire, P. E., 1989. Improved biocompatibility of contact lenses, Final Report for NIH/NEI Phase II Grant No. EYO5152-03.

37

Ultrasound for Modulation of Skin Transport Properties

Samir Mitragotri
University of California, Santa Barbara, California

Robert Langer
Massachusetts Institute of Technology, Cambridge, Massachusetts

Joseph Kost
Ben-Gurion University, Beer-Sheva, Israel

I. TRANSDERMAL DRUG DELIVERY

Transdermal drug delivery offers several advantages over traditional drug delivery systems, such as oral delivery and injection, especially in regard to protein delivery [1]. These advantages include:

1. *The complex route of a drug through the stomach, intestine, and liver is avoided:* In the case of oral drug delivery, the drug has to pass through the stomach, intestine, and liver, where it can be metabolized. Specifically, orally taken proteins may be highly susceptible to degradation in the digestive tract. On the other hand, in the case of transdermal drug delivery, the drug diffuses through the skin and is absorbed by the capillary network under the skin. Accordingly, the complex route of the drug through the stomach and the liver, and hence its potential degradation, can be avoided.
2. *Better patient compliance:* Transdermal drug devices are easier to handle and use than injections. In addition, the pain associated with the injection can also be avoided. These characteristics of transdermal drug delivery offer a significant advantage in cases where frequent drug doses are required, for example, in the case of insulin delivery.
3. *Sustained release of the drug can be obtained:* Transdermal drug delivery can provide sustained release of the drugs over a sufficiently long time (up to a week). Hence, it is possible to maintain a steady drug concentration in the blood, a feature that cannot be achieved using oral drug delivery or normal injections. This is especially important in the delivery of proteins which have a short half-life in the blood and need to delivered at a controlled rate in order to achieve a therapeutic effect.

Transdermal drug delivery, however, suffers from the severe limitation that the permeability of the skin is very low. Therefore, it is difficult to deliver drugs across the skin at a therapeutically relevant rate. This, in fact, is the main reason why only a handful of low molecular weight drugs (molecular weight < 400 Da) are clinically administered by this route today.

A possible solution to this problem is to increase the permeability of the skin using physicochemical driving forces, referred to as penetration enhancers, for example, chemical enhancers [2] and electric fields [3–5]. Chemical enhancers are compounds, which, when applied to the skin surface along with the drug, are believed to penetrate the skin and modify the transport pathways through it, thereby inducing higher drug transport. Electric fields can also enhance drug transport through the skin by two ways. First, electric fields can enhance the transport of a charged molecule across the skin by electrophoresis, a phenomenon referred to as iontophoresis [6]. In addition, application of short pulses of high voltage electric fields can also enhance transdermal transport through the formation of short-lived pores. This phenomenon is referred to as electroporation [3].

II. ULTRASOUND AS A PENETRATION ENHANCER

We and others [7–52] have shown that ultrasound can enhance the transdermal transport of a drug. This type of enhancement is termed sonophoresis (or phonophoresis), indicating enhanced transport of a molecule under the influence of ultrasound (a sound wave possessing frequencies above 20 KHz). Ultrasound waves are characterized by two main parameters: frequency and amplitude. Amplitude of ultrasound waves can be represented in terms of peak wave pressure (in Pascals) or in terms of intensity (in W/cm^2). Ultrasound can be applied continuously or in a pulsed manner. In the latter case, an additional parameter, duty cycle, is required to characterize ultrasound application. Duty cycle is the fraction of time for which ultrasound is ON.

Ultrasound is generated using a device referred to as a sonicator. It consists of an electrical signal generator which generates an electrical AC signal at the desired frequency and amplitude. This signal is applied across a piezoelectric crystal (transducer) to generate ultrasound. The thickness of the piezoelectric crystal is selected so that it resonates at the operating frequency. Sonicators operating at various frequencies in the range of 20 kHz to 3 MHz are available commercially and can be used for sonophoresis.

If a sonicator operating at the desired frequency is not available commercially, it is possible to assemble one using commercially available signal generators, amplifiers, and transducers. Such sonicators operating at frequencies of 10 MHz and 16 MHz have been assembled by Bommannan et al. [11, 12]. For a discussion of the relevant methods for making a custom sonicator, see Refs. 11 and 12.

For sonophoretic delivery, the desired drug is dissolved in a solvent and applied on the skin. Ultrasound is applied by contacting the transducer with the skin through a coupling medium to ensure a proper contact between the transducer and the skin. This medium can be the same as the solvent used to dissolve the drug or it can be a commercially available ultrasound coupling gel (for example, Aquasonic, Polar, NJ).

III. CLINICAL STUDIES OF SONOPHORESIS

Numerous clinical and nonclinical reports have been published concerning sonophoresis. The clinical technique involves placing the topical preparation on the skin over the area to be treated and massaging the area with an ultrasound probe. Some of the earliest studies done with sonophoresis involve hydrocortisone. Fellinger and Schmidt [53] reported successful treatment of polyarthritis of the hand's digital joints using hydrocortisone ointment with sonophoresis. Newman et al. [42] and Coodley [54] showed improved results of hydrocortisone injection combined with ultrasound "massage" compared to simple hydrocortisone injection for a bursitis treatment.

In addition to joint disease and bursitis, sonophoresis has been tested for its ability to aid penetration in a variety of drug-ultrasound combinations, mainly for localized conditions. The major medications used include the anti-inflammatories, e.g., cortisol, dexamethasone, salicylates, and local anesthetics. Cameroy [55] reports success using carbocaine sonophoresis before closed reduction of Colle's fractures. Griffin et al. [16–20] treated 102 patients with diagnoses of elbow epicondylitis, bicipital tendonitis, shoulder osteoarthritis, shoulder bursitis and knee osteoarthritis with hydrocortisone and ultrasound. Sixty-eight percent of patients receiving drug in conjunction with ultrasound were rated as "improved" demonstrating a pain-free normal functional range of motion, while only 28% of patients receiving placebo with ultrasound were rated as "improved." Similar effects were presented by Moll [56], who published a double blind study with three groups of patients receiving either lidocaine/decadron with ultrasound, a placebo with ultrasound or a placebo with ultrasound at zero intensity. Percentage improvements for the three groups were 88.1%, 56.0% and 23.1% respectively.

McElnay et al. [31] evaluated the influence of ultrasound on the percutaneous absorption of lignocaine from a cream base. Mean data indicated that there was a slightly faster onset time for local anesthesia when ultrasound was administered when compared with control values (no ultrasound). However, the differences were not statistically significant.

Benson et al. [8] reported on the influence of ultrasound on the percutaneous absorption of lignocaine and prilocaine from Emla cream. The local anesthetic cream formulation Emla was chosen because it requires a relatively long contact time with the skin before the application site becomes anesthetized (60 min). Enhanced penetration of anesthetic agent was observed after ultrasound application. Benson et al. [10] demonstrated that ultrasound is also capable of enhancing the percutaneous absorption of methyl and ethyl nicotinate. For the lipophilic hexyl nicotinates no effect of ultrasound on its percutaneous absorption could be detected. The pharmacodynamic parameter of vasodilation caused by nicotinates was used to monitor percutaneous absorption. McEnlay et al. [33] also evaluated the skin penetration enhancement effect of phonophoresis on methyl nicotinate in 10 healthy volunteers in a double-blind, placebo-controlled, crossover clinical trial. Each treatment consisted of the application of ultrasound massage (3.0 MHz, 1.0 W/cm^2 continuous output) or placebo for 5 min to the forearm, followed by a standardized application of methyl nicotinate at intervals of 15 s, 1 min, and 2 min ultrasound massage. Ultrasound treatment applied prior to methyl nicotinate led to enhanced percutaneous absorption of the drug. The authors suggest that the ultrasound affects the skin structure by disordering the structured lipids in the stratum corneum.

Similar experiments were performed by Hofman and Moll [57], who studied the percutaneous absorption of benzyl nicotinate. For recording the reddening of the skin a reflection photometry was employed. Ultrasonic treatment of the skin at levels lower than 1 W/cm^2 reduced the lag time as a function of ultrasound power. Repetition of the ultrasonic treatment confirmed the reversibility of the changes in skin permeability.

Kleinkort and Wood [25] compared sonophoretic effects of a 1% cortisol mixture to that of a 10% mixture. Although an improvement of approximately 80% of the patients receiving 1% cortisol was demonstrated, the group treated with the 10% mixture showed improvement in 95.7% of the patients, while treatment of 16 patients with subdeltoid bursitis showed 100% improvement. In all groups, the 10% compound was more effective.

Williams [51] developed an electrical sensory perception threshold technique for use with human volunteers in order to evaluate the effect of phonophoresis on three commonly available topical anesthetic preparations. Low intensities (0.25 W/cm^2) of 1.1 MHz ultrasound had no detectable effects upon the rate of penetration of either one of the three anesthetic preparations through human skin under conditions where temperature increases had been minimized.

IV. NONCLINICAL STUDIES OF SONOPHORESIS

A. Therapeutic Frequency Sonophoresis

The therapeutic ultrasound conditions correspond to a frequency from 1 to 3 MHz and an intensity in the range of 0–2 W/cm^2. Levy et al. [58] showed that 3–5 min of ultrasound exposure (1 MHz, 1.5 W/cm^2) increased transdermal permeation of mannitol and physostigmine across hairless rat skin in vivo by up to 15-fold. They also reported that the lag time typically associated with transdermal drug delivery was nearly completely eliminated after exposure to ultrasound [58]. Several molecules have been used in sonophoresis studies performed under therapeutic conditions, although the enhancement varies from molecule to molecule. Mitragotri et al. [35] presented a hypothesis for this variation of sonophoretic enhancement from molecule to molecule based on their mechanistic conclusion that ultrasound induces disorganization of the SC lipid bilayers, thus increasing drug diffusivity and hence permeability of the SC. This mechanism suggests that drugs which possess low passive diffusion coefficients through the SC bilayers [35] should be more significantly enhanced by ultrasound application compared to those which diffuse through SC relatively fast. A mathematical equation was developed [39] to predict a priori whether application of therapeutic ultrasound under a typical condition (i.e., 1 MHz, 2 W/cm^2) will enhance transdermal transport of a given drug:

$$E \approx \frac{K_{o/w}^{0.75}}{4 \times 10^4 P}$$

where E is sonophoretic enhancement of skin permeability (dimensionless), $K_{o/w}$ is the octanol-water partition coefficient, and P is the passive skin permeability (cm/h). A list of $K_{o/w}$ and P values for various drugs may be found in Refs. 59 and 60 respectively.

Although several attempts have been made to enhance transdermal drug transport using therapeutic ultrasound, a typical enhancement induced by therapeutic ultrasound is about 10-fold or smaller. This enhancement, may be sufficient for local delivery of certain drugs such as hydrocortisone, but may not be sufficient for the systemic delivery of most drugs. Accordingly, despite of significant attention dedicated to sonophoresis, there have been no commercially available sonophoresis system for systemic drug delivery.

B. Low-Frequency Sonophoresis

This region corresponds to a frequency less than 1 MHz. Tachibana and Tachibana [46–48] reported that application of low-frequency ultrasound (48 kHz) enhances transdermal transport of lidocaine and insulin across hairless rat skin in vivo. They found that the blood glucose level of a hairless rat immersed in a beaker filled with insulin solution (20 U/ml) and placed in an ultrasound bath (48 kHz, 5000 Pa) decreased by 50% in 240 min [46]. They also showed that application of ultrasound under similar conditions prolongs the anesthetic effect of transdermally administrated lidocaine in hairless rats [48] and enhances transdermal transport of insulin in rabbits. Mitragotri et al. [36, 38] have shown that application of ultrasound at even lower frequencies (20 kHz) enhances transdermal transport of various low molecular weight drugs including corticosterone, and salicylic acid as well as high-molecular weight proteins such as insulin, γ-interferon, and erythropoeitin across the human skin in vitro [36].

Transdermal transport enhancement induced by low-frequency ultrasound has been found to be much more significant than that induced by therapeutic ultrasound. For example, although application of therapeutic ultrasound has no effect on transdermal permeation of lidocaine, low-frequency ultrasound has been shown to significantly enhance lidocaine transport across hairless rat skin in vivo. Quantitatively, Mitragotri et al. [38] compared the enhancement ratios (ratio of the sonophoretic and passive permeabilities measured in vitro across human cadaver skin) induced by *therapeutic ultrasound* (1 MHz, 2 W/cm^2, continuous) and *low-frequency ultrasound* (20 kHz, 125 mW/cm^2, 100 ms pulses applied every second) in the case of four permeants: butanol, corticosterone, salicylic acid, and sucrose. They found that the enhancement induced by low-frequency ultrasound is up to 1000-fold higher than that induced by therapeutic ultrasound [38].

Low-frequency ultrasound has been found to enhance transdermal transport drugs which do not permeate skin passively, for example, large molecular weight proteins. Application of very low frequency ultrasound (20 kHz, 125 mW/cm^2, 100 ms pulses applied every second) has been shown to enhance transdermal transport of proteins including insulin, γ-interferon, and erythropoeitin across human cadaver skin in vitro [36]. It has also been shown that application of ultrasound under conditions the same as those mentioned above delivers therapeutic doses of insulin across hairless rat skin in vivo from a chamber glued on the rats back and filled with an insulin solution (100 U/ml) [36]. Insulin-ultrasound treatment (20 kHz, 225 mW/cm^2, 100 ms pulses applied every second) reduces the blood glucose level of diabetic hairless rats from about 400 to 200 mg/dl (the blood glucose level of normal rats) in 30 min. A corresponding change in the plasma insulin levels was observed during sonophoresis. Normal hairless rats possessed a plasma insulin level of 101 ± 31 picomolar while diabetic hairless rats possessed a value below our detection limit (34 picomolar). During sonophoresis, the levels of transdermally delivered human insulin in rat plasma reached a value of 77 (±28) picomolar after 30 min and a value of 178 (±84) picomolar after 1 h [36]. No significant change in the plasma concentration of indigenous rat insulin was observed during sonophoresis. These results indicate that very low frequency ultrasound delivers therapeutic doses of insulin across hairless rat skin in vivo.

C. High-Frequency Sonophoresis

This region of ultrasound corresponds to a frequency higher than 3 MHz. Bommannan et al. [11, 12] hypothesized that since the absorption coefficient of the skin varies directly with the ultrasound frequency, high-frequency ultrasound energy would concentrate more in the epidermis, thus leading to higher enhancements. In order to assess this hypothesis, they studied the effect of high-frequency ultrasound (2–15 MHz) on the permeability of salicylic acid (dissolved in a gel) through hairless guinea pig skin in vivo. They found that a 20 min application of ultrasound (0.2 W/cm^2) at a frequency of 2 MHz did not significantly enhance amount of salicylic acid penetrating the skin. However, 10 MHz ultrasound under otherwise identical conditions resulted in about a fourfold increase and 16 MHz ultrasound resulted in about a 2.5-fold increase in transdermal salicylic acid transport [11, 12]. They also investigated the effect of shorter (5 min) ultrasound exposures under similar conditions on transdermal salicylic acid transport and found that, while 10 MHz ultrasound enhances transdermal transport by 1.6-fold, 16 MHz enhances it by about 1.8-fold. Application of high-frequency ultrasound was also found to reduce the lag time associated with transdermal transport. For example, Bommanan et al. [11] found that transdermally delivered salicylic acid appears much sooner in the urine if driven by sonophoresis than by passive permeation. These researchers also found that an electron dense tracer, such as lanthanum, was driven deep into the dermis by a 5 min application of high-frequency ultrasound in hairless mouse in vivo.

V. MECHANISMS OF SONOPHORESIS

In order to understand the mechanisms of sonophoresis, it is important to identify various effects of ultrasound exposure on the human tissue, since one or more of these effects may contribute to the mechanism of sonophoresis. A brief description of the various biological effects of ultrasound is provided below.

a. *Thermal effects*. Absorption of ultrasound results in a temperature increase of the medium. Materials which posses higher ultrasound absorption coefficients, such as bones, experience severe thermal effects as compared to muscle tissues which have a lower absorption coefficient (α). The absorption coefficient of a medium increases proportionally with the ultrasound frequency, indicating that the thermal effects of ultrasound are proportional to the ultrasound frequency. The increase in the temperature of a medium upon ultrasound exposure at a given frequency varies proportionally with the ultrasound intensity and exposure time. The thermal effects can be substantially decreased by pulsed application. For a detailed discussion of the thermal effects of ultrasound, see Ref. 61.

b. *Acoustic streaming*. Acoustic streaming, by definition, is the development of time independent large fluid velocities in a medium under the influence of an ultrasound wave. The primary causes of acoustic streaming are the reflections and other distortions of the wave propagation. Oscillations of cavitation bubbles may also contribute to acoustic streaming. The shear stresses developed by streaming velocities may affect the neighboring structures [62].

c. *Cavitational effects*. Cavitation is the formation of gaseous cavities in a medium upon ultrasound exposure. The primary cause of cavitation is the ultrasound-induced pressure variations in the medium. Cavitation involves either rapid growth and collapse of a bubble (transient cavitation) or slow oscillatory motion of a bubble in ultrasound field (stable cavitation). Cavitation affects tissues in several ways. Specifically, collapse of cavitation bubbles releases a shock wave which can cause structural alterations in its surroundings. Biological tissues contain numerous air pockets trapped in the fibrous structures which act as nuclei for cavitation upon ultrasound exposure. Accordingly, a significant cavitation activity is known to occur in biological tissues upon ultrasound exposure [62]. The cavitational effects vary inversely with ultrasound frequency and directly with ultrasound intensity.

Significant attention has been devoted to understanding which of the above-mentioned phenomena plays an important role in sonophoresis. Below, we summarize the conclusions of the mechanistic investigations of sonophoresis performed in each of the three ultrasound frequency regions.

A. Mechanism of Therapeutic Sonophoresis

Mortimer et al. [41] performed sonophoresis of oxygen across frog skin in vitro. They found that the sonophoretic enhancement of transdermal oxygen transport depends on ultrasound intensity rather than pressure amplitude. Based on this observation, they hypothesized that cavitation cannot be responsible for sonophoresis. They hypothesized that the observed enhancement occurs due to acoustic streaming in the solution around the skin. Levy et al. [58] performed an in vitro investigation of the roles played by thermal effects, cavitation, and mixing in sonophoretic enhancement of urea transport across polymer membranes. They found that the observed enhancement cannot be explained by the thermal effects or mixing. In an attempt to elucidate the role played by cavitation, they performed sonophoresis experiments using degassed solutions. Since degassing a solution decreases the cavitation activity in the solution, they hypothesized that if a decrease in the sonophoretic enhancement is observed upon degassing, it would indicate the importance of cavitation. Indeed, they found that degassing procedure reduced the sonophoretic enhancement of urea permeation by twofold suggesting that cavitation may play a role in sonophoresis.

Cavitation occurs in a variety of mammalian tissues, including muscle, abdominal tissues, brain, cardiovascular tissues, and liver upon exposure to ultrasound at a variety of conditions [62]. As explained earlier, the occurrence of cavitation in biological tissues is attributed to the existence of a large number of gas nuclei. These nuclei are gas pockets trapped in either intracellular or intercellular structures. Simmonin et al. [63] hypothesized that cavitation occurs in the follicles of the skin upon ultrasound exposure and enhances transdermal permeation by convective velocities through follicles. However, no evidence was presented to support this hypothesis. Mitragotri et al. [35] presented results of the experiments indicating that cavitation inside the skin plays an important role in sonophoresis performed using therapeutic ultrasound.

In the first set of experiments, the known effect of static pressure on cavitation was utilized. It is known that cavitation in fluids and porous media can be suppressed at high pressures. This effect is believed to occur due to the dissolution or collapse of the gaseous nuclei under the influence of pressure. Sonophoresis experiments were performed using skin compressed at 30 atm (between two smooth glass plates soaked in water placed in a compression press for 2 h prior to sonophoresis experiments). They found that while application of ultrasound (1 MHz, 2 W/cm^2, continuous) enhances estradiol permeability of the normal human epidermis by 13-fold, the corresponding enhancement for compressed skin is only about 1.75-fold.

In the second set of experiments, the heat-stripped human cadaver skin was degassed (under a pressure of 0.05 mm of Hg) prior to the permeability experiments. The authors hypothesized that when a skin piece soaked in buffer is subjected to high vacuum, the resulting low pressures should reduce the dissolved gas concentration in the buffer, thereby forcing small gaseous nuclei in the skin to dissolve. When the degassed skin was exposed to ultrasound, once again the effect of ultrasound on the estradiol permeability was minimal (1.5-fold), compared to 13-fold across the normal skin. Based on these two results, the authors concluded that cavitation inside the skin plays a major role in enhancing transdermal transport upon therapeutic ultrasound exposure. They provided the following hypothesis for the mechanism of sonophoresis performed using therapeutic ultrasound.

Ultrasound exposure in the therapeutic range causes cavitation in the keratinocytes of the stratum corneum. Oscillations of the ultrasound-induced cavitation bubbles near the keratinocyte-lipid bilayer interfaces may, in turn, cause oscillations in the lipid bilayers, thereby causing structural disorder of the SC lipids. Shock waves generated by the collapse of cavitation bubbles at the interfaces may also contribute to the structure-disordering effect.

Since diffusion of permeants through a disordered bilayer phase can be significantly higher than that through a normal bilayer, transdermal transport in the presence of ultrasound is expected to be higher than passive transport.

B. Mechanism of Low-Frequency Sonophoresis

Since it is known that cavitational effects in fluids vary inversely with ultrasound frequency [64], it is likely that cavitational effects should play an even more important role in low-frequency sonophoresis. Tachibana et al. [46] hypothesized that application of low-frequency ultrasound results in acoustic streaming in the hair follicles and sweat ducts of the skin, thus leading to enhanced transdermal transport. Mitragotri et al. [38] hypothesized that transdermal transport during low-frequency sonophoresis occurs across the keratinocytes rather than hair follicles. They provided the following hypothesis for the higher efficacy of low-frequency sonophoresis.

Cavitation induced by low-frequency ultrasound may cause disordering of the SC lipids. In addition, oscillations of cavitation bubbles may result in significant water penetration into the disordered lipid regions. This may cause the formation of aqueous channels through the intercellular lipids of the SC through which permeants may transport. The occurrence of transdermal transport through aqueous channels across the disordered lipid regions may enhance transdermal transport as compared to passive transport because (1) the diffusion coefficients of permeants through water, which is likely to primarily occupy the channels generated by ultrasound, are up to 1000-fold higher than those through the ordered lipid bilayers [38], and (2) the transport path length of these aqueous channels may be much shorter (by a factor up to 25) than that through the tortuous intercellular lipids in the case of passive transport.

This hypothesis also explains why low-frequency ultrasound can induce transdermal transport of drugs which exhibit very low passive transport. Drugs possessing low passive permeabilities are either (1) hydrophilic, which makes their partitioning into the SC bilayers difficult, or (2) large in molecular size (for example, proteins), which reduces their diffusion coefficients in the SC. Low-frequency ultrasound may overcome both of these limitations by providing aqueous transport channels across the skin. Since these channels are filled with saline, hydrophilic drugs can easily partition into the SC. In addition, diffusion of drugs through water is much faster than that through ordered lipid bilayer regions, thus allowing drugs to transport across the skin at a faster rate. Therefore, molecules such as hydrophilic drugs or proteins, may permeate skin with relative ease in the presence of low-frequency ultrasound.

C. Mechanism of High-Frequency Sonophoresis

Bommanan et al. [11] performed sonophoresis of lanthanum tracers across hairless mice skin at an ultrasound frequency of 16 MHz in order to understand the transport pathways during high-frequency sonophoresis. They observed the skin under the electron microscope after sonophoresis and found that 5 min of sonophoresis results in penetration of lanthanum tracers to dermal levels of the skin. They further reported that the tracer was patchily distributed within the intercellular lipid bilayers of the SC. They provided the following hypothesis for the mechanism of high-frequency sonophoresis. The micronuclei (air pockets) present in the SC oscillate in response to oscillating pressure fields of ultrasound and eventually collapse. The oscillations of these bubbles result in enhanced skin permeation. They also hypothesized that the patchy distribution of the lanthanum tracer revealed in the micrographs corresponds to the location of oscillating air pockets in the SC. In a later report, Menon et al. [34] presented additional microscopic studies of the hairless mice skin after undergoing sonophoresis of lanthanum tracer. They reported the presence of long confluent channels in the intercellular lipids filled with lanthanum tracers in the hairless rat skin exposed to ultrasound. They presented the following hypothesis for the mechanism of sonophoresis: application of ultrasound opens and expands gas-filled cavities in the SC much like pumping air through a collapsed rubber tubing. Enhanced transport of drugs may than occur through these confluent channels across the SC.

VI. CONCLUSIONS

Application of ultrasound enhances transdermal drug transport, a phenomenon referred to as sonophoresis. Proper choice of ultrasound parameters, including frequency, intensity, pulse length, and distance of transducer from the skin is critical for efficient sonophoresis. Numerous attempts of sonophoresis have been performed over the last 40 years. These attempts can be classified into three categories, i.e., *therapeutic frequency*, *high-frequency*, and *low-frequency ultrasound*, among which, therapeutic frequencies represent the most commonly used ultrasound condition for sonophoresis. Recently, attention has been more focused on low-frequency ultrasound. Mechanistic experiments performed by several investigators suggest that cavitation plays a major role in sonophoresis. It has been suggested that cavitation disorganizes the lipid bilayers of the skin through which enhanced transport of drugs may occur. Various studies have indicated that application of ultrasound under conditions used for sonophoresis does not induce any permanent damage to the skin or underlying tissues, although more work is required before arriving at definite conclusions regarding the safety of ultrasound exposure.

REFERENCES

1. Bronaugh, R. L., Maibach, H. I. (Ed), *Percutaneous Absorption: Mechanisms–Methodology–Drug Delivery*, Marcel Dekker, pp. 1–12, 1989.
2. Walters, K. A., Hadgraft, J., *Pharmaceutical Skin Penetration Enhancement*, Marcel Dekker, New York, 1993.
3. Prausnitz, M. R., Bose, V., Langer, R., Weaver, J. C., Electroporation of Mammalian Skin: A Mechanism to Enhance Transdermal Drug Delivery, *Proc. Natl. Acad. Sci. USA*, 90, 10504–10508, 1993.
4. Prausnitz, M. R., Edlemean, E. R., Gimm, J. A., Langer, R., Weaver, J. C., Transdermal Delivery of Heparin by Skin Electroporation, *Biotechnology*, Nov. 12 (11), 1205–1209, 1995.
5. Bommanon, D., Tamada, J., Leung, L., Potts, R., Effects of Electroporation on Transdermal Iontophoretic Delivery of Leutinizing Horome Releasing Hormone, *Pharm. Res.*, 11, 1809–1814, 1994.
6. Green, P. G. , Flalagan, M., Shroot, B., Guy, R. H., Iontophoretic Drug Delivery, K. A. Walters, J. Hadgraft, *Pharmaceutical Skin Penetration Enhancement*, Marcel Dekker, New York, 1993.
7. Antich, T. J., Phonophoresis: The Principles of the Ultrasonic Driving Force and Efficacy in Treatment of Common Orthopedic Diagnoses, *J. Orth. Sports Phys. Ther.*, 4, 2, 99–102, 1982.
8. Benson, H. A. E., McElnay, J. C., Harland, R., Phonophoresis of Lingocaine and Prilocaine from Emla Cream, *Int. J. Pharm.*, 44, 65–69, 1988.
9. Benson, H. A. E., McElnay, J. C., Harland, R., Use of Ultrasound to Enhance Percutaneous Absorption of Benzydamine, *Phys. Ther.*, 69, 2, 113–118, 1989.
10. Benson, H. A. E., McElnay, J. C.., Hadgraft, J., Influence of Ultrasound on the Percutaneous Absorption of Nicotinate Esters, *Pharm. Res.*, 9, 1279–1283, 1991.
11. Bommannan, D., Menon, G. K., Okuyama, H., Elias, P. M., Guy, R. H., Sonophoresis. II. Examination of the Mechanism(s) of Ultrasound-Enhanced Transdermal Drug Delivery, *Pharm. Res.*, 9, 8, 1043–1047, 1992.
12. Bommannan, D., Okuyama, H., Stauffer, P., Guy, R. H., Sonophoresis. I. The Use of High-Frequency Ultrasound to Enhance Transdermal Drug Delivery, *Pharm. Res.*, 9, 4, 559–564, 1992.
13. Byl, N. N., McKenzie, A., Halliday, B., Wong, T., O'Connell, J., The Effects of Phonophoresis with Corticosteroids: A Controlled Pilot Study, *J. Orth. Sports Phys. Ther.*, 18, 5, 590–600, 1993.
14. Davick, J. P., Martin, R. K., Albright, J. P., Distribution and Deposition of Tritiated Cortisol Using Phonophoresis, *Phys. Ther.*, 68, 11, 1672–1675, 1988.
15. Fogler, S., Lund, K., Acoustically Augmented Diffusional Transport, *J. Acous. Soc. Am.*, 53, 1, 59–64, 1973.
16. Griffin, J. E., Touchstone, J., Ultrasonic Movement of Cortisol in to Pig Tissue, *Am. J. Phys. Med.*, 44, 1, 20–25, 1965.
17. Griffin J. E., Physiological Effects of Ultrasonic Energy as It Is Used Clinically, *J. Am. Phys. Ther. Assoc.*, 46, 18–26, 1966.
18. Griffin, J. E., Echternach, J. L., Proce, R. E., Touchstone, J. C., Patients Treated with Ultrasonic Driven Hydrocortisone and with Ultrasound Alone, *Phys. Ther.*, 47, 7, 600–601, 1967.

19. Griffin, J. E., Touchstone, J. C., Low-Intensity Phonophoresis of Cortisol in Swine, *Phys. Ther.*, 48, 12, 1136–1344, 1968.
20. Griffin, J. E., Touchstone, J. C., Effects of Ultrasonic Frequency on Phonophoresis of Cortisol into Swine Tissues, *Am. J. Phys. Med.*, 51, 2, 62–78, 1972.
21. Johnson, M. E., Mitragotri, S., Patel, A., Blankschtein, D., Langer, R., Synergistic Effect of Ultrasound and Chemical Enhancers on Transdermal Drug Delivery, *J. Pharm. Sci.*, 85, 7, 670–679, 1996.
22. Julian, T. N., Zentner, G., Ultrasonically Mediated Solute Permeation through Polymer Barriers, *J. Pharm. Pharmacol.*, 38, 871–877, 1986.
23. Julian, T., Zentner, G., Ultrasonically Mediated Solute Permeation through Polymer Barriers, *J. Pharm. Pharmacol.*, 38, 871–877, 1986.
24. Julian, T. N., Zentener, G. M., Mechanism for Ultrasonically Enhanced Transmembrane Solute Permeation, *J. Control. Rel.*, 12, 77–85, 1990.
25. Kleinkort, J. A., Wood, F., Phonophoresis with 1 Percent versus 10 Percent Hydrocortisone, *Phys. Ther.*, 55, 12, 1320–1324, 1975.
26. Kost, J., Levy, D., Langer, R. *Ultrasound as a Transdermal Enhancer*, R. Bronaugh, Maibach, H. I., *Percutaneous Absorption Mechanisms Methodology Drug Delivery*, Marcel Dekker, New York, 595–601, 1989.
27. Kost, J., Langer, R., *Ultrasound-Mediated Transdermal Drug Delivery*, Maibach, H. I., Shah V. P., Plenum, New York, 91–103, 1993.
28. Kost, J., Ultrasound for Controlled Delivery of Therapeutics, *Clin. Mater.*, 13, 155–161, 1993.
29. Kost, J., Pliqueet, U., Mitragotri, S., Yamamoto, A., Weaver, J., Langer, R., Enhanced Transdermal Delivery: Synergistic Effect of Ultrasound and Electroporation, *Pharm. Res.*, submitted, 1995.
30. Machluf, M., Kost, J., Ultrasonically Enhanced Transdermal Drug Delivery. Experimental Approaches to Elucidate the Mechanism, *J. Biomat. Sci.*, 5, 147–156, 1993.
31. McElnay, J. C., Matthews, M. P., Harland, R., McCafferty, D. F., The Effect of Ultrasound on the Percutaneous Absorption of Lingocaine, *Br. J. Clin. Pharmacol.*, 20, 421–424, 1985.
32. McElnay, J. C., Kennedy, T. A., Harland, R., The Influence of Ultrasound on the Percutaneous Absorption of Fluocinolone Acetonide, *Int. J. Pharm.*, 40, 105–110, 1987.
33. McElnay, J. C., Benson, H. A. E., Harland, R., and Hadgraft, J., Phonophoresis of Methyl Nicotinate: A Preliminary Study to Elucidate the Mechanism, *Pharm. Res.*, 4, 1726–1731, 1993.
34. Menon, G., Bommanon, D., Elias, P., High-Frequency Sonophoresis: Permeation Pathways and Structural Basis for Enhanced Permeability, *Skin Pharmacol.*, 7, 3, 130–139, 1994.
35. Mitragotri, S., Edwards, D., Blankschtein, D., Langer, R., A Mechanistic Study of Ultrasonically Enhanced Transdermal Drug Delivery, J. Pharm. Sci., 84, 6, 697–706, 1995.
36. Mitragotri, S., Blankschtein, D., Langer, R., Ultrasound-Mediated Transdermal Protein Delivery, *Science*, 269, 850–853, 1995.
37. Mitragotri, S., Blankschtein, D., Langer, R., *Sonophoresis: Ultrasound Mediated Transdermal Drug Delivery*, J. Swarbrick, Boylan, J., Marcel Dekker, 1995.
38. Mitragotri, S., Blankschtein, D., Langer, R., Transdermal Drug Delivery Using Low-Frequency Sonophoresis, *Pharm. Res.*, 13, 3, 411–420, 1996.
39. Mitragotri, S., Blankschtein, D., Langer, R., An Explanation for the Variation of the Sonophoretic Transdermal Transport Enhancement from Drug to Drug, *J. Pharm. Sci.*, 86, 10, 1190–1192, 1997.
40. Miyzaki, S., Mizuoka, O., Takada, M., External Control of Drug Release and Penetration: Enhancement of the Transdermal Absorption of Indomethacin by Ultrasound Irradiation, *J. Pharm. Pharmacol.*, 43, 115–116, 1990.
41. Mortimer, A. J., Trollope, B. J., Roy, O. Z., Ultrasound-Enhanced Diffusion through Isolated Frog Skin, *Ultrasonics*, 26, 348–351, 1988.
42. Newman, J. T., Nellermo, M. D., Crnett, J. L., Hydrocortison Phonophoresis: A Literature Review, *J. Am. Pod. Med. Assoc.*, 82, 8, 432–435, 1992.
43. Pottenger, J. F., Karalfa, L. B., Utilization of Hydrocortisone Phonophoresis in United States Army Physical Therapy Clinics, *Milit. Med.*, 154, 7, 355–358, 1989.

44. Pratzel, H., Ditrich, P., Kukovetz, W., Spontaneous and Forced Cutaneous Absorption of Indomethacin in Pigs and Humans, *J. Rheumat.*, 13, 6, 1122–1125, 1986.
45. Skauen, D. M., Zentner, G. M., Phonophoresis, *Int. J. Pharm.*, 20, 235–245, 1984.
46. Tachibana, K., Tachibana, S., Transdermal Delivery of Insulin by Ultrasonic Vibration, *J. Phamr. Pharmacol.*, 43, 270–271, 1991.
47. Tachibana, K., Transdermal Delivery of Insulin to Alloxan-Diabetic Rabbits by Ultrasound Exposure, *Pharm. Res.*, *9*, 7, 952–954, 1992.
48. Tachibana, K., Tachibana, S., Use of Ultrasound to Enhance the Local Anesthetic Effect of Topically Applied Aqueous Lidocaine, *Anestheiology*, 78, 6, 1091–1096, 1993.
49. Tyle, P., Agrawala, P., Drug Delivery by Phonophoresis, *Pharm. Res.*, 6, 5, 355–360, 1989.
50. Kost, J., Levy, D., Langer, R., *Proc. Int. Control. Rel. Bioact. Mater. 13*, *Control. Rel. Soc.*, 1986, 177–178.
51. Williams, A. R. Phonophoresis: An in Vivo Evaluation Using Three Topical Anaesthetic Preparations, *Ultrasonics*, 28 May, 137–141, 1990.
52. Wing, M., Phonophoresis with Hydrocortisone in the Treatment of Temporomandibular Joint Dysfunction, *Phys. Ther.*, 62, 1, 32–33, 1981.
53. Fellinger, K., Schmidt, J., Klinik and Therapies des Chromischen Gelenkreumatismus, *Maudrich Vienna, Austria*, 549–552, 1954.
54. Coodley, G. L., Bursitis and Post-Traumatic Lesions, *Am. Pract.*, 11, 181–187, 1960.
55. Cameroy, B. M., Ultrasound Enhanced Local Anesthesia, *Am. J. Orthoped.*, 8, 47, 1966.
56. Moll, M. A., New Approaches to Pain, *US Armed Forces Med. Serv. Dig.*, 30, 8–11, 1979.
57. Hofman, D., Moll, F., The Effect of Ultrasound on in Vitro Liberation and in Vivo Penetration of Benzyl Nicotinate, *J. Control. Rel.*, 27, 187–192, 1993.
58. Levy, D., Kost, J., Meshulam, Y., Langer, R., Effect of Ultrasound on Transdermal Drug Delivery to Rats and Guinea Pigs, *J. Clin. Invest.*, 83, 2974–2078, 1989.
59. Hanch, C., Leo, A., *Substituent Constants for Correlation Analysis in Chemistry and Biological Sciences*, Wiley, New York, 1979.
60. Flynn, G. L., T. R. Gerrity, Henry, C. J., *Principles of Route-to-Route Extrapolation of Risk Assessment*, Elsevier, New York, 93–127, 1990.
61. Wells, P. N. T., *Biomedical Applications of Ultrasound*, Plenum Press, 1977.
62. Suslick, K. S., *Ultrasound: Its Chemical, Physical and Biological Effects*, VCH Publishers, New York, 1989.
63. Simmonin, P., On the Mechanisms of in Vitro and in Vivo Phonophoresis, *J. Control. Rel.*, 33, 125–141, 1995.
64. Gaertner, W., Frequency Dependence of Acoustic Cavitation, *J. Acoust. Soc. Am.*, 26, 977–80, 1954.

38

Properties and Clinical Applications of Shape Memory Alloys

F. J. Gil and J. A. Planell
Universidad Politécnica de Cataluña, Barcelona, Spain

I. INTRODUCTION

The first steps toward the discovery of the shape memory effect were taken in 1938 by Greninger and Mooradian [1] when they observed that a martensitic phase was formed and disappeared when the temperature was decreased and increased respectively in a Cu-Zn alloy. However, this basic phenomenon of the shape memory effect governed by the thermoelastic behavior of the martensite phase was first extensively studied 10 years later by Kurdjumov and Khandros [2]. The great breakthrough came in the early 1960s, when Buehler [3] at the U.S. Naval Ordnance Laboratory discovered the shape memory effect in an equiatomic alloy of nickel and titanium. This alloy is known as nitinol (Nickel Titanium Naval Ordnance Laboratory) and the use of NiTi as a biomaterial is steadily increasing because of its properties of superelasticity and shape memory effect. NiTi alloy combines such properties with an excellent corrosion and wear resistance, mechanical properties and a good biocompatibility. These properties make it an adequate biomaterial, especially for orthopedic surgery and orthodontics [4].

The shape memory and superelastic effects are related to a thermoelastic martensitic transformation (Parent phase or austenite → Martensite) which can be produced by cooling or stressing. Martensite formation can be initiated by cooling the material below M_s, defined as the temperature at which the martensitic transformation begins. M_f is the temperature at which martensitic transformation finishes. The transformation is reversible, A_s being the temperature at which the reverse austenitic transformation (Martensite → Parent phase) starts upon heating, and A_f the temperature at the end of the reverse austenitic transformation.

When a stress is applied to the parent phase material above A_f, a mechanically elastic martensite is stress induced in alloys which exhibit thermoelastic behavior. That is, the deformed material reverts to its original shape when the stress is released. This effect is called pseudoelasticity or superelasticity [5–7]. Figure 1 shows the stress-strain curve for a superelastic alloy. The stress necessary to induce the formation of stress induced martensite (SIM) is a linear function of temperature. The critical stress $\sigma^{P \rightarrow SIM}$ increases with increasing temperature while the yield stress of the parent phase decreases with increasing temperature. Hence, above a certain temperature plastic deformation of the austenite occurs before stress induced martensite can be formed [6–8]. The size of the hysteresis (i.e., the difference between

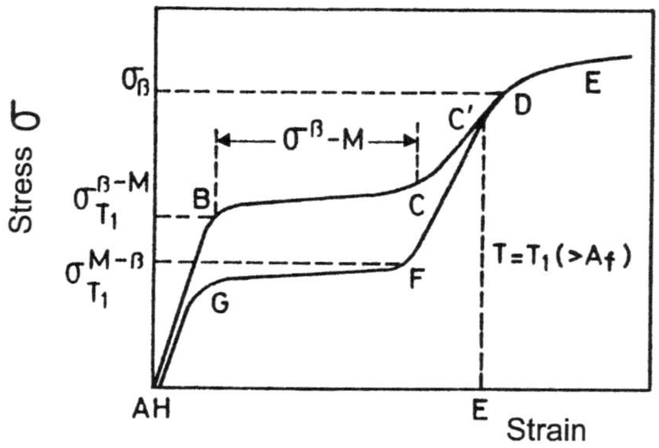

Figure 1 Stress-strain curve for a superelastic material.

$\sigma^{P} \rightarrow {}^{SIM}$ and $\sigma^{SIM} \rightarrow {}^{P}$) reflects the amount of irreversible process taking place during the formation of stress induced martensite [8–10].

Superelasticity occurs when a stress induced martensitic transformation can be produced in a range of temperatures in which the SIM becomes unstable when unloaded. In this case the transformation strain obtained during loading is recovered on unloading when the SIM reverts to the parent phase. We may identify a yield plateau along which the deformation grows without any significant increase of the load and a recovery plateau along which the deformation decreases at an essentially constant load. This effect allows for the application of constant forces to the dentition or bone over a long activation period, which results in a desirable biological response.

When the applied stress is released below A_s, the produced shape change persists since the material is martensitic and the reverse rearrangements of twins and martensite variants has not occurred. However, upon heating through the A_s to A_f temperature range the material regains its original shape by a reverse transformation from martensite to austenite. Thus, the original shape of the parent phase is obtained and the "deformation" of the martensitic phase is recovered; this is known as the shape memory effect (Fig. 2). It should be noticed that when the alloy is strained above a certain critical stress, the martensite is plastically deformed by slip processes and this plastic deformation is not recoverable by subsequent heating.

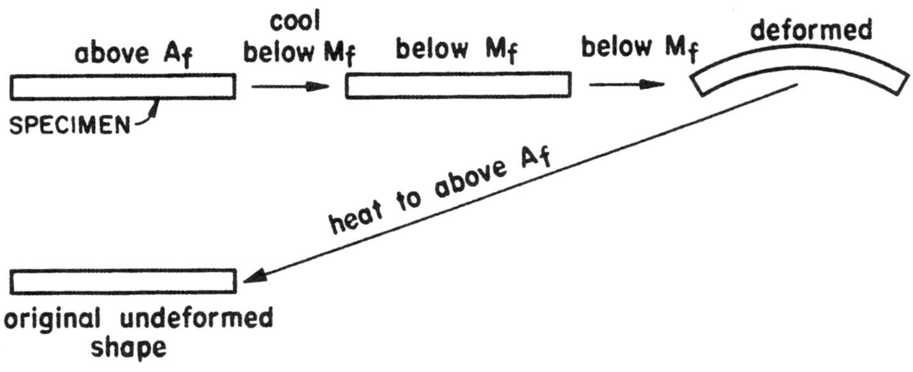

Figure 2 Shape memory effect [2].

II. BIOCOMPATIBILITY OF NiTi

There are few reports of NiTi biocompatibility due to the fact that the NiTi alloy is a relatively new material in the medical field. Cutright et al. [10] implanted wires of NiTi (equiatomic) into the subcutaneous layers of skin of 45 rats for up to nine weeks. The results after three days showed that healing at the wound site was beginning, Edema was prominent and loose fibrous connective tissue stroma had formed around the implant. The connective tissue stroma had become dense after one week and the proportion of collagenous tissue to fibrous connective tissue had increased in proportion to observation of three days. Swelling around the implantation area has almost disappeared. After three weeks, the dense fibrous connective tissue with large amounts of collagenous fibers had formed around the NiTi implant. From four to nine weeks, surrounding tissue grafts were stabilized and fibrous connective tissue became dense, and only a small number of chemically inflamed cells remained. Consequently, NiTi seems indistinguishable from stainless steel at similar time periods, and it appears that equiatomic NiTi histologically compares favorably with stainless steel and could be used in deep tissues [10].

Castleman et al. showed that the NiTi alloy implant was at least as good, if not better, than the Cr-Co alloy implant [11]. Plates were attached with screws to the unfractured femoral bone of 12 dogs. Cobalt-chromium alloy bone plates and screws were used as a control in four dogs. The animals were sacrificed after 3, 6, 12 and 17 months. Castleman et al. were able to conclude that:

- There were no signs of corrosion that could be attributed to the reaction between the metallic surfaces and adjacent tissues for either type of implant.
- In all implanted dogs a well-defined fibrous capsule covered the implant plate.
- Gross clinical and radiological observations indicated that there was no discernible difference between "sham" operated dogs and implanted dogs.
- There was no significant histological difference between tissue and bone samples taken from the adjacent area to either implanted or "sham" operated dogs.
- No significant histological difference was found between the "sham" operated dogs and implanted dogs with either alloy in tissue taken from liver, spleen, lung, kidney and brain.
- There was evidence by neutron activity analysis that no contamination occurred due to the implant in the liver, lungs and brain. For studies with Co-Cr alloy implant contamination with chromium can be observed in the bone adjacent to the implant after exposure of 17 months.

Ohnishi et al. [12] and Prince et al. [13] carried out studies with NiTi vena cava filters showing that they are biocompatible and sufficiently nonthrombogenic to avoid occluding the vena cava. However, nickel and nickel salts can cause contact dermatitis if exposed to skin. Some papers [14, 15] discussing the effects of nickel on the skin commented about eczema caused by the element or its salts and warned about the problems caused by nickel, advising against its use. Randin [16] studied the corrosion behavior of various nickel alloys and concluded that the NiTi alloy was very resistant to pitting and hence the likelihood of contact dermatitis was low.

III. CORROSION OF NiTi ALLOYS

The corrosion resistance of the NiTi based alloys is provided by a naturally formed thin adherent passive oxide film. This film is very stable, and, therefore, NiTi alloys are resistant to corrosive attack. However, under aggressive conditions containing highly acidified chloride solutions, breakdown of the passive film can occur [17–19]. In general terms NiTi shows a very high corrosion resistance (similar to the 300 series of stainless steels). Lu studied the behavior of a NiTi alloy in seven kinds of media simulating the conditions in the mouth and the human body [19]. The weight loss was measured on equiatomic NiTi coupons. The good results of corrosion rates are shown in Table 1.

The corrosion behavior of a NiTi alloy that was mechanically strained in the as-received condition and after heat treatment has been studied by Montero-Ocampo et al. [17]. They observed that straining led to a significant improvement in corrosion resistance and suggest that the thermomechanical processing might be an important factor to take into account in the manufacture of high performance surgical implants with improved corrosion properties.

IV. SOME MEDICAL APPLICATIONS OF SHAPE MEMORY ALLOYS

A. Harrington Rods for Scoliosis

A normal spine is not absolutely straight, it curves in the anteroposterior direction, but should still remain vertical. In scoliotic backs, the spine is seen to have one or more lateral curves. These curves may vary in severity and, in some cases, cause disability or malfunctions of some internal organ. Besides, this disease is progressive and induces large and painful deformations in the spine, making corrective surgery necessary [20–22].

The conventional distraction rod technique initially involved the attachment of hooks under the laminae of vertebrae above and below the curve to be corrected. The spine is then longitudinally straightened by an external device and a distraction rod (Harrington) attached to the hooks. However, measurements have shown that the correcting force falls to approximately 30% of its original value after 10–15 days and to reestablish the initial force after this period of time requires a second operation.

A new type of distraction rod made from shape memory alloy that could overcome the need for surgical correction was proposed by Schmerling et al. [23]. Implantation of the device would occur in the usual fashion with the rod's memory being straight. On placement, the same reduction of forces as previously described, would occur. However, after a given period of time the alloy may be externally heated above A_f and hence transform into its preset shape and exert the same, or nearly the same, force as originally applied. The research group found difficulties with heating the alloy. Matsumoto et al. [24] implemented radio frequency induction heating to induce complete recovery and the spine could be gradually corrected using NiTi

Table 1 Corrosion Rates of a NiTi Alloy in Oral Environmental Media

Environmental medium	Mass corrosion rate (mm/year)
Synthetic saliva	2.9×10^{-5}
Synthetic sweat	2.8×10^{-5}
Hank's solution	0
1% NaCl sol.	5.5×10^{-5}
1% Lactic acid	5.7×10^{-5}
0.05% HCl	0
0.1% $NaHSO_4$	6.9×10^{-5}

rods with a correction rate accurately controlled by controlling the temperature increase of the alloy.

B. Staples

In cases of osteotomy, bone fracture or bone fusion, the hospitals involved used "wavy" staples with 60° spikes at each end. The staples were elongated at 0°C along their U section and spikes placed within the predrilled holes in the bones. Upon warming by a hot saline pack (40°C), their "wavy" section returned to its original shape and thus produced a compressive force between the two bone fragments (Fig. 3). This compressive force encouraged osseous tissue growth and hence quick fracture recovery [25, 26].

C. Vena Cava Filter

Pulmonary embolisms still remain an important cause of mortality in hospitalized patients. While most patients can be treated with anticoagulant or fibrolytic therapy, there are some cases where this fails and mechanical interruptions must be made. These interruptions forced the venous blood to return to the right-hand side of the heart by means of the many collateral venous pathways, which are too small to carry dangerous emboli. Ochsner et al. [27] reported on the use of vena cava ligation to prevent such emboli. However, all these interruption techniques still require surgery, even though the patient may already be seriously ill from previous pulmonary embolism or underlying phlebothrombosis [28]. To bypass this major drawback, many investigations were set up to look into a device that would cause less trauma, by gaining access to the vena cava through a peripherial part of the venous system.

The shape memory vena cava filter consisted of a locking system and a filter mesh. The locking system incorporated two elements: sharp tipped leading and trailing wires which penetrated the endothelium to a depth of 1 or 2 mm, and were prevented from further penetration by small metal studs. The filter wire was straightened in ice cold water and placed within its delivery catheter with the rear stud of the locking system placed within a special notch. This enabled the rear end to be held until all the filter had been guided out of the catheter. Placement took place using the perfusion of cool saline to avoid forming the filter within the delivery system [29, 30].

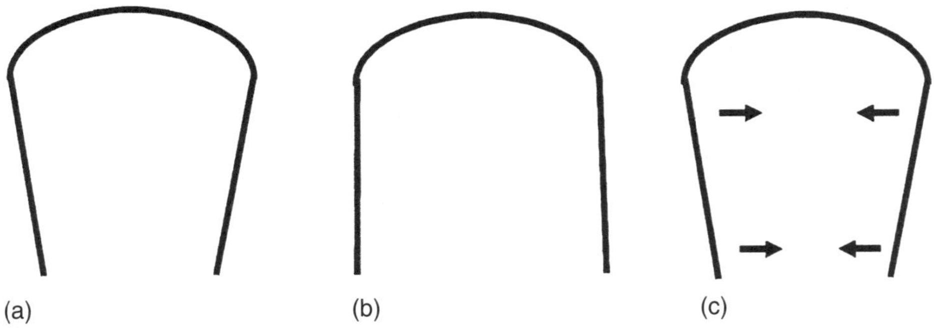

Figure 3 Scheme of the NiTi staples as used in bone fractures: (a) Before deformation, (b) after deformation and at time of implantation, (c) return to original shape, causing a compressive force.

D. Intravascular Stents

All vascular stents work on the same basic principle: they all hold up the walls of the artery by mechanical means, ensuring that a clear pathway is produced. Stenting prevents elastic recoil exhibited by arterial walls and may trap plaque and other debris against the vessel walls, thus allowing a reduction in the possible thrombus generation at the dilation site [31, 32].

There are two main deployment techniques used to position nickel-titanium stents in the intravascular system [33, 34].

a. Cold saline technique. This involves a NiTi wire whose transformation temperatures is around body temperature. The stent is transformed at a temperature well below its transformation temperature and inserted into the delivery catheter where it is kept cool by the flow of cold saline around it. In reaching the desired percutaneous transluminal angioplasty site it is pushed out of the catheter and regains its original shape as it warms up to body temperature.
b. Hot saline technique. This involves a NiTi with a transformation temperature just above body temperature. It is inserted using a catheter into the placement site, where warm saline is flushed over the stent and warms the shape memory alloy above its transition temperature, hence returning it to its original shape.

Other types of vascular endoprosthesis have been investigated in recent years, including biliary [35], ureteral [36], and rectal [37], although the tracheobronchial is frequently used in hospitals [38–40].

E. Spacers

Spinal surgery is another area where superelastic materials may exhibit their potential. Specifically it is well known that in a successful vertebral arthrodesis, two factors play a key role: the insertion of bone clips between the vertebrae which have to be grafted and the absolute absence of relative movement during the consolidation phase of the graft. The latter factor is paramount since the relative motion of the two vertebrae is the main cause of failure arthrodesis. The use of NiTi components bracing together the two vertebrae shows a great potential for success. When the compressively deformed spacer is implanted in the disk space, it changes its shape, due to body temperature, into a more circular one, [41, 42], as can be seen in Fig. 4.

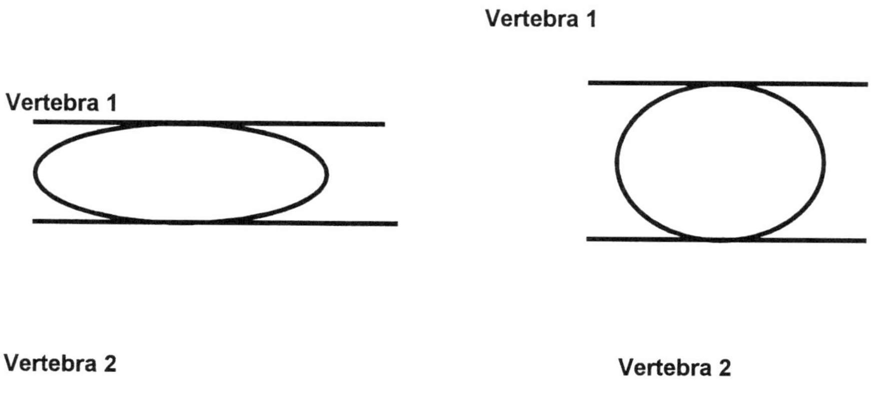

Figure 4 Spacer for bone-chip arthrodesis of a spinal column before and after heating.

F. Jaw Plates

Jaw plates, which are used in fractures of the lower jaw, have functions similar to the osteosynthesis plates. In addition to the function of fixation, any gaping of the fracture space when the lower jaw undergoes stress by mastication must also be prevented.

Figure 5a shows a scheme of a jaw with the memory plate screwed in place but not yet activated, the fracture gap is clearly visible. The memory plate was subsequently heated with hot water and Fig. 5b shows how the fracture gap is closed. The fracture gap is opened by mastication but the shape memory jaw plates at body temperature keep the fracture gap completely closed [41, 42].

G. Orthodontic Archwires

Tooth movement during orthodontic therapy is achieved by applying forces to teeth which result in bone remodeling processes. The elastic deformation of an orthodontic wire and the subsequent release of its elastic energy over a period of time gives rise to the correcting forces. It is generally assumed that optimal tooth movement is achieved by applying forces that are low in magnitude and continuous in nature. Such forces minimize tissue destruction and produce a relatively constant stress in the periodontal ligament during tooth movement. The superelasticity of NiTi archwires allows the orthodontist to apply an almost continuous light force with larger activations that result in the reduction of tissue trauma and patient discomfort, thus facilitating enhanced tooth movement.

In contrast, forces that are high in magnitude encourage hyalinization of the periodontal ligament and may cause irreversible tissue damage such as root resorption. The NiTi archwires produce teeth movement with greater efficiency and in a shorter time when compared to other orthodontic alloys, and they are specially adequate in situations requiring large deflections, such as the preliminary bracket alignment stage in the load deflection stage in orthodontic therapy.

The stress-strain curve for a superelastic material shows that the stress induced martensite $\rightarrow$ parent phase transformation upon unloading takes place at a fairly constant stress over a

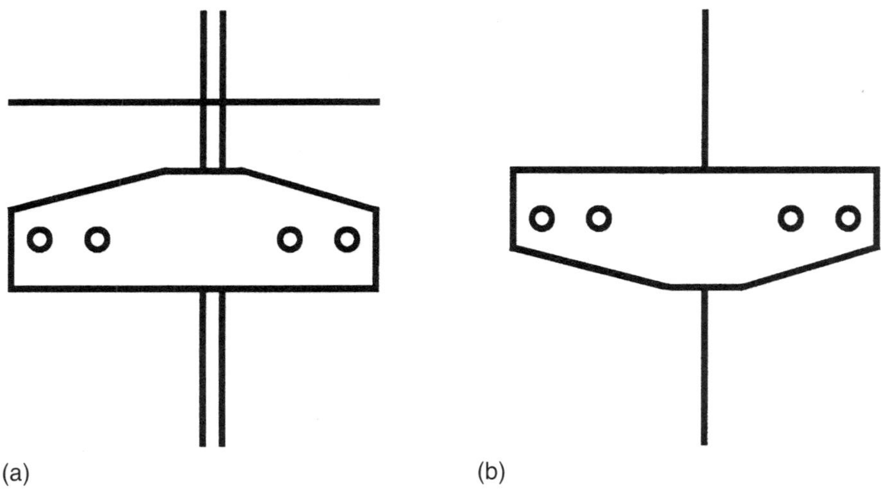

Figure 5 (a) Jaw NiTi plate and fracture gap before heating. (b) Jaw NiTi plate and fracture gap after heating.

significant range of wire activation with a desirable biological response. In Fig. 6 the different elastic behaviors of conventional alloys and NiTi alloy for orthodontic applications can be observed [43–49].

H. Partial Dentures

Shape memory attachments for partial dentures enable the partial dentures to be easily attached to and detached from the remaining teeth at room temperature. On the other hand, they recover their original shape at a higher temperature, they are suitably fixed to remaining teeth, and they do not detach easily in the oral cavity [50, 51].

I. Clinical Instruments

The stiffness of the stainless steel alloy is not suitable for instrumentation which has to run in curved canals. Walia et al. and [52] and Cuschieri [53] discussed the need for minimal access surgery to be performed for laparoscopic operations. Minimal access surgery allows a reduction of access trauma, encourages speedy rehabilitation and reduces the risk of wound infection, compared to conventional surgery, and thus is of great benefit to patient and surgeon. Shape memory alloys are used in laparoscopic spatulae that can achieve variable curvatures. The shape memory used is memorized into an arch with given dimensions; the instrument is then restrained within a standard laparoscopic cannular and introduced into the peritoneal cavity. Once in the correct region the spatula is extended out of its cannular until the required curvature is achieved (Fig. 7).

Another important application of shape memory alloys in clinical instruments is as guidewires. A guidewire is a long thin metallic wire that is introduced into the body through natural openings or small incisions. It serves as a guide for safe introduction of various therapeutic and diagnostic devices. The special mechanical properties of a guidewire provide the necessary balance between controllability and flexibility needed to safely and effectively negotiate the circuitous paths of the body's inner core. Once the guidewire is in place, a hollow tube or catheter is inserted over the guidewire and into the desired location. Guidewires are predomi-

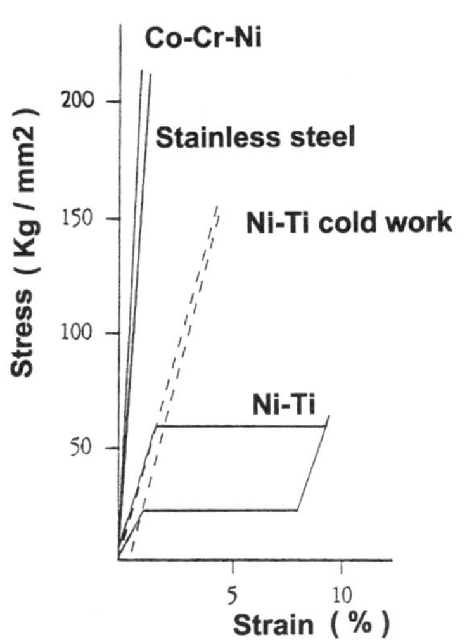

Figure 6 Stress-strain curves for different materials used as orthodontic wires.

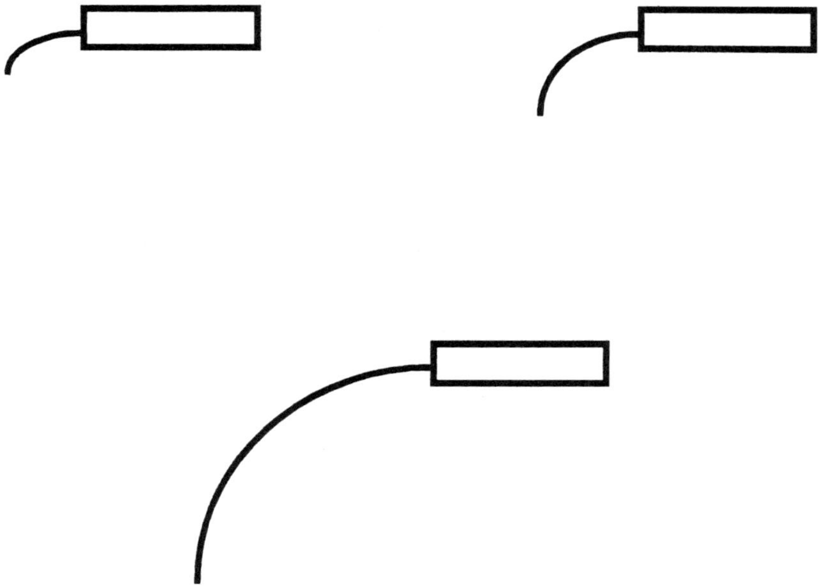

Figure 7 Shape memory spatula with variable curvature. The spatula was memorized into a semicircle and restrained within a standard laparoscopic cannula.

nantly used in cardiology and radiology and to a lesser extent in gastroenterology and urology. Stainless steel is now used almost exclusively for guidewires. Two anticipated benefits gained by the use of superelastic NiTi to replace stainless steel include diminishing the complication of the guidewire taking a permanent kink (which may be difficult to remove from the patient without injury) and increased steerability to pass the wire to the desired location due to an anticipated improvement in torquability (an ability to translate a twist at one end of the guidewire into a turn of nearly identical degree at the other end) [54, 55].

J. Concluding Remarks

Since its discovery, the shape memory phenomenon has been investigated on a vast array of fronts. Shape memory alloys exhibit a number of remarkable properties (superelastic response and/or shape memory), which open new possibilities in engineering in general and in biomedical engineering in particular. Until now, only some of the proposed applications in medicine have reached the commercial level, while others are still in the experimental state. However, there is no doubt that if all metallurgical problems are solved and reliable devices can be produced, shape memory alloys will become involved in the biomedical engineering field.

REFERENCES

1. Greninger AB, Mooradian VG. Strain transformation in metastable beta copper-zinc and beta copper-tin alloys. *Trans. AIME* **128**, 337 (1938).
2. Kurdjumov GV, Khandros LG. *Dokl. Akad. Nauk. SSSR* **66**, 211 (1949).
3. Buehler WJ, Gilfrich JV, Wiley RC. Effect of low temperature phase changes on the mechanical properties of alloys near composition TiNi, *J. Appl. Phys.* **34**, 1475 (1963).

4. Duerig TW, Zadno R. *Engineering Aspects of Shape Memory Alloys*. Butterworth-Heinemann. pp. 124–132 (1990).
5. Tosho S. *Shape Memory Alloys*. Volume 1. Ed. H. Funakubo. Gordon and Breach, Tokyo. p. 23 (1984).
6. Perkins J. *Shape Memory Effects in Alloys*. Plenum Press. New York. p. 12 (1975).
7. Iwasaki K, Hasiguti RS. *Martensitic Transformation*. Lovaine: The Institute of Metals, 198. (1982).
8. Guillemany JM, Gil FJ, Miguel JR. Ciencia y Tecnología de materiales con memoria de forma: Propiedades y Aplicaciones. *Rev. Técnica Metalúrgica* **5**, 213–224 (1986).
9. Kaufman L, Cohen M. Thermodynamics and kinetics of martensitic transformations. *Prog. Metal Phys.* **7**, 165 (1958).
10. Watanabe K. *J. Jpn. Soc. Dental Mater. Dev.* **2**, no 5, 594–602 (1983).
11. Purdy GR, Parr JG. Shape memory effect in NiTi alloys. *Trans. AIME* **6**, 23–25 (1981).
12. Cutright DE, Bhaskar SN, Pérez B, Johnson RM, Cowan GSM. Tissue reaction to nitinol wire alloy. *J. Oral Surg.* **35**, 578. (1973).
13. Castleman LS, Motzkin SM, Alicandri FP, Bonawit VL, Johnson AA. Biocompatibility of nitinol alloy as a implant material. *J. Biomed. Mater. Res.* **10**, 695–731 (1976).
14. Ohnishi H, Hmagnchi N, Miyagi T, Hamada S, Skida S. Proc. 3rd Conf. Japanese Soc. Biomat. **121**, (1989).
15. Prince MR, Salzman EW, Schoen FJ. Local intravascular effects of the nitinol wire blood clot filter. *Invest. Radiol.* **23**, 294–300 (1988).
16. Wells CG. Effects of nickel on skin. *Br. J. Dermatol.* **68**, 237–242 (1956).
17. Brandrup F, Larsen F. Nickel dermatitis provoked by buttons in blue jeans. *Contact Dermatitis* **5**, 148–150, (1979).
18. Randin JP. Corrosion behaviour of nickel containing alloys in artificial sweat. *J. Biomed. Mater. Res.* **22**, 649–666 (1988).
19. Montero-Ocampo C, López H, Salinas Rodriguez A. Effect of compressive straining on corrosion resistance of a shape memory NiTi alloy in Ringer's solution. *J. Biomed. Mater. Res.* **32**, 538–591 (1996).
20. Melton KN. NiTi based shape memory alloys in shape memory in engineering. *Aspects of Shape Memory Alloys*. TW Duering, KN Melton, DK Stöckel and CM Wayman, Eds. Butterworth-Heinemann. London, 21–35 (1990).
21. Lu S. Medical applications of NiTi in China in shape memory in engineering. *Aspects of Shape Memory Alloys*. TW Duering, KN Melton, DK Stöckel and CM Wayman, Eds. Butterworth-Heinemann. London, 445–446 (1990).
22. Lipscomb IP, Nokes DM. *The Application of Shape Memory Alloys in Medicine*. Mechanical Engineers Publishers. London, 96–97 (1996).
23. Pope MH, Stokes IAF, Moreland M. The biomechanics of scoliosis. *CRC Crit. Rev. Biomed. Eng.* 323 (1985).
24. Ghista PN, Vivani GR, Subbaraj K. Biomechanical basis of optimal scoliosis surgical correction. *J. Biomech.* **21 (2)**, 77–88 (1988).
25. Schmerling MA, Wilkor MA, Sanders AE, Wooseley JE. A proposed medical application of the shape memory effect: An NiTi Harrington rod for treatment of scoliosis. *J. Biomed. Mater. Res.* **10**, 879–902 (1976).
26. Matsumoto K, Tajima N, Kuwahara S. Correction of scoliosis with shape memory alloy. *J. Jpn. Orthop. Assoc.* **67**, 267–274 (1993).
27. Kuo PPF, Yang PJ, Zhang YF. The use of nickel-titanium alloy in orthopaedic surgery in China. *J. Orthop.* **12 (1)**, 111–116 (1989).
28. Dai KR, Hou XK, Sun YH. Treatment of intra-articular fractures with shape memory compressive staples. *Injury* **23 (10)**, 651–655 (1993).
29. Ochsner A, Ochsner JL, Sanders HS. Prevention of pulmonary embolism by cava ligation. *Annu. Surg.* **171**, 923–935 (1970).
30. Dolan JE, Alpert JS. Natural history of pulmonary embolism. *Progr. Cardiovasc. Dis.* **17**, 259–270 (1975).

31. Simon M, Kaplan R, Salzman E, Freiman DA. A vena cava filter using thermal shape memory alloy. *Radiology* **125**, 87–94 (1977).
32. Palestrant AM, Prince M, Simon M. Comparative *in vitro* evaluation of the nitinol inferior cava filter. *Radiology* **145**, 351–355 (1982).
33. Palmaz JC, Sibbitt RR, Reuter SR, Tio FO. Expandable intraluminal graft: preliminary study. *Radiology* **156**, 73–75, (1985).
34. Maass D, Zollikofer CHL, Largiader F, Senning A. Radiological follow-up of transluminally inserted vascular endoprostheses: An experimental study using expanding spirals. *Radiology* **150**, 45–46, (1984).
35. Grenadier E, Shofti R, Beyar M. Self-expandable and highly flexible nitinol stent: immediate and long term results in dogs. *Am. Herat J.* **128**, 870–878 (1994).
36. Hausegger KA, Cragg AH, Lammer J. Iliac artery stent placement: clinical experience with a nitinol stent. *Radiology* **190**, 199–202. (1994).
37. Shim CS, Lee MS, Kim JH, Cho SW. Endoscopic applications of Gianturco-Rosch biliary z-stent. *Endoscopy* **24**, 436–439. (1992).
38. Bethge N, Knyrim K, Wagner HJ. Self expanding metal stents for palliation of malignant esophageal obstruction—a pilot study of 8 patients. *Endoscopy* **24**, 411–415. (1992).
39. Van Arsdalen KN, Pollack HM, Wein AJ. Ureteral stenting. *Sem. Urology* **2 (3)**, 180–186 (1986).
40. Simonds AK, Irving JD, Clarke SW, Dick R. Use of expandable metal stents in the treatment of bronchial obstructions. *Thorax* **44**, 680–681 (1989).
41. Wallace MJ, Charnsagavej C, Ogawa K. Tracheo-bronchial tree: expandable metallic stents use in experimental and clinical applications. *Radiology* **158**, 309–312 (1986).
42. Haarsters J, Salis-Solio G, Bensmann T. The use of NiTi as a implant material in orthopaedics in shape memory. *Engineering Aspects of Shape Memory Alloys*. Eds. TW Duering, KN Melton, D Stöckel CM Wayman. Butterworth-Heinemann. London, 426–427 (1990).
43. Krousbrock R. *Shape Memory Alloys in Metal and Ceramic Biomaterials*. Volume 2. P Ducheyne and GW Hastings. Eds. CRC Press. Boca Raton, Florida. 82–86 (1984).
44. Andreasen GF, Morrow RE. Laboratory and clinical analysis of Nitinol wire. *Am. J. Orthod.* **73**, 142–149, (1978).
45. Andreasen GF. A clinical trial of alignment of teeth using a 0.019 inch thermal nitinol wire with a transition temperature range between 31°C and 45°C. *Am. J. Orthod.* **78**, 528–536, (1980).
46. Libenson C, Gil FJ, Planell JA. Untersuchung der Nickel-Titandrähte auf ihre Verwendbarkeit in der Klinischen Kieferorthopädie. *Orthodon. Kieferorthop.* **28 (1),** 135–154 (1996).
47. Miura F, Mogi M, Ohura Y, Karibe M. The superelastic japanese NiTi alloy wire for use in orthodontics. *Am. J. Orthod. Dentofac. Orthop.* **94 (2),** 89–96 (1988).
48. Burstone CJ, Qin B, Morton JY. NiTi archwire a new orthodontic alloy. *Am. J. Orthodont.* **87 (6),** 445–452 (1985).
49. Libenson C, Gil FJ, and Planell JA. Etude des caractéristiques des fils NiTi à usage orthodontique. *Rev. Orthop. Dento Faciale* **30**, 129–141 (1996).
50. Tosho S. *Shape Memory Alloys*. Volume 1. Ed. H. Funakubo. Gordon and Breach. Tokyo, 23 (1984)
51. Perkins J. *Shape Memory Effects in Alloys*. Plenum Press. New York, 12 (1975).
52. Miyazaki S, Fukutsuji S, Taira M. Application of TiNi shape memory alloys to partial dentures. Proceedings of the International Conference on Martensitic Transformation (1992), 1235.
53. Gil FJ, Planell JA. Spanish Patent (1992).
54. Walia H, Brantley WA, Gerstein H. An initial investigation of bending and torsional properties of nitinol root canal files. *J. Endodont.* **14 (7),** 346–351 (1988).
55. Cuschieri A. Minimal access surgery and the future of interventional laparoscopy. *Am. J. Surg.* **161**, 404–408 (1991).
56. Stice J. The use of superelasticity in guidewires and arthroscopic instrumentation. *Shape Memory in Engineering Aspects of Shape Memory Alloys*. Eds. TW Duering, KN Melton, D Stöckel, CM Wayman. Butterworth-Heinemann. London, 483–486 (1990).
57. De Haven K. Peripheral meniscus repair—three to seven year results. Third Congress of the International Society of the Knee. Gleneagles, Scotland, 256–257 (1993).

39
Chitin and Its Derivatives

Mutsuhiro Maeda, Yukio Inoue, Kazuo Kaneko, Tsuyoshi Sugamori, and Hideaki Iwase
Juntendo University Institute of Casualty Center, and Juntendo Izunagaoka Hospital, Shizuoka, Japan

Ryouichi Tsurutani
Unitika, Ltd., Uji, Kyoto, Japan

I. INTRODUCTION

Chitin, poly-β-(1,4) linked *N*-acetyl-D-glucosamine, is a substance that sustains and protects the body of crustaceans and microorganisms. Chitin is insoluble in water and in most ordinary solvents. This property of chitin has restricted its use in spite of its abundance in nature and the advantages claimed for the material. The products made by methods using strong acids or dichloroacetic acid were not always high in quality and featured a markedly lower molecular weight due to the use of strong acid and alkali during dissolution. However, since Austin (1975) suggested the possibility of dissolution method with amide-LiCl solvent, Kifune and co-workers (1983) found that mixtures of *N*-methylpyrrolidone (NMP) or dimethylacetamide (DMA) and lithium chloride (LiCl) are favorable solvents for chitin.

Chitin is a bioabsorbable and degradable material catabolized by a specific metabolic pathway [1]. It is a natural polymer that is attracting more and more attention as a novel material for medical use because it is not toxic and features low antigenicity, good biocompatibility [2–4] and bioactivating effects. Since Prudden [5] reported chitin's ability to promote the healing of wounds, some bioactivating effects based on its immunological adjuvant activity have become apparent. Based on their properties, braided fibers, nonwoven fabrics and porous sponges have been applied clinically (Figs. 1–3).

II. PRODUCTION OF MOLDED CHITINS AND THEIR MECHANICAL PROPERTIES

We obtained purified chitin powder (purity 99.9%, diameter 50–100 μm, molecular weight 10^6) from the two stage decalcification with acid treatment and deproteinization with alkali treatment of crude Japanese pink crab (*Chionecetes opilio*) shells. In the first purification, the shells were treated with 0.1 N HCl for about 1 h at room tempreture and with 1 N NaOH for 3 h at 80–90°C. After drying, the shells were pulverized to 30 mesh size. Moreover, they were

Figure 1 SEM finding of chitin braided fibers.

treated with 0.1 N HCl for 1 or 2 h at room temperature and with 1 N NaOH for 2 h at 95°C. The preparation was dissolved in a mixture of dimethylacetamide and lithium chloride to make dope.

The chitin fibers were obtained from the process of dissolution, filtration, making dope, transport, wet spinning, coagulation, stretching, winding, washing and drying. Using the solution consisting of DMA and LiCl, highly pure chitin was dissolved to make dope and was then wet molded with a coagulant such as water or middle alcohol. The coagulation velocity and the shape of filter size of spinning nozzle, reeling speed and fiber stretch in the wet spinning process are important factors for obtaining high strength in extension and stretch of fibers. The optimum size and physical properties of chitin fibers for practical use are molecular weight 10^5, fiber denier 0.8–0.9 d, tensile strength 3.5–4.5 g/d, remaining extension 15–18%, and water uptake up to 18.5% in weight. After these fibers were spun a braider was used to produce sutures of varying thicknesses.

The dope was poured into the frame to make flat film. As for wet molding of transparent film, the choice of coagulants is important because coagulated chitin readily loses its transparency. Butyl alcohol or isopropyl alcohol is suitable for obtaining high transparency. Uniform loading with tension is important to avoid serious contraction in the drying process. We are able to manufacture transparent and high quality film with a thickness of 30–100 μm, a tensile strength of 60 kg/mm^2 and an elongation of 14–15%. Thicker film requires gradual coagulation with additional solvent such as dimethylformide.

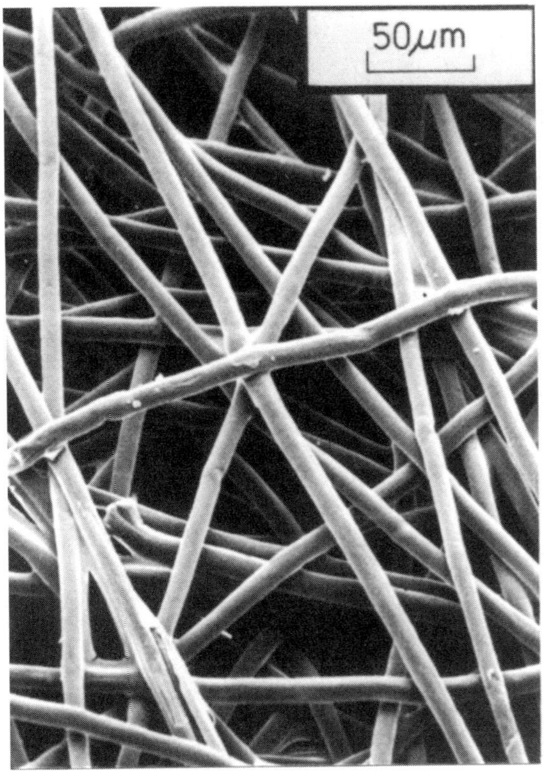

Figure 2 SEM finding of chitin nonwoven fabric.

The manufacturing process for rods was wet molding and gradual coagulation. The bending strength and elastic modulus were 19.5 kg/mm^2 and 638.9 kg/mm^2 (mean), respectively, in the dry state. In the wet state, the strength decreased to approximately one fiftieth. The water uptake was more than 100% in weight [3].

In order to produce nonwoven fabric, chitin fibers were cut to a length of 5 mm and PVA fibers (SML) were added as a binder with a ratio of 1 to 10, and the fibers were dispersed in water. Then the gel-like product was placed between filter papers and was pressed and treated with a thermal roller at 150°C. Because of its function as a temporary wound covering material, the fabric was manufactured at a thickness of 0.1–0.18 mm (Fig. 2). We designed the sheet to make sufficient room at maximum fiber swell. The sheet tensile strength of this fabric in the dry state was about 1000 g per 1 cm width, which was sufficiently tough and resistant for clinical use. The gas permeability is 120–150 $cm^3/cm^2 \cdot s$.

As for sponge production, additional treatment to make pores is to coagulate a mixture of chitin dope and water-soluble substance powder. PVA powder was used as a water-soluble polymer to make pores, so the pore size and rate could be adjusted by selecting the amount and size of the powder. After removal of PVA with hot water, lyophilization enabled the manufacture of a high quality sponge. Figure 3 shows the SEM findings for chitin sponge. The mechanical strength of chitin sponge (90% in porosity) was 6.2 g/mm^2 in the dry state and 3.8 g/mm^2 in the wet state.

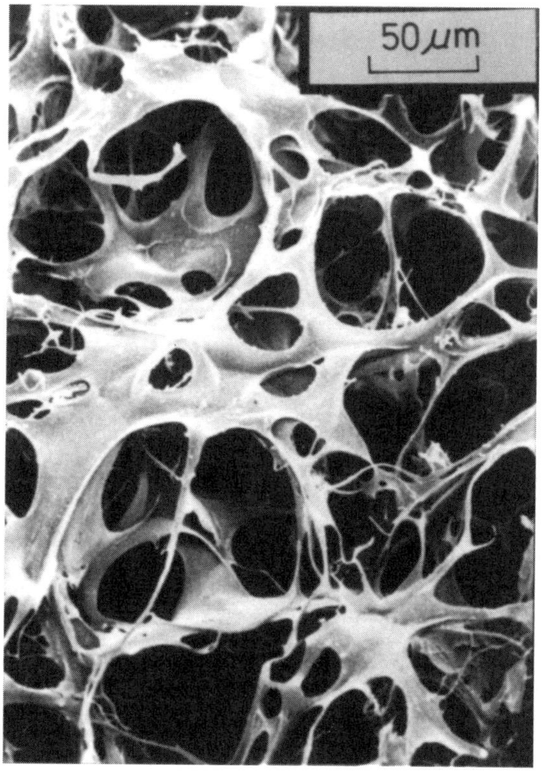

Figure 3 SEM finding of chitin porous sponge.

III. CHEMICAL MODIFICATIONS OF CHITIN

A. Deacetylation

Chitosan is the highly deacetylated product of chitin. Although chitin is insoluble in most solvents, chitosan is readily soluble in acidic solutions, but insoluble in water. However, chitosan with low molecular weight (less than 2000) is soluble in water. Chitosans are available mostly for industrial and biomedical applications.

Partially deacetylated chitins (DAC) were prepared by deacetylation of chitin or *N*-acetylation of chitosan. Sannan [6] prepared water-soluble DAC-45 to DAC-55 (45–55% of deacetylation) by homogenous deacetylation of chitin. On the other hand, DAC with heterogeneous deacetylation above 60% or below 40% was insoluble in water. The difference in water solubility is presumed to be due to the distribution of the acetamide group in the DAC molecule by Kurita et al. [7]. These results were helpful in understanding the mechanism by which the immunoadjuvant activities of these DAC molecules occur. In fact, immunological activity of 70% DAC was superior to the other deacetylation degree [8]. Peluso et al. [9] demonstrated 85% DAC has in vitro stimulatory effect on macrophage nitric oxide and chemotaxis. In the histological study of chitin-chitosan films with HA made from different deacetylation degree, 60–80% DAC showed the nonbacterial abscess in early experimental period and fibrosis and calcification in the late period. On the other hand, there were no significant histological changes in 95–100% DAC [10].

B. Carboxymethylation

Carboxymethylation is achieved with monochloroacetic acid and sodium hydroxide. The reaction takes place preferentially in C-6 hydroxy groups. Crosslinked carboxymethyl-chitin or -chitosan shows a high capability of separating bovine serum fibrinogen and albumin. The absorption ability of blood protein and Ca and Fe ions was studied by Tokura [11]. 6-*O*-Carboxymethyl chitin showed high immunoadjuvant activity as a result of macrophage activation [12].

C. Sulfation

The sulfation of chitin and chitosan has been one of the most attractive modification fields because of the possibility of preparing anticoagulant polysaccharides in this manner owing to structural similarity with heparin.

Hirano [13] reported that sulfated derivatives of chitosan such as *O*-sulfated-*N*-acetyl chitosan showed the anticoagulant effect with respect to activated partial thromboplastin time, thrombin time and antithrombin activity.

Murata [14] reported on the inhibitory effect of tumor induced angiogenesis by sulfated chitin derivatives, into which the 6-*O*-sulfate and 6-*O*-carboxymethyl groups were introduced.

IV. BIOACTIVATION EFFECTS

A. Promotion of Wound Healing

The specific chemical agent derived from cartilage that is responsible for the striking biological adjuvant effect of promoting the healing of wounds was reported by Prudden [5] and was identified as chitin. However, few studies concerning this activation effect have been undertaken because of the difficulty of producing pure molded chitin.

Our group [15] conducted reevaluation of their work from the aspect of adhesive strength found in incised wounds. From the application to the chitin powder to the incised skin of the backs of rats, the resistance to breaking on separation of the full-thickness strip of skin treated with chitin significantly higher than with absorbable polymer (PGA), nonabsorbable polymer (POM) or skin with no chitin treatment (incised only) on days 5 and 8. However, no difference was found between chitin and without chitin treatment on day 14 after surgery. In addition, Yano [16] had performed similar study to reevaluate the effect of topically administrated molded chitin fiber on tensile strength and the amount of collagen-hydroxyproline. The tensile strength of the incised wounds that were treated with chitin was significantly higher on days 3, 5, and 7 after surgery. The values of the incised tissues were higher than those of the control group, regardless of whether they had been treated with chitin, except for those with chitin treatment on day 3 and those without chitin treatment on day 5. However, the difference in the values between those with and without chitin treatment was not significant except for those on day 7. Paulette [17] documented an increase in the level of collagen in wounds in which "cartilage" was topically injected. In addition, our results indicate that the mechanism of such bioactivation effects might be sought in other events in the process of wound healing rather than simply in an increase in the amount of type I collagen resulting from collagen synthesis.

Kishimoto [18] immunohistochemically examined the healing process of experimental third degree burn wounds treated with a molded chitin dressing, and speculated as to its mechanism. Lysozyme positive cells, which are mainly histiocytes, were found 4 to 14 days after the occurrence of the burn in both the chitin dressing group and the group whose wounds were not dressed. In the no-dressing group, the peak number of these cells occured on the fourth day, but it lasted from the fourth day to the seventh day in the chitin dressing group. In the early (granulation tissue) stage, more fibroblastic cells and histiocytes were observed in the chitin dressing group than in the no-dressing group. Renewed collagen levels were also found to be different between the two groups. The collagen was found to be thick in the no-dressing group, but it was fine in the chitin dressing group. Even in the late (scar tissue) stage, fine collagen was observed in the chitin dressing group.

One mechanism of promotion of wound healing with unnecessary inflammatory reaction has been clarified. It is well known that chitin is a substrate of lysozyme (muramidase), which is found in histiocytes originating from monocyte macrophage. The biofunction of lysozyme is to decompose chitin to *N*-acetylglucosamine-*N*-acetylemuramic acid. According to Leiborich and Ross [19] the role of histiocytes in wound healing is biological wound debridement and the promotion of the proliferation of fibroblasts. These two kinds of collagen originate from different fibroblastic cells. Such speculation and the results of the study suggest that histiocytes might be induced by chitin and that they might promote the proliferation of fibroblastic cells that produced fine collagen in the process of wound healing. Kishimoto suggested these fine fibers might be collagen type 3. This speculation was confirmed by Matsue [20] and Sato [21].

Okamoto [22] demonstrated the increase of interleukin-1 (IL-1) and the fibroblast proliferation factor in the exudates taken from the surrounding areas where molded chitin was implanted in canine wounds. They found no difference between chitin and chitosan in terms of the wound healing process in animals. Such wound healing activation effects were also confirmed in the chitin derivatives such as chitosan, partially deacetylated chitin and *N*-carboxybutyl chitosan. However, the 35% DAC induced IL-1 and PG-E2 production by macrophage was less than lyophilized porcin dermal skin and collagen wound dressing [23]. Mori and co-workers [3] reported IL-8, a potent activator, and chemoattachment of neutrophils were induced in the supernatants of rat primary cultured dermal fibroblasts stimulated with chitin and its derivatives. Favorable angiogenesis and migration of neutrophils in order to have good granulation may be due to persistent release of IL-8 from fibroblasts.

B. Immunological Adjuvant Activity

Suzuki [24] and co-workers demonstrated tumor suppression abilities of chitin and chitosan in an experimental system among tumor bearing mice. They also observed that chitin and chitosan have an excellent ability to activate peritoneal macrophages and their enzymes. In a series of extensive immunological studies regarding chitin and its derivatives, they confirmed that *N*-acetyl-chitohexose, water-soluble homologue of chitin, is promising as an immunopotentiator. Nishimura [8,25] reported that 70% DAC displayed adjuvant activity for the induction of humoral and cell-mediated immunity including cytokine production. Such DAC was also found to be the most effective chitin derivative for activation of in vivo murine peritoneal macrophages for augmentation of host resistance against tumor growth and bacterial infection in mice. A close correlation was observed between the immunological activity and the degree of deacetylation of chitin. The degree of deacetylation was optimal at 70% for the induction of the adjuvant effect.

C. Bacteriostatic Effects

It is well known that chitosan is also used as a preservative in food processing. Allan [26] generally documented the growth inhibitory effects of chitosan solutions on bacterial and fungal infections. Suzuki and co-workers [24,27] reported good protective adjuvant effects of chitin and chitosan on *Staphylococcus aureus* and *Candida albicans* infection in mice. They also reported that *N*-acetyl-chitohexaose, hexamer of chitin, can be used as a biological response modifier for the opportunistic infection of microbes. Watanabe [28] reported that the high apo-transferrin inducible activity in partially degraded chitin resulted in the growth inhibition of *Candida albicans*. Iida [29] indicated that partial (70%) DAC effectively stimulated nonspecific host resistance against Sendai virus and *E. coli*.

Seo [30] showed the good antifungal and antibacterial activity of chitosan (deacetylation degree of 80%) and developed chitosan blended polynostic rayon fiber for underwear material.

Sano [31] and Tarsi [32] demonstrated good inhibitory effects of water-soluble derivatives of chitin, such as low molecular chitosan and phosphorylated chitin, on the adsorption of oral streptococcal strains, which causes tooth decay and possesses different hydrophobicities, to saliva treated hydroxyapatite.

V. BIOMEDICAL APPLICATIONS

A. Suture

Chitin is an absorbable suture material with suitable mechanical properties. The chitin suture was prepared from braided fibers (30–50 denier, 30–100 filaments) (Fig. 1). The biodegradation has been confirmed. Machnami [33] demonstrated that absorption of chitin filament occurred in foreign body giant cells. Chitin filaments gradually lost their strength when they were implanted into the back muscles of rabbits. One month after surgery, their tensile strength reached zero and their molecular weight varied from 10^5 to 10^4 over time (Fig. 5).

The straight pull tensile strength of the chitin suture of USP 4-0 size in a dry condition was 2.25 ± 0.05 kg, which is much the same as Dexon, and consists of commercially available polyglycol acid (PGA) and is much stronger than catgut. In a straight pull, the strength of chitin and catgut sutures was less in the wet condition than in the dry condition. However, in a knot pull, the strength of the chitin suture was not less in the wet condition than in the dry condition. The chitin suture had less elongation in a straight pull in a dry condition. In a wet condition, the elongation of the chitin suture was much the same as other suture materials. Nakajima [34] had measured the tensile strength of the chitin suture in the dorsum of rabbits or in incubated body fluids. In the dorsum of rabbits, the tensile strength of the chitin decreased over time and was half of the initial strength on day 14. Good tensile strength was maintained for a longer period than PGA and catgut sutures in serum, bile, pancreatic juice and urine. In contrast, the chitin suture was found to be weak against strong acid such as gastric juice with a low pH. These results suggest that the chitin suture has good resistance to hydrosis and digestive enzymes, and that care must be taken when chitin is exposed to a strong acid in the human body.

Nakajima measured the tensile strength of strips of rabbit skin sutured with chitin and PGA of a standard USP 4-0 suture size. They reported that the chitin sutures demonstrated significantly superior tensile strength after two and three weeks but that there was no difference in tensile strength between the two sutures four weeks after surgery. Its clinical application has already been established. The overall clinical results were acceptable; there were no abnormal tissue reactions or infections. There were no allergic reactions, and there were no untoward hepatic, renal, and hematological effects. The wound healing after the application of the chitin suture was reported to be uneventful, but clinically there were no apparent wound healing activation signs in comparison to the conventional sutures.

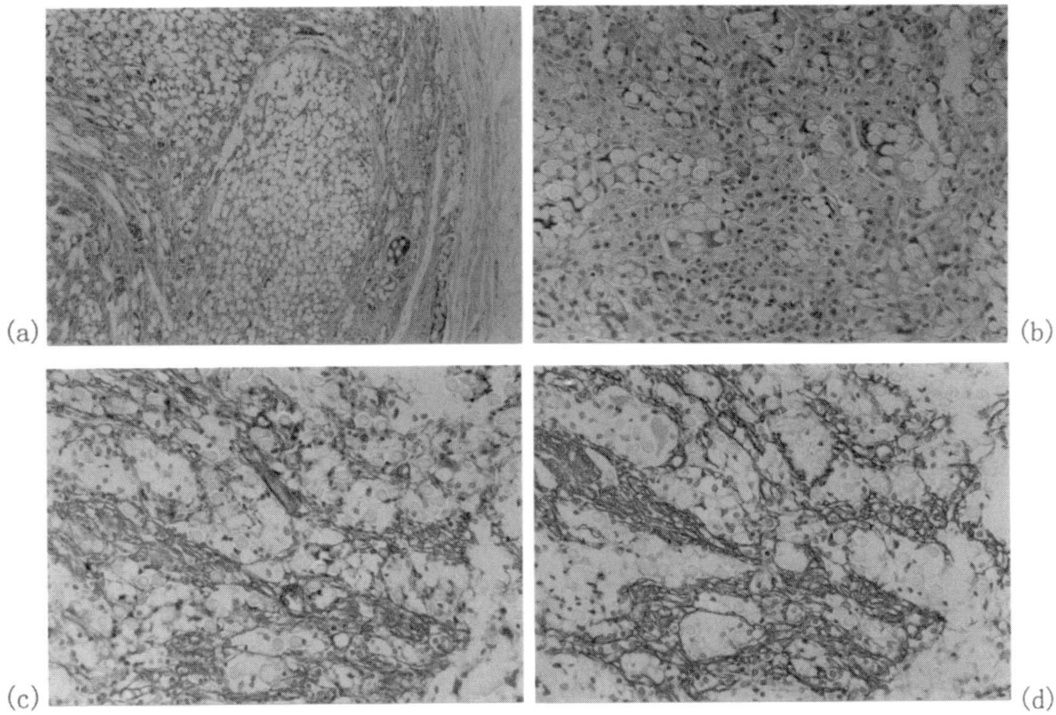

Figure 4 Collagen typing analysis of rabbit regenerated tendon after Achilles tendon reconstruction using chitin implant at 26 weeks (transverse view): (a) H-E stain at 9 weeks after surgery(×100). (b) H-E stain at 26 weeks after surgery(×200). (c) The fibrous tissue growing among the artificial tendon fibers is positive for type I collagen in a fibrous pattern at 26 weeks (streptavidin-biotin method, ×200). (d) Type III collagen in a fibrous pattern at 26 weeks (streptavidin-biotin method, ×200).

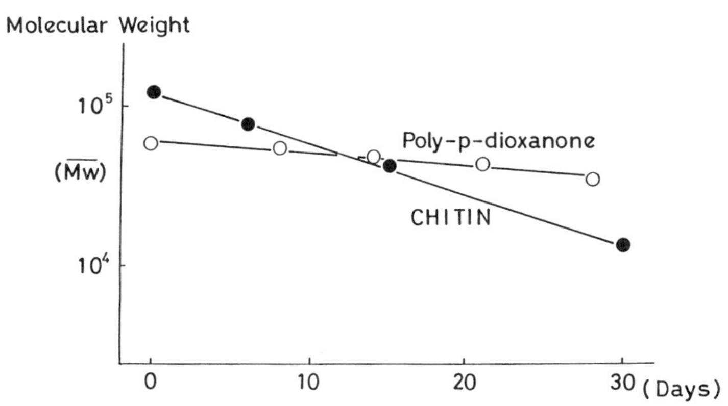

Figure 5 Time related alteration of molecular weight of chitin filaments (in vivo).

B. Wound Dressing

Temporary wound dressing materials have been produced by preparing chitin fibers, nonwoven fabrics, and sponges.

Chitin artificial skin, in the form of nonwoven fabric, is commercially available in Japan. A number of studies have already reported satisfactory results regarding its wound healing effects, histological reactions and wound management abilities [35]. The mode of application is similar to that of synthetic and biological dressing materials such as lyophilized dermal porcine skin (LDPS) and collagen fabric and membrane. Chitin artificial skin is superior to other dressing materials in terms of its analgesic, hemostatic effects, its liquefaction tolerance and reepidermization. It has also been noted that the surface of the healed wound with chitin treatment is smooth and normal. These aspects can be attributed to the bioactivation effects of chitin. According to Tokuhisa [36] and Nakamura [37], 35% DAC–induced production of inflammatory chemical mediators (IL-1, PG-E2) by macrophage on human peripheral monocyte culture was recognized but had lower capability. However, DAC-induced epidermal growth factor production of macrophage on human gingival fibroblast-like cell culture was higher than lyophilized porcine skin and collagen wound dressing. This results may show one of the mechanisms of favorable wound healing effect without unnecessary inflammatory reaction.

Sponge type materials were clinically applied for temporary wound coverage of severe soft tissue damage [38], for nasal and paranasal sinus treatment of intranasal surgery [39] and for postoperative treatment of oral surgery [40]. Such materials were used because of their ability in wound protection, pain relief, promotion of wound healing, and hemostatic and bacteriostatic effects. Cotton type materials were used for pressure sores and good results were obtained [41]. However, it is no use to apply molded chitins once infection has occurred.

Allan [26] documented the clinical application of chitosan membrane for burn therapy. Kratz [42] reported the wound healing stimulation effect of heparin-chitosan complex. Biagini [43] recommended *N*-carboxy butyl chitosan gel for skin tissue repair and speculated that the wound healing mechanism is similar to that of chitin.

Sashiwa et al. [44] developed 81% DAC povidone-iodine with PVP suspension as a wound lavage agent.

C. Sustained Release Drug Delivery

Pangburn [45] has shown the lysozyme sensitivity of partially deacetylated chitin and its possible use in self-regulated drug delivery system. We have reported the sustained release of cisplatin (CDDP) and amikacin (AMK) from capsules made of chitin [46]. By using CDDP-chitin complexes, Suzuki [47] found that they could release high concentration platinum under an implantation to mouse muscle for more than eight weeks. Also many researchers have made efforts to develop the new drug delivery systems using chitin and its derivatives such as 5-fluorouracil chitin and chitosan conjugates [48] and 6-*O*-carboxymethyl chitin gel [49]. As for application for drugs, Suzuki [50] reviewed recent studies briefly.

D. Topical Hemostatic

Olsen [51] reported the agglutination of blood cells by chitosan acetate added to blood. Malette [52] documented the hemostatic effect of chitosan from implantation of Dacron vascular prosthesis coated by chitosan solution.

We [53] have developed chitin hemostatic, which is a partially (58%) deacetylated chitin hydrochloride in the form of fibers and has high water absorption while readily changing to gel form in water. Chitin hemostatic showed significant reduction of blood loss from canine iliac bone hole. And significantly high amount of release of β-thromboglobulin and platelet factor IV in vitro indicated a platelet activation effect of chitin (Fig. 6). Also some clinical trials are ongoing in Japan [54]. We experienced its effectiveness in clinical application but had one case of insufficient wound closure probably due to residue of DAC-HCl in spinal surgery [55]. In our experimental study [56], noninfectious abscess formation was observed from the early stage of inflammatory reaction of 58% DAC.

E. Soft Tissue and Bone Substitute

The fibers and sponges of chitin were completely absorbed and disappeared five to six months later. The chitin sponge implanted into the bone of the rabbit's femoral condyle was found to be completely absorbed at the fourth month after surgery [15].

Our investigation [57] on the applicability of molded chitin has led to the development of a braided filamentous substitute for extra-articular ligaments and tendons. In an experimental model having a 1-inch (2.5-cm) gap in canine calcaneal tendons, it was shown that grafting with chitin ligaments provided superior reconstruction compared to polyester. Mean duration to filling of the gap and to induction of a neotendon composed of bundles of fibrous tissue was 6 weeks, and it took 12 weeks until mechanical properties were recovered to a level of normal tendon (Fig. 7).

There is a slight problem with chitin ligaments in terms of initial strength; nevertheless we obtained satisfactory results with respect to the process of absorption, replacement and tendon regeneration using chitin ligaments. Sato [21] showed the applicability of the scaffold type of chitin-other biodegradable polymer composite for tendon reconstruction surgery. Ohyabu [58] and co-workers implanted chitin sponge to the rabbit meniscal defect and observed the S-100 protein positive cells (chondrocytic cells) in regenerated tissue and suggested the possibility of meniscal repair by molded chitin. Maeda [59] reported the possibility of osteogenesis induced by chitin in the bone of rabbit femoral condyle. Borrah [60] documented

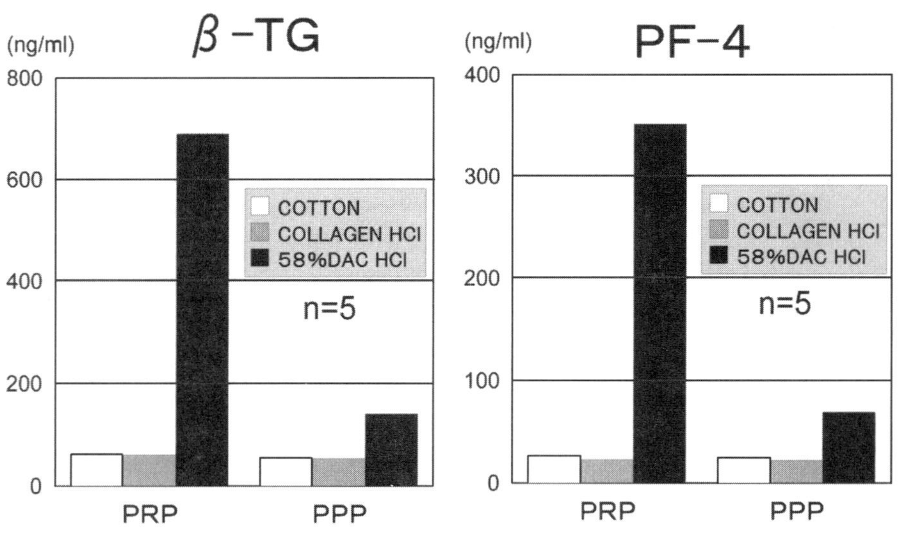

Figure 6 Platelet releasing factors (β-TG and PF-4) after a contact for 5 min with each samples (PRP: platelet rich plasma 60×10^4. PPP: platelet poor plasma 1×10^4).

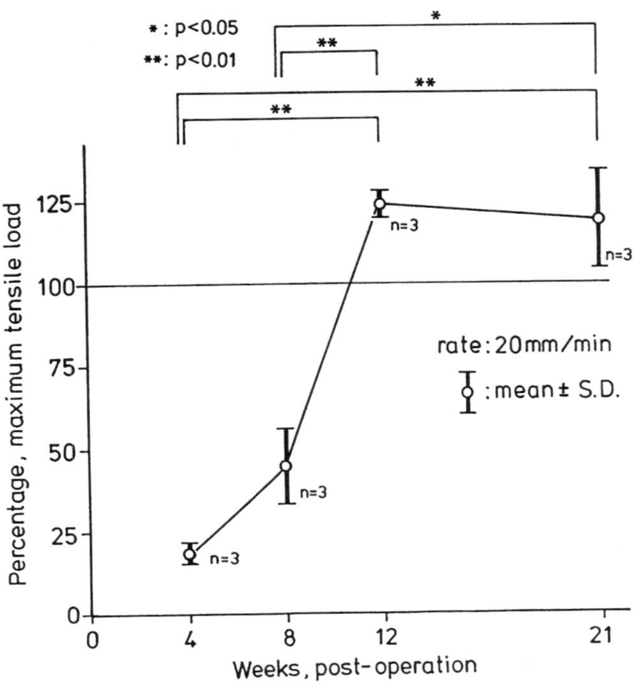

Figure 7 Time related changes of percentage tensile strength, compared to untreated control, of chitin regenerated tendon after surgery.

the osteogenic potential of *N*-acetyl chitosan regarding endochondral bone repair and osseous induction in the metacarpal and fibular bone defect model of the rabbit. Yoshihara [61] have attempted the medical application of freeze-dried carboxymethyl chitin and hydroxyapatite (HA) composite for bone repairing.

Maruyama and Ito [62] reported chitosan bonded self-hardening paste to hydroxyapatite granules (plus malic acid and polyphosphoric acid sodium) for bone substitute. Moreover, Hidaka and co-workers [10] demonstrated that its inflammatory reaction, implanted into the rat back, has changed by the degrees of deacetylation of chitin. Promotion of calcification on a water-insoluble carboxymethyl chitin disk was reported by Wan [63]. Calcium phosphate-polysaccharide composite cement for bone substitute is currently being developed [64]. This cement is self-hardening when both powder of α-tricalciumphosphate (α-TCP), tetracalcium phosphate (TeCP), dicalcium phosphate dihydrate (DCPD) and liquid (carboxymethyl chitin, succinic acid, disodium hydrogen phosphate) components are mixed and can be molded into a desirable shape.

REFERENCES

1. Kohn, P., Winzler, R. J. and Hoffmann, R. C., Metabolism of D-glucosamine and *N*-acetyl-D-glucosamine in the intact rat. *J. Biol. Chem.*, **237**: 304–308, 1962.
2. Maeda, M., Inoue, Y., Iwase, H. and Kifune, K., Characteristics of chitin for orthopaedic use. In *Chitin, Chitosan and Related Enzymes*, J. P. Zikakis (ed.). Academic Press, New York, pp. 411–415, 1984.

3. Mori, T., Okumura, M., Mastuura, M., Ueno, K., Tokura, S., Okamoto, Y., Minami, S. and Fujinaga, T., Effects of chitin and its derivatives on the proliferation and cytokine production of fibroblasts in vitro. *Biomaterials*, **18**: 947–951, 1997.
4. Schmidt, R. J., Chung, L. Y., Ancrews, A. M., Spyratou, O. and Turner, T. D., Biocompatibility of wound management products: A study of the effect of various polysaccharides on murine L929 fibroblast proliferation and macrophage respiratory burst. *J. Pharm. Pharmacol.*, **45**: 508–513, 1992.
5. Prudden, J. F., Migel, P., Hanson, P., Friedrich, L. and Balassa, L., The discovery of a potent pure chemical wound-healing accelerator. *Am. J.Surg.*, **119**: 560–564, 1970.
6. Sannan, T., Kurita, K. and Iwakura, Y., Studies on chitin 2. effect of deacetylation on solubility. *Mikromol. Chem.*, **177**: 3589–3600, 1976.
7. Kurita, K., Sannan, T. and Iwakura, Y., Studies on chitin 4. Evidence for formation of block and random copolymers of *N*-acetyl-D-glucosamine and D-glucosamine by hetero- and homogeneous hydrolyses. *Makromol. Chem.*, **178**: 3197–3202, 1977.
8. Nishimura, K., Nishimura, S., Nishi, N., Saiki, I., Tokura, S. and Azuma, I., Immunological activity of chitin and its derivatives. *Vaccine*, **2**: 93–92, 1984.
9. Peluso, G., Petillo, O., Ranieri, M., Santin, M., Ambrosio, L., Calabro, D., Avallone, B. and Balsamo, G., Chitosan-mediated stimulation of macrophage function. *Biomaterials*, **15**: 1215–1220, 1994.
10. Hidaka, Y., Ito, M., Yokoyama, K., Mori, K., Yamakura, K., Nakajima, M. and Igarashi, T., Rat subcutaneous tissue reactions to chitin and chitosan films and their mechanical prorerties. *JJSB*, **16**: 66–71, 1998.
11. Tokura, S., Nishi, N., Nishimura, S., Ikeuchi, Y., Azuma, I. and Nishimura, K., Physicochemical, biochemical and biological properties of chitin derivatives. In *Chitin, Chitosan and Related Enzymes*, J. P. Zikakis (ed.). Academic Press, New York, pp. 303–325, 1984.
12. Muzzarelli, R., Baldassarre, V., Conti, F., Ferrara, P., Biagini, G., Gazzanelli, G. and Vasi, V., Biological activity of chitosan: ultrastructural study, *Biomaterials*, **9**: 247–252, 1988.
13. Hirano, S., Tanaka, Y., Hasegawa, M., Tobetto, K. and Nisioka, A., Effect of sulfated derivatives of chitosan on some blood coagulant factors. *Carbohydr. Res.*, **137**:205–215, 1985.
14. Murata, J., Saiki, I., Makabe, T., Tsuta, Y., Tokura, S. and Azuma, I., Inhibition of tumor-induced angiogenesis by sulfated chitin derivatives. *Cancer Res.*, **51**: 22–26, 1991.
15. Maeda, M., Chitin as a biomaterial. *Bessatsu seikei geka*, **18**: 196–200, 1990.
16. Yano, H., Iriyama, K., Nishiwaki, H. and Kifune, K., Effect of *N*-acetyl-D-glucosamine on wound healing in rats. *Mie Med. J.*, **135**: 53–56, 1985.
17. Paulette, R. E. and Prudden, J. F., Studies on acceleration of wound healing with cartilage. *Surg. Gynec. Obstet.*, **108**: 408–415, 1959.
18. Kishimoto, S. and Tamaki, K., Immunohistochemical and histological observations in the process of burn wound healing in guinea pig skin under chitin membrane dressing. *Acta. Dermatol. Kyoto*, **82**: 471–479, 1987.
19. Leiborch, S. J. and Ross, R., The role of the macrophage in wound repair. A study with hydrocortisone and antimacrophages serum. *Am. J. Path.*, **78**: 71–100, 1975.
20. Ogata, T., Miyakawa, E., Matsue, M. and Matsue, I., The biological dressing effects of chitin membrane on the regeneration of palatal mucosa. *J. J. Periodont.*, **33**: 190–198, 1991.
21. Sato, M., Kurosawa, H., Inoue, Y., Maeda, M. and Iwase, H., Biomaterials for reconstraction of extra-articular tendons and ligaments. *Transactions of the 3rd combined meeting of the ORS of the USA, Canada, Europe and Japan*, **74**: 1998.
22. Okamoto, Y., Minami, S., Matsuhashi, A., Sashiwa, H., Saimoto, H., Shigemasa, Y., Tanigawa, T., Tanaka, Y. and Tokura, S., Application of chitin and chitosan in small animals. In *Advances in Chitin and Chitosan*, C. J. Brine, P. A. Sandford, and J. P. Zikakis (eds.), Elsevier Applied Science, New York, pp. 70–78, 1992.
23. Tokuhisa, M. and Tominaga, N., Effects of chitin on the production of chemical mediators by peripheral monocytes. *Jpn. J. Oral Biol.*, **36**: 591–597, 1994.
24. Suzuki, S., Watanabe, T., Mikami, T., Matsumoto, T. and Suzuki, M., Immunoenhancing effects of

N-acetylchitohexose. In *Advances in Chitin and Chitosan*, C. J. Brine, P. A. Sandford, and J. P. Zikakis (eds.), Elsevier Applied Science, New York, pp. 96–105, 1992.

25. Nishimura, K., Ishihara, C., Ukei, S., Tokura, S. and Azuma, I., Stimulation of cytokine production in mice using deacetylated chitin. *Vaccine*, **4**: 151–156, 1986.
26. Allan, G. G., Altman, L. C., Bensinger, R. E., Ghosh, D. K., Hirabayashi, I., Neogi, A. N. and Neogi, S., Biomedical applications of chitin and chitosan. In *Chitin, Chitosan and Related Enzymes*, J. P. Zikakis (ed.). Academic Press, New York, pp. 119–133, 1984.
27. Kobayashi, M., Watanabe, T., Suzuki, S. and Suzuku, M., Effect of *N*-acetylchitohexose against *Candida albicans* infection of tumor bearing mice. *Microbiol. Immunol.*, **34**: 413–426, 1990.
28. Watanabe, T., Hino, A., Ono, Y., Mikami, T., Matsumoto, T., Suzuki, S. and Suzuki, M., Apo-transferrin inducible activity in partially degraded chitin. *Chitin Chitosan Res.*, **4**: 25–29, 1998.
29. IIda, J., Une, T., Ishihara, C., Nishimura, K., Tokura, S., Mizukoshi, N. and Azuma, I., Stimulation of non-specific host resistance against Sendai virus and *Escherichia coli* infections by chitin derivatives in mice. *Vaccine*, **5**: 270–274, 1987.
30. Seo, H., Mitsuhashi, K. and Tanibe, H., Antibacterial and antifungal fiber blended by chitosan. In *Advances in Chitin and Chitosan*, C. J. Brine, P. A. Sandford, and J. P. Zikakis (eds.), Elsevier Applied Science, New York, pp. 34–40, 1992.
31. Sano, H., Matsukubo, T., Shibasaki, K., Itoi, H. and Takaesu, Y., Inhibition of adsorption of oral streptococci to saliva treated hydroxyapatite by chitin derivatives. *Bull. Tokyo Dent. Coll.*, **32**: 9–17, 1991.
32. Tarsi, R., Muzzarelli, R. A. A., Guzman, C. A. and Pruzzo, C., Inhibition of streptococcus mutans adsorption to hydroxyapatite by low-molecular-weight chitosans. *J. Dent. Res.*, **76**: 665–672, 1997.
33. Machnami, R., Kifune, K., Kawaide, A. and Tsurutani, R., A histological study of the fate of chitin suture material after intramuscular suturing. *Med. Sci. Res.*, **19**: 391–392, 1991.
34. Nakajima, M., Atsumi, K., Kifune, K. and Kanamaru, H., Chitin is an effective material for sutures. *Jpn. J. Surg.*, **16**: 418–424, 1986.
35. Kifune, K., Clinical application of chitin artificial skin. In *Advances in Chitin and Chitosan*, C. J. Brine, P. A. Sandford, and J. P. Zikakis (eds.), Elsevier Applied Science, New York, pp. 9–15, 1992.
36. Tokuhisa, M. and Tominaga, N., Effects of chitin on the production of chemical mediators by peripheral monocytes. *Jpn. J. Oral Biol.*, **36**: 591–597, 1994.
37. Nakamura, M., Tokuhisa, M., Tomonaga, N., Kuga, Y., Mizuno, A. and Umemoto, T., Effects of chitin fibroblast proliferation and production of epidermal growth factor. *JJOMS*, **43**: 157–164, 1997.
38. Maeda, M., Inoue, Y., Yanagihara, Y. and Iwase, H., Porous chitin (sponge) as a temporary deep wound dressing material. *Seikeigeka*, **43**: 1441–1446, 1992.
39. Manabe, Y., Saito, H., Takanami, N. and Matsuki, M., Application of chitin and paranasal sinus treatment. *Jibirinsyou Suppl.*, **25**: 65–69, 1988.
40. Furutani, M., Iida, M., Yamaguchi, Y., Mori, M., Fujita, S. and Sato, T., A new wound dressing of chitin sponge, its clinical application for oral surgery. *Shika J.*, **24**: 451–454, 1986.
41. Wada, H., et al., Treatment of pressure sore with sponge typed chitin membrane. *J. West. Jpn. Darmat.*, **52**: 761–765, 1990.
42. Kratz, G., Arnander, C., Swedenborg, J., Back, M., Falk, C., Gouda, I. and Larm, O., Heparin-chitosan complexes stimulate wound healing in human skin. *Scand. J. Plast. Reconstr. Hand. Surg.*, **31**: 119–123, 1997.
43. Biagini, G., Muzzarelli, R. A. A., Giardino, R. and Castaldini, C., Biological materials for wound healing. In *Advances in Chitin and Chitosan*, C. J. Brine, P. A. Sandford, and J. P. Zikakis (eds.), Elsevier Applied Science, New York, pp. 16–24, 1992.
44. Sashiwa, H., Okamoto, Y., Minami, S. and Shigemasa, Y., Antimicrobial and wound healing activity of partially deacetylated chitin-isodine suspension. *Chitin Chitosan Res.*, **4**: 18–24, 1998.
45. Pangburn, S. H., Trescony, P. V. and Heller, J., Partially deacetylated chitin: its use in self-regulated drug delivery systems. In *Chitin, Chitosan and Related Enzymes*. J. P. Zikakis (ed.). Academic Press, New York, pp. 3–19, 1984.

46. Iwase, H., Goto, K., Inoue, Y., Maeda, M., Arai, Y. and Kifune, K., Some characteristics of chitin for orthopaedic use—capsule of slow released chemotherapeutic drugs. *Trans. JSB*, **8**: 7, 1986.
47. Suzuki, K., Yoshimura, H., Matsuura, H., Katoh, T., Nakamura, T., Tsurutani, R. and Kifune, K., A new slow-releasing drug delivery system for chemically combined cisplatin with chitin for intraoperative local application: an experimental study. In *Recent Advance in Diseases of the Esophagus*, Springer-Verlag, Tokyo, pp. 865–870, 1993.
48. Ouchi, T., Inosaka, K., Banba, T. and Ohya, Y., Design of chitin or chitosan/5-fluorouracil conjugate having antitumor activity. In *Advances in Chitin and Chitosan*. C. J. Brine, P. A. Sandford, and J. P. Zikakis (eds.), Elsevier Applied Science, New York, pp. 106–115, 1992.
49. Watanabe, K., Saiki, I., Matsumoto, Y., Azuma, I., Seo, H., Okuyama, H., Uraki, Y., Miura, Y. and Tokura, S., Antimetastatic activity of neocarzinostatin incorporated into controlled release gel of CM-chitin. *Carbohydrate Polym.*, **17**: 29–37, 1992.
50. Suzuki, S., Trends in applications studies on chitic substances as drugs. *Chitin Chitosan Res.*, **4**: 1–11, 1998.
51. Olsen, R., Schwartzmiller, D., Weppner, W. and Winandy, R., Biomedical applications of chitin and its derivatives. In *Chitin and Chitosan*. G. Skjak-Break, T. Anthonsen, and P. Sandford (eds.). Elsevier Applied Science, London, pp. 813–828, 1989.
52. Malette, W. G., Quigler, H. J., Gaines, R. D., Johnson, N. D. and Rainer, W. G., Chitosan: a new hemostatic. *Ann. Thorac. Surg.*, **36**: 55–58, 1983.
53. Sugamori, T., Maeda, M. and Iwase, H., Local hemostatic effects of microcrystalline partially deacetylated chitin hydrochloride. *J. Biomed. Mater. Res.*, **49**: 225–223, 2000.
54. Miyahara, T., Clinical experience of chitin hemostatic. *Sinryou to Shinyaku*, **28**:1703–1710, 1991.
55. Maeda, M., Inoue, Y., Yanagihara, Y., Uta, S. and Iwase, H., Clinical experience of chitin hemostatic agent for spinal surgery. *Med. Consult. New Remed.*, **34**: 183–187, 1997.
56. Maeda, M., Iwase, H., Miyazaki, M., Inoue, Y. and Sugamori, T., How dose the residual topical hemostatic agent behave in vivo? *JJSB*, **16**: 196–201, 1998.
57. Maeda, M., Inoue, Y., Iwase, H. and Kifune, K., Experimental study on achilles tendon reconstruction using molded chitin. *Trans. Soc. Biomater.*, **17**: 96, 1994.
58. Ohyabu, N., Matsui, N., Ohtsuka, T., Taneda, Y. and Kato, T., Experimental study with chitin in the meniscal repair. *Cent. Jpn. Orthop. Traumat.*, :149–150, 1992.
59. Maeda, M., Inoue, Y., Iwase, H. and Kifune, K., Biocompatibility of molded chitin and its applicability as an implant material. *Trans. Soc. Biomater.*, **12**: 21, 1986.
60. Borrah, G. L., Scott, G. and Wortham, K., Bone induction by chitosan in endochondral bones of extremities. In *Advances in Chitin and Chitosan*. C. J. Brine, P. A. Sandford, and J. P. Zikakis (eds.), Elsevier Applied Science, New York, pp. 54–60, 1992.
61. Yoshihara, Y., Ishii, T., Nakajima, Y., Tojima, T. and Tokura, S., Study of carboxymethyl-chitin and hydroxyapatite composite for bone repairing. In *Advances in Chitin Science*. A. Domard, G. A. F. Roberts, and K. M. Varum (eds.), Jacques Andre Publisher, pp. 682–687, 1997.
62. Maruyama, M. and Ito, M., In vitro properties of a chitosan-bonded self-hardening paste with hydroxyapatite granules. *J. Biomed. Mater. Res.*, **32**: 527–532, 1996.
63. Wan, A. C. A., Khor, E., Wong, J. M. and Hasting, G. W., Promotion of calcification on carboxymethylchtin discs. *Biomaterials*, **17**: 1529–1534, 1996.
64. Yokoyama, A., Kamiura, Y., Yamamoto, S., Fujita, R., Iida, S., Komastubara, H., Takeishi, A., Kawasaki, T., Kohgo, T. and Amemiya, A., Possibilities of self hardening hydroxyapatite as a bone substitute material. *Abstract of IADR*, 1995.

40

Biological Activity of Hydroxamic Compounds

Constantin V. Uglea
University of Medicine and Pharmacy "Gr. T. Popa", Jassy, Romania

Daniela Spridon
Institute of Biological Research, Jassy, Romania

Formed in the "primary soup" as a second generation of organic compounds [1], the hydroxamic acid bond occurs in products resulting from fungi, yeast, bacteria and plants. The —CO—N(OH)— bond arises by oxidation of a free or bound amino group in a unit structure which is often closely related to conventional amino acids.

For the scientific community, the hydroxamic acids are century old compounds [2], but several million years earlier the living cells—especially those of microorganisms—seem to have discovered the need for such compounds in their metabolism of iron. The source and biological activity of some natural hydroxamic compounds are shown in Table 1. Most of these compounds reported result from fungi and actinomycetes but they occur also in yeast, bacteria and green plants. However, for mammalian cells, hydroxamic compounds remain only as "simple synthetic compounds," and they have not been accepted as partners involved in specific processes of metabolism.

As a result of their capacity of degrading the thioesters, the hydroxamic acids have been studied as early as 1964 [3] as potential biologically active agents and especially for their capacity of monitoring the process of cellular division. The above-mentioned reasons, along

Table 1 Source and Biological Activity of Some Naturally Occurring Hydroxamic Acids

Compound	Source	Structure	Action
Hadacidin	*Penicillium aurantioviolaceum*	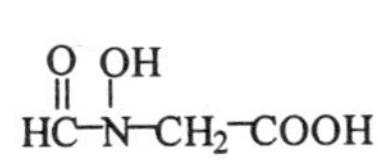	Antitumoral agent
Fusarinine	*Fusarium roseum*	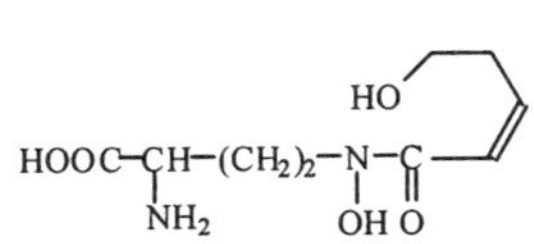	Antitumoral agent

Table 1 Continued

Aspergillic acid	*Aspergillus flavus*	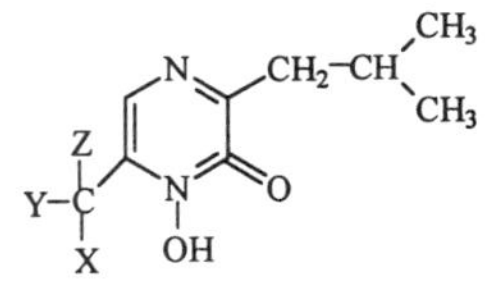 X=H Y=C_2H_5 Z=CH_3	Antibiotic
Hydroxyaspergillic acid	*Aspergillus flavus*	X=OH Y=C_2H_5 Z=CH_3	Antibiotic
Muta-aspergilic acid	*Aspergillus oryzae*	X=OH Y=CH_3 Z=CH_3	Antibiotic
Neoaspergillic acid	*Aspergillus sclerotiorum*	X=OH Y=$CH(CH_3)_2$ Z=H	Antibiotic
2,4-Dihydroxy-7-methoxy-1,4-benzoxazin 3-one (DIMBOA)	*Plant seeding*	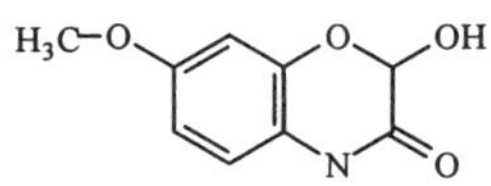	Fungistatic agent
Actinonin	*Streptomyces sp.*	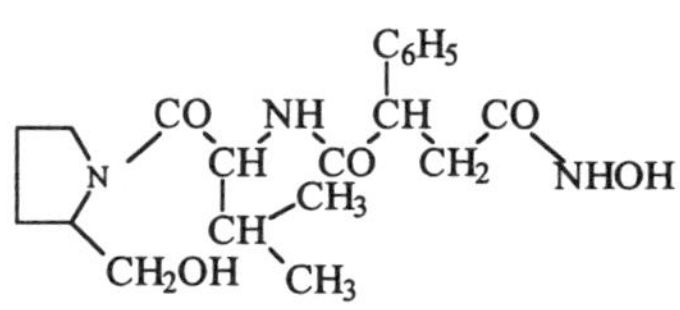	Antibiotic
Mycobactin P	*Mycobacterium phlei*	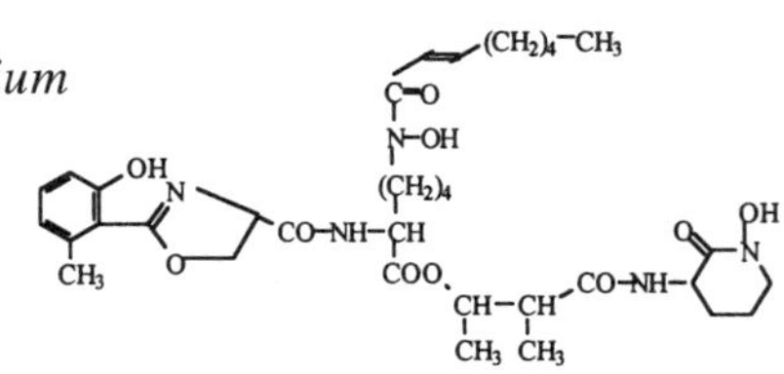	Growth factor
Mycellanamide	*Penicillium griseofulvum*	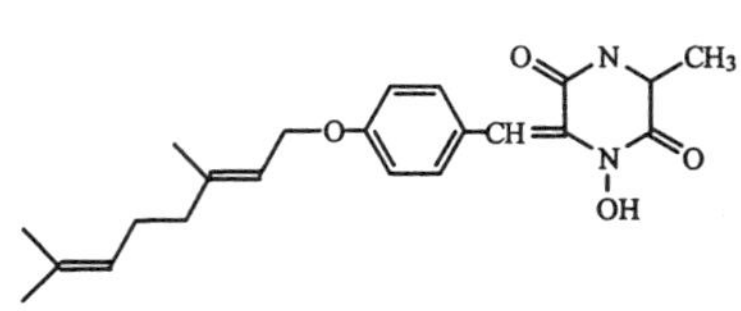	Antibiotic

with the most recently published results, as well as our own results on the antiviral and antitumoral activity of the hydroxamic compounds, urged us to write this chapter.

I. SYNTHESIS

The hydroxamic compounds may be obtained in laboratory conditions through numerous reactions, the most important of them being the following:

a. Condensation of esters with hydroxylamine [4]

$$R{-}COOC_2H_5 + H_2NOH \longrightarrow R{-}CO{-}NHOH + H_2O \qquad (1)$$

b. The reaction between anhydrides and hydroxylamine [5]

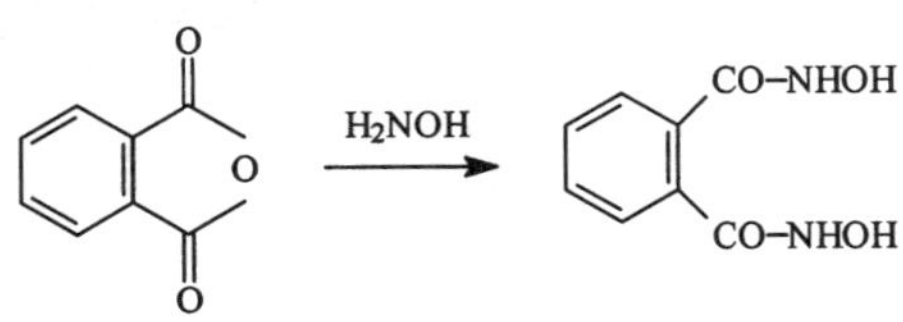

(2)

c. The reaction between amides and hydroxylamine [6]

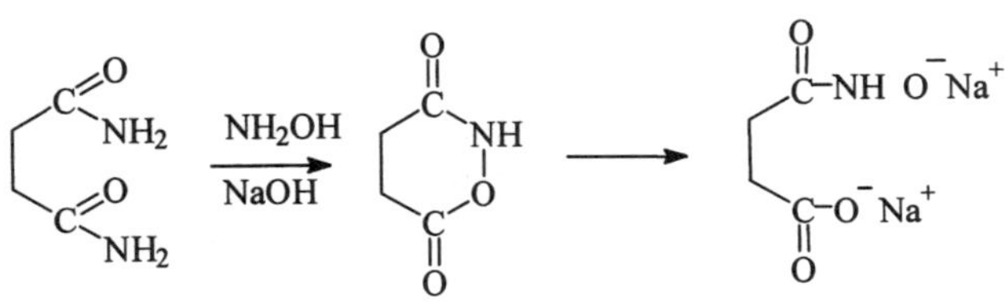

(3)

d. The condensation of carboxylic acids with hydroxylamine [7]

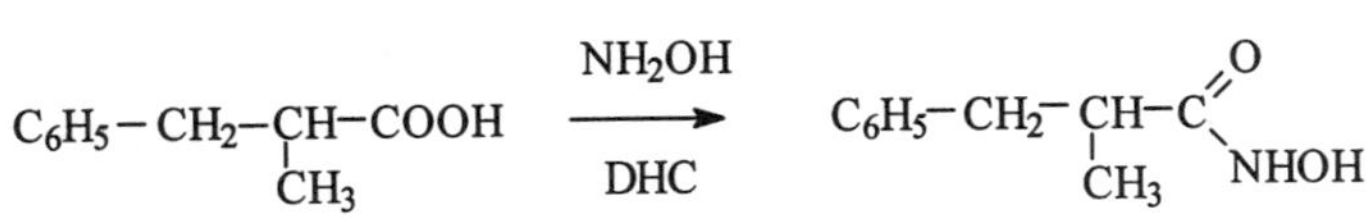

(4)

e. The reaction between aldehydes and hydroxylamine [8]

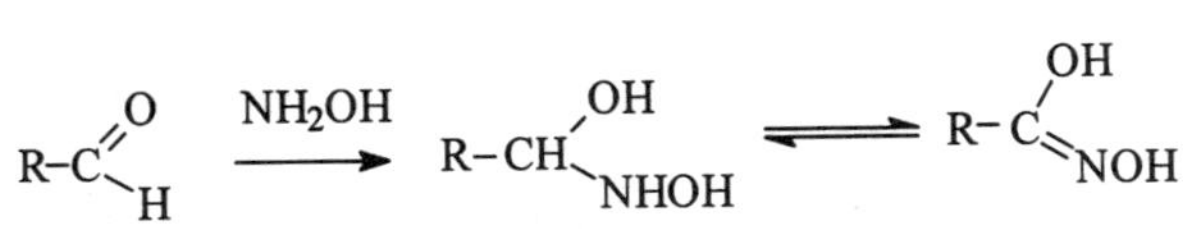

(5)

f. Finally, through the reaction between nitroderivatives and carboxylic acids [9]

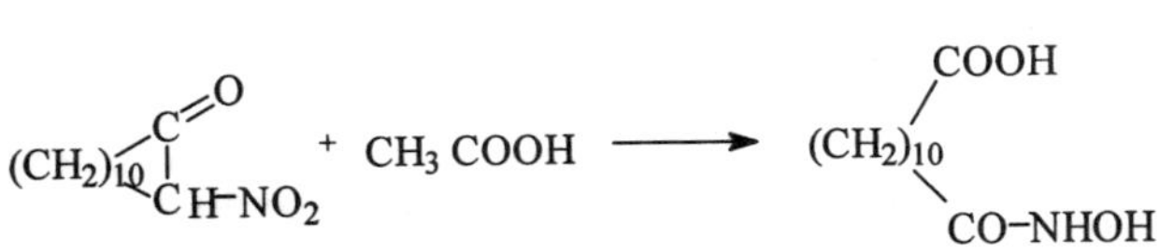

(6)

The synthesis of hydroxamic acids may be also performed through other reactions, special mention being made here of those between isocyanates and hydroxylamine [10], and oximes and boranes [11].

II. PHYSICAL AND CHEMICAL PROPERTIES

The most outstanding property of hydroxamic acids is their ability to dissociate a proton in slightly alkaline media ($pK_a \approx 9$). This allows attachment of a metal ion in a stable, five-membered ring. Binding occurs in a stepwise manner as the pH is raised.

Study of one of the most common and largely utilized representatives of hydroxamic acids—namely, hydroxyurea—showed that this compound is stable in a neutral medium, but it is hydrolyzed in an acid medium, hydroxylamine and a carboxylic acid resulting, according to the following reaction [12,13]:

$$H_2N-\underset{\underset{O}{\|}}{C}-NHOH + 2\,H_2O \longrightarrow NH_2OH + NH_4^+ + HCOO^- \tag{7}$$

Involved in the iron microbial metabolism, hydroxamic acids may be considered as iron transport agents. This role is compatible with the following observations. The ferric complex of the hydroxamic compounds is formed rapidly and it is remarkably stable related to the ferrous complex. This permits a mechanism for the pickup and delivery of the trivalent iron, the form encountered by aerobic organisms in nature. At the same time, the large discrepancy in stability between the two oxidation states rules out an electron-transfer function in the manner of them. The growth factor activity of some hydroxamic compounds is commonly replaced by much higher concentrations of them, and sometimes by the proper concentration of a synthetic chelating agent, but not by protoporphyrin. All of these suggest that the only role of hydroxamic compounds is to insert iron into porphyrin. Cells of *Arthrobacter sp.* which are starving for ferrichrome (an iron-containing nitrogenous pigment and precursor of cytochrome) are low in catalase even though the growth medium contains large amounts of inorganic iron.

Besides the biological processes, the hydroxamic link may participate, too, in various chemical reactions, such as alkylation [14], acylation [15], group isomerization [16], oxidation [17] and condensation with carboxylic compounds [18].

III. BIOLOGICAL ACTIVITY

Hydroxamic compounds apparently do not occur in animal tissues, and this has prompted the thought that it may be possible to build chemotherapeutic agents around this structure, the theory being that this would not interfere with natural processes in the host. Together with the property of these compounds of reacting with thioesters, the above-mentioned characteristics constituted the starting point in the development of some new chemical compound carriers of hydroxamic structures to be utilized in technology, agriculture, zootechnics, food industries and, last but not least, as drugs with considerable antimicrobial, anticonvulsive, antitumoral and antiviral effects.

Hydroxamic compounds have drawn the specialists' attention as to their possible utilization as drugs as early as 1950, when in Poland [20] and in the United States [21] the antitubercular activity of these compounds was shown. Even if the results obtained by the two teams did not permit a unique conclusion, the scientific community got the signal and, later on this category of compounds came to be considered as a potential carrier of biological activity and therapeutic effects.

In the following, thorough discussion on the influence of hydroxamic compounds on the enzymes' activity, as well as their antiviral and antitumoral effects, will be presented. Table 2 systematizes the application of hydroxamic compounds in other domains, such as agriculture, zootechnics, food, leather and paper industries as well as in cosmetics or medium environmental protection.

It is known that increase of urease's concentration in the urogenital system, as determined by certain microbial infections, increases considerably the risk of lithiasis, caused mainly by the decomposition of urea and formation of ammonium cations. The existence of these cations explains the increase of urine's pH up to 9.5, the decrease in solubility of certain salts present in the urine and, implicitly, precipitation of the calcium or magnesium ureates and oxalates.

Hydroxamates' capacity of inhibiting urease's activity [34] intensified the efforts made for the discovery of some new hydroxamic compounds, capable of inhibiting the activity of urease or of other enzymes. Thus, there have been obtained 2-ethyl-*n*-butyroylglycin hydroxamic acid and *n*-nonanoylglycin hydroxamic acid [35,36] with final application in zootechnics as deodorizers' [37]. The compound with structure **I** inhibits transaminase's activity, while that with structure **II** may control the intensity of the fermentative process through which the glutamic acid is transformed into phenylalanine [38].

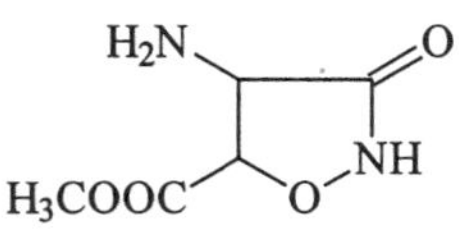

I

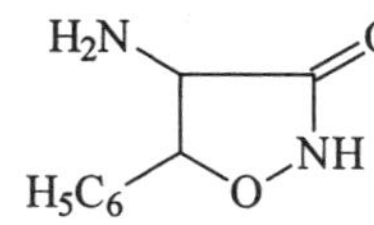

II

The phosphoglycolhydroxamic acid with structure **III** inhibits the activity of the enzymes' specific to the process of glycolysis, while the phenylhydroxamic acid inhibits activity of peroxidases [39,40].

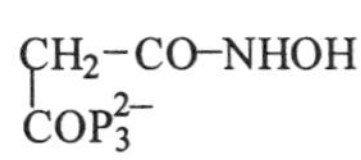

III

Table 2 Applications of Synthetic Hydroxamic Compounds

Structure	Applications	References
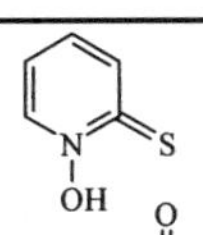	Antimycotic agent	22,23
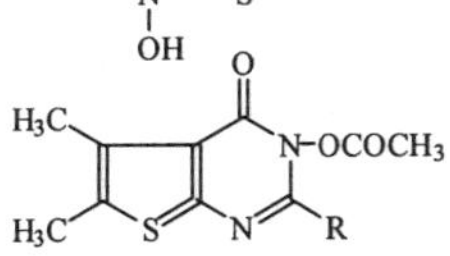	Antifungic and protective agent used in leather and wood industries.	22
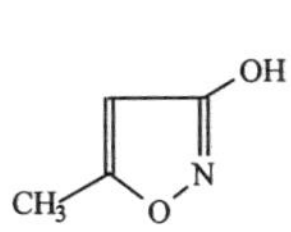	Used in agriculture as protective agent against the specific diseases of citric fruit, wine of noble stock and rice cultures.	20
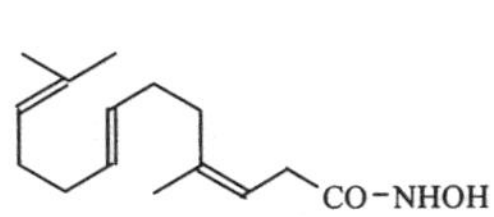	Fungistatic agent	20

Table 2 Continued

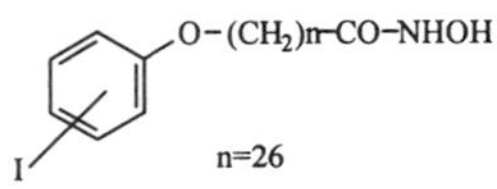	Fungistatic agent	20
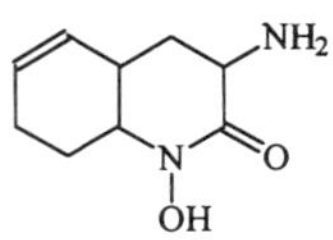	Bactericidal agent	25
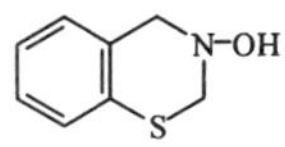	Bactericidal agent	25
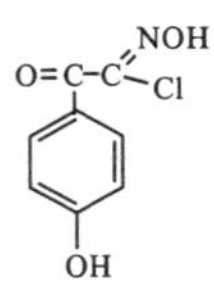 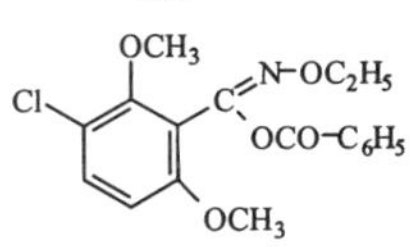 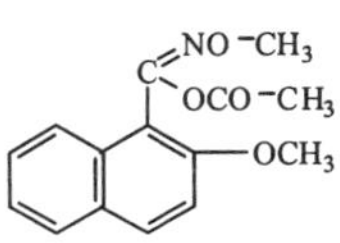	Bactericidal and preservative agent used in food industry.	25–27
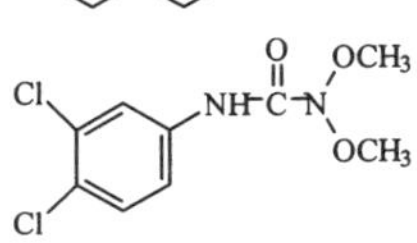 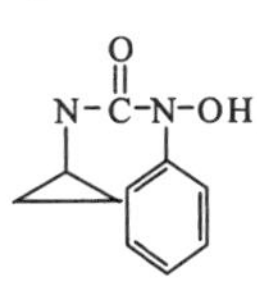 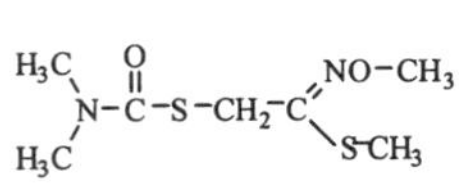	Used in agriculture as protective agent against insects and rodents.	28–33

The derivatives of the acetyl- and arylacetylhydroxamic acids, as well as those containing heterocycles in their chemical structure (compounds IV–VI), manifest a considerable anti-inflammatory activity; however, only compound IV came to be utilized in medicine as Parfenac [44,45].

The hydroxamic compounds with structures VII–IX evidence hypoglycemic effects [46–48], while the hydroxamic derivative of the γ-linoleic acid and the hydroxamic compounds with structures X–XV have anticonvulsive activity [49,53].

As early as the 1960s, studies [54,55] have been devoted to the antileukemic activity of hydroxamic compounds and especially of hydroxyurea. Following the observation of the favorable responses of a number of experimental tumors of hydroxyurea, a series of clinical trials was begun as a Phase I evaluation [56]. It was established that the maximum tolerated dose of

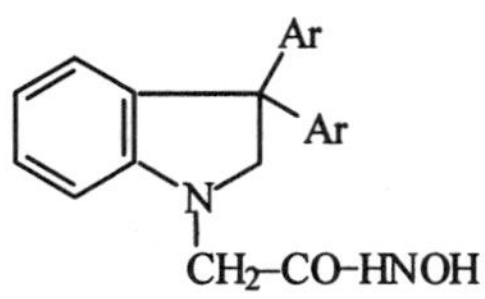

IV

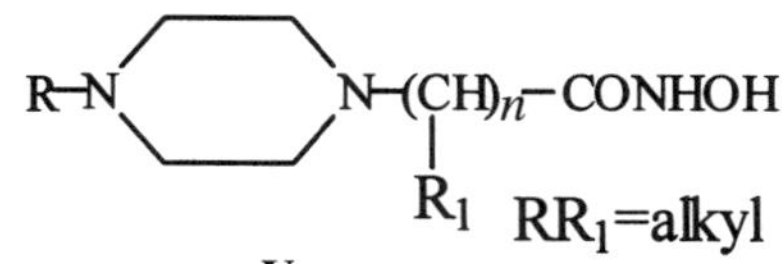

V

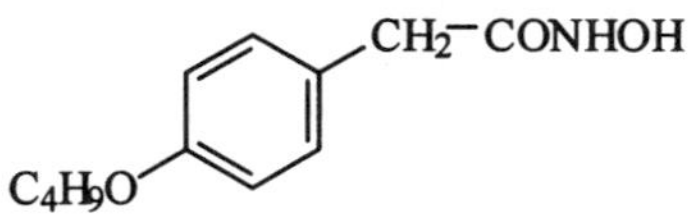

VI

HOS–NH–NH–CO–CO–NHOH

VII

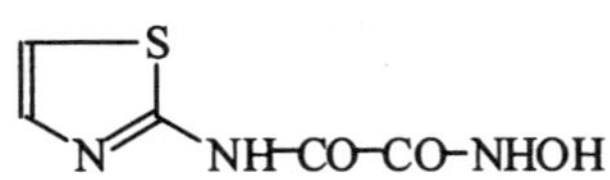

VIII

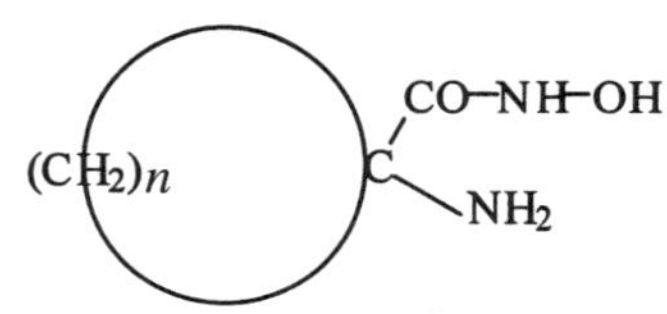

IX

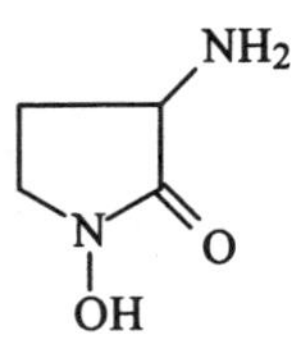

X

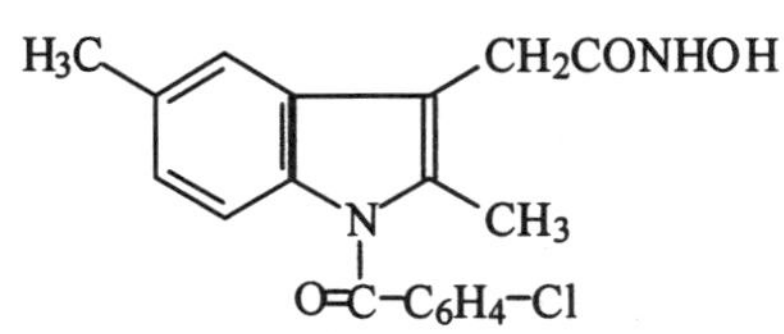

XI

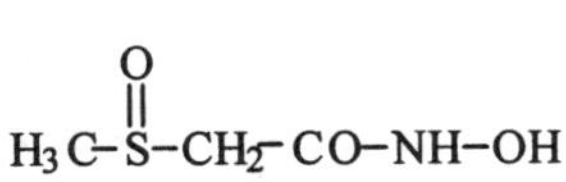

XII

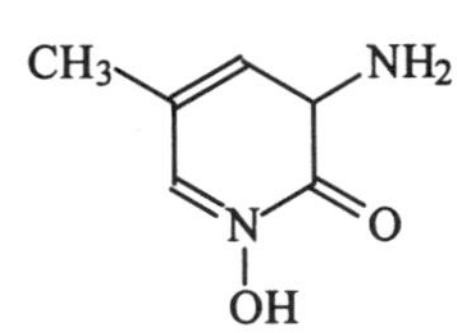

XIII

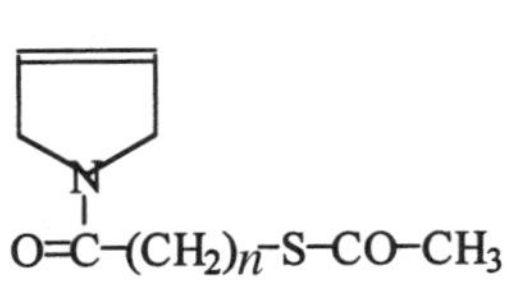

XIV

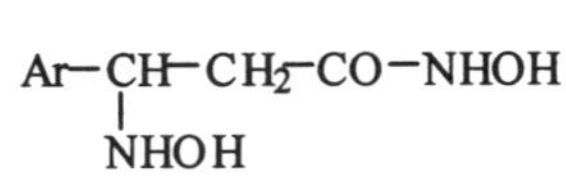

XV

hydroxyurea was 40–60 mg/kg body/day by the oral route. Toxic responses consisted of skin rashes, vomiting, alopecia, and cardiac arrhythmia, but the limiting factor in therapy was considered bone marrow depression. Evidence has been presented that 1–10% of the administered dose of hydroxyurea may be converted to hydroxylamine [57]. This was based on the discovery of acetohydroxamic acid in the blood of three patients receiving hydroxyurea. It was suggested that the acetohydroxamic acid detected was one of the products of the cleavage of acetyl coenzyme A by hydroxylamine formed from hydroxyurea, this action being proposed as at least one facet of the mechanism of leukopenic action of drug.

The mechanism by which hydroxyurea produces antitumoral effect is still unknown, but two proposals have been advanced: (1) Hydroxyurea is hydrolyzed in vivo and yields free hydroxylamine. This then cleaves the thioesters, in particular acetyl coenzyme A, disrupting oxidative phosphorylation and reducing concentration of cellular and mitochondrial adenosine triphosphate (ATP) [58]. (2) Hydroxyurea and a number of related derivatives of hydroxylamine cause fragmentation of the isolated DNA and induce chromosomial abnormalities in mammalian cells, cultured in vitro [59]. The postulated common mechanism is that oxidative transformation of these compounds occurs by forming the nitroxyl radical (HON=) which becomes dimerized as hyponitrous acid (HON=NOH), which induces cleavage of the main chain of cellular DNA [59].

However, two observations offer strong evidence against the concept that a disruption of oxidative phosphorylation is pertinent to the inhibitory effects of hydroxyurea on incorporation of ribonucleotides or thymidine into DNA. The former observation is that the ATP-generating source was not used in subcellular studies and the latter is that 2,4-dinitrophenol, a compound known to induce decrease in cellular ATP concentration, inhibited incorporation of leucine but not of thymidine in the cell system used by Young and Hodas [60]. The cellular and subcellular data [60] suggest that hydroxyurea alters the cellular synthesis of DNA by interfering with ribonucleotides (diphosphate) reduction.

The data of Frenkel, Skinner and Smiley [61] reveal the decreased conversion of cytidylic acid to deoxycytidylic acid by subcellular extracts of bone marrow taken from rats and patients after treatment with hydroxyurea. Their data suggest that hydroxyurea inhibits the conversion of ribonucleotides to deoxyribonucleotides in vivo as well as in vitro. However, since they did not observe any inhibition in the incorporation of thymidine into DNA of treated rats, the relation between the effects of hydroxyurea in isolated systems and in intact animals requires further clarification.

One of the most interesting aspects of the hydroxyurea molecule is its relative simplicity as compared with other compounds currently in use in the therapy of neoplastic diseases. It appears, however, that this simplicity of structure is not accompanied by chemical stability. This fact may lead to dissimilarities in the results obtained in some clinical test [56].

Some of the studies performed in our laboratory [62,63] showed that *N*-hydroxy-3,4-dihydroxybenzamide (Didox, compound **XVI**) manifests a significant cytotoxicity versus K562 chronic human erythroleukemy cell line and also an antitumoral activity versus the Walker 256 carcinosarcoma.

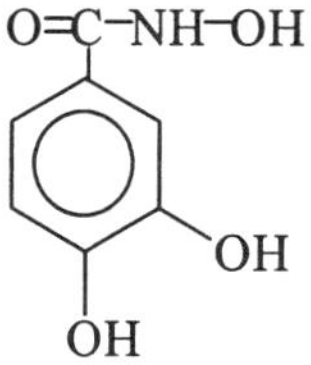

XVI

Similar to hydroxyurea, Didox is a substance with a simple chemical structure. A brief analysis of Didox's chemical structure evidences the presence of three structural moieties, namely the phenyl nucleus, the hydroxylic groups attached directly to the phenyl nucleus and the hydroxamic group. The major objective of our investigation is to determine the structure-biological activity relationships, i.e., to define the influence of each structural element on Didox's antitumoral activity. The problems to be solved were the following: Which of the above structural elements is responsible for the biological activities of Didox? To what extent does a synergism exist between them? Is the size of the molecule carrying these structural elements involved? Does it influence the intensity of their biological effects. Our intention was to find answers to these questions by evaluating the antitumoral activity of compounds **XVII** and **XVIII**.

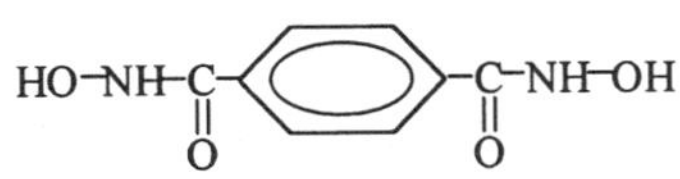

XVII

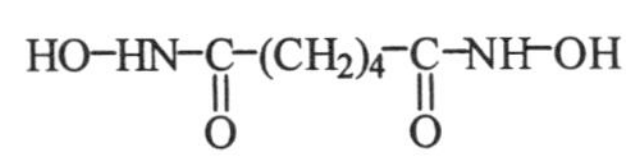

XVIII

Starting from the three structural elements forming Didox (the phenyl nucleus, the hydroxylic groups and the hydroxyamidic function), the observation was made that in compounds **XVII** and **XVIII**, two hydroxyamidic functions are present. Compound **XVIII** contains no aromatic structures, while compounds **XVII** and **XVIII** have no hydroxylic groups. The experimental data obtained by us show that such structural differences influence their biological activity [62].

Our results on the K562 leukemic line show that, in vitro, cytotoxicity is higher for compounds with an aromatic nucleus in the structure. As to selectivity, the data obtained permit us to assert that the hydroxamic groups are responsible for the cytostatic effect produced, while the hydroxyphenolic groups ensure selectivity.

The attempt of Uglea and Ottenbrite [63] to attenuate the side effects of the treatment with Didox (compound **XVI**) showed that, by this transformation, Didox becomes inactive in vitro, in parallel with a more intense antitumoral activity in vivo. Such a behavior suggests the hypothesis that the Didox-carboxymethylcellulose system is an indirect effector, its antitumoral activity occurring through the stimulation of macrophages.

Recently, we used a series of polyurethanes as carriers of compounds **XVII** and **XVIII** [Tinca Buruiana, Daniela Spridon, E. C. Buruiana, C. V. Uglea, V. Hefco, unpublished data]. The strucures of the obtained polymers are shown in formulae **XIX** and **XX**.

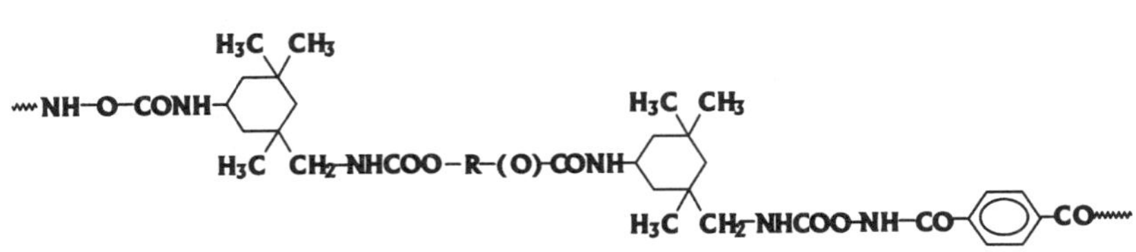

R : - [(CH_2)O $]_n$ - ; -[(CH_2 $)_4$ COO $]_n$-
PEO PCL

XIX (a) PUB1 (PEO: PCL = 1 : 1)
(b) PUB2 (PEO: PCL = 1 : 4)

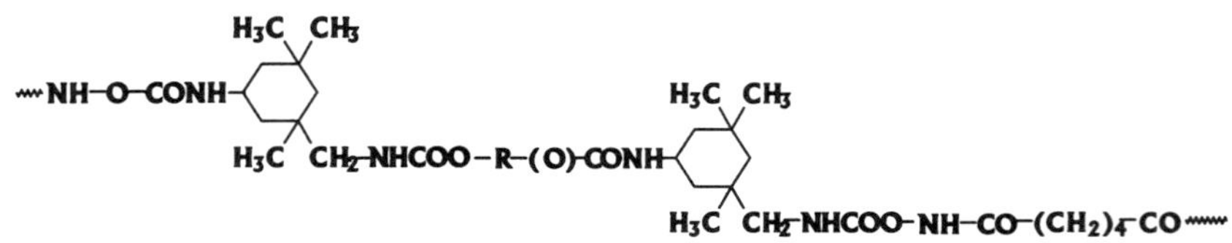

XX (a) PUB3 (PEO: PCL = 1 : 1)
(b) PUB4 (PEO: PCL = 1 : 4)

The in vivo evaluation of compounds **XIX** and **XX** on carcinosarcoma K256 bearing rats showed that, under these conditions, the aliphatic dihydroxamic compound manifests a more intense antitumoral activity than that determined for the aromatic compound. Our opinion is that such surprising behavior was determined by the modification of the hydrophylic:hydrophobic ratio, which influences the macromolecule's conformation.

Recently, a new treatment approach to HIV infection is based on the idea that the quantity of deoxynucleotides can be reduced, thereby impairing the synthesis of proviral DNA by inhibiting cellular ribonucleotide reductase.

This premise has gained recognition in several recent reports [64,65] that have demonstrated that hydroxyurea, a clinically used ribonucleotide reductase inhibition, inhibits HIV DNA synthesis in lymphocytes and activated macrophages in vitro. The in vitro trial reveals that splenomegaly, a marker for Rauscher murine leukemia virus, was suppressed or completely inhibited by Didox (500, 250, and 125 mg/kg body, qd), 3,4,5-trihydroxybenzamidoxime-Trimidox (250, 125, and 62.5 mg/kg body, qd) and hydroxyurea (1000, 500, 300, and 100 mg/kg body, qd). Trimidox was the most potent compound. Serum reverse transcriptase activity was inhibited at highest doses of Didox and Trimidox. It may not be necessary for these compounds to inhibit virus replication as monotherapies, since they have synergistic activity in vitro with nucleoside analogue antiviral drugs. These data support testing of ribonucleotide reductase inhibitors as a new treatment of HIV.

IV. CONCLUSIONS

The diversity of the biological activity of the hydroxamic compounds requires a very correct appreciation of the perspectives of chemotherapy, as well as of the objectives of drug administration engineering. There is a little doubt that drugs are intended to treat the cause of a disease rather than the symptoms.

Talking into consideration the limits of such a situation, an adequate strategy, which should include hydroxamic compounds in macromolecular systems capable of inducing their controlled release and their targeting, becomes absolutely necessary. In this respect we recommend the utilization of some macromolecular carriers to assist membrane permeability of the hydroxamic compounds and to modulate the release rate of the drugs. Such objectives assume essential modifications of our concepts. The first modification should to refer to the realization of a system capable of ensuring that drug release represents a multidisciplinary activity involving polymer and pharmaceutical scientists, chemical engineers, and a variety of biologically oriented scientists. A second modification refers to the necessity of employing new polymers, possessing specific properties, for the medium in which they should be applied.

Finally, the driving force for most of these changes is an expanding understanding of biology as it pertains to drug delivery systems. If the decades 1970–1990 represent the period to define what had been needed in controlled drug delivery and understanding, even at an organ level, of the disposition issues for the various routes of administration, the period 1990–2010 will represent the true biomedical polymer period. Of course, success of this period is dependent on continued economic success of drug delivery systems and the willingness of certain companies or entrepreneurs to invest in this future.

REFERENCES

1. Simionescu, Cr. I. and Denes, F., *The Origin of Life*, Academic Editorial House, Bucharest, 1983.
2. Dresler, W.F.C. and Stein, R., *Ann. Chem.*, 150, 242, 1869.
3. Neilands, J.B., Hydroxamic acids in nature, *Science*, 156, 1443–1447, 1967.
4. Gale, G. R., Kendall, S. M., McLain, H. H. and Du Bois, S., *Cancer Res.*, 24, 1012–1018, 1964.
5. Wise, V.M., *J. Amer. Chem. Soc.*, 77, 1058, 1955.
6. Dendon, M., Guinet, R., Carrez, M., FR Patent, 2353289, 1977; Cf. *Chem. Abstr.*, 89, 129285, 1978.
7. Staab, H.A, Lueking, M. and Duerr, P.H., *Chem. Ber.*, 95, 1275, 1962.
8. Smith, P.A.S. and Hein, G.M., *J. Amer. Chem. Soc.*, 82, 5732, 1960.
9. Kablaoni, M.S., US Patent 3998861, 1976, Cf. *Chem. Abstr.*, 86, 139426, 1977.
10. Pilgram, K.H., US Patent 4327033, 1982, Cf. *Ref. Zhur.*, 170, 364 P, 1982 (Russian).
11. Genem, B., *Tetrahedron Lett.*, 1976, p. 1951.
12. Deghenghi, R., *Org. Syn.*, 40, 60, 1960.
13. Davidson, J.D. and Winter, T.S., *Cancer Chemotherapy Rept.*, 27, 97, 1973.
14. Cooley, J.H., Bills, W.D. and Throckmorton, J.R., *J. Org. Chem.*, 25, 1734, 1960.
15. Mizurami, S. and Nagata, K., *Coord. Chem . Rev.*, p. 267, 1968.
16. Hurd C.D. and Bethune, V.G., *J. Org. Chem.*, 29, 916, 1970.
17. Sklarz, B. and Al-Sayah, A.F., *J. Chem. Soc.*, p. 1318, 1964.
18. Safir, R.S. and Williams, J.H., *J. Org. Chem.*, 17, 1298, 1952.
19. Hase, J., Kobaski, K. and Nakai, N., *Chem. Pharm. Bull.*, 19, 363, 1971.
20. Urbanski, T., *Nature*, 166, 267, 1950.
21. Gardner, T. S., Wenis, E. and Smith, F.A., *J. Am. Chem. Soc.*, 73, 5455, 1951.
22. Douglass, M.L., US Patent 3971725, 1976.
23. Wedig, J.H., Mitoma C., Howd, R.A. and Thomas, D.W., *Toxicol. Appl. Pharmacol.*, 43, 373, 1978.
24. Zhunget, G.I. and Artemenco, A.I., Hydroxamic acids, *Science*, Kishinev p. 105, 1986.
25. Bell, S.J., Friedman, S.A. and Leony, J., *Antimicrob. Agents Chemother*, 384, 79; Cf. *Chem. Abstr.*, 91, 746, 1979.
26. Wakeman, R.L. and Coates, J.F., US Patent 3427316, 1969; Cf. *Chem. Abstr.*, 70, 9637 t, 1969.
27. Noessler, H. G., Ger. Patent 1134387, 1967; Cf. *Chem. Abstr.*, 92870 I, 1972.
28. Simidzu, M., Kisawara, A. and Mathei, S., JP Patent 55-47008, 1980; Cf. *Ref. Zhur.*, 40, 89 P, 1982.
29. Mel'nikov, N.N., Novojilov, K. and Pilova, T.N., *Chemical Agents Used in Agriculture*, Kimia, Moscow, p. 135, 1980.
30. Kamikado, T., Jap. Patent 77-83823, 1978; Cf. *Chem. Abstr.*, 88, 22457, 1978.
31. Pilgram, E.H., US Patent 4299778, 1981.
32. Johnson, W.J., Can. Patent 1089763, 1980.
33. D'Amico, J. and Schafer, D.E., US Patent 4377366, 1982.
34. Rosenstein, I., *J. Antimicrob. Chemother*, 10, 159, 1982.
35. Jerusik, R.J., Kadis, S., Chapman, W. L. and Wooley, R.E., *Can. J. Microb.*, 23, 1448, 1977.
36. Kobashi, K., Takabe, S., Tarashima, S. and Hase, J., *J. Biochem.* (Tokyo), 77, 837, 1975; Cf. *Chem. Abstr.*, 83, 39342, 1975.
37. Kitai, K. and Arakaure, A., *Br. Poll. Sci.*, 20, 59, 1979.
38. Homutov, R.M., Severin, E.C. and Gulaev N.N., *Izv. Akad. Nauk. SSSR, Ser., Kim.*, p 1990, 1966 (Russian).
39. Lewis, D.J. and Lowe, O., *Chem. Commun*, 713, 1973.

40. Coletti-Previere, M.A., *Biochem. Biophys, Res. Commun.*, 107, 465, 1982.
41. Papadaki-Valiraki, A., Tsatsas, C. and Varanos, P., *Ann. Pharm. Fr.*, 32, 133, 1974.
42. Orzalesi, O., Ger. Patent 2400531, 1974.
43. Failer, M., A., A. Patent 277212, 1969.
44. Deli, D. and Borehan, D.R., *J. Phar. Sci.*, 60, 1268, 1971.
45. Schaefer, M. and Stuettgen, G., *Arzneimittel-Forsh.*, 28, 1021, 1978.
46. Petunin, P.A., Su Patent 535293, 1976.
47. Bezugli, P.A., Su Patent 740775, 1980.
48. Alburn, H.E., Clark, D.E. and Lapidus, M., US Patent 3845109, 1974.
49. Bonta, I.L., De Vos, C.J. and Grissen, H., *Brit. J. Pharmacol.*, 43, 514, 1971.
50. Tamietto, T., UK Patent 1570772, 1980.
51. Lafon, L., US Patent 4183951, 1980.
52. Jones, P.H. and Martin, Y.C., US Patent 3895114, 1975.
53. Petrillo, E.W. and Ondetti, M.A., US Patent 4284561, 1981.
54. Kabara J.J., Spafford N.R., *Advances in Tracer Methodology*, Rothchild S. Ed., Plenum Press, New York, vol. 1, p. 76, 1962.
55. Stearns, B. and Losee, K. A., *J. Med. Pharm. Chem.*, 6, 201, 1963.
56. Thurman, W.G., Bloedow, G. and Griffith, K.M., *Cancer Chemotherapy Rept.*, 29, 103, 1963.
57. Davidson J,D. and Winter T.S., *Cancer Chemotherapy Rept.* 27, 97, 1963.
58. Fishbeim, W.N. and Carbone, P.P., *Science*, 142, 1069, 1963.
59. Borenfreund, E., Krim, M. and Bendich, A., *J. Natl. Cancer Inst.*, 32, 667, 1964.
60. Young, C.W., Hodas, S. and Fennelly, J.J., *Science*, 146, 1172, 1964.
61. Frenkel, E.P., Skinner, W.M. and Smiley, J.D., *Cancer Chemotherapy Rept.*, 40, 19, 1964.
62. Spridon, D., Panaitescu, L., Vatajanu, A., Buruiana, E., Hefco, V., and Uglea, C. V., *Roum. Biotechnol. Lett.*, 2, 1310, 1997.
63. Uglea, C.V. and Ottenbrite, R.M., Polyssacharides as support for antiviral and antitumoral drugs, in *Polyssacharides in Medicinal Applications*, Severian Dumitriu, Ed. Marcel Dekker, New York, 1996, Ch. 26.
64. Eford H., *Retroviral Research*, Suppl., 17, 176, 1995.
65. Eford H., *AIDS Res. Human Retroviruses*, 11, Suppl. 1, 1995.

41

Development of a Modified Fibrin Adhesive

What Can We Learn from Biological Adhesion Mechanisms?

Frederick H. Silver, Russell T. Kronengold, Ming-Che Wang, Dominick Benedetto, and David L. Christiansen
UMDNJ-Robert Wood Johnson Medical School, Piscataway, New Jersey

I. INTRODUCTION

Tissue adhesives are needed to replace sutures and staples in applications where contact is required to prevent blood, lymph or air leakage or to join raw tissue surfaces. They have been used to repair a variety of medical problems involving the following tissues: musculoskeleton, skin, face, heart, lungs, ear, nose, throat, reproductive system, urinary tract, nervous system and eye [1]. Although tissue adhesives are used widely in Europe, in the United States they are used for only selected applications. The "approved" materials include fibrin glue made from cryoprecipitate, which is used to control bleeding in cardiothoracic surgery, and cyanoacrylates, which are approved for limited ocular use.

Both of these materials have major limitations, including tissue toxicity of cyanoacrylates and poor mechanical properties of fibrin glue. Cyanoacrylates have been the gold standard of adhesives to bond wet tissues together with high bond strengths. Unfortunately, they produce inflammation during degradation and hard masses when the degradation rate is reduced. Bond tensile strengths as high as 7 MPa [2] and lap shear strengths as high as 32 kPA [3] have been reported. Lap shear strengths for fibrin glue range from about 10 to 20 kPa depending on the fibrinogen concentration [1,4].

The purpose of this chapter is to examine the underlying physics and biology of adhesion in order to improve the design of tissue adhesives.

II. BACKGROUND

In theory the adhesion of two surfaces can occur by the formation of either mechanical or chemical bonds. Physical linkage occurs when a liquid adhesive is applied to two surfaces and penetrates the superficial pores forming a continuous bond. Liquid adhesives are solutions of short chain polymer molecules in a solvent that flow into the surfaces that they contact. Once the solvent evaporates, the polymer chains hold the two surfaces together by either electrostatic or van der Waals attractions. Unless crosslinks form between the chains on the two surfaces this type of adhesion is purely mechanical and is reversible. In wet systems, physical bonds are very weak and will not offer much resistance to shear failure.

In wet systems the only types of bonds with enough force to hold two surfaces together are covalent chemical bonds. Adhesives that form covalent bonds are liquids that are able to penetrate the pores on the surfaces to be joined and then crosslink to the surfaces. The requirements for penetration to occur include low viscosity of the adhesive and the ability to "wet" the surfaces. The adhesive viscosity is controlled by limiting the length and stiffness of the polymer chains before covalent crosslinking while the "wettability" of a surface has to do with the surface tension of the surfaces and the adhesive. An ideal biological adhesive has a low viscosity so that the material can flow easily into the pores of each surface. Once the adhesive is within the pores it should covalently crosslink to form a tough network with an infinite molecular weight. Both cyanoacrylates and fibrin glue are initially low viscosity materials capable of flowing into the surface pores. Fibrin glue self-assembles and crosslinks to surfaces during clotting, which physically and chemically binds it. "Wettability" is related to the relative surface tension of the glue and the surfaces. Surface tension represents the state of unfulfilled bonding of the surface atoms of a material. Materials with high surface tension are penetrated by liquids with lower surface tensions. Another way of saying this is that a surface will lower its surface tension by adsorbing a liquid with a lower surface tension. This is why oil with a surface tension of about 18 dynes/cm^2 will make a monomolecular film on water, which has a surface tension of about 72 dynes/cm^2. The critical surface tension of wet tissue has been estimated to be about 50 dynes/cm^2. Therefore the surface tension of any adhesive designed to wet biological tissue should be less than this value. In addition to having a surface tension less than 50 dynes/cm^2, a good adhesive will mix with material at the surface. This occurs when the solubility parameter of the adhesive and surfaces are similar. Although oil forms a very thin layer on water it does not mix with water because the solubility parameters are quite different. The solubility parameter for oil is 6.2, while that for water is 23.5.

A. Biology of Clotting

The basis for formation of a blood clot is the self-assembly of fibrinogen, a blood plasma protein formed in the liver. The blood concentration of this protein is about 3% and in the presence of other blood proteins, pieces of the fibrinogen molecule are cleaved and activated fibrinogen is polymerized to form a fibrin clot [5].

Fibrinogen is composed of six peptide chains—A, α, B, β and 2γ—giving the molecule a total molecular weight of about 330,000. The molecule consists of two terminal nodules (globular regions) attached to a central nodule by two three-stranded ropes. Dolittle proposed

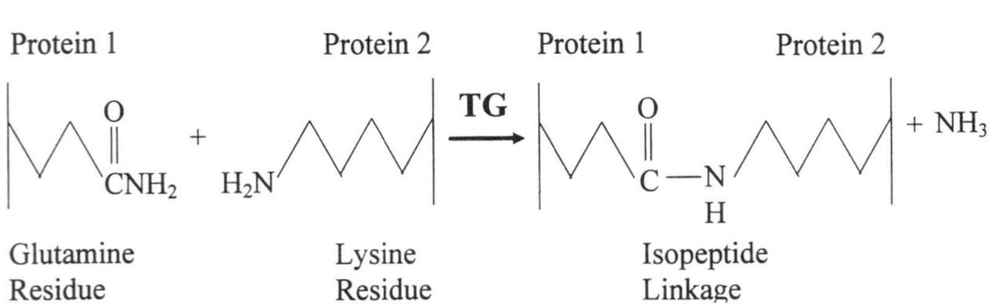

Figure 1 Diagram illustrating factor XIIIa crosslinking of fibrin, collagen, gelatin and fibronectin. Factor XIIIa is a transglutaminase which catalyzes the transfer of a glutamine from one peptide to a lysine on another peptide. This transfer is associated with the formation of a covalent linkage.

that the structure consisted of a central nodule containing the amino-terminal regions of all six peptide chains. Two terminal or distal nodules are attached to the central one by two three-stranded ropes in the form of coiled coils. Each of these coiled coils contains approximately 110 amino acid residues and extends about 160 Å in space. The central nodule contains 11 of the 29 disulfide bonds in the molecule. The molecular weights of the Aα, Bβ and γ chains are 64,000, 56,000 and 47,000 per chain, respectively (i.e., Aα are two chains and therefore have molecular weight of 128,000).

Fibrinogen is involved in formation of fibrin networks that make up the noncellular component of blood clots and is converted into fibrin via the intrinsic and extrinsic clotting pathways. In either of these pathways, prothrombin is converted to thrombin, which proteolytically removes fibrinopeptides A and B from the fibrinogen molecule. Removal of fibrinopeptides A and B (FPA and FPB) results in only a small decrease in molecular weight and allows assembly to proceed by the interaction of sites blocked by FPA and FPB. It has been proposed that removal of FPA leads to linear polymerization while removal of FPB leads to lateral growth. Initiation of self-assembly occurs by removal of FPA and dimerization of two fibrinogen molecules, which is followed by removal of FPB and linear polymerization. Further assembly occurs by lateral packing of elements that are formed during the previous step.

At the molecular level, splitting off fibrinopeptides A and B from the α- and β-chains of fibrinogen, respectively, occurs at specific arginyl-glycine peptide bonds. Finally, after fibrin monomer polymerizes, Factor XIIIa catalyzes the formation of covalent crosslinks between fibrin molecules. In addition, fibrin covalently crosslinks to fibronectin and collagen via Factor XIIIa [6,7].

Factor XIII, also known as fibrin stabilizing factor, is a plasma enzyme termed a transglutaminase that is responsible for catalyzing reactions that result in intermolecular crosslinks between several proteins found in extracellular matrix. The most prominent function of Factor XIIIa is crosslinking of fibrin monomers to produce insoluble, polymerized fibrin clots.

Plasma Factor XIII is a circulating blood protein that has a tetrameric a_2b_2-chain structure consisting of two a-chains and two b-chain subunits [8] with total molecular weight of about 320,000. Activation of Factor XIII occurs in two stages; first a peptide 37 residues long is cleaved from both a-chains by thrombin leaving the subunits noncovalently associated as $a_2'b_2$. Step 2 involves dissociation of $a_2'b_2$ in the presence of calcium, which binds to a_2', forming the active center.

Plasma Factor XIII is a transglutaminase that catalyzes a Ca^{2+} transfer between an ε-amino group on a lysine and the γ-carboxamide group of a glutamine residue on another peptide chain [9,10]. In the presence of active transglutaminase a glutamine and a lysine are covalently linked and ammonia is released (Fig. 1).

Activated Factor XIIIa catalyzes the formation of peptide bonds between the ε-amino group of lysine from one fibrin monomer and a glutamine residue on an adjacent fibrin monomer. This process creates roughly one peptide bridge per 500 residues in a fibrin molecule, which translates into about 5 to 10 crosslinks per molecule. Covalent crosslinking increases the strength of the fibrin polymer and provides increased resistance to lysis by plasmin.

Fibronectin is a another blood protein that is incorporated into a fibrin clot by both noncovalent and covalent interactions [10]. Fibronectin is incorporated covalently via a Factor XIIIa catalyzed reaction between a glutamine residue on fibronectin and a lysine residue on fibrin. The binding sites for fibrin on fibronectin are on the Fn1 modules, while the collagen binding modules on fibronectin are on the Fn1 and Fn2 modules. The collagen binding sites on fibronectin help to anchor the fibrin clot to damaged collagen of vessels walls and other connective tissue. In this manner, collagen and collagen derived products help to serve to regulate clotting and fibrin deposition. Since collagen also contains glutamine and lysine residues, theoretically fibrin is covalently crosslinked to collagen by Factor XIIIa. Experimentally there is some question of the evidence to support collagen to fibrin crosslinking [7]. Experimental evidence [11] demonstrates that transglutaminase crosslinks gelatin, the non-triple-helical form of collagen.

This background literature review suggests that the biological adhesion process associated with hemostasis includes the following steps:

a. Self-assembly of prepolymer (fibrinogen) into a noncovalent network
b. Inclusion of bifunctional linker molecules (fibronectin and collagen) to provide attachment to the surrounding tissues and damaged tissue elements
c. Covalent crosslinking of network (via Factor XIIIa) to provide mechanical reinforcement aswell as to slow lysis
d. Covalent crosslinking of network to linkers (via Factor XIIIa)

Based on this system, the fibrin glue system in theory should demonstrate improved adhesive properties if a higher level of linker molecules that may serve to better bind the fibrin polymer to tissue substrates was formulated into the system. A number of workers have added collagen to fibrin glue systems in the hope of improving the strength characteristics of the glue; however, in our studies we have chosen to use high molecular weight gelatin because of its high affinity for binding to fibronectin and its improved molecular chain flexibility over the rigid collagen triple helix. We hypothesize that torn or denatured collagen fragments generated during tissue trauma may crosslink more effectively to fibronectin and fibrin to limit bleeding in vivo.

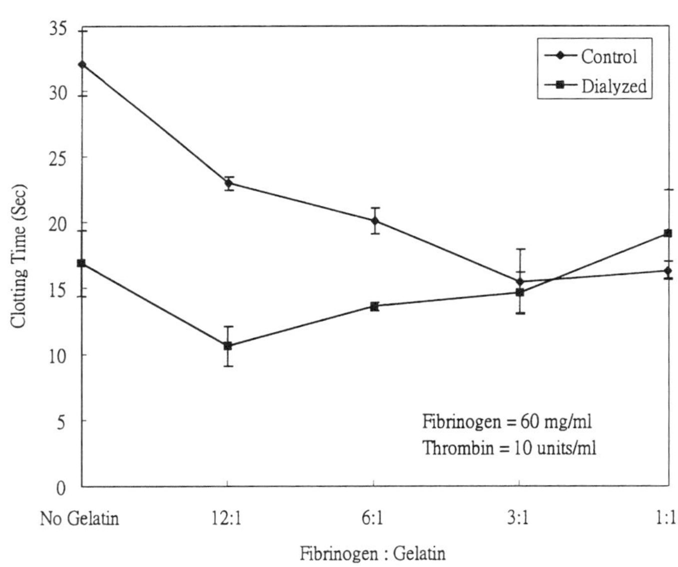

Figure 2 Effect of fibrinogen and gelatin on the clotting time of plasma. The clotting time for plasma is plotted versus fibrinogen to gelatin ratio at a fibrinogen concentration of 60 mg/ml with 10 units/ml of thrombin present. After dialysis to remove any traces of sodium citrate remaining in the plasma, the clotting time is independent of gelatin concentration.

III. PREPARATION OF MODIFIED FIBRIN GLUE SYSTEM

A. Preparation of Cryoprecipitate

Fresh adult bovine whole blood was obtained at slaughter and immediately mixed with cold sodium citrate solution, pH 7.4, to give a final concentration of 1% (W/V). The blood was kept on ice and then centrifuged at 600 g for 20 min at 4°C to separate the cellular components from the plasma. The plasma was removed using a pipet and then was centrifuged a second time. Aliquots of the cell-free plasma were placed at −15°C for at least five days and then thawed at 4°C overnight. The cryoprecipitated fibrinogen was pelleted by centrifugation at 1600 g for 5 to 10 min. The pellet was then resolubilized in up to 1 ml of distilled water, vortexed, allowed to stand, vortexed again, and then finally centrifuged at 1600 g for 5 min. Fibrinogen concentration was measured based on a the method of Ratnoff and Menzie [12].

B. Preparation of Fibrin Glue

Fibrin glue (FG) was prepared by combining 1 ml of cryoprecipitate and 1 ml of lyophilized bovine thrombin (ICN Pentex Bovine Thrombin, IVD Grade, Costa Mesa, CA) reconstituted in 20 mM $CaCl_2$ using a Micromedics Surgical Sealant Applicator (Eagan, MN). Modified

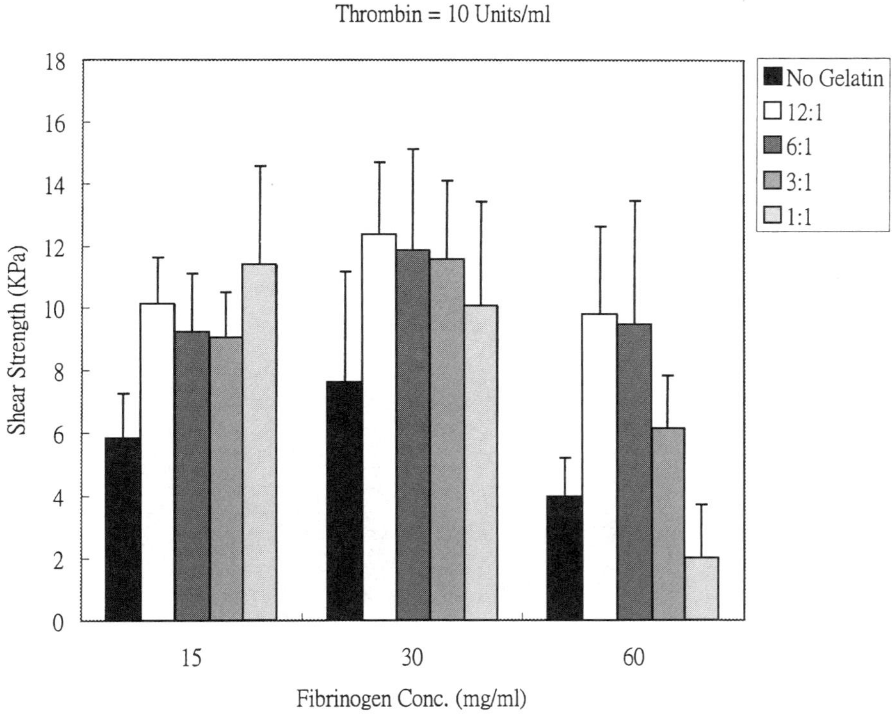

Figure 3 Lap shear strength of modified fibrin glue using 10 units/ml thrombin. Plot of lap shear strength versus fibrinogen concentration for different gelatin concentrations used to formulate fibrin glue. Maximum shear strength appears at a fibrinogen concentration of 30 mg/ml in the presence of gelatin up to a concentration of 10 mg/ml.

fibrin glue (MFG) was prepared by adding high viscosity fish gelatin (HVG, Norland Products, New Brunswick, NJ) solubilized in 20 mM $CaCl_2$.

Plasma fibrinogen concentration was determined by the clot collection method modified based on the method described previously [13]. The number of free primary amines that were removed as a result of crosslinking by Factor XIIIa were analyzed using a primary amine tag (fluorescamine) that was followed colorimetrically as described by Kronengold [14]. Clotting times and lap shear strengths were measured as described by Silver et al. [13].

IV. CLOTTING TIMES AND LAP SHEAR STRENGTHS FOR FIBRIN GLUE

Clotting time for fibrin glue made using 10 units/ml of thrombin appeared to be independent of gelatin concentration for solutions containing 60 mg/ml of dialyzed fibrinogen (Fig. 2). In contrast, the lap shear strength of bonds made using 10 units/ml of thrombin appeared to increase with fibrinogen concentration up to 30 mg/ml and up to about 3 : 1 fibrinogen to gelatin (Fig. 3). At 200 units/ml of thrombin the same trends appeared to apply especially at a fibrinogen concentration of 30 mg/ml (Fig. 4). The number of free primary amines decreased in the presence of gelatin over that noted in its absence.

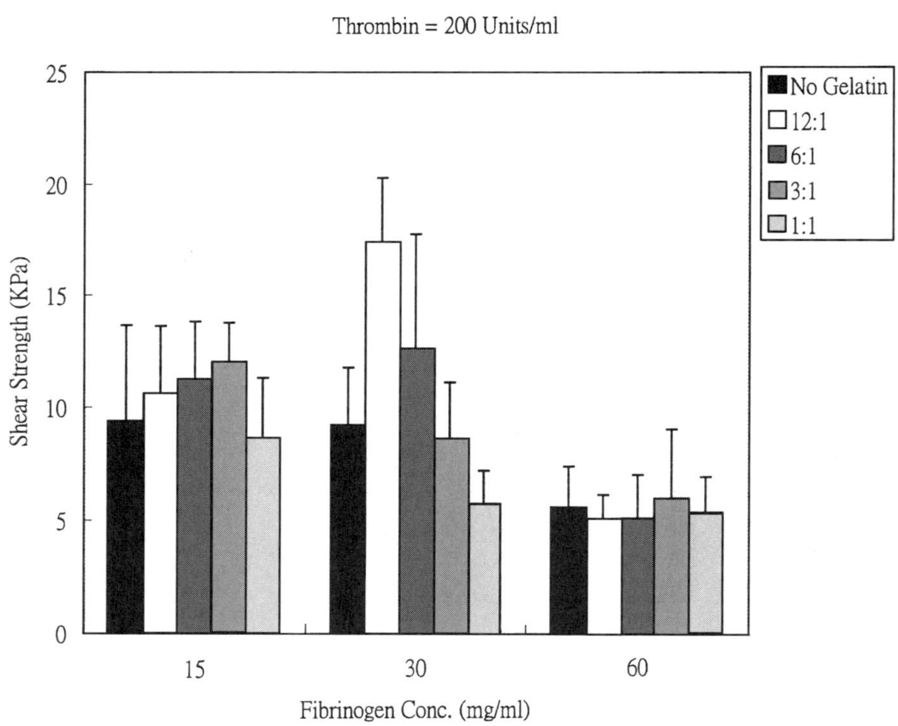

Figure 4 Lap shear strength of modified fibrin glue using 200 units/ml thrombin. Plot of lap shear strength versus fibrinogen concentration for different gelatin concentrations used to formulate fibrin glue. Maximum shear strength appears at a fibrinogen concentration of 30 mg/ml in the presence of gelatin up to a concentration of 10 mg/ml.

V. CORNEAL INCISION MODEL

Mechanical testing for determining the strength of fibrin glue sealed cataract surgical incisions was achieved using a burst test system (Fig. 5). Water from a 20 cc syringe was pumped through a 20 gauge needle into the eye to determine the burst pressure of fibrin glued incisions. A transducer (Omega Engineering Inc., model PX236-030GV, Stamford, CT) was inserted into the system to measure the pressure. The pressure transducer was calibrated with a physiograph meter and the infusion rate of the 20 cc syringe was controlled with a syringe pump.

Bovine eyes were obtained at slaughter (Moyer Packing Company) and transported cold. The eyes were tested using the corneal incision model and a thrombin concentration of 200 units/ml. Results obtained using this model are listed in Tables 1 and 2. The burst pressure was increased from 216 mm Hg for plasma clotted with thrombin to 289 mm Hg for eyes treated with FG containing 60 mg/ml fibrinogen. In the presence of 60 mg/ml gelatin and 15 mg/ml fibrinogen the burst pressure was increased to 365 mm Hg.

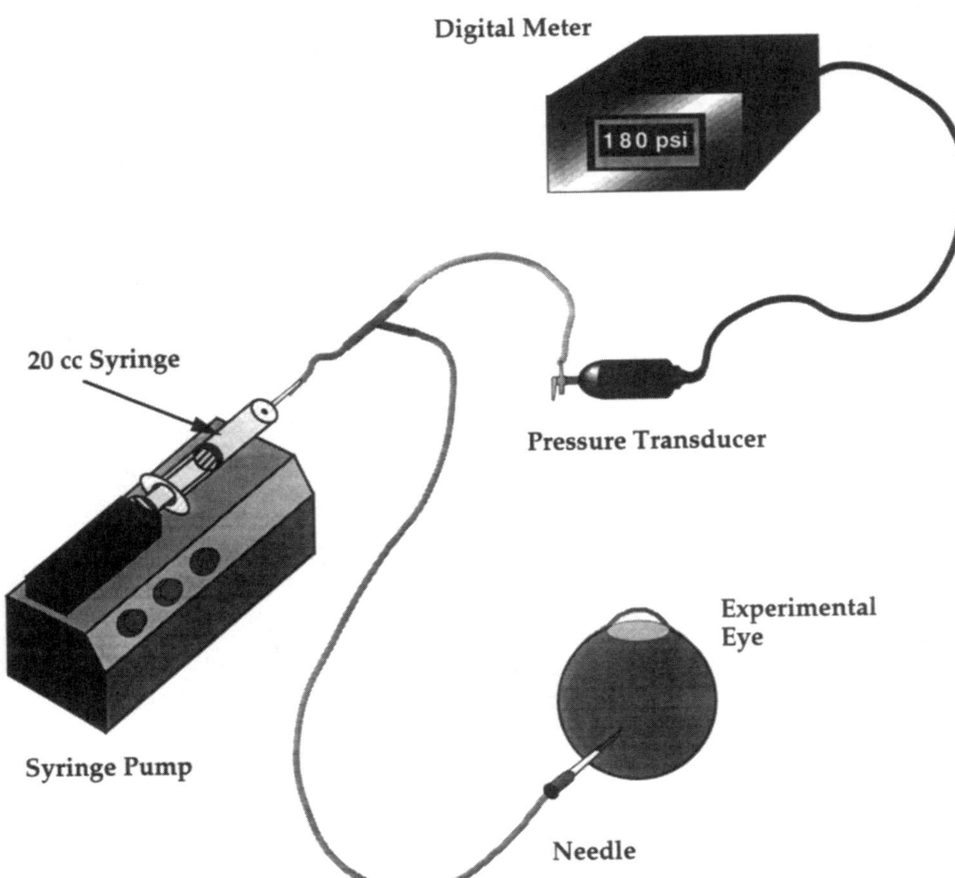

Figure 5 Burst strength test system. Schematic diagram showing components used to test the burst strength of ocular wounds. A syringe pump is used to pump water through tubing connected in parallel to a needle inserted into the test eye and to a pressure gauge. Water is pumped into the eye until it reaches a pressure sufficient to cause leakage from the wound. The pressure at leakage is recorded from the reading on the digital meter.

Table 1 Corneal Incision Model-Burst Pressures of Wounds Repaired with Fibrin Glue

Fibrin glue type	n	Mean burst pressure (mmHg)	p-value[a]
Plasma	12	216.1 ± 112.5	****
15 mg/ml[b]	14	209.5 ± 85.4	0.867
30 mg/ml	12	249.2 ± 66.2	0.392
60 mg/ml	13	289.4 ± 86.9	0.080

[a]Significance compared to plasma group.
[b]Fibrinogen concentration in mg/ml.

Scanning electron microscopy of control corneal wounds and corneal wounds repaired with fibrin glue are shown in Fig. 6. The samples were placed in modified Karnovsky's fixative and then postfixed with osmium tetroxide, dehydrated in a series of ethanol solutions (30%–100%), sublimed with Peldri II, and finally sputter coated. All samples were viewed using an Amray 1400 SEM at a 20 kV power setting.

VI. DISCUSSION

The strength of the fibrin glue system used to adhere two collagen sheets together is improved by adding high viscosity gelatin up to a concentration equal to that of fibrinogen. The addition

Table 2 Corneal Incision Model-Burst Pressures of Wounds Repaired with Fibrin Glue Containing Gelatin

Fibrin glue	n	Mean burst pressure (mmHg)	p-value[a]
15 mg/ml[b]			
+7.5[d]	9	247.8 ± 82.3	0.300
+15	9	247.8 ± 68.7	0.273
+30	8	337.1 ± 87.4	0.003
+60	8	365.4 ± 71.5	<0.001
+90	8	310.6 ± 125.8	0.069
30 mg/ml[c]			
+7.5	8	339.3 ± 106.7	0.057
+15	9	344.2 ± 113.2	0.044
+30	8	334.3 ± 91.1	0.026
+60	8	315.6 ± 71.5	0.047
+90	8	293.3 ± 95.5	0.278

[a]Significance compared to respective fibrinogen concentration alone.
[b]Prepared with 15 mg/ml fibrinogen.
[c]Prepared with 30 mg/ml fibrinogen.
[d]Gelatin concentration (mg/ml).

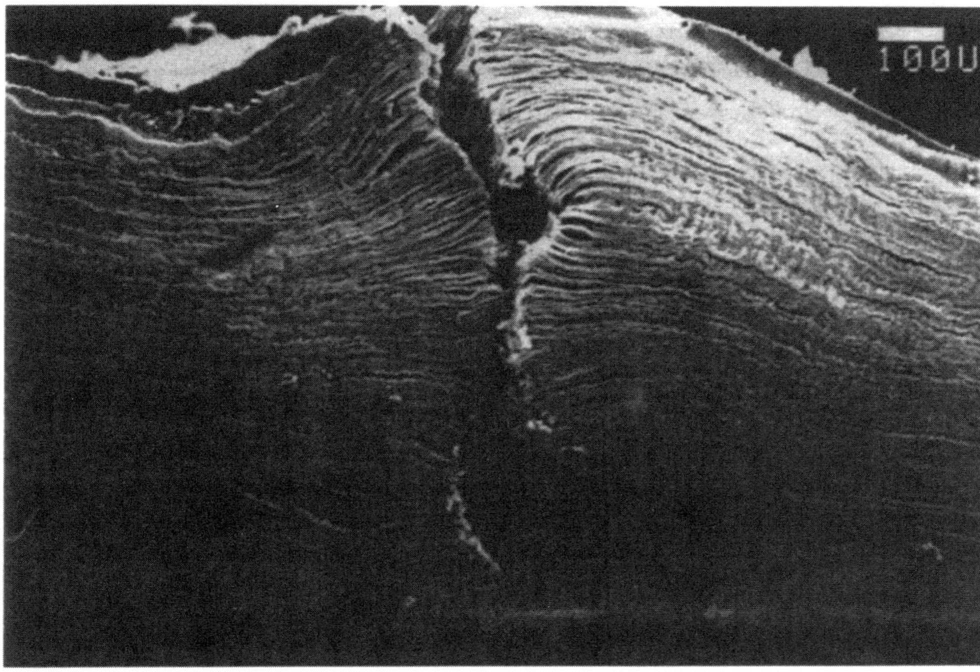

(a)

Figure 6 Scanning electron micrographs of corneal wounds. Micrographs showing separation between the wound surfaces in unrepaired corneal wounds (a and b) and corneal wounds repaired with modified fibrin glue containing 15 mg/ml of both fibrinogen and gelatin (c). Note presence of glue holding tissue edges in close proximity in (c).

of gelatin decreases the number of free amines suggesting that covalent crosslinking to collagen, fibrin and perhaps fibronectin occurs. Since gelatin and collagen bind to specific sites on fibronectin and fibronectin specifically binds to fibrin, then assembly of fibrin would appear to trap these macromolecules. Once fibrin assembly traps fibronectin, collagen and gelatin they are crosslinked by transglutaminase (Factor XIIIa) since all of these macromolecules contain glutamine residues. The net result is the improved strength of the fibrin network since there are more linkages between the macromolecular components.

The role of denatured collagen components in the regulation of fibrin gluing capacity may have a physiological role. When a vessel is cut during trauma, exposure of collagen fibers activates fibrin polymerization to enhance plugging of vascular leaks. The more extensive the trauma the more collagen is converted into a nonhelical form by the formation of breaks which would have a positive role in preventing further blood leakage. Increased collagen denaturation would increase the clot adhesive strength and prevent further blood leakage.

In summary, one mechanism by which the strength of fibrin glue can be increased is to mimic the biological clotting system. This is achieved by adding a linker molecule such as gelatin that has glutamine residues and adheres to fibronectin and fibrin. This strengthens the glue joint by providing more crosslinks to prevent slippage of fibrin fibrils.

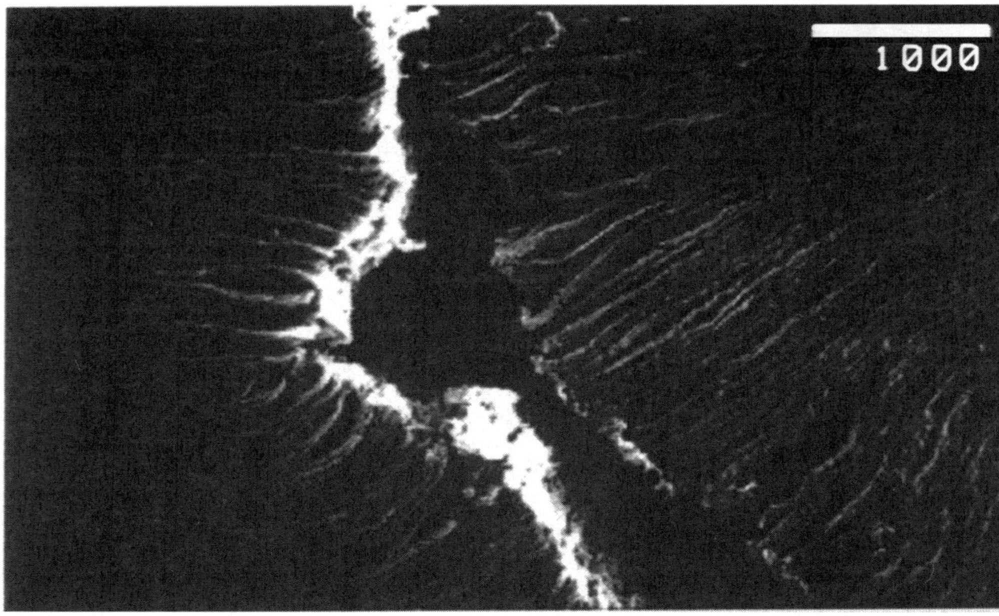

(b)

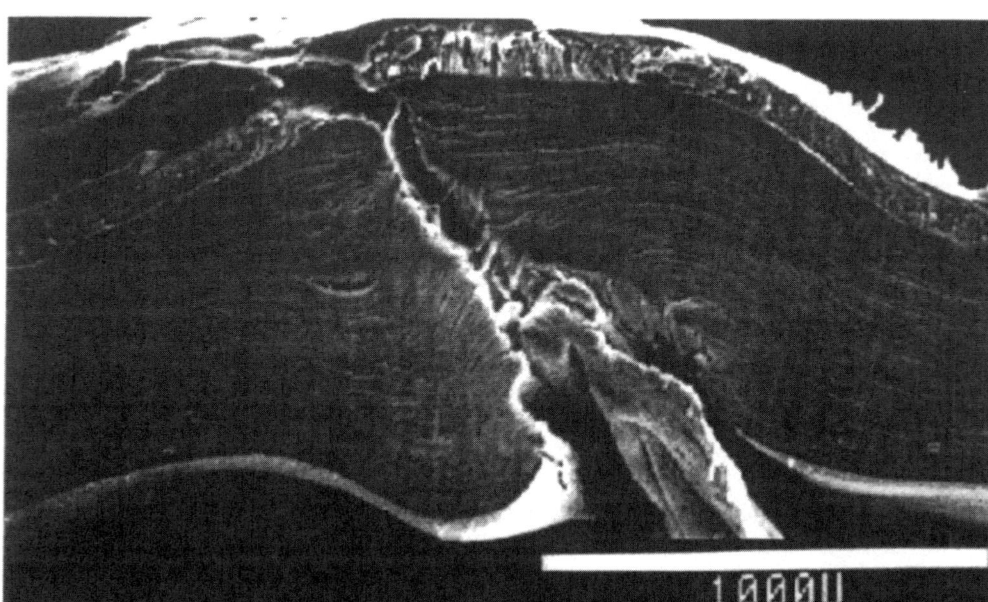

(c)

Figure 6 Continued.

REFERENCES

1. Silver, F.H., Wang, M.-C. and Pins, G. D., Preparation and use of Fibrin Glue in Surgery: A Review, *Biomaterials* 16, 891 (1995).
2. Brauer, G.M., Kumpula, J.W., Termini, D.J., and Davidson, K.M., Durability of the Bond Between Bone and Various 2-Cyanoacrylates in an Aqueous Environment, *J. Biomed. Mater. Res.* 13, 593 (1979).
3. Albes, J.M., Krettek, C., Hausen, B., Rohde, R., Haverich, A., and Borst, H. G., Biophysical Properties of the Gelatin-Resorcinol-Formaldehyde/Glutaraldehyde Adhesive, *Ann-Thorac-Surg.*, 56, 910 (1993).
4. Saltz, R., Dimick, A., Harris, C., Grotting, J.C., Psillakis, J., and Vasconez, L. O., Application of Autologous Fibrin Glue in Burn Wounds, *J. Burn Care Rehabil.*, 10, 504 (1989).
5. Hermans, J. and McDonagh, J., Fibrin: Structure and Interactions, *Seminars Thrombosis Hemostasis* 8, 11 (1982).
6. Buckert, F., Nyman, D., and Gastpar, H., Factor XIII, Fibrin and Collagen. *Thrombosis Hemostasis* 63 (Suppl) 391–401, 1978.
7. Mosher, D.F., Schad, P.E. and Kleinman, H.K., Cross-linking of Fibronectin to Collagen by Blood Coagulation Factor XIIIa, *Clin. Invest.* 64, 781 (1979).
8. Pedersen, L.C., Yee, V.C., Bishop, P.D., Trong, I.L., Teller, D.C. and Stenkamp, R.E., Transglutaminase Factor XIII Uses Proteinase-like Catalytic Triad to Crosslink Macromolecules, *Protein Science* 3, 1131 (1994).
9. Greenberg, C.S., Birckbichler, P.J. and Rice, R.H.,Transglutaminases: Multifunctional Cross-linking Enzymes that Stabilize Tissues, *FASEB J.*, 5, 3071 (1991).
10. Potts, J.R. and Campbell, I.D., Fibronectin Structure and Assembly, *Current Opinions in Cell Biology* 6, 648 (1994).
11. Fuchsbauer, H.-L., Gerber, U., Engelmann, J., Seeger, T., Sinks, C. and Hecht, T., Influence of Gelatin Matrices Cross-linked With Transglutaminase on the Properties of an Enclosed Bioactive Material Using B-Galactosidase as a Model System, *Biomaterials*, 17, 1481 (1996).
12. Ratnoff, O.D. and Menzie, C., A New Method for the Determination of Fibrinogen in Small Samples of Plasma, *J. Lab. Clin. Med.*, 37, 316 (1951).
13. Silver, F.H., Wang, M.-C. and Pins, G.D., Preparation of Fibrin Glue: A Study of Chemical and Physical Methods, *J. Applied Biomaterials* 6, 175 (1995).
14. Kronengold, R.T., Fibrin Glue and Biological Adhesion Mechanisms, Ph.D. Dissertation, Department of Biomedical Engineering, Rutgers University, 1998.

Index